The Routledge Handbook of Hazards and Disaster Risk Reduction

The *Handbook* provides a comprehensive statement and reference point for hazard and disaster research, policy making and practice in an international and multi-disciplinary context. It offers critical reviews and appraisals of current state-of-the-art and future development of conceptual, theoretical and practical approaches as well as empirical knowledge and available tools.

Organised into five interrelated sections, this *Handbook* contains sixty-five contributions from leading scholars. Part I situates hazards and disasters in their broad political, cultural, economic and environmental context. Part II contains treatments of potentially damaging natural events/ phenomena organised by major earth system along with categories of vulnerabilities to them and capacities to deal with them that exist in communities. Part III critically reviews progress in responding to disasters including warning, relief and recovery. Part IV addresses mitigation of potential loss and prevention of disasters under two sub-headings: governance, advocacy and self-help, and communication and participation. Part V ends with a concluding chapter by the editors.

The engaging international contributions reflect upon the politics and policy of how we think about and practice applied hazard research and disaster risk reduction. This *Handbook* provides a wealth of interdisciplinary information and will appeal to students and practitioners interested in Geography, Environment Studies and Development Studies.

Ben Wisner is a retired Professor who has worked on the interface between disaster risk reduction and sustainable human development since 1966. He currently conducts research and advises institutions such as the Global Network of Civil Society for Disaster Reduction.

JC Gaillard is Senior Lecturer at the School of Environment of The University of Auckland, New Zealand.

Ilan Kelman is a Senior Research Fellow at the Center for International Climate and Environmental Research–Oslo (CICERO).

The Routledge Handbook of Hazards and Disaster Risk Reduction

Ben Wisner, JC Gaillard and Ilan Kelman

Routledge
Taylor & Francis Group

LONDON AND NEW YORK

First published 2012
by Routledge
2 Park Square, Milton Park, Abingdon, Oxon OX14 4RN

Simultaneously published in the USA and Canada
by Routledge
711 Third Avenue, New York, NY 10017

Routledge is an imprint of the Taylor & Francis Group, an informa business

British Library Cataloguing in Publication Data
A catalogue record for this book is available from the British Library

Library of Congress Cataloging in Publication Data
The Routledge handbook of hazards and distaster risk reduction / edited by Ben Wisner, J.C. Gaillard, and Ilan Kelman.
    p. cm.
    Includes bibliographical references and index.
    1. Natural disasters–Risk assessment–Handbooks, manuals, etc. 2. Hazard mitigation–Handbooks, manuals, etc. 3. Emergency management–Handbooks, manuals, etc. I. Wisner, Benjamin. II. Gaillard, J. C. III. Kelman, Ilan. IV. Title: Handbook of hazards and distaster risk reduction.
    GB5014.R68 2011
            363.34'2–dc22

ISBN: 978-0-415-59065-5 (hbk)
ISBN: 978-0-415-52325-7 (pbk)
ISBN: 978-0-203-84423-6 (ebk)

Typeset in Bembo
by Taylor & Francis Books

Contents

Contents

Contents

Contents

Figures

Figures

Tables

Tables

Boxes

Author biographies

Jonathan Abrahams is coordinator, Risk Reduction and Emergency Preparedness, Department of Emergency Preparedness and Institutional Readiness, Health Action in Crises, World Health Organization, Geneva, Switzerland.

Bob Alexander is an independent disaster risk reduction (DRR) consultant with research and consultancy foci including community-based disaster risk reduction (CBDRR), food and livelihood insecurity, participatory vulnerability assessment methods, displaced persons and programmes for at-risk youth. Under the moniker 'barefoot bob', he's also a recording artist and award-winning performing songwriter.

David Alexander is Ministerial Contract Professor at CESPRO, Centre for the Study of Civil Protection and Risk Conditions, at the University of Florence in Italy. He is the author of several books, including *Natural Disasters*, *Confronting Catastrophe* and *Principles of Emergency Planning and Management*. He is co-editor of the international journal *Disasters*.

Tammam Aloudat is a Senior Health Officer in the International Federation of Red Cross and Red Crescent Societies (IFRC) Secretariat in Geneva, Switzerland, in charge of emergency and disaster health programmes. He has an MD from the University of Damascus and an MSc from the London School of Hygiene and Tropical Medicine. Since contributing to this handbook, Tammam has moved to work as an Emergency Health Adviser for Save the Children.

Margaret Arnold is currently a Senior Social Development Specialist with the World Bank focused on the social dimensions of climate change.

Agnes A. Babugura is a lecturer and head of the Geography and Environmental Science section at Monash University South Africa. She obtained her BA and master's degrees at the University of Botswana and her PhD in environmental science at the University of Witwatersrand South Africa.

Christopher M. Bacon is a political ecologist. His research explores how environmental governance relates to sustainable livelihoods, biodiversity conservation and environmental justice in the Americas. He teaches at Santa Clara University in California, USA.

Greg Bankoff writes on disasters, natural hazards, human–animal relation, development, resources and community-based disaster management. Among his publications are *Cultures of*

Disaster: Society and Natural Hazard in the Philippines (2003) and *Mapping Vulnerability: Disasters, Development and People* (2004). He is Professor of Modern History at the University of Hull.

Charlotte Benson is an economist, with some twenty years' research experience in economic, financial and budgetary aspects of disasters and related vulnerability in developing countries. Formerly of the Overseas Development Institute, London, Charlotte has been working independently for the last thirteen years.

Gregory Berger is a Mexico-based filmmaker, journalist and Professor of Film at the Art Department of the Universidad Autónoma del Estado de Morelos.

Mihir Bhatt has worked with the All India Disaster Mitigation Institute, India on action and learning around disaster risk reduction at community level since 1995.

Sálvano Briceño, Director of the International Strategy for Disaster Reduction (UNISDR) Secretariat, 2001–2011, is a lawyer. Prior to UNISDR, he was Coordinator of BIOTRADE and GHG Emissions Trading Initiatives at UNCTAD, having also worked at UNCCD, UNFCCC and as the first Coordinator of UNEP's Caribbean Environment Programme, 1987–1991.

Philip Buckle is a consultant in public sector policy development, strategic planning and operations management for disaster risk reduction and disaster response. He has taught university courses in vulnerability assessment, risk management and disaster risk reduction in Australia, Switzerland and the UK, and has contributed regular publications to refereed journals and conferences.

Jean Connolly Carmalt is Assistant Professor of Law, Politics, and Society at Drake University. She has a JD from Cornell Law School and a PhD in geography from the University of Washington. She has spent several years working professionally for international human rights non-governmental organisations (NGOs). Her current research focuses on the relationship between geographic analysis and human rights law.

David Chester is a graduate of the Universities of Durham and Aberdeen and is currently Reader (Associate Professor) in the Department of Geography at the University of Liverpool. For many years Dr Chester has carried out research on disasters, particularly those produced by earthquakes and volcanic eruptions, and in the last decade he has become increasingly interested in religious responses to extreme events of nature.

Lene Christensen is Knowledge Management Adviser for Danish Red Cross, previously Technical Adviser and Deputy Head of the International Federation of the Red Cross and Red Crescent Societies (IFRC) Reference Centre for Psychosocial Support. She has an MPhil in social anthropology (1999) and has since then worked with development issues and humanitarian assistance.

Ian Christoplos is currently a researcher in natural resources and poverty at the Danish Institute for International Studies. He has worked as a researcher and practitioner over the past twenty years, looking at how local institutions manage risk amid rural development efforts and humanitarian action, with assignments around the world.

Ana Maria Cruz is an international consultant in the area of risk management and emergency response practices, with areas of expertise including Europe, the USA, Turkey, Japan and China. Dr Cruz's research interests include modelling and assessment of flooding, storm, earthquake, tsunami and climate change.

Claudine Haenni Dale is a human rights and humanitarian policy adviser. She has worked as a political adviser to the Representative of the Secretary-General of the United Nations (RSG) on the human rights of internally displaced persons, and as an Interim Focal Point on Protection in Natural Disasters. She is engaged currently in re-drafting the Inter-Agency Standing Committee (IASC) Guidelines and Manual on Human Rights in Natural Disasters.

Zenaida G. Delica-Willison is currently the disaster risk reduction adviser for the Special Unit for South–South Cooperation in the United Nations Development Programme (UNDP). Prior to this she served as the Training and Education Director at the Asian Disaster Preparedness Centre. She is known to be one of the pioneers in advocating citizenry-based development-oriented disaster management at all levels.

Claude de Ville de Goyet, MD has dedicated his entire career to risk reduction. For 25 years he worked for the Pan American Health Organization (PAHO) regional office for the Americas of the World Health Organization (WHO) as director of the Emergency Preparedness and Disaster Relief Coordination Program. Since 2002 he has been an international consultant and evaluator for European Union (EU) and UN agencies. His contribution to risk reduction has been recognised internationally through a Certificate of Distinction from the UN Sasakawa Award (2006).

Chris Dibben is a lecturer in health geography at the University of St Andrews. He has worked on, amongst other subjects, epidemiological studies into recovery after myocardial infarction, the causes of low birth weight, the survival of drug misusers and the impact of air pollution.

Angus M. Duncan a graduate of the University of Durham and University College London, is Professor of Volcanology at the University of Bedfordshire and has worked on volcanoes in Italy, the Azores and Costa Rica for more than thirty years.

Salvatore Engel-Di Mauro teaches geography at the State University of New York at New Paltz. His main research interests include soil acidification, social causes and consequences of soil degradation, and gender and environment.

David Etkin is the Graduate Program Director of Disaster and Emergency Management at York University, Toronto. He worked for 28 years with the Meteorological Service of Canada in a variety of positions, including operations and research. He has edited several special volumes on hazard and emergency management, participated in several international projects and published over 60 papers.

Timothy Fewtrell received his PhD in hydraulic modelling of urban floods from the University of Bristol in 2009 and has been working as the Willis Research Fellow on novel methods for flood risk assessment for the insurance industry since 2008.

Maureen Fordham is Principal Lecturer in Disaster Management at Northumbria University, UK. She has worked on disaster and development issues since 1988. She is a founder member (since 1997) and long-term manager of the Gender and Disaster Network (www.gdnonline.org).

JC Gaillard is Senior Lecturer at the School of Environment of The University of Auckland, New Zealand. His research, policy and practical interests span a wide range of topics related to disaster risk reduction (http://web.env.auckland.ac.nz/people_profiles/gaillard_j).

Herman Gerritsen has been active in the modelling of hydrodynamic and transport processes in shelf seas since 1980. His recent work centres on structured improvement and validation of models and flood forecast systems by means of data assimilation. He is Senior Researcher in oceanography at Deltares, formerly Delft Hydraulics.

Pascal O. Girot is currently the Regional Programme Coordinator for the International Union for the Conservation of Nature (IUCN), based in San José, Costa Rica. He worked for 15 years as a geography professor at the University of Costa Rica, teaching courses in economic and political geography.

Delia Grace is a veterinary epidemiologist working at the International Livestock Research Institute, Nairobi, where she develops and manages research on applying risk-based approaches to animal, and especially zoonotic, diseases in developing countries. Her other interests include ecohealth, animal welfare and gender, and livestock.

Manu Gupta is the co-founder and director of SEEDS. His work includes advocacy, mobilisation of community-led efforts in recovery and risk reduction, training and research.

Katharine Haynes is a Research Fellow at Risk Frontiers, Macquarie University, Sydney and a visiting fellow at the Centre for Risk and Community Safety, RMIT University, Melbourne. Her research focuses on disaster risk reduction and climate change adaptation.

Annelies Heijmans is completing her PhD on community-based approaches to disasters in conflict-ridden areas at Wageningen University, the Netherlands.

Alexander Held holds a Masters of Science in Forestry and has been a fire ecologist at the Global Fire Monitoring Center (GFMC) since 2001. He is a Fire Manager with Working on Fire International, Burn Manager, Aerial Ignition Specialist, Incident Commander Type 3, and Fire Analyst. He serves as Fire Manager and Project Manager in Sub-Saharan Africa and Europe.

Kenneth Hewitt is a Research Associate from the Cold Regions Research Center, Wilfrid Laurier University, Canada. He was among the first researchers to criticise the deterministic, ahistorical and asocial concept of hazards and disasters and its dependence upon the use of choice and decision models.

Kaz Higuchi is an adjunct faculty member in the Faculty of Environmental Studies, York University, Toronto. He graduated with a BSc in physics from Carleton University and then joined Environment Canada as a weather forecaster-in-training, which he passed and was stationed as a forecaster at Arctic Weather Central in Edmonton.

Dorothea Hilhorst is Professor of Humanitarian Aid and Reconstruction at Disaster Studies of Wageningen University, the Netherlands.

Susanna Jenkins is a Research Associate at Cambridge Architectural Research. Current research focuses on probabilistic volcanic hazard, vulnerability and risk assessment, and the development of associated low-cost tools and methodologies.

Rohit Jigyasu is a conservation and risk management consultant from India and is currently invited Professor at the Research Center for Disaster Mitigation of Urban Cultural Heritage at Ritsumeikan University in Kyoto. His main research interest is in the area of disaster risk management of cultural heritage.

Cassidy Johnson is a Lecturer at the Development Planning Unit, University College London, and a specialist on urban and housing issues, disaster mitigation and post-disaster reconstruction.

Sebastiaan (Bas) N. Jonkman is an adviser and researcher in the field of flood risk management and has worked on projects in the Netherlands and other regions (New Orleans, Romania, Cambodia and Vietnam). His PhD was on loss of life estimation and flood risk assessment. He works for Delft University and Royal Haskoning.

Mark Keim is Senior Science Adviser at the Office of Terrorism Preparedness and Emergency Response, National Center of Environmental Health, Agency for Toxic Substances and Disease Registry, Centers for Disease Control & Prevention, Atlanta, USA.

Ilan Kelman is a Senior Research Fellow at the Center for International Climate and Environmental Research – Oslo (CICERO). His main research and application interests are disaster diplomacy and sustainability in island communities. See www.ilankelman.org.

Allan Lavell is specialised in regional development and disaster risk management, with a PhD in geography from the London School of Economics and Political Science. He currently works at the Latin American Social Science Faculty in San José, Costa Rica.

Joanne Linnerooth-Bayer is leader of the Risk and Vulnerability Program at the International Institute for Applied Systems Analysis (IIASA), in Laxenburg, Austria. Her current interests are investigating options for improving the financial management of catastrophic risks in developing countries, and exploring how these instruments can be linked to climate adaptation.

Cinna Lomnitz is Professor of Seismology at the Institute of Geophysics, Universidad Nacional Autónoma de México (UNAM). His most recent book is *Disasters: A Holistic Approach* (Springer, 2011), with Heriberta Castaños.

Alejandro López-Carresi is founder and Director of the Centro de Estudios en Desastres y Emergencias (CEDEM), a centre for studies on disasters, based in Madrid, which provides specialised training, research and consultancy on disaster management, international aid and development. He previously worked for several emergency medical services in Madrid and has worked on emergencies and disasters in Indonesia, Pakistan, Bolivia, Peru, Kosovo and Sudan.

Emmanuel M. Luna is a Professor of Community Development at the University of the Philippines-Diliman. He specialises in community-based disaster risk management and pioneered the integration of DRR in community development education in the Philippines.

Brian G. McAdoo is Associate Professor of Earth Sciences at Vassar College, New York. He began studying tsunamis after a landslide-generated tsunami in Papua New Guinea killed over 2,000 people. He has participated in post-tsunami surveys in Sri Lanka, Aceh, the Maldives, Java, the Solomon Islands and Samoa.

Michael K. McCall is a social geographer working in eastern Africa, South Asia and Mexico in community mapping and participatory geographic information systems (PGIS) applied to risks, vulnerability, environmental services and others.

Sabrina McCormick is a sociologist and currently a Science and Technology Policy Fellow of the American Association for the Advancement of Science (AAAS), Research Faculty at George Washington University, and Senior Fellow at the Wharton Risk and Decision Processes Center.

John McDermott is Deputy Director-General and Director of Research at the International Livestock Research Institute. He has worked in livestock development and animal and public health in developing countries for over twenty years as a professor, researcher and manager. He has published over 200 peer-reviewed articles and conference contributions and has supervised more than 30 postgraduate student theses.

Bill McGuire is an academic, science writer and broadcaster, and is currently Professor of Geophysical and Climate Hazards at University College London. He was a member of the UK Government's Natural Hazard Working Group, established in 2005 following the Indian Ocean tsunami. He has published over 300 papers, books and articles on volcanoes, natural hazards and climate change.

Marcel Marchand has been working for more than twenty-five years in the field of policy analysis, environmental impact assessment and integrated coastal zone management (ICZM) in various countries in Europe, Asia and Africa. He did his PhD on the modelling of coastal vulnerability for tropical storms and floods in India and Vietnam. He is specialist/adviser on ICZM at Deltares, formerly Delft Hydraulics.

Adolfo Mascarenhas is a Founder/Director and Vice-Chairman of LiNKS, an NGO in Tanzania dedicated to research into local knowledge and its utilisation. He was Founder/ Director of the Institute of Resource Assessment, and first Tanzanian Professor of Geography, University of Dar es Salaam.

Jessica Mercer leads the Climate Change Adaptation programme of Oxfam Australia in Timor-Leste. Prior to this Jessica worked within both academia and NGOs in the field of disaster risk reduction.

Ehren B. Ngo, MS, EMT-P, is Director of the Emergency Medical Care Program and Emergency Preparedness and Response Program, and an Assistant Professor of Cardiopulmonary Sciences and Global Health at both the School of Allied Health Professions and School of Public Health, Loma Linda University, USA.

Geoff O'Brien is an academic actively involved in research on vulnerability and resilience, particularly on building resilience for climate adaptation. He is active in public life and teaches at Northumbria University in Newcastle-upon-Tyne, UK.

Walter Gillis Peacock is Director of the Hazard Reduction and Recovery Center and Professor in Landscape Architecture and Urban Planning at Texas A&M University.

Mark Pelling is Reader in Geography, King's College London. He has written extensively on social vulnerability and adaptation to natural disasters and climate change with a focus on urban contexts, and acted as consultant for the UN Human Settlements Programme (UN-HABITAT), UNDP, the UK Department for International Development (DFID) and others.

Graciela Peters–Guarin is a Colombian geologist whose PhD applied participatory and geographical information system (GIS) tools to disaster risk assessment. She has research experience from the Philippines, Latin America and alpine Europe.

George Platsis has an academic background in business administration (strategy and international business specialisations) and disaster and emergency management (business continuity and terrorism focuses). Currently, he heads an interdisciplinary consultancy advising organisations of resiliency measures to counter current and emerging threats.

Carla Prater is Associate Director of the Hazard Reduction and Recovery Center, and Senior Lecturer in Landscape Architecture and Urban Planning at Texas A&M University.

Tim Radford is a journalist. He was, until 2005, science editor of *The Guardian*, London. He served on the UK committee for the International Decade for Natural Disaster Reduction (1990–99).

Nick Rosser is an RCUK Fellow at Durham's Institute for Hazard Risk and Resilience, specialising in landslide mechanisms. His research interests consider the failure of brittle slopes in mountainous and coastal environments, and more widely consider the social implications of the hazards posed by such failures.

Keiko Saito is Deputy Director of the Cambridge University Centre for Risk in the Built Environment (CURBE) and Cambridge Architectural Research and Willis Research Fellow. Her area of interest is in the application of geographical information systems and remote sensing to quantify and visualise the risk from natural disasters on the built environment.

David Sanderson has worked for twenty years in development and emergencies around the world. He is currently a professor at Oxford Brookes University.

Heather Sangster is a geography graduate of the University of Liverpool. She is currently a research student in the Department of Geography working on the pre-industrial responses to volcanic eruptions in Western Europe.

Wendy Saunders is a social scientist at GNS Science (Institute of Geological and Nuclear Sciences), Lower Hutt, New Zealand, where she specialises in land use planning for natural hazard risk reduction.

Hanna Schmuck is a social anthropologist (PhD) and has been working on disaster management issues for almost twenty years in Africa and Asia, mainly for the Red Cross.

Julio Serje is a software engineer with over twenty-five years of experience in both the public and private sectors, and supported by long-standing experience within the UN system mainly in Asia and Africa. He is currently responsible for risk and disaster information systems that support and contribute to the Global Assessment Report (GAR) in the UN International Strategy for Disaster Reduction (ISDR).

David Simon is Head of Department and Professor of Development Geography at Royal Holloway, University of London and is a member of the Academy of Social Sciences. He advises UN-HABITAT on cities and climate change, and has published widely on theoretical, policy and practice aspects of urbanisation and environment-development.

Aidan Slingsby has a background in geographical information science, specialising in geographical data. His research interests include the design, use and evaluation of interactive visualisation techniques for finding insights into large multivariate datasets, assessing data quality and uncertainty, presenting data to different user groups and applying these methods for the insurance industry.

Thomas A. Smucker is Visiting Assistant Professor of Geography and Director of the International Development Studies programme at Ohio University, and a visiting scientist at the International Livestock Research Institute (ILRI) in Nairobi, Kenya. His field research has examined rural livelihood change, land tenure systems and drought coping strategies in East African drylands.

Danang Sri Hadmoko lectures at Fakultas Geografi, Universitas Gadjah Mada, Indonesia and is a researcher at the Research Center for Disasters in the same university. His research interests deal with landslide hazard.

Jane Strachan is a post-doctoral researcher for the National Centre for Atmospheric Science Climate Division, and is based at the University of Reading. Jane works on high-resolution climate modelling, with a focus on the simulation of tropical cyclones. As a Willis Research Fellow, working closely with the insurance industry, she also looks at how these simulations may be utilised in storm risk assessment.

Martha Thompson is the Program Manager for Rights in Humanitarian Crises at the Unitarian Universalist Service Committee. She has lived and worked in Latin America for nineteen years as a journalist and humanitarian worker.

Roger Underwood is Chairman of the Bushfire Front in Western Australia. He is a former District and Regional Forester and for nine years was the General Manager of the Department of Conservation and Land Management in Western Australia. He is also the Director of York Gum Services, an independent consultancy practice with a focus on bushfire management.

Dewald van Niekerk is the founder and Director of the African Centre for Disaster Studies at North-West University, South Africa.

Juan Carlos Villagrán de León completed his PhD in experimental physics and a post-doctoral fellowship in the University of Texas at Austin. He worked then on emergency management, warning systems and vulnerability assessment throughout Central America. In 2009 he joined the UN-SPIDER (UN Platform for Space-based Information for Disaster Management and Emergency Response) Program of the Office for Outer Space Affairs in Vienna, Austria.

Ben Wisner has worked on the interface between disaster risk reduction and sustainable human development since 1966, when he lived in a Tanzanian village for two years. He is a retired professor who still conducts research and advises many institutions, including the Global Network of Civil Society for Disaster Reduction. His email is b@igc.org.

Foreword

This *Handbook* represents an important step forward in advancing and disseminating knowledge of risk and vulnerability related to natural hazards and, therefore, can contribute greatly to the implementation of policies and programmes for disaster risk reduction (DRR).

Understanding and implementing DRR has come a long way since the time of the United Nations International Decade for Natural Disaster Reduction (IDNDR) in the 1990s, when only experts and practitioners were involved. Indeed, for the first half of the IDNDR there was hardly any inclusion of social aspects. At the end of the decade in 1999, very few governments had any kind of policy or programme aimed at reducing risk and vulnerability to natural hazards. Ten years later, and following many tragic disasters, governments are now rapidly turning their attention to DRR. Almost all governments on earth are officially doing something in this field.

However, despite the advancement in government attention, there is still much to be done to give risk reduction the priority it needs. This becomes increasingly urgent as vulnerability grows rapidly, mainly due to a growing gap between rich and poor, rapid expansion of urban density, degradation of ecosystems and the impacts of climate change and other creeping environmental changes. These dangerous processes are taking place in an already unequal and unjust world, with many development policies based on economies of war rather than economies of peace.

There are obstacles to moving faster on DRR as well as progress in reducing human and social vulnerabilities. Here, I would like to address a few of them.

Short-term and scattered practices are more common than the long-term approach required for DRR. The implementation of the UN's master plan for reducing disaster, the Hyogo Framework for Action (HFA), could take five to twenty years in any given country, depending on existing vulnerabilities and capacities. However, long-term planning in government systems is still not a regular practice, with most governments in high-, medium- and low-income countries tending to focus on short-term interests. In the realm of international co-operation, short-term 'aid' remains the most common practice despite well-meaning policies recommended by the Paris Declaration and Accra Agenda for Action on Aid Effectiveness.

This short-term approach coupled with a lack of serious team efforts among donors and agencies, exacerbated by constant competition, does not augur well for implementing effective DRR policies and programmes. We have seen a deterioration of such team and long-term efforts when comparing the follow-up to the 2004 Indian Ocean tsunami and the most recent effort in Haiti, which was a much 'simpler' effort since Haiti is a single country (whereas the tsunami affected more than twelve) with a small population and is surrounded by much richer and more capable neighbours.

This *Handbook* makes a point in many of its chapters of linking temporal and spatial scales and, in particular, the long-term causes of disaster are traced to 'root causes' and the question of recovery and capacity building at all levels is seen as needing long-term commitments. The combined voices of its eighty authors add to those that have been advocating comprehensive and long-term, so-called 'joined-up', thinking. The chapters on history and religion are particularly good at reminding us of the long term, while chapters on planning on various scales provide practical advice about taking a long-term and comprehensive view.

Humanitarianism's excessive focus on disaster response sometimes turns into a narrow-minded 'delivery' business. It does not facilitate the long-term approach required for effectively reducing the risk of disaster, despite some very well-intentioned recommendations such as those offered by the principles of Good Humanitarian Donorship. Additionally, there is a tendency to pay greater attention to risks that have a high political profile, such as terrorism and conflict. Natural hazards affecting poor and marginal groups, especially slow-onset hazards such as drought, are neglected.

How to reduce the risk of disaster is well known and rather simple to achieve if the right policies are put in place. At the same time, reducing the risk of terrorism or some industrial and technological events still requires much study to understand the hazards and related vulnerabilities. However, in comparing international or regional conferences on the topics, those focusing on response and preparedness or wider-risk approaches develop faster and have greater resources than those focusing on reducing risk and vulnerability to natural hazards.

The editors and contributing authors address this obstacle well in the *Handbook*. While the editors do have a section on response and recovery, other sections of the *Handbook* are devoted to dealing with prevention and building capacity in facing natural hazards. I am delighted to see that Chapter 42, on emergency management, is written by a former emergency management technician from Spain, now turned researcher and trainer. Indeed, a refreshing thing about this *Handbook* is the large number of authors who have a base in practice and the large number of voices from the global South and authorship by young people, as well as balancing both genders. This gives me hope that DRR will advance as the next generation takes over.

In the end, working for development is the most humanitarian thing one can do. This is not to say that immediate relief and life-saving assistance are a low priority. Rather, it means that prevention has the potential to save lives as well, and improve the quality and security of lives, especially when DRR is linked to livelihood improvements and social mobilisation.

A continuing lack of ecological awareness is also a major obstacle to DRR. Awareness of the necessity and possibility of reducing risk related to natural hazards parallels in many ways the evolution of awareness of ecological, environmental and later sustainable development issues. Both failure to address natural hazard risk and failure to protect ecosystems make development a more expensive and almost impossible task. Both hazards and ecosystems are essential components of nature which without proper attention will simply become more difficult obstacles to human, social and economic development.

During the 1970s, following the awareness that development could not proceed without respecting nature, and that natural resources were finite, development policies underwent a major shift. Integrated ecological considerations were advocated as the starting point from which land use planning should be undertaken. At the beginning, such environmental approaches were recognised in development planning mainly as an add-on. The identification of economic and social objectives was usually undertaken first and then environmental considerations came later, aiming at avoiding the potential negative impact of development activities on ecosystems. They took the form of so-called 'environmental impact assessments' and, alas, were often pro forma and ignored. It was only later on that the full understanding of the essential role of nature in development led to using ecosystems, 'homogenous natural areas', bioregions, biomes and other territorial concepts as the starting point for land-use and development planning.

Wisner, Gaillard and Kelman have constructed a handbook that has a strong emphasis on ecology and environment. Various authors explicitly focus on ecological relations, for example in Chapters 29–32, the last being an excellent discussion of plant diseases, pests and erosion of biodiversity by Latin America's Director of the International Union for Conservation of Nature, Pascal Girot. However, in many other chapters the other authors also focus on society–environment mutual relations.

Natural science's and engineering's understanding of natural hazards developed in parallel with and separately from social science research on hazard risk and the DRR community's hands-on experience. Natural scientists and engineers were seen by governments as having privileged knowledge, and this reinforced the traditional perception of disasters as 'acts of God' – somehow inevitable and therefore wrongly labelled as 'natural' disasters. A natural consequence of such perception is a main focus on preparedness activities with less attention paid to prevention, mitigation and what one calls today DRR. Now I turn to this last, but all-important, obstacle.

The confused use of the phrase 'natural disaster' is still common. This misperception, currently encouraged by the climate change spotlight, has led to an excessive focus on understanding the hazards themselves from a physical and natural science point of view, and does not facilitate the much-needed attention to understanding and reducing human and social vulnerability. Despite efforts to avoid the use of the term 'natural disaster' in major international negotiations, such as those leading to the adoption of the Hyogo Framework for Action, there is still a tendency to use the wrong term when discussing DRR policies, legislation and programmes.

I am pleased to see that this *Handbook* specifically rejects the 'naturalness' of disaster, and spends time in many chapters discussing social vulnerability. Indeed, the lead editor specifically wrote about 'taking the "naturalness" out of "natural disaster"' in an essay published as early as 1976. The framework chapter, Chapter 3, provides a way of asking questions about a wide spectrum of social and human risk factors and different time and spatial scales, in addition to understanding the natural hazards themselves.

These are just a few of the key obstacles that need more intensive attention by governments and leaders. Government leaders are emphasised here because of their immediate potential for rapid policy advances. However, their efforts must be complemented, often led, by the work of parliamentarians, leaders in opposition, local and other sub-national government, civil society, the private sector and, above all, affected communities themselves and individuals who want to see change. DRR, like good environmental stewardship and management, could become a national priority above political divides if promoted by everyone. Reflecting on the experience of change in values, attitudes and behaviour on issues such as health, environment, transportation crashes and gender, among other issues, similar success can be imagined for DRR.

I find the *Handbook*'s chapters on education and communication, as well as its treatment of popular culture, very interesting in this respect. Media, early schooling, popular films and songs can have a big impact on mainstreaming DRR in the public mind, and that, in turn, can complement 'top-down' efforts to legislate, institutionalise, make resources available for DRR and remove obstacles that block the sustainability of livelihoods and the capacity of ordinary people to 'live with risk'.

This *Handbook* represents a major effort in compiling the most advanced knowledge on DRR. I congratulate its editors and authors for undertaking this valuable effort. I strongly recommend the wide dissemination of the *Handbook* as well as using extracts from it to raise the awareness of key and visionary leaders around the world. Avoiding the loss of lives and livelihoods due to natural hazards is not only possible, but in many cases is also rather simple, as long as the right long-term and team efforts are promoted by leaders and undertaken by everyone.

Salvano Briceño
Former Director, UN International Strategy for Disaster Reduction (ISDR), Geneva

Acknowledgements

We would like to thank our chapter and box authors for their patience and good humour during a long process and much back and forth of drafts. In addition, specific authors and others need to be singled out by name for special thanks.

Routledge's intrepid editorial team, led by Andrew Mould, for keeping up with a long stream of emails and for responding fuly to all queries, with answers packed with detailed advice and much needed encouragement. Jake Rom D. Cadag (Université Paul Valéry – Montpellier III, France) for enormous help with the bibliography and map of the world. Dr Shabana Khan (independent scholar) for help with the proofs.

Cedric Daep (Philippines), Geoff O'Brien (Northumbria University, Newcastle, UK), Greg Bankoff (Hull University, UK), Salvatore Engel-Di Mauro (State University of New York at New Paltz, USA), and Johann Goldammer (Gobal Fire Monitoring Center, Germany) for advice on drafts of certain chapters.

David Simon (Royal Holloway University of London, UK), Omar-Dario Cardona (National University of Colombia, Manizales), Allan Lavell (Latin American Social Science Faculty in San José, Costa Rica), Mihir Bhatt (All India Disaster Mitigation Institute, Ahmedabad, India), Ailsa Holloway (Stellenbosch University, South Africa), and Sonia Kruks (Oberlin College, Ohio, USA) and Andrew Mould (our Routledge editor in the UK) for encouragement and brainstorming at the beginning, and encouragement and advice along the way.

The editors also thank UMR 5194 Pacte – CNRS (France), University of the Philippines Diliman, University of Grenoble (France), The University of Auckland (New Zealand), the Center for International Climate and Environmental Research–Oslo (CICERO) (Norway), and Risk RED (Risk Reduction Education for Disasters) for institutional support.

Challenging risk

We offer the reader a left-foot book

The Editors

May we have this dance?

The late twentieth century and early years of its successor witnessed some terrible disasters involving natural hazards, including the 2004 Indian Ocean and 2011 Japanese tsunamis; earthquakes in Pakistan, Indonesia, China, Haiti and Japan; flooding in Pakistan, China and Mozambique; and deadly tropical cyclones hitting Asia, the Caribbean, the USA and Central America; along with a continuing series of droughts and epidemics amongst many others. Simultaneously, this same period saw awareness growing in many science, policy and practice circles on several major points regarding how to deal with disasters. These insights and even paradigm shifts coalesce around the words 'disaster risk reduction' (DRR), referring to the process of understanding, analysing and managing the causes and origins of disasters and the risks that accumulate and lead to disasters.

In a broad-brush summary, DRR (see also Figure 1.1):

- Must be a multi-sectoral effort, top down and bottom up, across scales from global to local to individual, and informed by knowledge in the present and short term to centuries past as well as projections into the future.
- Requires outside specialist knowledge from many professions, backgrounds, values and disciplines.
- Requires local knowledge that resides with ordinary people, including marginalised groups, in their homes, settlements, community spaces and work places.

No single person can possess the knowledge and skill to map out and successfully implement DRR. Teams are required that command many kinds of knowledge. Such teamwork needs to be complemented with partnership among communities, non-governmental organisations, sub-national and national government, regional and international organisations and the private sector. Therefore, a compilation of the state of such myriad components of DRR was needed: hence, this *Handbook*.

Yet DRR does not sit in isolation. It is impossible and fruitless to try to distinguish between human development and DRR. Whether one uses the term 'human', 'economic' or 'social'

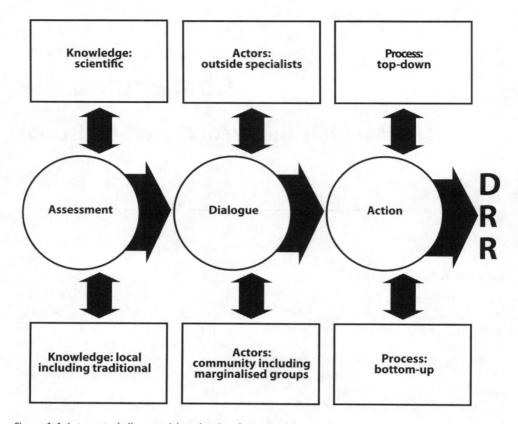

Figure 1.1 Integrated disaster risk reduction framework

development – or the trendier 'sustainable human development' or 'human security' – it is clear that no investment in development is risk neutral. Actions bring about reactions. They have consequences.

Investment in a new port, mine or variety of crop, or promulgation of a new local or national policy to close off a fishery or to create a nature or forest reserve, all affect the risks that people face. 'Good' development attempts foreknowledge of these consequences and aims to address the risks. Contrariwise, 'good' DRR is sustainable and accepted in the long run if it is woven into the fabric of people's day-to-day lives and livelihoods. People are happier with seismically safe houses that also are more comfortable to live in, respect their culture, are cheaper to maintain and offer space and amenities that make their livelihoods more productive. They accept drought-resistant grains that are easy to cook and taste good.

This confluence of development and DRR, within the overall ethos of sustainability, means that yet more topics are required reading (and understanding) for anyone who wants to understand either development or DRR, or someone who intends to participate in a team engaged in hands-on development or DRR-related work. At a stroke, the *Handbook* has just become fatter.

Indeed, the institutional silos that used to house 'humanitarian studies', 'natural hazards research and applications', 'disaster studies', 'development studies' and 'climate science' are now interconnected by many tunnels and bridges – an entire landform overlaps. Some denizens of institutional silos have even deigned to come up outside, like Plato's cave dwellers, and are blinking together in the sunshine. This *Handbook* seeks to produce a common space for those refugees from the silos, where practical and rigorous joint work can be achieved.

What is a handbook?

Each of the chapters to follow is a substantial, accessible, self-contained essay that provides a review of knowledge of and experience with one or another of the many components of DRR. 'Substantial' means 6,000 words, far longer than an encyclopaedia entry. 'Accessible' means that we have assured that most (not all) technical jargon and the specialist vocabulary of disciplines is avoided in favour of ordinary language, or is defined. 'Self-contained' means that each chapter presents a generalist, curious reader with an overview of background, methods, findings, experiences, debates, uncertainties and resources to have some grasp of the component or aspect treated.

'Comprehensive', 'exhaustive' and synonyms are not mentioned, because that would be impossible. No claim is made that everything is contained in this volume.

The editors admit to a great deal of cross-referencing and to vigorous exchanges with authors, resulting in many cross-cutting themes recurring in different forms. Examples are governance, local knowledge and community participation, and the centrality of livelihoods, vulnerabilities and capacities. Indeed, the common vision we as editors share and our own positions and practice cannot be divorced from the work we edit. That vision is set out briefly in Chapter 3 and will be refined in successive section introductions as well as in the *Handbook*'s conclusion.

Presence of a guiding vision of DRR might be reason to think of this book as more a set of guidelines, a compilation of advice or a starting point to discover the world of the topic. However, 'handbook' it remains because our attempt at rigour, accessibility and inter-disciplinarity means that the reader will, we hope, find more balance, dispute and uncertainty acknowledged than in most publications called 'guidelines'. Our vision guides the *Handbook*, but the *Handbook* is not meant to guide anybody, at least not directly.

The reader is introduced to a host of natural and social science areas, aspects of public administration, public health, engineering and the humanities, amongst other disciplines and professions. The reader is encouraged and, we hope, aided in asking critical questions about all these.

With all this in mind, we hope that the reader can understand why we asked Routledge if we could call the collection a 'Left-Foot Book', rather than a handbook – something to signal its difference. Our publisher was not amused, but the reader may find it helpful to think of this tome in that way. Using this phrase, we wish to suggest that there is more here than the usual technical 'handbook': something that invites critical thinking, along with departure from disciplinary norms and expectations, and the euphemisms and politeness of diplomatic language used by United Nations organisations. The editors remember an extraordinary 1989 film about a special human being who produced art in a surprising manner and made people think differently about art. The film was called 'My Left Foot' (www.imdb.com/title/tt0097937). We do not claim to bring as much freshness and innovation, courage and hard work to DRR as Christy Brown brought to painting, but we have tried.

Why we are the editors: Putting our collective left foot forward

We are not anonymous, grey people in dingy, cluttered offices who arrange semi-colons and worry exclusively about the page references in a journal citation. We have edited the *Handbook* because we are committed not only to understanding the world but to changing it – all the while learning and trying to improve ourselves. All three of us have worked not only as researchers and teachers, but also as trainers and advocates, with communities, governments and institutions at different governance levels. Working in such varied circumstances and with so many colleagues and team members, we have ourselves barely glimpsed the 'big picture' that explains the role of natural hazards in human life and what factors account for successful DRR – and, indeed, whether or not 'natural hazards' need be hazardous at all. These experiences in many parts of the world have led us to a shared vision that underlies both our desire to produce the *Handbook* and the *Handbook*'s architecture. This vision can be put down in three propositions.

First, we are convinced by our over 70 years of combined experience (heavily weighted towards experience of the first editor) and other available evidence that local people's knowledge and capacity is rich in potential for dealing with poverty, environmental (including climate) change and disaster risk, but it has too frequently been disregarded. Many *Handbook* chapters discuss what people believe they know, correctly or incorrectly, about natural hazards and the social construction of disaster risk. Our authors also explore from different standpoints how outside specialist knowledge – generally termed 'expert' knowledge – engages or fails to engage with local specialist knowledge. Intermediary institutions such as non-governmental organisations are often important in facilitating (or blocking) such engagement, hence many of the chapters discuss these mediating systems: education, civil society, local government, planning at and policy for multiple scales, religion and popular culture.

Second, we are convinced that a complex systems framework is required to understand fully disaster risk and adequately to inform policy and practice. 'Complex system' has become a cliché, which is unfortunate given the importance of the notion. In the words of Heylighten:

> Let us go back to the original Latin word *complexus*, which signifies 'entwined', 'twisted together'. This may be interpreted in the following way: in order to have a complex you need two or more components, which are joined in such a way that it is difficult to separate them. Similarly, the Oxford Dictionary defines something as 'complex' if it is 'made of (usually several) closely connected parts'.
>
> *(Heylighten 1996)*

In the case of our *Handbook*'s approach to DRR, one must deal with the constant interaction of a large number (n) of social and natural component processes producing an even larger number of interactions (n! or n-factorial). In fact, one could even question whether 'complex system' is a meaningful phrase, since a 'system' by definition seems to be inevitably complex. This is the same for 'complex emergency'. An emergency, by definition, is complex.

One does not have to descend to the quantum level to confront uncertainty and even indeterminacy with potentially catastrophic human consequences (Perrow 1998). The implication is not fatalistic. This position leads to scepticism about the value of heroic master planning and micro-management – yet it does not negate the possibility of planning and informed policy. It does require planning to be recursive and self-correcting. It also requires policy-makers to engage in iterative consultation with the risk-bearers. For too long people whose class position made them safe defined what 'acceptable risk' was for others, whose lives were quite different.

Third, we believe that research and action (and reflection upon action) are required at many scales from that of the locality and temporality of daily life to planetary space and its immediate surroundings over the long term. The physical and natural processes that may trigger a disaster are multi-scalar manifestations of planetary dynamics. Climate and weather, the crust of the earth, landforms, surface water and both useful and hazardous life forms all emerge from processes that have been going on for millennia: in the atmosphere, lithosphere, cryosphere, hydrosphere and biosphere. Those planet-scale processes and humanity's planet-wide impacts, increased through economic and cultural globalisation, anchor one end of the scale continuum.

National-scale legislation, policy, planning and politics – often geared towards results within a multi-year electoral or power cycle – are found towards the midpoint on the scale continuum. The private sector varies, from owner-operated one-shop firms as a family's lifelong livelihood, to multi-national corporations operating across the globe, attempting to yield month-to-month profits irrespective of long-term costs. At the other end – the micro-scale – many highly local processes are at work: the particular characteristics of people, households, livelihoods and communities that are shaped by biology, culture, history, resources, power and choices – and too often by lack of resources, power and choices. Many specificities of geology, hydrology, soil structure, ecology, weather and climate are also highly local and contribute to shaping hazard risk.

The three propositions just discussed recur in many *Handbook* chapters that draw on a wide range of examples and case studies. These case studies cover both recent and older disasters and DRR efforts from the vast majority of the world's countries (see Figure 1.2).

They provide ample evidence that 'emergency' and 'disaster', amongst other vocabulary, must be seen as continuous with 'normal' and 'routine' processes and practices. Twin-core concepts that will crop up in numerous chapters are 'vulnerability' and 'capacity'. 'Resilience' or 'resiliency' is often used too, amongst a host of other words and synonyms. Rather than generating confusion by sifting through too many vocabulary and definitional debates, we stay with 'vulnerability' and 'capacity', while recognising that many readers will be upset at this choice. We would ask those who are adamant about their set of vocabulary to interpret this *Handbook*'s work in the context of their preferences and we thank those readers for their patience and understanding in our choice.

Earlier, we wrote that actions have consequences and that no development investment or policy is risk neutral. Another truism stemming from DRR work is that people move about and occupy territory for reasons. If people live in a flood plain or on the side of an active volcano, it is for a variety of historical and contemporary reasons, combining choice and lack of choice. Social, economic and political processes that deny some people access to land, credit, technology and markets, amongst many other processes that make them poor and marginal, also make them less able to invest in self-protection in the face of hazards and vulnerabilities. The relations that govern daily life and the routine use of resources and space are the same relations that determine who in society will suffer death, injury, disruption and livelihood loss in hazard manifestations, and who will have the resources to recovery quickly and achieve even better circumstances than before (Hewitt 1997; Wisner *et al.* 2004).

How to use the *Handbook* and avoid two left feet

First and foremost, this brick of a book will make an effective door stop. In critically reviewing another large book published by one of the editors, a PhD student doing field work in El Salvador wrote that she found it excellent for smashing cockroaches.

Besides these practical applications (and many more, we are sure), the *Handbook* can be read in several ways. As noted earlier, much research and project work as well as evaluations and studies that inform policy ('lessons learned' – or, often, just identified and then forgotten) are carried out by teams. Therefore one typical reader we have in mind is the specialist in one topic (say, seismic engineering) who wishes to get up to speed with the rudiments of the disciplines of teammates from other fields (for example, a sociologist specialising in women in disasters and a public health researcher concerned with health care system emergency response).

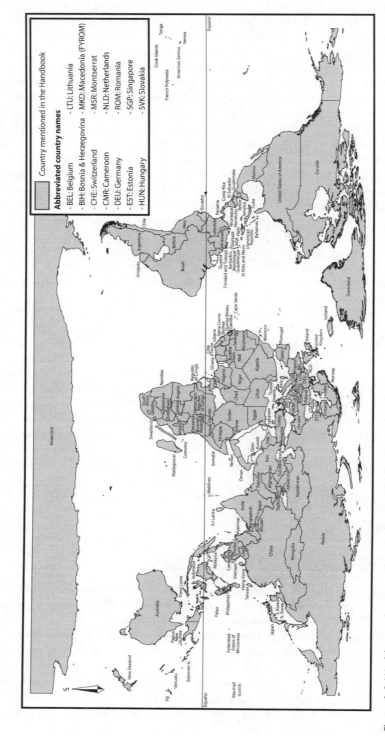

Figure 1.2 World map showing all countries mentioned in the *Handbook*

Alternatively, whole sections or sub-sections can be read by students at various levels who are puzzling through issues related to these themes in such diverse courses as environmental studies, geography, development studies, public administration, public health, climate affairs, history, environmental engineering and international law. Others with more operational and policy concerns can mix and match across sections by reading chapters that take various approaches to the same theme. For example, 'communication' is dealt with in chapters dealing with the media, education, preparedness and evacuation, and popular culture (amongst others).

Finally, for the marathon-running readership, we offer an invitation to read straight through from first to last page, making sure that you read the microscopically printed health advisory on the book's hefty spine. One might also read the chapters or sections in reverse or random order, just to be different from the linear presentation forced by a book's structure. After all, the *Handbook* was conceived as a space to invite a journey towards broad and deep understanding of natural hazards and DRR. There is a huge number of possible paths through the chapters and topics, although the journey is the same.

We hope that we will not end up as the only three ever to have read the whole *Handbook*! After all, any journey together can start with a left foot. Left ... right ... ; left, right, left ... ! Do join the dance!

Part I
Big picture views – hazards, vulnerabilities and capacities

2

Introduction to Part I

The Editors

Why a big picture view?

This *Handbook*'s first section of chapters provides a big picture view of hazards, vulnerabilities and disaster risk reduction (DRR). It introduces key macro-issues that are essential for understanding and reducing the disaster risk: a chapter that frames the issues and questions along with chapters that discuss the perspectives on disaster and DRR offered by history, politics, knowledge, culture, religion, urbanisation and sustainable development. Although embedded in local contexts, these issues are relevant beyond national borders and societies. They are increasingly recognised as crucial factors affecting vulnerability and capacity in facing natural hazards, while providing ways forward for DRR.

Such issues are not easy to appraise because they are fully enmeshed in the structure and evolution of contemporary societies. The initial chapter of this section, 'Framing disaster: theories and stories seeking to understand hazards, vulnerability and risk', provides a way of understanding these structural issues in the context of DRR. The *Handbook* agenda for research, policy and practice draws upon and elaborates the Pressure and Release (PAR) framework developed in the classic book *At Risk* (Blaikie *et al.* 1994; Wisner *et al.* 2004; see Figure 2.1 and Figure 2.2). The framework provides an analysis that traces unsafe locations and fragile livelihoods through dynamic pressures to root causes. It further ties vulnerability to access, marginalisation, capacity and recovery through an original suite of integrated diagrams.

The key issues addressed in this initial section of the *Handbook* deal with what the PAR framework identifies as root causes and dynamic pressures that shape people's vulnerability and capacity in facing natural hazards. Four decades of academic literature on disasters (e.g. Baird *et al.* 1975; Maskrey 1989; Oliver-Smith 1994), backed up by a profusion of practitioners' reports from the field (e.g. Anderson and Woodrow 1989; Heijmans and Victoria 2001), have shown that disasters deeply reflect failed or skewed development. Considering vulnerability to natural hazards through the sole lens of potential damage created by rare and extreme natural phenomena is a remnant of a paradigm that has been completely up-ended. In that context, sustainable DRR requires significant changes in the structure of societies, not short-term band-aid solutions, especially not those focused on only hazards.

A big picture view of these key issues is therefore essential. This introduction emphasises and brings together conclusions from each chapter of this section under three main themes: politics,

The progression of vulnerability

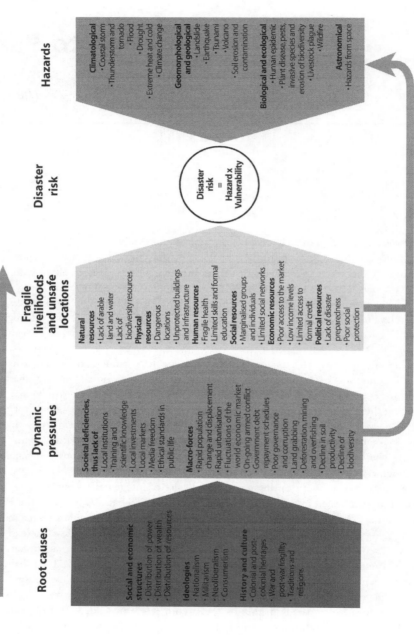

| Root causes | Dynamic pressures | Fragile livelihoods and unsafe locations | Disaster risk | Hazards |

Root causes

Social and economic structures
• Distribution of power
• Distribution of wealth
• Distribution of resources

Ideologies
• Nationalism
• Militarism
• Neoliberalism
• Consumerism

History and culture
• Colonial and post-colonial heritages
• War and post-war fragility
• Traditions and religions

Dynamic pressures

Societal deficiencies, thus lack of
• Local institutions
• Training and scientific knowledge
• Local investments
• Local markets
• Media freedom
• Ethical standards in public life

Macro-forces
• Rapid population change and displacement
• Rapid urbanisation
• Fluctuations of the world economic market
• On-going armed conflict
• Government debt repayment schedules
• Poor governance and corruption
• Land grabbing
• Deforestation, mining and overfishing
• Decline in soil productivity
• Decline of biodiversity

Fragile livelihoods and unsafe locations

Natural resources
• Lack of arable land and water
• Lack of biodiversity resources

Physical resources
• Dangerous locations
• Unprotected buildings and infrastructure

Human resources
• Fragile health
• Limited skills and formal education

Social resources
• Marginalised groups and individuals
• Limited social networks

Economic resources
• Poor access to the market
• Low income levels
• Limited access to formal credit

Political resources
• Lack of disaster preparedness
• Poor social protection

Disaster risk

Disaster risk = Hazard x Vulnerability

Hazards

Climatological
• Coastal storm
• Thunderstorm and tornado
• Flood
• Drought
• Extreme heat and cold
• Climate change

Geomorphological and geological
• Landslide
• Earthquake
• Tsunami
• Volcano
• Soil erosion and contamination

Biological and ecological
• Human epidemic
• Plant disease, pests, invasive species and erosion of biodiversity
• Livestock plague
• Wildfire

Astronomical
• Hazards from space

Accentuation of some (not all) hazards

Figure 2.1 The progression of vulnerability

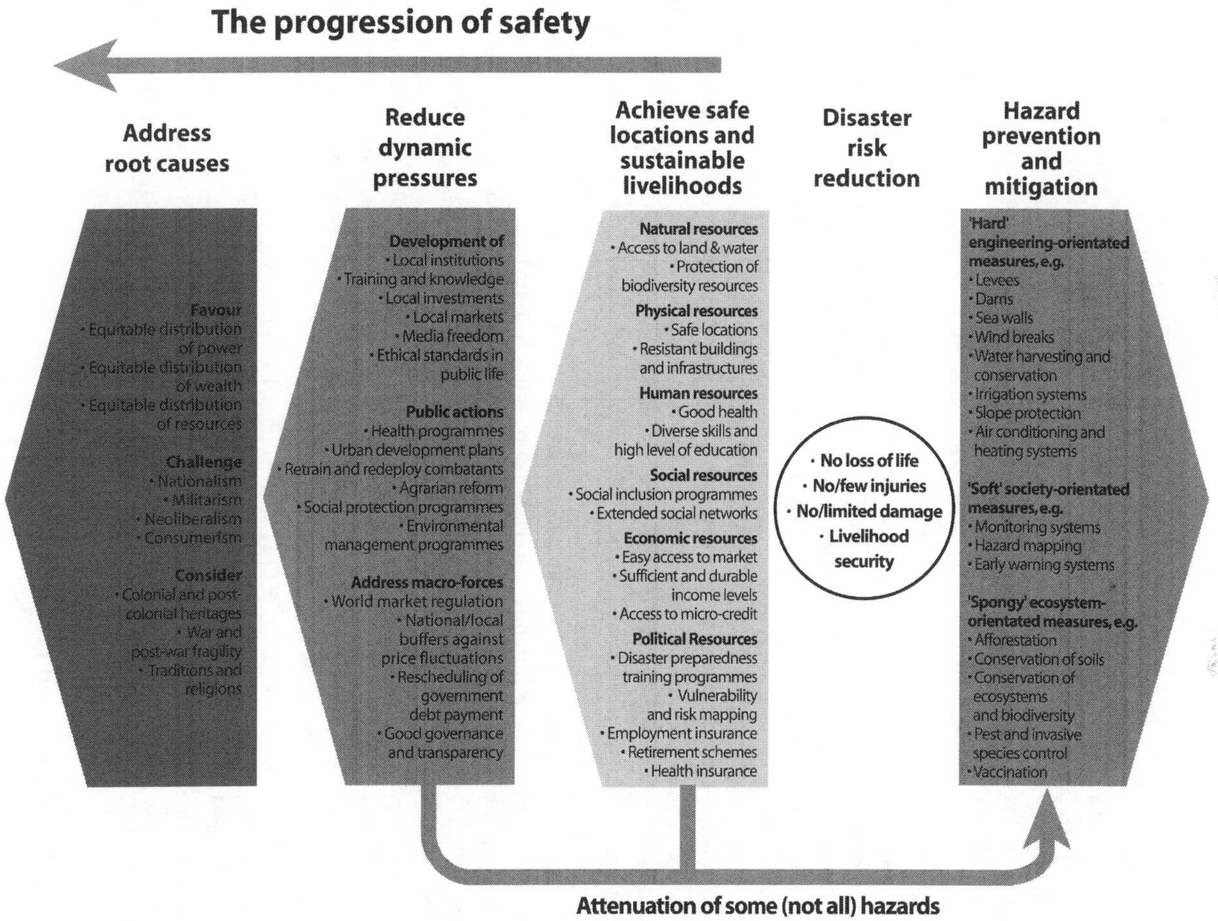

The progression of safety

Address root causes

Favour
- Equitable distribution of power
- Equitable distribution of wealth
- Equitable distribution of resources

Challenge
- Nationalism
- Militarism
- Neoliberalism
- Consumerism

Consider
- Colonial and post-colonial heritages
- War and postwar fragility
- Traditions and religions

Reduce dynamic pressures

Development of
- Local institutions
- Training and knowledge
- Local investments
- Local markets
- Media freedom
- Ethical standards in public life

Public actions
- Health programmes
- Urban development plans
- Retrain and redeploy combatants
- Agrarian reform
- Social protection programmes
- Environmental management programmes

Address macro-forces
- World market regulation
- National/local buffers against price fluctuations
- Rescheduling of government debt payment
- Good governance and transparency

Achieve safe locations and sustainable livelihoods

Natural resources
- Access to land & water
- Protection of biodiversity resources

Physical resources
- Safe locations
- Resistant buildings and infrastructures

Human resources
- Good health
- Diverse skills and high level of education

Social resources
- Social inclusion programmes
- Extended social networks

Economic resources
- Easy access to market
- Sufficient and durable income levels
- Access to micro-credit

Political Resources
- Disaster preparedness training programmes
- Vulnerability and risk mapping
- Employment insurance
- Retirement schemes
- Health insurance

Disaster risk reduction

- No loss of life
- No/few injuries
- No/limited damage
- Livelihood security

Hazard prevention and mitigation

'Hard' engineering-orientated measures, e.g.
- Levees
- Dams
- Sea walls
- Wind breaks
- Water harvesting and conservation
- Irrigation systems
- Slope protection
- Air conditioning and heating systems

'Soft' society-orientated measures, e.g.
- Monitoring systems
- Hazard mapping
- Early warning systems

'Spongy' ecosystem-orientated measures, e.g.
- Afforestation
- Conservation of soils
- Conservation of ecosystems and biodiversity
- Pest and invasive species control
- Vaccination

Attenuation of some (not all) hazards

Figure 2.2 The progression of safety

history and power; culture, knowledge and religion; and environment, development and sustainability. Challenges are identified for improving policy and practice.

Politics, history and power

History is replete with examples of disaster, such as the 1755 earthquake and tsunami that destroyed Lisbon, Portugal. However, the historical record is also full of narratives of survival, recovery and adaptation. Rebuilding Lisbon provided the opportunity for Europe's first master-planned city. Chapter 4, on history, goes even further to discuss 'historical concepts of disasters and risk'. Notions of acceptable risk are neither universal nor stable. They change over time and in relation to the rise and fall of powerful groups who usually decide on behalf of the less powerful what risk is 'acceptable' and what is not.

Chapter 5, on 'politics: power and disasters', delves even deeper to find root causes in the forms and distribution of power in societies. The recent history of colonialism, decolonialisation and globalisation have changed and transformed, yet conversely also partially preserved, patterns of discrimination and privilege that combine to allocate 'marginal people into marginal places' (Susman *et al.* 1983: 280). Power, however, is dynamic and also has the potential of being claimed and exercised from the bottom up. The politics chapter describes how to analyse power from different perspectives.

The twenty-first century is witnessing both enormous interest in and research on human rights. Meanwhile, around the world these rights are being eroded and denied, from Sudan and Zimbabwe to Iran and Palestine, from Myanmar to North Korea. Chapter 6, on 'human rights and disaster', reviews research on the links between disaster risk and human rights. A strong basis exists in documents such as the Universal Declaration of Human Rights for the assertion of a 'duty of protection' on the part of governments in order to address avoidable death and other harm to its citizens. People displaced by natural hazards, in common with refugees and those fleeing conflict, have a human right to protection and a right to assistance in recovery, including protection against arbitrary, premature return to hazardous conditions in the location of origin. Likewise, those sheltering have a right to protection from sexual abuse, exploitation and a right to at least minimum basic services (Sphere Project 2004).

The theme of conflict was just mentioned. It is not by accident that thinking about human rights and disaster leads to connections with conflict and post-conflict situations. This section devotes Chapter 7 to 'violent conflict, natural hazards and disaster'. Conflict makes relief and recovery efforts more difficult when a natural hazard occurs in the midst of a war or violent civil instability. Disaster risk reduction becomes more difficult under such circumstances as well. Furthermore, natural hazards may trigger or exacerbate conflicts, as opposing groups struggle over who controls relief assistance, as in Sri Lanka after the 2004 tsunami. The situation can also go the other way, as in the case of Aceh, Indonesia after the 2004 earthquake and tsunami, where opposing forces agreed to a ceasefire that has held at least as far as this *Handbook*'s completion.

Culture, knowledge and religion

The influence of culture may be positive or negative for DRR. Cultural patterns can strengthen people's ability to face natural hazards, while disproportionally high casualties amongst certain sectors of the society may be partially explained by cultural patterns that prevent those groups from accessing available means for tackling disaster risk. In the aftermath of disasters, culture may also explain why people decide to resettle in the same dangerous places in order to stay in the vicinity of sacred or familiar places that in part define their identify and security. Often, people follow a pattern of what they know, are used to and have previously been permitted to do. Nonetheless, cultures are seldom isolated. They interact, mingle and often clash with each other as detailed in Chapter 8, on 'culture, hazard and disaster'.

Big-picture forces affecting disasters and DRR can be deeply embedded within society. Chapter 9, on 'knowledge and disaster risk reduction', shows that local knowledge of vulnerabilities and hazards enables people to deal with them, perhaps entirely averting a disaster. Disaster risk reduction can consist of traditional ceremonies, communal labour to implement DRR measures, and efforts to expand and supplement knowledge from different internal and external sources as well as application of such knowledge. Hybrid knowledge or combined knowledge forms are needed for effective DRR.

Chapter 10, on 'religious interpretations of disaster', provides views of how religious beliefs, among other cultural patterns, contribute to shaping not only people's responses to natural hazards and disasters, but also DRR strategies from all sectors. Religion has both positive and negative roles in DRR, as discussed through established and evolving interpretations of disaster from existential, metaphysical and moral perspectives.

Cultural representations of disaster in films, songs and other art forms are powerful media for interpreting disasters and for teaching and learning about DRR, as shown by Chapter 11 on 'hazards and disasters represented in film', and Chapter 12 on 'hazards and disasters represented in music'. Unfortunately, those media are often overlooked by policy-makers and practitioners, who often continue to rely on the more traditional forms, such as authoritative posters, brochures, public service announcements and websites.

This sub-section shows that culture, knowledge and religion are dynamic and continually shaped and reshaped by internal and external forces – and also shape our understanding and experience of disaster and DRR. Predicting people's responses to natural hazards and disasters is a difficult task because culture, knowledge and religion interact with other factors, such as the physical environment, social organisation, economic constraints and political structures. People neither assess risks only according to the threat of hazard, nor through single filters. They always consider a large range of losses and benefits for their daily life. Consequently, it is crucial to consider DRR in its various contexts.

Environment, development and sustainability

In 1956 a landmark volume nearly the size of this *Handbook* was published: *Man's Role in Changing the Face of the Earth* (Thomas 1956). Other post-war scholarly collections followed, such as *The Careless Technology* (Farvar and Milton 1972), *Operating Manual for Spaceship Earth* (Fuller 1969), and a small pivotal narrative that combined lyricism and science: *Silent Spring* (Carson 1962). By the 1970s many states were enacting comprehensive legal regimes to regulate environmental management. The United Nations (UN) held its Conference on the Human Environment in Stockholm in June 1972. A decade and a half later, the Brundtland Report (WCED 1987) provided the baseline for defining 'sustainable development'.

In 1992 the venue for the twenty-year follow-up to the Stockholm Conference was Rio de Janeiro. International outcomes were the UN Framework Convention on Climate Change, the Convention on Biological Diversity, the Convention to Combat Desertification, and Agenda 21 to support local authorities and sustainable development. A decade later in Johannesburg (United Nations 2002b), disaster risk was incorporated as one of the consequences of mal-development.

Earth Summit 2012 or the Rio+20 meeting, entitled 'Sustainable Development – The Peace of the Future', will have continued this long series of environment and development initiatives.

Meanwhile, 2015 represents the deadline for major progress in DRR (UNISDR 2005b, 2009a; see also GNDR 2009, 2011), and the achievement of the Millennium Development Goals (United Nations 2000).

Such programmes and goals, and many more, have waxed and waned. One trend has been the increasing recognition and acceptance of overlaps and interlinkages amongst environmental management, development, sustainability and DRR. Each cannot be achieved without the other, in policy and in practice. The mutually beneficial characteristics of tackling them all simultaneously are amply demonstrated. For instance, sustainable ecosystem management contributes to DRR, while successful DRR must involve sustainable ecosystem management (Sudmeier-Rieux and Ash 2009). Development cannot work without embedding DRR processes in day-to-day activities, while DRR cannot be completed in isolation from other development processes (Lewis 1999; Wisner *et al.* 2004).

In parallel, scientific progress has been made to support policy and practice endeavours that link environment, development and DRR. Some of that science, and the link to policy and practice, are presented in Chapter 13 on 'hazards, risk and urbanisation' and Chapter 14 on 'disaster risk and sustainable development'.

Chapter 13 explains how urban areas have changed from locations of opportunity to encompass major disaster risk concerns, even as (and because) people flock to them for the opportunities that cities provide. Poor environmental management and inadequate development regulation have led urban risk to display dangerous differences from the forms of risk experienced by the majority of humanity in previous centuries.

Chapter 14 consolidates aspects of the history and possible future of sustainable development efforts and their application to DRR. The rhetorical invocation of 'sustainability' (and a good deal of corporate 'greenwash') is contrasted with reality. An integration of sustainable development and DRR is needed. Research and practice show how the two can be complementary, especially in the context of sustainable livelihoods.

Using the big picture for DRR

Reducing the risk of disasters requires all these big picture views to be factored in. Interactions between insiders and outsiders can dictate success or failure, often mirroring power relationships between dominant outside interveners and dominated local recipients. One example is the orientation of DRR policies which have long been framed by the interests of the most affluent countries, usually aiming to foster transfers of their experience, knowledge and technology to the less affluent regions of the world. Insider-versus-outsider interaction also occurs within a given country when dominant cultural groups or urban-based, centralised institutions and policy-makers fail to consider peripheral or historically subordinate cultures.

The *Handbook* aims at balancing such bias – as much as is feasible within the constraints of publishing an academic work in English – by including authors from all permanently inhabited continents (that is, not including Antarctica). Authorship is also balanced by age and gender. This meant that some key people in the field are not represented as chapter authors; however, others who will be key, or who should be known as key people, are among our chapter authors and the missing 'usual suspects' are cited in the review of literature.

Using these multiple perspectives and appraising the links and interactions amongst history, politics, religion, culture, knowledge, development, natural hazards, vulnerabilities, risks, disasters and DRR assist one to take actions to avoid disasters or to limit their impact on society. One continual frustration expressed by researchers, policy-makers and practitioners is that each subject has its own specialists who mark out their own territory. Silos are built and each group talks amongst themselves. Progress must continue on integrating and connecting diverse fields.

UNISDR (2005b: 1), for example, specifically states 'that efforts to reduce disaster risks must be systematically integrated into policies, plans and programmes for sustainable development and poverty reduction'. This aims to end the isolation of work on disasters and DRR – and especially to overcome the perception that dealing with disasters is about solely humanitarian response. As Part I shows, DRR can benefit from applying the lessons of other work – from environmental management to the study of knowledge systems and from an appreciation of religious interpretations of disaster to sustainable development and human rights.

The world continues to witness creeping impoverishment and inequities worldwide. Despite many successes, as discussed in this section and throughout the *Handbook*, disaster risk continues to threaten many, especially those mired in poverty and suffering inequity. Part I discusses large-scale societal forces that are root causes of these trends. Solutions will come from attempts to change these 'big picture' processes.

3

Framing disaster

Theories and stories seeking to understand hazards, vulnerability and risk

Ben Wisner

AON-BENFIELD UCL HAZARD RESEARCH CENTRE, UNIVERSITY COLLEGE LONDON, UK

JC Gaillard

SCHOOL OF ENVIRONMENT, THE UNIVERSITY OF AUCKLAND, NEW ZEALAND

Ilan Kelman

CENTER FOR INTERNATIONAL CLIMATE AND ENVIRONMENTAL RESEARCH–OSLO, NORWAY

Introduction

How to organise a seeming chaos of facts and ideas?

While not all people are curious about how children acquire language or why some animals hibernate, those who have witnessed a disaster or heard about the destruction and suffering involved often want to know why it happened. Millions of disaster survivors are especially keen to have answers. The drive to understand hazards, vulnerability and risk comes in part from the questions that ordinary people ask, especially when these questions take on political salience and governments begin to question in turn. In part, the desire to understand emerges from compassion for those who suffer. Practical steps to reduce disaster risk must be informed by knowledge and wisdom. Finally, also playing a role in piquing curiosity is the sense that one learns a good deal about human society and about planet Earth by studying disasters.

So there is a commonsense and practical side even to a chapter with a word in the title that might scare away most readers: the word 'theory'. The ancient Greek word *theorein* means nothing more demanding than 'to look about in the world' in the sense of the German *Welt-anschauung* (Jung 1989: 327). So looking about the world at flood disasters in South Asia, earthquake disasters in the Caribbean, drought disasters around the Horn of Africa or hurricane disasters in the USA, there is a lot to take note of and to ponder. What assistance can one seek in organising what one sees or has experienced? There are 'facts' about a wide spectrum of processes and events – physical and biological, political, economic, social, psychological and cultural. A framework assists in organising this welter of facts. It is a first step toward understanding that marshals, arranges and reminds one not to forget to ask certain questions.

Vade mecum: Reminder of good questions to ask

Young physicians can carry with them small, dense reference books that remind them of differential diagnoses. The *Handbook* framework serves a similar function to the medical companion, a printed friend that invites the hospital intern to 'follow me' (*vade mecum* in Latin). Quite understandably, someone trained as a civil engineer would be inclined to ask questions about structures and forces. S/he needs to be reminded by the framework also to ask about processes at work in society – and vice versa for the sociologist and other perspectives.

It is equally likely that someone trained as an economist will ask about losses and costs. The framework reminds her/him to ask also about the natural hazards (which could also be termed 'environmental hazards') themselves. Writing about post-Katrina attempts to plan for an even worse hurricane in New Orleans, Verchick (2010: 247) lists the various components of a disaster scenario for half a page and then asks, 'How will this mass of information be organized and communicated to legislators and the general public?'

This chapter provides an organising and nudging reminder framework, an *aide-mémoire*, for those seeking to develop an holistic view of natural hazards, disaster risk management and reduction. It also serves to extend remarks in this *Handbook*'s introduction by showing how the editors see the topics treated by individual chapters fitting together.

Theories and stories

Caveats are required at this point. Just as the *Handbook* makes no claim to being exhaustive, the framework presented in this chapter is only one possible way of organising the reality of disaster. It is the framework that the editors believe arranges a wide array of information to reveal key questions that lead to risk-informed, evidence-based decision-making for the long term.

All frameworks are grounded in and derived from generalisations about the world that the framers judge to be sound and reliable. On the whole, these generalisations help to answer the question, 'why?' They concern cause and effect. Thus, frameworks – and the one presented here – assume the validity of various theories that collect repeatedly observed causes and effects in nature, society and the arts. As Chapter 61 on university research shows, a large number of academic disciplines have applied their theories to various aspects of hazard, risk and disaster.

'Cause and effect' may sound too straightforward and deterministic. That is not intended. Uncertainty and contingency is rife in the study of disaster at all scales and within all disciplinary perspectives. Challenges continue: understanding the failure of welding in the steel frames of modern buildings in Northridge and Mexico City earthquakes; determining where the cholera bacterium survives in the environment in between outbreaks; and how to overcome the 'moral hazard' created by flood insurance if it perversely encourages people to live in a flood plain. These are amongst the many challenges to understanding hazards and vulnerabilities.

Also, importantly, the framework includes non-Western, oral and vernacular understandings of hazards, risk and disaster. This is the reason why the word 'story' appears in the chapter title alongside the word 'theory'. Ordinary people have a variety of ways to discuss among themselves the existential and practical questions that emerge when confronting hazards and taking risks. The framework welcomes and embraces local attempts to understand in collaboration with outside specialist efforts.

Myths and facts

Anyone who works on disasters as a planner, first responder, researcher, journalist, policy-maker or other role has heard statements that swallow or assume more general propositions about the world and people's lives and behaviours. 'Looting is common after disasters'; 'dead bodies must be disposed of quickly'; 'the poor are superstitious and fatalistic'. These kinds of statements are myths (Eberwine 2005). They are based on generations of anecdotal observation, or claimed observation, and reaffirmation, perhaps just repetition of what others said. They are not grounded in a coherent and consistent body of observations about the world.

Theories arise when these bodies of fact have been accumulated by following systematic methods, which allow generalisation and accumulation of new facts by asking questions guided by those methods and generalisations. The distinction between myth and fact is vital to clear thinking and good work in fields that address natural hazards, disaster risk management and reduction. That said, it is nevertheless one of the goals of disaster studies to explain why such myths persist.

Our framework and some key definitions

The following presents the framework in six variations, written around a suite of six diagrams. During the discussion, a series of key terms are introduced, clarified by their context and, in some cases, formally defined:

- Resource and hazard
- Vulnerability and capacities
- Livelihood and location
- Access and marginalisation
- Disaster and recovery

Social construction of resource and hazard

Figure 3.1 represents the physical and socio-political worlds in a highly schematic way. At one end is the natural environment (box 1). At the other end (box 8) are international- and national-scale political and economic systems. It is fair to suggest that, today, economic systems seem to influence many aspects of life and, consequently, what should be done to change the value system within which the globe currently operates.

The reader should note two aspects about the 'natural' environment. First, it is not entirely 'natural' but is influenced by human activities (economic decisions, land use, policy, etc.), represented by the four arrows that originate in boxes 5–8. Second, the natural environment (as dynamically modified and 'constructed' by human action) is the origin of both a series of possible opportunities and a series of possible hazards. This is the dual-faced character of nature that has been the focus of students of natural resource management and natural hazards since at least the 1950s (Zimmerman 1951). It is a well established element – perhaps even an axiom – of a major branch of human geography theory that treats society–nature relations.

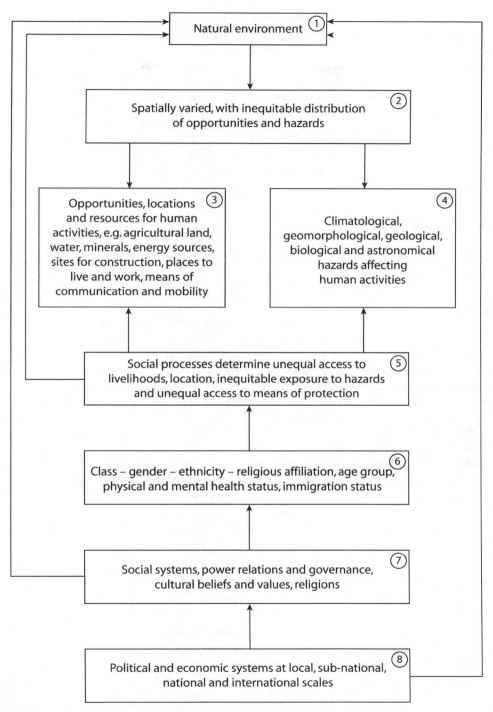

Figure 3.1 Nature's two faces: resource and hazard

The central significance of this version of the framework is to reveal disaster risk as a doubly contingent situation. Given the action of boxes 6–8, which contain nested political, economic, social and cultural processes, access enjoyed by a given household to 'nature-as-opportunity' or 'nature-as-hazard' varies (box 5). Access to natural resources upon which to build a livelihood is inequitably distributed, as is access to safe home sites, infrastructure enabling mobility (including evacuation), communications and marketing, spatial and temporal exposure to natural hazards and means of protection. Figure 3.1 is adapted from a figure developed by Terry Cannon, one of the co-authors of *At Risk*, where the dialectical character of disaster risk is elaborated (Wisner *et al.* 2004: 6–8): '[T]he natural environment presents humankind with a range of opportunities ... as well as a range of natural hazards ... But crucially, humans are not equally able to access the resources ...; nor are they equally exposed to the hazards'.

Defining vulnerability

Chambers was one of the first to introduce formally the term 'vulnerability' into the analysis of rural poverty. It came as one of five elements that interlocked with each other, producing what he termed a 'ratchet effect' or 'deprivation trap' (Chambers 1983: 112): a condition of 'integrated rural poverty' from which it is very difficult to extract oneself. The other elements were political powerlessness, physical weakness (ill health), isolation and income poverty.

Building on this pioneering work, and that of Blaikie and Brookfield (1987), the framework here uses 'vulnerability' to denote the degree to which one's social status (e.g. culturally and socially constructed in terms of roles, responsibilities, rights, duties and expectations concerning behaviour) influences differential impact by natural hazards and the social processes which led there and maintain that status. Thus, depending on the society and situation, social characteristics such as gender, age, physical and mental health status, occupation, marital status, sexuality, race, ethnicity, religion and immigration status may have a bearing on potential loss, injury or death in the face of hazards – or resources made to be hazards – and the prospects and processes for changing that situation.

Many other definitions of vulnerability exist (e.g. Wisner *et al.* 2004: 13–16; IPCC 2007a; Naudé *et al.* 2009b; Gaillard *et al.* 2010; UNISDR 2009c), which interpret the word from different points of view: social, economic, public health, climate change, amongst other sectors and topics. Nevertheless, the definition used in framing the *Handbook* overlaps sufficiently with others that its 'family resemblance' should facilitate mutual comprehension across disciplines.

The progression of vulnerability

Root causes

Figure 3.2 turns Figure 3.1 on its side and expands the short-hand descriptions in Figure 3.1's eight boxes. In the search for the answer to the question, 'why?', Figure 3.2 adds to the framework's nudging reminders a series of 'root causes' about which one should enquire. These overlap with the 'political and economic structures' mentioned in Figure 3.1 (box 8), but suggest that one should trace the origins of such structures historically and explain the ideological and cultural assumptions that give those structures perceived legitimacy. A number of the *Handbook* chapters in Part I deal with these root causes: chapters dealing with political power, history, religion and culture.

The progression of vulnerability

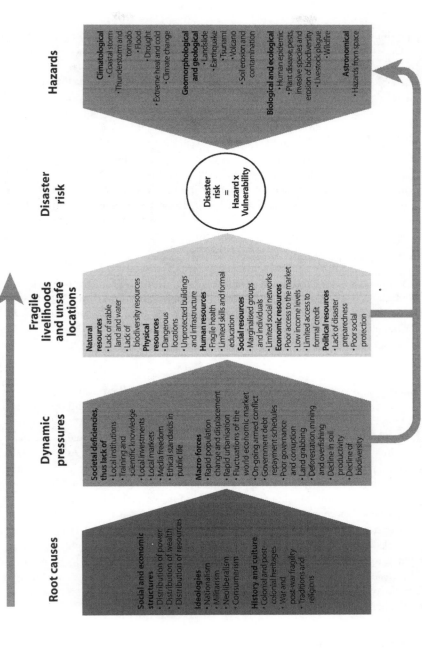

Figure 3.2 The progression of vulnerability

Disaster risk

A formal definition of disaster risk is contained at the core of Figure 3.2:

$$DR = H \times V.$$

As with everything else in the versions of the framework presented, this is meant as a mnemonic device, not necessarily a mathematical equation to be used for calculation. It is a reminder to enquire about both vulnerability (V) and hazard (H) – correcting a long-standing bias toward physicalist or hazard-focused research and policy. The definition has the appearance of a mathematical function, which has led to some confusion over the years since *At Risk* first appeared (Blaikie *et al.* 1994).

Disaster risk is a function of the magnitude, potential occurrence, frequency, speed of onset and spatial extent of a potentially harmful natural event or process (the 'hazard'). It is also a function of people's susceptibility to loss, injury or death. Also, some people are better placed to recover quickly from such losses than others. Taken together, susceptibility to harm and the process that creates and maintains that susceptibility to harm can be called 'vulnerability'. Vulnerability, in turn, may be counteracted either by individual and local capacity for protective action (C) or by protective actions carried out by larger entities such as government (M, which stands for mitigation and prevention). So, in fact, $DR = H \times V$ can be expanded and rewritten as the following mnemonic (Wisner *et al.* 2004):

$$DR = H \times [(V/C) - M],$$

where DR is disaster risk, V stands for vulnerability, C represents capacity for personal protection and M symbolises larger-scale risk mitigation by preventive action and social protection. Numerous definitions exist for all these terms, which are frequently ambiguous leading to contentious discussions and frequently disparate understandings.

Hazards

Many of the chapters with fine-grained focus in Part II of the *Handbook* deal with specific natural processes and events that are potentially harmful to people and their assets and disruptive of their activities. These are listed on the right-hand side of Figure 3.2.

As in Figure 3.1, hazards are not entirely free of human influence, although some hazards are difficult to influence at a large scale such as space weather and earthquakes. At smaller scales, the electromagnetic pulse experienced by specific components in satellites can be altered by shielding. Similarly, the peak ground acceleration experienced by a given building in an earthquake is affected by how the building and the land around the building are constructed.

The arrow labelled 'accentuation' at the bottom of Figure 3.2 is meant to suggest the influence of human activities, just as similar arrows showed that in Figure 3.1. One weakness of most frameworks is that they either focus on the human (left) side of Figure 3.2, making only a slight reference to natural hazards and the physical environment, or they focus mostly on the physical (right) side of Figure 3.2, giving only a nod to or brief treatment of the many underlying risk factors on the human side (for example Smith and Petley 2009; Turner *et al.* 2003; UNISDR 2004: 15; Burton *et al.* 1993).

Unsafe livelihoods and locations

A livelihood is an arrangement for making a living. Chambers and Conway (1991: 1) define sustainable livelihoods as follows: 'a livelihood comprises people, their capabilities and their means of living, including food, income and assets'. Livelihoods thus encompass all resources required to sustain durably people's basic needs. Basic needs refer to food, shelter, clothing, cultural values and social relationships.

The livelihoods concept is often associated with that of sustainability, especially in hazardous environments. Chambers and Conway (1991: 1) emphasise that 'a livelihood is environmentally sustainable when it maintains and enhances the local and global assets on which livelihoods depend, and has net beneficial effects on other livelihoods. A livelihood is socially sustainable which can cope with and recover from stress and shocks, and provide for future generations'. Natural hazards, as well as economic shocks and social disruption, may thus be of serious threat to people's livelihoods.

The concept of sustainability implies that basic needs are met on both an everyday basis and in the long term. It is therefore essential to consider everyday life when dealing with both the sustainability of people's livelihoods and their vulnerability to natural hazards. Social and economic threats to daily needs, especially to food security, are almost always more pressing than threats from rare or seasonal natural hazards. This is particularly true when hazards turn to resources, such as for fertile volcanic lands or flood plains, or dumpsites for those who have no alternative than scavenging rubbish to make a living (Gaillard et al. 2009). In common with various versions of the livelihood approach in development studies and practice, Figure 3.2 lists six categories of resources that are vital to dealing with hazard events as well as being central to sustainable livelihoods: natural, physical, human, social, economic and political.

Dynamic pressures

Figure 3.2 also shows a set of macro processes that 'transmit' the historic weight of root causes along the 'chain of explanation', as an intermediary between them and fragile livelihoods and unsafe locations and conditions. The list is not meant to be exhaustive but indicative of the large-scale, external drivers of significance. They fall into two parts.

First, there are societal deficiencies. The United Nations' (UN) road map for decreasing disaster impacts (UNISDR 2005b), on which a number of Handbook chapters have commented, lists numerous actions that national governments should undertake. Since 168 governments signed this Hyogo Framework for Action in 2005, one would expect that they would be getting on with the job. Yet serious lack of positive government action persists in building local institutions for disaster risk reduction (DRR), training and scientific research into hazards and disaster risk, credit and investment in households' economic resources, provision and maintenance of farm-to-market roads and other transport and market infrastructure, as well as failure of attention to such prerequisites for trust in government and good two-way communication as media freedom and ethical standards in public life.

Another dynamic pressure includes socio-political and economic processes as well as negative trends in conservation of land, water and biosphere. Many of these dynamic pressures are taken up in detail in Part I Handbook chapters, including rapid and unplanned urbanisation, population change and displacement, global economic conditions and violent armed conflicts.

Continued erosion of biodiversity and devastation of ecosystems, including deforestation, is a major factor. Globally, extinctions are proceeding at a rate that threatens keystone or 'backbone' species in many ecosystems, a fact acknowledged in 2010 when 193 countries met in Nagoya, Japan and negotiated a new Convention on Biodiversity (www.cbd.int), and by the International Union for Conservation of Nature (IUCN) list of threatened and endangered species (IUCN 2010). Soil erosion and fertility decline, waste and contamination of increasingly scarce fresh water, pollution of coastal waters and over-fishing all also weaken already fragile livelihoods.

Figure 3.2 is an elaboration and adaptation of the 'progression of vulnerability' framework developed in the course of two editions of *At Risk* (Blaikie *et al.* 1994: 21–45; Wisner *et al.* 2004: 49–86). However, it has a much longer history, as recounted in Box 3.1.

Box 3.1 Origin of the pressure and release framework

Ian Davis
Senior Professor in Disaster Risk Management, Lund University, and Visiting Professor in Cranfield, Oxford Brookes and Kyoto Universities

During 1976, while writing *Shelter after Disaster* (Davis 1978), I covered both post-disaster shelter needs and the vulnerable conditions that caused buildings to collapse under earthquake loading. I had been impressed with a paper in *Nature*, 'Taking the naturalness out of disasters' (O'Keefe *et al.* 1976). Its purpose was to define the characteristics of vulnerability. The paper seemed particularly relevant since it reinforced evidence that was coming from Guatemala following a devastating earthquake in February 1976. The earthquake had been described as a 'classquake' on account of the selective impact on poor families. I was present a few days after the quake and walked through largely undamaged streets of upper and middle class neighbourhoods in Guatemala City, then to arrive suddenly at one of the precipitous ravines that crisscrossed the city. There poor families had illegally perched their homes and suffered appalling casualties and damage from landslides and building collapse.

Considering all this, I began to sketch in an attempt to visualise vulnerability and hazard as a pair of converging arrows, meeting in a disaster, or meeting where there was the potential for a disaster. I decided to include the diagram in my book, and the rather crude first version of the 'crunch' diagram appeared. I placed 'types of hazard' on one side of the 'vulnerability to disaster interface' and on the other, I listed six 'dangerous conditions'. These were selected to relate to the construction of buildings, the theme of the book.

Later, while developing a series of slide presentations for the United Nations Centre for Human Settlements (UNCHS) (now UN-HABITAT) to accompany the *Shelter after Disaster—Guidelines for Assistance* (1982), as published by the Office of the United Nations Disaster Relief Co-ordinator, I used the same diagram. However, it suddenly dawned on me that the dangerous conditions needed to be further unpacked into root causes of vulnerability, leading to pressures that in turn led to dangerous (or unsafe) conditions.

In the late 1980s, I began to work with Piers Blaikie, Terry Cannon and Ben Wisner in the early development of *At Risk*. On one occasion (probably in 1992), I recall drawing the diagram on a blackboard during a discussion. My co-authors were enthusiastic to adopt the diagram. Within thirty minutes, one of my co-authors (I forget which) suggested we go further and develop a reverse of the pressure or 'crunch' diagram with the arrows pointing outwards, to symbolise the release of pressures away from a disaster. In that way we could introduce 'capacity' into the model. So we also used this 'release' diagram throughout *At Risk* in both its first and second editions. The twin diagrams became known as the PAR (Pressure and Release) framework.

The triangle of vulnerability

Figure 3.3 probes more deeply into the nature of vulnerability by highlighting access and marginalisation. All of the elements of the framework are here, but they are rearranged to provide such emphasis.

Root causes and dynamic pressures are rolled into the three large circles at the triangle's three apexes. These serve as structural constraints which determine the degree and reliability of different people's access to the six sets of resources familiar from earlier diagrams, depicted in the inner, smaller circles. Many chapters in this *Handbook* illustrate that these resources are often available locally, but many people are unable to access them because of their age, gender, caste, ethnic and religious affiliation, and physical ability or because of poor governance, patronage politics and inequitable distribution of wealth. Part II of the *Handbook* covers many of these as fine-grained processes, complementing those that appear in Part I.

Poor and unstable access to resources results in marginalisation in both daily life and in facing natural hazards. Access to resources defines how resistant, diverse and sustainable people's livelihoods are on an everyday basis and how much they are able to secure a decent daily living. The extent, resistance and stability of livelihoods also determine people's ability to avoid harm when dealing with natural hazards (Gaillard *et al.* 2009).

Vulnerability, located at the centre of the triangle, ultimately reflects people's position within society (not only poverty) as a consequence of their ability or inability to secure access to a large, resistant and sustainable set of resources. The triangle further shows that root causes of vulnerability are interacting and so are the resources that enable people to make a daily living and protect themselves in facing natural hazards.

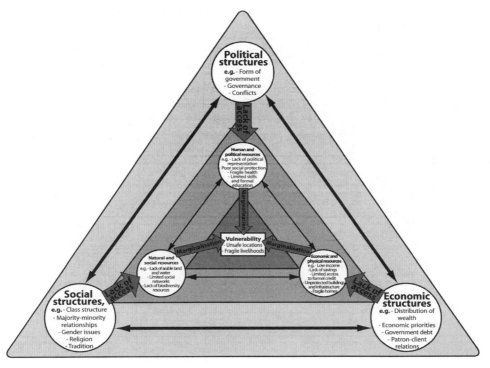

Figure 3.3 Triangle of vulnerability

Use of an access framework can become complex (Wisner *et al.* 2004: 98–112); however, for purposes of the *Handbook*, full detail is not always required and is not provided. While most of the access categories in the inner circles of Figure 3.3 are intuitive, 'political resources' perhaps needs a few words of explanation. Research has shown that people who are spatially isolated, living from the base of a poor or depleted ecological endowment and poor in terms of financial and livelihood resources also tend to have limited 'voice' or access to administrative officials and politicians. They are simultaneously spatially, ecologically, socially, economically and politically marginalised (Gaillard *et al.* 2010; Wisner 2010d).

The circle of capacities

Capacities refer to the resources and assets that people possess to resist, cope with and recover from disaster shocks they experience (Wisner *et al.* 2004; Gaillard *et al.* 2010). The concept of capacity also encompasses the ability to either use or access needed resources, and thus goes beyond the mere availability of these resources (Kuban and MacKenzie-Carey 2001).

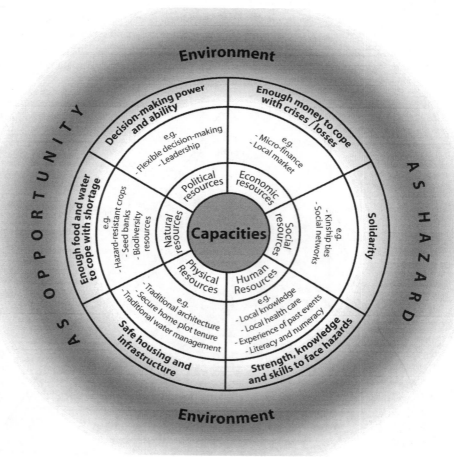

Figure 3.4 Circle of capacities

Capacities are not at the opposite end of vulnerability on a single, linear spectrum. Most people, including marginalised and vulnerable people, have capacities. These capacities fall within the same typology of resources used for assessing livelihoods and vulnerability, i.e. natural, physical, human, economic, social and political, as shown in Figure 3.4. People employ these in order to prevent, resist, cope with and recover from challenges wrought by natural hazards. No one is a helpless victim, nor should they be labelled only as such.

Figure 3.4 emphasises the fact that capacities are often, yet not exclusively, rooted in resources that are endogenous to the community facing hazards. By contrast, Figure 3.3 (the triangle of vulnerability) emphasises structural constraints on access, which are largely, yet again not exclusively, exogenous to the community, such as inequitable distribution of wealth and resources within the society, market forces, political systems and governance. People have more control over capacities, but they often have little purchase on external factors that create vulnerability (Gaillard *et al.* 2010). In practice, especially for those working at the community level (see Chapter 59), it is therefore often easier to enhance capacities than to reduce vulnerability. Capacities must therefore be recognised and used.

Enhancing capacities encompasses activities, often at the household or community level, which strengthen people's strategies to face the occurrence of natural hazards, such as agreeing on warning signals, infrastructure and livestock protection, meeting points, planning evacuation routes, vehicles and shelters, and preparing resources to cope with the disruption of daily life.

Marginalisation, disaster and failed recovery

In Figure 3.5 our framework takes yet another geometrical shape, but the basic concepts and logic behind it remain the same. The six livelihood and locational resources familiar from Figure 3.2 recur. Each descending wedge hits a crisis point, a disjuncture labelled 'disaster', as if pulled by the weight of the various aspects of vulnerability described in each case.

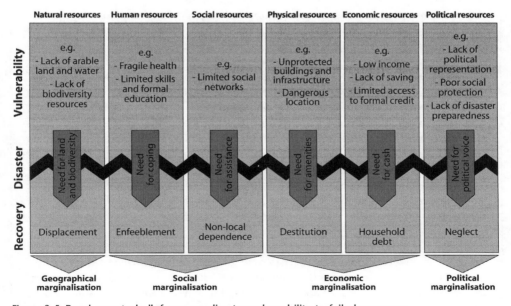

Figure 3.5 Road map to hell: from pre-disaster vulnerability to failed recovery

In the aftermath of a disaster, the affected have a series of needs indicated by the small central arrows. Failure to satisfy these needs leads to long delays or unsatisfactory 'recovery' and further marginalisation. This can become a vicious cycle (Chambers 1983; Susman *et al.* 1983; Gaillard and Cadag 2009). Instead of recovery, one finds displacement, continued weakening of human resources, perpetual dependence on anonymous, public or international charity, homelessness, indebtedness and unchallenged political neglect.

Those marginalised and vulnerable in facing hazards are often also those who struggle to recover in the aftermath of a disaster. Disasters increase the needs of resource-less survivors (Figure 3.5). Squatters settling in hazard-stricken areas need land to which they can relocate, although this is often a painful experience. People with limited skills and fragile health are often weakened when faced with changing social and economic environments. Survivors with fragile social ties and limited social networks need external assistance and thus increase their dependence on others. Those with poor economic resources often have to resort to high-interest, informal loans to provide for their need for cash to recover. They also often lose their sparse physical assets, including their house, thus leading to further destitution. Meanwhile, increasing needs of the most marginalised in the aftermath of disasters are frequently neglected by the authorities, for whom those survivors are often invisible. Disasters thus often further marginalise those who were already living at the margin before the events.

Box 3.2 Defining disaster and recovery

Disaster

Academics have spent a good deal of time debating the definition of 'disaster' (e.g. Quarantelli 1998; Perry and Quarantelli 2005). From a public administration and legal point of view, a distinction between disaster and emergency may make the difference in eligibility for outside assistance, and formal declarations of disaster by government authorities may have implications for some kinds of insurance.

Nevertheless, for purposes of this *Handbook* the editors suggest a common, simple definition: a situation involving a natural hazard which has consequences in terms of damage, livelihoods/economic disruption and/or casualties that are too great for the affected area and people to deal with properly on their own.

This situation is seen particularly for the uncountable small and 'neglected disasters' or 'invisible disasters' (Wisner and Gaillard 2009). Small, isolated communities often obtain no external assistance – or even recognition that a disaster has occurred there. The affected communities are forced to overcome the situation on their own, usually with unnecessary, extensive suffering. Such places could easily be given support to help themselves avert a disaster before it happens, or a little external intervention afterwards would avoid a small situation becoming a major disaster, yet that rarely happens.

In a sense, the true 'disaster' might be the failure to avoid suffering or to help when suffering occurs. For pragmatic purposes, rather than deep philosophical arguments, the definition of 'disaster' given above suffices.

Recovery

Recovery is an even more controversial and difficult term to define. Many policy-makers, donors, practitioners and researchers have broken it down into 'stages', which try to indicate clearly recognisable boundaries, but which cannot be partitioned so easily. The United Nations Development Programme's (UNDP) concept of 'early recovery' is an example (UNDP 2010a). Others try to distinguish between 'relief', 'reconstruction' and 'rehabilitation'. The *Handbook*'s chapter on shelter and reconstruction (see Chapter 46) will take up some of these points.

The term 'recovery' is cloudier for other reasons. First, references to restoration of normality or normality may be of little use if 'normal' was the situation of vulnerability for some of the population now affected. Returning them to the pre-disaster status quo will almost assure that they will be affected again by another disaster in the future.

Second, recovery has many aspects at a variety of scales. Many of these are covered by chapters in Part III of the *Handbook*. A return to fiscal stability in a country affected by a large disaster is one aim, which might or might not avert a future disaster. Revitalisation and strengthening of the livelihoods of households affected is another aim, at another scale, which has the potential for reducing vulnerability, if enacted appropriately. Beside economic recovery, there must be recovery and improvement of the 'life space' of home and community, along with the social and public space of the location, especially its treasured landmarks. Re-establishment of public services and infrastructure, transport and communications are other aspects of recovery. Finally, there must be psycho-social recovery, through which the mental trauma is healed or at least addressed to the extent that the affected people (including children) are not or do not become dysfunctional.

The progression of safety

Figure 3.6 shows how policy and practice have sometimes strengthened livelihoods and made locations and conditions safer, confronted and countered dynamic pressures and, occasionally, even addressed some of the root causes of vulnerability through legislation and the process of 'peace and reconciliation'. This figure is adapted from one with the same title that is used several times in *At Risk* to frame actual situations, such as in Bangladesh and Mozambique, and appears in a general form at the end of the book (Wisner *et al.* 2004: 344; Blaikie *et al.* 1994: 220).

The reader will note that the large boxes that converged on 'disaster' in Figure 3.2 move outward in Figure 3.6. This represents the impact of public policy and investment, good governance at multiple scales, increased awareness and preparedness, mobilisation and organisation at the village and street level, along with measures directed at some of the hazards.

Again, an arrow channels human action at various scales back to the box containing 'hazards'. One must remember the lesson of Figure 3.1: the natural environment is usually neither hazard nor resource until human action makes it one or the other (or both: hazard for some, opportunity for others – or hazard and resource at the same time, such as a flood that fertilises farmland while damaging the poorly constructed farm buildings). Thus, reforestation, conservation of biodiversity, water and soil, land use and forestry policy (and enforcement) can all help to mitigate at least some hazards (drought, flood and landslide, for example).

Prevention can be active or passive. Active prevention includes all efforts that aim at avoiding the hazardous phenomenon occurring, such as dredging rivers, deviating winds to prevent the accumulation of snow in the path of avalanches or triggering artificial rains in the event of insufficient rainfall. Active prevention also refers to hazards that are triggered under control, such as avalanches triggered early in the morning of ski days by using dynamite.

Passive prevention encompasses all actions that do not prevent the phenomenon from occurring, but rather focus on reducing its spatial or temporal extent, such as in the case of levees and dikes along rivers. Passive prevention also includes hazard monitoring and mapping of hazard-prone areas. Attempts at prevention can inadvertently create or exacerbate hazards,

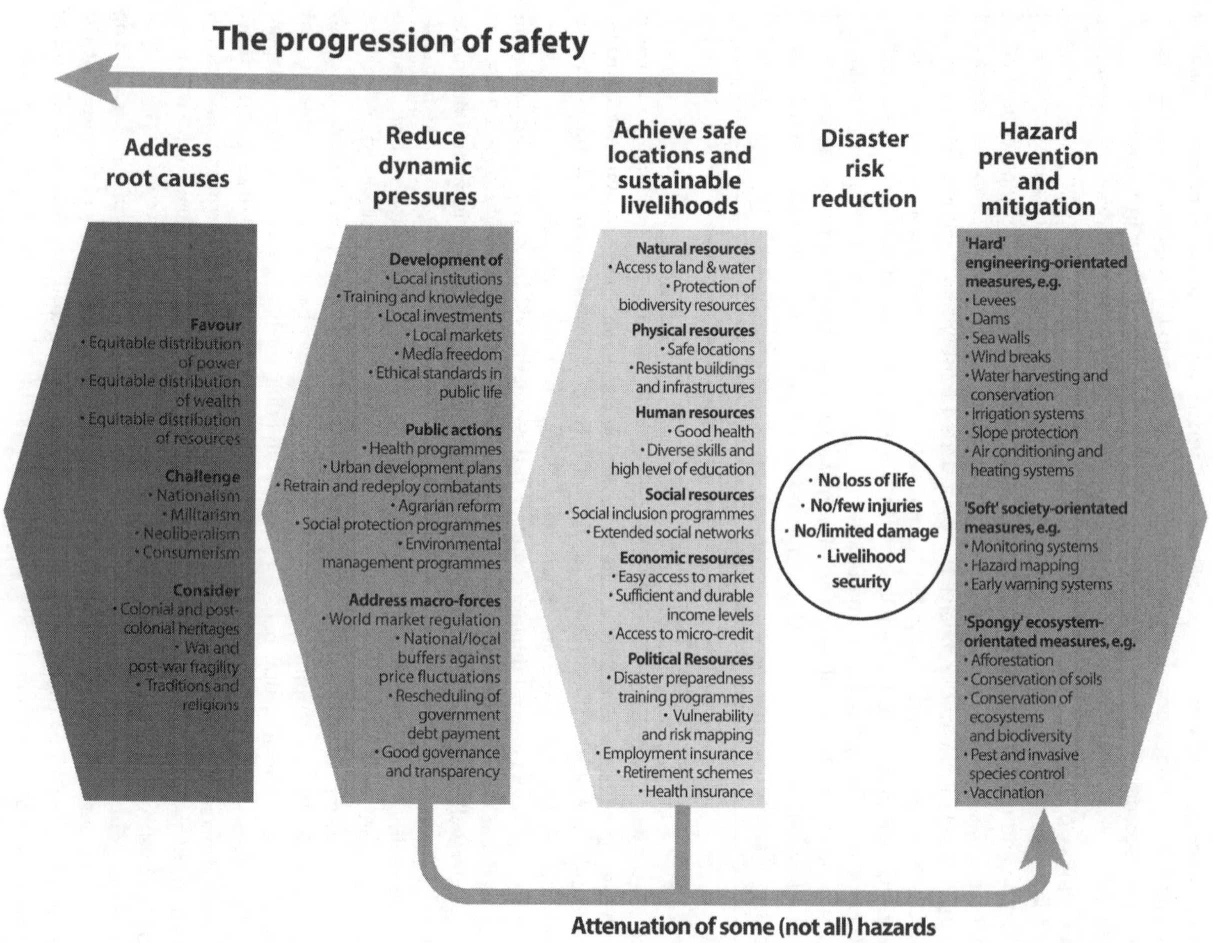

The progression of safety

Address root causes

Favour
- Equitable distribution of power
- Equitable distribution of wealth
- Equitable distribution of resources

Challenge
- Nationalism
- Militarism
- Neoliberalism
- Consumerism

Consider
- Colonial and post-colonial heritages
- War and post-war fragility
- Traditions and religions

Reduce dynamic pressures

Development of
- Local institutions
- Training and knowledge
- Local investments
- Local markets
- Media freedom
- Ethical standards in public life

Public actions
- Health programmes
- Urban development plans
- Retrain and redeploy combatants
- Agrarian reform
- Social protection programmes
- Environmental management programmes

Address macro-forces
- World market regulation
- National/local buffers against price fluctuations
- Rescheduling of government debt payment
- Good governance and transparency

Achieve safe locations and sustainable livelihoods

Natural resources
- Access to land & water
- Protection of biodiversity resources

Physical resources
- Safe locations
- Resistant buildings and infrastructures

Human resources
- Good health
- Diverse skills and high level of education

Social resources
- Social inclusion programmes
- Extended social networks

Economic resources
- Easy access to market
- Sufficient and durable income levels
- Access to micro-credit

Political Resources
- Disaster preparedness training programmes
- Vulnerability and risk mapping
- Employment insurance
- Retirement schemes
- Health insurance

Disaster risk reduction

- No loss of life
- No/few injuries
- No/limited damage
- Livelihood security

Hazard prevention and mitigation

'Hard' engineering-orientated measures, e.g.
- Levees
- Dams
- Sea walls
- Wind breaks
- Water harvesting and conservation
- Irrigation systems
- Slope protection
- Air conditioning and heating systems

'Soft' society-orientated measures, e.g.
- Monitoring systems
- Hazard mapping
- Early warning systems

'Spongy' ecosystem-orientated measures, e.g.
- Afforestation
- Conservation of soils
- Conservation of ecosystems and biodiversity
- Pest and invasive species control
- Vaccination

Attenuation of some (not all) hazards

Figure 3.6 Progression of safety

such as using levees or dams to stop water spreading out in slowly rising floods, which leads to the possibility of a flash flood from levee or dam failure.

Mitigation can be direct or indirect. Direct mitigation activities address symptoms of vulnerability in the 'progression of vulnerability' diagram (Figure 3.2). Examples include building sturdy houses to face cyclone or earthquake hazards, and facilitating the exchange of hazard-related knowledge. Indirect mitigation addresses the root causes of vulnerability and is geared towards better access to sustainable resources.

Conclusions

Understanding hazards and disaster risk draws on accumulated knowledge from many areas of human experience and enquiry. Equally broad knowledge is required for evidence-based and disaster risk-informed decision-making, practice and policy. This *Handbook*'s chapters contain a sampling of that wide range of data, information and knowledge – and, it is also hoped, at least a small amount of wisdom. The first three are the building blocks that allow one to attain the last (IFRC 2005: 13).

Since none of us in the twenty-first century are in the fortunate position of being brilliant crafter–artist–counsellor–scholars of the Renaissance, we need a way of organising this large array of information, bridging disciplinary boundaries and, above all, reminding ourselves of relevant questions outside our individual areas of expertise and experience. The framework presented is designed to do that.

The disciplines and professions studying disaster are numerous, as are the kinds of knowledge created, the language of these disciplines and criteria for relevance and confidence (no one should claim 'truth'). In addition, the processes, events and phenomena that comprise the content of 'hazard', 'vulnerability', 'capacity', 'risk' and 'disaster' are themselves complex, highly connected, cross time and space scales, and are contextual.

A 'framework', then, is only a first step in addressing the DRR process, which is a long journey from principles to data to wisdom – and then the reverse. Researchers, policy-makers and practitioners will find this rough path easier to travel if competent authorities take the following recommendations to heart:

- University authorities should encourage interdisciplinarity, non-disciplinarity and exploration among young scientists. It is wasteful to require narrow PhDs and discipline-focused publication until a person attains tenure or the equivalent.
- Researchers and practitioners should exchange more knowledge and experience – a truism, perhaps, but nevertheless important for generating and sharing wisdom. Moreover, arrangements should be made and funding provided for researchers to sojourn for periods in humanitarian and other disaster organisations and government, and vice versa.
- International, national and sub-national authorities should open up their planning institutions and processes to a full range of knowledge-bearers, including those from the arts and humanities, social sciences and lay people who represent communities and themselves. Too often planning for disaster is done by economists, engineers, some natural scientists, military and police experts alone.

These starting points for recommendations anticipate many that appear in the chapters throughout the *Handbook*. They will also be developed further in the Conclusion (see Chapter 65).

There is a long journey yet to go. This *Handbook* can be one stepping stone.

Politics, history and power

4

Historical concepts of disaster and risk

Greg Bankoff

DEPARTMENT OF HISTORY, UNIVERSITY OF HULL, UK

Asking why disasters happen depends much upon one's disciplinary perspective, but realising how they occur is always a question of history. Managing risk may be a matter of dealing with the political, social, economic and environmental dimensions of people and hazards, but understanding the particular nature of vulnerability and resilience in any situation is quintessentially an historical question: To recognise what makes people, households, communities and societies vulnerable or resilient in the present, you need to appreciate what made them that way over time. The old adage says that it is not earthquakes that kill people but buildings; actually it is not buildings so much as where they are situated, what they are made from, how they are built and why people use them that way that proves so fatal. Some people refer to the shared set of attitudes, values, goals and practices that inform all human activities as culture, and recently much more consideration has been accorded the role of culture in disasters; however, history underlies culture too, providing both its origins and the measure of its change.

A temporal perspective

As Susan Cutter writes, 'The temporal context of vulnerability is crucial yet the temporal dimension remains one of the least studied aspects of vulnerability' (Cutter 2006: 76–77) – not that history, or at least an awareness of the importance of time, has gone completely unnoticed in the existing literature. The importance of the temporal dimension is recognised but, with few exceptions, is largely 'de-historicised' in the sense that time is reduced to its component parts or at best applied to only the last few decades. In the Pressure and Release (PAR) model of disaster causality, the notion of history is implicit in the tripartite characterisation of the progression of vulnerability into root causes, dynamic pressures and unsafe conditions. In particular, root causes are depicted as 'temporally distant' (in past history), but in practice the discussion and examples refer to much more contemporary developments. The Access model, on the other hand, explains the significance of time in terms of frequency, the hour, season or speed of impact on affected populations (Wisner *et al.* 2004: 52–60, 106–9). These are all important factors but they are not historical per se. Kenneth Hewitt and Ian Burton, on the other hand, are primarily concerned

---

Box 4.1 Historical construction of risks in Mexico

Virginia García-Acosta
Centro de Investigaciones y Estudios Superiores en Antropología Social, Mexico

Mexican historiography, going back as far as the fifteenth century, includes a great amount of informative or descriptive material coming from archives, bibliographical sources (codices, chronicles, monographs, personal letters, old catalogues), newspapers (the printing of which began in the eighteenth century) and iconographical sources. Risks and disasters can be understood as processes by using these historical documents as ethnographic sources of information.

Diachronic and comparative studies of hurricanes, earthquakes, floods and droughts that occurred throughout Mexican history have provided important baselines. Reports of disasters associated with any of those hazards from the sixteenth to the twenty-first centuries show that there has not been more pluvial precipitation or more presence of hurricanes or higher intensity earthquakes, but that the accumulated risks and vulnerability and a growing lack of governance have led to disasters of greater magnitude. Two examples stemming from research carried out in this field follow.

Mexico City was founded in 1325 at the centre of the ancient lakes that made up a closed basin surrounded by mountains. In the colonial era (1521–1810), floods in this city were especially violent, largely because the new authorities, knowing practically nothing of dangers posed by floods and their consequent devastation, had ceased to maintain the hydraulic control works system that prevailed in the pre-Hispanic era. The 1555, 1580, 1604 and 1607 floods and the 1629 flood in particular, which left the city inundated for nine whole years, demonstrate this. Analyses of the socio-economic, cultural and political context before and after the phenomenon under which those events occurred have shown that these particular conditions, rather than the heavy rains, caused the terrible floods.

The Mexican republic is located in a highly seismic region. Since the pre-Hispanic era earthquakes have been recorded in pictographic codices (known as 'painted books'), due to their systematic occurrence. There are wonderful catalogues that go back as far as 900 BCE, which allow comparisons with similar phenomena in the same locations. In April 1845 and in September 1985, 8.1-magnitude earthquakes occurred, with the effects felt mainly in Mexico City. In this increasingly hazardous urban concentration, 17 people were reported as injured or dead in 1845; 10,000, according to official reports, in 1985. As Rousseau wrote to Voltaire in his 1756 letter in reference to the 1755 Lisbon earthquake, 'most of our physical ills are still our own work'.

Historical and anthropological disaster research has demonstrated that if disasters have become more frequent over time, it is not because there are more natural hazards, but rather that over time communities and societies have become more risky, more vulnerable. Nevertheless, research on disasters in Mexico has also shown that societies have not been simply passive actors in the face of them. People have historically formulated cultural constructions, adaptive strategies, to confront real and potential disasters. Discovering these is a new challenge for risk and disaster research.

---

with exploring what they call the 'hazardousness of a place', a vulnerability determined by the totality of interactions between the human-induced and biophysical factors in a given location. In particular, they recognise the importance of focusing on an all-hazards approach rather than on a single extreme event. In their study of London, Ontario, understanding the local context of hazardousness necessitates relating the predictability, frequency, speed of onset, intensity, extent of impact and duration of various hazards with traditional land use patterns, the state of civic preparedness, forms of native adaptation and the level of community resources (Hewitt and Burton 1971). Here, at least, is an implicit recognition of the importance of history, but time for most other scholars concerned with social vulnerability is still a very ill-defined notion, at best only implicitly acknowledged. In Jörn Birkmann's review of such literature, a small measure of recognition is accorded the 'revealed vulnerabilities in the past' as constituting one of the internal factors that lie at the core of what he terms 'intrinsic vulnerability' (Birkmann 2006: 14–17).

Even among scholars who recognise how the past generates social vulnerability, few have actually tried to apply it. There are two important exceptions, however, that give some intimation of the potential of this approach. The first is the seminal essay on the 1970 Yungay earthquake in Peru by the anthropologist Anthony Oliver-Smith. He is one of the few scholars to directly address the question of time as a primary unit of analysis that acts to either augment or diminish risk. He refers to this approach as an ecological perspective, one that focuses on the effectiveness of societal adaptation to its total environment. Culture and social organisation are affected by people's experience of their environment, but equally that environment is deeply shaped by the histories of those societies. To Oliver-Smith, the question of time is a crucial aspect in any consideration of vulnerability. As both an event and a process, it is people's adaptation (or mal-adaptation) to their total environment over time that determines their social vulnerability. In the case of the area affected in the 1970 earthquake, the social vulnerability of the region was as much a product of its historical underdevelopment, traceable back to the socio-economic and political consequences of the Spanish conquest, as it was of the physical hazard itself (Oliver-Smith 1994).

The other approach with an overtly historical dimension is James Lewis's notion of compound or derivative vulnerabilities. Vulnerabilities are compound in that the impact of an extreme event can only be understood as one in a line of intermittent and recurring disasters of all types that interrelate with the totality of socio-economic, cultural, political and natural factors, and derivative in the sense that they have accrued from the interaction of these factors over time. Lewis, an architect by training and focused primarily on the unique nature of island states, links hazardousness and place firmly together showing how the land's prior occupants influence what is done with it subsequently and so help create present-day vulnerabilities (Lewis 2009: 7–9).

The historian's contribution

What contribution can the historian make to a better appreciation of disasters? An important initial step was taken by David Alexander in defining the relevance of history to disaster research, stressing its impact, relationship to formative societal processes, lessons and message (Alexander 2000: 106). Alexander's emphasis, however, is more on how disasters lead to a better understanding of history, rather than on how history leads to a better understanding of disasters. Regarding history more from this latter perspective, there are at least six key areas, none of them the exclusive domain of the historian, wherein a better appreciation of the past can suggest how social vulnerability is generated and how disaster risk reduction (DRR) and management can be made more effective:

- A quantitative analysis of past disasters constitutes an invaluable aid in helping to model the frequency and intensity of present hazards and locating potential areas of risk.

- A qualitative assessment of how people coped with hazards and disasters in the past (i.e. their resilience) challenges our notions that contemporary ways are always better.
- A re-examination of 'court' histories or official accounts may uncover how people have been unfairly blamed for actions taken, not taken or still taken in disasters.
- Adopting a comparative historical methodology encourages a consideration of cross-cultural approaches, particularly of non-Western societies and their capacities.
- A fuller appreciation of the past shows how disasters are processes as well as events, indicating a continuum that stretches from the past through the present to the future.
- A greater recognition that disasters are also agents of change provides societies with 'windows of opportunity' to make needed reforms.

Using suitable examples of past disasters to illustrate these points, this chapter discusses how history and culture contribute to what can be termed the 'inherent vulnerability' of any society and why it is important that emergency and DRR planners and managers recognise what that constitutes before a disaster happens.

Patterns of the past

Disasters have two historical trajectories, one 'natural' and the other societal. They are both 'historical' in the sense that both force change over time. The fact that the nature of hazards varies over the years is perhaps less immediately apparent than that societies do. As Paul Edwards (2002) notes, this is largely due to differences between human, historical and geophysical scales of time. Human time scales are set by our natural characteristics – days, years, decades, in which the nature of hazards often changes too slowly for most of us to notice. Disasters are events that exist primarily in human-experienced historical time. Yet on a geophysical or even long-term time scale over centuries or even millennia, natural hazards and the disasters they give rise to are a predictable and measurable property of nature (Edwards 2002: 194–95). Climate, too, alters not only in the long term, over hundreds of thousands of years, but also within much shorter time spans that have affected human societies within recorded history. Variations in mean average temperatures associated with the Late Medieval Optimum (1100–1400), Little Ice Age (1600–1800) and Modern Optimum (since 1800) had (and continue to have) effects on agriculture, human nutrition and population density, as well as having an important bearing on the intensity and frequency with which people were subject to extreme events at any given location on the Earth's surface (Lamb 1977). More recently, scientific evidence suggests a strong correlation between floods, droughts and related hazards and higher temperatures, the release of heat-retaining gases into the atmosphere and variations in precipitation levels around the world. Even the magnitude of hurricanes and typhoons in the Caribbean and western Pacific has been linked to rising sea surface temperatures (IPCC 2007a).

The instrumental record, however, from which such patterns of the past are revealed, does not stretch back more than a couple of centuries at best, and in many countries is even less extensive. The frequency of a hazard, whether it is considered a one-in-whatever-year event and upon which industry standards and insurance premiums are calculated, is based upon these records without any certainty that the data series represents any longer fluctuations. Only by employing the historical methodology can such records be checked for their reliability. Impressive data series now exist based on descriptive documentary (archival) and proxy doc-umentary (indirect) sources for certain world regions that shed light on what occurred in the pre-instrumental age. Much of this evidence has been collated into large data series such as EURO-CLIMHIST, which contains 600,000 entries on weather-related events in Europe between 750 and 1850 CE. Other projects, such as CLIWOC (Climatological Database for the World's Oceans), are reconstructing the atmospheric circulation over the oceans through analyses of ship captains' logs and remark books held in major European repositories. Much more remains to be done, especially for the non-Western world, with Chinese-language sources promising a particularly valuable resource. By extending the record for all kinds of seismic and climatic events much further back in time, such data constitute an invaluable aid in modelling the frequency and magnitude of past hazards, create more informed industry standards and can help identify hitherto unrealised areas of risk.

Coping historically

The past challenges our notions that contemporary ways are always better and that techniques and practices developed by peoples and communities centuries ago to cope with the hazards that beset them have no bearing on how to deal with such events in the present. Indeed, science and technology are often wrongly credited as the product of the modern, largely Western age. A prime example of this is architecture and the historical design and construction of buildings in seismically active areas. Far from a recent innovation, evidence of seismic engineering dates back at least to Ancient Greece and Rome, where the classical temple façade of columns constituted a segmental (multi-block) rocking system for re-centring the structure's axial load during violent ground movements (Pampanin 2008: 118). Earthquake-resistant architecture has evolved in many cultures: the carefully bonded corners and alternate rows of headers and stretchers characteristic of Inca buildings in South America; the *hatil* (reinforcing beam) 'seismic culture' of houses in Byzantine and Ottoman Turkey; and the 'earthquake baroque' style of extensive buttresses, low body structures and squat bell towers of churches in the Spanish Philippines (Bankoff 2007a; Homan 2004; Oliver-Smith 1994). In Kashmir, an area of high magnitude earthquakes, three- to five-storey structures dating back to at least the eighteenth century are still in use today. There are two traditional construction methods employing techniques that render buildings more ductile and allow them to sway without falling: *Taq* consists of load-bearing masonry piers and infill walls laced together by wooden 'runners' at each floor level and *Dhajji-Dewari* is a system of patchwork-quilt walls of braced timber frames with masonry infill. Such techniques remained virtually unchanged for centuries and well-maintained structures built after this fashion performed well in comparison with more modern methods of construction in the earthquakes of 1967 and 2005 (Langenbach 1989).

A particular feature of this seismic adaptation is how past cultures without any obvious direct contact apparently found similar engineering solutions to building in earthquake-prone areas. Construction techniques shared the sophisticated notion that structures needed to respond to seismic disturbances as units, and were based on a compartmentalised wooden framework with masonry or rubble infill, often braced by diagonal crossbeams, special 'x' bracing to counter lateral forces and limitations on height. Spanish colonial architecture in the Americas, the eighteenth-century *casa baraccata* of Calabria and nineteenth-century housing on the Greek Adriatic island of Lefkás have much in common with the structures in Kashmir (Porphyrios 1971; Tobriner 1983; Walker 2008). Even more startling is how revolutionary new designs in contemporary seismic engineering now favour the introduction of jointed ductile systems involving precast 'hybrid' frame and wall units able to accommodate rocking motions utilising much the same principles as those employed historically (Pampanin 2008: 117). As in housing, so in many other features of the built environment: a fuller understanding and appreciation of past technologies combined with modern science can have useful application in reducing vulnerability in the present.

Re-examining 'court' histories

If history is written by the 'victors', then calamity is surely blamed on the 'vanquished'. Accounts of past disasters are largely 'court' or official histories that glorify the role of some, usually at the expense of others, often leaving a distorted record of the past in terms of capacities, responsibilities and culpabilities. Minorities, whether of a religious or ethnic nature, fared particularly badly: the great fire of London in 1666 was blamed on the perfidy of Catholics and, indeed, a commemorative monument to that effect still stands; the fires that broke out following the great Kanto earthquake in Japan in 1923 were popularly attributed to the nefarious activities of Koreans, who were massacred in their thousands as a consequence. Such prejudices manifest themselves in many different forms. The fire that consumed San Francisco in 1906 was anything but a social equaliser, even if the homes of rich and poor, native-born and immigrant alike were lost to the flames. Critical decisions as to what to save, when to abandon and how to rebuild mirrored the city's social stratification by ethnicity, race and class (see Chapter 38). Thus fire fighters fought the blaze in the 'best' parts of town but forsook Chinatown to its fate. Servants were left to mind the homes of the affluent while their owners sought safety out of town (Henderson 2008). The army were cast as the 'heroes' of the piece, called in to maintain social order and prevent anarchy, while the general citizenry were purportedly reduced to an intoxicated mob of looters and arbitrarily shot (Solnit 2009: 34–48). Much the same plot was written in the case of Hurricane Katrina when it devastated New Orleans on 29 August 2005. The media initially presented the mainly Afro-American population left stranded in the evacuated city as the victims not only of a pitiless nature, but of an unjust social system. Such a script, however, did not fit well with official versions of US society and these people were swiftly recast as dangerous, dissolute and disease-ridden, a situation that required the deployment of the National Guard who, like their predecessors a century before, showed little reluctance in turning their weapons on fellow citizens (Elliott and Pais 2006).

The multiple voices of disaster are easily lost over time, leaving posterity with only the official account of what took place. Such narratives need to be handled with care as they inevitably obscure how some use such events to their own profit. How in the rebuilding of San Francisco, for example, insurance repayments and rebuilding permits led to the further segregation of the urban area by class and economic function, or how the reoccupation and reconstruction of New Orleans is even now changing the ethnic composition of that city, 'whitening' and 'gentrifying' its residents (Elliott and Pais 2006; Henderson 2008). The historian has an important role in highlighting these processes, not simply to expose past wrongs but to ensure that those charged with disaster management and development projects in the present are fully aware of how issues of social justice are relevant and play out in such events.

Comparative perspectives

Comparing one society with another across time is a central tenet of historical methodology; looking to see what other cultures or communities did or do to tackle similar problems is educational to both. Western technological know-how and scientific expertise, in fact, have their limits, and developed countries may have as much to learn about disaster preparedness, management and recovery from non-Western, developing countries as the latter do from the former. In fact, a country's response to natural hazard may depend more on its social and organisational practices than on its wealth or resources, a realisation that until recently was sorely lacking in most discussions of disaster management. International recognition that all communities are resilient and that their capacities have relevance to DRR was finally accorded by member states of the

United Nations (UN) when they adopted the Hyogo Framework for Action in January 2005. The plan's emphasis on ensuring that DRR was both a national and local priority that utilised knowledge, innovation and education to build 'a culture of safety and resilience at all levels' was indicative of the need to think comparatively through time and across cultures (UNISDR 2005b). Just as the past has its own 'lesson' to disclose to the present, so communities and societies can learn from each other's experiences.

Qing China is one such historical example, with its empire-wide system of granaries that collected surplus grain after the harvest and released it to provide relief in times of shortage (Will and Wong 1991). A more contemporary case is Cuba. Through public education, national training exercises, a comprehensive early-warning system, an integrated civil defence structure and firm government leadership, Cuba is better able to protect its citizens and resources than most other states in the world. In particular, authorities are able to carry out large-scale evacuations in an orderly and timely manner without extensive loss of life, as when 750,000 and 1.5 million people were respectively removed from low-lying areas in the paths of Hurricane Michelle in 2001 and Hurricane Dennis in 2005 (Aguirre 2005: 66; Sims and Vogelmann 2002: 397). Hurricane Jeanne, which cut a swath of destruction through the Caribbean in 2004 leaving over 3,000 dead in neighbouring Haiti, passed without loss of life over Cuba (Bermejo 2006: 14). Cuba now serves as a model for the UN, and its government has even organised special medical brigades to provide assistance overseas in the case of disasters (Sims and Vogelmann 2002: 396). The Cuban model, however, has had no lack of critics who, while grudgingly acknowledging its effectiveness in certain areas of disaster preparedness and management, maintain that it depends upon an extensive system of social controls unrealisable in more 'democratic' societies (Aguirre 2005).

Nor is this capacity to face natural hazards only to be found at a state level; communities, too, are often largely self-reliant, dependent on their own resources in dealing with the hazards that confront them, as the government is often incapable of addressing the consequences of disasters adequately and there are only limited technological solutions to what are increasingly complex issues. In the Philippines, for instance, one of the most hazard-prone land masses in the world, there is a long history of a variety of formal and informal associations and networks committed to individual and community welfare that stretch back to 1565, when written records began (Bankoff 2007b). It is manifested nowadays in a strong commitment to community-based, development-oriented disaster management that has become an international model for communities to emulate around the world. The comparative approach used by historians emphasises the need to consider alternatives, that no one culture or state has a monopoly on 'best practice', and underscores the importance of dialogue between peoples and across cultures.

Process and event

A fuller appreciation of the past reveals how vulnerability is constructed over time, even if it only manifests itself in the present; how disasters are processes as well as events. An event is something that happens within clearly defined temporal boundaries that separate it from what came before or after: that at such and such a time, day, date or year something began; and that at such and such a time, day, date or year something finished. For example, the Great Tangshan earthquake that shook northeastern China took place at 3:42 am on 28 July 1976. A process, on the other hand, indicates a continuum that stretches from the past through the present to the future ad infinitum. It has no clearly delineated beginning or end because, at any point, the horizon stretches either before or away in either direction. Events occur along this continuum when trajectories or timelines of different actors or agents intersect at precise moments. That is to say, the Great Irish

Potato Famine of 1845 was the result of a collision between human society and environmental conditions: social and economic forces that made the population dependent on a single crop and vulnerable to anything that disrupted it; and agricultural management practices that made the potato vulnerable to the fungal pathogen *Phytophthora infestans* (Fraser 2003).

Understanding the dual nature of disaster as an event and a process is an important consideration in taking steps to reduce a community's vulnerability. Take the Hawke's Bay earthquake of 3 February 1931, measuring 7.8 on the Richter scale, which in tandem with fire laid waste to the North Island town of Napier in New Zealand. Quite clearly the earthquake was an event. There was a discrete physical trigger that tilted the immediate area upwards by over two metres and raised an additional 2,230 hectares of land above sea level. There was also a discrete period of response in which measures were taken to lend assistance to the injured and trapped, recover the dead and provide succour to survivors. How long this period exactly lasted depends on the criteria applied – hours, days, perhaps even weeks, but the event was finite and came to an end. Seeing the earthquake as part of a process, on the other hand, involves a wider perspective that encompasses not only the present event but its antecedents and consequences. In terms of history, it might begin with an examination of the manner in which Europeans colonised New Zealand: why certain sites were selected for settlement, the nature of the export economy, land usage, the transfer of British culture and how it adapted to local conditions in terms of the built environment – materials used, building codes and how they were applied, the spatial organisation of urban areas, the provision of emergency services and the social structure of society – among other considerations. Equally, however, there is a need to consider what happened subsequently: how the city fared after the earthquake. For a start, the scale of reconstruction work doubled the region's share of national building expenditure in the following years, creating new jobs and encouraging migration *into* the area as people came in search of work at the height of the Great Depression. Building codes were amended to ensure safer structures, wider streets, better public utilities including New Zealand's first underground power system, and a special fee attached to building permits that funded research into earthquake resistance (Grayland 1957: 124–32). Moreover, as Napier was all but destroyed, the city's centre was virtually rebuilt in the architectural style of the day, Art Deco, and enlivened with eclectic motifs drawn from local and international contexts. The result is a cityscape that resembles a 'period' film set. Nowadays, a good part of Napier's prosperity depends on its reputation as a major tourist attraction, drawing architectural enthusiasts from around the world, especially during Art Deco weekend every February. Thinking about a disaster as a process as well as an event underscores not only how vulnerability is constructed over time, highlighting mitigation measures that might have to be taken, but what its effects are on the future and the need to relate recovery with development.

Agent of change

Considering disaster as an historical process allows its impact to be evaluated as more than only a destructive event, at best a bane to be prepared for as well as possible, and at worst one to be simply endured. While not trying to minimise the destructive consequences of disasters in terms of human suffering, they are also simply agents of change in their broadest perspective. As such, they have recently received some recognition in the literature. They are variously referred to as a 'focusing event' or a 'policy window' to denote a sudden, relatively infrequent, localised, harmful natural hazard or industrial accident with the potential for causing even longer-term damage that has the effect of influencing the dominant issues in governance agendas, reopening channels of communication between stakeholders with the view to drawing up common mitigation strategies

Box 4.2 The Lisbon earthquake of 1755

Russell Dynes
Disaster Research Center, University of Delaware, USA

The 1755 Lisbon earthquake is an important historical case study for having evoked a co-ordinated state emergency response and a comprehensive effort at reconstruction to reduce the effects of future disasters.

At the time of the earthquake, Lisbon, with a population of 275,000, was perhaps the fourth largest city in Europe and one of the best known cities in the world. The earthquake presented a serious threat to Portugal's effort to modernise.

The earthquake occurred at 9:40 am on 1 November, when many of the residents were at Mass for All Saints' Day. A tsunami created additional damage and a fire lasting five or six days destroyed many of the remaining buildings. Estimates of death vary tremendously—perhaps 10,000 or four per cent of the population. The death cut across all social categories and it was suggested that many died 'unconfessed and unforgiven'. The major impacts were in the centre of the city. Some suggest that only 3,000 out of the 20,000 dwellings were habitable.

The young king, Jose I, deprived of his palace and leisure, had little interest in the consequences of the earthquake. He asked his ministers, 'What should we do?' Pombal, then minister of foreign affairs and war, was supposed to have answered, 'Bury the dead and feed the living'.

Pombal's initial action was to ask the Chief Justice to appoint twelve district leaders, with overarching emergency powers. He suggested to the Patriarch—the head of the Church—that the best way to dispose of bodies was to collect them on barges and sink them in the Tagus River. To ensure a continuing food supply, the military provided transportation for food from the countryside. The price of food was controlled. Fishing was encouraged and taxes on it were suspended. Ships were not allowed to leave port until it was determined that their contents could not be used in the emergency.

Housing was at a premium. By the end of November, plans for reconstruction were beginning. Land rents were controlled and laws passed that forbade landlords from evicting tenants. Debris was sorted to salvage materials needed for reconstruction. Fearing loss of population, a pass system was instituted to regulate entrance and exit to maintain some normality.

In order to make Portugal economically viable and politically strong, Pombal asked the Patriarch to stop alarmist sermons in parish churches suggesting that the quake was divine retribution for the past sins of Portugal. Warships were sent to Brazil, India and Africa to indicate that trade with Portugal was still secure.

Early in the emergency period, military engineers were charged with drawing up plans for a new city. Efforts were made to prefabricate and standardise materials and to make them more flexible in the event of future earthquakes. Plans were made to create a new square, placed on the old Royal plaza to be called Praça do Comércio. It was to reflect the direction Pombal intended for the future of Portugal. In June 1775 the reconstruction had progressed far enough to hold a dedication of the Praça, ironically adorned with an equestrian statue of Jose I.

and/or offering opportunities to push through needed policy solutions (Johnson *et al.* 2005). On an even more international scale, 'disaster diplomacy' entertains the notion that management can or cannot prompt long-term changes in international relations, that disasters caused by natural hazards can induce or not international co-operation amongst countries that have traditionally been 'enemies'. The case of the Indian Ocean tsunami of December 2004 and the peace accord reached in its aftermath between the Free Aceh Movement (GAM) and the Government of Indonesia is frequently cited as a classic example (Gaillard *et al.* 2008a).

As catalysts of change, disasters have long been transformative agents in their own right, causing political, economic and social adjustments, triggering needed adaptations in human behaviour and the built environment, as well as perhaps contributing to the overthrow of dynasties, economic systems and even civilisations. The rebuilding of Lisbon after the great earthquake of 1755 is just one such instance, as the Portuguese state became intimately involved in the recovery and reconstruction process, innovating policies for urban planning and developing seismic-resistant building codes, many of which continued to be standard practice until the 1920s (Mullin 1992). In the political sphere, the disaster relief campaign following the 1944 Argentine earthquake that reduced the town of San Juan to rubble allowed the aspiring Colonel Juan Perón to challenge the power of existing elites and launch his own rise to the presidency on a wave of social activism that opened up space for radical proposals (Healey 2002). Even in a modern industrial state like the United Kingdom, not known for its hazardous environment, disasters have at least 'punctuated' the contemporary planning process by offering periods of public interest and media scrutiny that have allowed (if not driven) changes in flood policy (Johnson *et al.* 2005). Once the historical importance of disasters as transformative agents is more fully appreciated, then policy-makers and planners are more likely to consider longer-term developmental goals as well as satisfy more immediate needs. The future reconstitution of the Haitian state as well as the rebuilding of Port au Prince following the January 2010 earthquake is a case in point.

Inherent vulnerability

Taking an historical perspective when considering disasters, then, has some very practical end results. In a more general sense, it can improve the setting of industry standards, provide modern technology with useful applications, highlight issues of social justice, prompt cultural comparisons of 'best practice', help reduce communities' vulnerability and link reconstruction work to developmental issues. As such, it is a form of 'applied history'; that is, applying an historical approach to analyse and evaluate present conditions to better inform policy decision-making in the future. In a very specific way, too, heeding local histories, a community's oral knowledge about its environment and the capacities of the people who live in it forms the basis of all effective DRR planning. Citizenry-based development-oriented disaster response and three-dimensional participatory vulnerability mapping, among the 'new tools' in the disaster management kit, depend for their successful implementation on local knowledge and an awareness of traditions – history by any other name. The good DRR practitioner of today also needs to be a good local historian of yesteryear.

More than simply a methodology, though, what happened in the past has a bearing on understanding social vulnerability in a wider context: the *gestalt*, so to speak, being greater than the sum of its parts. By emphasising the importance of culture, the historical approach to disasters can make apparent a community's vulnerability and resilience not just *after* a disaster but *before* it. It can be predictive. It reveals the *inherent vulnerability*, the background social vulnerability that has built up sequentially over time. With vulnerability, the emphasis is always one of degree, as all people are vulnerable in one way or another to some extent. However, some communities 'share' a common cultural and historical exposure to higher background levels of risk than others that requires special consideration before any engagement with more specific factors of vulnerability. Again, Haiti would be a case in point. There is a need, therefore, to explore the social and environmental relationships that precede disasters so that emergency planners and managers can be in a better position to predict what might happen: where disasters are more likely to occur; where they will have a higher impact on the population when they do; and where higher levels of intervention might be required both before and after a disaster. 'Disasters', claims Anthony Oliver-Smith, 'are seen to be more characteristic of societies than they are simple physical environments' (Oliver-Smith 1994). As such, disasters are very much a question of history.

Acknowledgement

This work is part of Economic and Social Research Council (ESRC) project RES-000-22-3070.

5

Politics

Power and disasters

Adolfo Mascarenhas

FOUNDER DIRECTOR OF THE INSTITUTE FOR RESOURCE ASSESSMENT, AND PROFESSOR OF GEOGRAPHY,
UNIVERSITY OF DAR ES SALAAM, TANZANIA (RETIRED)

Ben Wisner

AON-BENFIELD UCL HAZARD RESEARCH CENTRE, UNIVERSITY COLLEGE LONDON, UK

Introduction

One should consider afresh the relations between disaster and power in a world full of anomalies and contradictions. Much power is concentrated in the hands of relatively few individuals, institutions and nation states. This distribution of power affects the social distribution of risk and the resources dedicated to reducing risk. The distribution of disaster deaths is today skewed towards low- and middle-income developing countries. Why? Within these countries the rural and urban poor and marginalised are more heavily affected. Why? What has this distribution among nations and within countries to do with power and its distribution? Although human beings generally have to tread carefully and with great ingenuity on a restless planet, whilst the *hazards* dealt with by this handbook are natural, *disasters* are not natural; rather they are more a matter of power and social justice (Wisner *et al.* 2004).

Suffering in the aftermath of natural hazards such as coastal storms, flooding and drought have become part of the new normal, with a huge international industry built up around response and recovery. Much less progress has been made in addressing what the United Nations (UN) calls 'underlying risk factors' (UNISDR 2005b) and thus preventing loss and reducing risk. These include landlessness and highly skewed access to natural resources, absence of security of urban and rural land tenure, failure by governments to enforce land use, environmental and building code regulations, poor market access, gender and other forms of discrimination, among others. What is the role of economic and political power in shaping people's inability to cope with natural hazards and limiting initiatives to modify or remove root causes of vulnerability? A clue to the answer is the fact that even less discussed as 'disasters' are the day-to-day outcomes of political, social and economic policies in the disaster literature.

Unemployment, loss of resources and livelihoods, ill health and unaffordable health care are treated as collateral damage of systems that deliver economic growth. Only rarely has something as 'ordinary' as maternal mortality been labelled a disaster (IFRC 2006).

In the quest for economic growth and rolling out of market-based policies, people's livelihood, security and health have been eroded; many of them have been driven to live in dangerous sites and locations in order to make a living. All these factors increase the vulnerability and exposure of people to natural hazards. These impoverishing and displacing processes at work in the twenty-first century can be understood by considering the use and misuse of economic, political and cultural power over several centuries. For many formerly colonised people in countries that gained independence after WWII, misuse of economic and political power has negated whatever dreams and plans they had envisaged for a better life for the generations to come (Wisner *et al.* 2006), and the scientific discipline of development economics appeared to be a 'god that failed' (Yusuf 2009: 11).

A typology of power

Sociologist Max Weber distinguished between two basic types of power: that based on coercion and that based on authority (Weber 2004). As with all his work, he was presenting ideal types. In the real world these sources and expressions of power overlap and blur at the edges. For example, if after several warnings and perhaps ignored summons and unpaid fines a landlord's building is seized by the government for violation of the building code, there clearly is an element of coercion. However, authority is also involved. Weber distinguished between three sources of authority: charisma, tradition and law (rationality). In this hypothetical case, law is most likely the source of authority (see Chapter 53). However, in the early days of his tenure as President of Venezuela, Hugo Chávez was invested with great charisma and had the confidence of much of the population. This was a source of authority as well, although Weber warns that it is an unstable and often short-lived one. Nevertheless, Chávez was able to exhort the residents of Greater Caracas to learn from the flooding, landslides and building collapses that took lives in 1999. Charisma as a source of authority is definitely involved when the UN calls on former presidents and film stars to be its ambassadors for themes such as biodiversity or children's rights, or to function as a UN special representative for disaster recovery, a role that former US President William J. (Bill) Clinton served after the Indian Ocean tsunami.

Traditional authority remains a source of power at the local scale in many parts of the world, whilst in most cases legal rationality is at least the formal source of power in nation states in the early twenty-first century. At the local level, for example, the chieftainship system is still active in some parts of Southern and West Africa. While this is notably a rural phenomenon, in some Nigerian cities the traditional chiefs have a great deal of influence and have been crucial in small-scale micro-credit schemes that benefit, among others, market women (Mabogunje and Kates 2004). By extension, these investments benefit urban sanitation and fire safety, as markets are prime origins of disastrous urban epidemics and fires.

Politics is the formal and informal use of these varieties of power in order to promote the material interests of individuals and groups (Wrong 1995). Formal expressions include overt lobbying for legislation, such as a building code or land use regulations in a national or subnational rule-making body, or for financial support from a state institution for risk reduction such as hillside terracing or urban drainage. Informal politics may be involved in some of these same issues. For example, by dropping hints to elected officials that failure to deal rapidly with disaster recovery could lead to a *coup d'état*, a country's military exercises covert, informal power. Large landowners may also covertly explain to legislators that campaign finance may be withdrawn if proposed land use regulations are made law. Informal use of power at the local scale is very common. For example, Bosher (2007) found that informal networking with political leaders gave higher caste members more access to resources of importance for reducing the risk of cyclones on the coast of India's Andhra Pradesh state.

Power and disaster

The purpose of this chapter is to illustrate how the use or the misuse of the kinds of power discussed in the previous section is linked to disaster. There are many links but five are major avenues. While one can describe and discuss them separately, clearly they all overlap and interconnect.

Link one: disasters and power in the echoes of the colonial past

Many of the patterns established over centuries of colonial domination in Africa, Latin America and Asia are still present, albeit in modified forms, including: the dominance of export-orientation economic activity as opposed to production for the internal market; the persistence and growth of cities marked by smaller planned and serviced cores versus vast unplanned, under- or un-serviced peripheries; and, above all, the power of a small economic and political elite. All of this bears heavily on the social production and reproduction of disaster risk (see Chapter 4). Writing about Angola, an historian notes that 'colonial expansion uncannily syncopated the rhythms of natural disaster and epidemic disease' (Davis 2001: 12). The ways in which labour, land and other natural resources are exploited today by the elite may not be as direct and brutal as in Angola under Portuguese rule or in the Congo under Belgium's King Leopold (Hochschild 1999), but it nevertheless displaces people, undermines their livelihoods and increases their vulnerability to disaster.

Colonial occupation of Brazil by the Portuguese and the rest of South and Central America by the Spanish ended earlier than its counterparts in Africa and some of Asia. Still, patterns favouring landed and otherwise privileged elites continue. So strong are these patterns that Oliver-Smith has referred to the devastating earthquake in 1970 as 'the 500 year earthquake' (Oliver-Smith 1999). He argues that the Inca culture of prevention that included risk-averse architecture and land use had been destroyed by centuries of Spanish oppression. Furthermore, the reproduction of economic privilege by the post-independence land tenure system and system of rural production for export had weakened the livelihoods and coping ability of ordinary rural people.

The use of colonial and imperial power reached its zenith in the late nineteenth century. Disaster played a role. The consolidation of Britain's far-flung economic and politico-military control was assisted by large-scale climate disasters in China, India, southern Africa, Egypt and Brazil. All these calamities were linked to the warming of Pacific Ocean currents known as El Niño–Southern Oscillation (ENSO or, simply, El Niño). In his book, appropriately titled *Late Victorian Holocausts: El Niño Famines and the Making of the Third World*, Davis (2001) documents the full impact of imperialism when superimposed upon such natural events. The death toll was staggering – more than 12 million Chinese and over 6 million Indians perished of starvation during the years 1876–78. The callousness with which officials treated the people of the various colonial provinces was not lost on the Indians and it started the first nationalistic movement.

It would take almost half a century for a retrospective study to find that the reason for the millions of casualties in Bengal famines in 1943 was once again the British Raj (Sen 1981). Sen's seminal research revealed that the majority who died was not from scarcity, but simply that first the British moved food reserves from that part of India to keep it out of the hands of advancing Japanese troops, and then they failed to control the price that traders charged for the remaining food. Davis (2001) observes that the incorporation of the colonial world into the system of free-market economics amounted to cultural genocide. While few can defend an empire in moral terms, one has been encouraged to acknowledge its economic benefits; however, this, too, is questionable. There is little doubt that historically South Asia, in terms of its economy and trade, was one of the richest regions in the world. Between the first and the eleventh centuries this region had the world's largest economy. However, it shrank by two-thirds in the eighteenth century, to a mere twenty-four per cent of its pre-colonial maximum size a century later. South Asia's economy continued to decline and only reversed that trend well after independence (Maddisson 2003). The decline was due to the impact of disasters (Davis 2001; Mukerjee 2009), colonial manipulation (including that by the East India Company) and de-industrialisation.

In East Africa the situation was similar. Maasai pastoralists in the 1890s were a very much weakened people when the Germans (and later, the British) arrived. These feared warriors had suffered a drought, and also a new disease, rinderpest, had spread from the horses of the defeated Italian forces in Ethiopia and decimated their cattle. The disabling and painful infestation of human feet by parasites called 'jiggers' had accidentally been carried from Angola and had found its way to Zanzibar and then on to the mainland where it also affected the Maasai and others (Kjekshus 1996).

Link two: wealth, economic power and disasters

During the 'decades of development' following the end of WWII, the economic policy of many nations came under the influence of a 'modernising' ideology enforced by donor nations, many of which had been the earlier colonial powers, and by their allies. In the Cold War environment, elites prospered who supported the 'free world' and who gained wealth in this neo-colonial division of labour. International organisations including the World Bank and International Monetary Fund (IMF) reinforced this new order by setting conditions for their loans. By the 1980s a fully fledged neo-liberal world order had developed (Domhoff 2007). 'Structural adjustment' was mandated by the World Bank and IMF as the policy framework that would lead to prosperity. This required reducing the size of government and public expenditure, privatisation of many governmental functions and services and opening nations up to foreign investment. Structural adjustment programmes hit the poor very hard, eroding their ability to cope with additional stresses, including natural hazards. Alarmed by falling welfare indicators, the World Bank introduced 'adjustment with a human face' – a series of safety nets to soften the impact of the neo-liberal package of reforms on the poor (Mohan et al. 2000). Eventually, this morphed into a series of national Poverty Reduction Strategy Programmes (PRSPs). While many national PRSPs made references to disaster risk reduction (DRR), little implementation has followed, and the central trust of the neo-liberal model of development remains small government, privatisation, foreign investment and export-led growth.

Economic globalisation has led to the flow of investment in mineral extraction, urban real estate and construction, infrastructure and arable land that has enriched a narrow political and economic elite in many nations. Thus, there have been winners. On a global scale it is estimated that the top two per cent of the richest people own fifty per cent of global wealth, while ten per cent of the world's adults control about eighty-five per cent of global wealth (Davies et al. 2009). However, there have also been losers. These political and economic processes have driven many small farmers and pastoralists off their land either because they cannot compete with food imported into the country – the case for Mexican maize farmers and Haitians who used to grow rice, for example – or because their land has been annexed in one or another legal manoeuvre, or their access to forest products, pasture and other formerly common property resources has come to be regulated by the state (Bryceson 2009). Even the World Bank has become concerned about the rapid increase of what it euphemistically calls 'rising global interest in farmland' (World Bank 2010a) and what others would term 'land grabs' (Bretton Woods Project 2010). With rural livelihoods undermined, many rural people have swelled the population of cities and live in hazardous neighbourhoods exposed to flood, landslides, coastal storms, pollution and insect-vectored and water-borne disease.

Link three: state power and disaster

In many countries, the power of the state (or its proxies) has increasingly absorbed most rights to basic resources overriding tribal, clan or community interests. Despite World Bank-mandated down-sizing of government, the reach of some parts of the remaining oligarchic state has grown. Its attempts at control have skewed access to most resources in Africa, in significant parts of South America and now to a lesser extent in Asia. For example, in Bangladesh this is clearly seen in laws that regulate access to fresh water ponds and lakes (Khan and Haque 2010). As a result the livelihoods of millions of ordinary people have been compromised. Another example comes from Africa, where having lost two-thirds of their grazing lands during the colonial periods in Tanzania and Kenya, Maasai livelihood is under threat again (Homewood and Rodgers 1999). Since Tanzania adopted a free market economy in the mid-1980s, these pastoralists have witnessed annexation of lands by national parks and wildlife reserves, and concessions given for hunting and hunting reserves that provide income and recreation for wealthy individuals. Further undisclosed concessions and evictions in 2010 led to protests and several investigations including that of a Parliamentary Committee (FEMACT 2009). State power can be used at national and local scales to benefit some economically, to the cost of others.

Closely linked to the impact of state policy and legislation is sheer abuse and arrogant use of state power which leads to the gross neglect of governmental, professional and social functions, including the provision of basic services, enforcement of building codes, safety codes and the well being of its citizens. This includes corruption, something monitored by Transparency International, whose annual report for 2005 focused on corruption in the regulation of construction – associated with unnecessary earthquake deaths (Transparency International 2005).

A common buzzword among researchers and agencies concerned with development is 'governance'. There are functions that governments are expected to carry out as a matter of international norms and popular perceptions of legitimacy. Normally governments are expected to make decisions about 'acceptable risk'. These decisions are embodied in building codes, food safety standards, provision of water and sanitation, etc. However, how are these decisions made? Who is involved? In many cases economically and politically less powerful groups of people are not consulted, or their opinions and voices receive less notice. Governments also tax and raise revenues. Some part of these revenues should go to the management of risks. How are such resources allocated? Is the allocation socially equitable? Is it spatially (regionally) equitable? 'Decentralisation' is another buzzword (see Chapter 52), one of the hallmarks of 'good governance'. Yet one might ask if resources flow down from the centre at the same rate as mandates and responsibilities? Also, do local elites capture and use for their own ends resources that are available through decentralisation?

Link four: disasters and cultural- and knowledge-based power

In a 'wired' world, Western science and industrial country technology is hegemonic. A common, uncritical assumption is that all other knowledge is inferior. Some national authorities responsible for disaster reduction have little awareness or interest in, or patience with, local knowledge and skills that people utilise for self-protection against such hazards as drought, flood, animal and plant diseases (see Chapter 9). There is sometimes lip service given to 'indigenous technological knowledge' (ITK) and to 'participation' by farmers and urban dwellers, but the assumption is usually that the framework is provided by outside experts, whilst ITK is useful for filling in detail or for translating Western truth into vernacular languages. These assumptions are only contested by local non–governmental organisations (NGOs) and other militant groups. Ministries of education suffer lack of funding and have conservative educational philosophies often inherited from colonial days. Even 'memory' is truncated and modified by regimes of class and cultural power (Box 5.1).

Box 5.1 Disaster memoryscapes: how social relations shape community remembering

Susann Ullberg
Stockholm University

One hundred and sixty-one small wooden crosses symbolising casualties from the 2003 flood are stuck in the ground in one corner of Santa Fe's (Argentina) main square. The disaster is commemorated by groups of flood victims, who identify themselves as *los Inundados* (the Flood Victims), performing commemorative rituals as a means of protesting against the government's failed flood management practices.

One governmental and several 'civil' memorials also have been erected in different parts of the city. Testimonials and investigative journalism have been published. Several documentary films have been produced and publicly screened (see Chapter 11). Recently published school manuals inscribe the flood into local historiography. Memories from the 2003 flood are furthermore embedded in the urban landscape through street graffiti and the newly built flood embankments surrounding the northwest areas of the city. The new neighbourhoods built to relocate those disaster victims who could not return to their homes are a locus of flood memory. Every Santa Fe resident, young and old, vividly remembers the recent disaster either through being directly affected or through kin, neighbours or colleagues being affected, or as a disaster manager or political decision-maker.

Yet earlier floods of disastrous scope seemed to have left little trace in the Santafesinian memoryscape. Historical records include accounts of previous floods, but these events were not singled out as community disasters in the local memoryscape. Rather, public discourse depicted flood-prone and peripheral places as risky. The urban poor living in these risky areas were referred to as *los inundados*. *Los inundados* were in the public discourse either remembered as 'noble savages' of river life accustomed to coping with floods, or as astute beneficiaries of disaster assistance. The difference is that the 2003 flood-affected people (*los Inundados*, capitalised) included many affluent, middle-class people. Fieldwork in the city revealed that the 'forgotten' disastrous floods that had many times afflicted the poor in lowland districts were vividly remembered by the people actually living there, *los inundados*. They had not built any memorials and did not perform any commemorative rituals, but rather their flood memories were embodied in everyday living and extraordinary flood-coping practices embedded in local places and landscapes. Their memories were about coping as a fact of life, but also about loss and social suffering. Yet their memories dwelled in the shadows of the broader Santafesinian flood memoryscape.

Although disasters can be thought of as memorable events, ethnography of the flood memoryscape in Santa Fe showed that not all disasters are equally remembered. Memoryscapes are heterogeneous, shaped by social relations through different memory practices. Some memories become dominant. Such unequal remembering seems to add to conditions of social vulnerability more than enhancing capacities. Recurrent disasters are remembered as normal events that naturalise certain people as 'victims'. The effect of unequal remembering is that political efforts to reduce vulnerability to disasters are diminished or at best reduced to mere technological solutions that produce a sense of false security.

As noted earlier, authority is a source of power. Knowledge, in turn, is one of the sources of authority. Local officials and distant ministry experts 'know what's best'. Western knowledge cloaks incumbents with legitimacy. However, this common chain of attitudes can work the other way. Where state power has been repeatedly abused, mistrust develops, and the rural and urban poor may end up rejecting all that is associated with officialdom, including Western knowledge. Such a situation can be a grave obstacle to disaster risk awareness and reduction campaigns. If, as in Tanzania's Uluguru Mountains, there had been a coercive government programme to force farmers to terrace their fields, backed up with Western science, grandchildren of those who protested in the 1950s may be suspicious when a new generation of officials exhorts them to do the same thing, this time in the name of 'climate science' (Fosbrooke and Young 1960).

A contemporary example concerns climate change. The 'science' of climate change is a hegemonic English-language discourse. This body of knowledge is passed along in numerous vernaculars around the world, often by national governments that act as gatekeepers and interpreters. In Tanzania the first translation is into Swahili, the lingua franca of national broadcast and print media; however, further translation/interpretation takes place in 122 local languages (Bwenge 2010a). Political advantage may be taken by national and local elites by selectively interpreting climate science so that blame for disasters is diverted from them and their decisions. Box 5.2 gives an example of such a tactic at work in Venezuela.

Link five: disasters and violence as power

This chapter began with Weber's distinction between coercion and authority as sources of power. Violence and fear have been used for millennia as instruments to attain ends and interests against potential opposition. Wherever these instruments are in use today, ordinary people are either displaced or their ability to cope with natural hazards is reduced. Drug wars in Colombia have displaced hundreds of thousands of rural people, many of whom live in precarious conditions on the peripheries of Colombia's cities. Resource wars of long standing in the Democratic Republic of the Congo – a source of many valuable minerals (Essick 2001) – have killed 5.4 million people, and displaced over 2 million who live in refugee camps where they are at risk of cholera and, for women and children, sexual predation and violence (Shah 2010). With dozens of conflicts in the world and a long-term trend towards more civilians being killed and injured as well as displaced (Berkovitch and Jackson 1997), conflict has a major influence on vulnerability to disaster, as in the examples (see Chapter 7).

Box 5.2 Climate change: the 'perfect' political excuse

Alejandro Linayo
Research Center on Disaster Risk Reduction (CIGIR), and Latin American Network for Social Studies in Disaster Prevention (LA RED)

During the 1990s, just when global warming and its implications began to be recognised, another kind of 'warming' was taking place on the planet. This was the heating up of ideological confrontation over the destiny of humanity. The late 1980s had seen *Perestroika* and *Glasnost* and finally the fall of communist system. Some authors promoted the idea that we had reached the 'end of history' (Fukuyama 1993). The idea of liberal democracy (both economic and political) was trumpeted as the only option for society. Today, events have demonstrated how controversial this theory still is and how 'hot' the debate about 'socialism', 'neo-liberalism' and 'freedom' remains, particularly in Latin America (Douzinas and Zizek 2010; Raby 2006).

A barely studied phenomenon associated with these two types of 'global warming' is the invocation of 'climate change' as a political tactic. There seems to be a discursive game in which the responsibility for the occurrence of climate-related disasters is only attributed to greenhouse gas emissions of industrialised countries. In this way, the politicians provide an external 'excuse' for disasters caused by unsustainable development models.

Venezuelan scholars from the Research Center on Disaster Risk Reduction (CIGIR, www.cigir.org) have been studying political statements made after flooding and landslide disasters, against the background of situations such as that registered, for example, in the town of Santa Cruz de Mora, a small rural settlement located in the south of Merida state. In 2005 the town was affected by floods and landslides that killed almost one hundred people and destroyed much infrastructure. This research has observed contradictions between explanations for the disaster offered by political authorities and detailed studies that attribute flood impacts to deficiencies in local development policy (Linayo 2006).

In this case study researchers found that the Mocotíes River had been diverted during mid-1990s road construction by the regional and local government. Detailed damage assessment revealed that flood impact in 2005 was concentrated in the original river bed, where housing and other construction had been permitted and promoted by government after the road project (Laffaille, Ferrer and Dugarte 2005).

Blame, however, for such flooding was placed on foreign countries for their contribution to global climate change, at least from the point of view of local authorities, which used the language of climate change to explain the causes of the disaster (*Diario Frontera* 2005; *Diario de Los Andes* 2005). For example, Florencio Porras, Governor of Merida State, said: 'The amount of rain that caused this disaster is the highest ever seen in this valley … It is a signal of climate changes that we are experiencing' (Frontera 2005).

President Chávez said of the 2010 national emergencies: ' … [T]he disasters we are suffering because of these severe and prolonged rains are demonstration, yet again, that we confront a cruel planet-wide paradox: the most developed countries are destroying the environment' (Chávez 2010).

Power from below in DRR

The poor and marginal in society, including those who are most at risk from disasters triggered by natural hazards, have options. Historically they have sought and gained political power through the electoral system – for example, the remarkable increase in the voice of *adavasi* ('tribals'), low-caste members and 'untouchables' in India – or by 'speaking truth to power' through various forms of non-violent protest and civil disobedience.

In the context of DRR, John Gaventa's 'power cube' is a useful way to consider the ways that people are organising to contest the things that put them at risk and to gain control of resources that will make them safer (Gaventa 2007). Gaventa develops three sets of categories – hence the 'cube' diagram (see Figure 5.1). *Power* can be used in ways that are visible, hidden or invisible. Visible use of power takes place publicly and is open to general discussion and debate. Hidden use of power involves only certain people or groups. Power is invisible when some groups of people have internalised beliefs about their own powerlessness, incapacity or inferiority, as in the earlier example of local disaster reduction knowledge. *Places* where power is used can be global, national or local. *Spaces* of engagement where power is negotiated by different stakeholder groups can be closed, invited or claimed/created. Closed space is reserved for bureaucrats, elected officials and experts. This is often where 'acceptable risk' is determined – without consultation with those affected. Invited spaces are venues and situations where authorities invite certain stakeholders to take part. Most public consultations take place in this 'space', and usually marginal groups of people are not invited. Claimed or created spaces are those defined by organised groups of stakeholders themselves, using the media and often international networks of like-minded people to increase the influence of their voice (www.powercube. net/analyse-power/ways-of-visualising-power/powercube-powerpoint-2007).

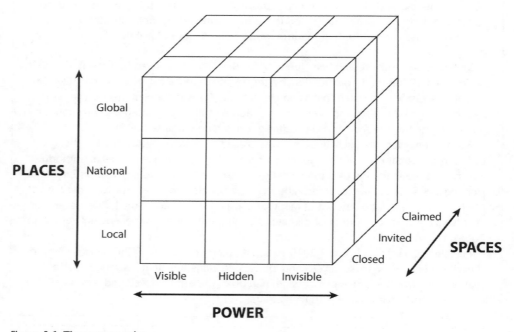

Figure 5.1 The power cube

Putting these notions into play in the field of disaster management and DRR, it is possible to identify worldwide a wide variety of 'spaces' in which different groups of people shape policy (see Table 5.1). Closed spaces abound. At the international and national scale there are many institutions either partly or wholly dedicated to DRR (see Chapter 50 and Chapter 51). There are also professional networks and organisations such as the International Association of Emergency Managers (IAEM, www.iaem.com), and some scientific unions and councils such as the International Geographical Union and the International Council for Science (ICSU, www. icsu.org/1_icsuinscience/ENVI_Hazards_1.html), which have developed programmes and projects that bring experts together in closed meetings. A question that will be taken up elsewhere in this *Handbook* is whether and how much of this activity 'trickles down' to the local level (GNDR 2009, 2011) (see Chapter 52 and Chapter 60). Generally, the national government ministries have branches and representatives at sub-national scale, including at local levels. Closed spaces at the local level include local government councils and their committees and sub-committees, as well as discussions among the local experts in civil protection, often

Table 5.1 Potential spaces of DRR policy – multiple but disconnected

	Official (closed) spaces	Invited spaces	Created/claimed spaces
External/ International	Bilateral donors UNDP UNOCHA UNICEF WHO World Bank Global Fund for Disaster Reduction International professional organisations (IAEM, etc.)	UNISDR ProVention Consortium (until 2010) Regional centres (e.g. ADPC) Donor–civil society dialogues	GNDR Duryog Nivaran LA RED INGOs (Oxfam, ActionAid, PLAN, etc.)
National	National DRR platforms Disaster management departments Various ministry departments and the Military National professional organisations (architects, engineers, etc.) Public (or private) utilities	PRSPs Top-down risk awareness campaigns, National Red Cross/ Red Crescent Society	National NGOs Unions of health workers and teachers, Rights organisations Social movements (youth, women, landless, etc.)
Sub-national	State/provincial civil protection authorities, State forestry, health, water, agriculture experts	River basin authority – water users forums	State/provincial networks of NGOs, rights organisations, social movements
Local	Local government councils Local civil protection expert (fire service, police)	Parallel structures: Dominant industries and plantations Faith-based leadership for DRR	Community-based organisations

Source: Authors, inspired by Gaventa (2007).

headquartered in the fire service or the police. Invited spaces have increased since donors and the World Bank have demanded more public participation in policy-making as part of 'good governance'.

Among the specialised institutions of the UN system, the UN International Strategy for Disaster Reduction (ISDR) led the way in broadening discussion beyond the national platforms set up after the 2005 international framework for disaster reduction (the Hyogo Framework for Action), reaching out to civil society and academia and creating a virtual common space for discussion (www.preventionweb.org). In principle, the World Bank-mandated participatory component of poverty reduction strategy programme (PRSP) monitoring provides a space for DRR issues to emerge. To a small extent this has begun to happen (see Chapter 51).

The same can be said of some donor–civil society forums at national level. Regional training centres such as the Asian Disaster Prevention Centre, located in Bangkok, organise policy discussions that involve selected stakeholders. National societies of the International Federation of Red Cross and Red Crescent Societies are also venues for invited representatives and experts to exchange ideas and experiences. Given the close social connection that the leadership of these societies often has with national political leaders, this turns out to be an important informal channel for ideas and opinions of people without access to closed policy spaces. Water users may be invited by river basin authorities to discuss flood control and related issues. At the local level, closed discussion is complemented by what is said within invited space organised by parallel structures that often have considerable informal influence over local government. These include locally active large industries and other major employers and large landowners, as well as religious leaders. Increasingly in some countries, large employers have their own disaster management plans and are concerned with business continuity, hence with the ability of their workers to survive and to suffer minimum losses so that they will return to work soon after a hazard event.

Those previously without a voice in any of these discussions may organise themselves as citizen-based organisations at the local scale, often supported by sub-national and national NGOs and rights organisations, or directly by international NGOs (see Chapter 60). Created policy dialogue space is least well developed at national and sub-national levels, where NGOs, rights advocacy organisations and social movements may take up DRR, but still tend to see it as a specialist concern and a distraction from their central issues. Also, woefully underutilised are the forums potentially provided by labour unions composed of workers at the 'front lines' of disaster and its prevention – health workers, teachers, sanitation workers and others working in public utilities.

The example of Bolivia under the leadership of Evo Morales pulls many of the issues of power and authority together. He has charisma and as the first indigenous Bolivian to be elected president he has brought the voice of the people excluded from policy space into the closed space of national decision-making. He has also rejected dominance by the IMF, donor nations and foreign corporations. His nationalisation of the natural gas industry is one demonstration of this. His government now pays more attention to new people-centred priorities: rights to land and resources and the needs of indigenous people. Morales had experienced life from below, including the impact of drought, not from an elite position. Every *El Niño* had for Morales a personal story of its own as he was growing up. He was aware of the cyclical warming-up of the Pacific's surface waters, and now he is highly involved with the issue of climate change, hosting a World People's Conference on Climate Change in April 2010, even before the ill-fated Copenhagen meeting. Living in an ecologically sensitive area, also prone to the impact of many natural hazards – ENSO, earthquake, volcanic eruption, landslide, flood, winter storm – Morales has led his government in mainstreaming concern about disaster reduction.

Bolivia's Ministry for Civil Defence and Co-operation of Integrated Development has welcomed collaboration with international and national NGOs in outreach to rural areas previously neglected. For example, in the drought-prone Chaco region, Care International has implemented a comprehensive community-based hazard assessment and action plan involving a variety of demonstration projects, also in co-operation with municipal civil protection authorities (CARE Bolivia 2008).

In the San Pedro River watershed, the German national agency for development co-operation, GTZ, partnered with municipal governments and community members to analyse hazards and develop strategies for reducing losses from drought, erosion and landslides that undermined livelihoods and food security, reduced the amount of available cropland and destroyed infrastructure (Brunner 2007). The approach to DRR involves livelihood enhancement and environmental protection as well (see Chapter 14). In Bolivia the government agency tasked with DRR is also responsible for environmental protection (see Box 5.3).

The change of priorities in Bolivia has improved several aspects of well being. Maternal and child mortality have declined and access to social services has increased (Center for Global Education 2010). Morales' policies might be dismissed as unsustainable populism that depends on the support of the poor. Yet in the recall referendum of 2008, the electorate had almost doubled to over eighty per cent of eligible voters, and he won the election with sixty-three per cent. This means that many who were not Amerindian voted for him. Morales changed the priorities, and took the poor into account as well as sustainable development and risk reduction. Yet the economy did not suffer. Radical changes and doing things differently need not be through violence.

Conclusions

Power has many sources and takes many forms. Most of these have direct or indirect impacts on the safety and sustainability of the built environment and the stability of people's livelihoods. Much of the world carries the burden of forms and distributions of power over decisions affecting those things that have been inherited from long decades of colonialism. The world now, ironically, has the 'new international economic order' that some political leaders in the global South demanded in the 1970s. Alas, structural adjustment of the 1980s and current economic globalisation is not what they had in mind. Countries of the global South are still integrated into a

---

Box 5.3 A village-level planning process in Bolivia

In its village development plan, a village decides to afforest two degraded sites of two hectares each. Since the funds available are only enough for one site, the village representatives and the Commission for Environmental Protection and Disaster Risk Management jointly discuss which of the two sites should be afforested. The Commission also considers whether the afforestation measures fit into the village's overall disaster risk management strategy. In order that the heavily overgrazed areas close to the afforestation site are not further degraded and the slope not destabilised, the Municipal Council passes a regulation declaring those areas out of bounds for grazing for the next two years. Implementation of the measure (e.g. selection of tree species, fencing-off, etc.) is being planned by the technical unit together with the village population, who will be carrying out the afforestation under the guidance of the technical staff.

(Brunner 2007: 14)

system of unequal exchange, and their political leadership has largely been won over by an ideology that values small government, privatised services and utilities, free trade, and export-oriented growth. Implementation of this ideology has led to land grabbing and marginalisation of poor farmers and pastoralists. Many of them have been displaced to urban squatter settlements rife with natural hazards. Violent conflict – an extreme manifestation of power as coercion – has displaced many others.

Yet power can also be manifested in the organised will of many small farmers, forest dwellers, herders and urban dwellers. Increasingly, social movements and various kinds of NGO are raising awareness that people can take control of their own lives. Community-based DRR is one example of 'power from below'.

6

Human rights and disaster

Jean Connolly Carmalt

DRAKE UNIVERSITY, IOWA, USA

Claudine Haenni Dale

INDEPENDENT RESEARCHER, GENEVA, SWITZERLAND

Introduction

International human rights law offers one avenue of inquiry into the relationship between ethical (and in this case, legal) obligations and the complex politics surrounding disasters. This chapter explores how human rights law relates to the disasters that arise in the aftermath of hazardous physical events such as earthquakes, floods and tsunamis. While it makes some reference to humanitarian law because of the potential interplay between that body of law and human rights in the context of disasters, its primary legal grounding is in human rights law, which applies both in times of conflict and in times of peace.

Human rights in the context of disaster

The arena of international law comprises a complex web of instruments and authoritative documents, the enforcement of which is rooted in political and social mobilisation strategies together with domestic implementation efforts, amongst others.

The sources of international law include everything from treaties, customs and state practice to a wide array of 'soft law' (non-binding) documents that provide authoritative interpretation, including guidelines or principles that pertain to specific situations (Meron 1986). In other words, while treaty rights provide crucial starting points for analysing the relationship between rights and disasters, they must be seen in conjunction with soft law and practice as it has evolved over the years.

For example, the Guiding Principles on Internal Displacement provides an authoritative set of guidelines pertaining to displacement, a problem that is particularly relevant to disaster situations (United Nations 1998). While these guidelines are 'soft' or non-binding law, they also have legal grounding in 'hard' law (such as that found in international treaties), which means that they illustrate the complex interplay that can emerge between different sorts of legal instruments in the international arena (see Box 6.1)

Box 6.1 Sovereignty and humanitarian intervention

In 2008 Cyclone Nargis devastated the country of Myanmar (Burma). The country is densely inhabited and very poor, with much of the population living in the low-lying Irrawaddy Delta. The storm was the deadliest in Myanmar's history. The humanitarian catastrophe that arose after the storm completely overwhelmed the country's limited capacities, and offers for aid poured in from around the world. However, the military junta running the country steadfastly refused international assistance, claiming it to be their right as a sovereign nation to refuse external assistance. Many people began to ask whether aid could be sent to people in Myanmar regardless of the junta's desires, using force to get it there if necessary. These calls were premised on the idea that humanitarian needs trumped concerns about sovereignty.

From the perspective of international law, the debate over whether it is acceptable to impose multilateral aid on an unwilling country is about the legality of intervening in a country, against that country's wishes, on humanitarian grounds. Article 2(7) of the United Nations (UN) Charter sets out the basic law regarding the principles of sovereignty. That Article states: 'Nothing contained in the present Charter shall authorize the United Nations to intervene in matters which are essentially within the domestic jurisdiction of any state or shall require the Members to submit such matters to settlement under the present Charter; but this principle shall not prejudice the application of enforcement measures under Chapter VII' (United Nations 1945).

The reference to Chapter VII, which deals with Security Council authority regarding threats or breaches of the peace and acts of aggression, makes it clear that the principle of non-intervention into sovereign affairs is limited. Chapter VII includes authority for the Security Council to use such force 'as may be necessary to maintain or restore international peace and security'. However, reading this provision together with the rest of the Article means that the only permissible breach of sovereignty under the Charter is when the Security Council authorises its use in the context of maintaining or restoring international peace and security. Since it is found in the UN Charter, this provision constitutes binding (hard) law on the countries that have agreed to be bound by it. This includes the vast majority of countries in the world. The military junta in Myanmar eventually allowed aid to enter the country from external sources, thus rendering moot that particular debate over whether aid delivery should trump concerns about sovereignty. To date, the authorised instances of Security Council interventions on humanitarian grounds have focused on situations of armed conflict, such as those in Somalia or Rwanda. There has been no precedent, therefore, for an authorised multilateral intervention against a country's wishes in the context of a disaster triggered by a natural hazard. However, the legal grounding for such an intervention remains the same as any other intervention: namely, that bringing uninvited military personnel into a country—for whatever reason—is only legal if the Security Council authorises it because of a threat to international peace and security.

Another, more constructive way of dealing with state sovereignty is to recognise it as a responsibility: 'In other cases, Governments may prefer, for a variety of reasons, to provide all necessary assistance themselves; and this is a legitimate exercise of national sovereignty and responsibility. However, when Governments refuse outside offers of humanitarian assistance, but at the same time are themselves unable or unwilling to provide adequate assistance to their own populations, they fail to discharge their responsibilities under international law' (Kälin 2005: 13).

Substantively, human rights are grounded in the idea that all people deserve to live in dignity and with respect (UDHR 1948). This basic principle is translated into a range of rights articulated by international treaties and expanded upon in soft law, including civil and political rights (e.g. the prohibition against torture or the right to a fair trial) and economic, social and cultural rights (e.g. the rights of housing, education and health) (UDHR 1948; ICCPR 1966; ICESCR 1966). Although human rights are considered universal and indivisible, there is a certain categorisation of so-called absolute rights that cannot be restricted in any manner, and those that can be restricted under certain conditions. The absolute rights are: prohibition of genocide and slavery; right to life; prohibition of torture; freedom of thought, conscience and religion; due process guarantees; non-discrimination; and the right to be recognised before the law (ICCPR 1966: article 4). Other rights, such as freedom of movement, right to housing, right to education or right to health, can be restricted provided that there is (1) an imperious reason, such as national emergency or a threat to public health; (2) a legal basis for taking the decision; (3) the measure restricting or suspending the right is proportional to the potential danger; (4) the decision is limited in time and periodically reviewed.

Each right, whether absolute or subject to restrictions or limitations, entails substantive and procedural requirements. The right to health, for example, requires (among other things) that health care be available and accessible, but it also includes requirements about how, procedurally, those elements should be implemented (UNCESCR 2000). Procedural rights typically include the prohibition against discrimination (i.e. the right must be implemented in a non-discriminatory manner), the rights to information, participation and remedy, and restrictions on retrogression that prohibit states from regressing in their efforts to implement specific rights. Some of these procedural requirements are entwined with one another. For example, the requirement for participation requires states to consult and include affected populations in processes of implementation, as well as in fashioning appropriate remedies for violations through redress and/or systemic change. The participatory requirement also overlaps substantially with the requirement for information, since effective participation is only possible if affected populations have relevant information about such processes.

One of the procedural rights particularly relevant to disaster contexts is non-discrimination. Under international law, the prohibition against discrimination includes those actions that do not intend to discriminate, but which nonetheless result in disparate negative impacts against certain groups of people (ICCPR 1966; ICERD 1965). For example, expanding an upstream housing development (protected by flooding by virtue of its location) could increase the risk of flooding to downstream communities. To the extent that downstream communities were already more vulnerable and therefore could not have access to the same choices that would mitigate their risk of flooding, and to the extent that the state chose not to take other measures that would mitigate that risk, this could represent the social and spatial transfer of social vulnerability (Wisner et al. 2004). If flooding later produced a disproportionate impact on those downstream communities, then that transfer of risk could constitute a prohibited form of discrimination under international law.

In addition, the prohibited grounds for discrimination can be broader under international law than they are in some domestic legal systems. The International Covenant on Civil and Political Rights prohibits discrimination on any of the following grounds: race, colour, sex, language, religion, political or other opinion, national or social origin, property, birth or other status (ICCPR 1966: article 26). This list is illustrative, not exhaustive, which means that it is possible to introduce additional categories (e.g. sexual orientation).

The substantive and procedural elements for particular rights entail corresponding duties, which in the context of human rights include the duties to respect, protect and fulfil

(UNCCPR 2006; Maastricht Guidelines 1997). The late twentieth century saw an expansion of the interpretation of human rights from the narrow state obligation to respect a right and to refrain from violating it to a broader interpretation that also includes the duties to protect and fulfil specific rights. The obligation to respect means that states have an obligation to refrain from violating the right through their actions. It is therefore a 'negative' obligation, meaning that it is framed in terms of what a state should not do, rather than in terms of what a state should do. For example, the prohibition against torture means that a state is legally required to refrain from engaging in torture: if the state engages in torture, it is taking an active measure which violates the prohibition, and it is therefore violating its duty to respect a person's right not to be tortured.

In addition to this basic duty to refrain from actively violating rights, however, governments must also protect rights and ensure that others cannot infringe or violate them. Using the example of torture, the obligation to protect means that governments have an obligation to train interrogators properly and to ensure that allegations of torture are investigated and followed through to completion. The duty to protect also extends to situations in which the government is not directly responsible for the treatment of victims; so, for example, in cases of domestic violence, governments are obligated to follow up and investigate allegations of domestic abuse. Importantly, the obligation to protect includes protection against indirect threats, including those arising from physical hazards (UNCCPR 2006; *Budayeva and others v. Russia* 2008).

Finally, governments have a duty to take the measures needed to implement particular rights through the obligation to fulfil. The requirement to fulfil includes obligations to both 'facilitate and … provide', which means, for example, that a state must enact legislation annulling confessions obtained under torture as proof in court. This 'positive' duty was once assumed to solely apply to economic, social and cultural rights, but contemporary law recognises how these duties also apply to civil and political rights (as demonstrated by the example of torture) (Eide 2001).

Each of these three levels of obligation – to respect, to protect and to fulfil – applies in the context of every right. Taken together with the procedural rights such as non-discrimination, information, participation and remedy that are outlined above, this means that the question of how human rights relate to disasters involves several steps of analysis.

The human right to life

The first paragraph of Article 6 of the International Covenant on Civil and Political Rights (1966) reads: 'Every human being has the inherent right to life. This right shall be protected by law. No one shall be arbitrarily deprived of his life.' That short phrase has been fleshed out through numerous interpretations and authoritative reviews, and through international and domestic jurisprudence (see, e.g. UNHRC 1984). As with other rights, a state's obligation to uphold the right to life includes the obligation to respect, protect and fulfil that right. In addition, the right to life entails the same procedural requirements that exist for other rights, such as non-discrimination and the right to participation. This section analyses the right to life in the context of disasters, with a particular emphasis on situations in which disasters are foreseeable and the extent to which states are obligated to implement measures of disaster risk reduction (DRR).

Legally speaking, the right to life should be interpreted broadly (UNCCPR 1982). Therefore, the state's obligations to respect, protect and fulfil the right to life apply even in situations where the threat to life is indirect (i.e. not from the state itself). Indirect threats to life include those arising from natural or physical hazards. As the European Court of Human Rights (ECHR) held in 2008, the right to life includes positive and procedural obligations in the context of threats arising from natural phenomena (in this case, mudslides) (*Budayeva and others v. Russia* 2008). In the Budayeva case, the Court found that the Russian government had failed to protect the lives of residents (including that of Vladimir Budayeva) in a mountain town with a long-recorded history of mudslides, despite being warned that it should at least establish observation posts to provide early warning. The Court ordered the government to pay substantial compensation to survivors, thereby establishing a legal precedent for the claim that the obligation to protect life includes positive measures to protect from physical hazards. The decision was significant for three reasons: first, because the right to life is one of the few rights that is universally recognised and has a fairly coherent and liberal interpretation throughout the legal systems; second, because there is relatively little legal precedent that develops the scope of the obligation to protect life in the context of disasters; and third, because the source – the ECHR – is a well-respected court with a relatively broad jurisdiction, the decisions of which are binding on its members.

In addition to this decision by the ECHR, the UN Human Rights Committee has also concluded that states have an obligation to protect life in the context of life-threatening natural hazards. In its 2006 review of the US report on implementation of the rights contained in the International Covenant on Civil and Political Rights, the Committee concluded that the US government should 'review its practices and policies to ensure the full implementation of its obligation to protect life and of the prohibition of discrimination, whether direct or indirect, as well as of the United Nations Guiding Principles on Internal Displacement, in matters related to disaster prevention and preparedness, emergency assistance and relief measures' (UNCCPR 2006: paragraph 26). While this conclusion was prompted by discussions about Hurricane Katrina specifically, as a legal analysis it applies to all disaster contexts. The Human Rights Committee is a treaty body (i.e. a group of independent experts mandated with commenting on a particular treaty's status and interpretation), which means that its conclusions about US practice are examples of soft law, and are only binding insofar as a given government chooses to abide by them. Nonetheless, the Committee is an authoritative body whose legal conclusions form an important part of the interpretive body of law that fleshes out the scope of human rights obligations. Therefore, its conclusion that the protection of life includes protection from physical hazards is legally significant to evaluating the scope of human rights obligations in the context of a disaster.

These conclusions by international human rights institutions provide authoritative statements supporting the idea that a state's obligation to respect, protect and fulfil the human right to life applies in the context of indirect threats to life, including those that arise from physical hazards. As was made clear by the Budayeva decision, these obligations are particularly important in the context of disasters that are recurrent. Legally speaking, the significance of a recurrent disaster is that it is foreseeable. This means that governments know or should know that there will be threats arising from physical hazards because they have witnessed those threats unfold in the past. Governments are therefore on notice that the threat to life exists and should take appropriate measures to respect, protect and fulfil rights in the context of foreseeable threats. The remainder of this section turns to an analysis of each of these levels of obligation.

The obligation to respect is fundamentally about making sure a state does not violate a right through its own actions. Therefore, in the context of a threat to life arising from a hazard instead of the state itself, the obligation to respect translates into an obligation to ensure that measures the state chooses to undertake are implemented in an appropriate way. In the case of human rights, the 'appropriate' way is one in which procedural rights – such as non-discrimination, participation, etc. – are respected. This means, for example, that if a state crafts evacuation plans, it should make sure that those plans are non-discriminatory in nature. Since the definition of discrimination under international law includes discriminatory effects, states must implement their policies (be they evacuation policies or policies that relate to security of tenure) in ways that do not result in a disparate impact for certain groups.

In other words, causal relationships between state actions and increased risks to life are relevant to the legal analysis of whether a state is violating its obligation to respect the right to life. Drawing again on the example of the procedural right to non-discrimination, this would mean, for example, that if a state's land tenure policies resulted in disproportionate threats to life for property-less individuals, those policies could violate a state's obligation to respect the right to life. In this way, the law provides a link between human rights and an understanding of disasters that focuses on the construction of pre-disaster vulnerability. The fact that discrimination is prohibited for a broad (non-exclusive) range of categories also means that states should take a more nuanced approach to ensuring their policies are not having disparate impacts – i.e. an approach which goes beyond simply analysing where a particular racial group lives, but which instead analyses the various elements that produce pre-disaster vulnerability (see Chapter 38).

Non-discrimination is only one of several procedural rights that apply to each level of state responsibility. In the context of disasters, another procedural requirement of particular relevance is the participation of affected populations. As the experience of the Sphere project has demonstrated, having affected populations participate in the implementation of post-disaster assistance programmes is directly related to the equity and effectiveness of those programmes (Sphere Project 2010). Similarly, the rights-based requirement for participation is related to the legitimacy and effectiveness of measures designed to prevent and respond to disasters.

The requirement that state policies do not themselves violate human rights is closely connected to the state's obligation to enact those policies in the first place. In the context of a disaster, this relates to the second level of state obligation, which is the requirement to protect against threats to life. As aptly formulated by the ECHR: 'This positive obligation entails above all a primary duty on the State to put in place a legislative and administrative framework designed to provide effective deterrence against threats to the right to life … [It] has been interpreted so as to include both substantive and procedural aspects, notably a positive obligation to take regulatory measures and to adequately inform the public about any life-threatening emergency, and to ensure that any occasion of the deaths caused thereby would be followed by a judicial enquiry …' (*Budayeva and others v. Russia* 2008: 129, 131). When there is a recurrent physical hazard that presents a foreseeable threat to life that would be mitigated by evacuation, for example, the state's obligation to protect life would include having an evacuation plan in place. The measures that will actually protect life in a given situation will always depend on the context (e.g. in some situations, evacuating might be more dangerous than staying, while in others the opposite could be true), but the obligation to protect life remains the same: a state must have a plan in place designed to minimise loss of life in the face of a foreseeable disaster. While there are guidelines that set out measures likely to protect life (such as the Sphere guidelines), the legal obligation does not dictate the details of particular policies. Instead, it sets the bar – protecting life – and requires states to determine how best to meet that bar. This obligation applies even in the context of a foreseeable risk that is not recurrent (e.g. a dense urban area on top of an active fault line, such as Port au Prince, Haiti or San Francisco, USA), since a government has an obligation to protect life when it knows or should know about an existing threat (*Öneryildiz v. Turkey* 2004) (see Chapter 41).

Within the legislative and administrative framework that states need to put in place to provide effective deterrence to the threat to the right to life, one could argue that this could also include a duty to implement measures of DRR (see Box 6.2). The extent of an obligation to engage in DRR measures is related to the type of disaster at issue, since those that are

Box 6.2 Tensions surrounding evacuations

Evacuations are a classic case of the tension and necessary balancing between potentially conflicting interests and rights, including the rights to life and freedom of movement. Evacuations are a restriction of the freedom of movement and right to choose
one's residence, in favour of the right to life.

In the context of natural disasters, evacuations can be ordered and, if necessary,
carried out forcibly, if:

- the authorities have a legal basis to do so (these powers have to be foreseen in a law);
- there is a serious and imminent threat to the lives or physical integrity/health of the
  persons that warrants their forcible evacuation; and
- all other, less intrusive, measures have been considered or tried but are insufficient
  to protect the lives or physical integrity/health of persons concerned.

The people concerned have been:

- informed, in a language that they understand, of the need to evacuate or be evacuated, of where they will be evacuated to and how this will be undertaken;
- consulted and provided with an opportunity to participate in the identification of
  suitable alternatives, evacuation routes and measures that need to be taken to
  safeguard their belongings (both those left behind and those brought along);
- evacuated in conditions that respect their dignity and safety;
- not discriminated against during the evacuation or in the place to which they are
  evacuated;
- taken to a place that is safe, does not put persons at further risk and allows for living
  conditions that respect the dignity of the people who are brought there; and
- informed, throughout the process of the evacuation, in a manner that is accessible
  to them and in a language that they can understand of the causes, duration and
  evolution of the situation.

If undertaken under these circumstances the evacuation may be forcible or involuntary,
but not necessarily a human rights violation.

foreseeable (because they are recurrent or easily predictable) mean that the government knows
or should know of the need to reduce risk. To date, however, there is no case law on this issue.

The obligation to protect the human right to life includes an obligation to implement DRR
strategies to the extent that such strategies will protect people's lives. This raises the complex
question of the extent to which pre-disaster vulnerability may be a causal factor to loss of life in
the context of a physical hazard. This is a factual question, and one which requires analysis on a
case-by-case basis in a particular geographic context.

The experience of the 2004 Indian Ocean tsunami raised the question of what a state needs
to do to protect people's lives in the aftermath of a hazard. As numerous examples of post-
disaster situations have demonstrated, threats to life continue beyond the end of the immediate
physical disaster, be it an earthquake, tsunami, hurricane or other form of hazard. In many
instances, for example, there is inadequate potable water, the provision of which can prevent
additional loss of life (Leitmann 2007).

For example, when Cyclone Nargis struck the impoverished country of Myanmar in 2008, the national government had far too few resources to protect people from the immediate post-hazard threats to life, such as dehydration, starvation or disease. Thus, the government's refusal of foreign assistance violated its obligation to protect people from foreseeable threats such as further flooding, disease and hunger in the post-cyclone period. However, the immediate aftermath of a disaster also raises questions about a state's third level of obligation under human rights law, which is the obligation to fulfil the right at issue. The requirement to fulfil the right to life includes positive efforts by the state to lessen the threat to life posed in a disaster situation, and to put in place structures that make it possible for people to fully enjoy the right to life. Thus, it would overlap with the requirement to protect life in the context of providing humanitarian aid in the immediate aftermath of a disaster. Providing humanitarian aid in the immediate aftermath of a disaster includes multiple different roles for a national government, including those which protect people from immediate threats and those which implement structures that make it possible for people to rebuild their lives in sustainable ways.

However, the obligation to fulfil the right to life extends beyond the provision of humanitarian assistance. As the Human Rights Committee has stated, the right to life is 'the supreme right from which no derogation is permitted even in times of emergency', which means it 'should not be interpreted narrowly' (UNCCPR 1982: paragraph 1). In terms of the obligation to fulfil, a broad interpretation of the right to life highlights the underlying structures needed for people to live their lives in dignity. As Wisner *et al.* (2004) have demonstrated, there is a direct correlation between the access that people have to different types of resources required for livelihoods and the level of risk they face from physical hazards such as earthquakes and floods. From a human rights perspective, this point highlights the interdependency of rights, since it implicates the right to an adequate standard of living – an umbrella right which encompasses housing, food and improvement of living conditions (ICESCR 1966: article 11). In many ways, these fundamental components of life required for an adequate standard of living are also necessary prerequisites for a person to live their life in dignity, which means that they are relevant to a broad interpretation of the obligation to fulfil the right to life.

The obligation of states to respect, protect and fulfil the human right to life has numerous implications when it comes to the threat to life posed by physical or natural hazards. There is a strong relationship between loss of life following physical hazards and the situation in which people live before the disaster (Wisner *et al.* 2004; Hewitt 1997; Lewis 1999). These vulnerability factors that turn hazards into disasters play into the legal obligations that states have to respect, protect and fulfil people's human rights. The analysis laid out above regarding the right to life only represents the obligations that pertain to one particular right. However, human rights are interrelated and interdependent. This means that the measures discussed above are also related to civil, political, economic, social and cultural rights ranging from the capacity people have to engage in advocacy to the rights that people have to enjoy the benefits of scientific progress (UNESCO 2009). The next section turns to a discussion of some of these rights.

Other human rights particularly relevant to disaster situations

The right to life represents only one of the many rights that are relevant to disaster situations. Two other rights that deserve brief mention are the right to freedom of movement and the human right to housing.

The human right to move freely within and between states and to choose one's place of living is articulated by Article 13 of the Universal Declaration of Human Rights, and is set out in treaty law under Article 12 of the International Covenant on Civil Political Rights (UDHR

1948; ICCPR 1966). It is essentially about making sure that a government does not prohibit people from choosing where they want to go or where they want to live. While the right can be suspended in times of national emergency that threaten the public order, there are still state obligations that apply even in the context of unavoidable resettlement or displacement. For example, a state may force people to evacuate from their homes in the context of a physical hazard that will threaten people's lives, or it may enforce temporary travel restrictions during an epidemic (see Chapter 30), but it may not implement these measures in an arbitrary or discriminatory fashion. In other words, any derogation of the right that does occur must be strictly limited to the necessity of the situation, and must be carried out in accordance with the principles of international law (ICCPR 1966: article 4).

Freedom of movement is one of the more pressing human rights challenges in the context of a disaster. This comprises both a particular dimension of a right to return, as well as forced or arbitrary displacement. One of the most pressing issues facing victims of a disaster is that of return or resettlement. From the Indian Ocean earthquake and tsunamis, to hurricanes in the USA, to the devastating earthquake that hit Haiti in 2010, the displacement of populations in the aftermath of a disaster often proves to be one of the most intractable problems facing disaster-affected people. Like other negative impacts that arise from the disaster, the ability to return to one's life or to voluntarily resettle elsewhere is closely tied to pre-disaster vulnerabilities: those with fewer resources, networks, and those who are consequently more at risk are also those who are more likely to be displaced and remain displaced long after the disaster itself. Therefore, the issue of displacement can be relevant to the right to life in terms of both immediate and full enjoyment of a life in dignity, and in terms of vulnerability to future threats from physical hazards.

From a legal perspective, it is doubtful whether a right to return could be included as part of an obligation to fulfil the right to life. For one, conversations about the legal status of the right to return become very political very quickly because of their application to people who experience conflict-based displacement (e.g. the Palestinians). In addition, however, there are factual questions about the extent to which a right to return to one's home is a prerequisite for the ability to fully enjoy the right to life. This is largely a factual question, since it depends on the circumstances of a particular situation. That being said, the same social vulnerability that makes a right to return so salient in the context of disasters can also heighten the risk that people confront in the face of natural hazards. Therefore, to the extent that a broadly interpreted right to life requires a state to put in structures that lessen vulnerability overall, the outcome of implementing that obligation could also contribute to solutions to displacement.

Similarly, the human right to housing must always be implemented in non-discriminatory ways, and with processes that involve the meaningful participation of affected populations (COHRE 2000). Legally, the right to housing is part of the broader right to an adequate standard of living, which appears in the Universal Declaration of Human Rights and the International Covenant on Economic, Social and Cultural Rights (UDHR 1948; ICESCR 1966). The right to housing has seen substantial legal development over the past two decades, and has been expanded upon through soft law instruments ranging from the Committee on Economic, Social and Cultural Rights to resolutions by UN bodies. Thus, in the context of a disaster, the legal scope of the right to housing includes issues ranging from the safety of housing and its ability to withstand known physical threats to the availability of post-disaster accommodation that is clean, safe and appropriate to the context. Housing is often a key component of the ability of populations to return to their homes after being displaced by a hazard, and its availability therefore speaks to more complex issues surrounding displacement.

In addition to housing and freedom of movement, there are many other (interrelated) rights that speak to the ability people have to live their lives in dignity despite the presence of physical hazards. The right to health, for example, requires states to ensure the 'highest attainable standard of physical and mental health' (ICCPR 1966: article 12). In the context of a disaster situation, this right implicates everything from the toxicity levels of floodwaters to the availability of humanitarian relief. In addition, rights related to fair trial (especially when the hazard is geographically widespread, so that legal infrastructure is destroyed), voting and education are often particularly threatened by disruption in the context of disasters arising from physical hazards. Rights pertaining to children and to gender issues are also crucial to consider in the context of disasters, particularly when post-disaster chaos gives rise to abuse and violation of those rights. Here again, there is a link to pre-existing vulnerability, since problems like increased trafficking of women and children are typically most prevalent for those who were also most vulnerable before the disaster. What all of these issues have in common is that they emerge in large part from the structures and systems that were in place before the hazard occurred. In this way, the evaluation of state responsibility to respect, protect and fulfil these rights should take into account complex approaches to understanding pre-disaster vulnerability.

Like the right to life, the rights of movement, housing, health, as well as those related to children and gender, involve three levels of governmental duty: the obligation to respect, to protect and to fulfil. Also, as with the right to life, these rights entail certain procedural requirements, including non-discrimination, information, participation and the right to remedy once a violation has occurred.

Conclusions

Contemporary international law holds that governments have an obligation to respect, protect and fulfil rights even when threats to those rights arise from indirect sources, such as physical hazards. Every level of obligation is relevant to decreasing the risk people face from physical hazards and subsequent disasters. There are many ways in which the relationship between human rights and disasters can be further explored, clarified and fleshed out by different actors:

- Governments should implement rights-based approaches to disaster risk reduction, with a particular focus on ensuring non-discrimination, both direct and indirect.
- Social scientists can provide a factual basis for claims about the causal relationship between loss of life and livelihood in the context of disaster and pre-disaster vulnerability that arises from things like disparate impact or structural discrimination.
- Human rights lawyers and scholars can expand the legal development of rights as they relate to disaster.
- Advocacy groups can bring cases, submit shadow reports to UN treaty bodies and prepare reports that analyse the causal relationship between structural vulnerability to disaster and human rights obligations in particular geographic contexts.

7

Violent conflict, natural hazards and disaster

Ben Wisner

AON-BENFIELD UCL HAZARD RESEARCH CENTRE, UNIVERSITY COLLEGE LONDON, UK

Introduction

Definitions and context

There are many ways in which violent conflict complicates, confuses and obstructs the efforts of disaster risk reduction (DRR). For example, civil war and the so-called war on drugs in the Latin American country of Colombia have displaced more than one million rural people, who have sought a more secure existence on the edges of large cities such as Bogotá (IDMC 2009). This influx of unemployed, poor people into highly dangerous locations where they squat in self-built houses in steep ravines adds a great deal to the challenge faced by emergency management planners in that country. Although some very important innovations in earthquake and landslide preparedness and mitigation have come out of Colombia in the past decade (see Chapter 51), the number of people displaced by violence threatens to overwhelm efforts to implement such innovative designs and programmes.

During the 1990s and 2000s, especially, one has seen more frequently the application of knowledge that could prevent loss from natural hazards blocked, deflected or diluted by war and its aftermath. After the tsunami that affected Sri Lanka and over ten other countries, the Tamil Tigers and the Sri Lankan government failed to agree an arrangement for sharing relief and recovery assistance (Izzadeen 2005). Thus, as ambitious as it might seem, a dialogue between the disciplines of peace studies and disaster research should be fostered, for the benefit of both sides.

Peace studies and disaster research have similar, and at some times overlapping, histories. Peace studies began as a discipline in the 1970s in part out of dissatisfaction with 'realist' approaches to international relations that take the necessity of war or the threat of war for granted in international relations. Instead, peace studies drew on a venerable, centuries-long tradition of concerns with social justice and non-violent conflict mediation to produce a positive notion of peace (Barash 2009). It sought the root causes of war in what are conventionally considered 'normal' economic and political power relations.

Peace studies approached a definition of violence and conflict from a broad perspective. Drawing from this research tradition, the analysis in this chapter treats several manifestations of violence:

- Organised activity intended to kill or harm others. Only one-to-one, interpersonal violence is excluded, although strictly speaking even such a thing as domestic violence has been shown to be correlated with disaster impacts (see Chapter 35). Organised violence takes the form not only of state vs. state war, but increasingly as the activity of war lords, urban gangs and mobs.
- Use of the threat of violence to displace or coerce others.
- Impact of 'structural violence'. This describes entire economic, social and political systems, the normal functioning of which produces and reproduces hunger, ill health and premature death.
- Impact of the historical memory of violence. Even when the actual use of violence or threat of violence has subsided, historical memory of violence may affect the ability of groups of people to generate the trust required to implement disaster reduction measures.

Contemporary thinking about DRR overlaps a good deal with notions common in peace studies, particularly the importance of history and root causes (see Chapter 3 and Chapter 4).

Magnitude of the challenge

The scale and human cost of war has so far been greater than the human loss from natural hazards. Despite the lethal reputation of earthquakes, epidemics and famine, a much greater proportion of the world's population has their lives shortened by events that are often unnoticed: violent conflict, illnesses and hunger pass for normal existence in many parts of the world, especially (but not only) in less developed countries (LDCs). If one totals deaths during the twentieth century (1900–99) from political violence, disasters involving natural hazards including epidemics, transportation crashes and industrial disasters, it is political violence that accounts for sixty-two per cent of these 424 million deaths. By contrast, rapid-onset hazards such as earthquakes and some volcanic eruptions account for only two per cent, while epidemics take twelve per cent as their share (Wisner et al. 2004: 3–6).

During the 1990s and 2000s many violent conflicts broke out around the world, and many civilians were killed, maimed (especially by landmines), injured, deliberately mutilated, starved, occasionally enslaved and displaced by the belligerent parties. Such conflicts continue. So great was the need for humanitarian relief in these conflict and post-conflict situations that some 'normal' development assistance was diverted, and opportunities for self-generated development delayed or destroyed, further worsening the position of marginal and vulnerable populations in the longer term. Furthermore, there was confusion in development agencies, including non-governmental organisations (NGOs), about how to act in regard to:

- civilian/military relations;
- relations with war lords, local elites and the army;
- ways to move from relief to recovery, and then to development;
- internationally acceptable standards of assistance; and
- mobilisation of international support for relief.

Conflicts have continued to exacerbate natural hazards such as flooding in Madagascar (2002) and Sri Lanka (2002), drought in Afghanistan (2002), epidemic disease in Côte d'Ivoire (2002) and the volcanic eruption in eastern Democratic Republic of the Congo (2002). In 2005 an example of the way in which conflict complicated recovery from the Indian Ocean tsunami in Sri Lanka is a vivid example, while in Zimbabwe capacities for dealing with drought had been undermined by chaotic years of farm nationalisations and famine relief denied to opponents of the ruling party.

Collapse of sanitary and health care systems combined with drought to produce a major cholera emergency in 2007–09. Drought and flood in Somalia have been difficult to respond to because of a protracted, ongoing and multi-sided civil conflict. In Timor-Leste violent conflict had destroyed seventy per cent of the new country's infrastructure in 1999 and the displacement of tens of thousands of inhabitants of the capital city, Dili, in 2006 combined with lack of market and transportation infrastructure to produce critical levels of hunger (Gonzales Devant 2008: 23–26). Displacement superimposed additional food insecurity on a 'normal' or baseline situation of seasonal hunger and poor harvest due to heavy rains (WFP 2006).

Interactions between violent conflict and disaster vulnerability

Violent conflict poses challenges to DRR. Efforts aimed at integrated mitigation, prevention and preparedness are made more difficult by past, present and possible future conflict. For example, early warning may be difficult under conflict conditions. Goma, a city of 500,000 in eastern Democratic Republic of Congo (DRC), had no public warning of a perilous volcanic eruption in 2002 (Wisner 2002). There was no municipal government since the city was under the control of a rebel army contesting the authority of the central government in Kinshasa.

Violence as cause of social vulnerability and institutional weakness

Violent conflict is often one of the main causes of social vulnerability. In conflict situations an increasing proportion of the casualties are civilians (Loescher 1993: 16–17; Murray *et al.* 2002). In addition to death and injury, the civilian population often finds its livelihood disrupted, leading many into more hazardous means of obtaining the necessities of life. Women and children are particularly affected by these stresses (see Chapter 35 and Chapter 36). In extreme cases famine may be the result, as in Bengal in 1943, Biafra (the Igbo-speaking breakaway territory of southeastern Nigeria) in 1969, Cambodia in the mid-1970s, Angola and Sudan in the 1980s and 1990s and Chad over the last three decades (Djindil and de Bruijn 2009).

Institutional weaknesses due to past wars combine with natural hazards to produce a downward spiral. This is evident in the case of Central America where most countries have societies shaped by wars. In the case of El Salvador, few of the elements of the 1992 peace accords had been implemented when Hurricane Mitch hit the region in 1998. Questions of land tenure and reform of the police and judiciary bear directly on social welfare and economic development. These issues were still not settled when two earthquakes hit in 2001, killing more than 1,000 people, injuring more than 8,000 and causing damage valued at US$2.3 billion. Forty per cent of the country's health centres were destroyed along with one-third of the schools. Some 150,000 homes were destroyed and another 185,000 were damaged (Wisner 2001a). Since then, the poorest rural people have suffered hail and drought, both devastating food crops, and the collapse of the world price for coffee. Small farmers and landless labourers have suffered most. Caught up in a similar vicious spiral, 500,000 poor rural Nicaraguans have crossed the border into Costa Rica seeking work. These immigrants are likely to live in places and in conditions that expose them to hazards such as flooding, landslides and disease (Bail 2007).

Such mass movement of poor people may be interpreted as motivated by a 'pull factor', namely economic opportunity, and not the 'push factor' or war. To some extent, that is true; however, one has to place the economic and institutional weaknesses of El Salvador, Nicaragua and other countries in the context of long histories of civil war.

Displacement of large numbers of people in war and other violent conflicts can lead to new risks. According to the United Nations (UN) High Commission for Refugees there were

15.2 million refugees in the world at the end of 2009. These numbers do not include internally displaced people, only those who have crossed a national border seeking refuge. The total of all people 'forcibly displaced' was 43.3 million (UNHCR 2010). Most of these refugees are fleeing violence. In many cases they face new risks that include exposure to disease and unfamiliar hazards in new rural or urban environments. Deadly outbreaks of cholera and other communicable diseases have affected displaced persons who fled the genocide in Rwanda and, earlier, the civil war that led to the creation of Bangladesh. In the densely populated neighbourhood of Alexandra Township in Johannesburg, South Africa, refugees from the war in Mozambique were among the poorest residents (Wisner 1995). When international refugees are finally repatriated to their home countries, they often end up in new locations – not their original homes. These new locations are sometimes hazardous or experience regular hazards with which the new population is unfamiliar.

Violent conflict often destroys infrastructure, which may intensify natural hazards such as flooding, the effects of droughts or epidemic disease. Among the infrastructure targets in recent conflicts have been irrigation systems, dams, levees, roads, bridges, water treatment plants, refineries, pipelines and electricity systems. Such destruction may rapidly erode public health and also throw large numbers of people into unemployment. Both these effects increase the population's vulnerability to future hazards. In the case of Iraq, the destruction of water treatment and distribution systems, drainage and sanitation facilities, and electricity supplies during the recent wars has contributed to health hazards that have cost the lives of more than 500,000 children (CESR 2003). During the first Gulf War, the US-led international coalition under a UN mandate destroyed electricity supplies, shutting off power to hospitals and water treatment facilities. This began a series of disastrous events that undermined public health. Transportation networks were also targeted so that distribution of food and other essential items to Iraq's primarily urban civilian population was disrupted.

Violence increases hazard frequency and intensity

Violent confrontations often wreak havoc on vegetation, land and water, and they undermine sustainable development. Chemical defoliants were used by the USA in South-East Asia during the Vietnam War. This caused long-lasting health effects and aggravated flood hazards (Austin and Bruch 2000). Unexploded ordnance and landmines make some agricultural land unusable in post-war regions. In 2009 there were sixty-six countries affected (ICL 2010). Deforestation of mountain slopes in Pakistan by the so-called 'timber mafia', said to include the Taliban, has been discussed as contributing to the devastating 2010 floods in that country (Walsh 2010).

Displacement of civilians also has a role. The presence of large numbers of displaced persons and refugees in dense concentrations can cause local de-vegetation and soil erosion (Black 1998). In Central Africa in the 1980s and 1990s areas from which civilians had fled grew wild, producing prime habitat for the tsetse fly, vector of deadly livestock and human disease, which returned.

Violence as an obstacle to effective disaster relief and recovery assistance

Violent conflict can interfere with the provision of relief and recovery assistance. The wars in Africa during the 1980s and 1990s often challenged the ability of humanitarian agencies to provide essential relief to the civilian population. In Sudan the UN Children's Fund (UNICEF) was able to negotiate 'corridors of tranquility' during its so-called 'Operation Lifeline Sudan' (Minear 1991). More commonly, arrangements for relief and recovery assistance have been ad hoc, unreliable and rapidly changing, as they have been more recently in Afghanistan and Iraq.

Worse than this, there is some evidence from case studies, mostly in Africa so far, that middlemen and war lords actually profit from and wish to perpetuate a 'relief economy' in which they are able to trade relief goods they steal or divert for guns, or use relief aid they come to acquire to 'buy' support among civilians (Keen 1994).

Violent conflict diverts national and international financial and human resources that could be used for development and for mitigation of natural hazard risk. During its war with Eritrea during the 1990s, Ethiopia let its national famine early warning system deteriorate. Resources were used for war and not for such social investments as maintenance of the food monitoring system that had been put in place following the famines of the 1980s. Subsequently the Ethiopian government was 'surprised' by a widespread food emergency that it should have been able to detect much earlier (Westing 1999). Eritrea has also suffered increased vulnerability to hazards as the result of the war and its societal militarisation (see Box 7.1). On the international scale, donor attention has been so fixated on Afghanistan and Iraq that insufficient attention has been given to a fulminating combination of HIV/AIDS, flood and drought in southern Africa, among other 'under-reported' humanitarian emergencies (Wisner and Gaillard 2009).

People fleeing violent conflict, such as illegal immigrants and asylum-seekers, who find themselves in large cities in Europe, North America and elsewhere, may be highly vulnerable to natural hazards, but they can be difficult for professionals to contact. This is because of language difficulties as well as a lack of trust. Cape Verde, a small island country off the coast of West Africa with a tradition of emigration, is now facing an influx of illegal immigrants from conflict-torn Guinea Bissau. Migrants primarily seek shelter in the capital Praia and have little choice but to settle informally in flash flood-prone gullies (*ribeiras*) (Gaillard 2010a) (see Chapter 38). In the aftermath of the 1994 Northridge earthquake in Los Angeles, illegal immigrants avoided hospitals and recovery services for fear of deportation (Bolin and Stanford 1998). Some people affected by the 2010 Haiti earthquake actually took advantage of having been brought to Guadeloupe for medical treatment to escape, and now live illegally and hidden on the slopes of Soufrière volcano, a hazardous location (Gaillard 2010b).

Violence as an obstacle to good DRR practice

Participatory methods meant to empower and engage socially vulnerable groups may be difficult or impossible during violent conflicts (see Chapter 51 and Chapter 59). An integrated approach to DRR has much in common with what the UN Development Programme (UNDP) calls a 'developmental' approach to humanitarian assistance (Smillie 1998). In both cases the goal is not only to address the specific crisis at hand – an earthquake or flood, in the first case, a violent conflict in the second – but to do so in a way that builds capacity to mitigate or to prevent a future occurrence. In both cases, the role of local knowledge and capacities is important, hence participatory methods based on trust are vital tools (see Chapter 9 and Chapter 64).

In conflict situations people are less inclined or able to take part in such 'bottom up' efforts, be they directed toward preventing damage from future natural hazards or toward peace-making. Some successes have been registered in Afghanistan with community-based DRR. However, these are likely to be an exception that proves the rule that under intensive violent conditions participatory and developmental approaches are very difficult (UNDP–Afghanistan 2010).

Application of existing knowledge for mitigation of risk from natural hazards is often difficult or impossible during violent conflict. Over the past four decades or more a large knowledge bank has grown as regards preparedness, mitigation, warning and response to natural hazards. Violent conflicts disrupt the communication necessary to make application of this knowledge effective. Advances in hydrological modelling and the use of current information and

---

Box 7.1 Eritrea: War and vulnerability to drought

Katuscia Fara
Environment and Disaster Risk Management Specialist

The humanitarian situation in Eritrea in 2010 was alarming. An increasing number of people were fleeing the country as a result of growing food insecurity. In the decades of the 1990s and 2000s the country had become increasingly vulnerable to the impacts of successive droughts and coping capacities of the population had been eroded. However, the causes behind Eritrea's increased vulnerability cannot be attributed solely to lack of rainfall.

One of the poorest countries in the world, Eritrea declared its independence in 1993, two years after ending a thirty-year liberation struggle with Ethiopia. With about eighty per cent of the population living in rural areas, the country has always been highly reliant on rain-fed agriculture and pastoralism for food security. While tradition-ally people adapted their practices to the predominant arid and semi-arid conditions, insecure land tenure, clearance of land cover during the years of fighting, increasing population densities and the expansion of agriculture into fragile areas have led to high erosion rates and resulted in lower yields.

Eritrea's inability to produce more food and its dependence on food imports was further aggravated by the two-years' war that broke out with Ethiopia in 1998 over the border zone. The conflict led to over 100,000 deaths and displaced over one-third of its 5 million citizens. War left many internally displaced persons, much destroyed infra-structure and landmines in Eritrea's most fertile lands in the southwest.

Since the end of the conflict the Eritrean government had become increasingly repressive and had progressively forced the majority of the population into national service. Both men and women enter into the service of the state from a young age, most of them serving in the military. This resulted in a critical shortage of agricultural workforce, thus disrupting food production, and reduced the range of household income opportunities and coping strategies available to people, such as livestock raising and off-farm employment. The result has been a progressive process of impoverishment.

The increasingly intransigent and controlling attitude of the government towards development organisations prompted the expulsion in 2006 of most of them, including the World Food Program, with only nine NGOs remaining operative in the country at the time of writing (2010). Even these had severe limitations to their activities. The government focus on prioritising defence spending and controlling the population, its tight control over the currency that effectively limits imports, together with severe fuel shortages, have led to rising prices of consumer goods (when available) and the further decline of the already weak economy. Strained relations with neighbouring countries including Yemen, Djibouti and Sudan have limited the coping strategies of pastoralists who can no longer follow rains across national boundaries and have restricted access to external markets. With very little food aid imported, due to the government monetisation policy, the already fragile food security situation could rapidly deteriorate.

communication technologies (ICTs) make management of large river basins feasible (see Chapter 17 and Chapter 21). However, tensions among the twelve countries within the basin of the Nile make common management difficult, even in the absence of overt warfare. It is not only current violent conflicts that complicate the DRR. A long history of conflicts, as, for example, in southern Africa, left behind weak infrastructure and institutional arrangements. Such a history may have played a role in the breakdown in communications between authorities in Zimbabwe and Zambia who released water from dams on the Zambezi River that took Mozambicans downstream by surprise during the floods in 2000 (Christie and Hanlon 2001) (see Chapter 21).

Implementing DRR in the context of violent conflict

Violent conflict affects each and every element in DRR. So, in the face of such blockage of efforts in times of war, situations of violent conflict and post-war conditions, one must ask what can be done.

Adapting DRR to fit humanitarian response in conflict situations

The UN has laid out the basic requirements for effective humanitarian assistance. These constitute a set of a dozen 'core protection principles' adopted by the Security Council in 2002 (Annan 2002). Bearing in mind the previous discussion of how conflict can obstruct response to and recovery from the additional humanitarian effects of natural hazards, some of these principles have practical and policy implications for DRR. These principles and others bear on DRR as well, including long-term prevention and mitigation of disaster risk.

Adapting disaster response and recovery assistance to conflict situations

Separation of civilians and armed elements – The UN principle is to 'maintain the humanitarian and civilian character of camps for refugees and internally displaced persons'. In practice this principle is difficult to enforce. Recent large international efforts to provide relief have inevitably involved civilian UN workers, civilian international NGO staff and local civilians working alongside military personnel from multiple countries.

Responding to the 2010 Haiti earthquake, a pre-existing UN peacekeeping force led by the Brazilian military played a role, as did UN peacekeepers after Hurricane Mitch in Guatemala. Towards the end of the civil war in Sri Lanka some civilians who had been forcefully 'separated' from separatist rebels and resettled in camps found themselves flooded. Such a situation of a civilian population fenced in and guarded by the military subsequently affected by a natural hazard is ambiguous from the point of view of the UN principle of 'separation'.

Access to vulnerable populations – The UN principle is to 'facilitate safe and unimpeded access to vulnerable populations as the fundamental prerequisite for humanitarian assistance and protection'. Another side of the 'separation' coin is the question of access to vulnerable populations. This has been a problem for humanitarian workers for decades – at least beginning with the Biafra civil war in the late 1960s in Nigeria. In conflict situations there and later in Sudan, Somalia, Afghanistan, Colombia and elsewhere, access to vulnerable groups may only be negotiated through war lords, rebels or the national military. In these cases people already caught in the crossfire of violent conflicts had been afflicted by secondary disasters with natural triggers such as drought, epidemic disease, flood, landslide and cold weather and blizzard.

Effects on women – The UN principle is to 'address the specific needs of women for assistance and protection'. Rape has become a weapon of war in many of the civil conflicts since the 1990s. State-of-the-art provision of post-disaster shelter already includes consideration of protecting women and children from sexual abuse by humanitarian workers and others, including attention to public lighting, the location and design of sanitary and bathing facilities, provision of specialist services concerned with women's reproductive health (see Chapter 35 and Chapter 44) and sensitisation. In addition, workers who were involved with supervening natural hazards in the course of a conflict should be highly aware that many of the women in their care may well already have been abused during the hostilities and provide counselling and reproductive health care.

Effects on children – The UN principle is to 'address the specific needs of children for assistance and protection'. What applies to women also applies to children. A good deal of work has been done on psychosocial recovery (see Chapter 36 and Chapter 47) of those affected by natural hazards and the integration of therapeutic exercises (via art, drama, song) into the 'school in a box' approach of UNICEF to re-starting education as soon as possible (Wisner 2006a). This is all the more important when children have already been traumatised by war. The situation and needs of former child soldiers are even greater and more complex. A shelter or camp following natural hazards during wartime could easily have former child soldiers among the displaced. They should be identified and provided with special assistance.

Safety and security of humanitarian and associated personnel – The UN principle is to 'ensure the safety and security of humanitarian, United Nations and associated personnel'. Many humanitarian workers have been killed during missions over the past two decades and the rate of workers attacked per 10,000 in the field more than doubled in 1997–2008 (ALNAP 2010: 26). On some occasions NGOs have temporarily shut down their operations and left an area or a country when egregious assaults on their staff have taken place. In some cases, such as Somalia, most UN and international NGO assistance takes place by remote control from Nairobi, in bordering Kenya, and it is local staff who take the risks. This raises difficult ethical questions. Most expatriate humanitarian workers have life insurance, medical evacuation insurance and solid training in emergency medical procedures, communication and avoidance of kidnap. Not all national staff may enjoy these benefits and safeguards.

Media and information – The UN principle is to '(1) Counter occurrences of speech used to incite violence. (2) Promote and support accurate management of information on the conflict.' The age of digital, satellite communication, 'citizen journalism' via blogs and social networking platforms makes it very difficult for an 'incident commander' to control information. Rumour is always rife in post-disaster situations, and it is likely to be compounded by a conflict situation. Did the rebel forces blow up a levee to cause the flood? Is the government army stealing drought relief food for its own use? These kinds of rumours need to be countered in keeping with the UN's wise general principle (see Chapter 63). In addition, within temporary settlements of the disaster (and war) displaced, citizen journalism and the new ICTs as well as old-fashioned printed news sheets can be used to inform people about practical issues like public health provisions and to boost morale (Wisner and Adams 2003).

Adapting DRR to conflict and post-conflict situations

Training of security and peacekeeping forces – The UN principle is to 'ensure adequate sensitisation of multinational forces to issues pertaining to the protection of civilians'. Mentioned above was the role of peacekeepers already in place in a conflict or post-conflict situation when an earthquake or hurricane occurred. The training of such military police and infantry troops from many nations has improved a great deal over the years. Their general training should include an overview of the past natural hazards in the location where they will be posted and what to do both to protect themselves and then to assist local authorities and the people in response and immediate relief. Without training it is easy for troops to mistake survival foraging for looting, and untrained military personnel can overreact. Whether and how these troops would be involved in a disaster involving a natural hazard while they are deployed should be written into the protocols agreed by the UN, regional bodies such as the Economic Commission of West African States (ECOWAS), particular countries providing staff and the national government.

Security, law and order – The UN principle is to 'strengthen the capacity of local police and judicial systems to enforce law and order'. DRR has institution and capacity building as a core value (see Chapter 51 and Chapter 52). A valuable contribution to all efforts at capacity building among the national and local police would be modules and material on their role in responding to the consequences of natural hazards. Such training should be on-going in the 168 countries that signed up to the Hyogo Framework for Action (HFA) (UNISDR 2005b). It is all the more important in conflict and post-conflict situations, not only for its practical utility, but also as part of the humanising and democratising of police forces. Those forces may have been themselves brutalised and desensitised to community needs in the course of prolonged, polarising, violent conflict.

Justice and reconciliation – The UN principle is to '(1) Put an end to impunity for those responsible for serious violations of international humanitarian, human rights and criminal law. (2) Build confidence and enhance stability within the host State by promoting truth and reconciliation.' Building the confidence that the public has in its country's judiciary, as well as building the capacity of judges, can benefit from raising awareness that no so-called 'natural' disaster is, in fact, 'natural' at all. There are always human causes that contribute, and sometimes malfeasance or corruption that are among the root causes. Not all judges may be aware of judicial proceedings against those allegedly responsible in these ways for, among others, school collapse in earthquakes in Italy and Turkey (Transparency International 2005) and a deadly landside in Hungary (see Chapter 6).

Disarmament, demobilisation, reintegration and rehabilitation – The UN principle is to 'facilitate the stabilisation and rehabilitation of communities'. Demobilisation provides a large number of potential reconstruction workers in situations where war has caused great destruction in the built environment, infrastructure, and environmental degradation. Such a workforce is all the more valuable where a natural hazard during conflict heaped destruction upon destruction. Deploying former militants from all sides of a conflict in this way also can provide bridging employment and income to ease the return to civilian life. Where training can also be included, all the better in terms of later livelihood recovery. A good example is the post-1992 ceasefire resettlement of former combatants from both sides of the civil war in El Salvador in the Lower Lempa River Basin (Lavell 2008; Wisner 2008). With the assistance of civil society organisations, the settlers carried out hazard and vulnerability assessments and worked with the Ministry of Environment to develop measures to reduce both flood and earthquake risk, while also improving pest control, crop storage and marketing, and coastal resource management.

Mine action – The UN principle is to 'facilitate a secure environment for vulnerable populations and humanitarian personnel'. De-mining must be central to DRR where there has been conflict because, if for no other reason, landmines deny access to so much farm land, and make journeys to fetch clean water and fuel wood hazardous in the extreme. Since effective DRR must be seen as integrally linked with livelihoods and achievement of the Millennium Development Goals (MDGs), removing this obstacle to access to vital resources has to be a high priority. That said, it should only be attempted by trained workers in a systematic and professional manner. Maps are important if those involved in the conflict kept such records, although they may not be accurate since landmines are easily moved by floods and mud flows. It should not be attempted by civilians from affected communities on their own or by DRR or other development workers. Mine awareness training should be an integral part of DRR outreach and activities and embrace the whole of affected communities, including children who have been maimed by mistaking such a metal object for a toy.

Natural resources and armed conflict – The UN principle is to 'address the impact of natural resource exploitation on the protection of civilians'. Control over such natural resources as fossil fuels (especially oil) and minerals such as diamonds, gold and uranium have contributed to violent conflicts in the past. More localised conflicts over pasture, water and arable land are not rare. Thus, post-conflict reforms involving legislation, the judiciary and ad hoc reconciliation bodies need to address access and ownership issues if a recurrence of conflict is to be avoided. At the same time, taking a livelihood approach to DRR, access to natural resources is also key to household and community ability to face natural hazards (see Chapter 14 and Chapter 58). This set of issues becomes all the more crowded when one adds mounting international and national programmes for forest resource management as a means of both climate change mitigation and adaptation, such as REDD (Reduction of Emissions from Deforestation and Degradation), often ignoring the needs of the local people.

Humanitarian impact of sanctions – The UN principle is to 'minimise unintended adverse side effects of sanctions on the civilian population'. Sanctions can weaken the ability of authorities to provide disaster response and can erode the DRR abilities of local communities. These negative consequences must be borne in mind by political decision-makers and monitored by DRR researchers and relevant NGOs.

Box 7.2 Disaster diplomacy

Ilan Kelman
Center for International Climate and Environmental Reseasrch, Oslo, Norway

As a contribution to understanding the politics of disasters, the website *Disaster Diplomacy* (www.disasterdiplomacy.org) examines how and why disaster-related activities do and do not induce co-operation among enemies. The key phrase here is 'disaster-related activities', covering pre-disaster actions such as prevention, mitigation and preparedness, along with post-disaster actions such as response, recovery and reconstruction.

Case studies examined cover three different categories. First, *Disaster Diplomacy* deals with a specific geographic region that experiences disaster, such as North Korea's international relations following floods, droughts and famines since 1995, along with an April 2004 train explosion. Second, the website considers a specific disaster incident. Tsunami diplomacy had potential in many countries following the 26 December 2004 Indian Ocean tsunami disaster. The most successful outcome was the peace agreement in Aceh, Indonesia, but negotiations had pre-tsunami origins and cannot be attributed solely to the tsunami's aftermath (Gaillard *et al.* 2008a). Third, this web-based collection includes more general disaster-related trans-boundary topics. Examples are international co-operation in identifying disaster casualties from many countries and regional seismic hazard assessments in conflict zones.

In examining the numerous case studies and attempting to develop conceptual models to categorise and explain the case studies, while potentially predicting outcomes of future examples, some patterns emerge. Overall, disaster-related activities do not generate entirely new diplomatic efforts, but they can catalyse diplomacy that has a pre-existing basis. That basis might be cultural or trade links connecting the parties involved or, as with post-tsunami Aceh, ongoing secret negotiations amongst the parties in conflict.

Disaster-related activities seem to influence conflict and peace-making over the short term, on the order of weeks or months. For longer timeframes, such as years, non-disaster factors tend to supersede disaster factors in influencing diplomacy, conflict and peace. These factors could range from a leadership change, to promoting an historical grievance, or to a decision that conflict has advantages irrespective of disaster-related costs.

For instance, when Cuba was led by Fidel Castro, both Cuba and the USA offered one another assistance after respective hurricane impacts. In all cases, the aid recipient put effort into ensuring that disaster diplomacy would not result from the aid offer. Mutual enmity bolstered the power base of Fidel Castro along with many anti-Castro politicians in the USA.

Many other reasons contribute to explaining why disaster-related activities sometimes have less diplomatic influence than might be desired, expected or assumed. International disaster assistance is not always needed, so in some cases there is no basis for disaster diplomacy. Even where disaster assistance is needed and is given, political gaffes can derail reconciliation. After over 26,000 people died in an earthquake in Iran in 2003, the USA provided aid. The US government tried to follow up with a high-level political visit that Iran declined, cooling any thoughts of long-term earthquake disaster diplomacy.

Overall, the political lesson from examining disaster diplomacy is that decision-makers frequently see priorities other than dealing properly with disasters or creating peace, even after a disaster or when goodwill is present.

DRR's possible contribution to peace-making

So far this chapter has dwelt on the impact of conflict on DRR. What about the reverse? Is it possible to implement and institutionalise risk reduction and disaster management systems in ways that address disparities and grievances that may lead to violent conflict? Income and power disparities are certainly among the causes of violent conflict. Thus, if one is in the situation of a consulting engineer, for example, giving advice about storm water drainage and a large city's water supply, is this not the perfect opportunity to suggest an extension of the drainage system and safe water supply into the low-income squatter settlement adjacent to the city centre? If one begins to look for opportunities to use risk reduction to eliminate disparities that are among the root causes of conflict, it is surprising how many opportunities there are.

Conclusions

War and violent conflict complicate the challenges of disaster management and DRR in a number of ways. These complications can be taken into account in order, at a minimum, to increase the chance that policy advice, programming, project planning, design and training activities will be robust enough to survive the chaos of conflict situations.

More ambitiously, DRR can contribute to the building of stable and effective institutions that contribute to the credibility of governments in the eyes of the poor, and even help to reduce the economic and social disparities in fragile states that can lead to violent conflict. Indeed, it should be possible not only to 'build back better' – a common buzzword since the Indian Ocean tsunami in 2004 – but to 'build back better and build peace at the same time'.

Culture, knowledge and religion

8

Culture, hazard and disaster

Kenneth Hewitt

DEPARTMENT OF GEOGRAPHY AND ENVIRONMENTAL STUDIES, WILFRID LAURIER UNIVERSITY, CANADA

Introduction

Culture is discussed as an influence on human affairs in many fields including disaster risk reduction (DRR) and management. However, in much of the latter, the primacy of scientific data and models is taken for granted and the focus tends to be on technical solutions, the work of professionally trained personnel, officials and non-governmental organisations (NGOs). There is often little awareness of, even resistance to, the idea that socio-cultural issues are key aspects of risk and disasters. If it is mentioned, culture is often seen more as an impediment to understanding and effective action. Indeed, it is widely assumed that disaster-related work is, or should be, simply driven by the 'sovereign facts': as close as possible to an exact mirror of environmental and societal realities.

It is, therefore, useful to begin by asking whether humans do view or can respond to environmental dangers, even injury and damage, as unmediated facts? Or are all people, including 'experts', always engaged with hazards through learned and shared responses plus collective, place- or institution-based views, values and beliefs? In fact, few of the actions that adjust and adapt society to given habitats seem to be simple reflexes, although some may become more or less routine. However, actions that are critical in adjusting to changing circumstances, while setting and achieving goals, are always preceded and accompanied by thought, messages and conversations — even when people are not actually following instructions, training manuals, mandates and protocols already laid down, which are, essentially, cultural constructs.

It seems unlikely that such thought and dialogue can be driven only by the 'sovereign facts'. Rather they will be influenced by conditions in and controls over how people exchange information and ideas, how they arrive at shared goals. There is always an on-going process of translation or mutual adaptation of what is observed in terms of education, the tools and priorities people share with their neighbours and the institutions in which they work. Experience suggests that people's responses to apparently similar facts can vary widely in different societies, times and places.

The foregoing supports the view that socio-cultural phenomena are always likely to be critical aspects of DRR. Knowledge and practical endeavours are mediated by shared languages, meanings, modes of communication, association, preferred technologies and the terms of collective action, i.e. by and through what is usually called the cultural context. Hence, culture would seem to be integral to understanding risk and responses to danger or loss. Whether these remarks seem self-evident or problematic, they refer to matters which have often been ignored in research and official work on risks and disaster, if with notable exceptions (Douglas and Wildavsky 1982; Oliver-Smith and Hoffman 2003).

To be sure, part of the problem is that notions of culture are contested if not chaotic, even among those who recognise their importance. However, to treat these questions as marginal or irrelevant concerns seems unwise when speaking of public, national or global threats, especially in actions involving more or less large and diverse groups or populations. Indeed, culturally aware research suggests that it is, above all, where socio-cultural conditions are ignored that they tend to frustrate safety goals the most (Mankiller 2009). It does mean that a critical and effective approach requires reflection upon the technocratic, corporate and governmental styles in today's world. These, too, comprise a 'culture'. Yet it not only differs in crucial respects from that of people most vulnerable to hazards and a majority of disaster victims, but also modern organisational cultures have an enormous influence on the fate of such people. We must confront reports of institutions dominated by technocratic culture being poorly prepared to assist people because of cultural difference and insensitivity or, not least, the assumption of cultural superiority.

Questions of definition

> Depending on who is doing the talking, of course, 'culture' may be made to signify any number of different material processes ...
>
> *(Adam and Allan 1995: xiii)*

Definitions of culture are generally unsatisfactory, whether from incompleteness, claiming too much or being overly influenced by the culture of whoever does the defining. Perhaps it is a mistake to attempt to isolate a cultural element or essence in the mix of multi-disciplinary contributions to risk and disaster. Here we will resist describing culture as 'a thing' – some separate and distinct entity or phenomenon that one can pick up and examine like an apple among oranges.

On one level, efforts to characterise cultures, or to distinguish groups of people, variously emphasise behaviour, context, artefacts or beliefs. Intellectuals and philosophers tend to argue that the underlying fabric or enablers of culture are symbolic, language and communication (Edgar and Sedgwick 1999). Intelligence or 'cognitive' skills are seen as basic: irreducible parts of any person's capacity to enter and share in a culture. Without them, one lacks the ability to receive and interpret information or initiate communication, to share in a belief system or set goals.

What is indicated in many definitions is how control of media, of the language and/or the message, offers powerful instruments and incentives to shape agendas (see Chapter 11 and Chapter 63). Choices and uses of media, or the settings of exchanges, can be made to serve some purposes or social arrangements better than others. Such is the cautionary tale of possible abuse or hijacking of cultural projects. Running parallel with it, however, have always been the possibilities to refine and open up any language, medium or knowledge field to novel ideas and greater inclusiveness.

An alternative path is to consider the sense of 'culture' in terms such as agriculture and silviculture, cultured pearls and cultured persons, with their emphasis on process rather than content, form or medium. This highlights the shaping or cultivating of phenomena, members of the group, or the self; something that can be constantly evolving. One must reflect carefully on

Box 8.1 A landslide story from Central Karakoram Himalaya

Long ago, Old Haji Ali said, where the waste of boulders and sand called Ghoro Choh now lies, there was a great city, its people rich and blessed with fine houses, fields and orchards. One day a travelling holy man came there and asked the ruler, the Rajah, for food and shelter. He was turned away, and none of the other wealthy folk would help him either. Finally, an old woman, the least well-off, gave him shelter and a share of her food. Next morning he instructed her to climb up to a place of springs above the valley, now the site of a tiny village called Mango. The holy man climbed up the opposite slope and smote the rock with his staff. A great part of the mountain came down burying the city, with all its wealth and pride.

This sounds like the morality tales told of disasters in many other places, including other landslides in the Karakoram. It may seem more like myth than fact. However, on the one hand, who may say that such tales do not work well in their communal ethos? No one has the technology to even predict, let alone prevent these great collapses. However, a local story with a 'moral' may help guide and in certain ways comfort people. Most of the inhabitants live rather harsh lives, are remarkably devout and communal consensus and reciprocity are basic to everyday survival. On the other hand, Old Haji proved to be a good geologist and landscape detective. He could teach things absent then from studies of Karakoram landforms.

For more than a century, the Ghoro Choh landslide had been interpreted by some of the foremost scientists as moraines dumped by glaciers during the Pleistocene, not Holocene landslides. Old Haji gives the main clue that would help in definitively separating these landslides from the many moraines that are indeed found near them, and for which many others had been mistaken. The ridges of Ghoro Choh consist of a single, green crystalline rock. The glaciers that once filled Shigar valley carried quite different rock types and a great variety of them, as the rivers passing through Ghoro Choh do today. Yet, highly influential studies of the Quaternary here were sure that they had found evidence at Ghoro Choh not just of ice, but the same series of three or four glaciations that had been identified in Europe. Whoever traced out the events for Haji Ali's story knew better! Science may be dedicated to the facts, but it has its myths too: conclusions that tend to suit the preoccupations and convictions of given times and disciplines.

This is not intended to be an argument against science, but first to recognise that it, too, needs careful, critical reflection as a cultural construct. That is not to suggest that modern knowledge has no worth in contexts where it is not being developed, only to illustrate the value of openness to other cultural constructs. More specifically, it brings us to the argument for attention to socio-cultural contexts.

the relations between culture and cultivation in the larger sense. Virtually anything with which humans come into contact is 'cultivated', not just gardens, clients or friendships. For instance, the methods employed in community-based DRR attempt to build trust and solidarity, apart from their more technical functions in hazard mapping, etc. (see Chapter 59 and Chapter 64). In this way, everything and anything is given meaning and value, not least 'nature', hazards and wilderness. As such, culture is seen to engage with the universal human activities of satisfying needs, seeking security and pursuing interests. It situates culture at the heart of matters as basic, changing and diverse as work, nurture, education, institutions and economy. If required to choose, a radical approach will surely adopt this one.

However, rather than seeing these as separate alternatives, cultural anthropologists and others regard the symbolic and the cultivating work of humans as interdependent and mutually supportive. Edgar and Sedgwick (1999: 102) draw together the themes explored above in suggesting that 'the two most important or general elements of culture may be the ability of human beings to construct and to build, and the ability to use language [to include] all forms of sign system'.

Some background to the place of culture in disaster studies

There is, of course, an immense literature dealing with cultural issues (Inglis 2004). By far the largest body of work comes from thinkers, ancient and modern, and in every part of the world, who have contributed to ideas of disaster and danger (Tuan 1979). They include philosophers, artists and theologians, the founders of religious orders, who have thought about the meanings of fear, trauma and untimely death, belief, caring and responsibility. In the modern world much of this is carried forward by a huge 'arts and humanities' literature, and by institutions committed, in one way or another, to a culture or ethic of safety, responsibility and compassion (see Chapter 11 and Chapter 12). It is likely that everyone is influenced, perhaps unknowingly, by such cultural debates and creative responses. Yet, except for persons and agencies whose work is subsumed by religious faith or an ethic of care, it rarely appears directly in the modern risk and disasters field – unless in quotations at the head of chapters (see Chapter 10 and Chapter 47).

Nevertheless, a good deal of modern professional and academic work addresses cultural matters, if it has played a fairly marginal role in DRR until recently. In particular, anthropologists have been well-placed to address these questions, based on their study of ideas and responses to danger among pre-modern peoples, usually small communities and sub-state populations (e.g. Oliver-Smith 1986). More recently, an important emerging field has looked at the role of what is termed local or indigenous knowledge of disaster risks (e.g. United Nations 2005). Questions of culture belong to the broader subject of how knowledge is produced and shared, and how some knowledge comes to dominate (Marglin and Marglin 1990) (see Chapter 9).

By contrast, sociologists and some of the more recent anthropological studies struggle with issues of cultural context in urban and industrial societies, and in the large, cognate field of 'accidents' (Kroll-Smith and Crouch 1990; Oliver-Smith 2006; Rodríguez and Dynes 2006; Squires and Hartman 2006; see also the blogs *Savage Minds* savageminds.org/2005/09/07/disaster-anthropology and *Anthropologi Info* www.antropologi.info/blog/anthropology/2005/the_anthropology_of_disaster_anthropolog).

Psychologists, psycho-historians and health professionals have acknowledged the cultural dimensions of suffering in disasters and war, incorporating challenges of meaning and mental distress for people with disabilities, including the growing field of post-traumatic stress disorder (Wolfenstein 1957) (see Chapter 47). Some scholars have had a considerable influence on contemporary understanding of disaster by addressing the question of culture (e.g. Anderskov 2004; Oliver-Smith and Hoffman 2003).

More radical approaches such as political ecology have challenged the whole idea that disasters are 'natural', rather arguing that they are caused by socio-economic or political interests and conditions (e.g. Hewitt 1983b). These approaches have had a preference for materialist interpretations in which culture is ignored. Only rarely is it included by political ecology as part of the framework of safety improvements (e.g. Wisner et al. 2004).

A failure to consider cultural matters begs the question of profound differences in the way different groups organise, control and adjust their relations to similar conditions and other peoples. What usually appear as technical or official 'materialist' viewpoints may not be helpful

Box 8.2 Māori and disaster risk reduction in Aotearoa/New Zealand

Rawiri Faulkner and Julia Becker
GNS Science, New Zealand

Māori are the indigenous people of Aotearoa (New Zealand). Māori organise and associate through tribal groups called *iwi* and *hapū* (tribes and sub-tribes), which are a base for political, economic, cultural and environmental discussions and planning. From the arrival of European settlers throughout the 1800s, Māori traditions, values and customs were negatively affected through colonisation and the imposition of a Western way of life. In recent times, Aotearoa has seen a revitalisation of Māori culture, with recognition of the contribution that indigenous knowledge can make to all New Zealanders.

Māori retain unique indigenous knowledge relating to natural hazards. This is developed through their interaction and adaptation with the natural environment over many centuries. Examples of this can be seen in the views that many Māori share regarding the relationship they have with natural earth processes. The Ngāti Rangi people, who live near Mt Ruapehu, see the need for communities to work with, rather than in conflict with the natural environment. As a consequence, many of the *marae* (community settlements) are located in places that mitigate the risks associated with living at the foot of Mt Ruapehu. For example, many *marae* are located away from the path of lahar (volcanic mud flow) and other natural hazard events. There are also numerous examples of where *marae* have been relocated due to hazard events of the past. One obvious example is the relocation of Waihi Marae in Tokaanu, which was moved away from the foot of the Hipaua landslide after slope failures in 1846 and 1910.

A current challenge for DRR in the Māori context is how to incorporate indigenous knowledge into contemporary planning and policy development and implementation. A legislative requirement exists in New Zealand for local government to consult with *iwi* and *hapū* on matters of environmental management, including natural hazards. However, such consultation rarely captures the extent of indigenous knowledge, nor is it often able to effectively incorporate indigenous knowledge into policy. Low levels of participation and collaboration between *iwi/hapū* and agencies such as local government are one barrier to achieving a desired outcome. Additionally, as knowledge is considered a *taonga* (treasure), many *iwi* and *hapū* groups feel that the wide dissemination of traditional knowledge can lead to the information being abused or used inappropriately.

Recent work with respect to flooding in 2004 in the North Island has shown how Māori communities can contribute effectively in responding to disasters. Local *marae* have the capacity to house, feed and provide support to victims during disaster events. While many *iwi* and *hapū* are willing to help with readiness, response and recovery for disasters, there need to be conversations with local government and other agencies to agree upon the role of *iwi* and *hapū*, and agreement made to provide resources and financial support where appropriate.

in dealing with different peoples, language groups, distinctive habitats, histories and technological environments. How could one appreciate their materiality, let alone mentalities, without cultural awareness and dialogue? It also seems mistaken to argue that 'culture' can be used to perpetuate the status quo, or to distract or deceive outsiders, but cannot be used to inform and assist in collective and creative adaptations – the other side of the coin of communication and meaning. Surely if it is relevant at all, it can do both.

Contexts of engagement with cultural issues

A concern with culture has emerged mainly where researchers and responsible agents, usually literate, formally educated and accredited knowledge workers, seek to address the predicaments and concerns of other national, ethnic, linguistic, gender and/or faith groups. 'Culture' is widely raised as an issue in encounters with people commonly, perhaps misleadingly, typified as 'developing', 'traditional', indigenous', etc., and more or less outside urban and industrial contexts. More recent studies of disasters in modern industrial and urban settings by anthropologists mentioned earlier remain the exception, not the rule.

It is certainly the case that a large fraction of those most vulnerable to disaster worldwide live outside centres where national or globalised wealth and power are concentrated. This includes informal or 'squatter' settlements in cities as well as rural areas (see Chapter 13). They may be at risk from, and their safety compromised by, metropolitan and modern developments, but they have little or no access to modern expertise and research. This situation increases the chances that socio-cultural problems will arise when formal, institutional efforts are made to assist them. Some compelling studies have shown how people are made more vulnerable through repression or exploitation by state and other forces, and may be poorly served in relief efforts. In such cases, cultural differences are sometimes used to excuse treating relief recipients as inferior, or to deny them rights and freedoms that others have (Kim *et al.* 2000). Cultural difficulties and clashes are widely reported in such contexts. At the minimum, culturally inappropriate or ignorant measures are found to magnify rather than reduce dangers (e.g. Bodley 1998) (see Chapter 6).

However, such problems are not necessarily confined to stereotypical cross-cultural encounters with 'pre-modern', 'exotic' and so-called remote peoples and places. As just noted, some of the largest groups of people living at greatest risk and reported disaster victims dwell in or near what are, otherwise, centres of political power and wealth. This emerges above all in studies of the most rapidly growing concentrations of vulnerable people in slum and squatter settlements in urban areas, although they hardly fit, or enjoy, the presumptive urban, civil life (Davis 2006). For them, cultural barriers and clashes arise within 'mass society' if inextricably mixed with economic and political differences and agendas. Again, however, the risk contexts underline the presence of socio-culturally distinct populations, and people more or less excluded from the services, discussions and, often, investigations and protocols that determine disaster management policy (see Chapter 53).

Disaster losses also reveal the existence of unusually vulnerable people within any given cultural group, even within otherwise influential and relatively safe groups. Disadvantaged and subordinate members in families and formal institutions as well as society at large can be at greater or lesser risk. It usually follows from how, and how well, people are treated according to gender, age, employment and religious or political conviction by their own cultural milieu (see Chapter 34, Chapter 35, Chapter 36, Chapter 37 and Chapter 38). Entrenched cultural underpinnings or reinforcements of status and risk profiles are commonplace, notably gendered divisions of labour and influence that reinforce different risks for men and women (Enarson and Morrow 1998).

Nevertheless, a case for addressing cultural concerns and influences in practice is seriously weakened or undermined if one assumes that they just apply to what popular Western or global stereotypes represent as underdeveloped, traditional, ethnic or 'backward' societies, particularly those with strong religious underpinnings, and the corollary that they have less relevance in

Box 8.3 Cultural vulnerability at Mt Merapi and Mt Agung, Indonesia

Katherine Donovan
Department of Earth Sciences, University of Oxford, UK

When Mt Agung erupted in 1963 on the island of Bali, Indonesia, hundreds of people died as they prayed in temples or ceremonially processed towards the hazards expelled by the volcano. Some bodies were found still grasping their traditional Gamelan instruments. The eruption had not begun unexpectedly, but the local people living on this volcano decided to greet the hazards rather than flee from them. To these people this eruption represented their deities descending from the volcano and out of respect many prepared to welcome their gods.

In 2006 Mt Merapi volcano, Java, began to show signs of unrest. Despite efforts to evacuate local residents, some reportedly refused to leave, believing that they had not received traditional warnings. It was feared that a tragedy similar to Mt Agung could occur at Mt Merapi, a volcano that is situated in one of the most densely populated regions in the world.

Although the residents' strong cultural ties to the volcano were extremely influential, they were one of many elements that appeared to be making the local population more vulnerable. For example, a reluctance to evacuate was driven mainly by the population's reliance on their livestock, which would starve if abandoned. Local livelihood pressures were exacerbated by the rich and distinctive culture of the volcano. If traditional precursors, such as premonitions and messages from the supernatural creatures called *Makhluk Halus*, were not received, then some local people found it difficult to believe the official warnings. It seemed that by relying on their traditional belief systems the local people were placing themselves at an increased risk from this extremely active volcano.

At Mt Merapi the local culture varies spatially due to differences in eruption experience, geographic isolation and acceptance of mainstream religions. In certain regions, for example in the more isolated northern areas, regular ceremonies are held to appease the creatures and the level of belief in traditional stories is high. These hotspots could be referred to as culturally intense areas. At Mt Merapi it appeared that these cultural hotspots corresponded with areas where the population was reluctant to evacuate, confirming that cultural beliefs were increasing local social vulnerability. Combining cultural intensity with conventional indicators of social vulnerability, such as population density, gender, age or disability, could provide a more holistic view of risk at this extremely active and dangerous volcano.

After the 2006 eruption, many people acknowledged that they would not evacuate in the future because they believed that their village was safe from Mt Merapi. Indeed, Mbah Maridjan, the *Juru Kunci* or spiritual caretaker for the volcano, explained that 'the creatures will protect this area'. Tragically, in October 2010 the volcano erupted again displacing over 200,000 and killing over 250 people, including Mbah Maridjan, who had remained in his home, refusing to leave. He believed that it was his duty to remain on the volcano in order to appease the supernatural creatures of Mt Merapi.

As the examples from Mt Agung and Mt Merapi demonstrate, it is vital that cultural vulnerability is better understood and taken into consideration during pro-active emergency planning.

urban-industrial and otherwise 'modernised' societies. If it is taken for granted that modern societies, or academic and technical actors, are not culturally constrained, the concern loses conviction where it is most likely to have influence. It also takes on a colonising, 'looking down upon' viewpoint, and is weakened if cultural critiques are considered irrelevant for governmental and corporate institutions (Wisner 2010b).

Indeed, if culture has any fundamental reality for risk and disaster, it must be relevant for understanding those who are best protected as well, not just an added problem of people at grave risk. Hence, it is important to also reflect upon the culture of the 'knowledge workers' themselves or, for want of a better term, the 'risk and disasters community'. In other words, it is not possible to treat human praxis and agency apart from culture. The culture debate in the disasters field must consider those who shape, bear and transfer knowledge, and their notions of risk, as well as those who produce and transform material conditions.

Cultural interactions and DRR

In accepting a cultural dimension in disaster work, one finds most places and institutions are already 'multi-cultural'. It may be as a result of migration, displacement, social upheaval, urbanisation, education or combinations of these. Few endeavours or problems anywhere are without contacts among people from diverse social milieus: persons who are or were residents of different regions, countries and continents, who profess different religions or political affiliations and have different first languages. Cultural issues affecting risk may then depend upon the degree of integration of various groups, as well as legacies of rejection and exclusion. Within contemporary societies cultural debates and interventions are a function of actual or perceived cultural rights and heritage.

Of equal relevance, if one conceives of 'cultures' as separate, tightly knit entities, is how most so-called remote places, ethnic groups and hostile environments are in more or less continuous exchange with the wider, globalised world. Indeed, while contacts and exchanges between different cultures are of key concern, we must resist the stereotypical idea that these are only about different geographical regions. Sometimes location or place is all-important in cultural forms; in other cases it seems irrelevant. A prevailing trend in the disasters community itself is for globalised dialogues. One encounters a remarkable uniformity of language and problem-formulation worldwide, at least in official documents and research papers. In most countries, adherence to the same terms and principles is found in dedicated institutions and mandates. An example is the so-called Hyogo Framework for Action (United Nations 2005). However, experience usually reveals differences in assumptions or interpretation that alter seemingly common notions, or their outcomes, in different cultural settings.

For a critical perspective, therefore, key questions are always about, 'Whose culture?' 'Culture for whom and to what end?' and 'Cultural influences and contacts between whom?' The approach taken here is intended to be a critical and precautionary one. Above all, there is a need to be alert for how 'culture' can be invoked to promote, avoid or obfuscate issues of danger and loss, notably when it is mixed or identified with political power and material interests.

In many ways, the more appropriate and positive aspect of culture for the risk and disasters field is how it frames the possibilities of human creativity and action beyond merely crisis responses to or 'coping with' hazards. Perhaps this goes to the heart of a common misreading. Many traditional cultures, or their vulnerability, are seen as 'cultural' because they are represented as, or presumed to involve, fatalistic beliefs about disasters. 'Fatalism' is widely attributed to the beliefs of other, non-Western or pre-modern cultures seen to attribute them to supernatural forces, in contrast to modern science (Dupree and Roder 1974; Kolawole 2007). In fact, people in modern societies often blame the powers of nature, accident, luck or human nature, and beliefs in spiritual or supernatural powers remain widely present. What should concern us is not the belief, but where it hides, or what it goes against, evidence showing that many of the deaths and damages in recent disasters could have been prevented with readily available measures. It should not ignore the equally compelling evidence of risk-averse practices and effective protection of the vulnerable in cultures worldwide otherwise regarded as religious. These are the grounds for suggesting that, not least in confronting environmental hazards, human intelligence, creativity and collective action are basic to security as well as of risk, and that 'culture' designates the realm in which this is possible.

The most difficult thing to grasp, perhaps, is that while all humans share the attributes and qualities necessary to and for culture, it is also a common source of difference or 'otherness' that can be very divisive, of insecurity, enmities and harm. Divisiveness and conflict are so often identified with different social, economic, political and institutional groups. Blame for disasters is sometimes shunted along re-established lines of difference. Rumours led to Koreans being blamed for the fires that followed the 1923 earthquake in Tokyo. Jews were often held responsible in gentile imagination for outbreaks of plague in the fourteenth century.

Thus, important as culture is, it is a mistake to assume that 'cultural' groups are ever only about belief, language or distinctive artefacts. Their predicaments or strategies can rarely be understood only in terms of their own culture because culture cannot be readily separated from material, social, spiritual and other domains. Once one accepts the importance of culture, one must be especially careful about exaggerating it or using it as a catch-all explanation. Rather, as these preliminary remarks indicate, culture raises questions not unlike those surrounding other general notions like objectivity, perception and reality, and more familiar ones in the risk and disaster arena like 'community', vulnerability or commitment. Each involves contested histories and struggles. They may be attempts to control or close off the discourse, or about more fundamental, alternative and progressive approaches. Perhaps 'culture' identifies the larger matrix in which all such questions arise. It is a topic that one is damned for ignoring, and likely to be damned for considering.

Contexts and the power of culture

> From its elitist connotations to its anthropological definitions, culture has always been political …
>
> *(Siapera and Hands 2004: ix)*

A radical engagement with culture in disaster work can hardly avoid politics, or the power of culture, and not only political cultures, but the politics of cultural phenomena and arguments (see Chapter 5). Risk and disaster are not just likely to bring people into contact who have different backgrounds, status or world views. They also involve contact between persons and groups of very different degrees of power and authority. Persons whose 'belief systems' have histories of conflict may be brought together in crises. The knowledge/power nexus can be, even more critically, a site of differential cultural power.

Professionals, officials and journalists often cite clashes of interests or concern with residents, groups or communities, perhaps over changes that official assessments and responsibilities seem to require, or where assistance is offered that has strings attached based on donor interests. The changes may involve protections that will alter land uses or the landscape, possibly the relocation of people and activities. In one way or another, such changes are likely to challenge and require modifications to pre-existing norms, taken-for-granted activities and relations to place – in effect, 'cultural' practices. They can highlight, and their success or failure depends upon, differences in the preoccupations and influence of different groups. In such cases cultural practices can be empowering or disempowering for different actors.

Knowledge workers and professionals are obliged to operate within the existing social order and cultural politics of the society or institution where they are engaged, and usually do so. However, in most communities, most countries and the world, the main institutions are largely controlled and directed by small, powerful elites, or by managers who benefit greatly from serving their need before that of others. Then again, roles are rarely or not necessarily absolute. Political alternatives and action can and do challenge the status quo. They may even be nurtured by a culture of progressive, public, equity-based values.

The importance of culture in risk and disaster lies, especially, in the way it is woven into the fabric of everyday life, not merely in issues of faith, art and ethics. It is here that cultural powers can exert a great influence on survival and loss in disasters. However, an increasingly critical aspect of today's world concerns the many ways in which states, or national agencies and international bodies, recognise and treat groups identified culturally, how the state intervenes in, or refuses to acknowledge, cultural matters. The risk and disasters field contains plenty of examples that echo James C. Scott's (1998) *Seeing Like a State: how certain schemes to improve the human condition have failed*.

Culture and the state

Much of the recent debate about culture or local knowledge, as of rights and entitlements, is at international levels and in multi-national forums. Globalisation has become an arena for these debates and decisions, not least about the fate of minorities, indigenous peoples and others defined culturally. Nevertheless, for most people, state power remains fairly decisive in the risk and disaster arena, at least as far as the laws, funding and expertise to improve safety are concerned. Moreover, many states have agencies or departments with special responsibilities for people identified culturally, whether set up for dealing with indigenous and minority affairs, or to promote multi-culturalism. Changes in public safety may involve land use regulation, building codes, official languages, school curriculum, the entrenching of given groups' customs in law, the treatment of highly valued places or labour mobility.

It is, of course, difficult to separate some of these issues from the state's role in developing and enforcing legislation or building codes, civil and human rights, or failing to do so (see Chapter 6). The debate is often couched in such terms rather than of culture or, indeed, risk per se. Moreover, it is usually inseparable from, and often in conflict with, state efforts to promote nationalism and the subordination of other identities, to make alliances or war with common ethnic or religious groups on the other side of state borders (Omenugha 2004).

What all this has meant for culture or cultural identities is that they become increasingly politicised. Rather than simply shared as givens, they enter and are sharpened, if not distorted, by struggles for recognition and status. (Cultural) 'identity politics' may be equally relevant for opposition to state, corporate or dominant group attempts to control peoples' land and labour, whether these place them at perceived or actually greater risk, or seem, from outside or technical perspectives, to be safer and wiser. Achievements in these struggles can play a major role in the material well being and safety of the groups concerned.

Such concerns are of special significance for minorities or disadvantaged majorities everywhere, but of special relevance in two settings: multi-ethnic states, and those comprising large, diverse immigrant or settler populations. The latter are increasingly identified, officially, as multi-cultural. In countries like Nigeria, for example, with more than 250 recognised ethnic groups, studies addressing risk and social vulnerability in disasters quickly become engaged with ethnicity and cultural difference. More particularly, they cannot escape the politics or political history of endangerment. It is hard to understand without resurrecting how given groups were treated under colonial rule and since independence (Watts 1983).

In many post-imperial countries with dominant settler populations, the indigenous peoples are particularly involved in such cultural struggles. Amerindian or First Nations communities in North America not only suffer from unusual vulnerability and harm in natural hazards and disease outbreaks, but also are deeply disadvantaged by centuries of abuse, displacement and employment problems. All of them have been subject to more or less constant government regulation and interference by agencies identified specifically with 'Indian' cultural affairs (Prucha 1984).

Wherever the state intervenes in everyday life and development, the exposure and vulnerability of given cultures to dangerous conditions are modified, for better or worse, according to how the state designates and treats them. States that privilege one or more linguistic, religious or status group over others can create the conditions for serious disadvantaging of others. This was seen both before and after Hurricane Katrina among poorer African-Americans and the elderly in New Orleans, and Native-Americans living in the Bayous near that city (Squires and Hartman 2006).

Disadvantage is something that applies especially where there are strong or lingering superiority–inferiority assumptions, whether based on visible biological features, gender or religion, but it is basically about culture – the culture of those who think they are better – and selected cultural differences of those treated as inferior. In most cases, governance in and by nation-states is the key to legal and principled developments and compliance with improvement in safety for any and all cultures within their territories. Unfortunately, although many states now accept a duty, or sign treaties, for the protection of minorities or disadvantaged majorities, most have poor track records in compliance, let alone progressive innovation. At least, all of this underlines the need for an awareness of how culture is part of a history of struggle, and engaged in on-going political strategies. It challenges the point of, or ability to distil, uniquely cultural signatures, but also why culture becomes central to the security of many, if not most, people.

Concluding remarks: Towards an ethics of engagement

The difficulties as well as benefits of cultural discourse are in highlighting arguments for a more inclusive view of disaster work, combining arguments for a professional voice in cultural matters with the need for the professional to listen to others and strive to understand predicaments and concerns from other people's points of view (Caughey 2006). If scientific and professional engagement with disasters is justified by a belief that safety can be improved and losses reduced, then, coupled with a desire to see that happen, cultural awareness requires greater efforts in the following areas:

- Openness: There must be a readiness to accept and actively engage with people of diverse background, to find mutually beneficial terms, concerns and agendas for risk reduction. An approach is required that brings DRR into the broader context of socio-cultural, economic and political issues, and conceives of risks as more than just environmental and technical issues. If, as suggested above, the struggle over political and economic power is often expressed in cultural terms, this is a key area in which DRR must look for the reduction of existing vulnerabilities.

- Partnerships: In a globalised world characterised by rapid change, growing uncertainties and complexities, partnerships between and among different groups and sectors acting at different scales are necessary to develop innovative solutions and allow exchange of resources for improved disaster risk reduction. Mutual cultural understanding and tolerance is necessary to make such partnerships work.
- Empowerment: It is necessary to find ways to give greater power and resources to initiatives that address the culturally specific concerns, whether local or international, of people in particular places or dispersed populations of common origins or identities. Where groups of people are undergoing rapid 'development' through outside forces, resources are needed to enable them to direct more of the tasks of vulnerability reduction and design and implementation of safety measures.
- Involvement: People professionally identified with the risk and disaster field require an ethic that guides contact, translation, facilitation and 'intervention'. This is especially needed where new initiatives are being introduced, supporting effective and sensitive 'cultural' respect and dialogue. Such measures will contribute to better understanding of communities' needs and priorities in order to reduce already existing vulnerabilities.

For all of this, a rights-based approach to knowledge and safety initiatives is required that would guarantee that local and outside knowledge will not be abused and that people at risk are able to participate fully in actions affecting their communities and livelihoods.

9

Knowledge and disaster risk reduction

Jessica Mercer

OXFAM AUSTRALIA

Introduction

Traditionally, knowledge of natural hazards has been a product of research within physical science disciplines. Volcanology, seismology, engineering and hydrology have, for example, been seen as the sites of knowledge production concerning volcanic eruptions, earthquakes and floods. Complementing this traditional approach, knowledge developed through social science disciplines has gained ground since the 1940s (White 1945; White *et al.* 2001). Disaster risk reduction (DRR) reflects these developments.

Both physical and social science disciplines have an important role in knowledge production for DRR. To some extent social science disciplines have made progress in involving 'at risk' people, yet vulnerability to natural hazards continues to increase (Wisner *et al.* 2004). Recognising this there have been initiatives that identify the importance of involving those most 'at risk' alongside other relevant stakeholders including government, academia, communities, business and civil society. These stakeholder groups tend to base their decisions and behaviour on different sources of knowledge. Without adequate communication existing amongst such stakeholders the potential impact of DRR knowledge is limited, with a tendency for each group's knowledge to operate in isolation of the other. This may be partly attributed to the invisibility of those at risk, who are consequently not reached or consulted in the decision-making processes. Another obstacle is mutual access to and intelligibility of each group's knowledge.

Traditionally, a top-down authoritative approach to DRR is favoured, in which solutions to natural hazards are often developed outside the specific context to which they will be applied. Yet, there is a wealth of knowledge existing within 'at risk' communities which could be utilised for DRR. This has been especially highlighted since the 26 December 2004 tsunami (e.g. Gaillard *et al.* 2008b; Shaw *et al.* 2008), when traditional memory of past tsunamis was employed by a few groups and lives were saved. However, accessing such knowledge and addressing the social component of disasters is undoubtedly more difficult than dealing with the physical components of the hazard. Such efforts often involve altering power structures within a society (Wisner *et al.* 2004). It is these very same power structures within society that influence the use, development or abandonment of various kinds of knowledge. Thus political, economic and social power in society determines 'whose knowledge counts'.

Types of knowledge

In very broad terms knowledge consists of information and/or skills acquired through education and experiences. Experiences may include an internally embedded awareness of a situation as a result of a direct experience. For example, children experiencing their community's survival strategies may internalise this knowledge for active use in later life, e.g. where to find famine crops in times of drought. This type of knowledge is more generally known as tacit knowledge, i.e. it is not necessarily articulated in written or verbal form but is implicit in the actions and practices of individuals or groups (Polanyi 1958). Such tacit knowledge is often part of a community or organisation and therefore becomes a normal part of everyday life. Holders of such knowledge are not necessarily aware of the knowledge they possess or how it may be valuable to others. This makes it difficult to verbalise or articulate such knowledge in written form for easy access by outsiders.

In contrast, explicit knowledge describes the things individuals know and are able to articulate, whether verbally or through some form of written or pictorial communication (Polanyi 1958). This type of knowledge develops through more formal types of education – observation, reading or group work – and can if necessary be outlined and disseminated verbally, in print and/or electronic form for others. For example, the National Economic and Development Authority (NEDA) within the Philippines government is mainstreaming DRR into local government decision-making and planning processes. This contributes to ensuring that there is explicit knowledge of DRR within the local government system, with opportunities for such knowledge to be transferred to other relevant stakeholders including local businesses and communities. Similarly, the education of young people by community elders on what to do in an emergency is also explicit knowledge, which is transferred down through generations. Explicit knowledge for DRR is often articulated through radio programmes, the introduction of DRR into school curricula and poster campaigns. The importance of the education system and of media and communication technology for DRR is explored in more detail in the sub-section on communication and education within this volume (see Chapter 61, Chapter 62, Chapter 63 and Chapter 64).

Often within DRR, explicit and tacit knowledge are referred to as scientific and indigenous knowledge, respectively. However, the defining of knowledge in this way for DRR is not strictly correct. Indigenous knowledge can be both tacit and/or explicit, whilst scientific knowledge based on Western scientific norms is nearly always explicit. Given this, a more appropriate and simplified terminology would be to view knowledge from an 'inside' and an 'outside' perspective. Outside knowledge is generally understood to involve Western technology or specific Western approaches which have been empirically proven or developed using methods that are judged by Western 'experts' of the time to be rigorous and in accordance with accepted practice. However, a consensus definition for tacit, indigenous or inside knowledge is lacking within DRR. Examples of ways in which this knowledge is often referred to include 'indigenous knowledge', 'local knowledge', 'traditional knowledge', 'indigenous technical knowledge', 'peasants' knowledge', 'traditional environmental knowledge' and 'folk knowledge' (Sillitoe 1998). Wisner et al. (1977) write about 'people's science', referring to people who have a large understanding of their own environment and who cope with natural hazards. However, this knowledge and the adjustment processes based on it are often ignored by outside stakeholders who prefer instead to use 'what they know', rather than incorporate 'what they don't know'. Glantz (2007) refers to the need for 'usable science', meaning that research findings are targeted directly to helping local people. However, for this to occur it must entail working with people and their knowledge on their own terms.

Collectively, inside knowledge is acquired by local people over a period of time through accumulation of experiences, society–nature relationships, community practices and institutions, and through passing it down through generations (Brokensha *et al.* 1980). Outside knowledge is considered global whereas inside knowledge is thought to be local. However, this distinction is also questionable for two reasons. First, communities share knowledge over time as, for example, five distinct Kenyan tribal groups were found to have shared many of the same drought coping mechanisms (Wisner 1978). Second, whilst some details of inside knowledge may be endogenous to a community or locality, culture or society, it is embedded within a global context, thus the term 'inside' is not strictly correct. Inside knowledge and behaviour are thus dynamic and reactive to global change. Such context further includes social and environmental changes and impacts. This is emphasised by Flavier *et al.* (1995), who state that [inside] knowledge systems are dynamic, continually influenced both by internal creativity and experimentation, and by contact with external systems. Inside and outside knowledge often combine to form a type of hybrid knowledge. Hybrid knowledge builds upon the strengths of both inside and outside knowledge. This occurred on Ambae Island in Vanuatu where Cronin *et al.* (2004) developed volcanic hazard management guidelines with the Island residents. These guidelines incorporate inside knowledge with risk management frameworks drawn from outside Western science. Richards (1986) uses the term 'opportunistic' for the way that local people may appropriate and modify outside knowledge. In this respect and given the many different terms outlined by Sillitoe (1998) the term 'local knowledge' better defines knowledge that has evolved within (inside) a specific community or area, but that has potentially incorporated or been shaped by outside knowledge in its continuous evolution.

Shaping disaster risk reduction through knowledge

Polanyi's (1958) original classification of knowledge has in many cases been removed from its original contextual analysis. This has contributed to the development of a substantial divide between knowledge identified as local (inside) and knowledge identified as outside (scientific) (Grant 2007). Indeed, Polanyi (1958) originally referred to all knowledge as having a local component and that local knowledge in some cases can be scientifically verified or made explicit. This clear divide has been reproduced within DRR.

Local (inside) knowledge in DRR

Communities have coped with change for centuries, utilising local knowledge passed down from generation to generation to ensure their survival. For example, rural communities living on the slopes of Manam, an active volcano in Papua New Guinea (PNG), have adapted their homes to ensure that the roofs have steep sloping slides. This ensures any volcanic debris immediately falls from the roof thereby minimising fire risk. In a different context, in Mongolia pastoralist communities have used their detailed ecological knowledge to develop a mobile system of pastoral land use with neighbouring communities which has developed and evolved over centuries. These are just two examples of local knowledge which has contributed to the survival of many communities worldwide, yet this knowledge is often ignored, disregarded or deemed insufficient for DRR.

Scientists and other stakeholders working to reduce risk in many cases prefer instead to base their recommendations on scientific analysis or Western solutions. Often the wider interrelating factors such as social, economic, environmental, political and technological issues contributing to a hazard becoming a disaster are ignored (Wisner *et al.* 2004). This is in favour of a focus upon specific elements of a hazard or a particular solution. Contributing to the problem is the propensity by Western experts to generalise strategies from the situation in which they were developed to other similar situations, without taking into account potentially important local differences. This has occurred in Bolivia, where indigenous farmers resort to a number of strategies enabling their communities to farm in a difficult mountainous environment. The strategies include knowledge of climate signs and indicators, an ability to manage soil diversity, diversification of crops, use of soil rotation systems, management of the traditional agricultural calendar and an integration of agricultural, forestry and animal husbandry systems (Regalsky and Hosse 2008). In addition, these strategies are all based upon communal control over equal rights of access to natural resources with a system in place for the reciprocal exchange of workforce amongst community members. However, such strategies are disregarded by 'outsiders' and the present Andean agricultural crisis is seen by indigenous communities as the result of a forced introduction of inappropriate technologies for the Andean ecology. These technological innovations have concentrated upon boosting mono-crop farming, introducing mechanical and chemical techniques for ploughing and treating soils, and replacing native plant species with exotic varieties developed for other ecological and social systems.

In some areas foreign 'expert' dismissal of local knowledge has led to its abandonment by communities themselves as they seek to 'develop' and to 'modernise' (Mercer *et al.* 2007). Some communities in Bangladesh have, for example, adopted corrugated iron roofing due to local perceptions of this Western material as a symbol of wealth. Given the potential cyclone hazard in Bangladesh, traditional bush materials in this case are far superior due to their easily replaceable and lightweight nature, as opposed to corrugated iron roofing which in itself presents a hazard during a cyclone event, with the potential to cause fatalities if ripped from rooftops. This inclination towards outside approaches and subsequent abandonment of local knowledge became particularly evident throughout the 1970s–1980s through the adoption of a Western approach to disaster management. The Western culture of relief aid provision after an emergency, rather than contributing to community development, has led in some cases to increased vulnerability of local communities to natural hazards due to a loss of local knowledge. In Vanuatu, for example, traditional communities abandoned the growth of famine crops and traditional cropping practices which had previously helped them to cope with hazardous events. Instead, in this case they came to expect and rely on external aid in the form of relief goods as a new 'Western' solution (Campbell 1990).

The dismissal of local knowledge by scientists dangerously ignores valuable resources which could greatly enhance the capacity of local communities to reduce the risk of disaster (Mercer *et al.* 2007). As with outside knowledge, local knowledge has also contributed to saving lives. Examples include islanders' response to Cyclone Zoe which hit the Solomon Islands in 2002 – they took shelter under overhanging rocks on higher ground as the cyclone struck (e.g. Anderson-Berry *et al.* 2003) – and the way some indigenous groups survived the 2004 tsunami in South-East Asia. In the latter case this included the people of Simeulue Island off the coast of Sumatra, who survived through a traditional folk tale. The story described a tsunami that devastated Simeulue on 4 January 1907 (Gaillard *et al.* 2008b). A Simeuluean word, 'Smong', was assigned to this type of event and according to the Simeulueans it consisted of three stages: (1) a strong earthquake; (2) a receding sea; and (3) a large wave and flooding. This resulted in their immediate evacuation and subsequent survival of the tsunami that occurred on 26 December 2004. In many cases local knowledge has struggled to establish itself as a global knowledge source due to its specificity to a particular community or locality and its intangibility to outside stakeholders. Although it is its very strength, the context-specific dimension of local knowledge makes it difficult to be raised in policy circles. However, this is not to over-romanticise local knowledge. Given the pace of change experienced today, some local knowledge may no longer be applicable or viable, whilst some may actually be increasing vulnerability. Local knowledge should be analysed carefully to ensure its applicability and effectiveness in addressing disaster risk (e.g. Tibby *et al.* 2007).

Outside (Western scientific) knowledge in DRR

Outside knowledge has made considerable progress in the field of hazard research for DRR, contributing to the saving of much life and property. For example, this knowledge has enabled us to forecast and provide hazard warnings, map hazardous areas and develop earthquake-resistant housing and cyclone shelters. The earthquakes in Chile and Haiti that occurred in early 2010 illustrate this point. The Haiti earthquake, measuring 7.0 on the Richter scale, caused complete devastation in the capital, Port au Prince, whereas the earthquake measuring 8.8 on the Richter scale that hit the city of Concepción in Chile caused much less damage. This has been attributed to strict building codes and an ability to implement these, which ensured that much of the city's infrastructure in recent decades was built using quake-resistant building techniques.

A particular strength of the outside (Western scientific) approach is the ability to invest time, effort and resources into one particular area of research, e.g. the lava flow paths of a particular volcano or the seismology of a particular area. This enables a wealth of data about a particular subject to be developed, which can subsequently be tried, tested, proven and utilised through standardised methods. For example, the mapping of lava flow paths for volcanoes with large populations living on their slopes, such as Mount Merapi in Indonesia, enables the identification of evacuation routes and safe areas if an evacuation were required.

Application of outside scientific knowledge and Western strategies can save life and property, but it can also be misused if always portrayed as superior. Following the 2004 tsunami, which impacted many communities across South-East Asia and beyond, the global community rushed to assist those affected, often without adequate consultation. The Sea Gypsies from the Surin Islands in Thailand all survived the tsunami as a result of their local knowledge of the sea (Arunotai 2008). They knew to flee to higher ground as they witnessed the sea receding from their homes along the shoreline. However, in rebuilding their homes 'experts' placed these out of sight of the sea above the shoreline, a distance which (1) would not protect them from another tsunami and (2) removed them from what saved them in the first place – their view of the sea. Similarly, homes rebuilt using advanced engineering techniques in response to the 2005 Pakistan earthquake were in some cases later abandoned by local communities, who were not consulted in their construction and deemed them inappropriate for local conditions (Sudmeier-Rieux *et al.* 2007). This is in spite of their earthquake-resistant properties. The failure in many cases of scientists to involve local communities and acknowledge the potential viability of local knowledge for DRR has resulted in similar scenarios to those above. Lack of involvement of those 'at risk' can also result in the limited or inappropriate uptake of outside risk reduction strategies by local communities themselves, as these are not communicated in an easily accessible and/or understandable format (Weichselgartner and Obersteiner 2002).

Whose knowledge counts?

Rather than allowing one form of knowledge to dominate, as in the case of outside knowledge, or weighing up one kind of knowledge against the other, e.g. local versus scientific, insider versus outsider or tacit versus explicit, there should be an effort to 'bridge the gap' (Wisner 1995) or to 'reconcile science and tradition' (Dods 2004). Whilst there are clear differences between local and outside knowledge, there are also clear similarities with regard to DRR. Both kinds of knowledge aim to reduce risk, and both bodies of knowledge are being constantly utilised, revised and developed. Despite these similarities and the benefits that both kinds of knowledge could bring to DRR, most often outside (Western scientific) knowledge has dominated.

Table 9.1 outlines the strengths and limitations of both kinds of knowledge. Outside knowledge clearly has considerable strengths in terms of globally accepted and proven methodologies, whilst local knowledge has developed from an understanding of a given context, and the subsequent development of strategies that fit within this. In addition, each body of knowledge could potentially be supported by the availability of other resources and assets, which may or may not be available to individuals, households and/or communities, e.g. financial assets, social networks, natural resources, tools, etc. In some cases these may balance out knowledge limitations in a particular area or location, thereby enabling an individual, household or community to continue functioning. For example, a household may not have the knowledge to mitigate the effects of flooding, but their financial assets and social networks enable them to cope and therefore recover. Conversely, people may have the knowledge to face natural hazards but lack the resources to take advantage of this knowledge.

Thus, it is not a question of 'whose knowledge counts?' but rather one of 'what knowledge counts?' As demonstrated in Table 9.1, the limitations of outside knowledge could be addressed through the strengths of local knowledge and vice versa. Subsequently, there are clear benefits to utilising the strengths of both kinds of knowledge to create a hybrid form of knowledge.

Hybrid (combined) knowledge in DRR

The integration of both local and outside knowledge to form a type of hybrid knowledge has been especially discussed in the field of natural resource management (Rist and Dahdouh-Guebas 2006). However, the integration of both kinds of knowledge has often only occurred in response to a specific problem at a specific time in a specific place, e.g. water contamination or soil stability (Briggs and Sharp 2004). The solutions have also often been of a fairly technical nature, using knowledge that is easily accessible and identifiable by both outside and local 'experts', such as soil management techniques, water preservation systems or use of medicinal plants. Hence, local knowledge is often only used when it is seen to offer solutions which fit within the existing outside scientific/development view (Briggs and Sharp 2004). As outlined above, local people are known to opportunistically integrate elements of outside specialist and local knowledge in response to changing environmental conditions. In El Salvador, for example, communities have started to recycle old car tyres, filling these with concrete and using the material to stabilise slopes at risk from landslides. This is in addition to more traditional techniques which utilise available wood resources to secure slopes and prevent soil loss.

Whilst the resulting outputs may be of benefit, at other times this has been to the detriment of communities concerned (Mercer et al. 2007). This has been seen amongst rural communities in PNG who have been introduced to the cash economy and the growth of new cash crops, e.g. coffee. This has resulted in the use of more modern, intensive farming techniques. The growth of coffee crops interspersed with their traditional vegetable crops has contributed to increased erosion and landslides in the area due to more intensive land use. Thus, what seems to be required is a more systematic approach to creating and understanding the function of hybrid knowledge. Not only should research be dedicated to this question, but the topic of hybrid knowledge should be highlighted amongst all relevant stakeholders including extension officers

Table 9.1 Strengths and limitations of local and outside knowledge

Outside knowledge:		Local knowledge:	
Strengths	Limitations	Strengths	Limitations
Proven using accepted scientific methodology.	Does not often recognise the wider socio-cultural context – danger of cultural incompatibility.	Local in nature – knowledge is specific to a given context.	Not necessarily scientifically proven.
Often global in nature and applicability therefore transferable across contexts and countries.	Often developed in isolation from the context in which it is applied.	Supported by local community.	Not easily accessible to outsiders.
Highly specialised – providing detailed information about specific elements of a hazard or hazard impact.	Very narrow in outlook – often failing to consider the wider context within which the strategy is to be applied	Culturally compatible and utilises local resources	Problem of over-romanticisation – not always the best solution especially in light of new risks, e.g. climate change.
Supported by international community.	Often not understood or accepted by end users – a result of its highly specialised nature.	Addresses specific issues identified within a community – focused on community needs.	Solely dependent on local resources (or, where family members have migrated, on remittances)

in various ministries and university students. This is in order for it to become an established, viable and accepted basis for DRR.

Several researchers have assessed the potential possibilities of integrating outside and local knowledge to develop more effective, culturally specific and sustainable DRR solutions (e.g. Cronin *et al.* 2004; Dekens 2007). Mercer *et al.* (2010) outline one such methodology, identifying how local knowledge may be successfully integrated with outside scientific knowledge for DRR in PNG. However, hybrid knowledge co-production and utilisation requires additional time and effort from all knowledge holders. This is in order to ensure that both bodies of knowledge are made accessible in a form that the other understands. For example, this could include the identification and documentation of local knowledge for scientists, and outside knowledge demonstrated and delivered in a form understandable and accessible to those 'at risk'. In integrating different kinds of knowledge, each needs to be assessed on individual merit with neither one identified as superior to the other. The most appropriate knowledge irrespective of its origins can then be integrated to form 'hybrid knowledge'.

Box 9.1 Disaster risk reduction and knowledge in Sierra Leone

Araphan J. Bayor, Richard Thoronka, Ahmed P. Samura
Caritas Makeni, Sierra Leone

Abass Kamara, Jessica Mercer
CAFOD, UK

In 2009 CAFOD in partnership with Caritas Makeni developed a community-based DRR programme with five communities in Bombali District, in the Northern Province of Sierra Leone. Bombali District experienced considerable death, destruction and displacement as a result of the civil war in Sierra Leone in 1991–2002. The communities are typical rural-based communities depending on rain-fed agricultural activities for their livelihoods. They are currently at risk from drought, wind storms and fires.

Caritas Makeni undertook a comprehensive baseline survey with each of the five communities. This was in order to identify existing knowledge utilised to address community disaster risk and to establish what, if anything, needed to be changed, developed and adapted to further reduce risk. The process used was one of guided discovery, whereby the communities themselves were able to explore their situation and identify potential solutions. All five communities identified actions that they could take themselves through local knowledge and actions that they felt they could take in partnership with others, such as Caritas Makeni, nearby villages and local government.

The communities undertook an assessment of the hazards they faced, and their vulnerabilities and capacities. This enabled the identification of existing capacities which they could build upon to address disaster risk. Local capacities included: strengthening community bye-laws to protect their natural resources; establishment and/or further development of village development committees; development of appropriate DRR and contingency plans supporting the inclusion of vulnerable groups; use of local herbs for medicine; use of manure for fertiliser; expansion of inter-cropping techniques to preserve soil fertility; continued use of communal land tenure systems; water harvesting techniques; and an ability to reforest areas that had been depleted.

The communities wanted to work in partnerships with others. They recognised the need to lobby nearby villages and local, provincial and national governments

---

concerning preservation of forests, and to ensure strict bye-law enforcement. Communities felt that they would benefit from working closely with government extension workers to identify and utilise hazard-resistant crop varieties, and to ensure implementation of sustainable farming practices whilst maximising production. The communities recognised that women and children were more vulnerable than men and wanted Caritas Makeni's support to ensure women's participation in policy formation, planning and implementation. In addition, they felt that they needed support to ensure access to sustainable water supplies through the provision of hand pumps and use of groundwater supplies. A combination of both local and outside technical knowledge is needed to effectively address disaster risk of the five communities.

Each community developed an action plan. Importantly, the communities clearly identified the necessity of involving nearby villages and the local government, thereby ensuring that activities were not carried out in isolation and that a holistic approach was taken to ensure the sustainability of initiatives in the long term. However, the available knowledge, both local and external, and appropriate solutions will differ among communities and across different contexts—what is clear is that a combination of knowledge types or a hybrid form of knowledge is needed to effectively address disaster risk.

Difficulties and benefits of integrating local and outside knowledge

Despite the obvious utility of hybrid knowledge, very few studies have attempted to integrate local and outside knowledge for DRR and it is rarely, if ever, achieved. There are two main difficulties preventing the smooth integration of outside and local knowledge. The first results from the perceived superiority of outside knowledge or the dismissal of local knowledge amongst policy-makers and scientists, and the uneven power dynamics that thus develop with those at risk. The second difficulty arises as a result of the context-specific and embedded nature of local knowledge.

Schmuck-Widmann (2001), in researching the knowledge of both engineers working to control the Jamuna River in Bangladesh and traditional *char* dwellers who live on silt islands (*chars*) in the river, identified that whilst the engineers had readily dismissed the knowledge of the *char* dwellers, the knowledge of both the engineers and local people with regard to the Jamuna River was very similar. Each party could have benefited from sharing each other's knowledge, but the view that 'the scientist knows best' and traditional power dynamics prevailed amongst the engineers to prevent effective knowledge sharing. This then also prevented the subsequent integration of such knowledge to reach the shared end goal of reducing risk. A process of co-discovery of solutions to particular problems is required that includes critical examination by everyone concerned (outsider and insider) of both kinds of knowledge and their applicability to a specific case. The community needs to be at the centre of engagement. Local people need to be able to understand and use outside knowledge that is relevant and applicable or adaptable to their local context. However, at the same time, the researcher or outside expert needs to understand the internal logic present in local knowledge and practice.

In many cases those working to reduce risk come from vastly different cultural backgrounds. Such cultural and language difficulties also contribute to widening the gap between the two kinds of knowledge. As in the case of the 2004 tsunami, traditional folk tales are often part of a community's risk reduction strategy (Gaillard *et al.* 2008b). However, such tales, whilst often correlating with scientific observations, are difficult to quantify and therefore integrate with outside risk reduction strategies. Similarly, other tales or actions may in themselves be inadequate risk reduction strategies, but they may be a precursor to more adequate DRR strategies as a community prepares for a potential hazard. For example, amongst indigenous communities in the Department of Cochabamba in Bolivia environmental signs such as the appearance of certain animals or changes in vegetation are used to forecast a dry year or a rainy year (Regalsky and Hosse 2008). The communities then adjust their agricultural strategy accordingly to cope with the expected hazards for the year. In Mexico, traditional Aztec ritual leaders are widely believed to be in tune with the spirit of the Popocatépetl volcano. Outside Western scientific prediction based on seismology and remote gas chromatography simply ignores the precursory signs used by the local shamans. Culture conflict results and evacuation orders are not well heeded. In dismissing such stories and strategies the 'outside' community is in danger of widening and perpetuating the already existing gap between local and outside knowledge for DRR. It would be far better to begin a dialogue between local and outside trained 'experts'.

Chapter 64 in this volume, which discusses 'participatory action research and disaster risk', presents one such option by which knowledge limitations and biases can be overcome. Participatory action research provides a facilitated process whereby at-risk communities and associated stakeholders are able to come together and identify DRR hybrid knowledge from both outside and local knowledge. Such hybrid knowledge needs to be applicable, appropriate and, most importantly, culturally sustainable given the local context (Mercer et al. 2007). For example, there is clearly no benefit to implementing an outside approach if this is at odds with the local cultural context. Rather, participatory action research provides a way in which an exchange of knowledge between all knowledge holders may be facilitated. This is explored in further depth within the community risk assessment toolkit (available at www.proventionconsortium.org). Community risk assessment uses participatory action research methods (see Chapter 64) to place communities at the forefront of actively planning, designing and implementing an action plan to reduce disaster risk. This provides an opportunity to identify local capacities and explore how these could fit with local government initiatives or other 'outside' expertise in order to reduce community risk further.

Box 9.2 Local knowledge and community-based DRR in Zimbabwe

Kuda Murwira
Rural Development Facilitator/Consultant, Zimbabwe

The success story of the Chivi community in southern Zimbabwe is an example of how rural communities can reduce their vulnerability and strengthen their capacity in facing drought and disruptive political conflict. After a serious drought in 1992, ITDG Zimbabwe (now called Practical Action) used participatory methods to develop a local plan for overcoming food insecurity. In the course of the work an effort was made:

- to revive traditional knowledge and skills and to build upon them with modern methods;
- to understand community priority needs;
- to involve all members of the community, especially the poorest and women;
- to create innovation platforms for sharing and teaching everyone in the community;
- to strengthen local organisational capacity; and
- to document experiences and to share them with other communities.

Food security was enhanced. Subsequent droughts did not cause hunger. The community continued to rely on their own crop and livestock varieties, selectively bred and shared with other communities in seed and livestock fairs. The community now has a broad crop and livestock diversity that enables them to counter the effects of drought. Also, the community has been able to withstand external threats which might have divided them and caused conflict. The political polarisation that affected Zimbabwe in the late 1990s and 2000s did not affect Chivi's working together, although people belonged to different political parties.

Throughout the process there was intermittent contact with government extensionists who offered technologies and learned to respect local knowledge of pest control, livestock and crop breeding, and seed storage, while the community accepted the superiority of outside approaches to vegetable irrigation and fencing. Also of note is that the Departments of Agriculture Extension and Research in Zimbabwe are in the process of publishing a 'participatory extension approach' promotional manual based on the experiences in Chivi and other communities.

Twenty years later, the community has continued to be a resource for neighbouring communities, the district, other district communities in Zimbabwe, as well as neighbouring countries in the region in terms of knowledge and skills in DRR and self-organisation. The level of self-organisation has grown and the values of the community have been passed on from one generation to the next. Leadership in the community is beginning to be devolved to the younger generation.

Two main factors contributed to success:

- The facilitation began with the most pressing priority. The drought Chivi had experienced was the worst for the last forty years. Selecting the right intervention point enabled them to rally the whole community.
- The project encouraged all stakeholders to contribute to identifying both causes and possible solutions to the problem. Its slogan: 'Nobody knows everything; everybody knows something'.

More information: Win (1996); Wedgwood *et al.* (2001); ProVention case study (2001): www.proventionconsortium.org/themes/default/pdfs/CRA/Zimbabwe.pdf and update (2007): www.proventionconsortium.org/themes/default/pdfs/CRA/Zimbabwe2_EN.pdf.

Conclusions

Local communities worldwide have relied on their best practices for millennia as knowledge is passed down from one generation to the next. However, given the rapid changes occurring today including increasing population, rapid urbanisation, environmental degradation and climate change, it is clear that in some cases local knowledge may no longer be applicable. Rather, there needs to be an integration of the most appropriate outside and local knowledge to address disaster risk.

Scientists and other stakeholders, including policy-makers, have a responsibility to identify with and listen to local knowledge when making decisions about DRR. There are valid and appropriate lessons that can be learnt from local community knowledge which has developed and evolved over centuries to cope with changing conditions. It is this knowledge that should form the basis of community risk reduction strategies. However, in light of the changes outlined above and increased societal impacts from natural hazards, scientists and other stakeholders also have an obligation to ensure that their 'science' is made accessible to the general public and to those it is intended to benefit. In developing a risk reduction strategy for a particular community or area there should be a responsibility placed on those working with 'at risk' communities to ensure that their knowledge is used wisely and integrated with appropriate and applicable local knowledge. Rather than one kind of knowledge dominating the other, outside and local knowledge should be mutually reinforcing one another and blended into a hybrid form of knowledge.

Traditionally, communication between scientists and local communities in terms of sharing DRR knowledge and expertise has been lacking. Scientists, researchers, government, non-governmental organisations (NGOs) and other associated stakeholders need to explore innovative means by which communities at risk can become more involved in developing appropriate risk reduction strategies. Similarly, outside knowledge needs to be transmitted in a format that local communities are able to understand. For example, innovations such as mobile phone communication technologies, which are constantly being developed, are increasingly becoming economically feasible for large sections of society, and are being utilised to discuss and share local knowledge. These need to be further exploited to enhance communication and therefore knowledge exchange between scientists, policy-makers and local communities. Communication tools such as these could potentially enable a more participatory dialogue enabling the voices of those 'at risk' to be heard, whilst facilitating DRR knowledge exchange between different sectors of society. This would enable enhanced risk reduction capabilities, as actions taken to reduce risk are able to effectively access and integrate relevant and applicable local and outside knowledge.

10

Religious interpretations of disaster

David Chester

DEPARTMENT OF GEOGRAPHY, UNIVERSITY OF LIVERPOOL, UK

Angus M. Duncan

RESEARCH GRADUATE SCHOOL, UNIVERSITY OF BEDFORDSHIRE, UK

Heather Sangster

DEPARTMENT OF GEOGRAPHY, UNIVERSITY OF LIVERPOOL, UK

Introduction: Archaeology and geomythology

Clinical psychology notes that people often appeal to deities when coping with severe stress, and this is also observed amongst those who are faced with disasters associated with natural hazards (Chester and Duncan 2009). Religious reactions to disasters are found within most traditions of faith, across societies located in different parts of the world and over the long span of human history. In some cases it is possible to extend the time-scale and interpret archaeological evidence to show how prehistoric societies used theistic frames of reference to make sense of the suffering triggered by natural events. Effigies found on the slopes of Popocatéptl volcano have been interpreted as evidence of the propitiation of divine wrath by the society that once inhabited this part of Mexico, while at Pylos in ancient Greece there are indications that the god Poseidon, who was believed to be the cause of earthquakes, was worshipped as early as the Mycenaean era: *ca.*1600 to *ca.*1100 BCE. For archaeological sites evidence of human reactions to disasters is usually either lacking or can be interpreted ambiguously. A wall painting (*ca.* 6200 BCE) at the town of Çatal Hüyük in Anatolia (Turkey) is, for instance, the oldest known example of the artistic depiction of a volcano. It shows a volcano located close to a town which could imply disquiet among the population about future eruptions, or may simply 'reflect the aesthetic sensibility of the artist' (Chester and Duncan 2008: 203).

In recent years oral traditions that preserve the memories of prehistoric disasters have been extensively studied (e.g. Harris 2000), and in some societies these myths continue to co-exist alongside written records. In Pre-Meiji Japan (i.e. before 1868) 'from prehistoric (times) an ensemble of beliefs and views on the Earth and on nature has survived' (Barbaro 2009: 26), while in parts of the Upper Nile, the Sudan and East Africa various indigenous groups continue to tell stories about the dead who reside in the underworld and are thought to be the cause of earthquakes. A note of caution needs to be sounded about oral evidence as a source of information because when myths are translated and written down by outside observers cultural presuppositions may influence accounts. One example occurred in 1986 during the Lake Nyos gas disaster in Cameroon, where a single word in a local language covered at least two meanings in the majority of European languages. There was one word for smell and taste, and the word for 'red' referred to all primary colours in the original language (Freeth 1993). In addition, fully understanding the explanations of disastrous events reported by indigenous peoples may be difficult, particularly when the literate group with whom contact has been made also has recourse to theistic explanations of their own. In southern Africa in the nineteenth century, Endfield and Nash (2002) have shown how the perceived 'heathenism' of tribal peoples was thought by some Christian missionaries as being responsible for severe droughts. Tribal groups, in their turn, not only professed a number of religious myths about their environment, but also perceived the missionaries as being either responsible for drought or, conversely, as 'harbingers rather than withholders of rain'. There is, however, overwhelming evidence that oral traditions, especially those found within pre-industrial societies, have preserved a range of theistic responses to disasters.

Within the Judaic/Christian and Islamic traditions perhaps the best known example is the flood myth that, although recorded in the Book of Genesis (Gen. 6–7) in the Christian Old Testament (Hebrew Bible), represents a much older oral and written tradition which was present across ancient Mesopotamia and was later recorded by the Greeks. In Europe and North America doubts over the veracity of the biblical flood were still part of scholarly discourse until the nineteenth century. Dundes (1988) has shown how flood myths occur throughout the world and, as well as in the ancient Near East, are also to be found in Mesoamerica, South America, Asia and Africa. In Germany North Sea floods in 1570, 1634 and 1717 were similarly ascribed to God, using a variety of theological explanations (Mauelshagen 2009), while during pre-industrial times in Finland summer frosts were assumed to be due to God's punishment of sinners.

Medieval Arabic texts contain reference to pre-Islamic beliefs concerning theistic responsibility for plagues as well as earthquakes. Many verses from the early part of the Qur'anic revelation continued to ascribe these phenomena to God (Akasoy 2009) and, indeed, a similar model of divine punishment was frequently invoked to explain plagues in Christian Europe during Medieval times (Smoller 2000). As far as droughts were concerned, not only was God's responsibility undoubted within the Hebrew Bible (e.g. 2 Chronicles 7: 13–14), but it was still invoked in the 1970s to explain a lack of rain in northwestern Nigeria by Christians and Muslims alike (Dupree and Roder 1974).

Today the study of myths about disasters is a major academic field and a full review is beyond the scope of this article, but a flavour of the range of responses may be gleaned from Table 10.1, which uses volcanic eruptions as an example, and shows how human reactions transcend place, time and culture.

Disasters and the major religions of the world

Religious interpretations of disasters are still to be found across all the major religions of the world. Within Judaism and Christianity attempts to reconcile the concept of an omniscient and loving God with the simultaneous existence of evil and suffering in the world is termed theodicy (Greek *theos* – God, and *dike* – justice), the word first being coined by Gottfried Wilhelm Leibniz

Table 10.1 Examples of pre-industrial societies in which volcanic eruptions have been interpreted theistically

Example	Nature of response
Southern Europe	During the 'classical' period the Etruscan god *Velkhan* was thought to be a destructive agent but was sometimes viewed as a god of productive fire. *Hephaestos* (Greek) or *Vulcan* (Latin) was normally viewed as a constructive craftsman, but volcanoes were also personified as the destructive god *Titan*. On Etna Lucilius Junior (first century CE) recorded that the people offered incense to appease the gods who were thought to control the mountain. In classical literature Thera (Santorini) is frequently mentioned and may have been the setting for Plato's *Atlantis*, which was created by *Poseidon*.
Northern Europe	In Icelandic mythology the god *Surtur* was the incarnation of eruptions.
Africa	Before European contact, tribal groups located in the vicinity of Nyamuragira and Niyragongo volcanoes (Central Africa) annually sacrificed ten of their warriors.
	In Ethiopia, some older inhabitants still lay offerings to the serpent-god, *Arwe*. The serpent-god was a god of terror who could be propitiated by the annual sacrifice of maidens. The god was killed around 1000 BCE, but worship is still evident in some remote areas.
Central Mongolia	In the Mongolian language, the name of the little-known Har-Togoo volcano means 'black pot'. There is a local belief that a dragon lives in the volcano.
Indonesia	Until recently in Java, human sacrifice to appease Broma volcano was practised. Chickens are now substituted. In Central Java the residents have their own perceptions about the dangers of Mount Merapi. They believe in spirits and these provide the residents with a subjective sense of security.
Japan	The *Oni* monster is a horned red giant, whose effigy is still to be found in souvenir shops. In Japan many volcanic features are called *Jigoku*. *Jigoku* is a term for hell introduced into Japan from Buddhism with its notions of an underground prison. Archaeological sites in southern Kyushu show evidence of spiritual activity. There is an example of a pot being offered to propitiate disaster during the accumulation of ash around 1300 BP.
Hawaii	There are many legends about the need to propitiate the Goddess Pelé, who was thought to control human fate. Offerings of fruit, bananas, pork and tobacco are traditional.
Vanuatu and Fiji	Volcano-initiated disasters are sometimes described in myths and legends [in Vanuatu] and are usually attributed to demons or spirits who wish to punish a breach of a social or cultural taboo. In oral history many eruptions were initiated by a sorcerer using a ritual song or incantation. Should ritual rules be broken, then eruptions could punish the people. Until recently, animal sacrifice was important. In Fiji there are legends about the two gods, *Tanovo*, who presided over Ono Island, and *Tautaumolau*, who ruled southwestern Kadalvu.
New Zealand	In the period before European settlement, each group within Māori society possessed its own sacred mountain. *Te Heubeu*, chief of the Ngati Tuwharetoa tribe in the Taupo district, told an Austrian geologist in 1859 that volcanic fire was sent from the mythical Māori homeland in response to a call from *Ngatoroirangi*, the high priest.

Continued on next page

111

Table 10.1 (continued)

Example	Nature of response
North America	Harris (2000: 1312–13) has proposed that Native-Americans may have preserved memories of the Mount Mazama eruption in Oregon (*ca.* 7500 BP). The Klamath tribe believed that this large eruption was a battle between the Mazama god, *Llao,* and *Skell* the sky god. Basaltic lava flows dating from the mid-eleventh century from Sunset Crater (northern Arizona) show corn casts on the surface, which have been interpreted as representing ritual practices of divine appeasement. Devil's Tower may also be interpreted deistically.
Central and South America	Propitiatory human sacrifices were a feature of Maya, Aztec and Inca responses to volcanic activity. In Nicaragua it was held that Coseguina volcano would not erupt if a child was sacrificed every 25 years. Virginal sacrifice was also a feature of the society who lived on the flanks of Masaya.
	In the years before the Spanish conquest, people living in the vicinity of Huaynaputina volcano in Peru sacrificed sheep, birds and personal clothing to the volcano, with some people even claiming that they conversed directly with the demons who supposedly controlled the mountain.

Note:
Modified from Chester and Duncan 2008, Table 10.1: 206–07 and the references cited. Further examples may be found in Harris 2000 and Sigurdsson 1999: 11–20. The locations of the places mentioned in the table are shown in Figure 10.1.

in 1710. Attempts to understand the reasons why people suffer have exercised the minds of philosophers and theologians for thousands of years, being notable features not only of the Hebrew Bible, the Christian New Testament and Christian theology, but also the sacred texts and theologies of other major world religions.

In Judaism and Christianity until recently two models of theodicy have been prominent: the 'free will' (Augustinian) and the 'best possible worlds' (Irenaean). The former emphasises that human beings have freedom and that suffering is not only a result of the operation of free will, but may also involve sinfulness because a person has acted against God's purpose. Divine punishment is sometimes stressed, being the predominant feature of most disasters recorded in scripture and many that have taken place subsequently. For instance, in Iceland in the Middle Ages, 'the fearsome noises that issued from some … volcanoes were … thought to be screams of tormented souls in the fires of Hell below' (Sigurdsson 1999: 75); in England in the 1380s pestilence, declining fortunes in war and finally an earthquake in 1382 caused many people to assume that these presaged a final divine reckoning; and in continental Europe during the sixteenth and seventeenth centuries disasters, especially earthquakes, floods and hailstones, were thought to be increasing in frequency and were held to foreshadow an imminent apocalypse. Although much reduced as an explanation of human suffering, especially since the eighteenth-century European Enlightenment, divine retribution is still an explanation used by some evangelical Christian groups and was held by a small minority following the Indian Ocean tsunami disaster of 2004, the British floods of 2007, the Australian drought of the first decade of the twenty-first century and the Haitian earthquake of 2010. Following the La Josefina landslide in Ecuador in 1993, some farmers still adopted either a Christian notion of divine wrath or one based on a syncretic understanding of the relationships between Christian and earlier animistic beliefs (Morris 2003).

Figure 10.1 The location of places mentioned in Table 10.1

A best of all possible worlds theodicy holds that it 'would probably be impossible to design any system of nature which did not have the potential to injure unsuspecting humans' (Murphy 2005: 345), and that God's purpose is to use disasters to enable a greater good to be achieved (Chester and Duncan 2009). Without earthquakes, for instance, there would be no global tectonics and without volcanic eruptions no planetary atmospheres, and within the literature there are many instances where a best of all possible worlds theodicy has been used. London was struck by two small earthquakes on 8 February and 8 March 1750 and, although many clergy preached of divine wrath inflicted on the people of Britain, some adopted a best of all possible worlds theodicy, a reaction that was also evident following earthquakes in Venice in 1873 and Colchester (England) in 1884. Both the free will, involving divine retribution, and the best of all possible worlds theodicies were also evident in the aftermath of the Lisbon earthquake of 1755.

The word Islam means submission to the will of God, and in the Qur'an, *Hadith* (oral traditions relating to the actions of Muhammad) and many statements of Muslim theology, suffering is perceived as a punishment for sin, as instrumental to the purposes of God and as a severe test of faith (Bowker 1970). In the Qur'an earthquakes 'are interpreted within two different, yet related frameworks and are viewed either as signs of a future apocalypse or as punishment of a limited duration for a specific group of people' (Akasoy 2009: 184). Disasters more generally either express God's anger at human actions, or act instrumentally as warnings of future punishments of sinfulness and/or the eventual Day of Judgment (Akasoy 2009). Although statements of Islamic theology regarding disasters are more heterogeneous than may be assumed from this summary, many historic disasters have been interpreted by those affected in this 'traditional' Islamic manner. For instance, the sixteenth-century polymath Imam Jalaluddin al-Suyuti, emphasised moral explanations of all manner of disasters including judgements for adultery and drinking wine; the 1576 Cairo earthquake was widely understood to be a reproof for the popularity of coffee houses in the city; the Krakatau eruption in 1883 was interpreted as punishment for sins including the acceptance of colonial rule imposed by the Dutch; and, in writing of drought in Yelwa, northern Nigeria, Dupree and Roder (1974) report that most of the people who were affected thought it was blasphemous to speculate on future droughts because human fate is in the hands of Allah. More recently, the eminent physicist Pervez Hoodbhoy (2007) has discussed the impact on an Islamic society of the earthquake that struck Pakistan in 2005 and killed *ca.* 90,000 people. In Pakistan no prominent scientist challenged the idea of divine punishment, while some Mullahs encouraged their followers to sell their television sets, which were thought to have provoked Allah's anger, and many of Hoodbhoy's students accepted an explanation based on divine wrath.

In Buddhism, Jainism, Hinduism and Shinto there is a common belief that a person's behaviour leads to an appropriate reward or punishment, and this has often encouraged fatalistic attitudes towards disasters. *Karma* means that current actions are the seeds of future happiness or suffering and there is no notion of innocent suffering. In Buddhism reincarnation is emphasised, sins committed in former lives are important and happiness together with suffering are both deserved. In Buddhism and Jainism there is no belief in a creator God and, hence, the issue of theodicy is far less acute than in Christianity, Judaism and Islam. In Buddhism the notion of *anicca* (impermanence) is a central element of faith and, following the Indian Ocean tsunami of 2004, many survivors including the bereaved perceived the event 'as yet another, if dramatic, example of *anicca*' (de Silva 2006: 284). Interviews undertaken amongst survivors in Sri Lanka indicated a correlation between belief in *karma* and a 'pessimistic explanatory style' on the one hand and, on the other, poor health and the occurrence of post-traumatic stress disorder (PTSD) (see Chapter 47) some six months after the disaster struck (Levy *et al.* 2009: 44).

In early *Vedic* Hindu texts the gods are powerful and can be appealed to, but in later Hindu thinking *karma* becomes more prominent and even the gods are subordinate to it, with suffering being embedded into the spiritual structure of the world. Interviews carried out following the Indian Ocean tsunami concluded that many Hindus accepted disaster as part of a cycle of creation and destruction which had to be passively – even fatalistically – accepted (Stern 2007a).

Shinto is a syncretic religion showing features similar to Buddhism together with some unique features. Shinto means 'the way of the *kami*', which is the way of the gods or spirits. The *kami* are the mystical figures and forces of the natural world. Ancestors have a special role to play in the common memory of the Japanese people and, following some decades in a cemetery, the ancestral spirits become helpful *kami*. The notion that *kami* reside in shrines and in some cases are permanently present within them is a feature of Shinto, and shrine worship of mountains – including volcanoes such as Fuji – is strongly developed in Japan (Chester and Duncan 2008).

Sometimes responses to disasters are syncretic and simultaneously show elements reflecting the theologies of different religions. For example, in their study of Merapi volcano, Lavigne *et al.* (2008) discuss how reactions to eruptions are complex, showing the influence not only of Hinduism, Islam and Buddhism, but also spirit cults, ancestor worship, spirit healing and shamanism. Examples of syncretism are discussed by Chester and Duncan (2008).

Academic interpretations of religious responses to disasters

Academic scholarship has been highly critical of the influence of religion on the human understanding of disasters. According to this reading of intellectual history, the period from the rise of Christianity as the official faith of the Roman Empire until the seventeenth and eighteenth centuries is interpreted as a long 'Dark Age', in which superstition largely replaced the naturalistic explanations of disaster that were embryonic in the classical era (Sigurdsson 1999: 71; Chester and Duncan 2009). It is further maintained that, following the eighteenth-century Enlightenment, religious explanations became less prominent and were progressively replaced by more scientific and social scientific explanations of disasters. The notion of the 'act of God' was superseded by a perspective in which 'natural catastrophes' represented interactions between vulnerable human populations and a demoralised natural world. This change first began in Europe and North America, later spread to other parts of the Christianised world and has more recently influenced disaster responses in countries where other religious traditions predominate. According to the hazard analyst David Alexander, 'the repetitiveness of impacts and forms of damage, the deliberate or inadvertent creation of vulnerability, and the gross predictability of the consequences of disasters all add up to human, not supernatural, responsibility' (Alexander 2000: 186–87), while for Dennis Dean, an historian of geology, 'theorising independent of the Bible began in America with the Enlightenment. In the mid-nineteenth century the danger that geology represented to religious orthodoxy was apparent and, although there were attempts at reconciliation, the mechanistic view quickly became dominant' (Dean 1979: 291). It is maintained by the academic consensus that where religious interpretations are still to be found, these places represent last redoubts of backwardness and superstition.

In order to test the veracity of this account of intellectual history we have compiled two archives. The first lists major volcanic eruptions that occurred throughout the world between 1850 and 2002 and, of the 49 events recorded, only 16 (*ca.* 33 per cent) show no evidence of interpretations being cast in religious terms (Chester and Duncan 2008: 213). The second archive was compiled to record religious reactions to major earthquakes and volcanic eruptions that occurred between 1900 and 2008 in countries with a predominantly Christian ethos

(Chester and Duncan 2009: 313, 315–19). Of the 61 discrete events, *ca.* seventy-two per cent show evidence of people responding using religious frames of reference, a figure that would probably be even higher if local records could be consulted in more detail. For instance, records of the 1902 and 1929 Guatemalan earthquakes show no evidence of religious responses, yet more detailed reports following the 1976 disaster show that religious explanations were both deep-seated and of long duration within Guatemalan society. The archives indicate that religious responses are not just confined to societies untouched by modernism and during the 2001 eruption of Mount Etna in Sicily, for example, the Archbishop of Catania celebrated Mass in the village of Belpasso in order to enlist God's help in protecting their homes and livelihoods against the lava flow that was threatening the village. The Mass took place in a well-educated society, involved community leaders and the high levels of participation (*ca.* 7,000) indicate that religious responses are still widespread across the Etna region.

Evidence from recent disasters confirms that religious interpretations are still important across a range of faith traditions. Losses in the 2003 Bam earthquake in Iran were widely interpreted in Islamic terms whilst, following the 2004 Indian Ocean tsunami, Buddhists in Sri Lanka tried to make sense of disaster losses using a religious frame of reference to their particular faith.

Changing attitudes

In recent decades there has been a major transformation, indeed a paradigm shift, in the ways in which hazard analysts have studied natural hazards, and this new perspective has been paralleled by a similar paradigm shift in Christian theodicy (see Chester and Duncan 2009). Until the mid-1980s hazard analysts used a methodology known as the dominant approach, in which extreme natural events rather than vulnerable human populations were held to be the principal causes of disaster. This approach was accepted implicitly by the small number of Christian theologians who addressed suffering caused by natural extremes because it could be used to support a best of all possible worlds model of theodicy (e.g. Farrer 1966). Two quotations from Austin Farrer illustrate how the two strands of the argument, one from theology and the other from science, are woven together:

> If an earthquake shakes down a city, an urgent practical problem arises – how to rescue, feed, house and console the survivors, rehabilitate the injured, and commend the dead to the mercy of God; less immediately, how to reconstruct in a way which will minimise the effects of another disaster. But no theological problem arises. The will of God expressed in this event is his will for the physical elements in the earth's crust or under it: his will that they should go on being themselves and acting in accordance with their natures.
>
> *(Farrer 1966: 87–88)*

> It is not, then, that the humanly inconvenient by-products of volcanic fire are cushioned or diverted; it is not that all harms to man are prevented. It is that the creative work of God never ceases, that there is always something his Providence does, even for the most tragically stricken.
>
> *(Farrer 1966: 90)*

More recently, many theologians have found Leibnizian models of theodicy increasingly unconvincing, arguing that they are a means of focusing evil and suffering on God and absolving humankind from any responsibility. John Wesley, founder of Methodism, interpreted the 1755 Lisbon earthquake as a divine punishment for the sins of the Inquisition, and the notion that

Box 10.1 Disaster risk reduction at the Parliament of the World's Religions

Andreana Reale
RMIT University, Australia

The 2009 Parliament of the World's Religions, in Melbourne, was an event bringing together hundreds of faith practitioners. During one forum, entitled 'faith, community and disaster risk reduction', various speakers converged to discuss the role of religion in disaster risk reduction (DRR).

A Western paradigm encourages us to set humans against nature. Natural events like floods and fires are seen as hazards that humans must contend with and conquer. In the ancient spiritualities of many indigenous traditions, humanity is understood to be inextricably linked to nature. One forum participant, Len Clarke, an indigenous Australian leader and educator, pointed out to participants that it is impossible to divorce people from environment. When humans do not respect Mother Nature, Mother Nature lets us know, with increased fires, floods and other disasters.

This theme was picked up by Deborah Storie of TEAR Australia, a Christian development, advocacy and relief organisation. She contended that many of the world's disasters are caused by overconsumption of the wealthy. This level of consumption has a knock-on effect throughout the world's economies, which eventually results in more floods, more cyclones and more droughts. This is felt disproportionately by the world's poor, and forces more people to live in disaster-prone areas. Deborah pointed to Christian scripture exhorting people to consume sustainably.

The religious worldview does not always encourage people to see disasters in this way. Hafiz Aziz ur Rehman, Assistant Professor at the Faculty of Shariah and Law, International Islamic University in Islamabad, explained that in the aftermath of the 2005 Pakistan/Kashmir earthquake, many Islamic leaders interpreted the disaster as divine retribution. This kind of thinking, said Rehman, does not help society to address the root causes of disaster. Fundamentalist Christians also attributed the destruction of New Orleans by Hurricane Katrina to divine displeasure with a sinful city.

Many Islamic non-governmental organisations (NGOs) came to the aid of Pakistan's earthquake victims, as well as volunteers from across the faith community. Without the religious component, the disaster response would have been much slower. Rev Chi Kwang Sunim, a Buddhist nun, meditation teacher and survivor of the 2009 bushfires in Victoria, Australia, explained how churches and Buddhist groups worked together to provide relief for bushfire victims. She also organised a Buddhist service to help survivors cope with the aftermath, and continues to use her position as a community leader to sit on committees to help her community recover.

Ruth Maetala, a researcher, community leader and government official of the Solomon Islands, emphasised the importance of engaging with religious leaders in any kind of DRR strategy. In the Solomon Islands nintey-eight per cent of the population belongs to a Christian church, and church leaders hold more clout than any other kind of authority.

Reflecting on this forum, one can see how important it is to consider the religious angle in DRR. As Rebecca Monson, an Australian National University PhD candidate, pointed out, religion sits alongside other factors to determine a community's overall vulnerability. Religion plays a role in disaster worldview, preparedness, response and rebuilding.

Box 10.2 The position and role of faith-based organisations in DRR

Bruno de Cordier
Conflict Research Group, Ghent University, Belgium

The importance of faith-based organisations (FBOs) in risk reduction and disaster response has long been underestimated. At first glance, when needs are most acute in the aftermath of a disaster, it does not matter much for those affected whether aid efforts come from faith-based or secular actors. In terms of preparedness and coping with the impact of disasters, however, identity and culture including religion and faith-based actors do matter. Many contexts in the global periphery highly vulnerable to natural hazards are societies where religion and faith-based actors are much more present and important than they are in strongly secularised Organisation for Economic Co-operation and Development (OECD) countries. Often, Christian, Islamic and other local FBOs and their volunteer networks provide a number of social services where the state is absent or has contracted in the social field. Although their values, discourse and modus operandi are often not those promoted by the international aid sector, they are often much more anchored in, and representative of, the context's real social tissue than the urban elite groups who are well represented in the international aid sector and often determine the latter's priorities and partnerships.

It is in that capacity of local social institutions that FBOs have a role and added value. Their long-term presence in the context, their volunteer networks and the social infrastructure like schools, clinics and welfare centres that religious organisations run form a solid local framework for rapid and effective aid efforts. In the Pakistani sector of Jammu-Kashmir, for example, long-present Islamic charities like the Al-Mustafa Welfare Society and the Al-Khidmat Foundation were bringing aid to displaced and earthquake victims in 2005. In the Indonesian province of Aceh, volunteers of the Islamic group Persyarikatan Muhammadiyya also carried out aid work immediately after the tsunami of 2004. These efforts have often been overlooked or even consciously minimised once large-scale international aid and the global media arrived. Another role for FBOs in DRR is to stress people's responsibility to mitigate hazards before they happen, for example through land planning and relief preparedness measures.

Religious infrastructure like churches and mosques can also serve as shelters and distribution points, a purpose they actually served in a number of emergencies. In emergency situations with refugees and internally displaced, religious actors and FBOs can, besides the delivery of practical and technical aid, facilitate the observation of religious festivals and duties among those who face difficulties in observing religious obligations because they have been uprooted from their usual environment, or because they lack the resources. Second, there is what Jean-François Mayer calls 'infuse tragedy with meaning' among communities and populations where there is a cultural reflex to make sense of a disaster or displacement or to find solace for psychological distress through religion. On the other hand, FBOs can also have a negative impact when they distribute relief goods to their confessional following only and, as such, create or exacerbate conflict.

---

human sinfulness resides at the heart of disasters is a more important insight than was realised at the time. When sectarianism is removed from Wesley's remarks then they are close to what many contemporary Christian theodicists and hazard analysts are proposing. For hazard analysts, disasters are increasingly viewed as outcomes of vulnerability for which human beings are largely responsible, because vulnerability often occurs as a result of global, regional and local disparities in wealth, poverty and power. More than nintely-nine per cent of volcano-related deaths in the twentieth century occurred in countries that may be classified as economically less developed. From the perspective of theodicy, what is sometimes called the liberationist approach argues that it is structural rather than individual human sinfulness which is reflected in disaster losses because these too are reflective of differences in poverty, wealth and power which occur both within and between countries. Structural sinfulness was first identified by Latin American liberation theologians of the 1970s and 1980s, such as Clodovis Boff, Leonardo Boff (Boff and Boff 1987) and Gustavo Gutiérrez (Gutiérrez 1988), where it was viewed as a process that kept the poor and disadvantaged in a state of subjection. Starting with the 1970 earthquake in Peru there has also been intense reflection on earthquake losses, especially by the theologian Jon Sobrino (2004).

With regard to other world religions, similar paradigm shifts have not taken place but there are many signs that these other faiths are becoming theologically more diverse in their approaches to disasters, and more accepting of measures to reduced hazard exposure. Islam has experience stretching back more than 1,000 years of accommodating differing political and social conditions across the world, and this is manifested in a great diversity of practice. One example of theological accommodation may be seen following the 1992 Dahshûr earthquake in Egypt, when the government produced a report entitled, *Earthquake Catastrophes and the Role of People in Facing Them*, which successfully reconciled scientific, planning and Islamic perspectives on disaster losses (Degg and Homan 2005). Such an approach promises much for other Muslim countries where Islamic believers and Enlightenment-inspired earth and social scientists have often viewed disasters with mutual incredulity. A further development over recent decades has been the growth in Islamic humanitarian agencies which provide relief aid to people affected by disasters, within contexts fully in accord with Islamic theology and faithful observance. Writing of the UK-based organisations, *Islamic Relief Worldwide* and *Muslim Aid*, which provide aid following disasters, de Cordier (2009: 612) argues that both charities find their justification in *zakat*, one of the five pillars of Islam and one which involves those with an income above a certain level making compulsory donations each year to the needy. Today in the UK and other economically more developed countries, voluntary charitable donations, or *sadaqa*, are usually made in cash, being frequently sent to Muslim countries that are suffering from the effects of disasters. In Iran, following the 2003 earthquake, the Supreme Leader of the Islamic Republic (Ayatollah Ali Khamenei) emphasised that the disaster represented God's testing of the population rather than divine punishment of individual sinners.

In the aftermath of the 2004 Indian Ocean tsunami in Sri Lanka, de Silva (2006) suggests that the more optimistic teachings of Buddhism, including *viriya* (energy and vigour) and *karuna* (compassion), may be employed in both counselling disaster-affected people and justifying active involvement in relief. In Sri Lanka numerous Buddhist monks also provided immediate help and support. In reviewing the Indian (Gujarat) earthquake of 2001, a survey of survivors found that those who helped others during the relief phase of the emergency often did so as a means of improving their *karma*.

Conclusions

Although there are a small number of examples where people have resisted relief efforts due to their religious faith, these are exceptions, and for the most part belief neither inhibits more practical measures being taken to reduce individual or group exposure to hazards, nor does it hinder people accepting help from the civil authorities. Believing at the same time in two mutually incompatible worldviews, or holding one opinion and acting contrary to it, is known in the literature as parallel practice, or more ambiguously as cognitive dissonance (Chester and Duncan 2009: 325). This has been discussed in detail with reference to countries with a predominantly Christian ethos, though it appears across the spectrum of world faiths (Chester and Duncan 2008), and this, combined with new perspectives that focus on helping disaster-affected people rather than blaming the disaster on their sinfulness, means that religious institutions and their leaders have the potential to become valuable resources in disaster reduction and mitigation.

'All over the world faith communities are already active in response to disasters and recovery activities … [and] these activities could increase' (Wisner 2010a: 129). Religious leaders, for instance, often know where people reside, can identify the missing and survivors, may provide psychological support and pastoral care, and not least act as community leaders and spokespersons. The world's major religions also have access to financial resources, may be established providers of education and medical care and are valuable 'assets' in the process of recovery as well as in prevention and risk reduction (Chester and Duncan 2008).

11

Hazards and disasters represented in film

Gregory Berger

UNIVERSIDAD AUTÓNOMA DEL ESTADO DE MORELOS, MEXICO

Ben Wisner

AON-BENFIELD UCL HAZARD RESEARCH CENTRE, UNIVERSITY COLLEGE LONDON, UK

Introduction

Humanity has been representing hazards and disasters for a very long time. From cave paintings to science fiction 'space opera', or the resigned yet risk-aware winter haiku of Issa and blockbuster films like *The Day After Tomorrow*, people express their anxiety in the form of art. Indeed, the human being could as well be called *homo metuens* (one who fears) as well as the better-known epithets *homo sapiens* (the knower), *homo faber* (the maker) and *homo ludens* (one who plays).

Do visual and other representations of disaster do more than 'keep alive our fascination with natural destruction' (Wallin 2005)? Some have even suggested that disasters are themselves 'readymade artworks', in the words of Scha (2005). That fascination is evident in Susan Sontag's novel, *The Volcano Lover*, for example. Yet there is more than mere fascination. Disasters are part of our shared history and pre-history (mythologised or chronicled). They are something about which we are puzzled. We search for their meaning, if for no other reason than to put the suffering and death of other people in perspective.

Pre-literate and pre-scientific people interpret nature through narratives and visually. These narratives not only help to make sense of the place of human beings in the world and cosmos, but also are sometimes bearers of morals. Where, as in the European Middle Ages, disaster was understood and communicated as the result of sinfulness, the moral is clearly not to sin (see Chapter 10). If the volcano spirit is unhappy or ancestral spirits withhold rain, and the cattle die, clearly one has to rectify relations between people and those higher powers. It is interesting that in many cultures the displeasure of gods is caused by ruptures and injustice in society. The tempests suffered by Odysseus on his homeward voyage from Troy were in part the result of the gods' displeasure with the human relations that resulted in the Trojan War.

The production, as opposed to consumption, of disaster art has also come to be understood as a tool in psycho-social recovery from the traumas experienced in an extreme natural event (IATO 2010) (see Chapter 47). Since the end of the nineteenth century this has also been true of films that depict disaster, the subject that this chapter takes up in what follows. Film, like the many representations of disaster, has all the functions just mentioned. It is entertainment that indulges our primordial fascination with calamity; it tells stories with morals and lessons; it seeks meaning; and it is also cathartic and a means of recovery.

Early history of disaster films

From the earliest days of motion picture production in Europe and the USA, the spectacle of mass destruction has been a fixation for filmmakers and moviegoers alike.

When Edison, the Lumière brothers and other pioneers of motion picture technology developed machines to capture and reproduce moving images at the end of the nineteenth century, the world was undergoing rapid and often unsettling changes. Industrial weapons of war were developed that would soon lead to the deaths of millions in the trenches of Europe. Accelerated industrial development in Europe and the Americas led to the pollution of the natural world on a scale previously unknown (Thorsheim 2006; Jenks 2010). Mass migration and the accompanying boom in urban populations led to a new urban working class for whom entertainment and spectacle served as comfort in the midst of an increasingly uncertain existence.

There was much material to work with and much to produce anxiety (Steinberg 2000; Svensen 2009). In the first decade of the twentieth century alone, tens of thousands of people were killed by a combination of seven major fires (including one in the Paris Metro), at least two particularly deadly shipwrecks and two highly lethal volcanic eruptions, including Vesuvius and Mt. Pelée, the latter of which destroyed the principal city on the island of Martinique. Further, there were three severe earthquakes including the 1906 San Francisco event and its fiery aftermath, and one that killed 70,000 in the Italian city of Messina. This period also saw the great hurricane that took 7,000 lives in Galveston, Texas and a typhoon that killed 10,000 in Hong Kong. Newspapers were abundant and news spread rapidly among newly urbanised populations (Nord 2006; Williams 2009) (see Chapter 63).

Just prior to the introduction of the Nickelodeon movie houses in the USA, theatrical recreations of disaster and catastrophe were in vogue in Coney Island and other urban amusement centres. Staged scenes of disaster, replete with pyrotechnics and cascading debris, attracted thousands of paying working-class customers every day, many of them recently uprooted immigrants. Floods and fires were particularly popular. On any given day, a theatrical recreation of the deadly 1889 flood of Johnstown, Pennsylvania could be seen in Coney Island just down the street from a show based on the storm surge that killed thousands of people in Galveston, Texas in 1900 (Stulman and Warnke 1990: 101–3).

In fact, the Galveston disaster was also the subject of one of the first films made for the commercial film industry, *Searching the Ruins of Broadway, Galveston for Dead Bodies*, produced by Vitagraph in 1900. Due to the tremendous success of this and similar films of the 'actuality' genre, the Edison company made a concerted effort to film as many fires and other calamities as it could between 1900 and 1903 (Stulman and Warnke 1990). By the time the first theatre dedicated exclusively to film projection opened in Pittsburgh in 1905, the disaster spectacle had established itself as a popular commodity in the new film industry.

After the film industry moved most of its operations to California, fiction and fantasy soon replaced the 'actuality' genre. However, as Hollywood produced ever more expensive films, it became clear that scenes of large-scale destruction, including the crumbling of Babylon in D.W. Griffith's *Intolerance*, could draw a dependable audience of people willing to pay to see catastrophe.

The first film to feature the full-scale annihilation of an entire modern city seems to have been RKO's 1933 film *Deluge*. The film opens with a series of spectacular natural events around

Box 11.1 Fortune, chance and risk

Ben Wisner
Aon-Benfield UCL Hazard Research Centre, University College London, UK

The libretto of Carl Orff's choral work 'Carmina Burana' invokes 'Fortuna, Imperatrix Mundi' (Fortune, Empress of the World):

> O Fortune
> like the moon
> you are changeable,
> ever waxing
> and waning;
> hateful life
> first oppresses
> and then soothes
> as fancy takes it,
> poverty
> and power,
> it melts them like ice …

Oddly, the notion that 'all that is solid melts into air' is not only a medieval reaction to plague, fatal childhood diseases and death in childbirth, and the violent coming and going of armies; it is also a very modern, even post-modern idea. Berman (1988) used these words as the title of a book that explores the self-destructive nature of modernism, an echo of Marx and Engels in the *Communist Manifesto* (1848), who wrote:

> Constant revolutionizing of production, uninterrupted disturbance of all social conditions, everlasting uncertainty and agitation distinguish the bourgeois epoch from all earlier ones … All that is solid melts into air, all that is holy is profaned, and man is at last compelled to face with sober senses, his real conditions of life, and his relations with his kind.

> *(Marx and Engels 1848)*

Returning to the old manuscripts found by Orff in Munich's Royal Court Library, and the secular songs recorded and preserved by Benedictine monks:

> I bemoan the wounds of Fortune
> with weeping eyes,
> for the gifts she made me
> she perversely takes away.
> It is written in truth,
> that she has a fine head of hair, but
> when it comes to seizing an opportunity,
> she is bald.

As popular imagination has shifted its idiom from fortune to risk, it is easy to assume that the former is superstitious and based on ignorance and emotion, while the latter is a matter of informed calculation: cool and dispassionate. Is this so? It is likely not true. In the twenty-first century trust in calculation and planning by experts is eroding. Many people have become aware that neither fortune nor chance govern one's risk of suffering and loss from natural hazards. Rather it is the action and inaction by governments and corporations that place different places and people differentially at risk (Wisner *et al.* 2004).

In Western society the transition from popular notions of fickle and unknowable fortune to calculable risk has eased the idea of chance. Early use of the terms 'risk' and 'chance' in the 1500s–1700s utilises numbers to bring order into a seemingly chaotic world (Luhmann 2002: 8–14). 'Chance made the world seem less capricious' (Hacking 1990: xiii). Yet in the twentieth century there was a revolt. Beck (1992) documents resistance against a rational calculus of risk that he called 'ecological modernisation', which seemed to hold out a false promise of protecting society. Once more, despite wider and deeper knowledge of nature and of the environment, the rapid growth of technology, of cities and the 'constant revolutionising of production' brought unforeseen consequences, at least for ordinary people.

the globe, unleashing a massive tsunami that overtakes and then crumbles New York City over the course of a meticulously rendered seven-minute sequence. *Deluge* is perhaps the first true example of the Hollywood disaster film. The sequence of images used to illustrate the destruction of New York – the ominous approach of the tsunami, panicked crowds running for cover, buildings crumbling one by one and a few detailed shots of some unfortunate souls as they meet their gruesome death – established the now familiar formula used ad nauseum to this day in the disaster film genre. Film director Roland Emmerich used the film for inspiration in his own depiction of New York City's demise after the melting of the polar ice caps in 2004's *The Day After Tomorrow*.

In the 1930s, rapid advances in the field of cinematic special effects accompanied a worldwide economic crisis that only deepened the collective anxiety of people in the industrialised world. A boom of other disaster-themed films followed *Deluge* in the same decade, including *The Last Days of Pompeii*, *In Old Chicago* and *The Hurricane*. From the 1930s onward, when moments of intense collective anxiety have coincided with innovations in special effects there has been an upsurge in interest in Hollywood disaster films at the box office (Brendon 2000).

Politics and existential anxiety

Although focusing primarily on the Japanese and US monster movies of the 1950s, it was the literary theorist Susan Sontag who first observed the connection between collective existential anxiety and the spectacle of Hollywood disaster films in her 1965 essay 'The Imagination of Disaster'. Disaster films, she noted, offer anxious viewers a kind of inoculation from the fear of annihilation by subjecting them to the cathartic experience of witnessing massive destruction from a safe and detached vantage point (Sontag 1965: 42).

While acknowledging the universal enjoyment of the disaster film, however, Sontag also mused about the potential negative impact of these films on our comprehension of disaster. After all, she noted, most Hollywood disaster films are devoid of social critique (Sontag 1965: 48). In most disaster films the root causes of the catastrophe at hand are not revealed, let alone changeable by human action. Disasters and their accompanying destruction are seen as the inevitable destiny of humankind. Most high-grossing films from the genre do not explain disasters as the end result of poverty and social vulnerability to extreme events (Wisner *et al.* 2004). The destructive force might be an atomic moth or an out-of-control colony of Brazilian ants, but the message is always the same: Disasters come from God, or at least from a force as inscrutable as God.

The Cold War and McCarthyism's hunt for 'disloyalty' in Hollywood was undoubtedly another reason why the root causes of disaster were not investigated in film (Seed 1999; Doherty 2003). Asking about the social, economic and political causes of vulnerability to disasters might seem seditious. As we shall see below, grappling with root causes often involves questioning the distribution of power and privilege in society (see Chapter 3 and Chapter 5), and such questions would have raised suspicions of 'disloyalty' during the McCarthy anti-communist witch hunt during the 1950s in the USA.

In the early 1970s, as the USA faced intense political and economic crises, Hollywood entered into its most prolific and profitable era of disaster film production. *The Poseidon Adventure*, *Earthquake*, *The Swarm* and dozens of other films drew millions of moviegoers to the theatres and helped to rescue Hollywood from a prolonged economic slump. A similar upsurge took place in the late 1990s, as computer-generated graphics reached new levels of sophistication and a cascading series of geopolitical crises brought on a new age of anxiety. With the notable exception of 2004's *The Day After Tomorrow*, explanations of the root causes of disaster were nearly always absent from these films.

Box 11.2 Comic books and DRR communication

Ben Wisner
Aon-Benfield UCL Hazard Research Centre, University College London, UK

Ilan Kelman
Center for International Climate and Environmental Research–Oslo, Norway

Comic books and cartoon strips are used to communicate disaster risk reduction (DRR) endeavours. Some examples are provided here.

The European Commission's Humanitarian Aid and Civil Protection Department (ECHO) uses a comic strip, 'Hidden disaster' by Erik Bongers, to tell the story of how it responds to disasters. An earthquake disaster strikes in the fictitious country Burduvia. The main character, a young woman, leads the reader through the deployment of humanitarian relief, including challenges such as intrusive journalists who publicise the story, and rebel leaders reluctant to accept aid. The approach is quite top down, but it highlights local innovation to deal with the aftermath of disaster as well as the role of civil society. It is significant that the protagonist is a young woman (see: www.euromedcp.eu/index.php?option=com_content&view=article&id=611%3Ahidden-disaster-comic-book&lang=en).

Wahana Lingkungan Hidup Indonesia (WALHI) is the Indonesian Forum for the Environment, founded in 1980. The group has a presence in twenty-five provinces with over 400 member organisations actively campaigning on environmental issues. They produced a comic book in Indonesian as part of a project in West Java on capacity building for community-based volcanic risk reduction. The comic book describes the main volcanic hazards and how communities can deal with them. It also treats social complexities such as difficulty in communication between urban intellectuals and rural people and between young and old (see the cover: www.accu.or.jp/esd/mt-static/ino/indonesia/WALHI%20-CoverPage-Intro.pdf; part one: www.accu.or.jp/esd/mt-static/ino/

indonesia/WALHI%20Part-1.pdf; part two: www.accu.or.jp/esd/mt-static/ino/indonesia/
WALHI%20Part-2.pdf; and annex: www.accu.or.jp/esd/mt-static/ino/indonesia/WALHI
%20Annex.pdf).

'Tokyo Magnitude 8.0' is an animated Japanese television series with eleven episodes,
at twenty-three minutes per episode. The series begins with an 8.0 Richter-magnitude
earthquake striking offshore from Tokyo, leading to the devastation of the city including the destruction of landmarks. A young schoolgirl and her brother, who were on a
day outing, must find their way back to their home (see: www.animenewsnetwork.
com/encyclopedia/anime.php?id=10704).

Historical examples exist as well. A 1956 16-page comic book from the USA called
'Mr. Civil Defense Tells About Natural Disasters' features a character like Li'l Abner, a
comic strip that ran for forty-three years and was often accused of being stereotypical,
elitist and sexist. Li'l Abner lives in a fictional place called Dogpatch, and he is the epi-
tome of rural idiocy. 'Mr. Civil Defense' emphasises that communities can and should
take charge of dealing with disasters themselves, but depicts the process as one in
which the authorities instruct the community how to act. It is ensconced in the racism
and sexism of the era, representing only white people as authoritative and action-
oriented, while suggesting specific and stereotypical roles for boys and girls to get
involved in 'civil defence' (see a high-quality version (11MB in PDF): www.ilankelman.
org/miscellany/mrcdhr.pdf; or a medium-quality version (7MB in PDF): www.ilankelman.
org/miscellany/mrcdlr.pdf).

This intellectually flawed approach to disasters in film has in no way been limited to Hollywood productions. Well-known European films, the stories of which revolve around catastrophes, have also equated disasters with God's will. One of the clearest examples of this is the
1950 classic *Stromboli*, from director Roberto Rossellini, one of the great masters of Italian
neorealism. Emerging from the post-war defeat of Fascist Italy, the neorealists rebelled against
the artificial spectacle and cultural hegemony of Hollywood by producing films featuring
stories of poor and working–class characters struggling to survive. Neorealist directors, including
Rossellini, seldom worked with professional actors, choosing instead to cast ordinary people in
their films.

Stromboli's story is primarily set on the island of the same name, home to one of Italy's active
volcanoes. A Lithuanian woman in a displaced persons' camp after WWII wins her freedom by
marrying an Italian fisherman, only to find a life of isolation and misery in a village on the island
of Stromboli. Rossellini cast well-known Swedish actress Ingrid Bergman as the lead, in order
to sharply contrast the character with the non-professional actors from the island who play
many of the other roles. *Stromboli*'s protagonist believes herself to be from a higher social class
than the simple, modest and religious people of the island. In the film's climax she attempts to
scale the volcano to get to a ferryboat, and thus her freedom in a bigger village on the other
side of the island. The volcano's toxic gases nearly overtake her, and after narrowly escaping
death she finds her ego, and sense of class superiority, humbled before the supreme power of
the volcano.

Stromboli is a masterpiece that showcases some of the best achievements of the neorealist
movement, including intricate sequences of fishing and other activities of the real-life islanders'
routine woven into the dramatic fabric of the story. As a political statement, however, *Stromboli*
is filled with contradictions. Rossellini himself said that he wanted to make a statement about
the dangers of egotism and assumed class superiority after the disastrous results of Nazism and

Fascism in Europe (Camper 2000). However, his social critique never extends to the relationship of the islanders to the volcano itself. In one of the most memorable sequences of the film, the volcano showers flaming projectiles into the village and its residents are forced to flee in fishing boats to a safer distance from the volcano, adrift in the water, where the village priest leads them in a prayer of the rosary until the eruption subsides.

The sequence is notable for its realistic recreation of evacuation proceedings, featuring poor, rural villagers in the role of themselves, a rare moment in cinematic history. However, in the final analysis of the film, we are left with the impression that the volcano is a mystic and inscrutable entity, an extension of God himself. Rossellini's social critique does not extend itself to a deconstruction of the institutions and human behaviour that have placed the villagers of Stromboli in harm's way (see Chapter 28). The question of how Stromboli's residents can lead a life without the continuous threat of an active volcano is never asked in the film. In this complex film are clearly resonances of the oldest theme in Western painting's treatment of disaster: divine will as its cause as a consequence of human weakness and moral failing. While in the Middle Ages these failings were seen as spiritual, secular society sees them as intellectual: the citizen does not make the effort to learn self-protection or does not pay attention to expert advice.

Disaster film of another kind

In the years since film was born, new social conditions have arisen that make more people vulnerable to disaster than ever before (Wisner *et al.* 2004). More people now live in crowded and poorly developed cities than in the countryside (see Chapter 13). Climate change exacerbates hazards and vulnerabilities threatening millions of people. New industrial development has polluted soil and water upon which millions depend for their life and livelihoods. In much of the world, continued government downsizing has limited access to health and infrastructure for the poor (see Chapter 5).

References to these underlying social conditions are absent from most disaster films produced in the USA and Europe, while the effects of these new global economic, political, demographic and environmental realities can be felt most acutely in the global South. No one doubts the effect of movies and television on the way people perceive the world and their place in it (Herman and Chomsky 2002) (see Chapter 63). Even though Hollywood is no longer the world's most prolific producer of films – India and Nigeria both produce more films per year than the USA – its products are still widely consumed in the global South, albeit in the form of pirated DVDs which can reach youth and others even in isolated rainforest enclaves during a violent civil war (Richards 1996). The entertainment delivered by these films comes at a price: the proliferation of misleading ideas about disasters in an increasingly dangerous world.

As counterpoint, the most thoughtful films to deal with disasters come from the global South. Though lacking in the resources and broad distribution reach of their Hollywood counterparts, films produced in countries such as Argentina, Iran and China have brought the social dimensions of disaster to the screen. Arguably the most influential director in the history of Latin American cinema, Fernando Birri's work built on and expanded the techniques and concerns of the Italian neorealists. Considered by film historians to be the 'father of new Latin American cinema', and co-founder of the region's most important film school, Birri is also the director of *Los Inundados*, a darkly comedic and astute observation of the political dynamics of disaster in Latin America, produced in 1962.

The film tells the story of a poor family living on the banks of the Rio Salado, on the outskirts of the city of Santa Fe, Argentina. After the river floods their homes, the displaced people are rescued by a flotilla of soldiers and then set up in a camp in the centre of the city (Ullberg

2009). The local wealthy elite are aghast at the presence of poor 'refugees' in their midst. Meanwhile, competing political parties attempt to use the refugees' cause to win votes in upcoming elections. One of the parties even goes so far as to designate a 'national day for flooded people'.

Once the elections are over, promises made to the refugees are forgotten and the police evict them. One of the families refuses to leave the abandoned boxcar where they have found shelter, only to awake one morning to find that they have been attached to a moving freight train. When stopped at the next station, railway officials cannot find the boxcar's serial numbers in their records. To officialdom, therefore, the family simply does not exist. Thus the family continues its journey along Argentina's railways, observing the industrial and agricultural development from the window of the boxcar as detached castaways for whom the modern advances they witness have meant nothing. Eventually they find their way back home and are sent back to their village, now dried out. The last sequence features the protagonist rebuilding his thatched roof and looking at the sky, joking in anticipation of having to repeat the whole experience again after the next rainy season.

Los Inundados hasn't been seen much outside Latin America, and yet its impact on the cinema and politics of the region is profound. The film sparked a movement, born in Argentina and systematised by other filmmakers soon afterwards, called 'Third Cinema'. Proponents of the movement combined elements of both documentary and fiction filmmaking, emphasised participation of workers and ordinary citizens in the production of films, and saw their films as tools for social change to be screened block by block, community by community, rather than as a commodity to be used for monetary gain or to obtain prestige at elite film festivals.

It is not by accident that Birri chose disaster as his subject matter. In Argentina, as in the rest of Latin America, disasters historically have been seized upon by politicians to pander to affected people while ignoring the root causes of vulnerability. In the decades following the release of *Los Inundados*, and the creation of Third Cinema, many Latin American countries fell prey to military dictatorships and authoritarian regimes (Funari *et al.* 2009). Yet popular movements also flourished despite adversities, intellectually nourished by films like *Los Inundados* and other works of Birri's contemporaries. Mishandled disasters, such as the 1985 Mexico City earthquake and the 1972 Managua earthquake in Nicaragua, became turning points for social movements which, unlike the hapless family of *Los Inundados*, challenged ineffective and cynical governments and began to radically transform their societies (Wisner *et al.* 2004; Olson 2010).

For all of the billions of dollars that Hollywood films have spent on special effects, Birri's simple film, therefore, has had a greater and more positive impact than any mega-budget apocalyptic disaster film. While it might be an exaggeration to claim that *Los Inundados* is single-handedly responsible for the revolution in thinking regarding disasters in Latin America, it undoubtedly has played a key role in the evolution in thinking regarding disaster risk reduction (DRR) and is well known among the scholar activists who launched the Latin American network of social science for disaster reduction (*LA RED*) in 1992 (LA RED 2010).

Thirty years after the production of *Los Inundados*, an unlikely revolution in cinema was taking place in the Islamic Republic of Iran. The best known of these filmmakers, Abbas Kiarostami, happened to shoot a film in the small village of Koker, in northern Iran, in 1987. Most of the actors of the film were local townspeople rather than professional actors. Three years later, in 1990, an earthquake struck the Rudbar-Manjil region and killed 40,000–50,000 people (MCEER 2010). Immediately after the catastrophe, a concerned Kiarostami set out with his son to learn the fate of the town of Koker and the cast of his film. Kiarostami's journey back to Koker is documented in his masterpiece *And Life Goes On*, an ambitious hybrid of fiction and documentary filmmaking that casts an actor in the role of the real-life director, stages thoughtful meditations on the meaning of the tragedy, but which features dozens of authentic interactions with actual survivors of the earthquake.

Kiarostami is not considered to be a particularly political filmmaker, yet his film insightfully captures many of the social dimensions of how humans deal with disaster. While many of the characters believe that the earthquake is God's will, Kiarostami questions that assumption in a series of memorable images and conversations. While on the way to the quake zone, the director's son points out a cement factory that, unlike many of the mud homes of the region, is still standing. Later on they find an old man whom the director recognises from the film, and he stops to give him a ride. 'No, this is not the work of God', says the old man. 'The disaster is like a wolf. God wants his servants alive.' Later on, a boy recounts the old man's story to a villager who has lost one of her sons in the earthquake. However, the boy also mentions something he read in a 'history book' recounting the story of Abraham and Isaac. The boy wonders aloud if the deaths of so many people weren't a sacrifice demanded by God (see Chapter 10). We immediately cut to a shot of the boy's father staring beyond the wrecked village at a herd of sheep. The meaning is clear. Whereas God is believed by Muslims to have sent angels to put a lamb in place of Abraham at the last minute, the sheep of Koker are living, while many children are dead. No, this is not the work of God, Kiarostami argues to us through his images. Kiarostami's images are both beautiful and radical.

Above all, *And Life Goes On* is a story of resilience that celebrates the inherent strength of ordinary people. Much of the film is a montage of villagers digging neighbours out of the rubble, trudging by foot along blocked highways to bring provisions to isolated family members and defiantly engaging in mundane activities necessary to keep on living. Notably absent are government rescue workers. Much like the survivors of the Mexico City earthquake or even many parts of New Orleans after Hurricane Katrina, it is the community itself that is responsible for its own rescue and recovery, for better or for worse. The heart of the film lies in a conversation that the director has with a young couple who chose to get married the day after the tragedy, despite losing sixty-five relatives. Their marriage is an act of persistence, as is the director's own quest to find survivors. Although the film does not pretend to speculate on how to avoid such a tragedy in the future, Kiarostami's film is still, in part, a celebration of popular power.

Throughout Asia and Africa, regional film industries are increasingly outpacing Hollywood in their film production output. Films like *And Life Goes On* and *Los Inundados* from these new centres of production could potentially enrich the dialogue on reducing people's vulnerability to disasters in some of the more hazard-prone corners of the globe. South Korea, for example, has given birth to a generation of cinematic masters over the last several decades. However, the country's first attempt to mimic the Hollywood disaster film offers few promises for innovative subject matter: 2008's *Tidal Wave* features a devastating tsunami that wipes out the beach city of Haeundae, leading to the misnomer of calling a tsunami a 'tidal wave'. Costing US$16 million to produce, and featuring Hollywood-grade digital special effects, the film was hyped nationally and internationally as 'South Korea's first disaster film'. In fact, the flood sequences in *Tidal Wave* are not much different to the images of New York City's destruction in *Deluge* 100 years ago.

Nigeria's first disaster film – a remake of James Cameron's *Titanic* – cost only a fraction of *Tidal Wave*, and at first glance is also devoid of commentary on the social aspects of disaster. However, while the film may not tell us much about how to protect vulnerable passengers from icebergs while on transatlantic passage, it does give us an interesting lesson on the impact of disasters on the construction of historical memory.

Masoyiyata/Titanic is a product of Nollywood, the nickname of the Nigerian film industry. Nollywood, which is less than twenty years old, has grown rapidly to become the second largest producer of films in the world (behind India) and by far the largest centre of film production on the African continent, producing an estimated 200 films per month (Saul and Austen 2010). The birth of this industry has become the stuff of urban legend. Supposedly, an entrepreneur found himself with a warehouse full of excess VHS videocassettes, so he set out to reuse the tapes for a film which he shot on video. He then used the remaining tapes to reproduce the film and distribute it himself for home use at open-air markets throughout the country. Many more such films followed, and an industry was born.

Digital technologies facilitated the Nollywood model of film production and distribution, drastically increasing the affordability and speed of producing DVDs for sale. Although most Nigerian films are produced in English, there are several regional centres of production that shoot films in local languages. *Masoyiyata/Titanic*, for example, is in the Hausa language from northern Nigeria.

Director Farouk Ashu Brown made his own version of *Titanic*, without permission from the director, intercutting a few wide shots of the ill-fated ocean liner from the original film with his own scenes, which make up the majority of the movie. Like the entire industry itself, *Masoyiyata/Titanic* is an astute, digital appropriation of the Hollywood production model, remade in an African image, by and for Africans.

Krings (2007), however, argues that the film is far more than a simple low-budget reproduction of a Hollywood hit. Krings believes that the filmmakers are keenly aware of the role that disasters play in history and human memory. Disasters, despite their negative connotations and tragic outcomes, punctuate moments of historical significance and are immortalised by historians. Because Eurocentric historians have systematically excluded African contributions to human achievement and the African experience in general, Krings notes, 'The Titanic, historically owned by the White Star line, is turned into a Black Star liner. There is no longer just a single black worker below deck ... but the whole ship itself which is Africanised ... What at first seems rather odd – the claiming of a disaster – on the other hand makes sense, since the film radically claims what black people purportedly have been denied' (Krings 2007: 8). What they have been denied is historical agency in history. They did cross the Atlantic historically, but as anonymous slaves crammed below decks. One also has to wonder if the film has meaning at another level – a warning to the African political elite whose 'ship of state' sails into peril while they enjoy luxuries.

Masoyiyata/Titanic and *Tidal Wave* may be the harbingers of a new internationalised disaster genre that adapts the familiar Hollywood motifs to local means of film production. Meanwhile, there seems to be no end in sight for the current wave of disaster films produced in the USA. On the contrary, the scope of destruction and the degree of havoc inflicted on planet Earth would appear to be ever more relentless.

Conclusions

The Earth may well be in a state of peril, yet the prognosis for the disaster film has never been better. The continued production of disaster films from Hollywood indicates more interest in the genre than ever before, while the ability of more and more local industries to produce them, particularly in the global South, will, one hopes, incorporate diverse perspectives into the production of a genre that is alternately escapist and contemplative. We may well see the day when thoughtful cineastes in the tradition of Fernando Birri and Abbas Kiarostami produce films with the broad popular appeal of *The Day After Tomorrow*. As was the case with *Los Inundados*, films can indeed inspire conversation, and perhaps even action. If such films lead to an authentic reduction of vulnerability to disasters, then 100 years of celluloid Armageddon will not have been in vain.

12

Hazards and disasters represented in music

Bob Alexander

INDEPENDENT CONSULTANT, RURAL LIVELIHOOD RISK MANAGEMENT CONSULTING, USA

Introduction and methods

Many popular songs have been written that reference hazards and disasters because of the compelling imagery that can be conjured with such references. Examples are avalanches of feelings, earthquakes of emotion, tsunamis washing over me, twisting like tornados, calm in the eyes of hurricanes. Though interesting entertainment, such songs do not entice listeners to proactively consider their own vulnerability.

After the Indian Ocean tsunami, articles asked what could be done to help raise awareness so that future events would not be as disastrous (Tsunozaki 2007). Some organisations responded by promoting materials including songs, since studies show that material with music is more effectively remembered (Snyder 2000; Anton 1990). Potential noted problems of using audio disaster risk reduction materials include the lack of visual stimulus and the language barrier to distributing such materials extensively (UNESCO 2007).

A literature review was conducted to determine how using songs and music for disaster risk reduction have evolved through history and how effectively such use overcame these problems to convey beneficial disaster risk reduction messages. As literature on this topic is limited, much insight was obtained from internet searches, discussions with academics and disaster risk professionals, and analysis of lyrics and descriptions of disaster-related songs and music. For the purposes of analysis, songs were categorised as having been used as education tools, as history, as coping and relief mechanisms, and as expressions of disaster risk capacities.

Results and analysis

As education tools

Music plays important roles in formal and informal disaster risk education. Songs in classrooms are beneficial for social and environmental studies because of their abilities to capture student interest, sensitise attitudes towards issues, draw from a rich data source, and make connections between theory and real situations (Ramsey 2002; Turner and Freedman 2004). For classroom science education, many videos with songs about hazards science are produced in the USA and are used in geography and science classes, including parodies of popular songs in Bill Nye the Science Guy's educational television show (e.g. 'Lavaflows', 'Earthquake Rumble') and videos from organisations like www.Time4Learning.com (e.g. 'The Hurricane Song'). Some teachers ask students to write lyrics or to make multimedia presentations that include songs to demonstrate what they've learned about hazards (Government of Australia 2010).

Disaster-related songs have also been used in videos and activities aiming to teach about socio–economic vulnerability and preparedness. Some of these videos continue to use songs with pedestrian, whimsical lyrics that perpetuate myths of vulnerable victims with no disaster risk reduction capacity. An example is using The Little Girls' 'The Earthquake Song' in videos for teaching about California earthquakes (see also Box 12.1). Others ask students to explore the dimensions of past events through the historical popular songs about them, for instance learning about effects of drought and famine through songs about the dust bowl.

Disaster preparedness is being taught to children in the USA with www.SongsForTeaching. com resources that include songs that teach tornado and fire safety, illustrated by 'get out–get in–get down' and 'stop drop rock and roll'. For Indonesia, the Education Development Centre developed a DVD that teaches primary school students in Aceh what to do before, during and

---

Box 12.1 From rap song to classroom ditty: the making of 'Grandpa Quake' in Turkey

Marla Petal
Risk RED (Risk Reduction Education for Disasters)

Prior to the 1999 Kocaeli earthquake, disaster mitigation and preparedness education had little salience in Turkey. The Ministry of Education had published a fine book on the subject after the 1998 Adana-Ceyhan earthquake, and the civil defence authorities held occasional seminars, yet the public had little awareness of their earthquake risk, and almost no idea what could be done to avert such disasters. When the 1999 Kocaeli earthquake hit the industrial heartland of Turkey, sixty kilometres from Istanbul, the earthquake risk to the entire Marmara region and other parts of Turkey became a subject of great interest. However, earthquakes by themselves do not teach people what to do to avert disasters.

Several months after the Kocaeli earthquake had taken the lives of some 18,000 people, a lively rap song called 'Earthquake's Coming!' was released on the radio. It featured well-known earthquake expert Professor Ahmet Mete Işıkara in the mix, his voice 'sampled' from a television interview. Professor Işıkara was the head of Kandilli Observatory and Earthquake Research Institute at the time, and was dedicated to educating children throughout the country.

The radio release caught him by surprise. At first the distinguished professor wasn't sure if the unauthorised song might be undignified, but the tune was catchy, and the chorus of 'earthquake's coming … get ready!' sent a good message. The content was 'accurate' and inoffensive and public educators rapidly recognised its power to spread the message of disaster prevention. Professor Işıkara embraced the song and its composers and sought a local animation company to set the rap to a content-rich instructive cartoon highlighting household disaster mitigation measures. This music video has been used extensively in Turkish schools to promote earthquake safety.

after an earthquake through song lyrics and dance steps in a traditional Acehnese form, while lyrics are being taught to students on Simeulue island about the relationship between earthquake and tsunami, and what to do in the event of an earthquake (EDCflix 2010) (see Figure 12.1).

Informally, songs and CDs, parodies and multimedia presentations are helping to spread awareness messages outside of schools. Songs written by or influenced by disaster risk reduction practitioners include those by this chapter's author about the need for sustainable adaptation (Alexander 2008) (see Figure 12.2). Other examples are from some of Indonesia's top musicians, who wrote, recorded and released a CD of 'science in music' songs. It was based on a workshop of disaster risk reduction presentations co-produced by the Indonesian Institute of Sciences. By enabling such fact-influenced song writing, the public might more easily access accurate myth-busting information about disaster preparedness.

Attention-seeking parody songs have been used to highlight many disaster risks. Examples include poking fun at the overdramatic drought forecasts in England in the parody 'Drought' sung to the tune of Tears for Fears' 'Shout'; the false early-warning system alarms in Hawai'i in Frank Delima's 'The Tsunami Song'; and the fears of no snow for the 2010 Vancouver Winter Olympics in Bodhi Jones's 'Hey, El Niño'. Songs from film soundtracks, such as 'Trouble the Water' about Hurricane Katrina, often successfully contribute to awareness. An illustration from the Pacific is the song about El Niño awareness on a video on Micronesia's Pohnpei, which became a local radio hit (Hamnett and Anderson 1999).

Travelling performances in rural areas of developing countries are helping overcome some language and technical issues of communicating messages. In Bangladesh school children are

Figure 12.1 Teaching what to do in an earthquake through song and dance in Aceh, Indonesia
Source: Photo courtesy of the Education Development Center.

learning and teaching about climate change and disaster impacts by singing songs for other schools and communities in plays written by the International Union for Conservation of Nature (IUCN) about cyclone preparedness and crop resilience during flooding (United Nations University 2010). Other Bangladeshi groups are performing Pot Songs, a popular effective type of travelling folk song performance combined with bright pictures. It creates mass awareness of a localised specific message, such as disease prevention through hygiene after cyclones Sidr and Aila, along with preparedness for major disasters.

Similarly, 'Wan Smolbag' performs travelling risk communication musical theatre in Vanuatu. It has recently begun online sale of DVDs and CDs of songs explaining human inability to control nature and the effects of disasters, for example 'El Niño, La Niña' and 'Nature's Power' (www.wansmolbag.org). In Indonesia, the Department of Oceans and Fisheries has launched a communications campaign to educate people on characteristics, signs, impacts, preparedness and prevention of earthquake and tsunami disasters using local entertainment media. They use shadow puppets, stage humour and singing to disseminate information in an easily under-standable local language and context.

As history

Broter Dege's song 'Battle of New Orleans', commemorating Hurricane Katrina and its after-math, was promoted with materials describing songs as archiving people's relationships with disasters in an oral history set to music. Such archiving began with oral histories and mythologies of people's or land's origins, creation, destruction or use. In some cultures, stories describe songs used to create the hazards themselves, such as the Native-American Lakota people's Creating Power singing to initiate flooding and earthquakes that created the Earth (First People 2010). In others, songs recount fabled events that destroyed land, like the earthquakes and flooding of 'Atlantis', which swallowed the island. Some re-enact creation events, like 'Coming of Pele' and 'Hawaiian Pele' depicting Pele's volcano flows and flooding to create Hawai'i (Flood and

Figure 12.2 Bob Alexander performing a song about sustainable adaptation at a disaster con-ference at University College London
Source: Photo courtesy of Carina Fearnley.

Kamalani 2002). Songs describe the *wandjina* ancestors of the Australian Aborigines causing the floods that enabled their society (Hare 2010; Advameg Inc. 2010).

Natural hazards faced in the migration stories of people are recounted in songs. For instance, an action song describes how those rowing from Samoa to Rotuma successfully survived the storms they faced. 'The Song of Amergin' recounts how a bard used a harp and the song's lyrics to stop the storms preventing Celtic boats from docking in Ireland.

Musical documentation after a disaster can serve as an historical reminder of what transpired and might manifest again. The Grammy-nominated CD 'People Take Warning! Murder Ballads & Disaster Songs 1913–38' includes twenty-four 'man vs. nature' songs of US disasters, serving to document that era's floods, droughts, tornadoes, earthquakes, hurricanes/cyclones, fires, mine accidents, flu epidemic and boll weevil infestations. Many Woody Guthrie songs serve as important archival documentation of the dates and geographic extent of the dust bowl storms (e.g. 'The Great Dust Storm', 'Dust Bowl Blues'), of the resultant migration (e.g. 'Dust Bowl Refugee'), and of the socio–economic effects on those affected (e.g. 'Dust Can't Kill Me', 'Dust Pneumonia Blues') from someone who survived it (Ramsey 2002). Though different in m, melody and lyric choice than Guthrie's songs, a 1973 Ry Cooder song reflects on the dust bowl events, not changing the stories but further propelling their impact (Mourits 2010).

In other areas, people who live near bodies of water are reminded of the history of dangerous sea and lake storms through Western songs such as 'The Wreck of the Edmund Fitzgerald' (Lake Superior) and poems such as 'The Wreck of the Hesperus' (Atlantic Ocean). Local artefacts contribute, too, such as oral history songs from Rotuma and a popular Palauan song.

Cape Verde's long struggle with recurrent drought is likewise documented in period pieces and reflective mournful songs of social commentary on coping with famine and migration (e.g. Djedjinho's 'Native Land', Tcheka's 'Rozadi Rezadu'). Modern songs about the Irish Famine of 1845–52 (e.g. Pete St. Johan's 'The Fields of Athenry', Luka Bloom's 'Forgiveness', Christy Moore's 'The City of Chicago', Primordial's 'The Coffin Ships') recall previous songs in lamenting famine and resultant migration. Some writers have speculated that these song-induced memories, along with the work of luminaries such as Bob Geldof and Bono, resulted in Irish people being among the highest per capita donors to famine relief elsewhere.

Songs with effects on modern consciousness include those reflecting on the 1907 tsunami on Simueleu island in Indonesia and on previous tsunamis affecting sea nomads in Thailand. Such songs were recollected well enough to save lives during the 2004 tsunami and to inspire more people to record songs about recent disasters (Kurita *et al.* 2007; LIPI 2010).

Many such songs can be used as reflective metaphors for contemporary vulnerability. Bessie Smith's 'Backwater Blues' was perfectly timed to become the most important song of reflection on the 1927 Mississippi River flood, even though it was written about a smaller Nashville flood the previous year and recorded two months before the big flood (Evans 2006). Because of its prominence in both the history and current issues of that area, many other songs have been written about the 1927 Mississippi River flood in its immediate aftermath, such as 'High Water Everywhere' by Charlie Patton and 'Mississippi High Water Blues' by Barbecue Bob. Songs were also written decades later, for example Randy Newman's 'Louisiana 1927' and Zachary Richard's 'Big River'. Kansas Joe McCoy and Memphis Minnie's 'When the Levee Breaks' remains popular and useful today, thanks to a recording by Led Zeppelin in the 1970s and its airplay after Hurricane Katrina.

Some such songs deliberately use historical metaphor to call attention to perceived contemporary problems. Music about the dust bowl experience, including dust bowl imagery in Bruce Springsteen's 'The Ghost of Tom Joad', illuminates subsequent agricultural and urban problems (Ramsey 2002; Mourits 2010). More directly, songs such as Great Big Sea's

'Fisherman's Lament' describe how storms used to be a way of life that is now lost because of government action (Ramsey 2002). Some 'batuko' songs in Cape Verde express social dissatisfaction about the threat of hunger from famine. More than thirty songs were recorded in the USA expressing political dissatisfaction with the George W. Bush Administration and the Federal Emergency Management Agency's (FEMA) response to Hurricane Katrina (Hurley-Glowa 2010). Sinead O'Connor's 'The Famine Song' accuses the history books and their writers of falsely teaching Irish children that nature, and not the English, should be blamed for the hardships of that time. The aim of such songs is not only to record history but to change how that history is remembered.

As coping and relief mechanisms

From the long history of local fundraisers to the globally known efforts of the past forty years, much has been written about the contributions of recorded and live music and their performers to post-disaster awareness and fundraising. After events like the Concert for Bangladesh and the Concert for Kampuchea in the 1970s, the Band Aid recording for Africa's 'We Are the World' raised more than US$60 million through record sales. Since that time there have been, among others:

- Many events and recordings for those affected by the 2004 Indian Ocean tsunami, including the 'Of Hands and Hearts: Music for the Tsunami Disaster Fund' compilation album, and the Indonesian and Japanese duet of the song 'Kokoro No Tomo'. Neither had lyrics related to disasters, but both subsequently became associated with fundraising and relief efforts.
- Over forty songs recorded to raise money for Hurricane Katrina survivors.
- Songs for the 2007 Australian Drought Appeal, for example Paul Dillon's 'Tears from the Sky'.
- The 'Promise: Earthquake Song' produced by Jackie Chan for survivors of the 2008 earthquake in China.
- Events and recordings for survivors of the 2010 earthquake in Haiti, such as 'Hope for Haiti Now: A Global Benefit for Earthquake Relief', including remakes of songs previously associated and not associated with disasters; a 'Helping Haiti' single reissue of REM's 'Everybody Hurts' along with a rock memorabilia auction; and a remake of 'We Are the World'.
- Events and recordings following the 2010 floods in Tennessee, such as Vince Gill's Flood Relief Telethon and over forty performers in 'Artists for Tennessee Flood Relief' singing 'City of Dreams' to help the Red Cross efforts.

Additionally, the non-profit Music for Relief (www.MusicForRelief.com) was established by top artists and music industry professionals, dedicated not only to disaster relief but also to proactive mitigation.

Using music and songs for coping and understanding directly is less documented. In addition to aforementioned contributions of songs to documenting history and mythology, songs are also used as a way for writers and affected listeners to process and frame their experiences.

Guthrie's dust bowl songs were not protest songs or change songs, but were coping songs in that 'songs reflect the need for people to turn their problems into a story to make sense of what is happening' (Mourits 2010). After Typhoon Ondoy in the Philippines in 2009, an original song and video was posted to YouTube with lyrics, translated into English, asking the typhoon, 'what is your message to us?' Accompanying comments said that the song is dedicated to all people affected by Typhoon Ondoy with the hope that they can express their post-typhoon emotions through singing this song. Many haiku and songs were written for local coping after the earthquake in Kobe, Japan. As well, the song 'Note From Kobe' was posted on YouTube as having been written by an elementary school teacher who survived the Kobe earthquake and was trying to encourage others to deal with the experience.

Sia Figiel started a poetry 'chain' with 'The Day After', a framing poem of encouraging coping from inside Samoa after the 2009 tsunami. Many people both inside and outside affected areas of Samoa and Tonga joined the chain with the next person starting with the last line of the poetic contribution of the person before them. In this manner, outsiders and insiders not directly affected often write and release songs or post YouTube videos to show solidarity and to try to help others cope. That includes many tributes to the survivors of that tsunami in Samoa and Tonga, for example Nifoloa's 'Tsunami Song'. Other examples are 'Zindagi: the Earthquake Song' as a tribute to survivors of the Pakistan earthquake; 'Citycell Flood Song' for those affected by the 2007 Bangladesh floods; and Brad Mossman's 'Water', one of many songs recorded for expressions of empathy with the Hurricane Katrina survivors (Mossman 2009).

As referenced in the lyrics of 'The Ballad of Springhill', also called 'Springhill Mine Disaster', written by Ewan MacColl and Peggy Seeger and referring to the 1958 disaster, 'We're out of light and water and bread, so we'll live on song and hope instead', singing songs has always been used as a way to smile through the pain in times of uncertainty as well. Similarly, songs can be used to help people to 'move on' once the event has passed, including expressive therapy of writing lyrics or using popular songs to express details of post-traumatic stress. Such expressive therapy can be guided individually, in groups, or, in larger communities, through radio and television shows dedicated to using the power of music for healing.

Inability to sing or to listen to music that is important to psycho-social recovery impedes the process. Consequently, post-event rehabilitation includes efforts to ensure radio access and ability to respect culturally significant annual or seasonal ceremonies. In post-tsunami Aceh, the Aceh Emergency Radio Network was established to provide information, entertainment and song requests. Other locations have adhered to post-harvest song rituals to honour those that need to be honoured. More symbolically for showing that people are 'moving on', a music-filled Mardi Gras was held in New Orleans after Hurricane Katrina.

As expressions of disaster risk reduction capacities

In some belief systems, such as the aforementioned Celtic 'Song of Amergin' and black magic in parts of the Pacific Islands, people were believed to have controlled storms and other hazards to their advantage and to others' disadvantage through using chants and songs. The Sundanese of West Java believe that the rock that controls earthquakes will keep the earth still if they chant to it that they are still alive on Earth. More modern approaches, as seen in significant structural mitigation over the past 100 years, seem not to have been rewarded with songs that are as prominent as the songs lamenting a major disaster when structures fail. Perhaps such songs have been missed in this analysis, or perhaps reasons lie in the appeal of songs of either great success or great misery, so that songs of steady performance aren't dramatic enough.

Many examples exist of songs expressing capacities of people to live with disasters. Agricultural people, such as those in Thailand, Laos, Cambodia and Zambia, have historically sung songs to support their religious and cultural beliefs of ensuring adequate rain. In Thailand, Buddhist monks continue to sing chants to Buddha during the Royal Ploughing Ceremony for blessings and protection of crops. To avoid drought risks, Cambodian, Thai and Zambian farmers have sung songs to the rice goddess or rain spirits when planting and harvesting, as

137

depicted in the Thai Rice Growers' Dance. In times of delayed monsoon at the time of sowing, Thai farmers would sing songs with coarse language while splashing water on a cat to induce the gods to send rain. In Laos and northeast Thailand, the two-day Bun Bang Fai 'rocket festival' includes music and dancing accompanying the firing of rockets to prompt the heavens to send rain so that farmers can grow crops (Tu *et al.* 2004; PANAP 2010; Von Kotze and Holloway 1999).

Islanders whose livelihoods have depended on the ocean, such as those of Samoa and Palau, have songs that depict their ongoing relationship with storms of the sea. Other songs celebrate the ability to balance their boats and, in effect, their lives and communities in the midst of calamities. 'Celtic Mass for the Sea' similarly celebrates the dynamics of how people need to live in harmony with disasters, because rewards for living with their ferocity are opportunities to thrive on their vitality (MacMillan and Brickenden 1993).

Other songs express such capacities more like an ill-fated hero. Rodney Crowell's much-covered 'California Earthquake' accepts the challenge to keep rebuilding every time an earthquake strikes. The protagonist understands that future earthquakes are going to be as devastating as, or worse than, the ones that have already destroyed so much, but accepts the challenge to keep rebuilding every time an earthquake strikes.

Natural hazards are also heralded as a conduit to opportunity. Songs from the Philippines include the aforementioned quest for a message from Typhoon Ondoy or the popular song 'Bagyo Bagyo' that calls for change in response to a typhoon. Traditional chants in Yogyakarta, Indonesia, honour Mother Nature for improving the land's fertility through the volcanic ash that comes to their fields from Mt. Merapi and the water and sediment that comes from ocean storm flooding.

Johnny Cash's song 'Five Feet High and Rising' is an example that, through the years, has filled roles of education, history, coping and expressions of capacity and opportunity. The song was used on the children's television education programme 'Sesame Street' to teach about living with flooding. It is a fun play-by-play, first-hand historical account of how farming families lived with the socio-economic effects of different levels of floodwater from the Mississippi River in the 1930s. It also receives contemporary airplay in times of flooding to help affected people cope, especially because of the message of opportunity and hope in the introductory words:

> We couldn't see much good in the flood waters when they were causing us to have to leave home. But when the water went down, we found that it had washed a load of rich black bottom dirt across our land. The following year we had the best cotton crop we'd ever had.
>
> *(Johnny Cash, 'Five Feet High and Rising')*

Discussion and recommendations

Like other means of communication, disaster risk communication via songs and music evolved through the years of oral tradition and recordings. Such gradual change has given way to a recent relative revolution in the ability of songs to spread messages across previous geographical, language and cultural barriers.

Historically, song sharing was limited to oral tradition. It was restricted to communicating about hazards and disasters under the categories of oral history and education (including informal travelling shows), documentation as reminder and temporal metaphors, individual or community coping, rituals for living with hazards, and sometimes local fundraising.

Recording allowed more extensive formal and informal education, metaphors that could be spread spatially as well as temporally, expressions of solidarity from neighbouring areas through song recordings and – as equipment for transportation, telecommunications, and performing and recording improved – even global-scale concert and recording fundraising events. During this time, Alan Lomax and others began documenting the everyday lives of places through recordings of their songs. As hazards and disasters were an integrated part of these lives, these archives provided both historical documentation and education for others.

The internet was still almost a decade from its takeoff by the time 'We Are the World' and Live Aid put songs about hazards and disaster action on centre stage for global awareness and fundraising. Nonetheless, radio and televised music videos were enabling people to have ever-increasing access to more music with more messages about hazards and disasters.

An increasing supply of communication sources does not necessarily imply more accurate messages. Although more international travel was enabling better understanding of language and cultural differences, most people were restricted to songs from their primary languages. The emergence of a mass market for selling songs resulted in less emphasis on the role of songs in education, history, relief and capacity. Instead, there was more emphasis on popular, danceable, catchy songs with no basis in fact, logic or storyline required.

Meanwhile, disaster-related work has evolved to focus less on paradigms of hazards, massive structural investments and relief. Instead, the emphasis is more on underlying capacities and vulnerabilities while integrating disaster risk reduction into sustainable development. Therefore, songs representing hazards and disasters seemed ill-suited to be the communication medium of choice, because the history of popular music cherishes the tragic and the heroic, not long-term good practices.

Much like newspaper reporters stereotypically disappearing after the initial post-disaster misery gives way to hope (see Chapter 63), songs had grown as a medium that commemorated either pain or triumph, but not avoidance and mitigation. Record company executives at song-writing workshops teach that commercially successful songs must do one of two things: tug at the heart strings by getting the listener to feel the pain, or let the heart soar through choruses of victory in the face of adversity. Songs of foresight and planning and the disaster that never happened are destined to have no soaring chorus or wailing screams. Thus, they are doomed in the eyes of commercial viability.

As evidenced previously, local cultural songs about hazard events can express how these events provide opportunities, require ritual or studied mastery, and provide lessons for reflection. However, these songs compete with popular music that tends to care less about the message to the brain and more about the message to the heart and pocketbook. Thus, the myth of the disaster protagonist who is the helpless victim with no coping mechanisms, buffering ability or choices on acting is bolstered through song, despite theoretical and practical advances in disaster risk reduction.

Home recording and video equipment increases song availability. In tandem, YouTube, MP3s, cell phones and internet translation engines increase access. Overall, digital technology has resulted in a direct distribution revolution with significant potential for dispelling the myths and conveying accurate disaster risk reduction messages. Songs of real activity, hope and calls to action can be spread quickly and accurately if the song is compelling enough to make people want to listen and learn.

Nevertheless, certain categories of such songs are likely to be more successful than others and false messages can be spread too. Relief songs and coping songs are the most likely to perpetuate the myth of the helpless victim. Although the funds raised can assist recovery efforts, relief songs and events need to portray people affected by disasters as victims in order to effectively reach donor wallets. Even though lyrics of many songs used for disaster relief are unrelated to hazards or vulnerability, images that accompany them in videos or live performances continue to portray the myth of the helpless victim.

Songs previously described as helping with coping in the aftermath or aiming to revitalise documentation of the original story also have an inherent problem. They are often created specifically for the purpose of convincing the writer or others that nothing could have been done to avoid tragedy: the fates were cruel and no proactive avoidance of misery was possible. Nobody questions the protagonist of 'When the Levee Breaks' about why the house was built in the floodplain. Instead, there is empathy.

Songs that do retroactively question and express dissatisfaction can be merely finger-pointing, blame-shirking or history revision for partisan victory. They can, however, also illuminate vulnerability, both directly and metaphorically, while dispelling such notions as government responsibility for protection and relief of those who do not proactively help themselves. As well, although some finger-pointing parodies are strictly comedy and are clearly inaccurate, they can help raise awareness of a perception that something is wrong that could be better addressed.

Songs of solidarity may help to break the myths if, like Figiel's poem, they initiate dialogue about true capacities and vulnerabilities that enable appropriate action. Songs for therapy might fall into the adverse selection trap of focusing these efforts on those whose capacities are already well developed. They can, however, be geared towards appropriate self-selection. Crowell's 'California Earthquake' and other 'living with disasters' songs might be ambiguously construed as communicating helplessness against all but building back and waiting to be hit again, but such songs could also be viewed as emphasising proactive disaster risk reduction, like 'Bagyo Bagyo', with the event as an opportunity to improve – not only improving disaster risk reduction but also living conditions.

Science education songs are still primarily focused on the hazard rather than on the risk that results from the combination of hazard and underlying vulnerability. Nevertheless, education material development focusing on preparedness, and the extent to which these materials are becoming accessible, helps disaster risk reduction initiatives. Travelling multimedia shows in Bangladesh and Vanuatu, plus the recordings in Indonesia, aim to ensure that accurate messages are being portrayed and received through these informal education songs. By utilising messages that combine the outside knowledge of disaster risk reduction professionals with perceptions of what will be received and experienced well by the performers, their future as disaster risk communication tools is promising.

An interesting parallel arises between the needs of the different elements of disaster risk reduction and the different types of uses of songs. Disaster risk reduction and related activities have elements of coping, relief, rehabilitation, reconstruction, mitigation and preparedness that might each be called upon to varying degrees to help manage the feelings of disaster risk at that time. Likewise, certain types of songs might be more helpful in processing or making decisions regarding such processes.

Generally, songs that help to document and frame what happened for coping, as well as to raise relief funds, should prove to be the most useful for coping, relief and rehabilitation. While early recovery, reconstruction and preparedness might reflect on songs highlighting opportunities and acceptable risks of past events, such songs would help most in reflecting how to implement disaster risk reduction. Educational songs about awareness should help in considering reconstruction and mitigation decisions, but should be especially useful in planning and preparedness. While songs of historical documentation of past events might be useful at all times, especially when trying to cope, songs that act as reminders or metaphors are probably most useful when considering reconstruction and mitigation.

This analysis and discussion provides examples of a few distinct types of the uses of songs related to hazards and disasters. These examples and categories are from limited information from a few countries and were not meant to be exhaustive. They show that the uses of different types of songs seem to be compelling enough to warrant development of a more inclusive database. This database could help to better understand location-specific and universal themes of songs as disaster risk communication and might contribute to training and disaster risk reduction practices.

The mechanisms for both formal and informal education and awareness programmes seem poised to best utilise the advances in technology and methods to dispel myths and to propel accurate disaster risk reduction initiatives (see also Glantz 2007). Thus, studies of the communication effectiveness of these programmes are recommended. Language barriers and culturally entrenched themes of types of songs continue to prevent utilisation of songs for some purposes. Yet an advantage of songs is that they are generally written by creative people who might rise to the challenge of co-determining ways to overcome these barriers effectively. Thus, the recent Indonesian Institute of Science initiative of enabling disaster risk professionals and musicians to collaborate on songs that are well-informed about disaster risk issues should also be promoted elsewhere.

Acknowledgements

Faruque Ahamed, EDC, Carina Fearnley, Vilisoni Hereniko, Sanny Jegillos, Saadia Majeed, Irina Rafliana, Richard Salvador, Erlin Sarwin, Supin Wongbusarakum, Etsuko Yasui.

Environment, development
and sustainability

13

Hazards, risk and urbanisation

Mark Pelling

KINGS COLLEGE, UNIVERSITY OF LONDON, UK

Introduction: from urban sanctuary to hotspots of risk

Humanity seems to be drawn towards an urban model of living. Throughout history civilisations have urbanised, with urban settlements prospering best when they meet three basic preconditions. First, they provide security for their citizens from external threats – from armed conflict to natural hazards. Second, they generate a mechanism for the sustainable extraction and concentrated use of social and ecological surplus – the basis for food and water security which under globalisation can involve chains of exchange that extend the reach of the city, and its dependencies, over large distances. Third, they are maintained through a social contract that balances legitimacy and power – not to be confused with equity and justice. Meeting these preconditions allows cities to offer security to citizens within political regimes that are environmentally sustainable and socially just. Of course, this is an ideal vision and cities rarely, if ever, live up to these standards. However, in coming close urban centres can offer stability to their citizens and surrounding political economy. This characteristic was illustrated well by Drèze and Sen (1989), who identified the advantages of urban centres that, in the midst of drought and famine in the Indian countryside, provided security to residents through the management of food stocks. In this case urban centres gained security through enhanced administrative capacity. This is only one facet of governance that lies at the root of determining who is at risk in cities.

Meeting the preconditions for urban security is not easy and can be put under pressure from competing demands, if capacity is undermined or new pressures increase human vulnerability and hazard risk. Historical evidence suggests that many urban civilisations have failed through a combination of environmental change and weakness in governance, leading to failed natural resource management. Amongst contemporary cities, even in those that appear secure at first glance, diversity in land use, socio-economic capacity and local governance produces a mosaic of risk and security where local – and sometimes extensive – insecurity can be found amidst apparent safety. Most critical in this respect are the rapidly growing numbers of people who live in urban slums and squatter settlements with limited access to basic services and political capital, but who are often highly exposed to risk in all its forms from crime and violence to economic exploitation and environmental hazard, but also those living in formal housing that has been quickly constructed ignoring building standards or with inappropriate land-use zoning (UN-HABITAT 2007).

In addition to those very human processes that drive internal inequalities and human vulnerability, urban risk can also be a product of overwhelming hazard. Hewitt (1997) describes the extreme case of place annihilation where urban disaster events are of such magnitude that they destroy not only the physical space – built infrastructure and settlement pattern in a city – but also the cultural meaning attributed to place. The possibility of place annihilation is interpreted more positively by Vale and Campanella (2005), who argue that in some cases this can be an opportunity to remake the city – to redefine its values through a reconstructed physical and cultural fabric (see Chapter 46 and Chapter 48). At the moment of writing it is uncertain whether, following the 2010 earthquake in Port au Prince, reconstruction will take the opportunity to strengthen institutions of governance as well as physical infrastructure. Certainly this opportunity exists when reconstruction is sensitive to the values of citizens as well as the economic demands of urban development. Too often, though, the opportunity is missed and financial interests dominate. This is most clear when the destruction of homes and communities of the poor in disaster opens up scope for redevelopment of high-value land, often close to the city centre, forcing original residents to relocate.

The nature of hazard and risk in cities – who the vulnerable are, the scale of vulnerability and scope for its amelioration – is then at its most basic level an outcome of competing values and visions of the city and the distribution of political, economic and social power, which in turn determine expenditure on physical and social infrastructure and the application of technological innovation. The observed increase in the number of people living in urban centres does not of itself generate risk, nor does the potential future increase in hazardousness associated with climate change and other natural phenomena. Both challenges are resolvable through appropriate local, metropolitan and national governance regimes. The challenge is to build on existing urban governance that in many cases falls short of meeting the basic preconditions of security while population increase and climate change means that the stakes are continuing to rise – cities are fast becoming hotspots of risk.

What makes urban disaster risk different?

Contrasting urban visions

What is a city? Cities can be viewed in many ways, and how one sees a city helps to order priorities for risk management. Table 13.1 presents five common visions of the city and for each identifies priorities for vulnerable objects, pathways for managing that vulnerability and provides a note on the academic traditions that are both drawn upon and help give substance to that particular vision.

No one vision is right or wrong, but the consequences for policy are profound, rooted as they are in economic and material interests (Kohler and Chaves 2003). Differences in urban vision go some way to explaining why it is that so many urban risk governance problems appear intractable. If different actors hold such fundamentally contrasting, and even conflicting, views then collaboration for risk management will not be easy. One way around this impasse comes from the literature and practice of sustainable urbanisation, which argues that the different elements of vulnerability are interdependent (Pelling 2003), making them less contradictory than Table 13.1 implies.

Urban risk management and sustainable development

The understanding of urbanisation that a sustainable development lens brings (see Chapter 14) is very close to that of disaster risk reduction. Both champion integrated or holistic approaches to

Table 13.1 Linking visions of the city to pathways for managing vulnerability

Vision of the city	Vulnerable objects	Pathways for managing vulnerability	Literature
An engine for economic growth	Physical assets, labour force and economic infrastructure	Insurance, business continuity planning	Econometrics of business continuity and insurance
An integrated system linking consumption and production	Critical/life-support infrastructure	Mega-projects connecting urban and rural environmental systems	Political-ecology, systems theory
A source of livelihoods	The urban poor, households, livelihood tools	Extending and meeting entitlements to basic needs	Livelihoods analysis and medical sociology
A stock of accumulated assets	Housing and critical/life-support infrastructure	Safe construction and land-use planning	Political-economy and urban sociology
A political and cultural arena	Political freedoms, cultural and intellectual vitality held in discourse or materially through museums, religious centres, etc.	Inclusive politics and the protection of human rights	Discourse analysis and public administration/ political theory

Source: Pelling and Wisner 2009

policy and the centrality of procedural as well as distributional justice in governance and decision-making. The major difference until now has been some within disaster risk reduction falling short in including non–human entities and future generations (too often the ecological and carbon footprints of risk reduction or reconstruction activities are given only superficial attention). Reorienting cities towards a vision of sustainability where environmental risk can be minimised places emphasis on the need for open and inclusive urban management. This will need to include the views of actors operating at a range of scales from the local to the international and able to contribute knowledge of risk–generating as well as risk–reducing processes at these scales. Inside cities, municipal government occupies a pivotal position in its varied roles of service provider, community resource mobiliser, regulator of the private sector, advocate and strategic planner. However, the capacity of municipal governments has very often been limited by financial and human resource scarcity and, especially in capital cities, by political competition with central government. In Georgetown, capital of Guyana, competition between political parties dominating national and municipal governments has slowed the disbursement of funds earmarked by international development institutions for urban renewal, increasing vulnerability (Pelling 2003). Disasters tend to weaken municipal government even further as their functions are overwhelmed by incoming international and national disaster response and reconstruction agencies able to pay high wages and with stronger management capacity (TEC 2006a). This is a critical challenge to sustainable urbanisation.

The character of urban places is traditionally offered as a contrast to the rural. Some indicators – such as urban administrative units – are spatially bound (yet can include rural land use), while others focus on economic characteristics (the preponderance of primary production as a defining character of rural economies) or the density of people and services, such as specialist health and education facilities. From a disaster risk perspective it is important to include the source and sink regions for environmental services upon which urban life support infrastructure is built, as interruption in the flow of water or electricity caused by a natural hazard outside the urban boundary can have significant impacts in the city. Damage caused to transport networks can interrupt trade, causing economic impact or restricting access to goods, leading to local price inflation. In 2001 Typhoon Nari hit Taipei, causing the city's most important traffic artery, its underground railway system, to close for several weeks. Knock-on financial losses for the city were estimated to amount to US$500 million (Munich Re 2004). Those cities with linkages to the global economy are similarly exposed to variability with scope for global contagion effects following a disaster in a city of the global economic core. This has not yet materialised, but scope for contagion was demonstrated by global oil price increases following damage to the oil fields off New Orleans associated with Hurricane Katrina (Klein 2007).

Urban vulnerability is shaped further by the high density of people and assets, which can lead to large losses from spatially concentrated events such as landslides, an horrific example being the mud and debris flow that buried the city of Armero following the Nevado del Ruiz volcanic eruption in Colombia in 1985. High density can also be a factor in the spread of risk, as in the case of urban fires spreading from shack to shack in tightly packed squatter settlements, all too common an event in South Africa, for instance. Mixed land use, especially industrial and residential, can lead to technological failures, for example when earthquakes trigger the release of toxic or flammable chemicals (see Chapter 56). Calcutta and Baroda are just two cities where the close proximity of manufacturing, hazardous materials storage and residential areas has been a cause for concern (ADPC no date a).

In essence, the production of disaster risk in urban centres can be distinguished from that in rural settlements by the greater separation between hazard and those at risk by layers of institutions and infrastructure. When they function, these layers can protect through flood drainage, social insurance, building standards, etc. However, poverty, political distortions and the uneven presence and capacity of civil society across a city mean that access to such protecting institutions and infrastructure is uneven.

The continuum of urban risk, its accumulation and redistribution

Our visions of urban risk are also influenced by available data. Data that are visible at the international level and have been integrated into assessments of disaster loss are very limited for cities. Where data can be disaggregated this is possible only for large events and large cities where national impact is very close to that of the disaster event. This bias in the scientific data is reproduced for popular knowledge through the news media, which prioritise large-scale events (see Chapter 63). Emerging work that uses local media and aggregates local loss data has shown clearly the significance of small events and of everyday risk that overlaps with public health concerns (UNISDR 2009a).

In rural and urban societies risk accumulates in the degraded infrastructure, dysfunctional institutions, eroded natural capital and constrained livelihoods of those at risk. Everyday and small disasters add to risk burdens through the incremental erosion of capitals and opportunity costs of living with risk. However, these risk burdens are not evenly distributed, geographically or socially. The urban poor suffer from a four-fold burden of environmental risk: local public health hazards, local and city region industrial pollution, global environmental change including the local impacts of climate change, as well as suffering vulnerability to natural hazards such as earthquakes and storms (Pelling 2003). These burdens come from a combination of local maldevelopment and the costs of development gains enjoyed elsewhere in the city or world.

Global patterns of urbanisation and disaster risk

Urban areas have long been the primary site of earthquake risk, with impacts largely a result of building failure rather than ground motion. In addition to this, as early as 1994 Green (in Wisner *et al.* 2004) commented that flooding, once a classic hazard of rural areas, was becoming a predominantly urban hazard as cities expanded into flood-prone areas. Failure of urban infrastructure and institutions to respond to environmental and demographic change has also generated new hazards of urban drought and temperature extremes (see Chapter 23). Given the overlapping geography of urbanisation and natural hazards, the riskiness of urban life should perhaps not be surprising.

The statistics for urban growth are impressive. The United Nations Development Programme (UNDP 2004) estimates that by 2030, twenty-seven countries will account for seventy-five per cent of the world's urban population, with all but seven in less developed countries, and that Africa and Asia will have more urban population than any other major area, with Asia alone accommodating over half the urban population of the world. Asia is also the most hazard-prone continent worldwide.

Urban settlements are becoming larger and more numerous through a combination of natural population growth and urban migration, with new urban landscapes of risk being created through a variety of motors (Mitchell 1999). Political stability and economic opportunity can lead to small rural settlements expanding into towns, as is happening in Central America, and taking on new social and environmental challenges and opportunities in which managers might not be experienced. At a larger scale, rapid expansion of urban corridors, such as that along China's seaboard, can reconfigure risk profiles at the regional level.

Large cities and megacities

Large cities with more than 5 million population, and megacities with excess of 10 million population, are becoming more common, especially in Asia and Latin America. In 2003, four per cent of the world population resided in megacities. By 2015 this share is expected to rise to five per cent. Almost three per cent of the world population in 2003 was estimated to live in cities with 5 million to 10 million inhabitants, rising to nearly four per cent by 2015. The concentration of large numbers of people and assets in these places generates a very high risk potential and creates new challenges for risk management (UN-HABITAT 2007).

Increasingly, the world's largest cities will be found in Africa, Asia, Latin America and the Caribbean. This trend looks set to continue, with the fastest growth rates of large cities in 2003 being recorded in per cent per year for Dhaka – 6.2, Lagos – 6.1, Delhi – 4.1, Mumbai/ Bombay – 3.1, Karachi – 3.7, and Jakarta – 3.3 (UNDP 2004). Urban populations follow economic investment, so that large cities also contribute substantially to their country's gross domestic product (GDP). Mexico City is responsible for around one-third of Mexico's GDP, for example. Combining concentrated population and economic assets with high hazard exposure (seismic as well as hydrometeorological hazards) is the urban region of Dhaka, Bangladesh, which is home to around 12 million people and also hosts activities contributing to 60 per cent of national GDP, although major engineering works have sought to mitigate riverine flooding. Not only large cities but also urban regions are made risky, because impacts can spread through their many economic interdependencies, as was noted in Mozambique following the floods of 2000 (Christie and Hanlon 2001) (see Chapter 21). Large cities, even in prosperous and relatively well-administered countries, can be caught out when there is a failure to adequately scan and prepare for risk. The heatwave that hit Europe's urban centres, especially Paris and London, in 2003 is a case in point, with little indication of learning from earlier heatwave tragedies in North American cities.

Small urban settlements

Small towns and cities with fewer than 500,000 inhabitants have been and will continue to be the type of urban settlement in which the largest share of the world's urban population resides. In less developed regions these settlements alone housed twenty-one per cent of the urban population in 2000; by 2015 this proportion is projected to be twenty-five per cent (UNDP 2004). The total population of medium and small urban areas exposed to environmental risk is likely to exceed the total at-risk population resident in megacities. Despite this, we know relatively little about smaller cities and it is tempting to project knowledge gained from larger cities (Cross 2001). There are, however, some important differences in governance: local government may be closer to citizens and more responsive, but possibly also have less capacity; civil society may also be less well developed and governance as a whole is likely to be oriented more towards rural than urban settlement concerns.

Social and demographic change

Planned and unplanned urban expansion leads to rapid urbanisation, which can generate vulnerability where this did not exist before. In El Salvador, free-trade zones in San Bartolo, El Pedregal, Olocuilta and San Marcos were promoted by the government without adequate concern for earthquake hazard. During the 2001 earthquake, large losses were reported from amongst migrant workers who supplied labour to foreign-owned enterprises in these new towns (National Labour Committee 2001). In Dhaka, Bangladesh, amongst the most vulnerable to flood risk are poor rural migrants whose lack of access to secure housing and livelihoods is compounded by the absence of familial support (Rashid 2000). Migration not only disrupts family support networks but can lead to demographic shifts and, in particular, disparities in gender and age that undermine social cohesion. In Cape Town, some of the settlements most at risk from fire hazard have a high proportion of young, male labour migrants living in guest houses with a subsequent loss of social capital believed to be a contributing factor to high shack fire incidence. In just one settlement, the Joe Slovo informal settlement in Cape Town, seven large-scale residential fires were recorded between March 1996 and January 1997, in which 153 shacks were burned and 498 people displaced (Mehlwana 1999, in Hardoy, Mitlin and Satterthwaite 2001).

If post-disaster legislation, policy and practice do not take account of demographic and social diversity, this can contribute to the production and reproduction of vulnerability. In Indonesia and India the registering of land solely through the male line has disadvantaged women survivors (Silverstein 2008), pushing female-headed households without family support closer to vulnerability.

Building standards and the limits to planning

The rapid supply of housing to meet rising demand, without appropriate regulation, is a principal generator of risk within the formal and informal sectors. Where government is active, building codes can be improved. For example, Jamaica's use of British building standards contributed to losses in hurricane events until reform backed up by enforcement following Hurricane Gilbert in 1988 (Pelling 2003). More important, though, is a general failure to implement existing building codes at the local level. Time and again this is flagged as a cause of losses to urban disaster, for example in reports on earthquake damage in Turkey (Özerdem and Barakat 2000) and in the collapse of multi-story buildings in the Armenian earthquake (Kreimer and Munasinghe 1992). Municipal authorities are normally charged with overseeing construction standards, but are prevented from fulfilling their duty for lack of resources and human skills compounded by institutional cultures that allow corruption to distort regulation and enforcement (see Chapter 52 and Chapter 53).

Outside of the reach of government, in slum and squatter communities, different approaches are needed. Some 924 million people lived in urban slums in 2001 (UN-HABITAT 2003) and the trend is upward. It is not unusual for the majority of urban residents in very large and small cities to be excluded from the formal housing market. In Manila, people in informal settlements at risk of coastal flooding make up 35 per cent of the population; in Bogotá, 60 per cent of the population live on steep slopes subject to landslides; and in Calcutta, 66 per cent of the population live in squatter settlements at risk from flooding and cyclones (Wisner et al. 2004). It is the speed of growth in these settlements, as well as their scale, that overwhelms the capacity of government to provide services. Urban development planning based on five-year cycles or decennial population censuses can offer only limited insight into the true needs of urban dwellers in rapidly growing cities and towns (see Chapter 53). Where government is weak and with limited access to market opportunities, social capital – both the underlying values that enable social collaboration and the organisations this produces – has become to key resource for development and risk management amongst the poor (Pelling 2003).

The formal involvement of civil society actors is most prominent in local risk management programmes and projects. Community-based organisations (CBOs) and local non-governmental organisations (NGOs) are common leaders or partners in local risk reduction through community health care, local hazard mitigation or livelihood strengthening work, as well as through the national societies of the Red Cross and Red Crescent, often active in schools and boasting many hundreds of trained first aiders and other volunteers (see Chapter 59 and Chapter 62). Because all development and risk reduction work affects local actors, civil society is potentially an active and leading partner in each of four areas of practice discussed below (see Table 13.2). Roles played by civil society vary from passive consultation to participation, through labour, to leadership roles as local actors contributing to plans and to regulation and management decision-making. For example, local citizen groups may be included in government-led upgrading of slum settlements.

Governance for urban risk reduction

Urban governance for disaster risk reduction requires co-ordination of four kinds of activities and actors (see Table 13.2): development planning, development regulation, risk reduction and emergency management. Such co-ordination has not been easy to achieve – particularly between urban development and disaster management professionals (Wamsler 2006) (see Chapter 53). Where disaster risk reduction works best, urban dwellers and their civil society organisations are involved.

Of the four areas of practice identified in Table 13.2, only one – emergency management – includes the emergency services, civil defence and disaster management co-ordinators. The remaining areas of practice have primary stakeholders from the development community. This highlights the significance of disaster risk reduction and management as a development concern. Each neighbourhood and city has its own balance of public, private and civil society involvement in the activities that comprise urban disaster risk reduction. This is determined by the legacy of past development policy and by present initiatives. In Dar es Salaam, Tanzania, for

Mark Pelling

Table 13.2 Urban disaster risk reduction: multiple activities and stakeholders

	Development planning	Development regulation	Risk reduction	Emergency management
Core activities	Land use, transport, critical infrastructure	Building codes, pollution control, traffic policing	Vulnerability and risk assessment, building local resilience	Early warning, emergency response and reconstruction planning
Primary stakeholders	Urban planners, city engineers, critical infrastructure planners, home-owners, private property managers, investors, transportation users, taxi drivers' associations, other professional associations, academia	Environmental regulation, law enforcement, contractors, factory owners, drivers' and transporters' associations, homeowners' and neighbourhood associations	Primary health care, sanitation and water supply, community development, local economic development, infrastructure management, waste hauliers' associations, water users' representatives	Environmental monitoring, emergency services, civil defence, disaster management co-ordination, fire fighters, police, military, Red Cross/ Red Crescent societies

Source: Pelling and Wisner 2009

example, there is a strong foundation of local government and civil society involvement built on post-colonial investments in adult literacy, democratic elections, a strong (although mixed) legacy of *Ujaama* (co-operative development) and traditional local leadership structures. This social heritage is enhanced by the active presence of many international developmental NGOs. The influence of the international private sector is also felt, with support from financial institutions such as the World Bank, which have promoted privatisation of potable water provision, albeit with limited success (Kiunsi *et al.* 2009).

Targeting urban disaster risk reduction

Tackling urban disaster risk requires efforts to reform institutions, policies and techniques. Table 13.3 outlines the focus and provides examples for each level of engagement. Each level of engagement supports the others so that, for example, it is easier to lobby for technological improvements in a city where there is a legal responsibility on the part of local government to facilitate risk management.

South Africa's Disaster Management Act 2003 is one of the most comprehensive national frameworks for disaster risk reduction worldwide (see Chapter 51). The act identifies specific responsibilities for municipal government for hazard mitigation and risk reduction (Pelling and Holloway 2006). New techniques for reducing vulnerability are also being applied across urban settlements. Some innovations are local, while others have been imported from experiences elsewhere, often through the work of international NGOs or South–South co-operation. For example, The Women and Shelter Network (www.hicwas.kabissa.org/memtray.htm), with partners in 30 countries, promotes women's rights to safe and secure housing, including safety from environmental hazards and disasters (Pelling and Wisner 2009). At the forefront of applying new techniques to reduce disaster risk are local actors: CBOs, local government, primary

Table 13.3 Institutions, policies and techniques for disaster risk reduction

Level of change	Focus	Examples
Institutions	Lobby for supportive legislation. Challenge cultural norms and received wisdom that assume disaster risk and loss are acceptable costs for economic growth. Integrate disaster risk reduction stakeholders into decision-making across city and national government.	Legislation enacting a national framework for disaster risk reduction with specified responsibilities at the municipal and local levels. Disaster risk reduction advocates invited to sit on committees for economic and policy planning.
Policies	Integrate the goals of disaster risk reduction into poverty alleviation, economic planning and environmental management policy.	Social policy including social safety nets, social housing, community health care and social insurance to target the reduction of vulnerability to disaster risk as part of poverty alleviation.
Techniques	Introduce disaster risk reduction techniques into everyday work practices for urban development.	Construction techniques, land-use planning, slum upgrading to include disaster risk reduction tools.

Source: Pelling and Wisner 2009

health care workers, local police and emergency services. A common challenge for all is the difficulty of building on local success stories so that they can spread to the wider city – or even to other cities.

Conclusion: approaching challenges

Each of the preconditions for urban security faces emerging challenges. Climate change is increasingly a factor exacerbating these challenges. Where economies are constrained and governance is weak, overcoming these challenges will be especially difficult. Living with all the problems produced by these pressures also raises challenges for the use of technology, the ways in which wealth is generated and distributed, and the balance of power in society. There are some big questions to be faced, three of which are outlined below (see also Chapter 18).

For those cities and neighbourhoods facing high levels of current and future risk, at what point does relocation become preferable to managing risk *in situ*? There has been some talk of moving major cities away from zones of earthquake risk, for example in Tehran where a large earthquake could bring the country to a standstill, and in the discussion of reconstruction of Port-au-Prince, Haiti. However, climate change will make many more cities face this question over the coming decades as risk profiles change, especially those associated with sea-level rise (storm surge and coastal flooding) that shift the economic balance between decisions to protect or move. There is extensive knowledge from past urban relocation projects driven by economic development, as well as disaster reconstruction, from which to learn (UN-HABITAT 2003). Most efforts, unhappily, have not enhanced the economic or social opportunities for the relocated, nor have they improved environmental sustainability at the urban level. This is despite considerable experience. In Bangkok alone, for example, between 1974 and 2001 close to 100,000 households are estimated to have been relocated with mixed outcomes for human well-being (Viratkapan and Perera 2006). The failure of developers to adequately consult with

Box 13.1 Social geography of urban disaster vulnerability

Juha I. Uitto
Evaluation Office, United Nations Development Programme (UNDP), New York

In the late 1990s the United Nations University (UNU) launched a project on Urban Social Disaster Vulnerability to systematically analyse how socially differentiated vulnerability may be better integrated into urban risk management. The project was conceptualised following a conference on Megacities and Disasters held in Tokyo, Japan, in 1994 (Mitchell 1999). Early research in Tokyo and Los Angeles informed the project (Uitto 1998). Tokyo is one of the best-prepared large cities in the world to respond to disasters, and the Tokyo Metropolitan Government has a long history of using sophisticated geographical information tools for disaster response planning. Still, they have focused primarily on data regarding physical environment, population numbers and economic vulnerability, although some social vulnerability dimensions (notably age) have been added.

The project was based on case studies in six megacities – Johannesburg, Los Angeles, Manila, Mexico City, Mumbai and Tokyo – where policy-oriented research was carried out in collaboration with local universities and research institutes. The research analysed whether vulnerability had been specifically included as a variable in the municipal disaster prevention and response plans.

The 'hazardscapes' vary significantly between the cities. Common hazards include earthquakes, floods and storms, as well as technological hazards, such as location of hazardous industrial facilities in densely populated areas. The composition of vulnerable groups also varies depending on the level of economic development. In Los Angeles and Tokyo the ageing population, with an increasing number of elderly persons, was identified as a specific group deserving attention. In Manila squatters and slum dwellers are particularly vulnerable because they are often located in the most hazardous places (Velasquez *et al.* 1999). Everywhere, socially disadvantaged people are particularly vulnerable.

Despite these differences, a number of common issues were found (Wisner and Uitto 2008). Early on, citizens' participation and strong links between municipal authorities and NGOs working with vulnerable groups were identified as critical factors determining success in vulnerability reduction. However, NGOs were frequently found to be too narrowly focused or formed in response to a specific disaster and did not sustain activities over a longer period or develop permanent capacity. They were also often seen as advancing their own political agendas and therefore not trusted by the municipal government officials. Similarly, problems with effective decentralisation of municipal disaster risk management and co-ordination among metropolitan authorities were found in virtually all cities. Both lack of training and lack of resources contributed to inefficiencies. Disaster managers tended to come from either engineering or law enforcement backgrounds, with little training in social sciences or ability to integrate social aspects into physical planning and risk analysis. Nevertheless, cases of innovative uses of neighbourhood groups and co-ordination between municipal authorities with NGOs were observed. Democratic participation in city governance, better education and training, more inclusive development involving women, minorities and youth, all were found to contribute to urban disaster risk management and reduced vulnerability.

Further reading: Uitto (1998); Velasquez *et al.* (1999); Wisner and Uitto (2008).

and plan for the needs of the relocated is compounded by the lack of tenure and large numbers of informal renters in most slum areas.

Cities are an important new operating environment for those humanitarian and development actors that are now beginning to consider cities as a site for disaster risk reduction. This builds on but is different to existing expertise in response and reconstruction. Existing approaches in urban disaster risk reduction too often focus on limited interventions around awareness raising or small-scale infrastructural investments. The real challenge is to find long-lasting solutions to everyday development challenges that generate local risks. This would, for example, be illustrated by a shift from attempting to reduce urban flood risk through one-off drain-cleaning exercises that are unlikely to be repeated by urban authorities, towards the promotion of community businesses for plastic bottle recycling or solid waste management that can provide livelihoods and an incentive for environmental management that also reduces risk. Risk reduction can also help more directly to lessen the impacts of disaster when it strikes. For example, it is important to register land claims before disaster hits, especially in unregulated and informal settlements, to contain land-grabbing post disaster. This is especially important for women and children, who may otherwise lose title on the death of their husband or parents. Post-disaster urban contexts also provide significant new challenges. Distinguishing disaster survivors is more difficult where populations can quickly become disbursed. Experience from the Indian Ocean tsunami suggests that wider targeting to include the poor as well as those directly affected is more realistic and prevents social tensions from arising (TEC 2006a).

International funding for adaptation to climate change is set to increase in the coming years. Without some strict safeguards, there is a danger that international agencies will be pulled towards projects that undermine, or miss opportunities to improve, local environmental and social sustainability as part of adaptation. The costs of managing multiple projects have in the past led to a preference amongst international agencies for support of large-scale, 'one-off' projects such as mega-dams and river engineering schemes, in preference to multiple local projects for risk management. Similarly, the awkwardness of being seen to interfere in local and national politics has made donors and development agencies wary of support for projects with strong governance reform elements. Those states with poor governance records have also been reluctant to propose projects of this kind. These concerns point to the increasing importance of local and international civil society to act as a watchdog on the evolving international adaptation funding architecture and as a source of, and champion for, projects that embrace local diversity for urban disaster risk reduction.

14

Disaster risk and sustainable development

Christopher M. Bacon

ENVIRONMENTAL STUDIES INSTITUTE, SANTA CLARA UNIVERSITY, CALIFORNIA, USA

Introduction

In late October 2008 Hurricane Mitch slammed into Central America. Vulnerable communities and eroded landscapes suffered disastrous consequences as more than 10,000 people lost their lives. The worst single tragedy occurred when the heavy rains fell upon the Casita volcano's deforested slopes, to produce a lahar consisting of mud, rocks and trees that sped down the steep slopes destroying more than 1,500 homes and killing an estimated 2,500 people including many living in the sleepy agricultural town of Posoltega (Gerulis-Darcy 2008).

What caused this hurricane to result in such disaster? Why were these smallholders and rural workers living in this hazardous landscape? Why had these slopes been deforested? The disaster could not have come as a complete surprise to high-ranking Nicaraguan officials because the government had participated in a World Bank-funded research project that had designated these slopes as high-risk danger zones (Luis Rocha 1999).

The presence of hungry farmers and farm workers living upon those steep slopes was a by-product of an economic model adopted by Nicaragua that was centred on agricultural exports. In the first half of the twentieth century, the town of Posoltega saw expansion of extensive cattle ranches (Luis Rocha 1999). This led to an agrarian structure composed of large livestock estates and peasant smallholdings. Farmers cleared trees to facilitate cattle production systems. During the 1960s and 1970s the government encouraged the spread of cotton production upon the valley floors surrounding the Casita volcano. Deforestation rates accelerated, land ownership concentration increased and smallholders were pushed towards – and eventually up onto – the volcano's slopes. The brief period of agrarian reform in the 1980s redistributed many large estates to rural workers, but in this case it did not convince the majority of small farmers to leave the volcano's slopes and resettle elsewhere. Once the Sandinista government had been voted out in the 1990s there was a renewed stage of land concentration. Properties once redistributed in the agrarian reform were sold off to larger estates.

This mainstream model of economic growth had created vulnerability that turned to disaster as death and economic loss followed Hurricane Mitch. Conventional wisdom holds that 'sustainable development' may offer the alternative. More than twenty years after Gro Harlem Brundtland and colleagues authored what became a flagship United Nations (UN) report titled *Our Common Future* (Brundtland 1987), this ambitious and ambiguous concept remains very influential and deeply contested. When launched, 'sustainable development' had been positioned as *the* response to both economic poverty and environmental degradation. Given the multiple meanings of the term 'sustainable development', this chapter delves deeper into definitions prior to exploring how this concept informs risk, vulnerability, hazards and disasters.

The concept of sustainable development

The Brundtland report articulated a commonly accepted definition of sustainable development: 'Sustainable development is development that meets the needs of the present without compromising the ability of future generations to meet their own needs' (Brundtland 1987: 8). The report criticised past economic growth models for their failure to eradicate poverty or ensure environmental sustainability. Brundtland blamed previous growth patterns for frequently damaging the environment and causing biodiversity loss and water contamination. The willingness to confront potential negative effects of economic growth is evident in phrases such as: 'where economic growth has led to improvements in living standards, it has sometimes been achieved in ways that are globally damaging in the longer term'. This attention to future generations would become the report's core contribution to development thinking. The report also called for, 'changing the quality of growth towards one focused on meeting essential needs, and merging environment and economics in decision making' (Brundtland 1987: 49). It also suggested a greater emphasis on human development, participation and equity.

The Brundtland report was based upon two concepts. First, it focused on *basic needs* and especially the basic needs of the poor and those living in the global South, as well as the needs of future generations. Second, the report emphasised *limitations* imposed by the ecological conditions, the state of technology and social organisation on the environment's ability to meet present and future needs. The report argued that the current path of economic growth would continue to produce risks; however, it failed to devote substantial attention to precisely how development processes produce risks and distribute them unevenly. Nor did the report engage the different cultural interpretations about which basic needs are to be prioritised, what is to be sustained, for how long, and by whom (Redclift 2006). This chapter will consider the rise of this term and the work of subsequent researchers who have sought to integrate more closely vulnerability with risk reduction and sustainable development.

The greening of development

Several factors contributed to the rise of the sustainable development agenda. The environmental movements of the 1960s in Western Europe and the USA profoundly influenced environmental policy and practice and opened the political space for this concept to emerge. These movements also intersected with a crescendo of criticism identifying the negative social and ecological consequences of the conventional economic growth models (Farvar and Milton 1972).

Prior to the Brundtland report, the connection between development and environment had been debated at the 1972 UN Conference on Human Environment held in Stockholm. In 1989 the UN General Assembly declared that the 1990s would be the International Decade for Natural Disaster Reduction (IDNDR). As the IDNDR was just beginning, the 1992 Earth

Summit on Environment and Development in Rio de Janeiro, Brazil provided a venue for international recognition of the connection between development, environment and disaster risk reduction (DRR). This connection was further emphasised at the third of the series of UN global conferences on environment and development, which took place in Johannesburg in 2002 (United Nations 2002b).

Gradually, the same greening trend associated with the sustainable development agendas spread into national economic planning, multilateral development banks and selected business associations. National governments in the North and South started publishing progress reports on sustainable development. Soon after the publication of the Brundtland report, the World Bank opened a vice-president's office for sustainable development and started actively promoting this agenda. However, civil society organisations such as the Bank Information Center and community groups in the global South have had to lobby the World Bank to put words into deeds by incorporating sustainable development into its full lending portfolio.

One study found that despite pledges to facilitate a global transition towards sustainability and a low-carbon economy, the World Bank's energy portfolio in 2008 remained highly imbalanced in favour of fossil fuel development (Mainhardt-Gibbs 2009). In some cases the World Bank has modified its practices. For example, the Bank was initially involved in financing China's Three Georges dam, which has displaced millions of people and may undermine longer-term ecological sustainability; however, after more research and social resistance it pulled its funding from this project.

Businesses, corporations and their different membership associations have adopted several strategies related to sustainable development. Although their shade of green is often of an even lighter hue than the negotiated language of UN documents, many firms use sustainability rhetoric in advertising and communications campaigns, and some have started to change their practices. For example, Nike Shoe Company claims that since 2001 it has developed an energy efficiency plan and worked systematically to reconfigure its supply chain and product design processes in favour of higher environmental standards and better working conditions. The company also announced plans to set climate reduction targets (see www.nikebiz.com/crreport for their self-reported achievements).

The diversity of business-based positions concerning sustainable development ranges from conservative corporate associations such as the US Chamber of Commerce, which rarely uses the term, lobbies against most environmental legislation and has funded global warming sceptic campaigns, to the much smaller Specialty Coffee Association of America (SCAA). The SCAA consists of a global membership including 3,000 roasters, retailers, farmers' associations and traders representing a retail market worth US$20 billion. The SCAA provides a platform for mainstreaming organic and fair trade certification, invests in voluntary climate mitigation projects and has made a pledge to help achieve the UN's Millennium Development Goals.

Despite initial corporate public relations efforts to suppress the sustainability agenda, the concept is now openly talked about by some of the world's largest firms, such as Toyota, Google, Nike and Unilever. Most firms frame the concept as part of their corporate social responsibility policy, which frequently includes donations to charity and small-scale demonstration projects. This public relations–oriented social responsibility response does not fundamentally alter the most common response by corporations to the sustainability agenda: business as usual (Utting 2008).

A small but growing number of firms are looking for sustainability-based strategies to redesign their products and processes, and transform their supply chains. The potential waste reduction and environmental benefits from these green innovation efforts should not be underestimated. There is little evidence that these strategies can effectively address current socio-economic inequalities and thus contribute to social justice sustainable development unless opportunities for marginalised communities are built into the plans from the outset, as proposed by some looking at sustainable urban redevelopment and green jobs (Agyeman and Evans 2003), as well as in more holistic efforts such as the earlier example of the coffee roasting association.

The main challenge is to get corporations to understand that no investment decision is risk neutral and all have consequences for sustainable development. In siting a plant, purchasing and moving minerals and other commodities in bulk, or offering employment that attracts people to a particular place, risk from natural hazards can be increased or decreased. Most corporations are not even aware of the disaster risk implications of what they see as straightforward business decisions (Wisner *et al.* 2004: 32–35).

Whose sustainable development?

Civil society and organisations based in social movements as well as researchers highlight the social justice dimensions of sustainability, as they analyse the current inequalities and deep-rooted processes of exclusion that are normally ignored by most sustainable development proposals (see Chapter 5 and Chapter 60). One should also interrogate the narrow spatial and short temporal scales that large corporations and national governments frequently use to frame sustainable development in limited terms. For example, Japan is well known for celebrating the preservation

---

Box 14.1 Strategies to shorten the seasonal hungry months

Seasonal hunger is so common in Central America that most people, including coffee-growing farmers, call these hungry periods *los meses flacos* (the thin months). In Nicaragua this generally occurs between April and September, before the corn and beans are harvested and after the money earned from the coffee harvest is spent.

In September 2009 The Community Agroecology Network, in collaboration with a progressive coffee roaster, a fairtrade coffee co-operative and a Nicaraguan-based non-governmental organisation (NGO), initiated a project focused on shortening the thin months. The project involves:

- Pilot testing community-based food banks that provide access to corn, beans and other crops during the thin months.
- Investing in kitchen gardens and the further diversification of planting of crops, such as bananas, cassava and the root crop *malanga (Colocasia esculenta)*, which are available for harvest during the thin months.
- Involving community youth to conduct the research, design and implement the project, and to re-think their agriculture and food systems in terms of human needs and sustainability.
- Leveraging awareness and investment from fairtrade enterprises and the specialty coffee industry.
- Investing in storage (modern and traditional) technologies to avoid food loss.

Although many of these strategies, such as improved storage, have demonstrated tangible results previously, the future impacts of this initiative are still to be determined (see www.canunite.org and aftertheharvest.blogspot.com).

---

and sustainable management of its national forest, but this obscures its dependence on high levels of wood imports and connections to deforestation in Asia and Latin America (Meyers 1989). Deforestation in these parts of the world is one of the factors leading to both the occurrence of landslides and vulnerability among economically poor farmers living on steep slopes.

A growing array of projects marketed under the 'green' and sustainable development banners directly undermine the goals that they ostensibly claim to promote. The so-called Mayan Riviera, located along the southeastern coast of Mexico, epitomises this contradiction. A closer look at the environmental history of coastal development in the region shows that large-scale hoteliers, who aggressively market the allure of 'jungle' landscapes and Mayan culture, constructed a five-storey, 1,000-room complex upon the ancestral lands of small-scale fisher folk, destroyed coastal vegetation and thus led to marginalisation and increased vulnerability to hurricanes and sea surges. Most of the traditional Mayan inhabitants were pushed off the land, occasionally returning to work in the new luxury hotels (Manuel-Navarrete et al. 2009).

In other examples, the perpetrators of unsustainable development are small-scale operations. For example, the many small-scale gold and diamond mining operations in central and southern Africa cause more environmental contamination and contribute to more worker injuries than their larger corporate competitors (Hilson 2006). These examples raise the importance of community participation and consideration of rights in the decisions about costs and benefits of sustainable development investments.

Some have argued for a 'just sustainability' (Agyeman and Evans 2003). Localising sustainable development in the nexus of power relations and cultural values of a place implies taking a people-centred development approach (see Chapter 5). It also implies giving attention to the multiple ways in which sustainable development is interpreted within different ways of life.

Linking environment, development and DRR in policy

The challenge of building DRR into sustainable development starts with the contested meanings of sustainable development just reviewed, and the way in which these interpretations intersect with different approaches to DRR. The current mainstream interpretation of sustainable development emphasises economic growth, albeit growth that does not undermine the ecological conditions for future growth. This might be argued to be an absurdity in a finite world where there logically must be some 'limit to growth'. Nevertheless, numerous UN documents assert that sustainable growth can have benefits because it extends basic services to formerly excluded people and allows investment in public infrastructure. However, this approach is still tied to the persistence and expansion of a global capitalist economy that is rooted in production and consumption practices (e.g. fossil fuel economy). Concern with climate change has begun to undermine confidence in such a business–as–usual scenario.

This sustainable growth approach may be coupled with disaster risk management based on techno–centric and engineering interventions. These approaches can have some important short-term impacts, providing early warning for a tsunami, for example, but they fail to change fundamentally the unsafe conditions or root causes that have sustained and perpetuated high levels of vulnerability among the economically poor and marginalised (Wisner et al. 2004) (see Chapter 3).

In contrast to economic growth-based 'sustainable development' and the technocratic approach to risk management, an alternative model departs from notions of 'just sustainability' and focuses on social relations and participatory planning. Such community-based efforts strengthen the capacity of local user groups and improve natural resource management. This holistic approach often seeks to address the root causes of social vulnerability from the bottom

Box 14.2 New Zealand's land-use and natural resource management legislation

Wendy Saunders
GNS Science, Lower Hutt, New Zealand

The Resource Management Act 1991 (RMA) has implications for how natural hazards are managed. The RMA is centred on sustainable management of natural resources, and thirty of its sections are relevant to natural hazard management.

Three other pieces of legislation have been enacted that contribute to integrated hazard management: the Civil Defence Emergency Management Act 2002 (CDEM Act), Local Government Act 2002 (LGA) and Building Act 2004. As shown in Table 14.1, these four statutes have common themes that promote integration: sustainable management or development; social, economic and cultural well-being; and health and safety.

Table 14.1 Common themes in key legislation that manage natural hazards in New Zealand (emphasis by editors)

Statute	Purpose
Resource Management Act 1991	'promote the sustainable management of natural and physical resources. *Sustainable management* means managing the use, development, and protection of natural and physical resources in a way, or at a rate, which enables people and communities to provide for their *social, economic,* and *cultural well-being* and for their *health and safety*'
Civil Defence Emergency Management Act 2002	'improve and promote the *sustainable management* of hazards … in a way that contributes to the *social, economic, cultural,* and *environmental well-being and safety*'
Local Government Act 2002	'provides for local authorities to play a broad role in promoting the *social, economic, environmental,* and *cultural well-being* of their communities, taking a *sustainable development* approach'
Building Act 2004	'provide for the regulation of building work, the establishment of a licensing regime for building practitioners, and the setting of performance standards for buildings, to ensure that: (a) *people who use buildings can do so safely and without endangering their health*; and (b) buildings have attributes that contribute appropriately to *the health*, physical independence, and *well-being* of the people who use them; and … (d) buildings are designed, constructed, and able to be used in ways that promote *sustainable development*'

These various pieces of legislation provide avenues for the implementation of risk reduction while providing for key well-being and health and safety. However, risk reduction is not being achieved to its full potential. Reasons for this include:

- While sustainable management is defined under the RMA, sustainable management and development are not defined under the other legislation, causing problems when assessing if a proposed development is sustainable.
- There is an assumption that the Building Act will achieve health and safety goals via construction standards alone; there is not enough emphasis yet on land use and other non-structural risk factors.
- In practice there is a focus on mitigation, rather than avoidance. Mitigation measures (i.e. engineering, warning systems and evacuation plans) may not achieve sustainable risk reduction.
- Slow bureaucratic procedures mean that new natural hazard and risk knowledge is difficult to incorporate into the land-use planning process.
- Many land-use planners are yet to fully realise their role and responsibilities in reducing disaster risk.

Further reading: Glavovic, Saunders and Becker 2010.

up. There is a rapidly expanding array of tools, case studies and effective strategies that promote community-level participation, and that can yield both sustainable development and DRR (see Chapter 59 and Chapter 64).

One example is the Green Belt movement, which began in Kenya but now includes Africa-wide participation of hundreds of thousands of community-based natural resource management groups involving more than 8 million members. They co-operate to promote more sustainable management of agricultural systems, watersheds and forests in ways that can improve livelihoods and reduce vulnerability to natural hazards (Pretty and Ward 2001). In Kenya, the government invested in supporting several thousand community groups to improve social and water conservation. The follow-up evaluation showed the following improvements when farmers in these groups were compared with non–group-based farmers: maize yields that were between fifty per cent and one hundred per cent higher, greater fodder availability, more trees and a higher diversity of crops grown, reappearance of springs due to groundwater recharge and a doubling of the amount of income earned by a unit of the farmer's labour (Pretty and Ward 2001: 213). The processes of building trust, co-operation and technical capacity among these land-management groups resulted in greater food security and conserved water, lowering vulnerability to drought.

An example of urban flood control illustrates the importance of land-use planning, government capacity and inclusive participation. Tornadoes, violent thunderstorms and frequent floods on the Arkansas River during the 1970s and 1980s established Tulsa, Oklahoma as the most disaster-prone city in the USA (Godshalk 2003: 138). After broad-based community dialogue, the city government developed a comprehensive flood-management plan, cleared and relocated 875 buildings from the floodplain, tightened watershed regulations, co-ordinated a public awareness programme, established open space and returned flood plains to wetlands, and mandated a storm water utility fee that annually generates US$8 million to help cover operating costs. This entire effort focused on urban economic development within spatial and other constraints that recognise environmental limitations, while also addressing a recurring natural hazard. The effectiveness of Tulsa's floodplain management won national recognition in the USA and some of the nation's lowest flood insurance rates. Tulsa's creative response also shows the benefits of integrating civic participation, land-use planning, public works and comprehensive disaster management in the move towards disaster prevention, risk reduction and sustainable urban development (see Chapter 53).

Mainstream policy tools: the precautionary principle and livelihood framework

There are several mainstream concepts that offer ways to link DRR to sustainable development. The precautionary principle is best known.

The precautionary principle

The precautionary principle moved onto the world stage through the 1992 Earth Summit, which brought the concept of sustainable development into prominence. The parties to this UN conference developed Agenda 21 as the core policy document to enable countries, firms and cities to plan and implement sustainable development. Agenda 21 recommends the swift adoption of the precautionary principle for sustainable development management and planning (Whiteside 2006).

The precautionary principle states that if an action, technology or policy contains a risk of causing harm to people or the environment, in the absence of scientific consensus that the policy or technology is harmful the burden of proof that it is *not* harmful falls on those proposing the action. It starts with the simple premise that society and science have limited knowledge about the future risks that accompany a technology or policy and that, in the face of uncertainty, action should proceed with caution. The starting point fits with notions of ecological limits and thresholds. The Earth Summit's declaration holds that 'where there are threats of serious or irreversible damage, the lack of full scientific certainty shall not be used as a reason for postponing cost-effective measures to prevent environmental degradation' (Whiteside 2006: x). Others prefer the simple phrase, 'first, do no harm'.

How does the precautionary principle inform strategies to promote sustainable development and DRR? The precautionary principle offers an important evaluative lens for screening proposed investments, projects and policies.

For example, the Chinese government often lauds the sustainable development benefits of the Three Gorges dam, claiming that it annually produces enough electricity to replace 50 million tonnes of polluting coal and thus reduces China's carbon emissions by 100 million tonnes. Civil society groups and many scientists see beyond the narrow focus upon a single environmental indicator (i.e. CO_2 reduction) and several economic benefits such as electricity generation and revenue gains. They question the fact that the dam has already contributed to the resettlement of 1.3 million people and could displace up to 5 million, many of whom were once rural smallholders (Bezlova 2007). The dam will also impact an estimated 58,000 square kiloms of land, fragmenting local ecosystems, increasing the risk of landslides and sedimentation, and accelerating both aquatic and terrestrial biodiversity loss (Wu *et al.* 2004).

Application of the precautionary principle by the Chinese authorities might have caused them to modify its design to mitigate these consequences. By contrast, the Caribbean Development Bank (2004) has published guidelines for integrating natural hazard impact assessment into the development project cycle as an integral part of the more standard 'environment impact assessment'. This amounts to a specific application of the precautionary principle to site-specific hazards such as landslide and hurricanes.

Despite its philosophical appeal, the deployment of the precautionary principle is often complex, and requires increasingly sophisticated and possibly speculative risk analysis. Also, the

Christopher M. Bacon

Box 14.3 Incentives for DRR

Mihir Bhatt
All India Disaster Mitigation Institute, Ahmedabad, India

Incentives for DRR at community level are scattered, uneven, often conflicting, unsustainable and top down. What communities want is consolidated, even, harmonised, sustainable and lateral (mutual) incentives. The experience of Concern Worldwide in India (CWI) over the past ten years in encouraging DRR activities after the 2001 Gujarat earthquake provides a very good example. CWI includes twenty-six local partners spread across six most-at-risk states in India, including Gujarat, Orissa, Tamil Nadu and Bihar.

The incentives for DRR are scattered across sectors and locations to start with. In a given village stone embankments are built by the flood management authority on river banks to stop flood water flowing into the village, but no incentives are available to individual villagers to enhance the plinth height of their homes or dig water-harvesting structures for the drought that visits the village with almost certain regularity. Often incentives are latent and not articulated across communities and interest groups. Incentives are there as hidden opportunities to be safer, or have more income, or build new assets, or protect a family. The scale of incentive is not known by the community, and also it is not known if the incentive will come in cash or kind. The unevenness of incentives is also due to differences in the size of incentives for the same or similar measures. One NGO pays around 12,000 Rupees (US$270) per flood diversion structure constructed on a river bank and in another case some 8,000 Rupees (US$180) for the same construction made of wire and stones in another location or community.

Incentives often conflict with other DRR incentives or development incentives. Incentives for using farm fertilizer clash with incentives to use drought-resistant local crops. Incentives for crop insurance displace the use of low-yield but heavy rain-resistant monsoon crops: insurance will pay for the loss! Incentives are almost always a pilot scheme and by the time the community comes to know about it, the incentive has changed, or modified, or been withdrawn. Social incentives to set up community-based disaster risk management committees after the Indian Ocean tsunami did not spread to non-tsunami states, or to non-tsunami districts within the affected state of Tamil Nadu, as if disasters only follow past disasters.

Finally, incentives always come from the top to the community. Lasting incentives are local, offered by one community to another, to displace the risk they face. In one example, two communities on either side of a river co-operated to grow food on both sides and agreed to mutual aid so as to avoid loss of food due to flood. This works for both communities. The communities provide each other with incentives.

principle itself is highly contested, in part because of difficulties in comparing risks and that misapplication of the principle could lead to paralysis in decision–making (Sunstein 2005).

Sustainable livelihoods and DRR

During the past two decades, the concept of sustainable livelihoods moved from the margins to the mainstream of development policy (see Chapter 58). The livelihood framework expands an approach to poverty reduction narrowly defined by monetary income, directing one to consider the multiple assets that individuals and households mobilise to satisfy their needs.

Assets include social relations (family relationships, group memberships, etc.), natural resources access (e.g. water, land for food, fodder and fuel, etc.), financial resources (loans, savings, etc.), and other assets that households can utilise to buffer exposure to a hazard and to cope with losses. Although much of this work has focused on rural livelihoods, some scholars have also used this framework to identify the most important assets for urban livelihoods and to design effective sustainable development strategies.

Livelihood-focused policy interventions seek to build the community-level assets that enable coping strategies in the face of natural hazards. For example, strategies for coping with drought are part and parcel of pastoral livelihood systems in Africa. Turner (2010) notes several coping strategies in a study from the Sudano-Sahelian region of West Africa. These include changes in the composition of the herd from primarily cows to smaller livestock, especially sheep and goats, and the increasing importance of mobility and access to changing landscapes and vegetation types at different moments during a changing seasonal cycle (see Chapter 22).

Coastal mangrove vegetation not only provides protection from tropical storms and coastal erosion, but is itself prone to increasing damage from storm hazards, especially in South-East Asia (Macintosh and Ashton 2003: 5). In Bangladesh, there has been a good deal of experience with mangrove restoration. Biswas *et al.* (2009) compare four restoration projects and conclude that the most successful ones involved not only ecological considerations but financial benefits for the community, took social relations into account and benefited from strong leadership and rigorous monitoring. From such examples one must conclude that incentives for ecologically sustainable practices at the community scale are very complex, yet increasingly necessary to ensure these practices though time.

A critical policy tool: environmental justice

An environmental justice perspective affirms the rights of all people to habitable communities that include environmental goods, such as clean air, water and soil. Environmental 'bads', including toxic waste incinerators and garbage dumps, have been shown to be unequally distributed along lines of race and income. According to the US Environmental Protection Agency, '[environmental justice] will be achieved when everyone enjoys the same degree of protection from environmental and health hazards and equal access to the decision-making process to have a healthy environment in which to live, learn, and work' (EPA 2010).

Research has found that those on low incomes and those who suffer racism in the USA pay a disproportionate sum of the environmental costs and harvest fewer environmental benefits when compared with national averages and their more privileged peers. The primary explanation for this is a history of exclusion and oppression contributing to a lack of power and access in these communities (e.g. Bullard 1994) (see Chapter 38).

Research into environmental justice has taken on global challenges, as the evidence accumulated that economically poor and socially marginalised populations worldwide suffered from environmental injustices (Schroeder *et al.* 2008). More consideration was given to pesticide exposure in the global South and the uneven resource flows among countries. The research and action agenda continued to evolve, moving to consider the cumulative impacts of multiple hazards (versus a focus on a single pollutant or a single natural hazard) and to incorporate indicators to identify social vulnerability. In the natural hazard context a corresponding concept and body of empirical evidence points to the impact of 'extensive' risk compounded by many small hazard events over time (UNISDR 2009a).

Hurricane Katrina is a powerful case study of how an environmental justice framework can inform strategies to promote sustainable development and DRR. Prior to the deluge in August 2005, the city of New Orleans had developed land-use patterns and a system of poorly maintained levees that favoured the reduction of small everyday hazards, yet left low-income and many African-American residents highly vulnerable to a large flooding event (Colten 2007). Small repairs to the levees could stop relatively small ocean surges, but more costly investment that would enable levees to withstand a large hurricane were deferred.

Linking environment, development and DRR in practice

A study conducted immediately after Hurricane Mitch on the slopes of the La Casitas volcano revealed that small farmers who had adopted agro-ecological management practices in Nicaragua's northern mountains were spared. These practices are of growing importance worldwide under various names including 'site-specific agriculture', 'permaculture' or simply 'sustainable agriculture' (Gliessman 2006). The agro-ecological practices in the Nicaraguan mountains include contour planting and use of soil-conservation barriers as part of a diversified farming system. The study compared farms active in the *Campesino a Campesino* (farmer-to-farmer) social movement with those who did not adopt these practices. They found that farms practising agro-ecology suffered less erosion and were more resilient to Hurricane Mitch (Holt-Gimenez 2002).

The genesis for agro-ecology can be found in an intercultural collaboration that links ecological science with insights from traditional indigenous farming systems, applying the resulting principles to the design and management of sustainable food systems (Gliessman 2006). Local knowledge (see Chapter 9) is evident in the many different varieties of corn that farmers cultivate on a single farm. Mesoamerican communities have contributed to the more than 20,000 varieties of maize cultivated globally. Some maize seeds are better for rainy years; other seeds are more drought resistant. This crop genetic diversity is complemented by species diversity since many farmers intercrop maize with beans and squash to create the essential 'Three Sisters' combination. They have learned that leguminous plants fix nitrogen from the air in the soil, thereby fulfilling corn's appetite for nitrogen. These systems have evolved over thousands of years and can also be observed in Asia, Africa and the Pacific, as well as the Andes, where thousands of varieties of potato are still grown.

Local knowledge is context- and place-specific and often acquired through lived experience, trial and error, passed down through generations. Such knowledge and practice extends far beyond agro-ecology. For example, a recent study sought to identify cases of local knowledge and coping strategies that can be useful in small island states seeking to mitigate natural hazards (Mercer et al. 2010). Many indigenous communities with generations living in the same island landscape site their settlements on high ground to avoid storm surges and floods. Local veterinary care practices from Asia, Latin America and Africa provide another example of the ways in which people cope with hazards to their livelihoods (McCorkle et al.1996) (see Chapter 31).

Local knowledge and practice, whether in building, drainage and water management, farming, care of livestock, management of heat and cold, or early warning, may be the starting point for strengthening DRR. Participatory action research allows the combination of local knowledge with outside specialist knowledge (see Chapter 64). Participatory action research starts with a commitment to connecting researchers, local organisations and community members. It is a cyclical approach that involves a wider diversity of stakeholders as active participants in a process of both knowledge generation and action for positive change. Participatory action research is deeply rooted in the recognition of the agency, aspirations and ideas of all participants, and is created with an explicit intercultural goal of uniting different forms of knowledge (indigenous, Western, practical and theoretical) (see Chapter 9).

Conclusions

Sustainable development may contribute to DRR. However, investments and projects under the 'sustainable development' banner may also generate higher risks. Where the dominant economic growth paradigm remains unchallenged and 'sustainability' is an add-on, DRR is not likely to be increased. The diluted, add-on version of sustainability does not consider social equality. Histories of uneven development combined with the lopsided and relatively narrow applications of sustainable development continue to perpetuate environmental inequalities and uneven distributions of risk. Acknowledgement of the links connecting environmental justice with sustainable development would begin to address these shortcomings.

Mainstream policy tools can be useful, but critical policy tools deserve more attention if the goals are to find longer-term solutions that avoid reproducing the same vulnerabilities that result from the current export- and growth-focused development model. Use of the precautionary principle for vetting proposed policies and investments could substantially reduce disaster risk. When coupled with a renewed focus on bottom-up participation, social equity and consideration of different forms of knowledge, cultural and community practice, breakthroughs to safety and enhanced livelihoods can result.

Part II
Fine-grained views – hazards, vulnerabilities and capacities

15

Introduction to Part II

The Editors

Why the need for fine-grained views?

According to Samuel Johnson's biographer (Boswell 1822), Johnson refuted a 'proof' of the non-existence of matter by forcefully kicking a large stone. Johnson's cry, 'I refute it thus!' must resonate over 200 years later in the minds of many whose field work puts them in direct contact with natural hazards and their consequences for society. The shift in dealing with disasters and disaster risk reduction (DRR), from a mainly hazards view to accepting that vulnerability must be considered, never denies the reality of hurricane-force winds with a destructive energy that is proportional to the wind speed cubed. No one who writes about the vulnerability of landless labourers or herders who have lost all their animals disputes the power of flood waters or the way the sun bakes the cracked soil.

It is important to understand the relationship between these very real, material hazards and vulnerability. To do that, studying the natural world is essential – as long as it is coupled with and not isolated from studying society. The chapters in this section do so, covering two aspects of fine-grained, on-the-ground views of disasters and DRR: natural hazards and the marginalisation of many people and communities, which makes them more vulnerable to those hazards, while reducing their capacities to cope and to recover.

Categorising natural hazards

The hazard-by-hazard chapters in this section explore how different categories of hazards bear on DRR policy and practice. A key component is the interaction of hazards and people's actions as they attempt to reduce their vulnerability to the hazards. People view and address hazards differently, suggesting the importance of understanding hazards from many varying perspectives, including modern science and relevant local understandings.

The authors of these chapters do not merely contrast top-down and bottom-up approaches to understanding, interpreting and addressing hazards. Instead, they seek to combine processes from different directions by finding a common space where external and internal understandings complement each other rather than clash. A single, isolated description of a hazard rarely provides the needed picture for successful DRR. Multiple perspectives must be joined.

Consequently, departing somewhat from more conventional treatments, the authors here deal with multiple 'we's' in the phrase, 'what we know and what we do'. People's experiences of natural hazards are a form of knowledge, as is Western science. The hazard chapters here cross disciplines, continents and worldviews to combine and distil humanity's collective perceptions, knowledge and understanding, giving illustrative rather than comprehensive summaries.

The first chapter in this section, Chapter 16, covers 'data sources on hazards'. It introduces examples of what we know and do not know about how hazards can be detected, monitored and interpreted. The data improves our scientific understanding of Earth processes. It can also be used operationally for such purposes as land-use planning, for siting of infrastructure and settlements, for issuing warnings when a threat might manifest and for indicating how human activities positively and negatively affect natural processes, sometimes creating or exacerbating hazards.

Chapter 17 is on 'tools for identifying hazards'. Hazards are considered from the point of view of methods available for assessing their frequency and magnitude in the short, medium and long term, as well as pinpointing post-event data. While some technology remains complicated and costly, the trend is towards more user-friendly, lower-cost methods that can be applied locally. This chapter also discusses visualisation methods such as the use of maps, graphics and other images for communicating risk and planning.

Individual chapters treat specific hazards in four overlapping categories, deliberately and necessarily including hazards over all space and time scales (see Table 15.1):

- Hydro-meteorological/climatological hazards: Separate chapters cover 'hazard, risk and climate change', 'coastal storm', 'thunderstorm and tornado', 'flood', 'drought', 'extreme heat and cold' and 'wildfire'.
- Geophysical hazards: Separate chapters cover 'landslide and other mass movements', 'earthquake', 'tsunami', 'volcanic eruption' and 'soil erosion and contamination'.
- Biological/ecological hazards: Separate chapters cover 'human epidemic', 'livestock epidemic', and 'plant disease pests and erosion of biodiversity'.
- Astronomical hazards: A single 'hazards from space' chapter summarises the topic, focusing on space weather and Earth impacts.

Table 15.1 not only summarises the categories but also indicates the limitations of categorising through the many overlaps and ambiguities that result. For example, wildfire is included in the hydro-meteorological category due to its link with weather, yet wildfire is also an ecological process and so is included there as well. Volcanoes produce many hydrological hazards including lahars and floods, along with many meteorological hazards such as lightning and vog (volcanic fog). Landslides can be rainfall- or earthquake-induced. Avalanches and glacial surges are frequently studied as geophysical phenomena, yet originate with hydrological processes. A separate chapter on tsunamis is included in the *Handbook* even though tsunamis always originate with other hazards and even though they represent a form of flooding. Storms are not treated as a single entity, but are divided amongst 'coastal storms' and 'thunderstorms and tornadoes'.

Lack of comprehensiveness is also a criticism of the choice of hazard categories and ultimately a limitation of the *Handbook*. Flash flooding can be a grave danger, yet is covered in only a box in the flood chapter. Haze, hail, fog and dust storms, amongst others, receive only brief mentions. Wind storms and tropical cyclones are not treated in their own category; instead, they are dispersed amongst several chapters.

These decisions do not indicate the perceived importance or lack of importance of any particular category or phenomenon. Nor is the number of chapters for each sub-category

Table 15.1 Summary of hazard categories

Hazard category	Origins	Examples (not a comprehensive list)
Hydro-meteorological/ climatological	Hazards from water, weather and climate	Avalanches, climate change, drought, floods, fog, glacial surges, hurricanes, icebergs, lightning, precipitation (e.g. freezing rain, hail, ice, rain, sleet, snow), storm surges, temperature extremes or fluctuations (cold and heat), tornadoes, waves, wildfires and wind.
Geophysical	Hazards from geology	Earthquakes (and associated hazards such as tsunamis and landslides), erosion, landslides/rockslides (and associated hazards such as tsunamis), poison gas, sandstorms, soil loss and contamination, volcanoes (and associated hazards such as fire, fumaroles (gas emissions), lahars (mudflows), jökulhlaups (glacial floods), tsunamis and vog (volcanic fog)).
Biological/ ecological	Hazards from living organisms or ecosystems	Wildfires, microbial pathogens, and poisonous, aggressive or otherwise dangerous plants and animals.
Astronomical	Hazards from outside the earth	Space weather and collision or near-collision of celestial bodies (e.g. asteroids and comets) with the Earth and associated hazards (e.g. tsunamis, wildfires).

indicative of the magnitude or intensity of each hazard type. Consider that a single meteorite strike could cause a global mass extinction, a consequence unlikely from any hydro-meteorological hazard. For a handbook, the objective is not to be comprehensive and exhaustively detailed, but to be illustrative and demonstrative of hazards, vulnerabilities, capacities, risks and disasters.

Hazard characteristics for whole categories are not provided here. The reason for this is that there would be a danger of superficiality and loss of detail. Hazards are often compared through characteristics such as:

- A physical, chemical and/or energy description of the hazard, e.g. rapid motion (from an earthquake or landslide), heat (from lava or temperature extremes) or mass (from hail or lahars, each of which also includes motion).
- Magnitude and intensity, e.g. moment magnitude for earthquakes or the Saffir-Simpson Hurricane Intensity Scale.
- Temporal characteristics, e.g. speed of onset and decay, duration (the temporal extent) and frequency (the temporal dispersion).
- Spatial characteristics, e.g. areal and volumetric extents and patterns of distribution.
- Predictabilities of the above characteristics and the quality of these predictions.

Yet disaster characteristics might not emerge directly from the hazard typology. For example, sometimes simultaneous or sequential hazards represent the disaster challenge. Southern California experiences cycles of devastating wildfires, sometimes during droughts, with the drought ended by wild storms that unleash ruinous mudslides on burned-over slopes. At the end of 2004 and then again at the end of 2009, the Philippines experienced a series of tropical storms and typhoons which layered disaster upon disaster, including mudslides and floods. Standard hazard typologies cannot always reflect the disaster that results from multiple hazards – simultaneous or sequential.

Some authors of hazard chapters describe selected typological traits explicitly, while others weave the material and interpretation throughout the text. A glimpse of the detail available for characterising hazards through various typologies (e.g. Alexander 1993; Wisner *et al.* 2004) is given, but not all the details, which would raise each chapter to the level of a textbook. Where useful, specialty material available for delving into further details is provided, from processes leading to volcanic explosivity (see Newhall and Self 1982 for an index), to a categorisation of how floods interact with buildings (Kelman and Spence 2004).

The overall lesson is not necessarily to avoid a universal typology of hazards. Instead, it is to recognise that 'hazardousness' (Hewitt and Burton 1971) is largely, but not completely, a human construct. Hazards are influenced by humans, but also exist in their own right. Hazards must be understood within the context of the limitations and inconsistencies in all approaches to categorising. The chapters on hazards in this *Handbook* provide a start through the baseline material, while connecting directly to the fundamental vulnerabilities often exposed by the manifestation of hazards.

Categorising vulnerabilities and capacities

The chapters in the latter half of Part II on fine-grained views reflect the argument that those people who are marginalised in society are the most vulnerable in facing natural hazards, but still have capacities. Yet the needs and abilities of marginalised people are usually less considered in actions to reduce the risk of disaster. Such material is not new, as it dates back at least to works by Wisner (1978) and Blaikie and Brookfield (1987). However, it is often overlooked or bypassed despite the growing gap between the poorest and the richest, between people regularly affected by natural hazards and those who have never been concerned, and between people and governments able to deploy an increasingly large arsenal of measures and those unable to resort to even the most basic strategies of self-protection. Such concerns militate for a thorough reinvestigation of issues of marginality and prejudice in the context of disasters and DRR.

Marginalised people are vulnerable because they lack or are deprived of access to available measures by the most powerful (Figure 15.1). Marginalisation is a matter of poverty, but not only that. Obviously, limited economic resources prevent people from choosing where to settle, which includes choosing safer areas which can be expensive. Poor people are also unable to afford DRR measures, although these are often available locally. Examples are wire ties to secure the roof against high winds or retrofitting to increase flood or seismic resistance. Yet vulnerable people and communities in the face of natural hazards also include those who are socially and culturally excluded from dominant policies and DRR activities. These encompass all neglected segments of societies, often including children, elderly, women, non-heterosexuals, people with disabilities, refugees, illegal immigrants, prisoners, homeless people and ethnic minorities. Some (not all) are physically weak, unable to move easily and dependent on the decisions of others – frequently because society does not factor in their needs even in the absence of hazards and disasters. Disasters often further marginalise those who were already living at the margin beforehand.

The chapters on marginality and prejudice in this section take up these themes. They focus on specific groups that share similar patterns of marginalisation in facing natural hazards. The chapters are 'disability and disaster', 'gender, sexuality and disaster', 'children, youth and disaster', 'elderly people and disaster' and 'caste, ethnicity, religious affiliation and disaster'.

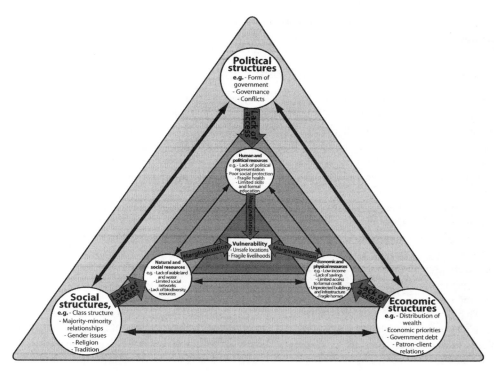

Figure 15.1 Vulnerability triangle

The scope of the chapters is necessarily limited as they do not intend to cover all marginalised groups. Refugees, prisoners, illegal immigrants and homeless people lack individual chapters, although they deserve to have them, and there is some literature that explores their vulnerability to disaster (e.g. Wisner 1998). For instance, more than 1,000 incarcerated prisoners died when the December 2004 tsunami flattened two jails in Meulaboh and Banda Aceh, Indonesia (Gaillard *et al.* 2008a). The homeless are among those who most frequently become ill or die in extreme heat and cold. Illegal immigrants avoid seeking official aid following a hazard event – even avoiding hospitals – due to the fear of deportation (e.g. Bolin and Stanford 1998). Refugees sometimes have no shelter except in exposed locations such as steep ravines or volcanic slopes on the outskirts of cities. Such consequences of marginality appear in passing in chapters elsewhere in the *Handbook*, such as 'extreme heat and cold' (Chapter 23) and 'violent conflict, natural hazards and disaster' (Chapter 7).

An obvious and persistent lack of available data and studies exists on many marginalised groups, which is indicative of their marginalisation (Wisner and Gaillard 2009). We hope that flagging these gaps will trigger further interest, studies and action by and for those people who most need attention.

Addressing marginality and disaster requires a fine understanding of local power relationships among different segments of a given society. Those marginalised in one context may be those with power in another; for example, the Hutu–Tutsi relationship in the Great Lakes region of Africa is not always a majority dominating a minority, but depends on the location and has shifted throughout history. The chapters in this section provide a starting point for appraising such power relationships, and their impact on people's vulnerability in the face of natural hazards and their ability to deal with disasters.

Using the fine-grained views for DRR

This section on fine-grained views discusses both hazards and vulnerabilities. It is rare that both subjects are brought together in a single space, providing *Handbook* users with basic information on hazard and vulnerability characteristics and dynamics side by side. Providing a primer on some aspects of social science and physical science, and their combination, exposes those new to the areas to basic information; it consolidates material for experienced researchers and practitioners – and, we hope, also inspires readers to seek out material on topics missing here or incompletely covered.

Hazards and vulnerabilities are each relevant in their own right, yet need to be understood together to realise the complex interactions that produce outcomes including disasters. Recalling Dr Johnston's refutation of idealism, one needs to acknowledge the physical forces and natural energies behind hazards. Yet the forces and energies that make up hazards cannot be divorced from their societal contexts. Bringing in fine-grained views of vulnerability alongside fine-grained views of hazards contributes to contextualising how individuals and societies experience nature.

That means never forgetting that nature can and does pose dangers for individuals and societies. Sometimes hazards must be curtailed directly. When an asteroid heads towards the Earth, deflecting (or possibly destroying) the asteroid is necessary rather than trying to reduce vulnerability to a planet-wide cataclysm. A sudden release of gases from Lake Nyos in Cameroon suffocated 1,746 people in 1986. A project to de-gas the lake makes sense, and has been implemented (Kling *et al.* 2005), rather than waiting for another lethal release and hoping that a warning system works, or moving the people away from their homes and livelihoods which could expose them to other difficulties or to other hazards.

Finally, hazard identification and understanding is meaningless for DRR without resultant action. In Haiti, for example, the seismic hazard was known prior to the January 2010 earthquake, but tunnel vision on the part of government and donors focused on climate-related risks, with tragic consequences (Mora 2010). By melding the fine-grained views of hazard and vulnerability as presented in this section, it is possible to begin to see steps that may be taken to change as many hazards as feasible into sustainable opportunities.

All hazards

16
Data sources on hazards

Julio Serje

GLOBAL ASSESSMENT REPORT TEAM, UNISDR, GENEVA, SWITZERLAND

Introduction

This chapter attempts to provide simple and practical guidelines to help those involved in disaster risk reduction (DRR) address their data and information needs, with a strong focus on quantitative data. DRR can be applied at many scales from community to global. Data and information needs vary accordingly. As practitioners work on more detailed scales, for example at community level, different sources of data will be needed. However, some needs are the same for everyone. Despite problems that are rather different at various scales, these common needs are the target of this chapter, given that in many cases these needs can be addressed using globally available data in the public domain. Additionally, researchers and practitioners should always seek other data produced locally and not necessarily available in electronic formats.

The amount of data available over the internet today is simply overwhelming. The purpose of this chapter is not to synthesise it all. Instead, it provides some starting points and makes suggestions regarding finding and using data. Box 16.1 provides some background to risk and vulnerability indices. Two main concepts are used to describe and give an idea of the scope and detail of data sources (UNDP 2004).

First, level of observation is the geographical extent covered by a specific data source. This chapter refers mainly to publicly available data sources at a global and country (national) level. Second, resolution of data is the type and size of the spatial unit to which collected data is associated. Event data sources are frequently associated with points or grids, while other data sources are associated with geometrical entities (polygons). Data such as demographics or disaster damage are usually attached to administrative boundaries like country, province or municipality. Environmental information is often associated with environmental units such as watersheds. Data coming from local-scale studies, such as community-based risk assessments, involve data pertaining to people's livelihood, choices and behaviour, so may be obtained at different and heterogeneous scales and spatial units.

Many of the data sources and documents mentioned here reside on websites that evolve over time, so an effort has been made to keep these as compact and general as possible and reduced to a minimum.

Box 16.1 Risk and vulnerability indices

The creation and use of an 'index' has been a popular methodology for evaluating relative levels of a status, such as development, poverty or something similar. In the area of DRR, in order to assess risk and other applications, it is common to find indices that attempt to capture levels of vulnerability to natural hazards, either physical or social, or combining multiple aspects.

Indices usually comprise a set of indicators and, through some mathematical combinations, an index number is derived for a community or geographic unit, which can be used to make comparisons with other communities or units. Indices are attractive because of their ability to summarise a considerable amount of technical information in a way that is easy for lay persons to understand (Davidson and Lambert 2001). Simpson (2006) conducted a thorough review of the process of building these indices, concluding that, in general, frameworks and indicator lists are either un-weighted or the weighting of the components is a subjective process. In many cases, these weights are adjusted based on expert panel reviews or iterative on-the-ground application, calibration and review.

Another key issue to consider with indices in general and in particular with vulnerability indices is that they take a group of indicators and produce a snapshot of reality. Indices are quantitative subjective measures, acting as proxies for the concept under examination. One considerable shortcoming of using composite indices for vulnerability is that there is no simple way to get scientific validation of a particular index (Davidson and Shah 1997), a process with opportunities for the interested parties to alter or adjust the indicator to suggest what they want it to do.

Vulnerability indices pose, in addition, several practical difficulties when used to assess risk. They cannot easily be applied generically to multiple types of elements at risk. Each element (e.g. human life, infrastructure and environment) must be calculated separately, adding complexity and subjectivity to the entire model. On the other hand, while indices are easy to understand and useful for conveying information, they are not necessarily suitable for conversion into vulnerability functions or factors that could be utilised to calculate expected losses.

Disaster risk data

Disaster risk reduction information needs can be very broad; it would make little sense attempting to describe them all given the wide spectrum of requirements and types of DRR applications and scales at which this work is conducted. Thus, sources are explored in relation to the most basic need: the knowledge of 'assessed risk', seen as the potential of future losses, and of 'realised risk' or past disaster data. Attention is paid to the commonly agreed components of risk, namely hazard and vulnerability, with exposure sometimes included as well. These categories comprise most of the publicly available data for DRR (see Chapter 3).

Describing risk quantitatively but simplistically at any point in space involves a two-dimensional variable. On one axis, the probability of events of varying magnitude (hazard) occurring is shown, and on the other axis, the potential losses that would happen given such events are shown. The probability is commonly expressed as 'return period' of an event.

Generators of risk information are faced with decisions on how to present this complex information in a practical and usable way. One common approach is the production of a 'risk index' providing users with a relative measure of risk, such as 'high, medium and low'. The majority of these indices 'flatten' the probabilistic dimension by taking a fixed return period for the index, or using a particular statistic such as historical loss rate. For example, the Natural Disaster Hotspots study calculated mortality rate as an estimate of the proportion of persons killed during a twenty-year period (Dilley *et al.* 2005). Another good example of a risk index is the Global Assessment Report on Disaster Risk Reduction (UNISDR 2009a).

A common compromise is to present a probabilistic risk assessment in the form of a set of 'risk scenarios' in which losses are calculated at various probabilistic values. Each risk scenario corresponds to a return period (25, 50, 100, 200 years, for example) in order to factor in most time scales. An event such as the 2004 tsunami in Aceh, Indonesia, has a return period of 200 years or longer (CCOP 2009).

Risk index-based maps and information are especially utilised for prioritising DRR efforts, focusing resources and formulating strategies. When risk is expressed also as the potential of losses, then it becomes even more usable in the decision-making processes of DRR, and all the more when the probabilistic dimension is also included. Justifying investments, for example, can only be done by measuring the amount and cost of losses and the probability of such losses happening. It is not the same to plan for yearly recurrent event types such as cyclones or spring floods than for an extreme earthquake, which may or may not happen in the next 400 years.

Global risk

GAR/PREVIEW (preview.grid.unep.ch)

One of the most comprehensive data sources found today, and probably the first place to go when looking for information, is the Global Risk Data Platform, hosted and developed as support to the Global Assessment Report 2009 by a consortium of United Nations (UN) agencies, for example the United Nations Environment Programme (UNEP), the United Nations Development Programme (UNDP) and the United Nations International Strategy for Disaster Reduction (UNISDR).

This platform is an effort to share spatial data information on global risk from natural hazards. Users can view, download or extract data on past hazardous events, human and economic hazard exposure, and risk from natural hazards. It covers tropical cyclones and related storm surges, droughts, earthquakes, biomass fires, floods, landslides, tsunamis and volcanic eruptions.

The site offers access to a set of pre-built risk, event and hazard maps, graphs and datasets. While most of the information is freely available, it is worth noting that some of the raw and processed data used to build the platform are not available because of copyright and commercial restrictions. However, in all cases, the platform specifies what data exists (metadata) and where to find it.

Natural disaster hotspots (www.ldeo.columbia.edu/chrr/research/hotspots)

Although most DRR work accepts that 'natural disasters' do not exist (see Chapter 3), this study uses the phrase in its title. This project produced a global analysis of disaster risks associated with six major natural hazards – cyclones, droughts, earthquakes, floods, landslides and volcanoes – accompanied by a series of case studies. The global analysis assessed the estimated spatial distribution of relative risks of mortality and economic losses. Risk levels were estimated by combining hazard exposure with historical vulnerability for two indicators of elements at risk: gridded population and gross domestic product (GDP) per unit area. Calculating relative risks for each grid cell rather than for countries as a whole allows risk levels to be compared at sub-national scales (Dilley *et al.* 2005).

The resolution of both these global analyses is relatively coarse. Their users must keep in mind that global datasets may inadequately capture important factors that affect local risk levels.

UNDP DRI Tools (gridca.grid.unep.ch/undp)

Another interesting tool, also developed by UNDP and UNEP, is the Disaster Risk Index site which features profiles for most countries, mostly based on historical occurrence of disasters. It can be used as a starting point, especially when few or no data are available.

Global earthquake model (GEM) (www.globalquakemodel.org)

GEM aims to establish a uniform, independent standard to calculate and communicate earthquake risk worldwide. In addition to risk, GEM plans also to develop a comprehensive Global Exposure dataset.

Regional risk

Pacific Disaster Net (www.pacificdisaster.net)

Pacific Disaster Net is the Virtual Centre of Excellence for Disaster Risk Management in the Pacific Region. It is a living collection and growing disaster risk information resource to research, collaborate and improve information and knowledge management, providing in-country information for distribution within the region and hosting material related to various sources like country governments, organisations and agencies at regional, national and international level. The website provides filtered, dynamic and fixed data and information with events, contacts, links and basic facts, available by country and organisation amongst other categories.

SIAPAD (www.georiesgo.net)

SIAPAD is the Disaster Prevention and Management Information System for the countries of the Andean Community (Colombia, Peru, Ecuador, Venezuela and Bolivia). Available only in Spanish, the site provides users with data and resources related to disaster risk in the area. Emphasis is placed on geographic information and disaster data (with links to the DesInventar disaster databases of Latin America; www.desinventar.net), and to a wealth of documents gathered in a searchable knowledge database.

DISCNet (www.pdc.org/osadi)

This initiative, started by the ASEAN Committee on Disaster Management to gather and exchange DRR information for member countries, promises to deliver high-quality, standards-based risk- and disaster management-related information to member countries. In addition to OSADI (Online Southeast Asia Disaster Inventory), a regional initiative to collect disaster data (described later), it also contains hazard maps and other spatial disaster-related data. It is expected to facilitate disaster management information sharing among ASEAN member countries, and to support and influence decision-making processes at all levels.

CAPRA (www.ecapra.org)

CAPRA (the Central American Probabilistic Risk Assessment) is designed primarily as a mechanism for exchanging, communicating, understanding and managing the risks in the region. It attempts a probabilistic approach to risk assessment, using cutting-edge methodologies in natural risk assessment and involving computer technology and advanced communications in the socialisation of natural hazards and their possible effects. Sponsored by the World Bank, the Inter-American Development Bank, the UN and others, the developing group plans to provide and make available to the general public the results of the assessments and its raw materials obtained in Central America.

Disaster impact and loss data

This category encompasses historical damage and loss databases available publicly. Damage and loss databases contain a set of indicators, such as mortality or number of houses or hectares of crops lost, describing the effects of disasters upon society and environment. The Hyogo Framework for Action (HFA) (UNISDR 2005b) states that compilation of disaster risk and impact information for all scales of disasters is essential to inform sustainable development and disaster risk reduction. The ultimate outcome of HFA and DRR is to reduce losses caused by disasters. The HFA advocates the 'systematic recording, analysis, summarising and dissemination of statistical information on disaster occurrence, impacts and losses, on a regular basis through international, regional, national and local mechanisms'.

Damage and loss data collected can inform about risk patterns and feed into risk assessments. It also serves as an indicator mechanism to monitor the dynamic nature of risk, helping to identify emerging trends and measuring the effectiveness of DRR interventions. Nonetheless, one discouragement is that damage and loss data are missing or under-registered for many countries worldwide (Dilley et al. 2005).

Compiling disaster loss databases, global or in developing countries, is a challenging job, as is maintaining those datasets up to date. As a consequence, there are limitations and constraints of which users must be aware. Global sources usually rely on few, second-generation information sources (such as the UN Office for the Coordination of Humanitarian Affairs (OCHA) and the World Health Organization (WHO), or media sources), which in turn rely on nationally produced data. National-level data sources vary, depending on data availability, ranging from media sources to purely official information to a mixture of these. In all cases, users must check with data producers to get a good idea of accuracy, coverage, possible biases and other limitations, especially when working on cross-boundary projects.

Global losses

There are several global disaster data sources. Probably the best known is EM-DAT (www.emdat.be), but there are many others with different levels of access, availability and coverage. Examples are SIGMA from Swiss Reinsurance, and NatCat from Munich Re. Unfortunately, the latter two, coming from private companies, are not available to the public, as only analysis reports are shared. These databases would be of tremendous help to numerous practitioners and many have requested that these databases, collected in many cases from public sources, be contributed to the public domain.

EM-DAT (2010) states on its website that it 'contains essential core data on the occurrence and effects of over 16,000 mass disasters in the world from 1900 to the present. The database is compiled from various sources, including UN agencies, non-governmental organisations, insurance companies, research institutes and press agencies.'

The EM-DAT database must be recognised as one of the most comprehensive publicly accessible sources of disaster data with global coverage. The Hotspots study, the UNDP DRI and GAR reports, among others, relied heavily on this dataset. Nevertheless, EM-DAT has been questioned on several occasions (IASC 2002) for being under-registered. It is easy to see that the 'core data' stated in their presentation are limited to only three variables available to the public: mortality, number of affected and economic losses. The 'affected' variable is difficult to define and in general terms is unreliable. The economic losses are evaluated by different partners that produce disaster reports with different and possibly inconsistent methodologies. There are also summaries available with homeless and injured per country or disaster type, but detailed data are not offered (Peduzzi *et al.* 2009).

Users of EM-DAT must be aware of its limitations: the nature of the information contained, made at a global level of observation and with a national level of resolution, makes it hard to use for sub-national purposes. The already mentioned under-registration problem is aggravated when working within a country, as only medium- and large-scale disasters are entered in this database. The database creators and managers imposed minimum criteria for disasters to be entered, of one of the following: ten or more fatalities, 100 or more people affected, a declaration of a state of emergency, or a call for international assistance. That leaves out many small and medium disasters that once accumulated, as it has been recently proved (UNISDR 2009a), represent a significant portion of the losses. As such, they should be taken into account by DRR practitioners.

An important related initiative is the GLIDE unique disaster identifier system (www.glide number.net). This initiative is building a fully searchable system by which information about disasters can be homogenised and integrated from multiple sources via a unique identifier. While the database does not contain indicators of damage and losses, it is an important point of reference as the number of sites using GLIDE identifiers grows continuously. Many GLIDE records contain usable links to pages which offer further information about the disaster.

Regional losses

There are few regional initiatives to collect disaster data. Several of them have been previously mentioned in this chapter, as they are part of wider initiatives, such as OSADI, DISCNet and Pacific Disaster Net.

Country losses

Efforts produced by a number of governments and independent groups from the academic sector, the UN and mostly Latin American non-governmental organisations (NGOs) have made important advances in the compilation of national disaster databases in both Asia and the Americas. More recently, Africa has been involved. These databases enable the exploration of loss patterns at the sub-national level.

Currently, there are as many as fifty national country databases identified, including countries from all continents and economic development statuses. Databases for the USA, Canada, Australia and many other developed countries are available. LA RED in Latin America and the UN in Asia and Africa have successfully applied the 'DesInventar' methodology and software (www. desinventar.net and www.desinventar.org) to build about forty national disaster databases. A number of other countries have also made independent efforts to build their own databases, such as Indonesia, Sri Lanka, Mozambique, the Philippines, Thailand, Vietnam, Papua New Guinea and Ethiopia.

These databases are normally collected with a national level of observation and a sub-national resolution, usually at a municipality level or equivalent (one level below provincial/state level). That results in much more detailed, disaggregated and less under-registered datasets. The owners of most of these databases typically do not impose any threshold for disasters to be included, or else impose lightweight ones which filter out a few small- and medium-scale disasters.

The size of some of these databases indicates at first glance their comprehensiveness. The Sheldus database for the USA (webra.cas.sc.edu/hvri/products/sheldus.aspx) contains more than 600,000 records of disasters of all scales. The EM-DAT (2010) database contains 705 records for the USA. In Colombia, the DesInventar database contains over 25,000 records, while EM-DAT contains only 146 for Colombia. Sri Lanka's DesInventar database contains over 10,000 records only on disasters involving natural hazards, as opposed to 73 in EM-DAT.

A study comparing these databases found some interesting results (IASC 2002). Despite the under-registration of disasters, the numbers of total mortality did not differ much between both types of datasets, because EM-DAT (and of course the country databases) contains records for the majority of large-scale disasters. One of the conclusions reached in UNISDR (2009a) was that, while mortality seems to be concentrated in large-scale disasters, a high proportion of the damage to economies and livelihoods is caused by a much higher number of small and medium disasters that, when added to the large ones, give a much more realistic view of the impact of disasters on society.

Another important aspect of national databases is a higher number of indicators collected. UNISDR (2009a), for example, used losses in different sectors (e.g. housing, agriculture, education) in both the main report and in many country case studies where the link between poverty and disasters was addressed. Among other facts, the study found evidence to show that communities in poor areas lose a far higher proportion of their assets, confirming that they have far higher levels of vulnerability, thus increasing inequality. Disaster impacts not only lead to reductions in livelihood assets, income or consumption, but can also negatively affect other aspects of human development, such as nutrition, education, health or the gender gap. Most of these national disaster databases contain extremely comprehensive indicators on damage to the agriculture, industry, education, water, communications, health and other sectors, for which there is no point of comparison against global datasets.

Exposure data

One of the major challenges faced by those working on disaster risk is to obtain an accurate picture of elements at risk. When looking into mortality risk, one could think immediately of demographics; many other elements are exposed to hazards: from infrastructure assets like homes, utility lines, roads and bridges, to livelihood assets such as crops, cattle, vehicles and any other material goods that may be lost in the case of disasters. In general terms, the first options for data sources are census offices and other national agencies (such as cadastral or agricultural agencies or databases).

However, a key question, extremely hard to tackle when doing DRR work, is the spatial distribution of these elements at risk. Most census and national data sources will present aggregates at certain resolution, mostly using administrative units.

Hazards and risk are not evenly distributed over space. The majority of hazard information, including its spatial distribution, is defined and calculated without any relation to administrative units. Consequently, a great difficulty comes when trying to match the spatial distribution of the hazard and the corresponding distribution of exposed elements. Several important efforts have been made to address this problem. Probably the most notable is the Global Population Database from Oak Ridge National Laboratory (ORNL) LandScan, which calculates population density for the year 2007. LandScan Global Population Database was created as part of ORNL's Global Population Project for estimating ambient populations at risk.

In LandScan, the population is calculated for each cell of a grid that can be as small as 30 arc-seconds, and is then divided by the area of the cell. Postings express population density in inhabitants per square kilom. Grids are also available for 2 arc-minute maps, 5 arc-minute maps and 20 arc-minute maps. This database is available at TerraViva! GeoServer: geoserver.isciences. com:8080/geonetwork/srv/en/metadata.show?id=237& currTab=simple.

With an initial distribution of population, many other element distributions can be estimated as a function of population (for example, housing). Nonetheless, finding the distribution of environmental, agricultural and infrastructural assets remains a challenge and is normally subject to the availability of local information and the funding of often expensive remote sensing data.

As mentioned above, the GEM project is working on building an exposure dataset which will, it is hoped, address part of such gaps.

The big hole: vulnerability data

There is no bigger challenge in the process of DRR than addressing the vulnerability component in the equation of risk. Just defining 'vulnerability' is a challenge. At a minimum, vulnerability must contain physical, economical and social factors (UNISDR 2009a), although some authors have identified up to eleven or more types of vulnerability (Wilches-Chaux 1993).

In the same way that the study of hazards has dominated knowledge generation about disasters, the study of physical vulnerability has dominated the much less formal study of disaster vulnerability. Many formulas, methods and measures of physical vulnerability can be seen in the literature, for example for earthquake hazards (Bommer *et al.* 2002) and other hazards such as landslides (Glade 2003), winds, etc.

As examples of the lack of agreement in the DRR community about what is 'vulnerability' and how it can be quantitatively estimated or measured, one can review the most current efforts on global and regional risk assessments. A reflection of the fuzziness of these concepts and methods is that there are few data available that could be seen as a 'vulnerability index' for generic DRR studies. That is despite the large number of studies and risk assessments aiming to calculate a vulnerability factor or index that can be used in a practical way to assess risk – especially in a quantitative manner.

There are several similar efforts in related areas such as the environment or economics. One is the Environmental Vulnerability Index (EVI), 'a vulnerability index for the natural environment, the basis of all human welfare, developed by the South Pacific Applied Geoscience Commission (SOPAC), the United Nations Environment Programme (UNEP) and their partners'. All indices and supporting data can be found at www.vulnerabilityindex.net.

Several characteristics of the EVI reflect the complexity of the topic of vulnerability. One is the strict focus of the index on the environment. It does not address vulnerability of other elements, including the population itself. The EVI description states that the index should be used with other social, physical and economic vulnerability indices in order to provide insights into the processes that can negatively influence the sustainable development of countries. The resolution of the index is at country level, which does not permit sub-national applications. That is a common problem in the few other data sources and case studies with data, such as the Commonwealth study on Vulnerability of Small States (Pelling and Uitto 2001).

At UN climate change talks in Bonn, it was felt that a need exists to devise a climate change vulnerability index (IRIN 2009). With the big money expected to come from the Adaptation Fund, this index would be used as a method to prioritise funding.

The answer for many of these questions may lie in the correct use of disaster data to generate vulnerability functions, rather than indices, based on empirical evidence and statistical processing. However, this may pose challenges when trying to determine and correlate the magnitude of events, given vulnerability's dynamic nature, which may be affected by the occurrence of disasters and which evolves with development processes (UNISDR 2009a). One of the most recent efforts, the already mentioned CAPRA project, is developing a series of vulnerability functions for several common risks – and types of elements at risk – in its area of study, in which damage statistics from previous events are one of the main components of the functions.

The two previously mentioned global risk assessments, the Hotspots study and the GAR/ PREVIEW study, have used disaster data (mostly EM-DAT) to calculate empirical vulnerability indices, called 'historical loss rates' in Hotspots. In addition to past disaster data and vulnerability functions with pre-calculated parameters, such as those published by CAPRA and others, many methods have been tried to model the social component of the vulnerability. These models involve statistical regressions with developmental and economic indicators such as GDP, basic unsatisfied needs and literacy indices (Peduzzi et al. 2009).

Hazard data (see also Chapter 17)

Earthquakes and other seismological data

Earthquake data exists in most countries of the world, usually gathered by local seismological institutions. Several global institutions provide free comprehensive information about earthquakes and associated geological data. The US Geological Survey (USGS, www.usgs.gov) provides worldwide data about earthquakes, landslides and volcanoes, in addition to many other useful resources on geography and geospatial information.

The Advanced National Seismic System (ANSS, www.ncedc.org/anss/catalog-search.html) composite catalogue is a worldwide earthquake catalogue created by merging local catalogues from contributing institutions. The catalogue currently contains earthquake hypocentres, origin times and magnitudes, and is searchable and downloadable by specifying a geographic window.

Another source of seismic information is the Global Seismic Hazard Assessment Program (www.seismo.ethz.ch/gshap), but unfortunately it ended in 1999. It still contains a useful downloadable earthquake hazard map as digital grid accessible. Detailed information for geological faults, peak ground acceleration and velocity and other geologic topics must usually be gathered locally.

Climate and weather data

Meteorological data are widely available from several sources, although they are not always free. The best sources of this type of data are the national meteorological services, which offer climate and weather data (such as temperature, precipitation, winds and atmospheric pressure) with different degrees of availability and level of detail. Airports can also be a solid source of long-term weather data, if records are kept.

The World Meteorological Organization (WMO) is probably the best starting point, as it offers links to existing national meteorological and hydrological services and in many cases can provide data directly from its website, www.wmo.int.

187

Floods

The Dartmouth Flood Observatory (www.dartmouth.edu/~floods) offers to the public domain a 'Global Active Archive of Large Flood Events'. The information presented in this archive is derived from a wide variety of news, governmental, instrumental and remote-sensing sources, documenting flood events from 1985 to the present.

Another possible source of flood information is PREVIEW, mentioned earlier, which contains a GIS model generated using a statistical estimation of peak-flow magnitude and a hydrological model derived from the HydroSHEDS dataset and the Manning equation to estimate river stage for the calculated discharge value. Interested users must contact PREVIEW to obtain the data.

Droughts

The Weighted Anomaly of Standardised Precipitation (WASP) was developed by the International Research Institute for Climate and Society at Columbia University in New York. It was computed on a 2.5° × 2.5° grid from monthly average precipitation data for 1980–2000. The WASP assesses the degree of precipitation deficit or surplus over a specified number of months, weighted by the magnitude of the seasonal cyclic variation in precipitation. Website at iridl.ldeo. columbia.edu/maproom/.Global/.Precipitation/WASP_Indices.html.

Landslides

The International Centre for Geohazards hosted by the Norwegian Geotechnical Institute in Oslo, working with UNEP GRID-Geneva, has developed a global landslide and snow avalanche hazard map that has been used for global analysis of these hazards. The map is based on a range of data including slope, soil and soil moisture conditions, precipitation, seismicity and temperature (Nadim *et al.* 2004). This index takes advantage of more detailed elevation data that recently became available from the Shuttle Radar Topographic Mission (SRTM) at 30-second resolution, compiled and corrected by Isciences, LLC (www.isciences.com).

Tsunami

The National Geophysical Data Center (NGDC) at the National Oceanic and Atmospheric Administration (NOAA) maintains a publicly accessible global Tsunami Database containing over 2,400 records of events dating from 2000 BCE to the present. The database is searchable by several criteria and provides indicators about the hazard (both earthquake and tsunami parameters such as magnitude, location and maximum water height), as well as the impact (mainly mortality, but several records contain damage indicators for the housing sector and estimations of damages in US dollars). Website at www.ngdc.noaa.gov/hazard/tsu.shtml.

Volcanoes

A useful information source for this hazard comes from the Smithsonian's Volcanoes of the World database as part of the Global Volcanism Program (www.volcano.si.edu/world). Volcanic activity from 79 CE–2000 CE was developed by UNEP/GRID-Geneva based on the Worldwide Volcano Database, and is available at the National Geophysical Data Center (NGDC, www. ngdc.noaa.gov/hazard/volcano.shtml).

The continuing search for useful new data

There are many data and information sources specific to risk and disasters. Only some have been presented here. Other areas of DRR data management, in particular those related to activities like community-based DRR, participatory action research and risk assessments, knowledge management and many others, have not been mentioned.

There is now a growing recognition that participatory research can generate numbers that are reliable, valid and representative of the communities involved (see Chapter 64). There is also awareness that data generated with participatory methods can be shown to be comparable and compatible with datasets produced elsewhere with the same methods, if careful standardisation takes place (Chambers 2007b). However, probably due to the highly localised nature of these types of studies, there is little available data. Given the relatively recent appearance of partici-patory numbers, there are still no recognised standards on applications to risk, despite the growing stream of literature in the form of case studies and theory.

Additionally, many other data sources exist for generic data management, in areas such as geography, cartography, environmental studies and economics. Many other areas are required for the complete understanding of risk, along with the processes of reducing and managing it. The many important advances in data generation, data standardisation, exchange tools, public domain availability and costs of spatial information could depict a roadmap for those facing similar challenges for the DRR practice.

Many barriers to a much wider dissemination of information about risk still exist, despite the tremendous advantage of the internet and other knowledge networks in which astonishing amounts of data are available. One barrier is the political and security issues of this information

---

Box 16.2 Public access to disaster and risk information

A couple of years ago, an extremely rigorous risk assessment was conducted in the state of Gujarat, India. Results were presented to the state parliament and, as a result, the entire study was declared 'top secret' due to security concerns and the recent confrontations with Pakistan. Unfortunately, not even a reference to the title or authors can be presented here, as it would pose a legal problem for them. The point is that the main recipient of this knowledge, the people from the state, will never receive it.

Several of the teams participating in the DesInventar project in Asia and the Middle East repeatedly found obstacles to obtaining and publishing information about disaster losses, ranging from public financial liability to political distrust to concerns about state security. Some of the extremely useful databases collected remain private and cannot be shared with the general public.

Stating that knowledge about trends and patterns of risk, or about past disaster losses, are essential needs for those involved in risk reduction and should be in the public domain may sound good in academic circles; however, politicians and govern-ments may see this knowledge rather as a liability in terms of their own inabilities, their mistakes in planning or their lack of actions to reduce risk. The word 'disaster' itself may involve financial responsibility for administrations that must provide relief by law.

Copyright and commercial interests are also in the way of public access to risk and disaster data. An example is the information that is put into the public domain by reinsurance companies, who only allow very discretionary access (or none) to extremely valuable and needed data. Only data that do not hinder their lucrative business will be released to the public.

(see Box 16.2). A few other barriers are the lack of standards in both data content and format, along with the prevalence of commercial interests over public well-being, not just in terms of costs of commercial data but also in terms of accessibility. Legal and copyright issues plus access to expensive software tools and the hardware to run them are other considerations.

Conclusions

More than a century ago, a group of visionaries started collecting weather-related data. The oldest time series of systematic climate data is a temperature record from central England beginning in the seventeenth century, although most instrumental records date back to the nineteenth century. Thanks to the vision and dedication of meteorologists and by means of mathematical analyses and computational resources that were rarely dreamed of then, scientists can analyse long-term climate trends while making available, in the comfort of everyone's homes every morning, accurate weather predictions that also feed early warning systems and help agriculture, navigation, transportation and many other sectors of society.

Wider disaster and risk data are a relatively new commodity, in an age where computing power and electronic storage costs little – for those who have access to the hardware and power systems needed. The DRR community has today an historical responsibility to follow the path laid down by meteorologists to systematically collect disaster and risk-related data, adhering to best practices, standardising it and, most importantly, making it accessible and available so that it fulfils its fundamental goal: to help save lives and reduce losses.

17

Tools for identifying hazards

Keiko Saito

WILLIS RESEARCH FELLOW, DEPARTMENT OF ARCHITECTURE, UNIVERSITY OF CAMBRIDGE, UK,
AND CAMBRIDGE ARCHITECTURAL RESEARCH, CAMBRIDGE, UK

Jane Strachan

WILLIS RESEARCH FELLOW, NATIONAL CENTRE FOR ATMOSPHERIC SCIENCE – CLIMATE DIVISION,
UNIVERSITY OF READING, UK

Timothy Fewtrell

CATASTROPHE RISK ANALYST, WILLIS ANALYTICS, WILLIS, LONDON, UK

Nick Rosser

INSTITUTE OF HAZARD, RISK AND RESILIENCE, DURHAM UNIVERSITY, UK

Susanna Jenkins

CAMBRIDGE ARCHITECTURAL RESEARCH, CAMBRIDGE, UK

Aidan Slingsby

WILLIS RESEARCH FELLOW, DEPARTMENT OF INFORMATION SCIENCE, CITY UNIVERSITY LONDON, UK

Katharine Haynes

RISK FRONTIERS, MACQUARIE UNIVERSITY, SYDNEY, AUSTRALIA

Introduction

This chapter uses illustrative examples to discuss critically some of the tools available for hazard identification at different space and time scales. This chapter complements Chapter 16, which describes data sources used for policy analysis by discussing how data, identification tools and presentation of data can be used directly for disaster risk reduction. The process of identifying hazards is an important part of disaster risk reduction processes, contributing to understanding the scales and specific places of the risks faced, assisting in characterising disaster risk reduction strategies.

Tools for hazard identification

Earthquakes

For background on earthquakes, see Chapter 26.

Event identification and short-term monitoring

Earthquake events are characterised by the amount of energy released (magnitude, also indicated by the maximum recorded motion) and the location, recorded by seismograph networks installed throughout the world. The earliest known seismograph was invented by Zhang Heng in 132 CE to serve the Chinese emperors, and it is said to have identified an earthquake six years later. In the mid-nineteenth century, Italian Luigi Palmieri invented a mercury seismometer that could record the time, intensity and duration of an earthquake.

Today, earthquake-prone countries such as Japan, Taiwan and the USA have dense networks of seismographs that uniformly cover much of the country in order to provide comprehensive monitoring for earthquake notification and emergency response. Seismic recordings from these stations are transmitted to operation facilities in real time and are generally made accessible to the general public. The individual seismograph stations, which record the ground shaking in the form of ground motion velocity or acceleration, provide basic information regarding the onset of shaking at the station and on-scale recordings of the ground velocity or acceleration that are used to produce location estimates for the epicentre, along with depth and magnitude. Data from at least three ground motion stations are required to provide an accurate location.

The United States Geological Survey (USGS) has been producing ShakeMaps (USGS 2010c) since 1997 wherein 'near real-time maps of ground motion and shaking intensity following significant earthquakes' are provided. These cover all earthquakes around the world above magnitude 5.5 and the information is distributed freely to subscribers within two hours, as well as published on the ShakeMap website. For earthquakes globally, the accuracy of the location of the epicentre as estimated in the first few hours can be off by up to 50 km (Saito 2009), when compared to the epicentre location estimated by the seismological institute monitoring in the local area. This is from differences between global and local velocity models. It is rare that locations are off by 50 km; typical global locations are within 10 km–20 km of the local network location.

However, many areas lack ground motion stations. In many cases, USGS estimates are produced much faster and in a more consistent manner, making ShakeMap attractive to emergency managers around the world. Similar systems have been produced by other organisations, such as the European-Mediterranean Seismological Centre (www.emsc-csem.org/index.php?page=home).

Short- to medium-term forecasting

Some early warning systems currently used in Japan and run by the Japan Meteorological Agency can issue warnings as soon as the primary (P) wave is detected. The early warning system takes advantage of the time lag of seconds between the arrival of the primary wave and secondary (S) wave at a location that depends on the distance from the epicentre. Although the time lag is not long enough to evacuate people, since the S wave causes the most destruction to the buildings and infrastructure (Coburn and Spence 2002), these early warning systems are used to automatically shut down critical infrastructure, such as nuclear power stations, public transport and elevators, thus reducing damage.

Similar systems are used in California, Taiwan and Mexico. However, in some cases the early warning does not reach the end user in time because the location is adjacent to the epicentre, so the S wave arrives almost at the same time as the P wave.

Long-term identification

Mapping of active faults has been ongoing globally to identify potential earthquake hazards. The ground's permanent movement is being monitored by a network of permanent GPS stations installed in some earthquake-prone countries, such as the USA, Japan and the European Alps region (Leonard 2004). At the monitoring stations, Kinematic GPS sensors are used, which measure movement in the range of millims. The drawback is that the cost of installing and maintaining such a permanent network is high. The implementation of one GPS station can cost between US$40,000 and US$70,000, and for maintenance a further US$9,000 per site annually is required, which is usually too expensive for less affluent countries.

Interferometric synthetic aperture radar (InSAR) techniques are also being used to measure the permanent displacement of the ground's surface. InSAR requires pre- and post-event imagery taken from the same orbit. Hence, depending on the orbit, there are cases where weeks are required after the event to acquire post-event images taken from exactly the same orbit as the pre-event image, in order to carry out the InSAR analysis. Radarsat, ERS-1 and 2, JERS-1, ENVISAT, ALOS-PALSAR, Cosmo-skymed and TerraSARX are some of the space-based sensors that can be used for InSAR analysis, and it can cost between approximately €500 and €7,000 per scene, depending on the wavelength and scene size.

Landslides

For background on landslides and other mass movements, see Chapter 25.

Historically, landslide-prone areas were identified through descriptions of previous events or via direct observation. When the field of geomorphology started in the late nineteenth century, interpretation of previous or future possible events was feasible through soil and slope analysis.

Pre-event identification

Pre-event identification is commonly based upon a deterministic assessment of past behaviour in any given area or region, aided by studying landslide deposits and scars that have been preserved in the landscape. The subsequent analysis of inventories of such events has been fundamental in defining future susceptibility to landslide occurrence and nature, a comprehensive review of which is found in Guzzetti (2006). Factors including seismicity, lithology, topography and land use are commonly found to be key variables in such analyses. This scale of assessment is normally undertaken at a regional administrative level to guide investment in mitigation.

On the macro scale, this type of regional susceptibility assessment can be married with constant monitoring of those conditions which in general lead to slope failure, including the potential for seismicity, from, for example, the USGS ShakeMap, or precipitation resulting from meso-scale climatic systems. Such monitoring systems include recent developments such as the

Tropical Rainfall Measuring Mission (TRMM), which uses satellites to measure precipitation in near real-time. At the local scale monitoring systems tend to be deployed after initial signs of movement or precursors to landslides have been identified. Two dominant approaches to in-situ monitoring of landslide behaviour are available: stress and strain. The former is well suited to assessment of general stability and likelihood of failure, whereas the latter is more appropriate for precise temporal prediction of failure and early warning.

Event identification and short-term monitoring

The majority of monitoring tools deployed assess the stress changes in a slope, which decrease slope stability. Examples include measuring pore-water pressure or precipitation, or peak ground acceleration during an earthquake, the level of which is then assessed using statistical approaches to develop exceedance thresholds after which failure is deemed probable. Such analysis varies in complexity, but can range from deterministic rainfall thresholds based upon data from similar sites, to simple infinite slope models, or to highly complex three-dimensional numerical simulations.

The local specificity of such approaches limits the value of transferring thresholds from one site condition to another. This challenge is compounded by the difficulty of collecting appropriate or widely representative monitoring data. Instruments can commonly cost many thousands of dollars per slope, only measure at single points, and rarely take readings actually at the failure surface which is the zone that dictates the deformation of the slope and behaviour of the ultimate failure. The levels of investment, maintenance and technical expertise, therefore, commonly restrict such approaches to high-value assets in developed countries. Often, the first attribute of landslide behaviour to be lost as a result of these compromises is an understanding of event timing, rendering such approaches limited in application as directly predictive tools.

Several tragic examples have been documented whereby the logical assumption that rainfall leads to failures, the type of assumption upon which these methods rely, has been well understood, as was the case at Guinsaugon, Leyte in the Philippines in February 2006. Here the potential for landslide was well known, resulting in the evacuation of the local population after five days of intense rainfall, only for them to return five days prior to the final catastrophic failure of the slope above, resulting in over 1,100 fatalities (Evans *et al.* 2007). This case demonstrates the limitation of often arbitrarily defined thresholds, and the presently limited understanding of the slope failure mechanisms in landslides.

Long-term identification

The second type of approach to assessing slope hazard is strain-based monitoring. This approach is ideally suited for temporal prediction and early warning. Such approaches are based upon the premise that a slope undergoes deformation and accumulates strain prior to failure, building upon the progressive failure model of Bjerrum (1967). Monitoring strain rate, which is used as the surface expression of subsurface deformation, researchers have identified consistent behaviour between different slopes, notably in the period immediately prior to final failure, termed the 'tertiary creep' phase. In this period of movement the deformation of the slope is controlled entirely by the progressive development of the landslide shear surface, the material characteristics of which define the bulk slope behaviour. Plotting inverse velocity against time yields a linear relationship in brittle materials and an asymptotic relationship in ductile or reactivated failures.

In brittle slopes, the extrapolation of this trend allows for a prediction of the timing of failure. As such, monitoring strain rate through time is a valuable tool for predicting exactly when a slope will fail. Notable examples include the retrospective analysis of movement data from the catastrophic failure of the slopes behind the Vaiont Dam in Italy in 1963, and observations from a range of cut slopes and embankments since (Skempton 1966). Such approaches are increasingly used as failure warning systems where slope failures threaten built infrastructure.

Measurement of the slope surface strain field for hazard identification is available via a range of approaches. Point-based measurements of displacement (vibrating wire extensometers with sub-millimetre precision), deformations (tilt meters; dGPS) or deeper-seated deformation (micro-seismics; acoustic emissions) are widely available, but are subject to the compromise between resolution, extent and precision balanced against the potential hazard and available resources. Costs for such approaches can be thousands of dollars per site. Unlike the locally specific stress-based approaches, the general behaviour identified by strain-based methods also allows relatively low technology methods for monitoring (e.g. repeat distance measurements across a simple stake network), and also have a relatively simple requirement for interpretation, permitting a more local interpretation in areas where available resource is more limited.

At a broader scale, recent technological developments have also become available which are now capable of obtaining the precision of point-based instruments from image-based or sensor networks which capture data from across the whole slope surface. Critically, these more recent approaches overcome the difficulties in locating instruments in 'representative' locations, limit the need to pre-define the limits of the area of interest and allow remote data capture from areas that are commonly difficult to access or dangerous. Such approaches include ground-based sensors, such as terrestrial laser scanning, which offers decimetre precision topographic images over ranges on the order of 500 m to 1000 m; ground-based interferometric radar, which resolves millimetre-level deformation at 1 km–2 km range at around 10 m resolution; airborne sensors such as repeat LiDAR or photogrammetry, which provide the ability to rapidly map swaths at high resolution (1–50 points/pixels per square metre at decimetre precision); and space-borne platforms such as interferometric SAR, which can again resolve deformation at millimetre precision over wide areas. Such techniques still remain prohibitively expensive to apply in all but the most high-risk sites, and often the quality of the data remains significantly greater than the fundamental understanding of the mechanics and predictability of failure.

The integration of such wide-area approaches with high-precision and short-frequency repeat measurement allows monitoring to become more predictive and to have the resolution of relevance to individual slopes with known instability problems.

Volcanoes

For background on volcanoes, see Chapter 28.

Identification of potential volcanic hazards used to be based solely on historical accounts of eruptions, passed down over many generations in the form of myths, legends and songs. Plato, circa 360 BCE, recounted the destruction of Atlantis 'in a single day and night of misfortune'. Māori mythology in New Zealand records a mighty battle between the volcanoes of Taranaki and Tongariro. Such historical accounts prove invaluable in aiding modern-day reconstructions of past hazards and form an important part of volcano knowledge, although care must be taken in interpreting them literally. Hazard identification today is commonly undertaken during two distinct volcanic regimes: (1) quiescence, and (2) precursory activity or the eruption itself.

During quiescent periods

Assessing past volcanic behaviour and using ongoing monitoring data to establish current volcanic behaviour offers insights into the likely nature of future precursory or eruptive activity. Key parameters of past volcanic behaviour to identify include the typical style and frequency of past behaviour and any patterns; the extent and magnitude of previous processes; and sectors around the volcano that are commonly impacted.

Long-term monitoring of volcanic behaviour began in earnest with the founding of the Vesuvius volcano observatory in Italy in 1841 to observe the volcano visually during a particularly active period in the volcano's history. Today, few active or potentially active volcanoes are regularly monitored, with significantly larger resources typically available for developed countries, e.g. the United States, than in low-income countries, e.g. Papua New Guinea. Long-term monitoring is costly. The Alaska Volcano Observatory, considered vital to aviation safety, receives total funding of approximately US$5 million a year (down from US$8 million in recent years) to monitor more than fifty historically active volcanoes.

The cost-effectiveness of such ongoing monitoring has been well proven during recent eruption crises. For example, the 1991 Pinatubo eruption in the Philippines incurred monitoring costs of around US$56 million in the lead-up to the eruption. Successful forecasting of the paroxysmal explosion was estimated to have saved 5,000 lives and prevented property losses of more than US$250 million (Newhall *et al.* 1997). Yet hazard identification, through monitoring and interpretation of geological records, started only after the volcano's first steam explosion, around two months before the most destructive phase.

Detailed, ground-based monitoring networks are rarely installed and operated on volcanoes that show no precursory signs of activity. In the late 1980s the USGS developed portable ground-based monitoring instruments that could be rapidly deployed to volcanoes exhibiting precursory or eruptive activity (see Table 17.1). In recent decades, the development of sophisticated remote sensing techniques for radar, aircraft or satellites has been used to supplement, and at some unmonitored volcanoes to monitor in the absence of, more traditional ground-based volcano monitoring techniques.

During precursory or eruptive activity

For unmonitored volcanoes, the first indication of a potential eruption may be the eruption itself. The first eruption in more than 9,000 years of Chaitén, an unmonitored volcano in Chile, in 2008 was initially thought to be from the neighbouring volcano Minchinmávida, which last erupted in 1835 (Smithsonian Institution 2008). For remote volcanoes, an eruption may only be detected by passing pilots or satellites. Like all monitoring techniques, precursory behaviour at a volcano, such as increased thermal output or ground deformation, can only be recognised when baseline, i.e. ongoing monitoring, data are available.

This may prove a problem for the long-term use of satellite data due to the large costs associated with data collection and the rapid interpretation of large quantities of data. The development of automatic data analysis and new satellites may help to reduce the costs and delays associated with monitoring eruptions in real-time.

Many types of monitoring equipment and methods are required in combination to gather baseline data, to detect precursory signals of an impending eruption and to monitor changes within the eruption itself. Monitoring tools can identify eruption activities such as explosions, pyroclastic density currents and rockfalls that are often obscured by clouds, steam or ash. Ideally, volcano monitoring, and therefore reliable hazard identification, does not rely upon any one tool, rather using a combination of geophysical, geochemical and geodetic techniques (see Table 17.1).

Over the past few decades, volcano hazard identification has evolved from field observations such as seismicity and thermal and gas outputs, to include sophisticated modelling and

Table 17.1 A synthesis of examples of monitoring tools used to detect and analyse major changes in a volcanic system

Signal	Monitoring tool	Description
Seismic activity	Seismometers	Seismometers record the types of and parameters from earthquakes. Seismometers require low background noise, such as from wind, waves, traffic and animal herds. A network of at least three stations is needed to pinpoint an earthquake's location.
Gas and ash emissions	Satellite	Satellite-based ozone mapping instruments, such as the Ozone Monitoring Instrument and the older Total Ozone Mapping Spectrometer, can measure SO_2 flux and ash emission from volcanoes in near real-time.
	Ground-based remote sensing	Ultraviolet spectroscopy can be used on the ground or attached to a moving object, such as a car or airplane, to measure SO_2 emission rates and concentrations in the plume. Infrared spectroscopy can be used to measure other volcanic gases, including CO_2, CO, HCl, HF and H_2O, as well as SO_2. Large releases of volcanic gases into the atmosphere are better measured by satellite-based rather than ground-based techniques.
	Direct sampling	Direct collection of gas is often conducted with later analysis in the laboratory. This allows detection of very low gas concentrations, but can be difficult and/or dangerous.
	Direct observation	Increased degassing through the central vent or fumaroles can be directly observed and recorded, weather permitting.
Visible signal	Satellite visible/ near infrared (VNIR) sensors	Optical remote sensing may be used, with clear weather, to supplement infrared measurements and confirm or identify volcanic changes such as dome growth, degassing, vent location and flow presence and extent. The sensors cannot penetrate cloud cover, but can be used to identify volcanic plumes and ash dispersal with 15 m resolution.
	Cameras and direct observation	Continuous or regular monitoring by cameras allows visual record of eruption precursors and eruptions, depending on clear weather and a good view of the volcano.

Source: After Scarpa and Tilling (1996), McNutt, Rymer and Stix (2000) and Francis and Oppenheimer (2004).
Note: Additional signals include ground deformation, thermal output, noise, electromagnetic disturbances and gravitation field change.

laboratory experimentation. The more recent development of digital technologies such as satellites allows automated analysis, data transmission and higher data resolution from remote locations. With increasing amounts of data, equipment and analysis comes much higher costs, both for initial purchase of the equipment and ongoing maintenance and data analysis.

Tropical cyclones

For background on tropical cyclones see Chapter 19.

Tropical cyclones are low-pressure systems that occur over warm, tropical or subtropical waters, with an intense convective activity, bringing high winds circulating around a relatively calm 'eye' and heavy rains. Tropical cyclones can also lead to storm surges. The WMO has designated Regional Specialised Meteorological Centers (RSMCs), which have the responsibility for monitoring and providing reliable forecasts of tropical cyclone tracks and intensities in tropical cyclone-prone regions, along with quantitative forecasts of storm surges and associated flooding (www.wmo.int/pages/prog/www/tcp/RSMC-TCWC.html). The RSMCs are often the first source of tropical cyclone information for local authorities.

Event identification and short-term monitoring

Tropical storm forecasting and identification has undergone huge changes during the twentieth century. Before the twentieth century, most forecasts were done by direct observation from weather stations. For example, during the 1870s Benito Vines introduced a forecast and warning system based on direct observation of cloud cover changes over Havana. However, direct observations were only available if the storm passed over land or coastal areas, which is often too late to prepare for the damaging winds and rainfall.

When radio technology was introduced, information from ships sailing in tropical waters could also be integrated into forecasts. During the 1930s and 1940s aircraft reconnaissance by the military was used to gain information about active tropical storms. Dropsondes were launched into storms to measure vertical profiles of temperature, humidity, pressure and winds. Identification of tropical cyclones improved as commerce and aircraft activity increased following WWII. Launch of the TIROS-1 weather satellite in 1960 led to a dramatic improvement in tropical cyclone identification.

Remote sensing, via the use of satellite data, in the form of visible and infrared data, has provided invaluable information used to detect and track potentially damaging tropical storms. Satellite programmes used to gain important tropical cyclone information include QuikSCAT (Quick Scatterometer), launched in 1999, with an initial budget of US$93 million. Using microwave radar, it provides high temporal and spatial (25 km) resolution wind speed and direction data. The Tropical Rainfall Measuring Mission (TRMM) was launched in 1997, with a total project cost to date of over US$750 million, using instruments such as the Precipitation Radar (PR) and the Microwave Imager (TMI) to monitor storm development, provide three-dimensional maps of storm structure and provide detailed information about the intensity and distribution of rainfall.

Current storm identification and tracking utilises analysis of surface and upper-air observations obtained from a combination of satellite, in situ and reconnaissance aircraft data, to provide storm position and intensity updates every six hours. Closer to land, land-based and airborne Doppler radar are employed to record changes in storms' location and intensity every few minutes.

Short- to medium-term forecasting

In terms of forecasting the movement and intensity of an identified tropical cyclone following detection, several 'guidance models' are currently employed by forecast centres (see Table 17.2), to provide forecast projections every six hours for 12, 24, 36, 48, 72, 96 and up to 120 hours into the future.

As an example of forecast improvement, errors for forecasts of North Atlantic storms track position, as recorded by the US National Hurricane Center, along with the variability of these

Table 17.2 Models used for short- to medium-term tropical cyclone forecasting

Model type	Details	Limitations
Extrapolation/ persistence	Based on speed, direction and intensity trends of current tropical cyclone.	Only has short-term applicability (often used as initial guidance for 12- to 24-hour forecasts).
Climatology and analogue	Utilising the average of behaviour of storms from historical storm records. Often restricted to a limited subset to reflect synoptic conditions.	Assumes a stationary climate system. Problems occur over data-sparse regions.
Statistical	Statistical regression based on relationships between historical storm behaviour and other parameters.	Only a limited period of reliable historical data available.
Dynamic	Based on physical equations governing the motion of the atmosphere, either globally, or over a limited area, which allows higher resolution. Storm vortices are tracked through the model output or output data is used to determine steering information.	Computationally demanding and expensive. Sensitive to inaccurate initialisation.
Statistical–dynamic hybrid	Statistical models that use both historical data and dynamic model output.	Same data limitations as the statistical model.
Statistical–synoptic hybrid	Takes into account current and forecasted synoptic storm information.	Same data limitations as the statistical model.
Empirical (e.g. Dvorak Technique, Dvorak 1974)	Pattern-recognition system based on visible and infrared cloud imagery. Used to subjectively analyse and forecast storm intensity in terms of maximum winds and minimum central pressure.	Relies heavily on forecaster judgement and experience.

errors, have been steadily decreasing over the last few decades. Forecast errors obviously increase with forecast lead time, but the gradual decrease in forecast errors has meant that current 72-hour forecasts have approximately the same error as 24-hour forecasts from the early 1970s: an annual average track position error of approximately 200 km. The 24-hour forecast now has an annual average track position error of approximately 90 km.

Seasonal forecasting

Seasonal forecasts of expected tropical storm activity for the upcoming tropical cyclone season have been issued for the last two decades, but are becoming more sophisticated with the development of seasonal prediction using ensemble forecast runs of dynamic global models (e.g. the UK Meteorological Office's GloSea model). Seasonal forecasting of tropical cyclone activity was pioneered by Professor Bill Gray from Colorado State University. Professor Gray's technique, developed for the North Atlantic Basin, is based on the fact that the global atmosphere and oceans have stored memory buried within them that can provide clues as to how active the upcoming season is likely to be. A combination of precursor signals, such as the El Niño Southern Oscillation (ENSO), sea surface temperatures, West African rainfall and the Quasi-Biennial Oscillation (QBO), is then used to form the prediction.

Seasonal forecasts will often include estimates of the total number of storms occurring in a season, expected number of higher category storms and a forecast of the Accumulated Cyclone Energy (ACE) Index. The ACE Index is often used as an indication of seasonal activity, taking into account the total number of storms over a season alongside the individual storm lifetimes and intensities.

Long-term monitoring

The Intergovernmental Panel on Climate Change (IPCC) has identified the need to quantify the variability in global cyclones, and in particular to characterise changes in cyclone tracks with anthropogenic climate change (see Chapter 18). An observed increase in the number of hurricanes in the Atlantic, for example, has been claimed to be associated with both human-induced global warming and natural climate variability. The most likely cause is a superposition of both factors.

High-resolution dynamic models of the global climate system now resolve sufficient detail to be able to simulate extreme weather events in a global climate context. Such model simulations are being used to investigate the relationship between modes of natural climate variability and tropical cyclones. This would not be possible by using historical observations alone due to the limited observational records available, which do not cover multiple periods of long-term modes of climate variability necessary for such studies.

Additionally, as a means of extending the limited historical storm record, a new discipline of palaeotempestology has emerged. It analyses palaeoclimate proxy data, such as sediment layers deposited behind barrier islands, coral chemistry and tree-ring patterns growing in coastal areas. Historical documents including ships' logs and newspaper accounts have also been used as a data source.

River floods

For background on river floods, see Chapter 21.

The information required for satisfactory flood identification includes the timing of flood water arrival and peak water levels, extent of inundation, instantaneous depth and velocity of flood water and the duration of flooding (Morris 2000). Historically, flood plains tended to be known through oral or written history describing past events. Local knowledge could often tell if a flood were imminent through examining rainfall patterns and river water behaviour. Current monitoring and forecasting systems tend to provide either 'at a point' temporal data (e.g. flow gauge data) or spatially distributed data that is zero-dimensional in time (e.g. synoptic satellite images).

Event identification and short-term monitoring

The identification, monitoring and forecasting of floods on a river network requires knowledge of the river discharge – the volumetric flow rate passing a given point at any given time. Real-time and short-term monitoring of waterways is conducted using flow gauging stations which provide direct or indirect measurements of the river channel discharge.

In the UK, the Environment Agency (EA) is responsible for maintaining and processing stage and discharge data from the network of gauging stations, which is archived and available to hydrology practitioners upon request. In order to supplement standard river flow data, the EA also operates an online portal specifically designed for evaluation of high flow conditions on river networks including the latest results of data reviews, site information and indicative suitability of data from each gauging station (Environment Agency 2010).

USGS operates a similar service in the USA, but delivers instantaneous stage and discharge data for the previous 120 days while archived discharge data for the entire record of the gauge are accessible through an online portal (USGS 2010d). Both services provide these data at an average of fifteen-minute intervals for the desired period. Such extensive networks are rare outside the more affluent countries, although improvements continue to be made, such as along the Ganges–Brahmaputra system (Hopson and Webster 2009).

Most river systems were generally instrumented for water resource or geomorphic change monitoring purposes, which often leads to poor siting for high-flow measurement purposes (e.g. near hydraulic structures). Additionally, gauging stations are often rendered inoperable during large flood events due to high water levels and communication disruption. Sensor networks placed along a river channel and on floodplains are being investigated to provide real-time monitoring of flood events, although these systems have yet to be implemented for operational purposes.

Short- to medium-term forecasting

In terms of forecasting flood events, most operational systems combine real-time gauge information with a numerical modelling framework that triggers warning levels if prescribed flood stage and/or discharge levels are exceeded. Numerical catchment hydrology models driven by meteorological parameters forecast catchment runoff which forms the boundary condition for a basin-scale or local-scale numerical hydraulic routing model to predict flood stage or discharge at specific locations (De Roo et al. 2003).

In the short term (~0–2 days ahead), forecasting systems use observed meteorological inputs from rain gauge and radar networks. For medium-term forecasting (~2–15 days ahead), numerical weather prediction (NWP) models must be used, especially where upstream river discharge data is unavailable (Cloke and Pappenberger 2009). Operational and research flood forecasting systems, run mainly by government agencies, are increasingly using ensembles of NWPs, rather than deterministic predictions, to drive the flood forecasting system.

Post-event identification and long-term monitoring

The Dartmouth Flood Observatory (Dartmouth 2010) uses satellite-based remote sensing to detect, measure and map river discharge. Specifically for flood analysis, the Observatory produces an archive of floods from 1985 to the present plus rapid-response inundation maps for current flooding episodes. Since the late 1990s, satellite and airborne remotely sensed data have been integrated into numerical modelling studies of flooding (Schumann et al. 2009). A number of research studies have demonstrated the utility of post-event surveys of wrack and water mark data for constraining hydraulic modelling studies (e.g. Neal et al. 2009). One- and two-dimensional hydraulic modelling tools are also used for mapping past flood events.

With respect to long-term monitoring, several satellites are currently operational that provide freely available altimetry measurements of river water levels (e.g. ENVISAT, TOPEX/POSEIDON). In addition, numerous recent and upcoming satellite missions (e.g. SWOT, ALOS-PALSAR, RADARSAT-2, TerraSAR-X, COSMOSkyMed and Sentinel-1) may provide sources for more accurate and timely visualisation of flood events and will increase the wealth of satellite data available for flood risk assessment. As a sample of costs, ALOS-PALSAR, RADARSAT-2, TerraSAR-X and ENVISAT cost roughly £2,000–£3,000 per image for commercial use, but

ENVISAT and ALOS–PALSAR images are free for academic use. Governmental and NGO users might need to negotiate regarding cost or might need to collaborate with academic institutions.

Visualisation of hazard data and its limits

Visualisation tools such as Geographical Information Systems (GIS) enable geographical datasets, including hazards, to be related and maps/images to be produced that convey these relationships. There are also community-based, participatory applications of GIS (see Chapter 64). Their cartographic heritage emphasises the static map for conveying a particular message. More recently, geovisualisation (Dykes *et al.* 2005) and visual analytics (Thomas and Cook 2005) emphasise the use of highly interactive graphics as integral to the process of knowledge discovery, where maps are considered interactive interfaces to data. General-purpose, off-the-shelf tools for this are fewer and are less well developed than in GIS, but several tools, frameworks and demonstrators are available.

A useful distinction can be made between maps and graphics for communicating a message, and maps and graphics for exploratory visual data analysis. Cartographic design is important in both cases. Appropriate design decisions depend on the data, the purpose, the target user and medium, and have implications for interpretation (Brewer 2005). Good cartographic design that produces cognitively plausible (Skupin and Fabrikant 2003) graphics needs to make effective use of visual variables (Slingsby *et al.* 2009). Some are suitable for conveying quantitative data (e.g. position, length, size, lightness), while others are better suited to nominal data (e.g. hue which has no intuitive order).

In cartography, position is used to encode geographical position, usually transformed through a map projection. The choice of map projection depends on the purpose for which the map is designed, and the extent of the region. Where hazard maps are intended primarily for locational and contextual purposes, the shape and layout of recognisable geographical features (e.g. rivers and coastlines) and labels help orientate the users, allowing them to identify places they know. Plotting hazards and their likely effects in the same spatial frame of reference allows these to be geographically related to places and other datasets.

Plotting large numbers of closely spaced, discrete data can result in high levels of occlusion and clutter. Such clutter can be reduced by generalising them into spatial clusters or area summaries and using visual variables such as colour (lightness) or size to represent counts or densities. Cartograms are techniques in which space is transformed such that units' sizes correspond to another characteristic of the area (such as population), magnifying areas where that value is high and reducing the area where it is low.

Hazards and hazard effects are also strongly temporal. A set of multiple small maps at evenly spaced time intervals can be produced in which the maps are arranged temporally in a grid or in a sequence, as a suitable means for revealing spatial patterns through time. Temporal trends in space can be conveyed using a single map, on which time-series graphics are placed on specific locations (e.g. Andrienko and Andrienko 2008). Animating through timesteps is another common technique for showing the evolution of hazard parameters or their probabilities of occurrence under different scenarios. Since only one time interval is visible at once, temporal trends are difficult to detect, but can form the basis of narratives (Robertson *et al.* 2008).

It is important to remember that all maps or data visualisations are incomplete simplifications of reality (Monmonier 1996) and thus have limitations. For hazard maps, in particular, an evolving situation such as a volcano or a drought is often a continual work in progress, and it is often difficult to convey this uncertainty on maps. Additionally, interpretation and use can be affected by socio-cultural and political factors, e.g. different colour associations (Moen and Ale 1998). Those producing hazard data need to be aware that different types of users may interpret and understand maps in different ways. Dymon and Winter (1993) found that this was often the case and suggested that a lack of funding for evaluating how hazards are communicated as part of hazard studies has inhibited research into this area.

Haynes *et al.* (2007) showed that for volcanic hazards on Montserrat, non-scientists had trouble interpreting simplified topographic maps, responding better to oblique aerial photographs and three-dimensional maps. In the maps tested, users found contour lines difficult to interpret and sometimes mixed them up with roads and exclusion zone boundaries. Decoding colours used to denote particular hazards was also difficult. People tried to match colours they saw around them with colours on the maps rather than recognising false colour symbols.

Conclusions

Regarding the future improvement of hazard identification tools, development and maintenance of current remote sensing observational programmes and tools is key in order to improve the data resolution, quality, timeliness and reliability. This would help to avoid placing people and expensive instruments in danger while providing needed data from and for low-income countries, such as covering ungauged river basins and unmonitored volcanoes. It should incorporate continuous gathering of baseline data and the development of rapid and reliable methods to collate, analyse and interpret that data when sudden changes are observed. Improved, verifiable and verified forecast models of hazard evolution, based on collectable data available, would result.

User friendliness and access continue to pose challenges as well. Many data from monitoring schemes, such as real-time earthquake and tsunami alerts, are easily accessible to anyone with internet access or mobile phones. Other remotely sensed data, such as for floods, require clunky software to find and interpret the images or models.

Improvements across hazard identification could be made by conducting evaluations and gathering feedback from those using visualisation tools, particularly if the intended audience is the general public. When possible, the users should be involved from the developmental stage in order to develop visualisation tools that suit their needs: ability to access data and ability to interpret material. Guidelines might emerge for presenting hazard information for different user groups, based on such empirical studies, thereby improving usability and access for everyone who needs and uses hazard identification tools.

Hydro-meteorological/ climatological hazards

18

Hazard, risk and climate change

David Simon

DEPARTMENT OF GEOGRAPHY, ROYAL HOLLOWAY UNIVERSITY OF LONDON, UK

Introduction

Climate change is undoubtedly one of the highest-profile environmental concerns of the early twenty-first century, and one with a profound impact on 'natural' hazards. Rather like sustainable development in the 1980s and 1990s, climate change has become the focus of heated debate, concerted scientific research, non-governmental organisation (NGO) activism, governmental commitments and international summitry in recent years.

Climate change can be caused by natural processes and human activity. The Intergovernmental Panel on Climate Change (IPCC) uses this definition, whereas the United Nations Framework Convention on Climate Change (UNFCCC) defines climate change as being directly or indirectly anthropogenic. This latter approach refers to the build-up of greenhouse gases (GHGs) in the atmosphere, leading to an increase in mean atmospheric temperatures and hence changes in wind and rainfall patterns and in prevailing climates around the world (IPCC 2007a: 21).

The broader concept of global environmental change (GEC) embraces not only these effects but also other changes in the biosphere or Earth system linked to them, such as sea level rise, land-use and land-cover change and urbanisation (Kalnay and Cai 2003), which reflect directly the impact of human activities and the increasing complexity of feedbacks among the various elements. As such, not all aspects embodied within GEC are governed predominantly by climate change. For example, land-cover change is influenced more by livelihood-related decisions than by climate change. This holds also for the bidirectional relationships between urban systems and the rest of the biosphere (Sánchez-Rodríguez *et al.* 2005; Sánchez-Rodríguez 2008; Simon 2007).

The family of GHGs – of which carbon dioxide, methane and nitrous oxide are the most important – trap solar radiation and albedo from the Earth's surface, thus triggering the temperature increases. While GHGs have always been present in the Earth's atmosphere, anthropogenic activities since the start of the industrial revolution have precipitated an unprecedentedly rapid increase in their concentrations. Pre-industrial GHG concentrations up to 1880 – the base year used for IPCC trend analysis – were about 280 parts per million (ppm), while current concentrations are about 435 ppm and rising by 1.9–2.0 ppm annually. However, numerous uncertainties remain in the climate science and hence projections, relating to under-lying drivers – for instance of decadal changes in tropical energy budgets (Hartmann 2002) – and the complexity of feedback loops.

However, because of lags in the complex feedback loops of the biosphere that will keep GHG concentrations rising for a considerable period even if emissions were capped at current levels, achieving the IPCC's (2007a) perceived 'safe' limit of 480 ppm would require substantial cuts in emissions. How to achieve this, what targets to set and how to divide the burden among different groups of countries that bear the bulk of historical responsibility for cumulative emis-sions or contribute differentially to current emissions, and how poorer countries will be assisted financially and technically to adapt, are the principal negotiating items in successive Conferences of the Parties (COPs) of the UNFCCC seeking to agree a successor to the 1997 Kyoto Protocol.

As a counterpoint to this approach, quaternary scientists searching for palaeoecological ana-logues to the current context of gradual atmospheric temperature increases over a century or two have demonstrated increasingly clearly that ancient catastrophic climate change has often occurred remarkably abruptly and rapidly, over the space of a few years or decades at most, especially during the last glaciation (Adams *et al.* 1999; International Geosphere-Biosphere Programme 2001). While very useful possible pointers, such studies cannot themselves answer the key question of to what extent current anthropogenic climate changes will mimic past natural changes.

Distinguishing between natural and anthropogenic climate change hazards

Natural hazards arise over all time scales, lasting from a few seconds or minutes (e.g. earthquakes and tornadoes), to hours (e.g. tsunamis and storms), to days (e.g. hurricanes and some floods), to years or decades (e.g. some volcanoes and droughts). During or immediately after many disasters, search and rescue commences and after the prospect of finding more people alive recedes, the emphasis shifts to cleaning up and rebuilding or rehabilitation as appropriate.

Not all natural hazards yield that pattern. For instance, droughts may last from a few months to several years, especially in semi-arid and arid zones, where they are endemic. However, since the landmark droughts and associated famines in the Sahel in the 1970s and Ethiopia in the 1980s, it has been appreciated that anthropogenic environmental changes, especially in terms of widespread deforestation and loss of other ground cover, affect microclimates and rainfall pat-terns. Over the longer term such local or regional environmental changes – including deserti-fication and other 'creeping environmental changes' (e.g. Glantz 1999; see Box 18.1) – contribute to climate change/GEC and the complex associated feedback loops (see Chapter 22).

Climate change/GEC comprises two complementary and mutually reinforcing elements. First, extreme climate-related hazards are predicted to be affected, with some increasing in severity and frequency and others diminishing. Hurricane Katrina in New Orleans in 2005 has become perhaps the iconic example of the destruction that an 'off-the-scale' event with surges of over 4 m can cause, although it is not clear how much climate change did influence the hurricane. Yet, it was but one of a series of severe storms that hurricane season. Data on recent Caribbean hurricane seasons attest to their increasingly damaging impacts, which have been linked to increased vulnerability to the combination of 'low elevation and relative sea-level rise (up to 1 cm/yr along parts of the Louisiana coast), only part of which is climate-related', while in Venice, Italy, there has been a marked increase in storm surge frequency since the mid-1960s (IPCC 2007a: 92–93). Their impact has been exacerbated by ongoing anthropogenic sub-sidence. Worldwide, tropical cyclone intensity, rather than increasing frequency, has been the

Box 18.1 Creeping environmental changes: why bother?

Mickey Glantz
Director, Consortium for Capacity Building, University of Colorado, Boulder, Colorado, USA

People today are concerned about rapid climate change. Since the late 1970s there has been a search for a 'dread factor' by the scientific community with the hope of prompting policy-makers to agree to the reduction of greenhouse gases emitted from human activities. Proposed 'dread factors' include a disintegration of the West Antarctic Ice Sheet, a shut-down of the oceanic currents called the conveyor belt and an abrupt climate change.

Yet, while researchers search for a real dread factor, slow-onset changes, called Creeping Environmental Problems or Changes (CEPs), are eroding the quality of soils, water and the atmosphere. Slow-onset, incremental but cumulative changes are diminishing the quality of life as well as the quality of the atmosphere. The family of CEPs that often become prominent include air pollution, acid rain, climate change, soil erosion, deforestation, desertification, depletion of groundwater and nuclear waste accumulation.

These incremental changes are happening each day but their impact on people and the environment is not discernable in the near term. Only after an extended period of time (depending on the specific characteristics of the environment undergoing change), when the incremental changes have accumulated to a visible level and considerable damage has been done, does society seek to address the changes. However, doing so is in response to a late warning not an early one.

The way in which individuals, societies and governments react to CEPs is not encouraging. CEPs are seen as problems that can be addressed in the future, because more pressing issues are viewed as demanding to be resolved first. Dealing with CEPs is constantly delayed until a threshold of change has been crossed.

At that time, the seemingly innocuous minor changes have accumulated into a major crisis. Those thresholds are not usually identified until the threshold has been crossed, even though it is easier and least expensive to deal with the CEP early in its process rather than waiting until it has crossed a threshold of change.

There are few political systems that have dealt with CEPs in a timely, least expensive, most effective way – before they have become a crisis. 'Why bother' to address a CEP is sometimes a first response. Yet bothering with such changes early in their development stages stops them becoming an environmental problem and then a crisis. 'Forewarned is to be forearmed.'

Sadly, being forearmed is apparently not enough to encourage correct decisions. A perfect example of an early as well as a late warning that was ignored, with awareness of the consequences, is the demise of the Aral Sea. It is sandwiched between two Central Asian deserts with a water supply from melting glaciers. A USSR plan in the early 1950s to double the amount of area devoted to cotton production by diverting large amounts of stream flow from the two rivers feeding the Aral Sea has led to its decimation.

In 1960 the Aral Sea was listed as the fourth largest inland sea in the world. Today, not enough of it remains to make the list of the largest inland seas.

dominant change thus far, although the Atlantic Ocean has experienced both trends (IPCC 2007a: 108).

Second, the growing severity and possibly frequency of extreme climatic events will increasingly be occurring on a rising base of increased atmospheric temperatures and higher sea levels, thus magnifying their impact. These latter changes represent the slow-onset events of which we have decades of advance warning and which increase slowly in magnitude, but herald long-term changes once they occur. Associated vulnerabilities and disaster risks will rise accordingly.

Where increasing severities and frequencies of extreme events are experienced, if underlying vulnerabilities are not addressed then not only will there be progressively shorter recovery times between events, but the impact of each event is also likely to increase as a result of greater prior vulnerability. Even a single disaster is known to affect the most vulnerable people disproportionately because they stand to lose a high proportion if not all of their possessions and assets (e.g. livestock, which in pastoralist societies serve multiple functions including as a source of wealth and status), and have few resources other than their labour power to assist recovery. One consequence is that economic differentiation and potentially social tension therefore often increase within disaster-affected communities. A series of such events in quicker succession greatly exacerbates this process. If through misfortune multiple hazards are experienced in close succession (e.g. an earthquake followed by drought or floods), the problems may be further exacerbated because of compounded impacts before recovery, with the different disasters potentially affecting different assets and livelihood activities.

Disaggregating climate change impacts

Globally, climate change/GEC is expected to increase both the number of vulnerable people and places and the extent of their vulnerability.

Coastal zones

Low-elevation coastal zones contain dense human population concentrations in both rural and especially urban areas, as well as a high proportion of littoral countries' key economic infrastructure, including ports and associated facilities. Rising sea levels and more frequent and severe storm surges will increase the scale of damage and the areas of temporary and permanent inundation (see Figure 18.1), as well as salinisation of freshwater aquifers underlying coastal dune fields or wetlands on which many people depend (Sánchez-Rodríguez et al. 2005; McGranahan et al. 2008) (see Chapter 19). Many estuaries and deltas are regarded as vulnerability hotspots through a combination of natural and anthropogenic subsidence, as well as other anthropogenic land-use factors such as ecosystem destruction and infrastructure development. For a non-delta ecosystem example, see Box 18.2. The delta regions of the Nile, Ganges/Brahmaputra and Chao Phraya (south of Bangkok, the Thai capital) are at extreme risk, and others like the Mississippi and Changjiang (Yangtze) are at high risk (IPCC 2007a: 327). The increased presence of shallow surface water, coupled with rising temperatures, is expected also to facilitate the breeding of mosquitoes and other disease vectors, with a consequent risk of malaria and other diseases (see below).

In many coastal cities, poor and vulnerable people have constructed houses or flimsy shelters on the banks of rivers, on seasonally inundated flood plains and wetlands, on stilt houses over water as on Lagos Lagoon (Simon 2010), or on steep hill or mountain slopes on the urban periphery. Where floods become more frequent and rainfall more intense, such areas become

Figure 18.1 Poor, displaced rural dwellers are swelling the population of cities like Dakar, leading to housing construction in inappropriate, flood-prone areas near the sea, which are at greater risk as sea level and storm severity rise
Source: Photograph by David Simon.

increasingly flood prone and, in the case of steep slopes, also unstable, heightening the risk of disaster. Examples of this phenomenon have occurred on the outskirts of Rio de Janeiro in Brazil and in Manila and other Filipino cities (see Chapter 25).

Inland areas

In many inland tropical and subtropical areas, the IPCC Fourth Assessment Report (IPCC 2007a, 2007b, 2007c) predicts increased prevailing temperatures and more variable rainfall, leading to desiccation, forest denudation, heatwaves, droughts and the expansion of arid and semi-arid areas. Water shortages, bush or grass fires and agricultural production deficits would then become increasingly severe, while livestock losses would rise. Conversely, high–latitude montane environments may be subject to more intense cold spells. Such changes also imply and will certainly affect existing seasonal variations in prevailing conditions (see Chapter 23).

Individually and cumulatively, these trends would undoubtedly have strongly negative impacts on the already precarious agriculturally based livelihoods of many rural and peri–urban poor households, and probably even currently reasonably secure smallholders. For instance, groundnut production in the Sahel has declined markedly for various reasons, but climatic changes towards warmer, drier conditions have seemingly played a catalytic role (IPCC 2007a: 106). In many tropical and subtropical regions, all available land is already being cultivated,

211

Box 18.2 *Páramos* and adaptation

Gustavo Wilches-Chaux
Latin American Network of Social Studies on Disasters and Risk Management (LA RED),
and Manager, WWW Ltda, Bogotá, Colombia

The *páramos* can be found in the northern corner of South America (Colombia, Ecuador and part of Venezuela), and in some parts of the Peruvian Andes. Those are very particular high-altitude ecosystems (approximately 3,000 to 4,100 m above sea level – temperatures around 0°C to 10°C), in which soils are poor in nutrients but very rich in both vegetable and animal biodiversity. One of the most characteristic and endemic plants of the *páramos* are the *frailejones* (friar-like plants) of the Asteracea family, of which the best-known are Espeletias. Peat mosses (genus Sphagnum) are not endemic but very common in the *páramos*. Each little moss plant is like a metaphor of the whole *páramo*: their structure enables them to absorb water in an amount equivalent to almost seven times their own weight – or more. During long periods of the year, fog (horizontal rain) is the permanent atmosphere in the *páramos*.

The *páramos* are enormous sponges conceived by nature to absorb, conserve and deliver water to lower territories and, as such, they are key factors in the resilience-immune systems of these countries. Unlike other Andean cities, Colombian high-altitude cities and villages do not depend directly on mountain glaciers for their water security, but on the conservation of the *páramos*. This is true under 'normal' conditions and it will be especially important in climate change scenarios. The *páramos* are key actors in climate change adaptation in the future.

The problem is that the *páramos* are being occupied, exploited and used for activities such as agriculture (legal and illegal: potatoes and poppies), cattle raising and mining, which are destroying their integrity and diversity and, hence, their capacity to provide the environmental services on which lives depend. In this case, the deep ecology approach (according to which the *páramos* have a sacred right to exist) and the anthropocentric approach (according to which we need healthy *páramos* for our own benefit) are not contradictory, but complementary. People living in the *páramos* may stay there if they become 'resources' for the *páramos* (part of their own resilience capacity), but those undertaking destructive activities must leave. Agriculture and cattle raising (and the people who continue practising them) must move to other lands and that will mean moving people from one place to another and land-tenure reforms. Social, political and economic changes that have been delayed for decades are becoming urgent not only in social terms but at the front of climate change adaptation priorities.

Deep ecology has the challenge to provide meaningful reasons that allow people to understand and accept that the many changes and sacrifices that must be made in their daily lives are not forms of forced impoverishment, but are necessary investment to make life possible for future generations. Many agreements must be made between governments, communities and private-sector leaders in order to reorganise people in their territories, but bearing in mind not only human priorities but also the priorities and interests of ecosystems. The more vulnerable *páramos* become, the more hazardous the territory and the future will be for human communities. Those agreements will not be reached without deep conflict. Conflict transformation is going to be a key tool – perhaps the most important one – for adaptation to climate change.

including marginal land that was settled or brought into cultivation during previous periods of hardship, such as structural adjustment in the 1980s and 1990s, and because of population increases. Seeking to expand landholdings or the total area under cultivation is therefore not an option for many. There may be some scope for increasing agricultural intensity and utilising improved hybrid or genetically modified varieties of crops to raise yields, but these usually require a complete package of new seeds, fertiliser and irrigation in appropriate amounts at specific times in the growing cycle, making them more expensive and often impracticable in the face of the variability, uncertainty and related stresses introduced by climate change.

For semi-nomadic and nomadic pastoralists, the impacts of climate change vary. In Tibet, warming in the high montane environment has improved livestock production, while in Mongolia, the warmer and drier conditions have adversely affected pasture biomass (IPCC 2007a: 106). In the Sahel, climate change is likely to represent a potent threat to pastoralist lifestyles and livelihoods, as inadequate access to water will either force them to range over ever greater distances (with all the associated costs and risks) or to restrict themselves to areas close to water sources. The latter is likely to exacerbate existing tensions and conflicts with sedentary cultivators over resource access. This problem first gained attention during the great Sahelian drought of the 1970s but has intensified in bad years more recently, providing a portent of what may happen as a result of climate change/GEC.

One strategy adopted for various reasons (of which climate change/GEC is just one) by several Sahelian governments and those in other semi-arid areas, like Botswana, is to promote the sedentarisation of such pastoral communities. This has usually met resistance from pastoralists, since it would threaten their identities, lifestyles and continued social existence. More common is for pastoralists and cultivators – whose livelihoods are being rendered unsustainable as a result of ongoing desertification and the other stresses referred to above – to abandon the land altogether and migrate to urban centres or coastal villages in search of alternative survival strategies and livelihoods, from wage labour or fishing, for instance. Disappointment in that search, coupled with declining fish stocks as a result of commercial overfishing and altered marine conditions due to climate change/GEC, has been documented as contributing to the flow of impoverished illegal migrants from Senegal and other West African countries seeking to reach the Canary Islands or Madeira – as the nearest parts of the European Union (EU) – in unseaworthy boats. Large numbers also struggle across the Sahara Desert and then attempt to cross the Mediterranean Sea into the EU in voyages that often end in disaster (Guèye et al. 2007). However, climate is not the only driver of such internal and international migration, which follows routes established by generations fleeing harsh economic and political conditions and violent conflict.

For inland urban areas (see Chapter 13), rising temperatures and declining rainfall patterns will increase environmental stress, especially in terms of heat island effects, damage to infrastructure unsuited to such temperatures and humidity, increasing energy consumption on air conditioning and refrigeration, and water shortages. Food supply from adjoining peri-urban and rural areas is also likely to be affected for reasons outlined above. Incidences of heat stroke and related morbidity will increase (see above) and the epidemiology of diseases will change. High-altitude cities dependent on snow and glacier melt for part of their fresh water supply are expected to face particular challenges, since these supplies are declining as some Andean and Himalayan glaciers retreat. For instance, the Chalcaltaya glacier feeding La Paz, the Bolivian capital, had declined in area from 0.22 km^2 in 1940 to under 0.01 km^2 by 2005 and was predicted by some to disappear by 2010 (IPCC 2007a: 87). By May 2009 the 18,000-year-old glacier had been reduced to just 'a few lumps of ice near the top' (BBC 2009). However, there may be more local variation than is sometimes implied in relation to Himalayan glaciers, and in the assertion by the IPCC (2007a: 493) – based on one grey literature source citing a few examples – that they are melting at an unprecedented rate and may disappear by 2035 if this trend continues, which triggered a major controversy. Some glaciers appear stable, others are retreating slowly, and others may actually be expanding (e.g. Resilient Earth 2010; BBC 2010a).

These impacts highlight the close interrelationship between climate change/GEC and human security, a broad concept that includes people's ability to meet basic needs and sense of well-being. Underlying this are factors or 'drivers' including power relations that determine the equitability or otherwise of the distribution of goods and services within a social group. In the context of climate change impacts, this also raises concerns regarding social and environmental justice, i.e. the need to tackle the causes of vulnerability to the impacts. Failure to do so increases the prospects of resource conflicts over access to water, land, grazing and food, in particular, both within and across political boundaries. Such conflicts during cyclical droughts or famines in semi-arid and savanna areas have already been increasing as a result of sedentarisation and increased land scarcity, but human displacements in the face of climate change/GEC impacts will almost certainly exacerbate such problems – as the Senegalese example above indicates (e.g. UNDP 2007: chapter 2; World Bank 2009: chapter 2).

Health hazards

In areas that experience increased mean temperatures and heatwaves, cases of heatstroke and related morbidity are expected to increase, along with mortality of elderly and other particularly vulnerable people. Precisely such abnormally high mortality rates occurred among the elderly, especially those living alone, in France and other parts of Europe during the unusually hot summer of 2003 (IPCC 2007a: 108).

Disease epidemiology is also very likely to change as a result of the interplay between two related factors: geographical shifts in the occurrence of disease vectors such as mosquitoes and ticks, and changing vulnerabilities due to associated or unrelated anthropogenic changes. Some areas in the Kenyan highlands that were malaria-free until the 1980s are now highly infected, but it is difficult to separate possible climate changes as a cause from others such as reduced spraying, increased drug resistance, changes to migration patterns and agricultural practices, and declining forest cover. More generally, however, as prevailing temperatures increase, it is expected that mosquitoes will both survive year-round where they currently cannot, and penetrate seasonally further towards temperate zones.

Similarly, increased incidences of tick-borne Lyme Disease due to milder winters in central and northern Sweden have also been documented but other anthropogenic influences may have contributed too. Along with the absence of adequately long time-series data, this difficulty in separating anthropogenic from climate-related changes has made it difficult so far to be clear on the role of the latter (IPCC 2007a: 107–9). Ongoing environmental health monitoring activities linked to disaster risk reduction (DRR) initiatives to help identify structural and other vulnerabilities (Songsore et al. 2009) can and should readily be adapted to monitor changing environmental/ climatic conditions and associated risks.

Economic impacts

Using a methodology akin to that of the IPCC, the 2006 Stern Review (Stern 2007b) concluded that the economic costs of not taking concerted action to address climate change would greatly exceed the costs of doing so. Moreover, taking action represents a huge opportunity to stimulate the economy through investment in new, clean or green technologies, to undertake the extensive retrofitting of buildings to increase their energy efficiency and to promote more sustainable development. While sceptics argue that securing jobs should take precedence over environmental issues, investing in important programmes of this sort could, overall, promote both objectives.

There is already growing evidence of the greatly increased recent cost of economic losses due to disasters. Measuring this over time is problematic on account of changes in wealth, population and population density and hence buildings and other infrastructure in the areas affected by disasters (see Chapter 54). While the extent of insurance cover in urbanised and higher-income contexts is normally reasonably high, among low-income and rural communities, especially in poorer contexts (particularly where monetisation is incomplete), insurance take-up is usually low. Instead, people rely on traditional risk spreading techniques such as field and crop diversification and using traditional, hardy crop varieties. Additionally, where government economists or loss adjusters seek to value subsistence production and 'traditional' rural dwellings and infrastructure for national accounting or post-disaster assistance purposes, these are usually undervalued relative to assets in the formal economy. All these factors greatly reduce the reported economic cost of any losses, although those losses are no less real. The direct corollary of this is that uninsured people are particularly vulnerable to the kinds and scales of losses that will arise through climate change impacts.

Coping with climate change

Efforts to cope with climate change are conventionally divided into two related and complementary categories, namely mitigation and adaptation. Mitigation is generally defined as human interventions to reduce the sources or enhance the sinks of GHGs. More simply, these are actions to reduce GHG emissions (i.e. pollution) and/or increase their absorption. Adaptation comprises deliberate changes in human or natural systems to cope with actual or expected climate changes/GECs by reducing their negative impact or exploiting new opportunities that arise (IPCC 2007a: 750).

In terms of hazards and DRR, the focus should be on avoiding conflict between mitigation and adaptation activities and sustainable development. If appropriately framed and undertaken, they can and should be mutually reinforcing (IPCC 2007b: 696) (see Chapter 14).

Understanding mitigation and adaptation

Each category comprises a spectrum of actions, both short and long term. Mitigation actions range from minor to major behavioural changes. Minor changes can be farmers reducing the burning of crop residues or felled trees and everyone with electric power switching off lights and appliances when not in use. Major changes might include technological innovations using less fossil fuel and producing fewer emissions per unit of economic activity (like more efficient aircraft engines) or avoiding emissions reaching the atmosphere (like catalytic converters on cars and carbon capture and storage at power stations). Some of these need careful assessment in terms of their cost-effectiveness and actual contributions, especially if they ultimately avoid or reduce the need to adapt our lifestyles and behaviour and thus do not promote sustainable development.

Switching energy supplies to 'greener' renewable sources is often promoted as important since it has both mitigatory and adaptive elements. Harnessing wind, solar, geothermal and wave power is providing a growing proportion of national electricity generation in many countries, both rich and poor. For instance, Kenya's reliance on fossil fuel for electricity generation has been reduced by the exploitation of the geothermal capacity on the edge of part of the Rift Valley, making it one of the world's leading geothermal energy producers. On the southern shore of Lake Naivasha, Oserian, one of the country's largest horticultural export farms, has exploited geothermal vents on its land to heat the greenhouses and provide energy, substantially reducing both its carbon footprint and medium-term production costs. Dependence on the interruption-prone national grid (a major source of vulnerability and risk) is thereby also being reduced, hence improving the farm's sustainability through more reliable and cheaper production and more secure employment.

As wind turbines and solar batteries have fallen in price, their uptake has increased, even in poorer locations. Sometimes they are helping to reduce the reliance of poor households in rural, peri-urban and urban areas on woody biomass for energy. In addition to emissions mitigation (reduction), this relieves some pressure on vegetative ground cover and contributes to environmental conservation (reducing run-off and soil erosion and hence flood risk, and maintaining the soil's profile and water infiltration capacity, hence conserving productivity). In other words, this contributes to DRR, environmental management, livelihoods and ultimately sustainable development. Access to such locally generated, small-scale power in areas remote from national electricity grids also enables dramatic lifestyle and livelihoods changes in villages and even individual households, promoting economic development and lifestyle adaptation through access to a range of electric equipment from simple grinding mills and refrigeration to more sophisticated plants as well as mobile phones and other information and communications technologies for development.

By contrast, large dams, construction of which is experiencing a renaissance in the name of development despite extensive accumulated knowledge about the vast economic costs and environmental and social problems created (World Commission on Dams 2000), are often regarded as a source of clean or green energy because water is a renewable resource. However, reservoirs in tropical and subtropical zones are important emitters of methane, one of the most damaging GHGs, from rotting submerged vegetation (International Rivers 2006). Hence, such schemes are causing direct environmental damage in often-sensitive localities as well as contributing directly to GHG emissions and inducing large-scale population displacement, thereby contributing to local livelihood destruction and impoverishment.

Large dams may also contribute to disaster risk in two other ways: the volume and weight of water behind large dam walls at full supply level are known to have the capacity to trigger earth tremors; second, in areas where snow melt or rainfall is expected to increase, or floods as extreme events increase in severity, reservoirs may overflow more frequently. Uncontrolled releases or releases timed to suit upstream conditions may cause disastrous flooding downstream, especially if adequate warning is not given. This was the case three times during the 2000s in the fertile agricultural lands of the lower Zambezi valley in Mozambique, necessitating large search and rescue and disaster relief operations. Although nutrient-rich silt was deposited on the flooded land, local livelihoods – especially of the poorest households – were severely disrupted, infrastructure destroyed and development set back (see also Figure 18.2).

Some interventions have indeterminate outcomes because of the complexity of the local interactions and feedback mechanisms. For instance, burning of grass residues at the end of the dry season in savanna and semi-arid zones, and of felled trees on tropical land being cleared for cultivation, is often undertaken as a way of returning nutrients to the soil and encouraging new growth – i.e. contributing to sustainable livelihoods. However, the burning also contributes to emissions and atmospheric pollution, while on a large scale such vegetative loss contributes to environmental change. Hence, reducing such burns is often advocated as an emissions mitigation measure, while allowing wood to rot naturally may be ecologically preferable and hence

Figure 18.2 Washout of the main trunk road through semi-arid Pokot Central District in north-western Kenya by river flooding following unprecedentedly heavy rainfall in early 2005 after a period of drought. Several such recent episodes may be linked to climate change
Source: Photograph by David Simon.

more developmentally sustainable. Local research would therefore be required in each case, but mitigation (emissions reduction) and developmentally positive adaptation could potentially both be achieved through this behavioural change.

The need to 'mainstream' mitigation and adaptation

Reducing risk from climate change/GEC requires mitigation and adaptation in numerous ways, including existing policies, planning and implementation actions. Treating adaptation as an additional burden or shopping list item on the agenda of public authorities and private companies is unlikely to be successful, since this will set adaptation in competition with other agenda items for funds. Only when climate change awareness and actions become embedded or 'mainstreamed' into normal ongoing DRR, development, planning and implementation activities will prospects be enhanced. In terms of mitigation, for instance, not only must existing buildings be retrofitted to improve their insulation and energy efficiency, but construction standards and regulations for new buildings need appropriate amendment.

Such changes will also contribute to reducing vulnerabilities and promoting development through economic stimulation and enhanced resilience. There is considerable scope for bringing the often separate discourses and actions of DRR, GEC/CC, poverty reduction and development closer together (Schipper and Pelling 2006; Wisner et al. 2004). Such agendas should not conflict with but reinforce one another. Moreover, it is only by integrating their concerns that the 'shopping list' approach and problems of conflicting pressures in a context of budgetary constraints can be minimised.

In terms of adaptation, upgrading existing infrastructure to withstand more severe and frequent extreme events – as, for instance, with the rebuilding and rehabilitation of New Orleans after Hurricane Katrina or re-housing disaster-displaced poor communities in the Philippines (Dodman et al. 2010) – is no less vital than ensuring that new construction work takes place to appropriately revised standards. There is no point constructing new low-income housing or upgrading an impoverished shantytown in a flood-prone, low-lying locality according to standards that take no account of expected climate change impacts, as that investment – and the expectations and efforts of residents – may well be washed away within a few years. Hence, the additional upfront expenditure to adapt new developments to climate change will be far cheaper than the cost of not doing so – which is precisely the logic of the Stern Review on the economics of climate change (Stern 2007b).

Similarly, new adaptation funds made available to poor countries by donors need to avoid creating conflicts between development and adaptation. A development crisis could emerge if people who depend on non-timber forest products are displaced in the course of implementing forest conservation under the new Reduced Emissions from Deforestation and Degradation (REDD) funding. Likewise, acquisition of large tracts of land for production of bio-fuels may seem a good mitigation and adaptation measure, but if this displaces poor farmers or reduces food crop production, the situation becomes more complex. Finally, the mental and institutional dichotomy between mitigation and adaptation is unhelpful; successful initiatives that address both simultaneously and thereby contribute to environmental sustainability and long-term development should be the objective. As with any other forms of development activity, bottom-up, community-led interventions targeting mitigation and adaptation are more likely to succeed when enabling structures, contexts and conditions are appropriate (IPCC 2007b: 713–14; Dodman et al. 2010).

Conclusions

Strategies for coping with changing severities and frequencies of extreme events – one aspect of climate change/GEC – have much to gain from lessons learned over decades of DRR work, not least in terms of prediction, vulnerability reduction and post-disaster recovery. Recovery from even short-duration disasters is not always quick since particularly bad floods, volcanic eruptions or earthquakes, for example, may have impacts lasting years or even decades. In extreme cases, the scale of destruction, risk of recurrence and/or cost-effectiveness of reconstruction has necessitated permanent relocation or evacuation – as from the Montserrat volcano.

However, slow-onset, (semi-)permanent changes like sea-level rise, increasing ambient temperatures and desertification also represent challenges that must be addressed now. In some areas, such as the southern margins of the Sahelian Sahara, long-term desertification as a form of environmental change preceded the recent climate change concerns. Lessons learnt there over recent decades from DRR-type environmental management interventions to promote water infiltration and retention by the soil, sand stabilisation and livelihood diversification will be relevant. However, if the rate and scale of desertification increase, sustainability thresholds may be crossed beyond which novel responses will be necessary. GEC/climate change science is highlighting the lags and feedback loops in biospheric systems which means that current emissions and polluting activities, and mitigation actions to reduce them, will have only gradual effects, so the impacts of those activities will endure long into the future.

This essay has exemplified the diverse climate change/GEC impacts already occurring or expected to occur in different agro-climatic, latitudinal and geographical zones. In policy terms, the most vulnerable groups of people (usually poor) in the most vulnerable areas in relation to particular hazards and risks should be prioritised; there is often a high degree of overlap between human vulnerability and the exposure of locations. While climate change/GEC often exacerbates existing vulnerabilities and threats, it is also likely to increase the areas and total number of people at risk.

Existing vulnerabilities have diverse origins, often historical and/or structural. Attempts to address climate change/GEC impacts will not in themselves, therefore, generally address these underlying inequalities without direct attention to them. In this respect, there is much in common with appropriate DRR policies, where assistance, relocation allowances, compensation or other interventions are targeted and calibrated in accordance with the extent of vulnerability and need rather than with existing resources, wealth or power. The ultimate objective should be to minimise disruption and dislocation and to maintain or enhance the sustainability of livelihoods and quality of life – an objective that underscores the consistency of appropriate climate change/GEC coping interventions with those of sustainable development, especially in poorer contexts.

Mitigation and adaptation actions are neither mutually exclusive nor in opposition. Smaller, incremental changes are relatively easy to undertake but thereafter the required effort and investment increase. Trade-offs become more likely in high-cost and large-scale interventions, but well-conceived and targeted actions can address both successfully. Importantly, too, conventional DRR and development activities should not be seen as in competition with mitigation and/or adaptation. Instead, only by 'mainstreaming' and embedding climate change/GEC awareness and appropriately modified standards, practices and technologies into normal, ongoing planning, development and risk-reduction activities by individuals, households, firms and both formal and informal institutions in these spheres are long-term vulnerability and risk likely to be addressed successfully.

19

Coastal storm

Sebastiaan N. Jonkman

DELFT UNIVERSITY OF TECHNOLOGY, THE NETHERLANDS, AND ROYAL HASKONING, THE NETHERLANDS

Herman Gerritsen

DELTARES, DELFT, THE NETHERLANDS

Marcel Marchand

DELTARES, DELFT, THE NETHERLANDS

Introduction: living at the edge

Disasters such as Hurricane Katrina, which struck the US Gulf Coast in 2005, and Cyclone Sidr, which struck Bangladesh in 2007, highlighted coasts as being hazardous places to live. On average, every year more than 100 million people are found to be exposed to tropical cyclone hazards worldwide (Pelling 2004). Countries with substantial populations located on coastal plains and deltas with a relatively high exposure to cyclones include India, Bangladesh, Honduras, Nicaragua, the Philippines and Vietnam.

These and other low-lying coasts experience disasters owing to their exposure to the dynamics of the environment, while at the same time they attract human occupation because of the richness of their natural resources, such as fertile soils, fish stocks and navigation facilities. This paradoxical situation is likely to be exacerbated owing to climate change in combination with population changes, especially migration to coastal areas, exposing more and more people to these natural hazards. Among these people, the most vulnerable are those with least access to resources and choices that would allow them to cope with the hazards.

Many of the world's megacities can be found in deltas and coastal plains, such as Jakarta, Mumbai, London and New York (see Chapter 13). A migration trend can also be found in many countries towards the coastal zone. Population densities in the coastal zone can be as high as 1,500 inhabitants/km^2 (Bangladesh), with growth rates of more than 2.5 per cent per year (Mekong Delta, Vietnam). Many small island states are entirely coastline. What makes people want to live at the edge of land and sea?

Formed by the interplay between rivers and sea, deltas and coastal plains have flat, highly fertile soils that are easy to till. They can be travelled across by water and are full of fish. However, as the sea gives, it can also take. Their low-lying topography makes deltas vulnerable to storm surges and river flooding. If not through surface flooding, the seawater enters from below: seepage of saline groundwater poses a constant threat to crops.

Hence, people choose to dig, drain and develop. Land reclamation, irrigation, soil drainage and embankments have made many deltas hospitable, under the assumption that a place to live safely has been built while living off the fruits of the land and sea. Several deltas have developed into the major granary or rice bowls of an entire country, such as the Red River Delta in Vietnam and the Godavari and Krishna deltas in India.

Besides agriculture, transport and industrial development, tourism also generates an increasing economic profit along the coast. For instance, the Mediterranean coast is the world's most important tourist resort. Around 250 million travellers came to this region in 2001 (Sardá et al. 2005). The value of beaches in Spain can be as high as €700 per m^2 per year (Azira et al. 2008). Coastal storms have not diminished that.

Of all potential long-term hazards, sea-level rise is probably the most important for deltas and coasts. Estimated current rates of sea-level rise are 2 mm–6 mm per year, or two to three times higher than for the previous century (IPCC 2007a, b). Densely populated and heavily industrialised urban areas in, for instance, the Netherlands are already located below mean high water level, making them extremely vulnerable to future rise in sea level.

Instead of looking at the absolute rise in sea level, though, it is more important to assess the rise relative to vertical movement of the land. While tectonic uplift exists in some coastal areas, which partly offsets sea-level rise, other coasts experience subsidence. This often originates through a combination of geotechnical processes, such as compaction, and chemical processes (e.g. oxidation of peat soils). Anthropogenic sources are also often involved, such as excessive groundwater exploitation.

Combined with subsidence of the coast (see Chapter 29), sea-level rise can lead to a series of changes in coastal environments. The change of the land height with respect to the sea increases the flood risk during storm events because more land is exposed to lower storm heights. It also increases coastal erosion, thereby threatening human settlements and enlarging the area at risk from coastal flooding. Furthermore, sea-level rise and coastal subsidence lead to landward movement of the tidal influence and salt wedge in rivers, jeopardising freshwater intakes for agricultural, industrial and domestic water supply systems.

Physical characteristics of coastal storms

Coastal storms can be divided into two main categories. First, extratropical storms: atmospheric pressure disturbances and associated winds result in intense energy transfer from the atmosphere to the ocean, which increases water levels (a surge). Second, tropical storms, which extract energy from the warm ocean water to grow in strength. Tropical storms are known under different names – cyclones (Indian subcontinent), typhoons (South-East Asia) or hurricanes (Americas) – but their physical characteristics are essentially the same.

Extratropical storms

As a typical example of extratropical storms, a European coastal storm is the result of a disturbance in the atmosphere over the Atlantic Ocean, leading to a local pressure low. The horizontal pressure difference gives rise to air flow, or winds, in the direction of the depression. Deflection due to the rotation of the Earth creates a storm with characteristic anti-clockwise rotating winds around the depression in the Northern hemisphere and clockwise rotating winds in the Southern hemisphere.

Over deep oceanic waters, the pressure difference around the pressure low results in an immediate rise of the water level below. A 1 millibar pressure decrease represents 0.1 per cent of standard atmospheric pressure and equals a 1 cm water-level rise. When the storm reaches the continental shelf, air pressure differences no longer result in immediate water-level rise, mainly due to friction along the sea bed, which reduces and slows down the response.

For these more shallow waters, wind dominates, setting the water in motion through surface drag. For strong depressions, which are associated with high and sustained winds, the surface wind drag leads to a water-level rise or surge. The surge propagates in the direction of the wind, also with a deflection due to the Earth's rotation. In decreasing water depths towards the coast, interactions with the sea bed and tide increase, resulting in a further enhanced wind-induced surge height along the coasts, as long as the wind maintains the effect. Depending on the presence of local sea bottom characteristics, plus constrictions such as river inlets and estuaries, severe storms may easily lead to peak surge heights that exceed 5 m. When the wind decreases, or changes direction, the surge height decreases as the water surface tends to regain its original state, such as normal tides.

Storm winds also generate and propagate surface waves referred to as 'sea state'. Waves propagating through a basin, which have been generated by winds elsewhere, are called 'swell'. For a severe storm in open sea, significant wave heights may reach levels of 8 m–10 m or more. In shallow water, moving to the coast, interactions with other water and land phenomena, breaking and further dissipation take place, leading to a decrease of wave height.

Both the storm-induced surge and wind waves cause hazards for navigation and port operations, along with potentially severe damage to coastal structures including flood barriers. Examples are dune erosion, along with dike collapses as a result of saturation due to sustained wave overtopping or pressure from surge and wave forces.

The 0–12 Beaufort Scale is used to indicate wind severity. Named after its originator, Rear-Admiral Sir Francis Beaufort (1774–1857), the Beaufort Scale is based on a combination of visually recorded state of surface waters and damage caused by the winds (Hsu 1988). Beaufort Scale 6 (40 km/hr–50 km/hr) corresponds to a strong breeze, while Beaufort Scale 10, 11, 12 correspond to winds of storm, severe storm and hurricane force (winds of 88–101 km/hr; 103–15 km/hr; >116 km/hr), respectively.

Tropical storms

Tropical storms are cyclones that originate over a tropical ocean. Their measured winds exceed 200 km/hr and are accompanied by torrential rains (Riehl 1979). They are generated in the band of the trade wind current of the tropics, just north and south of the equator, tending to move away from the equator initially, but then sometimes following complex pathways with multiple changes of direction.

Several conditions are needed for the formation of a tropical storm. The main ones are sea surface temperature above 26°C; below-normal air pressure in low latitudes (which means under 1,004 millibar) and above-normal air pressure in high latitudes; an existing tropical pressure disturbance; movement of the disturbance at a speed less than 6.5 km/hr; some special dynamic conditions in the upper air flow; and heavy rain or rain showers in the area (Hsu 1988). The effect of the low-pressure disturbance on local winds and pressure is such that the winds form a closed cyclonic or counter-clockwise circulation in the Northern hemisphere (anticyclonic or clockwise in the Southern hemisphere).

The warm water heats the air moving from the surrounding water toward the central low. The ocean feeds heat and moisture into the storm, providing energy that causes the warm air in the centre to rise faster. As long as the cyclonic centre remains over warm water, the supply of energy is almost limitless. As more and more moist air spirals inward into the low pressure centre to replace the heated and ascending air, more and more heat is released into the atmosphere and the wind circulation continues to increase (Hsu 1988). When the wind speeds exceed 119 km/hr, the storm has formally reached the level of a tropical cyclone.

The Saffir-Simpson Scale (see Table 19.1) divides tropical storms into five categories, based on their damage potential from wind only – surge height is calculated and is included, but damage from flooding, especially from rainfall, is not factored into the Saffir-Simpson Scale. Although Table 19.1 has limitations, it is an internationally used classification. For example, storm surge height at landfall also depends on local topography and bathymetry while the observed damage must always be influenced by vulnerability.

Moving into coastal shelf areas, the intense winds around the pressure low create large surges, which propagate with the storm towards the coast, where they sweep across the often low-lying coastal areas. The strong winds, surge waters and wave forces of even Category I and II storms may cause significant loss of life and destroy infrastructure and vegetation.

Damage and loss are a function of the magnitude of the hazard and the vulnerability. The latter differs considerably by economic status and access of the affected population to technical resources such as warning systems and means of evacuation, and financial resources such as insurance. The relatively mild tropical storm Agatha in May 2010 killed over 150 people in Guatemala alone, largely owing to the torrential rainfall and mudslides that occurred. Hurricane Katrina (2005) is a well-documented case of the damage to a country which is comparatively well-prepared for disasters (see Box 19.1). Hurricane Mitch in 1998 killed over 10,000 people when it was less powerful than a Category I cyclone – mainly through inland rainfall causing flooding and mudslides. The 1970 cyclone that struck Bangladesh and its low-lying coastal silt islands (chars) is estimated to have killed over 300,000 people.

Exposure, vulnerability and the effects of coastal storms

Exposure of low-lying coastal areas

Coastal storms will often lead to combined hazards of wind and flood effects, leading to casualties, societal disruption and infrastructure damage. Strong winds and low atmospheric pressure push water into the coast. Additionally, surface winds can generate large waves that can cause beach erosion and the destruction of properties in coastal areas. Some available coastal storm statistics from the twentieth century are summarised in Table 19.2.

Table 19.1 Saffir-Simpson damage potential scale for tropical storms

Scale number	Central barometric pressure (in hPa)	Wind speed (km/hr)	Storm surge height (at landfall/along coast)	Commonly observed damage
I	> 980	119–153	1.2–1.5 m	Minimal
II	965–979	154–177	1.8–2.4 m	Moderate
III	945–964	178–209	2.7–3.6 m	Extensive
IV	920–944	210–249	3.9–5.4 m	Extreme
V	< 920	> 249	> 5.4 m	Catastrophic

Source: Hsu 1988

Box 19.1 New Orleans and Hurricane Katrina

New Orleans is situated in the Delta of the Mississippi river. The city and its surrounding suburbs make up a metropolitan area that is largely below sea level and entirely surrounded by levees. As a consequence of its geographical situation, the area is vulnerable to flooding from hurricanes, high discharges of the Mississippi river and heavy rains. In the twentieth century the city experienced floods after hurricanes in 1915, 1947 and 1965.

In August 2005 Hurricane Katrina formed as a tropical storm in the Atlantic Ocean southeast of Florida. After making landfall in Florida, the storm entered the Gulf of Mexico and it began to take aim for southeast Louisiana. On 29 August Katrina made landfall in Louisiana, with Category 3 status and sustained winds of 200 km/hr. Katrina's storm surge caused massive flooding and devastation along a 270 km stretch of the US Gulf Coast. The entire coastline of the state of Mississippi suffered large-scale destruction due to surge flooding. The storm surge also caused massive overtopping and breaching of levees around New Orleans.

IPET (2007) estimated the flood damage to residential property in New Orleans at US$16 billion, and damage to public structures, infrastructure and utilities (like roads, railroads, water defences, electricity network, drainage, etc.) at US$7 billion. Apart from the economic consequences, the event led to loss of life (see below), physical and mental health impacts, pollution from industrial and household chemicals that mixed with flood waters, and several indirect and/or longer-term effects, such as the rise of oil prices in the USA and a decrease of the population of New Orleans in the years after Katrina.

This disaster gave insight into the potential public health impacts and loss of life due to coastal storms. In the days before the storm, about eighty per cent to ninety per cent of the population evacuated from the affected area. However, about 100,000 people remained in the flooded areas, either in their homes or in public shelters, such as the Superdome. The hurricane caused more than 1,100 fatalities in the state of Louisiana. Based on analysis of a dataset of 771 of these fatalities (Jonkman et al. 2009), it is estimated that one-third of the fatalities occurred outside the flooded areas or in hospitals and shelters in the flooded area. These fatalities were due to the adverse public health situation that developed after the floods. Two-thirds of the fatalities were most likely associated with the direct physical impacts of the flood and were mostly caused by drowning. The majority of the fatalities were elderly: nearly sixty per cent were over sixty-five years old. Similar to other analysed flood events, mortality rates were highest in areas near severe breaches and in areas with large water depths (Jonkman et al. 2009).

Since Hurricane Katrina about US$14 billion has been invested in the improvement of the hurricane flooding defence system, which is expected to be finished by 2011. The improved levees and a series of new storm surge barriers and gates will, in theory, stop the city of New Orleans from being flooded by a storm with a probability 1/100 occurring each year. However, several challenges remain, such as the continuing loss of wetlands around the city and the ever-existing threat of coastal storms.

The effects of coastal storms depend on the characteristics of the population and community, including demography, livelihoods and planning, along with coping capacities. An additional factor is infrastructure designed to provide protection in coastal areas along with its location and the elevation in relation to the nature of coastal hazards (wind, surge and waves).

In general, it is expected that flood effects will be the most significant cause of damage and loss of life (see Chapter 21). Although some fatalities may be associated with wind effects (and landslides in mountainous hinterland areas), most of the fatalities due to coastal storms are caused by the flood effects (Rappaport 2000). The number of wind fatalities will often be limited as most people will be able to find shelter during the passage of the storm. Low-lying coastal areas are especially susceptible to flooding, such as areas built on marshland in eastern England; river deltas, with examples being the Rhine (Europe) and Yangtze (China) rivers; and low-lying islands such as built-up barrier islands on the east coast of the USA and coastal areas of the Cook Islands in the Pacific.

Coastal floods are capable of causing large numbers of fatalities, as they are often characterised by severe flood effects (large depths, high flow velocities and powerful waves). In addition, coastal storms have sometimes occurred unexpectedly, i.e. without substantial warning. This allowed little or no time for warning and preventive evacuation, and resulted in large exposed populations. Bangladesh, in particular, has been severely affected by coastal floods (see also Table 19.2). Finally, flooding by saltwater can lead to substantial damage and difficult recovery of agriculture and ecology. Following the 1953 North Sea storm surge, flooded areas were still suffering from salt contamination for years afterwards.

Table 19.2 Overview of coastal floods

Date	Location	Hazardous phenomena	Fatalities[1]	People exposed
1 February 1953	Netherlands, southwest (Box 19.2)	Storm surge	1,836	250,000
31 January to 1 February 1953	United Kingdom, east coast	Storm surge	Approximately 307 on land	32,000
26 September 1959	Japan, Ise Bay	Typhoon	5,101	430,000
12 November 1970	Bangladesh	Tropical cyclone	300,000	
18 September 1974	Honduras	Tropical cyclone	8,000	
12 November 1977	India, southern	Tropical cyclone	14,000	9,000,000
25 May 1985	Bangladesh	Tropical cyclone	10,000	1,800,000
30 April 1991	Bangladesh	Tropical cyclone	139,000	4,500,000
29 October 1999	India, Orissa	Tropical cyclone	9,800	12,600,000
29 August 2005	USA, Louisiana and Mississippi	Hurricane (Katrina)	1,100[2]	100,000
15 November 2007	Bangladesh	Tropical cyclone (Sidr)	More than 3,000[3]	

Notes:
1 The reported numbers of fatalities may include considerable uncertainty. For example, for the 1991 floods in Bangladesh the estimated death toll ranges between 67,000 and 139,000 (Chowdhury *et al.* 1993).
2 The number of fatalities in Louisiana – most of these were due to flooding.
3 Estimate based on press sources, November 2007.
Source: EM-DAT 2010

Box 19.2 The 1953 flood disaster in the Netherlands and the Delta Plan

During the weekend of Saturday 31 January and Sunday 1 February 1953, a storm raged across the northwest European shelf. Compared with earlier major storms, the 1953 storm curved more strongly southward, and slowed down in its course, leading to a much longer period of sustained winds piling up water into the southern North Sea. Low-lying coastal areas of the Netherlands, England and Belgium were flooded. The resulting disaster in terms of loss of life and damage to infrastructure was enormous. In the Netherlands 1,836 people died in the flood, while in the UK and Belgium the casualties were 307 and 22 on land, respectively, but over 200 more died as ships sunk around the British Isles and across the North Sea.

The main affected area of the Netherlands was the Rhine-Meuse-Scheldt delta. Here, river arms and inlets run through the land, having created large islands, at the time only a few of them being connected to the mainland. The land itself lies largely below mean sea level and is separated from the sea by artificial dikes. The increasing and sustained high water plus wave overtopping led to weakening and saturation of many dikes. At over 150 locations dike stretches gave way, allowing water to rush into the low-lying polder through gaps that quickly started to erode further in width and depth (Gerritsen 2005).

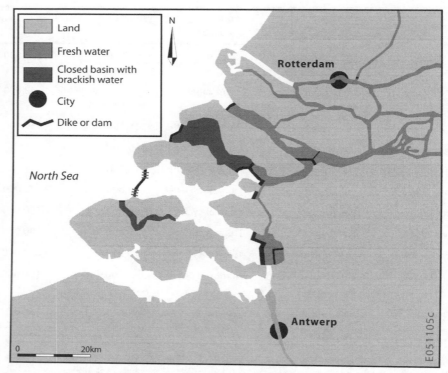

Figure 19.1 The Delta Plan after completion in 1986, with a storm surge barrier for the Eastern Scheldt, instead of the initially proposed fully closed dam

The relief effort to help and care for the affected people and livestock was of national and international scale. From the start, much effort was directed at closing the breaches, as tide flowed in and out twice a day, leading to continuing scouring and erosion of the dike and its foundation. It took nine months to close the last breach, a gap that had eroded more than 30 m deep. In many of the exposed areas, the soil was heavily contaminated with salt, and vegetation and trees died, so it took many years before agriculture was back to normal.

Only seventeen days after the disaster, the Delta Committee was founded by the Dutch parliament. Its task was 'to develop measures in order that such a disaster could not happen again'. The Delta Committee decided on a dike safety level that would, in theory, stop flooding from a coastal storm with a probability of 1/10,000 of it happening in any year, although some areas were given a probability of 1/4,000. It evaluated various options for major dike (re)construction. Economic considerations of improved connections, regional development and recreation tipped the scale in favour of the option of closing the main inlets with dams, which significantly shortened the length of the primary sea defences (CUR 2003).

Construction activities of this so-called Delta Plan officially started in 1956. In the early 1970s strong public concerns on the impact of a closed dam on the Eastern Scheldt ecosystem led to the adoption of an alternative, consisting of a storm surge barrier to be closed only during severe storms (see Figure 19.1). The completion of this Eastern Scheldt Storm Surge Barrier in 1986 was the last part of the Delta Plan.

The calculated level of safety from threats from the sea is now higher than anywhere in the world. With over fifty-two per cent of the population (more than 8.5 million people) living safely below mean sea level, and with ever-increasing value of infrastructure, the Netherlands has opted for optimal prevention, with minimal probability of having to deal with response (Van Veen 1962). At the same time, public awareness remains a priority, and much is invested to keep the technical and social infrastructure of flood forecasting and warning at state-of-the-art levels.

Consequences of coastal storms and floods

The consequences of a coastal storm encompass multiple types of damage. People can be injured or killed while their property can be damaged, such as residences, infrastructure and public facilities, contents in buildings, vehicles, crops and livestock. Other costs incurred include economic costs of rescue, evacuation, cleaning up and business interruption. Ecosystems can be damaged and livelihoods and social routines disrupted. Costs are also incurred through temporary accommodation, cleaning up, replacing items and moving back into repaired properties.

All these different types of consequences can be observed after large flood disasters, for example after the flooding of New Orleans due to Hurricane Katrina in 2005 (see Box 19.1). The damage is divided into economic and non-economic damage, depending on whether or not the losses can be assessed in monetary values. Another distinction is made between the direct damage, caused by direct effects of the wind or flood waters, and indirect damage that occurs outside the flooded area. For example, companies outside the flooded area can lose supply and demand from business and people within the flooded area.

In addition to these consequences, so-called chain reactions can occur. Floods can damage industrial installations, disrupt critical industrial processes or interfere with other infrastructure such as sewage treatment plants. Stored substances can be released and potentially undergo chemical reactions with water or air. Such effects can harm the environment, pollute drinking water and lead to additional fatalities (see Chapter 56).

Several methods and models are available to quantify types of damage due to coastal storms, including integrated methods for estimating flood and wind damage from hurricanes and storm surges. Methods for estimating direct economic damage due to flooding to physical objects (such as structures, houses) are well established (e.g. Dutta *et al.* 2003). Losses due to business interruption to the regional and national economy can be very significant and methods have been developed to assess such effects (Hallegatte 2008). Methods for estimating loss of life due to flooding have been developed that take into account information regarding the flood characteristics, an analysis of the exposed population and evacuation and an estimate of the mortality amongst the exposed population (Jonkman 2007).

Vulnerability and coping mechanisms

There are many mechanisms for avoiding or reducing losses and coping with them. These include individual household investments in wind- and flood-resistant housing, savings or assets that can be sold during recovery, remittances from family not affected, grants from governments, food/clothes/bedding, etc. aid from non-governmental organisations (NGOs) and faith communities; food for work programmes; migration; selling and pawning assets for consumption; reducing consumption; reciprocal labour; and using 'claims' (such as insurances) and loans. They also include community-based actions involving more than one household, such as the building and maintenance of storm shelters, micro-credit schemes, seed banks and mutual aid.

Evidence of the existence of these risk-reduction and coping strategies for floods and cyclonic storms is given by, for example, Winchester (1992) and O'Hare (2001). There are large differences in abilities to reduce risk and to cope with loss both within and among societies. For instance, see Box 19.3 for a case study from India. Even within households, women in India, Bangladesh and Pakistan are more likely to drown in a large coastal storm than men because they do not know how to swim, they wear confining clothing, and they are often attempting to protect small children while struggling to save themselves (see Chapter 35).

A household with good access to financial resources (via markets or via informal social arrangements) can borrow against future earnings to immediately rebuild asset stocks. Such a household might be expected to recover quickly. A household without this access may face a doubly slow recovery process, or even fall into the poverty trap (Carter *et al.* 2007). Because of these large differences in coping capacity, vulnerability is not equally distributed among people, even if they are exposed to the same hazard (Wisner *et al.* 2004).

Research findings from coastal floods occurring in central Vietnam revealed that flooding is an essential element for a coastal population, whose livelihood depends on productive functions of cyclical (freshwater) floods. The findings also revealed that floods, causing losses and damage, often inhibited economic development. The surveyed communities appeared to have evolved coping mechanisms to reduce the negative impacts of the floods, yet these coping mechanisms are under pressure due to environmental degradation (Tran *et al.* 2008).

The longer-term recovery process of a coastal flood disaster depends on many factors, both physical and social. Flooding with seawater affects soil fertility, which could take years to recover. High current velocities often result in significant displacement of sediments, leading to coastal erosion and blocking of waterways. Also the intensity of economic disruption plays an important role in the recovery process. However, the social and demographic changes are equally important and it is in this respect that we see big differences between countries.

Box 19.3 Coastal Andhra Pradesh: a history of cyclones

Coastal Andhra Pradesh belongs to one of the most cyclone-prone coasts of India, with more than sixty-five severe cyclonic storms in the last 100 years. A recent cyclone creating havoc was cyclone 07B on 6–7 November 1996, which caused 1,000 deaths and incurred heavy economic loss. Over 300,000 houses were fully damaged, 14,600 cattle lost and 346,000 hectares of paddy land destroyed. Many of the deaths were fishermen or their families, living in villages close to the shore (O'Hare 2001).

Despite the substantial losses incurred in cyclone 07B, there has been a general reduction in the number of casualties due to cyclones over the past decades. After the November 1977 cyclone that struck the Krishna Delta and left 10,000 casualties, most of the cyclones that made landfall along the Andhra Pradesh coast were characterised by much smaller human losses. This likely reflects an improved contingency planning system put in place by the government of Andhra Pradesh after the November 1977 disaster, in which the development of warning systems and evacuation procedures was made a priority. For example, early warning and the evacuation of 650,000 people at the onset of a cyclone in May 1990 restricted the number of casualties to 967 (Reddy *et al.* 2000). By contrast, economic damage showed an increase over the years, rising from US$38 million in 1977 to more than US$500 million almost twenty years later.

Poverty has been identified as one of the main factors influencing disaster vulnerability in India. Unfortunately, detailed statistical data on the distribution of human casualties and economic losses across social strata are lacking. However, housing conditions are a good indicator for direct susceptibility to damage and injury. Many poor rural families living in insubstantial village dwellings with mud walls and thatched roofs had their houses torn apart by the cyclone. Much less affected by house damage in almost every village were smaller numbers of higher-income landowning farmers and their families who live in more substantial brick and concrete dwellings (O'Hare 2001).

Based on qualitative interviews with several farming families, O'Hare (2001) found that a small number of farmers in the delta were bankrupted by the severe agricultural losses they suffered from cyclone 07B. However, the great majority were able to rely on savings and other resources to tide them over to the next harvest. Considerably more affected were the rural poor, especially landless agricultural labourers reliant on a meagre daily wage (O'Hare 2001).

This picture is corroborated by a longitudinal analysis over twenty years of economic development in the Krishna Delta in Andhra Pradesh, conducted by Winchester (1992). Whereas the rich and medium households were able to increase their assets (land, ploughs and animals), the poor households could not. The most successful households (the rich) were large and multi-occupational, and most of their family members were educated. These households had a variety of income sources and, over the years, had accrued enough surplus money to make them resilient to shocks such as from cyclones, mainly from assets and money lending. Almost the opposite was true in all respects for the least well off, who could barely help each other after calamities.

Cyclone impacts, while serious, do not significantly inhibit development of the coast and indeed are not the principal causes of poverty. Cyclones are only one of several hazards in a climatically drought-prone area that is subject to soil degradation and salt-water aquifer intrusion resulting in unsanitary conditions and the poor health of a significant percentage of the population.

Risk management for coastal disasters

People living on low-lying coasts have always invested in at least some sort of coastal defences, which they hope will reduce coastal flood risk. For instance, the construction of the first dikes in the Netherlands started as early as 1200 by the Cistercian monks. Around the same period the first dikes were constructed in the Red River Delta in Vietnam. Typical defence measures are breakers or groynes along the coast to dissipate the wave energy, embankments and sea walls that can resist the forces of the waves and remain intact during overtopping, sand nourishment for dune enforcements, and storm surge barriers across river estuaries. However important these traditional defence measures are, modern flood risk management also incorporates measures that aim to reduce loss of life in the coastal zone. These include safe havens, elevated construction on mounds, improved evacuation procedures and land-use planning.

Over the last decades, the capabilities to predict the occurrence of storms and cyclones, and to warn and evacuate the population, have improved (Schultz et al. 2005) (see Chapter 40 and Chapter 41). Contingency plans in combination with cyclone shelters and modern early warning systems have significantly reduced the number of casualties in, for example, Bangladesh and India. The warning time and thus the time available for evacuation depends on the characteristics of the storm and the presence and accuracy of warning systems. For example, for hurricanes in the USA the warning time is between two and four days, whereas it is one or two days for a storm surge in the North Sea.

The development of flood hazard maps for vulnerable coastal zones is an important tool for flood-conscious land-use planning. The EU Flood Risk Directive prescribes the issuance of flood hazard maps in the EU member countries, while the Flood Insurance Rate Map of the Federal Emergency Management Agency (USA) is used for insurance purposes. However important these sources of information are, it still proves to be very difficult to prescribe stringent regulations to keep risky places free from urban development. Hazard regulations that could promote risk reduction in the long run could turn into vehicles for short-term and short-sighted economic gain. Examples are coastal communities in the USA that rush to issue permits before the publication of new maps that show that the area to be built in is vulnerable to an increased or changed flood hazard (Armstrong 2000).

A complicating factor for coastal storms arises from the fundamental reason for living in this area: the connection to the sea is an essential livelihood asset, as exemplified by the harbour and related industrial activities, as well as the recreational value of the shoreline. This always makes structural measures a compromise. Traditional waterfronts are already a balance between tourists and livelihoods in many areas, such as the Mekong Delta and London. Storm surge barriers constructed across estuary mouths can be closed to prevent surges from entering the river, but need to be open under normal conditions to provide easy access for shipping and discharge of river water. Examples are the Thames Barrier near London and the Maeslantkering near Rotterdam. These apply mobile gates which can close or open the river mouth within hours. A similar barrier is currently being constructed for New Orleans.

Furthermore, the ability of coastal ecosystems such as coral reefs and mangroves to attenuate storm surges is increasingly being recognised (Mirza et al. 2005). Attention is given to the restoration of the natural capacity of coasts to mitigate storms, such as managed realignment of coastal embankments to create a buffer zone, rehabilitation of mangroves and artificial reefs. The advantage of these measures is that they often fulfil multiple objectives, including safety, valuing nature and recreational benefits.

The role of mangroves in reducing the sea waves is of particular significance. A rate of wave reduction has been reported of up to twenty per cent per 100 m of mangrove. For instance, a six-year-old mangrove forest of 1.5 km in width reduced 1-m-high waves at the open sea to 0.05 m at the coast (Mazda *et al.* 1997). Destruction of coastal mangroves in Orissa, India has reduced the buffer capacity of the coastal ecosystems and has facilitated storm surges and cyclonic winds in the region (Shiva 2002). The lack of protective forest cover also exacerbated flood inundation.

Conclusions

A large part of the world's population and many megacities are found in coastal areas that are susceptible to the potentially devastating effects of coastal storms. These storms can reach maximum wind speeds of more than 200 km/hr and lead to significant storm surge, wave action and extensive rainfall. Low-lying coastal areas that are often found in river deltas, atolls or areas that are or used to be salt marshes are especially susceptible to flooding and wave action. Sea-level rise is expected to increase the risks for those areas.

The consequences of a coastal storm encompass multiple types of damage, including loss of life, injury, property damage, livelihood disruption and ecosystem damage. Both wind and flooding cause extensive damage, but most casualties tend to result from floods, although a notable source of wind-related casualties is from iron sheeting being blown off roofs.

Coastal storm vulnerability should be reduced by a combination of approaches, such as investment in structural and non-structural measures for specific infrastructure, land-use planning, and action within communities and households. Top-down and bottom-up measures need to be implemented in tandem and combined to ensure that the people affected accept the measures, and the measures' limitations, without causing further problems. That includes considering, and being honest about the advantages and disadvantages of structural means such as dikes and storm surge barriers, ecosystem-based disaster risk reduction such as mangroves and other coastal vegetation, and long-term, community-based warning systems.

20

Thunderstorm and tornado

David Etkin

SCHOOL OF ADMINISTRATIVE STUDIES, YORK UNIVERSITY, TORONTO, CANADA

Kaz Higuchi

ENVIRONMENT CANADA, AND FACULTY OF ENVIRONMENTAL STUDIES, YORK UNIVERSITY, TORONTO, CANADA

George Platsis

SCHULICH EXECUTIVE EDUCATION CENTRE, YORK UNIVERSITY, TORONTO, CANADA

Introduction

Severe weather, including thunderstorms and tornadoes, contributes to disasters in many parts of the world. Extreme weather features such as hail, strong and gusty winds, downbursts, lightning, heavy rainfall and tornadoes can cause great damage to agriculture, homes and infrastructure, along with injury and death to people and livestock. As well, some of the major aviation disasters in the world have been associated with thunderstorm downbursts, which can produce tornado-like winds. This chapter describes some basic fundamental physical understandings and social responses to severe thunderstorms and tornadoes.

Thunderstorm hazards

A thunderstorm is an atmospheric phenomenon that is characterised by thunder and lightning, and, when severe, it is often accompanied by strong, gusty winds, heavy rain, hail and sometimes tornadoes. Climatologically, at any one time there are nearly 2,000 thunderstorms estimated to be occurring over the planet, resulting in nearly 16 million thunderstorms per year. These thunderstorms occur mainly over land, due to relatively low thermal capacity of the land, as compared to water, which results in quicker heating that gives rise to convection. For example, thunderstorms are 'in less than 10 per cent of the days over most oceans' (Dai 2001). Apart from the land-water bias in the spatial distribution occurrence of these storms, there is also a significant seasonal cycle (Dai 2001).

Thunderstorms occur mostly in the summer season (see Table 20.1). Although thunderstorms are most common in spring and summer, they can occur at any time of the year if certain meteorological conditions are met. All thunderstorms produce lightning and thunder, and are associated with cumulonimbus (meaning 'rain heaps' in Latin) cloud, colloquially called 'thunderhead'.

Severe thunderstorms are associated with deep moist convection (DMC) (Doswell 2001), something which is a major cause of weather-related property damage and loss of life. A typical DMC with a 5-km radius and 10-km depth can release energy nearly equivalent to a 25-kiloton bomb, and is associated with large hail, damaging wind gusts, heavy rainfall, and is quite often accompanied by tornadoes (Doswell 2001). Much of this energy is derived from the conversion of potential buoyant energy to kinetic energy. When the temperature of a rising air parcel is higher than the temperature of the surrounding air, 'positive buoyancy' occurs and forces the air parcel to rise. A vertically integrated value of positive buoyancy over a certain depth in the atmosphere results in what is called convective available potential energy (CAPE), a factor responsible for causing updraft.

Downdrafts are 'negative buoyancy' (or negative available potential energy) and are caused by evaporation and downward drag of air by heavy masses of falling precipitation (rain and hail), which are cooler than the environment. The leading edge of the downdraft is called the gust front, and sometimes these downdrafts can become quite intense, producing downward wind gusts (called downbursts) greater than those observed in some tornadoes. When these downbursts become intense in a localised area (around 5 km or less), they are called microbursts. Winds associated with these microbursts, sometimes more than 200 kilometres per hour (km/hr), resemble those produced by small tornadoes, and quite often can cause material damage that looks similar to that caused by a tornado. These microbursts are quite dangerous to aircraft and can cause major crashes, particularly during take-off and landing. Some of the major aircraft incidents, all of which had survivors, were crashes at Copenhagen Airport on 28 August 1971 killing thirty-two people; Dallas-Fort Worth International Airport on 2 August 1986 killing 135, including one person on the ground; and Phuket International Airport on 16 September 2007 killing ninety. They also damage buildings, infrastructure, and are the major cause of forest blowdowns in which trees are flattened. Dynamically, these microbursts are opposite to that of a tornado.

There are many different types of thunderstorms with a wide range of sizes and shapes, leading to many classifications of storm types. However, there are a few that occur quite frequently. Air mass thunderstorms, initiated by daytime heating, are one of the most common types, and usually occur somewhat randomly in the late afternoon/early evening. Hot, humid environments in the summer season are conducive to these thunderstorms. They are relatively short-lived (on the order of about an hour), small (a few kilometres) and are not usually associated with any major meteorological systems.

Table 20.1 Relative percentage of days of thunderstorm occurrences throughout the world

Hemisphere (summer season)	Percentage of days (per year) on which thunderstorms occur
Northern hemisphere	30–50 per cent (Southeast Asia, southeast USA, Central America, western Africa from 0° to 15°N)
	10–30 per cent (rest of USA)
Southern hemisphere	30–50 per cent (Africa, south of equator from 0° to 30°S)
	20–40 per cent (central South America)

Source: Dai 2001

Sea or lake breeze thunderstorms, which quickly dissipate in the evening, are caused by the conversion of air over land that results from the low-level circulation induced by the temperature contrast between land and water during the day. Squall lines are an organised line of deep convective cells and can extend for hundreds of kilometres. They usually occur along, or just ahead of, cold fronts and move relatively quickly. Quite often these squall lines contain supercells which are, spatially, relatively small, but can last for several hours, produce tornadoes, major hail storms, along with strong wind gusts on the order of 100 km/hr. There is another type of very destructive thunderstorm called meso-scale convective complex (MCC). These are non-frontal thunderstorms, spatially very large (hundreds of kilometres) and can last for ten to twelve hours, or even more. Many of these severe thunderstorms produce what are popularly known as the 'four horsemen' of thunderstorms: floods (which are not dealt with here – see Chapter 21), winds and tornadoes, hail and lightning. Each of these causes enough damage to be classified as one of the major causes of disasters.

Tornado hazards

A tornado is a dangerous, rapidly rotating column of air, usually associated with supercell thunderstorms, which becomes visible as a 'funnel cloud' when it picks up dust and soil (or any other debris) from the ground. When occurring over water, it is called a waterspout, and is made visible by the water it sucks up into the column. Air moves upward very rapidly around the tornado centre and is essentially opposite to what happens in a microburst, as described above (microbursts cause rapid downward movement of air that produces tornado-like damage).

The Fujita Scale (named after Professor T. Theodore Fujita, who developed the scale after many years of severe storm research at the University of Chicago) is commonly used worldwide as a measure of tornado strength and intensity based on observed damages (Fujita 1973). The scale ranges from F0 (weakest, little damage) to F5 (strongest, total destruction). One caveat, especially in worldwide application, is that the scale was developed based upon US data and can be difficult to apply where local building methods, materials and infrastructure differ from US building culture.

- Category F0: Light damage (with winds less than 73 miles per hour (mph)). Observed effects include signs torn off, small trees uprooted, branches broken off, minor damages to buildings.
- Category F1: Moderate damage (with winds 73 mph–112 mph). Observed effects include roof shingles blown off, mobile homes overturned or pushed off foundations.
- Category F2: Considerable damage (with winds 113 mph–157 mph). Observed effects include roofs blown off frame houses, mobile homes demolished, cars lifted off the ground, large trees uprooted or snapped, cars blown around.
- Category F3: Severe damage (with winds 158 mph–206 mph). Observed effects include roofs torn off well-constructed buildings, railway trains and heavy cars overturned, forest trees uprooted.
- Category F4: Devastating damage (with winds 207 mph–260 mph). Observed effects include cars thrown about, well-constructed houses destroyed, trees blown around and large missiles generated.
- Category F5: Incredible damage (with winds greater than 260 mph). Observed effects include car-sized objects thrown around at high speeds for hundreds of metres, well-constructed buildings lifted off foundations and thrown about and destroyed.

Seen by radar, the region in which the tornado occurs within the supercell is identified by a configuration that looks like a hook, and thus is called a hook echo. A smaller tornado is usually composed of a single vortex, but a larger, and more intense, tornado is characterised by several smaller 'suction vortices' (on the order of 10 m in diameter) that normally rotate counter-clockwise around the tornado centre (normally small meso-scale rotations are influenced by local circulation and not by Coriolis force). Some major damages are caused by these small, yet intense, suction vortices.

Globally, the USA by far has the greatest observed number of tornado occurrences (Dotzek 2003), in an area located mostly in the Midwest, and in Canada, into an area that extends into Ontario and prairie provinces. The popularly known 'Tornado Alley' in central USA is characterised, meteorologically, by the so-called Dry Line, where southeasterly moving dry and cool continental air overrides warm, moist and unstable air that is advected from the western sub-tropical Atlantic Ocean and the Gulf of Mexico, and is a situation arising mostly in the spring. The ensuing synoptic conditions are conducive to the formation of severe, deep, moist convections that lead to the creation of large supercells and squall lines. In many cases, tornado occurrences are a 'natural' outcome.

The USA, on average, has about 1,200 observed tornadoes per year, whereas the UK has about thirty-three per year (Holden and Wright 2004). Tornadoes in Europe are generally small and cause only minor damages. Canada experiences, on average, about eighty to one hundred observed tornadoes per year, but they are mostly in the F0 and F1 categories (Etkin et al. 2001); however, there are a few F2 tornadoes that touch down every year. Other areas in the world where frequent tornado occurrences have been observed include South Africa, Australia, New Zealand, Bangladesh (see Box 20.1), eastern China and southern Brazil. Bangladesh experiences the highest number of deaths due to tornadoes, at about 179 per year (Finch 2010), whereas the USA only has about fifty deaths per year, even though it experiences more F4 and F5 tornadoes than any other country in the world. The situation in Bangladesh is primarily due to high population density, poor tornado warning systems and inadequate construction.

The effects of climate change

Many climate models are suggesting a more variable climate in coming decades, with increased probability, and magnitude, of extreme weather events (see Chapter 18). It is practically impossible today to predict individual storms (thunderstorms and tornadoes) even a day ahead, let alone decades ahead under a warming climate. However, there is knowledge and understanding of regional-scale dynamics that are associated with deep moist convection and are exploitable in order to estimate the probability of change in both the frequency and magnitude of extreme weather events.

According to global climate model simulations, with increasing atmospheric carbon dioxide, surface temperature in many regions throughout the world will increase. Also, a warmer atmosphere will be able to hold more moisture and intensify the hydrological cycle, thus leading to an increase in moist static energy. For regions that presently experience severe thunderstorms and tornadoes, increased surface temperature will likely cause an increase in the number of days in which severe storms can occur, and when these storms do occur they will tend to be more intense. Overall, if our scientific understanding of climate change is generally correct, then the frequency and intensity of extreme weather events will increase, with an accompanying rise in disasters, unless society reduces vulnerability to these extreme events.

Myths and fallacies

Thunderstorms have played an important part in myth and religion since ancient times. For example, the Norse god Thor is the God of Thunder (also known as the God of Rain and

---

Box 20.1 Thunderstorms and tornadoes in Bangladesh

The Indian subcontinent is prone to numerous forms of disaster, with Bangladesh being one of the most susceptible areas, including thunderstorms and tornadoes. What is thought to be the world's deadliest tornado so far occurred on 29 April 1989 in Bangladesh, in which about 1,300 people were killed and 12,000 were injured. Because so much of the rural housing is made of mud, straw, bamboo or corrugated steel, communities are particularly vulnerable to tornadic winds. Corrugated steel and iron become deadly projectiles and affect people in two ways: first, via the initial impact to the person; and second, via wounds that can become easily infected. In the 13 May 1996 disaster, ninety-nine per cent of those injured had multiple injuries caused by flying sheets of corrugated metal that had been used as roofing. In eighty-four per cent of these cases, people needed antibiotics for infection control to stop the spread of sepsis (which ultimately killed at least seven per cent of those hospitalised).

Sadly, Bangladesh is no stranger to deadly tornadoes, with significant events reported in 1888, 1964, 1989, 1993 and virtually every year from 1995 up until the time of writing in 2010. Incomplete reporting suggests that the situation may be worse than what these data suggest, especially historically.

Apart from the immediate impacts of cyclones (see Chapter 19) and severe thunderstorms, there are longer-term consequences. Heavy rainfall results in soil erosion, soil moisture saturation and run-off, contaminated water, grain and fruit spoilage, and nutrient deficiency. These can result in increased disease, disruption of important socio-agricultural activities and thus lower food production, and transportation interruption. Additionally, the poor health care structure of Bangladesh makes much of the population highly vulnerable to the after-effects of these disasters.

An important factor in terms of how communities cope with these events relates to how deeply religious people are. Nearly one-half of the people in a post-disaster survey indicated that these events were acts of God, and accepted them on that basis. Of the remainder, only about ten per cent identified the tornado as a natural phenomenon. While that can work as a coping strategy for dealing with losses, it can also be a barrier to disaster risk reduction, contributing to the high level of deadliness and destruction that is witnessed in such events in Bangladesh.

Fertility), and his weapon was a hammer, which represented lightning. Ancient Greek and Roman stories tell of giant eagles or Thunderbirds ruling the skies, and that lightning was the result of their eyes blinking during thunderstorms (Frydenlund 1993). Other ancients thought that the gods, especially Zeus, sent bolts of lightning to Earth when angered by humans. Trees, or other places hit by lightning, were often used as sites to build shrines or temples. In the Old Testament the appearance of God was often accompanied by thunder and lightning, and this theme is also present in the Muslim Koran.

When severe storms threatened early Americans and Europeans, one of their responses was to go to church, where it was believed that God might protect them, and sometimes these churches were used for munitions storage (for the same reason of protection). Unfortunately, because of their height churches were more likely to be struck by lightning than many surrounding places. In 1856 the church of St Jean, Rhodes exploded when powder stored in its vaults was struck by lightning, killing about 4,000 people. Other lightning-related explosions occurred in Tangiers in 1785, Venice in 1808, Luxembourg in 1815 and Navarino in 1829. One French churchman, Nollet, believed that 'church bells by virtue of their benediction should scatter the thunderstorms and preserve us from strokes of lightning' (Frydenlund 1993) and disputed Benjamin Franklin's newfangled lightning rods.

Myths and fallacies about thunderstorms persist today. Examples include: getting hit by lightning is always fatal; buildings are lightning proof; surge protectors always protect; and rubber shoes or tyres will protect you. Lighting also exists in indigenous mythologies. For example, Iroquois stories of creation say that a turtle crawled around the sky collecting lighting as she went. She threw two balls into the sky, one that became the sun and the other that became the moon. People also have many misconceptions about tornadoes. For example, some believe that tornadoes are always preceded by hail, cars can be carried over tall buildings, mobile homes attract tornadoes or that opening windows will save a roof or a home from destruction.

Impacts of thunderstorms and tornadoes

Severe thunderstorm beneficial impacts

Impacts from severe thunderstorms are generally harmful. The main beneficial impact results from their contribution to agricultural systems and the ecosystem from their rainfall, especially in tropical regions. In some regions they contribute a majority of the required growing season precipitation, such as in southwestern USA, where they account for about seventy per cent of rainfall. Forest fires from lightning are also part of the natural ecosystem, and allow for the generation of new growth. The regeneration of some tree species, such as the Jack Pine (which cannot reproduce without fires), prevents the build-up of undergrowth that might result in larger catastrophic fires, and creates about twenty per cent of the global oxidised nitrogen, which then becomes available as a soil nutrient (see Chapter 24). Strong winds also blow down old trees, creating log ecosystems and opening up space for new growth. Thunderstorms play a role in the atmosphere's chemistry with updrafts (transporting pollutants aloft), nitrogen fixation in the soil and storm rainfall (scavenging vast amounts of pollutants from the air). The absence of thunderstorms, such as in the summer of 2005 in Illinois, led to US$2 billion in crop losses and is often a key factor in droughts (Chagnon 2006).

Severe thunderstorm harmful impacts

No accurate global damage database exists regarding thunderstorm deaths and damage, as reporting is highly variable amongst different countries. Events leading to large loss of life tend to be better reported and recorded more often than smaller events, which means that the cumulative impact of thunderstorm hazards is likely underreported in most parts of the world. Deaths from tornadoes and lightning in the USA are slightly fewer than from many other natural hazards such as drought, flood and heat waves, accounting for eleven to twelve per cent each. 'Based on the data provided by Property Claims Services, since 1949, tornado, hail and straight-line-wind losses account for more than 40 per cent of total natural economic losses in the USA' (Daneshvaran and Morden 2007). Thunderstorms in the Sahel in Africa, which tend to occur from May to July, cause soil erosion that is of serious concern to farmers through two mechanisms: dust storms associated with strong downdrafts and intense rainfall.

Lightning impacts

Lightning injures people through four main mechanisms that create either electrical currents or mechanical trauma: by direct strike; by contact where the victim is indirectly injured by touching an object that is charged by a lightning strike; by sideflash where the victim is injured when charge from a nearby object or other person flashes or splashes through the air to the victim; and by step voltage where lightning hits the ground or a nearby object and travels through the ground to injure the victim (Cooper 2000). Fires started from lightning can be a significant cause of casualties, but one that is often not reported. Over the past century the number of recorded injuries and deaths from lightning has been decreasing in urbanising and industrialised societies, from about five deaths per million in the early part of the twentieth century, to about 0.4 deaths per million by the end of the century. This occurred in parallel with the reduction of rural population (Holle 2008), as fewer people are exposed to this threat.

Data by country vary widely, ranging from near zero for Ireland and Lithuania, to 71.5 per decade per million people in Yemen (Holle 2008). Other countries and places with high lightning fatalities include rural South Africa, Zimbabwe and Hainan, China. In industrialised societies, men tend to be killed by lightning more often than women (Ripley 2008). At times, clusters of deaths can occur, as in Ndwana Village in South Africa, where there were fourteen deaths in less than thirty days (Dispatch Online 2010), though fatalities only occur in about nine to ten per cent of strikes (Holle 2008). Disability often results in the non-fatal cases, and consists of 'short-term memory loss, inability to process new information, personality changes, easy fatigability, decreased work capacity, chronic pain syndromes, sleep difficulties, dizziness, and severe headaches'.

Tornado impacts

Tornadoes cause injury and death, as well as damage to buildings and infrastructure (particularly power generation lines). Most of the deaths occur in mobile homes, which, though few in comparison to residential structures, are very vulnerable to strong winds. This results in greater proportions of deaths amongst poorer or marginalised populations. Underreporting again can be a significant problem, especially for weaker tornadoes. The USA has more reported severe thunderstorms and tornadoes (usually more than 1,000 per year) than any other country (Doswell 2003), and extreme events, such as the May 2003 outbreak in the USA (which spawned more than 400 tornadoes and 1,000 hail storms), resulted in losses of more than US$3 billion (Daneshvaran and Morden 2007). In the USA tornado deaths have been decreasing over time due to improved warnings and technology, being forty per cent lower in 1999 than in 1950. This trend appears to have occurred mainly after 1923 (Doswell 2003). In Canada, during an average year, approximately eighty tornadoes occur and, on average, cause two deaths and twenty injuries, plus tens of millions of dollars in property damage. The deadliest tornado known occurred in Bangladesh in 1989, and was estimated to have killed about 1,300 people. Bangladesh reports more than six tornadoes per year, a very small number compared with North America, but suffers disproportionately due to a lack of warning systems, poor construction standards and vulnerable populations. European tornadoes are generally weak, and therefore cause much less damage than stronger North American ones. Tornadoes in South Africa have caused hundreds of injuries and dozens of fatalities, while Australia and South America have had very few deaths.

Economic damage in the USA from tornadoes has not shown the same decreasing trend as mortality (Doswell 2003). Good statistical data are lacking from other parts of the world, so it is hard to generalise. One might infer, though, that countries without tornado warning systems are not experiencing the same decreasing trends. Warnings can save lives if people take appropriate action, but can do little to reduce property damage.

Hail impacts

The main impact of hail is the damage that it causes to buildings and crops, though deaths and injuries to people and animals also occur. In May 2010 a severe hailstorm apparently killed hundreds of deer and ninety per cent of pheasants and hares in Salzburg, Austria, resulting in a hunting ban. See Box 20.2 for hail in Africa. The amount of damage depends upon the size and density of the hailstones, and the angle with which they strike crops. Severe hailstorms can result in total crop loss and be devastating for farmers if they do not have hail insurance. Motor vehicles are more sensitive to hail density than size, but even small hail can cause building damage by blocking roof gutters and drains. The largest recorded hailstone was reported in Bangladesh during a storm in 1986 that killed over ninety people and 1,600 cattle – it weighed almost one kilogram.

Leigh (2007) found that in the period 1900–2003 hail accounted for about twelve per cent of building damage in Australia, after tropical cyclones, flood and wildfires, and accounted for five of the top ten, and thirty-one per cent of insured losses from natural disasters in the period 1967–2007. The most costly disaster occurred in Sydney in 1999 and amounted to over US$3 billion of insured losses, an amount much larger than in many other countries (in Switzerland for example, hail losses average nine per cent of insured claims).

Flash floods are covered in Chapter 21.

Box 20.2 Hail in Africa

Though statistics on hail in Africa are hard to find, there is ample anecdotal evidence to suggest that it is a significant problem, leading to damaged crops and houses, injuries and deaths, though it is often difficult to separate damage due to hail from that due to strong winds and heavy rain. The severity of hail is illustrated by the case of a storm in Klerksdorp, South Africa, in October 2009, in which the hailstones were 20 cm deep in some places.

In February 2007 a hailstorm in Cairo, Egypt killed four people, injured more than fifty and created traffic chaos. Storms in parts of Muzarabani, Zimbabwe in November 2009 killed two people, and destroyed homesteads and tobacco crops. Because of flood risk associated with hailstorms, the government has warned people to avoid living in flood-prone areas and to relocate to higher ground, but this has reportedly not happened, likely due to many factors, including maintaining livelihoods. Other hailstorms, such as that in the Amajuba District Municipality, South Africa, while not deadly, left over 1,200 people homeless.

In April 2009 in parts of the northern district of Amuru, Uganda a hailstorm resulted in lost crops, left over 10,000 people homeless and killed livestock. Because this came after a long dry spell, most residents were left with no source of food. Another storm in Kenya in October 2005 destroyed 561 hectares of tea crop.

Social vulnerability and coping strategies

The dynamic and interactive relationship between hazard and human vulnerability has been the ongoing concern of practitioners and academics for some time. Mid-twentieth-century disaster management approaches focusing on physical processes and engineering solutions, or rational decision-making, as a way to minimise risk have been broadened to include much more complex social/ecological/political paradigms which focus on unequal distributions of power, resources, wealth and knowledge (Hewitt 2007; Wisner *et al.* 2004), social traps (Rothstein 2005), adaptive and flexible management approaches, and human biases in the perception and assessment of risk (Tversky and Kahneman 1974). The degrees to which different paradigms are useful depend, in part, on the nature of the hazard and the intentions/interests of the user. Hazards that are more predictable spatially, such as floods (as opposed to tornadoes), offer more opportunities for people with more resources to mitigate or avoid risk, while those with fewer resources are often more exposed to the hazards. For example, richer people can choose to live outside of (or inside) flood plains, yet in many cases still remain in the same community, whereas choosing to settle away from tornadoes or thunderstorms would make residence impossible in several countries. However, at a micro scale, economic class and other social factors do appear to affect mortality from tornadoes. In the USA, for example, the chance of dying in a tornado is ten times higher for people living in pre-fabricated (mobile) homes than for those in permanent housing (Sutter and Simmons 2010).

Within the broad scale of atmospheric hazards, those related to thunderstorms lie at one end of the spectrum described by micro-scale processes. Though intense, they are of short duration and small spatial extent. They may cause disaster in a single community or, on very rare occasions, several communities, but larger social structures, such as provinces and states, are usually able to respond, and recover, using their own resources (unlike larger disasters associated with, for example, some extreme hurricanes or earthquakes).

For lightning, the primary exposure to health hazards results from rural livelihoods. Urban residents are much safer and, as with hail and tornadoes, lightning's ubiquitous nature makes people similarly vulnerable from a spatial perspective. Flash flood is very different, given its geographical specificity, and vulnerability to it is very much a function of where people are either forced to or choose to live, both of which are usually a function of social stratum and economic necessity. Vulnerability to tornadoes also depends strongly on wealth, as shown by the high ratio of deaths that occur in mobile homes mentioned above, and the large death tolls that have occurred in rural Bangladesh. In these cases, wealth equates to options for safer choices and capacity to reduce or share risk (such as buying insurance, government disaster assistance and high-tech weather warning systems such as Doppler radar), and is largely dependent upon individual and national wealth.

Burton et al. (1993) identified several coping strategies used by society to adapt to natural hazards, which include accepting losses, modifying hazard or vulnerability, and changing use or location. As noted above, not all people or societies have access to all of these strategies, or avail themselves of them even if they do. This set of strategies is thus presented with the caveat that successful disaster risk reduction largely depends upon social, cultural and political context – and changing those contexts so that people do have access to the strategies and can select amongst them is essential for disaster risk reduction (e.g. Hewitt 2007).

Accept losses

Each strategy contains both benefits and costs, which can mean that sometimes a combination of strategies may be best. The simplest approach is to bear the loss. This option does have advantages, namely that there are no up-front or mitigating costs associated with this choice. However, over the long run, even one disaster can result in total loss. If potential losses are catastrophic, a more rational choice, if affordable, may be to share the losses. With this strategy, the loss is spread over a group of people and is sustained by some sort of common fund (such as private insurance, government assistance or informal, kinship-based mutual assistance). Relief by non-governmental organisations (NGOs), faith communities and other civil society groups would also fit into this category.

Traditionally, in more affluent countries private insurance is commonly used. In Canada, the Insurance Bureau of Canada reports that between 1983 and 2006, about seventeen per cent of what they call 'natural catastrophe' payments were for hail, seventeen per cent were for flood and seven per cent were for tornado (IBC 2008). Data from Munich and Swiss Re suggest that global trends in 'natural catastrophe' payouts have been increasing over time. Flash floods, hail and tornadoes are referred to as secondary perils, which contribute about thirty per cent of total insured natural catastrophe losses. In less affluent countries, as well as Japan, insurance for these hazards tends to be seen as a luxury, cannot be afforded by most of the population or is not accepted as a cultural norm.

Government-based relief plans are common, generally based upon an exceedance threshold or by political decree, often have caps, such as A$130,000 in Australia (Emergency Management New South Wales 2010), or may only come in the form of tax relief, such as in Canada, in addition to direct payouts. These plans all have been critiqued as encouraging maladaptive behaviour, as they encourage development in risky areas without taking adequate measures or, in the cases of thunderstorms and hail, they can encourage disregard for the hazard on the assumption that direct payouts will cover any losses.

Reduce losses

Reducing losses can be accomplished by addressing either hazard or vulnerability. Historically there have been various attempts at hazard modification through cloud seeding in order to reduce hail activity (see also Box 20.3). Though there are still some active weather modification programmes in twenty-four countries, generally this sub-strategy is considered to be ineffective (WMO 1995; National Academy of Sciences 2003). No technological solutions are accepted for reducing the lightning (Zipse 1994) or tornado hazard prior to the hazards interacting with people or infrastructure.

With respect to vulnerability, Chapter 21 addresses floods. Hail impacts can be reduced through the use of hail-resistant roofs and protective coverings over windows and other glass openings. Lightning rods, air terminals and zones of protection can minimise lightning impacts. Buildings can be constructed to be highly wind resistant, and safe rooms can protect people against even strong tornadoes, though they may not be cost effective as there is a price point where the demand for mitigating measures becomes zero. For example, one tornado study conducted in Oklahoma showed that a price point of US$6,000 for a safe room would essentially drop demand to zero, but numerous financial mechanisms (see Chapter 54) exist that can change such models if society wishes to implement measures for which society would pay. Also, the use of Doppler radar, particularly at airports, can detect downbursts and strong wind shear that are dangerous to aircraft.

Rural people have also learned over the years to recognise the threat of hail and microbursts of wind associated with thunderstorms. In rural Tanzania there are low-cost techniques of anchoring roofing tin to rafters, and in the highlands of rural Lesotho older herdsboys teach the younger ones how to shelter under cows to protect themselves from large, potentially injurious or even deadly hailstones.

Changing use or location

In many cases, populations choose to settle in hazardous areas because those areas offer benefits, such as arable land or access to transportation routes. Using the arable land example, flash floods, a result of thunderstorms, often can move rich sediment deposits into an area or provide necessary

Box 20.3 Hail in France

Freddy Vinet
EA 3766 Gester, Université Paul Valéry – Montpellier III, France

France is prone to hail, and reducing the risk of disaster has long been a serious issue for farmers, who contribute significantly to the country's economy.

Regions affected by hail are those prone to thunderstorms. In winter hail falls are frequent but do not wreak serious damage because hailstones are smaller and crops less vulnerable. In summer (May–September), hailstorms associated with powerful convective cells are more harmful. Hail threatens buildings, cars and crops. On 15 May 2008 and 25 May 2009 two hailstorms affected the city of Toulouse. Some 90,000 vehicles were damaged and thousands had not yet been repaired one year later.

Agriculture is the most vulnerable economic sector. Every year, damage ranges from €100 million to €150 million. Crops such as wheat are less vulnerable, while others such as orchards are more vulnerable. Apples are the most crucial because they are exposed from May to September and are considered as highly liable by the insurance industry. Apples are also particularly vulnerable as they are consumed as fresh fruit, while grapes, for example, are transformed.

Thus, the geography of hail risk in France includes regions that both are exposed to thunderstorms during summer and concentrate vulnerable crops like orchards: the Garonne basin, intra-mountainous valleys in the Alps, and the Rhone Valley. In some areas of southern France the annual expected mean loss due to hail is thirty per cent, i.e. a hailstorm would destroy 100 per cent of the crop every three years.

Hail risk mitigation has been shifting from reduction of hail (weather modification) to protection of threatened crops. Until the mid-twentieth century in Western Europe, the favoured means of protection were cannons. The noise of the deflagration was supposed to split the hailstones. In the 1970s rockets proved insufficient, although local farmers' associations continue to resort to such means of protection. Today weather modification is widely practised in France. When thunderstorms are forecast, ground generators disseminate silver iodide into the atmosphere. The ice nuclei are supposed to facilitate the formation of many small hailstones that melt while falling.

Focusing on the hail hazard is not sufficient to reduce the risk. Indeed, the vulnerability of farms has increased because the agricultural sector has had to comply with the requirements of international markets. Until the 1950s, farmers produced fruit for their own consumption and were therefore able to overcome a temporary loss of production. Now arboriculturists who sell their fruit on the national and international market must provide a constant flow of quality produce in large quantities. Insurance can compensate for the losses, but it neither replaces the lost crops nor the loss of customers.

As an alternative, after the 1992, 1993 and 1994 stormy summers, farmers started covering their orchards with nets to protect from hailstones. In 2007 almost one-half of the apple orchards and twenty percent of the apricot orchards of southern France were protected with anti-hailstone nets. As a result, in some regions, e.g. the valleys of the southern Alps, orchards are now protected against frost (sprinklers), drought (irrigation) and hail (nets). Another way to reduce the vulnerability is for farmers to diversify their production to be less dependent on weather conditions and market constraints.

water for crops. In other cases, such as Bangladesh, relocation is impossible because the entire country is exposed to thunderstorms. Therefore, changing location might not be an option available due to lack of places to move to, or lack of social, political and economic context that permits the choice to be made. However, some rural people in Bangladesh do build their homes on elevated mounds of earth, providing some protection from flooding.

In areas subject to hail, lightning or tornadoes, changing location is generally not viable since the hazards are so widespread. Flash floods from thunderstorms are more spatially delineated, in that higher ground can be identified with less likelihood of flash flooding. Another option is to change use of the land, such as changing flash flood-prone land from housing developments to parks or wetlands. That has been successfully completed in Boulder, Colorado, USA and Toronto, Ontario, Canada, but there are complex societal issues that need to be considered, including land rights, compensation, power structures that exclude affected people from decision-making, and the use of hazards as an excuse to move people off land that powerful interests would like to acquire.

Conclusions

Many of the impacts from thunderstorms and tornadoes are preventable and predictable, yet much more could be done to prevent harm and disasters. In many cases it is suggested that catastrophe losses could be reduced by as much as eighty-five per cent for a minimal cost (Heffes 2008). Education (see Chapter 62) and public perception of hazards could be the key to implementing successful strategies. Lopes (1992) discovered that showing disaster images, including tornadoes, to a population in the USA resulted in little effect on how well the people prepared for disasters. Rather, the population must know that the disaster can happen to them and also be shown what to do in case of disaster. Some of the results from Ripley (2008) indicate that some such results are cross-cultural, but further investigations are needed, especially working with people from different cultures to learn what might work in each setting.

Similarly, government agencies designated as being the focus for disaster activities can provide good starting points for dealing with thunderstorms and tornadoes. Not all government jurisdictions, national or sub-national, have such agencies and the level of competence by and trust in government institutions varies substantially around the world (see Chapter 52 and Chapter 53).

Hazards related to severe thunderstorms can, and do, cause significant property damage, injury and death. Considering the prevalence of these hazards around the world, and the many examples of disaster risk reduction strategies that have worked and that have not worked, continued exchange of knowledge and actions can help to tackle people's and communities' vulnerabilities.

21
Flood

Hanna Schmuck

CONSULTANT, MAPUTO, MOZAMBIQUE

Introduction

Flooding is a natural part of the hydrological cycle and the life of rivers. It can be made more frequent and more severe by human actions such as deforestation or interference with the river channel, but people have been living with floods for centuries. Floods can be deadly and destructive, whilst they provide silt to nourish deltas and to fertilise crops and seasonal fisheries. At least 3,686 flood disasters occurred between 1900 and 2010. Each year at least 62,000 people die from floods and at least 28 million are estimated to be affected (EM-DAT 2010).

Economic losses are highest in Asia, followed by Europe and the Americas. This, however, does not mean that a family in Africa, for example, suffers economically less from a flood than a family in Europe. In fact, the loss of a basic house with household utensils, which may be worth about US$200 in monetary terms, is likely to be more disastrous for a family than the destruction of a family house in Europe. Whilst a family in Europe will get assistance from the government and to a certain degree also from insurance companies, a family in rural Africa can only hope to get temporary shelter in camps and some assistance from a non-governmental organisation (NGO) working in that area. If the harvest and seeds are destroyed, recovery can take years.

Flooding as a physical process

How harmful a flood is to human beings and their livelihoods depends on many factors. Apart from social and cultural parameters discussed below, the physical characteristics of the flood itself and the flooded area must be considered. Of great importance are the hydraulic characteristics of a flood: depth and velocity of the water, speed of water-level rise within a certain time, kind and height of waves and duration of the flood (Jonkman 2005; Kelman and Spence 2004). The kind of sediment and debris also make a difference to the damage a flood causes – whether the water contains only fine sand, or is loaded with uprooted trees, cars, parts of buildings and bridges, rocks or even ice. Also the causes of floods play a role: whether they are related to events in a river's catchment area (such as heavy rains, deforestation or even an earthquake or mudslide) or, in the case of coastal flooding, from a storm surge or tsunami (see Chapter 27).

244

There are many types of floods. Coastal flooding related to tropical cyclones or tsunamis involves seawater. Such floods are treated elsewhere in the *Handbook* (see Chapter 19 and Chapter 27). This chapter focuses on fresh-water flooding, which is mostly caused by exceptionally high rainfall, melting of snow or a landslide upstream that can cause an inland tsunami, or the breach or failure of a river training structure (e.g. dike or dam).

Kelman and Spence (2004) provide an explanation of the physical processes of floods and their impact on structures. Water contact with a structure and its contents – brick, wood, furniture and electronics – can result in irreparable damage by inundation or, if items are re-useable, they can take a long time to dry out and repair. As flood waters rise, the water leads to a pressure on structures. If the flood waters rise rapidly, then the difference between the outside pressure and the lack of water inside can cause structural damage or even lead to collapse. Flowing water adds pressure to structures, contributing to structural damage. Objects and structures in water experience an upward pressure due to the fact that objects less dense than water float. For example, although bricks forming a house are heavy, the entire house is mostly air. So it can be less dense than water and can be pushed up when it is inundated, making it easier for the pressure from flowing water to push a building off its foundations and then float down the river. There are many other ways in which floods damage structures, such as flowing water eroding around and underneath a structure; the force of waves going past, through or breaking on a structure; chemical contaminants; and debris within the water such as stones, cars and other broken structures hitting with a huge force.

As human beings naturally float, they are pushed upwards when the water gets deep and are pushed over by the force of flowing water or when hit by debris. Waves and debris cause physical trauma, cold can kill directly through hypothermia, or cold, panic and loss of consciousness from being struck by debris can lead to drowning. Another cause of death, especially in urban areas, is electrocution (Jonkman and Kelman 2005).

Drowning is the most common cause of mortality in floods and their immediate aftermath, but disease, electrocution, heart attack, snake bites, suicide, carbon monoxide poisoning and fire or smoke contribute to the toll. Analysing when flood deaths happen – before, during or after the flood – yields useful patterns. Deaths may occur during evacuation in traffic accidents. Heart attacks sometimes occur when people try to carry furniture upstairs or deploy sandbags. During floods, causes of death in rural areas frequently include snake bites, spider bites and crocodiles, amongst other wildlife. After floods, disease may kill weak individuals.

Seasonal floods may be expected and hence people are better prepared than for untimely ones. For example, if inhabitants of flood plains know that every spring snow and ice melt increase the river's water level and flow rate, they tend to be better prepared to protect themselves than if a breach of a dam upstream causes a flash flood (see Box 21.1). Indeed, some farming systems have evolved with close attention to seasonal floods, such as 'flood retreat' (or 'regression') farming along the Senegal (Horowitz 1995; Pratt *et al.* 1997) and Niger Rivers or rice farming and fishing in parts of South Asia (discussed below). Increased climate variability could complicate decision-making by farmers in such flood-dependent systems because their inherited knowledge is becoming less relevant (see Chapter 18). Also, the time of day a flood occurs may make a difference. If a sudden increase of the water level occurs in the night when people are sleeping, that is different than during the day when people can observe the event.

Flooding as a social process

Whether flooding has increased over the last few hundred years is controversial, but flooding affects many more people, at least in part because of increased numbers living in or near flood

Box 21.1 Flash floods

Eve Gruntfest
Colorado State University, USA

The simplest definition of a flash flood is 'too much rain; too little time'. As with all simplified explanations, there are numerous exceptions, so flash floods are difficult to define. The most catastrophic flash floods often result following days of steady heavy rain followed by additional, extremely heavy rains when the ground is already so saturated that major slope failures occur, bringing entire settlements with them. Flash flood impacts thus depend on topography. The most common flash floods result from heavy urban rainstorms that occur over intensively developed urban areas where the rainwater exceeds the capacity for storm water management. In many cities, storms that have twenty-five- to fifty-year recurrence intervals or longer can be deadly as motorists with no flood experience try to cross flooded roads.

Flash floods pose forecasting difficulties because they are so local in their impact. They also provide short lead times, combined with the possibility that the flash flood impacts can be felt far downstream from where the rain occurs.

In 1976, 145 people were killed during the night of the Big Thompson flash flood in Colorado, USA. Of the approximately 2,500 people in the canyon, most received no official warning that a catastrophic flash flood was imminent. Seven deaths seemed avoidable—initially these people reacted correctly: once they were aware of the flood's approach, they immediately went to higher ground. They then miscalculated the flood's actual moment of arrival and returned to lower ground to move a vehicle or collect something, and they were killed.

There have been advances in warning systems. These include automatic phone call systems and sirens linked to systems of stream and rain gauges, Doppler radar and satellite imagery. Unfortunately, flash floods often occur in catchments too small for the rainfall signal to appear on Doppler radar. However, even if heavy rain or its potential can be detected on radar, it may be difficult to notify campers and other non-residents (e.g., tourists) and cell phone reception can be poor in mountain canyons.

Flash floods are closely linked to other short-onset, deadly phenomena such as debris flows. In Venezuela, rainfall and flash flooding induced landslides in December 1999, causing property destruction and deaths. In this area, the alluvial fans are the only areas where slopes are not too steep to build. Rebuilding and reoccupation of these areas requires careful determination of potential hazard zones to avoid future loss of life and property.

After wildfires in mountainous areas, slopes have little vegetation, so rainfalls can cause serious mudslides and debris flows. Southern California has major post-fire debris flow events every few years, so communities there are now initiating mitigation efforts before the start of the winter rainy season, which helps to save lives and reduce damage.

plains and coasts (Wisner *et al.* 2004), as well as manipulation of rivers by engineering (Etkin 1999; Fordham 1998/1999). Rivers have always attracted human settlement. Ancient civilisations such as those of Mesopotamia and Egypt developed along rivers. Humans do not only settle along rivers to benefit from them for agriculture and transport, but also because of their beauty. People enjoy staying near rivers and watching their flow, and in some cities the riverfront has become a zone of entertainment and tourism. When the benefits of living close to a river are perceived to outweigh the negative impact of flooding, it is difficult to convince people to leave their land permanently for higher and safer areas.

In the USA, where national flood insurance has existed since 1968, rules were tightened in 1994 by the National Flood Insurance Reform Act. This followed the large-scale Mississippi flooding in the previous year and because thousands of people had claimed three or more times for flooded property, while refusing the offer of a government buy-out (FEMA 2004a; Kentucky 2010). Around the world, reasons for people refusing or being unable to evacuate in advance of a flood warning include desire to protect their homes or livestock; distrust of the flood warning or of the authorities' ability to provide appropriate evacuation centres; belief that they will not be harmed or will be protected by a deity; lack of resources to evacuate or to find accommodation elsewhere; or limited ability to understand the circumstances or to react in a way that the authorities direct (Handmer 2000; Twigg 2003; Thomalla and Schmuck 2004) (see Chapter 41). In such cases, governments have had to seek the help of the police or military to force people to evacuate. For example, in the 2007 floods in Mozambique, the government of the country deployed the army to evacuate inhabitants from the islands in the Zambezi river to camps on the mainland which were established by the government and aid agencies.

Whilst in non-industrial areas the main threat of floods is for homesteads, crops and livestock and infrastructure as well as people's health, river water can be highly contaminated and hence dangerous in cities and industrialised areas. This is especially the case where city planning is deficient or lacking (see Chapter 53). Flood water can be polluted with sewage, petrol, paint, household cleaners, industrial chemicals and pesticides.

The impact of floods on people also may depend on age and gender. Women and children tend to be more vulnerable to flooding than men (see Chapter 35). In much of the world, women are less likely to know how to swim. In some cultures, women's dress habits hamper free movement, for example in South Asia the sari, a long cloth wrapped around the body, can entangle a woman's legs in water. More than ninety per cent of the 150,000 deaths from the 1991 storm surge in Bangladesh were among women. The reasons included the customary practice that women were not supposed to leave the house without their husband's permission. When women received the cyclone warning, they were waiting with their children at home whilst their husbands had already taken shelter in safer places (Chowdhury et al. 1993).

A comprehensive description and analysis of the vulnerability of women in disasters can be found in Fordham (2004). However, the vulnerability of women cannot be generalised. A study on flood deaths from drowning in smaller-scale floods in Europe and the USA revealed that seventy per cent of the victims were male, mainly because males were more involved than females in driving through floodwaters, in emergency services rescuing people from floods, and in undertaking more risky behaviour around floodwaters (Jonkman and Kelman 2005). Children and the elderly are most likely to suffer from water-borne diseases during a long flood or when the waters recede (see Chapter 36 and Chapter 37).

Many other factors may affect individual vulnerability to floods other than gender, age and income. These include mental or physical disabilities, health status, use of a minority language (e.g. to understand warnings and to ask what to do) and belief systems. Situational factors include being inebriated or under the influence of drugs at the time of a flood and the activity in which the individual is involved at the time (e.g. rescuing, driving, sleeping).

Societal structures and political connections are another cause for different vulnerabilities to floods. Mustafa (1998) undertook a case study in five villages in the province of Punjab in central Pakistan, comparing the impact of flooding on communities and their social groups. He found that the main reason why landlords were spared from annual flooding was that they managed to manipulate the construction of a spillway. Small landholders and landless labourers were more heavily impacted by flood. On top of that, the Pakistani political system historically favours large landlords at the expense of small farmers and the landless. The case study shows that the vulnerability of communities and social groups depends on the level of their political and economic influence.

Mustafa writes, 'Vulnerability is largely a function of disempowerment vis-à-vis certain classes and institutions' (Mustafa 1998: 300). Power and the institutional relations that lead to this concentration in a few hands is the major structural contributor to vulnerable groups (see Chapter 5).

Pelling (1999) also showed with the example of urban Guyana that flood risks are not only related to physical processes, but also to political, social and economic system dynamics. About ninety per cent of Guyana's inhabitants are vulnerable to floods through rainfall, poor land drainage and impacts of sea-level rise and coastal erosion. Through an historical review, Pelling demonstrated that political power played the key role in the shaping of vulnerabilities. He traced back the origins of vulnerabilities to Guyana's colonial past and showed that institutional structures and cultural norms mould negotiations among political actors for control over urban development resources. In this manner the geography of vulnerability is produced.

However, floods are not only negative for people. For example, the Bengali language (spoken in Bangladesh and in the Indian state of West Bengal) differentiates between a harmful flood, *bonna*, and a beneficial flood, *borsha*. The usual flood, the *borsha*, is even necessary for fertilising and irrigating the fields and enabling fish to spawn. However, not even a *bonna* flood causes many deaths from drowning because it occurs slowly and people tend to know how to react.

Viewing things from the local inhabitants' various points of view, river training and control structures can provide short-term benefits, but create long-term challenges (Etkin 1999; Fordham 1998/1999). A flood can be harmful to some people, such as factory owners who are put out of business, while benefitting others, such as farmers who need the soil nutrients. How floods as a physical and social process are interlinked is clearly demonstrated in the following case study carried out in Bangladesh (Box 21.2).

Africa: Rural and urban flooding

Africans experience floods in rural as well as urban areas. Flooding of Africa's big rivers such as the Nile, Niger, Congo and especially the Zambezi affect hundreds of thousands of people every year. Usually not very well reported, but not less harmful for the people, are numerous small floods occurring all over the continent (Wisner *et al.* 2004; Pelling and Wisner 2009). These include smaller seasonal floods that displace people and disrupt livelihoods even in normally dry areas such as the Sahel of West Africa and northern Kenya. Not only people in rural areas are affected, but residents of cities. The urbanisation of Africa has increased significantly over the past years due to natural population growth and influx of people who have left their farms in hope of a better life or to flee civil war and conflicts (Pelling and Wisner 2009). Urban flooding in coastal areas is also common, for example, in Accra, Ghana (Songsore *et al.* 2009).

Rural flooding: the Zambezi River in Mozambique

One of the most flood-affected countries in Africa is Mozambique, mainly through the Zambezi River which enters into the Indian Ocean there after having travelled through eight countries. Apart from the Zambezi, a total of ninety-three rivers of various sizes and seasonal regimes cross Mozambique's coastal plain (Wisner 1979) (see Figure 21.2). Flooding of the Zambezi in the year

Box 21.2 South Asia: living with floods on the Brahmaputra River

The Brahmaputra, called the Jamuna in Bangladesh, is one of the largest rivers in the world. During the monsoon season it can be eight metres higher than in the dry season and transport fifteen to twenty-five times as much water (FAP 24 1996). Only eight per cent of its catchment area lies inside Bangladesh, so that Bangladeshis are affected by India's and China's activities in the watershed, including land use, dams and forest management (see Figure 21.1).

The Jamuna is a so-called 'braided river' consisting of several channels flowing around larger and smaller islands. The size of the channels varies, from 200 m to 5 km wide, and from 3 m to 40 m deep. They continually alter their courses, but especially in the monsoon season when the river erodes land, part of which it deposits at another place. The Jamuna transports up to 600 million tons of sediment per year,

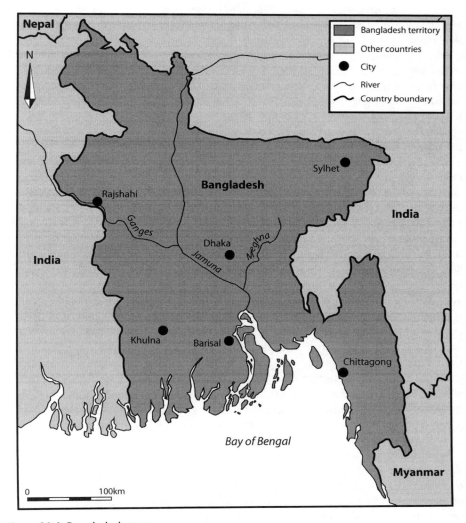

Figure 21.1 Bangladesh map

ranging from very small and fertile particles up to coarse sand, but no gravel or larger stones. The sediments are deposited as dunes in the river bed, which can be up to 600 m long. Flooding is usually slow, with the highest rate of increase of water level being about 30 cm within twenty-four hours (Schmuck-Widmann 2001: 121).

In Bangladesh, an island that consists of sediment deposits and exists only over a limited period of time is called a *char*. Although some *chars* are a few hundred metres long, they may disappear the following year, while others may develop over years into islands of more than 40 km² with dense vegetation. *Such* chars *remain stable for decades. At least 3 million people live on* chars, *or depend on them for their livelihoods. Their main income source is agriculture, supplemented by animal husbandry, small businesses and fishing. The inhabitants of the* chars *are especially exposed to the river's whims as their islands can disappear into the river within a few days. At the same time, new land accretes at another place. On average, a* char-dweller *aged forty-five has moved eight times because of this erosion (Schmuck-Widmann 1996: 22–4). From their experiences with the river, the inhabitants of the* chars *have developed stocks of knowledge and strategies for dealing with the Jamuna.*

Efforts to control the Jamuna and to stop mainland flooding and erosion date back to the 1940s, when Bangladesh was part of Pakistan. Other major efforts started ten to twenty years ago under the Flood Action Plan (FAP). European engineers worked in Bangladesh for the FAP to design structures that attempted to protect some banks from erosion. Potential knowledge of people living along the river and on its *chars* was not considered – and was even considered non-existent, as the local population dealt with the river by moving to safer places.

Yet *char* inhabitants have a profound knowledge about the river's behaviour, especially floods and erosion. To them, floods are not life threatening. It is rare to hear of a *char*-dweller drowning. Children falling into ponds is common, but that happens also during the dry season (Schmuck-Widmann 1996: 30–1; Schmuck-Widmann 2001: 120). Flood is described as 'inconvenient' when the flood level damages their homesteads so that they have to move to relatives living on higher ground or live on the roofs for several days or even weeks. Whilst too much flooding damages crops, too little flooding also results in a harvest failure and hence food insecurity. 'Floods are good' is a common statement of the *char*-dwellers. Floods also facilitate transport. Travelling by boat to other islands or to the mainland is much easier and faster than walking through the dry and sandy river bed in the dry season. In contrast, erosion is considered as 'the biggest problem' of life on the *chars* as it washes away villages and agricultural land. Erosion has its positive side, however, because new land is created from deposition.

By observing the interaction of width and curvature of a river channel, its depth, the velocity, the speed of water-level rise, amounts and kinds of sediments and debris, *char*-dwellers can predict erosion and flooding of *chars* and the river bank. Their knowledge is based on first-hand, daily experience and is transferred between generations through oral histories and growing up with the river.

In contrast to the Western engineers, *char*-dwellers consider their ability to influence the river, to prevent erosion and flooding, to be very limited. In order to protect their *char* from erosion, the farmers plant a reed grass locally known as *kaisha* (catkin grass or *saccharum spontaneum*). It grows up to five metres within one year and through its density of growth catches sediments so that after four to five years a *char* can develop. However, not every spot is suitable for this reed due to high flow velocity which will take it along with eroded soil (Schmuck-Widmann 2001: 85). Other methods for coping with the river are imbedded in the *char*-dwellers' way of life. For example, one can keep household items to a minimum and build houses in a way that they can be dismantled and transported easily for a move to a new *char*.

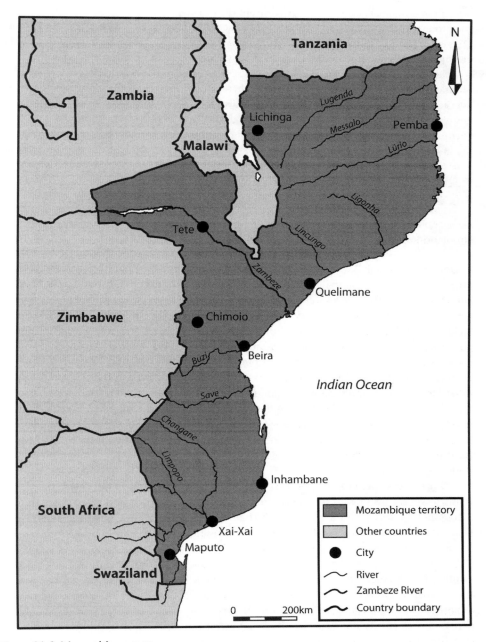

Figure 21.2 Mozambique map

Hanna Schmuck

2000 affected 4.5 million people (about twenty per cent of the country's population at that time); around 800 died (Wisner *et al.* 2004: 258). In the major floods of 2007 no more than 300,000 were affected, though water level was as high as it had been in 2000. Arguably, efforts by the government and national and international NGOs had paid off. Lessons learned from 2000 led to an improved warning system, establishment of protocols for disaster response, awareness–raising campaigns amongst the population, training of local government institutions and improved co–ordination among all stakeholders. Apart from that, many of those affected in 2000 had been resettled to higher and safer areas. Efforts to do so had been undertaken after previous floods since independence from Portuguese rule, but with only partial success. A study found that despite the government establishing resettlement areas and providing construction material, people who had lived in the risk zones preferred to return there after the flood waters receded because these areas had fertile soils, their ancestors were buried there and rituals took place in these localities (Universidade Eduardo Mondlane 2008: 11; see also Chambote and Boaventura 2008). Earlier studies in the Rufiji River delta of Tanzania during the period of government-mandated resettlement found that people had difficulty farming in their new environment and would return to distant riverside farms, and they suffered economic loss because at night the unguarded farms were ravaged by hippos (Mlay 1985; Sandberg 2010; see Box 21.3).

Box 21.3 Living with floods in Tanzania's Rufiji River delta

Audun Sandberg
Nordland Research Institute/Bodø Regional University, Bodø, Norway

Residing in the Rufiji River delta in Tanzania, the *warufiji* peasants had through centuries developed a robust risk-minimising agricultural system that could handle most flood conditions. From their homestead *madungus* (elevated houses on mangrove poles) they could command a fertile but risky environment. Their traditional ecological knowledge included oral history of past generations' experiences of large and small floods. These floods had resulted in a dynamic environment with micro-variations in elevation and a soil structure with crucial differences in nutrients and water retention capacity. Based on this knowledge an intricate agricultural system could be sustained – to some extent regardless of the magnitude of the floods. The *Masika* (flood farming) component of the system was based on floating rice that grew with the slowly rising flood water. With small or medium floods, this was usually a success. With large floods, this was a failure at low elevations, but the large amounts of silt laid the foundation for a successful *Mlau* (post-flood farming) component based on cotton, maize and peas. Even without rain during the growing season, the fresh nutrients and the slowly sinking water table would give a bumper crop before the onset of the next flood season, especially the clay soils. However, 'living with the floods' in this way was hard work, and often there were labour bottlenecks when a huge *masika* harvest coincided with a need for extensive field preparation of heavy soils ahead of a promising *mlau* season. Thus a low labour productivity was the risk insurance premium paid by the *warufiji* for their robust system.

In the course of the last 100 years, various ecology control measures have been instituted by a succession of colonial and independent government authorities and their 'experts'. However, none of these had fundamentally altered the river's seasonal regime. A large dam at Stiegler's Gorge on the Rufiji River for harnessing floods and generating electricity has been planned, but so far not built. Thus both the hazardous and the beneficial floods could continue unhindered. A large-scale prawn farming facility was planned in the delta mangrove forest, but it was shelved due to local and international opposition. Thus the delta forest protection from storm surges and tsunamis from the Indian Ocean is still intact.

A massive government resettlement programme of people into *ujamaa* (co-operative) villages on higher ground was carried out during 1968–72. This removed people from the physical flood hazard, but it also imposed difficulties in the form of long journeys to the fertile fields and increased vulnerability to animal attack on unguarded standing crops at night. This made the flood plain peasants vulnerable to hazards of quite another nature: the forces of modernisation, including individualised land tenure, market fluctuations, contract farming and land grabbing. The advantages of traditional micro-ecological knowledge are likely to become obsolete and large-scale mechanised and commercial farming will gradually take over from the complex risk-minimising agricultural system developed by the *warufiji*.

Further reading: Havnevik (1993); Sandberg (2010).

Urban flooding: Luanda in Angola

African cities are increasingly affected by floods, as are low-lying parts of many cities in Asia (e.g. Manila, Jakarta, Mumbai) and Latin America and the Caribbean (e.g. Mexico City, Guayaquil, Gonaïves, Georgetown). Most people in African cities live in a belt around the centre without any planned drainage or system for the disposal of waste. Where modest attempts at such infrastructure and services exist, they are badly maintained and poorly financed. Flooding, especially in the suburbs surrounding the colonial core of cities, is especially dangerous, as it may also expose people to toxic waste. Without drainage there is risk of diseases such as cholera and dysentery. One example is the capital of Angola, Luanda, one of the fastest growing African cities, with almost 5 million inhabitants. Cholera outbreaks in Angola occur every year, but peaked in 2006 and 2007 when 83,520 were affected and 3,140 died (IFRC 2008a). A twenty-seven-year-long civil war, which killed about 500,000 people, had only officially ended in 2002, and the country was just establishing a disaster management system that co-ordinated national and international humanitarian organisations.

The Angolan Red Cross has been officially mandated by the government to carry out disaster preparedness and response for cholera and other diseases, as they have the unique advantage of having volunteers within the communities. One focus is the suburb Cacuaco, where about 1 million people are estimated to live (see Figure 21.3). As a consequence of the 2007 cholera epidemics, the Angolan Red Cross in 2008 implemented a project of outreach and awareness-building in order to prevent another outbreak. The Red Cross informed people about the causes of water-borne diseases and how to minimise the risk. Measures recommended include cleaning water sources and latrines, dredging small channels, disposal of waste and building houses with affordable, accessible, flood-resistant material. Also action teams were established to conduct first aid, clean wells and latrines after flooding and to assist patients to reach the nearest health post. Apart from training, a basic kit was provided to selected families to assist in carrying out these activities (see Chapter 30).

Mitigation of flood loss and reduction of flood risk

A wide range of methods and techniques have been developed to reduce flood vulnerability and risk. Computer models have been developed but their results are widely varied. Some used in Bangladesh and northern Europe report forecasts based on rainfall and snow melt to provide

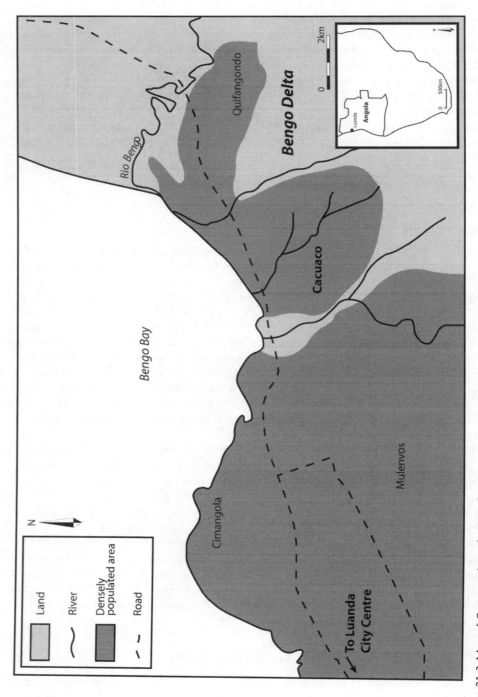

Figure 21.3 Map of Cacuaco, Luanda, Angola

flood warnings several days in advance and over several hundreds of kilometres. Others used in the American Rocky Mountains or for assessments of proposed dams include detailed flow patterns around structures based on a town being inundated in minutes by a flash flood or dam break.

Economic models and cost–benefit analyses are also used to analyse the impact on rivers of engineering, such as aiming for improved navigation or to open riverside land for development. Dams, dikes and levees are still the most prominent structures trying to control rivers, and they contribute significantly to irrigation, electricity production and river navigation. There are about 300 major dams worldwide that are at least 150 m high (International Rivers 2010).

Table 21.1 explains some structures for flood mitigation and river training. Some reduce floods at certain places, but may aggravate them at others, for example dredging, which is carried out to clear navigation. As the river is made deeper, the speed of water flow increases and thereby makes flood damage worse and can cause erosion at locations that had been stable before.

Relying on such structural approaches, without also addressing people's perceptions and long-term behaviour that affect vulnerability, may reduce flood risk in the short term, but usually increases risk in the long term due to steadily increasing vulnerability (Etkin 1999; Fordham 1998/1999). For example, structures usually require cost-intensive maintenance, which can be neglected due to lack of expertise and money or due to poor management. Meanwhile, a large dike makes people feel safe and reduces the awareness of flood risk and how to reduce flood risk.

Examples of alternatives to relying exclusively on structural measures include zoning and land-use planning. Land-use planning and zoning have to be done carefully. For example, building one-storey buildings and elderly care homes in flood plains should be considered carefully due to the evacuation challenges. Multi-storey businesses or apartments with the first few floors used for parking or as common rooms rather than for living could be considered.

Table 21.1 Flood mitigation and river training terms

Bandal	Frame of bamboo and sometimes also tree branches or jute stalks to influence the flow pattern and the sediment distribution for erosion protection and navigation support.
Breakwater	Protective structure of stone or concrete to break the waves.
Dam	A physical barrier constructed across a river or waterway to control the flow of or raise the level of water for flood control, irrigation needs, hydroelectric power production and/or recreation usage.
Dredging	Deepening or clearing the bed of a river, etc., of mud or sand to clear navigation.
Embankment	A man-made ridge of earth or stone that carries a road or railway or confines a waterway.
Floating screens	Panels installed in a river, anchored by ropes or cables to the river bed, to enhance sedimentation.
Groyne	Elongated structure made out of earth, concrete or steel pillars built as a spur dike almost perpendicular to the river bank, intended to stabilise the bank.
Levee or dike	An embankment raised to prevent a river from overflowing.
Reservoir	Natural or artificial pond or lake used for the storage and regulation of water.
Revetment	A facing of stone, concrete units or slabs to protect a dike or a seawall against erosion by wave action, storm surge and currents.

Any multi-storey building in which people would be expected to stay during a flood must not be in danger of collapse and must have adequate supplies and sanitation.

Often ignored is the fact that people have developed their own coping strategies and local knowledge to live with floods (see Chapter 9). As the example of the *char*-dwellers in Bangladesh has shown, their entire life is moulded around the annual climatic cycle and the behaviour of the river. Whilst some of these strategies are obvious, for example that homesteads are built on earth platforms and are easy to dismantle to facilitate moves, others are less visible because they are non-structural. For example, some *char*-dwellers purposefully arrange for their daughters to marry husbands living on the mainland in order to have a temporary refuge with in-laws in case of a severe flood.

For effective risk reduction, local, indigenous coping mechanisms should be explored in combination with outside, specialist knowledge. Participatory planning of flood protection measures involves the consultation and active involvement of all groups affected by floods. One successful example, in the UK, is described by Fordham (1998/1999). Instead of a project-by-project approach, the whole catchment area was examined and a wide-ranging consultation was carried out with residents on their preferred options for flood management. The new approach integrated flood defence with water resource management, pollution control and other development objectives.

Conclusions

Reducing flood vulnerability requires a balance of bottom-up approaches based on local knowledge and top-down measures and outside specialist knowledge. Too often it has been assumed that top-down, structural measures are sufficient by themselves. Even the act of warning of an imminent flood cannot rely alone on formal mechanisms because informal warnings and neighbourly gossip often supersede and are seen as more credible than official warning channels. Flood risk reduction requires working with communities to understand their needs in a participatory manner, which will cover vulnerabilities, existing capabilities, and capacities that are desired but that require improvement.

Analyses and actions must cover multiple time scales simultaneously, tackling many factors such as river engineering, climate change, water resources, changing population density and industrialisation. Ultimately, floods are normal river processes. They have benefits for ecosystems and for people. Many ways exist to ensure that floods do not become flood disasters.

Some of these techniques are known within communities, and sometimes they can be introduced from external sources, whether technical agencies or peer-to-peer exchange with other communities. Measures should be implemented by communities, but with external support where needed, especially to ensure that solving flood concerns in one location does not create or exacerbate flood problems upstream or downstream. That will assist in returning flooding to being a natural, normal and accepted part of the life of rivers and of river communities, rather than being surprising and disastrous extreme events.

22

Drought

Thomas A. Smucker

OHIO UNIVERSITY, ATHENS, OH, USA

Introduction

Drought is a complex hazard, the pervasive human impacts of which reflect the interaction of multiple physical and social processes. Unusual dryness may undermine a wide range of occupations and livelihood arrangements, endanger human and animal well-being and cause large economic losses. People have developed ways of living with episodic droughts, as they have also adapted to normally expected dry seasons; however, climate change, state policy, population growth and other factors are eroding the effectiveness of these traditional strategies. Nevertheless, because drought is both a cyclical event and a slow-onset hazard, there is great potential for progress towards drought risk reduction. Efforts to strengthen drought management and reduce risk will be most effective where proactive risk reduction measures take into account the multiple physical processes that contribute to drought impacts and the possibilities for building on established local strategies for managing landscapes and livelihoods impacted by drought.

Defining drought

Drought refers to an extended period of below normal or below expected rainfall which negatively impacts hydrological, biological or human systems. However, there is no universally recognised definition of drought, as the range of definitions used by governments and agencies around the world reflects a concern for different meteorological, hydrological, agricultural and socio-economic variables that can be used to identify the onset of drought and measure its severity. Thus, disciplinary perspectives on drought and peoples' vulnerability to drought abound. Drought as a physical and a human–environment phenomenon has been conceptualised differently through the lenses of meteorology, hydrology, agronomy and development studies.

Meteorological drought refers to a deficiency of precipitation for an extended period of time. Typically, it is measured as a magnitude of divergence from long-term mean rainfall. Thresholds for the identification of meteorological drought are often arbitrary rather than a reflection of likely drought impacts on society or hydrological systems (Wilhite 2000). Meteorological drought is a normal occurrence across the spectrum of climate types, not merely a characteristic of arid and semi-arid areas. Dry climates do experience a greater degree of temporal and spatial variability of rainfall and are thus referred to as 'drought-prone'. To be useful for the purposes of drought response planning, the magnitude and duration of below average rainfall used to identify the onset of drought should be relative to the sensitivity of hydrological, environmental or human systems to anomalous rainfall levels.

Meteorological drought is a product of large-scale drivers in the Earth's atmosphere, including global circulation patterns, monsoon systems and El Niño Southern Oscillation (ENSO) events. The interaction among these large-scale processes is responsible for patterns of temporal variation that produce seasonality and influence inter-annual and inter-decadal variability that prevails within the world's climate zones. For example, the warming of sea surface temperatures associated with ENSO results in wet and dry phases – differentially distributed across large areas of the globe – during which rainfall and temperature depart substantially from long-term averages for six-to-eighteen months. ENSO-driven variability is nested within longer-term drought cycles. This complexity of temporal variability is reflected in the modest progress that has been made toward seasonal prediction of rainfall (Ropelewski and Folland 2000: 22). Seasonal forecasting faces challenges in meeting the needs of end users. For example, whereas pastoralist herd mobility strategies may benefit from information about the timing of the onset of the rains, seasonal forecasts of total rainfall over the course of a rainy season is less useful (Luseno et al. 2003).

Drought monitoring often relies on indices that combine meteorological and hydrological variables to provide a composite view of moisture conditions (Hayes 2010). For example, the Palmer Drought Severity Index (PDSI) – which reflects moisture input, output and storage – has been widely adopted in the USA for drought monitoring. Byun and Wilhite (1999) have identified several weaknesses of drought indices, including insufficient temporal resolution to detect the precise onset and end of drought and inability to account for the interactions between meteorological and hydrological variables (e.g. runoff and evapotranspiration). Monitoring and early warning systems that use remotely sensed satellite data are increasingly capable of providing near real-time information to government and communities facing urgent agricultural and livestock management choices at the onset of drought.

Hydrological drought refers to deficiency in surface and subsurface water supplies relative to normal conditions. Levels of soil moisture, groundwater flow and stream flow are major parameters used in the identification of hydrological drought (Wilhite 2000). The relationship between meteorological and hydrological drought is complex as water use and modification of the landscape change flow patterns and may result in desiccation of the landscape, even if rainfall remains normal. For example, atmospheric conditions in upstream areas of major river basins may have major implications for hydrological conditions in distant downstream locations.

Agricultural drought is concerned specifically with the impacts of meteorological drought on crop and forage production. The impacts of agricultural drought reflect variation in soil characteristics, soil moisture and the properties of crops and forage plants. Because growth characteristics and moisture requirements vary by crop, assessments of agricultural drought impacts are often crop-specific (e.g., Mohamed et al. 2002). Agricultural drought assessment is thus concerned with the temporal distribution of rainfall relative to patterns of plant growth. Even under conditions of meteorological drought, rainfall may provide sufficient moisture during key growth stages of plant development, thereby limiting impacts (Passioura 1996). Assessment of drought impact on productivity within intercropped and agro-biodiverse systems of the humid tropics is particularly difficult (see Chapter 45).

Economic and social definitions of drought are concerned with the impacts on economic activities and wider social well-being. In economic terms, deficient rainfall or surface water may prevent the supply of particular goods such as hydroelectric power or agricultural commodities (NDMC 2006). Assessments have revealed a complex geographical distribution of winners and losers, as less affected firms and farmers take advantage of drought-induced price rises (Wilhite *et al.* 2007).

The decision to declare a drought emergency or to seek international assistance during a drought is a highly political one that may or may not take such technical, evidence-based definitions into account. Yet, at the other extreme, there are cases where the impacts of drought are evident to even the most casual observer. For instance, in early 2010 drought in Syria so deeply affected the farming sector that 300,000 people left the land to seek employment in cities, reducing the population of some farm villages by up to fifty per cent (UNOCHA 2010a).

Broader social development perspectives on drought range from underestimating or ignoring its role in development to its identification as a direct cause of underdevelopment (Glantz 1987). Social conceptions of drought should be flexible enough to incorporate vernacular, place-based and integrative perspectives. For example, oral drought histories in eastern Kenya reflect that people name drought periods and memorialise them based on impacts on the human condition (e.g. 'weakened joints', 'taker of relatives'), or the means by which people coped during the drought (e.g. 'beans' (consumed as famine relief), 'millet-tamarind porridge', 'planted cowpeas') (Smucker 2003; Wisner 1978). Popular discourses about drought also reflect an integrative perspective of drought as a human–environmental phenomenon. In Swahili, for example, drought is mostly commonly translated as *ukame*. In common usage among East African farmers and herders, *ukame* can refer to deficient rainfall, general barrenness of the landscape or a shortage of some kind. Moreover, the term is sometimes used interchangeably with *kiangazi*, or dry season (Bwenge 2010b).

Global climate change may result in more frequent occurrence of all drought types. According to the IPCC (2007a), regional climatic conditions resulting from global climate changes are likely to increase overall variability in precipitation (see Chapter 18).

Assessing the impacts of drought

Societal impacts of drought result from drought's impacts on biophysical processes, such as biological productivity or erosion (Wilhite and Vanyarkho 2000). Common direct impacts may include reduced crop productivity or failure, loss of wildlife habitat and forage, elevated livestock and wildlife mortality, reduced availability of water resources, and increased incidence of wildfires (see Chapter 24 and Chapter 31). Indirect impacts may include loss of income, occupational displacement, rural-urban migration, deepening socio-economic inequality, social and political conflict, and increased human mortality. Drought is perhaps best conceptualised as a 'trigger' event that causes damage by exploiting underlying (and sometimes longstanding) social vulnerabilities within populations exposed to its effects (Wisner 1978; Wisner *et al.* 2004).

Increased mortality in the wake of drought is one indicator of the severity of drought events. Insufficient water availability may lead to crop production decline or failure, and threaten established sources of safe drinking water, thereby increasing exposure to water-borne disease. However, assessments of the health and mortality impacts of drought must incorporate the social dimensions of food (and water) access, normal sanitary conditions and access to health care, not merely loss of productivity. The relationship between drought and famine is mediated by the diverse means through which people realise their entitlement to food through market exchange, social and kinship networks, their own food production and rights-based political access (Sen 1981) (see Chapter 6 and Chapter 45). Conventional explanations of the role of drought in famine mortality have often assumed that famine results from declining productivity or food availability decline (FAD), rather than the collapse of the means of food entitlement. There is now substantial empirical evidence indicating that food entitlement decline (FED) –

rather than FAD – is the essential precondition for famine. Even where FAD precedes famine deaths, mortality may reflect exposure to epidemiological threats that exploit weakened immune response resulting from malnutrition (Devereux 2001).

Global estimates of drought-related mortality vary widely. The Center for Research on the Epidemiology of Disasters (CRED) reviewed its 807 drought entries and 76 famine entries spanning 1900–2004 (Below *et al.* 2007). CRED used two criteria. First, drought losses must be attributed to drought in initial reports from governments, UN agencies, non-governmental organisations (NGOs) or the media. Second, 'drought-like physical conditions' must be associated with the reported losses, or confirmed by using the Weighted Anomaly of Standardized Precipitation (WASP) index to ensure that losses were concurrent with meteorological drought.

Sixty-seven of the 68 famine entries were classified by CRED as drought (as famine was re-categorised as an outcome of hazard rather than a hazard itself). CRED found 11.9 million deaths from drought between 1900 and 2004, with substantial regional differences. Asia (including southwest Asia) accounted for eighty-two per cent of drought mortality. Six of the ten drought events identified by EM-DAT with the highest mortality occurred in South and East Asia (two in China, three in India and one in Bangladesh). Taken together with the 1921 famine in the Soviet Union, Asia counts for seven out of the ten most deadly events.

Modelling of drought's economic impacts has taken a variety of forms, from simple regression analysis of time series data to assess the correlation between drought conditions and the economic performance of agriculture (Benson and Clay 2000), to process-based models that use historic or forecasted climate scenarios to simulate both crop productivity and economic consequences (Easterling and Mendelsohn 2000). In many cases assessments of economic impacts simply consider changes in income, employment, inequality and other characteristics in drought years relative to pre-drought economic conditions. Such quantitative approaches to drought impact assessment offer partial views of drought losses. They may rely on problematic assumptions about production conditions while ignoring environmental or social factors that interact with drought in reducing yields and farmers' incomes. Nevertheless, the valuation of drought-related economic losses may serve to convince policy-makers and the general public of the economic benefits of investment in drought risk reduction measures.

CRED attributes to drought seven per cent of disaster-related losses, totalling US$84 billion since 1900. Such accounting generally suffers from an inability to account fully for secondary impacts of drought, and there is a paucity of reliable quantitative information about the extent of drought losses borne by people themselves in low-income countries. For example, forty affluent country drought entries in CRED's EM-DAT database are associated with three times the economic losses of 228 African droughts (Below *et al.* 2007) (see Box 22.1). Despite bearing the brunt of drought impacts, the assets of people in the rural areas of the developing world have relatively low monetary value and their losses simply may not be recorded by governments, NGOs or insurance companies.

Western reporting on African droughts tends to emphasise numbers of malnourished and deceased more than economic impacts or the successful responses undertaken by ordinary people to cope. Such reporting sustains environmentally deterministic interpretations of drought as a common or necessary precursor of food shortage and famine. The 1992 drought in Southern Africa offers an example of pervasive drought impacts resulting in few deaths, despite drought conditions confronting nearly 20 million people. As in Australia, the drought cycle in Southern Africa is influenced by alternating wet and dry years associated with El Niño. Rainfall was suppressed in late 1991 and remained substantially below normal levels in much of the region. The Southern African Development Community (SADC) reported more than 86 million people affected in ten countries. Vogel and Drummond (1993) examine differential impacts

Box 22.1 'The big dry' in Australia

Assessments of Australia's recent drought period suggest new climatic factors and a wide range of direct and indirect impacts. At the onset of the six-year 'big dry' period in Australia, Nicholls (2003) suggested that the impacts of meteorological drought in the early twentieth century would differ from similar droughts of previous decades due principally to higher temperatures (and increased evapotranspiration). Since 2001 the annual inflow to the southern Murray-Darling basin has been the lowest on record due to the combination of meteorological drought and higher temperatures (Cai and Cowan 2008). Continued warming as a result of global climate change would significantly worsen the impacts of future meteorological droughts on the basin's hydrology and economic base, which depend on water-intensive land use.

In high-income countries the impact of drought is often measured in lost income, reduced economic output and occupational displacement. In the wake of the recent Australian drought, average farm incomes declined by A$29,002 between 2005/06 and 2006/07 alone (Edwards *et al.* 2009). Drought in southeastern Australia during 2001–07 was devastating to commercial agriculture production and, particularly, grain exports. A national survey of rural Australians found that 42.4 per cent of farmers and farm managers believed their farm was not economically viable under current weather conditions (Edwards *et al.* 2009). Beyond important economic impacts, political conflicts over water governance have pitted federal against state agencies and a sense of resentment has developed over the prioritising of urban water needs over those of rural communities.

between the commercial agricultural sector dominated by white South Africans and the livelihoods of black South Africans residing in apartheid-era Bantustans. In South Africa's commercial agricultural sector, agricultural subsidies and credit lessened the blow of drought impacts, though farmers' debts rose. As commercial maize production declined, large-scale importation was required.

Eldridge (2002) asks why there was no famine in Southern Africa given that drought conditions took a large toll on livestock and crop production throughout the region. In Zimbabwe's communal areas, maize production was less than one-tenth of average production of the previous seven years. Distress livestock sales and livestock diseases intensified. People throughout the region tried to make up for lost agricultural income by resorting to fuel-wood sales or searching for wage employment. Despite substantial food relief distribution programmes in several countries, drought-affected households purchased at least seventy per cent of staple foods that they required to make it through the drought. Thus, households themselves successfully bore much of the burden of coping. On the other hand, Southern African governments can be credited with pre-emptive actions to reduce the impact of drought (DeRose *et al.* 1999). SADC food reserves were not sufficient to deal with the scale of the drought, but by acting early to bring stockpiles onto regional markets and quickly implementing food-for-work and cash-for-work programmes, the governments of the region succeeded in averting famine (see Chapter 45).

Drought risk reduction

The United Nations (UN) master plan for disaster reduction (the Hyogo Framework for Action) emphasises the need for progress across the full spectrum of risk reduction activities, from early warning and community mobilisation to drought response and management (see Chapter 50). Gains made in one area may be threatened by lack of progress on other fronts. For example, technological advances have made possible more sophisticated early warning systems, but their efficacy depends on the ability of communities to act on the information that such systems diffuse (see Chapter 40). The greatest future challenge will be to establish and sustain political commitment at all levels to reducing social vulnerability to drought. International co-operation will remain important as societies across the globe cope with new patterns of drought risk and water stress in the face of global climate change (WWAP 2009).

Policies and governance

In many cases the implementation of drought risk reduction activities lacks sufficient resources and a clear focal point within government (see Chapter 51). Where this is the case, bureaucratic struggles or lack of clear division of labour in drought response may be a problem. In the long term, such struggles may hinder closer linkages between drought risk reduction and broader development strategies (UNISDR 2009b).

Following the 1992 drought in Southern Africa, countries in the region undertook reviews of national drought policies and response frameworks. In Namibia the contracting-out of drought response activities to NGOs was seen as an obstacle to the effective integration of emergency management into long-term rural development and risk reduction activities (Dumeni and Giorgis 1993). The National Drought Task Force (NDTF), an inter-ministerial task force, co-ordinated the efforts of ministries and agencies during the 1992 drought and gained considerable insights into successes and failures of the response effort. Because Namibia had been independent for barely two years before the onset of drought, the bureaucratic structures of the state were still in transition. The advent of a National Drought Policy in 1997 reflected a concerted effort to learn the lessons of the previous drought response efforts and to build on the institutional capacities that had been developed in the early 1990s (Government of Namibia 1997). Notably, the policy sought to reorient government activities towards risk reduction and to set stricter criteria for emergency relief provisions, which are to be enacted only during 'disaster droughts' that meet strict criteria related to rainfall, pasture and livestock conditions and crop productivity (Government of Namibia 1997: iii). While farmers in communal areas would gain access to relief vouchers for food and agricultural inputs under such extreme conditions, the policy reflects a concern that relief assistance for some private tenure farmers may act as a disincentive for investing in measures to minimise drought risk. This policy thus emphasises support for the drought risk management measures undertaken by farmers and pastoralists and these were integrated into rural extension services. Specific concerns addressed by the policy include diversification of crops and livelihood activities, improved water conservation and management, and range management by pastoralists. Priorities for drought mitigation activities identified by the policy include maintaining the reproductive capacity of livestock herds, managing potable water supplies, preventing environmental degradation and addressing drought-related health problems. Because drought is a frequent occurrence in Namibia, the focus on sustainable development is seen by policy-makers as imperative if the government is to avoid frequent and costly emergency interventions (see Chapter 14).

In contrast to the Namibian case, a different picture emerges in India where drought policy has been many decades in the making. Indian drought policies have been credited with greatly reducing or eliminating drought-related mortality through efforts to protect the assets of vulnerable farmers and public employment programmes that build infrastructure essential to drought risk reduction (e.g. expansion of irrigation, water storage facilities and terracing)

(McAlpin 1987). Due to the long history of state–organised famine relief in India, the country benefits from a clear (though complex) organisational structure for drought management that extends from the state to district and village government. Additionally, long–established programmes and institutions remain central to India's drought management policy. Perhaps most importantly, the Public Distribution System (PDS) administers a network of more than 45,000 Fair Price Shops which protect food access of low–income populations amidst rising food prices during drought. While the effectiveness of the PDS in alleviating hunger varies substantially from state to state, there is no doubt that it has served as a buffer against food price spikes for millions of Indians over the last fifty years. As substantial state resources have been invested in public relief, a challenge for Indian drought policy is to link relief to mitigation through support for people's coping mechanisms for managing drought–related threats to livelihood, especially where such strategies relate to common property or the management and use of sub–marginal lands.

Drought risk identification and early warning

The identification of drought risk requires an analysis of the interaction of the physical hazard relative to the social geography of drought vulnerability (see Chapter 16). Historical rainfall, soils and hydrological information and an understanding of regional climate contribute to risk analysis on the physical side of the equation. Equally important is the assessment of vulnerability and capabilities and how they are distributed within populations. The identification of groups who may be particularly vulnerable due to political, economic, cultural or geographical factors is an essential component of drought risk management. In post–disaster situations, drought impact assessment can contribute to prioritising action to mitigate future losses, though a general methodological approach to impact assessment is elusive and the potential for undercounting losses of the most vulnerable in society is great.

Increasingly complex systems of early warning have been implemented in various national and regional contexts over the last two decades, including China, the Horn of Africa, South Africa and Portugal (WMO 2006) (see Box 22.2). Early warning systems that provide alerts of impending drought–induced stresses can be a component of a more proactive strategy of disaster prevention and mitigation. While such technological systems contribute to drought preparedness and should not be ignored, the capacity of local actors to respond effectively to early warnings is equally crucial.

Thus, one of the concerns that critics of drought and famine early warnings systems have raised is the relationship between these 'expert' information systems and local systems of community-based drought mitigation and response. Walker (1989) asks 'Who should be warning whom to do what?'

Community drought knowledge and drought responses

The primary means by which populations cope with drought can be found in community, household and individual mechanisms that are buffers against drought–related losses. These responses are diverse and are components of more complex strategies that encompass agronomic, environmental management, economic and cultural practices. While such responses reflect local knowledge of environments, social and political networks, and other social conditions, expert information may combine with local knowledge-based responses to further buffer people against the impacts of drought. Responses to drought are dynamic and reflect wider changes in local livelihood conditions as well as the opportunities and limitations imposed by external factors.

Box 22.2 Early warning systems related to drought risk

The number of early warning systems related to drought risk continues to expand, offering opportunities for comparative assessment. Most early warning systems integrate meteorological data with information on remotely sensed vegetation condition, crop productivity, and prices of crops and livestock. Broader composite indicators incorporate health and nutrition indicators or local knowledge of factors associated with drought conditions that suggest the onset of drought. Effective early warning systems should trigger alerts for international, regional and national stakeholders and affected communities when critical thresholds for different types of drought are exceeded. The examples below reflect both regional systems of early warning that interface with national emergency response as well as national systems of early warning that are integrated, to varying degrees, into national response structures.

- Humanitarian Early Warning Service (HEWS) World Food Programme (www.hewsweb.org) is a multi-hazard 'one stop shop' that partners with agencies around the world to make early warning information, graphics and maps available to hazard management decision-makers.
- Global Information and Early Warning System on Food and Agriculture (GIEWS) Food and Agriculture Organization (FAO) (www.fao.org/giews) provides special reports and early warning alerts of impending food crises through rapid assessment and an established Crop and Food Security Assessment Mission (CFSAM) methodology, which draws on a wide range of data, from remotely sensed crop assessments to key informant interviewing. Food balance sheets and assessments of staple food import needs constitute the macro-level assessments of the CFSAMs. Micro-level analyses assess access to food on the part of crisis-affected populations through assessments of markets, current production, stocks and other sources of food entitlement. Specific national response recommendations are derived from these two components of CFSAMs.
- Famine Early Warnings System (FEWS) United States Agency for International Development (www.fews.net) is a network of early warning systems covering many countries and including regional centres and analysis for East, West and Southern Africa. FEWS NET provides monthly food security reports and emergency alerts, and alerts highlighting drought and other threats to livelihood and food security. FEWS NET reporting entails three major components: agro-climatic monitoring via remotely sensed data; markets and trade monitoring that draws on price, production and market information; and livelihoods mapping and reporting that interprets agroclimatic and economic information in light of baseline information on livelihoods and coping strategies. FEWS NET disseminates a suite of decision support products to inform national and international response.
- US National Integrated Drought Information System, National Oceanic and Atmospheric Administration, with participation of several federal agencies (www.drought.gov) is a federal government initiative founded in 2006 to consolidate current information on meteorological, hydrological and socio-economic aspects of drought in the USA. A web portal serves as the primary communication tool for drought monitoring and forecasting information. The initiative seeks to develop decision-support tools for drought response and information to inform planning to reduce sensitivity to drought and inform proactive drought mitigation planning.

- Ethiopian Disaster Risk Management and Food Security Sector, Disaster Prevention and Preparedness Agency (DPPA), Early Warning Unit (www.dppc.gov.et) has consolidated the collection of early warning information by government under the Early Warning Unit and decentralised data collection capacity to the *woreda* (district) level. Bi-annual assessments are conducted in conjunction with NGO partners such as Oxfam, Care and Save the Children, which conduct their own multi-agency assessments. Assessments incorporate meteorological information with local data on food availability and access. A major challenge is to maintain sufficient capacity for data collection at the local level and to incorporate non-agroclimatic factors into analysis, such as armed conflict, which disproportionately affect dryland zones.
- Beijing Climate Center (bcc.cma.gov.cn/en) is engaged in daily and monthly drought monitoring for China and East Asia. It produces daily drought maps and reports derived from data collected through remote sensing and an extensive network of meteorological and agricultural field stations. Assessments of the physical dimensions of drought are not complemented by livelihood and food security monitoring, limiting its potential as a more comprehensive drought early warning system.

While drought knowledge and response in agricultural communities reflects cultural knowledge developed in relationship with local environmental conditions, knowledge and practice of drought response interacts with political-economy and environmental changes that produce a varied landscape of drought vulnerability within and between communities (see Chapter 9). In agricultural communities, investments in soil and water conservation techniques are an important means of minimising agricultural impacts of meteorological drought, including the use of organic and inorganic inputs to change the topography and vegetative cover of cultivated areas so as to reduce evaporation and runoff. For example, in semi-arid western and northern China traditional pebble mulches have been used to conserve rainfall resources and expand crop production options for hundreds of years (Lightfoot 1994; see Box 22.3).

Additional drought adaptations are reflected in the mix of agricultural inputs that farmers select. For example, Mexican farmers have historically managed drought vulnerability and crop disease through the planting of traditional varieties of maize with varying phenological characteristics (Liverman 2000). Such strategies have been transformed since the mid-twentieth century through the diffusion of Green Revolution seeds and other inputs. Farmers with access to irrigation water have experienced increased maize and wheat productivity and less sensitivity to rainfall shortfalls. However, farmers who have adopted hybrid seeds for rain-fed agriculture have experienced a mixed record of productivity relative to traditional varieties, while also taking on debts that further constrain coping options. The growing importance of non-farm income sources – as a result of both rural diversification and circular migration between cities and villages – plays a rapidly increasing role in the strategies of farm households across the wealth spectrum.

Contemporary pastoralism is a livelihood system that has evolved through centuries of adaptation to great spatial and temporal variability of rainfall in various dryland regions throughout the world. A hallmark of pastoralist strategies to manage both normal environmental variability and drought conditions is based on mobility of people and livestock to take advantage of seasonally available fodder and water resources as well as trading opportunities in neighbouring agricultural communities. Movement to upland, riparian or wetland areas during

Thomas A. Smucker

Box 22.3 Indigenous drought-coping measures in China

Wenhua Fang
Academy of Disaster Reduction and Emergency Management, Beijing Normal University, People's Republic of China

In the arid areas of western China a variety of indigenous methods were developed by generations of local people through their experience of coping with drought hazard. Today some indigenous methods of long standing are still widely used and, in fact, are diffusing to other parts of China. These include the following (Ly and Chen 1955; Zhao *et al.* 2009; Ge 2009):

- Gravel mulch land: Known as *shatian* in Chinese, this is a farming method that originated more than 300 years ago in Lanzhou, Gansu Province. Gravel or pebbles are tiled into a 10 cm–20 cm-thick layer on the sandy soil to make a gravel mulch to increase water infiltration, decrease evaporation and stabilise soil temperature. At present there are around 200,000 hectares of gravel mulch fields in northwest China.
- Gravel mulch rainwater harvesting system: Crops like corn are planted in rows, and between every two rows gravel mulch is tilled into the soil. This helps to retain rainfall and to prevent soil salinity from developing. This method, called *paiyanlonggen* in Chinese, is very simple and easy to implement.
- Rainwater-harvesting system: The system, named *jishuixitong* in Chinese, is widely used in northwest China for providing clean water for both domestic and agricultural use. The system utilises a water storage tank to accumulate rainfall from small watersheds. The water is filtered by passing through an underground water tunnel through sand along the natural slope. The cleaned water is then stored in another water tank. Usually residential buildings and farmland are located downhill from the watershed so that provision of this filtered water can rely on gravity, and therefore does not require pumps.
- Sand barricades for preventing drought: A large area of desert is found in western China, and along its boundary there is a lot of farmland. In order to prevent desertification and retain rainfall and soil moisture, sand barricades, or *shazhang* in Chinese, are built with different easily found materials such as branches, hay, reeds, clay, rocks and living shrubs. The width of sand barricades is optimised with indigenous knowledge developed throughout history; this is an old system. Many barricades were built in the past, and many more are still being built.
- Water cellar: Water cellars, called *shuijiao* in Chinese, are rainfall micro-collection systems widely used in arid and semi-arid northwestern China since ancient times for domestic water and irrigation. These underground rainfall collectors are made up of a small catchment area, an excavation and sometimes a sedimentation tank, generally a concrete structure or made with clay. The water cellar is mostly used for collecting drinking water. At present, innumerable water cellars are still functioning in dry areas of China, benefitting millions of people. The Mothers' Water Cellars Project popularised this system in other parts of China on the basis of its great success in the northwest. By 2006 the project had invested over 330 million Renminbi and had constructed more than 100,000 water cellars. This campaign assisted more than 1.3 million people find a solution to water shortages in twenty-three provinces of China (see www.mothercellar.cn – in English and Chinese).

dry seasons and dispersal to drier savannah zones that are reinvigorated during the rainy season is a common pattern among pastoralists. However, where there is relatively little differentiation in the arid realm, movements may simply reflect opportunistic strategies based on local rainfall patterns (McCabe 2004). Because pastoralists throughout the world face extended periods of little or no rainfall on a regular basis, their patterns of movements are sometimes mistaken for drought-coping strategies.

The interface of pastoralist drought-coping strategies and state development policy in many developing countries has been characterised by conflict over mobility strategies. Many governments have pressed for sedentarisation of pastoralist communities in order to encourage the taking up of crop agriculture or modernisation, thus restricting the mobility strategies of groups such as the Maasai in Tanzania, Uigur in China and Raika in India. In other cases, pastoralist strategies are curtailed by strict limitations on livestock holdings, ostensibly so as not to exceed the carrying capacity of dryland areas. Thus mobility and holdings among livestock keepers in communal areas were strictly limited in South African-administered Namibia before independence. In other cases, pastoralist mobility strategies for coping with normal variability and drought are curtailed indirectly by rural development policies that favour the expansion of crop agriculture and exclusive property rights which restrict mobility within the savannah landscape (Campbell 1999).

Coping strategies of pastoral groups have evolved well beyond livestock-focused drought-coping strategies. A wide range of trading, resource extractive and wage labour activities now constitute normal livelihood, and these activities are also reflected in drought-coping strategies. Sedentarisation may be a temporary strategy to rebuild herds following drought losses or may be a more permanent transition.

Amidst this diversity and diversification, a few themes emerge that are important to the broader project of drought risk reduction. First, as the world's grasslands become more fragmented, mobility has become a luxury of the wealthy and well-connected livestock holders, who maintain access to seasonal pastures through social networks or pasture rental (Wangui 2008). Thus there is ever greater diversity at the community level in terms of the 'traditional' measures that can be relied upon. Little *et al.* (2001) suggest that the addition of supplemental drought-coping mechanisms is greatest at the extremes of the wealth spectrum, as the wealthiest take advantage of new commercial opportunities while the poorest resort to resource extractive activities and wage labour. Smucker and Wisner (2008) provide additional evidence of this dynamism among East African agro-pastoralists. They also point out a second key theme. The role of wider policy initiatives – from land privatisation to market liberalisation – may eclipse access to established means of coping. While millions of pastoralists around the world still rely on responses to drought that are based on knowledge of local landscapes, political and economic integration of the world's drylands into wider systems may block access to tried and true means of responding to drought. The study of change in pastoralists' drought coping thus reinforces the notion that drought risk reduction activities may have little impact when they do not confront the dominant thrust of rural development which contributes to the emergence of new forms of drought vulnerability.

Conclusions

Declining drought-related mortality should not obscure the fact that drought remains a fundamental threat to the livelihoods of hundreds of millions of people around the world. Research and policy discussions on the economic benefits of climate change mitigation may refocus attention on reducing underlying drought risk factors through investment in a suite of proactive drought risk reduction measures. Collaboration and dialogue among governments, UN agencies and international civil society has created enormous potential to learn policy and implementation lessons across borders. Of great concern is whether governments will maintain political and financial commitments to drought risk reduction activities and whether these will be implemented in a way which is sensitive to local aspirations and which builds on existing capacity for drought risk management and drought response.

23

Extreme heat and cold

Sabrina McCormick

GEORGE WASHINGTON UNIVERSITY, USA

Introduction

Early July, 1993, Sara (not her real name) was found in her apartment, head down on the kitchen table. She had visited a local health clinic the day before, and the physician discovered she was dehydrated. By the time the ambulance attendants found her the next day, Sara was dead. As they loaded her body into the ambulance, dense, hot air wafted from her apartment door. The windows were shut and the temperature inside had reached 49°C. Even after lying in the ambulance for a twenty-minute ride back to the hospital, her body temperature was 42°C. Sara was forty-eight years old when she died. The next day, a sixty-five-year-old man named Jimmy was discovered decomposing in his apartment. His neighbours had seen him three days prior. Since then, the heat wave had permeated Philadelphia, and both the concrete and the residents had absorbed its toll. The official cause of Jimmy's death was atherosclerotic heart disease, but heat was named as a contributing factor. Jimmy and Sara were not alone. One hundred and eighteen people died that month due directly or indirectly to a heat wave.

Why do people die of extreme heat and, even more often, of extreme cold? Nearly twenty years after the event that ended the lives of Sara and Jimmy, we are still struggling to understand how these events affect people and how to protect them. Although the physical sciences have traditionally provided information about temperature projections, weather patterns and environmental characteristics that affect humans, it is becoming clearer that social factors shape who is most affected and how they respond. This means that human behaviour, institutions and policies are moving into the spotlight to provide answers to the critical questions regarding how people are affected by extreme weather events. This chapter explores these factors, including the intersecting biological, sociological and environmental dimensions that determine vulnerability to extreme heat and cold.

Overview of heat and cold hazard

Extreme heat and cold cause deaths around the world, every year. These extremes also cause a great deal of damage to economies and livelihoods. Cold weather events have historically had great impacts on human populations. Global warming is increasing the frequency and intensity of extreme heat events, otherwise referred to as heat waves (Houghton *et al.* 2001) (see Chapter 18). In the USA between 1979 and 2002, total mortality caused by heat was higher than by floods, tornadoes, hurricanes, lightning and earthquakes combined. Death from cold events was almost double this number (CDC 2005).

There is a consistent relationship between heightened mortality and extreme heat. Approximately 400 people die of heat-related illness in the USA each year, and this number is projected to rise dramatically with climate change (Bernard and McGeehin 2004). Other countries have experienced similar rises in mortality as temperature severity and duration has increased. Of course, one must also bear in mind that over time the population has been increasing too and medical treatment has been keeping people alive longer, so there are more vulnerable people to die in a heat wave.

India and several other middle-income and low-income countries have similarly experienced recent increased impacts of extreme heat, especially in urban areas (Srivastava *et al.* 2007). Heat waves cause higher daytime and night-time temperatures in cities than in rural areas because buildings and asphalt absorb more heat than do trees and plants. However, even within cities, temperatures can vary significantly, by more than 10°C between sun and shade. Much of the heat variability within cities is due to variation in features of the built environment and the amount of green space – notably more in more affluent sections, less in poorer areas of a city (Ebi and Meehl 2007). Heat waves lead to poor health through two main pathways. Extreme temperature rise leads to heatstroke, while cardiopulmonary problems and respiratory illness are linked to heat-related shifts in air pollution concentration.

In many locations, extreme cold results in more deaths than extreme heat through direct effects such as cardiovascular stress and indirect effects like influenza (Huynen *et al.* 2001). These outcomes from extreme cold events are mediated by built environment and spatial determinants. For example, cardiovascular stress-induced mortality in Taiwan between 2000 and 2003 that followed cold surge events was shaped by geographic location within the city (Yang *et al.* 2009). Northern European mortality for people suffering hypothermia during cold waves ranges from twelve to forty-six per cent (Huynen *et al.* 2001). In highland Peru, mortality rates in children have increased due to extreme cold events (Barbier 2010). Some claim that increasing extremes in weather patterns will also exacerbate cold spells (see Box 23.1).

While much is known about the biological processes that produce negative outcomes to heat and cold, a consensus regarding definitions of when related morbidity and mortality occur has yet to be developed. The number of heat deaths published in any specific context often appears to reflect an objective and credible medical consensus. Diagnostic definitions are based on the political conditions in which a medical professional determines the existence of morbidity or mortality. There are a number of variables that have a distinct impact on whether or not a death is tied to heat or cold, including where and when the body is found, who first diagnoses the death and even the city in which the death is discovered (Nixdorf-Miller *et al.* 2006).

Impacts of extreme heat and cold

Disasters involving extreme heat and cold have a long history. In Beijing in 1743, 11,000 died in an extreme heat event (Levick 1859). One North American heat wave in 1936 killed almost 200 people in Canada alone. Two years later, a heat wave in Chicago, Illinois killed 800 people, mostly older persons (Klinenberg 2002). In 2003 a European heat wave resulted in roughly 70,000 deaths, in excess of the statistically expected number for that period of time (Poumadere *et al.* 2005). In the past decade, India has been affected by extreme heat, facing temperatures of 40°C for up to two weeks, resulting in higher mortality trends (Srivastava *et al.* 2007). One such heat

Box 23.1 How do heat waves and cold waves form?

Jostein Mamen
Norwegian Meteorological Institute

There are multiple definitions for heat or cold waves, but they are generally understood to mean longer periods of much warmer or much colder weather than normal. The deviation from the 'normal' is important in the definition, because the climate varies enormously across the globe. What is considered to be a heat wave in one area, e.g. temperatures above 30°C in Northern Europe, need not be so in another area, e.g. African or Asian lowlands.

Heat waves are always associated with high pressure systems, which can dominate the weather in an area for a few days to a few weeks. In summer, a high pressure system consists of warm air, light winds and clear skies. Because warm air can contain more moisture than cold air, heat waves also typically give sultry conditions. The warm and humid air makes us sweat more and the sweat does not evaporate from the skin as easily. Evaporation regulates the body's temperature, and when this is hampered most of us feel uncomfortable. For people in frail health, the condition can be dangerous.

Cold waves can occur in two ways. First, when a warm air mass is being replaced by a cold one, the temperature can drop considerably. The battle between the two air masses is often accompanied by strong winds and, in the winter, heavy snow. The second way in which cold waves can occur is during high pressure weather, namely at night time in the winter. The light winds and clear skies of a high pressure system give rise to substantial heat loss, especially from snow-covered ground. The coldest air is found in valleys and other low-lying areas. The stable nature of the high pressure system may let the cold weather dominate for weeks. During daytime, the sun is too weak to raise the temperatures significantly. North and South of each polar circle the sun never rises above the horizon for a part of the winter, leaving little opportunity for daily warmth to dispel a cold wave.

Prolonged exposure to cold, especially if one is not sheltered from the wind, can lead to death. As noted, 'hot' and 'cold' are relative. When the temperature drops down to 5–10°C in India, pavement dwellers and others without adequate shelter or clothing, especially those weakened by ill health and malnutrition, often die. This, too, is considered a cold wave in that context.

wave led to 2,600 deaths in the mid-1990s (Kumar 1998). Other cities around the world are also at risk of the extreme heat linked to climate change. Outside cities, livestock are vulnerable to extreme heat, and this impacts the livelihoods of the herders and farmers (Sirohi and Michaelowa 2007) (see Chapter 31).

Similarly, extreme cold spells strike communities around the world, causing mortality and putting vulnerable groups at higher risk. High-altitude indigenous communities have faced the loss of their herd animals upon which they depend for their livelihoods (Batjargal 2001). Such communities also lost staple food crops from frost damage. Likewise, herders in northern Asia have faced massive livestock losses in freak snowstorms and cold events. This phenomenon is common to many countries and deserves special attention since it tends to affect the poor, who lack access to other nutritional resources.

Cold also has serious economic consequences in more affluent countries (Kunkel *et al.*1999). For example, in January 1998 in eastern Canada, an ice storm brought down power lines and power line towers, leaving millions of people without power for days – and many without power for weeks. The ice storm itself was not an extreme temperature event. However, with temperatures dipping well below freezing while the power grid was being repaired, people were forced to deal with winter temperatures to which they were unaccustomed. After that disaster, a popular suggestion was to bury power cables to reduce the system's vulnerability to ice storms. A few weeks later, however, Auckland's buried power cables were affected by a heat wave, causing a power crisis in New Zealand's largest city for several weeks.

Vulnerability

The influence of weather patterns on human populations is mediated by social factors. Social determinants of disease include the social, economic and political conditions that define the existential situation of any given person. They include social status and networks, living and working environments, life skills, coping mechanisms, access to health services and many others. These factors interact with one another and shape biological processes.

The US National Academy of Sciences developed a framework for analysis of infectious diseases that can be adapted to the study of non-infectious conditions such as the impacts of extreme heat and cold. This so-called 'convergence model' brings together multiple factors that shape illness including physical/environmental, genetic/biological, ecological and social/economic/ political. In the case of extreme weather events the variables of importance include the physical environment as it is constructed in the built environment, related ecological variables, biological factors, and the social, political and economic factors (Smolinski et al. 2003).

Not all populations are similarly affected by extreme heat and cold. While some can protect themselves with cooling devices or appropriately insulated housing, others lack the necessary material resources. Additionally, there are social resources, which vary across populations, that play a critical role.

Biological factors in vulnerability

Biological factors intersect with these social contextual factors. Age is the most significant personal risk factor, with older persons being characterised by a number of physiological and social characteristics that place them at elevated risk (see Chapter 37). Ageing decreases thermo-regulation (the ability to effectively monitor internal temperature within a healthy range), and age-related decrease in fitness reduces the body's ability to adjust to outdoor temperatures. Pre-existing chronic disease, more common in the elderly, also impairs compensatory responses to these temperatures, as do certain medications, including those for cardiovascular disease, diabetes, as well as mental illness. For many individuals, a side effect of these drugs is that they may not be aware that high outdoor temperatures are making them ill. Many older adults tend to have suppressed thirst impulse as age inhibits thirst mechanisms. This inhibits ability to adjust to extreme heat. In addition, multiple diseases and/or drug treatments also increase the risk of dehydration. These vulnerabilities were reflected in the European heat wave in which people over the age of seventy-five constituted eighty-three per cent of deaths, and ninety-one per cent of those who died lived alone (Poumadere et al. 2005). The youngest groups are more at risk of heat death than cold. Babies and infants are at increased risk during a heat wave because they are more likely to dehydrate. Young children, especially in urban areas in some areas of the world, are also at risk of adverse heat affects – primarily due to their high prevalence of asthma (Weiland et al. 2004). Rising temperature also increases ground-level ozone smog production, which presents a serious threat to asthmatics, especially in combination with humidity.

Risks related to extreme cold events are similar to extreme heat, although cold events in the past have caused higher mortality. Like heat, extreme cold affects the very young, older persons and those suffering from pre-existing illnesses. Such deaths are also driven, in part, by cardiovascular, spatial and social factors (Yang *et al.* 2009). Cold events stress the cardiovascular system, such as the rise in blood pressure that occurs with exposure, while spatial location shapes temperature and exposure. Moisture and wind chill are some of the most important determinants of mortality in extreme cold events, and protection from them reduces mortality. Cardio-respiratory illnesses have the greatest link to mortality, especially in children and the elderly, and exposure to air pollutants has a greater impact on individuals at lower temperatures.

Socio-economic factors in vulnerability

Vulnerability to extreme heat and cold events is largely driven by the social determinants of disease. Biological, ecological, built environment and sociological factors intersect to shape who dies and who lives when temperatures suddenly drop or rise. Socio-economic status, even at the national level, determines access to resources and technological advances that can be used to address these challenges.

Individual-level vulnerability to heat and cold is shaped by social determinants, including built environment, land use, socio-economic status, social networks, social support and crime (Northridge *et al.* 2003). In the case of cold and heat, built environmental factors include housing quality and the combination of built structures in an area more generally. Built and natural features that contribute to increased vulnerability to extreme heat are: lack of canopy cover, preponderance of concrete and other surfaces that have low albedo (i.e. they absorb much of the heat from the sun), substandard building structures that cannot effectively insulate occupants from high outdoor temperatures, and buildings that lack appropriate passive and/or active cooling mechanisms (e.g. fans, air conditioning).

The structure of buildings themselves is important for cold risk. Lack of insulation makes it difficult for people to protect themselves from the cold. For example, the drastically higher winter mortality rate in Ireland than in Norway is largely due to poorer housing quality (Clinch and Healy 2000). Access to housing is also critical during cold events, and the homeless or those living in substandard housing are more at risk than others. During such events, snowfall can also damage weak housing structures and make them uninhabitable, sending residents into vulnerable conditions. Building and infrastructure design should consider the degree to which cold conditions can penetrate structures, pavements and soils. Housing conditions are often undermined by conflict situations when there is damage to structures and few resources available for timely rebuilding (Mann *et al.* 1994) (see Chapter 7). Extreme cold is more likely to affect groups living in regions at risk of these events, and who do not have the resources to protect themselves.

Built environmental conditions are often shaped by the social characteristics of inhabitants. People of low socio-economic status are more likely to have substandard housing and lack the social networks necessary to improve their conditions. Social isolation, and lack of social networks, plays a critical role in heat risk, in particular. Studies have found that sociological factors are associated with the impacts of extreme heat. For example, research in Madrid demonstrated greater impacts on the socially isolated and the less-educated (Díaz *et al.* 2002). Lack of neighbours, friends or family members to communicate risks and to check on vulnerable people has resulted in higher mortality.

These variables also interact with one another. For example, land use can shape social circumstances thereby creating or reducing vulnerability for communities. Neighbourhood-level

Sabrina McCormick

Box 23.2 Extreme heat and social isolation in Paris

Richard C. Keller
University of Wisconsin–Madison, USA

In early August 2003 a dangerous high pressure front stabilised over Western and Central Europe for nearly two weeks. It brought with it scorching temperatures and nearly windless conditions. However, its most devastating impact was the horrific mortality it left in its wake: some 70,000 excess deaths in Europe as a whole, and some 15,000 in France alone. In France, one of the most wrenching stories was that of the so-called 'abandoned' people affected by the heat wave: 100 people who had lived and died alone in Paris, buried at public expense when no relatives claimed their bodies. These people came to symbolise a welfare state in crisis: how, despite its top-ranked health system and its celebration of universal human rights, could France have allowed so many to die in such tragic isolation?

The heat wave was extreme in every sense. Daytime highs soared past 40°C for nearly two weeks throughout the country, and evening lows only dipped to the mid-twenties, giving bodies little respite from the heat. Ozone pollution reached record levels, making life in urban areas intolerable. The drought that France had experience since February of that year exacerbated the crisis, forcing shutdowns of nuclear power plants and the closing of public parks, in addition to fuelling the most extensive forest fires in French history.

Yet the heat was only one factor that structured risk and vulnerability for the French population. Mortality data reveal staggering concentrations of deaths in urban regions such as Paris and Lyon, and especially among the very elderly, the poor and the socially isolated. People over seventy-five years old comprised eighty per cent of the death toll, and those in the lowest socio-economic ranks constituted the most vulnerable group during the crisis.

The so-called 'abandoned' victims in Paris are therefore an important sentinel group, signalling some of the critical factors that have produced a social ecology of risk in contemporary France. Extensive field study of this group indicates that it was social factors, far more than a physiological vulnerability to extreme heat, that placed this group at risk. Conditions of desperate poverty, mental illness, addiction and poor health alienated these people from their communities and exacerbated the heat's already deadly effects.

Geographic vulnerabilities were also a critical factor that shaped risk, with those living on top floors of walk-up buildings and on the street – themselves important markers of poverty – bearing the brunt of the death toll. In conditions of potential rapid climate change, cities such as Paris can expect heat waves of increased intensity and duration, and must plan to protect such vulnerable groups as the homeless, the elderly and people with disabilities if they are to avoid such devastation in the future.

factors, like lack of access to social services, high levels of crime, and need for social services, have been cited as factors that increase risk in certain areas (Klinenberg 2002). More specifically, crime is both a reflection of and a contributor to lack of community cohesion and lack of proactive behaviours. In some areas, people are afraid to open their windows because of crime rates, despite the access to cooler air that this would provide. In addition, interactions between socio–economic status and environment are likely to be important factors in elevated risk among urban disadvantaged. Socio-economic factors shape risk of mortality linked to heat and cold. For example, countries in Europe with greater poverty, social inequality and deprivation suffer from higher rates of cold mortality (Analitis *et al.* 2008). Low-income groups are at higher risk of heat events, in part because they are less likely to have access to or utilise air conditioning (O'Neill *et al.* 2005).

Preparedness and mitigation of extreme temperature risk

Until recently there had been little official planning and few large-scale prevention programmes to address heat waves or extreme cold. In the eighteen cities in the USA reviewed by Bernard and McGeehin (2004), only one-third had a heat response plan, and most provided outreach only when a heat wave had already begun as opposed to prevention when heat was forecast, which is much more valuable. Additionally, in most of the world, especially in low-income countries, there are few watch/warning systems for extreme heat events.

Heat and cold are often managed differently and unevenly. There are very few formal plans to protect the public from extreme cold events despite the higher mortality rates in many settings. Heat warning systems are the most common plan for heat preparedness. They are based on the triggering of heat warnings and watches resultant from measurement of temperature and meteorological data. A part of this is often education about behavioural interventions such as seeking cool shelter, lowering stressful activity, drinking fluids and taking care of elderly relatives or neighbours.

Does preparedness work?

Heat preparedness programmes are growing around the world. To date, however, there are few, and those that exist need much work to be effective. Some appear to hold promise. Use of emergency medical services dropped by forty-nine per cent during a heat wave in the US city of Milwaukee, Wisconsin, in 1999 due in part to improved prevention efforts such as springtime education, telephone hotlines and improved co-ordination between government agencies and the media (Weisskopf *et al.* 2002). Initial evidence has shown that interventions in Philadelphia reduced mortality rates during heat events. An Italian intervention programme found that caretaking in the home resulted in decreased hospitalisations due to heat. Following the 2003 heat wave, France developed the 'Plan Canicule' focused on prevention, responsibility and solidarity. In 2006, when another heat wave occurred, the mortality rate in France decreased by eight per cent from the expected mortality for such an extreme heat event (Fouillet *et al.* 2008). While these developments appear promising, questions still remain about their effectiveness in many circumstances. Some have argued that lack of community-based intervention plans played a role in the 15,000 deaths in France during the European heat wave. This instance reflects existing questions regarding why or how programmes might increase protection from heat waves.

There is a similar lack of formalised preparedness for extreme cold events. This is particularly a problem in areas that are typically warm but that occasionally experience cold spells for which local populations are unprepared. Possible measures include cold wave warning programmes, increased awareness about cold risk, improved interventions for high-risk populations and public warm refuges. In addition, research has suggested that weather broadcasters and health specialists should collaborate with one another in the delivery of information to groups that are at high risk, such as the elderly, and young people who are outdoors continually (see Chapter 63). While much cold mortality could be prevented through protective measures, they are especially hard to implement in sudden, extreme events. For example, when Tajikistan experienced a

Sabrina McCormick

serious cold wave in 2008, water, fuel and electricity sources were compromised (see Box 23.3). Parts of the Andes also experienced cold waves in 1975, 2007 and 2009, with similar impacts.

Box 23.3 Cold extremes in Tajikistan

Charles Kelly
Aon-Benfield UCL Hazard Research Centre, University College, UK

In the winter of 2007–08 Tajikistan experienced unusually cold weather which had serious impacts on the lives of urban residents. While Tajikistan normally experiences cold winter weather, this cold spell was unusual for its duration and depth. Temperatures in the capital, Dushanbe, did not rise above freezing for close to forty days, quite different from the normal pattern of a few cold days followed by several days above freezing. Yet, the cold weather itself was not the crisis. The real crisis arose because most urban residents in Tajikistan rely on electricity for heating. The cold weather-induced demand for electricity overloaded a power system that had seen no investment in decades. Electricity generation was heavily reliant on hydro-electric power with limited reservoir capacity that could not meet increased demand. The 2007–08 cold spell affected all of Central Asia so that Tajikistan could not import electricity from neighbouring countries, as it normally does in winter.

As a result, electricity supplies were rationed with increased severity as the cold spell continued. There were extended cut-offs to residential users and almost a total elimination of electricity to industrial facilities, as well as restaurants and stores. As the volume of water in storage dropped (water stored from summer inflow provides most of the winter power in Tajikistan), cuts in electricity supplies became more severe, eventually focusing the limited electricity supplies on parts of Dushanbe and the aluminium factory (a major foreign income earner and employer for the country).

The human impacts of the cold weather and lack of electricity were varied. Aside from a lack of heat, there were problems with water supplies (reliant on electrical pumps to supply multi-story buildings), sanitation (no water meant it was hard to flush toilets or wash) and difficulty cooking (also reliant on electricity). There were also health consequences. An influenza-like disease circulated widely during the height of the cold wave and was likely associated with poorer-than-normal sanitation and crowded conditions as people sought to keep one room warm in a house. There were credible reports of carbon monoxide poisoning as urban residents attempted to use the traditional *sandelai* charcoal heater in apartment flats and homes to keep warm at night. While there is not a clear direct connection between severe cold weather and electricity shortages resulting in increased mortality, the cold weather did highlight the vulnerability of urban residents to acute electricity shortages and the country's reliance on a few hydro-electric facilities, as well as the lack of investment in energy infrastructure over several decades.

While urban residents in many countries are seen as receiving preferential access to resources and services, the cold wave in Tajikistan demonstrated that even this privileged access has little value when the natural events overwhelm infrastructure. Unless long-term planning takes into account the reality of extreme temperatures and invests in robust and redundant infrastructure, temperature extremes will likely contribute to more winters like the one in 2007–08 in Tajikistan.

Increased need for electricity drove rationing and shortages, which resulted in the increase in prices and lack of availability.

Social obstacles to preparedness

Even where there are programmes working to create preparedness for extreme heat and cold events, there are obstacles to engaging residents in them. A recent survey showed that elderly individuals in US cities with established heat preparedness programmes were not aware of appropriate preventive actions to take during heat waves (Sheridan 2007). Many knew about heat warnings, but did not consider themselves to be vulnerable. Few took the recommended pre-cautionary steps, such as increasing intake of liquids. Only forty-six per cent modified their behaviour on heat advisory days, and although many had home air conditioning, they did not always use it due to concerns about energy costs. Ethnic minorities in the USA have less access to air conditioning than others (O'Neill *et al.* 2005), which suggests that there are issues related to equity in access to prevention. In addition, in areas where air conditioners are used, high temperatures can be exacerbated by the release of hot air from air conditioners themselves. In Phoenix, although many residents know about heat warnings, only about one-half claim to take any action (Sheridan 2007). More generally, there is low public recognition of extreme heat as a hazard.

Educational messages for preparedness

One critical obstacle to uptake is developing appropriate educational messages about heat waves; yet, there is little evidence that such messages are being employed in preparedness strategies. Some evidence has shown that top–down educational messages achieve a very limited amount of resultant action. The receipt of information is not sufficient to generate new behaviours or the development of new social norms. Even when information is distributed through pamphlets and media outlets, behaviour of at-risk populations often does not change. It is important to note that more potential has been shown in influencing behaviour through community-level interventions than through individual-level directives at the population level (Kawachi and Berkman 2000). Those targeted by such interventions have suggested that community-based organisations be involved in order to build on existing capacity and provide assistance.

Research shows that heat and cold preparedness programmes should engage with communities in order to increase awareness (Smoyer–Tomic and Rainham 2001). Since social networks shape both heat and cold risks of at-risk populations, accessing these resources can lead to preparedness for extreme events. Community social resources such as networks, social obligations, trust and shared expectations create capacity to prevent, prepare and cope with extreme events. Scholars and policy-makers increasingly promote social networks as a long-term adaptation strategy. Social networks provide a diversity of functions, such as facilitating the sharing of expertise and resources across stakeholders (Crabbé and Robin 2006). Networks can function to promote messages within communities through preventive advocacy, or the engagement of advocates in promoting preventive behaviour.

Information about health risks has often been effectively distributed through a social network structure using 'opinion leaders' as a guide, and has promising application for changing behaviour regarding climate adaptation. Communities with stronger social networks are more likely to be ready for extreme weather because of access to information and social support. Indeed, social groups often are repositories of important traditional and other knowledge and skill in coping with extreme heat and cold (see Box 23.4).

Box 23.4 Economic and architectural adaptations to heat and cold

Ben Wisner
Aon-Benfield UCL Hazard Research Centre, University College London, UK

For millennia people have adapted their dwellings to ambient temperature and seasonality. Crops and livestock, too, have been bred for heat tolerance – the camel and Nubian goat, for example – and resistance to extreme cold – such as the yak in the Andes. Yet, extreme heat and cold waves can nevertheless overcome these defences, causing economic loss, illness and even death. An open question is whether these practices can serve as a basis for extending protection to higher and lower temperatures, especially as growing climate instability could increase the severity of heat and cold emergencies.

Architectural adaptations

Traditional shelter in many parts of the world utilised local materials and site-specific knowledge to provide cooling and heating. For example, in Iran the *Badgir* or 'wind catcher' is a mud brick chimney built over a fountain or underground water reservoir. The design provides cool, humidified air inside during the day, whilst the mud-brick radiates heat during the cold desert night (A'zami 2005). Similar systems existed in North Africa and the Middle East. Seasonally, in the mountains of the Korean Peninsula temperature can vary from -15°C to 30°C. Traditional architecture there uses adjustable features such as windows and doors to control ventilation. Attention is paid to the orientation of the house so that eaves block sunshine in the summer but allow passive heating in the winter (Kim 2006). Construction with high thermal mass – mud and stones, for example – provides protection from cold at high altitudes in the Andes, Nepal and elsewhere. Locally available materials such as dried grass are commonly used for insulation. Cross-ventilation is an important cooling practice in hot climates.

There are, in short, many architectural adaptations to heat and cold, but the challenges to building upon this knowledge base to protect against extreme heat and cold waves are numerous. First, local and traditional knowledge is often given low status and is disappearing, especially with the steady shift of population from the countryside to cities. Second, habitat changes such as deforestation mean that some of the traditional building materials are harder to find. Third, some of the people who suffer most from extreme heat and cold are homeless – for example, 'pavement dwellers' in India and Bangladesh (UNOCHA 2003).

Economic adaptations

In cold and hot climates crops and livestock have been selected for physiological resistance. The potato in the Andes is an example of a staple crop selected over 8,000 years by generations of farmers for cold, high-altitude conditions. Some 5,000 varieties of potatoes have been identified in this area of its first domestication (Cabieses 2010). Barley has similarly been bred to cope with such conditions on the Tibetan plateau. Heat tolerance and drought resistance have also long been bred into staple grains by farmers in semi-arid climates. Sorghums and millets grown in parts of Africa and South Asia are good examples.

Sorghum was likely domesticated in what is present-day Ethiopia around 7,000 years ago, and it remains a staple in hot, dry areas (CGIAR 2010). Livestock has also been genetically selected for heat and cold. Yaks, llamas, water buffaloes, fat-tailed sheep and zebu cattle (*Bos indicus*) are all examples. Management of these animals also takes into account extremes of temperature. Shelters are provided for winter in many herding systems, and fodder (including crop residues) may be stored as animal feed (Shrestha 1992).

As with traditional architecture, rural livelihood adaptations to heat and cold also face challenges in providing the basis for expanded protection against extremes of temperature. In particular, widespread efforts by governments to 'modernise' agriculture have resulted in the loss of many potentially valuable traditional practices. Changes in land tenure have also seen many smallholders displaced and now working for larger farmers who practise energy- and input-intensive agriculture.

The European and North American model uses fossil fuels to control temperature extremes. Examples range from large fans to stir the air and field heaters to guard Florida oranges from unseasonable frost, to energy-intensive hothouse complexes that allow production of high-value fruit, vegetables and flowers in Northern Europe even in winter conditions. However, as energy prices increase and the previously hidden 'external' environmental costs of oil drilling, coal mining and nuclear power become apparent, such technical solutions to heat and cold may turn out to have been a brief interlude. In addition, global climate instability and change will increase pressure on systems that have become dependent on such technical solutions.

Addressing social factors in preparedness promises to be critical for the protection of vulnerable populations (see Chapter 57). This includes incorporating communities themselves in understanding of and responses to extreme events. Top-down measures imposed by health practitioners that do not account for community-level needs and experiences are likely to fail. Greater attention to and support of community-based measures in preventing heat- and cold-related mortality can be more specific to local context, such that participation is broader. Such programmes can best address the social determinants of health outcomes.

Conclusions

Sara sat at her kitchen table on the afternoon of 8 July 1993. There was a knock on her door. As she stood to answer it, dizziness engulfed her and she fell. On the other side of the door, her neighbour heard a dull thud. After trying to get inside, he hurriedly called the heat hotline, which he knew would send a fireman to get in. Once Sara was found on the floor, unconscious, she was taken to a cooling centre where medical care was available to residents in the poor, urban area where she lived. Other neighbours were already assembled there, some relaxing and others beginning to implement the phone tree they had planned in order to make sure that their community members would not end up sick or dead from the heat wave. Meanwhile, city planners and a local grassroots organisation were collaboratively initiating work on the creation of green spaces in neighbourhoods where there were few trees and temperatures were chronically high.

This scenario is the alternative to what actually took place in Philadelphia in 1993, described at the beginning of this chapter. By putting existing knowledge of effective preparedness into place, mortality from extreme heat and cold events can be avoided. The risks of heat and cold are of critical importance in public health and city planning, especially as climate change alters historically familiar patterns making it difficult to predict and adjust to these events. Under-standing and acting upon the social determinants of vulnerabilities to such moments is a critical component of how deaths can be prevented. Putting social networks to use, addressing the limiting and catalysing aspects of social norms, improving resource constraints and acknowl-edging the need for further research are all a part of this agenda. By taking these actions, the most vulnerable populations can begin to combat extreme weather.

24

Wildfire

Roger Underwood

BUSHFIRE FRONT AND YORK GUM SERVICES, WESTERN AUSTRALIA, AUSTRALIA

Alexander Held

WOF INTERNATIONAL, SOUTH AFRICA

Wildfire (wildland fire, forest fire, bushfire) basics

The term 'wildfire' and its associated phrases refer to an unplanned fire in wilderness vegetation or bush, including grass fires, forest fires and scrub fires, i.e. any fire outside the built-up urban environment. 'Bushfire' tends to be used in Australia while the USA and Canada call it a 'wildfire' or a 'wildland fire'. In Europe and Asia it is usually called a 'forest fire'.

The three primary classes of wildfires are surface, crown and ground. Classes are determined by the types of fuels involved and the intensity of the fire. Surface fires typically burn rapidly at a low intensity and consume light fuels while presenting little danger to mature trees and root systems. Crown fires generally result from ground fires and occur in the upper sections of trees, which can cause embers and branches to fall and spread the fire. Ground fires are the most infrequent type of fire and are very intense blazes that destroy all vegetation and organic matter, leaving only bare earth.

There are three essential components of a wildfire: heat, oxygen and fuel. If any one of these three is missing, fires cannot start or keep burning. Fuel is the only element of the fire triangle that can be controlled by land or disaster managers. If bushland fuel levels are reduced before a summer wildfire starts, the fire burns less intensively, spreads less rapidly, causes less damage and can be more easily controlled by fire fighters. This is the principle on which the practice of 'prescribed burning' is based (Underwood *et al.* 1985).

Fresh oxygen is always being delivered to a bushfire by the wind and cannot be controlled, but it makes fire fighting much easier on less windy days. Heat cannot be usefully reduced in a wildfire even with helicopter water tankers and belly-lifter water-carrying airplanes, but rain helps. Using water on a fire is traditionally undertaken by fire fighters in urban areas to put out fires in structures where there is ready access to fire hydrants and mains water supplies. In a bushfire, however, the huge quantities of water needed to put out a fire are almost never available.

This principle is well understood and supported by science and field experience. In some parts of the world, such as Western Australia, the whole fire control strategy was designed

281

around fuel management, fuel being the component that can be managed by humans. However, in recent decades most countries have implemented a fire suppression approach, with the general message that fire and its effects are negative.

The three most widespread natural vegetation types in fire-prone ecosystems are grasslands, tropical and subtropical savannahs, and tall forests. Grasslands grow during the wet season, die off in the dry season and rot away or are eaten (and respired) by termites and/or herbivores during the next wet season. Each succeeding crop of grass is replaced by a new crop. If the grass is burnt during the dry season (and lightning-caused fires are a common feature of tropical areas) little changes, because the burnt plants would have decayed anyway.

Tropical and subtropical savannahs, areas of grassland dotted with trees, add a level of complexity. In terms of area, number and frequency, most wildfires occur in the savannahs of the tropics and subtropics (FAO 2007). Many savannah fires are 'natural', being started by lightning, but there is also extensive prescribed burning undertaken in these areas, especially savannahs used for cattle grazing and other forms of agriculture.

If fire is deliberately excluded from these areas, which is difficult but has been achieved in some small experimental locations, there is an increase in wildfire fuels over time. Dead material on the ground rots or is consumed by termites, other insects, or otherwise decomposes, but fire fuel accumulates as bark on fibrous-barked trees and in the woody shrubs that develop when fire is excluded. Fire exclusion experiments in Kruger National Park, South Africa, over the past fifty years have demonstrated the resulting bush encroachment dramatically (Shackleton and Scholes 2000). Essential grazing potential is lost for rural communities depending on cattle, sheep and goats. This increase in the fuel load means that late dry-season fires are more intense, causing death and damage to live trees and burning down dead trees. Intense fires will rapidly consume logs and branches on the ground which may otherwise have taken years to rot away.

Tall forests include the trees, woody shrubs and mid-storey vegetation along with the litter and accumulated organic debris on the ground. Eventually, all old trees begin to decay, and in the absence of fire the accumulated litter on the forest floor begins to rot away.

Given this variety of vegetation and vegetation ecosystems, wildfires vary in their size, speed and intensity. This variation is mostly determined by the amount of 'fine fuel' (defined as combustible material less than 6 mm in diameter). If there is no fine fuel present, then larger dead fuel (such as old logs and branches) or the living fuel in the trunks and canopies of the shrubs and trees will not ignite and burn. This is why even an intense fire goes out when it reaches an area that was burned recently, and carries no fine fuel.

However, the total amount of fuel consumed by a wildfire depends on the amount of moisture in the fuel. Dry fuels burn more intensely, and these intense fires dry out and burn the fine green fuels in front and above them.

Fuel reduction by prescribed burning employs low-intensity fires lit under mild weather conditions at a time when there is still some moisture in the fuel. This ensures that the flames are generally less than 1 m high and the fire is confined to the surface layer of fine fuel and the green material in the low shrubs.

Wildfire facts and situations

Statistics regarding wildfire numbers, casualties, costs and areas affected vary significantly by geographic region in terms of data availability and reliability, so a global overview cannot be given (Levine et al. 1999). Every year on average, Europe experiences devastation as forests and other rural areas are burnt by wildfires (see Box 24.1). Regions that experience even more fire, like Africa, do not have the capacity to provide statistics or damage assessment in terms of

Box 24.1 Wildfire in Europe

John Handmer and Joshua Whittaker
Centre for Risk and Community Safety, RMIT University, Melbourne, Australia

A combination of climate and demographic trends, often with local features, is increasing the wildfire threat across Europe. Food and Agriculture Organization (FAO) data indicate that Europe is seeing a growing number of wildfires and increasing wildfire risk. In 2003 Europe faced about 50,000 fires, with an annual average of 60,000 hectares burnt, which is more than double that of the 1970s (FAO 2007). Three-quarters of all forest fires and almost all the annually burnt area is in the Mediterranean region, with Greece having the most severe problem (Moriondo *et al.* 2006).

Historically, rural areas were relatively heavily populated with farm and timber workers and those that serviced these sectors. Many rural people historically worked to protect their livelihoods from wildfire as another natural hazard of farming, while also using controlled fire as a farming tool.

Changes in fire occurrence reflect the recent socio-economic changes underway in the European Mediterranean countries. Significant changes include depopulation of rural areas, increasing agricultural mechanisation, reduced grazing pressure and wood gathering, and increasing urbanisation of rural areas (Dimitrakopoulos and Mitsopoulos 2006), in addition to the expanding urban edge and increasing recreational use. These changes from traditional land use have led to the abandonment of large areas of farmland. In turn, vegetation has grown unchecked in these areas and accumulated as fuel for wildfires. The human changes parallel changes in the fire regime, with fires increasing in number and affecting a larger area every year (Dimitrakopoulos and Mitsopoulos 2006).

According to Xanthopoulos (2000), in Greece one of the most important fire-related trends that started in the late 1970s – of building secondary summer housing along the coast – accelerated in the 1980s. Although these areas were poorly planned, they were popular, driving up prices and enticing people to burn forests illegally in order to occupy the land. Evicting occupants was politically unfeasible, so the buildings were legalised. That in turn created a motivation for arson and building in fire-prone areas.

financial values. However, communities that live in severe poverty suffer the most from uncontrolled fires. These fires often not only burn the bush, but also crop and grazing land as well as houses built from natural materials. That affects the livelihoods of entire subsistence farming communities without any media or international agency attention and without any mitigation support from government or non-governmental organisations. An international agency that tries to keep track of fires is the Global Fire Monitoring Center (GFMC) (see Box 24.2).

In Europe, land-use changes during the last decades and the increase of mean temperatures and frequency of droughts due to climate change have accelerated this development. Southern Europe has experienced a number of catastrophic fires in recent years, as seen in Portugal, Spain and Greece, destroying large forest areas, properties and costing many lives. The Greek fires of 2007 consumed nearly 300,000 hectares of forest and other rural land, more than sixty lives were lost and a large number of homes destroyed. According to Greek media reports damages were estimated in the billions of euros.

Fires have even caused regional haze problems. In 1997 burning rainforests in Indonesia led to haze interfering with human health and activities across five countries. The crisis precipitated

---

Box 24.2 The Global Fire Monitoring Center (GFMC)

Johann Goldammer
Director, Global Fire Monitoring Center, Germany

The GFMC (www.fire.uni-freiburg.de) was established in 1998 as a contribution to the United Nations (UN) International Strategy for Disaster Reduction (ISDR) (and its predecessor arrangement, the IDNDR). It is located on the campus of Freiburg University, Germany, and serves as a global and public portal for wildfire documentation, information and monitoring.

The regularly updated national to global wildland fire products of GFMC are generated by a worldwide network of co-operating institutions. Web-based information and GFMC services include early warning of fire danger and near-real time monitoring of fire events along with interpretation, synthesis and archiving of global fire information. Supporting the development of wildfire-related policies, legislation and implementation strategies is part of the work along with liaison capabilities for providing assistance for rapid assessment and decision support in response to wildland fire emergencies under co-operative agreements with the UN and others. Training courses for international wildland fire management specialists are also run by GFMC.

GFMC is an Associated Institute of the United Nations University (UNU) and is serving as co-ordinator and facilitator of the UNISDR Wildland Fire Advisory Group and the UNISDR Global Wildland Fire Network, a global voluntary network providing policy advice and science and technology support for reducing wildfire impacts. Since 2008 GFMC has served as the Secretariat of the Fire Aviation Working Group, a consortium of countries with major aerial fire-fighting assets working under the umbrella of the UNISDR Wildland Fire Advisory Group.

---

a Regional Haze Action Plan that incorporated many aspects of fire prevention including public education, along with monitoring and enforcement of fire-related laws (Brauer and Hisham-Hashim 1998).

Calculating wildfire costs is challenging. In the 2003 Canberra wildfire disaster that killed four people, for example, the cost was estimated at over US$300 million (ACT Government 2003), yet the total cost of the fires including uninsured losses, the cost of suppression measures, and business interruption would be much greater than this figure. Added to the financial toll is loss of life, personal trauma, loss of invaluable and irreplaceable personal possessions, and the destruction of scientific equipment and data from the Mount Stromlo astronomy observatory. The implications do not end with the direct impacts of the fire. For example, considering only finances, businesses might not be rebuilt and insurance premiums might rise for everyone.

Normally, severe fire seasons occur every several years, when most of the damage is caused. Fire managers often suggest the potentially apocryphal rule-of-thumb that ninety per cent of fire damage occurs in ten per cent of the years. Considering the costs of catastrophic fires, the statistics suggest that reality might be even more imbalanced (e.g. ACT Government 2003; Butry *et al.* 2001). The wildfire issue cannot be confined to forest areas alone. It is a regional issue affecting public forests as well as farms, rural towns, rural infrastructure, private native forests and private plantations.

In order to confront such catastrophic events, new and innovative concepts for landscape management, including targeted management measures in forests, are needed in order to mitigate fire danger and intensity. This approach to fire prevention is often described by the term 'integrated fire management' (www.fireparadox.com).

Wildfires are getting worse

What do the statistics show? For the numbers available, the trend is towards larger and more intense and damaging fires (FAO 2007), with the three main contributing factors present in most fire-prone regions. First, there is a massive build-up of fuel on public and private land, which includes development encroaching into fire-prone areas sometimes with fuel management considered or attempted. Second, resources available to fight fires are declining. Third, there has been the loss of traditional fire practices through colonial rule, socio–economic factors, poverty and civil unrest, especially in the developing countries.

Rightly or wrongly, concerns are often articulated about increasing numbers of 'megafires', which are large, high-intensity and nearly unstoppable wildfires. Climate change is often suggested as being a major factor in more megafires.

If the current climate change models are correct, the possible temperature increases over the next century can have only trivial impact on wildfires and wildfire behaviour. Fire intensity is far more significantly affected by fuel quantity, fuel dryness and wind strength than it is by temperature. Climate change models also suggest significant changes in rainfall in many fire-prone locations. Reduced rainfall leads to increased fuel drying and increased fuel availability at lower temperatures. This is the same effect on fires as that of drought, a phenomenon which is common in many parts of Africa, Asia and Australia. Drought does result in more intense fires, but only if nothing is done to reduce fuels before the fire occurs.

For rainfall and wildfires, the critical factor is the seasonality of rain. In Australia's temperate regions, for instance, increased rainfall in late summer will generally lead to higher vegetation decay rates and generally lower fire dangers, while a corresponding decrease in winter rainfall would provide an extended opportunity for mild low-intensity burning. The opportunity for land managers is to get in first, in order to manage and reduce fuels before a potential megafire starts. In other words, the potential megafire can be forestalled simply by adopting a programme of fuel reduction prescribed burning under mild weather conditions.

The overall picture can therefore be complex. Flannigan *et al.* (1998) suggest that climate change is likely to contribute to reducing the fire threat in northern boreal forests. Meanwhile, Westerling *et al.* (2006) attribute worsening fires in parts of the American Rockies to climate change-related changes in spring and summer temperatures as well as an earlier snowmelt. It is clear that climate change is simply one of many factors influencing wildfires and there is plenty that fire managers can and should be doing to control wildfires, irrespective of climate change.

Although the principles are clear, needed data are still frequently lacking, even though practical observations in the bush are not lacking. Land management policies can be guided by decades of records of severe forest and water catchment damage by high-intensity fires along with long experience in forest management to meet ecological objectives, while protecting infrastructure and the public. Local knowledge alongside that data can and should be used for pre-fire management. Jakes *et al.* (2007) detail some successful examples from the USA.

Consequently, statements on worsening fire situations caused by climate change neglect the human capability to manage and mitigate the situation. Blaming climate change as culprit for more fire incidents is ignoring human intelligence and the societal duty to manage the fuels, and therefore the intensity and impact of a fire.

In fact, fire has always been a factor in the environment. Historical records indicate that it is only recently that frequent mild fires have been deliberately taken out of the forest, to be replaced by infrequent high-intensity fires, especially in the USA, Australia and Europe. Fire services and fire management regimes are suppressing small, low-intensity fires, but that leaves plenty of fuel for megafires. Instead, prescribed burning needs to be promoted because it reduces the size and intensity of wildfires.

Different fire management approaches

How could different elements in the fire triangle be managed? The weather is rarely controllable (see Chapter 20), and ethical concerns are raised about trying; nor can or should all possible avenues of ignition (including lightning) be eliminated. The main manageable factor is the large contiguous accumulation of fuel, meaning that broad-scale fuel reduction burning is the main defence against large wildfires. The idea is not to prevent wildfires, nor will fuel management prevent all wildfires, but it does ensure that wildfires are less intense, are easier and safer to control, will do less damage and are more related to the natural fire regime.

Western Zambia provides a good example, where the use of fire as a land management tool is anchored in the legal system. Fire use is regulated in Zambia's 'Forests Act' (Government of Zambia 1999). Fire can be used from January to May, while from May to December, which is the fire season, it is illegal. This legal setting encourages early or green burning, resulting in many, low-intensity fires. Meanwhile, community 'fire councils' decide when, where and how to burn. The Miombo woodland in western Zambia was recently assessed and found to be a stable, healthy ecosystem with appropriate fires, due to the fact that early burning is a regular tool. Miombo woodlands in other countries look significantly different.

The experience of Western Australian forest managers corroborates this approach. Numerous, large fires occurred in Western Australian forests from 1900 to 1960. After the 1961 Dwellingup fire disaster, the government's wide-scale fuel-reduction programme ensured that fuel accumulation was well controlled. After the burning programme diminished due to governmental cutbacks, the incidence of large wildfires began to climb again after about 1990 (see also Box 24.3).

Box 24.3 Securitising and militarising wildfire

John Handmer
Centre for Risk and Community Safety, RMIT University, Melbourne, Australia

Jason Flanagan
Faculty of Arts and Design, University of Canberra, Canberra, Australia

Unlike their emergency service counterparts, fire fighting has civilian and commercial origins and was driven in part by insurers. However, fire fighting is organised and pursued with increasingly expensive technology and equipment. There are suggestions that this is changing attitudes, the ethos and methods of work within sections of the fire-fighting industry, making them more like the military or security industry.

In the immediate aftermath of the February 2009 Victoria wildfires, Bergin (2009) published a piece in *The Age* newspaper entitled, 'Defending the home front is top priority', in which he declared: 'In the aftermath of Black Saturday, it's time to rethink the protection of the Australian people … We now need, as the Prime Minister noted last December, to face up to climate change as a fundamental national security

---

challenge.' Bergin went on to discuss how 'defending' against climate change at home might require different 'battle tactic and skills', outlining how Australian military equipment and personnel should play a direct role in addressing 'natural disasters'.

In parts of the USA wildfire policy is not only framed as a 'war' on fire, but is actually fought with military equipment and tactics. In some cases, fire fighters, for reasons of access, fuel reduction and attempts at containment, destroy some of what is being protected. Lines of cleared areas intended to contain fires especially may have several adverse environmental impacts, such as killing and removing vegetation, compacting and eroding soil, degrading water quality and fragmenting forest stands.

The growth of aerial fire fighting has contributed significantly to the securitisation and militarisation of wildfire. While aerial fire-fighting appliances are promoted by the media as a solution – and have probably become politically unavoidable – Australian research has shown that aerial suppression is generally no more effective in stopping a fire's forward spread than ground crews with tankers and bulldozers (Loane and Gould 1985). The use of aerial suppression measures is widely considered to be as much a public relations exercise as an effective fire-fighting measure, with fire officials in the US states referring to them as 'CNN drops'.

Fire retardants have a fleeting effect on wildfires, but a lasting effect on ecosystems, being particularly lethal to aquatic wildlife. Aerial approaches have some significant value, though. They can be used to attack fires as they start, especially in remote areas or difficult terrain, well before ground-based fire fighters can reach the areas. They can also protect assets as otherwise uncontrollable fire fronts pass.

The 'war on wildfire' approach has its winners: supporting large agencies, creating heroes, enlarging budgets and creating a clear mission. While individual fire fighters are celebrated as heroes, the organisations have also recently been subject to intense negative scrutiny as a result of escalating cost, issues of effectiveness and perceived commercial imperatives, especially in aerial fire fighting. In the USA the emergence of what has been labelled a 'fire-military-industrial-complex' has been seen, with suggestions that commercial interests are driving approaches that return profit rather than addressing fire risk management objectives (Ingalsbee 2006). Profits are best generated by the use of expensive technologies. They are not as easily generated by relatively low-cost community-based and risk reduction activities.

Yet no situation is ideal. In Zambia the national government has tended to aim for eliminating practices of shifting cultivation and pastoral pasture seasonal burns. Goldammer and de Ronde (2004) also document damage that has been done in western Zambia due to poorly managed deliberate fires. Madagascar has also witnessed a long and continuing conflict between rural farmers and the country's leaders with regard to how fire is used, indicating how most choices are politically based rather than using science or fully considering the ecological role of fire (Kull 2004). Fire management can rarely be decoupled from political interests.

In moving towards more evidence-based fire management regimes, many fire management lessons from indigenous people are still valid. 'Burning is our tradition' is one of the standard replies when talking to children in rural Namibia – except that burning was never done in October, which is very late in the fire season and would produce very intense fires. Instead, traditional use of fire started early in the season, producing low–intensity, patchy burns.

Too often, 'traditional' fire management is vague, idealised or misunderstood. Shepherds around the Mediterranean used to burn old grass. Today, with increased fuel loads, increased damageable infrastructure and air pollution laws, the Mediterranean shepherd's way might be traditional, but detrimental. Similarly, Germany's farmers traditionally used to burn agricultural residues to an extent that burning was declared illegal. Slash-and-burn agriculture, which can also be termed 'shifting' or 'swidden' agriculture, involves cutting down and burning forests and other vegetation to develop fields for agriculture. While it is a factor in destroying forests worldwide, techniques exist by means of which it can be sustainable if implemented properly (Kleinman et al. 1995).

Traditional fire practices should be looked at in the context of contemporary standards and interests, and adapted to them. The key is to let neither traditional nor non-traditional approaches dominate, but to use the best components of each to work together in a complementary manner (see Chapter 9). Traditional fire practices have greatest value in contributing to fire management solutions for today's problems. Land and forest managers can identify three potential options.

'Let-it-burn'

'Let-it-burn' assumes that nature knows best, so wildfires are left to burn – being extinguished by either running into last year's fire burn, at the onset of the rainy season, or by fire fighters at the edge of the bush if human assets are threatened. No government can afford (officially) to adopt the let-it-burn approach. Fires burning out of heavily forested country can be unstoppable when they reach the edge of the bush and their smoke can interfere with human activity far away. Additionally, under most legal systems, an effort must always be made by the land owner or manager, which is often the government, to suppress wildfires, because not to do so leaves them open to legal action.

All-out suppression

All-out suppression requires fires to be attacked immediately after detection, using emergency service resources, normally a 'fire brigade'. This approach originated in Europe's cities in the Middle Ages, being exemplified by the drama of the ringing alarm bells, galloping horse-drawn fire engines and magnificently uniformed and helmeted fire fighters. The current image is equally theatrical, with water bombers and helicopters sweeping the smoky skies, convoys of tankers filing along country roads and brilliantly uniformed Fire Chiefs being interviewed on television by breathless reporters.

All-out suppression is appropriate in cities, where permanent fire fighters are always on standby and can reach any fire within minutes. In earlier days in rural Europe, all-out suppression was implemented by volunteer brigades of farmers and bush workers, and was largely successful in well-managed farmland and country towns. In many rural areas today, such as in Australia, even in forest country, all-out suppression remains the dominant philosophy, even though it contradicts practical experience and vegetation fire science.

All-out suppression does not and cannot work in fire-prone ecosystems unless it is supplemented by other measures (discussed below), because many fires cannot be put out by humans. Even under relatively mild conditions, the intensity of fires burning in fuels over about 10 tonnes per hectare is too great to allow them to be attacked successfully. In scenarios such as commercial forestry plantations, all-out suppression is assumed to be mandatory if companies cannot afford to lose investment, even though the approach must fail in the long run. Insurance is available, but is comparatively costly, with some companies offering a reduction in premiums if prevention and mitigation measures are undertaken.

Forest and land managers are increasingly understanding this, leading to adoption of the third approach across Australia, North America, southern Africa and lately Europe.

Prescribed burning (green burning)

Prescribed burning (or green burning) recognises two points. First, vegetation fires cannot be prevented. Even if humans were banned from the bush, there would still be lightning strikes. Second, periodic, mild, patchy fires prevent the build-up of heavy fuels, so that when a fire does start, it is easier and safer to suppress and does less damage. A regime of green burning also produces a healthier and more vigorous forest and is better for biodiversity. This approach was applied rigorously in many Australian forests for several decades, with tremendous success (e.g. Fernandes and Botelho 2003). Pollet and Omi (2002) show similar successes in ponderosa pine forests in the western USA.

The green burning approach calls for a well-planned and professionally conducted programme of prescribed burning. It is not a scorched earth approach nor relentless firebombing. Prescribed burning does not mean that there is no need for fire suppression, because it deliberately does not prevent fires from starting. It involves less intense, trickling fires that burn quietly through the understory, burning away dry leaves and twigs, so that when a real fire comes in mid-summer, there is less fuel to burn.

Referring back to western Zambia's Miombo woodlands where the early burning or green burning approach has been practised over many years, the land is a patch mosaic of burned and un-burned fuels, and disastrous fires rarely occur.

Fire management decision-making

A combination of strategies is often needed for fire management (see Box 24.4). At times – with people, property or important heritage under severe threat – it might be necessary to mobilise all possible resources to extinguish a fire. At other times, the fire might be non-threatening while being deemed part of the landscape's normal ecological processes. It could be permitted to burn but monitored closely. Factors affecting the management decisions range from the weather to land-use plans. Two key messages emerge from operational experience in fire management.

First, once a fire 'crowns' – that is, starts to burn through the tree canopy rather than along the ground – fire fighters cannot stop it. Crown fires, especially in eucalypt forests, throw spot fires (burning embers) kilometres down-wind, so fire intensity and speed overwhelm human effort. Although eucalyptus trees are native to Australia, they have now been introduced to all other inhabited continents making the most dangerous crown fires a possibility around the world. There is only one certain way to stop a crown fire: reduce the fuel in front of the fire. When a crown fire runs into an area that has recently undergone a prescribed burn, the fire drops to the ground and then it can be tackled effectively.

Second, wildfire management systems must focus equally on the three basic elements of fire management: preparedness, damage mitigation and fire fighting. To focus only on the third element simply resorts to the usual paradigm of only emergency response, which frequently yields greater disasters than necessary. Experience frequently demonstrates that fires in heavy fuels under hot dry conditions are unstoppable. Fire management decision-making must start long before a fire starts and involve preparedness and damage mitigation.

Knowledge and experience in wildfire management is something that takes many years to attain. Fire behaviour is complex and the theory needs to be understood, but also observed and

Box 24.4 Prepare, stay and defend, or leave early

John Handmer and Joshua Whittaker
Centre for Risk and Community Safety, RMIT University, Melbourne, Australia

Concurrent with a strong emphasis on fire suppression is the 'rescue' or evacuation paradigm, according to which people at risk or otherwise in the way of a fire are evacuated. In many parts of the world, 'evacuation' is a conventional response to imminent wildfire threat – although far less so in rural areas. This may seem sensible until it is realised that often evacuation can mean travelling through a landscape and road network that is burning and made dangerous by smoke, flames, heat and embers. This approach has contributed to the perception that all fire is bad, but perhaps more importantly has likely reduced a sense of capacity within communities that live with wildfire risk by encouraging dependency on fire and emergency services.

In Australia, an approach known as 'Prepare, stay and defend or leave early' has been the official approach to community safety since 2005 (AFAC 2005; Handmer and Tibbits 2005; Handmer and Haynes 2008). The approach was widespread before becoming official and had been used in some rural areas for more than a century. Essentially, the approach is that people at risk of wildfire decide in advance whether they will evacuate before the fire makes it unsafe to do so ('leave early'), or stay and proactively defend the building that will protect its occupants from radiant heat, embers and smoke.

Any action requires considerable mental and physical preparation. People need to know the risk and have the skills, knowledge and confidence to ensure their safety. As developed in parts of Australia, the approach does not mean less emphasis on fire suppression or fuel management. Instead, it has meant a greater emphasis on self-reliance, with fire agencies calling for the public to 'share' the risk by taking greater responsibility for their own safety. The approach draws on evidence that, with preparation, houses in Australia can survive wildfires, and that an important factor in house survival is the presence of able-bodied occupants who proactively defend the structure against embers.

Evidence also shows that people are relatively safe inside a house during the passage of a fire front, as it offers protection from radiant heat and smoke – and conversely that the most dangerous place to be is outside in the middle of the fire. Research has shown that last-minute evacuations are very risky, with many deaths occurring in vehicles and on roads. The approach has been under critical examination since the bushfires of 7 February 2009 in the Australian state of Victoria. Staying to defend is now not recommended on very severe fire danger days. The approach is based on historical evidence, and needs further research in light of modern cars, housing, attitudes and capacities.

Elements of the 'Prepare, stay and defend or leave early' approach have been implemented, both officially and unofficially, in southern France, elsewhere in Europe and in some communities in the USA. It is likely that it is, or could be, used in many other parts of the world where fire intensities are relatively low, where people are committed to defending their livelihoods or other critical assets, or where evacuation is not a safe or practical option. Provided that they are properly implemented, these approaches that emphasise community action are generally low cost. This is in sharp contrast with the trend towards expensive technologies, in particular fire-fighting aircraft.

tested in real burning and fire suppression situations to allow a person to learn and gain con-fidence – without becoming overconfident in case they are suddenly faced with some form of relatively new wildfire situation.

Carrying out burning under mild conditions with experienced members of a burning gang is an essential prerequisite to understand fire behaviour at the low end of the scale, but also to be aware of the hazards involved and the safety measures that need to be observed. Similarly, when fighting wildfires, people need to be allocated a minor role at first and under the eye of someone with proven ability. Many practitioners suggest that fifteen or more years of practical experience is needed in fire fighting to become a 'fire boss', i.e. the person in charge of the resources and fire fighters in the field.

Over the years, tragedies have occurred where people with little practical experience in fire management have attempted to carry out a prescribed burn or to suppress a fire in forest con-ditions. Examples are the May 2000 Cerro Grande fire in New Mexico where a prescribed burn went out of control, and the September 2001 fire in Kruger National Park, South Africa that killed twenty-three people and the response to which was confused by ongoing back-burning. These unnecessary tragedies have sometimes led an agency involved to abandon pre-scribed burning as a means of reducing fuels and reducing the intensity of vegetation fires, which is unfortunate. Instead, the right approach for an agency to take is to do the research, train staff to translate the research results into field practice and then monitor the outcomes, adapting procedures as necessary while continuing with training.

Firewise principles being part of the solution

In places such as the American and Canadian Rocky Mountains and outside Australia's main cities, increasing numbers are inhabiting forests or forest edges. The quality-of-life advantages are tempered by the ecological process of wildfire becoming a major peril. Meanwhile, across Europe and southern Africa, rural exodus and changes in land use are producing high fuel loads, which make the rural communities fire prone where they never used to have a fire problem. Frequently, wildfires are not registered until they have become uncontrollable, because they started in remote locations at a small scale, such as from a single lightning strike, a smouldering cigarette butt, sparks from a motorcycle or an unextinguished trash burn or campfire.

Consequently, advocacy and fire awareness education are key components of integrated fire management. Ninety per cent of unwanted fires are caused by negligent human action, yet not all fires are bad. As noted, fire is required to maintain ecosystem function and, under controlled conditions, can be used as an efficient land management tool, as for instance in South Africa. This information, and the balance between needing wildfires while avoiding many human causes of wildfires, need to be communicated and understood in order to improve wildfire preparedness, damage mitigation and management.

It is particularly helpful for individual property owners to create a reduced-fuel zone around their properties. That is part of being 'firewise'. Principles are:

- Driveway accessible with address visible.
- Grass green and mowed within 9 m of the home.
- Vegetation mowed within 30 m of the home.
- Storage sheds located away from the home.
- 30 m of garden hose attached to the home.
- Chimney cleaned and screened.
- Woodpile, fuel tanks and other burnable materials are more than 9 m from structures.

- Coniferous trees are thinned and pruned.
- Outdoor burning is avoided, with recycling, mulching and composting preferred.

Translating this knowledge into action is frequently a challenge, so communicating the message must be considered. One study in Colorado found pre-conceived beliefs in agency credibility influenced whether or not residents adopted firewise behaviour, suggesting that cross-agency provision of the same, clear messages would assist in influencing behaviour (Bright *et al.* 2007).

Forest fire management system

In many countries the responsibilities and mandates for fuel management and fire management are unclear and are divided between numerous parties on national, provincial/state and local levels. Fire suppression is often dealt with by fire departments, whereas fire prevention tends to rest with the land owner and/or user. The situation needs to be rectified through the following policy and practice actions:

- Overarching legislation, dealing with all aspects of wildfires at the levels of all jurisdictions dealing with them.
- A national wildfire policy.
- An interagency agreement between the local, provincial/state and national authorities.
- A national-level agreement between the forest management agency and other key agencies.
- A single fire/land management organisation responsible for forest planning, forest management and wildfire management (i.e. a national Fire Protection Association – FPA).
- Preparation of a Fire Management Plan template by the responsible FPA.
- Adequate funding for the responsible FPA for all wildfire management operations, especially preparation and mitigation.
- Annual, independent monitoring and public reporting on outcomes.

Forming or designating (interagency) FPAs on local and provincial/state levels provides a focus for everyone involved. FPAs develop locally adapted fire management plans and regulations and then should report to a single national co-ordinating agency. In ideal cases, national co-ordinating agencies report to a single regional co-ordination body. The national umbrella FPA co-ordinates funds, assistance and resource-sharing, while providing guidance for a national fire strategy. New Zealand's Rural Fire Authority and the South African Working on Fire Programme are role models for this structure.

The FPAs would also co-ordinate the provision of fuel management services to landowners, governments and other land users. In this way, a single agency becomes the focus for fuel management and fire management, avoiding divided responsibilities and battles over control.

Conclusions

A good package of policies and activities-based practices on credible science, delivers community protection from destructive wildfires, minimises undesirable impacts on the environment, prevents needless costs to governments and communities, maximises fire-fighter safety and generates widespread political, community and media support. No approach can be a panacea, but it is necessary to continually learn and evaluate in order to improve, and in order to recognise differences amongst different contexts.

The most important lesson is to accept that fire is a normal and natural part of the wilderness and that human decisions can and should ensure that normal fires are given the space they need, without creating an artificial fire regime that hurts settlements and the environment. Nash (1985) attributes the desire to control all fires to the film 'Bambi', in which Bambi's forest is damaged by a human-caused wildfire (see Chapter 11), suggesting that 'Bambi' 'did more to shape American attitudes toward fire in wilderness ecosystems than all the scientific papers ever published on the subject', through its misguided messages that 'wilderness is good, fire is bad, and man [sic] causes fire … fire must be kept out of the wilderness' (p. 267). Instead, wildfire is part of the wilderness and must be accepted as such, but that does not mean accepting death and destruction from wildfires.

Geophysical hazards

25

Landslide and other mass movements

Danang Sri Hadmoko

FACULTY OF GEOGRAPHY AND RESEARCH CENTER FOR DISASTERS, UNIVERSITAS GADJAH MADA, INDONESIA, AND LABORATOIRE DE GÉOGRAPHIE PHYSIQUE – UMR 8591, CNRS-UNIVERSITÉ PARIS 1, FRANCE

Salvatore Engel-Di Mauro

DEPARTMENT OF GEOGRAPHY, STATE UNIVERSITY OF NEW YORK AT NEW PALTZ, USA

Introduction

Landslides are major natural and/or anthropogenic hazards that can result in major human suffering, livelihood and material losses, and environmental degradation. Worldwide, the total land area subject to landslides is about 3.7 million km^2, where nearly 300 million people live, or five per cent of the world's population (Dilley *et al.* 2005). Landslides can kill thousands of people. In the city of Huaraz, Peru, in 1941, a landslide transporting 10 million m^3 of materials travelled more than 23 km down the Cohup Creek valley, killing between 4,000 and 6,000 people and damaging twenty-five per cent of the city (Schuster and Highland 2007). More recently, on 1 March 2010, after days of intensive rain, a landslide buried 400 people and forced the relocation of more than 5,000 families in Bududa, Uganda (Uganda Red Cross 2010). Such destructive events make landslides a major concern for potentially affected communities, as well as scientists and government officials. The combined understandings of landslides from all these actors are essential to landslide risk reduction by improving the efficacy of such activities as hazard and risk assessment, vulnerability reduction and capacity enhancement.

Slope failure

Landslides (mass movement) results from slope failure. They are forms of downslope rock, sediment and/or soil movement, occurring from the surface of rupture – either a curved or planar slip surface – under the direct influence of gravity (Glade and Crozier 2005). Landslides can occureither along natural or artificial slopes and are frequently associated with the presence of water or ice, although these substances do not act as primary transportational agents.

Slope slippage may be due to earthquakes, some of which can even be provoked by human action (e.g. the 1963 Vajont disaster, in northeast Italy), or to human-made large-scale explosions, such as nuclear detonations. These intense vibrations trigger rapid and high-magnitude pressure. Stress is transferred to soil water instead of particles, which will start floating. The ensuing liquefaction can even unhinge the base of entire multi-story buildings, for example, and make them bob, slide and fall, like pieces of wood floating on water (see Chapter 26).

The initiation of movement can also be due to material added at the slope summit or removed at the slope bottom, giving rise to uneven weight distribution. Whether mass movement will result or not depends crucially on soil type as well. Added material can include rain water infiltrating and accumulating in a soil, the weight of which will then be increased. Another source of slope failure can be water saturation of a higher-positioned soil (or rock layer) having the same angle as the ground surface. This can also occur when less permeable soils are filled with water only in their upper horizon(s). Water weakens soil particle cohesion and the lubricated material will slide down the slope. More gradual forms of soil movement can be worsened or mitigated by human activity. For instance, soil creep, which is associated with wet–dry or freeze–thaw cycles, can be affected by human-induced regional climatic changes (e.g. urban heat island effects). It can lead to slope failure when the weight of accumulating material overtakes the capacity of a slope to hold it in place or slow down its movement. The complete removal of soils and its nearby dumping and accumulation create slope instability or accentuates pre-existing landscape susceptibility to slippage.

Landslide typology

No simple and ideal classification system exists for landslides because of their typically high level of complexity. However, slope failure can be classified according to mechanism, material type, slip surface morphology and movement velocity:

- Rockfall refers to materials falling, bouncing or rolling from a steep or even vertical slope along a surface on which little or no shear displacement has occurred. On 18 April 1991 one of the largest rockfalls occurred in Randa, Switzerland, where 30 million m³ of rocks fell and dammed the Mattervispa River. Between 10 cm and 40 cm of dust were deposited within a radius of approximately 1 km from the rockslide area (Glade and Dikau 2001).
- Topples include the forward rotation out of a slope of rocks around an axis below the centre of gravity of the displaced mass. There is no slip surface. On 27 September 1996, in Cowaramup Bay, Australia, a 14-m-high limestone sea-cliff toppled and killed nine people who were attending a primary school surf carnival. The disaster occurred while spectators were sheltering from rainfall under the cliff (Michael-Leiba et al. 1997).
- Sliding refers to all types of downslope movements of varied materials (rock, soil, debris or a mix of these) occurring on a slip surface, including translational slides which are characterised by a linear surface of rupture and for which the movement at the surface and near the failure surface shows relatively the same velocity; rotational slides, which are characterised by the spoon form of the failure surface. On 7 April 2010 heavy rainfall triggered dozens of rotational slides near Rio de Janeiro, Brazil, and killed at least 138 people, while 53 others were missing and 3,200 were homeless (Palermo and Engle 2010).
- Lateral spreads are horizontal extension of materials occurring on gentle, undulating or flat terrain usually accompanied by a vertical movement due to the subsidence of the fractured mass of cohesive materials. This process occurs usually with heavy and hard materials placed on weak and softer materials. Significant lateral spreads are occurring in Wadas Lintang,

---

Box 25.1 An uncommon type of slide: wasteslides in urban dumpsites

Franck Lavigne
UMR 8591 Laboratoire de Géographie Physique–CNRS, Université Paris I, France

Only a handful of contemporary disastrous solid wasteslides have been documented in the world, although such events often occur in squatter areas without any record. Wasteslides are often triggered by intense rainfall as for the 10 July 2000 disaster in Payatas, in Manila (Philippines), which occurred in the aftermath of two successive typhoons. Some events may also happen without any rain. For example, the increasing water pressure due to leachate circulation favoured the 8 September 2006 wasteslide at the Bantar Gebang dumpsite in Bekasi (Jakarta, Indonesia). Explosions due to biogas release may have triggered other wasteslides such as in Istanbul (Turkey) in 1993 and at Leuwi Gadjah (Bandung, Indonesia) in 2005.

Whatever the triggering factor, wasteslides always occur in unstable dumpsites or those with steep slopes. Understanding the quantitative parameters in waste dumps is challenging, suggesting difficulties in predicting them before an event. First, it is difficult to sample undisturbed waste deposits, which are typically heterogeneous and very fibrous, thus poorly cohesive. Second, the physical and chemical properties of the deposits rapidly evolve over time, due to the progressive waste compaction and decomposition processes. Third, the water pressure also varies over time within the waste deposits, as observed at the Doña Juana dumpsite in Bogotá (Colombia).

The volume of the largest wasteslides ranged from a few hundred thousand cubic metres (e.g. Bogotá in 1997, or Manila in 2000), to 1.2 million to 1.3 million m^3 (Istanbul), i.e. the equivalent of thousands of tons of waste. Half of the documented events were defined as slides (e.g. Istanbul, Jakarta, or Coruña, Spain in 1996), whereas the others were described as debris flows (e.g. Bogotá, Manila) or debris avalanches (Bandung). However, the process of waste motion remains poorly understood. Most of the existing data result from visual observations by eyewitnesses, whereas the internal structure of wasteslide deposits has rarely been analysed so far.

The main disasters occurred in the vicinity of large and quickly growing cities. In Manila, Jakarta and Bogotá, vulnerable, poor communities often struggle to make a living by scavenging an increasing amount of solid waste, although this constitutes a major hazard to their life. In these cities, waste disposal procedures are often uncontrolled and the slopes of the dump front often reached critical values (30°–45°) before failure. Several post-disaster surveys in Manila, Jakarta and Bogotá revealed that small wasteslides had occurred at the same dumpsites in the past, yet no preventive measures had been taken by local authorities. Additionally, people returned to the vicinity of the dumpsite because it provided for their daily livelihood and they had few other choices.

In that context, reducing the risk of wasteslides should consist of not only preventing the hazard from wasteslides, but also alleviating the poverty that compels poor people to settle and work on dangerous dumpsites. Long-term waste reduction measures are needed as well.

Indonesia. These are mainly caused by the presence of an expansive clay layer resulting from weathered volcanic materials. Lateral spreads rarely kill, but cause extensive damage to houses, buildings, roads and lifelines.

- Debris flows are continuous movement of saturated earth materials on a rupture plane. They usually occur with high to very high velocity within a short period of time and can travel hundreds of metres to several kilometres. A large debris flow occurred on 19 February 2007 in Yaka, Turkey, following a period of rapid snowmelt and heavy rainfall. It transported about 85,800 m³ of soil-like marl over 750 km. Debris materials formed a debris-dam lake which threatened 3,000 people (Özdemir and Delikanli 2009).

Physical characteristics of landslides

Landslide hazard refers to the threat posed by a potential landslide for local communities. Crucial questions to address include 'what', 'where', 'when', 'how strong' and 'how often' landslides occur at a given location. Large-magnitude events are rare, while small events occur more frequently. The relationship between magnitude and frequency can be visualised by a declining exponential curve or power-law function. High-magnitude landslides include the debris flows that struck Venezuela in September 1987. Unusual rainfall reaching 174 mm and pouring for almost five hours led to 2 million m³ of earth materials being transported (Schuster *et al.* 2002). In the same area a similar event mobilised 10 million m³ in the pre-Columbian era. Such large events are extremely rare and should not overshadow the smaller and more frequent landslides, the cumulative impact of which is higher.

Landslide hazard assessment includes the evaluation of the spatial and temporal probability of landslide occurrence. Assessing spatial probability consists in delineating homogeneous areas with the same landslide probability through either qualitative or quantitative methods. Qualitative analyses are based on experts' evaluation of different maps (e.g. geomorphological, geological and slope maps), and rely on the assumption that the relationships between the landslide hazard and the variables used are known, can be delineated on the map and indicated in models. Yet the use of models brings up questions about the reproducibility of results, the subjectivity of selection and the weighting of variables.

Quantitative analysis is done by relating a dependent variable (landslide occurrence) to several independent variables (e.g. slope, geology, land use, etc.). Values of landslide susceptibility refer to spatial densities of landslides calculated for each class of parameters. Hazard evaluation is based on the sole quantitative analysis of landslide data and maps, and does not involve experts' opinions. The temporal probability of landslides can be approached through an existing landslide database and long-term records of landslide-triggering factors (mainly rainfall and earthquakes).

Three main factors combine to cause landslides: predisposing, preparatory and triggering factors. Predisposition refers to static and inherent terrain variables not only affecting slope stability but also acting as catalysts that render the dynamic destabilising factors more effective. For example, a dip that is parallel to a slope will be more sensitive to triggering factors.

Preparatory factors consist of dynamic processes that reduce slope margin stability during a given period of time without starting any movement. In connection with predisposing factors, they can make a slope shift from more to marginally stable conditions. They can function at geological timescales, such as in weathering processes, climate change, tectonic activity, or on a shorter timescale, as in cases of deforestation or slope cutting for the construction of a road. Wildfires are also expected to affect shallow landsliding and debris flow significantly. Over burned areas evapotranspiration rates are reduced and lead to increasing soil moisture. Forest clearance through wildfire also removes forest canopies, which have an essential role in reducing raindrop energy.

Triggering factors consist of variables that directly cause the slope failure, as in a prolonged rainfall, an earthquake, slope cutting or over-burden on a slope. Major debris flows occurred in

Box 25.2 Multi-scale landslide risk assessment in Cuba

Enrique A. Castellanos Abella
Institute of Geology and Palaeontology, Cuba

Landslides cause a considerable amount of damage in the mountainous regions of Cuba, which cover about twenty-five per cent of the territory. Until now, only a limited amount of research has been carried out in the field of landslide risk assessment in the country. This research presents a methodology and its implementation for spatial landslide risk assessment in Cuba, using a multi-scale approach at national, provincial, municipal and local level.

At the national level, a landslide risk index was generated, using a semi-quantitative model with ten indicator maps, by means of spatial multi-criteria evaluation techniques in a GIS system. The indicators standardised were weighted and combined to obtain the final landslide risk index map at 1:1,000,000 scale. The results were analysed per physiographic region and administrative units at provincial and municipal levels.

The hazard assessment at the provincial scale was carried out by combining heuristic and statistical landslide susceptibility assessment, its conversion into hazard maps and the combination with elements at risk data for vulnerability and risk assessment. The method was tested in Guantánamo province at 1:100,000 scale. For the susceptibility analysis, twelve factor maps were considered. Five different landslide types were analysed separately (small slides, debris flows, rockfalls, large rockslides and topples). The susceptibility maps were converted into hazard maps, using the event probability, spatial probability and temporal probability. Semi-quantitative risk assessment was made by applying the risk equation in which the hazard probability is multiplied with the number of exposed elements at risk and their vulnerabilities.

At the municipal scale, detailed geomorphological mapping formed the basis of the landslide susceptibility assessment. A heuristic model was applied to a municipality of San Antonio del Sur in eastern Cuba. The study is based on a terrain mapping units (TMU) map, generated at 1:50,000 scale by interpretation of aerial photos, satellite images and field data. Information describing 603 terrain units was collected in a database. Landslide areas were mapped in greater detail to classify the different failure types and parts. The different landforms and the causative factors for landslides were analysed and used to develop the heuristic model. The model is based on weights assigned by expert judgement and organised in a number of components.

At the local level, digital photogrammetry and geophysical surveys were used to characterise the volume and failure mechanism of the Jagüeyes landslide at 1:10,000 scale. A runout model was calibrated based on the runout depth in order to obtain the original parameters of this landslide. With these results, three scenarios with different initial volume were simulated in Caují scarp at the scale of 1:25,000 and the landslide risk for ninety houses was estimated considering their typology and condition.

The methodology developed in this study can be applied in Cuba and integrated into the national multi-hazard risk assessment strategy. It can be also applied, with certain modifications, in other countries.

For further information: http://www.itc.nl/library/papers_2008/phd/castellanos.pdf and http://www.pacificdisaster.net/pdnadmin/data/original/2ndPRDRMmeetingNadi_Carlos_INSMET_09.pdf. Author contact: enrique@igp.gms.minbas.cu.

October 1954 and May 1998 in southern Italy after intense rainfall, which caused thin pyro-clastic layers to slide on steep limestone slopes, killing 300 and 160 people, respectively (Fiorillo and Wilson 2004). The 21 September 1999 Chi-Chi earthquake in Taiwan triggered more than 10,000 landslides which affected an area of about 11,000 km^2 (Khazai and Sitar 2003).

Anthropogenic aspects of landslides

The probability of occurrence of landslides is often increased by human activities both in rural and urban settings. In rural areas, deforestation due to commercial logging and erosion from intensive agriculture are major causes of landslides. In eastern Luzon, Philippines, decades of illegal commercial logging had weakened the slopes above the town of Real, which was buried by several huge landslides in late 2004, killing more than 400 people (Gaillard *et al.* 2007). Still in the Philippines, the devastating landslides that struck the town of Ormoc in November 1991 and killed 6,000 people originated in sugar cane plantations where the land was devegetated by decades of intensive exploitation.

Similarly, large mining activities contribute to significant erosion and subsequent landslides. In the case of south Wales, for instance, coal mining has resulted in greater frequency and magnitude of slope failure in an already landslide-prone region. As in other cases discussed in this chapter, such human-induced change cannot be explained by focusing only on local social dynamics. Mining is directly linked to national and world economies. In this case, a shift towards steel and other coal-demanding production promoted mine expansion in south Wales. The 1966 Aberfan landslide disaster is a well-known outcome of such confluence of profit-seeking concerns and pre-existing environmental processes (Bentley and Siddle 1996).

Subsistence agriculture may also lead to erosion and landslides through micro-topographic changes, such as in the case of terraces along steep slopes observed in many Asian countries. While terracing facilitates ploughing and irrigation and limits fertiliser losses, it also increases water infiltration and pressure by saturating surface and sub-surface materials, thus potentially increasing the probability of landslides. Landslides also frequently occur in areas planted with subsistence root crops, which are known for aggravating erosion in erodible soils. Farming and hunting activities may also indirectly lead to landslides. In Malawi, landslides result from hunt-ing through the use of fire which is effective in driving wild animals out but simultaneously leads to the destruction of slope-stabilising shrubs (Msilimba 2010). One must, however, be wary of claims imputing landslides solely, if at all, to such subsistence activities, as the case of Nepal, described below, demonstrates.

In urban areas, slope cutting and other engineering interventions weaken steep slopes and often lead to landslides. Alexander (1987) attributes the December 1982 landslides that affected a large area of the city of Ancona, Italy, to uncontrolled urbanisation and road building, which raised infiltration and disturbed surface sediments without the establishment of a proper drainage system. In September 2008, in Egypt, a massive rock fall entombed a section of Cairo's largest slum and killed eighty-two people. Shanties at the time were located both below and on top of a limestone cliff, with an inadequate sewage system, leading to the progressive dissolution of soft rock materials and their transformation into a flour-like paste (Makary 2008).

Landslides can cause enormous damage more indirectly through human-induced environ-mental changes, not only at the source of landslide materials in mountainous areas but also in places where transported materials accumulate. Landslides can reduce water quality and potability, lessen soil fertility and reduce stream and lake capacity, which can contribute to water shortages.

Landslides can damage forests for decades, especially in the tropics, where rainfall is intense. The productivity of a landslide-damaged forest was reduced to seventy per cent over the first sixty years in the Queen Charlotte Islands off the Canadian coast. Regeneration tends to be slower after a landslide. In Taiwan it took six years for a forest to recover up to ninety per cent of its former extent, after landslides in 1999 (Lin *et al.* 2006).

Landslides also lead to severe erosion and sediment supply. In the aftermath of the 1999 earthquake in Taiwan, landslides abruptly increased the Da-Chia River sediment content from about 98,363 tonnes/year with 6.14 mm of average annual erosion depth, to 353,088 tons/year with 22.07 mm of average annual depth (Lin *et al.* 2008). Large amounts of earth and organic materials were then transported through surface runoff and fed rivers, lakes and dams. It led to a decrease in water quality for local communities.

In California, USA, earthquake-triggered landslides in January 1994 led to an outbreak of the fungus *Coccidioides immitis*, which spread through airborne dust and contaminated soils in some semi-arid areas of Central and South America. It also caused 'valley fever' in affected communities (Geertsema *et al.* 2009).

In Padang, Indonesia, in September 2009, deep-seated landslides were triggered by an earthquake and dammed a river. Eventually intense rainfall led to the collapse of unstable materials and resulted in debris flow with very high sediment concentration, which buried large tracts of agricultural land, while a village was entirely buried by landslide materials.

The impact of landslides on people

Table 25.1 records recent catastrophic landslides that occurred worldwide over the last five decades. Most of these events were triggered by intense rainfall and earthquakes in both built and natural steep terrains.

Most of these events have occurred in developing countries where landslide-prone areas are often, but not always, settled by marginalised communities that cannot afford nor have the ability to claim safer settlements in gentler or flatter terrain. Exposure and vulnerability to landslides therefore intimately intertwine with people's political–economic status and livelihoods.

In Java, Indonesia, the 1997 economic crisis forced thousands of poor farmers to migrate towards upper volcanic slopes, considered to be the last agricultural frontier. In the years following, Lavigne and Gunnel (2006) documented hundreds of landslide events that killed tens of people. To sustain their daily needs, many people are compelled to settle and exploit steep and fragile slopes and face landslide hazard.

In La Paz, Bolivia, it is estimated that 500,000 out of a total population of 800,000 people are poor settlers who live in precarious houses along the dangerous slopes that tower above the city (Nathan 2008). As a consequence, landslides have washed away hundreds of houses over the last decades and killed tens of people.

In Colombia, thousands of poor families have fled their rural, conflict-torn villages over the last decades and flocked to cities such as Manizales where they hoped to find better livelihood opportunities in connection with the coffee and banana economy. Most settled on steep slopes, regularly affected by landslides, which stretch down from the upper flat ridge occupied by the richest communities (Chardon 1999).

In most circumstances, people are aware of the risk but lack the political power to put holistic risk reduction actions into effect. For example, before the September 2008 rock fall in Cairo, some of the residents reported the development of cracks in buildings to the local government, which ignored the situation (Makary 2008). In Tanzania, a major landslide occurred on 13 November 2009 along the slopes of Mount Kilimanjaro. At least twenty-five people were killed and hundreds of others were forced to evacuate in precarious conditions. The landslide was triggered by unusually heavy rainfall spanning four days. Although people were prepared to

Table 25.1 Selection of catastrophic landslides that occurred over the last five decades worldwide

Date	Location	Impact on people
07/04/2010	Río de Janeiro, Brazil	138 people killed, 53 missing, 3,200 homeless
02/03/2010	Bududa, Uganda	400 killed, 5,000 people displaced
09/10/2009	Benguet, Philippines	120 killed, 35,000 evacuated
06/06/2009	Chongqing, China	78 people missing
17/02/2006	Leyte, Philippines	1,221 killed, 19,000 people displaced
2005	Mumbai, India	500 killed
09/04/2000	Tsangpu Canyon, Tibet	130 killed
12/1999	Venezuela	30,000 killed, 400,000 homeless
04/12/1996	Guiyang, Guizhou, China	35 killed
1994	Cauca, Colombia	271 killed, 1,700 missing, 12,000 displaced
02/06/1993	Nepal	3,000 killed
23/01/1989	Tajikistan	10,000 killed
1988	Petropolis and Río de Janeiro, Brazil	4,263 homeless
27/09/1987	Medellin, Colombia	500 killed
03/04/1987	Cochancay, Honduras	2,800 killed
31/05/1970	Nevados Huascaran, Peru	18,000 killed
1967	Sierra des Araras, Brazil	1,700 killed
1966	Río de Janeiro, Brazil	1,000 killed
09/10/1963	Vajont, Italy	2,000 killed
10/01/1962	Nevados Huascaran, Peru	4,000–5,000 killed

Source: After EM-DAT 2010

cope with a prolonged drought thanks to the work of a local non-governmental organisation (NGO), they were unable to cope with heavy rainfall. This inability was worsened by the absence of support from local or national authorities.

It is not only the poor who are affected by landslides. In the European Alps there has been a significant increase over the past four decades in government subsidies to prevent settlement abandonment, in purchases of holiday or retirement homes by private individuals, and in largely private investments in tourist amenities and even housing speculation, such as in northern Italy (e.g. Val d'Aosta, Trentino Alto Adige, Tuscany). In consequence, landslides can threaten wealthier communities, as in the case of the wealthy municipality of Passy in the northern French Alps.

Farming on the steep slopes in Nepal

In tectonically active and often high-precipitation environments like the Himalayas, there is always great potential for slope failure. Certainly, soil erosion is generally enhanced, but the large amounts of yearly sedimentation have also been used to advantage. Material accumulating from up-slope erosion often contributes more nutrients and organic matter to land at lower altitude, among other positives such as increasing soil depth. In Nepal, the building of terraces is one major way in which people have benefited from high soil susceptibility. Owing to the constant, labour-intensive efforts of local inhabitants in creating and maintaining vegetated terraces, and in many cases even forests, erosion rates and slope failures can be less than they would be otherwise (Acharya 2005; Gardner and Gerrard 2003). Farming communities cope with the permanent threat of landslides not only by terracing, but also by ceasing to irrigate or even cultivate endangered terraces, thereby reducing the lubricating effect of water movement within soils.

Box 25.3 Avalanche deaths

Benjamin Zweifel
WSL Institute for Snow and Avalanche Research SLF, Switzerland

One type of slide is an avalanche, referring to snow and ice sliding down a slope. Avalanches tend to cause mostly disruption, through interfering with winter sports and damaging transportation routes, but less affluent countries have often experienced high-mortality avalanches.

The number of annual avalanche-related deaths worldwide is not precisely known. In a survey done by the International Commission for Alpine Rescue (IKAR), which represents twenty-one countries in Europe and North America, 160 fatalities per year from 1976 to 2002 were identified (Meister 2002). Additionally, snow-related tourism destinations in Australia, Eastern Europe, Japan, New Zealand and high-mountain regions – notably expeditions in the Himalayas and Andes – yield an estimated ten to twenty avalanche fatalities per year.

Avalanche deaths also occur in other mountain regions of the world – such as 102 avalanche deaths in four events in Pakistan during 11–18 February 2005 – but are often not reported or recorded. Settlements and traffic routes in remote mountain areas can suffer 100–200 avalanche deaths per year, with few avalanche safety precautions taken.

In more affluent countries, people in settlements and along land transport routes are generally well protected against avalanches because previous events have led to extensive safety precautions, including a well-developed avalanche warning service. In countries such as Iceland and Canada, eighty per cent to ninety per cent of avalanche deaths are related to tourism, including local tourism, even though extensive avalanche warning services are available. Most avalanche casualties are therefore winter sports enthusiasts who trigger the avalanches that lead to their deaths. In the USA more than three-quarters of the 440 avalanche deaths from 1950 to 1994 were related to sports or recreation.

The most frequent physical mechanism of death in avalanches is asphyxia: lack of oxygen from being buried in the snow, leading to acute suffocation. Most avalanche survivors kept their heads above the snow. The next most common mechanism of death in avalanches is physical trauma, potentially leading to shock and hypothermia. Regarding physical trauma, avalanche victims are frequently hurt by the impact of obstacles, for instance trees and rocks, or by snow pressure. Rarely a drowning can be avalanche related, if an avalanche sweeps people into a lake or river.

In such high-energy conditions, sudden terrace failure, mass wasting or large landslides are typical, regardless of human counter-measures, even if sometimes deforestation can magnify rates, depending on the type of ground cover left and intrinsic soil properties. Along lower-elevated contours next to river banks, terracing can be even more precarious and much vigilance and resources are often devoted to these areas. It takes years for terrace repair, its duration depending on economic circumstances. The likelihood of intensified erosion and terrace failure is magnified when basic resources are made scarce or unavailable as a result of international economic pressures, government policies and/or war.

For example, deforestation can be better understood in the context of land inequality, which renders poorer communities more dependent on forest resources. Population growth plays at most a minor role, and in some areas it has even been associated with greater slope stability

(Paudel 2002). The uneven geography of periodic disasters, due to combined and uneven effects of environmental and social forces, also affects some communities more than others. In any event, soil nutrient content may decline on some lands more as a result of cropping practices than erosion problems (Schreier *et al.* 1994). This can create conditions for widening existing economic disparities and greater landslide hazards in detrimentally affected areas (Johnson *et al.* 1982).

Understanding and assessing precursory signs of landslides

In order to reduce the potential impact of landslides, it is essential to recognise the signs that precede slope failure. These may differ from one area to another due to specific geological, geomorphological, climatic and human conditions. Understanding the precursory signs is essential to disaster risk reduction (DRR). These include processes that are easily identifiable in the natural and built environment (see Figure 25.1).

Extreme topographical change from very steep slope to flat terrain is common to landslide-prone areas. A head scarp at the top of a slope may reflect previous landslides. Ground cracking in linear form parallel to contour lines due to the slow slope movement may also serve as an indicator of landslide-prone areas and should be covered by earth materials to prevent rainwater infiltrating subsurface zones. Some slow movement may be accompanied by creaking, snapping or popping noises resulting from the friction of earth materials at the bedding plane. In times of heavy rainfall, special attention should be paid to muddy seepage and springs that may result from subsurface erosion on the potential slip surface of landslides. The uncommon increase or

Figure 25.1 Evidence of landslides in Central Java, Indonesia: (a) tilted coconut tree; (b) ground cracking on asphalted roads; and (c) cracking in a wall
Source: Photographs by Danang Sri Hadmoko

decrease of spring discharge is a common sign of hill–slope aquifer disruption. Tilting trees and withered plants are other signs of subsurface movement. Precursory signs are also evident in built environments such as with cracking in walls, foundations and drainage systems, or with tilted telephone poles. Visible space and sticking between doors and window frames are indications of building deformation due to slope movement.

Community-based susceptibility assessment

Community-based landslide DRR has been implemented in Manizales, Colombia. The *guardi-anas de las laderas* (slope guardians) are groups of young mothers who have been recruited, trained and hired by the local government to prevent and mitigate the effects of landslides. They are responsible for maintaining the drainage channels on steep and engineered slopes, and for training local residents in better garbage management (Hermelin and Bedoya 2008).

Community-friendly methods have also been developed to monitor slope instability. In the Philippines slope angle is measured by using folded paper (see Figure 25.2), while the strength of rock and soil materials is estimated with a carpenter's hammer. Visual observation enables the assessment of the status of vegetation cover, land use, drainage systems and artesian flow (Peckley *et al.* 2010).

a. To form a 45° angle, fold a square-shaped piece of paper into half, diagonally, forming a triangle of equal size.

b. To form a 30° and 60° angle, fold the square paper into three equal parts, diagonally. The corner with the smallest angle is the 30° while the next larger corner is the 60°.

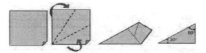

c. To form the 15° and 75° angle, from position b), fold the smaller angle (30°) one more time, into half. The smallest angle produced is the 15° while the next larger angle is the 75°.

d. To estimate the slope angle using this technique, find a spot outside the area being investigated where the slope angle *a* can be visually compared with any of the above paper-fold angles.

Figure 25.2 A non-expert approach to estimating angles of slopes by using folded paper techniques
Source: Copyright Peckley *et al.* 2010

Prevention, mitigation, protection and warning systems

Landslide DRR comprises measures designed to avoid (prevention) or reduce (mitigation, protection and preparedness) the undesirable impacts of landslides. The entire removal of hazards is unrealistic, so the reduction of key aspects of vulnerability should be conducted.

Strategies to be adopted to reduce landslide losses vary, depending on the geographical setting, climate and social conditions. Strategies should combine governmental decisions (top down) with the aspirations of local people (bottom up) in order to avoid inefficiencies in the implementation of risk reduction measures. These include physical protection and bioengineering solutions.

Physical protection in facing landslides

Physical protection comprises all the measures related to the modification of slope geometry, drainage, retaining structures and internal slope reinforcement, and large constructions (walls, dams, drainage systems) that can reduce or protect against landslide hazards. Large check dams are usually used in Japan to mitigate sediment-related hazards such as debris flows. A wide range of check dams are developed according to slope morphology, the sources of materials and the geological settings. Concrete check dams are the most common types, built both to capture the runoff sediment directly and to reduce the discharge of runoff sediments. The latter represents the so-called sediment control function (Popescu and Sasahara 2009).

Sophisticated and costly constructions are sometimes unrealistic for poor countries. Therefore, cheap, simple and low-technology means for homeowners are more effective for landslide protection. A traditional technique consists of covering soil cracks with ground coatings. This can be conducted by local inhabitants to avoid rainwater infiltration. Pipes made of bamboo sticks may also be introduced into the soil to drain slopes prone to landslides.

Similar community-based activities are also conducted in urban areas. The MoSSaiC (Management of Slope Stability in Communities) project has been implemented by governments and local officials in collaboration with the communities of four Caribbean countries to identify the potential causes of landslide on a house-to-house basis (www.mossaic.org). Community meetings and focus group discussions provided data and detailed maps on past landslides, drainage systems and instability factors. These data are used to identify landslide-triggering mechanisms and to develop simple, low-cost landslide risk reduction measures, such as the construction of drainage networks for optimum capture of surface water (roof water, grey water and overland flow of rainwater). These infrastructures were established, monitored and revised in collaboration with community residents, contractors and labourers from within the community, and local engineers.

Bioengineering solutions

Bioengineering techniques, such as tree planting, provide alternative and supplementary protection measures. Tree roots can penetrate deeply into the subsoil and parent rocks and thus stabilise steep slopes. Tree planting is relatively cheaper and may be conducted by local communities. It may be integrated into agro-forestry projects which provide protection from landslides as well as livelihoods to local people. For example, pines (*Pinus mercusii*) and tecks (*Tectona grandis*) are widely used species among mountain communities in South-East Asia to control erosion and landslides, and to support livelihoods in the long term (Razal *et al.* 2005). Between trees, local cultivators can also plant crops that can be harvested in three to six months (e.g. maize, bananas) for economic gains in the short term.

Warning systems

Warning systems for landslides range from very simple and cheap methods available to rural communities, to very sophisticated and expensive computer-aided systems. It is usually recommended that the most simple, applicable, easy-to-use and community-friendly system available be developed. In Java, Indonesia, several community-friendly warning systems have been developed for rural areas. There, people install their own slope-monitoring system by using traditional extensometers constructed with cheap materials including pins made of bamboo introduced into the slope and connected to a bell, which rings when the slope moves over a certain distance.

Simple but useful gauges made of cheap and easily obtainable materials, such as bottles, glasses and PVC pipes, help in monitoring rainfall in areas where the rainfall-triggering threshold is known. When rain occurs, inhabitants can regularly monitor the water level in the gauges, and if it reaches a certain volume, people can take the necessary steps in anticipation of potential landslides.

Conclusions

Many efforts to reduce landslide disaster risk have been conducted in many parts of the world. However, the occurrence of landslides is likely to increase with increasing urbanisation and development in landslide-prone areas. Acute competition for land, deforestation, changing climate patterns and the poor co-ordination between stakeholders responsible for DRR are obvious aggravating factors. Furthermore, most projects intended to reduce landslide disaster risk tend to be top down rather than bottom up in nature, leading to their repudiation by local communities.

In such contexts, governments, practitioners, scientists and local people should collaborate to integrate top-down and bottom-up risk reduction measures. Governments serve as the main actors in policy-making and in constructing infrastructures. Practitioners and scientists should develop and enhance cheap and community-friendly strategies to monitor landslide hazards. Finally, local communities should be regarded as actors rather than receptors, as they can participate actively in all activities related to landslide DRR. For this sort of involvement to be meaningful and effective, though, there must also be measures introduced to empower local inhabitants and to reduce, if not erase, economic gaps among them, as well as efforts at least to mitigate negative pressures from national and international forces and institutions that induce conditions of, or otherwise heighten vulnerability to, landslides.

Acknowledgements

The authors gratefully acknowledge the *Handbook* editors for their very useful comments and the Department of Environmental Geography, Faculty of Geography, Universitas Gadjah Mada, for the facilities provided to complete this chapter.

26

Earthquake

Cinna Lomnitz

UNIVERSIDAD NACIONAL AUTÓNOMA DE MÉXICO, MEXICO CITY, MEXICO

Ben Wisner

AON-BENFIELD UCL HAZARD RESEARCH CENTRE, UNIVERSITY COLLEGE LONDON, UK

A natural history of earthquakes

Earthquakes can produce many human casualties among vulnerable groups through collapse of structures and secondary hazards such as fire, landslide and tsunami. Earthquakes also can bring heavy economic loss. They can be destructive over large areas. People living in zones affected by frequent earthquakes have learned to live with them in a variety of ways. For example, in Japan there are neighbourhood-based volunteer fire-fighting groups dating from the 1700s, and traditional architecture in Japan tends to use light, flexible materials. Following the catastrophic earthquake that destroyed Lisbon, Portugal in 1755, the prime minister commissioned Europe's first city master plan. Lisbon was rebuilt with mandatory clearance between buildings to prevent the spread of fire, wide avenues and maximum building heights (Mullin 1992). Thus people have been learning experientially and adapting their way of life to sudden release of seismic energy in the Earth's crust. However, it was only recently that contemporary approaches to earthquake have taken shape.

The twin disciplines of seismology and earthquake engineering began to take their modern form after the 1906 San Francisco earthquake. Engineering had grown in sophistication during the period of rapid industrialisation in the latter half of the nineteenth century. Seismology did not yet know the precise cause of earthquakes, but it did have detailed knowledge of the kinds of ground motions (accelerations) that manifested on the surface (Howell 1990). It was assumed that homes and other buildings were destroyed by earthquakes because they were built to resist only the downward force of gravity, not lateral shaking. In the late 1920s and early 1930s Japan and California codified in building codes resistance to a *horizontal* (side-to-side) design force of ten per cent of the force of gravity. Engineered structures would have to be built to withstand a push from any direction amounting to ten per cent of the force of gravity. In 1967 a revolution occurred in understanding of the geological causes of earthquakes. This was the discovery of plate tectonics, following the general idea of continental drift put forward by Alfred Wegener. Previously it was believed that continents did not move. However, the position of continents was, in fact, the result of slow drifting in different directions over millions of years. The Earth's outer layer, known as the lithosphere, is like a puzzle of plates fitting together and moving away from or under each other at a rate of centimetres per year. About seventy-six per cent of all earthquakes are caused by relative motion (friction) between adjacent plates. Nevertheless, there are also intra-plate earthquakes that take place far from the edges (Stein and Mazzotti 2007). Examples of intra-place earthquakes include ones in Kenya and as far south as Malawi and Mozambique in the region of the Great Rift Valley, and the very large earthquakes associated with the New Madrid fault that affected a large area surrounding the Mississippi and Missouri Rivers in the USA in 1811–12 (Gunn 2008: 90–94).

Plate tectonics also explains how convection or temperature-driven circulation in the Earth's mantle causes upwelling of hot material at mid-ocean ridges and drags the plates along the ocean floor until they are forced downward (subducted) under the continents. There are fifteen major plates and many smaller ones (microplates), but by far the largest is the Pacific Plate, which underlies most of the Pacific Ocean. The active plate boundary around this plate and some of the adjacent plates is known as the 'Ring of Fire', which generates about ninety per cent of the world's earthquakes, including eighty per cent of the largest ones (USGS 2010a).

Coincidentally, about the time of the plate-tectonic revolution there was a serious worldwide reversal in the trends of earthquake disasters. The rate and severity of these disasters had been diminishing for half a century. Now they started to climb again. The reinsurance company Swiss Re calculates that more than 1 million people died in 360 major earthquakes in the period 1960–2010, with the decade 2000–09 being the deadliest, accounting for 450,000 fatalities (Swiss Re 2010: 10). Since the 1990s there have been some of the most expensive and deadly events in recent history: the 1994 Northridge, California earthquake (Magnitude (M)6.7, US$44 billion loss, although only seventy-two dead), the 1995 Kobe, Japan earthquake (M6.8, US$100 billion, 6,434 dead), the 2004 Sumatra-Andaman earthquake and tsunami (M9.3, damage estimates in the range of several billion US dollars, 229,866 dead), the 2005 Kashmir earthquake that killed 79,000 in Pakistan and 1,400 in India-administered Kashmir (M7.6), the 2010 Haiti earthquake (M7.0, 200,000 dead) and the 2010 Chile earthquake (M8.8, around 1,000 dead). Reported damage figures for these last two earthquakes run in the billions of dollars. Table 26.1 provides a worldwide overview of earthquake impacts.

Coinciding with this increase in deaths and financial loss was a worldwide rapid increase of the rate of urbanisation due to migratory movements from the countryside to large cities. How is this recent challenge being met (Castaños and Lomnitz 2011)? One approach is betting on significant advances in seismology and earthquake engineering. The other main approach has to do with new understanding of social systems. These two approaches are complementary, and are discussed in the following section.

Complementary natural and social science approaches

Advances in seismology and engineering

There have been important discoveries in earthquake physics. It was realized that plate motion must be driven by huge convection currents in the Earth's mantle. This discovery poses many unsolved questions, such as: Does convection involve the entire mantle, or only the outer part? How does plastic flow in the mantle induce brittle fracture in the crust? Why does brittle fracture take place at depths of up to 700 km below the surface? Why can't we predict earthquakes?

The problem is that we do not have direct access to the Earth's interior. Indirect knowledge of the interior comes from what seismic waves can tell us. The deepest holes drilled into the Earth go to a depth of 13 km – less than 0.3 per cent of the way to the Earth's centre.

Table 26.1 Significant earthquakes, 1910–2010 (by alphabetical order of countries)

Location	Year	Magnitude	People killed
El Asnam, Algeria	1980	7.5	2,590
Spitak, Armenia	1988	7.0	25,000
Chillán, Chile	1939	8.3	28,000
Southern. Chile	1960	9.5	5,700
Haiyuan, China	1920	7.8	200,000
Tangshan, China	1976	7.8	242,000
Wenchuan, China	2008	8.0	69,227
Chayu, China	1950	8.5	4,000
Napo, Ecuador	1987	6.9	1,000
San Salvador, El Salvador	2001	7.6	1,167
Kefalonia, Greece	1953	7.2	476
Motagua, Guatemala	1976	7.5	22,778
Port-au–Prince, Haiti	2010	7.0	220,000
Bihar, India	1934	8.4	10,653
Assam, India	1950	8.7	1,526
Latur, India	1993	6.4	30,000
Bhuj, India	2001	7.7	20,000
Flores, Indonesia	1992	7.5	2,500
Sumatra, Indonesia	2001	9.0	226,898
Sumatra, Indonesia	2005	8.6	1,313
Yogyakarta, Indonesia	2006	6.3	5,749
Sumatra, Indonesia	2009	7.6	1,100
Bou´in-Zahra, Iran	1962	7.1	12,225
Dasht-e-Bayaz, Iran	1968	7.3	7,000
Tabas, Iran	1978	7.8	15,000
Manjil-Rudbar, Iran	1990	7.4	40,000
Bam, Iran	2003	6.6	30,000
Messina, Italy	1908	7.5	83,000
Avezzano, Italy	1915	7.5	29,978
Irpinia, Italy	1930	6.5	1,883
Friuli, Italy	1976	6.9	989
Irpinia, Italy	1980	6.9	3,114
L´Aquila, Italy	2009	6.3	307
Tokyo, Japan	1923	8.0	130,000
Sanriku, Japan	1933	8.9	3,064
Tottori, Japan	1943	7.2	1,083
Nankaido, Japan	1946	8.1	1,462
Fukui, Japan	1948	7.3	3,895
Kobe, Japan	1995	6.8	4,034
Chhim, Lebanon	1956	6.0	136
Al-Marj, Libya	1963	5.6	290
Skopje, Macedonia	1963	6.0	1,070
Orizaba, Mexico	1973	7.1	539
Michoacan, Mexico	1985	8.1	10,000
Agadir, Morocco	1960	6.0	13,100
New Guinea	1976	7.1	6,000
Managua, Nicaragua	1931	5.5	2,450
Managua, Nicaragua	1972	6.0	5,000

Table 26.1 (continued)

Location	Year	Magnitude	People killed
Quetta, Pakistan	1935	7.5	35,000
Makran, Pakistan	1945	8.5	4,000
Muzafarrabad, Pakistan	2005	7.6	80,000
Quiches, Peru	1946	7.3	1,400
Santa Valley, Peru	1970	7.8	66,800
Mindanao, Philippines	1976	7.9	3,564
Dagupan, Philippines	1990	7.0	1,660
Vrancea, Romania	1977	7.0	1,387
Chi-chi, Taiwan	1999	7.6	2,416
Erzincan, Turkey	1939	7.9	32,740
Ladik, Turkey	1943	7.7	4,013
Gerede, Turkey	1944	7.3	3,959
Yenice, Turkey	1953	7.3	1,070
Varto, Turkey	1966	6.9	2,394
Gediz, Turkey	1970	7.2	1,086
Lice, Turkey	1975	6.7	2,370
Muradiye, Turkey	1976	7.3	3,626
E. Anatolia, Turkey	1983	6.9	1,346
Izmit, Turkey	1999	7.6	17,217
Ashkhabad, Turkmenistan	1948	7.3	110,000
Anchorage, USA	1964	9.2	131
Loma Prieta, USA	1989	7.1	68
Northridge, USA	1994	6.7	72
Caracas, Venezuela	1967	6.5	236
North Yemen	1982	6.0	2,800

Earthquakes originate mostly at a greater depth – up to 20 km is a normal depth along the San Andreas Fault in California, where earthquakes close to M8 can occur. Structural inferences from seismic waves must supply the missing direct evidence, mostly through refinements in data processing. A major advance was the use of precise satellite Earth measurement (geodesy) that employed a global positioning system (GPS). This has enabled us to observe plate motion in real time, thus confirming the predictions of plate tectonics (NRC 2003).

Yet such new knowledge has not slowed the increase in severe damage from earthquakes. In the USA and Japan direct economic losses have been rising exponentially since 1970. During the same period there have been major earthquake disasters in developing countries (see Table 26.1). Secondary hazards such as tsunamis, liquefaction and landslides account for a rising share of casualties. Thus the 1970 Peru earthquake (M7.8) killed 66,800 people. Almost one-half of these were due to an avalanche of ice and snow dislodged from the Mt Huascarán glacier, which buried the town of Yungay. There were no survivors in the direct path of the avalanche, and altogether there were only a few hundred survivors in the town (Oliver-Smith 1986; Castaños and Lomnitz, 2011).

Secondary hazards

Tsunami, landslide and liquefaction are common secondary earthquake hazards. Tsunamis are large ocean waves generated by major subduction earthquakes. A tsunami wave can travel thousands of kilometres across an ocean and preserve its shape (see Chapter 27). Tsunamis have a wavelength of more than 100 km and a wave height of about 1 m so they are hard to detect in the open ocean. As they approach a coast they grow and may attain amplitudes of more than 20 m, depending on the shape of the coastline.

Landslides and avalanches are failures on unstable slopes of soil, snow or ice (see Chapter 25). They may be triggered by earthquakes. Rivers blocked by landslides will dam up the flow for weeks and then suddenly release a large water wave that may flood the downstream area. The 2008 earthquake in Sichuan, central China triggered more than 15,000 landslides and debris flows, and some of these accounted for many deaths (Petley 2008).

Liquefaction is also often a result of deforestation and mismanagement of the environment. Soils are intermediate materials between solids and liquids: they may flow like liquids when vibrated, as in an earthquake. In the 2010 Chile earthquake some brand-new apartment blocks sank into the ground and broke apart due to liquefaction. Similar problems were experienced in the 2010 Haiti earthquake. Large-scale damage due to liquefaction was first documented in the 1964 Niigata, Japan earthquake (M7.5), but is now understood to be extremely common (Scawthorn 2008). It is one of the main causes of structural damage and fatalities in earthquakes. Since about 1850 a number of bayside cities, including San Francisco, Tokyo and Kobe, have expanded by way of landfills into their respective bays. The recent fills in the Bay Area and in Kobe (e.g. Port Island) have been better compacted and rolled with heavy machinery using clay or mashed rock so that they held better in the earthquakes. However, the old (1906) Marina fill in San Francisco and some of the older fills in Kobe fared poorly.

Understanding social systems and earthquakes

New thinking on disasters and sustainability is increasingly influential in shaping policy, particularly in seismic hazard analysis and risk reduction. This may be attributed to the insight that social systems are not only vulnerable but contribute to risk in mostly unintended ways. Some of these ways include land use and the management of natural resources. For example, deforestation may increase slope instability and hence hazards secondary to earthquakes. Approaches inspired by development studies focus on access to living space close to livelihoods and access to the natural resources and other requirements for making a living. From such points of view, marginal people with little political or economic power are forced by the functioning of the space economy and social system to live in marginal places – that is, sites that are often prone to the worst shaking by earthquakes or to the hazards associated with this shaking. The poor and marginalised live in homes that are usually not professionally engineered (Wisner et al. 2004).

Social science also provides insight into the perception of risk. A programme intended as protection against earthquake hazard may unwittingly increase risk by encouraging a feeling of false security. An example is provided by the 2010 Chile earthquake. Most of the casualties were caused by the tsunami, even though a system of tsunami alerts was in operation along the coast. People had a false sense of security because of the existence of the warning system. However, the system was either disabled by the massive power failure immediately following the earthquake, or it had been disconnected by an error of judgement.

The dominant free-market philosophy that has existed since the early 1980s tends to cause governments to decline responsibility for disaster risk management in view of rising costs. Risk management has been partly transferred to private enterprise such as insurance companies, and the role of government is limited to regulation and emergency response (see Chapter 5). In many countries a large part of lifeline infrastructure (water, electricity, telecommunications), as well as many social services (education, health care), have been privatised, and citizens are expected to take care of their own needs. The private sector is tacitly assumed to be more efficient than state organisations.

Economics and the role of insurance

Earthquake insurance has been available since before the 1906 San Francisco earthquake. Worldwide earthquakes accounted for about forty-two per cent of economic losses from natural hazard events from 1950 to 2008. This amounts to US$819 billion (Gao 2010). Since about 1960 the average cost of earthquake disasters has been increasing exponentially all over the world. This applies to both insured and uninsured losses (Wisner *et al.* 2004: 62–64). Swiss Re estimated that the world economy lost US$230 billion in 2005 due to disasters, but only US$83 billion were insured. Disasters triggered by natural hazards contributed by far the largest share. Hurricane Katrina accounted for US$135 billion of the losses in that year, yet the disastrous Pakistan/ Kashmir earthquake cost only US$5 billion because of low asset values (Swiss Re 2006). Almost none of this damage was insured. In another example, the 1985 Mexico earthquake, the worst disaster triggered by a natural hazard in Mexican history, killed more than 10,000 people but caused only US$4 billion in damages in Mexico City (see Chapter 54).

Prevention of earthquake risk

An engineer's design budget for the capacity of a given structure to resist earthquake forces is usually a function of resistance to a given percentage of the acceleration of gravity in the horizontal direction. In principle, all structures could be made earthquake-resistant as the peak horizontal ground acceleration rarely attains twice the acceleration of gravity. Thus there are design and engineering contributions to be made to the reduction of earthquake risk. In addition, there are contributions by planning, regulation, warning technologies and community-based action.

Performance-based engineering

After the 1994 Northridge earthquake, which caused US$20 billion of economic loss, there was increased awareness of the need to improve building practices in order to prevent surprises such as the failure of brittle welds in steel structures, especially in the joints between beams and columns. In 1997 the Federal Emergency Management Agency (FEMA) published guidelines known as FEMA 273 in which a new approach known as 'performance-based engineering' was proposed and developed (Building Seismic Safety Council and Applied Technology Council 1997). This means that engineering calculations, which earlier had relied on designing a structure to resist some static lateral force, would henceforth have to be subjected to computer simulations driven by an actual earthquake input derived from a seismogram. Also, input would be selected to drive the building in this computer simulation to the breaking point. If the engineer finds that the building is underperforming s/he must go back to the drawing board.

Hazard assessment

Forecasting earthquake hazard is 'still in a primitive stage' (National Research Council 2003). Earthquake-resistant design means additional investment, depending on whether it is planned at the outset or incorporated at some later stage. Seismic retrofitting is more expensive; therefore, adequate earthquake provisions may involve significant savings when taken at the right moment.

However, predicting the peak ground acceleration input at a site, and the vulnerability of a structure to this input, involves a wide range of assumptions. In the USA, FEMA designed a public-domain software package and database called HAZUS, built on a geographic information system platform (FEMA 1999). This system classifies a specific earthquake hazard computation according to thirty-six building types and twenty-eight occupancy classes, and it provides a database of populations and lifelines based on census figures. Critical facilities such as hospitals are provided separate treatment, and so are secondary hazards such as flooding and fire. Casualties and costs for repair and loss of business are also assessed. This approach is still being developed, expanded and improved (see Chapter 56).

Earthquake zonation and micro-zonation are established approaches for hazard assessment in urban areas where soil conditions may vary greatly from one location to another (Ansal and Slejko 2001). In Mexico City, for example, the downtown area was formerly occupied by a shallow lake, which was gradually drained. The mud layer is now totally paved over and built up (Zone III), while construction on the surrounding hillsides rises from more solid rock (Zone I). Shaking will be greater over the former lakebed, and the location of the former lakeshore is precisely known. In this way, it is possible to determine in advance where a prospective struc-ture will be located in relation to likely seismic forces and appropriate building code provisions can be enforced.

Planners in both the private sector (e.g. insurance companies) and the public sector have tools available to help them. Probabilistic seismic hazard analysis (PSHA) was developed in the 1970s as a methodology for dealing with uncertainty in earthquake risk for extremely rare events (Algermissen and Perkins 1976). Hazard curves showing the relationship between seismic energy and structural damage for various kinds of assets are computed by extrapolating available earthquake catalogues and observed peak ground accelerations. These results are used as a basis for interactive guesses by a panel of experts (Krinitzsky 1993). PSHA has been useful in providing quantitative hazard estimates where none had been available before.

Records of earthquake disasters have been kept for more than 2,000 years. These provide the basis for treating earthquake impacts statistically. Another method of hazard assessment used by the insurance industry is based on extreme-value theory, a branch of statistics dealing with rare events. Extreme values of hazardous events are not necessarily less predictable because their distribution has a regular behaviour in many geophysical systems (Chavez-Demoulin and Roehrl 2004).

Building codes

Building regulations are nearly as old as engineering, but building codes evolved rapidly after the 1933 Long Beach, California earthquake which destroyed 230 school buildings. The Field Act of the California State Legislature prevented future damage in K-12 schools and community col-leges in the state. No buildings of this type were severely damaged after 1940. The Japanese and US codes were first issued in 1924 and 1927, but have been updated regularly, as have other codes that were often inspired by Japanese and US codes. There have been recent attempts to work toward an international building code that would incorporate best practices (see International Code Council, www.iccsafe.org/GR/Pages/adoptions.aspx).

Building codes represent minimum standards for the engineering profession (see Box 26.1), and enforceability varies from country to country. Earthquake provisions in building codes are generally enforced by local authorities or by national governments. However, the quality of engineering service is usually maintained by the engineering profession itself, and building codes are regarded as guidance. Innovative engineering would be expected to exceed these standards.

Box 26.1 Boumerdes (Algeria) earthquake of 21 May 2003

Djillali Benouar
University of Bab Ezzouar, Faculty of Civil Engineering, Algiers, Algeria

This earthquake affected a densely populated, industrialised region of 3,500,000 people. It was one of the strongest recorded seismic events in North Africa. The depth of the focus was about 10 km. The magnitude of the earthquake was calculated at M = 6.8. Some 2,278 lives were lost (plus 1,240 missing, not confirmed), and more than 11,450 were injured. Some 182,000 people were made homeless as at least 19,000 housing units and about 6,000 public buildings were destroyed or seriously damaged. Economic loss was approximately US$5 billion. The severity of these events confirmed that Algerian buildings are highly vulnerable to earthquakes.

The province of Boumerdes, including the coastal cities of Boumerdes and Zemmouri and the eastern part of the capital city of Algiers, were most affected. Most cities and villages along the coast were damaged, from Algiers to Dellys, a zone 150 km long and 40 km wide. The epicentre was located at 36.89N–3.78E, about 10 km offshore from Zemmouri, located 50 km east of Algiers. Widespread liquefaction, rock falls, landslides, ground cracking and lateral spreading were reported in the surroundings of Zemmouri. The earthquake triggered a tsunami which was observed on the southern coast of the Balearic Islands (Spain). Seawater in coastal zones of Algiers and Boumerdes retreated by 200 m.

In Boumerdes civil protection teams began search and rescue operations six hours after the earthquake. However, efforts to search for victims were first attempted by the local population. Several countries sent rescue and first aid teams within twenty-four hours. The Algerian Red Crescent provided food, water, sanitation and health care to the victims within twelve hours.

Civil protection authorities, the Algerian Red Crescent, the armed forces and foreign non-governmental organisations (NGOs) established official 'tent-camps' a week after the earthquake. Some affected people refused at first to move into the government-supplied tents. Families displaced from damaged multi-story buildings preferred to stay in camps, but only those close to their buildings, for fear of looters. They also preferred to remain within their neighbourhoods, where they could support each other. Later, families slowly moved to official campsites where conditions were better than in makeshift camps. Schools and other educational institutions were closed for four months and only reopened after safety inspections by engineers and repairs. The earthquake occurred little over one month before the end of the school year, thus authorities closed the schools and postponed examinations. Final-year high school students who were preparing for the Baccalaureate were redirected to high schools in unaffected zones.

This earthquake raised the awareness of the government and the whole population. In its wake, the government improved the Algerian seismic building code (RPA99/ revised in 2003), made compulsory the implementation of the seismic code for public and private construction, introduced a compulsory natural hazard insurance scheme, made standard teaching about disaster risk reduction at all education levels, adopted a disaster prevention law (Law 04/20) and encouraged informal disaster education by the Algerian Red Crescent and the civil protection agency.

Rigid prescriptive regulations tend to be replaced by performance regulations, leaving it up to the engineer to ensure the desired performance of the structure in an earthquake. In Japan standard earthquake provisions are enforced by law since after the 1923 Tokyo earthquake. Partly as a result of strict earthquake regulations, the professional degree of architect-building engineer (*kenchikushi*) is a unified degree in Japanese universities. Thus a single professional individual is certified to be in charge of application of building code provisions, while civil engineers are trained mainly in the design of infrastructure and lifelines. Nevertheless, non-compliance with codes and outright corruption in construction practices are still present in many countries (Transparency International 2005).

Earthquake early warning systems

Some early warning systems for earthquakes are based on the delay time between the occurrence of the earthquake at the epicentre and the arrival of the seismic wave at an inland location. They usually rely on recording an earthquake in the epicentral area and transmitting the information to a target location by satellite, internet or radio. It takes longer for the energy of the earthquake to be transmitted through rock and soil than for the electronic warning to reach its destination. Utilising that small amount of time, warning signs on major roads can flash, telling drivers to pull over, fire fighters and ambulance teams can drive their vehicles outside depots that might collapse, gas and other inflammable pipelines can be shut off by automatic valves, etc. All this required a reliable automated system (see Chapter 40).

An early warning system based on an array of twelve stations along the Pacific coast of Mexico exploits the lag of around fifty seconds between the time of occurrence and the arrival time of the seismic signal in Mexico City. An experience of almost twenty years suggests, however, that the reliability of this system depends on the automatic identification of the signal as an earthquake. Spurious signals at a single station must be discarded. This means waiting ten to twenty seconds for other stations to record not only an initial pulse but some of the later wave train as well. Because of the inevitable trade-off between the risk of missing the alarm, as happened in the Chile tsunami, or emitting a false alarm as occurred several times in Mexico, it was finally decided to entrust the decision to a human operator. The trade-off represents a serious limitation of the overall utility of the system by adding delay.

Community-based loss prevention

While engineering and governance are important to earthquake loss reduction or even prevention, vital community-based and community-led activities are also required (see Chapter 59). One is citizen engagement in policy formation and implementation. There need to be demands from citizens for enforcement of adequate seismic building codes. This goes hand in hand with broader, cross-cutting issues of democratisation of decision-making, exposure and elimination of corruption, as well as the need for a well-educated citizenry, the existence of a legal framework that discourages malpractice and deviance from professional standards, and a government that respects the rule of law. An example of such activism is a group in Vancouver, Canada, called Families for School Seismic Safety (FSSS, at fsssbc.org). Awareness of earthquake hazard among lay people in Vancouver grew in 2002–03 after the Nisqually quake that rattled Seattle in the US state of Washington to the south, and the deaths of children in a school collapse in Italy in 2002. Families, high school students and teachers became concerned that many multi-storey, nineteenth-century schools in Vancouver were hazardous. They formed FSSS in June 2003, and teamed up with a seismic engineer at the University of British Columbia, further informed themselves, and launched a lobbying campaign at city, province and eventually national level in Canada. In the end, funds were appropriated to assess the safety of 864 schools and to retrofit or raze and rebuild 311 schools found to be unsafe (Monk 2005).

Communities in areas exposed to earthquake hazard also can do a good deal to prepare themselves. Neighbourhood teams can be formed and trained. Water, emergency supplies and tools can be stockpiled, rotated and maintained for use in emergencies. Volunteers in neighbourhoods can be trained in fire fighting, light search and rescue, first aid, transportation of the injured and communications. Many earthquake-prone cities have such trained neighbourhood groups. These include several cities in Japan, several on the west coast of the USA, and in Turkey. The national Red Cross or Red Crescent societies in many earthquake-prone countries train volunteers to respond and also provide education for children in school about earthquakes (IFRC 2010a; Mayer 2008).

Public awareness is also critical. This includes teaching school children and youth about earthquakes and what they and their families can do (see Chapter 62). Schools can become centres for diffusing innovative ideas and practices throughout communities. So, for example, there have been successes in having school children encourage their parents to develop family plans for earthquakes, to secure objects and furniture that could move, fall or shift and injure people, etc. (Wisner 2006a). The media are also an important vehicle for building public awareness (see Chapter 63).

Preparedness is essential in many venues. Not only homes, but public buildings such as schools and hospitals, and also factories, need non-structural measures to reduce earthquake risk in addition to having adequate structures. These include institutional plans and drills for earthquake, attention to objects and furniture that should be anchored, potential sources of fire or chemical spills in the case of violent shaking (FEMA 2004b) (see Chapter 41). Guidelines to non-structural mitigation have been developed, for example, for hospitals (SEARO 2006).

Conclusions

The challenges for earthquake risk reduction implied by the foregoing are well exemplified in the contrast between the 2010 earthquakes in Haiti and Chile.

Haiti

An earthquake of Magnitude 7.0 destroyed the capital, Port-au-Price, and two neighbouring cities, Jacmel and Léogâne, in Haiti on 12 January 2010. The epicentre was only 23 km away from Port-au-Prince. The population in the epicentral area was about 3 million. Casualty figures were highly unreliable, but the loss of life was unusually high, on the order of 220,000–230,000 dead (USGS 2010b). The Enriquillo Fault on which the earthquake occurred defines a microplate in the Caribbean–North America plate boundary. An earlier destructive earthquake on this fault had been in 1770. Unfortunately, social memory of this event was weak. Large earthquakes on this plate boundary are sufficiently rare to make the region especially vulnerable, as several generations may go by without having experienced a damaging earthquake. Lack of awareness of earthquake risk partly explains the absence of local earthquake regulations and resultant destruction of structures of every kind, especially on soft ground. There was no building code and no licensing requirements for engineers or builders. Fierro and Perry (2010) found that thousands of collapses were caused by the absence of earthquake detailing in confined masonry construction, which performed essentially as if reinforcement were absent, even in multi-storey buildings.

At the time of writing, discussions in Haiti about reconstruction and recovery includes proposals to disperse the formerly dense population of Port-au-Prince to a number of cities that would be expanded or built as 'new towns'. Such top-down visions fly in the face of years of experience of earthquake reconstruction and recovery summarised by such institutions as ALNAP (2008) and the World Bank (2010b), which finds that people are very reluctant to resettle far from their original homes and that jobs and livelihood considerations trump safety in the decision to relocate (see Chapter 46). In fact, it is the lack of viable livelihood opportunities that had caused so many people to congregate in Port-au-Prince in the first place. Obstacles to rapid provision of temporary housing in safe locations with the potential for transition to permanent settlements include:

- Lack of clear tenure to land.
- Insufficient consultation with either the potential host communities or the affected people, despite the fact that Haitian civil society is highly organised.
- Shortage of qualified engineers to inspect existing structures for damage.

Chile

The case of the Chile M8.8 earthquake of 27 February 2010 was apparently the opposite of that of Haiti. Chile had seismologists, seismic monitoring stations, building codes, an active and well-managed programme of earthquake research and, more importantly, a solid economy and an excellent educational system. Yet there were still around 1,000 fatalities. The economic losses may have been of the same order as in Haiti.

The coastal segment which ruptured in the Chile earthquake had been identified as a seismic gap, but the rupture was much more extensive than expected. The emergency response after the Chile earthquake was adequate, but the massive communication failure produced significant delays in terms of search and rescue and medical assistance in general. The tsunami warning system was not activated on time, which contributed to the loss of life in coastal localities. The response of the coastal population was complicated by massive power failures that plunged the entire region into darkness. With few exceptions, the building code ensured an acceptable performance of modern housing and other buildings. Two hospitals and one clinic in Santiago were evacuated because of structural damage.

Severe damage and casualties occurred in the coastal zone, which suffered the impact of the tsunami. This extended coastline has a low population density. A tentative explanation for Chile's death toll being two orders of magnitude smaller than Haiti's despite suffering an earthquake with an order of magnitude more energy is as follows. Consider the four basic rules of disaster prevention:

- Have a robust economy.
- Have an appropriate seismic building code and enforce it.
- Have a quality educational system.
- Don't neglect social and environmental factors.

The lesson of the Chile disaster is that a robust economy, acceptable enforcement of an appropriate building code and a quality educational system are not enough to guarantee seismic safety. Equally important is a balanced system of regional development with attention to pockets of poverty and social stress. The 4,000 km of coastline, and particularly its southern sector, is an unevenly developed, partly impoverished region. A major resource in this region was fishing, but the fishing industry, once involving many small fishing boats, is now dominated by a handful of large concerns that contribute little to creating local jobs or preserving biotic resources. Over-fishing and joblessness are the results. To fully grasp the connection between uneven development and vulnerability to the earthquake and tsunami, one needs to delve into the history of this southern coastal region.

For three centuries the indigenous Mapuche people were able to keep first the Incas and later the Spanish out of their ancestral lands. The Maule River in the epicentral region of the 2010 earthquake represented the natural frontier. Fierce Mapuche warriors waged intermittent guerrilla warfare against the white settlers. However, in 1881 the Indian Wars ended with defeat and the shift of the native population to reservations. Agricultural lands were turned over to European immigrants while systematic discrimination reduced the indigenous population. At present, fewer than 230,000 Mapuche are holding fast onto the remaining tribal land.

The Allende administration (1970–73) sought to restore tribal lands to the original owners, but under the Pinochet regime (1973–90) these policies were reversed. Many Mapuche were persecuted as leftists. The legacy of the dictatorship among the Mapuche was an extreme disparity of incomes, alcoholism, joblessness and new health problems such as diabetes, plus continuing racial discrimination. This process of marginalisation was not just a spatial and economic process, but a political one. The Mapuche have practically no voice and influence with central government, and the resources and infrastructure available in depressed coastal areas are much less than elsewhere in Chile. Young Mapuche emigrate from the former tribal territory and attempt to merge with the poor in the coastal villages and in the major cities.

Dozens of small beaches between Pichilemu and Bucalemu to the north and Puerto Saavedra to the south proved to be vulnerable to tsunami because of the flat topography and the shape of the bays. Originally, these were Mapuche fishing villages, which now cater to seasonal tourism. There were some casualties because of the erroneous information that the earthquake had not caused a tsunami. There was severe tsunami damage in housing made of materials that included wooden planks, zinc roofing and scavenged materials: these fragile structures were swept away by the tsunami. In the small coastal community of Dichato near Concepción there were fifty dead. The proportion of fatalities from the tsunami was 3:5 as compared to fatalities from collapsed homes in Concepción and cities inland. While some of these fatalities were tourists, no doubt economic development among the minority Mapuche population would have reduced the death toll.

27

Tsunami

Brian G. McAdoo

DEPARTMENT OF EARTH SCIENCE AND GEOGRAPHY, VASSAR COLLEGE, POUGHKEEPSIE, NEW YORK, USA

Introduction

Earthquakes, landslides and even meteor impacts can generate the giant sea-waves known as tsunamis. When these low probability hazards impact highly populated and vulnerable coastlines, they often evolve into significant coastal disasters that cause hundreds of deaths and considerable economic losses every year (EM-DAT 2010). The destructive power of tsunamis captures our imagination – from the stories of waves generated by the eruption of Thera (Santorini) that may have contributed to the decline of the ancient Minoans on Crete (Antonopoulos 1992), to the images of the devastating 2004 Indian Ocean tsunami that changed the way the world views tsunami risk. Scientists, engineers and exposed populations have a variety of tools to plan for and respond to tsunamis, and yet disparities in environmental and socio-economic vulnerabilities have a direct effect on how exposed communities fare. Long recurrence intervals and short-term unpredictability pose further challenges.

Tropical cyclones, another coastal hazard, hit certain coastlines during a four- to six-month period at somewhat frequent intervals (several events over the course of a 100-year period for a given region), and even without appropriate communication infrastructure, residents can usually recall the last major storm. However, there is no 'tsunami season', and they occur much less frequently than cyclones (large tsunamis occur every 300–1,000 years for a given region). Collective memory of the last major tsunami often fails, therefore planning is compromised. Yet not all tsunamis evolve into disasters. A massive tsunami, with runup heights exceeding 20 m, affected over 200 km of coastline in the sparsely populated Kuril Islands in 2006, yet it caused no deaths nor damaged a single building as none were exposed (MacInnes *et al.* 2009). The hearty coastal vegetation experienced the worst of the inundation.

While it is impossible to predict when a tsunami-generating event will occur (earthquake or landslide and, to a lesser extent, volcanic eruptions and asteroid impacts), scientists and engineers are becoming very good at predicting the timing and extent of the waves' impact, and are working on methods to incorporate interactions with exposed populations and ecosystems. By identifying the vulnerable and resilient elements from other disasters, exposed populations can work to indentify and strengthen critical institutions and structures (natural, social and built). Policy-makers, scientists, engineers, the private sector, non-governmental organisations (NGOs) and local communities can work together to mitigate against tsunami damage by encouraging and creating structures and mechanisms (social, physical or biological) that are meant to counter the impact of the waves. Special care must be taken, however, to ensure that these structures and mechanisms do not cause unanticipated short- or long-term problems that might ultimately make the communities more vulnerable.

Tsunami risk

Tsunami risk is a function of both the hazard and vulnerability. Hazard refers to the actual physical process involved, while vulnerability is the compounded effect of the exposure and susceptibility of people and their assets to the impact of the hazard as well as their capacity to anticipate, cope and recover (Wisner *et al.* 2004). Therefore, understanding tsunami risk reduction necessitates a multi-faceted approach.

Hazard

Tsunami (津波) comes from the Japanese words for 'harbour' (*tsu*, 津) and 'wave' (*nami*, 波), as the shape of natural harbours tends to amplify the wave. Tsunamis are a series of gravity waves that propagate in all directions from a large movement of water, usually in the oceans but also in inland seas, fjords or lakes (see Table 27.1). These mass movements of water can be triggered by landslides that either occur entirely underwater or fall from the land into the water, volcanic eruptions and even very rare meteor impacts. The larger the displacement of water, the larger the tsunami. Depending on where they hit and on how vulnerable coastal communities are, the larger the devastation on nearby and possibly far away coastlines. However, the vast majority of destructive tsunamis are generated by submarine earthquakes.

Subduction zone earthquakes around the Pacific Ocean, the eastern Indian Ocean, as well as the Caribbean and Mediterranean Seas generate the vast majority of tsunamis. These subduction zones exist where two of the rigid plates that cover the Earth's surface collide head on and one plate is forced to slide beneath the other. This sliding, however, is not smooth – friction at the contact between the two plates causes them to stick. When the stress at this interface builds high enough to exceed the friction between the plates at the boundary, the plates rapidly slide past one another creating an earthquake. The size (or magnitude) of the earthquake is related to the area of this stuck interface (or length of the subduction zone) and the amount of time that has passed since the last earthquake, and in general the bigger the earthquake the bigger the tsunami. While subduction zone earthquakes greater than Magnitude 7 on the Richter scale have the potential to generate damaging tsunamis, it is important to point out that not all large-magnitude subduction zone earthquakes produce tsunamis, nor is a large-magnitude earthquake required to generate a tsunami.

In the aftermath of a major earthquake and tsunami, people often fear that aftershocks will generate subsequent tsunamis. Aftershocks are smaller, stress-relieving earthquakes that follow major earthquakes. They drop off in size and frequency over time, and do not tend to generate tsunamis. Doublet earthquakes, where a large-magnitude earthquake triggers a second in an adjacent region, do have the potential to generate damaging tsunamis, as was the case in the Sumatran earthquakes in 2004 (M_w = 9.2) and 2005 (M_w = 8.7). In any case, if the ground in coastal areas shakes hard for more than thirty seconds, people should move quickly to higher ground.

While earthquakes are the dominant source, other phenomena can generate tsunamis (see Table 27.1). Landslides, both those that occur on land next to a large body of water or underneath the water surface, may generate a tsunami (e.g. Nisbet and Piper 1998). In 1929 a M_w = 7.2

earthquake triggered a landslide offshore Newfoundland, and as the failed material moved downslope a tsunami was generated that caused waves greater than 7 m high on the adjacent shoreline, killing twenty-eight people. Another relatively mild earthquake (M_w = 7.1) in 1998, offshore Papua New Guinea, presumably caused an underwater slump that generated a 15-m-high tsunami which killed over 2,000 people (Tappin *et al.* 1999). The largest tsunami ever recorded was the result of a landslide in Alaska in 1958, when a M_w = 7.9–8.3 earthquake shook loose a 30 million m^3 landslide that fell into the head of Lituya Bay, generating a 524-m-high tsunami (Miller 1960). These events can be particularly dangerous because the earthquake shaking is not indicative of the size of the tsunami.

Volcanic eruptions on subduction zones have also generated deadly tsunamis. The explosiveness of these volcanoes, as compared to those within oceanic plates that tend to have more fluid eruptions (e.g. Hawaii), can cause the caldera to collapse below sea level, and as the water rushes in to fill the void, a tsunami is generated. This was likely the case in the eruption of Thera (Santorini) in 1650 BCE, and Krakatau in 1883 BCE. Asteroid impacts can also generate

Table 27.1 Significant prehistoric and historic tsunamis

Name	Location	Cause	Effects
Støregga (7000 BCE)	Offshore Norway	Submarine landslide	Possible coastal Celtic population resettlement at St Andrews, Scotland
Thera/Santorini (1650 BCE)	Greece	Volcanic eruption	Possible contributor to the decline of Minoan civilisation
Cascadia (1700)	North-west USA, southwest Canada	Subduction zone earthquake	Relocation of Native American populations, recorded in Japan
Lisboa (1755)	Portugal	Earthquake	Added to the destruction of Lisboa, affected colonialist aspirations
Krakatau (1883)	Indonesia	Volcanic eruption	35 m high tsunami killed 30,000 in Sumatra
Sanriku (1896)	Japan	Subduction zone earthquake	'Tsunami earthquake' – low magnitude, big tsunami. Killed 22,000, started Japan's tsunami preparedness
Grand Banks (1929)	Newfoundland	Submarine landslide	Killed 29 in Canada's worst earthquake-related disaster to date
Aleutians (1946)	Alaska	Earthquake	Possible tsunami earthquake in which 165 people died. Led to establishment of Pacific Tsunami Warning Center
Sissano Lagoon (1998)	Papua New Guinea	Earthquake/landslide?	Landslide or tsunami earthquake debated – 1,200 killed
Indian Ocean (2004)	Sumatra, Andaman and Nicobar Islands	Subduction zone earthquake	250,000 people killed in a dozen Indian Ocean countries – led to establishment of an Indian Ocean tsunami warning system

tsunamis. The impact that defined the boundary between the Cretaceous and Tertiary Periods (60 million years ago) hit a shallow ocean offshore present-day Mexico, generating a tsunami that was 50 m–100 m high when it reached present-day Texas (Bourgeois *et al.* 1988).

Once a tsunami is generated, the displaced water radiates out in all directions, as when a pebble is thrown into a lake. The velocity of the wave increases with the water depth – a tsunami in the open ocean may be only centimetres high, but travels at speeds of up to 700 km/hr – about the speed of a commercial jet aeroplane. A tsunami generated in South America can take up to fifteen hours to reach Hawaii, over twenty hours to reach Japan, but less than twenty minutes to hit the coastline closest to the epicentre. As the wave approaches land and the water shallows, bottom friction slows the tsunami and it begins to build in height as faster moving water piles up behind it. Computer models use the ocean water depths to calculate when the tsunami will hit coastlines surrounding the region, and how big the wave will be when it strikes.

The travel time between the source of the tsunami and a given coastline is critical when considering risk reduction measures. It may take a tsunami as little as a few minutes to reach the 'near field' area closest to the epicentre that undergoes strong shaking. As the wave approaches these near field shorelines, it is often the trough of the wave that hits first, which manifests itself as the ocean rapidly receding, exposing the seafloor. Curious onlookers, unaware of the hazard, explore this newly uncovered seafloor and are often the first casualties when the inevitable peak follows. Destructive tsunamis often have wave heights at the coast of between 5 m and 10 m, and travel inland at velocities too high for people to outrun, especially when carrying personal belongings and helping the very old in heavy traffic with debris possibly blocking exit routes. Depending on the nature of the landscape (flat, heavily vegetated or developed, etc.), the wave can proceed inland for kilometres (inundation distance, see Figure 27.1).

While strong shaking on land near subduction zones is a natural early warning for the near field populations, those outside the region that felt shaking (far field) have no such warning. The tsunami often approaches the shoreline as a 'positive' wave (crest leading), which does not allow for the receding ocean to serve as a warning. Tsunamis that occur in countries in closed basins such as the Mediterranean and Caribbean Seas are particularly vulnerable because of the short near field travel times. Further away, like in Sri Lanka, where it took the 2004 tsunami some two hours to reach the coast, no shaking was felt and the only warning to those who happened to be looking out to sea was a strange white (or sometimes black) line on the horizon as the turbulent wave approached.

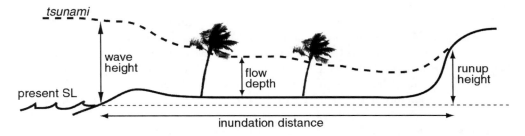

Figure 27.1 Tsunami terminology
Note: The tsunami wave height refers to the elevation of the tsunami above sea level at a given location. The flow depth is the height of the water above ground. The inundation distance is the distance the wave travelled inland, perpendicular to the coastline, and the runup elevation is the height above sea level at this point.

Vulnerability

Tsunamis affect people, settlements and ecosystems in a variety of ways. For instance, some populations rely on inshore fisheries for both food and income, and use mangroves for building material and fuel. Mangroves also stabilise the coast and protect it from storms and tsunamis. Even where people do not directly depend on coastal biological resources for their livelihoods, they rely on healthy coastal ecosystems; however, their dependency is indirect, and their vulnerabilities tend to lie in the built environment. Trends from recent disasters suggest that similar hazards produce larger loss of life but smaller monetary losses in low-income nations than in richer nations (EM–DAT 2010). Wealthier nations exposed to tsunami hazard such as the USA, Japan and New Zealand tend to have appropriate and enforced building codes, integrated warning systems and disaster management plans, along with resources available to support areas after an event occurs. Furthermore, in all nations it is often the wealthier, less vulnerable of the population as well as tourists who choose to live by or visit the coast for its aesthetic and recreational offerings.

Many communities, especially those in countries along the Pacific Rim including the Philippines, Indonesia and other Polynesian, Melanesian and Micronesian countries, rely on the land and sea for their livelihoods and food security. If ecosystems that support these activities are damaged either before or during the earthquake and tsunami, populations will be less able to recover. During the 2007 Solomon Islands earthquake and tsunami, fringing and barrier reefs on the island of Ranongga were lifted up to 3 m above sea level in places. This not only killed the coral, which will affect the reef fishery, but also adversely affected the recreational diving industry that accounts for up to sixty per cent of the region's income, and adversely affected the ability to deliver aid to affected communities (McAdoo *et al.* 2008). In the rural areas outside of Pangandaran on Indonesia's Java Island, the 2006 tsunami inundated low-lying rice *padi*, destroying crops and salinating soils, which affected food security. Coastal aquaculture industries damaged by the 2004 Indian Ocean tsunami are being rebuilt with the goal of providing local jobs and products for export; however, they occupy a sensitive niche in the intertidal ecosystem where mangroves could be established. In Thailand the owners of aquaculture schema are not always local residents, and the local populations realise few benefits from the operations (Sathirathai and Barbier 2001). The decision as to whether these aquaculture ponds are to be rebuilt or the intertidal ecosystem restored must ultimately be made at the local level with the costs and benefits analysis informed by data from experts in the field.

In both wealthy and less affluent nations, coastal ecosystems provide a variety of important protection services (Sudmeier-Rieux *et al.* 2006) (see Chapter 32). In the tropics, fringing and barrier coral reefs and lagoons reflect and diffuse tsunami energy. Unfortunately, villages tend to be located at breaks in the reef where natural or sometimes artificial channels focus tsunami energy, increasing the wave's amplitude, making it more destructive. Sandy beds of turtle grass that offer habitat for several important species also stabilise sediment that can buffer against the surges, and is the one coastal ecosystem that showed a positive correlation with lowered tsunami heights during the 2004 Indian Ocean tsunami (Cochard *et al.* 2008). Mangrove forests and other coastal wetlands offer a vast array of ecosystem resources, from nurseries for offshore fish species, to firewood for humans, and have evolved to absorb a certain amount of wave energy. But the rare tsunami that exceeds the height of the forest can rip up trees, turning them into projectiles that can cause additional damage to structures. The more robust the coastal ecosystem, the more effectively the local human population, and the ecosystem on which they rely, will be able to deal with the disaster.

Densely populated coastal cities that lie directly adjacent to active subduction zones are most vulnerable to the effects of tsunami. On 30 September 2009, a large-magnitude subduction

Box 27.1 Effects of the September 2009 tsunami on Samoan fishing communities

Joyce Samuelu Ah-Leong
Samoan Ministry of Agriculture and Fisheries, Fisheries Division, Samoa

On 29 September 2009 a powerful earthquake followed by a devastating tsunami struck the shores of the Samoan archipelago, killing 148 people and badly affecting the lives of coastal communities for which fishing is an essential livelihood. The loss of relatives and loved ones was the hardest, and the realisation by these communities that the sea that they depended mainly on for livelihood was also the destroyer.

Straight after the tsunami the people fled their coastal homes, or what was left of them, and made homes on their interior and higher lands. Fishing equipment was damaged and people's priorities were set on the rebuilding of their homes. Normal fishing activities were thus halted for almost four months, although most of these fishermen are subsistence fishers. In fact, during the reef assessments in tsunami-affected communities by the Fisheries Division in December 2009, there were no signs of any fishing activities, and the coastline and the marine area were still littered with debris from the tsunami.

The absence of fishing meant that people had to find food elsewhere. Most fishermen's families actually turned to food donated by the government or resorted to the very many donations from overseas Samoans that poured into the country after the tsunami. The few fishermen from the affected districts were not seen selling their catches after the tsunami, which is still the case today. However, this did not affect the overall landings of the inshore resources as most of the sellers are from non-affected areas.

The government, NGOs and international organisations such as the United Nations Development Programme, the Food and Agriculture Organization and the Pacific Regional Environmental Programme initiated a cleanup campaign which slowly won the people's support and the villagers became involved. This also encouraged people to access the sea and resume fishing. The government is prioritising the recovery of the affected fishing communities. The Samoan Ministry of Agriculture and Fisheries helps in restocking the main edible bivalves and other invertebrates, and in restoring marine habitats through the reconstruction of coral reefs and the building of artificial fish houses. The government also has distributed fishing gear and contributed to the restoration of damaged commercial fishing vessels.

zone earthquake devastated the Indonesian city of Padang (population ~800,000) on Sumatra's west coast, killing an estimated 1,100 people and displacing another 1.2 million from their homes and jobs (UNOCHA 2009). Fortunately, the earthquake was too deep (81 km) to generate a tsunami. Had the earthquake occurred closer to the trench, the damage in Padang may not have been as severe, but it could have generated a tsunami. Tsunamis can be more destructive in areas close to the earthquake's epicentre, where the shaking has weakened buildings that are then exposed to the tsunami, as was the case in Banda Aceh to the north.

Following the shaking of a large coastal earthquake, people often seek refuge inside their homes if they appear at least superficially sound. In the village of Gleebruk in Indonesia's Aceh Province, survivors of the 2004 Indian Ocean tsunami reported that a family was killed when, upon hearing the approaching tsunami, which was described as a very loud roaring with occasional 'booms', they ran into their house and locked the doors, thinking it was an escalation in the armed conflict between the Indonesian military and the Acehnese separatists. This may have been a reasonable response in a community that had been affected by this conflict for twenty-five years prior (Gaillard *et al.* 2008b), and where the last tsunamis were in the years 1907 and 1350 (Monecke *et al.* 2008). In the small village of Asili on American Samoa, school children had been educated on tsunami hazard by the US Department of Homeland Security some six months prior to the 29 September 2009 event. When they felt the large earthquake, the tsunami hazard they were facing did not register until they saw the lagoon emptying as the sea retreated. All the children and their families, save one, rapidly evacuated to higher ground, and none were killed. Unfortunately, a Korean family that had opened a shop in Asili only four months prior did not receive the information and ran back to their shop to protect their goods from feared looting. Two of the four family members lost their lives as their refuge was washed away.

Mitigation efforts

People in exposed communities must work hand-in-hand with scientists, development agencies and policy-makers to recognise the nature of the hazard and vulnerabilities before attempting mitigation measures. The first step is to work with geoscientists to understand the nature of the hazard: How often do tsunamis come, and how large are they when they hit? Then the community must identify the most vulnerable structures, people and resources, and decide how it will mitigate the risk. Will they spend money on advanced communication infrastructure to tie into a regional early warning system? Should they develop strict building codes and invest money in inspection and enforcement? Is the loss of land and access to resources that would result from implementation of a buffer zone worth the cost of protection? While the answers to these questions should be based on the available scientific data, the decisions ultimately lie with the local community that will have to live with and implement the solutions. It is also crucial to ensure that the means of protection are fairly available and accessible to all sectors of the community, including the most marginalised and usually vulnerable people.

Tsunami science

The frequency and magnitude of past tsunamis is one of the more helpful pieces of information that should be used to decide on tsunami mitigation measures. While the historical record of earthquakes and tsunamis is perhaps the most reliable, palaeotsunami data can extend the record to prehistoric times. A turbulent tsunami can pick up sand from offshore and deposit it in marshes, and sometimes on uplifted terraces, where trained geoscientists can determine the age of the deposit and sometimes how big the tsunami must have been. If conditions are right, several such layers of sand can be preserved, and a tsunami recurrence interval can be determined. A tsunami that occurs in a given location once every 200 years will elicit a different mitigation response than the same size tsunami that occurs once every 1,000 years.

Risk assessment

After the nature of the hazard is determined, an overall assessment of the risks contributes towards developing community awareness, and hence can reduce risk. This assessment can be done by a variety of people, from national and local governments and community-based organisations to academic researchers, NGOs and UN agencies. Interested parties from outside the community must work with exposed coastal communities to determine how they live within the coastal zone, and how well their structures might respond to the hazard, be they environmental (ecosystems, landforms), physical (buildings, roads, ports, communication, etc.), economic (agriculture, aquaculture and fishing activities, businesses, banks, insurance) or social (people's networks and interrelationships, religion, communication). A community that relies heavily on local environmental resources (fisheries, agriculture, tourism, etc.) will assess the health of their ecosystem to determine their vulnerability. Coral disease and bleaching, destructive fishing practices, mangrove fragmentation, sand dune modification, etc. affect not only the propagation of the waves, but the system's ability to recover after the event, and to provide resources for rebuilding after a tsunami hits. With the interconnectedness of these systems, it is vital that all outside actors work in concert with the exposed population to understand the linkages.

Box 27.2 TSUNARISQUE: an integrative project for reconstructing and reducing the risk of tsunami disaster

Franck Lavigne
UMR 8591 Laboratoire de Géographie Physique–CNRS, Université Paris I, France

Junun Sartohadi
Research Center for Disasters, Gadjah Mada University, Indonesia

The 26 December 2004 tsunami triggered an unprecedented disaster throughout the Indian Ocean. In Indonesia one-third of the city of Banda Aceh was flattened and 70,000 people died because of exceptional waves (up to 30 m high) which affected vulnerable coastal communities. The TSUNARISQUE project conducted in Banda Aceh and neighbouring areas between 2005 and 2007 involved French and Indonesian scholars from various disciplines, including tsunami modellers, geologists and geophysicists, geomorphologists, human geographers, historians and anthropologists.

The project encompassed: (1) the chronology and dynamics of the tsunami, through measurements of runup, wave height, flow depth, flow direction, bathymetry and topography; (2) the scope of structural damage; (3) the environmental imprint of the tsunami, including water and soil salinisation, beach erosion, destruction of sand barriers protecting the lagoons or at river mouths, bank erosion in the river beds, boulder transportation and deposition, and sand deposition; and (4) the identification of the underlying social causes for the disaster, based on questionnaire surveys among affected communities and interviews with key informants.

The project bridged some gaps in the scientific understanding of extreme-magnitude tsunamis. The project open-source database (www.tsunarisque.cnrs.fr) provides researchers with data for better calibrating numerical models. A careful analysis of these data shows a significant discontinuity in the tsunami flow depths along a line approximately 3 km inland, which is likely where the front of the wave broke. Damage to buildings reflects the propagation of the wave, with less damage beyond the line where the front of the wave broke. The project therefore enabled provision of a quantitative scale of tsunami damage intensity. In parallel, social scientists emphasised the importance of different local contexts, especially ethnic histories and the transmission of tsunami warning knowledge, in explaining the unequal scope of damage throughout the province of Aceh (see Chapter 38).

Beyond its scientific value, the project also strengthened the capacity of the Indonesian scientific community. High-tech measurement devices (e.g. an electronic theodolite and distance meter, laser range finders, GPS Geoexplorer, etc.) were turned over to local partners, i.e. the Centre for Disasters Studies (PSBA) of the Gadjah Mada University (UGM) and the Indonesian Meteorological and Geophysical Agency (BMG). As young, local researchers were fully involved in the field activities they acquired the skills required to handle these tools.

Finally, the project provided space for an NGO to collaborate with scientists. The NGO Planet Risk drew on the results of the project to develop a tsunami awareness campaign in Java. This included exhibitions of photographs and posters, distribution of leaflets and the creation of a permanent tsunami awareness centre in Parangtritis, Central Java. Another significant feature of the awareness campaign was the multiple screening of a documentary film made as part of the project. This movie featured locals acting as they had to face a pending tsunami threat. All these materials were prepared in close collaboration with local partners to ensure that they fitted within the local cultural context.

Buildings

The quality of the built environment is one of the critical factors that underlies the overall vulnerability of populations exposed to earthquake and tsunami hazard. Ultimately, the quality of buildings is an economic decision made by the builders at the local level, and these decisions must be made with an understanding of the hazards and vulnerabilities. Structures should be built to codes designed with the hazard in mind, with high-quality, reinforced concrete and a flow-through design that allows water to move through with fewer obstacles. In areas with appropriate and enforced building codes, structures from specially designed escape buildings and bridges can be used as refuges during tsunami inundation in regions with high, near field population densities. In Japan, special evacuation buildings have been designed to withstand the largest earthquakes and tsunamis, and similar structures have been built in Meulaboh and Banda Aceh in Indonesia's Aceh Province (see Figure 27.2). This level of quality requires substantial investment and their allocation must be informed by the frequency of recurrence – a structure must survive until the next tsunami, which could be 100–500 years distant (Monecke et al. 2008). The decision to design against these low-probability, high-impact events must be made not only with the hazard exposure in mind, but also the societies' willingness to accept a given degree of risk based on the probability of occurrence of a hazard and the impact on their livelihoods.

Coastal zone planning

Infrastructure planned, designed and located with informed awareness of risks is essential for contributing towards risk reduction. Approaches will vary between communities, from shoring up the docks at a fish processing facility to constructing new bridges along a coastal evacuation route. Should financial institutions and businesses be moved, or should they remain in place to better support daily economic activity? Where are the schools and how well are they built? Is the hospital better off in an exposed but central location, or in a harder-to-reach location on the outskirts of town, but away from the hazard? How are these elements spatially distributed? Tsunami mitigation policies put in place by the communities that are impacted by the decisions must include inputs from engineers, social and natural scientists, as well as international actors that may or may not be present to support mitigation activities.

Figure 27.2 Intertwined measures to face tsunami hazard in Banda Aceh, Indonesia
Note: This figure illustrates some of the multiple ways in which exposed communities reduce their risk of tsunami disasters. In the foreground, aquaculture ponds, stabilised with mangrove seedlings, will provide food and income for coastal zone residents. Fishing boats use the intertidal zone for safe harbours, as well as access to markets. In the event of another tsunami, residents can seek refuge in a tsunami evacuation structure (background), designed by the Japanese government to withstand very large-magnitude earthquakes and substantial tsunami inundation.
Source: Photograph by Brian McAdoo

One potential challenge facing local decision-makers and investors has to do with the recognition of short-term vs. long-term benefits realised by both the actors and the local populations. Outside investors and donors might have agendas that are at odds with the local communities. For example, many international NGOs had money earmarked for post-tsunami mangrove ecosystem restoration in Aceh following the 2004 tsunami. In some cases, the local communities sought to have the ecologically damaging, yet job-providing aquaculture industry restored. In their eyes, the short-term (and possibly long-term) benefits of fish and shrimp farming exceeded the long-term payoffs that would result from supporting healthy ecosystems, plus some locations felt that healthy ecosystems and small-scale shrimp farming were mutually compatible. Despite a well-voiced knowledge that healthy coral reefs offered some protection against the tsunami in Sri Lanka, post-tsunami coral mining progressed, driven in part by the need for building material. This foreseeable clash of values must be carefully negotiated.

Coastal zone setbacks designed to act as buffers against tsunami and storm surges are another controversial policy that emerged in many Indian Ocean countries following the 2004 tsunami. In Indonesia, Sri Lanka and Thailand coastal communities affected by the tsunami were relocated by the government for a variety of stated reasons. On Indonesia's Simeulue Island (Aceh Province), for example, the coastal community of Latiung (which was only minimally affected by the 2004 and 2005 tsunami) was moved several kilometres inland, ostensibly to protect them from future tsunamis, but in reality to provide a labour force for an oil palm plantation owned by a government official. Aceh, however, does not have the pressure from tourism and other industries that encouraged the setbacks in Sri Lanka and Thailand. In 2005 the Sri Lankan government hastily enacted a mandatory 100 m 'no build zone' designated as a 'green belt', but this was quickly reduced to 35 m after public pressure (Pattiaratchi 2005). In Thailand, however, coastal residents were removed from their property, designated as 'public land' by the government, so that tourism and access to natural resources could be pursued (Human Rights Center 2005). It is important to point out that in each case justification for the setbacks was based on politics and economics, not on the recurrence interval of the hazard.

Setbacks can also help protect the natural buffers such as coral reefs, coastal vegetation or sand dunes that tsunamis may encounter as they approach shorelines. This natural coastal environment may deflect the wave energy and cause it to be dissipated, thus providing some natural protection from such high-energy waves. While the ecosystems provide a valuable buffer to coastal communities, economic development that may be at odds with the ecosystems could ease poverty and hence vulnerability. Thus decisions are necessary concerning how best to manage development and ecosystems. The rural poor usually rely on natural resources for their livelihoods and therefore want to exploit coastal resources in ways that may decrease the natural vegetation. It is the urban poor who are usually most at risk from tsunamis. So a balanced approach to coastal land use has to be adopted. In principle such an approach can be informed by a set of tools for use in preparing for and mitigating the effect of tsunami. This includes local knowledge of past events and engineering structures that can withstand the stresses of both the tsunami and the earthquake (if near the source), and policies from the national to local levels that include land-use planning, risk awareness and agency co-ordination (Jonientz-Trisler et al. 2005).

Early warning systems

Working hand-in-hand with coastal planning and educational activities, early warning systems have proven effective in countries of the Pacific basin (Gonzalez et al. 2005). These systems are based on instrumental responses to large, potentially tsunami-generating earthquakes, but must be linked with careful pre-event planning at the country, regional and local levels (Kelman 2006). Evacuation planning is key to effective implementation: escape routes aim to be clearly marked, and communities are aware as to what to do during an evacuation to avoid congestion. Special care must be taken for those in the near field, where travel times are shorter, and in countries that do not have the resources to support the infrastructure of a permanent warning system. While a country like Japan exposed to frequent tsunamis can and must afford to make these investments, countries like Indonesia, the Philippines and small island states in the Caribbean and western Pacific must pick and choose based on what they consider acceptable risk and what they can afford, keeping in mind that the tsunami warning system was developed as an international public service with most information freely available to anyone, including national governments. As well, donors sometimes make decisions regarding which systems to implement and which not to implement, as demonstrated by the challenges in developing an Indian Ocean tsunami warning system (Kelman, 2006), meaning that the poorer countries often do not have a say.

Awareness

While coastal communities consider mitigation efforts focused on physical and ecological resources, perhaps the most important issue is awareness. In the near field, where the travel times of tsunamis are mercilessly short, populations must be aware of the hazard and advised of the warning signs, and not wait for an official notification from a central authority (Dengler 2005). Indeed, hazard awareness is the first step to reducing the risks posed by tsunamis, and forms the foundation of actions that can be taken should an event occur. As demonstrated in the case from American Samoa described earlier, the combination of awareness and an action plan saved countless lives. The community at Asili was bolstered by an outside agent, in this case the US Department of Homeland Security, to remind them of the hazard and offer advice on how to react. In some more isolated, less transient communities like those on the small islands of Thailand and Indonesia as well as the Solomon Islands, cultural memory is preserved, and the indigenous populations devise their own responses. It is imperative to assess local knowledge at the community level (as it relates to hazard awareness), so that information offerings from governments and NGOs may be integrated rather than imposed.

Conclusions

On average, every year tsunamis kill hundreds of people and cause millions of dollars of damage (EM-DAT 2010). The visceral media images from recent events like the 2004 Indian Ocean tsunami have heightened awareness of the hazard. Yet despite this increased awareness, hundreds of lives have been lost in subsequent events, and lots of work needs to be done to mitigate against future events. Tsunami scientists are also adding critical data to the discussion, by constraining the location, frequency and magnitude of potential events, and learning how the waves interact with the physical, biological and human environments. Communities, once aware of the hazard, work to determine their vulnerabilities and shore them up with well-informed mitigation efforts focused on the natural, built, economic and social environments. Early warning systems that link high-tech earthquake location and tsunami detection with in-country networks and local knowledge are revolutionising the way in which exposed communities deal with tsunamis, from communication networks to evacuation planning and awareness campaigns. By making connections between the hazard and the vulnerable parts within communities, tsunami risk reduction is tenable.

28

Volcanic eruption

Susanna Jenkins

CAMBRIDGE ARCHITECTURAL RESEARCH, UK

Katharine Haynes

RISK FRONTIERS, MACQUARIE UNIVERSITY, AUSTRALIA

Introduction

At present, volcanic and human activity co-exists in uneasy discord: approximately 5.5 million people were evacuated, injured or made homeless during the twentieth century alone (Witham 2005). Increasingly, the world's population and accompanying urbanisation, agricultural cultivation and industrial development are becoming concentrated in large conurbations that lie within reach of some of the most hazardous volcanic processes. The dense populations surrounding many volcanically active regions on Earth are testament to the benefits of volcanic eruptions, such as fertile land for agriculture, higher zones that capture rainfall for use in the surrounding plains, aggregate for construction, geothermal energy and even volcano tourism. With increasing aviation travel, explosive volcanoes without dense population settlements may still pose considerable economic and health risks to airborne populations and aeroplanes. Some of the busiest air routes cross the volcanically dense and active regions of South-East Asia and the north Pacific. The explosive eruption from Eyjafjallajökull volcano in Iceland in April 2010 caused major disruption to air travel across Europe with significant losses for the aviation industry.

Underlying vulnerabilities play a fundamental role in determining the extent of volcanic impacts. Social, economic and political factors determine who lives, works and has assets in the high-risk zone (e.g. agricultural workers), and also shape people's capacities to cope, recover and adapt. While nothing can be done to prevent the actual eruption, reducing underlying social vulnerabilities and improving understanding of volcanic processes can prevent volcanic disasters. This chapter will provide an understanding of the physical processes and mechanisms exhibited during volcanic activity alongside the social processes through which people are exposed to and are impacted by volcanic activity. The actions that people have taken to mitigate against and improve their capacity to cope with and adapt to volcanic impacts will also be discussed.

Physical processes and damage mechanisms

Most volcanoes on Earth are distributed along either divergent or convergent plate boundaries, home to the largest number of active submarine and active subaerial volcanoes, respectively.

Intraplate volcanoes form the third main zone for volcanism and can be found in oceanic or continental settings. Volcanoes born from intraplate locations or divergent plate boundaries produce the largest amount of magma but do so relatively quietly and without the explosive tendencies of those inhabiting convergent boundaries. Convergent plate volcanoes represent just ten per cent of the world's magma production but more than eighty-five per cent of the reported eruptions (Sigurdsson *et al.* 2000) due to their explosive nature and proximity to land and eyewitnesses. As may be expected, not all volcanoes fit neatly into their prescribed pigeonhole. Mount Etna in Italy, for example, lies at the junction of three converging plates, yet is a voluminous magma producer that erupts regularly with relatively low explosivity.

The explosivity of an eruption is governed by the magma silica content and viscosity (and therefore gas content), as well as by the presence of ground- or seawater: the increase of any or all of these factors results in more explosive eruptive behaviour. As a general rule of thumb, the more explosive a volcanic eruption the larger the impacted area and the greater the potential for loss of life, although there are of course exceptions to this rule. Volcanoes that exhibit typically explosive or effusive behaviour may act 'out of character' or exhibit the full range of volcanic processes throughout the course of one eruption. For example, degassed high-silica magma erupts domes of viscous lava non-explosively; however, such lava domes commonly then act as a cap to the underlying magma, preventing degassing and eventually resulting in explosive obliteration of the dome. Explosive eruptions typically comprise volatile release of energy through rapid-onset hazards such as pyroclastic density currents, ash falls, lahars, blasts and volcanic earthquakes and have a much greater impact on populations, infrastructure and agricultural activity. Such hazards may allow limited time for evacuation, sometimes rendering affected areas unsuitable for rehabilitation and human habitation for decades. Continuing ash emissions in the final phases of explosive events may result in repeated disruption over months or even years, causing disruption over wide areas (e.g. Sakura-jima, Japan 1955–present; Tungurahua, Ecuador 1999–2009; and Soufrière Hills, Montserrat 1995–2003 and still ongoing). Further remobilisation of ash fall by wind or rain can lead to disruptions extending for years beyond this; for example, in the years following the 1991 eruption of Pinatubo in the Philippines, the homes of more than 10,000 people were destroyed by lahars (Newhall and Punongbayan 1996). Effusive eruptions release energy more slowly than their explosive counterparts and can be characterised by low-viscosity lava flows and large gas emissions, e.g. Lake Nyos, Cameroon in 1986. While being more frequent than explosive eruptions, effusive eruptions commonly present less of a hazard to the population. Lava flows move relatively slowly and offer sufficient time for people and livestock to move to safety; however, land and physical infrastructure suffer extensive damage.

An important aspect of volcanic activity is that one eruption may produce a number of hazardous processes that may occur repeatedly or in combination throughout the eruption, with varying duration, spatial impact and intensity. For example, edifice failure, debris avalanche, lateral blast, pyroclastic density current, lahar and explosive ejection of ash all took place within the first thirty minutes of the climactic eruption of Mt St Helens on 18 May 1980. The eruptive event then persisted until 1986 with intermittent activity continuing to this day. Volcanic events may continue for years (Soufrière Hills, Montserrat 1995–2003, 2004, 2005 and continuing), or even decades (Kilauea, USA 1983 and continuing), with consequent repeated or long-term disruptions to social and physical infrastructure. As shown by the recent crisis on Montserrat, intervals between consecutive eruptive events may be on the order of months, allowing little respite for affected communities. Conversely, repose periods between eruptions can be on the order of hundreds or thousands of years, particularly for the more explosive eruptions. Volcanic eruptions can therefore take place at volcanoes or locations not previously thought to warrant study. For example 2,942 people were killed by an eruption from Mount Lamington, a forested peak not previously recognised as a volcano, in Papua New Guinea in 1951 (Blong 1984), and two of the largest eruptions of the last century, Pinatubo in the Philippines in 1991 and Chaitén in Chile in 2008, took place after around 500 and 9,400 respective years of quiescence. Rehabilitation of impacted areas post-eruption also poses significant problems. Areas covered by lava flows or large ash, pyroclastic or lahar deposits can prevent resettlement for years, or even decades, particularly for communities that rely upon agriculture.

Hazardous volcanic processes have been described in detail by a number of authors (e.g. Blong 1984; Sigurdsson *et al.* 2000) and we provide a simple summary here. Table 28.1 summarises the main range of physical processes associated with volcanic eruptions, their methods of generation, damage mechanisms and typical extent with historical eruption examples. Consequences and mitigation of such processes for human activities are considered in more detail in the following sections.

In addition to processes noted in Table 28.1, volcanic activity may result in ground deformation, lightning, acid rain, atmospheric sound and shock waves and, in larger-magnitude events, climatic variations such as global cooling. Some potentially damaging processes, such as lahars, gas emissions, acid rain, phreatic explosions, ground deformation, volcanic earthquakes and even pyroclastic density currents, may take place independently of an eruption. The spatial extent of volcanic processes varies widely, with ballistics commonly confined to within 5 km of the vent (Blong 1984) and pyroclastic density currents, lava flows and lahars expected to travel anywhere between 1 km and 100 km (McGuire 1998). Volcanogenic tsunamis or ash falls have the potential to travel thousands of kilometres from source. For example, ash falls exceeding 100 mm were deposited more than 2,000 km from the site of the 1815 Toba eruption in Indonesia (Ninkovich *et al.* 1978), while the caldera collapse of Krakatau volcano in Indonesia in 1883 generated a tsunami nearly 40 m high at source that remained large enough to temporarily strand small harbour boats in Sri Lanka, nearly 3,000 km away (Choi *et al.* 2003).

As well as varying widely in their potential impact over space and time, volcanic processes, damage mechanisms and consequences are extremely diverse. Immediate impacts include the violent destruction of people, livelihood assets and infrastructure through blast, burial and exposure to high temperatures within the path of pyroclastic density currents, lahars and debris flows. Longer-term impacts include a reduction in the health and socio-economic well-being of affected communities through a loss of income or livelihood, temporary and permanent relocation or prolonged exposure to gas, ash and contaminated water. In the years following the 1815 eruption of Tambora in Indonesia – the largest volcanic eruption in recorded history – more than 60,000 people lost their lives in areas thousands of kilometres from the volcano, mostly through indirect impacts such as famine and disease (Sigurdsson *et al.* 2000).

Hazard exposure and impacts – the who, why and how

The majority of volcanic disasters have social, economic and political root causes. These include the factors that lead people, their livelihoods and assets to be located near volcanic activity, in addition to the issues that govern their preparation for, response to and recovery from an eruption.

People may inhabit areas of volcanic risk for a range of reasons. For many, the benefits of living in a volcanic area – such as fertile soils, mineral wealth, geothermal energy and tourism potential – often outweigh the risks. For example, despite the area around Mt Etna, in Sicily, representing only seven per cent of the island's land area it is inhabited by twenty per cent of the population (Duncan *et al.* 1981). Even so, most people do not choose to live or work in

Table 28.1 Summary of the major volcanic processes, their generation and damage mechanisms and typical spatial extent from the vent

Physical process	Description	Generated by	Key damage mechanisms	Damage extent	Example (fatalities)
Pyroclastic density current	Fast-moving currents of hot ash, gas and rock	- Dome collapse - Column collapse - Interaction between hot material and water	- Temperature - Duration - Dynamic pressure - Presence and size of missiles	To 100 km	Mont Pelée, Martinique 1902 (29,025)
Ash fall and projectiles	Rock fragments 2 mm to more than 1 m diameter	- Explosive eruption of magma	- Vertical loading - Chemical content - Grainsize (respiratory) - Size, temperature and density of projectiles	- To 10 km (projectiles) - Potentially thousands of km (ash fall)	Mount Agung, Indonesia 1963 (163)
Lava flow	Streams of molten rock	- Non-explosive eruption of magma	- Viscosity/speed - Runout length	- To 50 km (low-viscosity basaltic flows) - To 15 km (high-viscosity andesitic flows)	Mount Nyiragongo, Democratic Republic of Congo 1977 (70)
Lahar and flood	Flowing mixture of hot or cold volcanic debris and floods	- Rapid melting of snow or ice - Lake breakout - Heavy rainfall - Infill of river channels with volcanic debris - Violent displacement of water, leading to dam overtopping	- Density - Temperature - Velocity - Presence and size of missiles - Sediment concentrations (range between 1% and 77%)	To 100 km	Nevado del Ruíz, Colombia 1985 (23,000)
Volcanic earthquake	Ground shaking	- Movement of magma - Conduit explosions - Readjustment of the edifice post-eruption	- Magnitude and intensity - Hypocentral depth	To 5 km	Sakua-jima, Japan 1914 (25)

Continued on next page

Table 28.1 (continued)

Physical process	Description	Generated by	Key damage mechanisms	Damage extent	Example (fatalities)
Gas emission	Release of volcanic gas	- Sudden release from crater lakes - Fumarolic activity and gas plumes from degassing volcanoes	- Composition - Concentration - Quantity	- To 10 km (direct exposure) - Hundreds of km (atmospheric dispersion)	Lake Nyos, Cameroon 1986 (1700)
Blast	Explosive discharge of gas, ash and rocks	- Edifice failure - Dome explosion	- Velocity - Temperature - Dynamic pressure - Presence and size of missiles	To 35 km for lateral blasts	Mt St Helens, USA 1980 (57*)
Edifice failure and debris avalanche	Sudden flank collapse and gravity-controlled flows of volcanic debris	- Eruption - Tectonic earthquake - Heavy rainfall	- Mass/volume - Runout length - Proximity to sea	To 100 km	Mt Unzen, Japan 1792 (10,000)
Tsunami	Series of water waves caused by displacement of the water	- Pyroclastic density currents, debris avalanche, lahars or atmospheric shock waves impacting water bodies - Volcanic earthquake - Major submarine explosive eruption - Caldera collapse - Edifice collapse or subsidence	- Runup wave height - Runout length - Inundation distance - Presence and size of missiles	Potentially hundreds or thousands of km	Krakatau, Indonesia 1883 (30,000)

Note: * Including those killed by pyroclastic density currents.
Historical eruption examples, and the approximate number of fatalities within the event that can be attributed to the process, are provided.
Source: Data from Blong 1984; Sigurdsson et al. 2000

these areas based on a rational assessment of costs and benefits. Political and economic factors often determine who lives and works in risky areas. For example, low-income or marginalised groups may be forced to take up cheaper, less desirable land in high-risk locations. Others inhabit land passed down through their family for centuries or follow settlement patterns directed by governments with short-term priorities other than long-term risk reduction.

Livelihood approaches are becoming increasingly common in the social sciences, with recognition of the relationship between livelihood insecurity and vulnerability to environmental hazards (e.g. Wisner *et al.* 2004). This relationship is often seen during volcanic disasters, where victims attempt to protect or continue their livelihood. Those killed on Montserrat in 1997, and the majority of residents who continued to enter evacuated areas, were those returning to look after crops and livestock or because they couldn't afford comfortable accommodation elsewhere (Haynes *et al.* 2008a). In 1999 residents returned to their properties around Mount Tungurahua in Ecuador, despite government warnings, in order to continue their farming and tourism-related livelihoods (Tobin and Whiteford 2002).

A volcanic eruption will often expose the social and political problems within a community that turn an eruption into a disaster. The management of a volcanic crisis is also complex. The examination of past volcanic crises has identified a number of key factors that often determine the outcome of a volcanic event (see Table 28.2). These include issues, which can be controlled and mitigated against, and natural aspects such as the nature and duration of the volcanic activity, which are beyond human control.

Occasionally a volcano will erupt with only one or two days' notice, presenting such an obvious threat that the population evacuates with full compliance, as occurred at Chaitén,

Table 28.2 Key factors that often determine the outcome of a volcanic event

Negative	Positive
No national or local capacity for disaster risk reduction	National and local capacity for disaster risk reduction
No eruption in recent history and no planning to reduce volcanic risk	A recent eruption and/or planning to reduce volcanic risk
A lack of baseline data resulting in poor knowledge of the volcano's past and potential volcanic behaviour	Good baseline data leading to greater confidence in potential activity
Lack of demonstrated violence by the volcano leads to doubt and loss of credibility	Obvious and demonstrated threat to the public leading to reduced scepticism and increased co-operation
Long drawn-out precursory activity	Rapid build-up and definite climax to activity. Justified inconvenience, minimal social and economic disruption
Poor communication and sometimes disagreement between scientists leads to a confused message	Co-ordinated monitoring efforts between scientists and institutions
Uncertainties used to fuel political gain and discredit others. Inappropriate reaction by authorities and by media (e.g. sensationalising or playing down), leads to increased public confusion and reduced credibility	Good communication between scientists and authorities, authorities and media Timely and appropriate information accessible to public and authorities (e.g. Mt Pinatubo), leading to improved understanding of the risks

Chile, in May 2008. By March 2009 the volcano was still active and the town of Chaitén was being relocated 10 km to the north of its current location: an example of how even a sudden-onset eruption can leave an area of considerable economic importance abandoned and unusable for many years while the threat of a further major eruption lasts. Deciding when an eruption is finally over is incredibly difficult and can elude scientists: a final layer of uncertainty that adds to the potentially large economic and social losses of a volcanic crisis.

Aside from the emotional desire of residents to return to their homes, the return of residents after an eruption crisis will depend heavily upon the availability of jobs and resources, the duration of the preceding unrest and eruption phases, the damage suffered and the hospitality of relocation areas. The economic deterioration of Basse-Terre in Guadeloupe during the crisis of 1976–77 prevented many residents from returning, which in turn inhibited economic growth and so on. Similarly, many Montserratians relocated in the UK and elsewhere would like to return, but are prevented by the ongoing crisis and lack of available jobs, homes or resources. The economic impact of the Montserrat crisis was considerable with gross domestic product (GDP) decreasing approximately twenty per cent for each of the two years following the crisis start (1996 and 1997), and the eventual destruction of Plymouth, the island's economic and administrative capital.

Structural mitigation of volcanic hazard

Methods for reducing, or mitigating against, volcanic risk include modification of the volcanic process itself. Engineering solutions have been used, with varying success, to modify the path, magnitude or impact of volcanic processes. Lava flow diversion attempts include earth dams (Etna 1983, 1992, 2001), aerial bombing (Mauna Loa, Hawaii 1935, 1942, 1975, 1976) and cooling lava flow fronts with seawater (Heimaey, Iceland 1973). While barriers and earth dams appear to be relatively successful at modifying or restricting the path of lava, the success of other methods remains unproven. Diversion of flows can be controversial where protection of one community may result in increased risk for another. The first historically documented attempt at lava flow diversion in 1669, for example, involved the deliberate (and temporary) rupture of a lava flow channel from Mount Etna, Italy that successfully slowed the flow towards Catania but resulted in lava being diverted towards the smaller town of Paternò.

Using engineering structures to modify or confine the path of lahars has had more success with the use of sediment retention basins and Sabo dams, for example around Usu, Unzen and Sakura-jima volcanoes in Japan. Unfortunately, such structures are expensive to build and maintain, and more simple levees or dikes constructed from unconsolidated earthen material are sometimes used with lesser degrees of success, e.g. following the Pinatubo, Philippines eruption of 1991. Artificial draining of crater lakes has been undertaken to prevent the generation of crater lake breakout lahars, e.g. at Kelut in Indonesia, although there may be associated social and cultural implications that prevent such action, e.g. Ruapehu, New Zealand. Degassing tubes were installed in the crater lake of Lake Nyos in Cameroon in 2001 as a first step towards preventing a repeat of the catastrophic release of CO_2 in 1986, which asphyxiated 1,700 people and 3,500 livestock (Baxter et al. 1989).

Modification of ash fall or pyroclastic density currents (PDCs) is, to date, close to impossible and protection must be taken in the form of evacuation, improved resistance of structures or, in the case of ash fall, face masks that prevent inhalation of fine volcanic ash. Nevertheless, training dikes with interior tunnel shelters have been proposed for dome collapse PDCs from Unzen volcano in Japan (Sigurdsson et al. 2000) and underground shelters were built on the slopes of Mount Merapi in Indonesia to offer protection from a PDC.

Box 28.1 The 2002 eruption of Nyiragongo volcano and the Goma disaster, Democratic Republic of Congo

Dieudonné Wafula Mifundu
Département de Géophysique, CRSN/Lwiro, Democratic Republic of Congo

Hazards from Nyiragongo volcano, in the Democratic Republic of Congo, have been studied since it erupted in 1977. Then, fast lava flows (60 km/hr) killed seventy people. In 1994 an international scientific team in collaboration with Zairian volcanologists issued the so-called 'Goma declaration', which called for increased risk reduction measures in the face of the potential threats posed by Nyiragongo volcano.

When the activity of the volcano resumed in 2001, scientists were able to identify precursory signs for a major eruption, i.e. changes in the lava lake's condition, shifts in the seismic activity pattern from intermittent small swarms of low-frequency earthquakes to tremors of larger amplitude, black smoke at the crater, increased temperatures along flank fissures and at the crater, increased gas emissions, and rumblings. On 17 January 2002 a major flank fissure eruption spewed high-speed lava flows down the slopes of the volcano.

Lava flows buried agricultural fields and houses in several villages nearby, then reached the city of Goma and Lake Kivu located 19 km away from the volcano. Some 145 people were reported killed while a further 250,000 were evacuated. Lava covered thirteen per cent of the city and damaged eighty per cent of the local infrastructure, including 5,000 houses and the entire business district. Thousands of families lost their livelihoods and faced soaring prices of basic goods.

The disaster's scope largely resulted from the region's political condition. For years, eastern Congo and neighbouring Rwanda had been torn by civil strife and armed conflicts, which resulted in permanent flows of vulnerable refugees back and forth. Most of these people sought refuge in the city of Goma, the population of which doubled in ten years without appropriate institutional support. By 2002 Goma was ruled by a rebel group, the Congolese Rally for Democracy, which failed to report to the population information about the eruption made available by the scientists.

There was no evacuation plan and further miscommunication amongst scientists, local authorities and the media, creating confusion amongst evacuees who fled in different directions, including to Rwanda. Many were afraid of seeking shelter in the official evacuation camps previously used for conflict-affected refugees, which had been swept by cholera outbreaks, although no epidemics were reported in the aftermath of the 2002 disaster. Shortly after the eruption many evacuees came back to Goma and many settled back on top of lava deposits.

The 2002 Nyiragongo eruption shows that volcanic monitoring and prediction on the side of the scientists, whilst crucial, is insufficient to prevent a disaster. There is a need for comprehensive risk reduction measures and contingency planning which consider the political setting and the difficulties inherent in the social and economic conditions of local communities. Since 2002 significant progress has been made with the strengthening of the Goma Volcanological Observatory, the activities of which include scientific research, hazard monitoring and warning, hazard and risk mapping, and awareness campaigns among local communities.

Improving the resistance of infrastructure (see Chapter 56) to volcanic actions can be an important alternative or additional method for mitigating volcanic risk. Assessments of building damage following the Pinatubo 1991 eruption (Newhall and Punongbayan 1996) and the Rabaul, Papua New Guinea 1994 eruption (Blong 2003) suggest that roofs with long spans (>5 m) or shallower roof pitches (<45°) are more susceptible to damage from ash fall loading. The construction type of the roof covering and support will also affect the resistance, i.e. reinforced concrete can resist larger ash loads than tiled roofs with timber supports. Upright 'props' that spread the ash load along roof support trusses and purlins are recommended to prevent collapse. Building houses on stilts or with some elevation above the surrounding ground can help to provide clearance from smaller lahars, while strengthened bunkers may offer some protection from projectiles, and possibly PDCs, for those caught in an eruption.

Research in other areas, particularly for floods, has shown that a reliance on structural mitigation measures often encourages more people to live with greater risk (Kelman and Mather 2008). Such mitigation measures are often expensive and do not address the underlying vulnerabilities and root causes of potential disasters (Wisner et al. 2004).

Volcanic risk awareness and communication of warnings

During periods of quiescence or low-level precursory activity

The majority of volcanic outreach campaigns are one-way; that is, information about the likely impacts and appropriate actions needed to reduce risks to life and property flow from professionals to those at risk/the public. Typically, outreach activities will be ramped up once precursory activity is detected. However, a range of factors determine the level and quality of volcanic risk communication. Risk communication is more likely to be developed where there is recent history of volcanic activity, and people and assets at risk. Also important, however, are the scientific and institutional capacities that enable risk information to be produced and disseminated. For example, Goma in the Democratic Republic of Congo has a large population at risk and a very recent history of volcanic activity; however, the lack of any functioning institutions meant that risk information could not be communicated prior to the eruption of Nyiragongo volcano in 2002. In comparison, Mt Rainier in the Pacific northwest of the USA has not had a lava eruption for approximately 2,200 years. However, the US Geological Survey and National Parks Service provide educational materials to the local community, government and emergency service with a particularly strong emphasis placed upon educational programmes in schools.

It is now well understood that one-way volcanic risk information does not often generate the changes in preparedness and behaviour that emergency managers and volcanologists would hope. Instead, successful volcanic education and communication campaigns are those that work with local communities, well before an eruption, taking into account their perspectives and livelihood commitments. Cronin et al. (2004) identified weaknesses in volcanic risk communication on the Pacific Island of Vanuatu stemming from considerable differences in perspective between the traditional beliefs of island-dwellers and external volcanologists and emergency managers. Cronin et al. (2004) conducted a participatory workshop in order to combine the islanders' local traditional viewpoints with science-based management structures. It was hoped that the alert system and volcanic hazard map created would be more readily accepted than earlier 'top-down' approaches, instigated by government and outside agencies. In addition, older members of the community provided information on their knowledge of precursory signs of activity such as sounds, smells, strange activity of ants and birds and even visions and dreams.

The exercise did not attempt to reconcile the different belief systems of the local actors but aimed to integrate all views to produce a shared understanding. Such participatory workshops can be crucial tools in integrating bottom-up and top-down approaches to volcanic management and providing dialogue between stakeholders who rarely work together (Gaillard and Maceda 2009).

During a crisis or precursory activity

Information on volcanic activity, such as real-time detection of volcanic hazards, volcanic unrest or the production of a new hazard map, must be disseminated effectively for it to be of any use. This is usually carried out through formal meetings between the scientists and officials, and through press releases from the authorities to the public via the news media, internet, radio, loudspeakers or even SMS to mobile phones.

Recent eruption crises have shown the importance of a structured and recognised communication framework, one in which the population are able to receive clear and straightforward advice and the scientists are able to concentrate on assessment of the eruption. Many improvements in the communication of important information between scientists and authorities and also between authorities and the public have developed over the course of the ongoing crisis on Montserrat. To aid decision-making through the provision of strategic scientific advice, the Scientific Advisory Committee (originally the Risk Assessment Panel) was set up in 1997 as an interim body between the scientists at Montserrat Volcano Observatory and the UK and Montserrat governments. In turn, Montserrat Volcano Observatory has developed a strong involvement in the information and training provided for the local populations about volcanology and the activity of Soufrière Hills.

During a volcanic crisis informal warnings among the population are widespread and often friends and family are among the most trusted sources of information (Haynes *et al.* 2008b). After the eruption of Mt Merapi, Indonesia in 1994, which claimed sixty-six lives, local people from different villages on the slopes of the volcano formed an early warning communication network (Sagala 2009). People are in contact with each other through radio communications and the system works in conjunction with official government warnings. Local people also assist with monitoring activities, which include observing volcanic signs alongside other indications such as animal behaviour, dreams and spiritual signs. Interviews conducted with residents after the 2006 eruption identified that the local monitoring and early warning networks were effective in reducing the risk (Sagala 2009).

Volcanic disaster risk reduction: moving forward

Traditionally, authorities have managed volcanic risk through monitoring and the provision of emergency services, notably evacuation, aid and rehabilitation. Increasingly, however, a more holistic and proactive approach to volcanic risk reduction, which integrates technical and community development based approaches, is emerging.

Due to the infrequent nature of hazardous eruptions and the variety and complexity of the processes involved, knowledge of and therefore preparation for the eruption risk is often limited. Risk assessments are valuable tools in these cases. The use of probabilistic analyses to characterise the uncertainties associated with assessment of volcanic eruptions has become increasingly popular. For example, event trees provide a range of possible hazard events, potentially incorporating exposure of local communities, livelihood assets and infrastructure, with probabilities of occurrence assigned to each node, or branch, in the tree. Volcanoes for which there is a good understanding of the breadth and associated probabilities of potential eruption impacts enable decision-makers to plan risk reduction strategies such as evacuation or long-term land-use planning and to estimate likely costs and times associated with rehabilitation.

Although often difficult to implement, modification of the exposure of elements at risk through long-term land-use planning, evacuation of people and assets, and access restrictions (ground, sea or airspace) can significantly reduce the hazard potential. In the case of Vesuvius, where more than 560,000 people live within the zone at risk from pyroclastic density currents, the creation of a national park around the summit in 1995, and more importantly the enforcement of no construction within its boundaries, has slowed previously unregulated construction on the slopes and consequently the increase in people and assets at risk. Unfortunately, although land-use planning is one of the most effective forms of risk reduction, it is also one of the most difficult to achieve. For example, following the almost total devastation of Montserrat by Hurricane Hugo in 1989, and despite scientific evidence that identified areas highly likely to be impacted in a future eruption, opportunities to reduce volcanic risks were not taken. Redevelopment was almost identical to that which had gone before, with all key infrastructure within easy reach of the volcano. There had been no eruption since the island had been settled and volcanic risk was not considered. In any case, given the inherent uncertainty, a judgement to make economically unviable changes to the island's infrastructural layout based on a highly uncertain event would have been unpopular and unfeasible. In addition, permanent or long-term resettlement can expose people to other or greater risks elsewhere (Kelman and Mather 2008).

Eruptions still occur unexpectedly at volcanoes not subject to active monitoring, for example the large explosive eruption of Chaitén in Chile in 2008. As a rule, monitoring of volcanic unrest based on a combination of geophysical, geodetic and geochemical methods, rather than reliance on any single technique, offers better chances of producing reliable physical models and an estimate of potential volcanic hazard. For those countries with limited resources, however, remote sensing and rapidly deployed teams of international scientists, like the Volcano Disaster Assistance Program, may play a greater role in monitoring unrest indicators than continuous in situ monitoring.

The losses associated with volcanic crises can be spread through disaster relief and/or insurance, known as 'risk transfer' (see Chapter 54), although the long drawn-out nature of volcanic crises can inhibit such action. For example, property insurance was withdrawn on Montserrat approximately two years into an eruption crisis that has so far lasted more than fifteen years. Reducing an individual's risk through insurance cover, hazard modification or improved infrastructure resistance is most effective in small-to-moderate volcanic events.

That underlying social vulnerabilities contribute to disasters has led to greater recognition of the need to integrate social development with volcanic disaster risk reduction (VDRR). For example, although the management of the 1991 Pinatubo eruption was largely considered successful, due to the large number of people evacuated it remains a humanitarian disaster with more than 1 million people suffering through displacement and loss of livelihood. VDRR entails a holistic, long-term approach that aims to reduce vulnerabilities by eradicating poverty and strengthening local capacities (public, civil society, government and non-governmental organisations) to prevent disasters. This approach follows the international agreement for managing disaster risks, the Hyogo Framework for Action (HFA) (2005–15), signed by 168 countries in January 2005.

Poverty can be reduced and people will become less vulnerable if economic growth can be stimulated and livelihoods diversified and supported outside of the high-risk zone. In the short term if a population is confident that their livelihood will continue while they are evacuated, or post-eruption, they will be more likely to evacuate (Kelman and Mather 2008). Ongoing risk

Box 28.2 Vulnerability of agricultural communities: the August 1991 eruption of Mt Hudson, Chile

One of the largest eruptions of the last century occurred in August 1991 at Mt Hudson in southern Chile. An eruption column 12 km high was produced and ash was deposited over approximately 8,000,000 hectares of agricultural land in Santa Cruz province, Argentina. Owing to the dry semi-desert conditions within much of the Santa Cruz province the ash could not settle and become incorporated into the soil. Elsewhere it was continually re-mobilised over a period of months to years (Inbar et al. 1995).

Wilson (2009; Wilson et al. 2009) undertook an interdisciplinary study to examine how the agricultural community coped with and adapted to the heavy ash fall. The study found that:

The eruption occurred at the end of a cold, stormy winter overall and a period of drought in Argentina. Therefore grazing, feed stocks and animals were already in poor condition. Most farms were further overstocked due to poor meat prices in the previous season putting strain on grazing and feed supplies. Horticultural farmers were less exposed as crops had yet to be planted or were in dormancy.

The major impact was on the health of sheep herds. Many sheep died of starvation as ash covered feed and caused digestive problems and worn-down teeth when consumed. Ash also silted up irrigation systems, leading to stock dehydration. Some sheep and cattle became blind due to abrasion from the ash. Fifty per cent of stock was actually lost in the Andean region and irrigated valleys, and ninety per cent in the large steppe region. In parallel while the initial impact for horticultural farmers was limited, wind-blown ash caused problems during spring growing periods.

No farmers had livestock insurance. Some evacuated their stock, which was expensive, and many sold their stock at reduced prices. The rehabilitation of soils was aided by irrigation, cultivation, stripping ash, spreading grass seeds and adding hay to increase the organic content. The success of these techniques depended on the thickness of ash, financial and psycho-social capacity of the farmer, and the location of the farm (climate, water availability, etc.). Horticulture farmers had better access to equipment and financial reserves and were more successfully able to cultivate the ash into the soil. Government assistance helped with the evacuation of humans and livestock, providing feed, credit and advice on the cultivation of ash. In Chile the government bought farms within a 60 km radius of the volcano.

Many farms were abandoned for a period of months to years following the eruption. Long-term farm abandonment was related not only to ash thickness, but also the capacity of the farming system to cope and recover. Mono-agricultural sheep farming in the Steppe region, which received comparatively less ash fall, was already stressed prior to the eruption. These farmers found it harder to cope and adapt and had little option but to de-stock. In comparison, some farms closer to the volcano that practised a horticultural and pastoral mix were better able to diversify. Due to greater capital these farmers were able to access technological improvements to cultivate and reduce the impact of the ash. They were also aided by more favourable farming conditions including greater access to water.

reduction in the Philippines provides an example of where people have adapted to reduce volcanic risks while remaining productive (Usamah 2010). Approximately 10,000 families needed relocation after Typhoon Durian (locally known as Reming) mobilised lahars around

Mt Mayon in November 2006. The Philippines government, with assistance from international non-governmental organisations, built new homes in a safer, elevated area and provided assistance with new livelihood opportunities. However, people have not moved permanently, instead preferring to remain in their old houses during the safe season and leave during the more dangerous wet/typhoon season or during periods of increased volcanic activity. During periods of relocation, people will commute when possible to continue their livelihood near their original settlement. This dual occupancy is not what was intended and some risks – especially those to livelihoods and livelihood assets – in the danger zone remain. Nevertheless, the level of risk to life and property has decreased, some livelihood diversification has occurred and people remain productive.

If people are able to continue their livelihoods elsewhere, or have diversified their livelihood and have the capacity to withstand economic setbacks, they are likely to be less vulnerable to disaster. The case study of the Mt Hudson eruption in Chile in 1991 explores the impact of ash on an agricultural community. The differential recovery of farmers impacted by the eruption, which in many cases was irrespective of ash impacts and distance from the volcano, demonstrates the importance of the underlying vulnerabilities and the economic and social health of the farming system (Wilson 2009).

Mitchell (2006) used a deliberative Future Search methodology on Montserrat and St Kitts to develop plans for a future less vulnerable to volcanic hazards. The sessions brought a range of government and non-government stakeholders together to identify, first, desirable outcomes and, second, the measures and actions needed to be taken to attain these outcomes by 2020. The strategy was novel in that in addition to volcanic risks it considered all disaster risks and climate change impacts.

National and regional governments and institutions must take greater responsibility for volcanic risk reduction. This requires real political will, functional institutions and a comprehensive understanding of VDRR practice. VDRR does not need great technical knowledge or significant financing and often it is just a different way of viewing the issue.

Conclusions

Volcanic processes are diverse in space, intensity and time, and such diversity complicates risk assessment and risk reduction. Volcanic risk and disasters may be prevented, or at least reduced in their extent, through a series of measures taken before, during and after an eruption. As discussed in this chapter, volcanic risk reduction includes actions taken to (1) improve understanding of volcanic processes; (2) reduce people's exposure to hazards; and (3) increase their capacity to anticipate, cope with and recover from volcanic impacts. This necessitates a new approach to volcanic risk reduction that integrates technical and community-based development strategies. It is fundamentally important that technical assessments of hazard and risk are undertaken and communicated to decision-makers and the public. However, research has shown that methods that integrate lay (indigenous and non-indigenous) and scientific knowledge are most effective in engendering shared understandings and trust. The participation of local people in monitoring and communication networks can be highly beneficial. While engineering measures can help to mitigate hazards, volcanic risks will be reduced most effectively through proactive approaches that address underlying vulnerabilities. This may involve measures to support and diversify livelihoods in order to withstand shocks and stresses, or to actively encourage development outside of high-risk zones.

29

Soil erosion and contamination

Salvatore Engel-Di Mauro

DEPARTMENT OF GEOGRAPHY, SUNY NEW PALTZ, USA

Introduction

Often soils facilitate the provisioning of food, fibre and drinking water, support housing structures and contain or neutralise harmful substances, among other benefits. However, sometimes soils have disastrous implications, like sudden, deadly landslides, erosion–induced, wind-blown dust choking crops and surface waters, and toxic compounds showing up in food and water. That is, soil properties can combine with environmental forces, including human activities, to cause the sort of soil movement and/or contamination that imperils human survival.

Soils have context-specific characteristics that must be well understood to mitigate disaster risk. They are ecosystems primarily constituted by microscopic organisms embedded in a matrix composed of mineral and organic matter, water and air. They have separable, interrelated and internally mixed layers (horizons) and molecular-level binding of mineral and organic substances (clay–humus complexes). Most soil particles tend to be small (<2 mm) and differ in shape, composition and electrostatic charge in the case of chemically reactive grains <2 μm. The extent to which soils enable life-fostering or life-destructive outcomes depends on their characteristics, which vary because of differing constitutive materials, which are progressively altered by various forces, including climate, organisms and seismic events (Gobat *et al.* 2004; Schaetzl and Anderson 2005). What follows is an overview of selected processes that can transform soils into hazards, and examples of ways in which people have experienced and responded to such processes.

Soil movement

Particles can be detached, moved and deposited at higher than rates of addition and weathering (mineral and organic matter breakdown). Loosened material is known as sediment, which is transferred to and accumulates in other environments (sedimentation). Whether the process presents a hazard depends on the amount, moving speed and contaminant concentration, the first two being a function of the characteristics of forces and soils involved. Forces include the strength of wind, water, certain organism populations and cataclysmic weather and seismic events. Their relative capacity to move soil particles or entire soils depends not only on how well soils themselves can resist movement, but also on the degree of interplay between those forces, which often magnifies soil movement. Since erosion is included in soil movement, the terms 'movement capacity' and 'susceptibility factors' are used instead of 'erosivity' and 'erodibility'.

Movement capacity

Seismic and extreme weather events can rapidly move large quantities of soil, but some organisms' activities are no mean feat. People, especially when using machines, can quickly obliterate entire soils by removing thousands of cubic metres of material. Movement by non-human organisms tends to be gradual, but can be considerable. Mound-building termites in the tropics, for example, are known to erect earthen structures more than 10 m high (other organisms, like earthworms, also move much material, but largely within soils). Sediment resulting from extreme events and organism action is often highly prone to further movement by water and wind.

When precipitation exceeds soil infiltration, water may carry particles away from the soil surface. Water erosion occurs through splash (loosening of particles), sheetwash (particles transported through thin, continuous film over smooth surfaces), rill or gully development (concentrated overland flow in small channels), and subsurface channel formation (piping). The degree to which rainfall can dislodge and move soil particles depends on duration, intensity, raindrop mass and size, velocity at impact and frequency. Higher values for these rainfall erosivity parameters generally mean greater potential for soil particle detachment and movement. Erosion is not limited to rainfall. For instance, melting snow can be even more effective than rainfall through sheetwash erosion.

When wind speeds are stronger than soil particle aggregation (i.e. ~30 km/hr), soil particles will tend to be dislodged and carried away. Wind strength is related to frequency, magnitude, duration and velocity, with higher values increasing the likelihood of movement. Winds move soil grains in three ways. One is by saltation, when detachable aggregates (peds) are lifted up and bounce on the soil surface. This induces surface creep, which is the movement of sand-sized particles or, if sand-sized aggregates, their break-up into smaller particles. Such fine sand and silt- and clay-sized particles (<0.1 mm in diameter) are typically lifted up and diffused into the air (suspension), where they can be transported to great distances by prevailing winds, if they are not rained out. Most nutrients lost to wind erosion are in the clay-sized fraction (mineral and organic). During a drought, particle suspension is largely what composes dust storms.

Susceptibility factors

Susceptibility is generally given by the degree of soil component aggregation, consistency (how well soil components stay together under pressure) and shear strength (degree of resistance to pressure applied along or on a soil or part thereof). These are mainly dictated by surface cover (height, structure and density of plant cover, along with surface unevenness), internal soil characteristics (e.g. percentage distribution of particles of different average diameter and degree of particle aggregation) and topography (e.g. slope angle and length). These parameters change over time, and increasingly as a result of human impact. Susceptibility estimations should be done regularly so as to correct for changes in the likelihood and extent of hazardous conditions (Bryan 2000).

Land cover plays a major role in restraining movement, even on steep slopes, with forests usually being the most effective. Yet relatively high erosion rates can occur in forests when there is little to no understorey vegetation. Particles can be loosened by splashing water drops falling from the canopy. Low or no vegetation cover usually leads to greater dryness and reduced cohesion, so rain droplets or winds can detach and pick up particles more easily and carry them down slope or aloft. In arid regions, though, where soil surfaces are mostly bare and often hardened, wind erosion can be minimal. The effectiveness of water erosion will conversely increase, even if precipitation is rare, depending on slope characteristics.

Susceptibility to rainfall additionally depends on organic matter and clay content, as well as root density. These affect soil-particle aggregate (ped) size distribution, particle cohesion and degree of soil moisture retention. Soils with low organic matter and/or low clay content tend to be movable because of looser peds, lower particle cohesion and less water-holding capacity. Something similar may happen when soil organic matter is broken down faster than it is replenished (e.g. when soils are ploughed and left exposed for months). This is because organic matter binds soil particles together into larger peds. When they fall apart, wind and water can detach and carry soil particles more easily. Otherwise, changes in the level and type of aggregation of soil constituents can also contribute to movement. For instance, frequent treading by heavy equipment or by large animals in confined spaces can lead to compaction, especially in clayey soils. Downward pressure turns peds into plate-like structures with little space in between, impeding both water and air movement. Compaction lowers water infiltration rates and so increases sheetwash and rill erosion (Håkansson and Voorhees 1998).

Topography provides a crucial, yet complex setting for overall movement. Long and steep slopes may promote high movement rates, even with plenty of vegetation cover, but they are geometrically irregular. The rate and fate of moving sediment is further complicated by surface unevenness. There may be a succession of different vegetation, rock outcrops and landforms, among other factors, that may break or enable further movement. These features are altered or shift in position over time due to the combined impacts not only of cataclysmic events, but organisms as well. For instance, holes from burrowing and fallen trees will trap some of the sediment, while logging will tend to facilitate overall movement. What is more, soils along a slope can have differing degrees of susceptibility, so the amount of material added during downward movement varies. Differences between slope portions and in slope characteristics help explain the general lack of correspondence between stream and coastal sediment and soil erosion upslope. Soil movement must therefore be measured at the source and not deduced from downstream data (Schaetzl and Anderson 2005). The dynamic properties of slopes over space and time also make for high variability in the extent and magnitude of hazards resulting from soil movement.

Human impact on and social responses to soil movement

Soil particle loss (fresh sediment) is actually necessary for making future soils and improving the quality of existing soils. Nevertheless, much justified concern has arisen for the unprecedented levels of human–induced acceleration of soil movement over the past several centuries, and their ecological and social consequences. The acceleration may be almost imperceptible in the case of subsidence and lateral shifts, which can threaten people with flooding and/or building collapse. It is sudden in the case of slope failure (see Chapter 25). Soils may also be made more susceptible to wind erosion, causing extensive and frequent dust storms.

Whether gradual or sudden, removals can increase risk for societies reliant on local soils for livelihood. Indirectly, it can pose problems further away through upward price pressures. In the former case, there are instances of sheet and gully erosion leading to nutrient depletion and crop yield problems, such as in Africa (Lal 1995). Much caution must be exercised, however, in imputing productivity decline to soil erosion because of the dearth of data on actually cultivated plots and the large number of confounding variables involved (Scoones 2001). Soil type and historical impact differences also lead to diverse effects, even under the same conditions.

Resulting eroded material can eventually build up along coastlines or in rivers and bury high-quality soils and even inhabited areas. Sedimentation can make bodies of water more dangerous as a result of shallower, in-filled beds (e.g. it will take less rainwater for river banks to be breached). Such soil movement increasingly endangers human life, but, as with sites affected by erosion, some people are endangered more than others because of social inequality and pressures exerted by far-away social institutions (Hewitt 1983b; Wisner *et al.* 2004: 279–80).

Nevertheless, soil movement prevention features in many communities and often without outsider intervention. Greater attention to indigenous soil conservation practices can improve soil movement mitigation efforts. In parts of Nepal, for instance, local inhabitants create and maintain vegetated terraces, thereby markedly reducing erosion rates and slope failure frequency (Gardner and Gerrard 2003). The case of the Machakos area, Kenya, also demonstrates the effectiveness of community-based technical innovations, including terracing, in reducing soil erosion (Brookfield 2001: 212–13). In other places, such as northern Shewa, Ethiopia, local practices, such as ditch construction, have had mixed results and could benefit from introduced conservation methods (Reij *et al.* 1996).

Subsidence and lateral movement

Soils may slump (subside) because of human impact, such as aquifer depletion, wetland drainage, underground mining and increased weight above underground cavities. Soils may collapse, dragging downwards whatever lies on top, or gradually sink, rendering an area more prone to flooding. The latter has, for instance, affected the Po floodplain (northern Italy) as an outcome of decades of groundwater over-pumping by industrial processors and farms (Carminati and Martinelli 2002).

Similarly subtle movements are due to the occurrence of clays that swell when watered and shrink when dry. The expansion and contraction cause cracks in buildings and roadways that can lead to their rupture. In semi-arid and arid regions, periodic rainfall can saturate air spaces and dissolve bonding agents (e.g. calcium carbonates, clays) between coarse grains, which collapse under their own weight (hydro-compaction). The soil volume shrinkage destabilises whatever is on the soil surface and can lead to their fall. Areas with silt- and clay-rich soils prone to freezing can experience frost heaving problems in winter and liquefaction problems in spring. Locating residences without any regard for soil type is one way in which people have been put at risk through housing structure instability due to such lateral movements, as in suburban Montevideo, Uruguay (Musso and Pejon 2006).

Wind erosion and dust storms

Wind erosion can compound the above-described effects. It is particularly pronounced in valley bottoms and large plains, where wind speeds rise in connection with major, unimpeded air pressure differentials. Wind erosion can lead to dust storms and, in affected areas, permanent soil loss. Often, wind impact has to do with prolonged water shortage occasioned by periodic droughts combined with human-induced susceptibility (see also Box 29.1). Droughts can also be exacerbated by the resulting atmospheric dust (Cook *et al.* 2009).

Box 29.1 Desertification

Covering about forty-seven per cent of planetary landmass, dryland ecosystems occur where the ratio of precipitation to potential evapotranspiration (P/PET) is less than 0.65. Excluding hyper-arid zones (P/PET <0.05), about thirty-nine per cent of the world's landmass is susceptible to desertification, defined as 'degradation of land in arid, semi-arid, and dry sub-humid areas ... caused primarily by human activities and climatic variations' (UNCCD no date). It is claimed to be a hazard affecting at least 250 million people, largely in poor countries.

Among the major obstacles to identifying desertification is context-dependence related to dryland ecosystem and land-use diversity, both of which change and at different rates. Another is in defining the main indicators. High rainfall variability makes vegetation unreliable for discerning trends. Soil quality indicators, though increasingly used, suffer from sparse and inadequate information, especially with respect to surveys and long-term monitoring. Consequently, clear differentiation of climatic oscillations from land-use effects is rare.

More fundamentally, the conventional definition is misleading, since dryland degradation occurs even without human activity. Indicator values also depend on the sorts of livelihood promoted (e.g. high-quality soil for hunting and gathering may not be suitable for export-oriented farming). Furthermore, desertification claims, based on tenuous evidence, have been used to impose measures such as livestock destocking and large-scale irrigation systems. Some of these projects have actually made people more vulnerable to hazards, like famine and soil acidification. This is not to deny or minimise desertification as hazard; rather, it is to caution against facile claims and their possibly dire consequences through policy implementation.

Land use intensification in the North American Great Plains

The 1931–39 Dust Bowl features among the most notable cases of wind erosion. Decades of aggression against Native Americans opened up grasslands to intensive land use in the drought-prone and windy North American Great Plains (winds can exceed 100 km/hr). Reduction in plant cover and organic matter, along with aggregate break-up, raises soil susceptibility where even conservation tillage may be insufficient in preventing mass deflation (Merrill *et al.* 1999). As in the 1880s, such practices turned drought into catastrophe. The erosion of bare, desiccated soils led to large and frequent dust storms, affecting especially sandier soils. Airborne dust became a major respiratory hazard, especially among the poor. The displacement of millions, mostly because of farm foreclosure, exacerbated a coincident massive economic debacle.

Under much pressure from below, governments responded by introducing social welfare, massive farm subsidies (which worsened rural inequalities) and soil conservation programmes. Such policies and the development of large-scale irrigation and pumping technologies averted calamity during the even more severe 1950s drought, but less formidable subsequent droughts have at times been harsher because of high-yield–oriented farming. In the mid–1970s farm businesses capitalised on high grains prices by destroying protective features (e.g. hedges) or abandoning conservation techniques (e.g. strip cropping) to expand productivity.

Such human-accentuated droughts continue to generate periodic dust storms, which present major driving and respiratory hazards. In contrast, the drought-mitigating effects of irrigation systems have robbed water from many communities, while the Ogallala aquifer, one of the largest in the world, is now threatened by depletion and, in some areas, salt accumulation, known also as salinisation (Brooks and Emel 1995; Woodhouse 2003).

Farmland expansion in northern Kazakhstan and western Siberia

Similarly devastating was the 1954–64 USSR government policy to expand grassland cultivation in northern Kazakhstan and western Siberia. The so-called 'Virgin and Idle Lands Programme' (VILP) involved further expulsion of pastoralists and gatherer-hunters from areas characterised by wide-ranging inter-annual precipitation levels, dry and high spring–summer winds (90 km/hr– 110 km/hr), and recurring droughts. A primary incentive for the VILP was a combination of inter-imperial rivalry with the USA (especially military, requiring large capital outlays), and a need to raise grain production and augment meat and dairy availability (e.g. wheat in the steppes allowing fodder crop specialisation in other regions) to pacify an increasingly urban and wage-dependent population.

Eventually, more than 40 million hectares of grassland came to be cropped with the help of 650,000 colonists, despite specialists' reservations based on major crop failures in the 1930s. Worse, the VILP featured an anti-fallow campaign, eradication of pre-existing vegetation (especially with mouldboard ploughs), machinery-induced compaction and constant pressures on state and co-operative farms to maximise yield and forsake field rotation, contour-ploughing and other soil-protection techniques. Matters were little helped by limited irrigation, and agrochemical distribution and application difficulties.

A prolonged drought (1961–63) led to lower productivity and more frequent dust storms. Approximately 7 million hectares of newly established cropland had to be retired. The VILP also led to land concentration under fewer, larger state farms, mostly at the expense of remaining pastoralists and co-operatives. The calamity turned the VILP into a net financial loss for state coffers, spelling more dependency on North American imports, and contributed to the removal of Khrushchev by rival forces. Government response included the institution of a special soil erosion programme in the Ministry of Agriculture. Importantly, continued expansion of farmland contributed subsequently to reducing grain import dependence, when drought hit other regions. Impacts on grassland soils were reduced with the introduction and spread of minimum tillage, spatial alternation of annuals and perennials, and re-establishment of grasses, among other soil conservation methods. Previous practices were also reinstated, such as crop rotation, contour ploughing and tree planting for shelterbelts, but the regions remain nevertheless highly vulnerable to wind erosion (McCauley 1976; Zonn et al. 1994).

Soil contamination

Soil movement is inseparable from contamination. Detached and relocated material can contain toxic levels of compounds and/or pathogens. Dust from eroded soils, for instance, has been found to carry hantavirus, tuberculosis and anthrax (Griffin et al. 2001). Such losses can also greatly modify soil properties and lead to decreasing contaminant neutralising capacity, especially with organic matter reduction (Fullen and Catt 2004: 152–53). Toxins diffused through soil movement are often now the result of bombing raids, spills, leaks and other military-industrial hazards, but they may also be from non-human sources (e.g. volcanic eruptions). Natural hazards like hurricanes can dislodge large quantities of contaminated soil. Contaminant diffusion can affect not only disaster-stricken communities, but end up in places further away, where, sometimes, disaster survivors are resettled.

The politics of industrial, military or landfill relocation, to name a few processes, can be just as effective in spreading contaminants. For instance, wage-dependent communities largely chained to a polluting factory for employment are economically eviscerated (but health may not necessarily improve) by factory business migration to another community where unionisation rates and taxes are lower and/or environmental policy enforcement is weak or absent. Such occurrences reduce human welfare and long-term health in one community and introduce new sanitary threats in another without even offering much economic improvement. Soil contamination is then intimately tied to issues of health and livelihood well beyond the immediate sites affected.

Conventionally, contamination refers to human impact inducing concentrations of substances or pathogens above expected levels. Contaminants become pollutants when crossing established critical loads, beyond which harmful effects start to occur for a specific function in an ecosystem (e.g. plant growth rate) or human health. Organisms intolerant of high contaminant levels are often used as bio-indicators of soil pollution (Gobat et al. 2004: 143–46). Nevertheless, distinguishing between soil contaminants and pollutants is hindered by the absence of universal standards and the complexity of chemical interactions (Jennings 2008). One should be alert to the motivations or consequences related to the use or questioning of standards.

Among the main contaminants are heavy metals (elements > 6 g/cm^3 density), metalloids, organic compounds (e.g. biocides, petroleum and its by-products), inorganic compounds (e.g. synthetic fertilisers), pathogens and radionuclides. Sites most exposed to contamination tend to be industrialised urban settings, processing and energy-producing plants and environs, military installations and war-affected areas, mining and oil drilling operations, agro-chemically intensive farming and ranching zones, and high vehicle-traffic corridors. Once discharged onto or into soil, a substance or pathogen may remain a contaminant or become a pollutant depending on a variety of factors that can lead to differing rates of breakdown (or persistence), flow and storage within soils, and to diverse pathways and levels of concentration from soils to other environments, like groundwater and air.

These factors can be generally divided into the amount and properties of a substance or pathogen, the properties of a soil (including soil-dwelling organisms' characteristics) and prevailing environmental circumstances (e.g. timing and duration of rainfall). Some contaminants, like non-polar compounds of low molecular weight, may be volatilised or washed away instead of entering a soil. Contaminants that do infiltrate become hazardous to people through direct contact or assimilation (e.g. epidermal absorption or breathed-in dust), or by consuming food and water derived from such soils (Singh 1998).

Human-induced soil contamination and social responses

Contamination can be initiated or exacerbated by human impact and threatens people in highly uneven ways. It has become especially pronounced and widespread over the past two centuries, with often long-lasting effects that are also somewhat independent of human action. For instance, the 1986 radioactive release from nuclear reactor meltdown in Chernobyl, Ukraine, featured windborne radionuclide diffusion to areas in northwest Europe, affecting thousands of hectares, but it especially affected soils containing relatively high levels of reactive clays and organic matter, to which such radioactive substances bind more easily (Singh 1998). Soils, alongside other environmental factors, play crucial roles in determining whether and to what degree the introduction of contaminants threatens people's well-being. At the same time, it is social processes, through often highly asymmetrical power relations, that are behind contaminant production, diffusion and differential social impact.

Hazardous chemical production and disposal

Waste management provides a salient example. In 1982 North Carolina's government elected to transfer mostly PCB-laden soil from polluted roadsides to a landfill in Afton, a poor community with an African- and Native-American majority. PCBs are persistent organic compounds because of their resistance to breakdown and tight binding to organic matter, but molecular weight and chlorination level, among other factors, also affect their leaching and volatilisation (Whitfield Åslund et al. 2007). The decision sparked local residents' indignation and they organised against the potential threat of groundwater contamination. After prolonged struggle, involving direct action protests, mass arrests, co-ordination with other activists (including scientists) and legal recourse, the inhabitants succeeded, by 2003, in closing the landfill, having neutralisation occur in place and receiving public investment in local employment. Still, there have been no reparations awarded, nor any investigations on repercussions of soil removal from affected roadsides.

What sets this case apart from preceding ones, such as the soil contamination-induced protests and evacuations at Love Canal (New York, USA), was the eventually worldwide reverberation of an initially localised response. Though many other communities suffering from similar socially imposed hazards had already articulated the linkage between environmental and social justice issues, the Afton struggle gained enough notoriety to spur other systematic studies and consciousness-raising campaigns from below. Eventually, besides spurring the development of movements against environmental racism, it even led to US government formulation of environmental justice policies, to the international adoption of the concept among both policy-makers and activists (contributing to the 1992 Basel Convention), and struggles against international toxic waste transfers to poor countries (Adeola 2000; McGurty 2007).

As protests were continuing in Afton, one of the worst industrial calamities in history occurred in Bhopal, Madhya Pradesh, India. Between 2 and 3 December 1984, tonnes of methyl isocyanate, a compound used to produce carbaryl pesticide, were released, among other reaction products, from a Union Carbide plant, killing thousands of people and permanently debilitating the health of hundreds of thousands of survivors and their progeny (Mishra et al. 2009). The plant was closed shortly after the disaster. Years of prior wastewater discharge and post-closure leaking stockpiles have contaminated local soils and groundwater with carbaryl, hexachloro-cyclohexane and other persistent organochlorines, alongside chromium, mercury, nickel and lead above safe levels. Nearby residents, already suffering from exposure to toxic gases in 1984, remain at risk for neurological, hepatological, reproductive, endocrine and gastrointestinal damage, among other health effects, through soil contact, dust inhalation, water use and bioaccumulation (Johnson et al. 2009). Sharply contrasting the Afton case, the struggle continues for clean-up, reparations and sentencing.

Indirect contamination

Human activities can lead to raising levels of pre-existing or introduced substances or pathogens. Salt build-up (salinisation), for instance, often leads to lower soil permeability, toxicity and water shortage for most plants, along with calcium and other deficiencies. It heightens soil susceptibility to both water and wind erosion, as all but the most salt-tolerant species are driven out, and can yield potential water contamination, among other hazards (Szabolcs 1998).

The main source is water-soluble salts from soil-forming material and additions, especially in areas with shallow groundwater. Though mostly alkaline, salt-affected soils can actually reach extremely low pH (< 4), especially in coastal wetlands, where oxidation of pyrite and other sulphur-rich materials generates sulphuric acid. Salinisation generally occurs under a wide range of environmental conditions (including tropical and mountainous areas) as part of soil formation, but it can be accentuated if not caused by human activities.

In arid regions, salinisation can result from irrigation and/or groundwater extraction, or from the drying out or interruption of an inland freshwater source through fluvial diversion or damming, or lake and wetland drainage. Salt-affected areas have expanded with the diffusion of large-scale irrigation projects over the past decades, including the above-cited Great Plains. This undermines crop production in ten per cent of world irrigated cropland, with particularly deleterious consequences in terms of livelihoods in places like the Indus Valley, Pakistan (Fullen and Catt 2004: 37–38).

In coastal wetlands, soil drainage, often due to construction, will raise acid sulphate levels such that, for instance, heavy metals can contaminate water supplies and materials in building foundations can be corroded. In fact, any activity affecting soil water dynamics or composition in such environments, such as wetland drainage, deforestation and industrial emissions, can lead to salt build-up, including that of acid sulphates. For instance, large-scale irrigation schemes imposed through foreign economic pressures along the Gambia River have resulted in salt-water encroachment (salinisation). Attempts to reduce the impact by building tide-blocking dams in the 1980s have activated local acid sulphate soils, threatening long-term crop yield (Carney 1991).

There are many other ways in which soil contamination results from indirect forms of human impact, but few studies explicitly link these two processes and even fewer analyse their social determinants. Perhaps this relates to the often formidable logistical difficulties involved and the usual exclusion of such phenomena from mainstream discussions. At the same time, it might be useful to consider social responses like those in Bhopal and Afton (see also Box 29.2) as struggles exemplifying the tight connection between soil protection and health risk. This could be one

---

Box 29.2 Soil contamination in the Niger Delta, Nigeria

The Niger Delta, among the world's largest coastal wetland and mangrove swamp areas (ca. 26,000 km^2), is home to millions of people belonging to more than 100 minority ethnic groups. Their livelihoods depend primarily on subsistence farming, gathering and hunting. Soil types include highly weathered clay, organic, and clayey and sandy alluvial. On-shore drilling, piping, transport and refining of oil and gas have contaminated soils for decades. Continuous gas flaring and thousands of spills are major sources of surface contaminants, including polycyclic aromatic hydrocarbons and heavy metals tens of thousands of times World Health Organization (WHO)-recommended levels. Among the immediate health consequences have been child deformity, liver damage and dermal diseases, but carcinogens and heavy metals will contribute to making many lives short and painful for current and future generations. Soil pollution is also undermining water quality and agricultural productivity through crop destruction and soil nutrient decline, sometimes on land prone to acid-sulphate soil activation. Health hazards compound mass pauperisation and underdeveloped infrastructure resulting from livelihood destruction, heightening economic dependency on external institutions and militarised appropriation of most oil and gas profits by transnational corporations and central government. Such injustice, recently recognised in international courts, has been met since the 1970s by mass protests, legal action and armed struggles. These have often been brutally suppressed by the central government, as oil exports provide the bulk of state revenues and feed the economies of influential, arms-trading countries like the USA and China.

way of raising awareness about both direct and indirect forms of contamination more effectively into the realms of, for example, activism and policy-making.

Conclusions

Soils should matter a great deal to those interested in disaster risk reduction. Soil movement and contamination result from evolving interactions among soil properties and environmental forces, including human activities. As illustrated above, soils can be inherently susceptible to movement or can be altered to become so. This exacerbates disasters like earthquakes, even of minor scale, but soil movement can itself turn into calamity, such as dust storms. Some soils, such as acid-sulphate soils, release toxic compounds with often human-triggered changes in environmental factors. Some soils offer little or no buffer against anthropogenic contaminants and so pose greater health risks through toxin bioaccumulation. Sometimes deleterious consequences endure, as in the case of abandoned, leaking industrial sites. Contaminated soils may also become susceptible to movement and spread contaminants far and wide. All such potential soil-borne or soil-contingent hazards should be considered when evaluating disaster risk for communities.

Abating the role of soil as disaster risk would require, at a minimum, the development of infrastructure and institutions to inventory soils and monitor their status and properties. This would have to be supplemented by information gathering on factors related to movement susceptibility and capacity as well as contaminant sources. Resulting databases and reports would then have to be integrated with other data on risk so as to make evident the links between what goes on in physical environments and social settings, and avoid compartmentalisation of knowledge, which can be lethal (e.g. raising productivity and maybe income levels among smallholding farmers by providing technologies and incentives that eventually turn soils into a source of harm).

Such data and monitoring abilities, and therefore warning systems, do not exist in most places. Building such capacity entails a massive redirection and redistribution of resources within and among places, necessitating much greater international co-operation and co-ordination than currently exists. The development of such capacity would greatly help in establishing early warning measures or improving on existing ones. An example would be in assessing, in advance, the likelihood and probable direction of a landslide with data on changes in soil susceptibility and movement capacity magnitude (e.g. precipitation intensity and duration). Another example would be, aside from existing applications like determining contaminant travel time from soil surface to groundwater, the thorough examination of potential soil contamination risk prior to construction, so as to avert the introduction of soil contamination disaster risks.

It is equally important to understand soil movement and contamination as linked to both locality-specific and broader social processes. Human impacts from some societies over the past couple of centuries have increasingly contributed to turning soils into hazards. Social institutions, especially globally influential ones, sometimes create the conditions for unprecedented impacts on soils and, when inducing and reinforcing economic inequality, simultaneously tend to impose human-accentuated risks on the least powerful. At the same time, there are many cases of indigenous practices that help reduce or prevent soil movement and contamination, but these depend on what happens within and between households and communities (e.g. Brookfield 2001; Carney 1991; Reij et al. 1996). Hence, aside from analysing interconnections across places that impose risks on some societies more than others, there must also be close scrutiny of class relations and of social status processes (ethnicity, gender, sexuality, age and physical ability differentials) within affected communities. Identifying the most vulnerable and the most responsible for creating or aggravating situations of vulnerability (intentionally or not) is hampered by not knowing who does what and where, who pressures whom to do what, and who is made more vulnerable where, to what, and under what social conditions.

For these reasons, it is imperative that those most directly affected by the risks imposed by soil hazards be empowered (e.g. wealth redistribution, livelihood options and participatory processes), so that they effectively partake of decision-making processes with those involved in hazard analysis or monitoring and policy formulation or enforcement. Simultaneously, funds must be made available and legal provisions introduced that defend people whose lives are threatened especially by soil contamination. This might be politically difficult, but if military-industrial hazards are ignored, then the effectiveness and even credibility of disaster risk reduction as an approach is much diminished. As a concrete example, imagine how effective risk reduction would be if one were to dedicate efforts to anything but such hazards in the case of communities devastated by Hurricane Katrina, in a zone where soil-embedded carcinogens and all sorts of other industrially produced contaminants abound and have been diffused through the landscape by the catastrophic force of wind and water.

Biological/ecological hazards

30

Human epidemic

Chris Dibben

UNIVERSITY OF ST ANDREWS, UK

Introduction

Epidemics and disasters have many connections. First, an epidemic may be considered a disaster in its own right, creating mortality, economic loss and social disruption on a scale comparable to other grave emergencies triggered by natural hazards. Second, epidemics may contribute to secondary events such as food shortage when disease-affected people are unable to farm or work. Third, some disasters such as floods may be followed by epidemics of water-borne diseases, and displaced people sheltering in unsanitary conditions have been known to suffer epidemics of cholera.

As with other forms of disasters, there is considerable inequality in the impact of epidemics and striking differences in the responses available for and played out upon different groups. The vividness of particular unfamiliar or devastating epidemic diseases – the Black Death, Ebola or 1918 flu – may hide this fact, leading to epidemic diseases being seen as equally destructive across populations; no respecters of class, culture or wealth. This chapter will examine epidemics and societies' responses to them. It will take a critical perspective, discussing not only the accepted science and dominant modes of response but also the extent to which an epidemic is a socially produced phenomenon.

The biology of an epidemic

An epidemic is usually defined as an unusually high incidence of disease within a population. It is unusual because it is relatively new to a particular place or has not occurred for some time. Infectious diseases are the most usual form of epidemic disease, although the term is not solely restricted to them. It is their ability to 'infect' and therefore to suddenly affect a population that means they produce unexpected illness and death. It is natural, therefore, to focus on the nature and spread of infectious diseases in order to understand epidemics.

Pathogens or infectious agents have a fairly simple life cycle of release from one host, transport in the environment and then entry into another host; however, the nature of that transport varies greatly. In general, two main forms of transmission can be identified. First, spread can occur through propagation or movement along a chain of infected hosts. Propagation between humans can occur either directly through acts such as kissing, touching or sexual intercourse, or through the air in microbiological aerosols such as dust or droplets. Second, it can spread through a common vehicle of infection. Water, for example, can carry cholera-causing bacterium *Vibrio cholerae*. Common vehicle transmission is normally either through a common material or a vector such as insects.

There are a variety of types of pathogens; the most significant for human populations are bacteria, viruses and parasites. Other important agents include rickettsias, fungi, chlamydias and prions. Bacteria, unicellular independent micro-organisms, spread through asexual division. This division is in many cases rapid and this can lead to bacteria adapting relatively quickly to changing environments through a process of mutation and selection. This can allow, for example, strains of antibiotic-resistant bacteria to evolve over time.

Viruses are much smaller than bacteria. They are really only a simple bundle of infectious genetic material. Unlike other pathogens, viruses are incapable of independent reproduction. Instead they rely on invading the cells of a host and using those cells to reproduce themselves. The need for the virus to invade a host cell means that to a certain extent viruses are species specific. They need to be able to bind with a host cell and are therefore unable to make this connection to other types of cells. However, as with bacteria, mutation makes it possible for certain viruses to species jump. It also appears possible for intense sustained contact to allow viruses to cross species barriers. This species jumping is highly significant for epidemic disease, as viruses for which humans may not initially have a great deal of immunity are able to enter into populations and therefore cause sudden rises in morbidity.

Parasites are more complex organisms than either viruses or bacteria. They have various stages to their lifecycles, importantly some within hosts, others outside. Malaria is a highly significant parasite for human populations. It is hard to treat parasitic infections in humans because the complexity of their cell walls makes it difficult to find drugs that are toxic to the parasite but will not cause significant harm to the infected person.

An understanding of the development of an infection within the human, from invasion, to infectivity, through morbidity to recovery or possibly death, is key to disrupting diseases that spread through propagation. What is particularly important is the period of infectivity and the extent of inter-personal mixing during this period. If infection can occur during the period of incubation (i.e. the time between invasion and onset of illness), then it is often harder for the individual or other authorities to halt spread because they will be unaware of the threat unless a process of screening or testing has been implemented. Influenza and HIV are examples of two viruses that can be transmitted before the infected individual is exhibiting signs of illness. On the other hand, tuberculosis cannot be spread during the incubation period. Diseases that become infectious only when a person becomes ill are to a certain extent self-limiting. For example, the severe acute respiratory syndrome (SARS) outbreak in 2002 was constrained because the virus appeared only to be able to spread once people had become ill, at which stage they naturally reduced their contact with others, and similarly public health authorities were able to identify cases. In contrast, the 1961 'El Tor' cholera outbreak spread very widely across the world because of the high number of carriers who were infective but also asymptomatic. The overall spread of a contagious disease within a population is then determined by the number of infective contacts that can take place based on the 'density' of populations (i.e. the spatial and temporal concentration of possible contacts) and the extent of immunity or barriers to infection existing.

Epidemics resulting from diseases that require a common vehicle for transmission are usually a result of an environmental change in that vehicle, the contamination of water with the bacterium *Vibrio cholerae* or a change in climate allowing the malaria-carrying mosquito to survive in a different climate. In the case of *Yersinia pestis* bacterium or plague, it is the movement of the rat carriers. The spread is as a result slow, some 15 km–20 km a year (Cohn 2008). Changes in the environment can result in previously unknown diseases affecting areas of the globe. After exceptionally mild winters in 1998 and 1999 the West Nile Virus appeared to have infected birds in New York and then the population of the city, causing cases of encephalitis and meningitis. Since then the virus has spread to the rest of North America. Major anthropogenic changes to the environment, such as dam building, can also provide new habitats for disease 'vehicles'.

Important dimensions of epidemics

Poverty and power

The unequal distribution of power across societies produces vulnerabilities that infectious diseases can exploit. The bodies of the poor and powerless, some weakened by malnutrition, other diseases or unable to resist the necessity for risky behaviours, allow infectious diseases to thrive, providing spaces where they can spread.

Even with a highly virulent and pathogenic virus such as the 1918 flu, which killed perhaps 50 million people, there is strong evidence for its socially patterned impact. There was a very large thirty-fold difference in excess mortality attributable to the flu experienced in countries around the world (Murray *et al.* 2007). At least one-half of this variation could be explained by differences in per capita income: if you were relatively poor, your chances of survival were very much worse than if you were relatively wealthy. In India it was likely that malnutrition and pre-existing illnesses were implicated in the very high mortality rates suffered amongst the poor (Schoenbaum 2001).

Frequently dominant discourses, whether they are in the media, amongst policy-makers or even in the research community, exaggerate the extent of the personal agency of the poor and powerless. Their 'risky' behaviour is often attributed to either a lack of knowledge or recklessness. During the famines of the nineteenth century in India, which decimated the population as a result of starvation and the infectious diseases attacking weakened bodies, the colonial authorities blamed the fecklessness of the afflicted for their own plight rather than colonial mismanagement and worse. Blaming the individual for their own action allows societies to avoid taking the more fundamental and therefore painful steps of carrying out significant reform.

The recasting of the poor as part of the problem is a natural part of the reinforcing of disease discourses. The use of new protease inhibitor drugs for HIV/AIDS with the poor in New York in the 1990s was discounted in dominant discourses because it was argued that non-compliance, due to chaotic lifestyles, would be high and there would, therefore, be a growing resistance to these new drugs (Farmer 2001). In other words, the act of giving these therapies to the poor was recast as a threat-increasing action.

The intersection of poverty, gender and ethnicity is an important juncture for understanding epidemic disease (see Chapters 35 and 38). The majority of women around the world infected with HIV/AIDS are poor. Poor women are particularly vulnerable to infectious disease, due to a combination of gender politics in their societies that deny them access to what capital exists in their situation of poverty. They may also be vulnerable to sexual violence and therefore sexually transmitted diseases. In one study in South Africa, thirty per cent of women said that their first intercourse was forced, seventy-one per cent had experienced sex against their will and eleven per cent had been raped (Whelan 1999).

In many countries particular ethnic groups experience a vulnerability to infectious diseases due to the racism to which they are subjected, and also because of the consequent poverty they experience. The institutional racism they experience can be exemplified in the reporting of their deaths, so for example Briggs (2004) assessed that there had been some 500 deaths from cholera between 1992 and 1993 in the Orinoco Delta in Venezuela. However, only thirteen deaths were reported to World Health Organization (WHO) officials – the other deaths were effectively 'erased'. This institutional response was simply part of a process of blame for the cholera deaths that was placed particularly on *indígenas* (indigenous women) for failing to save their family members.

Globalisation, modernity and epidemics

One of the key features of the modern world is the extent to which space, as a barrier to and consumer of time within systems of human communication and exchange, has been increasingly weakened by technological advancement. The resulting increase in interconnectedness of people across the world is often referred to as globalisation. The motivation for this 'compression' of time and space is the benefit it brings to capitalist systems of production and in particular the extent to which it allows an increasing circulation of capital. This benefit has certainly been very important in the economic growth seen particularly within industrialised countries in the last 200 years. However, because space and distance is an important protective factor at a number of different levels to infectious diseases, space–time compression has also as a by-product created new spaces of vulnerability to epidemic diseases. It in effect brings reservoirs of disease (place specific – i.e. particular livestock settings) close to everyone (see Chapter 31). It makes possible the sudden occurrence of previously unknown and, as they have become known, 'emerging diseases'.

Although the increased interconnectedness of people across the globe has accelerated in the last 100 years, this increase has been a fairly constant feature of human history. The Plague of Justinian (541 CE–542 CE), for example, occurred within a collapsing Roman Empire (Marks and Beatty 1976). Similarly, it is likely that the speed with which the 'Black Death' was able to devastate European populations in 1348–1400 was a result of the increasing trading contacts, particularly within the Mediterranean basin but also to isolated places such as Iceland and Greenland.

Epidemics as agents of change

Epidemic disease has led to dramatic transformations throughout history. Crosby has argued that the domination of European powers over the Americas after the arrival of Columbus and the Spanish in 1492 was not the result of superiority (cultural, intellectual or racial), or simply to do with technological or economic factors, but was vitally linked to ecological factors. 'Where imperialists have been successful not simply in conquest but also in settlement, they have done so with the indispensable assistance of the life forms they brought with them' (Crosby 2004: 368). Crosby argued that empire led to the creation of 'neo-Europes' – temperate colonial regions dominated by those of European descent and European diseases. It is estimated that of a pre-contact native population of about 54 million, by 1600, after only 100 years of contact with the Spanish, this population had dwindled to 5 million–6 million (Denevan and Lovell 1992).

Europeans, unlike the native populations of the Americas, had become used to living in unsanitary towns and cities in close proximity to livestock and other reservoirs of disease (Crosby 2004). When the populations of South America were exposed to these European diseases, they in contrast had little or no immunity. This happened in other but not all places touched by European imperial expansion. So although part of the 'conquest' of the plains of North America owed something to military might, a very significant role was played by flu,

---

Box 30.1 Neoliberalism, Washington Consensus and vulnerability to epidemic diseases

It could be argued that the economic policies followed by the leading financial institutions of the world, the World Bank, International Monetary Fund (IMF) and World Trade Organization (WTO), and supported by the leading industrial nations over the last thirty years, known as the Washington Consensus, have done more to (re)produce spaces of vulnerability to infectious diseases than remove them. Structural Adjustment Programmes (SAP) initiated in response to the debt crisis affecting countries economically battered by the various financial crises, oil shocks and recessions of the 1970s and early 1980s have been seen as particularly problematic. In the 1950s and 1960s many countries of the global south had been experiencing economic growth and were developing health, social and educational sectors along the lines of Western welfare states. This was affordable because of the healthy trade balance that existed at this time, with a strong exporting sector. However, this trade balance was upset in the 1970s as recession in the industrialised world led to a fall in demand for goods. Many countries of the global South had instead to turn to lenders to finance, initially in the short term, their existing welfare provisions. As the economic turmoil of the 1970s continued, these countries found it increasingly difficult to obtain further loans from private sources, often necessary simply to finance existing debt, as the risk associated with such lending was seen as too high. Instead they had to increasingly turn to the IMF and World Bank, which insisted on economic and social reform or 'stabilisation programmes', later termed SAP, as part of their agreements to lend.

The forced cuts in public spending combined with the disruption of rural livelihoods associated with SAP meant new spaces of vulnerability to epidemic diseases were created. Health care systems were badly affected by cuts. User fees were introduced in many countries leading to falls in utilisation, and two-tier health care systems developed, with the middle class fleeing the public-funded health care system into the private health care sector. The disruption of rural livelihoods, associated in part with trade liberalisation and competition from highly industrial farming systems led, ironically, to less food security and more poverty amongst rural dwellers who were then forced to migrate to urban areas in search of work. The urban spaces they moved into were often socially and economically marginal and unsanitary. Within this context vulnerability to epidemic disease increased.

---

measles, typhus and smallpox. However, in the Philippines this did not occur because of the long history of domestication of animals and trade with Eurasia. In Asia, colonisation was much less successful, with the native populations of these countries fairly rapidly becoming able to resist European domination.

Epidemic disease has had transformative impacts at other times. Cholera and other infectious diseases were a central part of why a European late nineteenth-century sanitation revolution took place, when the social conditions of citizens particularly in cities came to the fore. The 1918 flu, though devastating in its impact, may have helped end WWI, as Germany, experiencing the second more virulent wave of the pandemic, simply ran out of soldiers (Quinn 2008). It was also a significant motivator for the formation of the League of Nations Health Organization, the predecessor of the WHO.

War, disasters and epidemics

War and other disasters typically intensify already existing vulnerabilities to infectious diseases. Women, for example, are particularly vulnerable to sexual violence in refugee camps or at the hands of armed groups. In refugee camps, the normal forms of protection from diseases are frequently absent, such as mosquito nets, effective sanitation or clean water. After the 26 December 2006 tsunami diarrhoeal illnesses, typhoid fever and cholera occurred amongst many of the displaced peoples in the affected region. In 1994 the Rwandan refugee camps located around Goma and Bukavu in the Democratic Republic of the Congo experienced a severe outbreak of cholera, resulting in about 70,000 cases and 12,000 deaths (Siddique *et al.* 1995). The area has since become one of the world's most active foci for cholera in the world (Bompangue *et al.* 2009).

The collapse of infrastructure that occurs in the midst of war means that normal public health practices such as vaccination and other public health processes halt. Since the invasion of Iraq in 2003 there has been a rise in tuberculosis, diarrhoeal diseases, measles and mumps (Dyer 2004), and therefore also in deaths (Roberts *et al.* 2004). Although generally causing hardship, wars do occasionally lead to a lack of population mixing and therefore limit some infectious diseases. In Mozambique and Angola the spread of HIV may have been limited by the civil wars in these countries (Strand *et al.* 2007).

Responses, reactions and prevention

The 'modern' public health response to epidemics

The dominant response to potential or actual infectious disease outbreaks typically involves forms of surveillance identifying threat and then the implementation of control over the host, vector, infected, environment or infectious agent. These controls are applied at different scales: some will involve the individual, others institutions, communities or even regions. Immunisation has been one of the most effective forms of control over infection in hosts, and the eradication of smallpox was arguably the most effective application of an active immunisation programme.

Chemoprophylaxis, or the use of chemical substances to prevent, halt the progression of or treat the symptoms of infection, is the other main form of medical control intervention. Antibiotics are a key tool in the fight against bacterial disease. The discovery of the potential of an extract from the penicillium fungi to kill bacteria in the 1930s opened the era of antibiotics. Bacterial infections such as syphilis and staphylococcus, which were serious and for which there were no effective interventions before the development of the antibiotic penicillin, could now be treated. The use of chloroquine to prevent the development of malaria, specifically to prevent malarial parasitaemia, is another example.

Other responses attempt to control behaviour. This may include, for example, attempts to change the way individuals have sex, work, eat or use drugs recreationally, in order to prevent infection. A significant number of infectious diseases are spread through sexual contact, including syphilis, HIV and *Chlamydia trachomatis*. The aims of public health interventions may focus on trying to encourage abstinence, limiting the number of sexual partners an individual may have or the use of various forms of protection from infection such as condoms. Amongst intravenous drug users the goal may be to encourage users to stop sharing needles. An important assumption often made within these types of public health campaigns is that the target of the policy has individual agency, for example, that a sex worker has sufficient power, within their relationship with a client and in the context of their working environment, to insist that a condom is used, even if they understand the reduction in risk of infection that it offers. This assumption is often questionable, particularly for the least powerful and most vulnerable in society.

Control is also applied routinely to the infected with the aim of halting the spread of disease. This control may be reinforced by legal sanction, not simply requests to behave in certain ways. It can take various forms, from the restriction of activities so that individuals with diarrhoeal diseases should not be preparing food, to forms of isolation. Enforced isolation or quarantine into hospitals and other institutions can and has been used to ensure that the infected do not come into contact with the non-infected; however, other forms of isolation – for example, not entering public places – are more common.

Though, of course, often effective in limiting spread, quarantine does involve the suspension of human rights. This may be ethically acceptable if it applies equally to all members of society; however, it has often been the case that the most powerful can resist these types of sanctions more easily than the least powerful in society. Even if there appears to be a clear and legitimate public health argument for enforced quarantine, the implications can be troubling. The case of 'Typhoid Mary', a cook in New York during the early twentieth century, is a notorious example. She was an asymptomatic carrier of typhoid and her status and also her refusal to stop working as a cook led eventually to her enforced quarantine for some twenty years until her death. Her position as a poor Irish migrant at a time of intense discrimination and, therefore, her treatment by the public health authority (as well as the fact that the idea of a 'well' carrier of typhoid would have been extremely odd to the public at this time) suggest that even in this case the ethics and social justice of the public decision-making might be questioned (Wald 1997).

Control can also be applied to the environmental 'vehicles' for diseases, with control over water through effective sanitation and the provision of safe water probably being two of the most significant for human populations. Direct protection can be provided from insects by using bed nets, sometimes treated with an insecticide; however, while this offers protection against malaria-carrying mosquitoes, it offers no protection against insects that bite during the day. The tsetse fly, a vector for the sleeping sickness, is active during the day. Instead, direct controls against the insects have to be used, for example insecticides or biological controls such as the introduction of sterile insects.

In order for controls to be implemented there needs to be a system of surveillance and then a structure that can implement responses. Within countries this is typically carried out by local public and environmental health authorities under the direction of central disease control centres (e.g. the Centers for Disease Control and Prevention in the USA). These control centres communicate internationally and are also aided in this work by the Global Alert and Response system of the WHO under the International Health Regulations of 2005. Other bodies, such as the European Centralised Information System for Infectious Diseases, collect, analyse and present data on infectious diseases for their region.

The pharmaceutical industry and infectious diseases

Vaccines and drug therapies represent some of the most useful tools with which to fight epidemic diseases. They are particularly useful, it might be argued, because their use does not require the need to substantially transform society in order to remove the vulnerabilities that would otherwise lead to large inequalities in exposure and illness. However, despite this clear benefit, it is notable that there is as yet no vaccine for the prevention of HIV infection. It has been argued that this is not necessarily due to scientific difficulties, but rather to the state of the pharmaceutical industry. Scientific infighting, inadequate funding and a lack of co-ordination have all acted as a deterrent to the effective process of vaccine development (Cohen 2001). More broadly there appears to be a greater impediment to development in the relationship between the market-driven impetuosity of medical research in the twentieth and twenty-first centuries and the economic inequalities that exist across the globe (Craddock 2007).

This is a situation that not only affects the search for a vaccine for HIV but also new medical products aimed at tackling epidemic diseases affecting economically poorer societies. The market-driven pharmaceutical industry has, not unexpectedly, turned away from research and development into potentially unprofitable medicine and vaccines and is instead concentrating on products that will guarantee a profitable return. Since the 1980s it is estimated that there have been some 180 new drugs developed for the treatment of cardiovascular disease but only three for tuberculosis (Kaufmann 2009). Research is focused on diseases of the Western world, where the health care systems can be expected to pay for the expense of developing new medicines. For drugs that may be used against infectious diseases, often there is a significant and unresolved dispute between pharmaceutical companies exercising their rights to exploit the intellectual capital they have in the patents for these drugs and attempts by countries where they may be most usefully used to reduce cost by using that scientific knowledge to produce cheaper generic versions of these products. It centres particularly on the contrasting entitlements of a country to protect its citizens' health against the intellectual property rights of companies, both of which are supported in the 1995 Trade Related Aspects of Intellectual Property Rights agreements of the WTO and discussed at its 2001 Doha meeting.

The politics of response

Responses to epidemics are of course not apolitical. The HIV/AIDS crisis illustrates this well. The responses of political leadership to the crisis around the world have varied greatly. The political leadership in China, for example, grossly underestimated the extent of the crisis in their country for many years (Bor 2007). When in 2004 Premier Wen Jiabao finally made a more realistic assessment of the extent of the disease, the central government felt able to triple the funding available to tackle the problem. In South Africa, despite the scale of the crisis, the African National Congress (ANC) leadership felt able to resist pressure to extend the programme of antiretroviral drug (ARV) therapy, despite concerted campaigns by civil society organisations and other political parties (Bor 2007). More generally, harm-reduction approaches, the use of condoms or the provision of clean needles, have become highly contested within the international response to the HIV/AIDS crisis. Conservative institutions have lobbied against these types of programmes, significantly affecting their funding (Pisani 2008).

Epidemics, blame and the oppression of vulnerable groups

The 'top-down' approaches to epidemics of surveillance and control may seem rational. However, it is important to critically examine the presumptions on which they are based, as well as their intended or unintended consequences. This is important because within the fear engendered by epidemics the pernicious forces of discrimination and various types of oppression have frequently been able to play out. Indeed these forces can be seen to become easily embedded in public health policy. This is due to a tendency to link epidemic diseases to ideas of deviance and marginality associated with particular groups in society and therefore for the responses to epidemics to become tools of the dominant forces of social control and the goals of fixing social positions (Rosenberg 1992).

Box 30.2 Facing pandemic threat in France

Claude Gilbert
UMR 5194 Pacte–CNRS, France

Experiences from how the French government faced the recent H5N1 and H1N1 pandemics allow two conclusions. First, preparedness activities have been important since the mid-1990s and a detailed plan, including the purchase of important stocks of vaccines, was drawn and revised several times. Such a plan resulted from the strong involvement of the government, eventually joined by other stakeholders in the health and public safety sectors. It primarily focused on the health and medical impact of a potential long-term crisis, but also considered the social and economic dimensions of pandemics. France was thus considered as one of the best-prepared countries in the world.

Second, although they were prepared, government authorities failed to avoid a nascent political crisis triggered by the outbreak of the H1N1 pandemic. Public detractors alleged that the government hastily purchased vaccines and that experts who recommended such a measure were closely connected to major pharmaceutical firms. France was following recommendations from the WHO, as did other countries, and, in fact, the WHO was also caught up in controversy over experts' alleged links with large pharmaceutical firms and the profits earned by these companies by producing the vaccine.

The H1N1 crisis is, however, rooted in a deeper tangle of structural causes that are evident in the inability of France, as with other countries and international organisations, to consider solutions outside existing frameworks. Tools and procedures for monitoring and mobilising stakeholders and appropriate resources are at stake. Each pandemic is different and it is crucial to develop frameworks that are flexible enough to consider multiple scenarios in terms of knowledge and multiple solutions in terms of action, not limited to medical and health-related measures. Surveillance systems should also be flexible enough to monitor both health phenomena and social, economical and organisational responses.

Epidemic disease is particularly useful in reinforcing public discourses of discrimination and oppression because it transforms the emotional language of fear, hatred or scorn into the legitimated language of medicine. So, for example, during the 1918 flu pandemic in many countries 'foreigners' were blamed for the spread of the disease. In parts of Eastern Europe, in particular Poland, the Jewish population was singled out for blame (Quinn 2008). In this case the anti-Semitism of these places was easily reinforced by the fear engendered by the disease, and this facilitated oppressive 'public health' interventions and worse. Similarly, in San Francisco in the nineteenth century smallpox allowed the widespread view that the Chinese community was undesirable to be legitimised within a public health context (Craddock 2000). Chinatown became a particular focus for this transformed view of difference, and the place and its inhabitants were subjected to oppressive public health sanctions including restrictions on external employment and building. This in effect imposed severe restrictions on where the Chinese could go, what they could do and how they might sustain their livelihoods. The oppression acted out through the public health responses was not new, but simply transformed what had existed. Smallpox had not changed the Chinese community into an undesirable one: this had

369

Box 30.3 Resistance in the face of epidemics

The treatment of excluded groups during periods of epidemic disease has had important consequences for their political evolution. The response of society to the epidemic may objectify the oppression that the group experiences and therefore provide a focus for political activism. Something around which it was difficult to motivate action is transformed by the disease and a society's response to it into something more tangible. It has been argued, for example, that the HIV/AIDs crisis and in particular the failure of the state to provide an appropriate response to it, while it was seen largely as a 'homosexual disease', solidified the political community of homosexual identity (Altman 1988). AIDS deaths cut across divisions of social class or ethnicity and therefore provided a unifying cause which eventually allowed the group to act politically against this and other forms of discrimination.

This possibility for transforming the discrimination, embedded in public health responses, into libratory action may only be possible at certain times or within specific contexts. For example, although the Chinese in San Francisco fought successfully for the repeal of certain public health-related discriminatory policies, they were unable to throw off the mantle of 'pathological deviance' attached to them in common discourses (Craddock 2000). Nor were they able to turn the tide of tough immigration laws linked very closely to this discourse. It is also possible that difference can separate resistance movements that might be expected to act on new issues of discrimination. The differentness of women infected with HIV, their ethnicity, class and nationality, may have separated them from the Western feminist movement, leading to these 'other' women being isolated from media for protest and resistance available to Western feminists (Farmer 2001).

---

pre-existed the outbreaks. In a similar way the homosexual community was not transformed in the eyes of society by HIV/AIDS into a 'deviant' group (Craddock 2000); it had always been seen as sexually deviant. Rather it was now also or predominantly seen as a diseased and dangerous group.

Alternatives: bottom-up approaches to epidemics

Top-down responses to epidemics are typically unable to respond to the highly variable social and environmental situations that often exist in different places. This is important because it is in these different spaces and places that elements come together to produce situations of vulnerability to epidemic disease. Certain diseases such as HIV/AIDS, cholera, plague and malaria, because their mechanisms of spread are highly embedded in the unique contexts of specific places, are particularly unsuited to nationally planned public health responses.

The alternative, bottom-up approaches instead focus on locally initiated and run schemes. They allow health messages to be site specific and therefore to take into account the constrained condition of many people's lives. They may use local experts to encourage behavioural change, using sex worker peer HIV/AIDS educators, for example (Walden et al. 1999), or villagers to form water and sanitation committees to oversee sanitation improvements in their villages (Pradesh 2001). They may focus on early diagnosis of diseases such as malaria, not in village health posts and dispensaries but by members of households and in particular women (Tanner and Vlassoff 1998). Sustainable vaccination programmes, for example, aim to increase coverage through gaining an in-depth understanding of the local contexts (Streefland 1996).

However, although a focus on the local is important in generating effective public health policy, it is equally important not to lose sight of broad, fundamental structures that lead to the inequalities upon which vulnerability to infectious disease is sited. Otherwise there is a danger that blame is shifted onto the individual and the potential for social change is reduced.

Conclusions

Globalisation means that most people across the world are now 'closer' to potential infectious disease reservoirs than at any time before. At the same time there are also very effective tools with which societies can respond to the threat posed by epidemic disease. However, human history is replete with examples of the differential use of available resources and their unjust application. To be effective and just, responses to epidemics must be based both on an understanding of the epidemiological expression of disease biology and also on the social transformation of this process as it plays out across unequally positioned individuals.

Prevention is more cost effective and sometimes far more critical than treatment in the case of epidemic disease. However, preventative interventions need to be critically scrutinised to avoid the risk of simply reinforcing existing situations of inequality. The limited agency of the poor also needs to be taken into account. Although prevention is vital, treatment and particularly its availability to the poor must not be overlooked. Too many of the epidemic diseases of the poor are under-researched, underdeveloped, with few treatments available. Where they are available, for a variety of reasons often they are not available to the poor. Bottom-up approaches to epidemics do produce effective local preventative measures. However, unless they are also acting at other scales, as processes of resistance, they risk being simply a 'sticking plaster' for deeper structural vulnerabilities. As with most types of disasters, the reduction of structural inequalities, and specifically poverty across the globe, would do much to reduce the destructive impact of epidemic diseases.

31

Livestock epidemic

Delia Grace and John McDermott

INTERNATIONAL LIVESTOCK RESEARCH INSTITUTE, NAIROBI, KENYA

Introduction

Since the widespread domestication of animals in the Neolithic era, 10,000–15,000 years before the Common Era (CE), human livelihoods have been inextricably linked with the livestock they keep. Domesticated animals must have been among the most valued assets of ancient humans: walking factories that provided food, fertiliser, power, clothing, building materials, tools and utensils, fuel, power and adornments. Inevitably, the innovations of crop cultivation and food storage that allowed people to settle and live in high numbers and densities also increased the number of animals kept, density of livestock population and the intimacy of human–animal interactions. Pathogens responded, undergoing intense genomic change to seize these dramatically expanded opportunities. Epidemics of highly contagious and lethal disease emerged, as livestock and people reached the critical population sizes needed for acute infections to persist. Diseases also jumped species from animal to humans: the lethal gift of livestock.

This chapter discusses which livestock epidemics are likely to constitute a disaster and why.

Livestock epidemics as disasters

When animal epidemics constitute disasters: livestock plagues

Epidemics are usually defined as occurrence of a certain disease above expected levels in a population. A few cases of a rare disease may constitute an epidemic, as may the gradual increase of chronic or benign disease; epidemics may also be non-contagious (e.g. bovine spongiform encephalopathy, commonly known as 'mad cow disease'). Increasingly the term may be used when the aetiology is non-biological (for example, an epidemic of lameness associated with concrete flooring). However, the word 'disaster' likely refers to those epidemics caused by rapidly transmitting pathogens that produce acute and serious disease in a large number of hosts. In livestock, rinderpest (cattle plague), Newcastle disease (fowl pest) and classical swine fever (hog cholera) are archetypal examples.

Historically these lethal, highly contagious diseases were known as murrains, pestilences and plagues: words still evocative of disaster. The former List A of the World Organisation for Animal Health (which retains its historical acronym of OIE) comprised sixteen of the most important livestock epidemics, chosen because of their potential to spread rapidly, to cause large socio–economic losses and to interrupt trade. The current OIE list is longer and the criteria for inclusion have been expanded to include animal diseases that can affect people (zoonoses) or that are emerging. In rich countries, the most serious livestock epidemics have been controlled and, as a result, many highly contagious and serious epidemics are labelled 'exotic' or 'foreign' diseases. Global organisations prefer the term transboundary animal disease (TAD), as 'foreign' is a matter of perspective and most diseases are 'at home' somewhere on the globe. Generally, these diseases are notifiable, that is, there is a legal requirement of reporting to veterinary authorities (or, curiously, to a police constable in the UK).

Many serious livestock epidemics also fit into the category of Diseases with High Externalities (DHE), a term used by the European Union (EU) to indicate that they pose a large threat to the wider economy and hence their control justifies public intervention. What these definitions have in common is recognition of high infectiousness and potential for major negative impact. This chapter refers to these as livestock plagues, to distinguish them from non–contagious, slowly spreading, chronic or benign livestock epidemics that are less likely to constitute disasters, and the zoonotic diseases which constitute disasters but for different reasons. Table 31.1 provides a rapid profiling of some important livestock plagues based on the former OIE List A.

These diseases are absent from, or controllable in, rich livestock-keeping countries. This is sensible given that if a country has eradicated a disease it will not wish to re-import it and is entitled to put it on its notifiable diseases list. However, once on the 'scare list', a disease becomes guilty by association and more feared than it might be on its own merits. This has implications for poor countries, the role of which is still too often to accept standards rather than set them. For example, lumpy skin disease is arguably neither more deadly nor less manageable than orf, a similar disease causing skin lesions in sheep (and a zoonosis to boot).

Table 31.1 Features of some important contagious epidemics afflicting livestock

Disease	Species affected	Lists	Brief description
Rinderpest	Bovine	AGEO	The most serious livestock epidemic and the first to be eradicated. Characterised by the 3 Ds (discharge, diarrhoea and death), it is highly lethal and contagious.
Contagious bovine pleuropneumonia (CBPP)	Bovine	AGEO	A respiratory disease with high mortality when first introduced, later disease is mainly chronic. It has been eradicated from most of the world but is now expanding in Africa.
Foot and mouth disease (FMD)	Multiple	AGEO	The third historical cattle plague (with rinderpest and CBPP). Primarily a disease of trade; mortality and morbidity are low when the disease establishes in a country.
Swine vesicular disease	Swine	AO	Little impact on productivity but clinically identical to FMD. If it became widespread in free countries then FMD could 'hide behind it' so is often notifiable.
Peste des petits ruminants	Sheep and goats	AGO	A rinderpest-like disease of sheep and goats causing high mortality and morbidity, now expanding.
Lumpy skin disease	Bovine	AO	A skin disease caused by a pox virus of high morbidity but low mortality. Currently spreading in Africa. Readily controlled by vaccination.

Continued on next page

Table 31.1 (continued)

Disease	Species affected	Lists	Brief description
Rift Valley fever	Multiple	AGEO	A feared but uncommon zoonosis (haemorrhagic fever). Most of Africa is endemic and epidemics are associated with El Niño events. Disease is mainly seen in sheep and mortality is relatively low.
Bluetongue	Sheep and goats	AO	Spread by midges; recent incursions into Europe have been associated with climate change.
Sheep pox and goat pox	Sheep and goats	AO	A smallpox-like disease of sheep and goats.
African horse sickness	Equine	AO	Highly fatal disease present in South Africa and spread by insects. An outbreak occurred in Spain in the last decade.
Equine encephalitides	Equine	GO	Three related viruses found in the western hemisphere and transmitted by insects. A zoonosis.
Classical swine fever (CSF)	Swine	AGO	A serious and highly contagious disease of pigs.
African swine fever	Swine	AGEO	The fourth 'Africa only' disease on the OIE former List A. Highly contagious with no vaccine. Clinically similar to CSF.
Teschen disease	Swine	A	The more severe form of a polio-like disease, found worldwide. Most countries in the Western world erroneously made Teschen disease notifiable before it was realised that the virus was so widespread.
Fowl plague	Poultry	AGEO	A pandemic of fowl plague or highly pathogenic avian influenza is currently being experienced. Highly contagious and highly lethal to poultry; ducks are asymptomatic carriers.
Newcastle disease	Poultry	AO	The major killer of backyard poultry; effective vaccinations exist, but are not widely used by the poorest.

Note: A = former List A disease (OIE); O = current OIE list; G = disease on Global Early Warning and Response list of priority diseases; E = disease on list of priorities developed by Emergency Prevention System.

However, orf is present in the major livestock exporting countries of the developed world that built the international system of disease control on the model of their own systems. Could this partly explain why orf is less likely to appear on global disease lists of major epidemics than its exotic counterpart lumpy skin disease?

Impacts of livestock plagues

The epidemics described in Table 31.1 are among the most likely to constitute disasters. They have multiple and severe socio–economic, health and ecosystem impacts, including loss of animal assets through death, sickness or culling; increased cost of production, resulting in increased cost of livestock products and potentially compromised food security; loss of livestock genetic resources, some irreplaceable; restriction of livestock and livestock products export; loss to other agricultural sectors (e.g. feeds); in some cases threats to human health (zoonoses); in some cases spill-over to wildlife; disruption of other economic sectors (tourism); and loss of ecosystem services provided by livestock and wildlife victims. In richer countries, for which agriculture is usually a small percentage of the gross domestic product (GDP), the costs to other sectors may be greater than the costs to the livestock sector: for example, in the UK the 2001 foot and mouth disease outbreak losses to tourism were actually greater than the losses to the agriculture sectors (Royal Society of Edinburgh 2002).

One of the most powerful drivers of human interest in livestock disease and epidemics is enlightened self-interest. Many human epidemics of Eurasia (e.g. measles, smallpox, influenza) originated when pathogens of domestic animals evolved to become human specific (Wolfe *et al.* 2007). In recognition of their ancient animal origin, these are sometimes called the old zoonoses. Other pathogens remained adapted to domestic animals but took the opportunity to infect the humans who exposed themselves to infection by consumption of livestock products or contacting animals (e.g. the pathogens responsible for tuberculosis, brucellosis, rabies). These are called classical or established zoonoses. For another group of diseases, the sporadic or emerging zoonoses, human infection is rare, either because the pathogen is poorly adapted to humans (e.g. Ebola, avian influenza) or occasions of transmission are infrequent. As these pathogens evolve, they may become better adapted to humans, and this concern underlies the efforts to control avian influenza in birds before it gets the chance to evolve into a Spanish flu-type strain capable of killing tens of millions of people, as happened in 1918. Hence, understanding livestock epidemics is important not only because of their impact on livestock population and production, but because of their role in disease emergence. However, zoonoses are an area in their own right and this chapter concentrates on diseases of importance to livestock.

While few argue that disease control is a bad thing, recent experiences remind us that if livestock epidemics have negative impacts, so too can the action taken to control or prevent them. During the avian influenza pandemic, which started in 1997 and as of 2010 is continuing, there have been several calls to 'restructure' the poultry industry, which in effect meant getting rid of the backyard sector which included most of the poorest producers, many of whom are women with limited other options for income-generating activities.

The pandemic of H1N1 influenza declared in 2009, which originated in pigs but has escaped its swine host and is now maintained entirely by human-to-human transmission, gives another example. In response to the pandemic, the government of Egypt ordered all of the country's pigs to be slaughtered in a costly and (because humans can only get the new flu from other humans), epidemiologically pointless move. This had far-reaching and unintended consequences. Cairo's 30,000 garbage collectors used to feed the city's organic waste to pigs and so their livelihood became endangered while the streets of the capital filled up with trash (ABC News 2009).

Plague epidemiology

The epidemiology of livestock plagues has important implications for their behaviour that are unfortunately not always understood. Three are highlighted: the requirement for crowds, the illusion of epidemic control and the (partial) bonus of herd immunity.

The requirement for crowds

Many livestock plagues, as for their human equivalents, require large animal populations (and are therefore sometimes called 'crowd diseases'). Without a constant supply of fresh victims, or if too many hosts die or become immune, plagues burn out rather than propagate because hosts are too few and contacts too sparse. The actual threshold population needed to maintain an epidemic depends on pathogen factors (e.g. ease of transmission, survival in the environment) as well as host factors (e.g. susceptibility and contact rates), but historical records suggest human habitations of about 250,000 are needed for major epidemics. (Given that livestock-dependent households typically require several animals for each household member it is possible that livestock epidemics pre-date human ones.) Where crowds are absent, so are epidemics. For example, arguably, in the many African countries with low densities of chickens and few ducks, even if avian influenza is introduced it will not become established. Hence, the large amounts of money spent on preparedness in these countries may not have been the most efficient use of scarce disease-control resources.

The illusion of epidemic control

Plagues that result in immunity and/or widespread death are frequently cyclical in nature. When first introduced to a naïve population not previously exposed, called a 'virgin soil epidemic' in a 'naïve community', mortality is very high. As hosts are removed through death or the development of immunity, the rate of infection slows until it is no longer at epidemic proportions. After new susceptibles are added by birth or in-migration, another outbreak occurs. Even in the poorest countries, the introduction of a novel plague is followed by control efforts. Even if completely ineffectual, control efforts are often accompanied by a natural decline in cases. Politicians and technicians with what psychologists term an internal locus of control (i.e. a tendency to attribute success to their own efforts rather than good luck) may attribute declines in disease to their actions rather than the natural history of plagues. Arguably, the recent decreases in avian influenza owe more to natural decline than to the huge but often not very well thought out global and national responses to the pandemic.

The (partial) bonus of herd immunity

Herd immunity is an epidemiological phenomenon first described in livestock populations, which proved sufficiently useful to be transferred without name-change to the epidemiology of humans (Coleman et al. 2001). Herd immunity refers to the resistance of a group to disease attack to which a large proportion of the group is immune. This underpins population vaccination campaigns: not all individuals need be vaccinated to ensure protection of the group, and those who are vaccinated protect those free-riders who are not.

Herd immunity has a dark side: if generated to a level that is below the level needed to eliminate a disease, it can paradoxically perpetuate disease by creating a partially immune population in which either the disease persists at a low and difficult to detect level or is sufficiently suppressed for its effects to be tolerable. The widespread private use of vaccination probably allowed rinderpest to maintain itself for thirty years in India and for avian influenza to remain endemic to this day in Pakistan (Roeder and Taylor 2007).

Major livestock plagues and the lessons from them

History has been partly shaped by livestock epidemics, as it has by human epidemics (many of which originated in livestock). Chinese, Egyptian and Indian texts describe animal epidemics millennia ago and classical authors wrote of plagues leaving not a single ox in the land (Blancou 2003). Retrospective diagnosis of plagues is a popular pastime of medical historians. Some of the plausibly, if not always definitively, identified livestock epidemic disasters of the past include:

- Cattle plague (rinderpest) entered Europe with the Hun invasions of the sixth century and followed every major war until the last century (Barrett *et al.* 2006).
- Sheep murrain (probably sheep pox or mange) is reported to have killed sheep on every farm in England in the thirteenth century (Fleming 1871).
- Black bane (anthrax) epidemics resulted in massive animal mortalities throughout history and the concept of cursed earths or 'terres maudites'.
- Lung plague (contagious bovine pleuropneumonia) was first described in Germany in the seventeenth century and spread round the world in the globalisation of the steam age with disastrous effects. The USA was infected twice in the nineteenth century, and the post of Secretary of State for Agriculture was created specifically for the control of this disease (Blancou 2003).
- Glanders (farcy) is one of the first diseases to be fully described, reflecting the importance of horses as the mainstay of transport, tillage and war. Surprisingly, its zoonotic potential was often not realised; for example, Vial de Saint Bel, the first principal of England's first veterinary college, maintained that glanders was not contagious right up until his death from it in 1793 (Wilkinson 1992).

In the roll call of historic animal diseases, pigs and poultry are not salient. In the past, as with the present, these were often the less-favoured species and so kept by women and the poor. In the past, as in the present, their owners' voices are hardly heard. While large-scale die-offs have been reported from antiquity and historical times, the detailed description of symptoms and course of disease that would allow tentative identification is rarely present.

While less is known about historical livestock epidemics in other regions of the world, especially those without a written literature, it is plausible that livestock epidemics have been one factor in the vulnerability of American and African cultures to European colonisation. For example, in Africa the Great Cattle Death of the 1860s (contagious bovine pleuropneumonia) was followed by the African Cattle Plague of the 1890s (rinderpest), which killed eighty to ninety per cent of cattle and susceptible wild ruminants; the result was famine, smallpox and unprecedented predation of people by carnivores. This overthrew the pastoralist hegemony in much of Africa; it has not recovered to this day (Tiki and Oba 2009). Box 31.1 looks more closely at some important epidemics to draw lessons on the drivers, impacts and control of animal plagues.

Managing livestock plagues

Past livestock plague management

The essentials of livestock plague control have been known for centuries. Quarantine, import bans, identification of suspicious animals and premises, duty of reporting (and punishment, sometimes capital, for failure to do so), isolation, compulsory slaughter, disinfection and compensation can be traced back to mediaeval times and before (Blancou 2003). As for the human epidemics, control attempts of the past were often ineffective in the face of ignorance and panic responses from frightened populaces. There are some exceptions that teach the important lesson that controlling livestock epidemics does not require modern technology or twenty-first-century institutions. For example, rinderpest was successfully controlled in the Papal States (1712–15) by movement controls and quarantine rigorously applied (Barrett *et al.* 2006).

Box 31.1 The emergence of epidemics: a hotter, wetter, sicker world?

Bluetongue is an evocatively named disease of ruminants resulting in severe disease in naïve sheep not previously exposed to the disease. Caused by a virus from the family that includes African Horse Sickness, it is widespread in the tropics and subtropics, and is spread by biting midges. For the last century, Europe was mostly bluetongue-free and brief incursions of the disease did not establish it. Since 1998, however, there has been at least one serotype of bluetongue virus (BTV) active in Europe every year with serious impacts. For example, two epidemic waves in Italy at the start of this century resulted in the death of around 100,000 sheep and an outbreak in Holland a few years later had net costs of €200 million (Velthuis *et al.* 2010).

There is a substantial body of evidence linking this emergence to climate change and bluetongue is often taken as the harbinger of the exotic diseases set to invade Europe as climate change creates new niches for nasty diseases. This may be alarmist, as bluetongue differs from most of the other livestock plagues discussed in this chapter in important ways: it is not contagious; it is not highly lethal; it is not easily detected; it has a wide range of hosts (including wildlife); and the midge vector is highly abundant. All these factors make bluetongue a worse candidate for control than other plagues long eliminated from richer countries (rinderpest, CBPP, sheep and goat pox, classical swine fever).

While climate change will undoubtedly bring changes in disease distribution, as the world gets warmer it also gets richer. From a centuries-long perspective, the overall trend is that the world is becoming richer and disease control better (albeit with local and temporary setbacks). Most diseases occur in areas that are hot, wet and poor. If they are not comparatively poor, then they tend to have disease levels comparable to non-tropical rich countries (e.g. Singapore and Hong Kong). Malaria, an old zoonosis, is also the most important climate-sensitive disease. However, studies show that while Malaysia became steadily warmer over the last fifty years, malaria has dramatically declined (Sian 2000). Development explains the difference. A series of malaria control programmes along with better diagnosis and treatment, changing environments and increasing wealth has led to a dramatic decline in cases. Of course, the helpful assurance that being richer in the future and hence healthier is little consolation for climate change-affected people today. The poorest countries, which have contributed least to the phenomenon of climate change, are most likely to suffer from climate-mediated change in disease distribution.

Developments and events of the nineteenth century improved the prospects for eradicating livestock plagues. Germ theory provided a scientific rationale for unpopular quarantine and culling, the emergence of a veterinary profession supplied human resources for the war against disease, the formation of state veterinary services allowed centralised and organised controls, and widespread public concern over livestock plagues and increasingly interventionist governments were all factors. Technological advance in the age of empire created a lot of the problem as massive numbers of animals moved by ship and rail around the world were responsible for a huge upsurge in livestock plagues.

In a nice example of finally getting rid of a disease long prone to troubling incursions, cattle plague was eliminated from Britain in 1898 after an eight-year-long extensive, centrally directed campaign, involving ruthless tracing and destruction of infected cattle. The USA declared freedom from infection of contagious bovine pleuropneumonia in 1892, foot-and-mouth disease in 1929, babesiosis in 1943, screwworm in 1959 and classical swine fever in 1978, while similar successes were achieved in Australia, New Zealand, Canada, parts of Europe and parts of Latin America.

The achievements in the eradication of livestock plagues from more developed countries in the last centuries show that top-down hierarchies, operating military-style campaigns with minimal stakeholder consultation and lots of resources, can be quite effective at controlling disease. Fortunately (or not), veterinary services in rich countries no longer have the liberty to ignore considerations of animal welfare, environmental impacts and society's approval in their zeal to control livestock plagues.

Present-day livestock plague management

Plagues know no boundaries and modern management is increasingly transnational. At the global level, three organisations have mandates that cover livestock epidemics. The OIE has a global mandate to set standards for trade in animals and animal products (see also Box 31.2). More recently, it has expanded its mission to cover food safety, animal welfare, veterinary services and support to animal disease control. The United Nations (UN) Food and Agriculture Organization (FAO) has had a long involvement in livestock epidemics. Its programme, Emergency Prevention System (EMPRES) for Transboundary Animal and Plant Pests and Diseases, aims to minimise the risk of emergencies developing and focuses on five livestock plagues (indicated in Table 31.1). The World Health Organization (WHO) is concerned with livestock plagues that are also zoonoses or have potential to evolve into human pandemics. Together, the so-called 'three sisters' of WHO, FAO and OIE operate the Global Early Warning and Response System (GLEWS), which has the objective of improving co-ordination for identification and management of major animal diseases and zoonoses (twenty-five in total; see Table 31.1 for livestock plagues).

These global organisations are supported by reporting and/or information systems. OIE maintains the World Animal Health Information Database (WAHID), which covers just seven species (including bees) and 117 infectious livestock diseases, many of which are livestock plagues. These are notifiable; that is, there is an obligation for Chief Veterinary Officers of member countries to report to OIE. FAO has developed the Transboundary Animal Disease Information System (TAD-Info), which covers seven important livestock epidemics.

Most regions have a specialised organisation for animal health, for example InterAfrican Bureau for Animal Resources of the African Union, while at national level public veterinary services have the responsibility for the management of animal plagues.

At the national level, veterinary services have traditionally been responsible for livestock plague management. A useful distinction is between preparedness, prevention, surveillance and response. Table 31.2 summarises some of the activities under these rubrics. In the case of notifiable livestock plagues the initial response is usually to stamp it out. The rationale is that vaccination may not be completely effective but will keep disease at such a low rate that it can spread widely and establish in a country. Moreover, because vaccination and disease both lead to an immunological response it is not always possible to differentiate between vaccinated and infected animals and this can interfere with trade. Recently, the veterinary dogma of stamping out has been challenged but it remains the preferred option as a first approach to small outbreaks of exotic disease. Stamping out involves quarantine of affected farms or areas and the destruction of infected and in-contact animals. Because contacts can be difficult to determine, a

Box 31.2 The impacts of epidemics: trading our way out of poverty with livestock. Or not?

Many livestock-rich African countries are excited about the prospects of export to the high-value meat markets of rich countries. The Sanitary and Phytosanitary (SPS) regulations that govern international trade are seen at best as a barrier to be scaled and at worst as protectionism through the back door. The fear is that countries which are members of the World Trade Organization (WTO) can no longer exclude imports simply to protect their own producers and so have created the fear of livestock plagues to ban livestock and livestock products from poor countries.

More recent research suggests that while some developing countries are hugely successful exporters (e.g. Brazil), most have little competitive or comparative advantage. In particular, most African countries have little competitive advantage in production for high-end markets. It appears that meeting SPS requirements is not the major roadblock, but rather costs of production and ongoing quality assurance. Indeed, some southern African countries, including Botswana, Zimbabwe and Namibia, export or have previously exported beef to the EU under a highly favourable trade agreement for developing countries. Despite favourable conditions – relatively good veterinary services, many cattle and few people, a high price and a sure market – none was able to produce enough meat to fill the quota.

A sectoral approach to livestock export has also left unanswered questions about its equity and environmental implications. An economic assessment in Zimbabwe showed the direct impacts that foot and mouth disease (FMD) had on the poor and the measures for controlling it are very limited. Although most of the direct costs of FMD control are met by the public sector, the greater part (eighty-four per cent) of benefits is captured by the non-poor commercial sector. Many of the rural poor keep cattle but these are mainly used for asset accumulation and only two per cent are traded (Perry *et al.* 2003). A study from neighbouring Botswana found that the veterinary fences that criss-cross the country to control livestock diseases block the migratory pathways of wildlife and contribute to their decline in Botswana (where tourism now contributes more to the economy than beef export) (Mbaiwa 2006).

More encouragingly, recent studies have also underlined the large potential of the domestic, and to a lesser extent regional, markets. In Kenya, for example, domestic beef prices approach the world price and demand is so great that one-third of beef consumed comes on the hoof from Tanzania and Ethiopia (Aklilu 2008).

cull zone (of up to several kilometres) around the index case is usually recommended. If this proves ineffective then milder control means, such as vaccination may be considered.

Indigenous knowledge and community-based animal health care

While only a few decades ago many scientists and administrators in developing countries thought of farmers as ignorant and erroneous, now there is general acceptance that livestock-keepers can possess a vast storehouse of detailed knowledge about health and indeed every aspect of the animals on which they depend (Wanzala *et al.* 2005). Numerous examples exist of farmers' ability to identify and diagnose disease, often recognising signs such as the taste of milk or the smell of an animal that may be missed by Western diagnosticians. Livestock-keepers also have a wealth of

Table 31.2 Control of livestock plagues at national level: functions and sub-activities

Preparedness	Develop a list of notifiable diseases
	Conduct risk analysis: estimate negative impacts and their likelihood
	Develop a diagnostics manual and reference laboratories
	Develop contingency plans: plans to be carried out in the event of an epidemic
	Conduct simulation exercises: rehearsals for epidemics
Surveillance	Establish case definitions: what signs indicate what level of epidemic suspicion
	Build traceability system: allowing livestock to be traced back to farm of origin
	Conduct active surveillance: active searching for evidence of disease
	Conduct passive surveillance: routine gathering of information
	Establish emergency disease reporting system
	Establish alert system: showing different stages of concern (e.g. green, amber, red)
Prevention	Conduct import risk assessment of the risk of bringing disease into a country through imports
	Put in place import controls including quarantine
	Conduct border inspections and controls
	Ensure safe disposal of ship and aeroplane waste
	Conduct education and awareness raising
	Promote biosecurity: the actions that keep disease out of production enterprises
	Fund disease control in countries where it is endemic
Response	Establish national co-ordination group of all stakeholders involved in epidemic control
	Trace back and trace forward: where an infected animal came from and what it may have contaminated
	Depopulate and repopulate: kill all animals and restock
	Cull affected and in-contact animals
	Cull animals in a protection zone
	Dispose of culled animals safely and disinfect premises
	Put in place animal movement controls

traditional treatments, mainly plant-based, some of which have been shown to be effective in clinical trials (Mathias 2007).

However, there were no effective remedies against most major epizootics in the pre-modern era. Trypanosomosis was managed by keeping out of the tsetse-infested regions and rinderpest could only be combated by taking the entire herd into a remote area. Livestock keepers managed risk of herd wipe-out by developing elaborate systems of loans, gifts and animal exchange (Blench 2001). An interesting exception is preventive inoculation against CBPP, a traditional practice in west and southern Africa. Diseased lung tissue is inserted subcutaneously on the bridge of the nose resulting in a keratinous nasal excrescence. Ignorance of this advanced indigenous technique led a French physician in the late nineteenth century to incur the ridicule of anatomists by reporting that he had discovered a new breed of three-horned cattle (Blancou 2003). Treatments for CBPP are of more dubious value: for example, Fulani pastoralists burn cattle over the ribs thinking this may ease breathing and on the nose to prevent foot and mouth disease: painful and useless treatments (Grace 2003).

Most African countries, following the guidelines of the OIE, require animal treatments to be under veterinary supervision. These countries have typically a few hundred veterinarians, millions of livestock-keepers and tens of millions of animals. Consequently, most diagnoses and treatments are made by non-veterinarians, as shown by field studies (Grace *et al.* 2009), so community animal health has been promoted since the 1970s. Experts in livestock are selected by their communities and given from a few days to a few months of skills-oriented training in diagnosis and drug use. Evaluations have repeatedly shown the effectiveness and positive impact of this approach. Grace (2003) collates some examples: in conflict-ridden south Sudan community animal health workers (CAHWs) vaccinated more than 1 million animals a year; in Cambodia five years after training, ninety-five per cent of CAHWs are successfully treating animals; in Indonesia, training one CAHW cost US$15 and the benefits from improved animal productivity were worth US$170 per farmer reached. Community animal health programmes have fulfilled only a fraction of their potential, as public veterinary services lack resources and interest in supporting them, and private veterinarians oppose them as actual or potential competitors (IDL 2003).

Future livestock plague management

Extrapolating the trends of the past can give insights into the future (at the risk of missing the major discontinuities more likely to shape it). A key trend of recent decades has been the greater integration of human and veterinary medicine. One Health (OH) is a growing movement built around the premise that the health of humans, animals and the environment is inextricably interlinked and that disease is best managed in broad and inter-disciplinary collaborations. An obvious positive development from the 1997 avian influenza pandemic has been a visible need for better co-ordination between livestock, wildlife and human health services, and more support of the OH concept. Ecohealth is another integrative framework covering human, animal and ecosystem health, and with a strong emphasis on links between scientists, communities and policy-makers.

Another noteworthy trend is the democratisation of disease control. Increasing participation from a wider range of people has led to novel perspectives being introduced to livestock plague control such as the need to ensure animal welfare and to consider impacts on women and poor farmers. For example, over the last decade, hundreds of thousands of dogs in Chinese cities have been clubbed to death in attempts to control rabies. However, this traditional and very ineffective form of rabies control is now evoking non-traditional responses: in Beijing, over 500 people protested on the city's streets, and a petition of over 60,000 signatures was presented to the government, while the draft of China's first animal welfare legislation received over eighty per cent online approval when it was released in September 2009 (China.org.cn 2010).

There has also been a surge in novel surveillance and reporting tools that draw on a far wider range of field reports, from the traditional state veterinary officers, for example the Program for Monitoring Emerging Diseases (ProMed, www.promedmail.org), GeoChat (instedd.org/geochat) and HealthMap (www.healthmap.org). However, old diseases continue to thrive in the face of new technologies. The resurgence of CBPP, peste des petits ruminants and African swine fever in and out of Africa, the breakdown of livestock plague control in Zimbabwe, the failure to control avian influenza in poor countries where circumstances are propitious to its endemicity, the spread of climate-sensitive diseases such as Rift Valley fever and bluetongue – all these should prompt a rethink of animal disease control. Bottom-up approaches such as community-based animal health have been highly successful in getting animal health services to poor farmers at prices they can afford. Yet global as well as national veterinary policy still too often discourages these appropriate and inexpensive alternatives to scarce and expensive veterinarians.

Conclusions

The struggle with epizootics continues and has even intensified in recent times. Population-decimating animal plagues, such as contagious bovine pleuropneumonia, peste des petits ruminants, swine fever, Newcastle disease and avian influenza, continue to have lethal and devastating impacts on livestock and livelihoods. Livestock plagues are also shifting and emerging, while climate change, urbanisation, migrations, genetically modified crops and rapid land-use changes are examples of wild cards that could alter the present distribution of the disease dramatically for the worse. The declaration of an era of epidemics, though, might be premature. In richer countries, dependence on livestock is low, resources exist to effectively control disease and non-communicable diseases associated with modern farming systems (such as lameness and reproductive problems) pose the greatest problem to animal health.

In the developing world the situation is different. Many people depend on animal agriculture: 700 million people keep livestock and up to forty per cent of household income depends on livestock. Animal and human disease outbreaks are far more frequent, both for infections well-controlled elsewhere and for emerging diseases. In the poorest countries in Africa, livestock plagues that were better controlled in the past are regaining ground. Paradoxically, the fear of epizootics is much higher among the worried well in rich countries, who are highly concerned about the diseases of which they are very unlikely to fall sick or die. Thankfully, this enlightened self-interest is providing more support for control of epizootics in poor countries. However, it appears that while the centralised control of livestock plagues is effective (albeit, at high cost) in richer countries, it struggles in the poorest. New approaches are not only needed but need to be rapidly tested and made available. What is required now is the vision and courage to transcend sectoral and conventional veterinary approaches and apply innovations to these urgent problems.

32

Plant disease, pests and erosion of biodiversity

Pascal O. Girot

INTERNATIONAL UNION FOR THE CONSERVATION OF NATURE (IUCN), SAN JOSÉ, COSTA RICA

Introduction

Biodiversity is the variety of life, in all its forms. It provides the foods and medicines used around the world today, the fibre for clothes, many of the materials for shelter and houses, and a plethora of other goods and services such as soil nutrients, clean water, disease and climate buffering, energy and much else, which sustain the livelihoods of millions of rural poor and on which all people are ultimately dependent. As such, biodiversity is a critical element of human security.

Biodiversity loss particularly affects the poor, who are most directly dependent on ecosystem services at the local scale. Loss of access to biological resources, and the increasing erosion of agrobiodiversity, limit future options for development, resulting in many cases in the exacerbation of vulnerability of the poor. Ultimately, biodiversity loss represents a source of risk as unfraying ecosystems impact livelihoods and human health and provide less protection in extremes involving wind, ocean waves, flooding, drought and movement of slopes, amongst others. The mitigation of biodiversity loss can be achieved through in situ and ex situ conservation, through protected areas or botanical gardens, seed banks and other top-down approaches. Meanwhile, bottom-up approaches are increasingly emerging as an important option for limiting biodiversity loss, through community forestry, collaborative management of protected areas and other activities that contribute to the sustainable use of biodiversity, and for creating resilient livelihoods that reduce vulnerability through incorporating biodiversity into production landscapes.

The role of biodiversity in livelihoods and in reducing disaster risk

Basic concepts reviewed: biodiversity, agrobiodiversity and ecosystem change

Biodiversity is the variety of genes, species and ecosystems in which life forms are currently expressed on planet Earth. These three components of biodiversity are intimately entwined, as ecosystem dynamics depend on populations of species and their interaction with the abiotic environment, while genetic variation reflects the natural history of species. As defined in the

Millennium Ecosystem Assessment (2005: iii), an ecosystem is 'a dynamic complex of plant, animal and micro-organism communities and the non-living environment interacting as a functional unit. Humans are an integral part of ecosystems. Ecosystems vary enormously in size; a temporary pond in a tree hollow and an ocean basin can both be ecosystems'.

Conservation of biodiversity and the delivery of the associated ecosystem goods and services provide the foundation for human survival, providing food, fuel and building materials, in addition to soil nutrients, clean water, disease and climate buffering, energy, pharmaceuticals and much else.

A particularly important ecosystem around the world is the agro-ecosystem, which is composed of domesticated and managed plant species along with interacting wild biodiversity. An analysis of the importance of planned diversity for human welfare cannot be separated from the concept of agrobiodiversity. Agrobiodiversity can be defined as 'all crops and livestock and their wild relatives, and all interacting species of pollinators, symbionts, pests, parasites, predators and competitors' (Wood and Lenné 1999). At the heart of agrobiodiversity policy is the International Treaty on Plant Genetic Resources for Food and Agriculture, even as those resources continue to be eroded (see Box 32.1). These global common property resources are the subject of recent international agreements, but continue to be a source of controversy.

Consistent with evolutionary theory, ecosystems do not disappear, but are transformed. Domestication is a classic example of ecosystem transformation into agro-ecosystems. The co-evolution of ecosystems and human societies has also helped to forge natural and cultural landscapes which reflect complex interactions between natural ecosystems and human society. They also reflect changes in population densities, technical and economic models of agriculture and governance systems, as civilisations emerged from the control of water, agricultural surpluses and economic specialisation (Wittfogel 1964; Mazoyer and Roudart 1997).

The main agent of plant domestication has historically been the traditional farmer, who over time sought to maximise the natural diversity of their traditional crops, in order to increase yield stability and adaptability to new environments and to decrease vulnerability to disease. These

---

Box 32.1 Erosion of plant genetic resources for food and agriculture

- The Republic of South Korea referred to a study which showed that seventy-four per cent of varieties of fourteen crops being grown on particular farms in 1985 had been replaced by 1993.
- China reported that nearly 10,000 wheat varieties were in use in 1949. Only 1,000 were still in use by the 1970s. China also notes losses of wild groundnut, wild rice and an ancestor of cultivated barley.
- Malaysia, Philippines and Thailand reported that local rice, maize and fruit varieties are being replaced.
- Ethiopia noted that native barley was suffering serious genetic erosion and that durum wheat is being lost.
- Large-scale erosion of local varieties of native crops and crop wild relatives was noted by Andean countries. Argentina point to losses of *Amaranthus* and quinoa.
- Uruguay stated that many landraces of vegetables and wheat had been replaced. Costa Rica reported replacement of maize and bean varieties.
- Chile commented on losses of local potato varieties, as well as oats, barley, lentils, watermelon, tomato and wheat.

(Moore and Tymowski 2005: 20)

landraces have been the result of centuries of experimentation by local farmers seeking to improve their seed stock. Carl O. Sauer in his seminal work, *Agricultural Origin and Dispersals*, suggested that the origins of agriculture are probably most closely associated with woodlands and tropical ecosystems. As put by Sauer:

> The hearths of domestication are to be sought in areas of marked diversity of plants or animals, where there were varied and good raw materials to experiment with, or in other words, where there was a large reservoir of genes to be sorted out and recombined. This implies well-diversified terrain and perhaps also variety of climate.
>
> *(Sauer 1952: 14)*

However, the domestication of cultigens has been far from homogenous. Vavilov (1926) – recognised as the foremost botanist and plant geographer of his time and extensively cited by Sauer (1952) – rightly pointed out that some of the main centres of plant domestications coincide with areas prone to recurrent and cumulative natural hazards. For example, it is often among traditional seed stocks that the drought-resistant strains have been found.

Scientific plant breeding today has at its disposal a wide range of techniques to induce variation, to select desired traits and to propagate and reproduce new varieties through tissue cultivation. Still, both traditional and modern plant breeders have in common that they need reliable access to a wide range of plant varieties in order to ensure their genetic robustness. This in turn requires an adequate balance between traditional landraces, hybrid and genetically engineered plant material and their wild relatives, which are key for maintaining resistance traits, and for developing new traits in the face of a changing climate and local environment. As Moore and Tymowski (2005: 20) suggest, the International Treaty on Plant Genetic Resources for Food and Agriculture contributes to two main functions: (1) 'Sustainable production of food and other agricultural products'; and (2) 'Meeting new and unforeseen needs and conditions'.

Chronic famines, such as the Irish Famine of 1845–48 due to the Potato Blight, have been correlated with the combination of biological vulnerability, through the limited genetic diversity of potato landraces brought to Ireland 200 years prior from the Andes, which produced a lower resistance to pest and disease, and social vulnerability, through enclosure and exclusion from access to land. In this case the connection between agro-ecological resilience and social vulnerability is a direct one (Wisner *et al.* 2004).

Ecosystem services and livelihoods

Ecosystems are also a critical source of livelihood for millions of rural poor. Biodiversity has been enhanced by sound management and wise use by local communities. Increasingly, the role of rural communities as custodians of forests and biodiversity is being recognised. Some 250 million people live in communities and depend on forests for their livelihoods. It is estimated that twenty-two per cent of the forests found in developing countries are managed by local communities (Bray and Merino-Pérez 2004). Non-timber forest products and bushmeat are major sources of livelihood security for poor rural communities throughout the world. The trade in wild plants and animals and their derivatives is estimated at nearly US$160 billion annually (Millennium Ecosystem Assessment 2005: 53).

Ecosystems provide human populations with renewable goods and services that constitute the basis for adaptation to the changing conditions that are sure to characterise the coming decades. Agrobiodiversity continues to be the most reliable source of food security. Nonetheless, dozens of endemic varieties of cultigens are being replaced by commercial seeds or are simply lost to future generations. The past is still a source of important lessons on the importance of maintaining food security (see Chapter 45) in the face of a changing environment, as it provides more options for a society to choose from.

Livelihood strategies often adopt direct land-use practices and resource use to modify risk conditions and thus to lay the foundations for successful adjustment strategies. Examples are windbreaks, soil conservation and rainwater harvesting.

Local livelihoods

Given the reliance of the poor on ecosystem services for their livelihoods, a central element of this approach should be ecosystem management and restoration activities such as watershed rehabilitation, agro-ecology and forest landscape restoration. By protecting and enhancing the natural and managed ecosystem services that support livelihoods, vulnerable communities can maintain local safety nets, increase the buffering capacities of local ecosystems and expand the range of options for coping with disruptive shocks and trends. Some ecosystems regulate water flow and can be critical in maintaining the hydrological cycle by storing excess water, as in the case of riparian wetlands, and maintaining base flow during dry spells.

Research on the hydrological role of Tropical Montane Cloud Forests has confirmed their importance as flow regulators in upper watersheds (Bruijnzeel 2001). Major cities in Central America and some of the larger Caribbean islands such as Jamaica rely on cloud forests for their water supply. For the Andean highlands, the cold and wet *páramo* ecosystem is of vital importance. As groundwater extraction is scarce and difficult, the *páramo* is the most important water provider for major cities as well as most of the agricultural area in Colombia, Venezuela and Ecuador (see Box 18.2 on *páramos* in Chapter 18). The water is used for urban purposes, for irrigation in the drier, lower areas, and for electricity supply through hydroelectric power plants. The *páramos'* hydrological properties hinge on the soil–plant interaction. However, these properties are extremely vulnerable to irreversible degradation when the soils are disturbed. Therefore, their high mountain ecosystems guarantee a uniform and reliable water supply, and their conservation and sustainable use is the subject of considerable policy debate.

There also has been recent debate about the role of forests in flood control. While there is general acceptance of the role of cloud forests and mangroves (see Box 32.2) in water conservation and coastal protection, there have been discussions about the scale at which these benefits can be felt in the larger watershed. While there are numerous examples of successful micro-watershed management, the larger the size of the watershed and the more variable the precipitation, the more diffuse are the effects of forest cover on large-scale flood and drought events (Kaimowitz 2000 and FAO/CIFOR 2005), although other functions, such as control of soil erosion, are less debated.

Historically, there have been recorded shifts from forest to savannah ecosystems, due to anthropogenic use of fire (Moran 1981). Most often, the transformation of landscapes is the result of livelihood strategies, by which useful plants are nurtured and cultivated. Ecosystem change can also result from wholesale shifts in structure or function as a result of nutrient loading, invasive species or recurrent hazard such as wildfires (see Chapter 24). The resulting ecosystem, though stable and more resilient, is usually less productive and less diverse, and thus offers a diminished carrying capacity. Ecosystems can irreversibly flip into different form and function, as when a coral reef ecosystem shifts from a coral-dominated to an algal-dominated coastal ecosystem. Such phase shifts in coastal ecosystems have been documented in Jamaica and Belize, and in the Indo-Pacific Coral Reefs (Millennium Ecosystem Assessment 2005: 22).

Box 32.2 Mangroves and the mitigation of coastal hazards

In coastal ecosystems, it has been shown that restoring mangroves in cyclone-prone areas not only restores degraded ecosystems and increases physical protection against storms, but also boosts fisheries production, which generates much-needed income for local communities. In the tsunami-hit areas of South Asia in December 2004, some Sri Lankan communities with healthy mangrove forests and settlements protected by shelter belts tended to be less impacted than those with few natural sea defences (Dahdouh-Guebas *et al.* 2005). India and Bangladesh have come to recognise the importance of the Sundarbans mangrove forest in the Gulf of Bengal, not only as a source of livelihoods for fishing communities, but also as an effective mechanism for coastal protection. Vietnam is also investing in mangrove restoration as a cost-effective means of increased coastal protection. Similar benefits can be derived from healthy coral reefs, which are less brittle and which form better protection against storm surges and hurricanes (Dudley *et al.* 2010).

How is it possible to distinguish what is predictable (yet nonetheless uncertain) from that which is emergent, incremental and inherently unpredictable? How far does the current extinction spasm of species go beyond 'normal' historical episodes? Many of these surprises will bring benefits and opportunities to some, and will spell catastrophe and chaos for others. All directly impact the most vulnerable segments of the population: the poor and those who rely most heavily on natural resources for their livelihoods. The ultimate human insecurity is found in the unravelling of the web of life represented by the world's ecosystems.

Main drivers of biodiversity loss

As global climate change and pervasive changes in the world's biosphere take place at an accelerated pace, the range and quality of goods and services provided by ecosystems seem bound to dwindle. Combined with rapid habitat transformation, climate change will exacerbate the loss of biodiversity and the degradation of ecosystem services. Similarly, healthy ecosystems are also increasingly recognised as crucial buffers against several forms of extremes, as carbon sinks and as filters for waterborne and airborne pollutants.

While the International Union for Conservation of Nature (IUCN) Red List (www.iucn-redlist.org) provides an indication of extinction threats of a small proportion of the diversity of plant life on the planet, it does not reflect the wealth of agricultural biodiversity, and intra-specific diversity of cultivated plants. Lack of regulation also leads to overexploitation and destructive extractive techniques, especially of forest and marine species.

Policies that can address both the risks and opportunities posed by rapid environmental changes will require a combined focus on ecosystem management, sustainable livelihoods and local risk management (Sudmeier-Rieux and Ash 2009). Improved management of water resources and non-structural mitigation of weather-related hazards can reduce disaster risks by enhancing landscape restoration, mangrove forest management and local conservation and sustainable use initiatives (e.g. Box 32.2). In the face of the disaster of climate change, adaptation must be rooted in reducing vulnerabilities of communities which most depend on ecosystems for their livelihoods. Significant opportunities for this lie in ecosystem management and restoration activities. By protecting and enhancing natural services, it is possible to help to secure the livelihoods of the world's most vulnerable communities and to improve their capacity to deal with the impacts of climate change and simultaneously other disasters (Sudmeier-Rieux and Ash 2009).

Habitat loss

Today, one of the major drivers of biodiversity loss is habitat loss due to conversion to agricultural and other uses, as well as overexploitation. This is expressed by the loss of species, as land-use change, land degradation and pollution through overexploitation and overharvesting by industrial farming.

To date, some 1.8 million species have been described, of a total number of species most commonly estimated to range from 5 million to 30 million (Vie *et al.* 2008). The IUCN Red List provides a comprehensive assessment of almost 50,000 species, and shows trends in the overall extinction risk of sets of species. It offers a timely early warning mechanism for monitoring biodiversity loss.

Over the past two decades, the IUCN Red List has consistently shown a worldwide deterioration in the status of wild species. However, some regions have undergone steeper declines and have more threatened fauna. In Oceania, for instance, birds are substantially more threatened than in other realms, largely due to the impacts of invasive alien species. Amphibians are most threatened in the Neotropical region, covering most of Latin America and the Caribbean, where they are also most abundant, due to fungal diseases, airborne pollution and other habitat disturbances (Vie *et al.* 2008).

Invasive species

With the increase in international communication and travel, the spread of undesirable species has also tended to increase. The introduction of invasive species is another source of increasing pressure on fragile ecosystems such as small islands and inland water bodies. Insular ecosystems are most at risk from invasive species, due to greater opportunities to invade into a particular niche (or limited so-called 'redundancy') in species and ecosystem functions in many island systems.

In freshwater habitats, the introduction of alien species is the second leading cause of extinction. Introduced invasive species can trigger shifts in an ecosystem's structure and function, producing cascading effects on its productivity and therefore adversely affecting those livelihoods that depend on it. Historically, there have been numerous accounts of disastrous outcomes of intentional introductions such as that of the Nile perch (*Lates niloticus*), which resulted in the extinction of more than 200 other fish species in the Great Lakes in Africa. Another example has been the introduction of the Comb Jellyfish (*Mnemiopsis leidyi*) in the Black Sea, which has caused the collapse of twenty-six major fisheries and has contributed to the subsequent growth of an oxygen-deprived 'dead zone' in this inland water body (Millennium Ecosystem Assessment 2005: 22). Other invasive species include aquatic plants such as the Water Hyacinth (*Eichhornia Crassipes*), a South America native, which has spread to over fifty countries across five continents. Its rapid growth has choked inland waterways and deprived other species of oxygen and habitat. This requires constant investment in dredging and clearing of waterways for regional transport.

Emerging risks and biological hazards

More difficult to control are parasitic or viral infections of wildlife, such as the case of Avian Malaria (*Plasmodium relictum*) which is decimating Hawaii's native bird populations. Unfortunately, potentially damaging introductions continue unabated, in spite of concerted efforts on behalf of many countries. Careless behaviour leads to unintentional introductions. So-called 'accidents' now account for the majority of successful invasions. Another emerging biological hazard is animal-borne (zoonotic) diseases such as the West Nile virus and the Nipah virus. Warm winters, spring droughts and summer heat waves amplify the bird–mosquito cycle of West Nile virus. The disease has spread to 230 species of animals (including horses) and 138 species of birds in the USA, and is becoming a major public health concern (Harvard University Medical School 2004). Mortality of birds of prey could have ecological ripples, contributing to rodent-borne diseases and reductions in bird populations. That, in turn, can affect mosquito predation, with knock-on effects on insect populations, pollination, health and agriculture.

The Nipah virus is a newly emerging virus that is carried by fruit bats. Extensive fires in South-East Asia accompanying the drought associated with El Niño in 1997–98 removed food sources for bats, which led to their displacement onto pig farms. Over 100 people died and the pig industry was devastated. Nipah virus re-emerged in Bangladesh in 2003 and 2004 (Harvard University Medical School 2004). Similarly, outbreaks of avian flu are being reported in Asia, with devastating consequences from both domesticated fowl and wildlife, and an increasing public health concern for the Mediterranean and European region.

Incidences of pests and diseases in natural forests are on the increase. Bark beetle infestations in Canada and the USA affect pine forests from New Mexico and Arizona up through the west to British Columbia and into Alaska. In British Columbia nearly 22 million acres are infested, which is enough timber to build 3.3 million homes and to supply the entire US housing market for two years (Harvard University Medical School 2004). Similarly, following the 2001 drought in Central America a bark beetle (*Dendroctonus frontalis*) plague infested over 60,000 hectares of pine savannahs of Honduras and Nicaragua, illustrating the complex interaction between temperature and other stress factors (CCAD 2002). Drought and water stress encourage beetle infestations by drying out the resin that drowns the beetles, while warming encourages their over-wintering, reproduction and migration to new heights and latitudes. The dead stands

Box 32.3 Examples of ecosystem service provision

- Global: over 100 studies in protected areas have identified important crop relatives in the wild.
- Global: 33 of the world's 105 largest cities derive their drinking water from catchments within forest protected areas.
- Global: 112 studies in marine protected areas found that they increased size and population of fish.
- Papua New Guinea: in Kimbe a locally managed marine protected area network is being designed to protect coral reefs, coastal habitats and food security.
- Colombia: the Alto Orito Indi-Angue Sanctuary was set up explicitly to protect medicinal plants.
- Trinidad and Tobago: the restoration and conservation of the Nariva wetlands recognises their importance as a carbon sink, a high-biodiversity ecosystem and a natural buffering system against coastal storms.
- Sri Lanka: the Muthurajawella protected area has flood protection valued at over $5 million per year.
- Australia: management of Melbourne's forested catchments (almost one-half of which are protected areas) is being adapted in the face of climate change scenarios to minimise water yield impacts.
- Switzerland: seventeen per cent of forests are managed to stop avalanches, worth US$2 billion–US$3.5 billion per year.

contribute to wildfires, with losses of life and health, property and timber; harm to watersheds and water quality; and increased risk of avalanches (Harvard University Medical School 2004).

Mitigation of biodiversity loss, livelihood resilience and disaster risk reduction

The mitigation and reversal of biodiversity loss can contribute significantly to disaster risk reduction as healthy ecosystems naturally buffer human societies and livelihoods from pervasive hazards (see Box 32.3). The strategies adopted to conserve, restore and nurture biodiversity can be separated as two main categories: top-down measures and bottom-up measures. The first refers to the in situ and ex situ conservation policies adopted by governments to preserve unique and representative samples of the Earth's biodiversity. Top-down approaches include those efforts to collect and preserve ex situ, that is outside their areas of natural occurrence, seeds, germ plasm and plants in collections and botanical gardens throughout the world. Bottom-up approaches refer to the ancient practice of plant and seed selection by local farmers throughout the world, as well as innovative local practices that continue to be developed and implemented, to conserve and sustainably use biodiversity in the landscape. These local agricultural practices are still at the heart of maintaining agrobiodiversity, in spite of increasing outside pressures for their conversion to commercial agriculture, cash crops and industrial hybrid seed stock.

Top-down measures

The role of protected areas

Maintaining plant genetic resources over time requires a similar approach to the one adopted by wider biodiversity conservation measures. The United Nations Convention on Biological Diversity (CBD) identifies in its text two major strategies to conserve biological diversity: in situ conservation (Article 8) and ex situ conservation (Article 9). In situ conservation has led to the creation of protected areas systems, aimed at conserving genetic material and species within their original setting as functional ecosystems. Protected areas worldwide amount to approximately 120,000 state-run parks which encompass 13.9 per cent of the Earth's land surface. They contribute to storing considerable volumes of carbon, estimated at being on the order of 312 gigatons, amounting to fifteen per cent of the total terrestrial carbon stock (Dudley *et al.* 2010: 9). That does not include an unknown number of protected areas outside the state system, including indigenous and community conserved areas, which provide other significant social and spiritual value. These protected areas are also a source of wild relatives of agricultural crops, while contributing to coastal management including for settlements and providing water for urban populations (see Box 32.3).

In many countries the combination of a secured natural resource base, reduced exposure to natural hazards and diversified livelihood activities has supported disaster risk reduction for present and future threats, including the disaster of climate change, as in the case of Costa Rica where a long-standing tradition of in situ conservation has helped to provide ecosystem services in the face of climate-related hazards. In fact, a sustainable livelihoods approach in dealing with ecosystem management for disaster risk reduction, including climate change adaptation, has the advantage of meeting immediate development needs while contributing to longer-term capacity development that will create a basis for avoiding or reducing future vulnerabilities. Therein lies the importance of integrating productive landscapes, in and around protected areas, as part of an overall strategy to reduce risk.

Ex situ conservation

Ex situ conservation relates to all the strategies covering salvaging and conserving species and genetic material outside their original setting through the creation of seed banks, botanical gardens and zoos. In the case of the International Treaty on Plant Genetic Resources for Food and Agriculture, in situ conservation implies protecting farmers' rights, and protecting traditional agriculture from the impact of modern agriculture. However, these approaches are losing ground in the face of the introduction of hybrid seed stock and other genetically modified plants.

On the other hand, ex situ conservation of plant material is made possible by the creation of seed banks, plant collections and botanical gardens. Most ex situ plant collections are to be found in national or local seed banks, amounting to eighty-eight per cent of the total world collection. As much as twelve per cent of the world's ex situ collections of species identified under the International Treaty on Plant Genetic Resources for Food and Agriculture are held by International Agricultural Research Centers (IARC), such as the International Maize and Wheat Improvement Center (CIMMYT) in Mexico for maize varieties, as part of the Consultative Group on International Agricultural Research (CGIAR) (Moore and Tymowski 2005: 36). Particularly in the case of the management of seed stock, traditional farmers have historically managed seeds from one planting season to the next. They represent by far the largest portion of total world collection of plants and seed, and are thus critical to the future of agrobiodiversity. This should put into perspective the overrated dividend from applied research in genetic engineering, as the issue of collective rights over living components of biodiversity is at the heart of the current controversy over genetically modified organisms.

Bottom-up measures

Agrobiodiversity is the result of thousands of years of interaction between farmers and wild crop relatives, and the other wild biodiversity on which they depend – for example for pollination by insects, or nutrient uptake through fungal symbioses. Its conservation and expansion over the long term is at the heart of current debates surrounding food security, sustainable livelihoods and, more recently, adaptation to climate change with disaster risk reduction. The long-term management of a diverse intra-species diversity depends to a large degree on multiple strategies for the conservation of local landraces and seed stocks. These local, bottom-up agrobiodiversity conservation initiatives often hinge on broader processes at the community or landscape level. There are numerous examples of successful linkages with local community-based natural resource management and traditional agriculture, in the restoration of landscapes and watersheds (Girot 2000). These bottom-up approaches can involve community-based organisations, producers' associations and other local non-governmental organisations (NGOs) in community forestry, eco-agriculture and landscape restoration through soil conservation practices, sustainable forest management through no-timber forest products, ecotourism and other related activities (Pasos 1994). They can involve a varied combination of grassroots groups, community-based organisations and local small businesses, although both community-based organisations and small businesses require enabling environments for their rights to be recognised and for the opportunities to be harnessed.

These bottom-up sustainable livelihoods can also contribute to maintaining and restoring ecosystem services by managing wetlands, forest landscapes and coastal ecosystems. The effects of these local initiatives contribute to disaster risk reduction, but must be integrated into larger policy and practice frameworks, so that the ideas are implemented elsewhere and so that errors elsewhere do not undermine the successful local initiatives. There are numerous examples of regional fisheries in West Africa and community-based forestry in Mesoamerica, which have successfully scaled up and scaled out to have a larger ecosystem-wide effect (e.g. Pasos 1994; Girot 2002). These successful examples point to the need for adequate governance structures at multiple scales plus the recognition of local resource management organisations, through clear access and arrangements to share benefits, secured tenure rights and other key aspects for successful natural resource management to take place. The emergence of rights-based approaches for conservation of biodiversity tends to reinforce these local governance aspects, which are crucial to risk management and vulnerability reduction (Campese *et al.* 2009).

Conclusions

The linkages amongst biodiversity, livelihoods, disasters and disaster risk reduction are complex, bearing witness to the intrinsic relationship between societies and their natural and managed environments. If humanity seeks to avoid the impact of critical pervasive threats to human well-being, including disasters, it is clear that the unravelling of world ecosystems is arguably one of the greatest threats to life on Earth along with large-scale disasters including climate change. Global concatenation of changes at various scales might still have surprises in store. In this sense, a heightened awareness of all forms of current and emerging risks is necessary, as is the setting up of early warning systems for detecting biological hazards and invasive species.

In this chapter, the origins of agriculture and plant domestication have been reviewed, underlining the importance of biologically diverse landscapes and ecosystems as centres of plant diversity, and therefore as centres of the origin of domesticated plants. The evolution of agriculture is also inseparable from the surrounding ecosystems with which most farmers interact. In this sense, the future of agriculture is intimately linked to the maintenance of ecosystem services. The impact and success of the International Treaty on Plant Genetic Resources for Food and Agriculture will therefore depend on in situ conservation and sustainable use of plants by farmers as they have done for millennia, and on how agricultural areas are managed in relation to and with the wider landscape.

Others argue that ex situ conservation through zoos, botanical gardens and seed banks will provide the long-term solutions needed to feed the world. An ongoing controversy pits two world views against each other. On the one hand, there is a view based on cultural ecology and political ecology that places farmers from around the world at the centre of the long-term management of plant genetic resources. On the other hand, there is a view centred on genetic engineering and technical expertise, which empowers the technicians and the scientists in charge of creating seed banks, exploring genetic modifications and hybrids, and which claims to offer some promise of a world free of hunger and poverty.

These world views are clearly grossly simplified and are here used as a rhetorical recourse. The real world is more complex, and clearly both approaches are required to conserve and sustainably use biodiversity, and reduce risk, but the tension that emerges between locally controlled biological resources and globally controlled biological resources is very real and dramatic. It has a bearing on the way in which protected areas management, as well as the management of invasive species and other plagues and diseases, is pursued. The political ecology of the future of biodiversity will play out in the corn fields of Chiapas, Mexico, in the highlands of Ethiopia and in the foothills of the Himalayas. This is where local biological innovation, which has produced and maintained a stable and diverse source of useful plants and crops, is thriving.

These new geographies of conservation, as Zimmerer (2006) describes them, are not the wilderness of remote pristine landscapes to be preserved, theoretically unspoiled by human hands despite the challenge of that reality, but rather those landscapes composed of manicured

Pascal O. Girot

Box 32.4 Facing locust swarms in the Philippines

Ma. Florina Orillos-Juan
Dela Salle University, Philippines

Locusts have always been considered a grave problem in the Philippines. The majority of the country's crops – like rice, corn and sugarcane – are the basic sustenance of Oriental Migratory Locusts (*Locusta migratoria manilensis Meyen*), which cause the most destruction whenever they attack agricultural fields.

In pre-colonial times people used to invoke the help of a priestess to make offerings in order to appease the gods and deliver the crops from these voracious insects. Under the Spanish colonial government, locusts often devastated agricultural lands, leading to starvation. For ecclesiastical officials, swarms of locusts were punishments sent by god, so that the people would realise and repent for their sins. Consequently, they prescribed solutions that were anchored on religion: processions of images of saints in the rice fields, saying mass in the middle of the farmlands, blessing the fields by using holy water and appointing San Agustin to be the patron saint against locusts.

Civil officials had a more secular view and considered locusts as part of the natural world. The colonial government created local boards against locusts, whose primary duty was to undertake scouting and extermination works in the field through obligatory labour, ploughing of fields where eggs were laid, driving hoppers into pits and the manual catching of flyers. A bird, the martin, was also imported from China and released so that they could feed on locusts. Native Filipinos used a different strategy, eating them using a variety of preparations and recipes.

At the turn of the twentieth century, the Americans faced two different agricultural menaces: rinderpest and locust. Rinderpest was prioritised because draft bovine animals were dying by the hundreds every day, leaving farmlands uncultivated and barren, a condition that was favourable to the locusts. The American government in the Philippines enforced the legal compulsory rendition of service of all adult males to work in order to kill the locusts. Research and experiments also led to the use of biological methods (e.g. parasites, fungus and predators) and artificial techniques (e.g. soap solutions, baits laced with poison and chemicals like arsenic).

From the 1960s to the 1980s locust infestations occurred after each prolonged drought. During these years of severe outbreaks, chemical warfare was used to contain the locusts but the pesticide used to control them proved to be effective in killing the flyers but not the hoppers or nymphs. Eventually, the ashes spewed by Mt Pinatubo in 1991 and settling across hundreds of square kilometres around the volcano provided an ideal breeding ground for the locusts. Four years after the eruption, all six provinces surrounding the volcano were constantly invaded by locusts, destroying fields of rice, corn, sugarcane and minor crops.

At present, poisoned bait is still used to kill hoppers. Incipient swarms are controlled by manual, motorised or aerial spraying of pesticides and insecticides that are biodegradable and with claimed low toxicity to humans.

fields, forest patches and wetlands managed and preserved over generations by industrious and innovative farmers and fisherfolk. This romantic world view of resisting disasters and building resilient and sustainable livelihoods is often confronted by the stark realities of corporate expansion, monoculture and agri-business. Huge swaths of the Pantanal and Grand Chaco ecosystems of South America have been converted into soybean plantations, most of which are from genetically modified seed stock. Asian entrepreneurs are currently buying up large concessions of land in Africa such as in Madagascar for industrial-scale food and fibre production. The scramble for the production of biofuels has displaced food crops and smallholder agriculture in Central America. These are real and significant threats to food security and biological diversity. This is where the agricultural frontier reaches its final expression, and requires new approaches to understand.

As species continue to be lost, of which many still do not have full information regarding the complete natural history or their functions in the ecosystem, the cornucopia of life continues to unravel. Consequences take the form of destructive events, plagues, toxic algal blooms, fish stock depletion or massive forest fires – often claimed to be unexpected, yet usually foreseeable and with precedents (e.g. Box 32.4). Beyond the creative destruction of ecosystems, there is the risk of positive feedback loops and the concatenation of biological effects, as food chains unravel (Smil 1993). Ultimately, the final sufferers will be, as usual, the most exposed and vulnerable segments of human societies: those who depend directly on ecosystems for their livelihoods, and who have demonstrated, over time, flexibility, resourcefulness and innovation. The test of time will tell whether these new geographies of conservation can withstand this tide of change.

Astronomical hazards

33
Hazards from space

Bill McGuire

DEPARTMENT OF EARTH SCIENCES, UNIVERSITY COLLEGE LONDON, UK

Introduction

Hazards from space take the form of periodic collisions with asteroids and comets, powerful magnetic storms triggered by solar turbulence, and exceptional bursts of high-energy cosmic rays arising from exploding stars in the neighbourhood of our solar system. While having the potential to disrupt global communications and energy networks, or to obliterate a city, the associated day-to-day risk from hazards sourced beyond the atmosphere is small compared with more ubiquitous terrestrial hazards, such as earthquakes, floods and windstorms. As far as can be fully verified, no hazard from space has yet claimed a human life (but see later for some unconfirmed historical possibilities), and even for an impact large enough to have global consequences, the long-term, time-averaged mortality rate is just a few hundred deaths a year: a figure that will fall to fewer than 100 in a few years once detection of all such large objects is essentially complete. Nevertheless, religion, mythology and popular culture alike indulge in speculation about cataclysmic, other-worldly events that might affect the whole Earth (see Chapter 8 and Chapter 10).

For smaller land impacts, equivalent to a very large nuclear explosion, or for a tsunami-generating object that strikes an ocean basin, the long-term, time-averaged death rates are even smaller – fifty and fifteen, respectively, the latter being just three times higher than the annual number of deaths due to shark attacks (Chapman 2007). Despite the low average mortality rates, and from a disaster risk reduction perspective, it is worth noting that the probability of a land impact by an object in the 70 m–200 m size range in this century, killing around 100,000 people, is more than one per cent. While less in the public eye, the ramifications of a dose of extreme space weather also have the potential to be extremely disruptive, perhaps on a global scale. Notwithstanding the appearance of a supernova in our sector of the galaxy, the greatest threat arises from major solar outbursts causing geomagnetic storms capable of disrupting global communications and commerce and knocking out power grids.

The nature of the impact hazard

The Earth is under constant onslaught from space. Most incoming debris is efficiently incinerated as it enters the atmosphere, and only exceptionally does an object survive relatively intact to explode violently in the lower atmosphere or impact upon the surface. For this to happen, the original object needs to be at least 30 m–40 m across if it is mainly rocky, or smaller if it is a rare iron impactor. In modern history, only one major impact event has been verified, occurring at Tunguska (Siberia) in 1908. On this occasion a rocky body, generally regarded as having had a diameter of around 60 m, exploded 5 km–10 km above the surface causing a powerful blast that flattened around 2,150 km^2 of forest, destroying an estimated 80 million trees (Longo 2007; Mignan 2009).

The Tunguska object appears to have been a small asteroid and a member of the group of Earth-threatening rocky bodies known as near Earth asteroids (or NEAs), which, together with comets having short orbital periods, make up the population of near-Earth objects (NEOs). NEAs are bodies the orbits of which approach or cross that of the Earth or, more strictly, those of which the perihelia (closest orbital distances to the Sun) is <1.3 astronomical units (1 AU = the mean distance of Earth from the Sun). It is estimated that around one-fifth of NEAs currently follow orbits that enable them to approach within 0.05 AU of the Earth (Chapman 2007). These are termed potentially hazardous objects (PHOs), on the basis that orbital fluctuations, perhaps as a consequence of close encounters with our planet, may result in a near-term future impact. At the time of writing (January 2010), 1,086 potentially hazardous asteroids have been logged, together with 84 potentially hazardous comets.

NEAs, the major threat, arise primarily from collisions between larger bodies in the main asteroid belt between Mars and Jupiter, and progressively work their way into the inner solar system over millions of years. Collisions with the Sun and terrestrial planets (Mercury, Venus, Earth and Mars) act to reduce NEA numbers, but the population is continually being replenished from the asteroid belt, and sometimes by defunct comets from the outer solar system. As of end-May 2011, 7,961 NEAs had been discovered, including 828 with diameters of 1 km or greater.

Because the Earth's surface is subject to dynamic tectonic forces, its preservation potential in relation to impact craters is poor. Consequently, impact frequencies are primarily based upon observations of the numbers, sizes and orbital dynamics of near-Earth asteroids and comets. At the bottom end of the scale, one out of around 1 billion 4 m objects strikes the Earth every year, detonating high in the atmosphere with the force of around 5 kilotonnes of TNT (Chapman 2007), but having no effect on the surface. Tunguska-sized objects are expected to collide with the planet every several hundred to 1,000 years, although the actual frequency remains hotly debated. Such objects explode with the force of around 15 megatonnes of TNT and are easily capable of obliterating a major city.

A 300 m body, capable of wiping out a small European nation or US state, is expected to hit, on average, every 50,000 years or so, while a 1 km object – at the low end of the size range capable of having global environmental ramifications – is likely to strike the Earth every half a million years or so. Collisions with giant (10 km or more) asteroids or comets are extremely rare, and have return periods of perhaps 100 million years or more. The frequencies for a range of impactor sizes, together with probabilities, potential damage and required responses, are shown in Table 33.1 (adapted from Chapman 2007).

As noted by Morrison (2006), the risk of impact by an unknown asteroid is reducing progressively as surveys shrink the number of undiscovered 'global threshold' (capable of worldwide environmental effects) objects 1 km or more across. Partly as a consequence of this, attention has become increasingly focused upon the threat from smaller impactors capable of local or regional destruction. Accordingly, a National Aeronautics and Space Administration (NASA)-sponsored study (Stokes 2003) concentrated on two classes of impact event involving sub-kilometre objects: (1) a land impact or air-burst explosion over land; and (2) a marine

Table 33.1 Energy equivalent, probability, potential damage and required response for various sizes of impactor

Impactor diameter	Energy (TNT equivalent)	Probability this century (world)	Potential damage	Required response
<10 km	100 million MT	>1 in 1 million	Mass extinction; potential eradication of human species	No effective response possible at current technological level
<3 km	1.5 million MT	>1 in 50,000	Global, multi-year climate/ ecological disaster; civilisation destroyed (a new Dark Age); most people killed in aftermath	Probability of effective response remote; mitigation extremely challenging
<1 km	80,000 MT	0.02 per cent	Destruction of region of land impact or ocean rim if marine impact; potential global climate shock; approaches civilisation destruction level	Consider deflection or plan for unprecedented global catastrophe
<300 m	2,000 MT	0.2 per cent	Crater 5 km across; devastation of region the size of small nation or regional tsunami	Advance warning or no notice equally likely; deflect if possible; globally co-ordinated disaster management required
<100 m	80 MT	1 per cent	Low-altitude or ground burst larger than biggest ever thermonuclear weapon; crater 1 km across; regionally devastating	Advance warning unlikely; post-event national crisis management
<30 m	2 MT	40 per cent	Devastating stratospheric explosion; trees toppled and wooden houses destroyed within 10 km; many deaths if populated region	Advance warning very unlikely; all-hazards advanced planning would apply
<10 m	100 kT	6 per century	Extraordinary explosion in the sky; broken windows; little damage on the ground	No warning; no response required
<3 m	2 kT	2 per year	Blinding explosion in the sky; could be mistaken for atomic bomb	No warning; no response required
<1 m	100 t	40 per year	Explosion approaching brilliance of the Sun for a second or so; harmless	No warning; no response required
<0.3 m	2 t	1,000 per year	Dazzling, memorable fireball; harmless	No warning; no response required

Source: Adapted from Chapman 2007

impact with the potential to generate tsunamis. Trading off frequency against consequences, Stokes and his colleagues conclude that the greatest hazard from sub-kilometre impactors is associated with asteroid land impacts in the 100 m–200 m size range. As a consequence of tsunami propagation, sub-kilometre marine impacts have the potential to affect much larger areas and a higher number of people, to the extent that the total hazard is greater than that of sub-kilometre land impacts by a factor of five. On the other hand, as a consequence of the peak tsunami hazard being associated with larger (200 m–500 m) impactors, the frequency of such events is significantly lower.

In comparison with NEAs, the hazard presented by comets is regarded as small, contributing to around one per cent of the impact hazard (Stokes 2003). No known 'live' comet is regarded as providing a serious threat in the medium term, and defunct comets that have lost their volatiles and become captured in NEA-like orbits are known to be structurally weak and 'fluffy'. Such bodies would, therefore, be less able to penetrate the Earth's atmosphere intact than a more robust asteroid of equivalent size.

The degree of fascination and concern surrounding the hazard presented by large impacts is interesting, given the very low level of threat relative to other geophysical events capable of disruption on a global or near-global scale (McGuire 2006a). Although a 1 km asteroid strike might be expected on average every 500,000 years, the time-averaged return period for a comparably devastating volcanic super-eruption is just 50,000 years (Mason *et al.* 2004), with eruptions large enough to significantly perturb the climate for a year or two happening every 250–1,000 years. Lateral collapse of an ocean island volcano, capable of generating an ocean-wide tsunami (McGuire 2006b), occurs on average every 10,000 years. One argument is, therefore, that while real and important, both the overall threat from space and our response to it have been somewhat exaggerated.

Space hazards in the media: myth and reality

Notwithstanding endless accounts of alien invasion, natural hazards originating beyond the atmosphere have also long held a fascination for the media (Hartwell 2007) (see Chapter 63). One disaster blockbuster from Hollywood – '2012' – follows a sequence of films that includes 'Deep Impact' and 'Armageddon', which imagine the Earth at the mercy of incoming asteroid or comet impacts, or, in the case of '2012', brought to its knees by cataclysmic solar outbursts (see Chapter 11). This interest is mirrored in literary fiction in novels such as 'Lucifer's Hammer' by Jerry Pournelle and Larry Niven, and Arthur C. Clarke's 'The Hammer of God'. Sometimes the Earth survives pretty much intact as a consequence of the actions of a hero figure or figures ('Armageddon', 'Deep Impact') or, more rarely, through leadership ('Hammer of God'). At others, global society is knocked back to the Dark Ages resulting in a bunch of like-minded and honourable individuals coming together against a background of post-apocalyptic chaos, in order to endure against the more unpleasant elements of humanity ('Lucifer's Hammer').

Popular interest in the impact threat over the last twenty years has been driven noticeably by two events in science. First, the discovery of the giant (180 km) Chicxulub impact crater beneath Mexico's Yucatan Peninsula and its identification, in the early 1990s, as the probable 'smoking gun' in relation to the demise of the dinosaurs and the end-Cretaceous mass extinction (e.g. Alvarez 1997) – a hypothesis that has recently been supported by a major, international research study (Schulte *et al.* 2010). Second, the successive impacts of the twenty-one fragments of disrupted Comet Shoemaker-Levy 9 on the planet Jupiter in 1994. These events also spurred increased scientific research into the impact threat and raised the awareness of national governments.

Most importantly, in the latter context, this resulted in the US Congress directing NASA, in 1998, to detect and track ninety per cent of NEOs of 1 km diameter or larger by 2008. This task is now nearing completion, with close to 900 of an estimated large NEO population of 950–1,000 or so now located and monitored. Recently, interest in space hazards has been maintained by the identification of a 270-m-diameter asteroid, Apophis (e.g. Binzel *et al.* 2009), which will make a very close approach to the Earth on Friday 13 April 2029, passing within the geosynchronous orbits of communications satellites.

The raised profile of the contemporary risk presented by asteroids and comets also promoted a re-evaluation of enigmatic events in human history and recent prehistory that might conceivably be explained in terms of bombardment from space. Masse (2007), for example, subscribes to the idea that the great flood 'myth' present in many cultures has its roots in a marine comet impact occurring around 5,000 years ago in the Indian Ocean, perhaps with associated fragments striking elsewhere. The resulting tsunamis, together with torrential rains caused by water-loading of the atmosphere, are presented as underpinning the legends of Gilgamesh, Noah and others.

Impacts have also been implicated in the near simultaneous demise of several Bronze Age societies in the Near East, Africa and Asia. Notwithstanding mainstream thinking embracing the idea that the impact hazard remains broadly constant over time, a minority 'coherent catastrophist' viewpoint holds that impact events are clustered (e.g. Steel *et al.* 1994; Baillie 2007a). The corollary of this is that during some periods in recent Earth history, most notably the Bronze Age, collisions with extra-terrestrial bodies were far more common than they have been in the last millennium (see Box 33.1).

Looking further back in time, there is serious and growing scientific interest in the possibility of a significant impact having occurred in North America around 12,900 years ago, coinciding with a charred layer containing nanodiamonds, metallic spherules, iridium and other indicators suggestive of the arrival of a body from space. The proposed event is charged with triggering fires across North America causing the extinction of many large mammals, together with the demise of the palaeo-native American peoples who made up the Clovis culture. Enticingly, from an environmental impact viewpoint, the debris layer also immediately predates the Younger Dryas, a 1,000-year-long period that marked a temporary return to deep cold as the Earth was gradually warming following the last ice age. Kennett *et al.* (2009) propose that the nanodiamonds and other material provide evidence of multiple airbursts and possible surface impacts as a consequence of the Earth's collision with a swarm of rare, carbonaceous chondrite asteroids or comets.

Throughout human history, events in the heavens have been associated with the gods, and most often with their wrath. Comets have long been regarded as harbingers of doom, while preserved fragments of meteorites have always been held in high esteem within primitive cultures. Even today, the aurorae are rightly regarded as one of the world's natural wonders, fostering a niche tourist business and flagging a positive side to geomagnetic storms.

On occasion, an impact event or its legacy can become an important part of a culture's mythology. This is perhaps best demonstrated in Estonia where an impact around 2,700 years ago resulted in a Hiroshima-sized blast that flattened the local forest and left behind nine craters, the largest more than 100 m across. Now lake-filled and known as Kaali (Veski *et al.* 2004), this has long been regarded as having special significance. During the Iron Age, the crater was completely surrounded by a high wall and there is also some evidence for ritual sacrifices being held at the site. In Estonian mythology, Kaali is regarded as the place where the Sun went to rest, while in neighbouring Finland, the 2,000-year-old saga 'Kalevala' appears to incorporate the impact event within a heroic tale about the loss and rekindling of fire.

Box 33.1 Coherent catastrophe and clustered impacts

The coherent catastrophist school holds to the view that the impact rate over the last several thousand years has not been constant and that, instead, impacts are clustered. It proposes that a giant comet entered the solar system as recently as 20,000 years ago and broke up soon afterwards to form a trail of debris known as the Taurid Complex. Most years, the Earth passes through the less dense part of this debris train, resulting in the Taurid meteor shower that brightens the sky a few weeks before Christmas. According to the coherent catastrophists, however, every 2,500 to 3,000 years the Earth intersects a denser part of the debris train, resulting in a volley of rocks the size of Tunguska or larger. Such a barrage of hundreds of super-Tunguska objects, advocates the school, was responsible for the synchronous demise of Bronze Age civilisations across the planet.

The last time our planet is alleged to have passed through the dense part of the Taurid Complex was around 400 CE–600 CE, in the Dark Ages, following the collapse of the Roman Empire. The coherent catastrophists also claim an increased number of impacts at this time, partly based upon tree-ring evidence for a severe environmental downturn (Baillie 2007b) and partly upon evidence gleaned from legends and contemporary texts. Some even suggest that fireball storms formed as numerous boulder-sized rocks burnt up in the atmosphere may have promoted the social upheavals in Europe that led to the collapse of an already crumbling Roman state. Going further back in time, the Australia-based astronomer Duncan Steel has proposed that the great megalithic monument of Stonehenge was built to predict meteor storms associated with a much earlier Taurid Complex-related bombardment during the fourth millennium BCE.

There does seem to be a tendency, amongst some of the coherent catastrophists at least, to explain virtually every war, riot or natural catastrophe in terms of increased numbers of impacts, and it must be stressed that their ideas remain very much in a minority within the impact studies community. If, however, they prove to be correct, then a significant increase in the frequency of impacts may be expected in around a thousand years.

Because the sky and events occurring in it have always held great importance for societies and cultures, there are countless written descriptions and oral traditions of happenings in the heavens that may be interpreted as air bursts or fireballs heading towards the surface. While such accounts are clearly important when attempting to evaluate the nature and scale of the impact hazard in human history, there is a temptation to make assumptions based on rather flimsy historical or mythical accounts that are unsupported by scientific evidence. Convincing accounts of impact events supported by evidence are few and far between, with the 1908 Tunguska collision heading the list.

Nothing anywhere near as large as the Tunguska object has entered our atmosphere since, although a mini version of the event took place in 1947, also in Siberia, when a brilliant fireball hurtled across the Sikhote-Alin Mountains to explode with a deafening sound about 400 km north of Vladivostock. The iron object, estimated at about 10 m across, scattered 136 tonnes of fragments across the impact site that excavated more than 100 craters, the largest more than 20 m wide. In addition to the Sikhote-Alin event, perhaps a dozen airbursts are known resulting from small bodies exploding in the atmosphere. These include a blast of up to 1 megatonne over Brazil in 1930 and a 1 kilotonne–2 kilotonne explosion above Sudan's Nubian desert in 2008 that scattered the surface with small fragments. This object (designated 2008 TC$_3$) has the distinction of being the first to be identified and tracked prior to its impact twenty hours later – not much of an early warning, but nonetheless impressive given that the object was only 1 m–2 m across.

Consequences of future impacts

The ramifications of a future impact event are broadly dependent upon the size of the impactor. Most important in this regard is whether or not its diameter is sufficiently large to have global environmental consequences as a result of severe planetary cooling (cosmic winter) due to the loading of the atmosphere with massive quantities of dust. As indicated in Table 33.1, a 1 km impactor would be big enough to trigger a worldwide climate 'shock', but a true 'global threshold' object capable of engendering a global, multi-year cosmic winter is likely to require an object around 2 km across. Such an event is expected only every couple of million years and with most potential impactors larger than 1 km now accounted for, the chances of such an object appearing on the scene in the near-to-medium term are vanishingly small.

When and if a global threshold impact occurs, it is widely predicted that it would herald the complete failure of the global harvest, the loss of a substantial fraction of the human population and the arrival of a new 'dark age' (see also discussion in Bobrowsky and Rickman 2007). As addressed earlier, most concern is focused instead on the higher probability land or ocean impacts in the 100 m–500 m range. At the top end of the range, these are sufficiently large to kill millions to tens of millions, either as a direct consequence of the resulting blast and/or due to associated tsunamis in the case of a marine impact. While far from large enough to trigger a cosmic winter, a land impact approaching 500 m could raise sufficient dust to temporarily and detrimentally cool the global climate, in the manner of a large volcanic eruption. The broader ramifications of an impact in the 100 m–500 m size range would depend upon its location, with an impact in the UK, Europe, Japan or North America likely to incur losses of many trillions of US dollars, and having serious subsidiary consequences for the global economy.

The risk from (c. 50 m) Tunguska-like objects is regarded as small, primarily because the destructive footprint of such an impactor is tiny compared to the Earth's surface. The chances of such an object scoring a direct hit on a major urban centre, therefore, is extremely small. Nevertheless, it is real and its consequences have been considered.

To commemorate the centenary of the Tunguska event, Mignan (2009) modelled the consequences of a similar airblast occurring over New York City. They concluded that this would result in 3.2 million fatalities, close to 4 million injured and direct property losses totalling US $1.2 trillion. This does not, of course, take account of the major knock-on effect such an event would have on the US economy, nor the global financial mayhem that would follow the obliteration of Wall Street. While the probability of a direct hit on New York, London or Tokyo, the world's three key financial centres, remains very small indeed, a new piece of research suggests that the general hazard presented by Tunguska-like objects may be greater than previously thought. Boslough and Crawford (2008) have suggested that rather than 60 m, the Tunguska object may have been just 30 m–40 m across. If this is the case, then objects capable of a comparable level of destruction may strike somewhere on Earth as frequently as every few hundred years.

Space weather and its effects

Concern has been growing for some time that the more immediate threat from beyond the atmosphere comes from space weather rather than from space debris. Generally speaking, the 'weather' in near-Earth space is pretty benign; periodically, however, violent outbursts on the Sun's surface make it more interesting and potentially disruptive. At such times, streams of high-energy particles generated by the Sun strike and interact with the Earth's magnetic field (magnetosphere) causing so-called geomagnetic storms capable – amongst other things – of knocking out satellites in orbit and playing havoc with communications and energy networks on the surface. In 1997 a serious storm knocked out AT&Ts Telstar 401 satellite, while a year later another disabled PanAMSat's Galaxy IV satellite, affecting ATM machines, credit-card handling services, news-wire feeds, airline weather-tracking facilities and most US pagers. During another powerful storm in 2003, re-routing of high-latitude flights to avoid elevated radiation levels and radio blackout areas cost airlines up to US$100,000 per flight, while loss of the ADEOS-2 satellite, probably as a result of the storm, cost US$640 million. Most famously, in 1989 a geomagnetic storm caused by another violent solar outburst caused the catastrophic failure of Canada's Hydro-Quebec power grid, resulting in a blackout that affected 6 million people and lasted nine hours.

Looking further back, other big geomagnetic storms occurred in 1859, 1921 and 1960. The first of these was a 'superstorm', three times more powerful than any in living memory, which could provide a foretaste of future events. In its comprehensive 2008 report on the space weather hazard, the US National Research Council (NRC) (NAS 2008) recognises and warns of a catalogue of potential effects that could result from such a storm, including: interference with high-frequency (HF) radio communications and navigation signals from GPS satellites; degrading or loss of HF communications along high-altitude aviation routes; serious disruption and damage to electric power grids; corrosion of oil and gas pipelines due to induced currents; and damage to satellite electronics, solar arrays, imagers and tracking systems. The NRC report noted that a repeat of the 1921 storm would black out more than 130 million people in the USA alone, while an extreme solar storm could be massively disruptive on a global scale. The cost of such an event could be as high as US$1 trillion–US$2 trillion in the first year alone, with the world taking between four and ten years to fully recover.

Managing and mitigating space hazards

Based upon average return periods, the probability in any single year of a 30 m–40 m impact is about the same as that of an extreme geomagnetic storm. Unless the impactor strikes a major urban centre, however, the ramifications of an extreme geomagnetic storm are likely to be far greater and more widespread, with the potential for deleterious global consequences. Coping effectively with an intense geomagnetic storm is reliant upon critical infrastructure – in particular power grids and communications networks – being sufficiently robust; appropriate community preparedness measures being in place, in particular to survive an extended period without power; and sufficient early warning.

Typically, there may be two-to-three days' notice that a solar outburst capable of generating a geomagnetic storm is on its way, although in 1859 the lead time was less than eighteen hours. Forewarning of the onset of the geomagnetic storm itself may be just fifteen minutes or even less. In developed nations, the insurance industry is taking increasing interest in space weather and its effects (e.g. Swiss Re 2000). Satellites and aircraft that may be affected usually have all-risks insurance cover, while fires on the ground and business interruption associated with damage due to fires or induced currents would also normally be covered. Nonetheless, as a consequence of its potential to cause serious losses, insurers are working to reduce risk by putting pressure on the insured to take appropriate measures including building access to geomagnetic storm early warnings, developing greater robustness to the likely impacts of such storms and improving redundancy in critical systems.

Notwithstanding this, the increasingly high voltages of power grids are making them more vulnerable to being knocked out by geomagnetic storms, with a new 1,000 kV grid being developed by China highlighted as especially at risk. As noted by the NAS (2008), the USA – alongside other developed nations – is poorly placed in relation to mitigating and managing the effects of a severe geomagnetic storm. In the USA, for example, it could take more than a year for transformers to be replaced and new ones constructed, leading to a significant proportion of the country being without power for months. Unlike most hazards, the effects of space weather are likely to be felt less in the developing world than in developed nations, the primary effect being loss of power where previously available. Measures to cope with such an eventuality would be broadly the same as those required in response to power loss due to terrestrial hazards, for example the availability of back-up generators and alternative access to water that does not require electrically operated pumps.

One difference, however, may be the length of time required for transformers to be repaired and new ones supplied, due to greatly increased global demand. As a consequence, power may be unavailable for several months or even longer. This would impact, particularly, on hospitals and schools, which would require alternative power sources to be in place beforehand in order to continue to function during an extended period off-grid.

The progressive reduction of the impact risk is a consequence of being able to see potentially threatening objects coming well in advance. Once they have been detected, and their orbital parameters determined, therefore, their paths can be projected centuries into the future to see if they present a genuine near- to medium-term threat. In this context, the degree of hazard is indicated by the categorisation of the object on the Torino Scale (Binzel 2000) (see Table 33.2 and Box 33.2) or the Palermo Technical Impact Hazard Scale (see Box 33.2).

NEO discovery and tracking initiatives such as LONEOS (Lowell Observatory Near-Earth Object Search), NEAT (Near-Earth Asteroid Tracking), Catalina Sky Survey, LINEAR (Lincoln Near-Earth Asteroid Research) and Spacewatch have allowed NASA to close in on its goal of finding virtually all objects larger than 1 km. In 2005, however, the US Congress mandated the agency to detect, track and catalogue at least ninety per cent of all potentially hazardous NEOs greater than 140 m across (NAS 2009). This goal will be tackled with the aid of new search instruments, such as the 4.2 m Discovery Channel Telescope (Arizona), the 8.4 m Large Synoptic Survey Telescope (Chile), the Panoramic Survey Telescope and Rapid Response System (Pan-STARRS, Hawaii), the satellite-based NEOWISE (Wide-field Infrared Survey Explorer), together with Canada's NEOSSAT (Near Earth Object Surveillance Satellite), due for launch in 2011, and Germany's AsteroidFinder payload, slated for launch a year later.

Despite this impressive list of new instruments and sensors, the task remains a particularly difficult one that is unlikely to be completed by the target date of 2020. Nevertheless, as the orbital details of an increasing number of smaller and smaller NEOs are characterised, so the likely need for developing a technology to deflect incoming bigger objects becomes subordinate to the importance of constraining the timing and location of a future impactor of relatively small dimensions.

This is not to say that the need for deflection technologies will not be required at some future time, as NEO orbits evolve and impact probabilities change, and there is still a drive to develop these. With the Earth taking just seven minutes to move its entire diameter (12,672 km) in space, the intention is not to blast an object into pieces, but to infinitesimally alter its orbit so as to transform a certain hit into a near miss. Numerous intervention schemes have been proposed, including close-proximity 'stand-off' nuclear explosions; attaching low-thrust, long-lived motors, such as ion-drives; using foil or 'paint' to change the heat conducting properties or reflectivity of an object, leaving the Sun's rays to accomplish the required degree

Table 33.2 Public description for the Torino Scale

No hazard	0	Likelihood of a collision is zero, or is so low as to be effectively zero. Also applies to small objects such as meteors and bolides that burn up in the atmosphere, as well as infrequent meteorite falls that rarely cause damage.
Normal	1	A routine discovery in which a pass near the Earth is predicted that poses no unusual level of danger. Current calculations show the chance of a collision is extremely unlikely, with no cause for public attention or public concern. New telescopic observations very likely will lead to reassignment to Level 0.
Meriting attention by astronomers	2	A discovery, which may become routine with expanded searches, of an object making a somewhat close but not highly unusual pass near the Earth. While meriting attention by astronomers, there is no cause for public attention or public concern as an actual collision is very unlikely. New telescopic observations very likely will lead to reassignment to Level 0.
	3	A close encounter meriting attention by astronomers. Current calculations give a one per cent or greater chance of collision capable of localised destruction. Most likely, new telescopic observations will lead to reassignment to Level 0. Attention by the public and by public officials is merited if the encounter is less than a decade away.
	4	A close encounter meriting attention by astronomers. Current calculations give a one per cent or greater chance of collision capable of regional devastation. Most likely, new telescopic observations will lead to reassignment to Level 0. Attention by the public and by public officials is merited if the encounter is less than a decade away.
Threatening	5	A close encounter posing a serious, but still uncertain, threat of regional devastation. Critical attention by astronomers is needed to determine conclusively whether or not a collision will occur. If the encounter is less than a decade away, government contingency planning may be warranted.
	6	A close encounter by a large object posing a serious, but still uncertain, threat of a global catastrophe. Critical attention by astronomers is needed to determine conclusively whether or not a collision will occur. If the encounter is less than three decades away, government contingency planning may be warranted.
	7	A very close encounter by a large object, which, if occurring this century, poses an unprecedented but still uncertain threat of a global catastrophe. For such a threat in this century, international contingency planning is warranted, especially to determine urgently and conclusively whether or not a collision will occur.
Certain collisions	8	A collision is certain, capable of causing localised destruction for an impact over land or possibly a tsunami if close offshore. Such events occur, on average, between once per fifty years and once per several thousand years.
	9	A collision is certain, capable of causing unprecedented regional devastation for a land impact or the threat of a major tsunami for an ocean impact. Such events occur, on average, between once per 10,000 years and once per 100,000 years.
	10	A collision is certain, capable of causing global climatic catastrophe that may threaten the future of civilisation as we know it, whether impacting land or ocean. Such events occur on average once per 100,000 years or less often.

Note: At the time of writing (January 2010) only a single object scores above zero.
Source: Revised after Binzel (2000) to better describe the attention or response merited for each category.

Box 33.2 The Torino Scale

The Torino Scale (Binzel 2000) was developed by Richard Binzel at Massachusetts Institute of Technology (MIT) in order to answer a need to categorise the impact hazard associated with NEOs. Its name derives from an international conference in Turin (Italy) in June 1999, at which a version of the scale was presented. A more recent version of the scale was launched in 2005, which incorporated rewording so as to make the scale and its meaning more accessible to the public. The Torino Scale has eleven categories, numbered from zero to ten, with the lowest indicating an object that has a negligible probability of striking the Earth or is too small to penetrate the atmosphere, and the highest reserved for a certain collision with an object that is large enough to trigger a global environmental catastrophe and 'threaten the future of civilisation as we know it'. More precisely, every object is allocated a category on the basis of its collision probability and the amount of energy (expressed in megatonnes of TNT) that would be released if a collision occurred. Virtually every NEO so far discovered registers as zero on the Torino Scale, although Asteroid Apophis was briefly categorised as 4 before more detailed observations saw it downgraded to 0. At the time of writing (January 2010), only a single object that presents a potential threat this century merits a 1 on the scale: 130-m-diameter asteroid 2007VK184, which has a one in 31,300 probability of striking Earth in 2048.

A more complex logarithmic scale is the Palermo Technical Impact Hazard Scale, which is used primarily by astronomers and other professional scientists with an interest in the impact hazard.

of deflection; attaching a giant 'solar' sail of wafer-thin reflective foil, again allowing the Sun to do the work; and many more. Of the growing portfolio of deflection schemes, the 'gravity tractor' is probably the simplest and most promising. This involves 'parking' a spacecraft adjacent to the object for an extended period of time, with the deflection accomplished simply as a consequence of the gravitational pull exerted by the spacecraft.

Given that the primary impact hazard is from 100 m–500 m objects that may arrive with or without warning, a more appropriate coping strategy involves Earth-based mitigation and management. Where advance notice has permitted identification of an impact location, mass evacuation would be the primary response. Where warning time is insufficient, there would be no option but to advise residents of the area likely to be affected to seek appropriate shelter and ensure they have emergency supplies to hand to cope with the immediate aftermath. Dealing with the potential panic and attempted self-evacuation in such circumstances would inevitably provide an additional, possibly insurmountable, problem for emergency authorities,

For any measures to be enacted effectively, the impact hazard ideally has to be incorporated into emergency response and disaster planning at national, regional (e.g. European Union) and global levels. Currently, it is highly unlikely that the hazard is explicitly addressed in any emergency response plan. It is not, for example, specifically mentioned alongside terrestrial hazards in the Hyogo Framework for Action (UNISDR 2005b), nor is it incorporated within the all-hazards approach of the US Homeland Security Presidential Directive #5 (Chapman 2007). In principle, insurance may have a role to play in the context of extreme events (Smolka 2006) and in particular in post-event recovery from a small, damaging impact, with coverage provided by a typical property insurance policy. As noted by Kovacs and Hallak (2007), however, an object larger than a few dozen metres, which strikes a major urban centre such as New

York, London or Tokyo, is likely to overwhelm the insurance industry. Insurance, therefore, can hardly be viewed as a safety net in this instance, as it is more often than not for floods, storms and earthquakes. Furthermore, insurance cannot be used to lever policies or behaviour that might reduce the risk of damage resulting from impact, as it can for building in areas of high seismic or flood risk.

Failure to take account of the impact hazard as part of an all-hazards emergency response and disaster risk reduction policy, alongside floods, storms, earthquakes, tsunamis and other Earth-based hazards, is a clear reflection, first, of the relatively low return periods associated with impacts large enough to trigger a major catastrophe, and second, and more critically, of a continuing tendency amongst governments, emergency responders and disaster managers to view the impact threat as esoteric and (almost) apocryphal.

Two ways are suggested of overcoming this unhelpful viewpoint. First, to provide comparisons between return periods of variously sized impact events and low-frequency terrestrial hazards. For example, the return period for a 30-m impactor, capable of causing major damage and loss of life should it explode above an inhabited area, is less than the frequency of great, tsunami-generating earthquakes in the Indian Ocean, or the occurrence of a climate-perturbing volcanic eruption such as Pinatubo (Philippines) in 1991. Second, to ensure that all-hazards approaches incorporate global geophysical events (GGEs) (McGuire 2006a) and their potential consequences.

This matches one of the recommendations of the UK government's Natural Hazard Working Group (2005), established in the wake of the 2004 tsunami, which advocated identification and characterisation of 'potential natural hazards likely to have high global or regional impact' alongside other natural hazards. To date, this has largely failed to materialise. Until it does, Earth-based measures to cope with future impact events will remain of the 'make-do' variety, adapted from more generalised emergency response and disaster risk reduction plans, if and when required.

A great deal is now known about the impact hazard. The risk has been progressively reduced to a degree where it is fairly certain that the principal threat, in the near-to-medium term, comes from objects in the 100 m–500 m size range. While the physical consequences of future such impacts are well established, its location may or may not be known in advance. There is a moderate probability, however, that identification by a sky survey would provide sufficient warning and orbital determination to accurately fix the timing and location of the impact, enabling prior evacuation to take place and for other mitigatory and recovery measures to be put in place.

There remains the small possibility that an undetected object, three times or more larger than Tunguska, will strike a major urban centre unannounced, leading to destruction on a massive scale. Should this happen, it will behove the national government concerned to have an all-hazards emergency response plan in place that incorporates the impact hazard and is, therefore, flexible enough to effectively address the consequences and anticipate response and recovery requirements of a catastrophic air blast or surface explosion that is many times the size of the largest thermonuclear weapon.

Vulnerabilities and capacities

34

Disability and disaster

David Alexander

CENTRE FOR THE STUDY OF CIVIL PROTECTION AND RISK CONDITIONS, UNIVERSITY OF FLORENCE, ITALY

JC Gaillard

SCHOOL OF ENVIRONMENT, THE UNIVERSITY OF AUCKLAND, AUSTRALIA

Ben Wisner

AON-BENFIELD UCL HAZARD RESEARCH CENTRE, UNIVERSITY COLLEGE LONDON, UK

Introduction

It is a fundamental principle in the modern world that people with disabilities be given the opportunity to participate in modern society with as few impediments as possible. There is no justification for relaxing this principle when emergencies and disasters occur. Nevertheless, people with disabilities may encounter physical barriers or experience particular difficulties of communication that prevent them from reacting effectively to crisis situations and stop them from using the facilities and assistance made available to people who do not live with disabilities. In the aftermath of disaster there may be systematic, or systemic, discrimination against people with disabilities. In addition, in some, indeed perhaps many, situations people with disabilities can make valuable contributions to planning risk reduction and disaster response. However, to tap this potential and broaden participation, authorities need to take proactive steps discussed later in this chapter.

The whole question of how to assist people with disabilities in emergencies, let alone how they might help themselves, help each other and assist planners, has been roundly overlooked. The body of academic literature on this subject is small (e.g. Parr 1987) and recent attempts to renew it (e.g. Fjord and Manderson 2009; Kearns and Lowe 2007) have been neither numerous nor copious. Moreover, the subject is seldom a theme at emergency management conferences. Despite this, there are some useful initiatives both amongst the less and most affluent countries, such as the Verona Charter on the Rescue of Persons with Disabilities in Case of Disasters, some study centres (e.g. the Centre for Disability Studies, University of Leeds, UK) and some manuals of best practice by the US Federal Emergency Management Agency (FEMA), the

Shanta Memorial Rehabilitation Center in Orissa, India, and the non-governmental organisation (NGO) Handicap International.

Whereas populations threatened or afflicted by disaster usually receive aid that is supplied in an aggregate and generalised way to the entire group assisted, people with disabilities require individual assistance that is specifically tailored to their needs. This radically changes the meaning of disaster aid, but if such assistance is not offered, then the result can easily discriminate against people who, through living with disabilities, are among the most vulnerable in society – and through reduced earning capacity are often the poorest.

In the context of disasters and emergencies, what is disability?

People with disabilities are often severely affected by natural hazards. For example, only 41 of the 102 residents of a home for people with disabilities in Galle, Sri Lanka, survived the December 2004 tsunami as most could not leave or failed to understand in time the need to evacuate (IFRC 2007). Similarly, it seems that none of the 700 people with post-polio paralysis on an island of the Andaman archipelago in India were able to survive the tsunami as they were unable to run to the top of surrounding hills (Hans et al. 2008). A few hundred kilometres southwards, half of the 145 children with disabilities who were enrolled in schools run by the Indonesian Society for the Care for Children with Disabilities in Banda Aceh were killed in the disaster, which was twice the mortality rate in the area (CIR 2005).

The classic popular view of disability is that of a person in a wheelchair who must be man-handled away from danger. In reality the issue is far more complex than that. To begin with there are many forms of disability, including paraplegia, quadriplegia, deafness, blindness and defects of vision, mental illness and retardation, cerebral damage, stroke, senility, dementia and Alzheimer's disease.

There are also numerous forms of dependence on personnel, equipment and supplies, such as medicines for support to the vital functions that sustain a person's life. Although old age is not in itself a disability, many very old people are frail and lack mobility, and they may also be ill or susceptible to various diseases and conditions. Their cognitive abilities may be impaired by dementia. Except in some countries affected by warfare, or where there are many landmines, disability is more common in old age than among younger people.

The question of age-related disability requires special attention (see Chapter 37). In a study carried out in North Carolina, USA, Van Willigen et al. (2002) found that the effects of age were not perfectly correlated with those of physical disability. Fernandez et al. (2002) argued that age per se does not make a person more vulnerable to disaster: rather, it is the relationship between age and infirmity that generates special needs. They classified the needs into the categories of appropriate transportation, health care access, aid distribution, warning design and recovery needs. They found that in the USA, when frail elderly people were evacuated in advance of hurricanes, many were dropped off at shelters and nursing homes without care instructions or medical records. They also noted that, predictably, infirmity tends to vary considerably over the age range of the elderly (i.e. from sixty-five years onwards). In much of the global South, extended families care for the frail elderly, so the question of isolation or the strengths and weaknesses of institutional care in emergencies does not arise. However, with increasing urbanisation and long-distance wage migration of younger adults, such problems will appear.

Age-related disability topics are not just about the elderly. Peek and Stough's (2010) review of children with disabilities in disaster contexts identifies fundamental social processes that heighten the vulnerabilities experienced by the youngest disaster survivors. Furthermore, they highlight how re-establishing social networks and education can help to reduce disaster impacts.

In synthesis, a person with disabilities is one whose ability to move, think, perceive or express himself or herself is compromised by injury, illness or societal limitations. Clearly, the severity of disability is highly variable from one person to another. Countries with highly developed health and social care systems tend to set thresholds for the administrative classification of disability, which usually means the registration of individuals in official records of people assisted. Thus in the USA, 19.1 per cent of the population is covered by the federal Americans with Disabilities Act. Proportions in Western Europe vary from sixteen per cent to twenty per cent. One in five South Asian Indians lives with disabilities. In Laos, where twenty-five per cent of villages are contaminated with unexploded ordinance, an estimated 45,000 people have physical disabilities such as amputated limbs (Murray 1998). Around the world many more people have less serious cognitive or mobility problems. People with particular pathologies, including, for example, alcoholics, are not classified as having a disability, but to all intents and purposes some of them do. In short, at least one-fifth of the population has some form of disability status.

Box 34.1 Children with disabilities in disaster

Laura Stough
Department of Educational Psychology, Texas A&M University, USA

Lori Peek
Department of Sociology, Colorado State University, USA

According to the United Nations International Children's Emergency Fund (UNICEF 2007a), over 200 million children – representing about eleven per cent of the global child population – have some type of disability. Disabilities in children include physical disabilities, such as paralysis or orthopaedic impairments; intellectual disabilities, such as Down syndrome or foetal alcohol syndrome; sensory disabilities, such as blindness or deafness; cognitive disabilities, such as autism or learning disabilities; or psychiatric disabilities, such as depression or anxiety disorders.

Conservative estimates suggest that 7 million children with disabilities are impacted by disasters each year (Peek and Stough 2010). Millions more acquire disabilities during childhood as a consequence of disaster. For example, in Haiti, a country where people with disabilities are commonly known as 'kokobés' ('good for nothings'), hundreds of children lost their limbs due to crushing during the earthquake, while others were forced to undergo amputations as a result of secondary infections.

Historically, children with disabilities have been overlooked by disaster researchers and professionals. As such, children with disabilities may be among the least prepared and most poorly served, ultimately experiencing amplified physical, psychological and educational vulnerability.

Children with disabilities are more likely to be poor and to live in low-quality housing, which increases their exposure to hazards. When a disaster strikes, children with disabilities may have a more difficult time taking action, escaping from or withstanding the event. Children with mobility-related disabilities may be placed even more physically at risk if they are unable to evacuate in a timely manner. Even when evacuation is possible, children with disabilities and their families may be less likely to leave the threatened area.

Children with disabilities also encounter barriers during emergency response and recovery. They may find that public shelters do not accommodate wheelchairs or that announcements are not translated for those who are deaf. Disasters can be particularly hazardous for children with medical disabilities who rely on electricity for their medical support or who need medical care while they are sheltering away from home.

During disaster, parents are also impacted and this may limit their ability to respond appropriately to their child's needs. Following Hurricane Katrina in the USA, some children with disabilities were relocated to nursing homes or other institutions. Placement in these facilities sometimes meant separation from their families or sheltering with only one parent, creating further distress for these children and their caregivers.

To best protect children from the effects of disaster, children with disabilities and their caregivers need to be actively involved and considered in all disaster risk reduction activities. Evacuation procedures should accommodate children with disabilities; shelters must be made physically accessible; and school systems should offer academic modifications such as providing home instruction when schools are closed for extended periods to avoid regression of learned skills. Some children with disabilities may need the community infrastructure to be substantially rebuilt before they can return to their pre-disaster level of independence.

Disabilities fall into numerous categories with principal examples being:

- difficulties of personal mobility;
- inability to see (with possible need to use a guide dog);
- deafness;
- problems of communication and articulation of words (as with stroke-affected people);
- cognitive disorders;
- various medical problems, including those that require constant or frequent use of life-support systems;
- intolerance of chemical or environmental substances;
- psychiatric disorders and panic attacks; and
- infirmity associated with old age.

Smart (2009) discusses categories that are less medical than the approach taken in the list above.

The list is long and diverse and, of course, individuals may have more than one form of disability. Clearly, the different categories should be associated with a varied range of provisions during emergencies, including transport for people with reduced mobility, specialised means of communication for those with cognitive or speech difficulties, provision of portable or substitute equipment for those who depend on life-support systems and psychiatric support for those with mental health problems. The degree of autonomy of people with disabilities is highly variable, according to the nature of their disability and the availability of facilities that support them.

Finally, a study of disability in disasters should also consider those who have permanent injuries in the aftermath of catastrophic events. For example, 1,200 people among those 3,500 reported injured after the 1963 earthquake in Skopje, Macedonia, had permanent disabilities (UNDRO 1982a). Following the earthquake in Haiti, many amputees and people experiencing spinal cord injuries had very little support (Iezzoni 2010). Similarly, some early estimation by the United Nations (UN) and the World Bank suggested that the number of people with disabilities may have increased by twenty per cent in the countries affected by the December 2004 tsunami in South and South-East Asia as a result of the disaster. Such an influx of people with specific needs constitutes a real challenge for organisations in charge of recovery programmes. The situation was particularly critical in Indonesia after the December 2004 tsunami as revealed by field work conducted in the province of Aceh by JC Gaillard and the Tsunarisque project team in 2006 (see Box 27.2). People who had permanent injuries were joined by those who had suffered from the armed conflicts that had been ravaging Aceh for more than thirty years. Hundreds of people mutilated by acts of torture saw in the relatively peaceful environment that succeeded the disaster an opportunity to get health care. They flocked to the health centres and joined people who were directly affected by the earthquake and tsunami, thus brutally increasing the number of patients.

Marginalisation, disability and disaster risk reduction (DRR)

Unfortunately, discrimination against people with disabilities is more the rule than the exception. Coleridge (1993: 71–73) described three ways in which people with disabilities are perceived and treated by the rest of society: in the traditional model disability is considered to be a form of punishment, perhaps divinely imposed, and is thus a reason for social exclusion; in the medical model it is regarded as a departure from normality that needs to be cured; and the social model concentrates on removing barriers, rather than encouraging normalisation, treatment or care. Whereas the last of these three models is more benign than the first two, it is by no means able to solve the problem. Coleridge (1993) wrote passionately, and with a wealth of examples, about people's despair at being excluded, and he argued that we need to replace the three models with a rehabilitation model that redresses the balance and favours the reintegration of people with disabilities into mainstream society throughout the world.

Unfortunately, disasters tend to increase the level of discrimination against people with disabilities. For example, in earthquakes people in wheelchairs cannot take refuge under desks and tables, and neither can they rapidly exit a building down stairs (Rahimi 1993). Indeed, in both fire and earthquakes they may be trapped in buildings because elevators are not to be used in such emergencies. People who are deaf or have visual disorders may not hear verbal orders to evacuate or see emergency lights (Kailes 2002). Furthermore, people who depend on electrical apparatus (such as dialysis machines, ventilators, or simply electronic means of communication) may find themselves in difficulty when there are power cuts during emergencies. In China after the Sichuan earthquake in 2008, people with disabilities who were affected by the disaster were reported to search desperately for their radio devices in the rubble in order to get information on what was going on (Fu et al. 2010). Finally, all the services offered to people in disasters and crises – transportation for evacuation, shelter, counselling and so on – need to be accessible and intelligible to people with disabilities.

Accounts of such discriminating practices abound. In Bangladesh, the NGO Christian Blind Mission (CBM) reported that people with disabilities were not able to get their ration of food relief because they were required to queue with the other people affected by Cyclone Aila in May 2009. In the same country, elevated flood shelters are usually not accessible to people using a wheelchair. In the aftermath of the 2004 tsunami, Kett et al. (2005) reported that some people with mental illness were denied basic health services and treatment for serious injuries in public hospitals in Sri Lanka. In India, people with disabilities affected by the tsunami could not access latrines in evacuation centres. Similar patterns of discrimination were actually reported throughout the other countries affected by the tsunami, where a very limited number of specialised NGOs considered the particular needs of people with disabilities. Most of the organisations involved in emergency and recovery operation have, on the other hand, not fully considered people with disabilities (CIR 2005). The IFRC (2007) also reported that in the immediate aftermath of the 1994 Northridge earthquake in the USA, many people with cerebral palsy were denied access to shelters because they were thought to be intoxicated with drugs or alcohol. The same organisation observed that people with mental disabilities were also marginalised in the relief effort which followed the 1995 Kobe earthquake in Japan.

The problem is not an insignificant one. In India the disaster management law of 2005 makes no mention of the country's 90 million people with disabilities. In the USA 54 million people are classified as having a disability. At the time of Hurricane Katrina in the Gulf states of the USA, there were 155,000 people with disabilities in the cities of Biloxi (Mississippi), Mobile (Alabama) and New Orleans (Louisiana). Many found themselves in dire straits when the hurricane struck. In fact, as seventy-one per cent of the 1,330 known fatalities in the hurricane were people over the age of sixty, disability being correlated with age may have been a factor in the high percentage of elderly amongst Katrina fatalities (White 2007). Moreover, surveys confirmed the needs of people with disabilities. As McGuire *et al.* (2007) noted:

> many older adults may have required assistance evacuating before Hurricane Katrina ravaged the New Orleans–Metairie–Kenner, LA MMSA [Metropolitan and Micropolitan Statistical Area]. Indeed, as we now know, many of these disabled persons were left to fend for themselves. Some died; some may have had their primary disabling conditions untreated for several days; some went without prescribed medication, proper food and fluids; some were exposed to the elements; and some may have had their primary disabling factors complicated by hurricane-related secondary conditions, dehydration, infection and injury.
>
> *(McGuire* et al. *2007: 54)*

The detailed picture in the USA as a whole is disquieting. Considering just mobility-related disabilities, some 1.1 million Americans are affected by paralysis and 3.5 million have difficulty in walking. Poverty rates tend to be higher for households with disabilities, whose members tend to be less educated and have lower rates of health insurance than national averages. When Hurricane Bonnie struck North Carolina in 1998, households with people with disabilities incurred damages equal to eighty per cent of monthly per capita income, more than three times as high as the damages pertaining to households without members with disabilities (Van Willigen *et al.* 2002).

In the context of extreme poverty, sustaining food needs may be prioritised to the detriment of the needs of people with disabilities. In Kenya, CBM notes that the 2009 drought forced many parents to look for food and water instead of taking their children with disabilities to rehabilitation services.

Despite the size of the contingent of people with disabilities in the USA, a survey (NCD 2005) revealed that eighty per cent of the country's emergency managers had not made any special provision in their plans for those with disabilities, despite indications of the obligation to do so in national legislation represented by the Americans with Disabilities Act. Indeed, fifty-seven per cent of the managers did not know how many people with disabilities there were in their own planning jurisdiction, and only twenty-seven per cent had taken a course offered by FEMA on dealing with people with disabilities in disaster. Another survey of local emergency management authorities in the USA (Fox *et al.* 2007) found high levels of indifference to the needs of people with disabilities in emergency situations. Respondents cited costs, lack of staff and shortage of resources as the main reasons for not improving provisions for people with disabilities. However, lack of awareness was at least as important a reason.

The contribution of people with disabilities to DRR

We can conclude that major emergencies may put people with disabilities more at risk than other members of the general population and may impede them with new barriers that threaten their safety. Despite this gloomy picture, there have been more optimistic assessments. For example, Rooney and White (2007) found high levels of activism among people with disabilities – at least, those who were physically and mentally capable of it – and the formation of spontaneous helping networks among family, friends, neighbours, colleagues and even strangers. Douglas Lathrop suggested that, 'In some ways, disabled people who manage to live with a certain degree of independence are more able to face disaster than people who are not disabled. They have a "psychological advantage"' (Lathrop 1994). That may be so, but we cannot and should not solely rely on the resourcefulness of people with disabilities to get them through calamity.

The issue here is empowerment to break down isolation. Empowerment starts when people with disabilities are able to locate the root causes of their marginalisation within the larger society rather than as a physical inability in themselves (Wisner 2002). Eventually empowerment should be geared towards allowing people with disabilities to have a voice and a role in the decision-making process in both everyday life and DRR. When empowered, marginalised groups such as those with disabilities are able to engage in the assessment of their own vulnerabilities, needs and capacities to face those needs. As an illustration, in the aftermath of the 2004 floods in Bangladesh, Handicap International resorted to focus group discussions, community mapping, transect walks and peer discussions to involve people with disabilities in post-disaster recovery and DRR (Handicap International 2005).

Such participatory evaluation is possible because empowered people are more aware and critical of their social environment, and thus more open to partnerships and collaboration with other segments of society. They often form associations that are much more able to voice their needs in the face of national governments, international organisations and NGOs. Such associations prove to be valuable partners in DRR.

Most of the NGOs working with people with disabilities, such as Handicap International and CBM, worked closely with these associations in the aftermath of the December 2004 Asian tsunami. In China, in the aftermath of the 2008 Sichuan earthquake, associations of people with disabilities conveyed disaster-related information and social support to their affected members through mobile phone networks (Fu *et al.* 2010). In Sweden, some deaf and hearing people formed the Deaf Crisis Group, which collaborates with national agencies to reduce the risk of disasters. Volunteers received psychological and psychiatric training to assist people with hearing disabilities in time of emergency (IFRC 2007).

Box 34.2 Disasters and disabled people's organisations

Maria Kett
Leonard Cheshire Disability and Inclusive Development Centre, University College London, UK

John Twigg
Department of Civil, Environmental and Geomatic Engineering, University College London, UK

Disabled people's organisations (DPOs) vary enormously in size, scope and capacity; however, they have increasingly begun to play a larger role in responding to disasters.

As representative organisations, given the necessary recognition, support and resources, DPOs can play a major role, and one that can go beyond just focusing on persons with disabilities. For example, in Sri Lanka in the aftermath of the tsunami, DPO members helped with community food distribution and other inputs. In Pakistan after the 2005 earthquake, and more recently in Haiti, DPOs have played a key role in locating persons with disabilities, highlighting their needs and ensuring links with international agencies.

Studies of DPOs in emergency situations, however, show that the role they play can be complex. In Sri Lanka in the aftermath of the 2004 tsunami, research showed that even if DPOs were included in recovery efforts, they often lacked connections or even common ground with many other civil society organisations. As a result, disabled people were often excluded from broader discussions about political, social and economic issues that, whilst not about disability exclusion per se, directly affected their lives. For example, the entire debate on water privatisation, which had enormous repercussions for poor people across Sri Lanka, had no input from the disability community (Kett *et al.* 2005).

Furthermore, while the tsunami in Sri Lanka provided a wealth of opportunities to raise the profile of disability issues, as well as avenues in which to attract necessary funding and support, many of these opportunities were not pursued by local DPOs because they lacked awareness of and networks within the development community. Much of this funding went to larger, more resourced agencies. Moreover, it was found that DPOs often lacked capacity to undertake larger programmes, which led to a vicious cycle of exclusion and strain between DPOs forced to compete for the relatively small amount of money that remained.

Increasingly, however, DPOs and international agencies are beginning to work together to support inclusive responses. Some of the best examples of this come from post-conflict countries, possibly reflecting the increased opportunities that arise in a post-conflict or post-disaster phase – including funding, focus and resources. This may also reflect increased awareness of disability issues as a result of the visible 'wounds of war'. For example, many ad hoc self-help groups of people with war-related injuries (such as amputees) and those with other disabilities (though often impairment-specific) have formed registered DPOs in Sierra Leone.

Finally, the new UN Convention on the Rights of Persons with Disabilities (UNCRPD) has increased attention to and funding for disability issues. An example of this focus comes from Liberia, where the Human Rights and Protection Section of the UN Mission in Liberia have established a strong disability component in their work in collaboration with Liberian DPOs.

Challenges of increasing DPO involvement in conflict and disaster responses remain, especially those of DPO governance and accountability, but these are problems faced by many small organisations as they grow and professionalise.

Including people with disabilities in DRR

In an emergency situation it is comparatively easy to fail to recognise the type of handicap experienced by a particular individual, and thus to offer the wrong kind of assistance. Moreover, the organisations that work in DRR tend to be accustomed to thinking in terms of providing assistance to large groups of people, whereas people with disabilities have individual needs that may differ from those of the average person in a population. In fact, assisting people with disabilities in disasters requires not only particular procedures but also special preparations and plans tailored to their needs.

In emergency planning and management (see Chapter 41 and Chapter 42) it is feasible to take into account the problems, special needs and points of view of people with disabilities. For example, most evacuation plans require the ability to walk, drive, see and hear. They can be adapted to the needs of people who cannot do one or more of these things (Gerber in Fjord and Manderson 2009; Kailes 2002). As Enders and Brandt (2007) noted, the Geographic Information System (GIS), now commonly used to enhance emergency planning and management, can be a potent tool for locating people with disabilities and ensuring that their special needs are taken into consideration. More generally, Kailes and Enders (2007) suggested that planning could be more focused on people's needs in disaster in relation to health care, transportation, communication, supervision and the maintenance of functional independence. Mechanisms for support, leadership training and the delivery of services can and should be overhauled.

The experience of living with disabilities in disaster situations highlights certain needs. For example, there is a question of how to ensure continuity of services for people who depend for life support on electricity, telephones, water supply or other basic services. People with disabilities need to know how to manage when disorder and debris are present at home, and what transportation and mobility assistance will be available in disaster situations. They must be informed about how they can obtain basic necessities in emergency situations and how to manage the needs of guide dogs.

When cautiously integrated within the social context, technology is frequently considered a serious way to involve people with disabilities in DRR. Technological devices such as wheelchairs, hearing and computer aids, and remote controls are indeed very common implements for those living with disabilities. Therefore many researchers and agencies have explored satellite and mobile phone technologies and amateur radios for improving warning systems and keeping track of people with disabilities in time of emergency (Hans and Mohanty 2006; Fu *et al.* 2010).

Kailes (2002) notes that there are few if any empirical data on the efficient and safe evacuation of people with disabilities during emergencies and crises. Moreover, in many places there is a lack of integration and co-operation between the various organisations that work with people with disabilities and the community of people who plan for and manage emergencies. It is important to start the dialogue, for the issues are complex. No single emergency response strategy is valid for all types of disability. Moreover, the question of how best to assist people with disabilities in disaster is related to other issues such as providing help to ethnic minority groups, single mothers and people with special dietary and medicinal needs (see Chapter 35 and Chapter 38).

Basic assistance to people with disabilities in disaster should include the following:

- Procedures and services should be accessible equally at normal times and in disaster.
- Emergency communications should be available, comprehensible and reliable.
- Associations of people with disabilities should be involved in emergency preparedness activities and should be consulted as part of the emergency planning process.
- Where there is a significant risk of disaster, appropriate preparation, education and training should be provided for the benefit of emergency responders and people with disabilities who are at risk. For example, after the 1999 earthquake which struck Turkey, an academia-based NGO included the deaf in its recovery programme. A core group of six to eight deaf people who were affected by the disaster were then given a basic disaster-awareness programme, which they eventually passed to more than 2,000 other deaf throughout the country (Wisner 2002).
- Finally, efforts should be made to sensitise the mass media to their potential role as purveyors of emergency information to people with disabilities.

This last point raises the issue of advocacy. Since the 1980s associations of people with disabilities have been very successful in advocating for their global rights in everyday life on the international scene. The disability rights movement contributed to people with disabilities being seen as a minority with rights, rather than as a group with special needs which should appear on a technical checklist for planners (Wisner 2002). These rights obviously expand to DRR and emergency management. As an example of such advocacy, the International Disability Rights Monitor, a grassroots international research organisation, is henceforth including disaster-related activities in its current watchdog operations (CIR 2005). In Bangladesh, a country where 12 million people (or six per cent of the population) live with disabilities, the National Forum of Organisations Working with the Disabled (NFOWD) is a consortium of NGOs that advocates for the rights of these people, especially in times of disasters.

Some researchers and agencies have come up with guidelines for practitioners. Rowland *et al.* (2007) gave prescriptions for training programmes and Christensen *et al.* (2007) offered a framework for assisting people with disabilities in disasters, together with considerations on associated environmental, behavioural and organisational factors. Handicap International (2005) and the Shanta Memorial Rehabilitation Center in India (Hans *et al.* 2008) have also prepared some very useful toolkits for agencies and personnel in charge of DRR, and management who deal with people with disabilities. Another guide to emergency preparedness written specifically for people with disabilities by FEMA makes three recommendations. The first is for people with disabilities to assess the types of hazard that are present in the workplace and at home. Second, they should endeavour to create a support network of at least three people for each site that they habitually frequent. Third, they should estimate their own capacity to respond with self-protective actions in the event of a crisis. In addition, for kinds of disability that are not immediately apparent, it may be helpful to wear a brooch or bracelet that identifies the handicap in question. The IFRC (2007) further recommends that DRR and humanitarian organisations come up with small personal support groups, self-help networks or 'buddy' schemes. These groups gather three people who personally know someone with a disability and who agree to assist him/her in times of emergency.

In some countries, recent disasters have also contributed to an increasing recognition of the particular needs of people with disabilities. For example, the 1995 earthquake which struck Kobe in Japan was a particularly painful event for people with disabilities who were both severely affected by the disaster and eventually suffered severe neglect during the recovery period. Yet they organised into associations which led to the most powerful and influential advocacy for disability rights in decades and resulted in significant favourable changes in national government policies (Nakamura in Fjord and Manderson 2009). In Bangladesh, a national dialogue on disasters and people with disabilities was initiated in the aftermath of Cyclone Aila in 2009. In Indonesia, Handicap International is helping people with disabilities to give voice to their situation and needs on the national scene and is lobbying the government for a better integration of disability in DRR. In Europe, the Republic of Ireland has issued an innovative book on evacuating people with disabilities and designing buildings to ensure that they are warned of hazards (National Disability Authority 2008).

At the international level, the Sphere standards for humanitarian actions also integrated, although to a limited extent, some of the needs of people with disabilities. In the European Union (EU) the publication and signing of the *Verona Charter on the Rescue of Persons with Disabilities in Case of Disasters* was a milestone in the official recognition that there is a problem that must be tackled (www.eena.org/ressource/static/files/Verona%20Charter%20approved.pdf). The Charter is the culmination of a project that has investigated the condition of people with disabilities in disaster situations in various European countries and has thus contributed to the formulation of a clear picture of the problem and its potential solutions.

These are examples of good practices that can go a long way to improving the safety of people with disabilities in crisis situations. In contrast to common perception, the fundamental needs of people with disabilities in disasters are the same as everyone else (e.g. shelter, water, sanitation, food), even if supplying those needs might need to be tailored to the characteristics of specific disabilities. What matters is how support is channelled to those in need. Similarly, evidence suggests that including the needs of people with disabilities in DRR is not expensive. It usually costs less to build accessible buildings than to have to rehabilitate buildings later on to conform to laws and regulations (IFRC 2007). However, this form of pragmatism is dependent on developing an appropriately positive attitude in the DRR community toward people who live with disabilities. Much work remains to be done on creating the appropriate set of attitudes – and translating those attitudes into effective action.

Conclusions

As noted above, the academic literature on people with disabilities in disasters is sparse. This appears to be a sign that the problem is being neglected in terms of both research and applications. Yet disability in disaster is undoubtedly an important issue – morally, ethically and practically.

The solution to this problem of neglect lies in political engagement and strategies of inclusion of people with disabilities, rather than leaving them in a 'cocoon' of isolation. It also requires policy-makers to realise that people with disabilities are not a homogeneous category but a disparate collection of people with highly varied needs and capacities. The solution should involve a radical overhaul of attitudes and approaches. In many countries, more links are required between the disability and development communities, as well as between civil protection agencies and organisations that work with and on behalf of people living with disabilities.

We live in a world that is increasingly dominated by a strain of free-market capitalism that encourages individualism and egotism to the detriment of the principles of welfare and collective responsibility. Civil protection and voluntarism go against this grain, but in so doing they reaffirm values of civility, compassion, social participation, self-sacrifice and dignity. These reach their apogee in disaster situations. These are events that reveal the inner workings of society. Perhaps no aspect of disaster response is more diagnostic of social values than how the disadvantaged are cared for, and disability is one of the most important sources of social and economic disadvantage. Redressing the balance in favour of positive action is thus a reaffirmation of civility, not simply a form of welfare. It thus has great significance for social values in general and for the acceptance of people with disabilities as full members of society.

35

Gender, sexuality and disaster

Maureen Fordham

NORTHUMBRIA UNIVERSITY, UK

Introduction

In the aftermath of the Haiti earthquake, *The Independent* newspaper reported occurrences of the rape and sexual abuse of women and girls (Nguyen 2010). This is the most recent evidence at the time of writing that cases of gender-based violence are frequent and still at an unacceptably high level in disasters. This raises questions about why women, as compared with men, continue to be disadvantaged, abused or made vulnerable in disasters, and highlights the importance of recognising gender as of vital consideration in disaster management and in instigating measures for disaster risk reduction (DRR).

Fundamental to an understanding of disasters, and thus of how DRR should be maximised, is the fact that women and men experience disasters differently. Often – but not always – women suffer more in the event and bear much heavier physical and psychological burdens during the recovery period (Cannon 2002). Yet their needs and interests are overlooked; their knowledge, skills and capacities are not utilised for DRR and prevention; nor are their voices heard in the crafting of policy that affects disaster risk management.

When considering how to address such inequalities it is important to distinguish between the category 'women' and that of 'gender'. 'Women' was the primary analytical category in the work of development scholars, practitioners and activists (many from the global South), and before that in the work of feminist social scientists and activists (many of these from the global North). While acknowledging the importance of power relations between women and men, it was some time before a gender analysis was carried out in what had been assumed to be the gender-neutral context of disasters.

Whether we discuss 'women' or 'gender', both terms are problematic, suggesting homogeneity, essentialism and exclusively heterosexual relations. Women are not a homogenous category but display degrees of vulnerability and capacity according to situational context. Women and girls are not essentially weaker or more vulnerable than boys and men but are made so by socio-political structures and processes. Furthermore, while 'gender' is used most often as a surrogate for 'women', it is fundamentally an inclusive term and goes beyond the typical binary category. However, when gender is considered, it typically reflects an ideology of 'compulsory heterosexuality' (Rich 1980).

While gender is the term used most frequently, the majority of scholarship and practical intervention has been concerned with women and this social group represents the largest to be both marginalised and yet addressed in disasters policy: thus, this will be the group discussed most (but not exclusively) in this chapter. The chapter elaborates on some of these issues but the main body of the discussion draws on examples of the presence and absence of a gendered disaster sensibility before, during and after disasters. Lastly, the chapter identifies gaps and key messages for different audiences.

The place of gender in development and disaster

A gender perspective on disasters grows out of work in the development field. An easy shorthand for understanding the way issues of women and gender were addressed and progressed in development theory and practice is through the Women in Development (WID) and Gender and Development (GAD) conceptualisations. These had a strong influence on development policy and practice in the field.

WID was strongly influenced by European and North American liberal feminist approaches in which women first needed to be made more visible and their exclusion from the public domains of productive work, decision-making and political representation redressed. The scholarly rationale for this was to redress the imbalance between a dominant masculine focus on the public domain and abstract theorising, and a subordinated feminine concern with the practical and private dimensions of domestic life.

The WID approach, although useful in explicitly identifying women, tended to focus on their short-term, practical needs without radically changing the structures and processes that had kept them invisible in the first place. For example, while it might support women's greater employment opportunities, these might simply reinforce their reproductive role through only presenting options such as sewing but not options such as working with machinery or in mainstream businesses.

GAD approaches emerged as a reaction to the failure of many WID policies and projects to address women's long-term, strategic interests. This approach focuses more on gender relations – i.e. differential power relations between women and men. A GAD approach to employment might challenge existing fixed gender roles and seek to ensure that women had equal access to positions of power in the economy and the workplace.

The development of a separate discipline, activity and policy field called 'gender and disasters' (also GAD) can also be traced to the influence of feminist theory and feminist political movements in the late twentieth century, but it emerged later. Gender and disasters is as fragmented as are the other cognate areas of disasters research: hazards, humanitarianism (including conflict and refugee studies), disaster sociology and many more. It took time for the disasters field to catch up with development in terms of gender inclusion. It was an easier transition for the humanitarian field because of its closer links to development work in the global South. However, humanitarian research and practice involves largely separate groups of people, disciplines, and meetings and other kinds of communication than those in the hazard and disaster fields.

When the new *Disasters* journal appeared in 1977, social scientific and political approaches began to intervene in disaster studies (e.g. Wisner *et al.* 1977); nevertheless, women's separate needs and forms of experience received only tacit recognition within the general categories of 'the people' or 'the community'. The *Disasters* journal is referred to here because it was the first of its kind, but others followed a similar path to eventual gender inclusion. When specific women-focused papers appeared (Peel 1977), they treated women (pregnant or lactating) as medical case histories (positioning them and their children as passive receivers of medical aid).

This early work relied on a biomedical model of women with separate parts to mend, rather than on one that treats the whole and separate experience of women in disaster situations. There was seldom any study that addressed the separate and specific needs of women. Even discussion of the siting and design of latrines recognised only the needs of 'families', and the most salient appearance of 'women' is when they are photographed characteristically as victims or as people in need of medical aid (Peel 1977).

The journal articles reflect practice in humanitarian and emergency management systems, which in most countries have traditionally been, and continue to be, dominated by command-and-control approaches. These are primarily directed by men, often trained in the conventionally patriarchal institutions of the military and the emergency services, and do not discriminate between men and women in their organisational strategies.

Thus, we have a staged appearance of gender in disasters: (1) beginning in the development and humanitarian fields with a focus on developing countries and the work of grassroots groups and non-governmental organisations (NGOs); (2) appearing later in the scholarship of North American/European sociologists, learning from the global South; and (3) much later filtering into the policy and practice of developed world emergency managers. There is a continued near-absence in the work of those operating from within the hazards paradigm for whom the triggering, gender-neutral, 'natural' hazard continues to be of prime interest and regarded as the dominant causal agent of disasters. All of this must be seen as reflective of an historical and, despite many advances, still dominant, global patriarchy.

Gender discrimination – before, during and after disasters

Before: gender discrimination in the everyday

There is a wealth of evidence of gender discrimination in the everyday and thus inequity in disasters should not come as a surprise. Embedded patriarchal values in a majority of societies are at the root of much inequality. At the most fundamental level, preference for sons in many cultures accounts for why many women go 'missing' (Hausmann et al. 2009) – a polite euphemism for female infanticide and feticide – each year.

Girls' unequal access to education has been recognised in Millennium Development Goals 2 (achieve universal primary education) and 3 (promote gender equality and empower women). When girls do get access to education it often favours stereotypical 'feminine' skills training, which reinforces their location in the domestic environment, restricting them primarily to reproductive work.

Even in countries where equal access has been achieved, women still do not have equality of opportunity in the workforce (Hausmann et al. 2009). The International Labour Organization (www.ilo.org) reports that women have more access to employment now than ever before, but they still earn one-third less than men and it is too often insecure.

The inevitable consequences of their being economically disadvantaged are that women are more likely to fall into the traps of poverty and hunger. The chronic poverty of women and girls is perpetuated by their being subject to a 'confluence of gender-based vulnerabilities' (Ambler et al. 2007: 1). In both customary practice and law, for instance, women are not granted the same benefits and protection as men. They have less or no access to decision-making processes, and worldwide they constitute only a minority of elected representatives to government.

Having fewer financial resources and confinement to domestic duties and responsibilities often results in social isolation, making women further subject to threats or acts of violence: this is particularly the case in the event of widowhood. Women in poor families bear the greatest burdens of responsibility, which they assume when they marry in early teenage years. As a result of early childbirth, they and their infants are further subject to the dangers of ill health. In addition, women are the primary health care providers within the family and, in the case of those living with AIDS, it is they who must replace the labour lost in the event of a death.

These and other pre-disaster vulnerabilities mean women lack economic and social security and all the resources that go with those (safe houses, financial support, access to decision-makers, health care provision, etc.). Thus, women are less well placed to bear or recover from disasters when they occur, or to take up disaster-related work and livelihood opportunities: 'Women's economic insecurity increases, as their productive assets are destroyed' (Enarson 2000). The Gender and Disaster Network (www.gdnonline.org) provides resources related to many of these aspects.

During: disasters discriminate

Recent studies of disasters have shown that the suffering of women and girls is often more severe and yet, paradoxically, less visible in reports, policies and scholarship. On the other hand, their capacities and capabilities are even less in evidence. For example, the research of Stehlik *et al.* (2000) on Australian farm families living through extended drought clearly articulates the differences between women's and men's experiences and constructions of the drought. There were clear distinctions between 'inside and outside' (what is also referred to as 'public and private' in other gender writings). For women, their major commitment was to the inside domain of the home whereas the men rarely spoke of inside but referred to their outside domain on the land.

This is a familiar dichotomy to be found in many published examples. In Australia, women also had a lesser role in decision-making, for example in connection with banks and insurance companies, but were expected nevertheless to answer related telephone enquiries and to carry out the bookkeeping tasks – an essentially subordinate, secretarial role (Stehlik *et al.* 2000). Where women did find some employment or training 'outside' this was regarded by them with a mixture of self-fulfilment but also guilt – for the impacts such work might have on the running of the home. The ongoing nature of the drought meant that women's roles were expanded to assist in maintaining the properties (pumping water, etc.) and many of their partners came to recognise this important role. However, women seemed also to carry a multiple burden of not only rebuilding their own lives after the disastrous drought but also feeling the need to 'carry on' for the sake of their families and the wider community. Thus, although women were associated most closely with the home, they actually made a much wider contribution. This is the classic 'triple burden' that women bear: the (inside) reproductive role, the (outside) productive role and the (wider outside) community role.

Similar themes can be found from the global South. Delaney and Shrader's (2000) analysis of post-Hurricane Mitch reconstruction in Honduras and Nicaragua showed some of the more positive, transformative potential of disasters. The collaborative work that also took place between women and men created the opportunity to transform traditional gender roles and responsibilities. Women were carrying cement, digging wells and constructing latrines, and there were examples of men cooking. The economic and social empowerment of women changed household power relations; on the other hand, and perhaps more typically, there were examples that put women back in their traditional reproductive role.

A faith-based agency in Honduras prohibited women from participating in housing construction because they wanted them to dedicate time to their 'natural role' of childcare and domestic responsibility. Other cases produced role conflict; for example, the needs of women-headed households could not be met by some disaster committees which organised around

427

traditional gender divisions of labour in which women's responsibilities were to protect children and go to shelters and men's were to protect the family's assets such as land and animals. This meant that women-headed households had to choose between their children and their assets.

Rasheda Begum's (1993) account of her experiences as a relief worker in the 1991 Bangladesh cyclone revealed the highly gendered nature of the excess female cyclone deaths. Cultural norms (still in operation) requiring women to seek the protection of male family members restricts their mobility and thus their ability to escape. They lack appropriate social capital which makes them less likely to receive information and early warnings; when they do receive warnings, they do not evacuate for fear of blame and punishment for any loss to the home.

Such findings often have to be searched out because of a scarcity of gender-disaggregated data despite calls for its inclusion by many individuals and groups, over many years (Fordham et al. 2006). Where statistics are available, however, they are very revealing. Analyses across a range of disaster events reveal particular vulnerabilities of women. Neumayer and Plümper (2007) in an analysis of disasters in 140 countries, between 1981 and 2002, found that women's life expectancy was lower than that of men's and that the larger the disaster, the larger was the gender gap. This they explained by socio-economic patterns of gendered vulnerability rather than innate (biological) differences between males and females (the unpacking of single categories is discussed further below).

Of fatalities suffered by people over sixty years old in the 1995 Kobe earthquake, those of females were almost double those of men (Tanida 1996) – a fact which was primarily related to the susceptibility to collapse of wooden-frame houses located in poorer areas, which were occupied to a large proportion by elderly women living alone and tending to sleep downstairs because of difficulties with mobility.

What about men?

It is certainly true that men suffer in disasters, but men – as men – do not face the same formidable barriers and inequalities as women, who suffer simply because they are women. In any situation, men generally start with greater advantages. However, men face their own socially constructed roles and expectations which may also place them at risk. Men's lives are as profoundly shaped by gender relations as women's, since just as women are subject to conventions of femininity, so too are men to those of masculinity. Gender expectations also put boys and men at risk in disasters. For example, in search and rescue, they are socialised to be risk-takers while women are typically risk-averse (Fothergill 1996).

Sexual orientation and disaster

Beyond the more conventional categories of feminine and masculine, it is rare for sexual orientation to be considered in the disaster context. However, there has been a more recent inclusion of issues of sexual orientation in discussions of gender and disaster. Wisner (2001b: 2–3) noted the 'absence of sexuality or sexual orientation in check lists and post-disaster audits'. As part of a study of urban vulnerability he observed 'a very substantial population of male to female transsexuals' in West Hollywood which has a history of 'difficult relations' with various authorities, including the police, and which 'face very special needs in medical emergencies and situations requiring mass shelter'.

In India, the transgender Aravanis were almost completely excluded from the 2004 tsunami relief process. Aravanis (also known as Hijras or Jogappas in different parts of India) are an increasingly recognised but still marginalised transgender group who do not identify as either

Box 35.1 Integrating non-heterosexual groups into disaster risk reduction in the Philippines

JC Gaillard
School of Environment, The University of Auckland, New Zealand

Non-heterosexuals are amongst those groups marginalised within most societies and as a consequence almost systematically neglected in DRR. Yet lesbians, gays, bisexuals and transsexuals (LGBTs) have specific needs in times of disasters. On the other hand, LGBTs also display particular resources which are valuable additions to local initiatives to face natural hazards, but which are hardly considered in official disaster policies.

In the Philippines, non-heterosexual groups, especially gays, openly claim their identity and are recognised for their leadership and initiatives when it comes to community activities. Yet they often suffer from mockery and discrimination when in the presence of heterosexuals. On the larger political scene and despite several legal requests, same-sex marriage has not been recognised and LGBT congressional party lists have only recently been authorised to file candidacy. Within the family, LGBTs are frequently marginalised and tasked for demanding house chores.

This situation continues in times of disaster. In Irosin, in Southern Luzon, young gays are asked by the parents to clean their house in the aftermath of flash floods although they too are the ones who spontaneously go around the town to collect relief goods among their neighbours. When evacuated in crowded churches or public buildings gays suffer from the lack of privacy, being uncomfortable with both women and men. Their personal grooming needs are also the object of jokes from men in male comfort rooms where they are assigned. Nonetheless, they spontaneously care for babies and young children. For most of the young gays of Irosin, disasters, therefore, turn out to be even more stressful than for heterosexuals.

In the aftermath of the September 2009 typhoon disaster which killed more than 1,000 people, organisations militating for the rights of LGBTs raised support for the survivors without discrimination on the basis of their gender or sexual orientation (diversityandequality.ph) and then persisted as an advocacy group for the consideration of non-heterosexuals in DRR named LGTBI Pinoys for Calamity and Disaster Victims (groups.to/lgbtipinoysforcalamityvictims).

As part of a recent community-based DRR initiative in Irosin, young gays have been involved as both a singular group and members of the larger community. A special focus group discussion has enabled the assessment of their particular roles and needs in the face of natural hazards. They then participated in participatory mapping activities with heterosexuals. Gays identified their houses on the map in order to delineate specific areas where each of them would collect relief goods in times of emergency. Identifying their location points to the development of a trust relationship within their communities. Their potential contribution to the life of the community while evacuated in public buildings was also discussed. These activities conducted in the presence of heterosexuals contributed to the recognition by the larger community of the contribution of gays to DRR. This should help in reducing discrimination and mockery during disasters.

male/men or female/women, but prefer to wear women's clothes. They were not given places in temporary shelters, had no place where they could change their clothes and many would sleep in the open with a known risk of gender-based violence (Pincha and Krishna 2009).

There is a clear need for groups marginalised because of sexual orientation to be included within disaster risk reduction and disaster management planning. However, in many parts of the world homosexuality is illegal, and such laws would require considerable political advocacy to overturn. Thus, it is particularly problematic for formal disaster management institutions to respond to their needs and interests.

After: 'First the earthquake, then the disaster'

In disasters, there is both crisis and opportunity, but distributional effects mean that these are not always experienced by the same person or group. For many, the real disaster starts after the immediate crisis period, reflected in Anthony Oliver-Smith's reference to post-earthquake graffiti: 'first the earthquake, then the disaster' (Oliver-Smith 1999: 86). Gendered disadvantages in disaster situations are a direct consequence of already existing social inequalities and forms of gender discrimination, which are globally widespread. They lead to the dynamic pressures and unsafe conditions made concrete in formal and informal disaster management and recovery processes, across all sectors (Wisner *et al.* 2004).

Shelters and camps

The design and construction of relief and refugee camps often put women and girls at risk. In the aftermath of the 2009 droughts in Kenya, women who were compelled to go outside the protection of the camps to collect firewood were risking sexual assault. In the same year, it was reported that sexual and gender-based violence cases had increased by thirty per cent (in Kolmannskog 2009). Sarah Martin (2005), in her report on sexual exploitation and abuse in United Nations (UN) peacekeeping missions in West Africa and Haiti, attributes such increases to 'a hyper-masculine culture that encourages sexual exploitation and abuse' in the male–dominated response to disaster situations. It is a 'boys will be boys' attitude that has primarily to do with 'problems of abuse of power' and only 'secondarily [with] problems of sexual behaviour' (Martin 2005: 5). While Martin's report is specific in its focus on peacekeeping missions, the sentiments are applicable much more widely.

Despite Sphere Project guidelines and other recommendations on various aspects of shelter/camp design, they are still constructed in gender-blind ways. Too frequently official shelters and camps lack appropriate gender-sensitive services and conditions. Latrines are located at a distance and in the dark; camp-management committees have few women, some of whom face hostilities (TEC 2006a); there are insufficient safe spaces for women and children; and women's livelihood opportunities are absent.

These deficiencies are also evident in the global North. Thornton and Voigt (2007: 44) note that 'each phase of the Katrina disaster, ranging from the warning phase to the reconstruction phase, created conditions and opportunities for the victimisation, particularly the sexual assaults of women'. In another study, conducted by the International Medical Corps, women in post-Katrina trailer camps for the displaced communicated their perception of a lack of safety for themselves, especially at night, and also for their children even during the day. The study found that only one-half of the Gulf Coast camps had any kind of security provisions in place. However, even with such security, the even more frequent intimate partner violence would remain a risk (IMC 2006).

Health

A gender-blind approach means that women's needs and interests are unrecognised or dismissed (see Chapter 44). It is perhaps an irony that while women and girls are constrained within their reproductive role, their reproductive health needs (including those related to pregnancy, child-care, gynaecology, family planning, sexually transmitted infections including HIV/AIDS, and gender violence) are unmet in post-disaster situations.

In the 2005 earthquake in Kashmir, Pakistan, many women health workers were killed and their homes (in which they worked) were destroyed. This meant that many women lacked medical care because it could only be delivered by male doctors, who were not the accustomed providers and who were located at a distance from their families and villages. The absence of health care came not only via the physical difficulties of travelling to where it was available, but through cultural constraints such as loss of honour in being seen by males (including doctors) outside the family (Fordham *et al.* 2006).

When speaking of reproductive health it is important not to focus solely on women and girls as if they were the problem. Policies and programmes that deal with gender-based violence and challenge dominant definitions of masculinity are equally as important. Moreover, men are known to have a tendency towards unhealthy post-disaster coping strategies, such as alcohol abuse (Fordham *et al.* 2006). When they are unable to be physically active, men can be badly affected psychologically by disasters, particularly when the event undermines their socially constructed role of family and home provider (Fordham 1998) (see Chapter 47). Men might appear to cope in the short term, but can present significant health effects later, as was found in North Wales in the floods of 1990 (WCC 1992). There are many instances of men and boys crying in post-disaster despair, unable to act to repair the damage to themselves, their families and their communities (Fordham and Ketteridge 1998; Doppler 2009).

Land and property rights

The UN has declared that the basic rights of shelter – of land, housing and property – are fundamental to human existence, but 'almost a third of the world's women is homeless or lives in inadequate housing. This is due not only to systematic discrimination against women in inheritance laws and practices but to women escaping violence in both developed and developing countries' (UN-HABITAT 2008). These everyday realities for many cannot be avoided in disasters either.

Widows in Pakistan lost the land that they had previously occupied and worked on with their husbands to male relatives (Fordham *et al.* 2006). In Pakistan, efforts by Pattan (a grassroots NGO) in a housing project following the 1992 floods led to women's joint ownership of post-disaster housing. This was first resisted by men in the villages concerned, but was overcome by arguing that, as women are generally regarded as an economic burden, reducing women's vulnerability and building their capacity would lead to a reduction in this burden. Following the joint ownership agreements, evaluations noted decreased domestic violence and familial conflict as women became more empowered (Bari 1998).

Work and livelihoods

There is often a stereotypical view of women's livelihood options. The Tsunami Evaluation Commission (TEC 2006a) identified major problems for women who, along with other poor people, were marginalised by NGOs that tended to deal only with village officials. Women had greater problems than men did in accessing livelihood and other recovery programmes. Pressure on international agencies to spend relief funds meant that much money went to easily distributed items, such as fishing boats for men, without corresponding efforts to support women's status. Women small business owners (who are rarely officially registered) could not prove that they had lost their livelihoods and so did not qualify for assistance. In the aftermath of Hurricane Katrina it was much harder for women to get employment in rebuilding work and harder still if they were black women (Jones-DeWeever 2008).

Despite these typical limitations, there are examples of ways in which women have overcome barriers and achieved livelihood advances. The Emergency Committee of Garifuna in Honduras was begun by a small group of eight women and four men after the devastation of Hurricane Mitch. It focused on collective efforts to rebuild livelihoods and homes. They have set up community tool and seed banks; bought land for relocation to higher ground for future safety; and organised the training of local residents in building techniques for hurricane-resistant houses, advised for example by the Jamaican Women's Construction Collective. Women from the marginalised Garifuna community have played the lead role in these initiatives (Yonder *et al.* 2005). Other positive examples have been compiled and shared by GROOTS International, a network of grassroots women's groups which advocates for and facilitates women training other women in local organising, reconstruction, livelihood enhancement and disaster risk reduction skills and methods (www.groots.org).

Box 35.2 Women's employment after disasters

Elaine Enarson
Independent scholar, Lyons, Colorado, USA

Economic insecurity is a key factor increasing the impact of disasters on women as caregivers, earners and community actors. The gendered division of labour in households and in the global economy makes most women less able than men to control economic resources that can mitigate the effects of disasters. Women's high levels of pre-disaster poverty, secondary status in the labour force, extensive informal-sector work, lack of land rights and domestic responsibilities increase their exposure to economic loss long before a disaster occurs. Yet, in the aftermath of disaster, families become even more dependent on women's income. Women are significant economic actors whose time, efforts and income sustain life for others, and disasters cause these responsibilities to increase, especially among women heading households.

Post-disaster studies document that women's economic insecurity increases, as their productive assets are destroyed and they often become the sole earners in the household. In addition, their small businesses are weakened or destroyed, and they lose jobs. Gender stereotypes also limit their earning opportunities.

Yet after a disaster, women's workload increases dramatically. They strive to find more waged labour or income-generating work, and many engage in new forms of unpaid (voluntary) 'disaster work', including emergency response and political organising. Most women also face expanded obligations caring for family and kin who need them more than ever.

Women's working conditions in the household and paid workplace also deteriorate, through increased exploitation, lack of child care or other social protections, and increased work and family conflicts.

Women usually recover more slowly than men as they are less mobile, slower to be able to locate or return to paid work again and may also fail to receive equitable economic assistance in the aftermath. Many governmental and non-governmental reconstruction plans target men's work and overlook women's. 'The fisherman and his boat' was a common stereotype in the minds of international donors seeking to help after the Asian tsunami.

In the aftermath of the 1991 eruption of Mt Pinatubo in the Philippines, loss of the usual income sources gravely affected women (Delica 1998). Losing their harvest, as well as their backyard gardens, women fed their husband and children first. Before the disaster they had washed their family's clothes, but afterwards they eventually accepted outside laundry as informal employment. They not only cooked for their families, but also cooked to vend on the side. Sometimes they even worked as domestics, extending their responsibilities to others' homes. They sought out relief agency food or took on cash-for-work or slavish sub-contracts just to earn a little more to fend off hunger. Many women thus became breadwinners while their husbands were farmers with no more land to till. In addition, many relief agencies sought women's voluntary assistance packing relief goods, listing beneficiaries or delivering health assistance. All this adds work when women had even fewer resources than before the disaster struck, e.g. poor shelter, very limited water and few toilets. Yet they were expected to carry out their traditional responsibilities, and more.

It is imperative to increase women's economic security and to integrate gender analysis systematically through all sectors and phases of DRR. DRR must include the promotion of women's working rights and fair employment and earning opportunities.

Gender-specific capacity and DRR

There is a need to address the inequitable impact of disasters on women, based on an understanding that gender-based inequalities are not innate or biological but socially constructed. As many women have demonstrated, such action needs to be taken at a political level by challenging embedded social values and government systems that deny them the opportunities to actively manage their own situations.

What also needs to be overcome is the stereotypical view of women and girls as dependent and subordinate, reinforced by media images of them as weak, passive and in need of rescue. Women themselves, in many countries, are working to change both institutional practices and ideological attitudes, most often at community and grassroots levels. Female visibility and political representation are having a considerable impact on DRR: women and girls are challenging the socio-cultural norms of patriarchal military-based disaster response and management systems by becoming active rescuers, managers of emergency shelters and business and community leaders.

In Bolivia, the organising capacity of the Centro de Mujeres Candelaria involves over 1,000 women in many communities to reduce their vulnerability to drought, frost, hailstorm, snowfall and thunderstorm hazards, which can destroy up to half their annual crop and livestock production. Their community organising, working closely with organised rural unions and agricultural co-operatives and associations, provides food security for their families by using their traditional knowledge and skills to observe early signs of weather and climate changes to plan collective food crop, water and seed storage (www.groots.org and www.huairou.org).

In several communities in El Salvador, women and girls of all ages, supported by the work of Plan International and its local partners, are challenging socio-cultural norms and are actively

433

involved in planning for and responding to disasters (Fordham 2009). Girls and boys worked together to run local shelters after Hurricane Stan in 2005. Older women have surprised themselves at their abilities to carry a man on their backs while training to rescue people.

In Turkey, Kadin Emegini Degerlendirme Vakfi (KEDV) is a women-led NGO established in 1986 to support poor Turkish women. It supported the women who were active in the tent cities after the 1999 Marmara earthquake. KEDV volunteers provided emergency aid, arranged secure spaces for women and children and negotiated income-generating initiatives to enable women to support their families and benefit from a semblance of normality (Yonder *et al.* 2005).

In Zimbabwe, a Practical Action (formerly ITDG) project made a systematic effort to include women in focus groups and, after discussions to identify relative wealth and poverty, to reach out to the poorest women in the village. Practical Action wanted to work with existing community groups, one of which was the Garden Groups that were ninety per cent female. In gender-specific discussions, it became apparent that men and women saw the problems and solutions for their village differently. The men prioritised dryland farming while the women prioritised vegetable growing in their garden plots. Dealing with drought and water resources required a gendered analysis. Had they not done this then Practical Action (like so many others before them) might have seen the solution as a dam or irrigation system, which would have helped the men's interests but would not have supported women's livelihoods. Listening to the women resulted in the production of a low-cost clay pipe for subsurface irrigation of vegetables, which supported the work of the women's gardening groups (Murwira *et al.* 2000).

Conclusions

For researchers

Research needs to be gender specific and gender sensitive. There are still many disaster studies that do not tell us whether the people they discuss are women or men, old or young, or represent any of the other categories of social stratification. Furthermore, it is important to move beyond a one-dimensional checklist and look at double, treble and intersecting vulnerabilities. While recognising the need to ensure that no one overlooks the socially constructed vulnerability of many women and girls, we need more studies of the capacities of women and their contributions to building the capacities of their families and communities.

For disaster and development practitioners

All over the world there are grassroots groups working to improve the lives of their communities. Many of these groups are led by women and/or have a majority female membership. Disaster and development practitioners should make it their priority to work through such groups and not seek to set up new distribution and operational networks that rarely have the benefit of local knowledge and experience.

For disaster managers

In Pakistan during the 2005 earthquake, gender was treated as a cross-cutting theme in accordance with the IASC cluster concept and was supposed to be a component in the work of all the clusters. However, as Strand and Borchgrevink (2006: 18) note, 'treating gender as a cross-cutting theme has resulted in no one taking gender seriously'. This is a dilemma for disaster managers: how best to incorporate a gendered approach?

While this will be different in different contexts, it requires an active commitment to, and engagement with, gender equality issues at every level of organisation and through every stage of a disaster. More women are now joining disaster/emergency management professions (albeit often starting at and remaining in lower-level posts), but this does not guarantee a more gender-fair environment unless there is a change in the professional culture, which fully accepts the importance of gender.

While gender mainstreaming has become a familiar exhortation, it is clear that it too often fails in practice. Gender mainstreaming and gender as a cross-cutting theme risk being every-where and nowhere unless they are also supported by specific gender initiatives and a gendered oversight of all activities. The presence of gender equality targets and gender-focused management accountability processes will aid a truly equitable gendered approach.

36

Children, youth and disaster

Agnes A. Babugura

DEPARTMENT OF GEOGRAPHY AND ENVIRONMENTAL SCIENCE, MONASH UNIVERSITY, SOUTH AFRICA

Introduction

Disasters have devastating consequences, especially for children and the youth who make up one of the most vulnerable populations (Jabry 2005). In 2005, for example, 17,000 youth died in Pakistan due to an earthquake while more than 200 schoolchildren in the Philippines were buried alive in February 2006 due to a mudslide (ADPC 2007). Overall, between 1991 and 2000, about 75 million children under the age of fifteen and living in less affluent countries had their lives severely disrupted by disasters (Jabry 2005).

The vulnerability of children and youth to disasters is further compounded by poverty and HIV/AIDS (Babugura 2008). For example, based on the 2009 UNAIDS report, 14 million children have already been orphaned by HIV/AIDS, thus increasing the number of child-headed households. Millions of children were reported to be living in communities heavily burdened with HIV/AIDS, where they have lost a parent, both parents and caregivers to AIDS. With the loss of primary caregivers and other safety nets, children are left without those on whom they typically depend for security, support and protection during disasters. Moreover, children living in poverty lack protection from disasters as their caregivers do not have the resources to take preventative measures to deal with disasters (Anderson 2005).

Given that children and youth are heavily affected by disasters, it would be expected that their particular vulnerabilities would take priority in disaster risk reduction (DRR) policies and practice. However, they are often the least listened-to members of society especially in voicing their concerns and experiences with disasters. Often very little is done to address children's vulnerabilities and capacities during times of disaster (Anderson 2005). In most cases, children are simply mentioned as vulnerable groups and their issues limited to food security, malnutrition, HIV/AIDS and humanitarian response. Recently there has been much lobbying for more attention to be given to issues concerning child protection and unaccompanied children in disasters.

Save the Children (2008) argues that in cases where families are forced to evacuate from their homes, the poor usually have no choice but to turn to emergency shelters. In such situations, children's needs are not adequately provided for, thus increasing their vulnerability to injury and abuse in the shelter environment. For example, children are only sometimes counted separately from adults in shelter facilities, making it difficult to provide services that meet the specific needs of children. There is a great need to address children's issues beyond mere physical survival related to safe water, food, shelter, clothing and primary health care. More attention is required to attend to other needs such as protection from abuse and harm, education, rest, leisure and the right to participate freely in matters that affect children's lives, health and well-being (Jabry 2005).

Understanding children, youth and their vulnerability to disasters

Apart from being one of the most vulnerable populations, when a disaster occurs children and youth often depend on significant adult members in their family or community for guidance on how to manage and cope during and after the event. Given that disasters usually affect an entire community socially, the children's sense of security and normality tends to be undermined. As a result, the circumstances in which children and youth find themselves during and after a disaster present various concerns and challenges. The concerns include death, injury, displacement, exploitation and abuse, all consequences of the loss of family livelihood, socio–economic status and social networks. In addition, one must consider children's and youth's emotional reactions and their own coping mechanisms (e.g. ADPC 2007; Peek 2008) (see Chapter 47).

A clear understanding of the daily life of children and youth is essential as their normal situation can either pose a threat to their well-being when a disaster occurs or act as a protective factor that could buffer them from threats (Engle and Menon 1996). This is especially true for children who often experience challenges and hardships. For example, children and youth who have an accumulation of protective factors such as security, good health, social networks and support are likely to be less vulnerable and recover quickly from disasters. On the other hand, children and youth lacking such protective factors are likely to be at higher risk when faced with hazards and poorly able to recover after a disaster (Babugura 2008).

International organisations such as Save the Children, the United Nations Children's Fund (UNICEF), World Vision and Plan International have indicated that children who live in extreme poverty are vulnerable when faced with hazards. As a consequence of poverty, they are already at risk of being exposed to exploitation, abuse, violence, discrimination and exclusion. In cases where children and youth separate from families to seek work, these risk factors are all the more acute. Alternatively, their caregivers may migrate in search of employment, leaving the children to fend for themselves. Given their already fragile lifestyle, coping with disasters becomes a challenge.

Disasters usually leave children and youth feeling anxious, sad, confused, stressed and fearful. This is likely to have a negative impact on their development and emotional well-being if appropriate support is not provided. Very often, their emotional responses are exacerbated by seeing their parents, guardians and community members also feeling anxious (e.g. Babugura 2008; Osofsky *et al.* 2007). This leaves children and youth vulnerable to several other threats. For example, in circumstances where homes are destroyed because of a disaster, families are forced into temporary shelters. According to media reporting, such situations have proved to be traumatic, affecting a family's ability to protect children from abuse and exploitation. Bhalla (2009) reports on how traffickers took advantage of disaster-hit children in eastern India after the 2009 devastating floods. Vulnerable children were being trafficked to work as bricklayers, domestic servants and even sold as brides. Many girls were being sent to work in brothels.

Preparing and assisting children and youth to cope and recover from disasters

There is a growing body of literature, training tools, manuals and programmes providing information on how to prepare and assist children and youth to cope with and recover from disasters (e.g. Tanner *et al.* 2009; Wisner 2006a). A number of child-centred organisations such as UNICEF, Save the Children, Plan International, World Vision, Action Aid, the International Federation of Red Cross and Red Crescent Societies, among others, advocate and support efforts to assist children and youth to prepare for and recover from disasters. Due to the efforts of such organisations, researchers and DRR programmes are starting to advocate for children and youth as active agents of change instead of seeing them as passive victims. Though much more remains to be done, innovative risk reduction projects in which children and youth are supported in their efforts to claim their rights to safety and to campaign for DRR are already underway (e.g. Back *et al.* 2009).

The role of schools

Reviewed literature clearly shows the importance of schools in promoting and enabling DRR (e.g. UNISDR 2007b; Wisner 2006a) (see Chapter 62). Mainstreaming DRR into the school curriculum has proved to have several benefits, such as raising awareness and providing a better understanding of disasters for children, teachers and communities. Schools can offer a safe place for children and youth during and after a disaster. Deaths of school children in collapsed school buildings in Haiti, China, Italy and elsewhere show the great need for parents, teachers, children and youth to demand safe locations of schools, design, construction and maintenance.

Schools provide a platform that gives students of all ages the opportunity to actively study and participate in safety measures, as well as work with teachers and other adults in the community towards reducing risk before, during and after disaster events (Wisner 2006a). Suggested activities include raising awareness within school communities and making school building safer (UNISDR 2007b). However, for schools to achieve success in their role in DRR, active participation and support are required by various stakeholders such as government, international organisations, non-governmental organisations (NGOs) and local communities and families. The UNISDR (2007b) provides an account of how different stakeholders can participate in ensuring that schools effectively exercise their role in DRR. Wisner (2006a) documents some good practices that could be replicated.

The role of parents, guardians and communities

Literature has shown that the most important source of help for children and youth in disasters comes from parents and other adults who are most centrally involved in their lives. In cases where families have been separated, children orphaned and abandoned, community-based care has proven to be an effective way of providing the much-needed emotional and physical security. A sense of a caring community for such children and youth will lessen their stresses and fears, hence giving them some sense of security. Tolfree (2003) acknowledges that children in community-based care are at an advantage, as they have the opportunity to continue being cared for by familiar adults and they remain within their own communities. Allowing children to remain within their own communities not only retains a sense of belonging and identity but also provides a continuing support of networks within that community.

On the basis that children and youth will reflect the anxieties of their parents and adult caregivers, their ability to cope with disasters will also depend on the way in which parents and adult caregivers cope (Tolfree 2003). Parents, adult caregivers and community members therefore need to first manage their own reactions and take care of their emotions and stress in order to effectively assist children and youth. Once this has been done, parents, adult caregivers and

Box 36.1 Teaching teens to save lives

Haley Rich
Alliance for Empowerment, Pueblo, Colorado, USA

Ilan Kelman
Center for International Climate and Environmental Research–Oslo, Norway

A programme that originated in Pueblo, Colorado, USA has involved youth in emergency response and has linked that to disaster risk reduction and wider community sustainability. Haley Rich from Pueblo, on an entirely voluntarily basis, developed and implemented a programme at Pueblo West High School that has now been adopted by secondary schools across the USA. Based on the USA's Community Emergency Response Teams (CERTs), she developed Teen SERT (School Emergency Response Training) for teenagers to practise responding to multiple casualty scenarios.

The students learn basic and advanced first aid techniques, such as assessing the situation and making the scene safe, rescue breathing and CPR, the recovery position, and controlling bleeding and dealing with fractures. Leadership first aid skills are an integrated component, such as scene management, triage and dealing with bystanders.

The full-scale multiple casualty exercises are held with adults watching and taking notes and photos. However, only the teenagers are involved in the incident command and emergency management. So if an emergency happens and no adults can assist, the teenagers already have experience in managing the situation without any outside assistance.

The teenagers also act as 'casualties', so that the responders are dealing with both friends and peers whom they might never before have met – all of them screaming with pain or eerily silent and unresponsive. Many of the student 'casualties' particularly enjoy moulage, the act of using make-up and props to create fake injuries. These include penetrating wounds such as wood or glass blown into a 'casualty' by an explosion along with various forms of bleeding and fractures. After each exercise, full debriefings are held with adults.

As a direct result of the first aid training received through the Pueblo West High School's Teen SERT programme, at least ten Pueblo citizens' lives were saved by teenagers in the first four years of the programme. The programme's additional benefits have emerged as leadership skills, empowerment and community dedication for the students involved, along with a newfound awareness and respect for first responders and government services.

That translates into as simple an action as pulling over to the side of the road when a response vehicle with flashing lights is behind. Other outcomes have seen the youth more engaged in community activities, from picking up litter to spending time assisting elderly with day-to-day chores. The positive and proactive impacts of this programme reach far beyond the basic preparedness and response training for disasters.

Youth are the adults of the future. They will be running society, making decisions and teaching the next generation. By letting them learn and teach emergency response and disaster risk reduction as part of their basic education, society will perpetuate its own disaster vulnerability reduction.

community members are encouraged to establish a sense of control as well as build confidence in children and youth.

Parents, adult caregivers and community members can help children and youth cope with disasters in many ways. For example, in a family setting, parents and adults could start by preparing a family disaster plan and making sure that children and youth are involved in the preparations. Children and youth need to be informed about the disaster in age-appropriate language: the information should be simple and direct. Depending on the type of hazard, parents need to explain what is happening (e.g. impact of hazards on community and household). Having explained the hazards and related impacts, children and youth will need reassurance from parents and guardians about safety and the well-being of the family. Reassurance could include comforting, a lot of cuddling (especially for the very young children), verbal support and letting them know that they are safe. For older children and youth, parents and guardians could discuss issues such as the vulnerability of communities and the ability of families and countries to work together to help each other cope with the disaster.

Given that disasters tend to cause anxieties and various fears (e.g. fear that the event will happen again, someone close will die or be injured, that they will be left alone or separated from family, loss of hope in the future, belief that the world is unsafe and unpredictable) among children and youth, parents and guardians need to reassure them that it is normal to experience different reactions to disaster (including anger, guilt and sadness). It is also important that parents and guardians understand what is causing their anxieties and fears. This means allowing children and youth to openly discuss their fears and concerns. Children and youth, for example, can be encouraged to express their feelings through talking, drawing or playing. As children and youth express their feelings, parents and guardians need to listen, acknowledge concerns, observe and be prepared to answer any questions asked about the event. This provides clarity for any misunderstandings of risk and danger that children and youth may have about the event. In addition to answering questions, parents and guardians will need to instil faith and hope in the children and youth.

Studies have shown that children recover faster from disasters when they feel they are contributing to other people's recovery. Children and youth therefore need to be given specific age-appropriate tasks to let them know that they can help restore family and community life after a disaster. For example, some literature suggests that older children could help with clean-up tasks or reach out to those in need (e.g. Plan International and World Vision 2009; UNICEF 2007b). Engaging children (especially preschool and early elementary level) in physical activity is said to be a good way to relieve tension and anxiety. Various games or activities serve as a distraction, which in turn aid children to cope better.

For parents, guardians and community members to effectively exercise their role when a disaster strikes, they need to be guided by the young people's best interests. These interests can only be known by communicating with the children and youth. With proper guidance and support from parents, guardians and community members, not only will children and youth be equipped to cope with disasters and prepare before disaster strikes but also negative coping mechanisms that the youth engage in will be minimised or prevented. For example, in Botswana, Babugura (2008) found that in facing droughts girls were prostituting themselves to meet their own needs and the needs of their families. Alcohol and drug abuse, begging and stealing were also being used among the youth as a means of coping during drought disasters. In line with these findings, the literature acknowledges that reactions to disasters among children and youth are generally age-related and specific. Due to feelings of helplessness and guilt over not being able to take on full adult responsibilities when disasters strike, the youth tend to become disruptive or begin to experiment with high-risk or illegal behaviours such as alcohol and drug use (e.g. Schoeder and Polusny 2004).

Though the magnitude of sexual behaviour as a means of coping with disasters has not been adequately documented, a few studies such as Wiest *et al.* (2004) have reported cases of girls engaging in prostitution as a survival strategy following a disaster. Given that in most cases the girls are already marginalised when disaster strikes, the loss of security and protection forces them to prostitute themselves with the hope of gaining some income to sustain themselves and their families. Relief and social workers reported similar situations in countries devastated by disasters. In 2006, for example, a killer drought forced an alarming number of poor Kenyan women and children into prostitution in desperate bids to keep their families alive. Even girls below the age of thirteen were observed standing at the roadside selling their bodies so as to contribute to the welfare of their families (Oxfam 2006). Such coping mechanisms are very risky as the girls become more vulnerable to physical injury, transmission of HIV/AIDS and other sexually transmitted infections (STIs), and unwanted pregnancies (UNOCHA 2007).

The proactive role of children and youth in DRR

Despite their dependency on adults for protection and support before, during and after a disaster, children and youth should not be treated as helpless victims. The United Nations Convention on the Rights of the Child (UNCRC) clearly states that children and youth have the right to freedom of speech as well as the right to partake in decision-making processes that are relevant to their lives (see Chapter 6). The value of children and youth in DRR cannot be overemphasised (e.g. Mitchell *et al.* 2009; Wachtendorf *et al.* 2008). This is not only fundamental to policy-making that is responsive to children's needs and well-being, but also vital for their self-esteem and a means of empowerment.

Wachtendorf *et al.* (2008) state that, due to the connection of children to a large social network and support system through their schools, they have the potential to serve as conduits for disaster mitigation, preparedness, response and recovery information dissemination, both among their peers as well as to other household members. According to Mitchell *et al.*'s (2009) experiences with youth volunteer teams in El Salvador, Haiti and the Philippines on community risk mapping and mitigation activities, it is clear that children and young people have a much greater ability to participate in DRR than assumed.

There are several roles that children and youth can play to reduce disaster risk within their communities. These include taking part in activities such as managing evacuation centres, advocating and influencing policy-makers, protecting their families, creating disaster risk awareness, becoming informants, taking part in identifying disaster risks and learning how to deal with them, and participating in mapping hazardous areas in their communities.

Children and youth clubs/groups and safe child-oriented community centres could be put in place to create awareness about disaster risks and to build capacity. This allows children and adults to engage together in learning about and discussing disaster-related risks. Some effective methods used to create awareness include theatre, song, games and community meetings or debates. Various local and international organisations are already working with various communities in several countries (e.g. the Philippines, El Salvador, Bangladesh, India, Nepal and Indonesia) to set up children and youth groups (Back *et al.* 2009). The children and youth are engaged in various activities that enable them to acquire knowledge about hazards as well as build their capacity to cope with disasters. Activities include mapping hazards in their communities, preparing risk and resource maps, drawing up timelines and seasonal calendars, preparing disaster matrix ranking diagrams, and prioritising the responses to the most likely hazards and then undertaking their own DRR plan (see Chapter 64). In Mozambique, linking community projects with skills development and awareness raising was found to be particularly important for children and youth participants, in order to maximise empowerment and reduce the potential for feelings of helplessness (Back *et al.* 2009).

In Thailand children are involved in interviewing and mapping their communities (Benson and Bugge 2007). They also conduct assessments of risk and vulnerabilities, are able to educate peers and communities about disaster risks, advocate for government understanding of issues relating to children in disasters, lead the community in developing action plans to mitigate risks and conduct programme assessments to measure the impact of the work. Children in Sri Lanka also take part in formulating their communities' preparedness plans, help in the rebuilding of schools to ensure a child-friendly environment, and ensure that district plans incorporate issues affecting children and how their needs are met. In Vietnam children take part in community meetings to assess hazards and risks, are involved in creating risk maps as well as action plans for schools, advocate for mitigation measures, teach community members about responding to disasters, assess community preparedness, teach decision-makers about the impact of disasters on children, and educate communities about child rights and child protection (Benson and Bugge 2007).

Good practices have also been identified by Plan International and World Vision (2009) in countries such as El Salvador, Malawi and the Philippines. In southern Leyte in the Philippines, for example, children use theatre productions to educate communities about disasters. The

Box 36.2 Child-friendly participatory research tools for DRR

Grace Molina and Fatima Molina
Center for Disaster Preparedness, Philippines

Thomas Tanner and Fran Seballos
Institute of Development Studies, UK

Children form a significant group that is often overlooked by research and practice at the community level, in part because of a lack of appropriate action research tools.

Recent tools aim to serve as mechanisms for the children to have the opportunity to creatively express themselves, considering their strengths and limitations. Also, it is easy to comprehend that these tools will further encourage children's involvement. Most importantly, children were consulted during the process of the tools' modification, as the need arose, in order to adapt them to the children's level of comprehension.

These tools include:

- mapping of risks (hazards, vulnerabilities and capacities);
- stakeholders and communication pathways;
- ranking of risks, risk management actions;
- drawing of envisioned communities;
- interpretation of emotions and visualisations of motivation of participation;
- transect walks for risk identification and preparation of action plans;
- acting and theatre to present the impacts of disaster events and responses, as well as for advocating behavioural and policy change by others;
- message pyramids for the visual presentation of pathways from the community concerns and issues, with the people's interventions to address them; and
- participatory video.

Ice-breakers are also needed between sessions to keep the children energetic, develop their self-confidence and be able to introduce the tools and methods. In general, child-friendly DRR is most successful when:

- cultural norms and age range of participants shape the research design;
- research methods are focused on having fun;
- activities are carried out in small groups so that individual children feel confident enough to participate;
- methods are iterative, allowing children themselves to shape and change them;
- researcher intervention is limited to an explanation of the tool or method;
- a mix of oral, visual and written activities is used, helping children to express their perceptions, experiences and ideas concerning hazards, vulnerabilities and capacities; and
- children also gain from the experience of participating in the research.

These methods provide an avenue for the children to have the freedom to identify their priority issues in the community that need to be addressed. Thus, children are able to consider the full range and impacts of their DRR activities, consider their long-term plans and discuss initiatives that might help in dealing with risks and be able to recognise relevant stakeholders from community to national level. All in all, these tools help to foster a two-way learning process for researchers, local authorities and young people in the field. Their participatory and interactive nature allows each participant to share his/her thoughts and at the same time gain awareness from others' experience and insights. It also provides further opportunities to continuously strengthen and sustain efforts to improve safety and sustainability of communities.

For more information consult: www.iied.org/pubs/pdfs/14573IIED.pdf

children write the scripts, which are then directed and choreographed by youth leaders. Discussions are held after the performances, allowing people to share their views and opinions about issues raised during the performance. The children of the Petapa Emergency Committee in El Salvador have made a vast difference to their community's safety by being able to identify and address key underlying risk factors. One example given is how the children managed to identify the risks (erosion and higher risk of flooding) posed by the unregulated extraction of rocks and stones from a river in their community. This resulted in local leaders erecting signs that prohibit extraction of rocks and stones for personal use.

Policy issues

Vital to policy development and implementation is accurate knowledge regarding children and youth in disasters. Disaster literature (e.g. Osofsky *et al.* 2007) delineates important policy gaps in addressing specific needs of children and youth. Existing evidence such as experiences from Hurricane Katrina, the earthquake in Haiti, the Myanmar cyclone, and drought and famines in various countries, now make it apparent that every nation should consider the vulnerability of its children and youth to disasters. For example, in recognition of the risk and challenges experienced by children and youth during and after disasters, the USA established a National Commission on Children and Disasters to conduct a comprehensive study that examines and assesses the needs of children independently, in relation to the preparation, response and recovery from all emergencies, hazards and disasters. The study provided knowledge to guide effective policy and strategic planning to address the needs of children in disasters (Keegan 2010).

It is important that national governments provide the needed policy-level guidance and leadership in promoting child- and youth-sensitive disaster policies. Youth and childcare emergency plans need to be on the agenda of national financial DRR policies. Policy issues such as the development and enforcement of child- and youth-protection laws, education, evacuation plans, reunification of children and youth with family, establishment of national disaster preparedness standards for childcare centres and schools, establishment of national offices for children's advocacy, and development of monitoring and evaluation systems need great attention if nations are going to effectively invest in and take a leading role in the strategic planning and management of child- and youth-focused DRR.

National governments can provide an important supportive role by encouraging initiatives undertaken by regional and local governments, NGOs and private organisations. Some good practices that can be replicated at both national and regional level have already been demonstrated by international NGOs that have operated in previous disaster zones. One good example is the development of child- and youth-friendly spaces that are designed to address both physical and psycho-social needs of children and youth in a stable environment that invites trust. These friendly spaces are set up after a disaster has occurred. The child-friendly spaces allow children and youth to play, learn competences to deal with the risks they face, be involved in some educational activities and relax in a safe place.

Child- and youth-friendly spaces, for example, were successfully set up by World Vision in countries such as Bangladesh after the 2007 cyclone, in Indonesia after the 2006 earthquake and in the Philippines after the 2006 mudslide. Children and youth in these friendly spaces were able to participate in sports and games, counselling, cultural events, and received education and health care. Generally, the children and youth were able to cope better with painful experiences caused by disasters (World Vision 2010). Such initiatives can be enhanced as part of a holistic, top-down and bottom-up national framework and strategy for action and co-ordination of child- and youth-friendly spaces.

Countries such as Ecuador, Cuba, Nicaragua, Peru, Venezuela, El Salvador and Panama are known for their good practices in mainstreaming DRR into their national curricula. Nicaragua, for example, began implementing its policy to make risk management part of the nationwide curriculum in 2005. Since implementation, nine guidebooks for teachers and nine workbooks for students have been developed and used in schools. Since 2003, Costa Rica has developed extensive hazards and safety teaching in schools and 6,000 teachers have been trained to develop lessons based on local hazards and patterns of vulnerability. Cuba is also known for its strong history of reducing risk through its education system. The national curriculum covers disaster preparedness and response to hurricanes, which are the most significant local natural hazards. The Cuban Red Cross produces teaching materials, and the safety messages that children get in school are reinforced by what parents hear in training courses and drills in the workplace (Wisner 2006a).

In partnership with UNICEF, several countries (e.g. Kenya, Ethiopia, Burundi and Madagascar) are now adopting the cluster approach in disaster policy and strategic planning at national level. The cluster approach is a strategic framework that permits various partnerships to engage collaboratively to achieve shared objectives. One such objective has been to advocate for policy change to ensure that children's rights are promoted and that they receive assistance and care before, during and after disasters. The approach allows for improved predictability, accountability and leadership in humanitarian action. The approach has helped to strengthen the capacity to meet the needs of children and youth in a humanitarian setting. As a result, the needs of the most vulnerable children in Madagascar were addressed during the 2008–09 cyclones. Over 8,600 children identified as suffering from acute malnutrition were treated, thirty temporary classrooms were constructed in various schools to allow affected children and youth to return to school, 408 separated children were reunited with their families and child-friendly spaces were established to respond to the growing needs of unschooled children. Furthermore, the education and protection clusters were able to help children deal with negative experiences and psychological distress (UNICEF 2010).

At the international level, there is a need to focus DRR policies on strengthening the capacities of nations to prepare for, assess and respond to disasters. Policies must enforce regular monitoring, analysis and knowledge-sharing, with a special focus on the situation of children and youth for effective response to disasters. Although the Hyogo Framework for Action 2005–15 offers the opportunity to promote a strategic and systematic approach to reducing vulnerabilities and risks to hazards, it does not adequately address specific issues of children and youth. Apart from mentioning the need to include DRR in relevant sections of the school curricula and using other formal and informal channels to reach youth and children with information, other important issues are not addressed. These issues, for example, include the inclusion of children and youth in planning, assessing, implementation and decision-making; child protection to reduce vulnerability to abuse (important especially for children separated from families) and exploitation; and mainstreaming children's rights into DRR programmes.

Policy and decision-makers at international, regional and national levels have a responsibility to ensure that the views, voices and priorities of children and youth are taken into consideration if DRR policies are to be effective, fair and equitable. It is important to bear in mind that these very children and youth being excluded from participating in policy and decision-making processes will in future be the ones expected to continue the implementation of policies. It is therefore vital that they are allowed to familiarise themselves with the processes through meaningful participation from an early stage. Working with children and youth at an early stage grooms a knowledgeable and skilled generation of decision-makers.

Conclusions

It is now widely accepted that children and youth are highly vulnerable. Despite this acknowledgement, very often specific vulnerabilities of children and youth are overlooked (e.g. child trafficking, exploitation, neglect, rape and sexual abuse, psychological problems).

Given that children and youth are differently impacted by disasters, it is critical that their vulnerabilities and needs are well distinguished and comprehensively integrated into all disaster planning activities and operations. This includes addressing both immediate and long-term vulnerabilities. The right to special protection to curb kidnapping, abuse and sexual exploitation when disaster strikes requires special and urgent attention. It is everyone's obligation to respect the rights of children (UN Convention on the Rights of the Child, Article 12). Taking action to ensure the protection of all children and youth during and after a disaster will therefore require a collective responsibility (e.g. governments, schools, private and public sectors, communities, local leaders, parents and guardians, NGOs, etc.).

The value of children and youth's contribution to DRR is now being recognised. Children and youth can act as powerful agents of change. Hence, respecting the right of children to participate in making decisions on matters that affect them is critical both for the good of these young people and the good of the community. Educating and making children and youth aware of disaster risks and how to deal with these risks allows them to influence family and community actions. Schools have great potential to ensure that children and youth are knowledgeable about DRR. Several organisations have made it their priority to empower children and youth to become involved in their community's preparedness and mitigation plans. Parents, guardians and communities also have a major role to play in aiding children cope better during and after a disaster. This can be done by providing emotional support and reassuring the children and youth. It can also be done by lobbying for safely located and built and properly maintained schools that do not collapse in earthquakes and are not swept away by mudslides.

Having acknowledged the need to reduce the vulnerability of children and youth in and to disasters, it is also important to note that all efforts will be in vain if underlying factors that exacerbate their vulnerability are ignored. These factors include poverty, child labour, HIV/AIDS, exclusion, lack of access to education and inequality. DRR will not be achieved successfully in isolation from these underlying factors.

37

Elderly people and disaster

Ehren B. Ngo

LOMA LINDA UNIVERSITY, USA

Introduction

Disasters impact communities throughout the world in significant and devastating ways. Populations such as women, children, the elderly and the disabled are known to suffer more and in different ways during disasters. The elderly, in particular, have physical, economic and social vulnerabilities that result in unique challenges and subject them to greater harm, loss and difficulty in recovery from a disaster. Yet, despite the vulnerabilities of the elderly to disasters, their considerable life experience often means they possess the knowledge and skills to contribute meaningfully to disaster preparedness and recovery. Thus, the global growth of the elderly population presents both challenges and opportunities, particularly as it relates to disaster response, recovery and risk reduction.

The global elderly are a diverse demographic that defy a singular definition. Within the social structure and culture of a community, numerous factors including livelihood, social position, family role, functionality and health status play a significant role in defining the elderly. These socio-cultural perceptions, and even the self-perception of older age, are significant enough that the actual role of chronological age in defining the global elderly often becomes secondary. However, chronological age is most frequently used by governments and policy-makers in defining the elderly within a population. This 'legal' definition for the elderly is significant, as the chronological age of sixty-five is often chosen as the official transition to elderly status and the age at which pensions or entitlement benefits begin in many of the countries that provide these benefits. By comparison, many researchers, as well as the United Nations (UN), utilise a lower chronological age of sixty years or older to mark the transition to elderly status – in part recognising the lower life expectancy observed in many developing countries. Where age grades are culturally important, as among the Maasai pastoralists of East Africa, a man may achieve the status of 'elder' while still in the prime of his strength and economic activity.

Despite the widespread assumption that the elderly are a vulnerable population, it is imperative to recognise that not all elderly are vulnerable, or vulnerable in the same way. There is as much diversity among older adults as between the young and the old. Interventions to reduce the vulnerability of the elderly in disasters must take into account both individual and group differences. This requires the use of a holistic definition of the elderly based on more than an assignment of chronological age – a definition that also takes into consideration the many interactions between chronological age and individual attributes such as economic resources, relative health, and access to health care, along with broader socio-cultural factors such as social roles, community integration and even the prevalence of age prejudice and marginalisation.

Vulnerability of the elderly in disasters

Vulnerabilities of the elderly in a disaster exist for many reasons and fall into broad categories of physical, psychological, social and economic vulnerability. Vulnerabilities may be pre-existing conditions present before the disaster event, such as physiological processes of ageing, limited or fixed incomes, social limitations on the pursuit of livelihood, or social marginalisation and discrimination faced by the elderly. Vulnerabilities may also emerge for the first time among an otherwise healthy elderly population during or after the disaster when basic needs are not met as a result of neglect or well-intentioned but miscalculated interventions. Such was the case in the 1995 Hanshin-Awaji earthquake of Japan. There, a group of elderly were at first neglected by the disaster response effort, but were subsequently given priority in temporary housing; unfortunately, this created a community of the elderly and disabled living alone, and as many as eighty-three died unnoticed in the temporary housing (Tanida 1996). Finally, disaster vulnerability often exists as a complex interaction between the unique, personal characteristics of the elderly individual and an evolving, post-disaster social interface, requiring a clear recognition of the diversity among individual elderly.

In a disaster the vulnerabilities of the elderly often emerge in two ways: (1) disaster events magnify pre-existing individual and social characteristics and differences; and (2) disaster events accelerate existing risk factors in the population (Ngo 2001). These concepts are illustrated through observations made in numerous disasters.

For example, following the 2004 Indian Ocean tsunami, relief workers failed to take into account the limited mobility of the elderly, which made it difficult for the elderly to reach camps and shelters; once at the camps, those with disabilities were further challenged by a lack of assistive devices such as wheelchairs or walkers (HelpAge International 2005). Limited mobility and physical strength mean that the elderly are at significant disadvantage when post-disaster resources are located at a distance or require long waits in lines (Pekovic *et al.* 2007). Situations like these magnify pre-existing physical attributes of the elderly to a level of threatening vulnerability when the elderly are additionally forced to compete with more able-bodied survivors for limited resources (HelpAge International 2002).

Similarly, disasters can accelerate existing risk factors. For instance, research studies following the health impact of the 1994 Northridge earthquake in the USA and the Hanshin-Awaji earthquake in Japan found an increased rate of cardiovascular-related deaths immediately following the earthquakes, which dramatically accelerated the rate of cardiovascular deaths in the initial aftermath of the disasters and even contributed to an excess cumulative cardiac mortality in Japan (Ngo 2001).

Physical vulnerability

Physical vulnerability among the elderly can take on several dimensions. Short-term threats to life and health are primarily dictated by the ability to resist trauma and injury resulting from the physical forces of the disaster impact and to survive the immediate post-disaster environment. The physiological changes associated with ageing, which include a weaker musculoskeletal frame, decreased cardiovascular capacity, slower responses within the nervous system, diminished thermoregulation and loss of acuity in sensory function, compromise the physical well-being of the elderly in disasters.

Differential morbidity and mortality is common, as noted in the observations of medical response teams in China's 2008 Sichuan earthquake, where forty per cent of the patients in relief sites were over the age of sixty years of age (Chan 2008). In the USA's 2005 Hurricane Katrina and Japan's Hanshin-Awaji earthquake, the elderly comprised well over one-half of the fatalities, even though the elderly made up only fifteen to twenty per cent of the total population (Hutton 2008; Pekovic et al. 2007; Tanida 1996). Similar observations were made in Taiwan's 1999 Chi-Chi earthquake where the proportion of fatalities among the elderly (twenty-seven per cent) greatly exceeded their proportion of the population (8.1 per cent) (Lin et al. 2002). Reports from relief workers responding to the Indian Ocean tsunami concluded that, 'most of the older people could not run fast enough to escape the waves or swim to safety' (HelpAge International 2005: 7).

The elderly are especially vulnerable to heat waves and cold weather. An analysis of US mortality from twenty-six years of data revealed nearly exponential age-related differential mortality due to heat waves and cold weather (Thacker et al. 2008) (see Chapter 23). Likewise, the 2003 heat wave in France heavily impacted the elderly, with seventy per cent of the 14,800 deaths occurring in persons over seventy-five years of age (Hutton 2008).

The post-disaster environment can be especially challenging for the elderly, as the process of ageing affects basic mobility and can result in difficulties with essential capacities, such as activities of daily living. Additionally, the elderly have had more time to develop chronic diseases and degenerative processes. These physical changes often mean a need for assistance with walking, specialised diets, limiting exposure to temperature extremes and ensuring ongoing medical management of chronic conditions. In disasters and emergencies where food aid is being distributed, the unique needs of the elderly are rarely taken into account, forcing the elderly to travel long distances to distribution points, wait in lines for extended periods and carry heavy loads of rations (Wells 2005). The failure to account for the unique needs of the elderly resulted in gastrointestinal problems among the elderly who undercooked maize and beans in Turkana, due to a lack of familiarity with the food; likewise, the elderly in Darfur resorted to selling the sorghum they received, as it was too coarse for them to consume (Wells 2005).

Findings on the specific long-term impacts of disasters on the physical health of the elderly are less consistent, and there is some debate over how long physical health measures, such as general health, medical conditions, physical symptoms, functional impairment and fatigue persist after a disaster. Some studies suggest that physical health issues return to near baseline after one year (Fernandez et al. 2002; Phifer et al. 1988). Yet others have suggested that the elderly will never recover, as in the 2010 Haiti earthquake, which has left a vulnerable population of amputees and disabled to struggle for the rest of their lives (Goss 2010). A partial explanation for differences in long-term health impacts may be rooted in the intensity of the disaster experience. Numerous studies support an assertion that higher intensity disaster exposure is associated with more severe mental and physical symptoms (Kohn et al. 2005; Phifer et al. 1988; Ticehurst et al. 1996).

Psychological vulnerability

Psychological vulnerability among the elderly arises from a complex interaction of risk factors. For example, higher socio-economic status, being married and having access to adequate resources are recognised as protective factors for psychological health. However, the elderly often trend toward lower socio-economic status, widowhood and limited material resources due to diminished earning capacity or total loss of livelihood. The prevalence of age-related dementias and mental impairments also increases with age, thereby increasing the potential for vulnerability among the elderly.

Common psychological impacts of disasters across all populations include anxiety, irritability, restlessness, sadness, despair, depression, insomnia, loss of appetite and post-traumatic stress disorder (PTSD). Numerous studies have sought to determine the prevalence of the psychological impacts of disasters on the elderly and to establish whether or not a differential exists between younger and older disaster victims.

The resultant findings on the psychological impacts of disasters on the elderly have been mixed. One study found a greater PTSD risk and impact among elderly women (Ticehurst *et al.* 1996). The majority of studies have found no age-related differential in psychological impact (e.g. Kohn *et al.* 2005), while several studies have supported a lower psychological impact (e.g. Hutton 2008; Knight *et al.* 2000). In one study that conducted pre- and post-disaster measures of the elderly in Taiwan's 1999 Chi Chi earthquake, it was found that the elderly whose houses completely collapsed had more favourable psychological outcomes as opposed to the elderly with partially collapsed houses – a finding that may be attributed to the additional social support extended to the elderly who lost their homes (Lin *et al.* 2002) (see Chapter 47).

Factors such as maturity, wisdom that comes with age and prior experience with disasters have been noted to moderate the psychological vulnerability of the elderly, who handle life's transitions and stresses with a greater ability for coping and adaptation (HelpAge International 2002; Ngo 2001). Two hypotheses seek to explain the protective effect of age on psychological vulnerability. The inoculation hypothesis suggests that prior exposure to specific disaster experiences results in a psychological tolerance to subsequent, similar exposure. The maturation hypothesis suggests that the elderly are less reactive to stressful events such as disasters through the acquisition of mature coping styles and adaptive mechanisms, allowing them to reconcile these stresses through constructive outlets such as altruism, anticipation, humour and sublimation (Knight *et al.* 2000; Vaillant 1995: 80).

Despite limited indications that the elderly may be less psychologically vulnerable than younger disaster victims, caution should be used in extrapolating the results of prior research studies in this area to future disasters. First, very few published research studies have examined the psychological impact in populations with low human development, let alone under disaster conditions where the magnitude of community and personal loss may exceed the protective thresholds inherent in any particular group. Second, it is generally accepted that the severity and duration of psychological impacts are proportional to the magnitude of loss or other trauma experienced in a disaster, for both the young and old alike (Kohn *et al.* 2005; Phifer *et al.* 1988). Consequently, the psychological needs of the elderly should always be evaluated following a disaster (see Chapter 47).

Social and economic vulnerability

Early sociological studies on the elderly in disasters found that distinct patterns emerged along the lines of age. One of the first observations made by Friedsam (1961) was that the elderly were twice as likely to report their disaster losses as greater than other victims despite evidence that damages were relatively equal, a finding he attributed to relative deprivation. Subsequent studies mostly supported Friedsam's observations of relative deprivation, but also offered another consideration: the relative deprivation reported by the elderly was often accompanied by injury, disablement or moves to temporary housing (Ngo 2001). In examining social support for the elderly in disasters, researchers Kaniasty and Norris (1995) found that when the elderly suffered harm (injury or threat to life), a 'pattern of concern' emerged, prompting social support equal to younger victims. However, in the absence of harm, the elderly suffered from a 'pattern of neglect' and subsequently received less social support than younger victims.

The relationship between the post-disaster needs of the elderly and subsequent delivery of social support is influenced by a complex interaction of social patterns, perceptions, biases, ignorance and even prejudice against the elderly. How this interaction plays out in a disaster will vary based on social norms and culture, as well as the individual attributes of the elderly. Social vulnerability of the elderly emerges as a combination of intrinsic, individual attributes, as well as broad, external social attitudes and actions directed towards the elderly.

Individual attributes increasing vulnerability include limited mobility, illiteracy, low educational attainment and limited financial resources. The connection of the elderly to their community diminishes as peers begin to die off and functional limitations make it difficult to engage in their usual social groups, such as family and friends, local organisations or churches. Social vulnerability may also arise along gender lines. Older women are more likely to be widowed, as they generally outlive men. For women living in areas with patriarchal laws, this often results in the loss of their property and home, along with reduced social status from their widowhood. Older men may lose status and respect, and even face rejection by family and society, once they stop fulfilling the traditional male role of economic provider (HelpAge International 2002).

The economic impact of disasters interrelates closely with other social vulnerabilities. Often, the marginalised in society are relegated to live and work in the least expensive areas, such as coastal areas, on flood plains or on unstable hillsides, where the risk is greater. Additionally, economic forces at the individual or societal level often dictate the quality of housing and construction in an area, which affects vulnerability to hazards such as earthquakes and windstorms. Economic factors such as a lack of insurance, fixed income, non-liquid assets, limited

Box 37.1 Collapse of family support structures during the 2008 Sichuan earthquake, China

Liu Tingjin sits in a tent with other elderly survivors at a government camp following the 8.0 magnitude earthquake that shook Sichuan Province. She worries about her son, who carried her to the top of a hill a few days after the earthquake and left her there with other villagers without saying anything, not even goodbye. At eighty-two years of age and blind, Liu must now depend on her tent mates to help her with simple tasks, like guiding her to the toilet outside, which sits several hundred yards away.

Prior to the earthquake, urban migration left a greater concentration of elderly in some of the rural areas, as their children left for better economic prospects in the cities. Despite a culture that traditionally reveres its elders, attempts to reunite the elderly with their adult children have been largely unsuccessful. In some cases, adults struggling to care for their own offspring abandoned their elderly parents, leaving them to fend for themselves in congregate elderly tent villages.

With few people offering to take care of the elderly, the Chinese government stepped in to provide food and shelter, and pledged to build additional old-age homes. Despite the assertive role that the Chinese government has taken to ensure the basic needs of the elderly are being met, Liu and many of her tent mates lament that nothing can truly replace family (Chan 2008; Fanuthor 2008).

credit and outright poverty all contribute to a higher relative need of the elderly, making recovery a more difficult process for the elderly than younger disaster victims.

Economic vulnerability ranks as a leading concern for the majority of elderly. A loss of income or livelihood quickly threatens independence and can precipitate a range of consequences including threats to basic needs such as food and water, shelter, protection and health care. Following a disaster, the elderly are often overlooked in workforce assessments due to an assumption that they are no longer part of the workforce (Hutton 2008; Wells 2005). Such was the case in Sri Lanka following the Indian Ocean tsunami, where the elderly were not only overlooked in workforce assessments, but systematically denied the opportunity to participate in skills training and microcredit opportunities (Duggan *et al.* 2010). The loss of livelihood and income for the elderly represents a multilevel misfortune, because the elderly play an increasingly critical role in the intergenerational support of dependents, particularly where urban migration, conflict and HIV/AIDS have left older adults to raise children in the absence of parents and the middle-aged (Wells 2005).

Marginalisation, prejudice and discrimination based on age result in significant social vulnerabilities for the elderly, often impacting key areas of health and economics. For example, social policy may inadvertently deny the elderly access to health resources when limitations in mobility are overlooked and suitable transportation to care is not provided; alternatively care may be priced out of reach for the majority of elders without regard for their limited means (HelpAge International 2002). Prejudice about the suitability of the elderly for employment means they take low-paying jobs and informal work opportunities rejected by younger workers (Hutton 2008). During disaster events the elderly may face increased marginalisation, prejudice and discrimination as competition for resources mounts.

The interplay between the vulnerabilities of the elderly is significant. For many elderly, the interaction is simple but devastating: poverty results in social isolation and, in turn, social isolation feeds poverty (HelpAge International 2002). However, the elderly may also experience a more complex sequence of events, where physical vulnerability from illness results in an inability to pursue work or other livelihood, and a subsequent drain on already limited financial resources ensues. Alternatively, a lack of economic resources can limit access to health care, resulting in complications of chronic disease states or even disablement, thereby threatening the older person's access to social interaction. These complex interactions of multiple vulnerabilities among the elderly are common and tend to appear as a 'constellation' of risk factors, with the presence of one indicating the presence of other vulnerabilities.

Response and the role of the elderly in disaster risk reduction

Effective disaster risk reduction (DRR) requires actions that address the vulnerabilities of the elderly in dealing with disaster, and include disaster preparedness, response, recovery and mitigation (Hutton 2008). A growing awareness of the unique needs and potential contributions of the elderly during disasters is driving a global re-evaluation of how the elderly are recognised and included in disaster plans. Despite this growing body of knowledge, observations made at recent disasters indicate that the majority of disaster-response personnel and relief agencies have yet to incorporate an awareness of the elderly, let alone best practices, into their programmes.

Response to impacts on the elderly

Specific vulnerabilities, such as physical and psychological dimensions inherent to individual elderly, such as diminished strength, chronic disease, functional limitations and pre-existing

Box 37.2 Being affected again following the 2004 Indian Ocean tsunami, Sri Lanka

Like so many disasters that affect those with limited strength and diminished mobility the most, the Indian Ocean tsunami resulted in higher death rates among females and the elderly. Tragically, the elderly who survived faced a secondary threat in the ensuing relief operation in the form of exclusion, discrimination and marginalisation.

Conversational interviews with elderly survivors of the tsunami revealed several social problems associated with the relief operation: there was a disconnect between non-governmental agencies and the needs of the elderly, who were not given an opportunity to provide input on how their age-related issues, such as mobility, could be accommodated in the replacement housing that was provided; the elderly were systematically excluded from rehabilitation programmes, such as microfinance and skills training, which would have enhanced their ability to maintain their independence; inequities in the system for distribution of goods and resources were common, often resulting in younger family members receiving the full distribution of aid while the elderly received nothing; and frustration existed over the constraints placed on restoring their livelihood, with little consideration given to how the elderly sustained themselves through work such as fishing, tourism services and childcare, which were outside of the formal economy (Duggan *et al.* 2010; HelpAge International 2005).

Work by HelpAge International in the tsunami-affected areas found that consultation with the elderly in the post-tsunami environment would have assessed the needs of the elderly more accurately. However, just as importantly, engaging the elderly would have allowed relief organisations to understand the skills and contributions offered by the elderly, even in critical areas such as livelihood (HelpAge International 2005).

conditions such as dementia, are difficult to modify during or after a disaster. Therefore, effective intervention focuses on prevention and managing hazard exposure. Pre-disaster activities can and should include disaster preparedness training (Fernandez *et al.* 2002). Consultation with the elderly to adapt disaster preparedness training to meet their needs helps to ensure appropriate content, generates empowerment (Duggan *et al.* 2010) and identifies both limitations and capacities. Post-disaster prevention should include rapid assessment of the elderly to identify basic needs and those requiring immediate intervention.

The Seniors Without Families Team (SWiFT) triage tool is one such assessment tool designed to rapidly identify elderly who need intervention, and was pilot tested at the Houston Reliant Astrodome Complex shelter following Hurricane Katrina (Dyer *et al.* 2008). Prevention activities may also take a programmatic approach, such as instituting mass tetanus vaccination programmes when it is recognised that the population is lacking protection, as was the case in the Sichuan China earthquake, and again in the Haiti earthquake where teams were quickly deployed to vaccinate patients, even as they waited for further medical care (Chan 2008; Goss 2010).

Evacuation can play a key role in managing the risk exposure faced by the elderly; however, the elderly may resist evacuation due to inability or unwillingness (Pekovic *et al.* 2007). As Hurricane Katrina approached the USA in 2005, some of the elderly failed to evacuate because they were unable to walk to the buses leaving the city of New Orleans, while others refused to leave based on prior experience with hurricanes; this resulted in additional family members and care providers remaining in the affected area to care for those elderly (Eisenman *et al.* 2007). Ironically, evacuation is not without significant risks for the elderly. This is especially true of the frail elderly whose medical conditions can be aggravated by the loss of medications or essential medical devices, and separation from care providers. For others, the cost of evacuation may represent an economic hardship and the move to an unfamiliar environment may quickly threaten mobility and independence – especially with the loss of essentials such as eyeglasses or hearing aids.

Finally, providing the elderly a voice in post-disaster assistance, such as the type of shelters and housing in which they are placed, empowers the elderly and helps to avoid living conditions that are hazardous or exacerbate existing health conditions.

In the dimension of psychological health, while there is a lack of consensus on the differential vulnerability of the elderly, consistent themes have emerged and form the basis for evaluating the psychological needs of the disaster-affected elderly and planning appropriate interventions:

- Pre-existing mental-health conditions are a good predictor of post-disaster mental health vulnerability.
- Psychological distress frequently manifests itself in non-traditional symptoms among the elderly, making it sometimes difficult to recognise the underlying root cause as psychological.
- The more significant the impact on the individual and (to a lesser extent) the community, the more likely it is that the disaster-affected elderly will exhibit psychological symptoms.
- Psychological risk factors of the elderly may be mitigated or exacerbated depending on patterns of concern or patterns of neglect that may emerge post-disaster.

When instituting psychological intervention programmes, it is beneficial to remember that the elderly are less likely than younger disaster victims to utilise traditional counselling and other psycho-social supports typically provided following a disaster (Ngo 2001). Therefore, ideal interventions are those that integrate with existing psychosocial support structures and culturally appropriate mechanisms already familiar to and accepted by the elderly, such as church or community groups. Programmes utilising older adults to deliver peer-counselling and outreach to the disaster affected are particularly effective in addressing the psycho-social needs of the elderly (Myers and Wee 2005: 56) (see Chapter 47). Furthermore, the elderly respond psychologically to support in other areas; for example, mitigating the social and economic impacts of a disaster will help to minimise the psychological impact (Fernandez et al. 2002; Lin et al. 2002).

Among the myriad social and economic vulnerabilities, income is the number one priority identified by elderly in disasters and crisis, as it enables their independence and their drive to support themselves (HelpAge International 2000; Wells 2005). Prior research has shown that after a disaster, the elderly have an aversion to programmes perceived as 'welfare', despite need and eligibility, instead preferring 'hard services' such as housing assistance, low-interest loans, health care and transportation (Ngo 2001). Economic interventions on behalf of the elderly should be principled on the concept of empowerment. Offering the elderly an opportunity to extend their productivity and independence through a livelihood has proven benefits; it has been noted that retention of livelihood and self-sufficiency contributes to better health outcomes and well-being. Additionally, as the elderly play a crucial role as care providers for children within families, even small amounts of cash income find their way towards supporting the essential needs of children and grandchildren, including food, clothing and education (HelpAge International 2002; Hutton 2008; Wells 2005).

Social and economic vulnerabilities related to marginalisation, prejudice and discrimination remain challenging to address in the post-disaster environment. Pre-existing social and cultural conditions affecting the elderly are unlikely to be changed in the context of a programmatic response to a disaster. Therefore, interventions to these social and economic vulnerabilities should focus on the basic guidelines for creating effective engagement and partnership with the elderly, and utilise a robust programme evaluation process to ensure effectiveness.

Guidelines for initial interventions to the general vulnerabilities of the elderly may be built around shared principles from best practice recommendations that address the elderly in disasters. They include:

- Locate and recognise the presence of older people and identify them as an at-risk population (HelpAge International 2000; Ngo 2001).
- Identify and address problems in consultation with the elderly, identify their capacities and support empowerment (Duggan *et al.* 2010; Fernandez *et al.* 2002; HelpAge International 2000).
- Utilise existing resources, agencies and community organisations, and connect them with the elderly; ensure the elderly are represented in community-level decisions (Fernandez *et al.* 2002; HelpAge International 2000; Ngo 2001).
- Employ basic assessment tools and checklists to assess elderly vulnerabilities; these data can be used to evaluate effectiveness of programmes and interventions (HelpAge International 2000; Ngo 2001).

Building on the capacity of the elderly and reducing vulnerability

A common myth of the elderly in disasters is that they are passive beneficiaries and have little to offer as resources (Wells 2005). However, research on the role of the elderly in disasters and observations of the effectiveness of programmes that actively engage the elderly support the assertion that the elderly can and do play pivotal roles in DRR (HelpAge International 2009; Hutton 2008; Wells 2005).

The contributions of the elderly in the social dimension are significant, but often overlooked and underappreciated. Roles that the elderly play within the family include helping to maintain the household, cultivation and agriculture, and providing a presence to guard property (HelpAge International 2002). The elderly also fill essential support roles in many families, serving as caregivers for children, the sick and the dependent. In Pakistan, the establishment of child-friendly spaces by the UN Children's Fund (UNICEF) was modified to incorporate an inter-generational approach, where older people joined the children and found that caring for the children and organising their activities provided importance and meaning (Day *et al.* 2007). The impact of HIV/AIDS in many countries has changed the role of the elderly from a contributing caregiver, to one of primary and perhaps only caregiver for young, orphaned children, as entire adult generations disappear (HelpAge International 2002). The importance of the elderly in the lives of children is particularly notable, as the elderly will make enormous sacrifices to ensure that children have adequate food, clothing and even school fees, even if it means that they themselves go without. Consequently, supporting the elderly through a modest social pension or household grant not only reduces the economic vulnerability of the elderly, but can have a stabilising effect on the entire family unit (Wells 2005; HelpAge International 2002).

Within their communities, the elderly are also well positioned to share wisdom and guidance, due to their years of life experience, maturity and, it is likely, prior experience with similar disasters or emergencies. For example, in communities throughout Kenya, Bangladesh and India that have been devastated by drought, the elderly are often the only ones who possess an extensive, historical knowledge of past weather phenomena. Their knowledge of traditional practices on sustainable land use, reading climatic changes through animal and plant behaviours, as well as practices that helped them survive previous droughts, uniquely informs local

Box 37.3 Cross-cutting issues and challenges

Jo Wells
HelpAge International, UK

Special interest groups or vulnerable groups are defined by the humanitarian community: older people, people with disabilities, children and ethnic minorities. Consequently, organisations have evolved to 'champion' their causes, raise funds and deliver interventions on their behalf. However, issues such as age and gender concern us all and cut across programmes, as reflected within the UN cluster system. The result is a humanitarian system that both promotes specialist and dedicated response to individual groups of people, and recognises that protection and assistance of these groups of individuals should be 'mainstreamed'.

This contradictory approach can result in further marginalisation. Compartmentalising people into groups based on age, gender or physical status can be problematic because:

- it tends to ignore the family, household and community structures in which individuals live and which are central to their well-being in times of crisis;
- it assumes that if you belong to one or other group then by definition you are vulnerable (and, by implication, helpless), ignoring both individual skills and capacities and the fact that in particular circumstance others may be vulnerable, e.g. middle-aged men targeted in violent conflict;
- it ignores the complex interplay of social, generational, geographic, economic and political factors that may determine other vulnerable groups or individuals in particular contexts;
- it assumes that the needs of specific groups are covered by mandated organisations, yet when this is not the case there is no clear mechanism to fill the gap;
- it can result in a competitive funding environment where agencies are averting their gaze to priority issues because they fall outside their organisational focus; and
- at worst it can result in the discrimination against one group to the benefit of another.

How to improve upon the current situation?

1 Organisations that represent cross-cutting issues need to work together much more closely and collaboratively. All desire equality of:
   - access to humanitarian services;
   - access to information about their entitlements; and
   - participation and consultation in matters that affect their lives.

Put simply, this is good humanitarian practice.

2 There must be common assessment amongst key stakeholders of context, vulnerability and local capacity both prior to and at the onset of a disaster.

A common vulnerability assessment should look at how different groups will be affected by a crisis in different ways, because of their:

- pre-existing vulnerability (e.g. being very poor or discriminated against);
- exposure to various protection threats (e.g. gender-based violence) during and following the disaster;
- varying capacities for coping;
- ability to recover;
- access to social, legal and emotional support; and
- the various physical, cultural and social barriers they may face in accessing services.

A common vulnerability assessment helps to identify where there are protection and assistance gaps for particular groups or individuals, rather than focus on those who are most visible, vocal or who have agencies championing their cause.

communities and complements the scientific body on climatic change (HelpAge International 2009). Other areas where the maturity and life experience of the elderly can benefit communities include facilitation of problem solving and conflict resolution (HelpAge International 2002, 2007).

Formalising opportunities where the elderly can share their knowledge and expertise toward community problems is essential in bringing the elderly out from obscurity and into a forum where DRR and capacity building can be achieved in partnership with the larger community. Older people's associations (OPAs) is a model that has been used successfully around the world to enable the elderly to promote their own dignity and quality of life, increase their skills and ability to undertake activities that benefit themselves, their families and communities, reduce isolation and vulnerability among members, promote broader community involvement and cohesion, and increase the awareness of the elderly as productive and active members of society (HelpAge International 2007). OPAs contribute to DRR and capacity building through multiple avenues. In Rajasthan, India, village elders have shared their expertise and contributed traditional knowledge of water-harvesting techniques, while in Bangladesh, specific OPA members have been trained to go house to house and conduct damage assessments following disasters, and in Cambodia, OPAs have formed rice banks to improve food security and mitigate rice shortages during the year (HelpAge International 2007). Applied to DRR and capacity building, the OPA model provides an established template and roadmap for developing a sustainable, elder-based organisation for addressing the unique needs of the elderly as well as contributions in times of disaster.

Conclusions

The knowledge required to recognise at-risk elderly and effectively address their vulnerabilities in disasters exists today. However, translating this knowledge into co-ordinated action will require that policy-makers, academicians, practitioners, community members and even the elderly create the will to see this knowledge through to application.

For this to happen, policy-makers must foster accountability for the results of disaster response and risk reduction as it pertains to the elderly. They must create political will and be open to allowing a top–down and bottom–up approach. As purveyors of information, academics must continue to support and conduct research and development of best practices. Second, academicians must distribute and share this knowledge through formal education and non-formal means of knowledge dissemination (see Chapter 62). Practitioners with response and relief organisations must begin to recognise the unique vulnerabilities and contributions of the elderly, and apply the known body of knowledge into programmes with the same vigour as is seen for other vulnerable populations. Community members and the elderly must work toge-ther to organise and provide the elderly with a voice that can be heard clearly. Collectively we must cross organisational and social boundaries to work together collaboratively and pool our knowledge, gain understanding and achieve outcomes. Finally, to truly be effective in DRR for the elderly, we must look into ourselves and identify and begin to challenge our own biases and prejudices about older age and ageing.

38

Caste, ethnicity, religious affiliation and disaster

JC Gaillard

SCHOOL OF ENVIRONMENT, THE UNIVERSITY OF AUCKLAND, NEW ZEALAND

Introduction

This chapter deals with how people of different ethnicity, caste and religious affiliation face the risk of, cope with and recover from disasters. Typical questions this chapter asks are why Indian lower castes continue to lack access to basic resources to face natural hazards in spite of the country's significant progress in reducing the risk of disasters. Why were the black communities of Miami more severely impacted by Hurricane Andrew in 1992 than their white neighbours? To answer such questions, the chapter links up marginality and prejudice with the concepts of ethnicity, caste and religious affiliation, and with access to resources, capacity and vulnerability.

Castes often refer to hereditary groups distinguished from one another and connected by endogamy, division of labour and hierarchy (Dumont 1980). The concept of ethnicity is highly controversial (e.g. Jenkins 2008). In this chapter, ethnic groups are initially defined as those groups of people whose members identify with each other through a set of shared cultural practices which other groups consider as distinct. Obviously, drawing boundaries between such groups is often difficult.

Ethnic groups and castes are usually assumed to differ from each other by common kinship, language, religion, beliefs, customs, traditions and territory, all of these being often referred to as culture. Most of these distinct features are inherited from ancestors and history. Along that view, sense of ethnicity and caste, but also religious affiliation, are thus largely defined by the members of the group themselves. This is the endogenous dimension of caste and ethnic groups. Yet, castes and ethnic groups only exist in comparison and most often in interaction with other groups within a larger society (Barth 1969). Inter-group/caste interactions involve economic, social and political contacts and processes which shape the larger society. Inter-group/caste interactions further serve as powerful external forces which compel the members of ethnic groups and castes to constantly reshape their own identity (Nagel 1994).

Both endogenous cultural features and external forces actually explain why some groups are more or less vulnerable to, more or less able to cope with, and well or poorly equipped to recover from disasters. Study of ethnicity, caste and religious affiliation in relation to disasters thus requires both inside and outside perspectives, or a view from the bottom and a view from the top.

Ethnicity, caste, religious affiliation, marginality and disasters

Ethnicity, caste and religious affiliations are most often neglected by those collecting and aggregating disaster-related data. Therefore, there is no extensive and useful dataset pertaining to the disaggregated impact of disasters on different ethnic groups, castes and religious communities on a global scale. Available evidence only reflects the scope of research conducted in the field, which so far has been mostly in the USA (e.g. Bolin 2007; Bolin and Bolton 1986). There are only sparse, consolidated data from case studies in Asia, Africa and Latin America. Yet, the few existing materials are consistent in pointing to ethnicity, caste and religious affiliation as significant factors contributing to the differentiated impact of disasters.

Some examples across continents include the areas of Louisiana, USA, devastated by Hurricane Audrey in 1957, where Bates *et al.* (1963) noticed that death rates were higher among black than white communities (thirty-two per cent versus four per cent). In Slovakia, between 1995 and 2005, sixty Roma people were killed by floods, while only three ethnic Slovaks perished (United Nations Special Rapporteur on Adequate Housing, 2005). Similarly, the tsunami that battered the southern coast of the island of Gizo (Solomon Islands) in 2007 killed thirteen people in the village of Titiana settled by Gilbertese migrants, while nobody died in neighbouring Pailongge, which is the home of Simbo people (McAdoo *et al.* 2008). Seventy-four per cent of the people recorded in evacuation centres set up in the aftermath of the 1991 Mt Pinatubo eruption in the Philippines were lowlanders, yet ninety-two per cent of those who died from diseases in these shelters were from the Aeta indigenous minority (Sawada 1992).

The impact of disasters mirrors the everyday condition of the affected society. Therefore, ethnic groups, castes and religious communities' ability to face natural hazards reflects their position within the larger society. Those with social, economic and political power count less among those affected by earthquakes, landslides and floods because they often enjoy the choice of living in hazard-safe areas or, should they live in dangerous places, frequently possess the means to protect themselves from the harmful effects of natural hazards. In that sense, disasters most often detrimentally affect marginalised segments of society, e.g. children, elderly, women, people with disabilities, prisoners and refugees along with specific ethnic groups, castes and religious communities (see Chapter 34, Chapter 35, Chapter 36 and Chapter 37). Ultimately, the differentiated impact of disasters thus reflects the unfair interactions between people and groups within society (Wisner *et al.* 2004).

Some further observe that in the Indian context it is not exclusion from society that affects poverty, but rather inclusion in a caste-based hierarchical society (Bosher 2007). Barth (1969: 27) also notices that a 'stratified poly-ethnic system exists where groups are characterised by differential control of assets that are valued by all groups in the system'. Issues of access to those assets are therefore crucial in the differentiation of ethnic groups, castes and religious communities. Those resort to different arrays and combinations of exogenous and endogenous resources to sustain their livelihoods. These resources are unequally diverse, resistant and sustainable in the face of natural hazards (see Chapter 58). Furthermore, available means of protection are unequally accessible to different ethnic groups, castes and religious communities according to their position within the society. Some groups, however, display endogenous resources and coping strategies which are intimately embedded in their cultural identity and which often prove to be highly useful to face the threat of natural hazards. In contrast, some endogenous features may endanger some members of the group too.

Endogenous resources and coping strategies

Ethnicity, caste, religious affiliation and hazard-prone locations

People's decision to live in or move from hazard-prone areas is constrained by a large range of factors including ethnicity, caste affiliation and religion. Ethnicity may involve the attachment to native places. People often decide to settle in a given location because of personal and community histories, whether it is dangerous or not. This may be the place where they and their ancestors were born. It might also be the region around which the identity of their ethnic group or caste was shaped through experience of territorial markers, cultural heritage and other symbolic community activities (religious practices, folklore, festivals, ceremonies, etc.). Such places bear a strong and tangible sentimental value which largely overcomes the potential threat of natural hazards in people's decision to face such threats.

In the case of the 1976 earthquake that struck northeastern Italy, Barbina (1993) described how the destruction of houses and heritage buildings made local communities realise that the Friulan identity might disappear with the disaster. As a consequence, the survivors firmly opposed relocation in areas that government officials and scientists considered safer. Instead, they insisted on being temporarily sheltered in tents in the immediate vicinity of their native villages. Similarly, following the eruption of Mt Pinatubo in 1991, several lowland Kapampangan communities deliberately chose to face the seasonal threat of lahars. Some who were relocated tens of kilometres away went back to their native town on the banks of lahar channels, in part because of attachment to native places (Gaillard 2008). In both cases, the threat to an individual's identity through a loss of cultural heritage weighed heavier than the earthquake and lahar hazards.

Ethnicity, caste and religious affiliation in risk reduction

Techniques of facing natural hazards may reflect endogenous cultural patterns of ethnic groups, castes and religious communities. These include traditional settlement patterns, housing structures, gender relations, social networks, farming, fishing and forestry activities, etc., which make people more vulnerable or less vulnerable in the face of natural hazards. Some ethnic groups have learnt to live with natural hazards and developed indigenous strategies to avoid or reduce harmful consequences both in the long and short term (see Chapter 9). Kel Adrar Tuaregs have learned to live with the Sahel's harsh environment. To cope with uncertain and limited rainfall, they moved regularly over hundreds of kilometres in search of pasture for their herds, they stored limited food surpluses and they kept more animals than usually necessary for subsistence so that they may be shared among members of the group in time of scarcity. Kel Adrar communities also practised hunting and gathering and intensified trade activities when herding did not provide enough food (Swift 1973).

In Latin America, Oliver-Smith (1994) has similarly documented how Pre-Columbian Andean communities had adjusted their way of life to the threat of earthquakes, droughts, floods and landslides among other threatening phenomena. This meant farming a large variety of terrains at different altitudes, settling in small villages, building houses with earthquake-proof materials and techniques, storing grains in communal warehouses in the event of food shortage and even coping with disasters through 'positive' explanation for harmful events (see Chapter 10).

Box 38.1 Integrating ethnicity in disaster risk reduction around Mt Kanlaon volcano, the Philippines

Jake Rom D. Cadag
Université Paul Valery–Montpellier III, France

Mt Kanlaon is located on the island of Negros and is one of the most active volcanoes in the Philippines. The surrounding villages are exposed to volcanic and other natural hazards. The region is home to three major ethnic groups, namely Ilongo, Cebuano and Bukidnon.

Each ethnic group relies on different agricultural resources. Presently, Ilongo communities are tied to the sugarcane monocrop system. In the 1850s the sugar industry was introduced on Negros by the Spaniards, who replaced the subsistence farming practices of the natives such as production of rice, corn and edible root crops. The formerly self-sufficient natives were reduced to mere daily wage workers. On the other hand, the Cebuanos have encroached recently on the slopes of the volcano within the volcano's permanent danger zone. Free from sugarcane landlords and despite apparent threats of volcanic hazards and forest degradation, the Cebuanos produce rice, vegetables, corn and livestock and are thus more self-sufficient. They appear less vulnerable in the event of a food shortage due to the occurrence of natural hazards or disadvantageous market trends.

Diets also mirror cultural preferences. Although rice is considered as the staple food for all ethnic groups, the Cebuanos are also known as corn-eaters. In the event, again, of a shortage of food, Cebuanos thus might have a larger range of options.

Political status and interactions between ethnic groups also affect people's ability to secure sustainable livelihoods and access means of protection. In 2001 Mt Kanlaon was declared a national park, and this prohibited the slash-and-burn farming practice of the Bukidnons, thus affecting their livelihood. Presently, they are claiming their ancestral land in order to regain use of the newly proclaimed national park land. However, officials of the national park, who question the identity of the Bukidnons as indigenous people, and the more dominant Cebuanos, who are now intermixed with the Bukidnons, are against the claim. The uncertainty of land entitlement of the Bukidnons coupled with a discordant relationship with the Cebuanos threatens the entire livelihood security of the Bukidnon community. As a result, disaster risk reduction (DRR) programmes and policies become difficult to advocate specially for the Bukidnons.

Differences between ethnic groups are also evident in decision-making processes. Decisions at the village level are the sole jurisdiction of local officials in the Ilongo and Cebuano communities. On the other hand, Bukidnons are often subject to prior consent of the tribal chiefs regarding issues within and beyond the household. Yet, tribal leaders are seldom involved in the official DRR programmes.

Volcanic mythology is also unique in each ethnic group. The story of Suta, the spirit that controls Mt Kanlaon, is told in Ilongo literature but is rarely known to Cebuanos. Bukidnons have numerous beliefs that are more animistic in nature, keeping them culturally attached to the volcano. To what extent these ethnic groups perceive and believe mythology is a crucial factor in DRR strategies.

In time of emergency, different ethnic groups, castes and religious communities may also resort to particular endogenous resources to cope with natural hazards. These are deeply embedded in the communities' historical and geographical context and refer to norms, values, beliefs, knowledge, technology and legends. In Indonesia, about 170,000 Acehneses and Minangkabaus died out of approximately 1.3 million people living in the most affected districts of the northern tip of Sumatra during the 2004 tsunami, while only forty-four Simeulue people died out of a total of 62,000 people on the neighbouring Simeulue Island located near the earthquake epicentre. On Simeulue, the experience of past tsunamis allowed development of some specific cultural knowledge and practice to cope with the tsunami hazard. Immediately after the initial earthquake and the subsequent withdrawal of the sea, people quickly met at pre-agreed locations and eventually sought refuge in defined areas in the surrounding mountains for several days with rice and other foodstuffs (Gaillard *et al.* 2008b).

Ethnicity, caste and religious affiliation in recovery

Coping strategies may also be long term. In the aftermath of cyclones, the small island community of Tikopia in the Solomon archipelago copes with the destruction of natural resources and subsequent food shortage through temporary adjustments in their traditional way of life agreed upon by the whole community during daily public assemblies. These include the temporary abandonment of fallow periods, the redefinition of agricultural rights, stricter crime and thief repression, the omission or abbreviation of agricultural rites of lesser importance, the adjournment of wedding ceremonies and a reduction in size and even temporary suspension of mortuary ceremonies (Gaillard 2007: 528).

Religious affiliation also proves to be a valuable resource for disaster-affected people who often seek assistance from co-religionists and faith leaders or institutions in time of disaster. After the 2004 tsunami hit the southern coast of Thailand, many survivors approached Buddhist monks for ceremonies, rituals, counselling and temporary ordination, which all helped them in overcoming the disaster (Lindberg-Falk 2010). When disasters strike, international faith-based non-governmental organisations (NGOs) also play an essential role in raising resources. At one end, religious networks in regions extending help enable the quick and widespread collection of relief goods and other donations, whilst at the other end, the same networks in the affected areas allow for fast delivery of assistance (see Chapter 10).

International faith-based NGOs also reflect ethnic affiliation and migration patterns. Some of the huge Muslim NGOs in the UK were founded by Bangladeshi and Pakistani immigrants who aim to collect *Zakat*, a responsibility for Muslims to give part of their wealth to charitable purposes as promulgated by the *Qur'an*.

People's ability to adjust to post-disaster resettlement also depends on endogenous cultural patterns that reflect ethnicity, caste and religious affiliation. Following the 1991 eruption of Mt Pinatubo, the dissimilar responses of the people relocated in uniform resettlement sites resulted from the unique history and culture of their ethnic group. The Kapampangans living on the east side of the mountain were strongly attached to their ancestral territory and struggled to adjust to the relocation sites. On the other hand, the Ilokano living on the western side of the volcano proved to be loosely tied to their native villages and overcame the social uprooting induced by their resettlement (Gaillard 2008). After the 1963 Skopje earthquake, Davis (1977) also indicates that most of the Macedonian people relocated in other regions of the former Yugoslavia went back to their native towns less than two months after their resettlement because their children could not speak the local language.

Challenges emerging from caste, ethnicity and religious affiliation in disaster

Endogenous cultural patterns do not always empower ethnic groups, castes and religious communities with means of protection in facing natural hazards. Norms and values may lead to some members of the group being more vulnerable than others, although they face the same threat. In contemporary India, Pakistan and Bangladesh, gender discrimination to the disadvantage of women is due to ethnic, caste and religious reasons (see Chapter 35). It materialises in the daily distribution of household chores and/or in poor access to means of protection. Women who were confined at home to care for their children and did not know how to swim and to climb trees, proved to be more vulnerable than men in the face of earthquakes, floods and tsunamis (Chowdhury *et al.* 1993). Such unjust social structures are of concern for aid agencies in facing disasters.

Indeed, respecting local culture and channelling support through traditional social structures can mean depriving some community sectors (see Chapter 6). Religious affiliation also affects people's ability to face disasters. In the Muslim province of Aceh, Indonesia, the 2004 tsunami swept away people's meagre savings and capital in the form of gold, jewels and accessories kept at home rather than in interest-based bank accounts, which are prohibited by the *Qur'an* (Gaillard *et al.* 2008b).

Inter-group relations, access and vulnerability

Inter-group relations in risk reduction

Often the vulnerability of an ethnic group, caste or religious community reflects the lack or failure of access to sustainable livelihoods and means of protection that are available to other groups. People's ability to face natural hazards thus reflects inter-group relationships.

Wisner *et al.* (2004: 239) suggest that caste affiliation may bear on vulnerability to floods and cyclones through segregation in the location of homes in rural and urban India. In the Ganges River delta, villages are often structured around central elevated grounds where the more powerful castes usually settle. In contrast, less powerful castes occupy low-lying sites in the periphery. Similarly, many of the Senegalese and Guinean migrants who head towards Praia, Cape Verde, end up in the local flash flood-prone *ribeiras* (ravines) wherein settlement is prohibited because it is dangerous. Settling there is their effort to sustain their livelihoods, since many of them are illegal migrants who cannot afford to reside in less hazardous areas because they lack political legitimacy and financial ability. Language is a further limitation for these migrants who hardly speak Portuguese or Creole and thus have difficulty in advocating for their rights.

Often, access to land may reflect client–patron relationships amongst ethnic groups, castes and religious communities. In Angeles City, the Philippines, local Kapampangan leaders tacitly allow migrants from other ethnic origins to settle in flood-prone areas in exchange for political and economic allegiance. It therefore results in an ethnic segregation in the face of natural hazards with locals being safely located with stronger livelihoods and migrants threatened by floods with weaker livelihoods (Delfin and Gaillard 2008).

Access to means of protection also depends upon the interactions of ethnic groups, castes and religious communities. In Andhra Pradesh, India, Bosher (2007) shows that lower castes, which often are the poorest, have limited access to private and public resources that prove essential to avoid the harmful effect of natural hazards, i.e. land, house, livestock, savings, credit, farming and fishing implements, and vehicles. Access seems to be particularly difficult in multi-caste contexts where upper castes control political, economic and social assets to the detriment of the powerless marginalised castes. Similarly, sixty per cent of the poor black households (against thirty-five per cent of white households) of New Orleans lacked a vehicle, which would have helped them to evacuate quickly (subject to traffic jams) when confronted with the pending threat of Hurricane Katrina in August 2005 (Zack 2010). In rural India and urban USA, means of protection are obviously available locally, but are limited to those groups and castes with the financial and social ability to access them. Hence, the interplay between vulnerability and ethnicity is not only a matter of endogenous identity and resources but also reflects power relationships within larger society.

Inter-group relations in relief and recovery

Different ethnic groups, castes and religious communities may also have unequal access to resources to cope with the adverse effect of natural hazards. In the aftermath of Hurricane Andrew, which swept Florida, USA, in 1992, black households proved to be poorly covered by insurance schemes. When they were covered, they had seldom resorted to larger national insurance companies, but instead to rather obscure local corporations which were often unable to cover all rebuilding expenses. Peacock and Girard (1997) further report that to gain access to insurance funds, black communities frequently had to overcome the lack of street signs and house numbers by spray painting the name of their insurance company and their address on the walls of their house, while in rich neighbourhoods, adjustors had no difficulty finding their clients.

In Peru after the 1970 Huascaran debris avalanche which buried the socially stratified town of Yungay, the distribution of relief goods and aid two weeks after the event reawakened the ethnic tensions between Indians and mixed-heritage *mestizos* (Oliver-Smith 1979). Politically and economically dominant *mestizos* who lived in the town and who had lost most of their assets claimed to be more deserving of aid than rural Indians who had less to lose in the rural surroundings and who were allegedly less affected by the debris avalanche.

Similar discrimination and uneven distribution of resources among ethnic groups, castes and religious communities affect long-term recovery. Unequal access to insurance schemes and different incomes between Hispanic and white communities greatly affected the capacity of disaster-affected people to afford a permanent resettlement solution following the 1983 Coalinga earthquake in the USA. Indeed, Hispanics were living in the most severely damaged houses, which they were renting and which were unlikely to be repaired. Were these houses to be repaired, their rent after rehabilitation would be too high for their former dwellers (Bolin and Bolton 1986).

The story of the people of Niuafo'ou, Tonga, who had to evacuate their tiny island due to the restless activity of the local volcano in 1946, provides another powerful account of unequal inter-group relationships (Gaillard 2007: 526). An erratic journey led the people of Niuafo'ou to initially evacuate to the capital city of Nuku'alofa. There they suffered from the hostility of the locals with whom they had had little contact before and who expressed resentment that the displaced benefited from government food aid. Niuafo'ouans were eventually resettled on the neighbouring island of 'Eua, where their only means of livelihood at first was to work for local residents for very poor compensation. Antagonism further grew up when social contacts between the two communities increased as the people of Niuafo'ou gained access to public facilities and services.

By contrast, in some cases inter-ethnic, inter-caste and inter-religious relationships may be extremely valuable, such as when Hurricane Stan swept Guatemala in October 2005. Hinshaw (2006) reports that Stan led rich and poor, Mayans and non–Mayans, Protestants and Catholics to collaborate in evacuation efforts, relief provision and in seeking government assistance. On the international scene, a number of Christian and Muslim NGOs are also active in times of disasters, whether the affected area is Christian, Muslim, Hindu or Buddhist.

Discriminatory state practices and disasters

The foregoing picture may suggest that people's response to natural hazards and disasters may be determined by ethnicity, caste or religious affiliation. This would be oversimplifying. A larger view should encompass power projected by the state as reflected in development policies. Some development and disaster-related policies and practices deliberately force marginalised ethnic groups, castes and religious communities to settle in hazardous areas without proper means to protect themselves, to exclude them from the decision-making and delivery processes, and to erode their endogenous ability to face natural hazards.

Intentional discrimination and disaster vulnerability

The state and those monopolising power may act deliberately in disfavour of an ethnic group, caste or religious community. It is often evident in times of emergency, when shelters and relief goods are unequally distributed to the benefit of the most powerful. The adverse effect of discriminatory development policies is also obvious in South Africa where decades of apartheid compelled 'non-white' communities to settle in hazard-prone and precarious townships surrounding major urban centres. Since the end of the apartheid, these poor settlements have been a primary destination for political and war refugees in provenance of neighbouring countries such as Mozambique and Zimbabwe. Both non-white South Africans and foreign migrants prove to be particularly vulnerable in the face of natural hazards. In February 2000 some 500 shacks of the huge, poor township of Alexandra in the periphery of Johannesburg were washed out by flash floods. Alexandra is also regularly swept by fire and epidemics, but lacks proper water and sanitary facilities which would serve as protective resources (Wisner *et al.* 2004: 122–23).

In Japan, the Burakumin traditional untouchable caste suffered badly from the 1995 Kobe earthquake along with foreigners (Vietnamese, Koreans, Filipinos) who live in the district of Nagata, which was the most affected by the disaster. This district was composed of thousands of ancient wooden houses and apartments which had been left out of the city rehabilitation programmes. Furthermore, the city government suspended the water supply although this district hosted many small factories manufacturing rubber shoes which burned in the aftermath of the earthquake. Without proper hydrants fire fighting turned out to be difficult and it took more than thirty hours to extinguish the fire using sea water. In the aftermath of the disaster, Burakumin and other minorities were still camping in open areas such as parking lots while most of those affected were accommodated in public buildings, churches and tents (Wisner *et al.* 2004: 294).

In Slovakia, the Roma minority has long been marginalised by the state and ethnic Slovaks. Neglect often compels Roma to settle in flood-prone areas such as in Jarovnice, where 4,000 families used to live in shacks along the Mala Svinka River whilst ethnic Slovaks dwelled in the centre of the village that towers above the valley (United Nations Special Rapporteur on Adequate Housing 2005). As a consequence, in 1998 a powerful flood washed out 600 Roma shanties and killed fifty-three people. Ethnic Slovaks showed little solidarity with the affected Roma in the face of the disaster, and such events actually often trigger anti-Roma sentiments among local communities. Slovak politicians were further accused of exerting little effort to relocate Roma in safe places and limit temporary resettlement in rudimentary containers which eventually turned into permanent homes.

Unintentional impacts of state policy and practice on endogenous coping

Government policies and international interventions may also lead to the indirect erosion of endogenous coping strategies and increasing dependence on outside support. The ability of Kel Adrar Tuaregs to move with their cattle had been progressively constrained by colonial policies, the erection of national borders and post-colonial government programmes intended to ground them at definite locations. Savings in the form of living animals were heavily taxed by central authorities and hunting and gathering was disappearing as children attended formal schooling, to the detriment of indigenous knowledge. Finally, the traditional caravan trade was threatened by border policies and modern communication facilities (Swift 1973). Discontent with government eventuallly led to armed uprising and continuing deterioration of Tuareg livelihoods (Giuffrida 2005; Øni et al. 2009) (see Chapter 7).

A similar erosion of traditional coping strategies has been observed on the island of Mota Lava, Vanuatu, which is often visited by powerful cyclones (Campbell 1990). Encouraged by the colonial government, local residents have progressively converted land from subsistence crops (yam) to commercial crops (coconut for the production of copra) at the expense of their food security. Simultaneously, Western building techniques including utilisation of nails, masonry and particle board have progressively replaced indigenous materials, although they eventually proved to be much more vulnerable to powerful typhoons.

Sometimes, governments also take advantage of the reconstruction process to exert stronger control over some ethnic groups, castes or religious communities. In the aftermath of the eruption of Paricutín volcano in central Mexico from 1943 to 1952, Tarascan communities were relocated by the government in an urban centre, the dominant population of which was not Tarascan. The resettlement project was part of a programme of modernisation of the Mexican nation which had already been going on for years. In that context, the government has been accused of using the resettlement policy and the associated social services programme to 'civilise' the Indians (Gaillard 2007: 536). In a more recent case from Mexico, small farmers in Chiapas who suffered heavy flood and landslide losses in 2008 were offered resettlement but only if they agreed to grow tropical oil palm, which was part of a regional development plan being imposed from outside (Berger and Wisner 2008).

Ethnicity, caste, religious affiliation and disaster risk reduction (DRR) policies

Eroded endogenous resources and strategies for facing natural hazards are often replaced by standardised DRR policies that are culturally insensitive. Dominant DRR policies are indeed shaped around command-and-control top-down policies which give little attention to the plight of marginalised groups (see Chapter 51).

A quick review of the existing legislative instruments in South Africa, India and the Philippines, which count amongst the most progressive countries in the world for DRR, yields poor results. In India the Disaster Management Act of 2005 only states that 'while providing compensation and relief to the victims of disaster, there shall be no discrimination on the ground of sex, caste, community, descent or religion'. The Philippines Disaster Risk Reduction and Management Act of 2010 only recognises in its definition of terms that vulnerable and marginalised groups are 'those that face higher exposure to disaster risk and poverty including, but not limited to, women, children, elderly, people with disabilities, and ethnic minorities'. Finally, the Disaster Management Act enacted in South Africa in 2002 makes no mention at all of ethnicity-related issues.

On the other hand, in Australia, the Queensland Department of Emergency Services is fostering the participation of Aboriginal communities in the design and set-up of emergency planning. Aboriginal people are progressively integrated in fire fighting and medical teams in charge of managing emergencies. They are thus able to share their indigenous knowledge and to participate in crafting culturally sensitive DRR and management policies (Hocke and O'Brien 2003).

Box 38.2 NGO activities in rural India: inclusive or exclusive?

Lee Bosher
Loughborough University, UK

NGOs should ideally be available to anyone, irrespective of caste, class, religion and position of respect and responsibility, but in reality the beneficiaries of NGO activities can be restricted to a narrow social base.

Research conducted in villages in coastal Andhra Pradesh, southern India has found that marginalised communities (who are typically the lower castes) in multi-caste villages invariably have low access to assets, public facilities and political networks and therefore tend to be the most socio-economically vulnerable (Bosher 2007). Despite this fundamental issue, the same study found that many of the NGOs undertaking DRR activities in this region did not operate in multi-caste villages, apparently because they prefer to operate in relatively homogeneous single-caste villages. An executive-secretary from one of these NGOs explained the rationale behind this approach:

> Running a NGO is a business and I have my duties to secure the employment of my staff and to feed my family. Consequently, if funding is available to work with the 'fisherfolk' for example, this is what we will endeavour to do, even if those particular people are not the most needy people in the region. Also 'fisherfolk' villages tend to be single-caste villages and if we are to achieve the targets expected of us to obtain future funding, it is these single-caste villages that we will work with because we do not then have the inter-caste related barriers to contend with, that you would find in many mixed caste villages.

This example focuses on a specific region of Andhra Pradesh with the consequence that the findings are potentially very context specific. Nonetheless, the findings highlight a fundamental flaw in the way in which many NGOs operate in this region, through the targeting of perceived 'easy cases', but more importantly highlighting that the most vulnerable members of multi-caste villages do not necessarily receive the assistance they need from NGOs. For that reason, the means through which NGOs involved with DRR activities target their recipient villages will need to be improved, pointing towards an increased focus on improving the ability of the most vulnerable communities, irrespective of the caste or religious composition of a village, to face natural hazards. Central to this, it is important for NGOs to make explicit what they actually mean by 'community' or 'community-based participation': which communities are being included and which communities are being excluded. The recommendations made at the end of this chapter provide important starting points for more appropriately targeted DRR activities that are less about exclusivity and more about inclusivity.

Conclusions

Since marginalised ethnic groups, castes and religious communities are amongst those social groups that bear the brunt of disasters, it seems odd that the same groups are most often the last and least consulted in the drafting of disaster-related policies, and the last and least involved in actual DRR activities. Recommendations for better integration of ethnicity, caste and religious affiliation reflect the larger concerns expressed elsewhere in this *Handbook* for greater participation of local communities. DRR and management should be geared towards both integrating endogenous resources into policy and practice, and granting equal access to livelihoods and means of protection to all ethnic groups, castes and religious communities. Integrating community resources and coping strategies is a question of local political will which may be gained over the short term. On the other hand, access to protection is a matter of governance and is often dependent on global and structural constraints that require time to be changed.

To achieve this delicate balance of top-down and bottom-up actions, the following recommendations are given:

- DRR should be inclusive of all threatened ethnic groups, castes and religious communities. No one better than those concerned can assess the value and scope of endogenous resources, which are often difficult and time-consuming to appraise from outside because they reflect people's own identities.
- Because ethnicity is also a matter of inter-group relationships, it seems essential to foster a dialogue between these groups as part of DRR activities. Such a dialogue may be difficult but is even more crucial when inter-group relationships involve conflict or are unfair.
- Endogenous knowledge, norms and values should be recognised and integrated into DRR planning.
- To meet such objectives on a wide scale, there needs to be a legal recognition of the role of all ethnic groups, castes and religious communities in DRR policy documents at national and international levels.
- Government policies should further aim at facilitating fair access to means of protection that are available locally and that are culturally acceptable to those who are supposed to use them. This includes equitable access to sustainable livelihoods.

Only in that context will it be possible to mitigate durably the impact of disaster on marginalised ethnic groups, castes and religious communities. Such an objective seems to be a crucial issue in our contemporary world as interactions amongst those groups as well as regional and international migrations constantly erode endogenous resources and continuously reshape issues of access to livelihoods and means of protection in facing natural hazards.

Part III
Preparedness and response

39

Introduction to Part III

The Editors

If disaster strikes

This part of the *Handbook* is about preparedness for and response to disasters. It covers early warning, emergency management and the assessment of damage and needs following a disaster, including chapters on health and food. Recovery is also discussed in chapters devoted to settlement and shelter reconstruction as well as psycho-social and socio-economic recovery.

For millennia, individuals, families, communities, and wider social and political structures have coped with risks and uncertainties. Some of this activity involved material life, where storage and transportation innovations spread assets over time and space. Some involved ritual activities such as offerings and petitions to divine entities believed responsible for environmental events. An echo of such rituals might be heard in the routinised and formalised systems that exist today that are meant to prepare society for hazards.

For instance, Handmer and Dovers (2007: 52) schematise the process for dealing with disasters in a sequence of boxes labelled, 'What are we concerned about?' leading to 'How serious is the problem?' to 'What can be done?' and finally 'What's left over?' Chapters in this section review elements of these boxes, examining some of those routinised and formalised systems – and especially challenging the status quo in many instances. How does routinisation and formalisation set in? Is that appropriate? What myths and inertia emerge? Where problems result, how can change be effected? Where the system works, how can it be retained? The chapters are clustered into two sub-sections: preparedness and response, followed by recovery.

Candles in the dark and helping hands: preparedness and response

A people-centred approach is at the heart of Chapter 40, on 'early warning principles and systems'. This chapter covers studies and implementation of the warning process at scales ranging from small communities to global networks. Despite decades of literature on the topic, sadly it took a major disaster – the 2004 Indian Ocean tsunami – to put warning systems high up on the contemporary political agenda, yet still with mixed success.

Since the Indian Ocean tsunami of December 2004, a good deal of expertise and money has been invested in a tsunami warning system for this region of the world. Yet the system as a whole failed when a tsunami hit Indonesia's Mentawai islands (BBC 2010b). Even when the technical parts of warning systems are created and maintained properly, and the technical system functions, it is not assured that warnings reach isolated communities. That was further demonstrated by studies of the Dominican Republic and Haiti during the 2004 hurricane season (Wisner *et al.* 2005). Furthermore, many warning systems take for granted that people have evacuation options which they may not have, as in the case of Hurricane Katrina, when thousands of people without private motor transport had no way to leave the city (Verchick 2010).

Even where warning systems properly balance technical and social aspects, the system must be placed within wider response contexts. Chapter 41, on 'preparedness, warning and evacuation', links warning to action. Misconceptions are overturned, particularly the standard assumption that systematic and orderly preparedness leads to systematic and orderly warning, and that systematic and orderly evacuation is the inevitable result, which in turn leads to happiness amongst everyone involved. Instead, preparedness, warning and evacuation must be seen as social processes involving a wide range of people, each with knowledge and capacity but also capable of confusion and mistakes, not run like clockwork by a top-down technocracy.

Next 'emergency management principles' are discussed. The author, an emergency medical technician and logistician with experience in many countries, distils lessons and describes the roles of professionals, political decision-makers, the media and citizens. Topics covered include planning and management, communications, search and rescue, and medical triage, covering examples of operational emergency management systems used around the world. Emergency management has benefited a great deal from various forms of scenario building and modelling – with or without public discussion and input (Verchick 2010; Alexander 2002b). This chapter supports Albert Einstein's famous comment 'imagination is more important than knowledge', by emphasising that emergency management must involve a variety of people, ideas and approaches, focused on innovation and flexibility. Emergency management requires improvisation balanced with rule-based activity, and the combination calibrated with real-time feedback.

Damage and needs assessment is a vital part of real-time information required as part of emergency management – and must continue even as emergency management gives way to continuing relief and recovery. Chapter 43, 'from damage and needs assessments to relief', draws on the author's years of work with the Pan American Health Organization. Asking the right questions and seeking the correct information are the key. To do so accurately and efficiently, partnerships with community groups and local governments are essential. Locally available skills, information and materials are often overlooked, when they should be balanced with what is needed from external sources. Yet despite this wealth of experience, in many cases, a standard 'top down' methodology is still used.

One area of need concerns health and health care. Chapter 44, on 'health and disaster', deals with the immediate health care needs following a hazard, but goes beyond, linking it to longer-term public health monitoring. The special needs of women and other groups are also addressed. This chapter frames 'health' and health care more broadly than many practitioners tend to prefer, to ensure that the role of a strong pre-disaster primary health care system and healthy population is evident in providing the basis for solid health-related response and recovery (see Figure 39.1).

Paralleling the importance of health care is food security. Chapter 45, on 'food security and disaster', lays heavy emphasis, as do other authors in this section, on the often-ignored capacities of people to provide food for themselves following a disaster. There has been a vigorous debate over the pros and cons of externally provided food assistance (Crilly 2008), which this chapter's author sorts through. The four fundamental characteristics to monitor with regard to food security are highlighted: availability of food, access to food, stability of food supply and safe use of food.

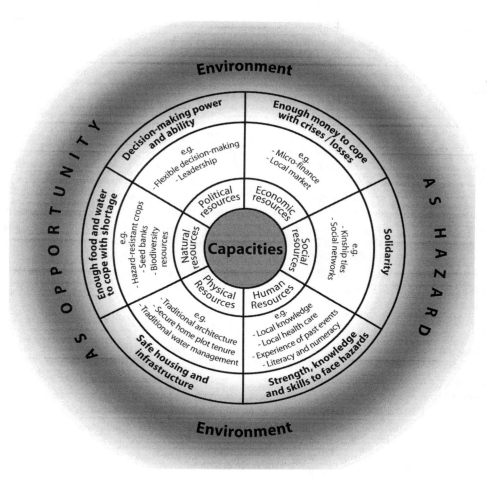

Figure 39.1 Circle of capacities

Settlement and shelter in the immediate aftermath of a disaster are addressed in Chapter 46, 'Settlement and shelter reconstruction', a chapter that bridges response and recovery. Debates about resettlement and shelter are reviewed. These are eerily similar to the discussions over externally provided food assistance. In the sector of settlement and shelter, a shift of opinion is documented towards fuller appreciation of local capacities and resources. The recommendation is that external interventions take a supporting role and provide enabling conditions for post-disaster reconstruction rather than direct provision of housing.

Each of the chapters introduced so far is based on orientating disaster preparedness and response towards the people who are affected. Repeating the recovery figure from Chapter 3, it is easy to see how preparedness and response may be inhibited without a robust food system with storage capacity, good health status and nutrition, and competence in construction of shelter (Figure 39.1). Pre-existing conditions influence post-disaster relief efforts especially in terms of supporting locally driven preparedness, warning systems, reaction to warnings and post-disaster assessments, which assist in avoiding further post-disaster marginalisation.

That does not mean excluding professionals, external experts or outside assistance for disaster preparedness and response. Instead, it means co-operation. It is well established that a significant proportion of immediate rescue and first aid is provided by family members, neighbours and others nearby a disaster-affected site. High levels of social solidarity and pre-disaster engagement and training in planning, preparedness and decision-making by community members will make them more valuable partners (again, see Figure 39.1).

Co-operation also means identifying and incorporating aspects that are not given extensive attention in this *Handbook*. For example, animal roles in preparedness and response, rescuing animals during disasters, where people keep pets and emergency care for livestock, which often represent people's key livelihood assets, require particular skills and actions. Despite the appropriate emphasis on people helping themselves, professional technical rescuers are still needed for preparedness and response. Public safety divers, fire fighters, swift water rescuers, law enforcement officers, emergency medical technicians, helicopter pilots, emergency room physicians and nurses, and many more professionals – including and beyond technical rescuers – play a vital role.

As Alexander (2002b: 1) puts it: 'we are all part of the civil-protection process. Emergencies can be planned by experts, but they will be experienced by the relief community and general public alike. Therefore, we should all be prepared for the next disaster as remarkably few of us will be able entirely to avoid it.'

The contradictions and cruelties of recovery

Many of the preparedness and warning chapters link explicitly to rebuilding after a disaster. Rebuilding is not just about providing four walls and a roof for each family, but also covers significant community services – such as water, food, health, schools, public safety and social support – along with communal places such as market places, civil buildings, territorial markers, and places of worship and recreation. Livelihoods and community interaction need to be continued in addition to the physical rebuilding. The strong message from connecting preparedness and warning to recovery is that the affected population should be enabled to lead the process of rebuilding and not be simply 'beneficiaries' (with or without consultation).

People may be traumatised by personal loss and grief, by injuries, by witnessing great suffering and damage, and often, in the case of women and children, by abuse after the event. People may also be traumatised, or empowered, by their experiences by helping in a disaster's aftermath, whether or not they are part of the disaster-affected community. Chapter 47, on 'psychosocial recovery', addresses such trauma and opportunities. It describes means by which people can re-establish lives of dignity where they do not enjoy the luxury of affluent countries with large numbers of highly trained clinical psychologists and psychiatrists. The chapter discusses a range of assessment and treatment methods that are simple, low-cost options, applicable to everywhere affected by disaster. Echoing a theme that emerges from all these chapters, they show that lay people can and should be trained to provide counselling and support.

Livelihoods also need support to recover. Chapter 48, on 'socio-economic recovery', continues a theme that runs throughout the *Handbook*: livelihoods are a locus of household and community capacity as well as a potential weak point where various aspects of vulnerability express themselves. The chapter discusses how indirect and household-level economic losses are often missed in damage assessments and recovery plans, and how 'winners and losers' emerge. Small businesses and artisans are often neglected, so economic recovery must be socio-economic, balancing 'macro with micro'.

These recovery-related chapters again demonstrate how pre-existing vulnerabilities must be attended to in the recovery process if disaster impacts are to be healed and if creating the

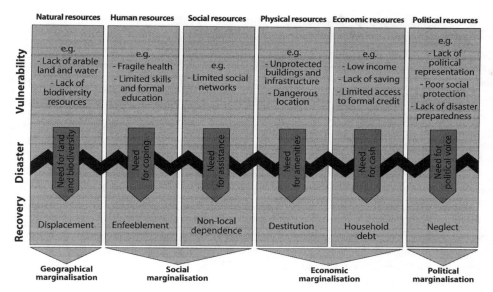

Figure 39.2 Road map to hell: From pre-disaster vulnerability to failed recovery

conditions for a new disaster is to be avoided. Figure 39.2 suggests that merely re-establishing the status quo leaves people no better off than they were before. Consequently, they are still vulnerable to the next hazard and might even be worse off. The end states at the bottom of each column in Figure 39.2 are unfortunate: households or whole communities displaced, enfeebled, dependent, destitute, indebted and suffering neglect. There is much rhetorical exhortation these days to 'build back better' (UNISDR 2010a). Unless recovery is tied to DRR and development and specifically to policies that address the root causes of vulnerability, improvement is unlikely.

Improving preparedness and response: overcoming myths

The chapters in this section of the *Handbook* cover topics that are particularly prone to myths that persist despite having been repeatedly and thoroughly debunked (Lopez-Caressi 2011; Eberwine 2005; Schoch-Spana 2005). Some examples of myths regarding behaviour in preparedness for and response to disasters are:

- People do not know how to prepare.
- People are not interested in preparing.
- People cannot help themselves in a disaster.
- People panic.
- Law and order break down.
- Social networks are abandoned.
- Any and all outside assistance is necessary and useful.
- Foreign life-saving medical and rescue assistance is required.
- Dead bodies are an acute hazard to health and need to be disposed of quickly.
- Temporary settlements away from people's homes are best.

- Recovery takes only a few weeks or months.
- The disaster can be put behind people and forgotten.

Of course, some of this behaviour and these conditions have been documented, but such instances are quite rare. The statements constituting these myths tend to be exceptions rather than rules in terms of disaster-related behaviour and conditions.

Popular, populist, journalistic and even professional assumptions about disaster risks and consequences are as dangerous as these common myths. Oddly, false assumptions are clustered around two extremes. First, some assume that science and professional use of technology can and should protect society from any and all risks, so that should be the focus of preparedness and response. Second, others assume that the future is so uncertain that planning and preparation are bound to fail, so one must simply respond to a hazard event and hope for the best. Truth lies in the middle.

Against the techno-optimists, one can cite a large disaster-related or disaster-applicable literature on complex systems, non-linearity and at least sixteen kinds of ignorance (e.g. Alexander 2002b: 137; Dovers et al. 1996; Perrow 2006; and Smithson 1989), illustrated in Table 39.1. Against the catastrophist-pessimist extreme, one need only reference enormous strides in dealing with disaster risk over the past century and, in particular, the effectiveness of good DRR and emergency management policies and practices in a variety of countries.

Many of these points are articulated in the chapters in this section. They show how preparedness and response need modern science and technology but should not rely exclusively on them. Those components are essential for dealing with natural hazards, but do not suffice in themselves. The chapters further demonstrate how combining many different approaches, including modern science and technology, can lead to solid preparedness, ensuring that responses to disaster threat and disaster are not influenced by the myths and extreme assumptions discussed above. Instead, disasters can be dealt with through preparedness and response, which contribute to long-term sustainability.

Table 39.1 Smithson's typology of ignorance

Ignorance					
	Irrelevance	Untopicality			
		Taboo			
		Undecidability			
	Error	Distortion	Confusion		
			Inaccuracy		
			Absence		
		Incompleteness	Uncertainty	Vagueness	
				Probability	
				Ambiguity	Fuzziness
					Non-specificity

Source: Smithson 1989

Preparedness and response

40

Early warning principles and systems

Juan Carlos Villagrán de León

UN-SPIDER PROGRAM OF THE OFFICE FOR OUTER SPACE AFFAIRS IN VIENNA, AUSTRIA

Introduction

In generic terms, early warning constitutes a process whereby information concerning a potential disaster is provided to people at risk and to institutions so that tasks may be executed prior to its manifestation to minimise its detrimental impacts, such as fatalities, injuries, damage and interruptions of normal activities. Early warning systems (EWS) have been designed and implemented to target sudden-onset hazards such as floods, landslides, tornadoes, tsunamis and earthquakes; for slow-onset hazards such as drought; and for hazards that can appear at a variety of time scales such as extreme temperatures, volcanic eruptions and epidemics/pandemics. In addition, warning systems have targeted complex issues beyond hazards such as famine and food insecurity.

In recent decades, efforts at the local, national and international levels have been conducted to change the paradigm concerning early warning systems. These efforts promote a transition from the common, technically driven systems to more participatory, people-centred approaches. While acknowledging the importance of understanding and monitoring the dynamics of potential disasters to make the most accurate forecasts possible, people-centred approaches introduce the notion that it is equally essential to involve those who would be affected by, and those who have to respond to, such disasters as participants in the design and operation of the system. After all, any early warning system will often be evaluated in terms of its capacity to save lives and reduce losses, rather than on how quickly it issues a warning. This issue has been underlined by former US President William J. Clinton in his statement during the Third International Early Warning Conference held in Bonn, Germany in March 2006 (PPEW-ISDR 2006a):

> We know the most effective early warning takes more than scientifically advanced monitoring systems. All the sophisticated technology won't matter if we don't reach real communities and people. Satellites, buoys, data networks will make us safer, but we must invest in the training, the institution building, the awareness raising on the ground.

Technically oriented systems are designed with a view to monitoring natural phenomena with a high degree of accuracy and precision in order to forecast potentially catastrophic hazards as quickly as possible. In contrast, people-centred systems are designed differently, beginning with a view to those communities which may be affected, and designing the system based on the characteristics of such communities.

The term 'last mile', which gained popularity soon after the 26 December 2004 tsunami in the context of early warning, includes all the notions regarding technically oriented systems: design the system with a top-to-bottom approach, ensuring that those at risk can then be incorporated at the end. In the context of people-centred early warning, it could be stated that such a 'last mile' is the most important mile of the system and, therefore, should not be left for the end (see Box 40.1).

This chapter focuses on the importance of ensuring that early warning systems are designed, implemented and operated with a view to empowering those who need them the most. Empowering these people with the proper information concerning the risk they face and how to minimise losses in case a damaging event is forecast.

People-centred early warning – basic notions

The best examples of people-centred early warning systems may be found in remote villages. There, people and their community leaders rely on their own knowledge for forecasting and on their own capacities for the required tasks, such as moving themselves to safe areas. For example, indigenous groups in Indonesia, Thailand and in the Solomon Islands have been able to link the process of ocean water receding along a coastline to an impending tsunami's arrival, using that for early warning purposes, and they have been able to pass this knowledge from generation to generation (UNISDR 2006, 2008c, 2008d). Similar experiences regarding the effectiveness of local or traditional knowledge in the case of tropical cyclones in the Solomon Islands have been documented by Yates and Anderson-Berry (2004).

In the last decade, efforts have been conducted within Central America with the support of the Organization of American States (OAS), the German International Co-operation Agency (GIZ), the United Nations Development Programme (UNDP) and many non-governmental organisations (NGOs) to establish community-operated warning systems for floods (Villagrán de León 2003). In these systems, volunteers in the middle and upper segments of the basin monitor and report the accumulated rainfall using simple rain gauges along with the river level using hand-made scales or simple electronic devices. They then transmit this information to communities at risk, where members of the local emergency committees and municipal officers assess the data to forecast potential floods. When a flood is forecast, the community evacuates itself to pre-identified temporary shelters in safe areas.

Initial notions concerning people-centred early warning were presented by Anderson (1969). He outlined the need to ensure that communities at risk become aware of the degree of risk they are facing, work together with disaster agencies to manage warning-related communications effectively and timely, and should be able to respond in case a warning is issued. All of these elements should be institutionalised into a plan known to all and enhanced through feedback gathered from previous episodes.

A decade later, Gruntfest et al. (1978) confirmed the elements highlighted by Anderson (1969) in their article focusing on the flood that took place on 31 July 1976 in the Big Thompson River Canyon, Colorado. They comment on the need to consider behavioural patterns characterising people who are exposed to such hazards as a basis for communicating a warning more effectively to ensure proper response.

Sorensen (2000) conducted a review of progress regarding early warning efforts in the USA while Handmer (2002) conducted a similar review of such efforts in North America and

Box 40.1 The first mile of warning systems

Ilan Kelman
Center for International Climate and Environmental Research – Oslo (CICERO)

When discussing warning systems, a plea is frequently made for 'the last mile'. The argument is that, despite the wealth of relevant material available for, and the extensive efforts put into, warning and evacuation, a gap exists in reaching the right people at the right time with the right material and information regarding warnings, in order to yield successful actions.

That gap must be filled to ensure that knowledge is not only developed, but also reaches the people who need that knowledge. The idea is that it would close 'the last mile' between where the knowledge comes from and where the knowledge is needed and used.

The last mile has been critiqued as placing last the people who should be considered first. Instead, 'the first mile' is suggested as the paradigm. The ethos is retained of trying to overcome the gap to reach the right people at the right time with the right information, but the first mile emphasises that connecting with the people who experience disaster, and who want and need the warnings, should be the primary step and focus, not the last endeavour.

The past decades of research and practice on warning systems from around the world, as discussed in this chapter, demonstrate that a top-down, externally imposed system is not likely to induce the desired actions in response to warning information. That is the case even if it is technically perfect, including perfectly closing the last mile.

In contrast, if the entire system is set up by starting with the potentially affected communities, i.e. the first mile, then people are more likely to accept the external warning information and to act appropriately. The value in the first mile is in explicitly starting with the people who will reap the rewards of the system so that the system is accepted from the beginning, for information (e.g. warning of an impending hazard) and action (e.g. evacuation), rather than these people being an add-on at the end.

An important characteristic of the first mile, and how it differs from the last mile, is the process of creating the warning system. In the end, the technical and management structures of the warning system might even be the same, but by having people's participation from the beginning, as in the first mile, the warning system tends to be more socially acceptable and socially robust, even if the non-social aspects appear superficially to be similar to the last mile.

Furthermore, people's knowledge (see Chapter 9) demonstrates not only the rich depth and breadth of understanding that communities have, sometimes even for distant and rare phenomena, but also how melding that knowledge with external specialist approaches yields significant advantages for warnings. That combination of knowledge bases tends to occur most effectively by starting with the communities and then considering what external technologies and management systems might be useful – to add on to what the communities can themselves provide. Again, that defines the first mile rather than the last mile.

Europe. Both agree that most developments are found in the areas of monitoring hazards, in forecasting potential extremes and in technological ways of communicating warnings. They note little advancement in the context of people-centred early warning.

In 2006, in the report entitled 'The Global Survey of Early Warning Systems', the Platform for the Promotion of Early Warning of the International Strategy for Disaster Reduction (PPEW-ISDR) took notice of these and other relevant issues and introduced the four elements of people-centred early warning systems (PPEW-ISDR 2006b, 2006c):

- Risk knowledge
- Monitoring and warning service
- Dissemination and communication
- Response capability

Risk knowledge targets the assessment of hazards and vulnerabilities, which may include coping capacities or deficiencies in preparedness. For example, flood hazard assessment could involve elaborating hydrologic models and cataloguing historic events to construct a probabilistic model characterising the extent of flood inundation with particular return periods. It could also involve compiling historic oral accounts of floods and qualitatively estimating updates due to landscape or climatic changes. Vulnerability assessment targets the identification of people, communities, infrastructure, services, livelihoods and processes to be affected by hazards and changes to those over time. As Handmer (2002) comments, 'When considering warning priorities we should focus on where they are likely to have the most impact'.

The monitoring and warning service targets the more technical aspects of early warning, focusing on the routine hazard vigilance and the capacity to forecast extremes. Hazard monitoring and forecasting procedures must have a sound scientific basis while benefiting from and incorporating local or indigenous knowledge. As already recognised by Anderson (1969) and more recently by Metzger et al. (1999), input from scientists and experts helps to verify forecasts and warnings issued. Meanwhile, incorporating indigenous knowledge captures what science cannot capture and may facilitate ownership of the system by communities at risk.

Dissemination and communication efforts target the transmission of the warning message using as many parallel communication channels as possible, and ensuring that people understand the message. As Mileti and Sorensen (1990), Gruntfest et al. (1978), Handmer (2002) and Sorensen (2000) stress, warning messages should provide information on the forecast hazard including locations to be impacted and the time of impact, along with instructions and guidance on what to do.

Finally, response capability implies that people at risk possess the knowledge and resources to move themselves and their key assets to safety. However, it is important to recognise the difference between being capable of responding and executing this action. As Mileti and Sorensen (1990) comment, 'warnings are more likely to be responded to with some protective action if they are understood, believed, and personalised'. A lack of awareness concerning how a catastrophic hazard manifests may deter people from taking effective action to minimise its impacts (Kelman 2006).

This element is also introduced to ensure that staff in civil protection agencies and authorities, as well as those at risk, know how to respond quickly and efficiently to minimise losses, and that they possess the resources to do so. This requires a high degree of interaction amongst the agencies that need to co-ordinate the evacuation and those to be evacuated (Chapter 41). This also requires a high degree of awareness on behalf of the people at risk concerning the measures they need to implement to minimise losses before the hazard manifests (Handmer 2002; PPEW-ISDR 2006b). Within government agencies, co-ordination can be achieved through the chain of command as described by Metzger et al. (1999). Organisations from civil society can be involved and co-ordinated through standard operating procedures or response plans, which should be tested and well-practised (Chapter 42).

A people-centred approach highlights the need to design the system ensuring that people who may be affected by hazards are aware of their risk level, and are able to respond in a timely and effective fashion to warnings with the support of a variety of agencies. This notion of early warning introduces the need to look beyond the scientific and technical aspects related to the monitoring of hazards and the dissemination of top-down warnings. It stresses the fact that the design, implementation and operation of any people-centred early warning system requires the contributions and co-ordination of a wide range of individuals and institutions, especially the people being affected. As expected, co-ordination is viable when strong and effective governance mechanisms are in place to promote early warning activities. Such governance is the basis for institutional agreements and for the allocation of the resources required to operate the system on a permanent basis.

Additionally, there is a need to recognise that societies are dynamic and that there are continual changes in the urban and rural settings, which implies the need to continually update the situation concerning vulnerability. Furthermore, migration trends such as from rural to urban areas, along with population growth, demand awareness concerning risks faced by the population, as well as of the various aspects of the early warning system. Then those at risk can react adequately, efficiently and in a timely fashion when a warning is issued.

Designing people-centred early warning systems

Preconditions for designing people-centred EWS

Designing a people-centred early warning system requires a successful blend of bottom-to-top and top-to-bottom approaches. It is based on strong governance, institutional and social arrangements, and it demands technical resources and reliable communication mechanisms. Three preconditions are essential when designing the system:

- A governance model that recognises the value of conducting activities in the context of disaster risk reduction, including early warning practices, as opposed to a governance model that only recognises the need to respond after the event has impacted a community (Chapter 51).
- Acceptance that the system is being designed to empower communities at risk in order to minimise losses through the provision of strategic information in a timely fashion (Glantz 2001). Such a conviction implies that all stakeholders, including communities at risk, share a responsibility in the design, implementation and routine operation of the system.
- Acceptance that the system can only be operational if it involves the active participation of a variety of individuals and institutions, from the local to the national level in many cases, and the international level in selected cases.

The need for a proactive governance model, rather than a reactive one, is essential for securing the participation of government institutions that are required to design and operate some of the components of the system (PPEW-ISDR 2006b). Experience shows that a reactive model does not lead to the establishment of early warning systems. For example, in five Central American countries, several hurricane and earthquake disasters struck between 1969 and 1976. These disasters promoted the enactment of policies targeting disaster response, which materialised through the establishment of national emergency committees or commissions. In the following decades earthquakes, volcanoes and hurricanes continued to impact the region. However, the institutional mandates dictated their activation only after a disaster had occurred.

It could be stated that the United Nations International Decade for Natural Disaster Reduction, 1990–2000, convinced some countries about the paradigm of risk reduction, highlighting the need to focus on vulnerability, while Hurricane Mitch catalysed the modification in legislation in these five countries, transforming the national emergency commissions into more modern disaster risk reduction and management agencies (Villagrán de León 2008). Hurricane Mitch also paved the way in this region for the establishment of many community-operated early warning systems targeting floods, particularly in small basins in rural areas, incorporating all the elements of people-centred early warning (CEPREDENAC 2004). Villagrán de León (2005) showed how these systems have been successful.

The second condition is essential for ensuring the participation of community organisations and communities at risk in the design, implementation and routine operation of the early warning system. This condition also implies that the operation of the early warning system is not only a responsibility of government agencies, but also of everyone involved (PPEW-ISDR 2006b).

The third condition is based on the design and operation of a people-centred early warning system targeting the four elements presented in the previous section. Risk assessment should be conducted in a participatory way by communities at risk with support from experts covering many disciplines and interests. Such an approach facilitates the awareness process concerning such risks by communities at risk and supports the dissemination of warnings and communication regarding response capabilities.

Key issues in designing people-centred EWS

In addition to the pre-conditions discussed in the previous section, the design of people-centred early warning systems should take into consideration further key issues.

Recognising the responsibility of a single authority of some kind

A single authority needs to be recognised for purposes of the EWS regarding the issuance of warnings (e.g. community leader, government officer, research institute). Considering that a warning constitutes a call to action for a community at risk to mobilise itself, it is imperative that the warning is issued by a single authority of some kind, who not only takes responsibility for issuing the warning, but who also has the credibility to assure people regarding the accuracy of the forecast in order for people to take action. That helps to avoid different 'official' warnings coming from multiple sources. One of the best examples concerning such a single-authority approach is the early warning system for hurricanes in Cuba, which has continually displayed its effectiveness, saving numerous lives (Thompson and Gaviria 2004).

Community-based hazard mapping and vulnerability assessment is critical

The assessment of risk conducted by communities at risk supported by experts using participatory approaches should lead to the identification of the different types of vulnerable groups, their geographical location, the level of exposure, strategies for the communication of warnings and strategies for post-warning actions, such as evacuation to safe areas where appropriate. As noted above, both hazard and vulnerability assessment require input from those in the community and those outside the community. Those different forms of knowledge (see Chapter 9) need to be combined, further contributing to communities' awareness concerning the risks that they face and ways in which they could minimise such risks through local interventions. Participatory approaches also permit communities to take ownership of the warning system, increasing the chances of proper maintenance. An innovative example stemming from the Buzi river in Mozambique is the use of satellite imagery and participatory GIS techniques to facilitate the assessment of vulnerability to floods and droughts by local community leaders in areas exposed to these hazards (Kienberger 2007).

Public understanding of warnings

This concerns the need to improve understanding of warnings and their implications. A comprehensive review of many relevant issues was undertaken by Mileti and Sorensen (1990). According to these authors, it is important to understand how people receive and perceive warnings. A need emerges to focus not only on issues related to releasing and disseminating the information, but also on how such information is presented to individuals and communities to ensure that they act adequately (see Box 40.2). In particular:

- The need to involve community members in identifying the best communication mechanisms and strategies. That might be the use of a particular local radio station which is popular within the community, rather than the government-run, national-level station which may not be equally popular or credible.
- The need to involve community members in designing warning messages to tailor the text of such messages to local customs to ensure that those receiving the message understand it and the instructions regarding how to act once a warning is issued.
- The elaboration and use of different types of messages for different target audiences according to their own needs and responsibilities. For example, in Mexico City, the seismic alert system targeting earthquakes includes specific instructions to drivers of trains in the subway system, requesting them to drive at slow speeds and to stop at the next possible station in case of a warning (Lee and Espinosa-Aranda 2003). The same system has the capacity to target schools independently, as well as public radio stations, which have different instructions regarding how to proceed once a warning is received (Espinosa-Aranda et al. 2003).
- In countries with several ethnic groups, providing warnings in different languages or dialects is needed. A case in point is the warning concerning El Niño episodes, which span entire regions and which manifest through different types of hazards – droughts in some areas but floods in others (Glantz 2001). Locally contextual warnings, for both language and specific risks, need to be considered. In cases where communities may be exposed to several hazards, communication channels and strategies should be used that have been identified by the local communities.
- The use of parallel communication systems to strengthen the warning message, but focusing on the same warning that was issued by the single credible source. For example, Anderson (1969) comments that in Hilo, Hawaii, sirens sound an initial alarm in coastal areas exposed to tsunamis, local radio stations transmit the warning and police officers are dispatched to critical areas to supervise evacuation and to ensure that no people return to hazardous areas prior to the all-clear signal.
- The need to ensure that warnings reach all those at risk. Once a warning has been issued by the appropriate authorities, it is up to the media and other networks to disseminate such warning to everyone who needs it. As mentioned earlier, the efficiency and effectiveness of the early warning systems will be enhanced if warnings reach the target audience as quickly as possible. It is therefore important for early warning system operators to keep informed

---

Box 40.2 Acculturation as part of the warning process

Laurence Creton-Cazanave
Laboratoire d'étude des Transferts en Hydrologie et Environnement, and PACTE-CNRS,
Grenoble, France

A project in France tested an approach to 'warning processes' that aims to understand how those involved in warnings and dealing with warnings recognise environmental clues in order to guide action. Using the flash floods warning process in the Vidourle catchment in southern France, the study focused on 'acculturation'.

Acculturation refers to continuous and direct contact between two groups with different cultures, leading to exchange and adoption of each other's identity, behaviour and values. The acculturation process allows developing and exchanging skills, ideas, knowledge and wisdom, linked to specific fields of activities and/or locations. These skills may be practical knowledge and heuristics, and rely on a long-term and collective practice of a job or a country. This study paid special attention to the acculturation process that occurs with non-human elements, such as technical and policy tools.

Riverside inhabitants unconsciously integrate knowledge about the Vidourle river's behaviour into their daily local life. As one resident said, 'Now, I understand how it flows, the Vidourle becomes a torrent as soon as it rains ... According to local people, and it appears to be true, when rain comes from the mountains, we can be sure there will be a flood, at least a little one.'

As another example, the weather forecasters express reservations about the potential benefits associated with new models: 'The new model is great, but, you know, each time they provide us with a new one, experience has to be re-built ... Each time, we need to learn again, how it works ... Actually, we need to use it a lot, to make it run many times before we can rely on it to do the forecast.' They are indicating the need for inherent knowledge and experience for the models to be useful.

A private company that helps municipalities with their warning and evacuation plans is involved as well. A mayor explained that this company is helpful for the municipality, thanks to long-term co-operation: 'When they give us information about the weather and water situation, there is no need to talk for three days ... Five minutes are enough; they know exactly what kind of information we are expecting ... And they also know what we can understand, what we can use, and what are our real issues.'

Acculturation is thus a resource for all who are involved in the warning process, occurring through unintentional assimilation more than through any formal education. Three main lessons result. First, acculturation supports the warning process and could be deemed essential for community-based warning processes over the long term. Second, acculturation results from daily, local, collective activities, not being feasible by prescription or through formal teaching. Third, acculturation is a long-term process that often does not fit into typical policy-making schedules.

Thus, warning policy-makers should accept and work with the acculturation resource that exists but that they cannot create, recognising the importance of local experience.

---

about advances in communication technologies as well as the communications systems that people are using and prefer.

The sustainability of the EWS and its implementation

This concerns the need to consider issues of sustainability when designing and implementing the system (human, technical and financial resources). While sophisticated equipment to monitor hazards is a useful contribution to warning systems, and while the most advanced tele-communication systems to communicate warnings potentially increase communication efficiency, before using the latest technology it is important to assess the community's capacity to use and permanently sustain and upgrade such high technology (Glantz 2001). To this end, it is important to design the system taking into consideration capacities and limitations in terms of budget and resource access for operating, maintaining and upgrading the system. For example, the Bangladesh community-operated early warning system developed by the Bangladesh Red Crescent Society (IFRC 2009) makes use of volunteers equipped with bicycles, sirens, megaphones and transistor radios to alert rural communities at risk in case of cyclones.

Social sustainability is as important as technical sustainability. That includes sustaining awareness campaigns, and the campaigns' impacts, and regularly testing all the system components, including operating procedures. One strategy could be to establish the system in conjunction with the civil protection agency, and linking early warning to preparedness (Villagrán de León 2001).

Levels of alert

This concerns the analysis of the use of different levels of warnings or alerts, depending on the hazard. Early warning systems may use different levels that could indicate time until hazard impact or expected severity of the hazard. Examples are three colours (such as blue, yellow and red used in Mozambique for cyclones), five colours (white, green, yellow, orange and red used in Europe by METEOALARM for a variety of hydrometeorological hazards) or two levels such as watch and warning levels used in the Netherlands, Norway, Jamaica and the USA for floods. It is important for community members to be involved in the system design to aim for a consensus on the number and type of levels. That would help to avoid problems, such as for colour blind people or where a colour might have cultural significance, such as green in many Arab countries. A balance must be sought between a unified system, so that people do not have to remember peculiarities for each system, and local contextuality including cultural sensitivity and inclusiveness of everyone who will use the system.

Incorporation of traditional knowledge

For the incorporation of traditional, indigenous knowledge used by communities for hazards and vulnerabilities, the critical challenge to overcome is the gap that exists between the local and the scientific communities regarding accepting each other's knowledge for early warning. A gap exists in the way in which knowledge is generated and accepted as valid and useful (see Chapter 9). In the scientific community, the peer-review process is essential for knowledge stemming from research to be accepted by the scientific community. In contrast, indigenous knowledge does not require such a formal peer-review process, although it is informally peer-reviewed by people accepting the knowledge and passing it down through generations. Examples are the shedding of leaves by special trees in Africa in the case of dry spells, or the observation of the level of nests built by particular species of birds above the ground as a sign of a probable level of floods.

In an attempt to move forward, the author proposed two potential approaches to validate indigenous knowledge (Villagrán de León 2005). The first approach is to identify the root causes that lead to changes in the environment that are the basis of indigenous knowledge and link them to the scientific theories of the earth sciences, so that the scientific community will recognise the value of such indigenous knowledge. The second one is to systematise indigenous knowledge via the Western scientific methods, such as its testing using statistical methods to assess its degree of accuracy and reliability.

Standard operating procedures

This concerns the elaboration and implementation of standard operating procedures (SOPs), which take into account the local context and are tailored to the needs of local communities, including a degree of flexibility when needed (Chapter 42).

Early warning systems involve conducting specific activities, some of them in a sequential order and others in a parallel fashion, so that the activities allow the system to function in a timely fashion. In selected cases, such as flash floods and earthquakes, time may be critical, so it is important to be able to proceed at a moment's notice. The use of SOPs allows those operating the system to act under the authorisation of decision-makers, even if the decision-makers cannot be reached immediately. Previously approved and practised SOPs allow such decision-makers to pre-authorise specific activities, including issuing a warning (see Chapter 42).

These procedures should include segments on the following topics:

- The roles and responsibilities of all agencies and vulnerable groups in the different activities contemplated in the plan. In particular, the emergency operations centre should be activated in case of any disaster. In places without an adequate emergency operations centre, one should be established.
- The information flow in case of warnings, indicating sources of information and agencies to which the information needs to be sent.
- Mechanisms to be employed to issue and disseminate the warnings to the target audience, including vulnerable groups.
- Mechanisms to ensure that agencies which conduct response activities are activated, such as search and rescue teams and fire fighters. If these agencies do not exist, then communities should have support to establish them.
- Mobilisation of strategic resources to where they may be needed to facilitate the evacuation of vulnerable groups to safe areas.
- The issuance of an 'all clear message' that signals the end of the potential threat. This is particularly useful in the case of tsunamis, for example, which may last several hours, or floods which may last several days.

Frequent testing of the early warning system

Once all the elements of the system have been implemented, it is essential to conduct tests to validate their functionality and to make improvements, especially as the technical and social aspects of warning systems evolve and as lessons are learned from other case studies. Testing in the form of table-top simulations assists designers of the operating procedures to test them in a controlled environment; to detect gaps and critical issues that need to be reviewed and corrected; and to make them locally contextual for even the remotest communities. For example, in Japan, on 1 September of every year, a drill is conducted to test earthquake preparedness measures and to keep alive social memory concerning earthquakes.

Conducting exercises involving all components and all stakeholders of the system serves three purposes. First, all those involved in or using the warning system can identify additional gaps and critical issues in all components of the system, whether or not they would appear in table-top exercises. Second, all those involved, both within and external to the communities, become aware of the functionality of the system and interact with each other. Third, all those involved in the operation of the system can maintain the required skills and knowledge.

Handling near misses and false alarms

An example of this is the Famine Early Warning System for Africa. As Verdin *et al.* (2005) comment, the resource limitations of local meteorological agencies in terms of a thin network and delays in reporting of climate variables add uncertainty to the forecasts. Coupled with the inherent unpredictability of weather and climate, the capacity to forecast rains then becomes even more difficult. Nevertheless, as Sorensen (2000) and Barnes *et al.* (2007) comment, false alarm concerns may be overemphasised and speculative. The public is frequently willing to tolerate such false alarms so long as the system is able to demonstrate that it has learnt from the experience and has adapted to improve its forecasting capacities. Costs of the false alarm to the individuals affected also make a difference.

Barnes *et al.* (2007) comment on cases where a warning characterises a hazard's probable parameters, but the hazard appears outside those forecast parameters. In this context, it is important again to learn from such experiences and to conduct an awareness campaign not only to explain the limitations of the system in its capacities to forecast, but also to discuss strategies and to gain input regarding how to deal with such issues in a participatory fashion.

An example was the false 'cancellation' of a tsunami warning in Chile following the strong, shallow earthquake that took place on 27 February 2010. The then Chilean President Michelle Bachelet went on the air to cancel a tsunami warning based on advice from the National Emergency Management Agency, ONEMI, only to realise later that there was indeed a local tsunami. Unfortunately, the cancellation of the warning led to the death of civilians who decided to return to their coastal area based on her false statement. The event has brought to light the difficulties associated with tsunami warnings.

Conclusions

In recent decades, government institutions, non-governmental organisations and citizens have promoted the adoption of people-centred approaches in the context of early warning. Such approaches widen the focus from the monitoring of hazards and the dissemination of warnings. They expand to the need to ensure that those who can benefit from the information provided by the system are empowered to do so through being involved in all stages of development of the system (design, testing, routine operation and evaluation). Those who must be involved include civil protection agencies or emergency committees, which also have a crucial role to play by ensuring that once a warning is issued, all tasks related to response are conducted in a co-ordinated fashion.

In addition, it is important to stress the view that such people-centred approaches imply the need to consider the system's permanent functionality through continuously conducting awareness activities and through the frequent assessment of risks, especially vulnerabilities. In this sense, the new view moves away from considering early warning systems as dormant until an event occurs, to one where such systems are permanently active through frequently conducting activities. For example, as part of the activities conducted within the seismic alert system in

Mexico City, schools conduct earthquake drills roughly once a month (Espinosa-Aranda *et al.* 2003).

This chapter has presented the basic elements of people-centred early warning. The implementation of these elements remains a challenge in some early warning systems. In other systems, progress is being achieved with the support of the international community and through communities taking initiatives. With people-centred warning systems supporting the notion of warning as a process (see Box 40.2) for the people being affected, warning systems will continue to improve, contributing to disaster risk reduction.

41
Preparedness, warning and evacuation

Philip Buckle

RESEARCH ASSOCIATE, CENTRE FOR DEVELOPMENT AND EMERGENCY PRACTICE, OXFORD BROOKES UNIVERSITY, OXFORD, UK

Introduction

This chapter focuses on the process of evacuation and its links to preparedness and early warning. Much conventional wisdom on this evacuation needs to be re-thought. Reading the available plans and manuals on planning for and conducting evacuations suggests that it should be planned and managed in an orderly, systematic and considered manner. In fact, this view reflects agency perceptions but evacuation planning or management rarely proceeds as smoothly as emergency service planners would like (DEFRA 2004). Therefore, this chapter reviews and describes commonly practiced and often (but not always) successful processes for planning and managing evacuation and for preparedness activities.

Is evacuation effective? The answer is not unequivocal, but evacuation is generally effective. Do evacuation plans work? They generally do, but only partly. Equivocal responses to the efficacy of evacuation planning and management arise because planning is usually well-intentioned but unhappily short-sighted, focusing on agency priorities and perceptions, limited by agency traditions, hierarchies of authority and resources. Evacuation planning and management are also constrained by the limited engagement of communities in the process.

Paul and Bhuiyan (2009), for example, found that many home-based and local preparedness measures, such as stockpiling food and medicines, are not hazard-specific but relate to disaster risk and the needs of everyday life. That is, managing one's own safety in a considered fashion is a non-specific preparedness measure. Yet even high-risk scenarios, such as an earthquake in Dhaka, may not prompt adequate preparedness. Preparedness may not occur for many reasons including different perceptions of the risk, focusing on other risks as priorities, lack of adequate resources to prepare properly or inadequate training and availability of training materials to prepare properly.

The tension between theory and practice

Two particular issues can be flagged. In many instances, most people self-evacuate and do not rely on emergency services or local authorities for transport and accommodation. This limits the

493

applicability of most evacuation protocols which do not plan for either limited or mass self-evacuation. When Hurricane Rita affected Texas in 2005, self-evacuation led to too many private vehicles on the road, clogging the evacuation routes (Litman 2006). Second, evacuation itself is not without risk, including the risks associated with leaving one's home and property, being separated from family and friends and moving into the 'unknown' in an environment with hazards. Evacuation itself may move people into danger where the hazard is unpredictable or moving rapidly, such as tropical cyclones and wildfires. Thus, although evacuation is a risk mitigation activity, it is itself not risk free. Examples are transportation crashes, stress on medically frail individuals, dealing with prisoners who try to escape, loss of livelihoods by being away (balanced with new livelihood opportunities created by the evacuation) and not having adequate resources for or supplies of food, water and medicine.

While it is by now well accepted that disasters are derived from and are a part of daily life (Lewis 1999; Wisner *et al.* 2004), governments and agencies still have difficulty publicly acknowledging this causal chain for reasons that concern vested interests, political accountability, public perception of government performance and public confidence in agencies. Plans and arrangements for early warning systems (EWS) (see Chapter 40) and evacuation therefore often appear naive and optimistic when compared with the real world they attempt to model. They also tend to place unwarranted emphasis on easily controllable elements such as technology and to correspondingly neglect the 'difficult' issues of managing, supporting and learning from people and their perceptions, needs and behaviours.

The meaning and process of evacuation

Definitions and ideal models of evacuation

Definitions of evacuation and preparedness are discussed in Box 41.1. One conclusion is that emergency services and governments see evacuation as a planned, managed and controlled activity that can be led by the emergency services.

Evacuation starts with the first warning or alert when people can begin getting ready for moving, but it does not end with the passing of the hazard. Evacuation ends when people return to their homes or home sites. That should not occur before the area is safe. Needed actions might be debris removal, control of disease vectors, removal of dangerous structures, trees and buildings, and restoration of essential utilities, notably water and power.

Much early warning and evacuation planning focuses on rapid-onset hazards because time and opportunity are so constrained by the speed of onset. The consequent lack of warning and preparedness time limits individual actions. Instead, systems need to be in place before the event to ensure that movement to safety can be achieved in an orderly and timely way. Such principles should then also be applied to slow-onset hazards.

Nonetheless, evacuation is not migration, which typically involves long-term, long-distance relocation without prospective return, and where, during relocation, daily activities and livelihoods are resumed. Instead, evacuation is relatively short-term, typically assumed to be hours or days (sometimes weeks), and includes plans for returning home. Beyond this evacuation may flow into more or less permanent relocation or migration.

On Montserrat in the Caribbean, the volcano that makes up the island started erupting in 1995. As the main population centres became dangerous, people were evacuated 'temporarily' to the north of Montserrat, to other Caribbean islands and to the UK, which is Montserrat's governing state. This evacuation became permanent for many of them, especially as their homes and communities were destroyed by volcanic activity and remain continually affected by lethal

Box 41.1 Defining evacuation and preparedness

Emergency Management Australia (EMA) defines evacuation as:

> The planned relocation of persons from dangerous or potentially dangerous areas to safer areas and eventual return.
>
> *(EMA 1998: 43)*

In its manual on evacuation planning EMA (2005) states:

> Evacuation is a risk management strategy which may be used as a means of mitigating the effects of an emergency or disaster on a community. It involves the movement of people to a safer location. However, to be effective it must be correctly planned and executed. The process of evacuation is usually considered to include the return of the affected community.

UK Resilience defines evacuation as:

> Removal, from a place of actual or potential danger to a place of relative safety, of people and (where appropriate) other living creatures.
>
> *(Cabinet Office 2009: 16)*

Often, evacuation is not defined. For instance, in the Pacific, the national disaster plans for the Marshall Islands (NDMC 1997a) and Samoa (NDMC 1997b) include discussions of evacuation but do not define or describe it. Often, evacuation is included as a subset of preparedness. Benson *et al.* (2007) define preparedness as:

> Preparedness is activities and measures taken before hazard events occur to forecast and warn against them, evacuate people and property when they threaten and ensure effective response (e.g. stockpiling food supplies).
>
> *(Benson* et al. *2007: 16)*

Similarly, internationally, UNISDR does not define evacuation, but defines preparedness as:

> The knowledge and capacities developed by governments, professional response and recovery organisations, communities and individuals to effectively anticipate, respond to, and recover from, the impacts of likely, imminent or current hazard events or conditions.
>
> *(UNISDR 2009c)*

The accompanying comment to this definition states that preparedness includes evacuation amongst other activities. Many agencies around the world adopt these definitions verbatim. Consequently, as an example, the Caribbean Disaster Emergency Management Agency also does not have a definition of evacuation, instead enfolding it within the definition of preparedness.

volcanic hazards. People still on the island continually enter the hazard zone for livelihood-related reasons due to factors such as confusing messages from and distrust of the authorities (Pattullo 2000).

Where evacuation lasts for more than hours or a few days, arrangements have to be put into place to ensure that evacuation centres or temporary shelter and settlements are sustainable and that the livelihoods of the evacuated people can be sustained. While this may in part be addressed by financial or other supplements, these measures are often not fully adequate or appropriate over a long period. For example, after evacuations due to volcanic activity around Mount Tungurahua, Ecuador in 1999, many residents ignored official advice and returned to their homes in order to avoid losing their tourism-related livelihoods (Tobin and Whiteford 2002). Other livelihood assets at people's original homes also frequently require work, such as animals or crops being tended to and infrastructure, equipment and machinery being maintained and monitored to avoid theft.

The evacuation process includes a set of activities that need to be managed:

- Providing warnings and evacuation support on the basis of a risk assessment.
- Providing services on the basis of need and vulnerability.
- Providing services equitably.
- Providing services in a timely manner.
- Maintaining services over the full span of the evacuation process.
- Managing the return of affected people.
- Managing the security of homes, assets and livelihoods as far as this is feasible (e.g. looting prevention).

The warning and evacuation process is made more complex by temporal factors. Differences arise whether the hazard occurs during the day or night, during winter or summer, during rest and holy days or work days and during rush hour or not. Any warning process needs to take this into account by being flexible and adaptive to changing circumstances.

Consequently, evacuation management is not a simple process that is concerned solely with technical service provision but must include much more. Managing movement and support to people in transit is important, ensuring the movement to safety and the return home. Managing confusion and uncertainty can be completed through effective planning and multi-faceted and unambiguous communications. Managing continuity of daily life and livelihoods, where possible, assists in minimising secondary disruptions to people's well-being.

Government agency approaches to planning – and equally to more general preparedness planning and often to early warning systems – tend to be just that, approaches that conforms to agency needs rather than to local needs. In contrast, effective evacuation management includes numerous community-based elements (MCDEM 2008). That starts with the planning process to identify risks, resources and options for addressing the risks and includes community engagement in leading and managing evacuation needs to be integrated with agency plans and EWS technologies. Successful examples include the Townwatch programme in Asia (Ogawa et al. 2005), the StormReady and TsunamiReady programmes in the USA and Community Disaster Volunteer Training Programmes in Turkey and the USA (see also Box 41.2).

Critical issues in evacuations

Many misconceptions exist concerning evacuation. Most prominently, evacuation is often assumed to be an agency-managed process. While it may be planned, most evacuations are

Box 41.2 Public awareness in New Zealand

David Johnston
Joint Centre for Disaster Research, GNS Science/Massey University, New Zealand

Julia Becker
GNS Science, New Zealand

Douglas Paton
School of Psychology, University of Tasmania, Australia

Michele Daly
GNS Science, New Zealand

New Zealand is exposed to a wide range of potentially devastating impacts from a variety of natural hazards. Recent research has showed that levels of preparedness are less than those desired by emergency management agencies despite enhanced public awareness initiatives over the last decade. Public awareness programmes must address the wide spectrum of psychological and social factors which determine levels of acceptable risk, and which govern mitigating action. Traditional mass-media means of public awareness is a necessary tool to initially motivate people to prepare. It helps to create an awareness of the hazardscape, initiate discussion and thinking about hazards and their repercussions, and form an understanding of how certain actions will be of benefit. However, mass-media public awareness alone does not lead to an uptake of preparedness activities.

Community development and awareness programmes have been found to effectively increase preparedness and foster a sense of community through empowering communities. These programmes focus on the ability of people to face natural hazards, such as self-efficacy, using action coping mechanisms and a sense of community, which cannot be achieved through media communication. There is much inherent strength within communities and this can and should be harnessed to improve local capacities and coping strategies. The active involvement of formal and informal community networks in hazard awareness and other mitigation activities has been shown to be a key predictor of preparedness. Recent practice in New Zealand has shown that community-based initiatives need to be integrated within community development initiatives. Such an approach is more effective than stand-alone, one-off programmes. Several New Zealand studies also suggest that school programmes need to be one of the centrepieces of a sustained, community-based effort.

Public awareness about natural hazards and disaster risk reduction in New Zealand is currently delivered both at a national and local level. The New Zealand National Public Education Strategy sets out the strategic framework for public awareness for the emergency management sector in New Zealand. The national awareness programme has multiple elements including: (1) media advertising (television, radio and print); (2) advertising in the 'Yellow Pages' regional directories; (3) dedicated website (www.getthru.govt.nz); (4) printed brochures; (5) household mail drop (with emergency plan and checklist); (6) promotional display stands and drink bottles; (7) a 'disaster awareness week'; (8) school resources ('What's The Plan Stan?' at www.whatstheplanstan.govt.nz); (9) public relations, sponsorship and promotional activities; (10) online Civil Defence Emergency Management (CDEM) public awareness toolbox. At the regional and local level, CDEM Group Plans outline a range of public awareness initiatives to be delivered. Current efforts are based around a mix of media advertising, printed material, public outreach and community-based programmes in schools and community groups. Evaluating the effectiveness of programmes presents a number of challenges but is an essential element of successful programmes.

There are many opportunities to improve New Zealand's capacity to respond to future disasters, requiring a mix of awareness education, social policy, exercising and empowerment strategies.

conducted by local people themselves. Only those with limited transport options and special sites such as prisons, nursing homes and hospitals may require directed and managed evacuation by the military, police or civil authorities. Estimates of self-evacuation run from about sixty per cent of the at-risk population to up to ninety-five per cent. Where government agencies are weak or absent, self-evacuation may account for the greater proportion of the evacuated population. Droughts (exacerbated by government mismanagement and war) have led to over 1 million people being displaced in Eritrea, the vast majority of them having evacuated their communities on their own without assistance.

Even with self-evacuation, planners often mistakenly disregard the fact that most people want to confirm in some way the warning they receive before evacuating. The 2009 wildfires (see Chapter 24) in Australia demonstrated the reluctance of many people to evacuate on the advice of the emergency services and to wait to confirm the imminence of the hazard with their own eyes.

Another way in which people's behaviour often diverges from planners' assumptions is that during the evacuation process people make a series of risk assessments to balance their own immediate safety against the need to protect homes, livelihoods and assets. If, in their judgement, the risk to the latter is greater than the former, then they may refuse to evacuate (as noted for Australian wildfires), they may try to enter the hazard zone (as noted for Montserrat) or they may try for early and pre-emptive return (as noted for Mount Tungurahua volcano in Ecuador). Road blocks around hazard zones are sometimes set up to keep people out of the area, which includes visitors as well as tourists, but law enforcement resources are usually limited and circumventing the roadblocks can be straightforward.

Governments and agencies typically underestimate the drive to return home and to restore livelihoods even before it is fully safe to do so. After the 2004 tsunami in Aceh and Sri Lanka, many evacuation shelters and later the temporary or transitional settlements and shelters were located at a distance from the coast. Consequently, they were not fully accepted or used by some who depended on fishing for their livelihood because they naturally preferred to be close to the shore, the sea and their boats.

Personal safety is a critical aspect of the evacuation process. This involves security at evacuation centres to maintain order and to prevent crime, from petty theft to assault. Requirements include adequate lighting, privacy for sanitary facilities and other measures to ensure safety from sexual predation or harassment such as of non-heterosexuals (see Chapter 35 and Chapter 36). It also involves maintaining order and property protection at the evacuated sites.

Evacuation is not cost free. Substantial costs in logistics, transport and personnel may be incurred in addition to the cost of establishing and maintaining evacuation centres. Costs are also incurred in the area to which evacuees have moved, such as in terms of disrupting the community's daily routines and in temporarily losing facilities such as schools and community halls used as evacuation centres. Meanwhile, the influx of personnel and aid can boost the local economy while creating a shortage of supplies.

Direct human costs are also associated with evacuation. While being evacuated, people cannot maintain their jobs or tend crops and livestock which may lead to losing short-term income, clients over the long term or entire livelihoods. Numerous studies in the USA have examined patterns of small business success and failure following disasters such as the 1994 Northridge earthquake, the 2001 Seattle earthquake, the 1993 floods in the Midwest and Hurricane Andrew in 1992. In many cases, business interruption leads to the end of the business.

There may also be psychological costs where the evacuation process is disruptive and worrying. Being confined with many others during periods of stress and uncertainty can generate psychological and emotional stress. In some cultures, the act of relative confinement, mixing with strangers or being in proximity to members of the other gender may be troubling. Operational field manuals for transitional settlement and shelter (see Chapter 46) emphasise the need to factor gender sensitivity into any form of evacuation support being provided, but field evidence shows that this is often not done.

Finally, evacuation may become permanent. Where a person or family is not tied to their original home site through weak tenancy arrangements, weak land tenure, having jobs remote from home or being concerned about future safety, they may choose not to return. This is a dislocation for them and also for their community which, after the disaster, may find itself depleted of members. This consequence has been noted in several wildfire disasters in Australia where the disaster acts as an agent of changing the community. Once people are evacuated and their lives are disrupted through the destruction of their home, many chose to move to a location that they perceive to be safer or one that offers them more, and often more immediate, livelihood and employment opportunities, rather than waiting for their community to be rebuilt. Few studies have specifically mapped these changes and baseline data are often lacking, but local and anecdotal evidence from people who stay behind in the communities is that population depletion may even continue several years after a disaster, with their neighbours citing the wildfire as the catalyst that made them decide to relocate.

Disaster preparedness

What is preparedness?

Some forward-thinking practitioners and policy-makers (e.g. Gabriel 2000) have argued that preparedness is not a separate management activity, but is what governments, agencies and communities do, or should do, to ensure that they can meet their responsibilities in community safety. Preparedness covers developing and maintaining appropriate technologies, gaining political support and agency mandates, acquiring resources, developing personnel skills, expertise and knowledge, managing information and managing linkages and networks with partners.

There are many definitions of preparedness but the IFRC's definition is useful:

> Disaster preparedness refers to measures taken to prepare for and reduce the effects of disasters. That is, to predict and – where possible – prevent them, mitigate their impact on vulnerable populations, and respond to and effectively cope with their consequences.

> Disaster preparedness is best viewed from a broad perspective and is more appropriately conceived of as a goal, rather than as a specialised programme or stage that immediately precedes disaster response …

> Disaster preparedness is a continuous and integrated process resulting from a wide range of activities and resources rather than from a distinct sectoral activity by itself. It requires the contributions of many different areas – ranging from training and logistics, to health care to institutional development.
>
> *(IFRC 2000: 6)*

Preparedness can appear to be difficult, complex and fuzzy. However, it may be more manageable, and perceived to be more manageable, if it is taken to be based in everyday life and partly framed as community safety, health and well-being along with adequate education, gender equity and sustainable livelihoods (see also Wisner *et al.* 2004).

Learning from and advising communities of the risks they face, their vulnerabilities and their capacities is a first step in preparedness. Awareness, training and education programmes are an integral component for local people, disaster management agencies and organisations (see Chapter 62).

Resources for preparedness

Beyond this first step of raising awareness is the need to acquire funding and resources, which includes evacuation centres, transport, communication and warning systems, stockpiled resources for medical aid, food, water, shelter and cultural support, as well as systems for power. Stockpiling is frequently used, although it is usually more efficient if sources of reliable supply during evacuation are identified, planned for and monitored. This avoids product deterioration, the risks of loss or theft and the substantial outlay of funds for materials that then, in effect, lie dormant. The types and levels of resources, as well as the location of evacuation centres, transport systems and other life- and health-sustaining measures need to be planned.

In the Dominican Republic, flood hazard mapping and preparedness planning in the mountainous north has identified flood risks to the local population. That action has been complemented by local community groups, mainly run by women, who have taken risk assessment even further to identify vulnerable groups and vulnerable livelihoods before a disaster has occurred. They have also detailed the type and level of emergency service support that will be needed for post-impact recovery and rehabilitation.

Resources for planning and preparedness decline slowly as time passes from a disaster. Other competing priorities tend to capture resources that may be better used to support disaster preparedness and community safety. This is not necessarily a cynical response, but one that recognises that governments must manage competing priorities, and competing interest groups, with finite resources. This reality can be turned to being an advantage by astute managers and community leaders. Resources for preparedness and government commitment will be highest immediately after a large event and this situation can be capitalised upon by managers and leaders able to promptly press their case to government. This applied noticeably on the Sri Lankan coast affected by the 2004 tsunami where prompt advocacy by local communities and humanitarian agencies secured resources for rebuilding damaged and destroyed houses.

Planning preparedness

Planning is an activity best conducted in a participatory manner in collaboration with local people, local agencies, disaster management officials and government officers. The final product is not the written plan in itself, but is the process of planning that engages all the participants. This ensures that they have a common understanding of the risks, understand and accept each other's capabilities and accept a set of protocols for initiating, continuing and withdrawing from emergency activities.

The planning process establishes credentials and credibility which are a prerequisite to rapid and effective action. Planning identifies risks, along with the resources and systems necessary to manage those risks, and sets out protocols for the interaction of local people, agencies and governments. No plan can ever guarantee success. Instead, a plan tends to limit the potential for failure. Plans, in turn, need to be supported by periodic and regular review and updating, training, promulgation, testing and refining. Testing many occur through desk-top exercises, simulations and full-scale, real-time field exercises. These are expensive and time-consuming, but they are the closest alternative to an actual event.

An essential part of planning and preparedness is post-disaster review where the plans and actions of communities and agencies are reviewed against agreed criteria. These may take the form of 'hot' debriefs immediately after the event, 'cold' debriefs well after the event, agency and inter-agency reviews, coroner enquiries and judicial and government reviews. These should feed into the planning process.

Planning for early warning and evacuation should involve pre-planning evacuation centres, liaising with suppliers of medical and health services, food, water, bedding, clothes, hygiene services, and security and safety procedures in order to ensure that these can be rapidly mobilised as soon as needed. Unfortunately, many emergency management agency planners often still assume that the heat of the moment will give them sufficient authority and priority to arrange these facilities and services at the time at which they are needed. Obtaining buy-in for preparedness from other agencies, government, civil society and the community is seen as being increasingly important.

It is also becoming more commonly agreed that planning should be inclusive and should engage with local people. Once people are engaged and once the emergency services have established trust and credibility, local people will be more responsive to preparedness, warning and evacuation arrangements. Local engagement – in addition to being a right since evacuation directly affects the safety, security and freedom of movement of people – can provide evacuation planners with useful local knowledge, local resources, a sense of local priorities and consequently local support for the work and a source of local resources and volunteers.

Evacuation planning must also consider the needs of people who cannot self-evacuate or who may have difficulty in fully managing their own lives through lack of resources, lack of choices, lack of mobility, poor health or reduced decision-making skills. These may include the elderly, disabled, children, people without their own transport resources or choices and people who do not speak or read the dominant language who may not understand warnings. Other special groups may require their own dedicated transport, support and security arrangements with examples being hospitals, prisons and tourists and other travellers. Hurricane Rita's evacuation included most of these groups needing assistance, especially since the widespread evacuation was spurred on by the catastrophic images from Hurricane Katrina having affected Louisiana a month earlier (Litman 2006). McGuire et al. (2007) showed that vulnerabilities may pile upon each other so that age, disability, ethnicity and other factors may compound to make otherwise ordinary tasks, such as leaving one's home, much more difficult.

Ideally, evacuation occurs well before the hazard is estimated to arrive and may even be arranged if the hazard is considered likely. That may be particularly useful in the case where evacuation routes may be threatened by the hazard. Routes may be damaged or blocked by the hazard, such as smoke from wildfires reducing visibility to zero or storms flooding or washing out routes. Evacuation routes may also be blocked by evacuees. Unplanned, uncoordinated or late evacuations can cause routes and the evacuation processes to be disrupted by congestion and lack of fuel. When Hurricane Rita threatened Texas in 2005, many evacuation routes became blocked with traffic from people evacuating, some of whom then ran out of fuel as they waited.

Evacuation planners and local communities further need to consider how evacuation will occur when happening at different times (as mentioned earlier), during inclement weather or when people may have a reduced capacity to receive warnings or to physically manage their evacuation, such as being inebriated, with examples being during sporting events or holidays. Similarly, the return from evacuation needs to be managed and balanced between safety and people's desire to return home.

The role of centralised planning may be to set out certain broad priorities and performance criteria and to ensure that central funding and resource allocation is adequate to meet planning objectives. Outcomes and performance standards need to be set for evacuation as they often are for emergency response (see Chapter 42). This is difficult to do and is lacking in most areas of disaster risk reduction, but performance criteria in terms of training and awareness raising, call-up of logistic and transport assets and time to take people to safety need to be established as a guide to action, as a means of effectively estimating the scale of needed resources and for purposes of accountability. Examples are SEEP (2009) providing standards for post-disaster economic recovery and Sphere Project (2004) aiming to set minimum standards in disaster response, both generally and for specific sectors such as water, health and shelter.

Testing, drills and simulations

Planning (and the plans that result) needs to be supported by applied and practical testing and training. This can take a variety of forms including 'hypothetical' or discussion exercises around the implementation of the plan or sub-plans. A step above these are scenario-based activities. These involve the designation of a 'real world' situation which is used as a problem to be resolved over a real or simulated time scale with a suite of 'players' representing or drawn from the agencies and groups who will be involved with any response activity. These scenarios can be desk-based, in effect discussions around a tabletop with props such as white boards, maps and planning documents as support materials. They could also be larger-scale, involving the actual deployment of agency personnel and resources in response to the scenario. Such deployment-based or field exercises may run for several days. All scenario-based exercises should be followed up by debriefing sessions from which lessons learned are drawn out and consolidated in agency operating principles and practice. ADPC (2008) is a good source of guidance on these matters.

More locally, evacuation drills, most commonly seen in the form of building evacuations, need to be regularly undertaken to test plans and to ensure that local people and agencies understand their tasks and roles.

Simulations may also be carried out through computer modelling activities, particularly using Geographic Information Systems (GIS) that can combine hazard data, demographic data, topographic and other environmental information, and data about transport routes and the location and availability of buildings and material in support of evacuation. These newer simulation methods can be powerful on the computer, but are best complemented by ground-truthing activities to ensure that the model meets reality.

All testing and simulation activities, as well as real disaster events, need to be followed up by debriefs. Two types of debriefs exist: 'hot' debriefs occur immediately after the event or exercise; 'cold' debriefs occur some time after the event or exercise, permitting a period of reflection by those involved.

Warning systems

Speaking of EWS (see Chapter 40) or evacuation and their component parts as separate entities is misleading. The focus should be on the goal of moving people to safety and then returning them to their homes while supporting and working with them in the intervening period. An EWS has no value if it is not heeded and if people do not move away from the hazard to safety. The evacuation process is of limited value if people are moved towards the hazard or are moved to safety but are not supported and cared for during the evacuation process (Betts 2003).

Warning systems technology is continually becoming more sophisticated, but brings with it unstated assumptions that the technology is at least as important as the warning message and that the warning message is as important as achieving the movement of people and the essential assets and possessions to safety. In fact, this approach is an inversion of the actual process that is needed. Instead, the process should work backwards from (1) what is needed to move people to safety, to (2) what systems and information are required to achieve warnings that are acted upon, and (3) what resources may be needed to help people move to safety and to live in an evacuation centre for what may be an indeterminate time.

An optimum EWS may not be the cheapest, as it has to take into account its efficiency in achieving its outcomes. Ideally, traditional and existing technologies are used as much as possible, ensuring that they can be maintained and repaired, and that the technologies are socially agreed and acceptable. Table 41.1 reviews some examples.

None of these technologies has intrinsic merit above others, although personal communication from another person who is known to and trusted by the message recipient tends to carry the greatest weight in influencing a person's decision to evacuate. Of course, this is also the slowest way to warn a large population and the one most likely to put the adviser at greatest risk or to change the message in the process of disseminating it.

Other forms of communication also have limitations. Sirens, gongs and bells have a limited range, as do flags. Audio and visual signals are inappropriate for people with certain disabilities. Mobile phones, radio and television have a rapid and widespread coverage, although again certain disabilities may limit their reach and effect, but they are costly and are subject to disruption by the hazard and to access limitations from people without the resources to afford them. Emergency service advisories fall between these two modes of communication in terms of robustness and effectiveness.

Any early warning system needs to rely on multiple systems and to have backup and redundancy built in to optimise speed and success of dissemination, despite the expense. In the majority of instances, local people will seek confirmation of the message from a separate source whatever the source of the original message.

Conclusions

Preparedness and evacuation are social processes. They do not exist outside the socially constructed risks that communities face. They are best planned for and managed in a participatory manner where the expert local knowledge is held in equal esteem to the expertise and formal knowledge of agencies and expert practitioners. Evacuation and preparedness are concerned with dealing with people and with managing their safety and the continuity of their livelihoods. Assigning time, knowledge, skills and resources to these rests upon making a judgement about the value of these outcomes *vis-à-vis* other local and national priorities.

As a result of a narrow, technical approach, many complexities and problems of evacuation are overlooked. Evacuation should be the managed exit and re-entry of people. Evacuation

Table 41.1 Examples of EWS technologies and their characteristics

EWS technology for rapid-onset events	Socially acceptable (integrated into daily life, able to reach all people, credibility)	Robust (cheap, sustainable, able to withstand hazard impact, locally repairable)	Cost/benefit	Efficacy (people will heed the warning and will move to safety)
Personal communication	High	High	High	Low (range of the warning is restricted)
Traditional warning systems (e.g. gongs, bells, flags)	High	High	High	High
Emergency services advisories	Medium (if planned and well-managed)	Medium	Medium	Medium (emergency services are often reluctant to put themselves at physical risk)
Sirens	High	High	High	Medium (limited range and a blunt message)
Flags, bells	High	High	High	Low (limited range and a blunt message)
Mobile phones	Medium	Low	Low	Low but increasing (the mobile phone network is susceptible to hazard impact and is easily overloaded)
Radio	Medium	Low	Low	Medium (of use only if the radio is on)
Television	Low (many people do not have televisions)	Low	Low	Low (of use only if the television is on)
Internet	Low but increasing	Low	Medium	Low (of use only if the system is on)

often has the potential to be movement towards danger, so the risks of evacuation should be considered. Warnings and messages are often ignored or misunderstood (see Chapter 40). The active engagement of local communities in planning and conducting evacuations needs to be encouraged in an equal partnership with governments and emergency service agencies (Maskrey 1997).

Yet local people can underestimate risks and can act contrary to their own safety. Responsibilities and resources should be devolved to the lowest possible level consistent with financial prudence, accountability and consistency and equity across communities. A balance is needed, but it should be a collaborative process which engages local people as equal partners in the risk assessment and planning processes for preparedness and evacuation.

42

Emergency management principles

Alejandro López-Carresi

CEDEM (CENTRO DE ESTUDIOS EN DESASTRES Y EMERGENCIAS), MADRID, SPAIN

Introduction

The objective of any emergency management system is to manage and control a major emergency in order to reduce the negative impact to life, property, infrastructure and livelihoods (Moore 2008). It is frequently argued that effective emergency management is virtually unachievable for two reasons. First, because the huge increase in urgent needs cannot be satisfied quickly. Second, because the unknown facts in the acute phase of the emergency make it impossible for responders to have the necessary information to make the appropriate decisions. These arguments are based on a valid premise: there are numerous factors in large-scale emergencies that cannot be controlled or known. However, those factors cannot explain or justify the lack of control over the responding resources.

The number of casualties and their location will only become clear over time. However, the number of responders and their location can and should be known at all times. Lack of knowledge about the emergency is understandable. Lack of knowledge about the response is not. Table 42.1 provides a summary of some facts which can or cannot be known or controllable in the initial stage of a major emergency.

For example, the management of an emergency can be labelled as poor not when an earthquake-induced fire is bigger than expected and the fire crews cannot control it, but when there is poor control of the responding fire crews, their number or location is unknown to the emergency managers, responding units are uncoordinated or there is no real command over the units that are intervening.

Table 42.1 Controllable and uncontrollable factors at the start of an emergency

Uncontrollable/unknown	Controllable/known
Number of casualties	Number of respondents
Location of injured	Location of responding units
Incident evolution and development	Response evolution development
Spontaneous, unrequested citizen collaboration	Spontaneous, unrequested institutional collaboration

The initial goal should be to achieve an effective control of the responding resources. It is through efficient management and co-ordination of all elements involved in an emergency that the ultimate objective of any emergency management system – control of the incident – will be achieved.

Stakeholders in emergency management

Emergency response involves multiple actors, and the range of affected people and institutions is large. Some of the main stakeholders in emergency management are described here.

Emergency services

Fire, rescue, emergency medical services and law enforcement all represent the first institutional response. They and other emergency responders may be involved in tackling the emergency on site, warning and evacuation (see Chapter 41), security of evacuees and evacuated locations, transportation of evacuees and goods, caring for the evacuees and communications to evacuees.

Authorities

Public administrations have a prominent role at all levels. Public servants or elected officials have the ultimate responsibility for making top-level decisions. The first line tends to be local authorities, but that can quickly scale up depending on the situation (see Chapter 52).

Other public services

Utility companies (e.g. electricity, water supply and communications) are often involved in the response to restore services. Some of these institutions have emergency procedures but many improvise their response, despite the opportunities available for training and practice, such as through drills and tests.

Non-governmental organisations (NGOs)

Local, national and international NGOs may also be involved in different aspects of the response. They may provide emergency and transitional settlement and shelter, water and sanitation, logistics, health services or psychological support. This contribution may be planned with skilled personnel, but many organisations not previously included in the emergency plan, and with or without proper training, often take part.

The community

The community's role in emergency management is often neglected because it is considered to be a passive recipient of help, instead of acknowledging the people's capacities and strengths (see Chapter 59). Over past decades, the development of community-based disaster preparedness projects in many countries promoted the establishment of local disaster committees and teams (e.g. Ogawa et al. 2005). These locally based disaster organisations have proved to be essential not only in disaster risk reduction but also in organising immediate disaster response and emergency management.

For example, parts of the USA and Turkey (amongst other countries) have developed community emergency response teams for undertaking immediate tasks after major emergencies in their own communities. Groups of citizens trained in basic but lifesaving rescue and medical assistance provide crucial and swift response after major emergencies. Additionally, they can serve in leadership roles, such as knowing where particularly vulnerable people live and ensuring that those people are taken care of during evacuations. The importance of communities as part of the solution and as a resource in emergency management must always be emphasised, especially in terms of neighbouring teams assisting areas that are experiencing an emergency.

Armed forces

In some countries, response to crises and disasters triggered by natural hazards is managed by civil defence or civil protection departments dominated by armed forces personnel. The involvement of the military in emergency response has led to many debates regarding ethics and credibility, especially when the armed forces of one country work in emergency response in another country, such as the US military assisting Aceh after the 2004 tsunamis and Haiti after the 2010 earthquake.

Another example is Spain's UME (Emergency Military Unit) which was created in 2005. Although it is part of the armed forces, weapons are not included amongst its equipment. UME intervenes in fire-fighting operations, search and rescue and floods, providing operational and logistical support to civilian teams in major emergencies. Armed forces are also often used to support law enforcement, in securing evacuated areas, maintaining curfews and protecting the locations to which people are evacuated.

Again, ethical issues arise, especially if the military is not trusted in some locations. There could be tensions if, for example, evacuees or their property are suddenly surrounded by armed forces as part of 'protection' or 'securing the area'. In some emergency management situations, involvement of the military and civil–military co-operation might not assist the situation (e.g. Weiss 2005) (see Chapter 60).

The tendency observed over the last decades in industrialised countries and in some other parts of the world is to assign disaster management to civilian authorities rather than the military. Conversely, many countries in Latin America and Asia still assign disaster management partially or completely to the military, and even where civilian authorities are in charge, it is often not full civilian control (Alexander 2002a).

The organisational bases of emergency management

Emergencies at different scales

Organisations responding to emergencies on a daily basis can be divided into two broad categories regarding their approach to large-scale emergencies. Some organisations address the problem as statistical, constituted by a sudden increase in needs. Subsequently, the response should be based on a corresponding increase in resources, but the organisational structures do not change; they apply the same system that is used normally. That is, large-scale emergencies are considered as a quantitative problem. For example, many emergency services plan their disaster response thinking only to obtain and send massive numbers of ambulances, rescue units, civil protection resources, etc., regardless of specific needs for special co-ordination or communication.

The other category includes organisations addressing large-scale emergencies and disasters as special situations, events that need something more than a mere increase in the responding units. Under this approach, major emergencies present distinctive problems and therefore require a different organisational structure of the responding institutions than that employed in more common emergencies. Disasters are thus considered as a qualitative problem (Quarantelli 1983). For example, London's emergency plan (LESLP 2007) covers, amongst other issues, co-ordination with the military and different government agencies which do not regularly respond to emergencies. It considers not only the increase in the resources that may be needed during an emergency (the quantity), but also the different type of organisations that may respond and the specific characteristics of large emergencies (the quality).

Another example of this differentiation is found in the case of the serious riots in north London, UK, in 1985, which led to the death of a police constable. This event made senior police officers realise that their usual rank-based command system was inappropriate for managing larger emergencies. A mere increase in the responding resources was not enough to guarantee satisfactory management of the situation. A specific structure, a qualitatively different structure, was designed – the gold, silver and bronze system – that is still used today (see Box 42.1).

Table 42.2 summarises the main differences between day-to-day emergencies and larger, less common emergencies, illustrating the necessity of different emergency management approaches (see also Moore and Lakha 2006). A small-scale emergency might be a landslide hitting a single house or several houses, or a single street being flooded. A large-scale emergency would have hundreds or thousands of residences affected.

Emergency management systems design factors

An emergency management system could be defined as a structure built of principles that guide decision-making and resource management in emergency response. It is a simplified overarching working procedure that is materialised, developed and implemented though more detailed emergency plans (e.g. LESLP 2007; Government of Samoa 1997). While many countries, cities and institutions have emergency plans, not all of them have an emergency management system.

Table 42.2 Comparing emergencies at different scales

Small-scale emergency	Large-scale emergency
Interaction with known individuals	Interaction with strangers
Familiar tasks and procedures	Unfamiliar or unusual tasks and procedures
Adequate radio communications	Radio communications saturated
Predominantly internal communications	Internal and external communications
Communications predominantly with local media	Communications with local, national and international media
Adequate organisational infrastructure to manage the responding resources	Organisational structure too overwhelmed to manage the responding resources; organisational resources and other assistance from outside required
Familiar organisations	Unfamiliar or new organisations
Limited citizen participation	Broad, generalised citizen participation
Services responding within their normal geographic remit	Services cross boundaries

Source: Partially based on Auf der Heide 1989

Box 42.1 Decision-making: the gold, silver and bronze system

The gold, silver and bronze system establishes three levels of decision-making for each emergency service that responds to large-scale emergencies: strategic (gold), tactical (silver) and operational (bronze) (Cabinet Office 2003). This system delivers command (vertical management objective) and co-ordination (horizontal management objective), describing a functioning, generic structure with general rules being applied. Variations emerge in the specific deployments of the system which do not affect its essence and main purpose: to achieve control and co-ordination of the response, and ultimately of the emergency.

The strategic level (gold) is the highest level of management, activated only during the most serious emergencies. It consists of those at the highest level of responsibility for the services involved in the response. The mission at this level of command involves overall, integrated management of the emergency and post-emergency recovery. Each service has a gold manager who exercises the highest level of responsibility for decision-making. The meeting place for managers at this level should be in a safe location, distant from the emergency. A manager at this level should never be present at the location of the emergency, since at that location they will neither be able to meet their responsibilities nor co-ordinate with other gold managers.

The tactical level (silver) is the next level of responsibility and applies the strategy designed by the gold level. There is only one silver leader for each service and for each incident (where there is more than one ongoing incident, such as multiple locations flooded by the same storm). At this level, different variations may be found in the actual deployment of the system.

The operational level (bronze) refers to management of operations during the emergency and involves executing instructions dictated by the tactical level (silver). Bronze-level commanders are deployed at the emergency scene since they are the operational team leader in close and constant contact with responders from their services. It also involves adequate co-ordination with the other bronze leaders of the different services involved.

An additional level of decision-making can be added on top, and that would be the political level. Sometimes it is called the platinum level. In the UK this layer of command is developed by COBRA (Cabinet Office Briefing Room A) and it is usually chaired by the prime minister. The idea of political management, above and separated from the operational, tactical and strategic management, should be equally applicable to other countries.

The political-level functions should be focused on those decisions that exceed the capacity of the professionals involved. Examples are whether all air and rail traffic outside the affected area must be interrupted, whether other institutional or public activities not directly affected by the incident should be interrupted, and whether it might be necessary to mobilise further financial resources in addition to those previously tagged for emergencies. It is not uncommon that politicians and elected officials are involved in the actual management of the emergency, sometimes even at the scene of the incident, breaking the barrier of the political management, interfering with tactics and operations and, contrary to their intention, causing problems for professional managers (Alexander 2002b: 32–3).

This is made clear when a jurisdiction has numerous emergency plans for different hazards, all describing different emergency response structures and management systems. In fact, countries with commonly agreed and widely accepted emergency management systems are scarce, while an internationally agreed system is not in place.

Additionally, Drabek (1986) differentiated between emergency planning as a product and as a process, yet many locations still view it as a packaged product rather than as a long-term, ongoing process. According to Drabek (1986), some put the emphasis on producing plans as paper documents which are perhaps valid for some time, but usually never rehearsed or reviewed. Others emphasise creating a process – a system of periodic reviews and scheduled meetings of appointed planning committees which test plans, analyse performance in real events and implement changes and necessary modifications.

Irrespective of the emergency management system selected, its design should include several characteristics that are detailed here: simplicity, shareability, flexibility and inclusiveness.

Simplicity

An emergency management system is usually applied in confusing, complex and uncertain situations. Therefore, any system must be as simple as possible. Structures need to be clear and simple to follow. Roles should be well-defined. The command relations amongst the different responders must be seamless. A lack of clear protocols on how the elements relate to each other, or a complex system where it is difficult to ascertain roles, contributes to the confusion in an already complex situation.

Shareability

Some services organise their response to disasters independently of other services that will almost certainly also respond to the situation. Some refer to this as 'Robinson Crusoe syndrome' (Auf der Heide 1989) as they behave as if they were alone on a desert island. Any emergency management system should take into account the best way to co-ordinate and share services and organisations, considering which elements are common to other institutions. This will help to ensure that they are understood and shared by other organisations.

For example, a triage system is a procedure to prioritise casualties for treatment and transportation to more advanced medical care, such as hospitals. If used by only an emergency medical service, but not shared by other emergency responders, then errors and confusion can result regarding the meanings of colours and tags that indicate the degree of severity of injuries.

Similarly, before an emergency it is essential to have agreements regarding the common use of radio frequencies or any other communications system, along with coded signals. Otherwise, the sudden convergence of diverse emergency services with distinct radio communication systems will cause additional organisational problems amongst those seeking to assist. Even if there are common frequencies or technical systems, the lack of an agreement regarding the use of common terminology or codes amongst different services can lead to messages being misinterpreted.

Flexibility

An emergency management system must be flexible and adaptable to circumstances while maintaining a basic, functional structure. Individual incidents should not be seen as static, but rather as part of a dynamic process. During the course of an emergency, conditions change and the response must subsequently be flexible.

For example, in the initial stage of an emergency involving mass casualties, most medical resources will be devoted to classifying the victims according to their injuries and condition (i.e. triage) and providing life-saving first aid. After that task is completed, resources must be redeployed to on-site treatment and decisions about transportation to hospitals. A similar pattern occurs in the medical care facility, redeploying resources from the reception of casualties to surgical treatment and ward care. Reassigning fire-fighting units from rescue to fire suppression, or the opposite depending on the changing needs of the situation, is another example of the dynamic characteristic of major emergencies. Management systems must be designed to respond according to this changing scenario.

Inclusiveness

As mentioned regarding stakeholders, a disaster will involve numerous actors, not just emergency services. The more prolonged the emergency, the greater the possibility for increased participation of other parties, especially those from farther away. The emergency management system must have the capacity to include details concerning later arriving responders so that overlapping activity is avoided and good communication is maintained.

Existing emergency management systems

Incident command system or incident management system

The incident management system (IMS), also known as incident command system (ICS) (Christen and Maniscalco 1998), was born in 1970, following a faulty and chaotic response to a series of forest fires in California, USA. This led to the launch of one of the first modern emergency management systems specifically designed to respond to large-scale emergencies.

Initially designed to co-ordinate fire fighters' response to forest fires, its simplicity and versatility led to its application to most emergency services throughout the USA (Irwin 1989). Mexico, Australia, New Zealand and the UK have since implemented similar systems (Hodgetts and Porter 2002), but, despite local variations, they all apply principles of a unified command system to avoid duplication and confusion. In spite of the system's capacity for adaptation and its application in several countries, it cannot be said that it has been fully accepted as an international standard or that it is widely used throughout the world or across large regions like the European Union (EU). Many countries and organisations still rely on improvised emergency response systems or just keep the structures that their emergency services use on a daily basis to respond to large-scale emergencies.

The basic structure of an ICS/IMS includes an incident commander and different sections into which response tasks are divided, namely operations, logistics, and planning and administration (see Figure 42.1). Operations cover 'what needs to be done?' The operations section assumes responsibility for most of the on-the-ground tasks related to the emergency, such as rescuing and evacuating people, suppressing fires, dealing with hazardous materials and cordoning off areas. Logistics deal with 'What is needed?' This department is responsible for meeting the human and material resource needs presented by the operations department, such as people with certain skills, radios, water, foam, vehicles, fuel, bandages and triage tags.

The planning section is responsible for collecting precise information about the emergency, determining which units and resources are involved in the response, the state of those resources and data about the emergency. If possible, emergency evolution, plans and maps should be covered along with control over all administrative and financial documents generated by the

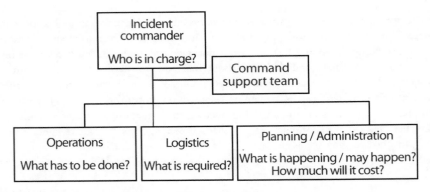

Figure 42.1 The basic structure of an incident command system

incident. Planning can answer the question 'What has happened, what is happening and what might happen?' Finally, the administration department, activated only in large-scale operations, will manage the files, documents, receipts, contracts, payments and all administrative arrangements. It covers the questions 'How much will it cost? What human resources are needed? What administrative arrangements are needed?'

In directing these sections, the incident commander's main objective is to direct and assign primary functions, such as specific operations and specific logistics. Her/his mission can be summarised in response to the question 'Who is in charge of the situation?' The incident commander must establish the intervention's overall objectives and designate the priorities, in addition to ensuring that the different sections work together in a well co-ordinated manner. An additional task could be managing a co-ordination team that will have responsibilities such as:

- Safety: responsibility for work safety for all those working on the emergency.
- Public information officer: the link with all communication media and the public.
- Liaison: links with agencies and organisations in the field other than emergency services, such as the armed forces or private institutions.

Due to its versatility, ICS/IMS has also been applied in other areas like hospital major incident response, where it is known as the Hospital Incident Command System or HEICS (San Mateo County Health Services Agency 1989; Arnold *et al.* 2005).

Other common systems

Variations of all systems are found around the world. New Zealand, for example, uses the Co-ordinated Incident Management System (Government of New Zealand 2005). Nepal has been in the process of developing a National Integrated Disaster Response System (NSET 2008). In Canada, the province of British Columbia emergency management structure has developed and adopted the British Columbia Emergency Response Management System (Provincial Emergency Program 2000). They have adapted the fundamental principles of ICS/ICM to their specific characteristics and needs to produce a more suitable system. For example, while some countries may have several police departments, others will manage several ambulance services. In each case, local variations will be observed but their emergency system still operates under the same general principles of avoiding confusion and lack of control of resources.

The deployment of the gold, silver and bronze system is also not standardised. Some services deploy their silver commanders at the scene, while others argue they should stay in the control rooms of their respective services, in frequent contact with other silver managers also located at their respective control rooms. These differences should be regarded as minor, and the general understanding of this level as underneath the gold level and supervisor to the bronze level remains.

Emergency operations centre

The emergency operations centre (EOC) is an essential element in emergency management. The EOC description developed in this section is generic since the exact nature and structure of EOCs will vary. This is an attempt to describe it as representative of what is likely to happen in a typical EOC, under this denomination or under a different name.

The EOC is a multi-organisational meeting point where top decision-makers gather and where all elements, organisations and resources can be co-ordinated. In some systems, the EOC is the decision-making level below the incident commanders in the field (Alexander 2002b). An EOC is meant to be established by the local government as the first responder to an emergency, and should be the secure hub of communication and decision-making. In the case of large-scale emergencies, EOCs of further responsibility could also be established at provincial, state or national levels. Although a strict parallelism may not be appropriate on all occasions, the EOC usually hosts the equivalent of the gold command level.

An EOC may or may not be included in an emergency plan, but most responses to major emergencies end up creating a meeting point – a location in which top senior officials from the responding services make decisions, to be communicated to their respective services. The difference between a properly planned and structured EOC – with drills, disaster exercises and discussed procedures by everyone involved – and an improvised EOC tends to be that the latter could end up being incomplete, where not all organisations involved in the emergency are included. Even worse, multiple, ad hoc EOCs may emerge, all of them managing part of the resources, but without a clear picture of the whole situation and without co-ordination.

An example is the improvised EOCs, as witnessed by this author, in many cities in Indonesia after the 2004 earthquake and tsunami. Controlled by the Indonesian military, they were attempting to include all involved in the response. Unplanned and improvised in adapted buildings, the difficulties of the task were compounded by lack of planning and inadequate resources for such an emergency, especially given that much of the damage was in Aceh, which was experiencing conflict and thus had previously been closed to many external interventions. Despite those problems, EOCs fulfilled an important role in attempting to organise the response of groups in the area, providing a daily meeting point for everyone involved.

EOC characteristics

The EOC is usually located in a designated building or other location. The EOC receives information from the emergency and elsewhere in order to assign and co-ordinate existing resources.

The EOC is frequently confused with the field command post, but they are different. The field command post is the location in the immediate vicinity of the incident in which emergency services officials deployed in the emergency gather and share information. Bearing in mind that rigid comparisons between systems may not be applicable in all occasions, the field command post, sometimes called the joint emergency services control centre (JESCC), usually hosts the equivalent of silver commanders. In contrast, the EOC is the core of decision-making where the top (gold-level) commanders are housed. The EOC is where the field incident commander reports to and where requirements beyond the capacity of the ICS/ICM organised in the field are requested, channelled and co-ordinated.

The EOC should have a tactical and strategic focus for making decisions about the emergency, whereas the field command post is focused on a more tactical and operational approach. The EOC may deal not only with the emergency itself but also with decisions that go beyond the affected area, such as continuing or cancelling municipal activities, reorganising public transport services to meet new needs, reassigning non-emergency personnel to different activities or communicating with the media (see Chapter 63).

The EOC is also different from an emergency call centre (e.g. the place to which emergency numbers such as 112, 911 and 999 are routed: see Box 42.2). In a widespread emergency the type of assistance sought is usually different to the assistance required in the case of an individual calling. In this case, the required services or resources are not normally found on site. They might include blankets, potable water, heavy machinery, vehicles, lighting equipment, tools and debris containers. The typical emergency number centre will not be able to answer those requests.

The EOC may be established in a fixed location, in a designated room within a building or in a specific building that is or is not normally used for emergency management. That does not detract from the importance of the EOC as a focus for the co-ordination and management of emergencies. In most towns and cities with an EOC, it is not a special room or reserved space. Instead, it is a plan – a shared idea about emergency management that is developed in a location that is predetermined but that can change if necessary. This particularly helps in case the EOC building or location itself becomes part of the emergency, such as being damaged in an earthquake or inundated in a tsunami or flood. That happened due to Hurricane Katrina in 2005 when several EOCs along the US Gulf Coast were rendered inoperable due to flooding (Smith and Simpson 2005).

The designated EOC building might be inaccessible, so improvisation may be needed. If the concept and plan do not exist beforehand, they are difficult and dangerous to create during an emergency. Nonetheless, permanent EOC headquarters frequently exist, often with permanent staffing. When an emergency occurs at night or during holidays, many municipal staff may be a long way away from their desks or they might find their route blocked. As well, EOC staff not in the EOC could have become casualties in the emergency. Consequently, a fully staffed EOC with duty staff who are expected to remain there throughout an emergency will need sleeping places, food, water and hygiene products.

EOC organisation

To achieve efficiency, the EOC must accommodate representatives of all those involved in the emergency management. That includes wider institutions such as the military, the Red Cross/ Red Crescent, NGOs and specialist rescue groups. In terms of internal organisation, several organisational models exist. Virtually all of these coincide in that they have a basic structure that is divided into two 'rooms': operations and political management.

The operations room is where those who are primarily responsible for responding services have a specific workplace with robust communications for contacting their various co-ordination centres and services. Those who are working should share the same physical space and the same room to better co-ordinate operations. This function consists of receiving requests from the emergency site for services that are represented in the room and trying to respond to these needs.

Box 42.2 Three-digit emergency numbers

Ilan Kelman
Center for International Climate and Environmental Research–Oslo (CICERO)

The first three-digit number to activate emergency services was the 999 number in London, UK, which started being operational in 1937. In the late 1960s the USA and Canada implemented co-ordinated emergency services, all of which could be accessed through the single, free emergency phone number of 911. The three-digit number system for emergency and other services has since been extended to much of the rest of the world.

Throughout the EU emergency services are now accessed by dialling 112, although some countries still keep their traditional emergency numbers, such as 091 or 061 in Spain and 999 in the UK; these numbers therefore co-exist with 112. The 999 number also influences former British colonies such as Zambia and Botswana, where 999 is retained as an emergency number. Other countries use their own system. For example, in New Zealand the standard emergency number is 111, while Australia uses 000 and many East Asian countries use 119. In Barbados, fire is 311, police is 211 and ambulance is 511. Although the numbers are different, the final objective is the same: to channel all emergency calls through a free, simple, quick number.

Simplicity has brought its own slew of problems with the switchboard receiving calls for reasons of loneliness, cats stuck in trees, or to ask directions. Over the past several years, many North American municipalities have been implementing 311 for non-emergency calls requiring municipal services, to try to differentiate these services from life-threatening or time-critical calls covered by 911.

Three-digit emergency numbers also experience problems during large-scale emergencies when the system can be overloaded with calls. For major emergencies, such as a city-wide earthquake, many jurisdictions advise their citizens not to use the emergency number, because these services will not be available. Instead, they suggest that everyone should be prepared to be on their own for seventy-two hours after a major event, although sometimes that is extended to one to two weeks.

The advent of mobile phones brought another challenge to three-digit emergency numbers: location identification. When an emergency number is called from a landline, the address associated with that line is usually available immediately to the operator. That way, even if the caller is impaired, panicked or does not know their location, the emergency services can still be dispatched to the correct address.

Mobile telephones do not have a fixed address. This can lead to emergency calls being routed to an emergency service far away, sometimes even in another country, especially when near an international border. As well, if the caller does not know where they are, delays can occur in the emergency services finding the appropriate site. Some advocates suggest installing GPS trackers in mobile phones which are activated only when an emergency number is phoned. Even that is too much for some extreme civil libertarians who object to any form of tracking of their mobile phone as an invasion of privacy, even if it might save their life.

The political management room is the room where those who are politically responsible make decisions that go beyond the emergency's immediate response. These decisions have an impact on sectors that are not immediately affected by the emergency but require immediate decisions.

An essential part of the EOC is to provide public communications to all those not affected by the emergency. The interest among the media and the general public will be extreme, and a designated public information person should channel that information. This will also assist in keeping media away from places in the emergency site where they might be in danger or might endanger others.

Multiple EOCs

If more than one emergency occurs in the same timeframe, there may be several command posts in the field, one for each emergency. Yet for co-ordination, there can be only one EOC within the same territorial unit. When several jurisdictions activate their EOCs simultaneously, there must be an overriding EOC that co-ordinates the activities of all the other EOCs. Thus, the launch of various EOCs in different locations would result in the launch of an EOC at the county, provincial, state or national level (depending on the size of the jurisdictions).

The higher-level EOC co-ordinates requests for assistance received from the EOCs across the affected area, along with offers of help from the other EOCs (see Figure 42.2). For even larger-scale emergencies, if various county/provincial/state EOCs were activated, then it would be necessary to activate an autonomous or regional EOC and so on until a national EOC is reached, if necessary. National-level EOCs should be co-ordinating with other national-level EOCs in the case of a multi-country emergency.

That sequence is not always followed. Chan et al. (2006) describe how on 21 September 1999 an earthquake disaster struck Taiwan killing several thousand people. Within minutes, the national government activated the National Emergency Operations Centre and the National Disaster Prevention Centre. The following day, EOCs were activated in most affected areas. Chan et al. (2006) go on to detail co-ordination, communication and resource allocation problems throughout the emergency response.

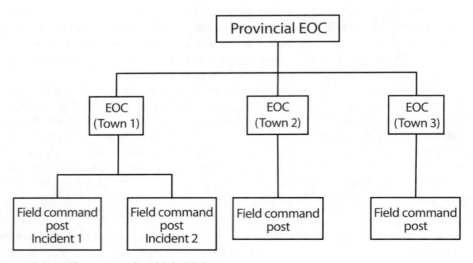

Figure 42.2 An illustration of multiple EOCs

Public relations preparedness and community liaison

The will to help after major emergencies is a natural reaction (Wenger and James 1994), but spontaneous efforts must be channelled and co-ordinated to avoid problems caused by well-intentioned actions that become counterproductive. Self-deployment into emergency sites is particularly discouraged, but that is also under the assumption that a full and competent response is being carried out by local and wider authorities, which is not always the case. In fact, in many forest fires, villagers are seen helping the fire brigades in their operations or providing support through food and water. After earthquakes, most of those rescued from buildings are brought out by people in the immediate vicinity.

Self-deployment can be troublesome because people untrained in rescue or medical skills can put themselves and others in danger. Unwanted donations of goods require human or material resources to manage them. Survivors and friends and relatives of those affected require privacy that can be invaded by onlookers or media. Dealing with all those aspects can decrease the effectiveness of emergency management.

One way to overcome such challenges is to involve communities in the emergency planning process, indicating the roles that they should and should not play. Being part of regular emergency exercises, evacuation drills and citizen training in basic emergency skills increases the sense of belonging, ownership and responsibility of emergency procedures before, during and after an emergency. Blood donations, logistical support, preparing meals and arranging sleeping facilities or shelters for evacuees and emergency management personnel are examples of activities that may be included in the emergency planning and management. Such previous work should assist in providing properly channelled aid from the public during emergencies. Good communications management from the EOC during the emergency will increase the chance of public co-operation by working through the planned arrangements.

Conclusions

While the quick control of a major emergency can be difficult to achieve, the adequate management and control of responding resources is not. Internal control of the response is the compulsory stage to achieve the next step of bringing the emergency under control and that should be the objective of emergency management systems.

Emergency planning must be implemented in order to obtain good results in emergency response. As noted by Drabek (1986), the emphasis must be placed on the creation of a process, not on the delivery of a product. Planning and preparedness must be the basis for adaptation, flexibility and improvisation in emergency response.

43

From damage and needs assessments to relief

Claude de Ville de Goyet

RETIRED, DIRECTOR, EMERGENCY PREPAREDNESS AND RELIEF CO-ORDINATION PAN AMERICAN HEALTH ORGANIZATION, WHO REGIONAL OFFICE FOR THE AMERICAS, BELGIUM

Introduction

The term 'relief' (or humanitarian response) covers the immediate response as well as the early recovery. Since the hazard has occurred, the term 'risk' (a probability) will not be used in this chapter, nor will there be consideration of secondary hazards such as rain and flooding or winter storm, which might complicate needs assessment and relief, or put relief workers and disaster survivors at increased risk.

The impact, still unknown at the early stage, will depend on the nature of the hazard as well as the community vulnerability and level of preparedness. When the impact exceeds the capacity of the community and requires 'external' assistance, the term 'disaster' is used. Under this relative definition, an incident or minor crisis in a well-organised and developed country may well be a major disaster in a small developing nation.

The main challenge in organising relief is often the lack of accurate and reliable information. Information management is a prerequisite and even the essence of effective disaster management. Several days or weeks may pass before a detailed view of the impact, and populations who are most at need, is known. That is particularly the case in disasters resulting from sudden-onset hazards which cause immediate life-threatening damage. Assessing the impact of an earthquake or cyclone is a matter of greater urgency and is a more demanding challenge than evaluating the result of a slow developing drought, flood or locust infestation.

This chapter will focus particularly on assessment and relief following sudden-onset hazards requiring massive external assistance, a situation occurring most frequently in least-developed countries. Conflict disasters will not be covered, but are covered in Chapter 7.

Types of assessment

There are many types of assessments following natural hazards according to their objectives and the information required. The main ones are:

- Damage (or impact) assessment that estimates the immediate physical and social impact of the hazard.

- Needs assessment that determines what is required or missing to assist the different groups of affected population.
- Economic valuation that places a cost on the direct and indirect impact on the various sectors.

Damage assessment

A damage (or impact) assessment is the most basic and common form of assessment. It offers an indicator of the severity of the impact: Is it a minor disaster or a major catastrophe in human terms?

Why?

Estimating the severity of the impact informs humanitarian actors and decision-makers in order to mobilise resources according to the magnitude of the disaster and the reaction of public opinion. Preliminary figures will immediately trigger appeals from the major relief agencies.

How?

The nature of data varies little according to the type of hazard (see Box 43.1). The basic indicators are the number of persons affected, dead, injured and homeless or displaced. The number of dead is the most influential data to trigger the immediate response.

Box 43.1 Main indicators for damage assessments

- **Number of persons affected:** Either those directly impacted by the hazard or more frequently the population resident in the administrative or geographical areas affected by the hazard event.
- **Number of dead:** May be the compilation of actual body counts or the result from local subjective estimates including missing persons.
- **Number of missing:** In organised societies, based on reports from relatives or neighbours; otherwise, based on the number of people believed to have lived or to have been present in the destroyed area or building. The practice of pooling dead and missing statistics is misleading. Duplicate counting is increasing as most bodies recovered after a few days cannot be identified.
- **Number of injured:** Most useful indicator when broken down according to type and severity of injuries. Figures are often inflated due to reporting from a large number of sources and levels of care (from first aid to hospitals). Earthquakes often cause the highest number of wounded with a high proportion of severe injuries, while floods, storm surges and tsunamis may kill many with few injured survivors.
- **Number of homeless and displaced:** Usually by estimating the number of houses destroyed and/or the size of temporary settlements. The two groups may be quite different.
- **Number of houses collapsed or damaged:** From local reports and generalisation of remote sensing estimates.
- **Number of health facilities destroyed or damaged (not operational):** An indicator increasingly used as a result of the global campaign for safer hospitals.
- **Other indicators:** Many sectoral indicators (e.g. water and sanitation, public health, food and agriculture, communications, energy, logistics, and collapsed infrastructure such as schools, bridges, roads, ports and airports, etc.) may be included.

The value of these indicators is in their comprehensiveness (an authoritative compilation aiming to present the full picture of the disaster). In large-scale events, the reports from administrative units will be the primary source of information to be complemented by satellite or aerial surveys and field visits. When government structures totally collapse, as in the Haiti 2010 earthquake, the international community is mostly unable to provide early data.

Who?

The national civil protection or disaster management agency with the technical support of line ministries is the prime and official source. Few other institutions can offer comprehensive and authoritative damage assessments. However, the national authorities might not always be entirely straightforward in their reports or have access to the most up-to-date or relevant information. The United Nations (UN) and Red Cross/Red Crescent systems offer a validation mechanism thanks to their cross-sectoral expertise or widespread community presence. The best official figures are usually the result of a consultation with the input of various sources. Non-governmental organisations (NGOs) and community organisations will contribute to ensure that their constituency is duly included in the tally and that the vulnerable groups or minorities are covered.

Issues and lessons learned

1   Reliable and geographically inclusive damage assessment takes time. Reported figures of losses increase over time as the magnitude and extent of the disaster is assessed. The public and the donor community often express impatience with the length of the reporting process. Estimates or guesses including those from 'scientific' models or worst case scenarios are, however, not a good alternative to assessment. They are also usually grossly overestimated.
2   In poor countries, the losses are high in terms of lives lost: More than ninety per cent of all deaths caused by disasters have been occurring in developing countries (UNISDR 2004), while in more developed countries the losses have been mostly economic (Kenny 2009). In democratic and decentralised systems, the number and autonomy of institutions (private, public, community) may complicate the assessment. In others such as Haiti, weak local institutions and poor governance are a major obstacle.
3   'No data' does not mean 'no impact or damage'. Early figures may well underestimate the gravity of the impact.
4   There are no universally accepted and applied definitions or criteria for collecting data. For instance, injuries may cover only those treated by the health care system or may include anyone who is transiting through a first aid station. Where those definitions exist, they are often not known locally.
5   The highest (most impressive) numbers, regardless of their reliability, will usually prevail. Consolidated (and usually lower) estimates resulting later from a systematic census are rarely published in less-developed countries. A notable exception is the official number of deaths following the Bam, Iran earthquake in 2003. Following a census conducted to determine the exact number killed, the number decreased from 41,000 to 26,271.
6   The number of dead is a powerful trigger, but a poor guide for relief decisions. Public response is conditioned, in great part, by the number of sudden casualties. The dead, however, do not require assistance. The Nevado del Ruiz volcanic eruption in Colombia in 1985 killed over 23,000 people, leaving approximately 10,000 survivors who were mostly uninjured. Similarly, the 2004 Asian tsunami caused an impressive number of dead. Both triggered a disproportionately large influx of life-saving relief and emergency funds leading, according to many, to overextended relief and 'early' recovery phases.

Needs assessment

Carrying out a needs assessment

Why?

Damage or impact assessment gives an idea of the magnitude of the disaster, not what will be needed where and when. A more action-oriented information tool is required. Needs assessment identifies critical requirements (life saving and others) that cannot be addressed locally, and can prioritise action and monitor the external response to avoid duplications or gaps. External response may be from neighbouring communities, provinces (or other sub-national jurisdictions) or national or international levels. The multi-sectoral scope of the potential needs, the competing interests and priorities of the actors (each with their own mandate or niche, be it a sector or a particular group of beneficiaries) and the fast-evolving situation make these objectives particularly important but elusive.

How?

To identify specific unmet emergency needs to which external actors may respond, the assessments should:

1 Differentiate between chronic shortcomings due to poverty and hazard-caused needs. Costly and short-duration emergency measures are not often best suited to addressing long-standing water shortages or endemic levels of diseases that were not directly affected by the natural event. Assessment requires knowledge of local conditions prior to the triggering hazard, a skill that is not always present in most of the assessment teams. Pre-established 'minimum standards', such those developed by Sphere Project (2004), are no substitute for the lack of baseline data. Reliance on those global standards for assessment leads to unsustainable levels of emergency response and crying disparities between attended victims and those not directly affected but living permanently far below those 'minimum' standards. This is not to advocate that relief and recovery objectives should be limited to a return to the unsatisfactory pre-disaster status, but that standards be adapted and made locally compatible with sustainable recovery goals.

2 Compare health and welfare status/needs of affected versus non-affected communities – an essential but rare practice. This act can lead to surprising results. Following the storm surge in Bangladesh which killed around 500,000 people in 1970, Sommer and Mosley (1971, 1972) uncovered a significantly better health condition a few weeks after the cyclone among survivors than in those villages not affected by the disaster. This was explained by the fact that surviving a storm surge (or a tsunami) is not merely a matter of luck (random), but also of fitness. Those who were too weak, old or sick to swim or cling to a tree for days died.

3 Take into account local capacity: external actors are supposed to complement or palliate acute shortcomings of local resources. Assessments should not merely report the humanitarian caseload but must estimate the critical gaps, taking into account the status of the local services and the influx of national volunteers. Failure to do so, which occurs too commonly in high visibility disasters, results in humanitarian assistance competing with, rather than complementing, the local efforts.

4 Consider the minimum time needed for external response when assessing short-lived needs. Primary care of trauma is a glaring example. Undoubtedly, rapid assessment may identify a

need for first aid and primary surgical procedures following an earthquake. However, in the case of recent disasters in Iran (2003), Haiti (2004), Indonesia (2004) and Pakistan (2005), foreign field hospitals (FFH) arrived between three and eight days after the impact (Von Schreeb *et al.* 2008), which is too late for emergency trauma care, especially when, after the Iran earthquake, nationals were capable of providing initial medical assistance and evacuated over 12,000 wounded before the arrival of externals (Abolghasemi *et al.* 2006). Most foreign medical personnel also leave too soon to serve as substitute for the collapsed facilities. PAHO-WHO (2003) developed guidelines for an effective use of FFH. As noted by Von Schreeb *et al.* (2008), none of the forty-three FFH surveyed complied with those guidelines. Still, those facilities are routinely requested and remain often underutilised. The lessons learned as early as in the Guatemala earthquake in 1976 (de Ville de Goyet *et al.* 1976) are unheeded. The cost-effectiveness (the daily bed cost is over US$2,000) and the immediate benefit for the affected population of belatedly rushing first aid supplies or ambulances across the globe are, at best, debatable.

5　Anticipate needs. Looking at yesterday's short-lived needs is not useful. Tomorrow's needs can, however, be projected or anticipated as disasters follow a rather predictable course (PAHO-WHO 2000), from emergency medical care to water/sanitation, food and psycho-social assistance, to temporary housing and concern with livelihoods and full recovery. This means that the assessment team should be familiar with this pattern of changing priorities over time.

　　At times, anticipating needs has been abused. Predicting massive secondary effects attracts attention and funds. Unfortunately, examples abound in the field of communicable diseases. Although the risk of major outbreaks of disease following disasters is demonstrably low, unsubstantiated prediction of imminent catastrophic epidemics leads to unjustified mass vaccination (TEC 2006c), improper handling of dead bodies (Morgan 2004; Morgan and de Ville de Goyet 2005) and wasting resources. The challenge is how to keep the projection objective and honest when generating fear is so profitable.

6　Balance emphasis on multi-sectoral and disciplinary assessments. Need can be extremely varied and technically detailed. Every discipline (from the sanitary engineer to the fishing industry expert) will require specific information to prioritise its response. Inter-agency needs assessments that force compromise on the total number of indicators may be of limited operational value for specialised NGOs or UN agencies. There is no surprise that agreeing on a universal standard protocol (a list of indicators) for rapid needs assessment is a difficult challenge. Therefore, agency or topic-specific surveys will complement inter-agency needs assessments. In practice, the narrower the topic or discipline (e.g. a food/nutrition survey or communicable diseases surveillance), the more likely the information will be effectively used. Each actor seems to trust only its own data, with the inconvenience of a multitude of simultaneous surveys (TEC 2006b).

7　Consider the time dimension. The answer is: the earlier, the better. A 'quick and dirty' assessment is often required. However, as needs are being attended to and evolve, new priorities or threats emerge and gaps (e.g. groups of populations or unattended needs) are identified. Assessment of needs is, in fact, a continuing process if it is to be used for co-ordination of the response. The agency with the most comprehensive and updated vision of the unmet needs is the best equipped to co-ordinate.

8　Be clear whose needs and priorities are driving assessment. In principle, the external response is driven by the needs and priorities of the affected households, but also those of the host communities in the case of displacements. In practice, assessments of needs by humanitarian actors and their resulting priorities are strongly influenced by what the specific agency can offer. An assessment by an NGO specialised (or funded) to build houses will not consider unrelated needs of the targeted beneficiaries. Similarly, assessments of needs by bilateral donors or UN agencies will accommodate a limited range of supplies and services. Specialised assessments are offer-driven, leading to a degree of frustration among intended beneficiaries.

9   Consider the needs of mobile populations. A condition for effective use of findings is some stability of the population surveyed. Following Hurricane Katrina, the high mobility of the affected population greatly limited the utility of the assessments (Rodriguez *et al.* 2006). Following the bombing of Afghanistan in late 2001 and early 2002 and the collapse of the Taliban regime, mobility issues also arose regarding assessments due to Afghans moving internally and returning due to the conflict, overlaid with people who were displaced and returning due to the drought (Ashmore *et al.* 2003).

10   Carefully monitor the logistical chain. Identifying end users' needs may not be sufficient to determine the best course of action. Lack of medical supplies at primary health care facilities does not mean necessarily that requesting or donating medicines is required. Excess supplies may already be clogging the airports and ports of entry or the warehouses of government or agencies. Assessment of local needs should be completed by an analysis of the logistical chain as the problem and the solution may then be in streamlining the distribution (see Box 43.2 on SUMA/LSS).

Translating needs assessment data into action

How?

Following the 2004 Asian tsunami, an international evaluation was conducted to assess the effectiveness of the post-tsunami needs assessments. It produced a profusion of data, figures and statistics (TEC 2006c). The amount of data testifies to the commitment of the international actors to assessing needs. Nevertheless, the evaluators were puzzled by the lack of a clear sense of direction conveyed by the reports. The question 'so what now?' was usually left without answer.

Box 43.2 Supplies management system: SUMA-LSS

After a few days, relief equipment and supplies often pour into the affected country and areas. Local or international donations of dubious usefulness compete with critical supplies for scarce logistic capacity. Requests and needs for new supplies are not checked with existing inventories, for lack of information on what is available or in the pipeline.

The Pan American Health Organization (PAHO) developed a tool called SUMA to address this information and co-ordination gap. SUMA is an inventory and classification system for all incoming or pledged supplies regardless of source and ownership. SUMA was adopted rapidly by national relief authorities. An inter-agency successor, the Logistic Support System (LSS), is now used globally in major disasters.

Although developed to match identified needs with existing stocks, SUMA-LSS is often adopted more for its accountability value (transparency in local management of donations) than for its contribution to the assessment of needs and co-ordination of response.

Available at www.lssweb.net (accessed 17 January 2010).

Experts and assessors seem to be too careful to venture far into 'speculative' conclusions by suggesting or recommending specific priorities or courses of action. By doing so, they share the responsibility when decision-makers choose the most popular path rather than the most effective. This shortcoming, observed in most large-scale disasters, can best be illustrated by the example of the otherwise effective monitoring of communicable diseases in the aftermath of the major earthquakes or cyclones in the 2000s. Weekly tables of statistics of cases reported by each affected community, camp or humanitarian actor are impressive. They offered a sense of professionalism, but little indication of whether any particular statistics required specific concern or action. Lack of control groups (people in unaffected areas) and, above all, absence of interpretative analysis from experts limited the effectiveness of this data. For instance, an increase of the number of cases reported may be the result of an exponential increase in medical coverage and scrutiny (up to tenfold) with the dispatch of international teams, the adoption of new criteria for reporting (symptoms-based rather than laboratory-confirmed) or over-reporting due to the fear of catastrophic outbreaks.

Missing information is also a problem. Field reports, appeals and press releases generally emphasise what is needed or wanted by the humanitarian actors. When actual needs greatly exceed what is available to meet them, these 'shopping lists' contribute to an effective response. However, the challenge is twofold: to get as much of what is needed locally while also minimising the disruptive flow of inappropriate donations and unsolicited personnel. LSS was developed partly for this objective. Its impact on stemming the flood of unsolicited donations has been limited, suggesting that lack of information is not the key factor.

Only occasionally do reports indicate without ambiguity what is not wanted or needed. Discouraging unwanted donations is perhaps the simplest and most effective change that could be brought to current practice of needs assessments. Discouraging certain types of assistance (inappropriate or in excess of needs) is not an easy task. Field assessment teams may be overruled by headquarters more sensitive to donors or public demands than to impact on beneficiaries.

If possible, a census of beneficiaries should be made. Estimating the aggregated number of affected persons (the task of initial damage assessment) is essential for quantifying needs. Some of those needs of certain groups or households are rapidly attended. Disaggregating the list according to those still in need is a formidable challenge. Duplications and gaps become almost unavoidable. Sectoral government agencies and relief organisations usually maintain separate (and incompatible) census records of their target groups (e.g. farmers, fisherman or a selected community). Knowing who needs what (e.g. at the family or community level) is impossible. The assessment of needs following major disasters could learn from the experience gained by the registering and issuing of identification by the United Nations High Commissioner for Refugees (UNHCR), World Food Programme (WFP) and cash-providing agencies. Sometimes retinal scanning is used to avoid fraud.

Who?

While initial damage assessment is usually the responsibility of one national institution, needs assessments are carried out by almost every sizable humanitarian actor. Teams crisscross the affected area, interviewing overworked authorities and stressed beneficiaries. In addition to bilateral specialised assessment capacities (e.g. Disaster Assistance Response Teams – DART), formal inter-agency mechanisms have been established. The most important are the UN Disaster Assessment and Coordination team (UNDAC) and the Red Cross Field Assessment Coordination Team (FACT).

As suggested in the acronyms, the UN and Red Cross clearly identify assessment as closely linked to co-ordination. Each developed procedures, protocols, standard forms and training programmes for their volunteer members.

UNDAC, the leader in this field, has impressive achievements in validating local needs assessments in medium-scale disasters, particularly when staffed by volunteers from the affected region. Its performance in massive disasters with too many actors and a highly sensitive environment is debated (TEC 2006b, 2006c). Lacking substantial human resources, the managing agency, the UN Office for the Coordination of Humanitarian Affairs (OCHA), often allowed the demanding task of liaising with hundreds of organisations and convening myriad co-ordination meetings that distract UNDAC from its main task of generating new information through ongoing assessment of the needs of the beneficiaries.

FACT, an efficient Red Cross instrument, has a less ambitious objective: to guide the Red Cross System response rather than that of the entire international community. Findings and reports are generally not shared outside the system.

Both FACT and UNDAC are supported by donors and tend to validate and justify external appeals and interventions rather than screen out unnecessary initiatives from their partners. After the 2004 Asian tsunami, FACT's attempts to discourage proposed contributions from participating national Red Cross and Red Crescent societies were largely ignored.

One effective global mechanism designed to assess one single need is worth mentioning. The International Search and Rescue Advisory Group (INSARAG) is managed under the UN's aegis and provides teams specialised in rescue of victims trapped in damaged buildings. Established after the co-ordination fiasco of the search and rescue response to the 1985 earthquake in El Salvador, where too many teams were competing in full view of the media for access to the few collapsed modern buildings, INSARAG adjusts the offer of assistance to an inelastic demand.

Most UN agencies and NGOs also have their own assessment capacity. Staffed by experts with often considerable familiarity with the country and the type of disaster, they focus on their own institutions' expertise and constituencies. Their role is to document sectoral or thematic needs and to promote their agency's priorities, be it water, public health or rural development in the competitive world of humanitarian assistance.

At home, the disaster management agencies of large developed countries (e.g. the USA, Canada, UK) have established sophisticated assessment capacities. These teams are often heavily specialised and especially oriented to technological or terrorism disasters of high complexity but rather limited geographical scope. When faced with a large-scale natural event such as Hurricane Katrina, the systems in place suffered from the same lack of co-ordination, competitive interests, and lack of sharing of data noted internationally.

One key actor, not officially in the business of needs assessment, holds the most influential position on decision-makers. Its assessment of the situation may trigger or derail key funding decisions. It is the international mass media (see Chapter 63). Often with their own access to the disaster site and to the affected population, they will determine whether the event is worth the public's attention and generosity. Political authorities will follow suit by allocating funds to known relief agencies. Only exceptionally have humanitarian actors included independent mass media representatives (as opposed to their own public relations officers) in their early assessments. This is a strategic error which contributed in part to the lack of impact of the inter-agency findings on the process of decision-making of major countries and funding agencies.

Finally, those in need of assistance are often the last ones able to present their own priorities. They are learning, in some cases, to master the new facilities offered by the internet. Local agencies and sometimes affected communities in developed and developing countries are opening websites, listing their perceived needs and actually by-passing the heavy layers between themselves and the public. Conversely, relief agencies are increasingly but still modestly empowering affected people to make their own choices by shifting to a cash or voucher approach instead of the mismatched panoply of kitchen utensils, food items, water container and clothing that often appears. These trends should be encouraged.

Issues and lessons learned

Are policies or funding decisions by governments or donors influenced by the findings of an external needs assessment report? Agency-specific needs assessments play an indispensable role. In addition to justifying participation by the agency, they determine operational details (e.g. where, what, when). During this process involving discussions with partner agencies, potential duplication is usually identified and avoided. General information is shared at co-ordination meetings, but detailed findings will often remain internal.

The inter-agency needs assessments are useful to validate the magnitude of the disaster and to justify the forthcoming appeals for assistance or funds, but that may already have been done by the media, with the balancing consideration being that the media can fuel rumours, unscientific estimates and exaggeration. UNDAC assessments alone are unlikely to influence the type of response and political commitment that emotional coverage by large media organisations can achieve. An opportunity to co-opt the mass media and to work jointly is often wasted.

There are also lessons for local communities. Local authorities are usually marginalised in the needs assessment process. One reason among others is that the bulk of human and budgetary resources usually resides at the national or international level. In non-Anglophone locations, it is also a problem of language: English is the disaster co-ordination language.

Being dependent on external assistance which is not always appropriate to local needs brings up the importance of local preparedness. Many urgent needs are short-lived. Communities are often on their own for the first twenty-four to forty-eight hours or longer after an event. The obvious solution is developing local preparedness, maintaining simple but effective mobile medical units (or field hospitals) and stockpiling the most critical elements (see Box 43.3 on disaster stockpiling). Given the powerful incentive of self-expansion of the world humanitarian community, affected communities will need to re-evaluate the benefit of self-reliance and assert more vigorously their priorities for recovery and reconstruction.

A single centralised database of beneficiaries and their pending needs is imperative, but is difficult to concretise due to repetitive surveys of affected families by different institutions; multiple interviews with representatives from UN agencies or NGOs, of which few may ultimately bring real assistance to the interviewees; incompatible forms and databases; and narrowly specific offer-driven assistance provided. Short of a binding high-level political decision in the affected country, no relief agency is likely to reshuffle its data collection to the format of a central system and share data even with an impartial UN agency. Integrating local needs and perspectives with a centralised database also has challenges.

Simplifying the recovery assistance through increasing reliance on cash donations (when local conditions and markets permit) may contribute to simplifying the assessment of needs and, most importantly, repatriate the decision-making to the affected communities and families. This may lead to a downsizing of external assistance by eliminating those who offer services judged not worthwhile by the beneficiaries who are now in charge of deciding what is in their best interests. That alone would be a positive development.

Box 43.3 Disaster stockpiling

Pros:

- Critical items pre-positioned near vulnerable sites are available immediately.
- Response is local, saving lives in the immediate aftermath.
- Supplies are standardised (emergency kits).

Cons:

- High cost of maintenance, especially when infrequently used.
- Offer-driven: the objective becomes to use the stock and promote the items.
- Politically vulnerable: stimulates demands and misuse.
- Ethical and practical concerns in countries where essential supplies are in short supply in normal times.
- Chance of vandalism and theft.
- Assumption that the stockpiles will be available when needed without contingency planning for their deployment and use.

Practice:

- Bilateral stockpiles in co-operation with donor agencies (OFDA, JICA, etc.).
- Food pre-positioning by WFP.
- Agreements with manufacturers and suppliers (virtual stocks).
- Borrowing from military forces.
- Operational stockpiles by relief agencies: UNICEF, WHO, MSF, ICRC, etc., to be used in their own operations.
- Common logistical stock by the UN Humanitarian Response Depot (UNHRD).

Economic valuation

Economic valuation is basically a damage/impact assessment with a single indicator: the financial cost. Every impact, physical or social, is translated into direct or indirect monetary value and percentage of gross domestic product (GDP).

Why?

During immediate emergency relief, cost is a non-issue because millions of dollars can be spent in search and rescue or airlifted field hospitals to save a few lives, while a few dollars could not be found for life-saving immunisation or essential drugs prior to the sudden-onset hazard. Generous 'minimum' standards (e.g. Sphere Project 2004) are rigidly enforced regardless of the general poverty context and the exorbitant cost per capita of emergency operations.

Cost-effectiveness and sustainability are, however, guiding principles for the recovery and rehabilitation process. Friction between the humanitarian community and long-term development institutions is not uncommon as emergency phases tend to overextend for months or years (as long as humanitarian funding is available). Meanwhile, planning or assessments for recovery/rehabilitation are now starting as early as the second week after the impact. Assessments for both humanitarian response and recovery/rehabilitation have merits and need not be mutually exclusive.

In recovery/reconstruction, sectors compete on the basis of their contribution to the total loss of income and capital. The health and social sectors, the favourite in the human-focused relief response, are often the losers in reconstruction. Dollar values of health or social losses are often 'negligible' compared with those from infrastructure and business. Priority is on 'productive' sectors.

How?

The Economic Commission for Latin America and the Caribbean (ECLAC 1991) has developed a methodology for economic valuation of the impact of hazards. This methodology has been used in every disaster in the Americas in the last decades and is now the global standard. In the health sector, for instance, it assesses the cost of reconstructing a damaged facility with reference to current health care and risk reduction standards. If a collapsed hospital with thirty beds needs now to be replaced by a 100-bed modern facility to accommodate the growing population and the national guidelines or standards, then that is the selected value. In addition to the physical losses (equipment, etc.), it will estimate the income lost by the health services and by businesses (untreated sick personnel or relocation to a place with better health coverage). Economic valuation of deaths is still under study as the topic is both sensitive and complex.

Operational approaches and resources mobilised for the assessment of needs and this economic valuation are strikingly different. While UNDAC is grossly understaffed for the task – it can barely register the existence of the many reports or documents produced by the main actors – the economic valuation project sends inter-sectoral field teams and recruits large numbers of national graduates to scrutinise all reports of damage and to seek out contradictions or gaps for further field study. The result is an authoritative cost valuation which greatly influences the process of reconstruction.

Who?

As can be expected, the international and national financial institutions lead this process. These include the World Bank and the regional development banks. All sectors (public or private) are deeply involved in ensuring that their losses are properly assessed. The challenge for the social ministries, health above all, is tremendous. The survey generally takes places after a few weeks, while the health sector is still overwhelmed with public health and medical demands. Few social sector (health, education, social welfare, youth affairs, women's affairs, etc.) officials are familiar with the concept and methodology of the economic valuation, but weak or incompetent participation will lead to under-evaluation of the economic cost of social losses and a further disadvantage in reconstruction.

Issues and lessons learned

Cost-effectiveness, a factor rejected by the purists of humanitarian response, should be considered as early as possible to provide as much assistance as possible to the greatest number. Participation of social sectors in economic valuation is important for the beneficiaries.

Conclusions and the way forward

The importance of information in disaster management is not questioned. All organisations active in response have invested in intelligence gathering to help them in their operational decision-making. In small or large disasters, such technical findings from their own assessments guide them effectively, when not overruled by politically convenient considerations. Effectiveness is due in great part to the specific and manageable focus (health, school, food situation, etc.) of their data collection and a willingness to act upon their own experts' findings.

The impact of inter-agency, comprehensive assessments attempting to guide other agencies in prioritising their work and allocating funds has been much more limited in large-scale disasters. In the immediate aftermath, coverage by the mass media is much more effective, if not always as objective. A closer collaboration leading even to joint international missions is worth considering.

Is a detailed and updated inventory of unmet needs possible? It is already a considerable challenge in developed countries with strong institutions and may be impractical in least-developed countries with the current disorganised patchwork of international relief organisations. It should, however, be the ideal goal of the relief community. To get closer to this goal, some conditions are required:

- Strong leadership and participation of local institutions (see Chapter 52). Preparedness and capacity building are critical. National authorities (see Chapter 51) are the main ones in a position to demand that the actors take into account the findings of assessments.
- A more substantial investment and commitment from donors, since the human resources allocated to needs assessments are grossly inadequate. Whether diverting a large amount of funding from actual relief to assessment/co-ordination would ultimately improve the situation of the beneficiaries remains to be studied. However, this is likely if the assessment makes response more demand-driven and less emotional and political.
- Data should be disaggregated at the level of affected individuals or households. Disparities in aid are often at this level. A centralised register of affected persons or beneficiaries may seem far-fetched, but given that the international community donated an average of US$5,500 per person affected by the 2004 Asian tsunami (TEC 2006b), a register would have been a negligible investment.
- Relief assistance (and the list of potential items needed) should be simplified, when possible, through the use of cash or vouchers. Households should be empowered, as they are in the developed countries, to prioritise their own needs. Of course, this does not apply to services of collective interest (e.g. public health, water, waste management, etc.).

In summary, improving needs assessment cannot be dissociated from the overall rationalisation of humanitarian assistance. Many expatriate actors are highly professional and offer value for money to the beneficiaries, but some could hardly justify the cost of their presence if their beneficiaries were offered the option to choose. A 2004 Red Cross Report called the relief community 'the largest unregulated industry' (IFRC 2004). Indeed it is, and one where the 'customer' is presently offered very little choice. Without addressing this issue, massive investment in needs assessment may not result in better relief and recovery.

44

Health and disaster

Mark Keim

CENTERS FOR DISEASE CONTROL AND PREVENTION, USA

Jonathan Abrahams

HEALTH ACTION IN CRISES, WHO, SWITZERLAND

Introduction: health and disaster

Saving lives and the protection of people's health are among the most critical objectives of disaster risk management and reduction. 'Saving lives' is often mentioned in the mission statements of community safety organisations, but this phrase runs the risk of narrowing the attention to actions that treat the *symptoms* of the event but do not adequately address the actual *causes* of the health emergency. Disaster risk management as applied to health, hereafter referred to as health emergency management (HEM), optimises health outcomes through actions to assess, avoid and treat risks that occur as a result of the interaction of hazards and health vulnerabilities in the community. Health is not just the responsibility of the health sector: there are many in other sectors, as well as the community, who contribute to the management of health risks and to these health outcomes.

For the most part, the world's health sector's interventions in disasters remain reactive and focused on providing health services in response and recovery. HEM is more than just treating people affected by these events. It is also necessary to explore how these health effects were caused, what strategies can be employed to minimise the health risks and who is responsible for making decisions and taking these actions. A combination of measures is required to manage these risks. Wherever possible, priority should be given to a preventive approach over the more traditional curative approach in order to promote health and avoid the risk of death, illness and injury before health is threatened. It has been said that all disasters follow a cycle including the following stages: prevention, mitigation, preparedness, response and recovery. This depiction of disasters as cyclical in nature may be seen to inadvertently imply an inevitability of disasters. This approach, to a degree, under-emphasises the opportunity for risk avoidance. Thus, the ultimate goal of HEM is to put much more emphasis on the 'prevention' part of the cycle and to break the disaster cycle (i.e. to prevent future disasters altogether).

The health sector could do more to strengthen the health system and the capabilities of personnel for their vital roles in HEM. The health sector could also be more proactive by working with other sectors to reduce health risks before hazards strike a community. There is a substantial challenge in integrating the health sector in mainstream disaster risk management and reorienting the health sector towards a multidisciplinary risk management approach which not only provides the necessary health care and services, but also addresses the actual causes of risk. A multidisciplinary approach is all the more important in the face of future challenges including the impacts of global climate change.

Statistics focusing on how many people have died or the number of people affected do not describe the full extent of the health impact or the human suffering. People also experience injuries, disease and long-term disabilities (Lawry and Burkle 2010) (see Chapter 34). They suffer the emotional loss and psychological trauma from these events. Disasters can stall and set back social development towards, for example, the Millennium Development Goals, including hard-earned achievements in health status. On the other hand, they also provide opportunities to catalyse action on health emergency management, as well as other areas of health policy and practice. When the 2004 tsunami struck Sri Lanka, mental health services were concentrated in Colombo and were extremely limited in the disaster-affected areas. Subsequently, the Sri Lankan government agreed to a national mental health policy which has resulted in psycho-social training for health workers and more mental health units in health facilities (Mahoney et al.2006) (see Chapter 47).

While disasters involve the destruction of lives and property, they also create something important: opportunities to improve safety, enhance equity and rebuild in new or different ways. Ideally those opportunities would be used to produce safer communities with more equitable and sustainable livelihoods for people. By focusing on human vulnerability and the ability of individuals and communities to recover, HEM places the individuals at risk at centre stage and tasks the responsible authorities with enhancing social equity and promoting community cohesiveness, alongside a heightened sense of individual responsibility.

Health impacts of disasters

Communities may be affected by hazards resulting in the following general effects on people's health:

- Increased number of deaths and injuries;
- New cases of disease and disability;
- Increased number of social behavioural problems and psychological disorders;
- Interruptions to the management of chronic diseases;
- Loss of shelter and population displacement, including refugees, missing persons and unaccompanied children;
- Loss of clean water, sanitation and hygiene;
- Possible food shortages and nutritional deficiencies; and
- Illness or injury among response personnel.

Hazards can also impact on the capacity of the health system to provide both routine health services, such as primary health care services and disease surveillance, and those needed specifically to respond to disasters, such as emergency medical services (see Chapter 42). The impact on these services is exacerbated when health care facilities and other health infrastructure are damaged, and when health workers are killed or injured in disasters, or leave areas wracked by insecurity and violence. After the 2004 Indian Ocean tsunami, seventy-five per cent of health workers in Banda Aceh either died or were displaced from their homes (Widyastuti *et al.* 2006). When floods in Burkina Faso in September 2009 closed down the country's main hospital, valuable medical equipment was damaged resulting in interruption to kidney dialysis treatments and diagnostic services (WHO 2009a). Widespread flooding in the USA in 1993 presented multiple challenges to the six metropolitan medical centres when these hospitals lost all public utilities. Health care leaders implemented extraordinary measures in order to maintain adequate amounts of water for laundry, fire protection, cooling, instrument sterilisation, renal dialysis, physical therapy and dietary services (Peters 1996).

The epidemiology of disasters

Disaster epidemiology is the study of the patterns of injury and illness associated with the impact of hazards on communities. It provides useful information on the common causes and effects of different hazards on health. This understanding of hazard impacts helps to anticipate, plan for and develop capacity to manage the health effects of disasters. It provides the ability to predict what types of injuries and other problems could be expected from particular hazards and, importantly, what is unlikely to happen. This understanding will help to avoid wasting resources which might otherwise be diverted to lesser priorities, with the potential for further loss of life and health. While diligent disease surveillance is important, it is also important to recognise the natural history of such disasters and thus prioritise interventions on more common and urgent needs, such as provision of trauma care, safe water, sanitation shelter, psycho-social support and restoration of the primary health care system.

Natural hazard-specific health impacts

Injuries are the leading cause of death in disasters associated with natural hazards, as opposed to deaths from disease, poisoning, suicide or other causes. Geophysical hazards tend to create higher injury/affected ratios, as compared to meteorological hazards. For geophysical hazards, burns, asphyxia and other forms of traumatic injury are the main causes of death.

For hydro-meteorological hazards such as cyclones and floods, the main cause of death is drowning (Malilay 1997). Most injuries (such as nail puncture wounds, lacerations, falls, burns, electrocutions and carbon monoxide poisonings) are sustained during the disaster recovery clean-up phase of the disaster. The main cause of death in tornadoes and landslides of hydro-meteorological origin is traumatic injury sustained during disaster impact. The main causes of death caused by heat waves are heat illness and exacerbations of chronic respiratory and cardio-vascular disease (Bailey and Walker 2007). Drought-related deaths are generally mediated by agricultural, economic and health effects such as malnutrition, poverty, poor sanitation and hygiene, unsafe water, infectious diseases and conflict (Wilhite 1993). The main cause of death during wildfires is injury, mostly burns and smoke inhalation. The public health impact of wildfires may include burn injuries, exacerbations of chronic obstructive pulmonary disease and asthma, and temporary population displacement that results in a need for emergency assistance (Sanderson 1997).

Infectious disease epidemics following disasters are rare and differ according to the community's development level and the endemic diseases present (Toole 1997). Geophysical hazards are rarely associated with epidemics of communicable disease (Floret *et al.* 2006). Hydro-meteorological disasters are uncommonly associated with epidemics of communicable disease. However, outbreaks of endemic diseases have been known to occur, particularly in low-income settings where risk factors include contamination of water and overcrowding leading to transmission of disease (Keim 2010a). Leptospirosis, an infectious disease that affects humans and animals, is a common zoonosis with a variety of clinical manifestations. In the Philippines after tropical storm Ketsana and Typhoon Parma in 2009, cases of leptospirosis surged after flooding, resulting in 3,000 cases and 240 deaths (WHO 2009d).

Preventing the health impact of disasters

HEM encompasses the functions that are needed to prevent, mitigate, respond to and recover from health emergencies and disasters. As such, it combines the elements of health systems and disaster risk management. In addition, the health sector has specific concerns such as safe health facilities and delivery of health services from the community level to the national level.

HEM shares some tenets with preventive medicine (Sidel *et al.* 1992). When disasters are seen as the outcome of accumulated risk produced by years of underlying vulnerability combined with hazard, the case for preventive action can be made more plainly. The word 'prevention' is common to the health sector and disaster management. Health, however, distinguishes between 'primary', 'secondary' and 'tertiary' prevention. Primary prevention, which may be equated with hazard avoidance, relies on structural and non-structural measures to prevent human exposure to hazards. Examples include locating housing and infrastructure away from hazard-prone areas and early warnings of hydro-meteorological hazards. Disease surveillance linked to an early warning system also serves as an effective means for primary prevention of infectious disease disasters. After the 2005 Hurricane Katrina in the USA, clusters of diarrhoeal disease were reported in evacuation centres in four states, and gastroenteritis was the most common acute disease complaint among evacuees in Memphis, Tennessee. This surveillance system and subsequent warning allowed for shelters to implement effective prevention and control measures (Ivers and Ryan 2006).

Secondary prevention is the health equivalent of vulnerability reduction. Persons located in close proximity to the hazard have a higher risk for injury and illness as compared with those less exposed. Populations are not equally susceptible to the same health hazard. Differences among persons are due to such factors as sex, age, genetic predisposition and health status. For example, elderly persons living in temperate climate zones are more susceptible to heat wave disasters than are young adults living in the same location (see Chapter 37), while vaccinated people are less susceptible to infectious disease disasters.

In general, healthy people are less likely to become ill or injured as a result of disaster. Socio-economic factors such as access to health services, socio-economic status, social networks, political influence and behaviour are determinants of both risks of health and to disasters (see Chapter 3). Programmes that endeavour to address these factors serve a dual purpose in that they also help to improve health status and build individual and community capacities to cope with the health risks of disasters. For example, nutritional programmes that address the high rate of vitamin A deficiency among Micronesians also make the population less vulnerable to other disaster hazards such as measles or food insecurity caused by sea-level rise (Keim 2010b).

Finally, tertiary prevention activities involve the post-impact phase of disaster, when actions are undertaken to minimise loss of life and damage and to recover to a better state than before. Tertiary prevention is applied when primary and secondary prevention do not effectively reduce the risk, leaving a residual risk which is managed by the health emergency response. The challenge for disaster risk reduction as applied to health is to broaden the focus of disaster risk management from that of tertiary prevention (response and recovery) to emphasise primary and secondary prevention

Key issues for health emergency management

Primary health care

Primary health care (PHC) describes an approach to delivery of health care and basic health services at the community level. PHC aims to increase accessibility of the population to health services and provide a community-centred approach which emphasises community participation in identifying problems and solutions which affect their health, including the promotion of self-reliance (WHO 2008). As in community-based disaster risk management, primary health care calls for a comprehensive, integrated, multidisciplinary and multi-sectoral approach to address the causes and effects of public health. Table 44.1 lists typical PHC services and examples of each.

As disasters often disrupt the ability of the health system to provide these services, a key objective of response is providing priority primary health care services and restoring the primary health care system. Primary health care workers at community level respond to the immediate health care needs and provide the first level of contact with the health system. In remote and rural areas with limited health resources in particular, PHC workers attend to both local medical emergencies and detect and report disease outbreaks, referring serious cases to district hospitals. PHC health workers are also the front line of the health emergency management system (see Box 44.1). In Bangladesh, in addition to multi-functional education and primary health centres which provide evacuation shelters for people exposed to natural hazards, training of all health workers in the prevention, management and control of diarrhoea has enabled communities to significantly reduce fatalities from diarrhoea following cyclones and floods (Ofrin and Nelwan 2009).

Successful HEM and the delivery of PHC are mutually dependent. PHC is an effective means for secondary and tertiary prevention of the health effects of disasters. This practical application of both risk reduction and treatment is critical for the sustainability of PHC, and to ensure that health development gains are protected during disasters.

Box 44.1 The role of local health workers in HEM

- Integrating health risks into local risk assessments;
- Identifying and strengthening local health resources that can be utilised for risk management;
- Representing the health sector and advocating for health in local multi-sectoral emergency management systems;
- Ensuring that health is addressed in community capacity development;
- Ensuring the safety and preparedness of community health care facilities;
- Co-ordinating health response at the local level with referral of patients to district and other levels of the health system;
- Providing community training and education in health (for community members, health workers, schools); and
- Maintaining emergency preparedness (i.e. plans, exercises, early warning) at the community level.

Table 44.1 Typical primary health care services

Primary health care services	Examples
Community health education	Community information to raise awareness of health risks (e.g. diarrhoeal diseases from unsafe water) and available health services, and to promote healthy behaviours, home-based treatment and self-referral to health care workers.
Prevention and local control of communicable diseases	Vector control such as impregnated bed nets and spraying of insecticide to combat malaria, disease surveillance, early warning systems, and treatment of diseases such as acute respiratory illnesses.
Immunisation programmes	Ensuring routine immunisation against national target diseases with an adequate cold chain in place.
Maternal and newborn health	Supply of clean delivery kits to pregnant women, antenatal care, essential newborn care, basic emergency obstetric care, post-partum care such as support for breast-feeding, and abortion care.
Sexual and reproductive health	Family planning, availability of condoms, counselling and testing, management of sexually transmitted infections, and clinical management of rape survivors.
Provision of essential drugs and medicines	Drugs, medical supplies, diagnostic agents and equipment for home-based and centre-based management of acute and chronic diseases, such as oral rehydration salts to treat diarrhoeal disease, antiretroviral treatments to maintain HIV/AIDS therapy, safe blood supplies and provision of emergency health kits.
Promotion of good nutrition	Community education on the safe preparation of food, mass-feeding programmes which provide adequate energy and nutrient intake, assessments of malnutrition, and supplementary and therapeutic feeding programmes.
Treatment of injuries and non-communicable diseases	First aid, injury care and life-saving essential surgery, mass-casualty management, treatment of diabetes and hypertension, and rehabilitation and management of people with disabilities.
Promotion of safe water and basic sanitation	Education to promote hygiene, such as handwashing after using the toilet, provision of adequate supply of clean water, emergency latrines and waste management.
Promotion of mental health	Psycho-social support, self-management of stress and identification and care of people with severe mental health problems.

Source: WHO 2009b

Mass casualty care

During a disaster, health care systems are confronted with increased demands and decreased availability of health care resources. Following a mass casualty incident, there are numerous critical steps that must be followed to ensure appropriate management and care of victims. Mass casualty management (MCM) refers to the medical response to an event where the number of casualties is large enough to disrupt the normal course of emergency and health care services. MCM should be seen as a chain of survival that extends from the field to the hospital and then back to the home of the survivor (see Chapter 42). Medical management begins with emergent field-based search and rescue, first aid, evacuation, and then extends to definitive care at the hospital and then back again to the community to include physical and psycho-social rehabilitation and recovery in the long term. As was the case in the 2010 Haiti earthquake, injury management includes the provision of long-term rehabilitative care for disabilities incurred as a result of traumatic brain and spinal injuries, as well as thousands of limb amputations (Iezonni and Ronan 2010).

Because mass casualties place such profound demands on the health care system, advance preparedness must ensure that ca\re will be delivered to the greatest possible number of casualties (see Box 44.2). The aim of preparedness should be to keep the health care system operational, flexible and functional, and to deliver acceptable quality of care to preserve as many lives as possible.

Public health and medical officials preparing for the 1996 Olympic Games in the USA worked to establish a standardised MCM system for use throughout Atlanta in the case of a disaster incident. This MCM system was then called into use when a bomb exploded in Olympic Centennial Park during the Games. The unified incident command system comprised participants from local, state and national levels (Sharp *et al.* 1998). Twenty-four local metropolitan hospitals were co-ordinated in advance of the attack along with five separate companies, providing emergency medical services in the pre-hospital setting. A system for geographical

---

Box 44.2 Mass casualty management (MCM)

The medical response to a mass casualty event operates at two broad locations: the field and the hospital. These spheres should be connected tightly by a critical and vulnerable link – the process of effective distribution of the casualties by emergency medical services.

A standardised and well-rehearsed incident command system (see Chapter 42) is paramount for proper operations during an actual disaster. This commonly involves establishment of an incident command system and allocation schemes for scarce resources (e.g. medical triage).

An MCM system relies on a number of key components including:

- Co-ordination mechanisms between local, sub-national and national levels, and with neighbouring sub-national and international jurisdictions.
- Co-ordination and communication between pre-hospital services (such as ambulance services) and health care facilities, and with other sectors (such as emergency services, law enforcement, transport and communication) to arrange for the transport and distribution of patients.
- A triage system that enables personnel in the field and in facilities to determine priorities for treatment of patients, taking into account the severity of their condition, chances of survival and the available resources.
- Emergency operations plans that provide for a rapid and organised surge of resources and logistics to support the influx of patients, health care personnel, equipment and supplies.

triage was put into place. This system was designed to triage disaster victims on site and then transport those least severely injured to facilities farthest from the disaster scene, allowing for the more severely injured to seek care at the closest hospitals. Trauma centres closest to the Olympic venue developed emergency operation plans in advance of this disaster. As a result, surge capacities were rapidly implemented and all victims of the terrorist bombing were processed through the emergency department and admitted to the hospital within only several hours after the attack (Brennan *et al.* 1997).

Communicable disease

The risk of infectious diseases after disasters is often specific to the event itself and is dependent on a number of factors. In low-income countries, outbreaks are most commonly associated with water-borne illness and, less commonly, to vector-borne illness. Flood-related outbreaks are normally due to diseases that are endemic to a given population, such as leptospirosis. High-income countries tend to suffer few flood-related outbreaks, but instead have a higher proportion of flood-related, non-communicable diseases such as injuries, mental illness, cardiovascular disease and chronic obstructive pulmonary disease (Keim 2010a).

The relationship between disasters and communicable diseases is frequently misconstrued (Keim 2010c). The risk of epidemics is often presumed to be very high in the aftermath of disasters. The risk of outbreaks after disasters continues to be greatly exaggerated by political officials, some health officials and much of the media (Noji 2005). Despite public concern to the contrary, non-endemic diseases do not spontaneously emerge after disasters. However, disasters may exacerbate diseases that are endemic to the affected populations. The risk factors for outbreak after disasters are primarily associated with factors such as displacement of vulnerable populations,

---

Box 44.3 Control of communicable disease during health emergencies

The World Health Organization (WHO) describes the fundamental principles of communicable disease control in emergencies as follows:

- Rapid assessment: identify the communicable disease threats faced by the emergency-affected population, including those with epidemic potential, and define the health status of the population by conducting a rapid assessment.
- Prevention: prevent communicable disease by maintaining a healthy physical environment and good general living conditions.
- Surveillance: set up or strengthen disease surveillance system with an early-warning mechanism to ensure the early reporting of cases, to monitor disease trends and to facilitate prompt detection and response to outbreaks.
- Outbreak control: ensure that outbreaks are rapidly detected and controlled through adequate preparedness (i.e. stockpiles, standard treatment protocols and staff training) and rapid response (i.e. confirmation, investigation and implementation of control measures).
- Disease management: diagnose and treat cases promptly with trained staff, using effective treatment and standard protocols at all health facilities.

(WHO 2005)

---

overcrowding, inadequate shelter, poor water and sanitation, malnutrition and low nutrient intake, insufficient vaccination coverage, exposure to disease vectors and lack of or delay in treatment (WHO 2005). The provision of shelter, water, sanitation, food and basic health care will help to protect health and minimise risks of communicable disease outbreaks in emergency settings.

For example, during the 1994 humanitarian crisis that occurred in Goma (in the east of present-day Democratic Republic of the Congo) as a result of refugees fleeing Rwandan genocide, there occurred a deadly cholera outbreak that claimed thousands of lives in the Hutu refugee camps. This infectious disease outbreak was initially detected by rapid needs assessment performed by non-governmental organisations (NGOs) responding to the crisis. Medical clinics were established and the disease was managed on a case-by-case basis. Non-medical interventions (including water transport and drilling of safe wells) were also put into place in order to avert the growing cholera outbreak. Unfortunately, disease surveillance systems revealed a trend of increasing mortality due to cholera. Further study revealed that non-hygienic water storage practices were also causing continued contamination of potable water sources. The outbreak was finally effectively managed after a broad variety of medical and non-medical interventions were in place (Roberts and Toole 1995). This experience provided an important lesson regarding the need for a comprehensive approach to management of the communicable diseases during a disaster.

Reproductive health

Women of childbearing age (from fifteen to forty-nine years) in disasters are at risk of death and illness associated with pregnancy and childbirth due to obstructed labour, infection, unsafe abortions and hypertension. Malaria, HIV/AIDS, malnutrition and anaemia can also contribute to maternal deaths. In disasters, women are at increased risk due to lack of access to emergency obstetric care and sexual violence. The Minimum Initial Service Package (MISP) focuses on priority activities to reduce reproductive health morbidity and mortality in the acute phase of an emergency. Activities include:

- Co-ordination of reproductive and sexual health programmes;
- Clean birth delivery kits for mothers, birth attendants, midwives and clinicians;
- Referral system to communicate and transport women to health centres and hospitals for management of obstetric emergencies;
- Prevention and management of sexual violence (e.g. provision of emergency contraception), pregnancy testing and treatment for sexually transmitted diseases, treatment for HIV/AIDS and support for rape survivors;
- Reduction of HIV transmission through blood safety, application of standard precautions for infection control and provision of condoms; and
- Plans for integrating reproductive health into PHC services.

(WHO et al. 1999)

In February 2005, two months after the impact of the Indian Ocean tsunami, the Women's Commission for Refugee Women and Children conducted an assessment of the implementation of the MISP in Aceh province, Indonesia. The assessment pointed to the value of implementing MISP from the beginning of an emergency to ensure that affected populations have access to life-saving reproductive health services, as well as the importance of strong participation of community organisations in the co-ordinated response to reproductive health needs (Women's Refugee Commission 2005).

Emerging issues and future directions

Safe and prepared hospitals and health facilities

Protecting heath facilities and health workers is a critical issue for the health of people at risk or affected by disasters, the investment in health infrastructure and health systems, for emergency response and for community recovery. When health facilities have been destroyed or damaged, there are huge health consequences. Hospitals, other health facilities (e.g. health centres, laboratories, aged care facilities, pharmacies, blood banks) and health workers are among the major casualties of disasters. When health care facilities are built safely and staff are trained and prepared, there are significant benefits to health.

There are numerous examples where hazards, including conflict, have damaged or destroyed health facilities, killed and injured patients and health workers, and affected the delivery of health services when they are most needed in times of emergency (see Chapter 7). These include the 11,027 medical institutions that were damaged in the Sichuan earthquake leading to the medical evacuation of some 10,000 casualties to over 300 hospitals nation-wide (China Planning Group of Post-Wenshuan Earthquake Restoration and Reconstruction *et al.* 2008). The 2004 Indian Ocean tsunami resulted in damage to sixty-one per cent of health facilities, while seven per cent of workers and thirty per cent of midwives were killed in Aceh province, Indonesia (Carballo *et al.* 2005). Conflicts around the world have disrupted HIV and primary health care services, such as feeding and maternal health services, and vital health personnel have left their posts due to insecurity and violence. Conflict has also disrupted vaccination of children despite the occasional success of ceasefires declared for purposes of vaccination of children, as in Afghanistan in 2000, southern Sudan in 1997 and El Salvador in 1985.

The safety and preparedness of a health facility involves structural, non-structural and functional factors (see Box 44.4). Structural issues relate to the planning and location in relation to hazards, the underlying soil structure and building standards and techniques. The non-structural elements include non-weight-bearing walls, lighting and the safety of equipment, material or supplies, which can be displaced or flooded in a hazard event. The functionality of a health facilities also relies on transport access to them and the availability of the critical infrastructure upon which health facilities depend, such as power, water and waste disposal (see Chapter 56). Health facilities and workers can also be prepared for their role in emergencies through emergency response planning, training of staff and exercises to test the readiness.

Countries are taking action to manage risks to health facilities. Bangladesh has invested in safely built, multi-functional facilities for health, education, agriculture and other community services. These have provided shelter and protection for communities in cyclones and floods. In 1990 in Costa Rica five major hospitals were being retrofitted when the country was hit by a 6.8 magnitude earthquake. The areas of the hospitals that had been retrofitted were in excellent condition after the earthquake, while other parts, which had not been reinforced, suffered extensive damage. The money saved from the prevention of damage far exceeded the cost of retrofitting the facility (PAHO 2004).

International health response

In 2005 humanitarian reform was formulated to address the changing environment of humanitarian crises, such as the increasing role of NGOs, and to fill significant gaps in international humanitarian response identified by an independent review commissioned by the UN Emergency Relief Coordinator. The 'cluster approach' (see Chapter 50) is one of the pillars of the

---

Box 44.4 Hospitals safe from disasters

The Hyogo Framework for Action 2005–15 promotes the goal of 'hospitals safe from disasters', by ensuring that all new hospitals are built with a level of resistance that strengthens their capacity to remain functional in emergency and disaster situations, and implementing mitigation measures to reinforce existing health facilities, particularly those providing primary health care. The United Nations International Strategy for Disaster Reduction (UNISDR) system recognised the issue by focusing the World Disaster Reduction campaign in 2008–09 on the theme of 'hospitals safe from disasters', and was accompanied by 2009 World Health Day, which drew global attention to the key actions that could be taken to make health facilities safer in emergencies. Assessments of the safety of health facilities have been conducted throughout Latin America and the Caribbean, Nepal, Oman, the Philippines, Sudan and Tajikistan, using assessment tools such as the Hospital Safety Index (PAHO 2008).

At a political level, alongside the longstanding commitment of ministers in Latin America and the Caribbean to safe hospitals, eleven ministers of health from the WHO South-East Asian region committed their governments to protecting health facilities from disasters by agreeing to the Kathmandu Declaration. The Global Platform for Disaster Reduction in Geneva proposed that by 2011 national assessments of the safety of existing education and health facilities should be undertaken, and that by 2015 concrete action plans for safer schools and hospitals should be developed and implemented in all disaster-prone countries.

---

reform, alongside strengthening humanitarian financing and a humanitarian co-ordinator system and strong partnerships between UN and non-UN agencies. WHO is the lead agency for implementing the global health cluster (WHO 2009b).

The cluster approach aims to:

- Improve the efficiency and effectiveness of thehumanitarian response to crisis;
- Increase the predictability and accountability in all of the main sectors of international humanitarian response, including health; and
- Ensure that gaps in response do not go unaddressed.

At the country level, clusters work closely with national and local authorities and provide support to the host government's efforts. The health cluster approach has been implemented in several disasters in Haiti (earthquake), Indonesia (several earthquakes), Pakistan (earthquake and flood) and the Philippines (several typhoons). It is critical that international and national actors alike are aware and trained in the cluster approach in order to plan for and co-ordinate international humanitarian response to disasters.

The health cluster approach has met with variable degrees of success. One of several significant accomplishments is making the lead agency (WHO) or co-lead agencies accountable for the performance of their cluster by clearly stipulating their responsibility to ensure adequate co-ordination of activities by partners involved in its specified area. Second, the cluster system aims for a common analysis and a commonly agreed strategy, since this was not always the case with the sector co-ordination of the past. Third, the cluster approach seeks to deliver predictability in tackling emergencies and crises (see Box 44.5).

Still, there have been challenges in putting the concepts adopted by the UN Inter-Agency Standing Committee into action on the ground. According to an independent evaluation

Box 44.5 Responsibilities of the health cluster

Health sector co-ordinating mechanisms involving the UN and other international organisations, international and national NGOs, donors, private-sector health services and donors often involve the following:

- Mapping of health actors, services and activities;
- Assessments and information on the health situation and needs, situation reports and health bulletins;
- A joint health response strategy with clear priorities and identified responsibilities;
- A joint contingency plan for response to future events;
- Agreed standards, protocols and guidelines for basic health care delivery, and standard formats for reporting;
- Training materials and opportunities available to all partners;
- Resource mobilisation strategies, such as joint appeals; and
- Joint monitoring, evaluations and lessons learned activities.

commissioned in 2007 of the entire humanitarian cluster system, 'there has been no observable increase in ultimate accountability'. The 2007 report also noted that, 'results of the global cluster capacity-building effort have not materialised in major ways in field operations' (WHO 2009c). A subsequent evaluation conducted in 2010 noted improvements to the cluster system and found that the benefits of implementing the cluster approach were slightly greater than its costs and problems. The report concluded that the investment in supporting and strengthening the cluster system should be continued. The performance of the health cluster is being addressed by building the skills and competence of individuals on the ground through training of health cluster co-ordinators, international and national health actors and provision of guidance and tools for implementation of health cluster activities.

Conclusions

In order to protect health from disaster risk, a systematic and comprehensive approach is required that integrates health aspects into many areas of activity at national, sub-national and community levels: health aspects of policy and legislation, risk assessment, hazard reduction, vulnerability reduction, capacity development, response and recovery. The development of such an integrated HEM approach involves the UN in partnership with national and local actors, key NGOs and international organisations from the health and non-health sectors. A stronger investment in the preventive approach will ensure that fewer people are exposed to hazards and that health impacts of disasters will be minimised. The health sector should also work with other sectors to promote cost-effective investments in disaster risk reduction at community, national and international levels. These measures should include risk sharing and risk transfer, structural and non-structural measures, including the protection of community infrastructure such as health facilities, water supplies and sanitation systems, and community-level planning for risk reduction. Simultaneously, the health sector has responsibility for providing the necessary health services to improve basic health status, reduce health vulnerabilities, strengthen health system capacity to focus on a primary health care approach and address the myriad health needs in response to and recovery from disasters.

Research activities should be directed towards effective disaster risk assessment and development of good practices for reducing human health vulnerability to disaster hazards. The health sector should be represented in multi-sectoral policy, planning, response and recovery mechanisms such as national disaster institutions and national platforms for disaster risk reduction. The international community also can support these efforts by facilitating HEM platforms at global and regional level to develop and share evidence-based practice and research.

Health professionals are well-placed to support a disaster risk management approach that places a greater emphasis on preventing illness and injury, while recognising the important role of the health sector in providing curative care particularly in response to emergencies. HEM should be recognised as an essential public health function integrated in health systems and health development plans.

45

Food security and disaster

Ian Christoplos

DANISH INSTITUTE FOR INTERNATIONAL STUDIES, COPENHAGEN, DENMARK

Introduction: food security and livelihoods

Food security has four main dimensions: food availability, access to food, stability of supply and safe and healthy use of food. The situation is deteriorating in all four dimensions. These dimensions demonstrate that risks are multiplying due to commercial speculation, population increase, climate change and other factors:

- *Availability* of food is decreasing due to scarcity arising from land degradation and decreasing yields, increasing cost of agricultural inputs, population pressure, worsening climatic conditions, demands for meat and dairy products (which increases use of basic grain for animal feed rather than human food) and shifts from food to biofuel production.
- Poor people's *access* (entitlement) to food is declining due to worsening terms of trade between wages and food costs. Speculation is driving these declining terms of trade and also (perhaps) reducing the access of the rural poor to land for subsistence farming.
- *Stability* is threatened due to increasing prevalence of disasters, erratic climate conditions, uncertainty regarding food prices and national protectionism.
- *Safe and healthy use* of food is deteriorating as particularly the landless rural poor switch to more monotonous diets, which lack essential micronutrients, and as both flooding and scarcity of potable water increase the prevalence of diarrhoeal diseases and malaria, which in turn affect assimilation and metabolism of food.

Success or failure in maintaining food security along these various dimensions is a fundamental indicator of whether or not the most basic aspects of risk are being effectively managed. Hunger is the bottom line for those who are chronically vulnerable due to poverty or acutely vulnerable to disasters. People affected by disasters, their politicians and the organisations that claim to protect them generally agree on the centrality of food security, but they do not always agree on what this implies for humanitarian and development action. This chapter aims to unpack why this is so and suggests where a new, more 'coherent' agenda on food security might emerge.

The issue of food security attracted considerable attention during the late 1980s due to the impacts of droughts in Africa and concerns about the nature of famine. At that time food security was being compromised by many of the factors that may have been early forms of climate change. Researchers such as Amartya Sen (1981) and Alex de Waal (2005) began to draw attention to how famine resulted not from a lack of food in a given region, but rather from political decisions, market failures and the structures of vulnerability that together resulted in a lack of 'entitlements' by marginalised populations to gain access to the food they need. Such entitlements were recognised as consisting of income, access to social protection or smallholder production. The ability to buy food, as influenced by local/national/international policies, markets and trends, came to be recognised as more important than household or even national food production levels. A new food security paradigm began to emerge.

Partly as an outcome of these new perspectives, during the early 1990s analyses of food security began to be subsumed under the broader concept of *livelihoods* (see Chapter 58). The complex struggles of poor people to survive in increasingly diverse and complex economies were recognised as a more appropriate starting point for preventing hunger than a narrow food production perspective. Today it is widely recognised that attention must be given to livelihoods in order to understand why people are hungry: i.e. to take into account the policies, institutions and processes that determine how well poor people are able to access food amid the context of vulnerability in which they live.

In disaster response, there has been an increasing focus on programming that is commonly referred to as 'doing livelihoods', i.e. creating short-term employment through cash or food for work, or providing cash, credit, seeds, tools, fishing boats or other assets to enable people to (re-)start their own livelihood activities. 'Doing livelihoods' has generally been promoted as a way of reducing dependence on food aid through 'early recovery'.

In the past few years the focus of attention has partially swung back toward food security and increasing yields. Food prices skyrocketed during 2007–08, causing hunger among the poor and sending shockwaves into the political systems of countries with large vulnerable populations, as well as wealthy countries dependent on food imports. The assumption that food aid would always be available was shaken when countries (and aid agencies) scrambled to buy food on international markets. Food aid suddenly became a quite expensive form of aid, rather than a way of disposing of surplus production, at the same time as the ability of the poor to buy food was declining due to increasing consumer prices. The outcome of this event was a reminder that the prime directive of the livelihoods of the poor is being able to afford enough to eat, and that food production does indeed matter. This price spike led to huge new commercial and aid investments in food production, but it is too early to assess what the outcomes of these investments will be in terms of household food security.

Food security and natural hazard impacts

Natural hazards have impacts on food security, but given the complex mix of factors that determine availability, access, stability and use, simple attribution is almost always misleading. Confusions arise due to a tendency to focus on either the *hazard* – be it climate change, an earthquake or a hurricane – or the *impact* – in this case hunger. What is missed is how any of these hazards alone is not the problem. Each hazard multiplies the existing problems facing a given population in accessing food or maintaining livelihoods. Even a devastating earthquake does not automatically result in a serious food security crisis if there are institutions and a socio-economic structure in place to replace lost food production with safety nets. In a disaster such as that which affected Haiti in January 2010, the range of other risks related to weak institutions and socio-economic structures combined with that of the earthquake to create an extremely fragile situation. On the other hand, there are many examples of famines that were expected by the international community but did not actually happen, due to unrecognised capacities of people to find their own ways to access food, to stabilise supplies and to use the food they have more effectively.

Context, in terms of the full range of prevailing risks, is therefore key to understanding if and how food security may be threatened by any given hazard. Events such as the 2004 South Asian tsunami, and many coastal storm surges, cause severe salinity problems in coastal agriculture, but many coastal low-lying areas are relatively fertile and producers may be able to manage occasional losses relatively well (see Chapter 27). Poorer farmers cultivating less fertile sloping lands may even benefit from higher grain prices due to their neighbours' losses. People crowd flood-prone coastal plains, treacherous but fertile volcanic slopes and other high-risk areas because they see these risks as being offset by the greater chance of good harvests than they would have if they farmed areas affected by chronic loss of soil fertility and vulnerability to less dramatic long-term and slow-onset droughts. Furthermore, in relatively highly populated coastal areas, river banks or in the usually densely populated slopes near volcanoes, their chances are better of finding alternative employment if a crop or business is lost.

Simple assumptions about causes and effects may also lead to misplaced strategies that respond to a given hazard, but ignore underlying vulnerabilities and capacities. After the South Asian tsunami there seemed to be a serious loss of availability of fish in local diets due to the destruction of fishing boats. The overwhelming aid response led to an oversupply of small coastal boats (with detrimental impact on fish stocks) without addressing the underlying factors that create or limit access to food. The livelihoods of many of the most vulnerable tsunami-affected people in Sri Lanka were actually more dependent on employment on large boats and other commercial enterprises than they were on fishing themselves.

Box 45.1 Famines that didn't happen

Humanitarian agencies regularly appeal for funding to prevent famines when crops fail due to droughts or when livelihoods are disrupted by conflicts, market disturbances or earthquakes. Some of these appeals receive a generous response, but others receive little or nothing. There is not necessarily a direct correlation between funding levels and famines that didn't happen. There are clear cases where early warning resulted in pre-emptive mobilisation of aid resources. One notable case was the response to the 1991–93 El Niño-related drought that affected Southern Africa. In this massive drought the aid response was generally viewed as exemplary, which meant that famine was indeed averted (Holloway 2000).

However, there are also many disasters that receive a far more meagre response. During the same 1991–93 drought there were some large areas of Angola that remained inaccessible due to the conflict and related factors and therefore received a small fraction of the food that was 'needed'. Nonetheless, there was no reported famine as a result. Are there many such non-famines where the aid didn't arrive? We do not know since records are rarely kept of what happens when the aid does not arrive. It is obvious that in most cases people have survived far better than predicted.

This is not to say that relief response is superfluous in contexts such as these. The coping strategies used when food does not arrive may involve selling off vital household resources such as land or draught animals, or going deeply into debt. Even if people are able to manage acute food security crises, the means by which they do so may increase their chronic vulnerability.

Food security amid changing patterns of risk

Food security in rural areas is dependent on managing many risks and opportunities. This is primarily related to the institutional capacities of disaster-affected people, farmers, governmental agencies and private firms to respond to weather and market uncertainty and unpredictability. The shifting landscape of risks to household, local, national and global food security are being managed as part of how these actors are already dealing with volatile markets and pressures on natural resource management regimes (see Chapter 22 and Chapter 31). Climate change is likely to include both extreme weather events and gradual declines in global and national production levels per capita, which in some regions and countries are expected to be severe. There are some areas where harvests are expected to increase as well, but the general picture is rather grim.

Different types of risk have very different implications for what to do about food security. Today food security is sometimes seen as being primarily related to fluctuation in aggregate global food supplies. On the other hand, there is a realisation that for households and individuals, supply is not the main limiting factor; it is access and stability of affordable supply. Food security relates to their access to food, and this is mostly related to risks inherent in a given household's livelihoods (see Box 45.2). This dichotomy between the inter-related but still very different issues of household versus national/global food security is often glossed over in discussions of future risks and ways to respond.

The trade-offs and mix of strategies that would be required to combine these national/global food security goals and those of rural households struggling to find livelihoods are rarely confronted in policy declarations, but are inevitably part of how food security and long-term risks, including those from climate and climate change, come together at local levels. The return of attention to food production and productivity has been beneficial in ensuring that the broader and long-term possibly catastrophic impacts of a range of hazards in global agrifood systems are in the spotlight. It is unclear, however, how this attention will ultimately influence approaches for dealing with disaster-related livelihood and food security risk at household level. A danger exists that the response to macro food concerns may ignore or undermine the strategies of the poor in dealing with extreme events (natural, socio-political and economic/market-related).

There is significant scepticism from many humanitarian and development actors involved in food security efforts about whether the sudden attention to climate change requires new perspectives and modalities, or if it is just another fad requiring insertion of new catchwords to meet funding windows (Kelman and Gaillard 2008). Indeed, many current efforts at operationalising climate change adaptation efforts draw very heavily on the community-based natural resource management efforts (Ensor and Berger 2009) and the disaster risk reduction toolbox (UNISDR 2008e) of recent decades. Such repackaging and re-naming is to some extent necessary and inevitable. However, protecting and promoting food security in light of climate risks will create demands on institutions involved in agriculture, disaster risk reduction and environmental governance. They are likely to be expected to both do much more of what they have done before and also do things differently (Christoplos et al. 2009). The surprises of the food crisis of 2008 were a reminder that the current and future landscapes of risk suggest the need for conceptual and operational frameworks that are significantly different from those of the past.

With respect to disaster risks, there are particular institutional dynamics and divisions that need to be considered. Disaster trends are ever more intertwined with chronic decline in food security due to reduced precipitation, glacial melt and heat stress on crops and livestock. The divide between acute humanitarian needs and the factors that generate chronic poverty and food insecurity are becoming blurred. The view that simple structures should exist whereby some agencies provide relief in disasters and that social safety nets will respond to address

Box 45.2 Land grabbing and modernisation for national food security?

Paradoxically, the need to raise aggregate food production has led many countries to pursue strategies that may have negative impacts on access to food by the rural poor. 'Land grabbing', whereby countries or entrepreneurs buy or appropriate large tracts of farmland for industrial food production, is an example of how some states intend to mitigate risks to national food security through 'modernisation' in the form of mechanised food production. Food importing countries with access to capital for investment but which are increasingly concerned that speculation and protectionism might reduce their access to sufficient food for their citizens in the future are looking beyond their borders for ways to manage potential future crises. Agricultural land has suddenly become a commodity of great interest. Even poor countries such as Ethiopia, which are heavily reliant on subsistence farming, see these investments as a way to break out of existing systems that have failed to maintain either household or national food security. Aid has failed them, so now they are eager to see if trade can yield better results.

This is a high-risk strategy, however. Such attempts to increase national food stocks may not have a positive impact on rural household food security. The livelihoods of rural, food-insecure households are usually reliant on diversified ways of accessing food and livelihoods. The seasonal employment they may access in industrial agriculture may not be enough to offset the loss of other entitlements. Subsistence agriculture may be a significant part of their livelihoods, but this is not a national priority in food-insecure countries that see continued reliance on subsistence farming as a symptom of failed poverty alleviation, and 'modernisation' as an obvious solution. A common approach is to encourage greater reliance on commercialisation and off-farm employment. There is an assumption that the food insecure may be better off leaving farming than in dealing with ever more frequent floods and droughts on their diminishing plots of land.

Both strategies include growing levels of risk. In most rural areas around the world a rapidly growing proportion of the population are no longer farmers. For the rural landless, working for the 'land grabbers' may be a better opportunity than not working at all. This so-called 'deagrarianisation' has generally not led to livelihoods that can be described as 'sustainable' (Bryceson 2009), but there are few other opportunities available. The situation of agricultural labourers is starting to be recognised as being just as important for food security as subsistence farming (de Schutter 2009).

chronic food insecurity is out of touch with these new realities. In the new landscape of risk related to chronic decline, unpredictability and uncertainty, it is in many respects pointless to try to differentiate between acute and chronic food security. This has major implications for rethinking institutional mandates and operational structures (Parry *et al.* 2009).

Finally, there are situations when gradual decline turns into collapse of livelihoods and agro-ecosystems. These so-called 'tipping points' have not been on the radar screen of most governments or of many in the aid community, but they should be, as the collapse of livelihoods and consequent acute/chronic food security emergencies in fragile states as diverse as Somalia, Haiti and North Korea are clear examples of the difficulties of understanding how to respond in places where risk is no longer manageable.

Ian Christoplos

Food aid and food security

In the past the solution to food insecurity was largely assumed to be food aid. This was due largely to a combination of the symbolic values of providing food, the desire of grain surplus nations to market and dispose of their grain stocks, a distrust in disaster-affected people to decide themselves what they need, and inertia within humanitarian bureaucracies (and perhaps mentalities) that tended to treat food aid as the default response to any disaster. Food aid decisions were often made without analysing the actual food insecurity problem in a given disaster, and were driven by the relatively easy availability of food compared to other response options (Levine and Chastre 2004). For these and other reasons, during the past decade the automatic association of food aid with food insecurity gradually began to be profoundly questioned (Barrett and Maxwell 2005). Cash-based responses have started to be recognised as a flexible and cheaper alternative that can provide positive incentives for local food production and local economies more generally since they put cash – rather than imported food – into the pockets of the poor (Harvey 2007; FAO 2006).

However, the 2008 food price spike and the diminishing value of cash for food purchases have drawn attention to food aid again. More efforts are now being made to purchase food locally, since there are no longer pressures to dispose of grain surpluses. The World Food Programme has recognised that their 'procurement footprint' may be able to contribute to both smoothing and stimulating demand and calming markets. It is still uncertain if or how this can be managed given the logistical constraints and intense competition from commercial buyers for grain in acute emergencies.

In addition to disaster response, in the past food aid was also used within development programming, partially as a form of social protection for the chronically poor. As food reserves shrunk, this form of food aid gradually diminished to a rather insignificant component of the social protection toolbox. Yet the need to manage chronic food insecurity has increased. While food for work, school lunch programmes and similar initiatives still exist in some countries, the range and scope of other social protection programmes (see Chapter 57) is likely to further displace them given the increasing need for safety nets and the disappearance of food surpluses.

Food security and fundamental choices in agricultural development

Food security opportunities and risks in relation to rural livelihoods have very often been ignored in discussions of response to the 2008 food price crisis. These factors are not ignored by the food insecure themselves as they look for work amid sudden shocks and increasing livelihood stress. It is also apparent to local governments that are trying to attract private and public investments. At local levels, the interplay between jobs and risks is a more concrete and immediate concern compared to the relatively abstract projections regarding climate change or likelihood of earthquakes or other hazards. This is important to stress, as it is within this interaction between private-sector actors considering investments, local governments trying to attract them and the strategies of the food insecure in search of livelihoods that the most important decisions impinging on household food security are likely to be made.

This is not new, but the lessons regarding rural livelihoods that have been learnt in the development community in the past decade have been slow to filter into interventions in response to disaster risk. Livelihood support (see Chapter 58) in relation to disaster recovery has been a growing component of humanitarian programming and has become an important component of early recovery programming to jump-start local economies. However, in a perspective of addressing livelihood-related risks, these projects have proven problematic. The livelihood programming in response to the South Asian tsunami was characterised by failures to analyse the conditions that must ultimately sustain the livelihoods – namely, markets for specific products and labour, and the carrying capacity of the natural resource base (Christoplos 2006a). Experience from Hurricane Mitch showed that the quick impact interventions of the humanitarian phase had little long-term impact, but that development actors were able to learn from this experience with better programming later (Christoplos *et al.* 2010). These findings suggest that livelihood-related support in response to disasters must reflect what has been learnt in other livelihood support efforts.

The 'yeoman farmer fallacy', wherein subsistence and semi-subsistence agriculture was expected to be a basis for rural poverty alleviation, has long been taken for granted among humanitarian agencies but has largely been debunked in development circles in favour of promoting a gradual shift to market-oriented farming. However, market integration is not a guarantee for the livelihoods needed to improve food security. Engagement with markets is part of how the rural poor find ways to spread their risks, but it is not a panacea. In recovery efforts after Hurricane Mitch investments in strengthened market chains primarily benefited those households with sufficient assets to take advantage of new market opportunities (Christoplos *et al.* 2010). Market-oriented post-disaster programming is usually focused on 'picking winners', i.e. those who are able to take advantage of markets. Vague assumptions that the development process that these winners undergo will later trickle down to poorer farmers and the landless are rarely followed up, and there is little evidence that such a process actually occurs.

Internationally, a range of new initiatives is being piloted that is intended to reduce risks and entry barriers related to market-oriented agriculture. Weather-indexed insurance, warehouse receipt systems and other interventions that increase access to storage and credit, together with relevant market and climate information, can perhaps help the rural poor to manage combined market–climate–livelihood risks. Insurance initiatives have received particular attention in recent years, but as yet these types of interventions have rarely been sustainably institutionalised and scaled-up within national and local structures (Warner *et al.* 2009). Where insurance is not available, people at risk often use micro-credit (see Chapter 54) to access money to deal with livelihood and food security shocks, even if this was not the intention of those designing these programmes.

Even this 'pro-poor growth' paradigm, which assumes that smallholders could or should be effectively supported to retain benefits from the new landscape of agribusiness, is now being criticised as being 'romantic' (Collier 2009). In this critique of so-called romanticism the entry barriers and risks in 'modern agriculture' are portrayed as being too great for smallholders, who are instead expected to be better off pursuing livelihoods elsewhere. Some ministries of agriculture (in Latin America and Eastern Europe in particular), long eager to promote 'modern' industrial agriculture, have taken on this narrative, effectively declaring the livelihoods of the most vulnerable rural poor unviable, and therefore unworthy of further support. Some donors have also endorsed this view, perhaps due to the lower transaction costs and greater and more measurable (national food security) outcomes that can be achieved with better-off farmers. There is a growing gulf between household food security policies and those focused on agribusiness and trade.

Much of the food-insecure rural population still rely on subsistence agriculture for a significant proportion of their livelihoods, not because they are 'romantic', but because they see this as the lowest risk alternative. Long-term trends suggest that subsistence agriculture is declining in importance and that even those households that are still partially subsistence-oriented are trying to improve their livelihoods through diversification (World Bank 2007b). However, in a context of high risk there is a growing acknowledgement that subsistence retains an important role, at least as a buffer in dealing with variability and uncertainty (Trivelli *et al.*

2009). Awareness of the implications of climate change and the impacts of the 2008 food price crisis has begun to lead to a reassessment. The importance of at least a partial subsistence buffer for those who cannot deal with market risks is beginning to be recognised. Programmes to subsidise and support subsistence production are being initiated in some countries, and in parts of Latin America (Nicaragua, Brazil) the role of subsistence agriculture in alleviating hunger is taking central stage. It is too early to determine the extent to which this trend will be sustained and become a significant departure from past policies.

It is also uncertain whether these new anti-hunger programmes are reaching the most food-insecure. Food security measures based on provision of seeds requiring controlled environmental conditions or livestock dependent on reliable access to feed may address some constraints to food security without taking into account the political, natural hazard and market-related risks experienced by the most food-insecure subsistence farmers. These farmers often lack land with access to irrigation or reliable rainfall and cannot manage livestock production where the 'feed security' of the animals competes with the household for food. The need to sell at least part of their production to cover the costs of inputs may paradoxically leave the chronically poor more vulnerable to markets. These food markets may even be flooded due to expanded production stimulated by anti-hunger programmes.

The social contract in food security and disaster risk

The entry point for many food security interventions is not in agriculture, but in efforts to reduce the risk of disasters. It has been widely acknowledged that disaster preparedness and risk

Box 45.3 Agricultural rehabilitation beyond seeds and tools

Many post-disaster agricultural rehabilitation programmes based on seed distribution are promoted as being targeted to the 'most vulnerable groups', despite the fact that the chronically poor are probably landless or near landless and therefore lack anywhere to plant the seeds that they receive (Longley et al. 2007). Agricultural rehabilitation programmes will only reach vulnerable people if these programmes take into account the access these groups have to the broader range of assets (other than just seeds) that need to be orchestrated to produce food and generate livelihoods. Supply-driven programming (see Chapter 43), be it seed distribution or 'doing livelihoods', often fails to recognise this context. When looking at these factors it will generally become apparent that the supply of inputs and tools is not the main problem; indeed, research shows that seed markets are far more resilient than humanitarian agencies think (Longley and Sperling 2002). Weaknesses are instead found in the institutions related to land tenure, credit, agricultural extension, etc., which impact on food and livelihood security. There is rarely a quick fix available to address what is broken in these institutions, especially (but not only) after a disaster.

Perhaps 'supply-driven' programming is to some extent inevitable in humanitarian agricultural rehabilitation. At the least it would seem that attention should be given to ensure that these programmes 'do no harm' with respect to increasing risks to would-be beneficiaries. New seed varieties and crops have been promoted in Afghanistan (and elsewhere) without much-needed agricultural research and extension services to verify that these seeds were appropriate and to ensure that farmers could use them effectively (Christoplos 2006b).

management require a combination of natural resource management, land use, food security and climate adaptation initiatives. Managing and responding to disaster risk is part of the social contract between states and citizens (Pelling and Dill 2009a, 2009b). The failures of the state to live up to minimal responsibilities after Hurricane Katrina have been described as a breakdown in the social contract (Ignatieff 2005). This refers primarily to the responsibility to respond to acute human suffering, with hunger an obvious aspect. Within the current discourse related to climate change there are claims that this social contract needs to shift from one of responsibility to respond to disasters to one of preventing them (see Chapter 6). Such a shift may seem logical at the outset, but is less self-evident when it comes to the expectations and decisions about the 'humanitarian imperative' (the responsibility to respond to human suffering) when a disaster strikes and people are hungry.

The prevention versus response dichotomy is frequently described in an either/or framework. It is claimed that, due to the effects of climate change, demographic pressures, etc., disaster risk must be the focus instead of (rather than in addition to) humanitarian response. Relief is increasingly portrayed as a dependency-creating and wasteful relic of the old aid architecture, which must be replaced by a convergence between disaster risk reduction and climate change adaptation investments. Part of this has to do with claims that a shift is needed from humanitarian response to food insecurity (epitomised by food aid) to developmental approaches (food production). The disaster risk reduction–humanitarian interface is thus described as a zero-sum game (deciding who gets the money), rather than a search for synergy. This is problematic in that decisions to put aside the immediate humanitarian imperative of addressing hunger and famine may carry with it a new form of triage in terms of working with those whose risks can be reduced (often those with the land and other resources to possibly produce more food), rather than addressing the consequences of risks for the most vulnerable (e.g. the landless). The fact that the most food-insecure populations are usually not able to rapidly increase their production levels tends to be ignored in calls to move from humanitarian response to agricultural development. Sweeping rhetorical claims tend to overshadow this triage. It is likely that front-line actors dealing with food security may have a very different perspective on ethical trade-offs since they directly see the consequences of these policies.

Risk-aware social protection for food security

The old divisions between humanitarian and development interventions are being rethought within the concept of social protection. Social protection may be defined as the public actions that address socially unacceptable conditions of vulnerability, risk and deprivation within a given polity or society (Norton et al. 2000). A holistic approach to social protection (see Chapter 57) is contingent upon ownership and political will at local and national level. If social protection is to become part of the social contract, it is likely that food security will need to be a central component.

Significant progress is being made in situating social protection at the centre of the new global aid agenda. A number of pilot initiatives have been launched to test how a holistic approach to social protection could be established. Wide-scale implementation, however, remains elusive. Social protection projects that have received high levels of praise have been discontinued after donor funding has been phased out (Bradshaw and Víquez 2008). Institutionalising social protection in response to multiple risks would partly require building on the local structures on which food-insecure people rely. Structures are needed to flexibly respond to and address the causes of the many small disasters that are occurring more frequently. Increasing challenges related to the cumulative impacts of such hazard events will create demands for major engagement from national structures as well. Multiple risks at multiple scales will require better co-ordination (or at least coherence) vertically between central and local authorities, and horizontally across different sectors.

Thus far harmonisation of response to food security among national and local institutions and across sectors is extremely weak. Responsibility for nutrition is usually placed with ministries of health, food with ministries of agriculture and social protection with ministries of social welfare. Emergency response institutions manage acute food insecurity, while other organisations deal with chronic insecurity. Some of these institutions have decentralised responsibilities to local actors while others retain vertical structures or lack field presence altogether. Food security is too often the responsibility of everyone and no one. Civil society organisations advocate for action among all of these parties, but can do little to bring them together. Non-governmental organisations (NGOs) may themselves be struggling with internal conflicts between humanitarian and development divisions.

Conclusions

Beyond humanitarian assistance, little will happen and nothing will be sustained unless the response to chronic and acute food insecurity is integrated into ongoing national and local governance strategies and processes. The umbrella concept of social protection may encapsulate what the new landscape of food security risk and vulnerability means for national policies. If social protection and food security are seen as part of the social contract that states have with their citizens, a range of intervention forms may become a part of institutionalised frameworks for addressing risk. This needs to go beyond short-term humanitarian projects, while at the same time concretely addressing the pressing issue of hunger. Social protection may provide a way of transcending national–local gaps, as well as the divisions between humanitarian and development institutions, and between those actors dealing with economic development and those concerned with food security and the livelihoods of the chronically poor. It might also provide a basis for aligning these efforts with the household strategies that have always integrated food security, natural resource management, disaster resilience and livelihoods.

46

Settlement and shelter reconstruction

Manu Gupta

SEEDS INDIA

Introduction

Reconstruction of shelter and settlements following disasters has remained a challenging task for humanitarian agencies, governments and the people affected. In recent years shelter recovery has become particularly daunting in terms of scale and complexity. The website of the United Nations (UN) Office for the Coordination of Humanitarian Affairs (OCHA) provides detailed data for each event; see www.reliefweb.int. The Haiti earthquake in January 2010 destroyed 97,000 homes. The 2008 Sichuan earthquake in China saw more than 5 million people homeless. In 2004 some 470,000 houses were destroyed by the Indian Ocean tsunami. In Myanmar 375,000 houses were damaged by Cyclone Nargis in 2008, while 200,000 houses were lost due to the 2005 Pakistan earthquake. In 2010 some 188,000 houses were destroyed in the earthquake in Haiti.

While shelter is a universal basic need, the needs for surviving communities are contextual and rooted strongly in local governance mechanisms, social and cultural practices and economic status. Humanitarian agencies involved in shelter recovery face the dilemma of timeliness of action versus quality in delivery and of addressing shelter needs in locally acceptable forms.

Concepts of post-disaster shelter

Shelters for surviving communities in the aftermath of disasters or other emergencies are thought about in different ways in accordance with their intended purpose and the point of time of the intervention. In addition to the usual terms such as 'temporary', 'interim', 'transit' and 'permanent', there are many other terms based on nuances in language, culture and governance structures. Distinctions are often made between shelter and housing in a post–disaster scenario, in which 'sheltering' denotes the activity of staying in a place during the height and immediate aftermath of a disaster, while regular daily routines are suspended. 'Housing' denotes the residential locus of a return to normal activities such as work, school and so on. Based on this distinction, four stages of shelter and housing are usually defined as:

553

- Emergency shelter, which may take the form of a public shelter (schools, stadia, etc.) or a shelter under a plastic sheet. The stay is expected to be short.
- Temporary shelter, which may be a tent or a mass shelter used for a few weeks following the disaster and accompanied by basic supply of food, water, health services, etc.
- Temporary housing, which is the initial step towards normal life with possible return to work and school. Communities live in temporary housing while awaiting a permanent solution.
- Permanent housing is the return to the former home after its reconstruction, or resettlement in a new home.

(Quarantelli 1995)

From a theoretical perspective, these distinctions imply a continuum of changing needs of disaster-affected communities and the pragmatic response by humanitarian agencies. In reality these distinctions are usually blurred. The temporary shelter and temporary housing are often merged and in recent years have given way to the terms 'interim' or 'transitional'. The three re-organised categories, corresponding to three stages of response process, may thus be identified as emergency, transitional and permanent (Corsellis and Vitale 2005). Primarily these are meant for individual households and thus exclude other forms of intervention such as mass evacuation centres – cyclone/flood shelters and community centres which offer refuge against sudden displacement of communities. The three stages follow an approximate, though distinctive, time frame as evident from collated evidence of post-disaster shelter recovery (Johnson 2007). The emergency shelter phase is active immediately after the disaster and may last up to thirty to forty-five days. The interim/transitional housing phase can last up to twelve to eighteen months. Work on permanent housing overlaps with the transitional housing period wherein community consultations can take place.

Emergency shelter

Emergency shelters are often tents or similar improvisations, and are useful only in immediate disaster relief. They have been unsuccessful in meeting community needs especially in protection against extreme weather, comfort and cultural preferences (Leon *et al.* 2009). The Sphere Project (2004) minimum standards on shelter and settlements recommend that people be given an opportunity to return to their own land and dwellings and such a provision be made by repair of damaged dwellings. In case return may not be possible, families should be hosted by another community where additional shelters may be added or extended. Only as the last resort, where such opportunities do not exist or there are no local materials available, should tents be used. Prefabricated shelters, too, have proved ineffective due to high unit costs, long shipping times, long production and their inflexibility in terms of design and use.

Transitional shelter

After the 2004 Indian Ocean tsunami agencies working on shelter recovery began to promote the concept of 'transitional shelter' to replace 'temporary shelter'. A transitional shelter offers a basis for up-grading or reuse as permanent housing. Transitional shelters mark a preliminary restoration of normality for surviving communities during which livelihood activities, especially for poor households, and other community practices can be resumed. They should be able to offer a basic sense of protection and safety against subsequent hazards in the same area and meet basic preferences of communities with respect to their socio-cultural and economic practices. Risk reduction practices such as use of disaster-resistant building technologies, as well as community redevelopment strategies, are therefore an essential part of the transitional shelter phase. Communities can be actively engaged in construction activities at this stage. Options and preferences can be negotiated and their participation in the process can lead to faster, more efficient and acceptable solutions. In Kashmir, India, in the aftermath of the 2005 earthquake, communities were encouraged to use salvaged material from damaged houses, thus enabling large-scale reconstruction in a very short time (Rawal 2009). For humanitarian agencies, transitional shelters offer time to sort out vexed issues of land tenure and relocation that are often associated with permanent housing. Emergency shelters alone, due to their limited useful life, can create undue pressure on early delivery of permanent housing. Following Hurricane Katrina in the USA, many thousands of mobile homes were brought in as 'temporary' housing. However, people housed there found it difficult to move to rental accommodation in the city because of increasing rental prices. Nearly four years after the hurricane, 3,400 people were still in trailers (Linthicum 2009). There were also health hazards associated with the materials used to build and to insulate these units (FEMA 2008).

Permanent housing

The normalisation of living conditions to the satisfaction of local communities marks the permanent housing phase. People's livelihoods are restored and community services are functional. Safe and sustainable housing acceptable to the community are other basic requirements that must be met at this stage. Lately, the evolution of design for permanent housing has moved away from the design studio into the community. It involves a community-led process that facilitates upgrading transitional shelters and recognising indigenous construction practices as part of upgrading. For example, in Pakistan, as part of the 2005 earthquake recovery efforts, the traditional construction practice of *Dhajji-Dewari* structures was recognised and with minor technical inputs was promoted proactively by the government and humanitarian agencies (Jha *et al.* 2010). *Dhajji-Dewari*, Persian for 'patchwork quilt wall', is a traditional form of Kashmiri construction recognised for its sustainability and resistance to earthquake damage. Ultimately, the final form and efficiency of permanent housing is a function of the entire recovery process that started immediately after the disaster hit the communities.

The unbundling of the shelter reconstruction into stages helps humanitarian agencies understand and prioritise action. However, the community perspective may be quite different. As Sudhanshu Shekhar Singh, an Indian humanitarian worker, told the author of this chapter, 'jargon such as transitional shelter, intermediate shelter, permanent shelter may make sense in academic papers and high-end proposals, but for poor families on the ground who lost everything, it is a house'. Also, since people are often exposed to multiple hazards, shelter and housing and the building location should all be secure against all the hazards common in an area. Some emergency and transitional shelter in Haiti following the earthquake turned out to be exposed to flooding and wind hazards (Gell 2010). Also, people understandably are concerned with security of property against theft from their homes as well as crime against persons, such as sexual assault where sanitary facilities are not adequately lighted and located for night time use.

Resettlement: the challenges of relocation

Agencies involved in reconstruction are often faced with the crucial question of relocation. Disaster events reveal the exposure of certain locations to particular hazards. It therefore might seem logical to locate reconstruction at sites that are safe from hazards. However, human settlements have historically evolved on particular locations for many economic, political and cultural reasons. Relocating settlements solely on the basis of their exposure to natural hazards generally faces resistance from communities. Past experience reveals that relocation is costly both in financial terms and for the social upheaval to which it leads (Davis 1981).

A few success stories of relocation provide useful lessons (Oliver-Smith 1991). The earthquake that struck Guatemala in 1976 left thousands of people displaced, living in sub-standard housing in both outlying and internal neighbourhoods of Guatemala City. In some of the working class neighbourhoods, in situ reconstruction was impossible due to the instability of the terrain. The leaders of the neighbourhoods in co-operation with the local church organisation and a group of students at a local university organised a 'land invasion' of over 1,000 families. The local authorities were forced into action. They bought land for a new settlement and set up basic infrastructure in the new site. In return, the affected community was required to contribute at least three weeks of labour in housing construction and pay a mortgage of US$8–US$10 per month. Land titles were transferred to the recipient households after a year of occupation and use of the property. In this case, the resettlement was voluntary and facilitated by organisations that communities could trust. Thus factors that have influenced success or failure of resettlement policies include the physical environment of the new settlement, the relationship to the old settlement and its social institutions, and the capability of the affected community to rebuild their livelihood.

Following the 2004 Indian Ocean tsunami, coastal topography was significantly altered along areas suffering direct impact. Villages and land were left permanently submerged or highly prone to flooding and unsuitable for reconstruction or agriculture. In Indonesia, public policy encouraged voluntary resettlement on land purchased by communities themselves or by the public agency. However, it turned out that relocation worked only in areas where the livelihoods were not too far detached from their original locations (da Silva 2010).

Several elements combined to allow people to live farther away from dangerous zones without leaving the old settlement area: a genuine need for relocation felt by the community, the involvement of facilitating organisations and a flexible approach adopted by public agencies. Above all, communities were allowed to make choices and decisions about land use and infrastructure. It is also helpful to see relocation not only as a means to 'compensate' for property losses but as a way to pursue the restoration and enhancement of income-generating capacity and livelihoods of affected people (Cernea 1997).

Policy vehicles for implementing reconstruction

The degree of success of reconstruction programmes around the world has been largely influenced by the institutional mechanisms employed. In recent years, setting up special government institutions usually headed by very highly placed government officials has become a normal practice: BNPB (*Badan Nasional Penanggulangan Bencana*, or national board for disaster management) in Indonesia after the 2004 tsunami; ERRA (Earthquake Reconstruction and Rehabilitation Authority) in Pakistan after the 2005 earthquake; and GSDMA (Gujarat State Disaster Management Authority) in Gujarat, India after the 2001 earthquake. Earlier on, existing institutions were specifically mandated to carry out reconstruction, for example SENA (*Servicio Nacional de Aprendizaje*, or national learning service) in Colombia, after the 1983 Popayan earthquake (Wilches-Chaux 1984). The creation of special institutions for carrying out reconstruction has proved to provide strong leadership for the implementation of the reconstruction programme. The leadership is reflected in clear policy, guidelines on the ground and continuous

Box 46.1 Post-disaster resettlement/relocation

Camillo Boano
Development Planning Unit, University College London, UK

Some disasters leave little alternative but resettlement, making the disaster-resettlement/relocation nexus at the centre of the debate on post-disaster practice. Relocation can be promoted as being the best option to reduce the risk of future disasters and to lessen site-specific vulnerabilities such as those inherent in slums on unstable hillsides and communities in flood-prone regions. Governments can propose that relocating people by compensating them with new land and new homes in physically safer places is considered a just solution to the post-disaster housing shortage. The debate about dispersing much of the pre-earthquake population of Port-au-Prince, Haiti to other parts of the country is a case in point.

Yet resettlement is not always the best solution as it can introduce new risks or exacerbate existing ones. Finding adequate sites for resettling disaster-affected communities can be an enormous challenge. Unsuitable sites can lead to lost livelihoods, lost sense of community, cultural alienation, poverty and the abandonment of the new site for the previous one. In post-tsunami Sri Lanka the rigid grid layout hinders a sense of community and prevents alterations, hampering appropriate provision of infrastructure and increasing community tensions. The economic, social and environmental costs of resettlement should be carefully assessed before the decision to resettle is finalised, and ways to mitigate these costs should be considered, including the use of transparent and participatory methods.

Despite great efforts to make resettlement more humane and affordable by devising standard guidelines (see www.housingreconstruction.org/housing), one of the main reasons for resettlement failure is de-emphasising the well-being of the population as a criterion for the selection of the relocation site. Inappropriate land may be chosen for a resettlement project because it can be acquired quickly, is owned or controlled by the government, or is easily accessible with topography that favours rapid construction. A lack of affordable land in areas close to sources of employment and appropriate livelihoods often necessitates resettlement to peripheral peri-urban areas where land is less expensive, distancing the new site from vital resources (grazing land, food sources), established social networks, livelihood opportunities and markets. In addition, unresponsive housing design and poor construction are often to blame for the rejection or failure of post-disaster resettlement projects, creating non-viable alternatives for the relocated population such as severe livelihood adaptation or abandonment of the new housing.

When resettlement is chosen, it is more likely to be successful when communities fully participate in well-planned and adequately financed programmes that include elements such as land-for-land compensation, livelihood generation, food security, improved access to health services, transportation to jobs, appropriate housing designs and settlement layouts, proximity of natural habitat, restoration of community centres and support for community and economic development. In other words, there is an increased chance of success when resettlement is conceived as a sustainable development programme that includes disaster risk reduction (DRR). Devising alternatives helps to protect communities from what could potentially become a 'land grab' or new form of 'forced eviction' under the false banner of risk reduction, which has contributed to eviction–relocation becoming the rule rather than the exception.

monitoring of the programme. Following the 2001 earthquake in Gujarat, one of the early tasks undertaken by the GSDMA was to produce a simple policy document that laid out the path for recovery (GSDMA 2001). The policy outlined a differential implementation strategy creating clear roles and responsibilities for the government, non-governmental and community institutions. In effect, an enabling policy environment was created thus opening space for a range of public–private partnerships which facilitated the recovery process.

In contrast to the case of Popayan, recovery following Hurricane Katrina in the USA is one in which government agencies remained paralysed. Vital time lost in the recovery process may have caused irreversible damage. For example, planning and implementation of post-Hurricane Katrina housing was fragmented among several agencies: the federal Department of Housing and Urban Development (HUD), the Federal Emergency Management Agency (FEMA) and the Louisiana Housing Finance Agency. There was much confusion and delay (US Congress 2009), and some experts worried that post-Katrina housing policy would perpetuate racial segregation (Seicshnaydre 2010). As noted earlier, thousands of affected families were still in 'temporary' trailers years later, and as late as 2010, FEMA's plans to sell off 120,000 trailers at a huge financial loss were widely criticised (Hsu 2010).

Clearly outlining the scope of recovery intervention, backed by non-negotiable principles of action, forms the backbone of a recovery programme. Continuous peer-level learning through sharing of experiences, networking, training and advocacy within governments and humanitarian agencies provides the necessary foundation for comprehending the problem much in advance and outlining the strategy at the time of crises.

Training and technical capacity building

In Iran, after the Bam earthquake in 2003, quality of rebuilt housing was assured by a system of 'demonstration houses', supported by a cascading system of technical and skill-based training (Jha et al. 2010). In Gujarat and in Pakistan, technical guidelines prepared and mandated by the government institutions provided the guidance on design, technology and building material parameters needed in reconstruction. In Ecuador, indigenous knowledge was showcased with as many as ten prototype demonstration houses using bamboo (Jha et al. 2010). The 'housing facilitators' approach followed in Aceh, Indonesia after the 2004 tsunami used four major players in building capacity. First, the programme employed housing facilitators who were young professionals, mainly engineers and architects, who worked with local communities to facilitate the housing construction process, especially ensuring construction quality. Second, there were community contractors who were the representatives of local residents. In Aceh, each neighbourhood was divided into a cluster of ten-to-fifteen houses. Each cluster selected one person as their representative. That representative was the community contractor whose responsibility it was to ensure the proper construction of houses. Third, masons and manual workers were employed by the community contractors for constructing houses. Finally, house owners had to be intensively involved in the construction process (Petal et al. 2008).

These lessons from Aceh are similar to ones apparent in a similar experience of post-earthquake reconstruction in Colombia after the 1983 earthquake, which damaged the city of Popayan. The task of reconstruction was assigned to an unlikely agency, SENA, a governmental organisation responsible for professional training and social development. The agency could, however, leverage its own training skills to facilitate a successful self-help construction programme. SENA's instructors were, regardless of their specialty, trained in a methodology based on 'learning how to learn' or acquiring methodological tools so that each person could, from their own individual or community work, extract knowledge required for specific objectives.

The methodology also emphasised learning how to acquire skills to enable a person to carry on a trade, and learning how to develop positive attitudes towards oneself and towards one's community. The objectives of the reconstruction programme were focused on the 'process', as SENA was not a housing reconstruction agency as such. The process, however, successfully embraced non-negotiable technical guidelines for construction, so that members of local communities, including men, women and even children, could learn seismic-resistant construction as they were progressively involved in the rebuilding of their new houses. SENA's instructors were divided into two groups – technical and social – and the community organised into modules, with internal 'monitors' and volunteers who would act as links with SENA instructors. Concurrent with the self-help construction programme, SENA encouraged enterprise building and training for employment-promotion activities in construction that further strengthened supply of housing material by tapping skills and resources from within communities. As a result, 1,153 shelters were built in about eighteen months (Wilches-Chaux 1984).

Community-led shelter recovery

The Popayan experience in Colombia introduces the general topic of community-led shelter recovery. Often the potential of surviving communities to take the lead in the shelter recovery process is underestimated. However, repeated experiences have shown that an owner-driven reconstruction process has been far more successful, sometimes with surprising and large spinoffs and side benefits. 'Owner-driven reconstruction' means that the process involves a number of individuals and households that lead the shelter reconstruction enterprise rather than leadership coming from an external agency. Of course, that does not mean that no external agency is involved at all. Numerous instances of completely independent, spontaneous self-help shelter recovery exist. Indeed, in the case of smaller hazard events this is quite common. However, the process referred to above as owner-led or -driven is one in which the external agency provides an enabling environment, but the siting, design and construction is led by affected people themselves.

Owner-led reconstruction has been practised in various forms around the world. The work carried out by the coffee growers' association in Colombia after the 1999 Eje Cafetero earthquake, in which 14,000 individual housing and related projects were completed within eighteen months, is an example (Lizarralde et al. 2009). In Gujarat after the 2001 earthquake, owner-driven reconstruction took two forms. In the first one, the government provided financial and technical assistance, and the survivors were given full freedom to construct on their own. In the second form of initiative, non-governmental organisations (NGOs) were encouraged to provide building material and technical assistance with financial assistance provided by the government. The collaboration of government and NGOs created an enabling environment in the form of access to safe land with secure tenure, material for building construction, appropriate knowledge and skills for disaster-resistant construction and an efficient system for grievance resolution (Barenstein 2006).

The success of owner-driven reconstruction strategies has led to the formation of a network in India called the Owner Driven Reconstruction Collaboration (ODRC) (www.odreconstruction.net). The ODRC is currently engaged with the government in preparing standard guidelines of practice on housing reconstruction. On the ground, the ODRC is working with the government to lead a post-flood housing recovery programme that complements government efforts in an ongoing social housing programme. This complementary activity includes enhanced financial assistance to house owners in order to enable them to add flood-resistant features in their reconstructed houses. The ODRC has added a further financial incentive if the work is completed in time. A key to success in this case is the presence of a generous supply of trained manpower, well informed about appropriate technologies and design in disaster-resistant construction to assist the reconstruction process at the individual household level.

Bridging the knowledge gap: from research to practice

Not all situations are as fortunate as the ones described above. There is sometimes a lack of appropriate construction knowledge. Thus, integration of disaster-resistant features in shelter recovery programmes often requires concerted efforts in transfer of knowledge (see Chapter 9). While at the international and national level knowledge of low-cost, disaster-resistant construction has grown exponentially, the communities at risk may remain unaware of these techniques and the availability of materials and assistance. Recovery programmes offer opportunities to bridge this knowledge gap.

Skills-based training

Post-disaster reconstruction creates a surge in demand for skilled and semi-skilled labour. There can also be a mismatch between chosen type of reconstruction and local capabilities (da Silva 2010). Given the challenge, mobilising human resources and building skills within existing communities may be the only option. Doing so, however, requires a well-designed training programme that focuses on providing basic inputs on construction technology and good quality control. With the urgency of reconstruction driving demand, quality can be compromised.

Creating a cadre of master trainers (even if they have to be brought in from elsewhere) supported by 'hands-on' training can jump-start capacity-building efforts. Hands-on training, however, has to be closely supervised. Lessons in theory, if any, would have to overcome barriers of language and illiteracy. In Colombia (Wilches-Chaux 1984) and in Gujarat (Petal *et al.* 2008) construction workers were mobilised within the community. As a result, the skills developed remained with the community.

Involving the community in skills-based training raised house owners' awareness of 'building back better', while preserving their own interpretation of what their home should look like. In Nepal, mason training programmes were accompanied by public awareness programmes using innovative methods such as small model houses for shake table demonstrations of what actually happens in an earthquake (Petal *et al.* 2008). Awareness leads to demand and provides an enabling environment for communities to volunteer for training in construction.

Local knowledge and co-production of design

Communication of new techniques to community members including specialised construction workers (carpenters, masons, etc.) via skills-based training should not neglect local construction methods and materials as a starting point. There is often considerable potential for local knowledge to become the basis for upgrading and use in contemporary settlements. As noted earlier, in Pakistan a traditional building technology that has built-in features of earthquake resistance was recognised and used in recovery. In the Indian state of Rajasthan, traditional designs and material technology of communities living in arid regions were considered by engineers and modified so that they could be incorporated into the shelter recovery process. In this case the use of the dried stalk of a local plant with unique properties was used for roofing. The stalk of the jowar plant (*Sorghum bicolor*) has a high surface tension that allows water to run off, yet is porous enough to allow ventilation. This material was retained in the reconstruction designs as other available alternatives such as corrugated iron sheets could in no way provide the double benefits of ventilation and protection against occasional light rains (Gupta 2008). In Afghanistan, the relative sophistication of traditional designs followed by host populations provided inspiration to the design and choice of materials and skills in transitional shelter for the internally displaced people (Ashmore *et al.* 2003).

The urban challenge

In urban reconstruction, many of the prevalent strategies such as community-led shelter recovery may turn out to be counterproductive. Urban housing may be more complex than in rural areas, where the built environment is dominated by individual, low-rise shelter units owned by households. While such individual housing is found in cities in the global North and self-built, low-rise residential units are common in informal settlements of the cities of the global South, rental is also very common, as is higher-density, high-rise accommodation. Thus there are often multiple families living on a single plot of land (in multi-storey apartment blocks) and there are landowners and tenants, with the latter not having land title. Furthermore, there are squatter settlements that do not have any legal right over lands they occupy (see Chapter 13 and Chapter 53).

Experiences in Mexico City and San Salvador highlight the complexities of urban reconstruction. Differential strategies that cater to communities belonging to a particular housing type have to be supported by local government intervention, economic incentives and microfinance schemes. In Mexico City, following the 1985 earthquake and due to considerable political protest by affected people and their support groups, the government eventually provided choices of locations and reconstruction approaches to the affected communities. This was in contrast to the government's initial intention simply to move all the affected people to new settlements on the extreme edge of the megacity region. Also, in the end, some 50,000 affected people were employed as paid construction workers (Kreimer and Echeverría 1991). In El Salvador, after the 1986 earthquake, the experience was quite the opposite, with uncertainties of land tenure, interruptions and ambiguity in the decision-making process, which led to a long delay in the reconstruction programme (Cuny 1987).

Overall, both rural and urban reconstruction has been successful when equity was considered in the approach towards house owners and tenants, when the government provided physical redevelopment plans that clearly offered better infrastructure and services and when these plans were well understood by the affected population. The latter is critical. An active facilitative mechanism that promotes extensive community consultations in setting local priorities has proven indispensable. Also successful urban reconstruction has included mechanisms for redressing grievances and disputes as well as a robust and transparent financing programme to provide accessible credit to affected communities and strong support by NGOs and volunteer groups in anticipating needs, problems and aspirations. Box 46.2 describes such a constellation of positive and crucial factors in place in both rural and urban situations in India. Box 46.2 is based on the personal experience of the chapter's author.

Principles of action

The 2010 World Bank document, *Safer Homes, Stronger Communities* (Jha *et al.* 2010), outlines ten principles of shelter recovery for any decision-maker framing a policy on shelter recovery. Similarly, the report *Shelter after Disaster*, first published by Davis (1978) and then released in 1982 by the Office of the United Nations Disaster Relief Coordinator (UNDRO, now UN-OCHA) (UNDRO 1982b), outlines twelve principles (republished as ten principles by Shelter Centre 2008). The two sets of principles were published with a gap of twenty-six years between them.

Box 46.2 Learning from experiences in Patanka, India

The 2001 Gujarat earthquake left more than one-half of the 253 houses in the village of Patanka totally destroyed. An owner-driven recovery programme was introduced. Affected Gujarati villages received financial assistance from the government for repair and reconstruction of housing. An NGO called SEEDS (www.seedsindia.org) facilitated recovery by providing an additional amount of support in the form of building material, and provided training to masons concerning disaster-resistant construction practices. An awareness and mobilisation programme was carried out concurrently to generate a consensus on the goals of the programme. This communication strategy was similar to the ones discussed earlier in the cases from Colombia and Pakistan.

The fifteen-month project provided some important lessons:

1 Building trust is vital. In order for communities to participate and lead the recovery programme, the external agency needs to make efforts so that they can be trusted. Programme objectives, along with the commitments of the agency, should be made clear. Staff of the external agency should be able to build rapport with local residents, and local leaders must be taken into confidence. In Patanka, a series of confidence- and trust-building meetings were organised. Locally preferred cultural activities, sporting events and other team-building activities helped to integrate local community lifestyles with programme activities and staff.

2 Social inclusiveness is essential. In Patanka, the SEEDS team volunteered to build the house of the only widow in the village. Similarly, other smaller groups of poorer households were given extra support. These groups were identified by the village government. Instead of a 'one size fits all' approach, a disaggregated strategy to meet specific requirements of vulnerable groups within communities is an essential recipe for success. Recovery programmes should allow some flexibility for community-level variations to be adopted. A comprehensive damage and needs assessment can support such an approach.

3 Mutual learning and teaching is required. Instead of set ideas on design and technology, the SEEDS team carefully studied local traditional practices in architecture and design. The external intervention was limited to addition of elements that ensured safety and sustainability. This brought about greater ownership and acceptability by communities. External agencies should be open to ideas and suggestions from the communities. A common meeting point must be reached on the basis of shared ideas and values. In architectural terms, the common meeting point can be a 'demonstration house' that is able to translate the ideas and aspirations of communities and matched by principles and guidelines to which external agencies need to adhere.

4 There needs to be joint monitoring. In Patanka, each house owner was given a 'smart card'. The card could be used to record progress made by the house owners in the stages of construction of their houses. This was checked periodically by government engineers and the SEEDS representatives. Based on progress made at individual house level, further follow-up action could be taken up such as release of a subsequent instalment of financial assistance, additional technical support, etc. Such a joint system of monitoring provided almost a real-time feedback to local stakeholders. When collated at community level, this provided a real picture of progress made on the ground. In owner-led reconstruction programmes,

communities cannot be passive implementers of the programme. A joint monitoring system such as the smart card system, or other means for real-time review of progress, creates transparency and efficiency in programme implementation without compromising on quality.

5   Recovery programmes need to be open ended. Communities are not concerned with project deadlines, donor reports and other constraints. In Patanka, local festivals, weddings in homes and family reunions were important events when construction activities would cease. For the project managers these can be serious setbacks to project timelines; however, for communities this is their way of life.

Interestingly, they reinforce certain common issues but also reflect a changing trend. The common issues relate to dangers of relocation versus in situ reconstruction and the need for incorporating DRR into recovery activities. The perceptible shifts in focus concern the role of humanitarian agencies as facilitators, on institutions and policy, and most importantly on the increased role of surviving communities as active participants.

Based on the analysis of available literature and experiences of the shelter recovery process, five factors emerge as having influenced the new, largely successful practice of post–disaster shelter recovery:

1   Government provides a conductive policy environment that recognises and empowers community-led action yet creates adequate checks and balances so that the process is anchored suitably in technical and legal rules.
2   Humanitarian agencies play a facilitator's role, taking on functions that positively influence (but do not interfere with) decisions made by house owners and communities. The approach is based on building mutual trust and local capacity, and it recognises existing knowledge, yet creates capacity for up–scaling efficient and timely action.
3   Communities lead a process which transcends mere consultation and actually has the community in the lead in reconstruction. This recognises the right of people to self-determination and to make independent decision on use of financial material and human resources (see Chapter 59).
4   The reconstruction process optimises timeliness of intervention, choice of building technology and cost.
5   Institutional arrangement ensures co-ordination among affected communities, humanitarian agencies and governments.

After the 2004 Indian Ocean tsunami, the phrase 'build back better' was popularised as a common goal for recovery efforts (Clinton 2006). In the context of shelter recovery, the word 'better' has been interpreted in a way that mirrors the high stakes and community aspiration involved in the recovery process. Surviving communities sometimes interpret 'better' as something more 'modern'. In some situations in Aceh, Indonesia following the tsunami, such 'modern' housing was pursued at the cost of safety because people removed key structural components from their new houses in order to save materials and money. They then used these components to extend the building or for fancy finishing so that they would appear more affluent (Kennedy *et al.* 2008). This produces a dilemma for outside agencies. Self-determination and community leadership are to be highly valued and have been shown to be essential for long-term success in recovery; yet safety might suffer. The answer is a more concerted and focused continuing dialogue with the community about these trade-offs.

Box 46.3 Corruption and post-disaster reconstruction

James Lewis
Datum International, UK

Corruption is present, in varying degrees, in all countries and in communities, both urban and rural, comprising the bad as well as the good. In construction and reconstruction, not only does corrupt practice pervert execution but, by siphoning of finance at its source, it denies the opportunity to build or, at least, reduces the amount and the quality, and increases the cost of new building and infrastructure. Bad construction practices may not be revealed except by the next earthquake or hurricane many years after builders and original owners have moved on. In many countries, corruption will have increased since that perpetrated years before.

In China, for example, in 2007, the year before the Sichuan earthquake, the Chinese Communist Party (CCP) secretary in Janwei county of Sichuan province was reported as having acquired US$5 million, while an anti-corruption chief of another province collected bribes worth more than US$4.4 million. Despite more than 1,200 laws, rules and directives against it, corruption in China is concentrated in those sectors with extensive state involvement, such as infrastructure projects and government procurement (Pei 2007). Strong public allegations were made after the earthquake when so many schools were destroyed, with some buildings collapsing adjacent to others that did not. China is not alone: the abuse of public office for personal gain in Italy, for example, is country-wide, regularly involving elected officials.

Extensive and persistent corruption in any country is not a phenomenon isolated from its broader political environment but one that 'involves a non-benevolent principal rather than bureaucratic or institutional slippage from a benevolent one' (Lewis 2005).

In the 2010 Haiti earthquake, thousands of survivors were made homeless, and national recovery was impeded by destroyed government infrastructure. Funding is required for reconstruction of clinics, schools, hospitals, police and fire stations and port facilities, for example, in addition to resettlement and re-housing of the homeless. Large amounts of money are being made available for a country with a recent history of severely corrupt administration and for a construction industry which, worldwide, is the most corrupt sector.

Construction and reconstruction of all kinds requires transparent governance and effective management against cartels and collusion in bidding, fraudulent allocations of aid, and against bribed consents and permissions. Measures for sound administration, made before construction commences, serve towards ensuring that processes are immune to bribery and backhanders and towards a consistent quality able to resist future earthquakes. For the avoidance of 'slippage from benevolence', donors and recipients of reconstruction funding need to comply with the same governance measures.

Any construction, inclusive of reconstruction and whether localised, national or international, requires frequent, consistent, informed and independent inspection to ensure accurate execution of its specifications. Inspection needs to be inclusive of contractual authority to facilitate the stopping of work, the uncovering of suspect work and the taking down and removal of work or materials shown to be inadequate. These are expensive procedures and awareness of their potential implementation will be a first measure towards compliance.

The ideal is difficult to attain. For communities, satisfaction is drawn from the timeliness of shelter delivery, the security of life and assets, privacy, adaptability of design to their cultural preferences and comfort. The latter may well include design features that symbolise the status of the household and affirm its identity.

Conclusions

Post-disaster shelter programmes are subject to a continuously changing context, and therefore successful practices have been found to have a number of things in common. First, they exist as part of a comprehensive policy framed at the outset. Second, they are implemented rigorously and respond to challenges that emerge from time to time. Third, they are supported through an enabling environment that includes driving institutions, capacity-building orientation and community-based approaches. Community involvement is especially important. With affected people involved in design and implementation, the concept of shelter must be interpreted in more fluid terms, so that it moulds itself to local needs and preferences. Thus the approach to shelter has shifted to a focus on 'process' rather than merely on 'product'. The successful shelter process is participatory, co-ordinated and optimises the use of resources.

There are still barriers where governance is poor or weak and where there is little history of community participation. However, increasingly there is a conscious shift towards an approach and official policy that empowers surviving communities to lead the reconstruction process while supported by a conductive framework. Such an enabling environment includes decentralised governance and a techno-legal framework that allows diversity of materials, codes and flexibility in designs without compromising on safety and environmental sustainability. This framework must also include adequate capacity in the form of technical skills and locally available building material. Consensus is also emerging around the notion of 'building back better'. This goes along with a growing awareness that every disaster is also an opportunity to bring about transformation in society. Vulnerabilities can be reduced and seeds of economic prosperity can be sown. With improved focus on the most vulnerable, shelter projects have contributed to economic and social security, as the experiences from Patanka in Pakistan, Gujarat in India and in many other parts of the world have shown.

Recovery

47

Psycho-social recovery

Tammam Aloudat

INTERNATIONAL FEDERATION OF THE RED CROSS AND RED CRESCENT SOCIETIES

Lene Christensen

DANISH RED CROSS

Introduction

When disasters strike they cause physical destruction and loss of life and property that affects individuals and communities. Disaster relief has traditionally focused on saving lives and ensuring basic needs of those who have survived the disaster by giving access to food, water, shelter and health care in the immediate aftermath of a disaster. In addition to physical damage, disasters cause significant psychological and social suffering to affected populations. Due to multiple losses, disasters and their consequences impact human well-being and ability to function. There is loss of life or ability, property and prospects of future livelihood. The psychological and social impacts of emergencies may be acute in the short term and can undermine the long-term mental health and psycho-social well-being of the affected population and threaten peace, human rights and development (IASC 2007).

It is commonly recognised that surviving or witnessing crises or disasters may produce a known set of biological stress reactions (Bromet and Havenaar 2002). Extreme stress reactions often occur when such events are (a) outside the range of human expectation, (b) take place in a sudden and unexpected manner and (c) threaten human life and create intense feelings of fear, helplessness and horror (Norris and Elrod 2006). Not everyone who is exposed to such life-threatening experiences reacts, nor do they react in the same way. Known reactions include physical symptoms (e.g. headache or stomach-ache), emotional reactions (e.g. feelings of numbness, detachment or strong emotions such as anger or sadness) and behavioural reactions (e.g. apathy, estrangement from family and friends, and change in temperament). Most people regain their balance again within a few weeks or months; however, some will require specialised medical attention (Salem-Pickartz 2009). The diverse psycho-social burdens and possible interventions leave space for different approaches for what needs to be done to handle them. Interventions range from highly specialised medical-psychiatric ones, to others that are based on the communities' ability to cope using their own resources with external support setting the condition for such coping mechanisms to be effective. Over the past years, and with more emerging evidence, direction has moved from much emphasis on the former approach to more balance between the two.

Many people who experience stress reactions are not aware that such reactions are a normal way for the human body to react. Baron (2002: 188–207) reports how a group of southern Sudanese refugees living in exile in northern Uganda complained of loss or inability to function. The complaints included feelings of fear and anxiety, physical pain (e.g. head, neck, joints), shortness of breath, disturbed sleep or nightmares, and loss of interest in care for self and family. Not being aware of the normal emotional and physical stress reactions, some refugees 'were distressed by their feelings of stress and became further distressed by the fear of not under- standing their own symptoms'. This led to the development of psychosomatic aches and pains, a common complaint after disasters (Baron 2002).

Psycho-social effects of disasters

Social effects of disasters

Normal routines are severely and uncontrollably disrupted by disasters. Family functioning is affected if there is physical separation; in fact, one of the greatest sources of human distress is not knowing whether relatives or significant others are dead, hurt or in other ways affected by what has happened. People who are distressed or suffering from acute stress reactions are less able to take care of themselves and each other than they normally would. Social disruption affects the ability of individuals, families and communities to function, and upsets the everyday routines in which family members normally engage. In many societies women have responsibility for caring for children and may not be able to retain this role even if the family is forced to relocate to a camp or temporary shelter if their dwelling has been destroyed. Men, who are often the breadwinners, may find it difficult to cope if they have lost their means of a livelihood. Such disruption to social norms creates increased tension and abuse rates are often seen to rise in post-disaster settings (IFRC and Johns Hopkins University 2008).

Societies, both before and after a disaster hits, are dynamic environments. Community members experience adversity differently, based on their pre-disaster experiences and way of life, the impact that the disaster itself has had on their life situation and the way in which the humanitarian response following the disaster is organised. In post-disaster settings, traditional support mechanisms may be affected or cease to operate because individuals and groups are no longer able to engage or interact as they normally would. A broad variety of issues may be experienced at the individual, family, community or societal level. The problems experienced may be predominantly psychological or social in nature, yet often interconnected as the com- posite term psycho-social suggests. Pre-existing social problems include extreme poverty and belonging to a group that is discriminated against, marginalised or being politically oppressed, whereas pre-existing psychological problems include severe mental disorder and alcohol or other types of abuse. Disaster-induced social problems encompass family separation, disruption of social networks, destruction of community structures, resources and trust, and increased gender-based disadvantage or violence, while disaster-induced psychological problems include grief, non-pathological distress, depression and anxiety disorders, including post-traumatic stress disorder (PTSD). Problems encountered as a consequence of poorly organised humanitarian aid encompass the *social* side – undermining of community structures or traditional support mechanisms and exclusion due to lack of access to services – while *psychological* consequences may be anxiety due to a lack of information about food distribution or other services delivered (IASC 2007).

Loss and grief

Disasters and crisis events are invariably characterised by loss: loss of life, destruction of property, loss of livelihood, physical injury and breakdown of social networks. In the case of sudden-onset disasters, e.g. associated with an earthquake, violence or major and mass-casualty accidents involving many people, reactions may be intensified. Grief is the natural, and often painful, process that sets in to release the affected person from what has been lost.

The grief process may be conceptually captured in four areas of adjustment: emotional recognition of the loss, living through feelings of grief, making practical adjustments, and turning towards the future by learning to live with the memory of what has been lost (Baker *et al.* 1992). While their manifestations will differ by culture and over time among human beings, on the whole ordering one's understanding of grief according to a framework that uses these four broad categories has been found to be useful.

People, on an individual level, respond differently to such life-changing events, and reactions also vary among different cultural settings. The cultural context in which people are brought up shapes the way they understand, act and make sense of the world that surrounds them; in a way, different cultures can be said to 'suffer differently' (Watters 2007). Cultural differences are also found in beliefs and practices of how people heal after a crisis. Cross-cultural differences in distress, grief, healing and recovery can potentially hinder an effective outside response. Utilising Western ways of categorising mental states and disorders in different cultural contexts may have the effect that certain culture-specific ways of expressing and interpreting the world are mis-understood or misinterpreted. It may also be that certain recognisable symptoms are identified, but that these have a different meaning or are not considered important in the local setting. Failure to recognise culture-specific characteristics of the community and their loss and grief can result in inappropriate approaches for the targeted community or parts of it. Researchers and counsellors arriving in Sri Lanka after the Asian tsunami did find PTSD symptoms in the affected population, but the nightmares, flashbacks and other items on the PTSD checklists were less important to the affected Sri Lankans. Their deepest psychological wounds concerned the loss or disturbance of one's role in the group. The idea of leaving the group setting and healing through individual interaction with a counsellor was not only a foreign concept, it also seemed to intensify the fear of social isolation (Watters 2007).

One example of a cultural practice that helps the grieving process and thus facilitates psycho-social recovery is appropriate burial rituals. In all cultures and religions across the world, respect for the dead is a deeply ingrained value. Conducting funeral rituals according to local tradition facilitates both the acceptance that the death has happened and furthers the grieving process. After large-scale disasters, people are commonly buried in mass graves, although both the World Health Organization (WHO) and the Pan American Health Organization (PAHO) advise against this practice. One important reason is that not knowing what has happened to a loved one deprives those left behind of the ability to mourn their dead, fosters continued speculation of their whereabouts, and leaves relatives in a limbo that may create further distress and impact negatively on the ability to recover (de Ville de Goyet 2004).

Psycho-social needs of vulnerable groups

During and after disasters the general population should have access to psycho-social support services, such as psychological first aid, social support and practical and emotional assistance. In every society, particular groups of people are at a higher risk of experiencing social and/or psychological problems. Many sub-groups of a population are potentially vulnerable; these need to be identified through specific psycho-social needs assessment performed by psycho-social support specialists supported by trained community volunteers. Vulnerable groups include women, some men in particular situations (e.g. ex-combatants, idle men, teenagers at risk of detention, abduction or being targets of violence), children, the elderly, extremely poor people, refugees, internally displaced persons (IDP), migrants in irregular situations, and people who have been exposed to extremely stressful events/trauma (e.g. those who lost close family members or their entire livelihoods, were raped or tortured, or witnessed atrocities) (IASC 2007: 3–4).

It is important to pay attention to the needs of potentially vulnerable groups; however, not everyone belonging to a vulnerable group lacks resources or is unable to take care of themselves. All disaster-affected community members should be seen as active survivors rather than passive victims, while at the same time care must be taken to ensure that they are protected from further harm. Within each group there are diversity and potential resources that may be tapped into when organising a psycho-social response for disaster-affected community members. Some groups (e.g. combatants) may be simultaneously at increased risk of some problems (e.g. substance abuse) and at reduced risk of other problems (e.g. starvation) (IASC 2007).

Box 47.1 Stress: different reasons in different contexts

Tammam Aloudat
International Federation of the Red Cross and Red Crescent Societies

On 26 December 2003 an earthquake measuring 6.3 on the Richter scale hit the town of Bam in Iran. The epicentre was close to the densely populated city centre and an estimated 31,000 people died and 17,000 were injured. An international humanitarian operation assisted the Iranian authorities in reaching those affected by the earthquake. A few days following the earthquake a large group of men was roaming the streets, showing violent behaviour and breaking into medical facilities. It turned out that the city of Bam lies on the Silk Road and is therefore a hub for handling and smuggling heroin and opium. As a consequence, the city has a high number of drug addicts who were experiencing abstinence symptoms as their drugs had been buried under the rubble of collapsed buildings. Until the usual supply routes had been re-established, the group demanded quite a lot of attention.

In Aceh, after the Asian tsunami, the usually disempowered women were put in a situation where they had to care for their families in the extreme situation in which they found themselves. They expanded their traditional role, took care of the young-sters, fed their families and helped find shelter. In contrast, men, who are the usual breadwinners and family protectors, were in a situation where they could not prac-tise their normal roles, work or protect their families. The situation reversed the 'normal' power balance, empowering many women, and put many men in a vulnerable situation.

The two examples above show that vulnerabilities and vulnerable groups change from one context to the other, and during and after the disaster in the same place. Such changes have to be understood, scrutinised and considered when planning and implementing mental health and psycho-social support programmes.

Emergence and development of mental health and psycho-social support (MHPSS) as a field of humanitarian assistance

Origins

Humanitarian aid, despite having existed in modern form for over 150 years (Stoddard 2003), is still a largely evolving field. The past three decades have seen a massive evolution from interventions such as indiscriminately airdropping food over Ethiopia during the 1984–85 famine and non-specialised celebrities leading on-the-ground relief work as in the Armenia earthquake (Keller 1988), into a better-regulated field of work moving swiftly towards evidence-based practice (Bradt 2009).

During those developments in the field of humanitarian aid, the scientific community was still debating whether disasters and crises caused psychological distress and, if so, what was its nature (Gerrity and Flynn 1997). The debate continued despite detailed studies categorising factors that increase vulnerabilities to psychological problems developing after crises and linking them to the characteristics of the disasters, the response, the individuals and groups affected. While programmes tackling psychological consequences of disasters started and evolved from the early 1990s, the debate on those consequences and programmes extended for another fifteen years until wider consensus was seen on what entails good public health practice in respect of mental health (Van Ommeren et al. 2005). Such widely accepted interventions take place away from vertical PTSD-focused programmes into social interventions integrated into the general health care.

A decision on the nature and providers of psychological support in emergencies was needed from the beginning. The question to answer was whether the psychological needs of populations affected by disasters would be most efficiently tackled by mental health professionals and psychologists in vertical programmes or by other means involving more community participation, local rituals and norms, and culturally relevant coping mechanisms. The former approach was widely known as disaster psychology, the latter as psychological first aid which is provided by family, friends and neighbours but not by mental health professionals (Jacobs and Meyer 2006).

The evolution of MHPSS programmes

The year 2004 marked a major turning point for humanitarian aid and MHPSS programmes. This was largely due to the Asian tsunami, one of the biggest disasters in recent times. The tsunami, with its unprecedented proportions and effects on populations in a dozen countries, led to one of the largest emergency response operations in history, and required and forced new thinking and practices of humanitarian aid. This was notable in many ways, including extending relief programmes to include recovery and rehabilitation that went on for several years.

The recognition of MHPSS was achieved and standardisation took steps before the tsunami through the Sphere Project (2004), which focused on information for affected populations, community participation, care for isolated persons, provision of schooling and shelter, maintaining cultural and religious events, appropriate burials, and provision of both psychological first aid and community-based psychological interventions alongside care for urgent psychiatric complaints. The Sphere minimum standards, among more information and research about MHPSS previous programmes and experiences, helped guide better planning and programming in the tsunami and other operations such as the 2005 Pakistan earthquake. However, many of the minimum standards were unattainable in many cases in the midst of the devastation, leading to burial in mass graves for example.

As the relief operation went on, however, the MHPSS gathered steam and MHPSS programmes were implemented by varying organisations ranging from Médecins Sans Frontières' individual and group counselling (de Jong *et al.* 2005) to wide-scale psycho-social support programmes by the Red Cross and Red Crescent National Societies, including support groups, community restoration and reconciliation activities, play activities for children, and commemoration and remembrance activities. The lessons learnt from the tsunami and later operations have guided a major change in approach, more mainstreaming and better understanding, which are demonstrated in later intervention guidelines (IFRC Reference Centre for Psychosocial Support 2009) and the revision of the 2011 Sphere handbook.

Box 47.2 Psycho-social programmes: hero book

Lene Christensen
Danish Red Cross

One example of a psycho-social approach that has been implemented in multiple settings with very positive outcomes is that known as hero books. The approach consists of a process whereby children or young people are guided to become authors, illustrators or editors of a personal hero book. Developed in sub-Saharan Africa, the approach builds on storytelling, which can take place wherever safe spaces are created for an individual to tell their story. The story can be told in words to other people, by making a map, drawing a picture or writing it down.

The hero book approach draws on the psychological body of knowledge known as narrative therapy or practice. One aspect of this practice highlights the concept of 'externalising conversations', which means that the problem (and not the person experiencing it) is the problem. It is a non-pathologising approach that centres people as experts in their own lives. The approach has proven successful in many different settings where psycho-social support care is needed and has helped children affected by HIV and AIDS, poverty or conflict.

Hero books are notable because they are seen to help young people communicate and deal with the difficult circumstances in which they live their lives. They have helped develop self-esteem and have empowered young people to tell their stories in positive ways. They have also been used to draw attention to abuse situations and harmful practices to which children and young people have been subject. It is an empowerment process where children are reassured that a problem they experience is not their fault, and nor is it exclusively their responsibility to solve it.

The impact of the hero book approach was investigated in 2006 by the Regional Psychosocial Support Initiative (REPSSI), University of Western Cape, University of Oxford and Cape Town Child Welfare. The research design did not allow for making causal interpretations about the impact of the approach; nevertheless, seventy-nine per cent of respondents interviewed reported that there had been an improved change in some or all of their problems and ninety-one per cent stated that the intervention had contributed to the change. The qualitative part of the study found that children found an ability to 'communicate inner feelings' by having been part of the hero book process. A selection of hero book stories is found online at web.uct.ac.za/depts/cgc/Jonathan/index.htm.

Community-level experience and action

While disasters have obvious human consequences and may leave people with increased vulnerability compared to their pre-disaster situation, many individuals and families cope relatively well in the face of adversity, show remarkable strengths and are able to use the resources at their disposal to get on with their lives and create a sense of normality under abnormal circumstances. The ability to cope has been treated in psychology and is used in a variety of ways. In relation to psycho-social recovery, coping describes both the competence of individuals who are put under stress and are seen to cope better than expected, and their ability to recover from distressing events. Understanding both the vulnerabilities and capacities of disaster-affected communities is essential for putting in place action that enables communities to get involved in responding to the disaster, while at the same time ensuring that people with special needs are looked after (Ungar 2006). Two examples of coping behaviour come from Sudan's Darfur region. Here a woman living in an area severely affected by famine preserved the family millet seed by mixing it with sand in order to prevent her children from eating it. In another location during the same famine families abandoned food distributions in order to return to their area of origin in time for the planting season (Regel 2007). This shows how behaviour, which at one level seems counter-productive, in fact has a well calculated rationale (see Chapter 58).

Community-based psycho-social interventions

Community-based psycho-social support uses the community as its starting point when an intervention or programme is being planned. Community-based psycho-social support may imply many different activities, from support groups for young mothers with children to livelihood activities for men who have lost their source of income, or establishing child-friendly spaces where children may have the opportunity to play and relax (Chapter 36). Finding out which type of activities are the appropriate ones in a particular community requires an understanding and ability to act on the information provided by community members. Facilitating genuine community participation requires an understanding of local power relations, patterns of community interaction and potential conflict, working with different sub-groups and avoiding the privileging of particular groups. Engaging community members in discussions about their particular problems and the resources that are available is an empowerment process in itself. As people become involved, they are likely to become increasingly hopeful, better able to cope and more active in rebuilding their own lives and communities. At every step, post-disaster relief and recovery efforts should support participation, build on what local people are already doing to help themselves and avoid doing for local people what they can do for themselves.

The starting point: psychological first aid

Disasters can strain individuals' and communities' ability to cope, leading to distress, grief, the placement of great stress on social norms and, in some cases, further psychological consequences such as PTSD. Immediate post-crisis interventions have been under debate and in use for several decades, ranging greatly in nature and practice. The currently accepted practice of immediate psychological intervention is psychological first aid (PFA), which is quoted as best practice by both the Sphere Project (2004) and the Inter-Agency Standing Committee (IASC) guidelines (IASC 2007).

Important issues need to be addressed when discussing the merits of psychological first aid, including the diversity in definitions, components and applications, and the evidence base for the use of PFA in the immediate aftermath of disasters. Several definitions have been in use since the use of PFA spread. While all the basic concepts are shared among different definitions, some variation is worth examining. The IFRC Reference Centre for Psychosocial Support defines PFA as 'providing basic human support, delivering practical information, and showing empathy, concern, respect, and confidence in the abilities of the individual. It is offered to individuals immediately after a critical event' (IFRC Reference Centre for Psychosocial Support 2009). Different definitions include that of the Sphere handbook (Sphere Project 2004: 293): 'basic, non-intrusive pragmatic care with a focus on listening but not forcing talk; assessing needs and ensuring that basic needs are met; encouraging but not forcing company from significant others; and protecting from further harm'; while yet another definition describes PFA as an 'evidence informed model utilised in disaster response to assist those impacted in the hours and early days following … disaster' (Uhernik and Husson 2009). In attempting to synthesise such definitions, common elements can be found. Psychological first aid can be defined as a basic psycho-social intervention (Jacobs 2007) that targets individuals affected by disasters in their immediate aftermath; respects the context, culture and specific audience; uses standardised techniques in a flexible and pragmatic way; and thus provides psychological relief, practical guidance and prevention of further harm.

Describing the components of PFA interventions is another aspect where opinions vary, while keeping the same basic spirit. Different specific components lists of what a successful PFA intervention entails were put forward by the IFRC (2009), the National Center for Child Traumatic Stress Network and National Centre for PTSD (2006), the North Atlantic Treaty Organization (NATO, 2009) and IASC (2007). Instead of listing and contrasting such documents, the varying components will be grouped in fewer categories, which are common to all approaches. The first set of components of PFA is related to the psychological needs and techniques, which include components of PFA interventions such as active listening, contact and engagement, genuine compassion, non-judgement, voluntary sharing of experience, gathering information about the individual, family contact, informing the individual about coping, and understanding people's concerns. The second set is concerned with the physical needs and their fulfilment. This is not necessarily done by the provider of PFA but it is their responsibility to ensure the connection of people they are serving with sources of physical assistance. Such components include ensuring the safety and comfort of disaster victims, protection from further harm, company, help with shelter, services and information on other further assistance. The third set of components is in regard to vulnerable groups. PFA providers should pay attention to the specific conditions and needs of children, adolescents, the elderly, separated family members in need of reuniting with their families, and people with psychiatric conditions in need of specialised attention.

Evidence from applications of psychological first aid

PFA differs from other psycho-social interventions in its use of non-specialised providers, its context and its cultural specificity as well as sustainability (Jacobs 2007). PFA is designed to be provided by non-specialists. The approach, in a similar manner to first aid, is given to individuals by others in their family or community. In the Red Cross/Red Crescent case, community volunteers are trained to provide PFA as part of their disaster relief activities. This consideration is specifically important to make up for the lack of specialist counsellors familiar with the specific context in sufficient numbers after many emergencies. Jacobs (2007: 936) also argues that 'if you have only one tool for providing psycho-social support, you may tend to do the same thing for everyone, everywhere, every time. Given the breadth of individual differences, this is unlikely to be effective. The saving grace of PFA is that it is adapted and shaped for each community in which it is implemented, so that the tool takes on many different shapes in different cultures.' The provision of PFA by local community members and volunteers who have only basic training makes it possible to implement it with minimal cost. This is essential for preparedness where in disaster-prone communities such training and sustainability can be ensured before emergencies when material resources are usually scarce.

A recent systematic review of PFA commissioned by WHO acknowledges the lack of systematic evidence for the effectiveness of PFA, while others call it evidence-informed (Uhernik and Husson 2009) rather than evidence-based (Bisson and Lewis 2009). While this might raise doubts about PFA, other evidence focuses on the negative effects of no intervention and people's feeling of lack of social support, which on its own has been associated with the development of PTSD (Brewin et al. 2000). Additionally, another review states that 'PFA is different from [the discredited] psychological debriefing in that it does not necessarily involve a discussion of the event that caused the distress' (IASC 2007: 119).

Psycho-social assessment: turning needs into action

In order to know what the psycho-social needs and resources are in a particular post-disaster setting, a psycho-social assessment must be conducted. This data-collection exercise provides the basis for deciding whether a psycho-social intervention is needed and, if the answer is yes, creates the basis for designing the programme or set of activities that addresses the psycho-social needs identified. Assessments provide a detailed understanding of the emergency situation and analyses the threats to and capacities for psycho-social well-being of disaster-affected groups.

Depending on the time and resources available, assessments may be conducted in a quick manner right after a crisis event. Such rapid assessments can last from only a few days to a few weeks and are often conducted while initial relief efforts are being put in place. Detailed assessments are comprehensive investigations, often conducted in co-ordination between different agencies and organisations. They usually take place some time after a crisis event has occurred as they must be well planned and co-ordinated. To make sure that all aspects of community life are investigated thoroughly, detailed assessments should be made with full community participation. Local volunteers, health workers or other key resource persons within a community are not only important providers of information, but may help facilitate community discussions and in other ways assist the data-collection process.

In summary, key features of psycho-social needs assessments are:

- Linking to rapid provision of effective support and services;
- Collecting information on how local people understand and experience their situation and how they are able to cope with it;
- Analysing how psycho-social impacts and access vary according to gender, age, ethnicity and other stratifications;
- Realising not only the deficits but the local resources available and how groups (e.g. women's or youth groups) may contribute to programme delivery and support;
- Analysing the situation and programme approach on an on-going basis;
- Mapping local power structures and gender relations, and identifying the most vulnerable, invisible groups and those not included in regular community discussions; and
- Recognising that Western concepts and tools may not apply in the local context and that especially spirituality may be a significant factor to be considered.

(The Psychosocial Working Group 2004)

Tammam Aloudat and Lene Christensen

Mainstreaming MHPSS

Psycho-social support programmes, despite the efforts to standardise and provide the tools for them, have been implemented differently in different countries, by different actors and in diverse crisis settings. While this seems appropriate given the diversity of cultures, coping mechanisms and social norms, it defies attempts to standardise interventions. Also, in what should be considered an early stage in the evolution of such programmes, evidence of outcomes is still being built.

Despite the incomplete evidence to validate and specify the most effective interventions, observations, data collected and well-documented programmes in previous disasters suggest guidelines to direct the implementation of MHPSS programmes as listed below. Continuous revision of available evidence from implementation in disaster operations, academic research, as well as flexible adjustment of the guidelines to the evidence should be adhered to. In the meantime, quantitative and qualitative research is being undertaken to ascertain the effectiveness of MHPSS programmes, and establish causal links between activities and outcomes and impact while gradually eliminating assumptions and proxy indicators. This happens in close co-ordination between academic institutions and implementing agencies.

Programmes ranging widely, from interventions such as community-based psycho-social support conducted by volunteers to psychiatric clinical interventions to religious activities have been labelled psycho-social support. While there is hardly a clear-cut set of criteria for including or excluding such interventions from the general MHPSS stream, it is necessary to carefully examine each of them for a set of criteria before making such decisions. Such criteria could include:

- That the aim of the intervention is to relieve the psycho-social consequences of crises and enable survivors to go on with their lives;
- MHPSS interventions adhere to best practices documented by Sphere and IASC guidelines, as well as to the standards promoted by the major actors in the field such as the Red Cross/ Red Crescent, WHO and UNICEF;
- Cultural and social appropriateness;
- Similar to all humanitarian interventions, MHPSS should be applied according to need, with impartiality and equity for the recipients of services; and
- Co-ordination of any intervention with humanitarian stakeholders, in a cross-cutting manner that allows for the interfacing with other sectors.

Another debate, both theoretically and practically, is whether MHPSS interventions are most effectively implemented as stand-alone interventions or integrated into other interventions, e.g. in the health or relief sectors. Both schools have their arguments. Stand-alone MHPSS interventions enable independent identification of psycho-social needs and implementation that is not limited by the scope or resources of the bigger intervention. The integration approach focuses on the benefits of providing psychological support to complement other relief, health, and water and sanitation intervention. The other school points to the inappropriateness of addressing one aspect, the psycho-social needs, without meeting other needs such as food, medical care, water and sanitation.

Other factors that intervene in a decision about the modes of implementing MHPSS programmes are the situation, the context and the capacities available. It is difficult, and indeed as yet premature, to put rigid rules in such situations before the body of evidence is expanded enough.

Conclusions

Mental health and psycho-social support is a relatively new field in humanitarian response to disasters. The lessons learned from previous and current interventions indicate that MHPSS should cover the wide continuum from specialised clinical mental health interventions to focused support for basic services and security. This implies organising a layered system of complementary supports that address the needs of different groups.

While needs are apparent to professionals working in disaster response, and some tools are available including guidelines and established approaches, all those are dynamic, changeable and subject to the best practices identified and evidence obtained. The evidence base that guides the implementation of effective programmes as well as the further evolution of the field is still being built.

To move on with effective provision of MHPSS to disaster survivors, rigorous co-ordination is a prerequisite to build up the body of knowledge and establish best practices, not merely the co-ordination among implementing agencies, but wider collaborative efforts with communities, governments, academia and other interested groups. An important part is active collaboration between aid agencies and academic institutions. The two entities are mutually dependent if proper research is to be conducted and evidence obtained that is inevitably needed for the validation and evolution of the discipline.

While implementation of MHPSS action is necessary both at services level and for the ability to conduct further research, this cannot happen without active support from donors. Such support has been improving but still requires more sustainable funding to enable programmes to proceed. A strengthened evidence base and enhanced documentation of the results achieved contribute towards being able to access funding for programming.

Another prerequisite for proper implementation is acceptance, from both the community members who participate and receive services and the governments that adopt and promote MHPSS at the policy level. Such acceptance can only be achieved by active advocacy for global and national policies that recognise the needs and effectively implement policy changes on all levels to enable active programme implementation.

By working simultaneously on the needs above, MHPSS can evolve as a strong, standardised, evidence-based and effective component of health and other interventions in international and local disaster response, filling gaps and helping survivors go beyond their immediate needs and into recovering their lives and restoring their livelihoods.

48

Socio-economic recovery

Rohit Jigyasu

RESEARCH CENTER FOR DISASTER MITIGATION OF URBAN CULTURAL HERITAGE, RITSUMEIKAN UNIVERSITY, JAPAN

Introduction: economic impacts of disasters

The increasing occurrence of disasters has resulted not only in a higher toll on lives but also in rising economic damage. Over the two recent decades up to 2003, direct reported economic losses from disasters have multiplied fivefold in real terms to US$629 million (IFRC 2003b).

The economic impact of disasters in the most affluent countries is much higher than in the global South, obviously because the concentration of industries and infrastructure is very high in the global North. By contrast, more lives are lost to hazard events in the low-income countries. Eleven per cent of all people prone to disasters live in the global South but these account for more than fifty-three per cent of the total deaths recorded between 1980 and 2000 (UNDP 2004). Nevertheless, low-income countries bear the heaviest economic burden in terms of average annual damage relative to gross domestic product (GDP), and also the impact of hazards on their citizen's livelihoods is severe.

Disasters are not merely sudden catastrophic events but are phenomena that result from physical, political, economic, social and cultural/attitudinal vulnerability conditions. These conditions are not only active as the context for disasters but strongly affect the recovery process that is initiated afterwards. Therefore, the crucial questions that need to be considered are:

- What are the basic premises and key indicators that determine the economic impact of disasters? How is the ability to recover at different scales – from individuals to countries – linked to the pre-disaster vulnerability situation? Do economic impacts affect all aspects of the economy of the affected region and all those who are affected (e.g. people and households whose livelihoods are not part of the formal economy)?
- What are the key challenges faced in economic recovery in post-disaster situations and what are the underlying reasons for these?
- Can successes and failures of previous recovery initiatives tell us about the essential considerations for sustainable economic recovery at various scales?

Assessing the economic impacts – identifying the gaps

Various tools and methodologies are available for assessing the economic impacts of disasters. There are measurements of direct and indirect impacts. The direct costs come in the form of damages to stocks of physical and human assets and include the costs of relief, rehabilitation and reconstruction. Examples include damage to buildings, crops and economic and social infrastructure. Indirect costs are secondary impacts, which take the form of lost output and investment, macro-economic imbalances caused by disasters, debt and increased poverty (ERM 2005), and cost of damage to means of production, etc. Some of these impacts are more visible and tangible (impacts to which a monetary value can be assigned) than others (which may not be expressed easily in monetary terms). For example, physical damage to houses, businesses and infrastructure is more visible than lost income through being unable to trade, bankruptcy and business closures (EMA 2002). Death of a loved one is a loss with many complex dimensions and cannot be collapsed into something like 'lifetime earning potential' (see Chapter 47). Some of these less tangible impacts may take a much longer time to manifest than physical damage, and therefore they are not generally accounted for in economic assessments that are made immediately after the disaster with the goal of informing reconstruction and insurance processes.

Most economic assessments are based on macro-economic analyses that take into account the GDP of a particular country. It is estimated that from 1999 to 2000, large-scale disasters resulted in damages constituting between two per cent and fifteen per cent of an exposed country's annual GDP (EMA 2005). The damages represent the costs of replacing physical assets at current prices. GDP, however, represents the flow of goods and services in the economy, and not the stock of assets, while economic impacts are experienced through damage to or destruction of assets or 'stocks' resulting from the disaster itself, or from events in the aftermath of a disaster (ERM 2005). Post-disaster reconstruction booms are common and reflect large-scale reconstruction projects of damaged infrastructure along with an inflow of foreign aid. However, in GDP terms this positive increase does not reflect an increase in national physical assets. For example, overall macro-economic impact on growth in Indonesia after the 2004 Indian Ocean tsunami was limited, largely due to increased investment in construction to replace assets destroyed, much of it financed from abroad.

Economic impacts may vary a lot among various sectors. Consider the case of the Maldives after the 2004 Indian Ocean tsunami. The economic impact on the Maldives was substantial due to grave damage to 14 of about 200 inhabited islands, causing around five per cent of the population to lose their homes, closure of one-quarter of tourist resorts and damage to eight per cent of fishing boats (ERM 2005). However, impact on one sector may influence other sectors, thereby causing higher cumulative economic loss. For example, the economic impact of the 1999 earthquake in Turkey was high, as it affected the industrial heartland of the country. The immediate surrounding districts were affected indirectly by their close economic linkages with the affected area, e.g. industries and small businesses supplying services and material inputs to each other's production processes. Taking all these districts together, the wider earthquake region accounted for thirty-five per cent of national GDP and almost one-half of the country's industrial output (Bibbee et al. 2000).

These larger macro-economic indicators seldom reflect damage and recovery at the scale of the micro-economics of local markets and livelihoods. This suggests the necessity of an alternative approach for examining the impacts of disasters in terms of who and what is affected. Economic impact of disasters may not just vary across sectors, but also among groups, especially those who are most vulnerable, e.g. socially marginalised, urban poor, people with disabilities, single parents and children. It is a crucial challenge for economic impact assessments to take into account variation among such groups. Also many livelihood activities of marginal groups, such as labour for construction, selling fruit and vegetables, etc. belong to the informal economy, so it is difficult to assess disaster impact on them through frameworks and indicators designed for formal economic sectors. This is not a trivial, technical point. There are strong functional connections between the formal and informal economies – the myth of 'dualism' being put to rest decades ago – and poor recovery in informal urban manufacturing, trade, urban gardening, and rural food, fibre and fish production can seriously delay full national economic recovery. In addition, such livelihood activities are critical for the immediate, medium- and long-term survival and well-being of the poor, many of whom may not receive much official relief for months, as one saw in Haiti after its 2010 earthquake.

Economic impact assessments also may not take into account urban–rural dynamics. Although urban and rural economies have their own characteristics, disruption in urban economies due to disasters may negatively affect the demand for goods and services from fringe and outlying regions and reduce the flow of remittances to rural areas. On the other hand, disasters in peri-urban and rural zones may stimulate an increased influx of people into cities because rural people who were already experiencing livelihood stress choose to rebuild where they saw better prospects for their children (see Chapter 13).

Economic recovery processes – key issues and challenges

Following a large disaster, the recovery process is often beyond the capacity of national governments, which are occupied with the provision of emergency services. The international community supports recovery through provision of aid both in cash and kind, and various international and local NGOs act as mediators or facilitators in carrying out the recovery process (see Chapter 50).

Winners and losers

Because most governments understand 'development' as identical to macro-economic growth, more attention is usually given during recovery to large-scale business enterprises at the cost of programmatic attention to the livelihoods of local communities, especially the vulnerable and most affected groups. Moreover, recovery initiatives seldom adopt a comprehensive view that addresses economic impacts of disasters on various sectors and on various sections of the community. Since many livelihood strategies operate at the micro level and work within complex social relationships (see Chapter 58), it is difficult for governments and agencies to intervene successfully without collaboration with communities and civil society organisations. Also, governments and agencies have a limited understanding of, and trust in, the coping capacities of poor households (Beck 2005).

In the case of reconstruction following the 1993 Marathwada earthquake in India, most of the reconstruction was undertaken by building contractors from outside the region, which imported state-of-the-art-technology, while the local skills of stone craftsmen and carpenters were not utilised. As a result, many crafts-people shifted to other occupations and even migrated to other regions. On the other hand, the hotel industry received an unprecedented boost from 'disaster tourism', owing to the visit of many rescue and recovery teams as well as researchers from around the world. Conversely, following the 1985 Mexico City earthquake, much of the labour was hired among affected working-class people due to popular protest against importing labour from outside (Kreimer et al. 1999).

There are several other examples to demonstrate how post-disaster reconstruction can adversely affect marginal groups. Following Hurricane Katrina in 2005, the state legislature in Louisiana passed revisions to its gaming law that allowed casinos to build on land rather than on water. However, funnelling of federal relief funds to rebuilding casinos, hotels and chemical plants and to wealthy districts of New Orleans threatened to replace the low-income housing with newly built mansions and condos in gentrified New Orleans (Birch and Wachter 2006; Klein 2005). Clearly, in disaster economic recovery there are winners and losers.

In Malawi, the recovery process following droughts in 2002 could not help small-scale farmers because they did not have physical access to sites of sale nor the means to purchase agricultural inputs such as seeds and fertilisers, and they lacked markets where they could sell their crops. Seeds and fertilisers were available, but credit was still mostly inaccessible to these farmers (Gabre-Madhin 2002).

The challenge of dependency and perverse economic effects

The recovery process carries a risk of increasing dependencies through external aid if local coping mechanisms and institutional systems for recovery are not strengthened. Also a crucial question is how long these economic initiatives can be sustained in the long run. When the reconstruction process is externally driven, the effects of economic recovery are often short-lived and depend on a steady supply of external economic and human resources. Moreover, their sustainability also depends on local enabling mechanisms that in turn require effective governance and community engagement. Often the interface between external and internal mechanisms is missing, and therefore the external support fails to build upon the local capacities and available opportunities.

Following the 1993 Marathwada earthquake in India, ten 'building centres' were set up by the Housing and Urban Development Corporation to promote construction activity and generate employment through training programmes for construction artisans, unskilled labour and unemployed youth. The centres supplied building materials to construction sites. In less than five years, all these centres had to be shut down (Jigyasu 2001). One of the reasons was that the centres were established through outside financial support without internalising the process by linking with local industries and delivery markets.

Experience has shown that large amounts of aid may provoke a rise in prices and damage private-sector activities not directly related to reconstruction, such as the export sector. The classic case is food aid that reduces the demand for local farmers' products (see Chapter 45). Therefore, it is important to make an effort to preserve jobs in the local manufacturing sector that has been temporarily halted due to disaster. In the case of Haiti, exports are small (some ten per cent of GDP), but were growing (at twelve per cent in 2009) and are highly concentrated in assembly industries including garments (some ninety per cent of exports are assembly goods). Following the January 2010 earthquake, it is important to ensure that the potential of job creation and growth in these sectors is not put at risk (Cavallo et al. 2010).

Recovery policies may not be effective if they do not take into account the nature of economic growth in a particular country or region. The economic package announced after the 2001 Gujarat earthquake in India turned out to be grossly inadequate when the actual construction started in 2003–04 because, by then, due to inflation, the cost of building materials had substantially risen. Often recovery policies do not create an adequate shield from such powerful market forces.

Another aspect of economic distortion can be seen in the Gujarat case. After the 2001 earthquake, abundant foreign materials such as cement and steel were made available by the government at subsidised rates, thereby competing on the market for traditional building materials such as stone, wood and mud. Concrete blocks achieved popularity on the basis of strong marketing and communication programmes by cement companies, and thus became the favoured building material. In fact, the commercial interests of these companies played a significant role in determining which materials were used. In spite of promotion by various non-governmental organisations (NGOs), the market availability of low-cost, locally available alternative building materials such as compressed soil blocks went down sharply from the initial phase of reconstruction. Modern construction materials were readily available and proved to be cheaper compared to traditional or low-cost materials. One reason for this was lack of encouragement of local entrepreneurship for manufacturing traditional materials. Another reason was that there were not enough builders trained in using these materials and technology (Jigyasu 2010).

Social and spatial effects of recovery policies

Post-disaster recovery policies have a significant and often unpredictable bearing on urban–rural dynamics. Following the 2001 Gujarat earthquake, the government imposed a blanket ban on any construction in the urban area of the town of Bhuj until the preparation of a master plan, but the plan took nearly two years to produce. As a result, the owners whose houses were damaged decided to opt for relocation by surrendering their damaged property to the government. This left tenants in these houses homeless. They were from poor or lower-middle income groups and now found themselves homeless. Over the years many such vacant lots and houses acquired by the government were occupied illegally by people migrating from rural areas in search of livelihood opportunities that became available during reconstruction (Jigyasu 2010).

Rural people in Gujarat benefited from the various post-earthquake interventions and economic packages announced by several organisations and by the government. For example, although the initial days were bad for craft workers engaged in handicrafts, this sector recuperated rapidly since it was not infrastructure-intensive. Also, the exposure to the outside world actually helped these craft workers market their skills and various products to an international client base. The rural people could also resume their traditional occupations of agriculture and animal husbandry relatively easily (Jigyasu 2010). The example shows that the post-disaster recovery process may have a different impact on people depending on their skills and assets and the opportunities provided by the recovery process.

In most cases, the recovery process affects marginalised groups in different ways, sometimes reinforcing or even increasing their vulnerabilities. The 1994 eruption of Mt Rabaul in Papua New Guinea severely impacted the Tolai traditional communities living on the volcano. Relocated dozens of kilometres away from their native villages, Tolai communities had to deal with the depletion of Tambu (traditional shell money with commercial and ritual values used to evaluate wealth) habitually accumulated from agricultural harvests on the slopes of the mountain. The lack of Tambu progressively resulted in the abandonment of the Tolai's traditional leadership system based on the accumulation of this shell money. Post-disaster relocation triggered changes that affected the culture, society and economy of the Tolais (To Waninara 2000) (see Chapter 8).

Sometimes disaster creates new vulnerable groups. After the 2005 northern Kashmir earthquake, up to forty per cent of the affected population migrated to some other areas in Pakistan (FAO 2005). Due to this labour migration, the proportion of women-headed households increased significantly to approximately twenty per cent of households in Pakistan-administered Kashmir (World Bank and ADB 2005). Poorer households temporarily headed by women became more vulnerable if migrating members could not find work (Beck 2005).

Livelihood-focused economic recovery interventions may miss some disadvantaged groups who may use altogether different livelihood strategies. In the case of Mozambique, agricultural livelihoods were supported by the provision of seeds and tools and the introduction of specialised crops; however, more complex livelihood strategies of urban households, semi-rural and fishing communities were not as well catered to (Wiles et al. 2005). Agencies did not have a logistical system that could cope with the needs of varied livelihoods. In the Kathmandu Valley in Nepal, land of the Guthis has been nationalised, thus depriving people of their primary asset. Indeed, for Guthis land provides a sustainable source of their livelihood as well as the basis for social and religious activities that are crucial for sustainable recovery (Jigyasu 2002).

These examples show that disasters frequently lead to more marginalised people, as those who lost their livelihoods are often unable to recover (Walker 1989; Wisner 1993). This is not only linked to the availability or unavailability of livelihood options but also concerns access of the marginalised to available resources. These examples demonstrate the close link between pre- and post-disaster vulnerability, which affects people's ability to recover in the aftermath of a disaster (Gaillard and Cadag 2009).

Recovery difficulties are foreshadowed by pre-disaster vulnerability

Post-disaster economic recovery programmes should be designed by studying pre-disaster vulnerability conditions. For example, in Turkey the process of urbanisation has been an integral part of a national policy of industrialisation. Industrial and urban growth has relied heavily on the

---

Box 48.1 Recovery from aridity in Iran

Faced with the aridity and high variability in the amount and distribution of rainfall, Qashqai communities of Iran have developed complex natural resource management strategies to exploit the uncertainties by recognising and working with the patterns that emerge in nature. This has led them to a livestock-related livelihood system based on migration in order to make the best use of precipitation patterns. Pastoralists divide their traditional drought-related activities into two main groups: activities before low rainfall periods that might lead to drought conditions and activities after the drought period. Preparedness activities were led by a variety of mechanisms for drought prediction (CENESTA 2004). This 'living with risk' approach has been rightly defined by Kelman and Mather (2008) as accepting that environmental hazards are a usual part of life and productive livelihoods, and therefore creating and maintaining habitats and livelihoods by using available resources without destroying them. These resources include environmental hazards, which might thereby become less of a danger and more integrated into day-to-day life and livelihoods.

However, in recent years their use of traditional knowledge and natural resource management systems has decreased tremendously, thereby drastically altering their response to drought, and thus affecting the potential of utilising these indigenous systems for sustainable recovery. One of the reasons is current government policies that severely limit their access to land. Land reform implemented in 1962 included nationalisation of all natural resources, including rangelands, forests and water. In the case of pastoral nomads, this meant that land that was ancestrally theirs, and which they managed sustainably and held as common property, was alienated from them and taken over by the state. Since then the nomads have had to obtain individual grazing permits based on a state expert assessment of the carrying capacity of the range. The present system of individual short-term permits means that the nomads are unable to work together to apply the principles of sustainable use. Unable to manage the rangelands, the government decided to give it to the private sector—but usually not to the traditional holders of right to the range. It was usually given to those with power and influence, who use it, more often than not, for speculation.

informal economy for cheap wage goods. The process has been so rapid as to put pressure on the regulatory, supervisory and governance structures meant to ensure public safety. The regularisation of illegal settlements was initiated by their residents, who used the process to secure investment in infrastructure and public services as municipalities incorporated new districts. Owners acquired certification of conformity to building codes for their buildings through amnesty, even if these still did not meet standards of the codes. All this resulted in catastrophic consequences during the earthquake that affected the country in 1999 (Bibbee *et al.* 2000). The process of economic recovery initiated after a disaster is short-lived if pre-disaster vulnerability is not addressed through structural changes.

Achieving sustainable economic recovery

Balancing macro and micro in recovery

Sustainable economic recovery requires a connection between macro-economic and micro-economic measures. Livelihoods of ordinary people are critical to recovery and are often overlooked or misunderstood. In some cases, it can be the other way round, so a balanced approach is the best. At the same time, development of large-scale business enterprises should not undermine the basic livelihood needs of those on the margins of society.

Targeted support to marginalised groups is needed to address their special needs. After Hurricane Mitch in Central America in 1998, the provision of seeds, agricultural inputs and – in some cases – cash helped farming families remain in their communities despite massive harvest, soil, housing and livelihood losses. Poorer groups that were landless could benefit from agricultural support only through increased labouring opportunities and potentially lower prices of subsistence foods (Beck 2005). Following Hurricane Ivan, which hit Grenada in 2004, the local Red Cross Society and British Red Cross developed a livelihoods project to help re-establish household food security and stabilise prices in the local food economy. The project provided seeds, tools and fertilisers to those engaged in small-scale production for local markets. With vegetable prices increasing in the aftermath of Hurricane Ivan, over eighty per cent of farmers anticipated selling the majority of their harvest in the market and were expecting to earn approximately US$150 per month (IFRC 2008b).

Livestock production is a major source of employment in Kenya, and makes a significant contribution to the economy. Over the past decade it has on average accounted for one-quarter of the country's GDP, and more than one-half of the income of small farmers. Therefore, the Kenyan drought in 2000–01 saw an unprecedented level of livestock-related interventions in pastoral areas. By September 2001 twenty-one livestock-related projects were under way or had been completed in ten districts. The livelihood sub-sector working group of the United Nations (UN) played an instrumental role in highlighting the plight of pastoralists and galvanising responses in this area (Aklilu and Wekesa 2002).

Market inter-connectedness also demands recovery strategies at a regional level that take into account the dynamics between cities and their hinterland. Diversification of livelihoods reduces future risk. Urban–rural linkages are an important part of diversification of livelihood as poor households may have members employed both as labourers or small farmers and also in urban petty trade or construction (Beck 2005). Communities relying exclusively on the natural resources available in their immediate environment are much more vulnerable in the event of the partial or total destruction of these resources. In the case of the 1991 Mt Pinatubo volcanic eruption, mountain-dwelling Aeta communities exclusively dependent on agriculture for their living were rendered jobless by the destruction of their fields by metres of pyroclastic deposits.

By contrast, the Aeta communities located on the foothills of the volcano, which had a more diverse livelihoods system, were less affected by the eruption (Gaillard 2007).

Economic recovery strategies at macro as well as micro levels should be broad-based and closely integrated with other sectors such as shelter, agriculture, health and infrastructure. Home-based enterprise and other informal income mechanisms are the single most important strategy for poor populations affected by disasters. The home is often also a place of work, for example a shelter for small and large livestock, or a base for petty trading and handicrafts. Location and design principles should take into account the need for secure places to store equipment and materials required for livelihood activities. When necessary, relocation should take place as close to their livelihood sources as possible (Beck 2005).

Reducing dependencies through local employment opportunities

As noted earlier, economic aid following a disaster carries the risk of increasing dependency. Local employment opportunities should be created during the recovery phase in order to reduce dependency. For instance, following the 2004 Indian Ocean tsunami, the International Labour Organization (ILO) developed a response strategy based on employment generation and new forms of livelihood. Labour-based technology was introduced in reconstruction to generate jobs quickly and to provide income while rebuilding basic infrastructure. The strategy also aimed at boosting the revival of local economies through a Local Economic Development (LED) approach which emphasised identification of economic opportunities, business promotion, employment-friendly investments, social finance, establishment of co-operatives, social dialogue and empowerment of local communities. Emergency public employment services were set up to provide training to help in the recovery of the labour market and to put job seekers in touch with available jobs. Technical advice and support for social safety nets and social protection was also provided as part of this strategy (ILO 2005).

After the Haiti earthquake in 2010, the UN Development Programme (UNDP) embarked on a cash-for-work programme in an effort to generate employment opportunities and income in the country. The programme helped some in the affected communities to earn a living and ensure basic necessities for their families. Included were short-term jobs clearing rubble and rehabilitating essential infrastructures such as streets and electricity. The programme also injected urgently needed cash into the economy, accelerating the resumption of small businesses and trade (United Nations Radio 2010).

Coping strategies, markets and recovery

Loss of material assets is typical after major disasters, and this usually has a disproportionate effect on the poor because of their relatively greater dependence on such assets. Sale of assets – for example small plots of land, grain, jewellery, handlooms, livestock and agricultural tools – is a key livelihood strategy for the poor. Beck (2005) notes that for many rural and urban people in the earthquake-affected areas in Pakistan, the buffalo is a form of bank, and people commonly own at least two buffaloes, worth approximately US$1,100 each. Distress sales and pawning by the poor are typical after disasters, often determined by male heads of households, with sales often below market value in a buyer's market. Once assets are gone, recovery for poorer groups is much more difficult. Asset replacement may be important to support livelihoods, for example by replacing livestock (Beck 2005).

In Jakarta, Indonesia, in a period of reduced rice production during the 1997/98 El Niño event, a novel programme was established to use commercial markets for aid delivery. Imported wheat was milled into flour by Indonesian flour mills, and Indonesian companies produced pre-packaged noodles, providing jobs for some of those recently made unemployed in the city. In addition, the noodles could be used by street-side cafes as well as households, ensuring that small food traders and vendors were not adversely affected by the provision of food aid. By using existing commercial networks to deliver noodles, the programme avoided the costs of establishing a parallel logistics system. The programme also used commercial marketing firms to identify clients and to provide advice on targeting and prices that would appeal more to people with low income and less to others, who had more discretionary income (ALNAP and ProVention Consortium 2009).

It is not merely important to strengthen markets. It is also important to improve the access that people have to credit and finance. Over the last few decades, micro-finance has emerged as an effective means of increasing access to credit, savings and other financial services in poor and vulnerable communities. Micro-finance can strengthen coping in the face of disasters by providing access to credit and other financial services to enable investment in higher-yield livelihood strategies, to diversify livelihoods and to enable investment in risk reduction measures (Cavallo et al. 2010). Besides, micro-credit can encourage local entrepreneurship to take advantage of the opportunities provided by the recovery programmes. For example, reconstruction in Gujarat after the 2001 earthquake led to several small-scale industries manufacturing compressed soil blocks, which were promoted by NGOs (Jigyasu 2010).

Working with or through existing organisations of long standing in the area, which already have an understanding of livelihoods and the measures needed to protect them, may assist effective post-disaster recovery. An example is provided by the Soni community after the Gujarat earthquake. Soni community groups received financial and material support from the business sectors with which the community had interacted before the earthquake. In turn, these community-based groups provided financial support for suffering families by providing livelihood kits consisting of basic tools. Field surveys and interviews with the stakeholders indicate that the Soni were the fastest recovering community in Bhuj (Nakagawa and Shaw 2004).

International NGOs can also work closely with community and local government to develop new livelihoods. For example, following prolonged drought in Timor-Leste in 2004, the local Red Cross Society (CVTL) decided to help in developing market gardens after a participatory risk analysis with the community. Through the process, the community's key priorities were identified and, on that basis, a one-year community action plan was drafted, specifying the respective roles of the community and the CVTL (IFRC 2008c).

Box 48.2 Community recovery from disaster: changing relationships between vulnerability and capacity

Etsuko Yasui
Applied Disaster and Emergency Studies, Brandon University, Canada

Nagata ward in Kobe city was the most severely affected area in the 17 January 1995 Kobe earthquake. Two of its neighbourhoods – Mikura and Mano – were equally vulnerable at the time of disaster because of shared historical, social and physical conditions – e.g. inner-city decline, economic reform from manufacturing to service industries, ageing population, a concentration of blue-collar and seasonal workers, issues of ethnic and racial minorities, high rate of social welfare, poorly maintained apartments and densely packed wooden housing.

However, the impacts and immediate responses following the earthquake were not the same. This was due in part to differences in community capacity between Mano and Mikura. Mikura had very little interaction among its residents in the pre-disaster period, whereas Mano had a long history of diverse and active neighbourhood associations dating from the 1960s. Furthermore, Mano had well-respected leaders, a devoted community planner, philanthropic partnerships with local small businesses, support by professionals and academics dedicating their skills and knowledge, and good relationships with the local government.

Seventy per cent of Mikura was burned by the earthquake fires, which caused substantial property losses and population relocation. Mano, on the other hand, responded quickly and effectively to the quake by recognising the needs of the most vulnerable residents in the community and setting up emergency headquarters to provide shelter and relief goods. The residents battled the earthquake fires themselves and prevented extensive damage to their community.

During Mikura's response phase, a volunteer group was formed that ultimately became a community-based organisation (CBO) for assisting in the recovery efforts. Its unconventional but progressive and bottom-up approach was well received by the survivors because many of them were disappointed by the government's top-down approach. While the Earthquake Restoration Project took almost ten years to re-zone Mikura's land area, the CBO assisted in the reconstruction process of Mikura with newly built apartments, wider streets, open spaces, two parks, a new community centre and seismically upgraded houses. The new look of the community was at first strange to its residents for its low-density housing and landscapes, whilst its population has gradually expanded with newcomers. The community has nevertheless strengthened itself with greater participation in local activities led by the CBO, has successfully taken collective decision-making processes, and accomplished effective use of external resources that resulted in building further community capacity. Mano's collective action to minimise losses from the earthquake further increased its capacity, but also paradoxically contributed to its vulnerability due to the continuance of fragile wooden housing, high building densities, ageing population and unresolved issues of ethnic minorities.

The relationship between vulnerability and capacity at the community level is not necessarily an inverse one. Rather, the interactions are highly complex and contingent on many contextual considerations. Two communities located in relatively poor areas of a city with high social and physical vulnerability may have differing community capacities, and therefore different responses and recovery processes. Although Mano and Mikura both experienced trade-offs that either exacerbated new vulnerabilities or interfered with existing capacities, the two communities are now safer and more comfortable places.

Conclusions

To mitigate the economic damage and loss from a disaster and to facilitate quick recovery, policy-makers mainly use four tools, i.e. increasing spending, regulatory and legal forbearance, tax incentives and risk mitigation (Birch and Wachter 2006). In addition, sustainable economic recovery requires consideration of long-term impacts, especially after external support is withdrawn and the affected region is left to manage with its own resources and management systems. Mere provision of economic aid is not sufficient unless enabling mechanisms are put in place or strengthened. These include good economic policies that take into consideration needs at a macro (national/regional) level as well as at a micro level (livelihoods of various sections of the community, especially the marginalised groups). Neo-liberal economic policies may not ensure equal opportunities for all. Therefore, local governance mechanisms must be strengthened and made accountable for equitable growth.

Key strategies should include reducing dependencies by creating and supporting local employment opportunities, by utilising local skills and resources, and strengthening and engaging market mechanisms that cater to local needs. Livelihoods of the poor should be supported by securing their assets, such as tools and livestock, and improving their access to credit and finance.

Another key aspect of sustainable recovery is engagement of various stakeholders including national governments in charge of policies, international agencies engaged in reconstruction, NGOs and, most importantly, the communities.

Rather than considering economic recovery in isolation, links with other sectors such as shelter, agriculture, social development, cultural affairs, education and health should be reinforced and a holistic approach adopted. This is crucial because all these sectors are intimately connected to the economic well-being of the affected society.

A post-disaster recovery process should not be aimed at merely restoring or building back the loss and damages suffered during the previous disaster. Economic recovery is also a significant opportunity to prepare for the next disaster. In this way, the recovery process should be seen as part of a continuum of disaster risk reduction and sustainable development (see Chapter 14). This implies that the post-disaster recovery process must be dovetailed with measures that address the 'normal' situation by reducing existing vulnerabilities and strengthening local capacities based on skills and knowledge systems that have evolved through experience over time.

Part IV
Planning, prevention and mitigation

49

Introduction to Part IV

The Editors

Can disaster be avoided?

The policies and practices discussed in the previous section are important and necessary, but they do not deal directly with the underlying causes of disaster risk. Preparedness per se and response are, by definition, palliative. Planning is somewhat more ambiguous. Emergency response planning assumes that a disaster will happen. Conversely, comprehensive planning for hazards assumes that action can taken before a hazard in order to prevent or mitigate adverse impacts from that hazard – perhaps even turning the hazard into an opportunity.

Yet prevention and mitigation of disaster impacts goes far beyond treatment of potential hazards. Prevention and mitigation of disaster impact must also incorporate reducing vulnerability and enhancing capacities. Capacities can be built and vulnerability reduced so that some hazards morph into resources, while planning for other hazards is integrated into people's day-to-day lives and livelihoods (Gaillard *et al.* 2010) (see Figure 49.1). Reducing vulnerability largely requires actions at the global and national levels to facilitate access to resources that make up livelihoods, ability to select disaster risk reduction (DRR) options for oneself and means of protection. On the other hand, enhancing capacities fosters resources that are endogenous to communities – as emphasised in Figure 49.1. DRR therefore necessitates a combination of top-down and bottom-up actions from a large array of stakeholders at different scales.

That does not mean throwing out post-disaster activities or hazard-focused approaches. Rather, it means understanding that an exclusive focus on reactive and hazard-focused approaches will perpetuate the disaster problem. To effect DRR, links amongst development, sustainability and DRR need to be forged, as one is reminded by Figure 49.2. Many people have knowledge, skills, expertise and experience; no single person or institution can possess what is needed for successful DRR. Instead, collaboration across fields and disciplines is essential, linking top-down with bottom-up, linking those outside communities with those in the communities, including marginalised groups, and linking all forms of knowledge.

These processes need routine funding, not just when a disaster occurs – which, of course, is too late. Four actions can be highlighted:

- Addressing the underlying social and human causes of disasters;
- Emphasising the contribution to prevention and mitigation that must be made by the people who bear the disaster risk and disaster vulnerability;

- Stressing the importance of place and context; and
- Continually exchanging information and ideas to create inspiration through the joint activities of teaching and learning.

This section of the *Handbook* details these topics. The first sub-section explores governance, advocacy and self-help for DRR. International, national and sub-national, including local, scales are covered. Activities include planning (covering different interpretations of that process and clearly showing the necessary overlaps between 'planning for disaster' and 'planning for hazard'), financing, government institutions and governance. The second sub-section involves different forms of communication and participation: teaching, learning, scientific publication, media publication and participatory processes designed to facilitate exchange.

The chapters particularly emphasise that disasters reflect or are a symptom of fragile, undiversified and unsustainable livelihoods. The policy implication is that DRR is a livelihood,

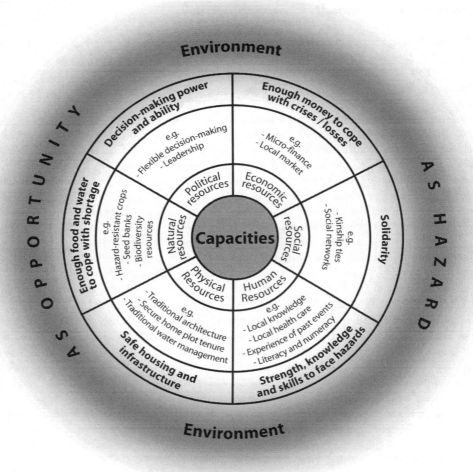

Figure 49.1 Circle of capacities

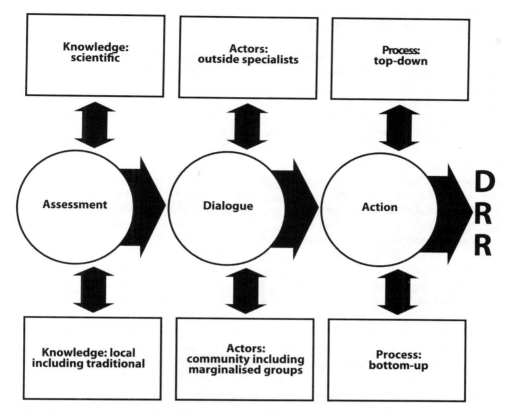

Figure 49.2 Integrated disaster risk reduction framework

development and sustainability concern. Research, policy and practice must go far beyond the top-down technocracy that often dominates disaster-related decision-making.

Governance, advocacy and self-help

International policy has experienced positive changes, especially concerning the importance of community participation and inclusion of social issues. In 1994 the so-called Yokohama Message of the 'Yokohama Strategy and Plan of Action for a Safer World' (UNIDNDR 1994) laid out some principles for DRR across scales, as eventually confirmed by the 'Hyogo Framework for Action' (HFA) (UNISDR 2005b). More recently, the United Nations (UN) 2009 Global Assessment Report on DRR (UNISDR 2009a) emphasised some underlying risk factors such as poverty, governance and ecosystem degradation.

Chapter 50, on 'international planning systems for disaster', gives a taste of this broad and rapidly evolving policy area. An overview of international humanitarian architecture and its history is provided and linked to a sketch of endeavours by UN and international development institutions to support DRR. There is much experience at this level and many recurring lessons, but often they do not seem to be learned. The international consensus advises establishment in each country of a DRR sector or system, while mainstreaming DRR into other activities. However, this does not go far enough. Instead, international planning approaches need to recognise that 'routine' development investments and programmes are also risk management approaches.

If international policies are slowly changing, their influence on national and sub-national policies and practice remain limited in some sectors as emphasised in Chapter 51 on 'national planning and disaster', Chapter 52 on 'local government and disaster' and Chapter 53 on 'urban and regional planning and disaster'. Many good practice case studies are given, from coastal management in Samoa and Mexico to dealing with volcanic hazards in an urban region in the US northwest. Many problematic areas are identified as well, such as poor pre- and post-Katrina planning in New Orleans and the fact that many Pacific island countries do not yet have formal land-use policies or national land-use plans.

Challenging questions that recur are: For what are different government authorities responsible with respect to DRR, for what should they be responsible and for what can they realistically be responsible? The answers vary widely, according to size – some megacities have more people than many countries – and national and regional contexts. It is particularly striking to note the variety of positive developments across Latin America, Africa and Asia compared to the intense problems still witnessed in Europe and North America. Of course, the less affluent countries nonetheless display severe vulnerabilities while the more affluent countries have demonstrated significant improvements in reducing disaster casualties. Good and bad practices appear everywhere, so everyone should be learning from each other while advocating needed actions.

Some powerful institutions such as the International Monetary Fund (IMF) have not yet accepted the need to address vulnerability and avoid the assumption of rich-to-poor transfer of help (Freeman et al. 2003). Other international institutions such as the World Trade Organization (WTO) have downplayed the negative effect of disasters on development. For instance, WTO (2006) states that 'while human suffering and localised damage can be very considerable, and the immediate effects on particular industries notable, the economy-wide impact of these events on trade and growth is short-term and generally minimal' (World Trade Organization 2006: xxi). Is that really saying that human suffering from disasters is acceptable because trade and growth are not affected?

Financial mechanisms can be useful tools for DRR. Two chapters on 'financial mechanisms for disaster risk' (Chapter 54) and 'economic development policy and disaster risk' (Chapter 55) demonstrate the plethora of initiatives possible to support DRR. These work at all scales. Financial mechanisms range from catastrophe bonds and conventional property insurance to micro-credit and micro-insurance. They involve the private sector as much as non-profit groups and governments. Economic development policy requires governance, advocacy and self-help elements such as legislation that is monitored and enforced, direct connections with development, government policies that are horizontally and vertically integrated, and accountable budgeting and expenditure.

How effective are international, national, financial and economic mechanisms at supporting sub-national scales for DRR? GNDR (2009, 2011) showed that the recommendations of the HFA rarely 'trickle down' to local communities threatened and affected by disasters. In fact, the international scale often has limited influence for fostering DRR at the national and sub-national levels. Meanwhile, national initiatives often negligibly and sometimes negatively affect local processes. This is particularly worrying since the implementation of DRR policies is defined by the UN's master plan for disaster reduction (the HFA) as the primary responsibility of national governments.

The story is not entirely bleak. A large number of good governance, advocacy and self-help practices have emerged at sub-national levels. In parallel, many non-governmental organisations (NGOs) have been building on past decades of experience by revising and formalising their policies and practices in order to address root causes of disasters and to advocate for the participation of those vulnerable to natural hazards. In some instances, feedback from NGOs and sub-national levels of government influences national and international policy.

Chapter 56, on 'protection of infrastructure', details some policy and technical approaches to avoid or limit damage and disruption that the built environment may face in a disaster. At times, infrastructure receives better treatment than the people who use the infrastructure. Decision-maker priorities are sometimes skewed away from those most affected by the decisions. An example is discussions of protecting so-called 'critical infrastructure' from disasters that do not include schools and hospitals, focusing instead on what is required for business continuity (telecommunications, transportation, etc.).

The human element comes back in with Chapter 57, on 'social protection and disaster'. Disasters can be seen as a reflection of chronic social system failures. This chapter explores a wide variety of ways in which robust social protection and social security systems facilitate DRR. Systems of pensions, subsidised health care and food, care systems for the elderly and many other kinds of safety nets provide a pre-condition for more direct self-protection actions and facilitation of DRR.

Stable and productive livelihood systems are also a vital pre-condition for DRR. Chapter 58, on 'livelihood protection and support for disaster', applies the sustainable livelihoods strategy to DRR by examining household assets, resources, incomes, and resource- and income-generating activities. The chapter is rounded out with a discussion of information needs for implementing livelihood support and possible objections to this approach.

Communities – not always well-linked to any government – are involved in DRR and advocacy. Chapter 59, on 'community action and disaster', demonstrates how community resources, solidarity and decision-making may establish the needed DRR baseline, irrespective of government activities. Yet there is much work to be done. Often, community approaches to DRR are done more in name than in practice and many efforts are led and sustained by outside players. Evaluation is often inadequate. Many good practices, from both rich and poor communities, lead the way and provide a solid baseline for continuing the work.

Chapter 60 – 'civil society and disaster' – illustrates the influential roles played in DRR by this sector at all governance levels. Civil society includes national and international social movements. At one end of the scale continuum, the international campaigning alliance Via Campesina embraces 250 million small farmer and landless farm labourer members worldwide (viacampesina.org/en). At the other end, there are small community-based organisations such as women's groups for raising improved poultry. International and national NGOs also fall under the umbrella term of 'civil society' and have among them great differences in their funded bases, technical capacity and links with governments and international agencies. This sector of society has seen many advocacy successes, and failures, in supporting DRR.

The common thread that binds all these governance, advocacy and self-help efforts can be seen through scale – spatial scales and time scales. DRR activities must span and connect the large with the small, and the long term with the short term. Good governance implies that activities are relevant at the local level.

Effective DRR is often achieved by making all sectors' actions relevant to people's day-to-day lives and livelihoods – yet DRR cannot be effective without thinking about decades and generations into the future. Ultimately, cross-scale collaboration is required for people to help each other while governing and advocating for themselves.

Communication and participation

How can the governance, advocacy and self-help detailed in the previous sub-section be enacted? Facilitating people's participation in helping themselves and in empowering those who are often left out of the DRR process requires exchange, support, teaching and learning from the top down and from the bottom up. For example, children should be in school (broadly defined) to be recipients of DRR knowledge passed down to them by adults, but should also provide their family, teachers, neighbours and the wider community with their own views on hazards, vulnerabilities, risks, disasters and possible solutions.

In many cases, too little trust and respect presently exists amongst governance scales and amongst scientists, government and agency officials, NGO workers and business people. Trust and mutual respect are required to implement DRR and to engage vulnerable people in fruitful and effective discussions. The chapters in this sub-section overview some ways for building and developing trust and fostering interaction and mutual support by focusing on communication and participation.

Much research occurs outside of universities, so Chapter 61, on 'university research's role in reducing disaster risk', describes only part of the picture. However, academic understanding of disasters and DRR has played a significant role in determining why disasters happen and what could and should be done about it. The cross-over between academics and practitioners has been particularly intensive and productive in some aspects of DRR as shown by the number of papers from practitioners appearing in top disaster-research journals. Universities also serve an important function in the cross-over of research and teaching, permitting students to learn from top academics while forcing top academics to continually go back to basics. DRR, as with all other fields of research, has suffered from the contemporary short-sighted devaluing and reduction of long-term research driven by curiosity.

Chapter 62 on 'education and disaster' links well with Chapter 63 on 'media, communication and disasters'. In both chapters, the core process or societal institution – education and media, respectively – is defined broadly, covering diverse modes of formal and informal exchange and conveyance of knowledge and ideas. Both education systems and the media have in the past set back DRR through distortion and misuse of information, but these chapters rightly focus on the creative, inspirational and agenda-setting aspects of education and media. Ensuring the safety of schools, keeping journalists free from threats and encouraging scientists to communicate with the media are amongst the priorities identified in these chapters.

As the penultimate chapter in this *Handbook*, Chapter 64, on 'participatory action research and disaster risk', brings together many themes for improving DRR through the techniques of participatory action research (PAR). As the chapter notes, Wadsworth (1998) defines PAR as 'research which involves all relevant parties … by critically reflecting on the historical, political, cultural, economic, geographic and other contexts'. As such, PAR connects theory, policy and action, across different time and space scales, while linking top down and bottom up.

Yet no technique is or can be a panacea. Communication and participation are always needed, but are open to abuse and error. Vigilance, critique and constructive evaluation must be a continuous part of all communication and participation.

Creating and supporting the DRR process

DRR should be context-specific. People and communities are different. No magic solution fits everywhere for DRR. Yet DRR policies and practices have often assumed that a clear distinction exists between 'safe' and affluent places in the global North in stark contrast to 'unsafe', poor, needy places in the global South. This assumption was congruent with wider development policies fostering top-down transfers of knowledge, technology and experience from rich to poor countries, because the poor countries were allegedly unable to cope without external assistance and could provide little themselves.

Such approaches have been shown to be obsolete as they failed to significantly reduce the occurrence and impact of disasters in the global South (Hewitt 1995; Bankoff 2001). Meanwhile, the richer countries discovered their own vulnerabilities from disasters such as Hurricane Katrina striking the USA in 2005, the 2003 heat wave across Europe and the 2011 earthquake/tsunami catastrophe in Japan. As this *Handbook* goes to press, the twenty-first century still awaits: a major earthquake along the US East Coast (not just the US West Coast) that could affect Boston, Massachusetts or Charleston, South Carolina, for example; a devastating North Sea storm surge; or an epidemic or volcanic eruption that stops air travel to and from Australia and New Zealand. The affluent are by no means exempt from natural hazards by virtue of their wealth.

We can and should all contribute to DRR. We can all learn from each other, exchanging experiences, creativity, proposals and concepts regarding DRR. That includes the aspects not covered fully in the *Handbook*. For instance, it would have been easy to have had separate chapters for each of professional training and informal learning. We must never forget our different contexts, but be willing to collaborate to determine what is and what is not transferable. Teachers always learn and learners always teach. We can and should be learning from while teaching each other, at all scales.

These approaches, with examples given in the chapters, of inclusive and participatory pre-disaster prevention and mitigation activities do not exclude more technical actions that directly address hazards or respond to disaster. The authors show that DRR can neither be solely hazard-focused nor solely vulnerability-focused, nor can it be solely bottom up or solely top down. It should result from the careful blending and respectful exchange of actions and interests at all levels, recognising the role played in disaster and DRR by the different elements.

This includes scaling out to the big picture and asking why disasters happen and what DRR needs in order to be effective. Participatory action research happens within historical, political and cultural contexts and is designed to share and extend knowledge. Communication and education use numerous media including film and music. Livelihood and social protection are linked to human rights discourses while governance and action beyond government also occur through these processes, being intertwined with politics, economics, culture, religion, knowledge and history. Sustainability and development emerge prominently in such a framing of disaster and DRR.

These connections complete the loop, because the reader has now been brought back to the beginning of the *Handbook* and the big picture views of Part I.

Governance, advocacy and self-help

50

International planning systems for disaster

Margaret Arnold

WORLD BANK

Introduction

Over the past several decades, a number of critical issues integral to but long neglected in development efforts became agendas to enter the mainstream of development policy and projects. These include gender (first called for at the Third World Conference on Women 1985) and the environment (Rio Declaration 1992). In a similar fashion, early work in the 1980s pointed out that misconceived and poorly planned development could be the cause of disasters in low-income countries (Hagman *et al.* 1984) and called for incorporating natural hazard risk assessment into development planning and projects (Organization of American States 1988).

One need only look at the many comparisons made between the earthquakes that occurred in Haiti on 12 January 2010 and Chile a month later to understand the many development failures that contributed to the tremendous loss of life and destruction in Haiti (see Chapter 26). The 2006 review of twenty years of World Bank lending for so-called 'natural' disasters undertaken by the Independent Evaluation Group (IEG 2006) recognised the foreseeable nature of hazard risk and found that those risks were infrequently considered in country programmes or project financing.

For example, of sixty-five projects in the transportation, urban development, and water and sanitation sectors approved between 2000 and 2004, the study found that in only three cases did the project documentation describe how a natural hazard might affect the project and how that risk could be addressed. The Active Learning Network for Accountability and Performance in Humanitarian Action (ALNAP 2010) reviewed some 700 evaluations by humanitarian agencies and concluded that co-ordination between the humanitarian sector and International Financial Institutions (IFIs) was not taking place. ALNAP also found that in recent years, given that more funding has been available for emergencies, there was a tendency in both camps to duplicate efforts. As a result, both humanitarian relief agencies and development institutions provided relief, rehabilitation and reconstruction. Here again donor co-ordination seemed important, but was largely missing.

Support for mainstreaming disaster risk reduction (DRR) picked up momentum in the mid-to-late 1990s, and since then there has been growing support and movement towards integrating disaster risk concerns into development planning and practice. In the process of staking its claim on the development agenda, some proponents of DRR went beyond the call for recognising and addressing the risk inherent in development activities to establishing DRR as a distinct field of study and practice. Disaster risk reduction (DRR) as a field is somewhat an orphan child. Promoted and still funded for the most part by humanitarian agencies, it has strived for its place in the development world. This chapter aims to provide an overview of the global-level dialogue and activities related to planning for disaster risk reduction, the current architecture of the DRR system and the progress and challenges of the DRR agenda as the awkward child of the humanitarian, development and now climate change adaptation fields (see Chapter 14, Chapter 18 and Chapter 51).

Overview of the international humanitarian system

To understand the current international DRR system, it is first necessary to understand the international humanitarian system and its relationship to development institutions. An assessment by ALNAP (2010: 18–19) of the humanitarian system describes a system of non-governmental organisations (NGOs) and the Red Cross/Red Crescent movement, donor states and United Nations (UN) agencies made up of approximately 595,000 humanitarian workers and with programming worth about US$7 billion in 2008.

Walker refers to the current system as more of an 'eco-system', and provides an insightful analysis of its history and competing agendas of containing crises to maintain order and governance (going back to the famines in India of the 1800s); promoting compassion (the birth of the Red Cross and the principles of impartiality and neutrality); advocating change (exemplified by the Save the Children Fund and its use of relief to promote the rights of children); and delivering welfare (or a 'soft' version of containment, with agencies such as CARE and the World Food Programme (WFP) seeking to 'ameliorate the collateral damage in the system but not to change it') (Walker 2010: 5–9; see also Walker and Maxwell 2008).

Walker states that as humanitarianism became big business, there were a number of related factors that brought into question the basic humanitarian principles of impartiality and neutrality (see Chapter 7). Many of these factors related to the growing complexity of conflict-related emergencies, including the political interests of the national donor governments funding humanitarian work, the use of for-profit and military actors in humanitarian response operations and the changing nature of contemporary conflict.

As a result of this changing landscape, a 'new humanitarianism' developed in the 1990s that recognised the political nature of all aid, and emphasised more human rights-based and developmental approaches to relief aid (Anderson 1999). Many viewed this as a necessary recognition of the realities of a post-Cold War world. Over time, however, experience revealed the dangers of stepping back from such principles as independence, neutrality and universalism. Fox (2001) and others (Terry 2002; Polman 2010) discuss the implications that new humanitarianism has for humanitarian action in the context of conflict, involving access to affected people, the safety of humanitarian workers, the distinction between 'deserving' and 'non-deserving' victims and the promotion of conditional relief aid based on a Western agenda. Fox (2001) distinguishes between the political manipulation of humanitarian aid and the 'conscious use of humanitarian aid by agencies to pursue political ends that is proposed in new humanitarianism' (Fox 2001: 288).

De Waal adroitly lays out the humanitarian's dilemma and the different cruelties inherent in humanitarianism. These include the individual cruelty of failing to do good in the margin (e.g., turning people away at a camp for internally displaced persons who do not meet the criteria for admission); the institutional cruelties that occur when humanitarian principles clash (e.g.

Box 50.1 Trust in the UN system

Ben Wisner
Aon-Benfield UCL Hazard Research Centre, University College London, UK

Ilan Kelman
Center for International Climate and Environmental Research – Oslo (CICERO)

(This box does not represent the views of the chapter's author or her employer.)

The effectiveness of international advocacy for DRR is to some extent linked to the credibility of the entire UN system. Scandals and controversies involving one specialised agency of the UN cast a shadow over the whole. Since 2005, several issues have arisen that might erode public trust in the system.

Recent controversies

Accusations were made in 2010 that the World Health Organization (WHO) had exaggerated the threat of the H1N1 virus pandemic, resulting in the waste of millions of US dollars by world governments, and that members of the expert group that advised WHO on the pandemic had links to large pharmaceutical companies who profited from sales of vaccines (Stein 2010). In response, the WHO's Director-General stated that 'potential conflicts of interest are inherent in any relationship between a normative and health development agency, like WHO, and profit-driven industry' (United Nations 2010). She noted that the WHO was in the process of establishing and enforcing stricter rules of engagement with the private sector, but stressed that 'at no time, not for one second, did commercial interests enter my decision-making'.

In December 2010 a French epidemiologist working for the French and Haitian governments announced results of tests that suggested that Haiti's cholera epidemic had originated at a UN peacekeepers' base outside Mirebalais, along a tributary to Haiti's Artibonite River. The strain of cholera matched one from South Asia (BBC 2010c). There had been suspicions and protests about this for weeks, with at least one protester being killed by the UN peacekeepers. The UN had steadfastly denied the allegations. Soon after the French scientist's announcement, the UN said it would organise a study of the issue.

These blows to the UN's credibility came on top of a long series of allegations of its peacekeepers engaging in illegal economic activities in some of their mission locations and also of engaging in sexual abuse (Polman 2010).

Longer-standing discontents

These recent instances came against the background of many years of criticism of the UN system from all points on the political and ideological spectrum.

For years some member countries and independent voices had criticised the UN system for being excessively bureaucratic, wasteful of resources and non-transparent. Conservative governments in the UK and USA in the 1980s insisted on reforms at a time when the UN's financial position was deteriorating. Criticism also came from the left, to the effect that parts of the UN system worked too closely with big business. For example, George (1976) documented how the Food and Agriculture Organization (FAO) of the UN promoted agricultural machinery made by large Western corporations and inputs from the chemical industry.

Some criticisms went further and were political. The USA left the UN Educational, Scientific and Cultural Organization (UNESCO) in 1984, and the UK left in 1985, over alleged anti-Western bias that they perceived in that specialised agency, although both countries have since rejoined. Other countries considered so-called 'structural adjustment' programmes enforced by the World Bank and International Monetary Fund (IMF) as a 'condition' of loans to be intrusive and heavy handed, and, in effect, working on behalf of Western banks and other financial institutions.

Thus criticised from many angles and dependent on member country financial contributions, especially those of the largest economies, the UN system must engage in a difficult balancing act.

Rwanda); and the escapable cruelties of 'failing to apply workable technologies in the right place, at the right time, or in the right way' (de Waal 2010: 134). De Waal points out that while research and learning from humanitarian action experience can, and have, improved humanitarian outcomes and reduced the number of 'escapable cruelties', there are certain elements of cruelty that are inherent to the humanitarian mission and with which we will continue to struggle (see Chapter 61).

Architecture of the international humanitarian system

The UN system

The Inter-Agency Standing Committee (IASC, www.humanitarianinfo.org/iasc) brings together all major humanitarian agencies, both within and outside the UN system. Chaired by the UN's Emergency Relief Coordinator, it develops policies and agrees on the division of responsibilities among humanitarian agencies. Besides the IASC, the UN's humanitarian system portal highlights the role of the following agencies (www.un.org/en/globalissues/humanitarian).

The humanitarian and disaster-relief efforts of the UN system are overseen and facilitated by the Office for the Coordination of Humanitarian Affairs (OCHA, ochaonline.un.org), led by the UN Emergency Relief Coordinator. Among its many activities, OCHA provides the latest information on emergencies worldwide, launches international 'consolidated appeals' to mobilise financing for the provision of emergency assistance in specific situations and co-ordinates disaster response (see Box 50.2). In war, the Secretary-General helps to negotiate 'zones of peace' for the delivery of humanitarian aid and UN peacekeepers protect the delivery of that aid (www.un.org/en/peacekeeping), whether provided by members of the UN system or such humanitarian bodies as the International Committee of the Red Cross (ICRC, www.icrc.org) (see Chapter 7).

The ICRC is part of the Red Cross/Red Crescent Movement that deals with conflict situations and serves as the guardian of the Geneva Conventions. The International Federation of Red Cross and Red Crescent Societies (IFRC) deals with non-conflict disasters. The two institutions, in addition to 186 national societies, comprise the Red Cross/Red Crescent Movement, which is the largest humanitarian network in the world. This division of labour was agreed upon in 1997 to address the complexity of humanitarian emergencies (www.redcross.int/en/history/movement.asp).

There are other actors in the UN system involved in disaster and humanitarian affairs, leading some critics to view the current system as suffering from sub-optimal coordination and effectiveness. A series of highly publicised events over the past few years (see Box 50.1) have also contributed to the view in some quarters that the UN may face a crisis of trust in its ability to carry out its assigned responsibilities. Irrespective of those views, shared neither by this chapter's author nor many at the UN itself, it is hard to imagine the world without the UN or its critical work in these areas.

Other international organisations and movements

The IFRC is a worldwide system in its own right, which began in 1919 and has grown to include 186 national societies and a long history of promoting and supporting DRR and community-based, local action. Its precursor, the Red Cross, was founded in 1863. The IFRC publishes the *World Disaster Report* annually, a publication that has consistently attempted to call attention to neglected aspects of vulnerability and DRR (www.ifrc.org/publicat/wdr/index.asp). The IFRC and a number of its national societies are active participants in policy discussions at international and national level, as well as providing expert technical assistance. For example, the Netherlands Red Cross hosts an IFRC-wide focal point for expertise in climate change and DRR, the Red Cross/Red Crescent Climate Centre (www.climatecentre.org).

Many other international professional, academic and civil society organisations and networks interact with the UN and IFRC systems on themes, projects and programmes that bear on DRR (see Chapter 60 and Chapter 61). While this form of collaboration is growing rapidly, still the most visible way in which many of these diverse international organisations interact is in disaster response, relief and recovery activity (see Chapter 42). Literally hundreds of different national agencies and groups may respond to a large disaster such as the 2004 Indian Ocean tsunami or the 2010 earthquake in Haiti. Co-ordination in these circumstances is crucial, and a good deal of energy has gone into developing mechanisms to do this (see Box 50.2).

Box 50.2 The cluster system

Ben Wisner
Aon-Benfield UCL Hazard Research Centre, University College London, UK

Margaret Arnold
World Bank

As part of the humanitarian reform process, an attempt to improve humanitarian outcomes came with the launch of the cluster approach in 2005. The IASC endorsed a proposal in September 2005 that assigned various UN agencies to lead sector 'clusters' intended to identify and fill assistance gaps in a predictable and accountable manner in all humanitarian operations. The cluster approach has five goals:

- to ensure that *sufficient global capacity* is built up and maintained in all main areas of response;
- to ensure *predictable leadership* in all the main areas of response and ensure a 'provider of last resort';

- to promote *partnerships* between UN agencies, the IFRC, international organisations and NGOs, and to avoid situations where governments have to deal with hundreds of uncoordinated international actors;
- to strengthen *accountability* at the global level and to beneficiaries; and
- to improve *strategic field-level co-ordination and prioritisation* in specific sectors/areas of response through the cluster leads.

(UNOCHA 2010b)

There are eleven clusters, each led by one or more organisation(s) (see Table 50.1). These include UN and international NGOs and other international agencies. There are also four cross-cutting clusters: age, gender, environment and HIV/AIDS. Membership of clusters incorporates dozens of bilateral agencies, NGOs and other actors in any given disaster situation.

Table 50.1 The cluster system and its co-ordinators

Cluster	Coordinated by
Agriculture	FAO
Camp coordination and management	UNHCR and IOM
Early recovery	UNDP
Education	Save the Children Fund and UNICEF
Emergency shelter	UNHCR and IFRC
Emergency communications	OCHA and WFP
Health	WHO
Logistics	WFP
Nutrition	UNICEF
Protection	UNHCR
Water, sanitation and hygiene	UNICEF

Although the cluster approach was designed in 2005, nearly all of the 160 key informants complained of lack of co-ordination to ALNAP researchers conducting the 2010 State of the Humanitarian System report (ALNAP 2010: 49). More than one-half of 250 respondents told ALNAP that the transaction costs of co-ordination on humanitarian actors were 'too high – not worth the burden on the organization' (ALNAP 2010: 42). Early on, in response to the 2005 earthquake in Pakistan, problems were evident in reaching out to local community leaders and development workers and getting them to participate in the clusters and sub-clusters (Street 2007). That has persisted and was observed again in Haiti in 2010.

Davies (2008) reports opinions that the cluster approach had opened up space for discussing difficult cross-cutting issues such as landlessness, disability and female-headed households – issues that are quite often neglected. It is still too early to judge if the cluster system is a recipe for too many meetings or if it will become a routine instrument for improved humanitarian assistance. An evaluation by the IASC in 2010 pointed to reduced duplication of efforts and better identification of gaps (VOICE 2010: 2).

The role of national governments in the humanitarian system

Contemporary national institutions that deal with disaster evolved from military and paramilitary institutions to deal with 'civilian defence' (revived in the US notion of 'homeland security' and still evident in some agencies, such as those in Latin America with the name *protección civil*). In former colonies and countries dependent on foreign economic assistance, these institutions were further structured around the need to co-ordinate as hosts the large amounts of overseas aid that would accompany disaster response and recovery (see Chapter 51). Little of any of this laid the foundation for prevention and DRR or established a connection with routine development planning. From the turn of the millennium onwards, there has been more emphasis on DRR at a national scale, and the various UN agencies working at country level have partners and focal points for DRR interventions, as well as often a variety of new laws to provide national frameworks for DRR.

There are also now a variety of regional institutions that support national efforts at DRR, such as the African Parliamentarian Group in DRR and Climate Change Adaptation, the Arab Disaster Risk Reduction Network, the Southeast Europe and Central Asia Disaster Risk Reduction Steering Committee, CDEMA in the Caribbean (www.cdema.org), the Pacific Islands Applied Geoscience Commission (SOPAC, www.sopac.org/index.php/sopac-programmes/community-risk-programme), the Asian Disaster Reduction Response Network (ADRRN, www.adrrn.net) and the Andean Committee for Disaster Prevention and Care (www.comunidad andina.org/predecan/contexto_caprade.html).

From managing disasters to managing risk

Many of the challenges facing new humanitarianism are more relevant to conflict situations, and their debate lies beyond the scope of this chapter (see Chapter 7). The DRR significance is that the principle of addressing underlying causes in addition to impacts, as is common with expressions of new humanitarianism, was taken up actively by some agencies working with disasters involving natural hazards. In the case of some agencies this could be considered 'mission creep' and in other cases a necessity because of the need to respond to more and more disasters. There was a growing realisation of the necessity to link relief and development efforts and to use complementary approaches to deal with both the underlying causes and impacts of disasters. Accordingly, there is more effort at the international level to deal with root causes of disasters before they occur.

The International Decade for Natural Disaster Reduction and the ISDR

During the 1990s the UN declared the International Decade for Natural Disaster Reduction (IDNDR) with the goal of reducing losses of life and property, and reducing the social and economic disruption caused by so-called 'natural' disasters (United Nations 1989b). While the IDNDR was never independently evaluated, it was commonly criticised for overemphasising scientific and technical approaches to risk reduction, to the detriment of understanding and addressing issues related to social vulnerability. The mid-term review of the IDNDR pointed this out, and an attempt was made to increase the focus on socio-economic vulnerability factors in understanding disaster risk (Alexander 1993: 617). Towards the later part of the IDNDR, there was also recognition of the importance of understanding local conditions of vulnerability – and the related need to involve and support local actors. This led to an emphasis on urban disaster risk reduction capacity during the last few years of the IDNDR (see Chapter 53).

At the conclusion of the IDNDR, it was replaced with the International Strategy for Disaster Reduction (ISDR), and a Secretariat was created to continue efforts towards reducing disaster losses. The specific goals of the ISDR are:

- to increase public awareness to understand risk, vulnerability and disaster reduction globally;
- to obtain commitment from public authorities to implement disaster reduction policies and actions;
- to stimulate interdisciplinary and inter-sectoral partnerships, including the expansion of risk reduction networks; and
- to improve scientific knowledge about disaster reduction.

(UNISDR 2005b)

The ISDR is simply a strategy framework which was adopted by UN member states in 2000. The strategy is served by a Secretariat which is meant to act as the focal point in the UN system for co-ordination of disaster risk reduction efforts. It is formally responsible to the UN's Office for the Coordination of Humanitarian Affairs (OCHA). In practice, the division of labour places relief and recovery co-ordination under the OCHA and DRR under the ISDR Secretariat. The UNISDR also refers to the ISDR system (or simply ISDR) of 'numerous organisations, States, intergovernmental and non-governmental organisations, financial institutions, technical bodies and civil society, which work together and share information to reduce disaster risk' (UNISDR 2010b).

With the challenging mandate to generate awareness, political will and public commitment for DRR, the ISDR has struggled somewhat to establish itself as a credible 'shepherd' of the international community towards commitments and action on DRR. With limited capacity, the UNISDR attempted to co-ordinate the diverse range of system stakeholders, many of which felt little ownership of the ISDR. The capacity of the ISDR is expanded by the efforts of specialists in other agencies, members of various bodies created to co-ordinate efforts focused on DRR. These include the Global Platform for Disaster Risk Reduction, many national platforms and regional platforms, as well as thematic platforms, an ISDR donor support group, an inter-agency group, a scientific and technical committee and a management oversight board.

A 2005 evaluation of the Secretariat (Christoplos *et al.* 2005) found that the Secretariat had applied a strategy of attempting new initiatives, but that these would require stable, long-term funding. Rather than harnessing and supporting the ongoing efforts of system partners, there was confusion around whether the UNISDR was an operational body or a non-operational entity focused on co-ordination and awareness-raising. The same evaluation found that this raised 'unrealistic expectations for more direct material support to national processes and a blurring of the Secretariat's role as a non-operational "honest broker" within the United Nations system' (Christoplos *et al.* 2005: 2). By the time of the ISDR's World Conference on Disaster Reduction (WCDR) in Kobe, Japan in 2005, support for the ISDR and for a costly international conference was waning. The occurrence of the Indian Ocean earthquake and tsunami disaster in December 2004, just before the WCDR in January 2005, reignited DRR as a priority, and 'In the view of many stakeholders, the WCDR was saved from near obscurity by the interest thus generated' (Christoplos *et al.* 2005: 10).

The result of the Kobe conference was the Hyogo Framework for Action (HFA) (UNISDR 2005b), a ten-year framework to reduce disaster losses which was signed by 168 countries. Several HFA principles and priority actions focus on integrating disaster risk considerations into development planning (see Box 50.3); however, the agreement is not legally binding. Despite this, the HFA has served as an important rallying point, which the UNISDR has used to build ownership and commitment to the DRR agenda.

Four years later, a follow-up evaluation of the UNISDR documented the progress of the Secretariat in becoming a relevant champion for DRR. The evaluation found that 'Since the evaluation in 2005, UNISDR has maintained its role as "honest broker" within and beyond the UN system and has increased its relevance in key areas, but has made little progress in clarifying roles and responsibilities' (Dalberg Global Development Advisors 2010).

The ISDR has established a system for governments to self-assess and to report progress of work under the HFA's five priority areas (see Box 50.3). The ISDR summarises progress on a biennial basis in its *Global Assessment Report* (GAR). The Global Network of Civil Society Organizations for Disaster Reduction (GNDR) undertook reviews published in 2009 and 2011 that provided a local perspective on progress towards the HFA. This is a bottom–up assessment that complements the top–down review undertaken in the GAR. The title of the 2009 report, *Clouds but Little Rain*, puts the findings in a nutshell (GNDR 2009). Some 7,000 local actors in forty-eight countries surveyed judged progress in implementing the HFA to have been stalled at the national level, with little trickling down. The GNDR carried out a second, ambitious 'reality check' in 2010–11, which focused specifically on the question of local risk governance.

Box 50.3 The five priority areas defined by the Hyogo Framework for Action

The Editors

The UN's framework for 'significantly' reducing disaster risk by 2015 calls on governments, among other stakeholders, to do five things. These are referred to as the five 'priority areas' (PAs) of the Hyogo Framework of Action (HFA) (www.unisdr.org/eng/hfa/hfa.htm):

- Ensure that DRR is a national and local priority with a strong institutional basis for implementation.
- Identify, assess and monitor disaster risks and enhance early warning systems.
- Use knowledge, innovation and education to build a culture of safety and resilience at all levels.
- Reduce the underlying risk factors.
- Strengthen disaster preparedness for effective response at all levels.

Development institutions

Towards the end of the IDNDR, development institutions also started paying more attention to addressing the underlying causes of disasters. For example, the World Bank and the Inter-American Development Bank began efforts to assess their disaster-related activities in a more strategic manner and to address disaster risk as an integral part of development. A number of bilateral donors, including the UK's Department for International Development, Norway's Ministry of Foreign Affairs, the Swedish International Development Co-operation Agency, and others that were facing increasing demands on their humanitarian action budgets, also began to shift attention towards ex ante action for disaster risk reduction.

Several disasters contributed to building the momentum of the disaster risk reduction agenda. Hurricane Mitch devastated Central America in October 1998, killing thousands and causing about US$6 billion in losses. At the donor conference held in Stockholm to raise support for the reconstruction efforts, the leaders of Central American countries called for a 'transformation' of the region rather than mere reconstruction, with a focus on reducing the region's vulnerability as an overriding goal. There was a growing realisation that the economic losses from disaster events were growing steadily, and increasing efforts to understand and document the developmental impacts of catastrophic events (Wisner 2001a).

Key milestones for the World Bank in this process were the establishment of the Disaster Management Facility (DMF) in 1998 and the launch of the ProVention Consortium in 2000.

With little existing capacity (or budget support) within the World Bank for disaster risk management, and recognising the cross-disciplinary nature of the issue, the DMF reached out to external experts in its formation of the ProVention Consortium. It established an informal network of organisations from development and humanitarian institutions, academia, civil society and the private sector. With funding support from a few bilateral donors, ProVention dedicated its efforts to documenting the long-term financial and economic impacts of disasters, demonstrating the linkages between disasters and poverty reduction, and proving that disasters were indeed a development issue. Risk analysis and modelling tools were also developed, as well as work with the private sector on alternative risk-financing mechanisms.

In early 2003 the ProVention Secretariat was transferred to the IFRC in Geneva. At that time, ProVention expanded its network of civil society and academic partners and initiated efforts to improve knowledge generation and sharing among researchers and practitioners in the global South. The relationship with the IFRC allowed ProVention to capitalise on the strengths of its new host to focus on community-based disaster risk management (see Chapter 59 and Chapter 64), and to forge closer linkages with humanitarian and local vulnerability needs. One of the innovations undertaken was collaboration with the development and solidarity organisation Grassroots Organizations Operating Together in Sisterhood (GROOTS, www.groots.org) in piloting a community disaster resilience fund (ProVention Consortium 2010).

A key ProVention initiative served as a watershed of sorts for the recognition of disaster risk management as a development issue within the World Bank. *Natural Disaster Hotspots: A Global Risk Analysis* was published in 2005 (Dilley *et al.* 2005). Where past risk maps identified where people were at risk, *Hotspots* also mapped economic assets that were at risk. Using a common geospatial unit of reference, *Hotspots* ranked countries in terms of highest risk potential, providing the scientific underpinnings to inform the World Bank and other donors where there was the greatest need for risk reduction.

Another key milestone at the World Bank was *Hazards of Nature, Risks to Development*, a review of twenty years of World Bank activities for 'natural' disasters since 1984. This covered 528 projects amounting to US$26.3 billion, or 9.4 per cent of World Bank commitments over that period (IEG 2006: 11). This study was the first comprehensive assessment of the Bank's disaster-related assistance, and served to inform efforts to update the Bank's emergency response policy. The study concluded that while the Bank had done a decent job at assisting client countries to rebuild infrastructure after disasters, it did not do as well at reducing vulnerability and addressing root causes of disasters.

In September 2006 the World Bank, together with UN/ISDR and bilateral donors, launched the Global Facility for Disaster Reduction and Recovery (GFDRR) as a partnership of the ISDR system to 'mainstream disaster risk reduction and climate change adaptation in country development strategies by supporting a country-led and managed implementation of the Hyogo Framework for Action (HFA)' (GFDRR 2010). The GFDRR builds on the work of ProVention, and represents an important effort to promote the application of knowledge generated at the country level. By providing technical assistance grants, the GFDRR serves as a sort of market development facility for risk reduction investment.

The international aid system will likely continue to sort out its architecture, shuffling funds between the development, DRR and climate change adaptation (CCA) pots. It is important to continue encouragement and pressure to invest more in reducing risk and preventing disasters as part of development. It is equally important, however, to encourage investment in the most appropriate, effective preventive measures rather than in prevention per se. This point is borne out very clearly in the joint World Bank–UN report on *Natural Hazards, UnNatural Disasters: The Economics of Effective Prevention* (World Bank 2010e) (see Box 50.4). The report emphasises that 'effective prevention depends not just on the amount but on what funds are spent on'. It goes on to say that while there are few specific DRR investments that governments need to make (e.g. weather forecasting technology, early warning systems, extra efforts to protect critical infrastructure), effective investments in disaster risk reduction will not come from DRR budget lines. Rather, effective reduction of disaster risk is about provision of basic services, maintenance of infrastructure, good governance, strong institutions, making risk information available and allowing for public involvement and oversight.

Are lessons being learned?

The cross-cutting nature of DRR means that efforts are often not considered as falling within a specific sector and not considered in a strategic manner. This also applies when it comes to evaluation. While many agencies assess their performance in individual interventions, more strategic assessments of agency roles in the response and recovery process are lacking. For development institutions, disaster operations are not looked upon as a sector and, therefore, not evaluated in a strategic manner. While this remains a problem, a shift towards a more strategic approach to evaluation, including for disasters involving natural hazards, began after the Joint Evaluation of the International Response to the Rwanda Genocide, when the international community realised the need to assess how interventions interacted with each other in a specific context (Sellström *et al.* 1996).

ALNAP (www.alnap.org) was established in 1997, following the multi-agency evaluation of the Rwanda genocide. ALNAP counts on participation of key humanitarian donors, NGOs, the Red Cross/Red Crescent, the UN and academia to improve humanitarian performance through the development of analysis and tools made accessible to all. ALNAP has developed a rich database of learning on humanitarian response to disasters.

In 2002 ProVention initiated one of the first inter-agency efforts to assess the long-term disaster recovery process, with a series of case studies on Bangladesh (1998 floods), Honduras (1998 Hurricane Mitch), India (2001 Gujarat earthquake), Mozambique (2000 and 2001 floods) and Turkey (1999 earthquake). ProVention also teamed up with ALNAP to publish a series of short notes on key lessons from relief and recovery efforts, each with a specific focus on urban disasters, earthquake, floods and slow-onset events. One of ProVention's final publications included a study of lessons from the recovery in Nicaragua ten years after Hurricane Mitch (www.proventionconsortium.org/?pageid=37&publicationid=170#170).

In terms of actual learning, it seems that there is a long way to go to the integration of the lessons into practice. In November 2006 a conference discussed disaster operations, highlighting aspects that facilitate institutional change and improve practices. A dramatic convergence of evaluation findings emerged. While the evaluations had been undertaken by diverse institutions (IFIs, bilaterals, humanitarian organisations, etc.), they bore out very similar findings (IEG

2008). A follow-up conference held two years later revealed that most of those lessons had not yet been applied. Among these one might highlight the following:

- to ensure long-term engagement for vulnerability reduction;
- to make prevention and mitigation a priority;
- to use disaster-resistant techniques in infrastructure reconstruction;
- to customise disaster response to a country's specific needs;
- to engage beneficiaries during interventions;
- to include local governments in decisions; and
- to strengthen line agencies' disaster risk management capacity.

Box 50.4 The economics of effective prevention

Bianca Adam
Global Facility for Disaster Reduction and Recovery, World Bank

In 2008 the World Bank and the UN embarked on a major report, which for the first time assesses the economics of disaster prevention in a comprehensive way. Launched on 11 November 2010, *Natural Hazards, UnNatural Disasters: The Economics of Effective Prevention* (www.worldbank.org/preventingdisasters) presents research by over seventy experts from over two dozen institutions around the globe, including economists, climate scientists, geographers, political scientists and psychologists.

The report delivers five key messages, four addressed to policy-makers and a final one to the donor community.

First, *governments can and should make information on hazard risks more easily accessible* to enable people to make informed prevention decisions.

Second, *governments should let land and housing markets work, supplementing them with targeted interventions when necessary.* When land and housing markets work, property values reflect hazard risks, guiding people's decisions on where to live and what prevention measures to take. In Mumbai, where rent controls have been pervasive, property owners have neglected maintenance for decades, so buildings crumble in heavy rains. The poor bear the brunt of the cumulative effects of such policies (tax structure, city financing arrangements, and so on), which produce only a limited and unresponsive supply of affordable, legal land sites for safer housing.

Third, *governments must provide adequate infrastructure and other public services,* ensuring that new infrastructure does not introduce new risk. Locating infrastructure out of harm's way is one way of doing so. Where that may not be possible, another way is to execute multipurpose infrastructure projects, such as Kuala Lumpur's Storm-water Management and Road Tunnel (SMART), which combines a roadway and a drain in one tunnel.

Fourth, *good institutions must develop to permit public oversight.* One robust finding of the report is that countries with well-performing institutions are better able to prevent disasters, including reducing the likelihood of disaster-related conflict. Fostering these institutions is difficult. Partly, it means recognising the role of the market. It means letting evolve a messy array of overlapping entities (the media, neighbourhood associations, engineering groups), which may not all have lofty motives but nevertheless allow divergent views to percolate into the public consciousness.

Fifth, *donors have a role in prevention as well.* Disaster aid can both help and hinder prevention efforts. It can increase prevention by improving the quality and quantity of public goods. However, at the same time, predictable ex post aid can reduce prevention through the Samaritan's dilemma – the tendency to under-prevent when external ex post support is expected. For example, Nicaragua declined to pursue a weather indexing programme after it had been priced in the global reinsurance market citing international assistance following Hurricane Mitch in 1998 as an indication of dependable alternatives.

The share of humanitarian funding going to prevention is small but increasing – from about 0.1 per cent in 2001 to 0.7 per cent in 2008. However, prevention activities often imply long-term development expenditure whereas the focus of humanitarian aid – already a tiny part of official development aid – is immediate relief and response. Donors concerned with increasing prevention in projects could earmark official development aid (rather than humanitarian aid) for prevention-related activities. Such aid, if used effectively, could reduce issues arising from the Samaritan's dilemma.

Conclusions

The international profile of the DRR agenda has increased in recent years, to the point of becoming a separate sector of research, business development and investment. This is due to a number of factors, not least of which includes a series of major catastrophes, Hurricane Mitch in 1998 and the 2004 Indian Ocean tsunami among them. The establishment of the DRR sector (some even call it an 'industry') has enabled important political and financial commitments on the part of the international community and national governments. These include the signing of the Hyogo Framework for Action, the creation or revision of DRR policies in several bilateral donor agencies and IFIs, the establishment of the World Bank's GFDRR and structural changes to strengthen the UNISDR system. While these developments represent important milestones, lessons from evaluations continue to point out that a key challenge remains to translate these commitments into change on the ground.

The mechanisms mentioned above also bring an increased flow of resources, particularly to national-level actors. Similar mechanisms have not yet emerged to channel funding support on a broad basis to local-level actors such as municipal governments and communities, where sustainable action for risk management needs to take place (see Chapter 52). Increasing decentralisation has given local authorities more decision-making authority; however, these efforts are often not combined with increased funding and capacity support. Moreover, mechanisms to get financial support directly to communities and households are needed.

Another priority is to make stronger links between the CCA and DRR agendas. While CCA is playing an increasingly prominent role in the DRR agenda, discussions thus far have often taken a more defensive approach, focused on institutional access and control of CCA funding rather than opportunities to leverage resources and expertise for reducing vulnerability, particularly of the poor.

In addition to climate change, other factors are contributing to patterns that intensify, accumulate and compound risk: urbanisation, increasing conflict and environmental degradation. The interaction of such risk drivers creates a complex and challenging environment for humanitarian and development actors where institutional barriers for understanding and co-operation are detrimental to providing effective support to poor communities. This highlights the urgent yet age-old need for inter-disciplinary and multi-stakeholder approaches to address the root causes of vulnerability.

So while governments and the international community need to be held accountable for the policies and investments made in DRR, it is not all about building a single DRR system, growing the DRR 'sector' or mainstreaming DRR into development. It is more about recognising that development involves risk; that development is risk management. The owners of that risk need to be informed and empowered to manage it. The most effective reduction of risk and of disaster impacts will come not from the DRR sector, but from the people, communities and institutions that bear the risk.

The field has yet to answer the question posed by Allan Lavell more than ten years ago at the conclusion of the UN IDNDR in 1999 (see also Lewis 1999; Wisner *et al.* 2004). He stated that 'concentration on the question of the impacts of disasters on development basically serves as a distraction from the fundamental question, which is the impact of development on disasters. Only by resolving this latter question will we ever get anywhere in terms of risk and disaster mitigation, and, consequently, in terms of reduced disaster impacts' (Lavell 1999: 2).

51

National planning and disaster

Allan Lavell

LATIN AMERICAN SOCIAL SCIENCE FACULTY, COSTA RICA

JC Gaillard

SCHOOL OF ENVIRONMENT, THE UNIVERSITY OF AUCKLAND, NEW ZEALAND

Ben Wisner

AON-BENFIELD UCL HAZARD RESEARCH CENTRE, UNIVERSITY COLLEGE LONDON, UK

Wendy Saunders

GNS SCIENCE, LOWER HUTT, NEW ZEALAND

Dewald van Niekerk

AFRICAN CENTRE FOR DISASTER STUDIES, NORTHWEST UNIVERSITY, REPUBLIC OF SOUTH AFRICA

Introduction

The notion of national planning is fraught with difficulties. In some low-income, highly indebted countries bilateral budget support makes up more than one-half of national expenditure. Assistance from the World Food Programme (WFP), the United Nations Programme on HIV/AIDS (UNAIDS), the World Health Organization (WHO) and other multi-lateral institutions has become routine, not a matter of sudden or unforeseen disaster. International non-governmental organisations (NGOs) provide many of the services that would be expected of a national government. Indeed, African Nobel Laureate, Wangari Maathai, suggests in the case of Africa that such dependency means that 'governments and individuals [in Africa] aren't the active partners in development' (Maathai 2009: 68). Crisis and emergency have become the new 'normal'. Furthermore, large transnational corporations are likely to have contracts covering such vital sectors

617

as water and electricity, and they control an increasing amount of farm land (World Bank 2010a). In such circumstances, national experts, parliamentarians and officials do play a role, and the appearance of sovereignty is closely protected. Yet both the words 'national' and 'planning' have to take on severely nuanced and limited meanings.

This chapter cannot take up all these issues. The focus here is on a cascade of questions that are no less vexed and difficult. In country X, is there national legislation that mandates disaster management and disaster risk reduction (DRR)? If so, what does the designated national institution actually do? How much of what they do is related to the reduction of disaster risk as opposed to either direct response to a crisis or co-ordination of external humanitarian assistance? Finally, if there is some disaster risk reduction activity, does it influence routine national economic, social, regional and infrastructural planning? In other words, how 'integrated' is the approach to development and DRR planning?

The idea of integrating reduction and management of disaster risk into national development planning is not new (e.g. Cuny 1983; Blaikie *et al.* 1994). From the 1990s researchers recognised the conceptual link between disasters and development; practitioners made efforts to bridge the two (Lavell 1999; Wisner *et al.* 2004). In part this is because of the high costs that low-income countries pay for disasters that destroy infrastructure and assets and reduce gross domestic product (GDP). Another reason is this: it is increasingly clear that disaster risk is the product of failed development – what one might call mal-development. Hurricane Andrew in Florida (1992), the Mississippi floods in the USA (1993) and the Great Hanshin Earthquake in Kobe, Japan (1995), all in countries with greatly developed engineering expertise, made this clear.

Disaster relief without development

The Latin American situation

Between the 1950s and 1970s most Latin American countries established civil defence structures to deal first with the perceived threat of internal civil disorder or military invasion, and second with disasters. These augmented the already existing Red Cross, medical services, fire fighting and military approaches to disaster response. The origin of major changes in institutions and laws governing disaster response and risk in Latin America can be traced to a series of key events in the first half of the 1970s. In 1970 a large earthquake off the coast of Peru killed 70,000 people and led to one of the largest disasters to affect Latin America during the last century. Shortly thereafter, in 1972, the capital city of Nicaragua, Managua, was severely damaged by an earthquake with an estimated 8,000 deaths. Then, in 1974, Hurricane Fifi hit northern Honduras leading to an estimated 6,000 fatalities. Subsequently, in 1976, Guatemala was affected by a large earthquake that led to the death of approximately 20,000 people and which was, due to the concentration of loss among poorer social groups, referred to by Alan Riding in *The New York Times* as a 'class-quake' (cited in Wisner *et al.* 2004: 9). These four events revealed in a dramatic way the inadequacy of the then existing response mechanisms and procedures (Cuny 1983).

In response to these events the Pan American Health Organization (PAHO) and later the United States Agency for International Development (USAID) Office for Foreign Disaster Assistance (OFDA) put together their first disaster preparedness training schemes. PAHO focused on hospital emergency plans (see Chapter 44) and response, including triage training; OFDA dealt with disaster response plans in general, including later efforts in the setup of incident command centres in civil defence organisations (see Chapter 42). Both initiatives, still ongoing in the 2000s, had major impacts in terms of disaster response training and preparedness

During the 1980s these response and preparedness structures were modified to a certain extent as disaster management widened beyond response to include predominantly structural prevention, DRR and post-disaster recovery aspects.

The African situation

In Africa, colonial authorities had used police, paramilitary forces and their metropolitan militaries to deal with civil unrest and mass-casualty incidents such as urban explosions, fires and transportation accidents. Decolonialisation in the period from the 1960s to the 1980s brought many foreign experts that were housed in different ministries. The planning priorities that these experts brought with them concerned infrastructure development ('spatial modernisation' in the 1960s), extending market relations into the countryside ('economic modernisation' in the 1970s) and finally in the 1980s the implementation of 'structural adjustment' and the safety nets that were supposed to come along with it (see Chapter 5).

In the midst of this busy planning activity led by outside experts, there was little space for national capabilities for planning to emerge, let alone innovations in the area of disaster risk reduction. Added to the colonial heritage of top-down planning and crisis intervention, Cold War 'aid' increased the prominence of Africa's militaries. Food emergencies and floods were dealt with during this period as anomalies unrelated to 'development'. The notion that disasters might be the outcome of mal-development was muted and rare (Wisner and Mbithi 1974; Wisner 1975).

Beginning with the 1984 famine in Sudan, African leaders and intellectuals began to question the relationship between the export-driven growth model of development and actual development that benefits all groups in society. Was something not systemically wrong if so many rural and urban poor were chronically at risk? The Sahel famine (1969–72) had caused questioning of this kind (CIS 1973; Franke and Chassin 1980), but these questions did not enter the mainstream. Asking such questions in the late 1980s, a cohort of African planners was ready to respond to the opportunities provided by the International Decade for Natural Disaster Reduction (IDNDR), 1990–99. Later, they responded to United Nations (UN) calls for 'national platforms', dedicated disaster-prevention departments and new legislation.

The Asian and Pacific situation

Frameworks for disaster management had long existed in the Philippines and Indonesia. In these countries disaster management had been guided by policies that dated back to the dictatorial regimes of the late presidents Marcos (1965–86) and Suharto (1967–98). These disaster-focused institutions were controlled by the military through their civil defence arms and included top-down, command-and-control institutional machineries that focused on disaster relief. This worked relatively well when responding to large-scale events that they treated as massive battle operations. In contrast, they were ineffective in the face of the more frequent small-scale disasters, which required local and sporadic resources. Since poverty alleviation was hardly considered as a prerogative of the military and civil defence organisations, disaster management also failed to address the social, economic and political causes of disasters. Finally, distrust between the military and civil society due to decades of dictatorships proved to be an issue when both parties interacted in time of disasters (Delfin and Gaillard 2008).

South Asia, especially India, Pakistan, Bangladesh and Sri Lanka, was more fortunate in its post-colonial experience. In part because of its neutrality during the Cold War, to varying degrees there was earlier and deeper involvement of civilian authorities in disaster management. The British colonial authorities had laid the foundations of a decentralised governance system, and it took hold, prospered and grew in capacity. For example, in India District Collectors have formal training in disaster management as part of their curriculum in higher education via the Indian Administrative Service (IAS). However, despite such 'civilianisation' (in marked contrast with Latin America and South-East Asia), only recently has disaster come to be seen as a priority in national planning.

Historically, in Pacific island countries there have been three key stages in disaster risk management. Prior to colonists and missionaries, Pacific communities maintained their own system of response to natural hazards, through co-operation (inter-island and inter-community), systems of food security, traditional knowledge and settlement characteristics, which all overlapped to provide ways of dealing with events, and enabled communities to be relatively self-sufficient (Campbell 2010). Many communities, particularly rural ones, still have adequate means for dealing with hazards. However, during the colonial era, communities were encouraged to move to the more hazardous coastal areas, where the new colonial government had easier access and control over communities.

There is a long history of external relief assistance in the Pacific, and this has now become relied upon by independent national governments. This has led to a decrease in self-reliance, with valuable techniques such as food storage and preservation no longer being perceived as required. This has made the Pacific island community more reliant on external assistance.

Legal and institutional mechanisms

Today, many governments find themselves in the midst of a paradigm shift from traditional disaster management towards DRR as part of development planning. This is evident in a number of new policies, plans and legislation, as well as ways of thinking about their impact and legal implications (see Box 51.1).

Latin American examples

In 1989 Colombia became the first country to create an inter-institutional and interdisciplinary disaster risk management system that embraced prevention and mitigation, preparedness and response. It was also decentralised and development-based (Ramirez Gomez and Cardona 1996). Such a move followed the impact of the 1985 destruction of Armero by a volcanic lahar with the loss of 20,000 lives. A variety of changes that reflect new ideas and demands have since been made. These changes, in turn, were stimulated by debate among academics and practitioners, international examples and guidelines, and perceived needs (e.g. Lavell and Franco 1996).

Although little real change in terms of widened risk reduction concerns would take place over the ten years following the Colombian experience in Latin America, with the impact of Hurricane Mitch in 1998 considerable impetus was then given to changes in legal and organisational structures and advance of the prevention and mitigation, disaster risk reduction argument in Central America and the rest of Latin America.

By the early 2000s Nicaragua and Bolivia had also created comprehensive systems. El Salvador created its National Service for Territorial Studies (SNET) in 2003 following its 2001 earthquakes, and this move helped to establish risk reduction and risk studies as something separate from disaster response. The most recent and most radical change has been seen in Ecuador, where the government has created a Risk Management Secretariat, substantially reducing the military's role and incorporating civil defence into this new civilian-run organisation. The Ecuadorean case is backed up by explicit mention in the country's 2008 Constitution of risk management and protection of the population against physical hazards

Box 51.1 Judging the effects of national disaster reduction laws

Jean Carmalt
Drake University, Des Moines, Iowa, USA

Under international law, the definition of discrimination includes intentional forms of discriminatory behaviour in addition to *discriminatory effect*. The latter arises when there is evidence of *disparate impact*, even when there is no intent to discriminate. In other words, if neutral laws result in an outcome that disproportionately impacts one group in a negative way, it is prohibited as a form of discrimination under international law.

This means that international law prohibits not just intended acts of discrimination, but it also prohibits the discrimination that comes about from more complex inequalities that emerge from structural forces.

In terms of disasters, prohibiting discriminatory effect is particularly important in terms of implementing new laws aimed at reducing the risk of disaster, such as those crafted following the Hyogo Framework for Action (HFA) (UNISDR 2005b) in a number of countries including South Africa, Peru and the Philippines. Like all laws, these laws must be implemented in a non-discriminatory way, which means that they must not discriminate in purpose *or in effect*. This includes a prohibition against leaving unaddressed the structural forces that, for example, result in residential segregation that disproportionately exposes one group to more hazard impacts than another group.

If, for example, a national disaster law mandates construction according to a seismic code, but this code is only enforced in the country's largest cities, then *in effect* the law is discriminatory and in violation of international law, although the framers of the law did not intend discrimination. Indigenous, isolated and marginalised groups in a country must enjoy the same coverage as all residents, which could, for example, equally require provision of timely and accurate information and warning, support in local preparedness for disasters, assistance if one occurs and resources for recovery.

Some of the international agreements that prohibit discriminatory effect include the following:

- UN General Assembly (1966) International Covenant on Civil and Political Rights. Resolution 2200A (XXI)
- UN General Assembly (1966) International Covenant on Economic, Social and Cultural Rights. Resolution 2200A (XXI)
- UN General Assembly (1965) International Covenant on the Elimination of All Forms of Racial Discrimination. Resolution 2106 (XX)
- UN General Assembly (1979) Convention on the Elimination of All Forms of Discrimination against Women. United Nations General Assembly Resolution 34/180.

Further reading: Carmalt (2011); UN (1989c).

African examples

In Africa, as in Latin America, there were triggering events of historical importance in preparing the way for new legislation and new government institutions. Although the situation was one of civil war and not natural hazard, the alleged and perceived inability of surrounding African countries and the rest of the world to respond to the humanitarian catastrophe in Biafra in 1968–71 provoked a profound rethinking of preparedness and response. In the course of the two and a half years of civil war after Nigeria's oil-rich southeast had declared independence, more than one million people, mostly civilians, died of injuries, starvation and disease (Forsyth 2007). The UN and the Organization of African States were paralysed and did nothing, despite early efforts by East African heads of state to mediate. This was the first large-scale humanitarian action in history dominated by NGOs, and they carried out an airlift of food surpassed only by the Cold War relief of Berlin (de Waal 2005). One consequence was that NGOs and their African partners began contingency planning for disasters elsewhere on the continent. The administrative structure and planning methods of African governments, by contrast, remained unchanged.

Table 51.1 provides an overview of institution building in Africa for disaster management and DRR. Early 5innovators were those countries experiencing major drought and food emergencies with accompanying heavy loss of life due to concurrent disease in the 1970s and

Table 51.1 Examples of institutional DRR development in Africa

Country	Date of legislation	Institution	Observations
Burkina Faso	1973	Ministry of Social Action and National Solidarity	Began as unit to co-ordinate food aid in drought emergency, and evolved
Ethiopia	1973	National Disaster Preparedness and Prevention Committee (Prime Minister chairing)	Ditto
Niger	1974	Prime Minister's Office	Ditto
Nigeria	1976	National Emergency Management Agency	Ditto
Sudan	1985	Humanitarian Aid Commission	Ditto
Cameroon	1986	Ministry of Territorial Administration and Decentralization	
Malawi	1991	National Disaster Preparedness and Relief Committee; and Joint Food Crisis Task Force (2002)	Response to recurring food emergencies
Ghana	1996	National Disaster Management Organization (under Ministry of the Interior)	
Burundi	1998	Ministry of Public Security	Response to internally displaced persons (IDPs) due to civil conflict
Mozambique	1999	Ministry of International Economic Co-operation	Expanded and revised national system after 2000 floods
Uganda	1999	Prime Minister's Office	Response to refugees and IDPs due to conflict
South Africa	2002	Department of Provincial and Local Government	Immediate priority after 1994 elections ending apartheid, and following Cape Flats floods
Angola	2003	Ministry of the Interior	
Swaziland	2004	Deputy Prime Minister's Office	
Tanzania	2004	Prime Minister's Office	

Source: Ben Wisner from national reports to the World Conference on Disaster Prevention (UNISDR 2005a)

1980s. Their institutions grew up in response to the need to co-ordinate large-scale external humanitarian assistance, and have evolved since. In the 1990s, responses to the needs of large numbers of internally displaced persons and refugees from civil conflict were a major cause of new legislation and creation of government institutions. Technical and financial assistance provided during the IDNDR was also a factor (see Chapter 50). In the case of South Africa, the government claims that it was flooding that affected informal settlement in the Cape Flats in greater Cape Town in 1994 which triggered a call by the first parliament of post-apartheid South Africa for legislation on disaster management. This was followed by legislation in 2002 (Government of the Republic of South Africa 2004). In the twenty-first century, bilateral donor and UN encouragement for the creation of national DRR platforms and policy increased. Thus a top-down process including some new legislation has recently begun, yet until large amounts of money and technical assistance became available for climate change adaptation, attempts to link up hazard vulnerability and mainstream development planning remained rare.

Asian examples

In Indonesia, a new law was enacted in 2007 in the aftermath of the December 2004 earthquake and tsunami which devastated the province of Aceh. In the Philippines, a new institutional framework emerged in 2010 after more than a decade of lobbying by civil society organisations. Regional and international organisations like the Asian Disaster Preparedness Center and the UN International Strategy for Disaster Reduction (ISDR) had provided impetus for these developments.

New legal mechanisms marked a significant shift from relief-oriented policies to disaster risk reduction and post-disaster recovery. Governments began to foster the participation of NGOs and community-based organisation (CBOs) in disaster risk reduction activities, although the military remained a prominent stakeholder. Another improvement was providing local governments with staff and financial resources for pre-disaster prevention and mitigation activities (BNPB and UNDP 2008; Disaster Risk Reduction Network Philippines 2010).

South Asia has also seen new legislation and institutional development, the drivers for which were the 2004 tsunami and the 2005 Kashmir earthquake. One or both of these two events seriously affected India, Pakistan and Sri Lanka. India passed a new Disaster Management Act in 2005 that established a National Disaster Management Authority under the Ministry of Home Affairs and a Disaster Mitigation Fund. Pakistan established a National Disaster Management authority in 2006. Sri Lanka passed its Disaster Management Act in 2005 that established the National Council for Disaster Management and a Ministry of Disaster Management and Human Rights, which has the authority to direct all other government ministries in the provision of disaster response and assistance (Hapuarachchi 2009: 55–6).

Pacific examples

With the exception of Fiji, the Cook Islands, Samoa and French Polynesia, the Pacific islands do not have land-use policies or national land-use plans. This is a major constraint to integrating DRR into routine land-use planning. In Fiji and the Cook Islands, where they do have land-use regulations, the consultation arrangements are too rigid and are not participatory (SPC 2008). In the late 1990s, SOPAC (the Pacific Islands Applied Geoscience Commission) included a DRR portfolio in its work. Whilst SOPAC predominantly provides scientific and geological services to the region, it has also established a community risk programme called CHARM (Comprehensive Hazard and Risk Management) (SOPAC 2005).

In response to the Millennium Development Goals, a Pacific Urban Agenda (PUA) has been developed and incorporated into the regional Pacific Plan. Among its three main objectives, the PUA seeks to integrate 'disaster management' into urban planning and governmental frameworks. To date, implementation has been slower than other elements of the Pacific Plan (Campbell 2010).

New Zealand is an exception, as it has been integrating DRR into national development planning through legislation and practice. The combination of the Resource Management Act (1991) and the Civil Defence Emergency Management (CDEM) Act (2002) has placed legislative mandates on planners to consider natural hazard risk reduction in all aspects of planning. In addition, emergency managers are required to develop consistent risk reduction options in regional CDEM Group Plans (Saunders *et al.* 2007).

Linking development planning and DRR

If it is likely true that 'no investment is risk neutral' (Wisner *et al.* 2004), then it is reasonable to ask to what extent DRR has been integrated into routine national development planning. Decisions to allow or even provide a subsidy for foreign direct investment, to build a dam or other mega-project, and to build a road, hospital or school should be made in an explicitly risk-aware manner. Each has the potential for decreasing or increasing disaster risk, or for shifting risk from one place to another or one group of people to another.

Evidence from Latin America

In Latin America the process of integrating development and DRR is discussed in language specific to the Hemisphere: 'corrective' and 'prospective' risk management (Lavell 2004b). These involve measures introduced to reduce existing risk or to control possible future risk, respectively.

In the 2000s Peru pioneered the use of public-sector financing decisions for reducing and controlling risk. Its Ministry of Finance and Economy made regulations and produced an operations manual. Elements of these have been gradually diffused into other Andean countries and Central America. The Costa Rican Planning Ministry is currently pursuing this line of action as well (see Chapter 54).

Colombia has been the regional pioneer in promoting land-use planning, regional planning and environmental plans as vehicles for the introduction of DRR at the local level in its more than 1,000 municipalities, which are the primary jurisdictions in this large country. The use of such mechanisms in the cities of Bogotá, Manizales and Medellín is now well developed. The Swiss Agency for Development and Cooperation in Nicaragua and the World Bank in Honduras and Nicaragua have promoted similar processes at the municipal level (see Chapter 53). The regional programme, PREDECAN (*Prevención de Desastres en la Comunidad Andina*), in the Andean countries promoted development of land-use planning methods at the municipal level based on pilot projects in all participating countries (Colombia, Ecuador, Peru and Bolivia).

Despite these positive steps, a limitation exists. This concerns the bureaucratic isolation of people and small units dedicated to merging DRR and development. In some countries risk reduction and management units have been inserted into existing ministries and departments such as agriculture, public works, planning and finance, education and health (for example Costa Rica, Peru, Colombia and Ecuador). This is a positive sign. However, some believe that such a sectoralisation or even 'ghettoisation' of the risk problem can only lead to its marginalisation. There is a danger that the rest of the institutional structure will believe that the problem is one for the risk reduction and management office and not for them. This is also a common problem in the African context.

Evidence from Africa

In Africa development has not yet been redefined to include DRR as a core component and cross-cutting concern. However, there are beginnings, and they cluster around two externally driven programmes – the World Bank's Poverty Reduction Strategy Papers (PRSPs), and programmes designed to monitor, project and plan for the impacts of climate change.

There are PRSPs in thirty-four African countries (World Bank 2010c). In some countries' documents (for instance those of the Democratic Republic of the Congo (DRC), Gambia, Malawi and Tanzania), language concerning DRR occurs once or twice. In its review of all PRSPs worldwide, the UNISDR found that most of them do not include any mention of DRR (UNISDR 2009b).

Although this is a small beginning, the PRSPs are important for a number of reasons. They are an attempt by the World Bank to assist about seventy countries worldwide (one-half of them in Africa) to plan in a comprehensive manner for poverty reduction. The countries concerned are those receiving debt relief under programmes for Highly Indebted Poor Countries (HIPC). Not only is the planning approach comprehensive and multi-sectoral, it is mandated to be consultative. Civil society representatives are part of the PRSP setup in each country and contribute to evaluation of it progress. The PRSP *Sourcebook* of planning techniques is extremely comprehensive, allowing national planners to tease out the poverty reduction implications of sectors such as mining and forestry, while dealing in other chapters with cross–cutting issues such as health, education and social protection (World Bank 2010c).

If used well, the PRSP planning process can increase the visibility of risk and focus resources on people who need assistance in reducing and managing risk (Naudé *et al.* 2009a). So, for example, Ethiopia has begun a large-scale rural employment guarantee scheme similar to India's National Guaranteed Rural Employment Act and has also expanded its network of strategic food reserve warehouses (Government of Ethiopia 2004).

Climate change is the second focus of activity in Africa that could lead to more consideration of disaster risk in national planning (see Chapter 14 and Chapter 18). This agenda is also externally driven, with money and technical assistance provided by the UN Development Programme (UNDP) and a number of bilateral donors. Many African countries are drought- and flood-prone to begin with, and some have suffered food price shocks and even food riots in recent years. Some have low-lying portions of coastal cities that will be affected by sea-level rise. Thus while climate variability has been the chronic burden of the rural poor in Africa from pre-colonial days, national elites are now beginning to understand that their country's social, political and economic stability may depend on intelligent long-range planning for drought and flood.

Evidence from Asia

Recently in Indonesia and the Philippines, DRR has been more fully integrated into development planning. In Indonesia, reducing the risk of disasters has been promulgated as one of the nine national development priorities. Discussions are ongoing for similar mainstreaming in the Philippines, where DRR is not yet a top priority in the national medium-term development plan. In both countries many government agencies are integrating disaster risk into their sectoral plans. A national hazard mapping programme is also being conducted with the ultimate goal of considering hazard-prone areas in provincial and municipal development plans (BNPB and UNDP 2008; Disaster Risk Reduction Network Philippines 2010).

In South Asia, Bangladesh and the Maldives come as close to fusing development and DRR as anywhere. Both are highly prone to cyclones, flooding and climate change, especially sea-level rise. In Bangladesh, a national Plan for Disaster Management guides the co-ordination work of the Disaster Management Bureau, in existence since 1993, and the Ministry of Food and Disaster Management. There has been good co-operation with local communities and civil society, and progress has been achieved in protecting lives from coastal storms. The Maldives enjoys a fully integrated planning and co-ordinating system which is the fruit of international technical and financial assistance after the Indian Ocean tsunami. Climate change concerns amplify political will, but the sustainability of this new administrative system once international aid is ended remains a question (Hapuarachchi 2009: 55–6). Other Asian countries are making progress towards comprehensive DRR from a more conventional starting point in disaster response. For example, Box 51.2 provides a brief sketch of DRR in Mongolia.

Box 51.2 National planning in Mongolia

Ben Wisner
Aon-Benfield UCL Hazard Research Centre, University College London, UK

Mongolia is a country with between 2.6 million and 3 million people, spread thinly, except for major population centres, across an area three times the size of France. Mongolia is prone to many hazards: severe winter weather, drought and wildfire, as well as human and animal epidemics (IFRC 2010c). Its capital, Ulaan Baatar, is located in a seismic risk zone. In 2009 a series of hazards including drought and severe winter weather (*dzud*) killed twenty-five per cent of the livestock. Many destitute herders migrated to cities, where they have few opportunities for employment.

The Mongolian government passed a 2003 Disaster Management Law, which is complemented by other laws that include risk-relevant aspects of public health, livestock, agriculture and the environment (Government of Mongolia 2004). However, a supplementary National Implementation Plan that sets out a series of activities over several years was still pending with the parliament in 2010 (Jeggle 2010). Meanwhile, there have been a number of efforts that broaden a more traditional emergency management approach in selected *aimags* (provinces), the capital city area and among rural herder communities.

There is an ongoing effort to develop a national information centre and also investment in connecting all centres by short wave radio (Government of Mongolia 2004). Many disaster response personnel receive in-service training.

With the growth of the capital city, an expanding urban earthquake risk has been addressed in recent years with assistance from UNDP that includes micro-zonation studies in order to map earthquake vulnerability.

There are more than 300 weather stations used for early warning of the highly variable and often severe winter weather and periods of intense rainfall and urban flooding. The government is keenly aware of the value of herder traditional knowledge and has historically sought to draw on its beneficial aspects. For example, Government of Mongolia (2004) cites herder knowledge of the pre-winter food collection activities of voles as a sign of the severity of a coming winter season.

Privatisation of land and state fodder reserves and dissolution of pre-1991 herding collectives have created difficulties such as over-grazing and inadequate availability of sufficient winter fodder reserves. Drought conditions in recent years also have weakened both pasturage and herds, increasing the animals' vulnerability to severe winter weather. Attempts to stabilise the livelihoods of herders include experimentation with agricultural insurance (Skees and Enkh-Amgalan 2002) and a vigorous programme of livestock protection that extends to most of Mongolia's districts (*soum*).

A remaining challenge is getting the balance right between market and government regulation. Price rises for basic food during 2009–10 (GIEWS 2010) suggest room for price controls or social protection.

Mongolia has an excellent plan and framework for DRR that extends beyond conventional 'natural hazards' to their recognised association with environmental management, public health and climate change impacts. It also has good practices underway, but these are not yet universal, nor has parliament yet passed the implementation plan that would encourage much wider and more sustained national commitments to integrating disaster risk issues in national development strategies. Democratic governance also seems to be well established (Hoover 2009) and has provided a basis for extending DRR.

Evidence from the Pacific

In the Pacific island state of Samoa, Daly *et al.* (2010) describe how planners partnered with communities in order to complete a vulnerability assessment of coastal infrastructure and to see how to reduce vulnerability using planning tools that combine land-use and DRR frameworks. Coastal infrastructure management plans were developed through local consultations within the context of a national planning strategy. That included intensive training of government staff involved in planning and DRR to undertake the village-based consultation work.

Increased self-reliance has been supported more recently by the Pacific island 'Framework for Action 2005–2015', the mission of which includes improved governance, mainstreaming of disaster risk management at all levels of governmental planning and decision-making, and strengthening partnerships among all stakeholders in disaster risk management (Campbell 2010). Samoa established in 2002 the Planning and Urban Management Agency (PUMA), with subsequent legislation passed in 2004 to enable land-use planning. This plan led to reducing the death toll from the 2009 tsunami due to nationwide evacuation training exercises held before the event (Campbell 2010).

In New Zealand national guidance is available for planners on how to implement and achieve risk reduction. Planning guidelines have been developed for landslides and active faults based on the Australian/New Zealand Risk Management Standards. The risk-based guidelines factor in hazard return periods along with a category describing the importance of infrastructure at risk (Saunders and Glassey 2007). The planning principles underpinning the guidelines describe collecting hazard information; trying to avoid the hazards before development is started; analysing and acting on risks for areas already developed; and ensuring that appropriate communication regarding risk is made to the people in the buildings affected. This model can apply to planning for risks for other hazards, and the Ministry for the Environment has a more general guidance note on all natural hazards for land-use planners available online (Government of New Zealand 2008).

Conclusions

This chapter has documented considerable progress in developing institutions and mandates to reduce and manage disaster risk in Latin America, Africa, Asia and the Pacific. So far, the impact of these institutions on routine national development planning and practice is small. Key events can be seen to have boosted national interest and political will, with successive rounds of legislation and policy development. Nevertheless, political will is eroded by day-to-day pressures of governance. There are major obstacles to integrating development with DRR. These obstacles include the following, and further progress will require that they be tackled.

- *Military legacy*: The legacy of earlier response-focused civil defence and civil protection systems is hard to overcome. These were dominated by the military, and in many countries they remain highly influential in disaster management as well as in society at large.
- *Aid dependency*: African and small island states depend on external aid for disaster recovery. There is a need to ensure that a percentage of this aid is invested in risk reduction initiatives with full participation of communities. A recommendation coming from the World Conference on Disaster Reduction (2005) was that at least ten per cent of relief and recovery assistance be given to prevention and DRR.
- *Narrow, donor-defined priorities*: Just as investment in HIV/AIDS swamped the limited health care systems in Africa, narrowly focused, donor-driven disaster management investment could also distort priorities and siphon off scarce human capacity. After the 1998 US Embassy bombings in Dar es Salaam, Tanzania and Nairobi, Kenya, US investment in disaster preparedness focused on the narrow issue of urban mass-casualty training that fitted into the context of its soon-to-be-formally named 'war on terrorism'. The political environment since the terrorist attacks on 11 September 2001 has served to intensify this focus. Climate change adaptation funding may also have a similar distorting influence.
- *Perception of budget conflict*: There is still a widespread perception of competition for budget resources between 'development' and 'DRR'. Development must be redefined before this false perception is eliminated, but its persistence as well as the silo mentality of conventional sectors and ministries make it difficult to redefine development. Sharing of effective practice on a South-South basis, more effective use of key disasters as 'teaching moments' and broader, deeper public and professional education are tools to help break this bottleneck.
- *Political will*: National governments have a crucial and unique role to play in ensuring the reproduction and systematisation of good practices observed among practitioners. Reproduction and systematisation require legislative instruments at the national level to make DRR compulsory at the local level. Unless strong political will from national governments emerges, laudable grassroots initiatives will remain scarce practices.

52

Local government and disaster

Geoff O'Brien

SCHOOL OF BUILT AND NATURAL ENVIRONMENT, NORTHUMBRIA UNIVERSITY, NEWCASTLE-UPON-TYNE, UK

Mihir Bhatt

ALL INDIA DISASTER MANAGEMENT INSTITUTE, AHMEDABAD, GUJARAT, INDIA

Wendy Saunders

GNS SCIENCE, LOWER HUTT, NEW ZEALAND

JC Gaillard

SCHOOL OF ENVIRONMENT, THE UNIVERSITY OF AUCKLAND, NEW ZEALAND

Ben Wisner

AON-BENFIELD UCL HAZARD RESEARCH CENTRE, UNIVERSITY COLLEGE LONDON, UK

Introduction: what is local government?

Local government can play an important role in disaster risk reduction (DRR). However, to understand this potential and why it is often not achieved, one must first be clear about the forms and characteristics of local government. A common definition of local government is state administration at a level that is closest to the population within its area of jurisdiction. This very simple description masks the diversity and complexity of local government across the globe. Ask anyone about their local government and they are very likely to describe political and administrative functions that both represent and manage their local area, and which is the first point of contact with government. Everyone will have some idea of their local government and its area of operation or spatial scale, and what it can do or its powers.

Three characteristics of local government bear directly and heavily on their role in DRR. These are scale, power and accountability.

Scale

A local government will cover an area that is smaller than the national footprint, except in some unusual circumstances such as in a single–island nation. In some cases there is a sub-national level of administration between local and national. Apart from that, spatial scale can be anything from quite large in sparsely populated rural areas to small and compact in the case of urban wards. The spatial scale of the jurisdiction of local government has historical determinants. These may have been imposed for administrative purposes such as revenue collection. They may have evolved over time from earlier forms of governance such as the city state as the unit of local government, or they may have coincided with the territory of an ethnic or linguistic group. They may be determined by physical characteristics such as a river, coastal or other physical boundary.

Finally the spatial scale of local government is dynamic. For example, electoral norms or rules may demand areas of equal population. As populations are dynamic this will require periodic adjustments to maintain balance. Local government boundaries change over time.

However, it must be noted that spatial scale does not always distribute power or wealth evenly. For example, in Japan the Tokyo prefecture is almost twice the size of others in terms of population and employment, giving it a significant amount of power. In Gujarat, India, the earthquake-affected district of Kutch, one of the twenty-five districts of Gujarat, is almost one-third of state land area. This is true of many of the world's megacities and even of smaller national capital city-regions in former colonial countries that function as primary cities. These political realities bear strongly on what resources are available for smaller cities, towns and rural jurisdictions for work on DRR (Freire and Stren 2001).

Scale and boundaries can have other serious consequences for DRR. This may mean that in all but the case of highly local hazards (e.g. one particular unstable slope or local flood), most hazards will have to be managed by drawing on cross-jurisdictional relations among more than one local government unit (LGU). Also, as an important element of preparedness, any given LGU needs to have mutual assistance agreements with neighbouring and even more distant counterparts, since the resources and infrastructure in the primary LGUs affected by a hazard event may be overwhelmed.

Power

In the context of local government's function in DRR, power can be thought of in two ways: access to financial and technical resources, and legal authority (see Chapter 5). Both are important for disaster risk reduction. The more resources and powers a local government has, the more it can act independently and, in principle, the more attentive and responsive it can be to local hazards.

The relationship between resources and authority is important, and the scope of legal authority enjoyed by local government differs by kind of state organisation. Broadly there are two forms of state: unitary and federal. A unitary state is a country that is governed constitutionally as one single unit, with one legislature. The political power of government in such states may well be transferred to lower levels such as regionally or locally elected assemblies, governors and mayors, but the central government retains the right to recall such delegated power. Lower-level governments in unitary states derive their statutory frameworks from parliamentary legislation or an executive order rather than from constitutional authority, and lower-level governments are not directly represented in national legislatures (Ansell and Gingrich 2003).

A federal state is characterised by a union of partially self-governing states or regions united by a central or federal government. India, Nigeria, Brazil and the USA are examples. The self-governing status of the component states is constitutionally entrenched and cannot be altered by a unilateral decision of the central government. Federalism is the creation of two layers of government, the federal government and the constituent states, which equally share the legal sovereignty of a country. Federalism is a multi-centred and non-centralised structure of government, where each centre is given a guaranteed portion of power which cannot be removed by the others (Baldi 1999). However, from the point of view of local government, the difference between unitary and federal systems may be academic. The important question from a DRR (and more general development) point of view is whether and how local government gains access to financial and technical resources and, therefore, what capacities LGUs have for DRR.

However, in federal systems, sub-national regions or states may have different approaches to DRR; for example, there is great diversity among Indian states, all of which take advice from the National Disaster Management Authority, but decide on their own approach and budget. Different regional or state policies will have impacts on what local government is able to do.

Power exercised locally and the availability of power are dependent upon what functions are passed from the central and sub-national to the local level. These institutional arrangements are referred to as decentralisation. Decentralisation is the transfer of authority and responsibility for public functions from the central (or sub-national) government administration to local government. The extent and actual implementation of decentralisation determines local power.

At the end of the day, the important issue is what funding, technical resources and trained staff LGUs are able to deploy in order to reduce disaster risk. Whatever legislative provisions might exist, it is still necessary to put LGU capacity under the microscope (see Chapter 51). Even when clear and adequate provisions exist in law, decentralisation may only be partly implemented. Money and technical resources may be 'delayed', diverted or simply disappear. Political power determines how implementation takes place. Local governments out of favour with the ruling national political power may find themselves starved of resources, as Wisner (2001a) documented in the case of recovery from El Salvador's 2001 earthquakes. Decentralisation may also mask the special interests that actually by-pass the legal, representative authorities, as Ribot (2003) has shown in Senegal.

Accountability

Accountability is a vital issue for DRR for two reasons. First, if local government is not representative of all the interests in its jurisdiction, DRR may proceed 'blindly' because the needs, skills and local knowledge of some residents will be invisible to authorities. Second, implementation of DRR may lack full support by residents who feel excluded. In many countries local governments are elected, as opposed to being appointed. Typically it is a requirement that those who stand for election live, or at the very least work, in the area covered by the local government to which they wish to be elected. Such elected officials often live in the electoral districts or wards they represent. Because of this proximity between elector and elected, coupled with often extensive local knowledge, local government is ideally placed to identify risks and vulnerabilities.

In Tanzania, for example, the 2002 Villages Act devolves a number of policy areas to elected village councils. This includes land-use planning and, by extension, reduction of risk from natural hazards. While this DRR function can be further developed at the village level, there is an institutional basis for it because of a history of work by committees within the village council that deal with such issues as maintenance of irrigation infrastructure and water users' rights, control of illegal tree felling and charcoal burning, and adjudication of herder–farmer conflicts. The elected village committees in Tanzania are complemented by the presence of government-appointed village officials (village executive officers). There is thus a mix of both appointed and locally elected officials and council members (Venugopal and Yilmaz 2010). Cutting across governance by both elected and appointed officials at the village level is the continuing authority of clan elders. These have no legal power but wield considerable influence. In the Indian state of Tamil Nadu local traditional councils of fishermen often have more influence than the elected village councils.

Local government and the practice of DRR

Routine local government functions

Though local government is generally recognised as having a duty to enhance the well-being of its citizens (and DRR is an important aspect of well-being), structural constraints may limit the ability of local government to integrate DRR into its daily functions. Luna (2007) and O'Brien (2006) note that there is often a dichotomy between the responsibility that local government has for the future of its jurisdiction and specific pre- and post-disaster planning functions.

In large LGUs (such as wards in large cities or 'cities' within megacities) pre- and post-disaster planning is often conducted separately by specialist units (Mitchell 1999). In smaller and more remote LGUs (which may well be among those with the least power) DRR plans may simply be done externally by a central government contractor or specialist government unit and be 'delivered' to the LGU for implementation. In both cases the core mission of local government is by-passed and with it also the potential for close collaboration with citizens and civil society, building on their resources and capacities.

For routine problems, such as road traffic crashes or even seasonal water supply or drainage problems, the specialist approach can be very effective. However, this approach ignores the wider issues of community preparedness. The barriers to broader engagement with hazard and vulnerability assessment and actual planning often lie in the political domain.

For example, pre- and post-disaster planning may be forbidden by legislation, which often requires a separate or stand-alone function and does not designate LGUs as the locus of such activities. Of course, some functions are quite specialised, such as fire and rescue services, and require ongoing training in the use of specialist equipment. Whilst such arrangements may work well in such instances, as they do within hospital emergency room contingency planning, there remains the challenge of broader organisational learning and integration of DRR into the full range of local government functions.

Disaster management functions of local government

Pre-disaster planning can be as limited as negotiation of mutual aid protocols with adjacent jurisdictions, or stockpiling equipment and goods likely to be required for response to a hazard event. It can also involve a wide variety of precautionary measures that include hazard assessment, awareness raising, land use and other regulations that reduce building and other uses in exposed spaces, and even measures to increase the stability of local people's livelihoods through micro-credit and encouragement of economic diversification (see Chapter 14). These more elaborate pre-disaster functions are most often carried out in collaboration between local government, sub-national or national government agencies and civil society organisations. The UN's 'Resilient Cities' programme has summarised a range of local government functions that ideally provide for much improved safety.

In New Zealand, community response plans (CRPs) are proving to be a successful tool in communicating local risks and encouraging communities to determine how they will reduce

Box 52.1 Ten-point checklist: essentials for making cities resilient (UNISDR)

1 Put in place *organisation and co-ordination* to understand and reduce disaster risk, based on participation of citizen groups and civil society. Build local alliances. Ensure that all departments understand their role in disaster risk reduction and preparedness.

2 *Assign a budget* for disaster risk reduction and provide incentives for homeowners, low-income families, communities, businesses and public sector to invest in reducing the risks they face.

3 Maintain up-to-date data on hazards and vulnerabilities, *prepare risk assessments* and use these as the basis for urban development plans and decisions. Ensure that this information and the plans for your city's resilience are readily available to the public and fully discussed with them.

4 Invest in and maintain *infrastructure* that reduces risk, such as flood drainage, adjusted where needed to cope with climate change.

5 Assess the *safety of all schools and health facilities* and upgrade these as necessary.

6 Apply and enforce *realistic, risk-compliant building regulations and land-use planning principles.* Identify *safe land for low-income citizens* and develop upgrading of informal settlements, wherever feasible.

7 Ensure that *education programmes and training* on disaster risk reduction are in place in schools and local communities.

8 *Protect ecosystems and natural buffers* to mitigate floods, storm surges and other hazards to which your city may be vulnerable. Adapt to climate change by building on good risk reduction practices.

9 Install *early warning systems and emergency management* capacities in your city and hold regular public preparedness drills.

10 After any disaster, ensure that the *needs of the survivors are placed at the centre of reconstruction* with support for them and their community organisations to design and help implement responses, including rebuilding homes and livelihoods.

United Nations International Strategy for Disaster Reduction (UNISDR),
www.unisdr.org/english/campaigns/campaign2010–11/
documents/230_tenpointchecklist.pdf

these risks and prepare for, respond to and recover from events. Local government emergency management officials work with the local community to develop CRPs, resulting in local ownership and acceptance of plans. While initially an emergency management tool, the CRP process provides the opportunity to raise awareness among communities about their risks, and their management options. The next step in this process will be seeing how the CRPs will influence future land-use planning decisions (Mitchell *et al.* 2010). In addition, New Zealand's Local Government Act (2002) provides for Long-Term Council Community Plans (LTCCPs), which are ten-year plans that include community outcomes and an annual financial plan. In spirit, this and other legislation places local land use and natural resource management, as well as disaster risk reduction, in the context of sustainable development, but implementation problems must be overcome (Saunders *et al.* 2007; Glavovic *et al.* 2010) (see Chapter 14).

Local governments are also in principle the hub of disaster response co-ordination; however, in very extensive and damaging events, the lives of many local government staff may have been lost and the capacity of LGUs severely reduced, as in the Haiti earthquake in 2010. However, in the small and moderate disaster events that are more common (and the accumulated effects of which have a very negative effect on people's livelihoods and life chances), local government remains the administrative centre of needed outside assistance (see Chapter 42). Whilst DRR begins at the local level, not all functions can be effected at that level. Some need to be bumped up to a higher level. For instance, not all local governments can dredge a river or manage an entire watershed. Thus the principle of subsidiarity should provide the basis for a melding of top-down and bottom-up governance of risk.

The same is true of recovery activities. Here, however, smaller LGUs can easily be swamped by donor professional capacity and levels of assistance. There have been both good and bad experiences regarding the role of local government in long-term recovery in recent years. On the positive side, local government took significant initiatives after the 2004 tsunami in India. The District Collector in Nagapattinam District, Tamil Nadu, India set up a non-governmental organisation (NGO) Co-ordination Centre that brought together local and non-local NGOs for relief and rehabilitation operations. The Centre also supported long-term recovery by working with the owners of houses that had been destroyed. This was a part of the owner-driven 'build back better' programme for long-term recovery supported by the United Nations Development Programme (UNDP) (Brusset *et al.* 2009). The outcome was safer houses. This programme is regarded as a flagship in tsunami-recovery projects in India. The owners of destroyed homes decided on what type of house, when, how and at what pace they would rebuild. Decisions were taken at the local level and if a deadlock occurred, problems were resolved through face-to-face dialogue. Both the house and the housing process were ultimately owned by the victim.

However, on the negative side, one can point to the failure of the City of New Orleans to lead recovery in the worst-affected, low-income area of the city affected by Hurricane Katrina. Years later, only a small percentage of the population of the Ninth Ward has returned, and plans for this area remain contentious (Wilson 2010; The Lens 2010) (see Chapter 48).

Good and bad local governance and DRR

Ideally governance at the local level is experienced by residents as being fair and equitable. It should also show leadership and encourage public debate, communicate clearly and co-operate with grassroots and NGOs. In short, local government has the potential of being a champion for wider civil society involvement in public affairs. Not only should it act, but local government should also encourage broader involvement by providing an enabling environment in which others can act (see Box 52.2).

For example, none of the 150,000 inhabitants of Dagupan City were killed when Typhoon Parma struck the Philippines in October 2009, thanks to an efficient and sustainable partnership between the city government, village authorities and local communities. An early warning system based on simple flood gauges enabled local residents in villages previously identified as the most endangered to evacuate in a timely manner to high-rise buildings and other previously established evacuation centres. The massive evacuation proceeded smoothly, as numerous drills and simulations had been conducted over the previous few years by the city government in collaboration with the village authorities (Luneta and Molina 2008). Village authorities were tasked with managing the village-based evacuation centres, while the city government managed larger shelters. Permanent communication between local authorities and the city government enabled the upper structure to fulfil needs at the local level adequately and rapidly. In Dagupan City, through the financial and technical support of the city government, village authorities routinely retailed subsidised rice and operated a community pharmacy. In the event of a crisis,

Box 52.2 Municipal government engagement with DRR in the Yucatán peninsula, Mexico

Emily Wilkinson
Aon-Benfield UCL Hazard Research Centre, University College London, UK

Municipal civil protection departments were first set up in the Yucatán peninsula in the late 1990s. Since then, significant progress has being made towards reducing the risks associated with hurricanes. Most measures are not carried out directly by municipal governments or even considered a government responsibility. Rather, decision-making to reduce disaster risk is pluralistic: fishing co-operatives, communities and households have developed a variety of measures to protect lives, reduce damage to assets and ensure continuity of economic activities after a disaster. Where municipal civil protection authorities are able to co-ordinate their activities with civil society, these actions are more efficient and effective.

In small, coastal municipalities civil protection directors have few resources of their own so they focus on DRR efforts with multiplier effects, including education and communication strategies to raise awareness of hurricanes and how to prepare. This, in turn, encourages autonomous actions by families to protect lives and property. Innovative municipal authorities have also learned to look elsewhere, liberating funds from higher levels of government and leveraging community resources. They have found that involving civil society directly in civil protection activities is advantageous in terms of efficiency and municipal autonomy. For example, fishing co-operatives and farming associations can now organise evacuations by themselves, so there is less need for state government or military intervention.

In the municipality of San Felipe, the municipal president used the federally backed Healthy Municipality Programme to create awareness of health risks such as dengue fever, and to promote responsible rubbish collection and disposal. The programme promoted a greater sense of pride in the community, so when Hurricane Isidore struck in 2002, citizens were anxious to clean up and repair damage to the main town as quickly as possible. As a result, economic recovery in San Felipe was faster than in neighbouring municipalities.

Because these governance reforms came from the 'top down', San Felipe residents now have high expectations of government. They assume it will provide efficient services, including disaster management, and expect it to co-ordinate efforts with civil society to protect lives and property. These good practices in disaster preparedness and response have been institutionalised as a result of popular demand.

Improving the efficiency of preparedness and response has gone a long way to reducing the physical damage and social disruption associated with hurricanes in the Yucatán peninsula, but coastal communities may have become over-reliant on these measures. Every time there is an early warning, the whole community has to be evacuated—along with valuable possessions including fishing equipment—even if the warning ends up being a false alarm. This is costly and, despite careful organisation, some assets will be left behind and may be damaged or stolen. Preparedness measures can therefore be seen as *necessary*, but not *sufficient*, to reduce disaster risk over the long term. Although municipal authorities have not yet devised ways to support longer-term adaptation, some families have taken the initiative themselves, building second homes inland where they can stay and safeguard their possessions during hurricane season.

rice and medicine supplies are therefore available as relief goods. Village authorities further provided pedi-cabs for a very cheap rent. Such initiatives enabled the poor to make a living on a daily basis while the village authorities earned additional funds for DRR and emergency operations.

A review of DRR work in the cities of Mumbai and Pune and the small town of Bhivandi in Maharashtra in India found that in cities where governance was better, for example where the mayor was elected, where there was an active standing committee, the annual budget review was transparent and there were transparent performance indicators:

- DRR was better integrated in the state policy level;
- There were operational guidelines for city and ward committees; and
- External relations had been developed with civil society, armed forces and a growing number of small and medium-sized businesses.

(Brusset et al. *2009)*

In fact, the challenges of mainstreaming DRR in city planning and the city development plan were better understood city-wide. 'Push decisions and the best people closer to where the action is', said the newly elected mayor.

Following the peace treaty and peace process in Aceh and Maluku Island in Indonesia, a review by the UNDP found that this had directly influenced DRR by making it more oriented toward partnership among stakeholders (UNDP 2010c). The local disaster community, comprising large and small community groups, businesses and leaders, were more readily able to come together and undertake specific tasks as governance improved. Issues of DRR governance such as community committees were better addressed, more decisions were taken at lower levels and direct participation of local civil society in the recovery process widened over time.

By contrast, poor governance is characterised by arbitrary policy-making, unaccountable bureaucracies, lack of enforcement of regulations or imposition of unjust legal provisions, the abuse of executive power, widespread corruption and, as a result, a civil society unengaged in public life and which mistrusts local government. Blair (2000) notes that there are often strong factors that can limit good governance at the local scale. There are huge pressures on some national governments, such as international debt, which also impact the local level. There can be significant tensions within nations such that gender, minority ethnicities and faith groups as well as the marginalised and poor are excluded from political processes (see Chapter 38). Changes in policy and leadership due to elections may affect the continuity and effectiveness of DRR.

Though the overall principles of response, recovery, mitigation and preparedness underpinning dealing with disasters have not changed, emphasis can often be more focused on institutional preparedness and security as opposed to public preparedness. The principles of disaster management have evolved from lessons learned from both natural and technological hazards, and often, particularly in the affluent, industrial countries, the responsibility has been devolved into a specialist agency or bodies that can be characterised as legally based, professionally staffed, well funded and organised (O'Brien 2006). This tends to reduce the role of communities and citizens in DRR.

The Global Network of Civil Society Organisations for Disaster Reduction (GNDR) has developed a checklist of twenty indicators of good local governance (see Box 52.3) that summarise much of the foregoing discussion. Implementation, however, is often more difficult than one would hope.

It is increasingly common for local government to have responsibility for disaster planning, but often this is not linked to the broader planning functions of local government (see Chapter 53).

Box 52.3 GNDR indicators of good local government

Inclusion and participation

1. Do local government policies and legislation explicitly recognise the rights of all people in society to security and protection from hazards, particularly vulnerable groups (e.g. children and youth, elderly, disabled, migrants) in high-risk areas?
2. Does local government support the active participation of vulnerable people and their representative organisations (including children and youth, the elderly, the disabled and migrants) in DRR decision-making, planning and implementation procedures?
3. Do local government disaster risk reduction practices take into account the different needs and priorities of men and women, and support their equal participation in decision-making, planning and implementation processes?
4. Do local government DRR practices take into account the specific needs of children and young people when planning and implementing measures to reduce disaster risk?
5. Does the local government actively promote the formation of partnerships between local authorities and other levels of government, civil society, academia and the private sector?

Capability

6. Does the local government have regularly reviewed disaster risk reduction policies and legislation in place at the appropriate administrative levels (e.g. municipal, district)?
7. Are there decentralised government processes with roles, responsibilities and authorities clearly defined and allocated to support the implementation of DRR measures within appropriate line-ministries and local administrative offices?
8. Does local government have sufficient expertise and technical skills to support the implementation of risk reduction measures within relevant line-ministries programmes, including major infrastructure projects?
9. Do local government risk reduction interventions build on indigenous knowledge, skills, experience and capacities (assets and resources)?
10. Does the local government have sufficient financial resources from appropriate budgets to implement disaster risk reduction measures within relevant work programmes and sectors?
11. Does local government provide training and learning opportunities in disaster risk reduction to enhance knowledge and skills of local authorities, community and civil society leaders?

Accountability

12. Does local government establish baselines and set geographically specific time-bound targets for disaster risk reduction with clear responsibilities?
13. Does local government regularly monitor and report on progress towards disaster risk reduction actions and relevant targets?
14. Does local government involve civil society and local communities in monitoring progress towards achievement of risk reduction objectives?

15. Does local government provide local mechanisms through which citizens can register complaints and seek a response for not meeting policy obligations and commitments?

Transparency

16. Does local government regularly collect, review and map information on local risk patterns and trends (e.g. climate change, hazards, capacities and vulnerabilities)?
17. Does local government connect traditional and indigenous knowledge and experience with external scientific and climate risk information to inform local action planning?
18. Does local government provide at-risk people and local organisations with regularly updated, easily understood information on disaster risks and risk reduction measures?

Coherence

19. Does local government coordinate DRR actions within local authority offices and across different line-ministries and departments (e.g. trade, industry, agriculture, health, etc.)?
20. Does local government actively support collaborative DRR actions between different locally active state and non-state stakeholders (local government, private, civil society, academia)?

This can be problematic as decisions taken, for example, about the location of a new development may be made without knowledge of the risks that may be produced. This is particularly challenging for slow-onset hazards driven, for example, by climate change. The time horizons between different planning functions can mean an area currently not vulnerable may become so in the future. Globally there is a surge in urbanisation and in many poor countries this is often unplanned (see Chapter 13). Many populous cities are located near to the coast, making them vulnerable to sea-level rises in the longer term. The enhanced social vulnerability created by unplanned urbanisation and poor governance in many parts of the world will be further amplified by environmental changes, including climate changes, land subsidence, deforestation, etc. Many issues will have to be tackled at the level of the city. Little work has been done in terms of vulnerability assessments, city disaster risk assessments or systematic responses for mainstreaming sustainable regulatory frameworks and codes into urban management practice (Parnell *et al.* 2007) (see Chapter 53).

Local government, DRR and learning

Learning has always been a key aspect of disaster management. However, in the disaster response community lessons learned are mainly focused on improving the response function. This is single-loop learning or technical error correction. It is certainly an important form of learning, but learning will need to be very different if local government is to improve its pre-disaster planning for DRR and recovery activities as well as build and consolidate partnerships with communities and civil society. For this, *social learning* is required – a process of iterative reflection that occurs when experiences, ideas and environments are shared with others (Keen *et al.* 2005).

Local government is ideally placed to facilitate these processes because of its proximity, local knowledge and access to forms of communication and dissemination. Social learning for disaster preparedness must recognise cultural factors and build upon accepted practice. For example, a study of two seismically active areas, Fukui in Japan and the San Francisco Bay area in the USA, showed that cultural differences shape the way in which learning is approached (Tanaka 2005).

Local government can encourage both the public and disaster professionals to work together, as was clearly shown above in the Dagupan City case study. Local government can also facilitate vulnerability and disaster risk assessment and is best placed for effective communication with the public as well as with local resident professionals (architects, engineers, etc.) and academics. Studies by Coppola and Maloney (2009) show the processes for developing campaigns for public education for disaster preparedness, and these show the importance of the co-ordinating role of local government. It is local government that has the capacity and authority, for example, to place signage for lifelines and to designate evacuation areas (see Chapter 41). In collaboration with more specialised civil society organisations, local government can have knowledge of vulnerable groups such as the elderly, immigrant groups and the disabled, where special evacuation measures may be needed, or those for whom English is a second language, where information may need to be produced in more than one language (Wisner and Uitto 2009).

In short, local government can promote new ways of learning and co-production of knowledge at the community level (see Chapter 9). At the same time local government can learn, particularly from previous experiences. In Bangladesh in 1970 and 1991 severe storms resulted in deaths of 500,000 and 138,000, respectively. Following the 1970 disaster the government and other agencies began to implement the Bangladesh Cyclone Preparedness Programme, a bottom-up programme aimed at communities reducing their vulnerabilities. This represented a different approach and a determination to learn from experience. An example of measures implemented is cyclone shelters. In the 1991 storm, fatality rates were 3.4 per cent in areas with access to cyclone shelters, compared with 40 per cent in areas without access to shelters. Because of improved preparedness, during another strong storm in 1994, 750,000 people were safely evacuated and only 127 died (Schultz et al. 2005; Akhand 2003).

Challenges for local governments attempting DRR

Local governments face challenges such as the impact of global recessions and stagnant or reduced financial resources. They also face growing demand for services, especially as higher levels of governance offload responsibilities to them while cutting funding. In addition, LGUs must face the long-term implications of climate change. Local governments in affluent, industrial countries are not exempt. Although any single given winter storm cannot be attributed to climate change, an example from the UK is instructive. The winter of 2009–10 in the UK was very severe and prolonged, and many local authorities came perilously close to running out of gritting materials for icy roads. Many were forced to target main routes at the expense of the wider road network. The shortage arose as many authorities had diverted resources from stockpiling gritting materials. Politically, it is much harder to defend contingency planning when there is a limited and even shrinking budget, and there are many other needs and demands, from education to fighting crime.

One of the most challenging issues for local government is how to invite and sustain the participation of people at risk in DRR action. The challenge, but also the potential benefit, of popular participation comes from the evaluation of the 'Linking Relief, Rehabilitation and Development' programme in post-tsunami Sri Lanka and the Maldives. The evaluation shows that the deeper and longer the participation of the communities in the governance of DRR activities, the stronger and more potentially sustainable the impact (Brusset et al. 2009).

Although there is an increasing consensus concerning the value of community participation in DRR at the policy level, there still remains a reluctance by donors and authorities to put resources into the hands of local governments and communities so that they can determine for themselves how, where, when and with whom they wish to reduce risk. For example, of the €43 million Disaster Risk Management programme of the Government of India, hardly any funds were offered to local communities to reduce risks as they saw fit. The participation of communities in deciding how money is spent was almost nil (ADPC 2009).

Another key challenge is the low level of investment in those who manage DRR. Managers of DRR locally (in government and also civil society) are more often than not handed ready-to-use guidelines and training in DRR project management. Very seldom are they helped to develop skills and capabilities that will help them analyse and deal with a complex and changing risk reality. Basic skills are lacking, such as organising communities, developing and motivating team members, understanding key financial statements, developing and delegating work effectively, setting goals for others or even managing their own careers. Yet getting such practical guidance is rare for those who govern DRR. Although the UNDP has useful guidelines for the administration of projects or programmes, and UNDP's Regional Centre in Bangkok and the Training Centre in Turin have designed useful training sessions, a direct focus on the governance of DRR is still nowhere to be found.

Conclusions

Improved co-ordination of DRR at the local level is needed. Good local government along with a culture of good governance is the most effective way of reducing disaster risk, where local political administration leads on building community capacity to cope with and respond to disasters. However, not all local government can be regarded as good or effective in being able to cope and respond. There is a need to identify the characteristics of good local government that support community capacity. This is a challenge for the politicians and DRR researchers. International development partners need to build bridges to local government and to direct resources for building local capacity. Similarly, local government needs to have support and trust from the national central government. Learning is a vital tool and should be used actively to encourage communities to be involved.

Local government in one form or another has been a feature of human life since the beginnings of society. It is tried and tested and sometimes it has been found wanting. Despite imperfections in local government, DRR can only really be effective if driven from the bottom up, as well as from the top down, and local government is the best vehicle for that.

53

Urban and regional planning and disaster

Cassidy Johnson

DEVELOPMENT PLANNING UNIT, UNIVERSITY COLLEGE LONDON, UK

Introduction

One of the impacts of having over one-half of our world's population living in urban areas is that disaster risk is concentrated in ever more densely populated places (see Chapter 13). This acceleration of urban growth is mostly concentrated in low- and middle-income nations, where an estimated 2.8 billion urban dwellers live, out of an estimated 3.5 billion worldwide (Satterthwaite 2007). While many people, communities and governments are aware of hazard risks and do what they can to prepare and cope, there is an important role for urban and regional planning in reducing disaster risk.

Urban and regional planning provide general guidance for managing development of a territory. It is concerned with balancing economic, social and environmental interests through guiding physical development of the area, i.e. the built environment, infrastructure and land use, as well as social and economic development of the society. Whereas *urban planning* concerns the territory of a city or town, including its smaller constituencies such as municipalities or boroughs, *regional planning* covers a wider territory, which may be mostly rural or could include several cities, towns or villages and their rural hinterland. The purpose of this chapter is to outline how urban and regional planning can contribute to disaster risk reduction and to discuss the underlying governance issues of planning and implementation of plans.

Urban and regional planning in disaster risk management

There has been an operational and professional separation between urban and regional planning and disaster management in the past. Disaster management has been the domain of government civil protection units at different scales and has generally been focused on contingency planning for response to disasters. More recently, disaster management has shifted focus from solely response and recovery issues to proactive risk reduction.

Regional and urban plans have generally been concerned with the use of space within their domains and usually employ regulatory frameworks such as land-use and building codes to achieve their goals. In the past, urban and regional planning has engaged little with aspects of disaster risk, except in the aftermath of a disaster, or perhaps when it has been imposed from higher levels of government (Burby 1998). However, government initiatives to introduce disaster risk reduction (DRR) into urban and regional development are becoming more commonplace. International agencies also play an advocacy role in spreading the message that risk reduction should be an integral element of planning (see Chapter 50).

For example, the United Nations International Strategy for Disaster Reduction (UNISDR) launched a 2010–11 campaign focused on urban development, 'Making Cities Resilient: My City is Getting Ready' (www.unisdr.org/english/campaigns/campaign2010–11). Non-governmental organisations (NGOs) have also become more aware of the need to integrate DRR into their urban development projects and programmes or to develop specific projects that focus on risk reduction (Wamsler 2008). Furthermore, the need to address the potential increased frequency of hazards brought on or exacerbated by climate change and the corresponding push for governments to develop adaptation plans have spurred more local-level planning for risk reduction (Bicknell et al. 2009).

At the same time, there has been a shift in recent years in the methods used for doing urban and regional planning as well as their substantive focus. Urban and regional planning used to concentrate only on the production of master plans, which are spatial or physical plans that depict on a map the state and form of a territory at a future point in time when the plan is 'realised'. However, this has been replaced by strategic spatial planning, which is more flexible and adaptable as it outlines in a conceptual way the desired future direction of urban development and the particular decision-making processes that are necessary to achieve these priorities (UN-HABITAT 2009).

Urban and regional planning used to focus solely on structural issues, such as land use and environmental quality, housing, transport and infrastructure and urban design. However, in many regions it has now become much wider in its orientation, and includes non-structural issues such as jobs and livelihoods, social programming and tackling issues of poverty and exclusion. The methods and process of planning have also changed. What were once static plans developed by technicians who gathered and analysed data and provided objective advice to decision-makers have been replaced with participatory methods for planning, where planners are responsible for facilitating all the people and organisations that are affected by the plans to become part of the process of generating information, discussing alternatives and reaching consensus on the objectives of the plans (Todes et al. 2010) (see Chapter 51).

The ideal of current practice today is that urban and regional plans will mainstream disaster risk throughout the planning process and in the daily application of the plans (Von Einsiedel et al. 2010). Thus, ideally, land development and building, as well as economic and social development, are all based on an understanding of risk, and decisions are taken to minimise risk where possible. Disaster mitigation plans may be stand-alone documents and thus separate from urban and regional plans, but ideally they should be part of the same planning process and mainstreamed into all elements of the urban or regional plan (Wamsler 2008; Godshalk et al. 1998). However, in practice in most constituencies, urban and regional planning is still not well integrated with DRR. In part, this is because the two activities of planning and DRR are still bureaucratically separated within local governments and the staff come from different professional backgrounds (see Chapter 52). This is also because planning staff in local governments have little or no training about how to apply DRR in planning. In time, though, these aspects will change: interest in climate change adaptation is helping to bridge bureaucratic and operational divides in some cities (see www.clacc.net and also Roberts 2008). Also, planning schools are now beginning to offer courses in DRR, which will give professionals an orientation in the subject as a basis of their education.

Any type of planning that includes disaster risk must be based on an understanding of the multi-hazard risks in the particular area. However, getting accurate data and producing maps about the vulnerability of people and about the level of hazard risk in a local area is expensive (Johnson 2010). Even though governments may understand the future benefits to having this accurate data, the expense can be too much given other urgent development priorities. Even for wealthy countries, the cost of hazard risk mapping can be immense for local-level governments (see Chapter 54). Thus, in the USA, in order to implement the national flood insurance programme in the 1970s, it was the national government that invested US$1 billion in flood hazard mapping (Burby 1998). In some places, universities have become important resources for creating data about hazards. For example, the National University of Colombia in Manizales has worked with the Manizales City planners to catalogue all small and large hazard events (Cardona 2008; see Box 53.1).

Community risk mapping and enumerations, in which residents are engaged to generate their own data about their neighbourhood, the risks, disaster events and current needs, have become an important source of information for planning, as well as a process whereby people can organise themselves and learn about disaster risk (see Chapter 59 and Chapter 64). For example, in Orissa, India, the women's savings groups in informal settlements have become involved with enumerating and mapping the risks in their communities. The information generated is used as a basis for negotiations with authorities for tenure and settlement upgrading (Livengood 2011).

Box 53.1 Science and urban planning for risk reduction in Manizales, Colombia

Omar-Dario Cardona
National University, Manizales, Colombia

Manizales (2009 population 360,000) provides an example of how a combination of technical and scientific work, political–administrative will and the community's acceptance can lead to successful initiatives. Manizales has invested heavily in the science and technology of DRR. The administration has encouraged the diffusion of knowledge about DRR and supported participation of citizens in planning. This work has contributed to improve the quality of life, vulnerability reduction, and protection of economic and social development. Disaster risk management in Manizales involved the following elements:

- A Municipal Emergency Plan directs the response in case of crisis and defines operational procedures and co-ordination mechanisms.
- A network of rainfall measuring stations monitors flood and landslide hazards and alerts the population and authorities in case of emergency.
- A network of accelerometers automatically sends a shakemap and damage scenario to the authorities by internet and cell phone a few seconds after an earthquake. Thus officials can see the probable effects of the earthquake and activate the emergency response plan, adjusting it accordingly.
- The Observatory of Volcanology and Seismology continuously watches the area's volcanoes and seismic activity.

- Teaching about disaster risk has been incorporated into formal education, and there has been continuous work with the media to disseminate DRR messages.
- The city promotes incorporation of risk in land-use planning.
- The Manizales building code incorporates earthquake-resistance standards developed since 1981. This was the first code of this sort to be issued in Colombia.
- Key buildings – main hospitals, fire-brigade stations, administration headquarters, several schools, university campuses and the Basilica Cathedral – have been retrofitted using state-of-the-art earthquake-resistant technology.
- The city has a specialised procedure of post-earthquake building damage evaluation. It has specific manuals, evaluation forms and a computer-based expert system for the evaluation of the buildings' safety.
- Seismic micro-zonation has served to optimise the municipality's financial protection strategies for public buildings and to evaluate the risk of all private buildings. Micro-zonation involves detailed study of soil, geology, slope and modelling of the forces at specific locations that are likely in earthquakes of different magnitudes.
- The city offers multi-hazard property insurance that can be paid with property tax. This insurance scheme allows full coverage of exempted properties, which benefits the city's poorest population.
- Public works to stabilise slopes have been carried out since the 1970s.
- The programme 'Guardians of the Hillside' involves single mothers in the maintenance of drainage works that stabilise steep urban slopes, for which they receive a payment.
- A large number of relocation projects have been implemented for people living in areas where the risk could not be otherwise reduced, such as on very steep slopes. The freed-up areas have been restored and protected, in part through community-based action.

Addressing vulnerabilities through urban and regional planning

Urban and regional planning should seek to address *vulnerability in the built environment* (i.e. the characteristics of buildings and infrastructure that affect their ability to withstand the impacts of a hazard) and also *social vulnerability*. Wisner and his colleagues define social vulnerability as the characteristics of a group that affect their capacity to anticipate, cope with, resist and recover from the impacts of a disaster (Wisner *et al.* 2004: 11). Social vulnerability can be addressed by urban planning through social and economic development, and by raising awareness at all levels of society about disaster risk. At the most fundamental level, urban and regional planning should address all kinds of vulnerability by making opportunities for people to have access to safe and affordable housing and infrastructure, and access to livelihoods, education, health and social services. Planning can also make special provisions to reduce social equalities and to specifically address the vulnerability of special groups such as women, the elderly and children.

In addressing vulnerability in the built environment, there are two major approaches: the first is the *location approach*, which is to use planning to limit the amount of occupation, use and development in areas that are deemed to be at risk of hazards. The second is the *design approach*, which is to make development in hazardous areas safer by altering the way in which buildings and infrastructure are built (Burby 1998) and by paying attention to issues of access for purposes of possible evacuation and ingress by emergency services.

Locational approaches to urban DRR

The *location approach*, which seeks to limit development in hazardous areas, is most commonly achieved through the process of land-use planning. Land-use plans must first identify areas that are at risk from hazards and then designate these areas, through zoning, for some low-intensity development. Land-use planning and zoning may also be used to set aside open areas that can be used for evacuation or emergency housing in case of a disaster and to plan for lifeline infrastructure that cities manage (e.g. water, drainage, transport and energy infrastructure). This approach can be very effective if it is adhered to over time, and if limitations to development are continuously upheld.

The preservation of this land requires support by consecutive local governments and other stakeholders because they must give up the economic benefit offered by developing the land (see Box 53.2). People may support the limitations to development in at-risk areas because the preservation of open space also has utility as parks, recreation areas, for agricultural use or the like. Typically, such areas include frequently flooded river margins, some exposed coastal areas and steep slopes. In addition, a comprehensive approach to urban risk would have to take into consideration the proximity of residences and other land uses to industrial facilities that might

---

Box 53.2 Reducing volcanic risk through urban and regional planning

Carina Fearnley
Institute of Geography and Earth Sciences, Aberystwyth University, Wales

Volcanic eruptions are often low-frequency, yet have the potential to be high-impact events. Consequently, planning for volcanic hazards presents a difficult problem in balancing between the benefits of a specific location and the risk of a catastrophic volcanic hazard that may occur once every millennium. Given how diverse and often widespread volcanic hazards can be (see Chapter 28), a regional planning approach is necessary in preparing for volcanic crises. Typically, hazard maps are developed, a risk assessment is conducted and a warning system is established; however, these methods often neglect vulnerability.

One such example is the town of Orting, on the flanks of Mt Rainier in the USA. Orting is built on the deposits of prior lahars that have historically occurred every 500–1,000 years. If Mt Rainier erupts, the inhabitants may be at severe risk from a lahar and have less than forty minutes to evacuate to high ground before it strikes. Although the City of Orting is located within Pierce County, the city has its own development regulations that enable it to purchase land adjoining its boundaries.

Given rapid growth in the American North-West and increasing pressure for affordable housing outside the main cities (Seattle and Tacoma), Orting and nearby communities are being developed and rapidly expanded in lahar danger zones around Mt Rainier. Since 1990, the population of Orting has grown over 200 per cent (according to the US Census Bureau), with a population of 6,319 in 2009. Financial gain to landowners and economic benefits to local communities drive this risky development.

So that people can safely live in the city, lahar hazard management strategies focus on establishing automated warnings and co-ordinating rigorous evacuation procedures. This requires maintaining consciousness over the long term, such as running regular drills so that all inhabitants can be prepared to move quickly, along with frequent awareness programmes. Additionally, Orting is working towards building a

---

'bridge for kids', so that school children can safely evacuate by walking over the Carbon River to the Cascadia Plateau.

Integrating volcano DRR into urban and regional planning is difficult because generally the hazards manifest on longer time scales than the tenure of those in political power or even those working at the institutions responsible for urban planning and development. One of the ways to reduce a volcanic crisis is to not develop new urban areas in vulnerable locations. This is often not practical. Orting remains one of the highest risk areas around Mt Rainier, yet its beauty, relative good quality of life for the cost of living and nearby cities for employment continue to place demand for more development. Integrating further political will into long-term urban and regional planning for volcanoes, covering both hazard and vulnerability, is an ongoing challenge.

suffer damage in a natural hazard event and produce secondary risks such as explosions, fires or toxic emissions (see Chapter 56).

The land-use planning and zoning approach may also be used to clear areas that are already developed but are at high risk of hazards. This may include areas that are repeatedly flooded or have been affected by an earthquake. Governments may mandate through regulatory measures that the properties are purchased and residents or businesses are relocated. If the relocation is against the wishes of the owners it may be a difficult task to convince people to sell and move.

Local governments have had little success in convincing residents to move in situations as diverse as the low-lying neighbourhoods of post-Hurricane Katrina New Orleans (Maret and Amdal 2010) and flood-prone fishing communities in St Louis, Senegal (Diagne and Ndiaye 2009). However, in some cases people may want to relocate if they are suffering repeated damages from disasters and this is negatively affecting their livelihoods. For example, after flooding in Iloilo, the Philippines in 2008, the Iloilo City Urban Poor Network negotiated resettlement to a nearby site that would be less vulnerable to frequent flooding (Dodman et al. 2010).

Design approaches to urban DRR

The *design approach* to urban and regional planning allows development in areas at risk of hazards, but seeks to control how buildings are designed and built. Designing safer buildings may limit some damage if a hazard strikes, but it does promote more development in the hazardous areas, and thus the potential for losses is greater. Levees and other protective structures along urban waterways have this effect. In cities, where the demand for land is intense, the design approach is more realistic than limiting development. This is because the land is valuable or people would rather live in centrally located areas, accepting the risk, rather than move further away. In addition, informal settlements in flood plains, on steep slopes and in other dangerous locations develop because of low-income people's need to be near sites of possible employment, petty trade and livelihood activities such as urban gardening. When relocated to the outskirts of a large city, these poor end up paying a large proportion of their daily wage for transportation. For similar reasons, fire- and earthquake-prone tenements in Los Angeles house many illegal migrant men per room, as rent sharing and proximity to work is vital to their livelihoods (Wisner 1998; Davis 1999: 93–148).

The most common method of regulating settlement design is through building codes, which specify how buildings should be built to withstand hazards. Building codes are usually developed at the national or regional level and then are adopted by local governments. Depending on the type of hazard, they may require buildings to be elevated above flood levels, designed and braced to withstand earthquakes or hurricanes, or to use fire-safe cladding materials, etc. Site design elements may also be included, such as retaining walls for hillside building sites at risk of landsides and larger drainage systems for local flooding.

Design approaches that try to impose structural changes for existing buildings (retrofits) have been difficult to implement. They tend to be politically contentious and thus not very effective unless accompanied by subsidies that cover the costs of retrofitting. However, for new development (rather than existing development), and in post-disaster recovery, the design approach can be quite effective if the building regulations are coupled with non-regulatory approaches to safe building, for example, public information and training programmes for builders and building owners about safe design and building techniques, and building manuals that describe in simple terms and with pictures how to build safely. Also economic incentives, such as low-cost loans and subsidies, might be needed to encourage people to use safer materials and methods for building. Post-earthquake training of thousands of people in the building trades in Pakistan and training of masons in low-cost school retrofitting are successful cases that show the feasibility of this approach (Petal et al. 2008).

For informal settlements, incremental improvements to the quality of the built environment can reduce disaster risk. As an example, for flooding this may include installation of drainage infrastructure, street paving that includes adequate culverts and drainage, small dikes, building retaining walls and raising buildings or furniture above flood levels (Wamsler 2007b). Widening streets and having more open areas can reduce high risk of fires in informal settlements and make it easier for emergency vehicles to enter the neighbourhood. Since buildings are often built without legal registration, more informal means of implementing safe building are necessary (i.e. non-regulatory means), such as making accessible information and training for builders and residents about how to build earthquake-safe houses or facilitating the organisation and financing for building communal infrastructure. It is also necessary to support existing coping strategies that people are already doing: for example, in Korail settlement in Dhaka, Bangladesh, people prepare for seasonal flooding and extreme heat by making design adjustments to their homes and the communal areas around them, as well as engaging in other coping strategies, such as household savings and building up assets that they can sell or use to rebuild (Jabeen et al. 2010).

Integral to addressing vulnerabilities in the built environment, planning can also provide the preconditions for neighbourhood organisation and information dissemination, in order to prepare for disasters and to reduce risk. Community development or social development departments in the city government may co-ordinate planning or work with local NGOs to run community centres that provide information and training on risk as well as a host of other social programmes. Local libraries can make available hazard maps that are free to the public. Disaster managers or fire brigades can do training about disaster risk, response and first aid for community members and also in schools.

The politics of integrating DRR into urban and regional planning

Developing citizen participation and a common vision

Ideally urban and regional planning is based on a common understanding shared by all of the stakeholders. Diverse residents' groups, the private sector and different levels and branches of government should be represented and participate in making plans for urban and regional development. It is through the process of planning that a common vision and goals are developed. This is true also of DRR planning and post-disaster recovery. Usually planning for DRR or post-disaster reconstruction in a city or a region is led by local government. As Burby (1998: 2) states, 'In the process of preparing plans, local governments engage in a problem-solving process that works to ensure that all stakeholders understand the choices the community faces, and that they reach some degree of consensus about how these choices will be made ... [P]lans require the systematic evaluation of alternative courses of action, so that the approach chosen to reduce vulnerability is an optimal solution given a communities' present circumstances, future prospects and goals and aspirations of its residents.' Participatory budgeting, such as that popularised in Porto Alegre, Brazil, where residents have an opportunity to decide on how public works budgets should be spent, is one such example of how people can participate in planning that can reduce risk (Vlahov *et al.* 2007).

While participatory planning processes are the ideal, it is not necessarily an easy process to reach consensus of a wide range of stakeholders, especially if decisions to be taken affect some people or communities more than others. Barangay Rizal is an informal settlement with legalised tenure in Makati City, Metro Manila, the Philippines, in which a planning project was undertaken in 2009. The goal was to reduce risk to earthquakes and floods through a process of participation involving local officials and local leaders in partnership with an international NGO, the Earthquakes and Megacities Initiative (www.emi-megacities.org/home). The project leaders found that engaging all of the important stakeholders from the outset was critical in forging consensus on the planning process for risk reduction. However, to implement the plan is a more difficult step. In order to address the physical vulnerabilities, the plan proposes to relocate, to nearby mid-rise housing, families that are living within five metres of the local fault line in very high-risk structures, and along the riverside. The idea is to create more open space in the neighbourhood both for parks and playgrounds, and also to create space that can be used for temporary evacuation in a disaster. The plan also included widening of access roads and designation of emergency routes. The difficulties are that households have very limited resources for relocation and need to be convinced of the project. As well, appropriate legal mechanisms are needed to administer the relocation. 'To achieve the redevelopment objectives the Makati City government as well as the city subdivisions councils will need to forge agreement among the project partners not only on the physical improvement plan but on a package of incentives as well, in order to provide a conducive environment for the proposed development to materialise' (Von Einsiedel *et al.* 2010: 41). Research in the USA by Stevens, Berke and Song (2010) studied sixty-five urban development projects, all within 100-year floodplain areas, and found that public participation was very important for the implementation of hazard-resistant design. They found that public participation in site plan review was important because elected officials can ignore hazard mitigation plans and policies if citizens or their advocacy and watchdog organisations do not draw attention to them, thus greater public participation increases the likelihood that hazard issues are identified and discussed, and ultimately implemented in projects.

In post-disaster situations things are more complex. There is a need for urgent action and there are often many external organisations involved on the ground in relief and rehabilitation (see Chapter 46). The local government involved may have lacked the capacity for multi-stakeholder planning before the disaster and is probably less able to do this after it has been affected by the event. The earthquake in 2010 in Port-au-Prince, Haiti, is an example of this, where urgent planning was needed to make land available for transitional housing and reconstruction. Multiple stakeholders, including civil society organisations (CSOs), international NGOs and donor agencies, are all involved in the reconstruction. The capacity for planning of the local government was weak before the disaster and made worse by the earthquake.

Civil society groups can make an important contribution to urban and regional planning, and this is especially so when it comes to disaster risk. In Windhoek, the capital city of Namibia with a population of 223,000, a network of urban poor groups, Shack Dwellers International, has been working with the local government as well as NGO professionals to develop improved regulations for land and housing policies. In Windhoek, twenty-six per cent of people live in informal settlements. The need for policy revision came about in part because of the recognition that former housing policies were not working for the very poor and thus were making health conditions worse for them. Under the new policy, the municipality recognises different levels of household income and makes allowances for types of tenure based on affordability. For example, changes to the tenure laws allow the very poor to access land by making it possible to have more than one family sharing ownership of a plot. Plot sizes have also been reduced so that people can hold titles to smaller plots, depending on what they can afford. One aspect that makes this policy possible is that there is no value placed on the land, and the municipality is only interested in cost recovery of the services it installs (Mitlin and Muller 2004). In this way, security of tenure can be spread more widely in urban space and with it a precondition for people's later investments in home improvements and upgrading is satisfied. Flood-proofing and other design measures fall into this category.

Constraints to linking the professional and political in urban DRR

Land tenure is vital to people's livelihoods, and it is hotly contested. It is not just a technical variable in a professionally produced plan. Thus one important constraint on the implementation of DRR as part of urban planning may create a situation that confirms in people's minds that governments are not to be trusted. Evictions are a prime example of the kind of administrative action that fuels mistrust. The right to occupy land that is at risk of a natural hazard or to return to an area affected by a disaster is a governance and planning issue, both for DRR and recon-struction (see Chapter 46).

A post-disaster situation may be an opportunity for people to gain access to safer lands. This was the situation in Iloilo, the Philippines, a city of 41,870 households including 16,754 households that live in informal settlements (in 2000). After Typhoon Frank inundated many coastal settlements in Iloilo in June 2008, the Philippines Homeless People's Federation was able to negotiate with the municipality to relocate almost 2,000 families to safer lands in San Isidro (Dodman et al. 2010). However, in other cases the reason for eviction may be to seize the opportunity to develop land for higher economic return. For example, in Thailand, after the Indian Ocean tsunami struck in December 2004, many of the villagers who had lived in fishing villages along the Andaman coast had to fight for the ability to return to their village because the land was being claimed by developers wanting to build hotels along the coastline (Bristol 2010).

As DRR anticipatory measures become increasingly adopted by local authorities before dis-aster strikes, neighbourhoods at high risk from natural hazards may be relocated against their wishes. For example, in Istanbul, Turkey, a city of almost 15 million people, there is a very high risk of a large earthquake happening in the near future. Due to the high density of housing and improper construction techniques, large portions of the built environment are highly vulnerable to earthquakes. Under the current municipality's law, neighbourhoods that are deemed to be at risk of earthquakes can be expropriated by the government and then redeveloped with new buildings that should supposedly be safer in earthquakes (Sengezer and Koç 2005; Yonder and Turkoglu 2011).

There are two tiers of interest that are driving the integration of DRR in planning. At the local level, municipalities and regions that are affected by disasters (either a large-scale one or repeated small-scale events) will be more likely to implement planning that addresses disaster risk, response and recovery. In the immediate aftermath of the disaster there tends to be a period of reform in which disaster risk is strongly considered in planning. Sengezer and Koç

(2005) outline how in Turkey many cities have integrated DRR into urban planning in the immediate period after an earthquake. Also at the national level, governments seeing a drain on national resources from disasters mandate that action be taken at local level to reduce the risk of disasters through planning and building. Governments are also beginning to think about how future risk will be shaped and how to adapt to climate change. In many cities and regions, it is work on climate change adaptation that is driving more local-level planning for risk reduction (see Chapter 18).

In theory, urban and regional planning can have a beneficial role in reducing risks of disaster, and in disaster response and recovery, because it facilitates the organisation of communities around common priorities, it regulates the production of the built environment and aids the production of healthy social and economic systems in regions, cities and towns. However, in practice, the use of urban and regional planning to address disaster risk has fallen short of expectations. Why is this?

First of all, the historical separation between planning and disaster management means that the two are not well integrated intellectually, professionally or operationally. Second, while intellectual debates in both planning and DRR indicate that planning should follow a partici-patory process, not all local governments know how to operationalise this, even if participation is mandated by law. Indeed, where there has been a history of mistrust between city govern-ment and some or all of the residents, participation may sound hollow and rhetorical to some stakeholders. Thus many plans are made, but since they are not based on consensus the plans become difficult to implement or in some cases can make vulnerability worse (e.g. leading to forced evictions). Issues of trust between residents and municipal governments mean that people are less likely to engage in meaningful discussions about development (Wisner and Uitto 2009).

Third, even if local governments undertake good planning and prepare plans that address disaster risk, local governments are also under pressure to develop their territories, so will sometimes ignore hazard risks. Furthermore, in rapidly urbanising areas, local governments do not have the capacity to manage the pace of building. People build, and much of it is informal and unregulated, so will not conform to codes. More soft (non-regulatory) mechanisms are needed, both to support community organisation and to promote safer and better building through education about risks.

Conclusions

The integration of DRR into urban and regional planning requires systematic efforts on the part of governments, civil society organisations and universities. Some key recommendations can be identified for each of these stakeholders.

On the part of municipal governments:

- Political will is necessary to prioritise disaster risk reduction in urban and regional develop-ment. This may come from a 'champion' at the local level, or from higher-level legislations or directives. While there may be interest to think about risk in the immediate aftermath of a disaster, the difficulty is to maintain these principles over time.
- In low- and middle-income countries, the achievement of wider pro-poor development initiatives such as tenure security, affordable housing, provision of water and drainage infra-structure, garbage collection and social services also helps to make neighbourhoods less vul-nerable to hazards. In part this is so because of the technical nature and specific consequences of these services, but also good services help to cement co-operative and trusting relations between residents and government, thus making further specific DRR measures feasible.

Planning and building regulations that allow people to build incrementally and on small plots of land help people to consolidate assets in housing. Of course, asset consolidation is best also protected by insurance schemes (see Chapter 54).

- One of the key aspects to reducing disaster risks, and something that continues to be an impasse for many governments, is the issue of trust. Transparent initiatives, which clearly outline the hazard risks and allow for multi-stakeholder decision-making, are needed to build trust between local governments, residents and civil society groups.
- Within municipal bureaucracies, there needs to be better integration of disaster management within the normal planning systems, for example mainstreaming disaster risk reduction into planning and not having it be only part of a special projects division or only within the emergency services.

On the part of higher education:

- Training of planners in disaster risk reduction is required. While large urban areas often have specialist people on staff, smaller municipalities or rural regions usually do not. This also should include a movement from traditional master planning to strategic planning, including the training of planners that is necessary to carry this out.
- More research is needed on the kinds of urban governance structures that lead to good DRR through planning, and why these work within different political economy contexts. As well, research is needed on a local level on disaster risk and its impacts on people and livelihoods, dynamics surrounding disaster risk reduction and evictions.
- More urban and regional planning programmes need to train planners about disasters and methods for disaster risk reduction.

On the part of civil society:

- Increasing urban residents' knowledge about disaster risks and what can be done about them is a critical need. This is also the role of governments, but CSOs are often well positioned to carry it out.
- It is also critical to develop the capacity of people to organise and thus to make demands on local government, to hold them accountable to people's needs.
- Lobbying, advocacy and watchdogs are other important functions of civil society. Organisations with interest in DRR can put pressure on governments to uphold planning regulations in the face of pressures to develop areas.

54

Financial mechanisms for disaster risk

Joanne Linnerooth-Bayer

INTERNATIONAL INSTITUTE FOR APPLIED SYSTEMS ANALYSIS (IIASA), LAXENBURG, AUSTRIA

Introduction

The human and economic toll from disasters can be greatly amplified by the long-term loss in incomes and health resulting from the inability of communities to restore infrastructure, housing, sanitary conditions and livelihoods in a timely way. Individuals and governments, especially those with limited savings or reserves, high indebtedness and low uptake of insurance, frequently cannot recover from major and recurrent events (Mechler 2003). Informal and formal financial arrangements made before disasters can help to ensure needed post-disaster capital. They can also contribute to development even before disasters strike because they provide the necessary security for households and firms to undertake high-return and high-risk investments. Barnett *et al.* (2008) have shown that the poor's ex ante risk management strategies commonly sacrifice expected gains, such as investing in improved seed, to reduce risk of suffering catastrophic loss, a situation perpetuating the 'poverty trap'.

The post-disaster capital gap experienced by many countries, combined with the emergence of sometimes novel financial instruments for pooling and sharing risks, has motivated governments, development institutions, non-governmental organisations (NGOs) and donor organisations to consider pre-disaster financial instruments as an important component of disaster risk management. Financial mechanisms that prepare individuals, businesses and governments for disaster response are not only important for the developing world, but also for wealthy countries, many of which do not have sufficient or reliable safety nets in place.

At the outset it is important to recognise that financial instruments are not neutral with respect to who pays or takes responsibility for disaster relief and reconstruction. Critics are concerned that market-based instruments, like micro-insurance and microcredit, shift responsibility from social institutions to the poor, while non-market mechanisms, like remittances or reserve funds, share responsibility. It is important to keep this distinction in mind – individual responsibility versus solidarity – throughout this discussion.

As another important caveat, the highest priority in risk management is to invest in cost-effective measures for preventing or mitigating human and economic losses. The residual risk can then be managed with risk-financing strategies for the purpose of providing timely relief and assuring an effective recovery. Disaster risk management thus consists of risk mitigation and risk financing. The two are interlinked, as discussed in the final sections of this chapter.

Disaster risk financial mechanisms

Table 54.1 gives an overview of financial mechanisms available for managing catastrophe risk on the part of households and small and medium-sized enterprises (SMEs) operating at the micro scale, microfinance institutions and donor organisations operating at the intermediary scale and governments as macro-scale operators.

In wealthy countries, most disaster-affected people finance relief and reconstruction by depending on their own savings, insurance, family support and also taxpayer solidarity. Because of a large tax base and favourable credit ratings, governments of wealthy countries are generally well prepared to provide backup capital to insurance programmes and finance even major disaster events, although the 2005 New Orleans hurricane disaster demonstrated the importance of political will and solidarity.

The picture can look different in low and lower-middle income countries,[1] especially if they are small and highly exposed to hazards. Lacking insurance and savings, and in the absence of reliable and solvent governments, those incurring losses may employ diverse non-insurance financial coping strategies, such as relying on international aid, remittances, selling and pawning fungible assets and borrowing from moneylenders. Likewise, low-income governments with high risk exposure rely on a range of post-disaster and inter-temporal strategies, like borrowing from international financial institutions (IFIs) to meet their post-disaster liabilities. Finally, donor

Table 54.1 Examples of financial mechanisms for managing risks at different scales

	Micro scale: Households/SMEs/ farms	Intermediary scale: Financial institutions/ donor organisations	Macro scale: Governments
Solidarity	Government assistance; humanitarian aid	Government guarantees/ bail outs	Bilateral and multilateral assistance; European Union solidarity fund
Informal risk sharing	Kinship and other mutual arrangements; remittances		Diversions from other budgeted programmes
Savings and credit (inter-temporal risk spreading)	Savings; micro-savings; microcredit; fungible assets; food storage; moneylenders	Emergency liquidity funds	Reserve funds; regional pools; post-disaster credit; contingent credit
Insurance instruments (risk transfer and pooling)	Property insurance; national hazard insurance; micro-insurance; crop and livestock insurance; weather hedges	Re-insurance	Sovereign risk financing; regional catastrophe insurance pools
Alternative risk transfer		Catastrophe bonds	Catastrophe bonds; risk swaps, options and loss warranties

Source: Adapted from Linnerooth-Bayer *et al.* 2010

organisations and insurers acting as intermediaries need protection against catastrophic risk, and they increasingly rely on alternative instruments, like catastrophe bonds, to transfer their financial risks to global capital markets.

The following discussion covers these insurance and non-insurance mechanisms and their role in managing disaster risks on the part of micro-, intermediate- and macro-scale actors.

Disaster risk financing at the micro and intermediate scales

Solidarity

Governments are heavily involved in reducing, absorbing and financing risks and losses from disasters. By acting as 'insurers of last resort', governments of high-income countries typically back up private insurers through government guarantees or bail outs, and provide relief and support for reconstruction to those affected by disasters in the private sector. It may be surprising that in the USA the average annual expenditure by the federal government for disaster assistance greatly exceeds the average annual loss borne by reinsurers on US catastrophe coverage (Froot and O'Connell 1999). The situation can be starkly different in low-income countries. As an example, after the devastating 1998 floods in Sudan, the government compensated those affected by only about fifteen per cent of their direct losses, and private flood insurance or even donor assistance was not available or forthcoming to make up the deficit (Linnerooth-Bayer and Mechler 2007). Solidarity extends beyond national borders and includes voluntary donations from the international community of individuals, NGOs and governments. After the 2004 Asian tsunami, pledged donor assistance was reportedly greater than direct economic damages (World Bank 2005), although pledges often do not translate into actual transfers (Walker et al. 2005). High-profile disasters in the future may continue to receive an unprecedented amount of donor assistance, even though overall international donor aid for disasters has averaged only ten per cent of direct economic losses and appears to be decreasing (Mechler 2003).

Informal risk sharing

At-risk individuals in low-income countries rely extensively on financial arrangements that involve reciprocal exchange, kinship ties and community self-help. Women in high-risk areas, for example, often engage in complex, yet innovative ways to access post-disaster capital by joining informal risk-hedging schemes, becoming clients of multiple microfinance institutions or maintaining reciprocal social relationships (see Box 54.1). Beyond sharing financial resources, whether through self-organised pools, exchanges or loans, households use informal networks to protect livelihood processes (Cox and Fafchamps 2006). Combined analysis of multiple surveys indicates that about forty per cent of households in low and lower-middle income countries are involved in private transfers in a given year, either as recipients, donors or both (Davies 2007).

A particularly important informal risk-sharing mechanism is remittances, or transfers of money from foreign workers to their home countries, which are substantial throughout the developing world. As demonstrated by the 2005 Pakistan earthquake, remittances are hugely important for disaster relief, often exceeding post-disaster donor assistance.

While remittances are simple in concept, their use can be complicated by associated transfer fees. Survey results show that the associated costs of an average transfer can vary widely, between 2.5 per cent and 40 per cent (Inter-American Development Bank 2007). Payments are usually sent through banks or professional money-transfer organisations, but often these channels break down and remittances are carried by hand (Savage and Harvey 2007). Some non-profit organisations disburse pre-paid cell phones to the affected population after a disaster, allowing easier communication with foreign relatives. Remittance transfers have been complicated across some borders due to initiatives to counter international money laundering and terrorism, as well as by stricter enforcement of immigration laws and a crackdown on illegal immigrants. This has contributed to a drop in remittance growth, which was six per cent in the first half of 2007 as compared to twenty-three per cent in the same period in 2006 (Carrasco and Ro 1999).

Box 54.1 Microcredit in India for reducing women's disaster vulnerabilities

Nibedita S. Ray-Bennett
Cranfield University, United Kingdom

The state of Orissa, India is highly prone to multiple hazards, including cyclones, droughts and floods. One vulnerability-reduction measure being undertaken is micro-financial services such as credit, insurance, savings and other risk-transfer mechanisms. After the 1999 cyclone, the NGO ActionAid (AA) Bhubaneswar, along with its partner organisation Bharat Gyan Vigyan Samity (BGVS), offered microcredit to women to facilitate recovery of livelihood assets and to reduce disaster vulnerability of women-headed households.

In the village of Tarasahi, Orissa, twelve women-headed households, six government officials and ten NGO workers were interviewed to determine the effectiveness of the microcredit efforts in the post-cyclone phase. This included the floods in 2001 and 2003 and the drought in 2002. The results were mixed. While the microcredit offered was instrumental in the respondents' recovery of their cattle lost in the 1999 cyclone, the offers were limited in providing only specific financial services. Instead, microcredit must go beyond credit, offering non-financial services such as disaster risk reduction measures. Four main suggestions emerged from the work in Tarasahi to improve the efficacy of microcredit in reducing women's disaster vulnerability.

First, comparative studies would assist in understanding the impact of microcredit on women- and men-headed households in multi-hazard locations, to see how gender affects microcredit programmes. In particular, the method for credit delivery requires further scrutiny because it can be instrumental in exacerbating or reducing disaster vulnerability. Another needed comparison is the advantages and disadvantages of grouping or separating women from the same social caste, class and status.

Second, while self-help groups are essential for successful microcredit programmes, the evidence from Tarasahi suggests that an external organisation needs to constantly monitor the programme and facilitate or catalyse specific activities beyond financial services. Pre-disaster services could include warning systems, health, financial management, lobbying to regulate food prices, and literacy. During disasters, women could be mobilised to work in emergency response and relief distribution. Microcredit programmes must be implemented within the context of these other activities.

This connects to the third point, that microcredit is important but cannot solve women's disaster vulnerability alone. Multiple disaster-related activities must be pursued simultaneously with microcredit neither dominating nor being neglected. For Tarasahi, this can be achieved through providing a primary health care centre, veterinary services and intervention in food management through the state's public distribution system. With such community building activities, pre-disaster microfinance programmes need to focus on supporting livelihoods, such as effective and affordable insurance and access to credit at low interest rates.

Fourth, livelihood diversification, essential for disaster risk reduction, can be supported by external interventions that support farmer-friendly research to improve productivity in disaster-affected areas; that improve warning systems for the community and link them to the community's livelihoods; and that reduce the progressive pressure on lands through creating self-employment opportunities in the secondary and tertiary sectors of the economy. Providing disaster-resistant vegetable seeds and tree seedlings, perhaps as microcredit, would be particularly useful to landless households so that they can grow food on their homesteads to have after a disaster.

Savings and credit

Literally, 'saving for a rainy day' might be considered the most obvious and self-sufficient financial management mechanism for financing disaster shocks. In the developing world, besides setting aside money, savings can be in the form of stockpiles of food, grains, seeds and fungible assets. Research in disaster-prone slum areas in El Salvador revealed that households spend an average of 9.2 per cent of their yearly income on risk management, including strategies for financing emergency relief and recovery. For example, one slum dweller deliberately fastened corrugated iron to his roof in a temporary fashion so that he could sell the iron in a post-disaster emergency (Wamsler 2007b).

Savings in low and lower-middle income countries are increasingly channelled through microfinance institutions (MFIs) and banks. Between 2004 and 2006 the value of global micro-savings deposits increased by twenty-four per cent per year, amounting to some 1.3 billion savings accounts (CGAP 2005). These institutions ensure that savings are accessible when needed, although smaller savings institutions themselves can be directly impacted by catastrophes, which can result in insufficient liquidity to handle a run on their accounts, as occurred during the 1998 floods in Bangladesh (Kull 2006). Savings can also be organised through less formal means, like community-based organisations (CBOs), which not only pool financial resources but often set up community grain and seed banks. Despite strides in enabling formal savings, the poor spend their limited income primarily on consumption and livelihood investments and many still lack access to safe, formal deposit services (CGAP 2005).

Lacking sufficient savings, many disaster-affected people take out loans to cover their post-disaster expenses. Case studies of seven major disasters impacting high-income countries in the 1990s showed that, on average, disaster-affected people themselves ultimately financed around one-half of their losses (Linnerooth-Bayer and Mechler 2007). For this, they relied heavily on post-disaster credit, sometimes from governments that offer low-interest loans.

In low and lower-middle income countries, the provision of small loans to the poor through MFIs has become mainstream. The 18–60 per cent interest rate charged on formal microcredit, although high compared to rates charged in wealthy countries, is generally far below the 120–300 per cent often charged by local moneylenders (Grameen Foundation 2008). Such 'loan sharking' is most common after disasters when demand is high. A recent CNN documentary exposed practices in India where farmers have been forced to sell their wives to pay back loans after repeated droughts (Sidner 2009).

Post-disaster microcredit is not without its own risks, however, as increased post-disaster demand can challenge the liquidity of microcredit organisations and tempt relaxed loan conditions or even debt pardoning. To deal with this problem in Latin America, development organisations and private investors created a novel intermediary institution, the Emergency Liquidity Facility (ELF), which provides needed and immediate post-disaster liquidity at break-even rates to MFIs.

Insurance mechanisms

While informal risk sharing and savings/credit arrangements as described above can work reasonably well for low-loss events, they are often unreliable and inadequate for catastrophic events. Faced with large losses, low-income households may be forced to sell productive assets at very low prices; post-disaster inflation may greatly reduce the value of savings; moneylenders may exploit their clients; entire families, even if geographically diverse, may be affected; and donor assistance rarely covers more than a small percentage of losses.

Anticipating a liquidity deficit, insurance becomes an option for risk-averse households and businesses; yet, typically, it is neither available nor affordable to the most vulnerable. As shown in Figure 54.1, catastrophe insurance density drops from around one-third in wealthy countries to less than one-tenth in higher-middle income countries, and it is almost negligible (one to two per cent) in low and lower-middle income countries (Munich Re 2005a).

Two examples can illustrate the stark difference in global insurance coverage. At the time of the 2010 earthquake in Haiti, insurance policies represented only about 0.28 per cent of the country's gross national product (GNP). Payouts to private Haitians by the largest insurers are expected to total below US$40 million (Carney 2010). This stands in contrast to Hurricane Katrina in 2005, which led to more than US$40 billion of insurance claims, making it the industry's most costly geological or hydro-meteorological disaster to date (Open CRS 2008).

Negligible insurance penetration in low-income countries is not surprising given the reluctance of insurers to enter these markets and the inability of households and businesses to pay high premiums. Unlike other types of insurance (e.g. life or health), catastrophes affect whole

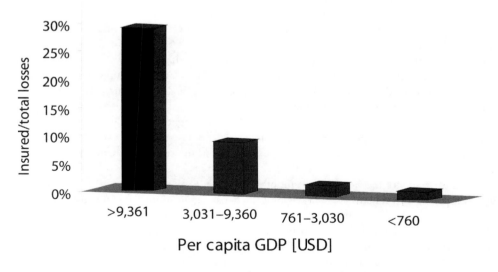

Figure 54.1 Catastrophe insurance density during the period 1985–99 according to country income groups (per capita GDP in 2000)

regions or countries at the same time (co-variant risk), which means that insurers must have costly backup capital or reinsurance. With insurance premiums sometimes many times the expected loss, it is understandable that low-income households and businesses rely on less costly, even if less reliable, alternatives.

Insurance penetration in wealthy countries is mixed and varies greatly across countries. Hazard insurance is almost non-existent in Italy, where after the 1997 Umbria earthquake the government compensated the affected people for almost one-half of their losses. This contrasts with the UK, where the government provides practically no assistance to the private sector after a disaster. After the disastrous 1998 Easter floods, insurers absorbed almost forty per cent of losses. Many developed countries, including Japan, France, the USA, Norway and New Zealand, have legislated formal public–private hazard insurance programmes. The US National Flood Insurance Program (NFIP) is unique in that the federal government serves as the primary insurer, and persons living in exposed areas should eventually bear their full risks. The aftermath of Hurricanes Katrina, Wilma and Rita in 2005, however, revealed large debts in the NFIP and its continuing dependence on taxpayer support. A different philosophy underlies the French all-hazard insurance system, which is backed by a public-administered fund and deliberately incorporates national solidarity through taxpayer involvement and cross-subsidies. To counter the problem of disincentives from the cross-subsidies, households must comply with specified risk-reduction measures. Moreover, their compensation will decrease with each subsequent disaster, which encourages at-risk households to relocate (Linnerooth-Bayer and Mechler 2007).

The Turkish Catastrophe Insurance Pool (TCIP) is the first of its kind to be legislated in a middle-income developing country. Premiums are affordable and the system viable because the World Bank reinsures two layers of TCIP risk in the form of a contingent loan facility with highly favourable conditions (Gurenko 2004).

Catastrophe insurance instruments are also emerging in low-income countries in the form of (usually pilot) micro-insurance programmes. Malawi offers a recent example. Groundnut farmers at high risk from droughts are participating in a pilot project offering seed loans that are insured against default with an index-based weather derivative (payouts are contingent on a physical trigger, in this case rainfall measured at a local weather station) (Hess and Syroka 2005). This programme provides the safety net necessary for farmers to plant higher-risk seeds that increase their productivity (in this case fivefold). Because of the physical trigger, there is no moral hazard (the propensity of insured agents to undertake riskier behaviour); to the contrary, farmers will have an incentive to diversify their crops and take other measures to cost-effectively reduce potential losses. This micro-insurance programme, like all other currently operating pilot programmes, is made affordable by international assistance.

Donors can also be strapped for cash after large-scale droughts and other natural hazards. For this reason, the World Food Programme designed and executed an index-based insurance system to provide capital to the Ethiopian government in the case of extreme drought (Hess et al. 2006).

Insurance provides the liquidity to smooth out disaster shocks, and by enabling productive investments has the added benefit of helping high-risk agents escape disaster-induced poverty traps. Insurance also has advantages to donors since providing support for these instruments will ultimately reduce their obligations to provide post-disaster assistance (Linnerooth-Bayer et al. 2005). Yet even with donor support, insurance is not appropriate in all contexts. Many countries lack the requisite governance and institutions, and low-income households, businesses and governments may have lower-cost alternatives for providing post-disaster liquidity. Finally, poorly designed insurance contracts can lead to moral hazard.

Disaster risk financing for governments

Governments generally have a large portfolio of public assets, and many are committed to providing post-disaster relief to the most vulnerable. Without pre-disaster financing arrangements in place, the post-disaster reconstruction of public infrastructure and provision of relief can be jeopardised. As shown in Table 54.2 (reformulated from Table 54.1), public authorities have access to a variety of financing mechanisms for this purpose. Wealthy countries rely heavily on reserve funds, issuing bonds on the domestic and international markets, diverting funds from other budget items and sometimes imposing or raising taxes. Typically, low-income country governments have less access to these sources because disasters significantly increase fiscal deficits and worsen trade balances. They become reliant on bilateral and multilateral assistance, credit from the central bank (which either prints money or depletes its foreign currency reserves) and from IFIs. Paying back loans and interest can cripple economies for years and detract from development measures.

Solidarity at the international level

Governments experiencing a post-disaster financing gap often turn to international donors and international development banks (Gurenko 2004). For low-income countries, bilateral donor assistance plays an important role, whereas middle-income countries more often turn to multilateral financial agencies for development lending. Pledges from the international community, however, often fall short of reality. For example, two years after the 2001 earthquake in Gujarat, India, assistance from a government reserve fund and international sources had reached only twenty per cent of original commitments (World Bank 2003).

Even wealthy governments rely on international solidarity instruments. As a case in point, the European Union Solidarity Fund (EUSF) was created in 2002 in response to massive flooding throughout Central and Eastern Europe. The purpose was to show solidarity with member states by granting post-disaster financial aid if losses exceeded the capacity of governments to respond (Commission Report 2004). Hochrainer et al. (2010) argue, however, that this fund does not meet its stated purpose, and suggest that it re-orient to provide backup capital for national and regional insurance pools.

Budget diversions

Resource-strapped governments typically divert funds from other budgeted projects to cover their post-disaster liabilities. In the developing world, these diversions are often from international loans for infrastructure projects. Based on anecdotal evidence, Lester (1999) cites a figure of thirty per cent of infrastructure loans from the World Bank diverted for this purpose worldwide.

Table 54.2 Examples of financial mechanisms for managing risks at the governmental scale

	Solidarity	Informal risk sharing	Savings and credit	Insurance instruments	Alternative risk transfer
Macro scale: Government	Bilateral and multilateral assistance; European Union solidarity fund	Diversions from other budgeted programmes	Reserve funds; regional pools; post-disaster credit; contingent credit	Sovereign risk financing; regional catastrophe insurance pools	Catastrophe bonds; risk swaps, options, and loss warranties

Whereas this response may be the least costly one for the government, it can be disruptive both economically and politically. Most countries require that budget reallocations have parliamentary approval, which can delay appropriation of funding.

Catastrophe reserve funds, debt instruments and contingency credit

To reduce dependency on debt financing, many countries (particularly in Latin America) have instituted a catastrophe fund (Charvériat 2000), which has a cost equal to the foregone return from maintaining liquid capital and a benefit in having the resources immediately available with less transaction costs. A major problem with a fund, however, is that it may be insufficiently capitalised, especially if the disaster occurs shortly after its creation. In principle, insurance companies also operate with a reserve to cover large outlays; however, private insurers are more concerned than the government that their reserves are sufficient to avoid insolvency, and for this reason they diversify their insurance portfolio. A second problem with a catastrophe fund is the political risk that it is diverted for other purposes in years with no disasters.

Bonds and other debt instruments, which transfer the burden to future periods and even future generations, are the most common sovereign post-disaster financing mechanism. Issuing bonds after a disaster is usually not a problem for wealthy countries where the hazard impact does not significantly affect the economy and thus the ability of the government to service its debt. The credit ratings of Japan and the USA were not affected by the Kobe earthquake in 1995 and Hurricane Katrina in 2005, respectively. This is not typically the case for low-income countries since domestic financial institutions may not be willing to lend on pre-disaster terms.

The World Bank and other IFIs fill this gap. Since the early 1980s, the World Bank alone has initiated over 500 loans for disaster recovery and reconstruction purposes for a total disbursement of more than US$40 billion (IEG 2006), and the Asian Development Bank also reports large loans for this purpose (Arriens and Benson 1999). IFIs and donor organisations, however, are greatly concerned about the dependence of developing countries on post-disaster capital grants and loans, which discourage governments from engaging in risk reduction activities (Gurenko 2004). Recipient countries, in turn, are concerned that international donations and loans for post-disaster reconstruction will continue to take an increasing portion of declining official development assistance (Mechler 2003). A major limitation is the growing discrepancy between the amount of reconstruction funds available from the international community and the growing funding needs of disaster-prone countries.

To partly counter this problem, governments can pay a fee for the option of a guaranteed loan at a pre-determined rate contingent on a disaster or some other defined event occurring. Colombia was the first country to secure contingent capital from the World Bank to provide immediate and less expensive capital to the government when it is most needed. Although a contingent credit arrangement can potentially provide a government with lower-cost capital relative to either a pure risk-transfer solution (such as insurance) or the accumulation of reserves, the major disadvantage is that it can exacerbate the country's debt burden. The appropriateness of this mechanism thus depends on the country's post-disaster financial profile, and more specifically on its post-disaster ability to service debt.

The World Bank is recently offering a new financial product to middle-income country governments called the Catastrophe Risk Deferred Drawdown Option (CAT DDO). Its purpose is to make financing immediately available after a disaster and is intended to fill the gap while other sources of funding, such as emergency relief aid, are being mobilised. Importantly, countries that sign up for the CAT DDO must have an adequate hazard risk management programme in place that is monitored by the World Bank (World Bank 2008).

Sovereign risk transfer: insurance, risk pools and catastrophe bonds

In theory, governments are not advised to insure public infrastructure damage and other post-disaster liabilities unless their infrastructure risks are highly correlated and losses cannot be absorbed by a large taxpaying public (Hochrainer and Pflug 2010). Many countries are not large enough to sufficiently diversify their risks through a national insurance programme, in which case they might consider purchasing insurance, as was the case in the Czech Republic after the devastating 2002 floods.

Governments can also form regional pools. An innovative recent example is the Caribbean Catastrophe Risk Insurance Facility (CCRIF), which provides the sixteen participating governments with immediate liquidity in the aftermath of hurricanes or earthquakes. Governments contribute resources to the pool depending on the exposure of their country, and the fund is reinsured in the capital markets. Many donor governments, and also the European Union (EU), have contributed to this pool, enabling poor countries to join.

Haiti, as a recent case in point, was given assistance to join the CCRIF, and its government received a payment following the 2010 earthquake. It is too early to assess how these funds will be administered. The Haiti example underscores the importance of good governance and reliable institutions to administer any sovereign risk financing programme, whether it is insurance, credit or donor assistance.

A second alternative to commercial sovereign insurance is a catastrophe bond, which is an instrument whereby the investor receives an above-market return when a pre-specified catastrophe (measured in terms of an index, for example earthquake intensity, or losses) does not occur in a specified time, but sacrifices interest or part of the principal following the event. Disaster risk is thus transferred to international financial markets that have many times the capacity of the reinsurance market. The first developing country government to issue a catastrophe bond was Mexico in 2006 in order to provide security to its reserve fund (Cardenas *et al.* 2007). One major advantage of a catastrophe bond is that it is held by an independent authority and not subject to credit risk or insurers defaulting on obligations.

Other similar instruments include catastrophe futures or options contracts, which are designed to provide insurers and reinsurers with an alternative or supplement to traditional reinsurance. They allow parties to hedge their catastrophic risk exposure through access to the capital markets (a hedge is any instrument that provides the contracting parties a payment if they experience losses). In 1992 the Chicago Board of Trade issued the first catastrophe insurance futures and options contracts based on a loss index that has subsequently been adjusted to better reflect insurer losses. Governments can also engage in risk swaps with another government facing non-correlated risks, an instrument that is used extensively by insurance companies (Cardenas 2008).

Linking financial instruments with disaster risk reduction

Financial mechanisms are not generally viewed as measures to prevent loss of life and property, and for this reason they are commonly regarded as an alternative or addition to investments in risk reduction. This view, however, overlooks the long-term preventative benefits of proactive risk-financing schemes. By enabling recovery, timely post-disaster capital can significantly reduce long-term indirect losses – even human losses – which do not show up in the disaster statistics.

The view that risk financing is an alternative to risk reduction also overlooks the propensity of well-designed programmes to provide incentives for physical interventions and lifestyle changes that reduce disaster risks. A few examples illustrate this. In Istanbul, apartment owners

---

Box 54.2 Financial mechanisms can inadvertently increase disaster losses

Much anecdotal evidence suggests that insurance can lead to an increase in disaster losses, such as the maxim, 'If you want a new couch, put your old one in the cellar when you hear the flood warning'. As this example illustrates, once an insurance contract is in place, households may be less inclined to take cost-effective measures like placing their belongings out of harm's way following the flood warning. Not only insurance, but any financial mechanism that does not reward preventative behaviour can lead to what insurers call moral hazard. Why should family members, expecting free compensation from remittances or international assistance, invest in costly loss-prevention measures?

Insurers guard against moral hazard by charging high deductibles or requiring co-insurance. As discussed earlier, index-based contracts avoid these disincentives since claims do not depend on losses. For example, Mongolian farmers participating in an index-based insurance programme have every incentive to protect their herds against adverse weather, since they will collect claims regardless of their herd mortality (Skees and Enkh-Amgalan 2002). Depending on its design, insurance may thus have less moral hazard than instruments that penalise protective behaviour.

---

who choose to disaster-proof their properties pay a lower insurance premium, thus making investments in safety more attractive. In Mongolia, herders who insure their livestock will face increasing premiums if climate change worsens weather conditions, giving them an added incentive to change practices or even livelihoods if animal husbandry becomes unproductive. Poorly designed insurance contracts, on the other hand, can discourage investments in loss prevention or even encourage negligent behaviour (see Box 54.2).

Finally, an innovative, but untested, idea is to link indexed insurance contracts with early warning to promote timely risk reduction. For example, if an indexed contract is triggered by a flood warning and not by the flood itself, this would provide timely cash to enable those at risk to take protective actions, such as sand-bagging or moving their possessions to higher territory (Suarez and Linnerooth-Bayer 2010).

Because of the potential to link financial instruments with disaster risk reduction, there is mounting interest on the part of development organisations and governments to include insurance in a climate adaptation regime. The UN Development Programme (UNDP), as a case in point, is taking steps to establish a Climate Risk Finance Facility, which would support financial instruments in low-income countries. Financial instruments also feature strongly in the climate negotiations. The UN Framework Convention on Climate Change and other international climate agreements specifically call for consideration of insurance and other risk sharing and transfer mechanisms for countries most vulnerable to climate change. In response, proposals have been put forth that call for solidarity on the part of wealthy (and high greenhouse gas-emitting) countries to support insurance instruments in countries that are highly vulnerable to climate-related disasters (for a discussion, see Linnerooth-Bayer et al. 2010).

Conclusions

Proactive financing for disaster response and recovery is an important and often neglected aspect of disaster risk management. This discussion has distinguished between market-based mechanisms, like microfinance and insurance, which place the burden on the at-risk community, and solidarity instruments, like remittances and international post-disaster assistance, which embed familial and international solidarity. Especially for catastrophic events, market-based financial instruments can have significant advantages over conventional solidarity instruments due particularly to their reliability and incentive effects, as well as their provision of safety nets needed for high-return investments. Yet, in addition to problems in implementing effective insurance schemes, they may place an unacceptable burden on the most vulnerable. The opportunity and challenge is to create international social protection programmes by building on the advantages of market instruments and rendering them available and affordable.

Governments, as well as the international development and climate adaptation communities, are responding to this challenge in a variety of ways: donors are supporting pilot programmes that test innovative ideas that enable access to financial instruments; governments of highly exposed countries are experimenting with novel risk-transfer mechanisms; UN organisations are developing new institutional arrangements that can provide pre-disaster financing support; and climate negotiators are considering proposals for including insurance instruments in a post-Copenhagen adaptation regime. The verdict is still out on the success of these international efforts, yet already existing programmes are demonstrating the large potential for smartly designed financial instruments to provide a more secure environment for those least able to cope with disaster risk.

Note

1 By low- and lower-middle income countries, this chapter refers to the World Bank classification of per capita income: low income US$975 or less; lower-middle income US$976–US$3,855; upper-middle income US$3,856–US$11,905; and high income US$11,906 or more.

55

Economic development policy and disaster risk

Charlotte Benson

CONSULTING ECONOMIST, VIENTIANE, LAOS

Introduction

Over the past fifteen years, there has been increasing recognition by government, civil society and development partners around the world of the need to 'mainstream' disaster risk reduction into development – that is, to consider and address risks emanating from natural hazards in national development frameworks and in legislation and institutional structures, as well as in sectoral strategies and policies, budgetary processes, the design and implementation of individual development projects and, finally, in monitoring and evaluating all of the above (Benson *et al.* 2007).

Mainstreaming requires analysis both of how potential hazard events could affect the performance of policies, programmes and projects and of the impact of those policies, programmes and projects, in turn, on vulnerability to natural hazards. This analysis should lead to the adoption of measures required to reduce vulnerability, treating risk reduction as an integral part of the development process rather than as an end in itself. It does not require a re-working of government objectives: instead, mainstreaming seeks to help ensure that these objectives are both attainable and sustainable.

This integral approach is considered essential in view of the fact that development initiatives do not necessarily reduce vulnerability to natural hazards but, instead, can unwittingly create new forms of vulnerability or exacerbate existing ones. 'Win-win' solutions for securing sustainable development and reducing poverty and vulnerability to natural hazards need to be explicitly and actively sought. In the longer-term framework, disaster risk reduction interventions, it is hoped, constitute minimum levels of response to a likely increase in hazard risks as a consequence of global warming and trends of rising vulnerability in many developing countries.

Some 168 nations and multilateral institutions are formally committed to mainstreaming disaster risk reduction into development as signatories of the Hyogo Framework for Action (HFA), 2005–15. This ten-year framework resolves to reduce disaster losses. It is centred on three principal strategic goals, the first of which targets more effective mainstreaming (UNISDR 2005b) (see Chapter 50). Mainstreaming has been identified as a central theme of a number of regional and sub-regional disaster risk management strategies (see Box 55.1). A number of

Box 55.1 Examples of regional and sub-regional strategies

Regional strategies include:

- The Africa Regional Strategy for Disaster Risk Reduction, approved by the Eighth Ordinary Session of the Executive Council of the African Union in December 2005 (www.unisdr.org/africa/af-hfa/docs/africa-regional-strategy.pdf).
- The Pacific Island Countries Disaster Risk Reduction and Disaster Management Framework for Action 2005–2015, endorsed by the Pacific Islands Forum Leaders (www.unisdr.org/eng/hfa/regional/pacific/pacific-framework-action2005-15.doc).
- The 2007 Delhi Declaration on Disaster Risk Reduction was adopted at the Second Asian Ministerial Conference on Disaster Risk Reduction by Asian and Pacific countries (www.ndmc.gov.mv/docs/declaration.pdf).

development partners have re-orientated their individual policies in this area to strengthen emphasis on disaster risk reduction and incorporate principles of mainstreaming.

This chapter identifies key features of an enabling environment for mainstreaming disaster risk reduction concerns into development and reviews practical progress to date. Despite high-level commitments, practical advancements have been ponderously slow thus far, reflecting a wide range of challenges implicit in this mainstreaming, not least its multi-stakeholder, cross-cutting nature across many sectors and all levels of government and society. Chapter 51 provided detailed country experience and discussed in more detail obstacles and opportunities for progress. As such, widespread awareness of disaster risk and commitment in tackling it is required across all levels of government and civil society. Along the way, the enactment and enforcement of appropriate legislation, the development and implementation of comprehensive disaster risk management strategies, the development of sufficient institutional capacity, the enhancement of intra-governmental horizontal and vertical integration mechanisms, the establishment of adequate budgetary arrangements and project appraisal and monitoring and evaluation tools, and the active engagement of local communities are all required – a tall order. In consequence, as so poetically voiced by Bishop Donald Mtetemela, a development worker with over twenty-five years' experience, there are 'many clouds – international initiatives and plans, but very little rain – actual change at the frontline' (GNDR 2009).

Key features of a supportive enabling environment

Awareness

The fundamental first step in mainstreaming entails awareness raising: before solid, sustainable progress can be made, there must be strong understanding and appreciation of the relevance of disaster risk reduction to sustainable development (see Chapter 14) and poverty reduction at all levels of government and civil society, from the highest ranking government officials to landless, marginalised households operating within the informal sector (Benson 2009). This requires knowledge of the hazards faced, the factors determining the nature and level of vulnerability within individual communities and broader society, potential developmental impacts of hazard events and the scope and opportunities for specific disaster risk reduction interventions. In practice, underlying hazard data is improving but so far remains patchy in many developing countries, often comprising a mix of broad-brush overviews and piecemeal high-resolution local maps, with little multi-hazard mapping (see Chapter 16 and Chapter 17). Physical exposure and vulnerability mapping is often even more limited in both developing and developed countries, and can be rapidly outdated in areas experiencing rapid socio-economic change. Ironically, although hazard data is often much stronger in more affluent countries, it is sometimes over-developed and given far too much emphasis.

Appreciation of the relevance of disaster risks to broad development goals has also remained relatively weak, despite formal commitments to mainstreaming. This in part reflects the strong, operational focus of post-disaster damage assessments on humanitarian relief and reconstruction needs, rather than wider socio-economic consequences. Data on the impact of small-scale localised events are particularly limited, a major oversight in countries with a high annual inci-dence of such events and thus where they may have a substantial cumulative impact, a point emphasised in the United Nations (UN) International Strategy for Disaster Reduction (ISDR) *Global Risk Assessment 2009* (UNISDR 2009a). There has been some improvement in recent years, in particular as comprehensive disaster loss assessment guidelines originally developed by the UN Economic Commission for Latin America and the Caribbean (ECLAC) almost four decades ago have begun to be applied by development partners in other regions of the world following major disaster events. Another tool developed in the Latin American and Caribbean context, the *DesInventar* methodology (www.desinventar.org), which records the impacts of highly localised, small-scale events, has also begun to be used elsewhere. Despite its aspirations, though, a 2004 review of *DesInventar*'s use in the Latin American and Caribbean context con-cluded that data on infrastructure, industry and services are not sufficiently complete or reliable (IDEA 2004), again hindering a build-up of understanding on the developmental impact of disasters unless concerted effort is taken to improve this aspect of data collection.

Reflecting inadequate appreciation of the relevance of disaster risks to broad socio-economic goals, disaster scenarios are rarely considered in government economic and budgetary planning and related investment decisions. In some countries this may constitute a significant oversight. For instance, the predicted growth and poverty reduction returns to investments in irrigation generated by a macro-economic forecasting model for Ethiopia were doubled by building inter-annual variations in rainfall, based on historical records, into the model (World Bank 2006a). There is also a very limited body of evidence on the economic and financial returns to individual disaster risk reduction initiatives. Absence of such hard data reduces the political will to invest in disaster risk reduction. Perhaps most surprisingly, there has been very limited sys-tematic analysis of the impact of disasters on the poor and near-poor anywhere, despite the oft-cited mutually reinforcing nature of poverty and vulnerability to natural hazards. Notable exceptions include research on the impacts of Hurricane Mitch (e.g. Morris *et al.* 2002; Carter *et al.* 2007) and the wider literature on food and livelihoods security, in particular pertaining to sub-Saharan Africa.

In theory, well-placed, high-level political champions with relevant expertise and knowledge can help to overcome some of the above constraints (see Chapter 5). It is relatively easy to garner this political interest in the immediate aftermath of major disasters, when memories are fresh. However, it is far more difficult to turn this interest into a longer-term commitment to reduce risk, particularly as the benefits of risk reduction investments may not be accrued for many years. Indeed, simply sitting back until the next disaster and then providing emergency relief can secure far more political favour than investment in much less visible ex ante risk reduction (Benson 2009). Moreover, potential champions need to understand and spearhead change across many areas and levels of government – a challenging task. As such, political champions for disaster risk reduction have been relatively few and far between across the developing world.

Legislation

Sufficient legislative arrangements for disaster risk reduction, including its mainstreaming into development, forms a further key component of an enabling environment, obliging and empowering actors and agencies to take action. In many developing countries, progress is being made in this regard, with increasing appreciation of the need to revise legislation outlining disaster-related responsibilities of government agencies and civil society to reflect the paradigm shift worldwide from an essentially post-disaster management to a more exhaustive disaster risk management approach, with much greater emphasis on ex ante risk reduction. In South Africa, for instance, this has resulted in the promulgation of the 2003 Disaster Management Act, which was 'applauded internationally as a path-breaking example of national legislation that promotes disaster risk reduction' (Pelling and Holloway 2006: 4). However, nine years of hard work lay behind this Act, 'demonstrating that reform requires long-term perseverance and commitment from those seeking to make change – sustained and supported by political leadership' (Pelling and Holloway 2006: 5). Similarly, in the Philippines at least ten years' worth of effort lay behind the 2010 passage of the Philippine Disaster Risk Reduction and Management Act.

The process was somewhat faster in Indonesia, where the urgency for legislative reform was acutely underlined by the devastating 2004 Indian Ocean tsunami, spurring on a reform process begun several years earlier to result in new legislation by 2007 (BNPB and UNDP 2008). However, the Indonesian case is far from the norm and legislative change is typically very slow, reflecting little sense of urgency on the part of the majority of legislators or their constituents to prioritise the passage of disaster risk management bills.

There are further issues relating to the enforcement of legislative and regulatory frameworks in many developing and some developed countries, for instance as regards building codes, land-use planning and environmental impact assessments. Even where sufficient legal requirements are in place, adequately addressing disaster risk concerns, actual practice may be far from satisfactory, for instance because inspectors are bribed to ignore building and planning regulations in order to cut construction costs. The extent of corruption in construction is greater than in any other sector of the economy (Transparency International 2005).

Disaster risk management strategies

A comprehensive disaster risk management strategy is required to implement the legislative framework and to provide co-ordination and monitoring mechanisms and arrangements (Benson 2009). This should include the identification of specific entry points and mechanisms for mainstreaming disaster risk reduction concerns both into the broader development agenda and into the design and implementation of individual development initiatives. In the past five years or so, there has been considerable progress in this regard in many developing countries as old, primarily ex post response-orientated strategies have been replaced with new broader disaster risk management ones (see Box 55.2). Development partners, particularly the UN Development Programme (UNDP) and the UNISDR Secretariat, have supported this process, including the development of related action plans.

Institutional capacity

Sound institutional arrangements and capacity for disaster risk management are a further essential element of a successful enabling environment for mainstreaming. In practice, lead national disaster management offices are typically located within relatively powerful ministries, such as the

---

Box 55.2 Disaster risk management in Nepal

A final draft National Strategy on Disaster Risk Management for Nepal was completed in March 2008 and approved eighteen months later. This strategy outlines a comprehensive, holistic approach to disaster risk management, organised around the HFA and covering ex ante risk reduction and preparedness as well as post-disaster response. Development partners have established a Nepal Risk Reduction Consortium to support the government in developing a related long-term Disaster Risk Reduction Action Plan. In the immediate-term donors are supporting five flagship programmes in areas of school and hospital safety, emergency preparedness and response capacity, flood management in the Koshi River Basin, integrated community-based disaster risk reduction and management, and policy and institutional support for disaster risk management. The development of legal and policy instruments to support effective National Strategy Disaster Risk Management (NSDRM) implementation is currently underway (UNDP 2010b).

Office of the President or Prime Minister, or Ministry of Home Affairs. However, they are commonly staffed primarily by technical disaster preparedness and response specialists and typically lack specific sectoral knowledge or economic and development planning skills. As such, they are poorly equipped to engage with other government departments, particularly ministries of finance and planning, on detailed discussions around disaster risk reduction and the implementation of disaster risk management strategies, particularly as regards the allocation of adequate resources and the mainstreaming of disaster risk reduction concerns into other development initiatives. Disaster risk reduction capacity is also often very weak within sectoral line agencies and local government, while local disaster management committees are commonly only constituted in the event of a disaster, rather than on a more regular basis to plan and co-ordinate risk reduction endeavours as well.

One possible way forward entails the establishment of disaster risk reduction focal points within individual line agencies to direct and co-ordinate sectoral initiatives, including the mainstreaming of risk reduction concerns into broader programmes of work; to identify and draw on existing disaster risk management expertise within each line agency; and to provide sector-specific technical support. For instance, such focal points have recently been established in key government departments in Nepal (Benson et al. 2009b), whilst a disaster management cell has been installed within the National Planning Department in Sri Lanka (Duryog Nivaran 2009). It is essential, however, to ensure that these focal points do not become lip service attempts at mainstreaming, allowing concerned governments to tick boxes on risk reduction capacity in relevant government agencies whilst, in practice, placing focal points in marginalised, peripheral units, rather than at the heart of thinking, planning and resource allocation within their respective agencies.

Well-placed assignment of specific mainstreaming responsibilities to ministries of economic development and planning would also be extremely constructive, preferably coupled with climate change adaptation mainstreaming responsibilities. Indeed, deliberate, urgent efforts are required more generally to bring disaster risk reduction and climate change institutional arrangements and capabilities closer together to exploit synergies, share experiences and maximise the effectiveness of their respective programmes and actions, including initiatives to integrate the two issues into national planning processes, strategies and budgets.

Integration of disaster risk into development planning

From a weak beginning, there has been steady – if slow – progress over the past few decades towards greater consideration of disaster risk concerns in poverty reduction strategies and national development plans (see Chapter 51). There has been a parallel growth in development partner commitment to integrate such concerns into their own programmes of work. Various guidelines have been developed in support of this process and the related mainstreaming of climate change adaptation concerns into development (e.g. Benson *et al.* 2007; OECD 2009a; UNDG no date).

In consequence, a recent desk-based survey of sixty-seven poverty reduction strategy papers (PRSPs) found that twenty per cent devoted a whole chapter or section to disaster risk; fifty-five per cent of the reports mentioned the relationship between disaster risk and poverty; and only twenty-five per cent included no mention of disaster risk at all (UNISDR 2009a). However, the nature of discussion of disaster risk in many PRSPs and national development plans remains very narrowly and traditionally conceived, and related commitments are often confined to plans to strengthen warning systems and disaster response capabilities, to target relief and rehabilitation assistance towards the poor, to invest in structural flood control measures and, in some cases, to reduce the vulnerability of the agricultural sector. Even the latter may sometimes be forgotten – a surprising oversight in countries such as Nepal, where the current national development plan both partly attributes poor historical agricultural growth to adverse climatic factors and places particular emphasis on increased agricultural productivity, yet fails to include any explicit initiatives to address the sector's sensitivity to climatic variability (Benson *et al.* 2009b). The Nepal example is particularly noteworthy in view of the fact that it includes a chapter specifically, and exclusively, on 'natural disaster management', located within the social development section of the plan, and an additional section on water-induced disaster prevention under the chapter on irrigation.

Few PRSPs or national development plans take that fundamental step further to integrate disaster risk management concerns into broader development strategies and programmes, and to tackle them holistically, taking a truly mainstreamed approach. A notable good practice exception – at least on paper – is Bangladesh, the 2005 PRSP of which includes comprehensive disaster management as one of sixteen policy matrices (GOB 2005). Various disaster risk management goals and actions were also included under other policy matrices, and disaster risk reduction concerns were firmly embedded within the aims and objectives of the section on agriculture. The extent to which the PRSP ensured comprehensive disaster risk management, environmental sustainability and mainstreaming of these concerns into the national development process was also identified as one of ten key goals on which the success of Bangladesh's strategy would be judged. The revised PRSP for FY2009–11 went a step further again, incorporating climate change adaptation needs as well as maintaining a multi-sectoral approach to disaster risk reduction. There seems to have been little effort to evaluate the actual achievements of the 2005 PRSP, though, or to build on lessons learned, a serious omission given that certain parties feel it achieved little with regard to its disaster-related goals. A recent report stated that 'despite having all the "policy documents" and "plans of action" [including a National Adaptation Plan of Action and a National Environment Management Action Plan] the risk reduction approach has not yet reached the point of becoming mainstreamed in the development process' in Bangladesh (PDRI 2009: 23). The few other PRSPs that appear strong on disaster risk reduction on paper could, similarly, be falling short in implementation.

In many developing countries, sectoral policies and strategies also continue blatantly to ignore disaster risk concerns, even where broad national development plans and PRSPs contain basic disaster risk reduction principles. Again, these shortfalls are often most startling in the agricultural sector, particularly in hazard-prone countries that depend heavily on rain-fed agriculture, such as Cambodia.

Intra-government horizontal and vertical integration

Horizontal and vertical integration of government at different levels is important in ensuring that principles of disaster risk reduction mainstreaming are reflected at all levels of government regardless of the original point of entry. Vertical integration is particularly critical in countries where considerable responsibilities have been devolved to local bodies. In practice, there are often significant breaks in the planning system between different levels of government, implying that locally identified needs are not necessarily reflected in higher-level plans and strategies, whilst there can be problems in implementing national policies and regulations at the local level. For instance, a recent study in the Philippines found that provincial investment plans and regional and national investment plans are formulated independently of each other, implying a break in the planning chain between regional and provincial levels (ADB 2007). There may be particular problems in implementing national policies and regulations at the local level in countries where local officials fall under the supervision of local chief executives rather than national line agencies and use of funding is largely determined locally.

Meanwhile, strong horizontal integration is required most critically because issues such as flood management and water resource management require careful co-operation and co-ordination between contiguous local units of government. Horizontal integration between national line agencies is also important given the cross-cutting nature of disaster risk and potential implications of decisions in one sector of government for vulnerability in another (Benson 2009). Current levels of horizontal integration are often rather patchy, however, particularly in countries that still lack a comprehensive national disaster risk management strategy.

Budgetary arrangements

The mainstreaming of disaster risk concerns into government budgets should be tackled from two angles, ensuring that:

- Levels of public expenditure on risk reduction are sufficient relative to the levels and nature of risk faced, the expected economic and social returns to risk reduction and the reasonable responsibilities and obligations of government; and
- There are adequate financial arrangements to manage the residual risk – that is, the remaining exposure to loss after risk reduction measures have been taken, as manifested in the form of post-disaster relief and recovery needs – based on a carefully formulated plan to cover different tranches of loss.

(Benson 2009)

The latter, although concerned with post-disaster response, is an important aspect of mainstreaming. Disaster events and related relief and rehabilitation operations place additional demands on public spending, diverting resources away from development activities and disrupting plans. As such, a mainstreaming strategy should explicitly advocate for the consideration of potential post-disaster financing needs as an integral part of the budgetary planning process in hazard-prone countries. This is particularly important in countries where there are still relatively limited levels of discretionary expenditure and thus limited budgetary flexibility, implying potentially high opportunity costs to the post-disaster reallocation of funding (Benson 2009).

In practice, although many hazard-prone developing countries make some budgetary provision for disaster-related issues, none have a comprehensive disaster risk financing strategy. Instead, arrangements focus primarily around limited regular budgetary allocations for humanitarian relief and disaster preparedness and, in a few cases, some use of risk-transfer tools. For instance, the Indian federal government operates a Calamity Relief Fund to meet immediate relief and emergency recovery expenditure on approved items of physical infrastructure and statutory personal compensation arising as a consequence of natural hazards (GOI 2005; GOI 2007).

However, in a number of countries even national and local contingency funding for immediate, humanitarian relief for small-scale hazard events occurring every single year is insufficient – as, for instance, in the case of Malawi (Benson and Mangani 2008). In many others, such as Vietnam, short-term humanitarian and early recovery needs may be met expediently, but reconstruction activities can take several years as their costs are largely met from the regular annual capital budget, thereby competing in terms of urgency and importance with development projects for fixed investment resources. Recent econometric modelling suggests that the economic impacts of disasters, as defined in terms of gross domestic product (GDP) losses, are much higher in such countries where public (and private) reconstruction resources are limited, and thus where reconstruction is spread over a number of years (Hallegatte et al. 2007). These authors suggest that this may partly explain why some poor countries that experience repeated disasters cannot develop and instead remain in a perpetual state of reconstruction, making it difficult to accumulate productive capital.

Meanwhile, budgetary resources for disaster risk reduction are almost certainly inadequate in many developing and some developed countries, although poor related tracking of expenditure makes it difficult to make this statement with absolute authority. Tracking issues in part reflect measurement problems relating to the cross-cutting nature of disaster risk reduction (see Box 55.3). More fundamentally, disaster risk mainstreaming is as much about an approach and even an attitude to development, as it is about expenditure on disaster risk reduction. Indeed, mainstreaming sometimes has little cost implication, resulting in an alternative approach to an issue but not necessarily a higher bill. This being so, the importance of high-level awareness of development–risk connections and existence of champions is again to be underscored. Nevertheless, an overview of risk reduction initiatives, related spending and how it compares with post-disaster expenditure is important in tracking progress and identifying gaps.

Box 55.3 Tracking disaster-related expenditure

Comprehensive systems tracking all disaster-related expenditure are critical in supporting structured, evidence-based decision-making around the appropriate balance and nature of risk reduction and post-disaster interventions, and in ensuring that any funding gaps are visible. Ideally, all information on such expenditure should be contained in a national database that also covers data on post-disaster reallocations, disaster losses and the economic returns to individual disaster risk reduction investments.

There has been some progress in the tracking of post-disaster expenditure over the past few years, particularly in the context of Indonesia following the 2004 Indian Ocean tsunami. The Indonesian system tracked resource flows from the government, international donors and the twenty largest NGOs, providing a powerful tool for reconstruction planning and monitoring (Goldstein and Amin 2008).

However, few, if any, countries have established a system for tracking disaster risk reduction. This partly reflects measurement difficulties as relevant initiatives may be scattered across a range of budget headings and, in some cases, form just one component – or are even simply an indirect benefit – of a wider development intervention, rather than a stand-alone risk reduction project. Depending how widely one defines disaster risk reduction, one could even argue that considerable effort, and related expenditure, is, in fact, already made in this area. For instance, one might choose to count spending on irrigation programmes, watershed management initiatives, food security initiatives, broad poverty-reduction programmes and even routine infrastructure maintenance, as poorly maintained structures are often more vulnerable to hazard events. There are no simple heuristics available on the incremental cost of, say, seismically strengthening new infrastructure or of incorporating knowledge on flood and drought management into training for agricultural extension workers. Moreover, many governments have limited experience in tracking any form of cross-cutting expenditure, even poverty reduction, although know-how is slowly building up.

In view of these challenges, a highly simplified system is required to track disaster risk reduction. A system that simply categorises all development spending according to one of the following seems the most plausible way forward:

- Explicit disaster risk reduction expenditure.
- Spending that incorporates disaster risk reduction features at some cost (e.g. construction of seismically strengthened schools).
- Spending that contributes to disaster risk reduction at no additional cost (e.g. irrigation).
- Other spending.

(Benson et al. *2009b).*

From a donor perspective, since 1995 the Organisation of Economic Co-operation and Development (OECD) Development Assistance Committee (DAC) has required donors to separate out spending on emergency aid in reporting aid flows and, since 2004, to draw a distinction between short-term reconstruction, relief and rehabilitation, and 'emergency and distress' relief assistance.

In 2005 it introduced a new sub-title on disaster prevention and preparedness and, effective 1 January 2010, went a step further, introducing a climate change adaptation marker against which all new aid activities must be assessed.

This marker will identify aid projects that have climate change adaptation as either their principal or a significant objective, complementing an existing DAC marker on climate change mitigation (OECD 2009b). The OECD notes that there is no internationally agreed methodology for tracking the exact share of aid activity expenditure that contributes to climate change adaptation or mitigation and that, until such a methodology exists, the markers will only provide approximate quantifications of aid in support of these issues (OECD 2009b). Nevertheless, the introduction of these markers represents a further step forward and will generate useful experience on which development partners and governments alike can build.

Well-defined lines of dedicated disaster risk reduction funding accessible by all relevant line agencies are required as well, set up in such a way that they can also support interdepartmental initiatives, for instance in the area of watershed management. The creation of such lines of funding may seem to fly in the face of mainstreaming, as it implies that disaster risk reduction requires specific stand-alone solutions instead; however, progress in risk reduction in many countries has been very poor, whilst mainstreaming is a long-term goal. As such, short-term incentives are urgently required to help move forward by tangibly demonstrating that investments in risk reduction really can 'pay'. In countries with considerable decentralisation of government, similar lines of funding should be established that local bodies can access if they put up matching funding, again providing solid tangible incentives to kick-start risk reduction initiatives on the ground, especially those in partnership with local communities and civil society.

Finally, it should be remembered that insurance and other financial risk-transfer mechanisms are not a panacea, magically resolving all disaster risk-financing problems (see Chapter 54). Over the past ten years or so, such mechanisms have received considerable interest and attention, even emerging as a new 'sector' of lending for the international financial institutions (Benson *et al.* 2009a). However, they are often implemented in the form of stand-alone tools far removed from any risk reduction initiatives or the proper integration of disaster risk concerns into financial planning. Financial risk transfer mechanisms can, indeed, play an important role, but careful effort is required to ensure that they reinforce risk reduction principles and are used in the context of broader financial risk management strategies to address medium- to higher-level tranches of residual risk (see Chapter 54).

Tools facilitating the analysis of disaster risk as part of the project appraisal process

Consideration of disaster risk concerns as part of the standard project appraisal process is an essential element of mainstreaming, helping to ensure that gains from individual development projects are sustainable, and highlighting related issues of responsibility and accountability for hazard-related human, physical and economic losses. However, current practice is generally poor in this regard. Disaster risk analysis is typically considered primarily the domain of environmental assessments, but related guidelines often provide little guidance on disaster risk analysis and many aspects of disaster risk may be ignored. Most obviously, environmental impact assessments are geared to considering the impact of a project on the environment rather than vice versa, and so the potential impact of a hazard event on a project and related risk reduction measures may go unexplored. Disaster-risk concerns are also commonly ignored in guidelines for other areas of project appraisal, such as economic and social impact analysis. In consequence, governments can struggle to demonstrate even the merits of disaster risk reduction projects themselves, as, for instance, in the case of Nepal where the Department for Water-Induced Flood Control consistently underestimates the benefit to cost ratio of proposed investments because it is not equipped to tackle certain measurement challenges, particularly as regards averted loss of life and reclaimed land (Benson *et al.* 2009b).

A simple, low-cost mechanism, starting by requiring just a few lines in the scoping document for all proposed projects outlining the relevance of disaster risk to the project outputs and objectives and indicating how, if relevant, disaster risk would be addressed, could play an extremely useful role in moving forward. Proposed projects that are vulnerable to hazard events – or that themselves could affect the level or nature of vulnerability in the project locality or wider economy or society – could then require further disaster risk-related investigation in the project appraisal (e.g. via a full environmental impact assessment, based on revised guidelines comprehensively covering assessment of disaster-risk concerns) and technical design stages, including examination of project design alternatives. The Inter-American Development Bank has already introduced procedures along these lines for its operations in Latin America and the Caribbean, offering interesting case experience (IDB 2008).

Monitoring and evaluation

Capacity to monitor and evaluate disaster risk reduction initiatives, generate hard evidence on related inputs, outputs, results and impacts, and learn lessons for the future is an essential component of the enabling environment for mainstreaming (Benson 2009). In practice, the use of benchmarks and indicators to monitor and evaluate disaster risk reduction initiatives is not very common anywhere (UNISDR 2008b). This partly reflects an inherent challenge relating to the fact that the *success* of a disaster risk reduction initiative is ultimately measured in terms of something – a disaster or a particular form or level of loss – that does *not* happen as a consequence of a hazard event. There are further complications relating to the fact that the design hazard event may not even occur over the life of a project, with the normal evaluation time frame (Benson *et al.* 2007).

With a little imagination, such problems are not entirely insurmountable. For instance, lead or process indicators can be selected that will at least provide some sign of progress towards the achievement of project objectives (e.g. the number of schools constructed to withstand a flood of a specified depth or an earthquake of a certain intensity; progress of a mangrove planting scheme intended to provide protection against sea surges in terms of rates of growth and survival of the trees) (Benson *et al.* 2007). However, such solutions only really work for structural risk reduction projects and, moreover, force reliance on measures of physical output. Less tangible achievements in areas such as institutional strengthening, capacity building and the level of regard paid to disaster risk issues in development planning remain much harder to capture, except perhaps as measured in input terms in the form of budgetary resources provided.

A global monitoring system has been established by UNISDR to track national progress in implementation of the HFA, including towards the achievement of a number of these less tangible goals. Such a system is very much needed. Unfortunately, however, the UNISDR system is self-reporting, leaving much open to interpretation in according scores and, somewhat inevitably, largely failing to capture practical progress on the ground rather than merely on paper. Some countries are apparently very honest in their HFA implementation reporting, but certain others appear to err somewhat on the side of generosity. Such apparent discrepancies make country comparisons difficult and, in some cases, create a false illusion that disaster risk is far better managed than is actually the case.

A further initiative launched by the Global Network of Civil Society Organisations for Disaster Reduction (GNDR) has generated more in-depth insights – and, in likelihood, a more honest reality – on practical progress to date. This initiative focuses on the local, grassroots level. It is modelled around the same five strategic areas of the HFA, aiming to support and complement the UNISDR national monitoring efforts. The first assessment was based on interviews with over 7,000 local government officials, civil society organisations and community representatives in forty-eight countries. The assessment revealed a significant gap between national and local perspectives, producing a global average score of 2.38 out of 5, compared with 2.95 out of 5 according to the UNISDR survey.

Conclusions

There has been disappointingly slow progress in disaster risk reduction and its mainstreaming into development to date. Political commitments and related national strategies are increasingly in place and efforts are being made to reform associated legislative frameworks and strengthen scientific hazard data and analysis. However, the continued shortage of funding for DRR – one of the ultimate tests of government commitment to any concern – clearly indicates that most countries are still a long way from genuine, sustained commitment to the issue.

Moreover, they are even further away from genuine, practical mainstreaming of disaster risk reduction concerns into sectoral policies and individual development initiatives, moving beyond the need for dedicated risk reduction budgets to the incorporation of adequate measures to address disaster risk concerns as part of standard good practice in the design and application of national and local development frameworks, sectoral strategies and policies, and individual development initiatives. In consequence, there has been limited practical progress towards achievement of the ultimate objective of reducing risk at the grassroots level. Indeed, vulnerability to natural hazards continues to rise in many countries.

There are, admittedly, significant challenges in reducing disaster risk and mainstreaming related concerns into wider development. However, there is also, unforgivably, very limited desire or effort on the part of those in the commanding seats to move beyond rhetoric and wholeheartedly tackle those challenges. Appropriate entry points and opportunities for disaster risk reduction and related mainstreaming into development, and mechanisms for tackling possible obstacles, urgently need to be identified; related financial resources, capabilities and supporting guidelines secured; and vulnerable people engaged in developing and implementing the participatory practical measures required to ensure that sustainable socio-economic development and disaster risk reduction move forward together, hand in hand.

56

Protection of infrastructure

Ana Maria Cruz

DISASTER PREVENTION RESEARCH INSTITUTE, KYOTO UNIVERSITY, KYOTO, JAPAN

Introduction

Society's lifeline systems (e.g. water, waste disposal, energy, transportation, communications) as well as public–private services housed in the built environment (e.g. hospitals, schools, essential industrial/commercial establishments) are critical to the well-being of a society and may be vulnerable to natural hazard forces or other types of external hazards. In a world that is more and more complex and interconnected, even a small disruption locally can lead to major upsets globally. Furthermore, extreme natural hazards can result in cascading failures. Thus, it is of vital importance to understand the vulnerability of the exposed systems in order to design and implement adequate protection measures to ensure that these remain operational during and after a major disaster event. Examples show notable failures in infrastructure protection as well as relatively successful experiences.

The failure example is Hurricane Katrina in the USA in 2005. Hurricane Katrina flooded large parts of New Orleans, destroyed thousands of homes and killed almost 1,000 Louisianans (Brunkard *et al.* 2008). Many lives could have been saved and people might have thought twice before constructing their homes in New Orleans had they known that they were not protected against Category 4 and 5 storms. The efforts to protect the city from hurricane-induced flooding were flawed from the very beginning because the levee system that was put in place had been designed for a Category 3 storm, although stronger storms were possible, creating a sense of 'false security' (see Box 56.1). The levees failed for several reasons, including lack of proper maintenance, but mostly because they were not designed to withstand the most severe hurricanes.

In contrast, flooding in Austria in 2002 (Preuss 2005) shows how success could be achieved. In August 2002 heavy flooding affected Austria, among many other countries in Europe. Although many parts of the country experienced severe flooding, Vienna fared well compared to other major cities in Europe (Preuss 2005). Vienna's floodplain management programme has included the construction of the New Danube Canal and Danube Island, which provided sufficient storage capacity for the 2002 event. The project included a multi-disciplinary effort to re-network the rivers and streams that had been isolated during previous years, ecological restoration of the floodplain, the delineation of recreational areas and hazardous areas, and a requirement that homes be elevated in those areas. The floodplain management programme involved an

Box 56.1 Relying on structural flood defences can increase vulnerability

Ilan Kelman
Center for International Climate and Environmental Research–Oslo

Structural approaches – such as walls, dams, dikes, levees and barrages – are frequently used for coastal flood management without fully acknowledging how they affect vulnerability. Structural defences stop small floods from happening, so that inhabitants behind the walls tend to become inured to the absence of regular, smaller floods. Because few extremes occur, flood risk reduction activities tend to lapse. There is decreased awareness of the potential flooding, decreased understanding of how to predict and react to floods, and decreased ability to cope with floods as and after they happen. As well, the structural defences are highly visible providing daily reinforcement of 'protection' while permitting vulnerability to increase.

Eventually, a large flood must occur that overwhelms the structural defences, with the tendency to cause damage that is far greater than would have occurred if the affected community were used to regular, smaller-scale floods. This phenomenon is termed 'risk transference' (Etkin 1999) because the risk is transferred onto future events, yielding potential short-term gain for definite long-term pain. The risk is transferred because the hazard usually changes little, but vulnerability has increased substantially.

That does not mean that structural defences should be avoided. Obvious benefits accrue over different time scales. The absence of regular, smaller floods permits people to live and work without that disruption on land that might otherwise not be available, including for agriculture, by avoiding regular contamination by salt water.

Yet the benefits are frequently articulated, and the gains in the short term tend to be overemphasised, with long-term costs being underemphasised. Terms such as 'flood control', 'flood protection' and 'flood defence' reinforce that view, rather than using the more realistic 'flood alteration'. Structural approaches for floods provide only some control, protection and defence. They do so through altering the water's behaviour to some degree in some circumstances, not through complete power over what water always does or could do.

The policy consequence is not to rely on only structural approaches when using them. Structural approaches are one option from amongst many and, as with all disaster risk reduction approaches, have advantages and disadvantages. Other measures are needed in conjunction with structural approaches, to seek a balance. Honesty is needed in describing what structural and other approaches do and do not achieve over different time scales. That might permit an informed decision knowing the full consequences, positive and negative, of each alternative – including issues beyond disaster risk.

interdisciplinary and co-operative undertaking involving hydrologists, geotechnical engineers, biologists, landscape architects, urban planners and others. By having a comprehensive floodplain management and restoration programme in place, Vienna was able to avoid damage.

Earthquakes provide another contrast. The built environment was catastrophically affected in the Haiti earthquake in early 2010, while the infrastructure fared comparatively well in a higher release of seismic energy in Chile (see Chapter 26).

These examples highlight the need to evaluate and plan for threats involving the impact of natural hazards on buildings and infrastructure that can result in potentially high death tolls and can severely affect people's livelihoods. The successes demonstrate that taking adequate measures to protect infrastructure and people from the impacts of natural hazards can greatly reduce loss of life and hardship on livelihoods while speeding up recovery. This chapter takes such lessons to discuss the challenges involved in protecting buildings and infrastructure and indicating how to overcome those challenges.

Challenges in the protection of infrastructure

The challenges in the protection of infrastructure and associated systems from natural hazards will vary by location depending on the particular hazards to which the territory is subject, and the availability of materials and resources for construction. Buildings and infrastructure can be protected in a number of ways. These include the adoption of stringent design codes and standards, the adoption of structural and non-structural protection measures, land-use planning, and disaster mitigation and response planning. Any codes, legislation or regulations adopted must also be monitored and enforced to ensure that they are complied with. While structural mitigation measures can help communities reduce disaster risk, structural mitigation measures alone may also, in some cases, add to the overall risk.

A typical example is building levees and floodwalls to protect communities from river flooding without considering the natural environment. That often leads to devastating impacts due to river degradation, biodiversity reduction, shrinking of the habitats of aquatic fauna and flora, degradation of water quality and changing of the water–soil cycles. This was the case for major flood-control projects in the USA during the first half of the twentieth century after the passing of the Flood Control Act in 1934. These measures did not protect the citizens from major natural hazards, particularly Hurricanes Betsy and Camille in 1965 and 1969, respectively. Later evaluation investigations proved that engineered flood-control measures alone, particularly structural mitigation measures, could often disrupt or destroy the natural environment, could be extremely costly and could create a sense of false security. Similar problems were documented following Hurricane Katrina in 2005 and during the 2000 Tokai flood and the 2004 Niigata/Fukui flood in Japan (Ikeda et al. 2008).

Thus, a comprehensive approach to infrastructure protection should be adopted in order to integrate structural and non-structural measures. It would strengthen the capacity of local communities to make their own informed disaster risk management choices and to promote the participation of all stakeholders, in particular community groups, non-governmental organisations (NGOs) and municipal governments, in all stages of disaster risk management.

For example, the adoption of building codes and standards and building code enforcement have worked fairly well in many wealthier nations with professional engineers and builders, well-trained inspector and well-educated users. Nonetheless, building code enforcement is more likely to succeed where there is social demand for safe construction; there are resources to educate builders on how to implement the standards; there are the financial resources to meet these standards; and well-trained and adequately paid licensed technicians and professionals who are able to respond rapidly to problems. These conditions tend to be absent in the majority of less affluent countries, making the need for a comprehensive, community-based approach very important.

The adoption of prevention and mitigation measures helps to reduce damage and losses. High-tech engineering solutions are often expensive, but the cost of rebuilding or repairing collapsed or damaged infrastructure could be worse. In low-income countries, the cost of frequent repair or replacement of infrastructure could be particularly burdensome from a financial point of view, and even more disruptive where there are no redundant transportation routes, communications, etc. Finding low-cost prevention and mitigation measures that take advantage of local expertise and materials is also very important.

Risk reduction options for the protection of infrastructure

Infrastructure and its associated systems are often located in areas subject to natural hazards, and thus may be at risk of damage or disruption if appropriate measures are not taken to prevent or prepare for such events. Water and wastewater treatment plants and certain industrial facilities that handle large quantities of hazardous materials deserve special attention due to the potential for hazardous materials releases. In general, most infrastructure and its associated systems and essential facilities are composed of a combination of buildings and lifelines. The vulnerability of buildings and lifelines will vary for each type of hazard, as well as the risk reduction measures available for their protection.

Buildings

Buildings may be affected by earthquakes, tsunamis, floods, high winds, fire and landslides and soil problems, among other hazards. Buildings vary considerably around the world in materials and complexity. However, in the thousands of cities on the planet with a few hundred thousand inhabitants or more, buildings are complex combinations of the foundation and structure, and the systems within them such as plumbing, electrical, heating, ventilation, air conditioning and ancillary systems (see Chapter 13). These may suffer damage when one or a combination of these systems fails. Building structures are used for office use and administrative functions related to the operation of the building or as part of an infrastructure system (e.g. communications, water treatment), as well as to store materials and house equipment. Control rooms are generally housed in building structures and may require special design to ensure that they are operational following a major disaster (see Chapter 42). Damage or collapse of buildings may result in human casualties or major upsets to an infrastructure system. For example, following the Gujarat earthquake in India in January 2001, water distribution systems were affected when water pump buildings collapsed, damaging electrical controls and emergency generators (Eidinger 2001).

Earthquakes

Large earthquakes pose significant threats to buildings and structures. The impact of an earthquake on a structure will depend on the magnitude of the earthquake, the depth of and distance from the epicentre, the geology and topography, the local soil conditions and the duration, frequency and acceleration of the impacts (see Chapter 26). The adoption of appropriate seismic building codes for new structures and the retrofitting of older buildings to updated seismic building codes can help minimise loss of life and building collapse during earthquakes. Many countries subject to earthquake hazards have adopted modern seismic building codes (see Box 56.2). However, mainly wealthier countries have been successful in implementing them, including monitoring and enforcement of the adopted code.

One example is Turkey, which adopted modern building codes in 1975, and had updated and strengthened them in 1998 prior to the 1999 Kocaeli earthquake. Nonetheless, the extensive damage to residential buildings (more than 215,000) and the hazardous materials releases from industrial plants during the earthquake were attributed to poor decisions, lack of clear housing and land-use policies, and lack of oversight of building regulations (USGS 2000).

Box 56.2 Bogotá, Colombia: using microzonation to protect a large city's infrastructure

Omar-Dario Cardona
Associate Professor, Instituto de Estudios Ambientales, IDEA, Universidad Nacional de Colombia, Manizales

Bogotá is a city of more than 8 million inhabitants located in the eastern mountain branch of the Andes mountain range in Colombia. The city was affected by earthquakes in 1785, 1826, 1827 and 1917, when the city had a population of fewer than 100,000 inhabitants. However, the vulnerability conditions of Bogotá have dramatically increased since the 1950s and at least seventy-five per cent of the city grew in a disordered way and without earthquake-resistant standards.

On the other hand, according to estimates by the Directorate of Prevention and Attention of Emergencies (DPAE), there are more than 450 unstable hillside zones and flooding areas in the south and the east of the city that have been occupied. A strong earthquake is not necessary to produce a huge earthquake-triggered disaster in Colombia's capital.

In recognition of this growing vulnerability, the city administrations of the last twenty years have been implementing a strong disaster risk management strategy through seismic microzonation, initiating vulnerability reduction measures for infrastructure, and mainstreaming risk reduction in the different planning and development sectors. Microzonation involves detailed study of soil, geology and slope, and modelling of the forces at specific locations that are likely in earthquakes of different magnitudes.

Bogotá began microzonation in the 1990s and has relied on its university and research sector. It became a model for other cities in Colombia seeking to reduce earthquake-related damage.

Using microzonation data, a study was made of the vulnerability of the bridges and the impact on city mobility in the case of an earthquake. This was the basis for detailed evaluation and retrofitting of all vehicle and pedestrian bridges and the airport terminals of the city in the second half of the 1990s.

All public services of the city have completed detailed vulnerability studies and the retrofitting of telephone and energy substations, natural gas and water pipelines, water storage tanks and pumps, and landfills. The vulnerability of public infrastructure is now low and the redundancy of services is high. In the last fifteen years, the vulnerability of all hospitals has been evaluated, and all of them have been retrofitted according to the national earthquake-resistant construction code.

In the last ten years, more than 200 public schools have been evaluated and retrofitted. More than US$460 million have been invested to reduce the seismic risk of children and teachers through reinforcement and construction of new facilities for the benefit of 300,000 students.

Using different scenarios of damage based on seismic hazard microzonation, DPAE has developed an earthquake emergency response plan for Bogotá and has conducted public information campaigns and simulations. In addition, the Finance Secretariat has developed a risk-transfer strategy for the financial protection of public assets and the promotion of insurance of private buildings.

There are nonetheless good low-cost building practices that can be adopted by low-income and rural areas in the developing world (Blondet *et al.* 2003; Coburn *et al.* 1995). For example, the use of improved seismic-resistant adobe mud building construction has proven to be an effective and low-cost solution in Peru and El Salvador (Blondet *et al.* 2003). Low-tech solutions of masonry construction for improved seismic performance have been used in Latin America and India, among other countries (Brzev 2007; Schacher 2009).

Because of the high seismic risk in Japan, it has been a world leader in seismic protection of buildings and other infrastructure systems. Japanese requirements are performance-based standards where the building is required to satisfy performance criteria with respect to materials, equipment and structural methods. Other regions of the world have adopted performance-based standards for certain infrastructure. However, many country codes are based on the International Building Codes, which require the design of buildings for the one in 475-year earthquake, ensuring that buildings remain standing following a design-level earthquake to allow its inhabitants to get out safely (such design is sometimes referred to as survivable collapse). The adoption of seismic building codes and good building practices have resulted in a substantial reduction in human casualties from building collapse during most earthquakes in Japan, the USA and other countries, with Chile as another example.

Tsunamis

Buildings located in areas subject to tsunami hazards may be vulnerable to tsunami wave impact and flooding. Site conditions in the run-up zone will determine the depth of tsunami inundation, water-flow velocities, the presence of breaking wave or bore conditions, debris load and warning time, and can vary greatly from site to site (see Chapter 27). The vulnerability of buildings to tsunami loads will depend on several factors including number of floors, the presence of open ground floors with movable objects, building materials, age and design, and building surroundings such as the presence of barriers.

The National Tsunami Hazard Mitigation Program (NTHMP) in the USA recommends four basic techniques that can be applied to buildings and other infrastructure to reduce tsunami risk, including:

- Avoiding development in inundation areas: This is, of course, the most effective mitigation strategy but not always possible, particularly for existing buildings.
- Slowing techniques: These include the use of specially designed forests, ditches, slopes and berms which can slow and strain debris from waves.
- Steering techniques: These are used to guide tsunamis away from vulnerable structures and people by placing structures, walls and ditches and using paved surfaces that create a low-friction path for water to follow.
- Blocking water forces: This technique consists of building hardened structures such as breakwalls and other rigid constructions that can block the force of waves.

Tsunami-specific building codes are not yet available. Current structural designs to protect buildings in tsunami-prone regions are generally based on loadings due to riverine floods and storm waves, providing little guidance for loads specifically induced by tsunami effects on coastal structures. Modern building codes aiming for unified building codes such as the International Building Code (IBC 2009) include design requirements and standards for fire, wind, floods and earthquakes, but they do not contain requirements for tsunami-resistant design. While a few communities in the USA, for example, have adopted tsunami-resistant building design standards, the vast majority of coastal communities have not.

Floods

Flood loads are similar to tsunami loads (see Chapter 19 and Chapter 21). Buildings located in river basins and near large water bodies may be subject to flood loads. As with tsunamis, flood protection measures include avoiding building in flood-prone areas, making buildings water resistant, and slowing, steering and blocking techniques. Elevation of buildings or important building components above a specific flood contour level can protect building functionality and contents.

The US Army Corps of Engineers (USACE) has done extensive work in flood mitigation and control for the USA, although the approach is highly technocratic and often exacerbates flood problems. The 1993 floods in the Midwest USA are an example, when approximately seventy per cent of levees in the upper parts of the river failed, leading to a review of the policy of relying on structural approaches to attempt to deal with river flooding (Tobin 1995).

Most wealthier nations (e.g. the USA, Germany, Italy, Spain, France, Japan) as well as many developing countries (e.g. Mexico, Colombia) limit or prohibit development in the 100-year flood plain. However, the law generally applies to new construction. Thus, existing buildings located within the 100-year floodplains are not protected. Furthermore, political pressure and corruption sometimes result in authorisation of building permits or illegal development in flood-prone areas. As well, despite its popularity, the choice of the 100-year floodplain is arbitrary. Other options are the historical maximum flood or the maximum expected flood based on geomorphological features.

The combined use of facilities to maintain the water-retaining and -retarding functions of river basins, the creation of incentives to use land safely and to build flood-resistant buildings, and the establishment of warning and evacuation systems for both tsunami and riverine flooding can help to protect communities from flood losses. Development in high flood-risk areas can be regulated through land-use planning controls and floodplain zoning. The above measures should be used in addition to other non-structural and structural mitigation measures such as increasing the flood resistance of engineered structures, and the construction of seawalls and other barriers to protect critical facilities and other coastal infrastructure from flooding, storm surge and tsunami waves – as long as the short- and long-term costs of benefits of engineering approaches are assessed honestly, rather than assuming that any engineering structure is better than not having it.

High winds

Building structures may be subject to wind damage, particularly storm-induced winds, hurricane winds and tornadoes (see Chapter 19 and Chapter 20). Engineering design codes are used to ensure that buildings and structures are constructed to withstand particular wind speeds depending on the characteristics of each region. In the USA, the American Society of Civil Engineers (ASCE) provides the guidelines for the design and calculation of wind loads in the design standard 'ASCE 7 Minimum Design Loads for Buildings and Other Structures' (ASCE 2006). ASCE 7 requires design for the fifty-year wind speed with an importance factor for critical infrastructures and industrial facilities containing hazardous materials. This results in the equivalent of a 500-year wind speed for these structures.

It is important to note that very often wind damage to building structures is due to failure of roofing materials, doors and windows. These failures, which are often less expensive to prevent or mitigate, lead to weather penetration and damage.

Landslides and other soil hazards

Landslides and other soil hazards often accompany natural hazards such as earthquakes, floods, hurricanes and volcanic eruptions (see Chapter 25). The Wenchuan earthquake in Sichuan Province, China, on 12 May 2008 triggered 5,117 landslides, 3,575 rock falls, 358 debris flows and 34 barrier lakes (Shi 2008). The slope failures were responsible for destruction of infrastructure, associated systems, industrial plants, roads and bridges. Hurricane Mitch in Central America in 1998 triggered hundreds of landslides, including re-mobilisation of older volcanic mudflows, which were responsible for damage to buildings and the majority of fatalities (Spiker and Gori 2003).

As with other natural hazards, landslide hazard reduction includes both structural and non-structural prevention and mitigation measures. Structural measures include construction of earth-retaining walls, construction of surface water drainage systems, slope surface protection such as hydro-seeding, sprayed concrete and reinforced concrete grids, and re-compaction of fill slopes.

Lifeline systems

Natural hazards have the potential to disrupt lifeline systems. These systems include roads, bridges, ports, power generation and distribution systems, communications systems, water distribution systems and waste management systems. Lifelines, infrastructure and associated systems were seriously affected by liquefaction and strong ground shaking during the 1999 Kocaeli earthquake in Turkey, resulting in disruption and damage (Tang 2000).

Damage to lifeline systems can delay or impede emergency response activities. For example, loss of water due to multiple pipeline-breaks delayed emergency response to several of the gas-caused fires following the Northridge earthquake (City Administrative Officer 1994). The loss of water and power outages following the Kocaeli earthquake hampered emergency response to earthquake-triggered hazardous materials releases (Steinberg and Cruz 2004).

Bridges and roadways

Liquefaction, ground settlement and slope instability can cause extensive damage to bridges, elevated highways and roadways during earthquakes. Transportation systems, including highways, bridges and roads, suffered damage during the Kocaeli earthquake. Typical damage included pavement openings, ground heaving, fissures, displacement and settlement, and road buckling due to compression damage (Tang 2000). In addition, the large amount of debris from damaged buildings, and an increased amount of traffic as rescue efforts and other traffic converged to the affected area, overwhelmed the region's road network. Extensive damage to transportation routes was also reported following the Kobe earthquake, which destroyed the city's main highway, several railroad tracks and much of its port (Dawkins 1995).

The structural integrity and performance of bridges and roadways can be improved with proper design and materials. Tunnels, although expensive, usually prove to be cost effective in the long term to avoid landslide hazard in transportation routes with slope problems – unless the landslide blocks the tunnel entrance. Spiker and Gori (2003) observe the need to establish standardised codes for excavation, construction and grading in landslide-prone areas.

Ports and marine terminals

Ports and marine terminals are affected by earthquakes and tsunamis, and liquefaction and soil problems during earthquakes (Tang 2000; Erdik 1998). Ground shaking, settlement and lateral displacement caused damage to port facilities in Izmit Bay following the Kocaeli earthquake (Tang 2000). Ground subsidence and/or submarine slides caused the loss of 200 m of pier at the AKSA chemical company in Yalova on the south shore of Izmit Bay (Steinberg and Cruz 2004). Liquefaction and permanent ground deformation devastated the Port of Kobe, Japan, in 1995, damaging more than ninety per cent of the port's moorings (Erdik 1998). Damage to ports can have severe economic impact on a region, as occurred following the Kobe earthquake, cutting Kobe off from the rest of Japan and the outside world (Cataldo 1995).

Ports and marine terminals are susceptible to storms, including tropical cyclones. Several ports in Central America were severely affected by Hurricane Georges in 1998 (Beam *et al.* 1999). Protection of ports and harbours from wave action and storm surge may include natural or artificial breakwaters and surge barriers. Several national and international bodies have published guidelines for the design and protection of ports and harbours from various natural hazards (e.g. Eskijian 2006; PIANC 2001).

Underground pipelines

Underground pipelines can be affected by earthquakes, poor ground conditions, liquefaction, flooding, storm surge, erosion and landslides. Earthquakes and flooding have caused extensive damage to gas, water, sewage, wastewater and oil pipelines. Damage to gas and oil pipelines can result in secondary hazards such as gas leaks, fires and explosions. Lindell and Perry (1997) reported nine petroleum pipeline ruptures during the 1994 Northridge earthquake in California. The spills caused property damage and an injury in connection with one spill when leaking crude oil was ignited.

Damage to water distribution lines can leave people without drinking water for long periods of time. Following the Gujarat earthquake, damage to water pumping stations and transmission lines left people without a drinking water supply. Restoration of potable water supply via pipeline took four-to-six months to restore (Eidinger 2001). Typhoon Aila in May 2009 damaged the few potable water distribution systems available in Sajnekhali, West Bengal. The high storm surge overtopped levees, killing many people and livestock, while causing salt water intrusion to drinking water reservoirs leaving victims without any source of water.

Loss of water can also hamper or impede response to disaster-triggered fires. Erdik (1998) reports over 2,000 water-line breaks during the Kobe earthquake, having a negative effect on fire-fighting capabilities. Steinberg and Cruz (2004) reported that damage to the main water pipeline, which provided service to several industrial facilities in Korfez, severely hampered emergency response to the multiple earthquake-triggered fires at Turkey's largest oil refinery following the Kocaeli earthquake.

Mitigation measures to reduce pipeline (as well as other lifelines) vulnerability to natural hazards include avoiding areas close to active faults or in areas susceptible to liquefaction, land-slides, flooding or other natural hazards, the use of proper materials, and the following of international standards for design and construction. Detailed guidelines on lifeline protection are available from ASCE's Technical Council on Lifeline Earthquake Engineering (e.g. ASCE 2009).

Electrical power systems

Electrical power is highly susceptible to natural hazards. Damage to power systems can severely hamper emergency response capabilities. Earthquakes, storms and high winds can knock down electrical power lines. Power outages have been reported during most major hurricanes. Often electrical power is shut off as a preventive measure before the arrival of hurricane-strength winds, and remains so until crews verify the integrity of electrical power lines and poles after the storm. Damage to electrical power systems during hurricanes is often caused by water penetration in power stations and toppling transformers and electrical power lines and posts.

The 1998 ice storm in eastern Canada knocked out power to several million people, with some not having power back for several weeks. Auckland, New Zealand also suffered a 1998 blackout during a heat wave that fried several power lines.

Most major earthquakes have resulted in electrical power outages of varying lengths. The most vulnerable components during earthquakes include generators and transformers, with damage often occurring due to improperly anchored equipment (Erdik 1998). Indirect damage to electrical power lines and poles caused by building collapse can also be extensive, as was documented by Tang (2000) following the Kocaeli earthquake.

Communication systems

Communication systems are highly susceptible to natural hazard impacts. Most often, communication systems fail during earthquakes due to poor seismic protection of backup power systems. During the Kocaeli earthquake communication systems suffered due to failure or lack of mitigation measures (Tang 2000). Damage to several communication towers and unavailability of back-up power affected both landlines and wireless networks following the recent Padang, Sumatra earthquake in September 2009, leaving the region without electricity-based communications for at least six hours (Tang 2009).

Protection and mitigation measures for communication systems include the adoption of appropriate building codes to ensure that buildings can withstand the forces of the natural hazard, the use of anchoring mechanisms for electronic equipment, batteries and back-up generators, and the raising above specific flood levels of sensitive equipment in flood-prone areas, with the added potential for making the equipment flood resistant.

Industrial facilities and water and wastewater treatment plants

Natural hazards can cause extensive damage to industrial facilities, and disrupt water and wastewater treatment plants. Damage to storage tanks and plant processing equipment may result in unintentional releases of hazardous materials. Of particular concern are water treatment plants because many store and use chlorine gas for water treatment.

Hundreds of hazardous materials releases triggered by natural hazards occur every year around the world. For example, Steinberg and Cruz (2004) reported more than twenty-one incidences of hazardous materials releases, some with severe offsite consequences to neighbouring residential areas, following the 1999 Kocaeli earthquake in Turkey. Van Dijk (2007) reported on hazardous materials releases due to the 26 December 2004 Indian Ocean tsunami in the city of Banda Aceh, Indonesia, including oil spills and release of other materials from two depots of fertiliser and pesticides. Krausmann et al. (2010) reported releases of ammonia and other hazardous materials with possible effects on nearby residents following the Wenchuan earthquake in China in 2008.

Industrial facilities and water and wastewater treatment plants can be protected against natural hazards through the adoption, monitoring and enforcement of design standards that account for natural hazard loads (e.g. wind, seismic, flood and tsunami loads) on buildings, steel support structures for processing equipment and storage tanks. Industrial risk management practices for the prevention of accidental hazardous materials releases (e.g. toxic gases, fires, explosions) which specifically address the potential impacts of natural hazards on their installations are needed. Additional risk management actions that can be promoted to make plants less vulnerable to natural hazards include the use of redundant safety systems, natural hazard-resistant designs, the provision of guidelines to inform industry about how to plan for natural hazards, and requiring the strategic placement of hazardous substances in less vulnerable areas.

Infrastructure interdependencies and disaster planning

Experiences from disasters have shown the need to better understand infrastructure failure interdependencies and their societal significance. One of the major concerns is that the existing interdependencies between infrastructure and their associated systems can lead to cascading failures. To illustrate, Menoni (2001) analysed the interactions and couple effects induced by the Kobe earthquake on systems such as lifelines, industrial facilities, hospitals and emergency response facilities, residential buildings and people. The study showed how the earthquake-induced hazards in each of these systems (e.g. gas leakages and toxic releases) led to effects in other subsystems such as economics, emergency services and social systems. Menoni (2001) noted the need to incorporate both structures in the physical environment such as lifelines and building stock as well as organisational, social and systemic factors into the analysis.

There is ongoing research to improve understanding of cascading failures, infrastructure and system interdependencies, and criteria for defining 'criticality' of facilities. Criticality refers to how essential infrastructure is for different societal functions. Definitions and criteria vary, without much agreement (Moteff et al. 2003). Some definitions have focused on economic and military importance, while others prefer to look at criticality in terms of large loss of life (e.g. office buildings or apartment blocks), affecting morale (e.g. monuments of national significance), essential societal services (e.g. schools and hospitals) or changing quality of life (e.g. water treatment plants).

As an example, the University of British Columbia in Canada runs the project entitled 'Analyzing Infrastructures for Disaster-Resilient Communities' (www.chs.ubc.ca/dprc_koa). The main objective of the project is to develop and disseminate knowledge that is needed to prioritise investments for fostering disaster-resilient infrastructures and, thus, more disaster-resilient communities. The European Community has set out non-binding guidelines for identifying and designating European Critical Infrastructure and for assessing the need to improve their protection (Bouchon et al. 2008).

Conclusions

This chapter has reviewed some options available for protecting infrastructure and their associated systems from natural hazards. Protection of infrastructure and their associated systems may include physical risk reduction measures such as those that can be included in the design, construction and maintenance of buildings and lifelines, but also human, organisational and operational measures must be considered. Thus, a comprehensive approach to infrastructure protection should be adopted that integrates structural and non-structural measures, strengthens the capacity of local communities and promotes the participation of all stakeholders in disaster planning.

Most importantly, when addressing the protection of infrastructure and their associated systems, it is necessary to analyse them not as independent entities, but as a part of a much larger, connected system. Disaster consequences may be greatly reduced with a collective effort to understand and prevent ripple effects from infrastructure failure.

57

Social protection and disaster

Walter Gillis Peacock

TEXAS A&M UNIVERSITY, COLLEGE STATION, TEXAS, USA

Carla Prater

TEXAS A&M UNIVERSITY, COLLEGE STATION, TEXAS, USA

Introduction

Disaster risk reduction (DRR) is fundamentally concerned with reducing the probabilities of sustaining harmful impacts associated with natural disasters (UNDP 2004: 36). Disaster risk is a function of the exposure to hazards and vulnerability. Recently a more comprehensive view of vulnerability has emerged that considers not only physical dimensions but also broader social, economic and environmental conditions (UNDP 2004: 42; Naudé *et al.* 2009a). Vulnerability can be generally understood as the lack of capacity to anticipate, cope, resist and recover from disasters (Wisner *et al.* 2004) and is shaped by an interaction among physical, social, economic and environmental factors. Of particular relevance for understanding the relationship between social protection (SP) and DRR are the socio-economic structures and processes that determine access to scarce resources such as income, wealth, social assets (see Chapter 58), power, cultural factors that shape belief and customs, and driving forces such as urbanisation and demographic change.

It is critically important to recognise that vulnerability and disaster risk are not uniformly distributed within a society even after holding hazard exposure constant. Economic, social, cultural and racial/ethnic factors can significantly predispose segments of a population to higher levels of vulnerability (Wisner *et al.* 2004; Enarson *et al.* 2006; Bolin 2007). Economic status, for example, is a determinant of access to opportunity structures, political power, health care, education, housing, communications, transportation, land and natural resources, credit and insurance. It should be no surprise that research generally finds poor individuals, particularly those who are also women, the elderly and minorities, have much greater difficulty responding to warnings and preparing for disasters, suffer disproportionate losses and deaths, and have much greater difficulty adjusting to and recovering from disasters. It is this linkage between poverty and higher levels of disaster vulnerability and risk that makes manifest the basic linkage between SP and DRR, for SP has fundamentally been focused on poverty reduction, which is also a fundamental factor shaping disaster social vulnerability.

Social protection, particularly in its early stages, was generally conceived as an approach to combat poverty. These programmes are usually traced to Bismarck's establishment of a social insurance programme in late nineteenth-century Prussia, which was intended to reduce the appeal of socialist political agendas through co-optation of their emphasis on poverty reduction (Midgley and Tang 2010). SP has been defined as 'public policies that assist individuals, households, and communities in better managing risks and that support the critically vulnerable' (Holzmann 2009: 1), where risk is generally with regard to falling into poverty. In practice, SP has most often consisted of two types of programme: social insurance and assistance. The former involves pooling contributions from workers to provide financial support in case of a loss of livelihood; the latter refers to income transfers to those deemed most needy and worthy of support. Thus, for example, old-age pensions and cash transfers to the disabled such as those found in the USA's Social Security programme are usually encompassed under the term social protection (Midgley and Tang 2010). However, with increasing development and democratisation, and in response to the labour movement, SP evolved to include policies and programmes associated with workplace standards such as minimum wage laws, child labour prohibitions and similar labour regulations (DFID 2006).

The SP movement has been characterised as a 'quiet revolution' (Barrientos and Hulme 2009), for it has expanded in both scope and breadth from traditional formal programmes to more inclusive programmes. For example, included under the SP rubric are programmes similar to pension programmes in South Africa and innovative programmes like India's Rural Employment Guarantee system, as well as rural health and nutrition programmes, anti-retroviral programmes for people living with HIV/AIDS, micro-insurance and microcredit, child protection programmes and disaster relief programmes. These expansions in part seek to address the important distinction between acute and chronic poverty.

When addressing acute or transient poverty, SP programmes focus on short-term assistance meant to help individuals and households cope with shocks (e.g. disasters, economic and political crises), while asset accumulation and asset protection programmes are better suited to addressing long-term, chronic poverty (Barrientos et al. 2005). The latter programmes are more likely to be useful in reducing high levels of vulnerability, thus increasing overall societal resilience. Yet even here there are structural exclusionary impediments to addressing chronic issues that have led many to call for *transformative* SP. While the evolution of SP has been dramatic, the history of that evolution and current state of SP for particular countries and regions has been quite variable, particularly when comparing older members of the Organisation for Economic Co-operation and Development (OECD) to non-members. Indeed, the SP movement must continue to evolve to better address DRR and unique issues facing many less-developed countries (LDCs).

The evolution of social protection

Different histories

In wealthier countries, including OECD members and some oil-rich and medium-income developing countries, SP developed along the lines of social insurance, that is, pooled contributions gathered by taxation or as premiums. These contributions were then used to provide unemployment insurance and old-age pensions to workers. There is also a history of privately funded organisations such as funeral associations, often limited to members of particular occupational groups that performed similar functions. Elaborate systems of public education, labour market policies, labour and job safety standards, child labour laws, job-training programmes, and even disaster assistance and aid evolved along with various forms of social insurance and in response to organised labour and other movements. Most OECD countries have also developed some form of publicly funded health insurance system, usually based on a system of taxation and with wide or universal coverage. The main exception is the USA, which is currently dealing with a variety of problems including overly high administrative costs and inadequate access because of the lack of such a system.

At one time it was commonly argued that overly generous SP programmes were a drag on development, providing disincentives to economic participation. It was noted that the USA appeared to respond more quickly to economic downturns, and this speed of adjustment was attributed to its low levels of SP. However, research into this relationship failed to support the hypothesis (Abdulai *et al.* 2005; Blank 1994), and the relationship between income supports and positive economic results has lately been acknowledged by scholars and funding agencies (World Bank 2001; Holzmann 2009; DFID 2006; Gentilini 2009). This was an important development, freeing national governments in LDCs and international agencies such as the World Bank and the International Labour Organization to investigate the uses of SP to promote more equitable and just economic development.

The dominant characteristic of SP programmes and policies in OECD nations is formalism; they are highly dependent on governmental involvement either directly through running programmes or indirectly through regulation and formal economic activities because the funding for these systems and their regulation is derived from employer and employee contributions and through efficient tax systems. While the distinction between 'formal' and 'informal' economic activities is far from clear and they are often highly interrelated and interdependent in any economy (Valodia 2008; Chen 2008), the ability to establish, fund and operate 'formal' SP programmes and policies similarly to OECD nations is impossible for much of the world.

As noted by Chen (2008), 'informal employment broadly defined comprises one-half to three-quarters of non-agricultural employment' in areas like the Middle East, North Africa, Latin America, Asia and sub-Saharan Africa, and in many areas is growing in response to globalisation. Furthermore, there can be distinct gender- and age-based bias in the coverage offered by formal programmes (Charlton and McKinnon 2001) and these same groups, in addition to ethnic/racial minorities, are more likely to be 'confined' to informal activities (Chen 2008). The net effect is that, as traditionally conceived, SP is not likely to reach those most in need of assistance and the gap in coverage is growing. In order to expand SP, non-formal approaches and other alternatives to 'formal' government-provided SP should be included as part of the mix, especially in LDCs (Norton *et al.* 2000).

More recent examples of social protection

Research suggests a clear link between levels of development and SP provided by governments (Midgley and Tang 2010; Charlton and McKinnon 2001). Gentilini (2009) shows the link between direct foreign aid as a percentage of national gross capital formation and levels of SP. These levels range from the absence of SP in very poor and politically fragile countries (i.e. Afghanistan, Somalia, Haiti), through chronically poor countries (i.e. Kenya, Mozambique, Nicaragua and Bangladesh) that still manage to provide some elements of SP, to emerging SP systems in countries with moderate capacities (i.e. Egypt, South Africa and Ecuador) (see Chapter 5). Finally, consolidated SP systems are found in Brazil, Russia, India, China, Mexico and other countries with more functional agricultural, market and governmental systems.

However, this assessment is highly focused on formal programmes, which along with access and coverage problems have at times been tainted by corruption. Indeed, the fear of corruption in LDCs has been a rationale for discouraging national SP programmes. Although sometimes associated with recipients misrepresenting themselves, corruption often results from mis-appropriation of funds by those charged with administering programmes. It has been argued that corruption itself was part of a general syndrome created in part by inflation and strict controls on government expenditures required by Structural Adjustment Programs (SAPs), resulting in poor oversight of SP programmes that had meanwhile lost public support as their benefit levels and coverage fell (Charlton and McKinnon 2001). The loss of government audi-tors and other supervisory capacity was an unintended side effect of SAPs. While corruption can be problematic, the major problems with formal SP remain coverage/access and addressing chronic poverty.

Of course, individuals and households have long employed a host of informal and 'tradi-tional' coping strategies ranging from curtailing consumption, attenuating/depleting savings, the sale of assets, increasing labour supply by pulling children out of school, migration of all or part of the household on a temporary or permanent basis, utilising extended family resources, as well as various community arrangements based on reciprocity (Holzmann and Jorgensen 2001). While some of these, particularly international remittances, can have benefits for maintaining consumption levels and enhancing livelihoods and assets, others like selling assets and with-drawing children from school can deepen poverty levels, thwart asset accumulation and limit, if not destroy, livelihood and human capital development (Holzmann and Jorgensen 2001). Fur-thermore, many of these forms are of limited utility for larger community and regional shocks as well as more pervasive and chronic poverty in large proportions of LDC populations. In light of these limitations, informal examples of SP are insufficient. Fortunately, a host of SP programmes and policies are being considered and implemented, including expanded social services in the form of health care and education and training programmes (cf. Goudge et al. 2009), social transfers (food/cash), asset transfers, public works programmes, social pension programmes, livelihood diversification programmes, social funds, and microcredit, micro–insurance and microfinance.

Social transfers in the form of food and cash are often seen as polar opposites and yet they can play complementary roles. The provision of food and other in-kind aid was once the major form of aid, particularly during crisis periods, to address starvation, malnutrition and avert famine. Similarly, government retail food subsidies were common early SP programmes but were often subject to cuts as part of SAPs (Killick 1995) during the 1980s and 1990s. While both forms of food aid remain, the former in particular has come under increasing criticism for accomplishing little in the way of addressing chronic food shortages or poverty; the provision of food alone does not promote agricultural production, enrich livelihoods or address poverty.

More recently, direct cash transfers have been promoted by many as a more viable option with a number of advantages including enhancing household choice and flexibility, stimulating market demand and supply, stimulating asset accumulation and diversification, and greater cost effectiveness (Harvey 2006) (see Chapter 45). Research suggests that ensuring predictable and timely cash transfers can enhance programme effectiveness, as can household planning and decision-making, targeting of recipients, and conditioning transfers to participation in health care programmes and school attendance. For example, Kenya's National Social Protection framework offers two targeted programmes linked to schooling and health care facilities (Davies et al. 2008). Similarly, the *Bolsa Família* of Brazil and *Oportunidades* in Mexico have been used to increase family income through cash transfers, and build social capital by keeping children in school (Midgley and Tang 2010). These programmes are means-tested, although it has been argued that uni-versal coverage would be more likely to reduce child poverty (Esser et al. 2009). Interestingly, Ethiopia's Productive Safety Net Programme utilises both cash and food transfers in conjunction with seasonal employment on public works projects to reduce household vulnerabilities, while enhancing food security and improving infrastructure (Davies *et al.* 2008).

Microfinance, microcredit and micro–insurance SP schemes seek to extend conventional market-based financial resources to poor households that rarely have access to credit and savings mechanisms. Microcredit is generally promoted as a mechanism for increasing the capacity of vulnerable groups by financing micro-enterprise and other income-generating activities, thereby increasing incomes, savings and food expenditures, and potentially diversifying household live-lihoods, as well as increasing administrative skills and social solidarity. Microcredit in particular has received a good deal of attention, since it can be targeted toward women and female-headed households. The literature has been somewhat mixed, when considering the full range of potential positive effects from stimulating savings, reducing poverty, increasing food expen-diture, diversification, etc. The most consistent finding is that these forms of SP are better targeted to less marginal households and not the poorest of the poor (cf. Kondo *et al.* 2008).

To maximise coverage of individuals engaged in informal economic activities, SP can be offered to the poorest and most vulnerable by incorporating them as non-contributory pen-sioners, or by creating a separate social assistance programme. For example, social pensions are cash-transfer programmes similar to more formal pensions, except that entitlements are not contingent on individual or employer contributions (see Chapter 37). Research suggests that even a modest provision to elderly household members can allow that person to retain their place as a respected and useful member of the household, continuing to offer very real services such as child care, contribute to the household's income maintenance and asset purchases, and free younger members of the household to pursue more lucrative employment or educational opportunities (Charlton and McKinnon 2001; Davies *et al.* 2008; Goudge *et al.* 2009).

Social funds are a unique form of SP that can stimulate community-based groups and orga-nisations to join together on projects or programmes to address local poverty issues and enhance their developmental capacity. Generally, social funds are programmes that fund grants for small-scale development projects such as schools and community infrastructure or for capacity-building programmes such as microfinance, literacy or training programmes. These programmes are generally planned and implemented by local community organisations, often without working through conventional governmental structures. Since the first social fund was established and funded by the World Bank in Bolivia in 1987, funds have spread throughout Latin America, Africa, Asia and the Middle East. The World Bank provides support for social funds in some forty-three countries around the world, and in a number of cases such as the earthquake in Pakistan has used the infrastructure of the national social fund as a vehicle for rapid dispersal of large amounts of funding for relief and recovery.

Social protection, vulnerability and marginalisation

As discussed above, on the surface the linkage between SP and DRR appears self-evident: SP was initially introduced to address poverty, or more expansively economic status, and is intimately associated with social vulnerability and hence disaster risk. However, this linkage remained tangential and obscure for two reasons. First, SP was specifically addressing poverty risk, not disaster risk, hence vulnerability was narrowly focused on economic vulnerability and issues of DRR were not explicitly addressed. Second, despite the emerging and more expansive view of SP attempting to address more chronic issues associated with poverty, these approaches often failed to address more fundamental issues associated with marginalisation. The problem is not simply that the poor do not have access to resources and assets allowing them to pull them-selves out of poverty or properly address DRR; in many societies exclusionary structures deny individuals and groups such as women and ethnic/racial minorities basic human rights, resources, safety and security. In these situations SP must move beyond considering traditional approaches and must be more comprehensive in order to contribute to DRR. Two recent movements in SP associated with social risk management (SRM) and transformative social protection (TSP) explicitly address each of these issues.

Social risk management

Holzmann and Jorgensen (2001) introduce SRM as a conceptual framework for SP and provide a more comprehensive concept of SP. Their SRM approach seeks to reposition SP's traditional instruments of labour market interventions, social insurance and safety nets 'in a framework that includes three strategies to deal with risk (prevention, mitigation and coping), three levels of formality of risk management (informal, market-based, public) for many actors ... against the background of asymmetric information and different types of risk' (Holzmann and Jorgensen 2001: 529). Not only does the SRM framework explicitly address natural hazards as an important factor in vulnerability and risk assessment, but it also explicitly addresses prevention, mitigation and coping, which, as Vakis (2006) notes, is highly complementary with what is conventionally thought of as the disaster cycle. Indeed, Holzmann and Jorgensen (2001: 544) offer an expanded set of examples of what they term non-formal, market-based and public-sector SP strategies for addressing risk reduction (prevention), mitigation and post-event risk coping. Essentially they discuss many of the expanded set of SP schemes discussed above as well as some additional forms, and explicitly point out how they potentially promote disaster prevention, mitigation and coping.

While focusing on poverty reduction and critical economic features of protecting basic livelihoods, and promoting economic risk-taking as well as noting that equity issues must be addressed, Holzmann and Jorgensen do not explicitly address broader social vulnerability issues, particularly those related to marginalisation. In a series of papers, Norton et al. move SP further toward addressing marginalisation, suggesting that SP should 'increase the "voice" of the poor and vulnerable' through the development of standards and entitlements that directly address human rights (Norton et al. 2000: 11).

Transformative social protection

It was not until the more recent emergence of a TSP perspective that marginalisation became a central focus of social protection. Sabates-Wheeler and Devereux (2007) offer a major critique of SP approaches. They point out that these approaches fail to address fundamental structural issues shaping vulnerability and as a consequence neglect chronic determinants of poverty and perhaps even underestimate the potential of informal strategies for addressing SP. Indeed, they offer a new more comprehensive definition of SP:

> Social protection describes all public and private initiatives that provide income or consumption transfers to the poor, protect the vulnerable against livelihood risks, and enhance the social status and rights of the marginalised; with the overall objective of reducing the economic and social vulnerability of poor, vulnerable and marginalised groups.
>
> (Sabates-Wheeler and Devereux 2007: 25)

Their concept of a TSP framework identifies four key elements or measures: protective, preventive, promotive and transformative.

Protective measures are designed to provide more short-term relief from critical 'deprivation' and fall along the lines of traditional safety net and social assistance approaches (i.e. food/cash emergency relief, disability benefits and pension programmes), while *preventive* measures seek to 'avert deprivation' and might include health insurance, savings clubs and unemployment benefits, as well as income and livelihood diversification (Sabates-Wheeler and Devereux 2007: 25). *Promotive* measures seek to enhance income and livelihood capabilities and include public works projects for food or wages, school feeding programmes and direct cash transfers (Sabates-Wheeler and Devereux 2007: 25). *Transformative* measures address more directly structural features associated with marginalisation and might include promoting minority rights, collective worker actions, anti-discrimination campaigns and various social funds to promote change.

The TSP framework has been rapidly promoted and discussed within the context of climate change and variability (cf. Davies *et al.* 2008), in part because of its ability to address fundamental structural change in reducing marginalisation and enhancing broad-based adaptation that will be required to address consequences of climate change that are neither predictable nor clear, particularly at the local level (see Chapters 18, 21 and 22).

While there are still major questions about both SRM and TSP, with both perspectives the linkage between SP and DRR is much more clearly articulated. First, the SRM framework directly addressed the need to consider DRR as a critical element when considering the nature of SP measures to be undertaken. Furthermore, TSP has taken a critical step in addressing more fundamental issues of social vulnerability and marginalisation as a critical element within SP – for without considering transformative change, problems of chronic poverty and social vulnerability (in particular gender and racial/ethnic issues), and even HIV/AIDS, cannot be addressed. Analytically this is a large advance over the rather superficial way in which the United Nations International Strategy for Disaster Reduction (UNISDR) Hyogo Framework of Action (HFA) treats what it calls 'underlying risk factors'. Because the HFA is a consensus document hammered out in diplomatic negotiation among 168 nations in 2005, it addresses neither rights nor marginalisation, nor gross disparities in wealth and power, nor the economic forces driving farmers off the land and urban dwellers into unsafe locations (see Chapter 50).

Social protection and community solidarity: recognising and enhancing participatory governance

TSP, with its direct focus on challenging fundamental processes and structures that shape social vulnerability and thereby thwart comprehensive DRR, has not been without critics. It has been suggested that TSP is much too holistic or ambitious, trying to change too much, demanding political involvement generally shunned by international agencies and NGOs, and begging the question of who actually counts when setting transformative agendas (cf. Aoo *et al.* 2007). Yet in many respects there is a great deal of commonality between TSP's transformative position and the vigorous ongoing debate found in the broader development and SP communities related to creating effective programmes through enhancing participation, creating spaces or places for local participation and responsive government and, more fundamentally, with the deepening of democracy (see Chapter 5).

TSP's call for addressing citizenship and rights is not new, nor radical, but rather harks back to the United Nations (UN) Declaration on the Right to Development (see Chapter 6). As noted by Gaventa (2002: 2), this declaration links not only 'the idea of development to the concept of rights, but also names the rights to meaningful participation and social justice as its inherent components'. In other words, the *transformative* elements of TSP are essentially the same arguments raised in the participatory governance or community participation literature for creating effective programmes and policies that might be associated with SP. While this literature is far too rich to be captured herein, it is worth considering some of its common themes and complementarity with TSP.

This literature goes beyond the simple declaration that participation is important, to examine the mechanisms and structures, both legal and social, that help ensure not simply participation but also responsive government and accountability. It draws heavily from examples and research in the developing world on the transformative possibilities arising from participatory governance within SP programmes. Some of the examples include: HIV/AIDS treatment campaigns in South Africa; farmers' management of irrigation systems in Andhra Pradesh, India; land reform in the Philippines; the Ugandan Participatory Poverty assessment process; the 2005 National Right to Information Law in India; and the Rural Integrated Project Support Program in Tanzania. Clearly many of these are directly concerned with more conventional as well as transformative SP issues. The lessons garnered from this work make it clear that there is not a set of guidelines that fits all situations. Indeed, it is quite clear that sometimes success stories are initiated from below, other times imposed from above, and in all cases the picture is far more complex than a simple top-down/bottom-up characterisation suggests.

A number of authors have offered general design properties or key lessons winnowed or distilled from these and other case studies (Blackburn et al. 2002; Gaventa 2004). For example, some of the lessons Gaventa identified for participation *and* responsive government include: (1) legal or statutory provisions that enable participation; (2) a recognition of diverse local and regional contexts; (3) simultaneously focusing on community empowerment and the capacities of local government to understand and respond to community empowerment; (4) developing clear guidelines for the rules and roles of engagement for community leaders, elected officials and government staff; (5) clarifying forms of accountability underlying different forms of representation; (6) creating incentives for quality representation and participation, ensuring the opportunities for real decisions over resources and strategies developed by locals; (7) building broad coalitions in civil society and alliances among civil society, progressive government figures, academics, professionals and grassroots organisations; (8) seeking a level playing field by identifying and addressing power relationships in the participatory process; and (9) recognising that fundamental change is only possible with sustained efforts over time.

While all of the above are important, legal or statutory provisions (point 1), which enable participation, are of course basic, for without the creation of a 'democratic space' to allow for participation, community involvement can be difficult if not deadly. This, however, does not mean that the process is always top-down. As Gaventa and others have noted, in many cases, were it not for concerted grassroots movements and civil society pressuring and shaping this space, legal and statutory provisions may not have developed in the first place (e.g. Chile, South Africa, the Philippines and Mexico). Furthermore, maintaining and expanding that space demands concerted efforts and pressure from community-based organisations and broader civil society, including faith-based organisations, unions and worker organisations. Hence coalition building (point 7) can be critically important for ensuring, maintaining and expanding space for participation, as well as for sustaining and reinforcing long-term efforts (point 9) for change. Interestingly, coalition building can extend both nationally and internationally when international agencies exert pressure and demand local participation and involvement, as in the current case of Haiti where organisations like Plateforme Haïtienne de Plaidoyer pour un Développement Alternatif (PAPDA) are demanding a voice for the Haitian people in reconstruction. Given Haiti's violent past, it will be important to ensure sustained involvement in this process. The aftermath of the Guatemalan earthquake clearly showed that introducing participatory

Box 57.1 A focused definition of social protection when self-protection fails

Terry Cannon
Climate Change and Development Team, Institute of Development Studies, UK

There are broad and narrow concepts of 'social protection' that are useful in under-standing DRR. The broad notion encompasses all the forms of social assistance and social security accessible in a particular society, such as old-age pensions, food assistance, fertiliser subsidies, free or subsidised health care, etc. These are often crucial in reducing poverty (and vulnerability to hazards), and may help in boosting self-protection. In addition, one can use the term 'social protection' to cover a narrower range of public actions: those specifically aimed at reducing disaster risk. This narrower use of the term can be defined as any action taken by actors other than the individual or household that helps to reduce vulnerability to a given hazard. It is often a replacement or sub-stitute for self-protection when people are too poor or give priority to other problems or livelihood opportunities.

Many people live and work in places that are hazardous, often in order to pursue livelihoods, or because poverty leaves them no alternative. People also have more immediate everyday needs, which means that they may not give high priority to hazards. They have a different attitude to risk than that of DRR specialists. So some people are unable or unwilling to protect themselves from known hazards. This failure of self-protection means that intervention of other actors through social protection is all the more important (Cannon 2008).

Social protection may also be essential because for some hazards the household or community is not the appropriate level of intervention for protection in facing natural hazards. Certain types of flood prevention cannot be undertaken by households or communities on their own. Building standards essential to earthquake, flood, storm or tsunami safety must be implemented at a higher level. People at risk of earthquakes are highly reliant on good social protection: adequate building codes, their proper imple-mentation and high-quality construction. Social protection for DRR may therefore be a substitute for self-protection and an essential component of risk reduction in situations where self-protection is never going to be sufficient, and where different scales of intervention are required.

Social protection is highly dependent on governance. Good governance will allow for proper social protection to be in place for known and anticipated risks. Bad gov-ernance is likely to lead to poor or missing social protection. Good social protection requires an active civil society and free media (as monitors of corporate behaviour and government action and inaction). Good social protection is unlikely without good governance, and DRR will be hampered by the failure to allow open media and debate about disaster policy, corruption and power.

Further reading: Cannon (2008).

forms of community involvement exposed nascent community leadership to violent and deadly reactions once international attention began to wane (Bates and Peacock 1987).

Of course, local or community-based involvement, particularly with respect to extant leader-ship and elites, does not always ensure effective or even responsive government or progressive change. In fact, the myth of community as harmonious, homogeneous and conflict-free, as well as any romantic notions of community solidarity working toward the common good in which informal SP will provide for all, must be dispensed with (see Chapter 59). As Gaventa (2004: 27) notes, ensuring effective participatory governance requires 'working on both sides of the equation'. In other words, training and building toward effective leadership and representation on both the community and governmental sides may well be necessary. Furthermore, enhancing community capacities through developing strong, democratic community and governmental organisations, fostering transparency and information sharing, and understanding and learning the roles and functions of representative processes and democratically elected leadership are but some of the issues that may have to be addressed. The point is that participation is not a one-way street; rather, it is critical to incorporate both the community and governmental sides of the equation for promoting effective participatory governance.

The above provides only a small glimpse into the literature on participation and participatory governance, reflecting its consistency with TSP. Both address the extension of fundamental rights and the removal of exclusionary structures that generate marginalisation, focusing on the right to participate and seeking to create space for a deepening of democracy. Indeed, the complementarity between these approaches is also evident. While TSP is focused on protective, preventive and promotive SP policies and measures to address DRR, these are extended to include transformative measures, addressing fundamental issues that generate social vulnerability and marginalisation, by introducing rights and promoting participation and anti-discrimination policies. It is precisely the latter issues, in addition to accountability, that the participatory governance literature addresses. It promotes a rights-based approach for ensuring effective policies and programmes development, including broad-based community participation and responsive government. Consequently, more traditional SP approaches, addressing DRR through protective, preventive and promotive measures, but basing these measures on a participatory governance approach, become, in effect, transformative.

Conclusions

Social protection has undergone a quiet revolution not only if one considers the evolution of SP since its nineteenth-century beginnings, but even more significantly if one considers the foolhardy attempts to eviscerate many SP programmes that accompanied the radical structural adjustments agenda promulgated during the latter part of the twentieth century. While the very nature of SP is in flux, the tool set of options is quite varied and diverse and the issues germane to DRR are clearly a central part of the mix. This is indeed an exciting time, and yet it is also a moment in which ideology may again be more of a guide than hard evidence and empirical outcomes. This is perhaps most evident in the often heated discourse regarding the relative effectiveness of cash transfers versus food aid, not to mention the debates between those in favour of transformative and more limited SP initiatives. Indeed, what is most clear at this point is that programmatic innovation is far ahead of the limited attempts to engage in systematic research and data collection. There is a need for systematic collection of data to test the assumed effectiveness of these programmes. For the time being, it is necessary to act with the best information available today, while it is possible to facilitate the systematic gathering of information so that tomorrow's programmes do better. Governments and civil society should continue their use of SP with open minds, assisting the collection of data and supporting honest programme evaluations in order to improve programmes and spread the most useful innovations.

58

Livelihood protection and support for disaster

David Sanderson

OXFORD BROOKES UNIVERSITY, OXFORD, UK

Introduction

An early, influential conceptualisation of livelihoods was developed by Scoones (1998) and, also building on work by Swift (1989), was concerned primarily with chronic rural poverty. Described by Chambers (2008) as 'a real breakthrough', the conceptualisation sought to combine rural economic concepts concerning food, access, supply and availability. It also sought to describe a relationship between tangible assets (described as stores and resources), intangible assets (claims and access) and livelihood capabilities. The conceptualisation comprised three overlapping circles, representing: claims and access; livelihood; and stores and resources.

The focus on livelihoods was given much greater prominence by Chambers and Conway (1991), whose influential paper 'Sustainable rural livelihoods: practical concepts for the twenty-first century' provided the basis for the development of the sustainable livelihoods (SL) approach in the following decade. The impact of this paper on development thinking was great. Soles-bury (2003) chronicled the development of livelihoods and found a significant impact among development researchers and practitioners. The shift in thinking came at a time when previous dominant theories and practices – particularly those associated with integrated rural development – were losing their intellectual and political attraction. To these ends SL offered a 'fresh approach' (Solesbury 2003: 14).

Within the 1991 paper Chambers and Conway proposed a new definition of livelihood:

> A livelihood comprises the capabilities, assets and activities required for a means of living; a livelihood is sustainable when it can cope with and recover from stress and shocks, maintain or enhance its capabilities and assets, and provide sustainable livelihood opportunities for the next generation.
>
> *(Chambers and Conway 1991: 6)*

Their paper described livelihoods as being based on three areas: capability (drawing in particular on Amartya Sen's 1981 work on famine), equity and sustainability. It also expanded on Scoones' earlier work by adding two significant areas. The first was the description of vulnerability.

Chambers and Conway (1991) argued that vulnerability has two manifestations: internal, relating to the capacity to cope; and external, relating to the ability to withstand shocks and stresses. Shocks were defined as 'impacts which are typically sudden, unpredictable, and traumatic, such as fires, floods and epidemics' (Chambers and Conway 1991: 10). Stresses were defined as 'pressures that are typically continuous and cumulative, predictable and distressing, such as seasonal shortages, rising populations or declining resources' (Chambers and Conway 1991: 10). What is missing from these early formulations are stresses such as chronic low-level corruption, market failures, and poor governance manifested as insecurity of land tenure, land grabbing by powerful elites, urban criminality and extortion by 'muscle men', etc. (see below).

Framing sustainable livelihoods (SL)

Between the mid-1990s until the early 2000s SL was adopted and promoted as an analytical and programming framework by several aid agencies. The adoption of SL was backed in particular by the British Government's Department for International Development (DFID), as well as other agencies including the United Nations Development Programme (UNDP), the Food and Agriculture Organization (FAO), and several non-governmental organisations (NGOs), including in particular CARE and Oxfam.

Each agency sought to develop its own interpretations of SL. Oxfam was one of the first, followed closely by DFID's influential model developed by Carney (1998). Research by Hussein (2002) for the Overseas Development Institute documented the SL frameworks of fifteen agencies (though even more existed), including those of international NGOs, donors and the United Nations (UN) organisations. Almost all frameworks developed during that period comprise similar elements. They were:

- People-centred, wherein households, communities and/or individuals are at the centre of the model;
- Multi-sectoral;
- Multi-level (macro–meso–micro); and
- They all identified a vulnerability context based, for example, on concerns of (rural) seasonality, trends (large-scale changes such as economic growth/decline) and shocks.

They were also:

- Dynamic and flexible (to reflect the dynamic nature of livelihood strategies); and
- They included 'institutions, organisations, policies and legislation that shape livelihoods' (Twigg 2001: 11), known in Carney's framework as Transforming Structures and Processes (TSPs).

Figure 58.1 presents a version of an asset-based sustainable livelihood diagram developed by the present writer while working for CARE International.

Within disaster studies, the access framework developed by Wisner et al. used the concept of livelihood. The access framework examines the causes of unsafe conditions in relation to economic and political processes that allocate assets, income and other resources in society (Wisner et al. 2004: 88–91). The framework describes the nature of livelihood strategies and their role in reducing vulnerability, and includes the household as a central feature, along with structures of domination and social relations.

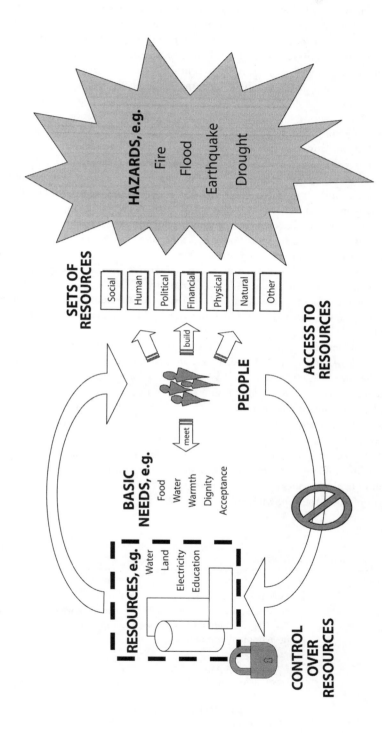

Figure 58.1 An asset-based livelihood framework
Source: Sanderson 2009

In this framework, as with other SL approaches, the starting point is people. People have basic needs, that is, the requirements to sustain life. They include 'food, health services, favourable habitat conditions (potable water and shelter), primary education, and community participation' (Frankenberger 2002: 1). To meet basic needs people access resources. Resources include food, land, water, as well as services, such as health care or electricity.

For a livelihood that seeks to be sustainable, meeting basic needs is not enough. Over time people build up and use assets. These range from belongings and goods to safely built houses and savings, as well as relationships, skills, abilities and political organisation. While there has been much debate regarding which asset types exist, consensus mostly exists on six asset types, often described as soft (or non-tangible) assets and hard (or tangible) assets (see Table 58.1).

Assets can fall under one or more headings – it is not a precise categorisation. For example, land could be considered a natural asset (e.g. for farming), a financial asset as a means of income (e.g. through renting it out), or a physical asset (e.g. for extraction of clay for brick-making). To these ends the fine delineations are not as important as what assets do for improving and stabilising livelihoods. Assets are thus both means and ends – for instance, earning an income is the means by which a livelihood can be more robust, and it is also the end result.

Assets play a pivotal role for reducing vulnerability to hazards and thus in disaster risk reduction (DRR). As Buchanan-Smith and Maxwell (1994) note in their comparison of relief to development approaches, 'the central idea is that household vulnerability is defined by the capacity to manage shocks: some households may be unaffected by shocks, some may recover more quickly, and some may be pushed into irreversible decline' (Buchanan-Smith and Maxwell 1994: 4).

This is the same understanding that lies at the core of other asset/vulnerability approaches, two of which include Anderson and Woodrow's Capacity and Vulnerability Analysis (CVA) (Anderson and Woodrow 1989) and Moser's Asset Vulnerability Framework (AVF) (Moser 1998). Within these approaches assets are the buffer between people and shocks. They are also the means people use to build stronger livelihoods. For example, members of more socially cohesive communities are more likely to work together to reduce the collective risk of a disaster than communities that are less cohesive. In such a case the non-tangible asset is social cohesiveness. This can be seen in the example from the Hoar region of Bangladesh where collective community action in building flood defences and stock-piling animal feed on higher ground reduce vulnerability to flood shocks.

Assets therefore perform the following two key functions:

Table 58.1 Non-tangible and tangible assets

Non-tangible assets	
Human	Skills and abilities, training, knowledge, health
Social	Networks, social groups, family ties, trust, knowledge sharing
Political	Political representation, organised groups for change, e.g. slum development committees
Tangible assets	
Physical	Productive instruments, personal belongings, goods, tradable items, livestock
Financial	Savings, cash, stocks, remittances
Natural	Water, air, land, forest, environment, biodiversity

1 They build capacity, enabling better access to resources. For example, education (human asset) increases the chance of getting a well-paid job and thus to accumulate savings against a shock; group invasion of land by a homeless group (based upon social and political assets) enhances the chance of securing land and thus expands livelihood opportunities.

2 They reduce vulnerability by acting as the buffer between people and external shocks and stresses; the stronger the assets, the less vulnerable people are. An obvious example would be a well-built house being less susceptible to damage in an earthquake. Stronger social ties (social assets) enable easier borrowing and lending of cash and goods in times of hardship.

Importance of social assets

The earlier list of assets comprises 'hard' (tangible) assets such as well-built buildings and 'soft' (intangible) assets such as skills, abilities and the formation of networks. Rural road building is an example of a hard asset (see Box 58.1). Projects and programmes focused on DRR often ignore soft assets, focusing instead on hard assets alone, such as belongings and income.

The significance of soft or intangible asset building can be seen in the following example. After the Kashmir earthquake of 2005 that killed well over 80,000 people, a large NGO relief-to-recovery programme in Pakistan sought to provide a mix of both goods and training (hard and soft assets). Seven programme components were implemented, which included the repopulating of rural areas with livestock, planting trees, rebuilding schools, and skills development. An independent evaluation of the thirty-two-month programme (Hasan et al. 2009) found that efforts to build soft assets (the cheaper programme components) were often more successful than those that concerned the provision of hard assets alone (the more expensive programme elements).

One soft asset component in particular concerned the building of skills of unemployed men, through the provision of Construction Trade Training Centres (CTTCs), which offered training in the building trades such as carpentry and plumbing. Trainees attended a forty-day training session in one trade, and at the end sat an exam. In the life of the programme nearly 2,000 trainees were awarded certificates. In interviews, graduates stated that they would not even have been given a chance to work by their employers had it not been for this training. Training helped them get employment and provided a basis on which more skills could be built through apprenticeship with senior craftsmen. The training helped to enhance the income of graduates and in turn their families. Such income was critical in recovery from the earthquake.

Mobilisation of social assets in the form of networks and well-established relationships can also save lives. Following floods in Catuche, a neighbourhood of Venezuela's capital city, Caracas, Manuel Larreal from the organisation Ecumenical Action–ACT described how:

> the organisation of the neighbourhood and the solidarity of the people saved hundreds of lives … as the flooding progressed, community members mobilised to assist one another. Neighbours who knew each other swiftly communicated the news of the rising water. Older residents were helped from their homes by younger neighbours. When a few were reluctant to leave because they didn't believe the threat or because they were afraid their few possessions could be stolen, neighbours broke down doors and carried people forcibly to safety.
>
> *(Jeffrey 2000, cited in Sanderson 2000: 100)*

In urban Bangladesh, research undertaken by Conticini (2005) with street children in Dhaka uncovered complex arrangements of social and political assets that reduced vulnerability. This included bartering for goods, lending and borrowing money, supporting one another and

David Sanderson

Box 58.1 Roads, livelihoods and landslides in Nepal

K. Sudmeier-Rieux
University of Lausanne, Switzerland

Rural roads in hilly and mountainous Nepal are livelihood links. A road to a distant village may reduce travel time from days on foot to hours by vehicle. They open opportunities for quicker transportation of goods, better access to employment, education, tourism and health opportunities. For better or worse, access to transportation may alter the local economy and type of cultivated goods from a largely subsistence, local market-based economy to cash cropping. Public transportation may provide nearby employment opportunities rather than oblige out-migration. However, roads in hilly areas of Nepal also increase the likelihood of landslides as they often cut through fragile geology, destabilising slopes and altering local hydrological conditions.

Rural road building is often initiated and conducted by communities, largely in co-ordination with local authorities and financed locally or through state funds. In Katahare village, near Dharan city in eastern Nepal a rural road was constructed using collective proceedings from years of membership fees from the community forest user group and local timber sales from the community forest. Unfortunately, the road collapsed during the first rainy season, destabilising slopes and dislodging large boulders which destroyed several sheds and houses.

Further north in Terathum district, near Basantapur of the middle hills range in eastern Nepal, a road built fifteen years ago by the Asian Development Bank connected the district headquarters, Manglung, to the main Dharan road. As a consequence, farmers began growing cabbage, cauliflower and cardamom as cash crops, dependent on transportation to local and regional markets. Fifteen years later, the road is beginning to collapse and a major landslide has already destroyed a number of houses, several hectares of fields and is threatening to destroy several others. Geologists had surveyed the road when it was initially constructed, but inadequate measures were taken to stabilise and drain the road. Farmers are increasingly concerned that they will soon no longer be able to bring their crops to market. Should the road collapse, their options will be limited to carrying out crops manually, which is a very labour-intensive task in steep terrain.

Landslides cause on average over 300 reported deaths per year in Nepal and severely impact rural livelihoods by destroying fields and houses, wiping out roads and damming rivers. The number of landslides is expected to increase with more intense rainfall patterns and the increased number of locally constructed roads. To minimise future casualties and economic losses, Practical Action, Nepal has initiated a gravity roping cable car initiative for transporting goods to villages in steep areas as an alternative to road building. It is also exploring how roads must be constructed to handle more extensive drought periods with heavier rainfall. Cost-effective bio-engineering methods such as bamboo plantations, gabion walls, drainage and geo-textiles have the potential to stabilise many newly constructed roads, yet unfortunately these techniques are often under-used locally due to poor dissemination of information and communication between the Department of Roads, communities and local authorities. In order for rural roads to remain effective livelihood links, improved collaboration with communities is required to ensure sustainable road construction and maintenance.

agreeing leaders by consensus. Conticini also asked street children to identify and prioritise their own asset list. They identified eight asset types that were important to them. These comprised: feelings of love and trusted friends; co-operation; money (including savings, remittances, debt and credit); work and play; food; health; knowledge; and use of space.

Accessing resources and tackling discrimination

To build assets people need to access resources. Broadly speaking, *resources* comprise those things necessary to produce their livelihoods and to meet basic needs. *Access* concerns how households acquire the resources needed. Wisner *et al.* (2004) present a detailed framework for analysing the factors that influence the access a household has to labour power, knowledge and skills, tools and livestock, land and other natural resources, as well as to marketing and re-investment opportunities. In urban areas the framework must be adapted to give prominence to the sale of labour to buy goods and services and is more likely to include using cash.

According to both urban and rural SL frameworks, two conditions help or hinder access to resources: local social discrimination in determining access and more distant (e.g. governmental or corporate) control over resources.

Discrimination in determining access

Discrimination relates to 'position in society', and it may be positive or negative. For those who are vulnerable it is almost always negative, and may be based on gender, age, disability or ethnicity. In essence, this concerns 'who you are', and how this affects ability to access resources.

Research by Narayan *et al.* (2000) asked communities in Dhaka for their perceptions of discrimination. They found that in some communities poor people are further divided into sub-categories that include the helpless poor and the hated or bottom poor. They found that the 'hated poor' were the most vulnerable: 'the hated poor ... were terms used to describe households, usually headed by women or elderly men that have no income-earning members. Disabled people are also among the hated poor. Members of these households often starve. Lacking land and other assets, they do not have access to loans, even from family or friends. In addition, they are not accepted as members of local organisations, and thus cannot benefit from group assistance as a last resort' (Narayan *et al.* 2000:120–1). The 'helpless poor' were 'identifiable by their old clothes and pained faces' and were classified as those unable to afford health care or education.

Resources are invariably controlled, e.g. by governments, local authorities, tribal chiefs, markets, mafia, private sector utilities. Forms of control include legislation, rule of law and, particularly in urban areas, organised crime. Famously, organised crime in the USA grew rapidly during the Prohibition years of 1920–33 as highly organised gangs managed and controlled the distribution and selling of illegal alcohol.

Governance and control over resources

National and sometimes local governments exert control over the use of urban and rural land and natural resources such as water, forests, fisheries and minerals. Resource access is governed both by legislation and its enforcement, and also by failure or intentional lack of enforcement. The latter takes place, for example, when a rich individual bribes a forestry official and hauls off lorry loads of timber despite laws against such a practice. Such controls exist also in urban areas, where space and location are themselves resources and their access is controlled. In addition, as Beall and

Kanji (1999) note, 'The urban poor are linked into structures of governance through their dependence on the delivery of infrastructure and services by city institutions, as well as through the impact of meso- and macro-level policies' (Beall and Kanji 1999: 3).

The qualities of governance and vulnerability are closely linked – almost always, the weaker the system of governance the more vulnerable the population. A stark example is provided by the Haiti earthquake of 12 January 2010, which at 7.0 on the Richter scale killed 223,000 people. Ranked at 149 out of 182 countries on the 2009 Human Development Index, Haiti before the earthquake was the poorest country in the Western hemisphere, with some seventy-eight per cent of its population living on under US$2 a day.

What made the earthquake so devastating is the complete lack of land-use planning and enforcement of building safety codes. Shortly after the disaster an even larger-magnitude earthquake occurred in Chile, but a very small number of people were killed because governance had insisted on safe building design and construction. Governance in Haiti had been better, but had declined over decades of despotism. Michele Wucker, executive director of the World Policy Institute, cited many of the causes of this decline in the ability effectively to govern:

> By the time the U.S. pulled out in 1934, Haiti's own institutions had atrophied ... The Duvaliers (who ruled Haiti from 1957 to 1986) left Haiti economically decimated. A large number of educated professionals left the country during the Duvalier regimes, and the period that followed was so unstable, it was hard to lay down roots and build infrastructure. International investment was limited because it was an unreliable business environment.
>
> *(Wucker 2010)*

Using the SL approach in disaster risk reduction

The SL approach provides a tool for undertaking analysis and for deciding on programming actions. A mantra often associated with livelihoods is the need to undertake holistic analysis followed by focused strategies.

Attempting a holistic analysis

SL provides a menu of issues to consider in any analysis of a specific rural or urban situation. SL invites an analysis of the range of issues and factors that contribute towards vulnerability (Wisner et al. 2004). Livelihood-based analysis has been used in a variety of instances. In Ethiopia the Livelihoods Integration Unit (LIU) within the Ministry of Agriculture and Rural Development uses an SL approach to map hazard impacts on vulnerable households (Boudreau 2009). The Ethiopian LIU coined the phrase 'people-centred risk assessment', and this approach is centred on the need 'to encapsulate an understanding of how households survive and, implicitly, what hazards will affect them'. The approach focuses in particular on food security and depends on the gathering of baseline information on how people obtain food, their food expenditure, their access to goods and services and their coping capacity.

A challenge to holistic understanding is the sheer complexity and dynamism of people's livelihoods. For example, having suffered from post-volcanic eruption debris flows, farmers in Comoros turned the new sand deposits into a livelihood resource (see Box 58.2).

In Kosovo the SL approach provided the basis for a participatory assessment undertaken for the UN Mission in Kosovo (UNMIK) in 1999–2000. Although this was a post-conflict situation, the SL approach used is nearly identical to those one may apply in post-disaster circumstances, and the Kosovo experience is so rich that it deserves to be studied by disaster professionals.

Participatory rapid appraisal (PRA) tools were used in the Kosovo case to gather information. PRA tools, including mapping and Venn diagrams to highlight relationships as perceived locally, were used to gather information on each livelihood element (see Table 58.2) (see Chapter 64).

The assessment was carried out by trained teams of Kosovans who undertook PRA exercises in fifteen villages and three urban settlements throughout Kosovo. PRA exercises for each settlement lasted three days and included a sequence of activities that included mapping, transect walks, wealth ranking and power analysis. The resulting analysis provided UNMIK with information to plan subsequent government and UN activities.

According to Westley and Mikhalev (2002), the SL approach was chosen as the methodological basis for analysing a conflict/post-conflict setting due to its dynamic nature, i.e. 'the framework allows an understanding of how household livelihoods change over time. Why and how do some households maintain assets and livelihoods during conflicts while others do not? How resilient are households to conflict and what makes them so?' (Westley and Mikhalev 2002: 6). The direct applicability to DRR and to post-disaster recovery is obvious.

Box 58.2 Lahar hazard and livelihood strategies on the foot slopes of Mt Karthala volcano, Comoros

Julie Morin
Université de la Réunion, France

JC Gaillard
School of Environment, The University of Auckland, New Zealand

Vouvouni is a village located on the foot slopes of Mt Karthala volcano, Comoros, where 4,000 people struggle to sustain their daily needs through limited resources from overseas remittances, small informal businesses, craftwork, construction work and subsistence farming.

As a consequence of the April 2005 eruption of Mt Karthala, lahars inundated Vouvouni, buried several houses and affected scores of other families (see Chapter 28). Despite the scope of the damage, many people realised that the large amount of sandy deposits brought by the lahar might easily turn into a valuable resource as the demand for construction materials is important in Comoros. Indeed many people face the social obligation to build sturdy houses to comply with the requirements of the 'Grand Mariage', a rite of passage considered a prerequisite to step up in Comorian society.

The owners of the two lots with the largest amount of lahar deposit initially prevented access to their field to secure control over the sand. In response, some villagers diverted permanent river channels towards their own property to get their share of the resource, although they were conscious that this might endanger their lives. As expected, the diversion of the rivers led lahars to invade the entire village and affected hundreds of people. Many had to abandon their houses while others lost precious crops buried by lahar deposits.

Yet sand was now plentiful and freely accessible in many private and public spaces. Furthermore, extracting sand requires few skills and utensils so that people who could not farm anymore were able to earn comfortable incomes, at least for a time. For those not directly affected, lahar materials became an important additional source of income. For example, students started to extract sand during school holidays. In most cases, people sold sand to wealthier local traders who eventually retailed construction materials with significant benefits. In late 2005 the entire community of Vouvouni was thus involved in sand extraction.

However, the increase in sand supply led to the market being quickly saturated. Simultaneously, pipes in toilets, bathrooms and kitchens were quickly filled up by sand deposits resulting in frequent dirty flooding in the house courtyards. Rubbish also spread as a consequence of flood and lahar events, and contributed to the outbreaks of gastro-enteritis. The economic and social cost of lahar diversion and sand extraction quickly turned into a serious burden for the local community and the number of households relying on sand extraction eventually decreased from fifty in 2006 to only twenty in 2009.

The shifting perception of lahars from resource back to threat became evident in 2008 and 2009 when villagers pooled their labour to build levees to protect the village. Yet the levees did not withstand the 2009 lahar season, which engulfed Vouvouni and brought serious damage again. As an ultimate alternative, people eventually decided to engage in small-scale personal protection measures such as sandbagging and the construction of small rock walls in front of the houses. Most of the households also raised their belongings onto small plots. The next rainy season will tell if these measures will suffice to bring safety to Vouvouni.

Attempting to focus actions

While SL is holistic, identifying and mapping complex issues and related factors, SL-based interventions and actions must also be focused. There is a tension here between theory and practice that one commonly observes in community development work. However, 'focused' does not necessarily mean limited; rather it is essential that actions be thoughtful, well-timed interventions that are made with an understanding of the likely impacts that the action will have on the many interconnected parts of a livelihood system. Local knowledge and analytical capacity are essential in order to apply SL concretely to specific situations (see Chapter 59).

An example helps to clarify this tension and its resolution. In 2008 CARE Bangladesh, a large NGO, identified as one of its four strategic directions the need to work with the most marginalised in urban areas. Following extensive consultation, the impact they wanted to achieve was that 'the most marginalised groups in urban areas have secure and more viable livelihoods and are increasingly treated as equal citizens by the state and society'. This was in response to the relatively neglected sector of urban poverty and burgeoning demand of rapidly increasing vulnerability of poorer people in urban Bangladesh.

To these ends CARE adopted an SL approach to undertake a holistic analysis, followed by the development of a focused strategy for interventions. A key issue concerned identifying the causes of marginalisation of people, two of which are:

1   *Weak instruments of governance delivery.* The rule of law was undermined by corruption. In particular *Maastans*, or musclemen, abused and exploited the urban poor without hindrance from the police. Slum-dwellers in Bangladesh enjoy little assistance from law enforcement agencies and they face extortion (from *Maastans*) and fear of physical harm or eviction if payoffs are not made (World Bank–Urban Development Unit South Asia Region 2006).
2   *Social and economic exclusion.* The urban poor were regularly excluded from society and from opportunities to earn an income. The marginalised suffer not only physically, through lack of shelter and regular beatings, etc., but also psychologically through abuse and dismissal as

Table 58.2 Information to be collected, tools used in relation to stakeholder themes

Livelihood component	Relevance of livelihood components	Principal PRA tool	Secondary PRA tools
Context			
Institutions	Presence and importance of community-level institutions; interaction of population with external institutions; control of resources by institutions	Venn diagram	Household interviews; focus group discussions
Natural resources	Food economy zone; presence of common property resources; availability and access to natural resources; access to land	Area mapping	Secondary data; key informants
Infrastructure	Availability of education, health, social services; water and sanitation infrastructure; roads and transport infrastructure	Area mapping	Venn diagram; interviews; secondary data
Culture	Ethnicity, religion and gender	Secondary data	Livelihood profile; interviews
Political context	Broader political context; access to voting; household and community decision-making	Secondary data	Venn diagram; interviews; key informants
Settlement patterns	Migration and settlement; perceptions of vulnerability and risk	Key informant interview	Mapping; interviews
Processes (rules, regulations, etc.)	Impact of rules, regulations and policies on households and communities; potential impact of taxation; access to passports; institutions at community level; participation	Venn diagram	Secondary data; interviews; key informants
Household Assets			
Social	Exchanges of goods and services; assistance to or from extended family networks; membership in community groups; gender	Household interview	Livelihood profile
Physical	Housing; agricultural implements; vehicles; machinery; shops; household-level water and sanitation facilities; water and sanitation; food security	Household interview	Livelihood profile
Human	Education level; ability to work; dependency ratio; education; health; gender; skills and uses	Household interview	Livelihood profile
Financial	Livestock; savings; remittances; access to credit; access to finance; household economy; remittances; pension; food security and agriculture	Household interview	Livelihood profile

Continued on next page

David Sanderson

Table 58.2 (continued)

Livelihood component	Relevance of livelihood components	Principal PRA tool	Secondary PRA tools
Political	Belonging to affiliations; linkages to formal and informal power structures	Household interview	Livelihood profile
Natural	Land; access to common property resources; distribution of poverty within communities; land holding	Household interview	Livelihood profile
Vulnerability to shocks and stresses	Susceptibility to hazards and/or other phenomena; strength and use of household and community soft assets, e.g. networks, affiliations, sense of identity, skills; possession and quality of hard assets, e.g. well-built buildings, savings, belongings to use/trade	Household interviews	Economic activity analysis; livelihood profile; interviews

Source: Westley and Sanderson 1999

being unimportant people. For them, their plight was consolidated through regular social confirmation that their position was at the bottom. Narayan *et al.* (2000) asked women their views on exclusion and the plight of the 'bottom poor'. They found that women in this marginal position said that they were disregarded and were never invited to social events. Furthermore, they explained that such neglect caused lack of attention by government functionaries so that the urban poor did not receive government assistance (see Chapter 38).

Three manifestations of marginalisation were an inability to meet basic needs, including 'food, shelter, health, education and freedom of speech' (Islam quoted in CARE Bangladesh 2007: 11); an inability to build and use assets, in particular soft assets such as social ties; and powerlessness, in particular the most marginalised being excluded from the freedoms and choices that others take for granted.

In developing its strategy, CARE identified three overlapping and mutually reinforcing sets of activities. The first was the immediate requirement of meeting basic needs. Programmes to meet these needs included soup kitchens, needle exchange for drug users, provision of medical assistance, provision of clothing and basic sanitation. The second set of activities tried to build up personal and group assets (in particular soft assets such as networking and development of skills). These activities included skills training, microfinance, education and community mobilisation. The third set of activities addressed the underlying causes of poverty such as faulty governance at all levels with its attendant unfairness, corruption, crime and violence, as well as its entrenched discriminatory attitudes. Activities included policy engagement, advocacy and attempts to influence the attitudes of government officials.

The CARE strategy is summarised in Figure 58.2. The horizontal axis indicates the graduation of interventions from meeting basic needs through provision of services to gradual empowerment. The vertical axis indicates the level of intervention, from meeting immediate basic needs to addressing the underlying causes, and engagement with government using success in provision of material needs as a means of gaining credibility in dialogue with officials.

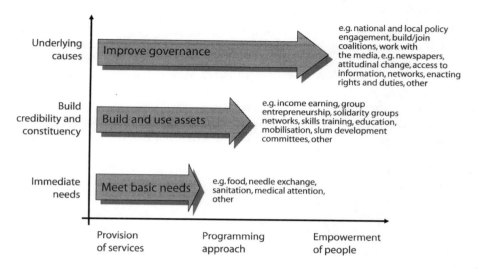

Figure 58.2 CARE Bangladesh programming strategy

The decline of SL approaches?

The SL approach dominated the development agenda from the late 1990s until the early 2000s. However, by the mid-2000s it had been dropped by many of its chief advocates, including DFID, UNDP, Oxfam and CARE (Pinder 2008). By contrast, among many field-based NGOs and academics, SL continues to be widely used as a tool for programming and analysis.

There are several reasons why support and interest in livelihoods declined among large institutions. A longstanding critique of SL is that it is thought to be hard to implement, especially by large, bureaucratic institutions. SL is holistic (recognising the complexity of interlinked factors) and multi-sectoral (beginning with people in the centre, rather than the structures of implementing agencies). So, some large institutions wondered how SL could be used effectively, when their agencies are organised by sector and are specialist by history or mandate. Pinder (2008) identified other criticisms of SL relating to implementation, that SL focuses on the micro level (that is, is people-centred) and this makes the scaling-up of projects to macro level and the use of market mechanisms difficult. This last point is significant because the decade of the 2000s witnessed a massive growth of pro-market, neo-liberal thinking among many governments and in the large, international NGOs that function as conduits and vehicles for bilateral development assistance.

Pinder (2008) concludes, 'the main constraint on the wider application of SL is its assumed micro-nature at a time when most multi- and bi-lateral development agencies have moved, and are continuing to move, towards support for international and national development frameworks, for example through direct/general budgetary support, and the Aid Effectiveness Agenda' (Pinder 2008: 3). Such a move toward general budget support, even for weak and corrupt governments, precludes use of SL analysis which usually includes assessment of governance, corruption and discrimination. Further, Westley and Mikhalev (2002) record that, 'NGOs seemed more at ease with the SL framework and able to see how it could apply to their own strategies and programmes. Other agencies may have preferred a more quantitative approach or better macro-analysis of social factors' (Westley and Mikhalev 2002: 6).

Resurgent sustainable livelihood approach?

The decade beginning with 2010 has seen increasing interest in bridging development, DRR and climate change issues (O'Brien *et al.* 2008). Such linkage is facilitated through the SL focus. So-called climate science desires to 'downscale' its analysis and recommendations from global circulation models to regions, and eventually to national and sub-national situations. However, without corresponding study and action from 'the bottom up', evidence-based climate change adaptation (CCA), DRR and economic/social development will remain an impossible ideal. SL would seem to have a natural niche in helping to facilitate such linkage across scales and among sectors. However, against this more optimistic view, research undertaken by Wamsler regarding urban disaster management led her to conclude that, 'Although the concept of sustainable livelihoods has the potential to bring together disaster and development people, the interviews revealed that, in general, disaster people do not apply this concept' (Wamsler 2007c: 155). Thus it may be that the use of SL in urban DRR and corresponding linkages with CCA and development will proceed faster in rural areas and must still develop more appropriate frameworks in urban areas (Pelling and Wisner 2009).

Conclusions

While SL has been 'off the agenda' in recent years for some larger international NGOs, inter-governmental agencies and donors, it is still used by many national NGOs, and continues to provide a powerful approach basis for holistic analysis and focused strategies that span both disaster management and developmental understandings. Most important of all, SL puts vulnerable people at the centre and reminds practitioners, policy-makers and researchers that, without remembering this, development- and disaster management-related interventions are ultimately of little effect.

For researchers, SL presents a good basis for analysis, forcing a consideration of the links between issues that reduce vulnerability and increase capacity, ranging from the micro, household level of everyday issues concerned with savings, relationships, skills and politics, to the macro, policy-level environment of governments and other bodies that hold the 'reins of power'. For practitioners, SL makes smooth and natural links between relief and development, and reminds programmers that all interventions have a knock-on effect, and are ultimately linked to efforts that either contribute towards or erode the goal of a sustainable livelihood – for example, in programming terms shifting quickly from meeting basic needs, which does not build a livelihood, to building assets and addressing governance and discrimination, which does.

Finally, for donors and governments there is the opportunity, in addition to considering the above, to reconsider and to readopt SL as a holistic analytical and programming tool that is deliberately not oriented towards institutional silo-based delivery, but rather is cross-cutting and based on the everyday needs and abilities of people themselves.

59

Community action and disaster

Zenaida Delica-Willison

SPECIAL UNIT FOR SOUTH–SOUTH COOPERATION, UNITED NATIONS DEVELOPMENT PROGRAMME–ASIA-PACIFIC REGIONAL
CENTRE, BANGKOK, THAILAND

JC Gaillard

SCHOOL OF ENVIRONMENT, THE UNIVERSITY OF AUCKLAND, NEW ZEALAND

Communities, participation and empowerment

This chapter is about communities and their roles in disaster risk reduction and management. The concept of community is used in very different, sometimes loose and often contested ways (e.g. Walmsley 2006). The concept's origin can be traced to the German sociological literature on *Gemeinschaft* in the late nineteenth century. Community then referred to a small aggregate of people, often located away from centres of power but sharing a common and continuous way of life, similar beliefs, close ties, trust and frequent interactions. In the twentieth century the concept spread to a large array of social sciences, which eventually had a tremendous influence on policy and practice in the areas of development studies and disaster risk reduction (DRR). Since the 1960s, development and disaster workers have often considered community-scale initiatives as an alternative or necessary complement to globalised top-down development and disaster management practices.

The concept of community has thus come to bear a positive meaning. Invoking community emphasises the local dimension of development and disaster reduction. It further suggests that ordinary people are capable of finding collective solutions to their problems. In that sense the concept of community is most frequently associated with that of participation (see Chapter 64). Since the 1960s many development and disaster workers have indeed been advocating for an increasing involvement of local people, especially those often marginalised, in the assessment of the problems they face, in the search for solutions to these problems (e.g. Burke 1968; Chambers 1983). Taking this view, the implication is that increasing communities' participation in development and DRR should ultimately lead to people's empowerment.

People's participation and empowerment have been seen as real only if development and DRR activities are community-based. This bottom-up alternative paradigm for development and DRR gained wide ground in the 1980s and 1990s with such paradigms as community-based natural resource management, community-based coastal management, community-based forest management and community-based disaster risk reduction and management (Adams 2001). Community-based activities and people's participation yet required the development of specific tools and methodological frameworks such as Participatory Rural Appraisal (PRA) and Rapid Rural Appraisal (RRA) (see Chapter 64). These tools and methodological frameworks have been crucial and symbolic of the emergence of community-based approaches to development and DRR.

The growth and increasing influence of this community-based, participatory paradigm did not go without critics. Some researchers and practitioners have emphasised the drawback of people's participation through the so-called 'tyranny of participation' or 'myth of community' discourses (e.g. Cooke and Kothari 2001; Guijt and Shah 1998). These critics stress participation becoming an end in itself rather than a means to an end, thus leading to populist approaches and community exploitation. Some organisations involved in development and DRR projects refer to community-based participatory projects to seek funding and please policy-makers for whom community and participation have become the key concepts. Yet many of these organisations either lack the ethical commitment, the ability or the time to run participatory activities with necessary respect for people's needs and cultural preferences. Some researchers actually question the relevance of participatory tools and methodological frameworks as they allegedly fail effectively to address issues of power and diversity within the community. Communities are indeed rarely homogenous, in gender, age, ethnicity, castes, religious affiliation, etc., and always include power relationships between people with resources and those who lack access to these resources, which are often difficult to appraise for outsiders (e.g. Guijt and Shah 1998).

The role of communities in disasters

Disasters are local events that first and foremost affect local communities. Since the local people are the ones affected when disasters occur, they become the first responders at the household and at the community level. Both the scientific and practice literature acknowledge the capacities of local communities to respond to natural hazards on their own, as long as they are empowered with adequate resources (e.g. Delica-Willison and Willison 2004; Quarantelli and Dynes 1972). Local people rescue, evacuate and provide assistance to less capable and differently able members of their communities. For example, eighty-five per cent of the survivors of the 1985 Mexico earthquakes were rescued by their friends, kin or neighbours, who were on the spot at the time of the event (Quarantelli 1986–87). Those rescued with large media coverage by international rescue teams coming over hours or days after a disaster are a very small minority. Local people can do more than react and respond to disasters. They also can prepare and, above all, plan and act so that disaster risk is reduced. The question is how to build the capacity of the local people so that they are able to assess the risk beforehand, and identify, prioritise, plan and implement risk reduction measures at the community level.

Shortly after a disaster, affected people also engage very quickly in recovery and reconstruction, often much faster than do aid organisations (Davis 1978). People are usually eager to recover as soon as possible and thus display a great potential for action. They usually gather fallen materials and start over, building shelters for their families and neighbours. They often care for their neighbours and kin. For instance, the Ashapura Nagar slum, in Bhuj, India, had been there for 100 years. It was devastated by the Gujarat earthquake on 26 January 2001. Immediately after the earthquake, people came together to establish a community kitchen and survived with it for fifteen days without external relief assistance. None of the external agencies and municipal authorities was aware of any damage or casualty in this community of 100 families, as this was an unauthorised settlement on government land (AIDMI 2009a).

Although people and communities are able to handle many tasks in responding to disasters, they often need external assistance too. Local communities are unable to undertake timely, long-range, massive evacuation, clean-up very heavy debris or undertake complex medical operations. The people thus have to partner with external institutions in the management of aid. When not impacted by the disaster themselves, local institutions are frequently better able to channel this assistance because their spatial proximity makes them more reactive. Furthermore, they usually better understand the needs of the people and the local social and cultural context. Unfortunately, providers of outside assistance frequently overlook the ability of both communities and local institutions to face disasters. They may provide the wrong support or provide unnecessary support, which prevents people's initiatives or becomes redundant, and leads to social disruption and economic market breakdown. It is sometimes hard for external agents to understand that local people are not merely recipients of external assistance, but also can be the channels for this help. They are not passive, but active participants in the implementation of DRR and post-disaster reconstruction activities.

Nobody is more interested in reducing the risk of disasters than local people whose lives and livelihoods are threatened. Local communities are in a better position and have more incentive to plan and implement DRR actions for their own safety than a central government. DRR is not just about physical structures, but includes livelihood interventions, legal assistance measures, proper urban planning and urban management. More often the expertise needed exists in the locality. Support from outside is necessary, but local control is important (Maskrey 1989). For example, in Raidhanpar village, India, after the Gujarat earthquake, many organisations were willing to provide assistance but most of them had their own plans and priorities, which did not match those of the people (AIDMI 2009a).

The progressive recognition of people's and communities' ability to face disasters along with a shifting view of the causes of disasters among practitioners and eventually policy-makers and scientists led to the emergence of the concept of community-based disaster risk reduction and management.

The emergence of community-based disaster risk reduction

For decades, institutional disaster management practice worldwide has been predominantly employing a top-down strategy. The assumption was that simple exposure to natural hazards constitutes risk. However, civil society organisations that were working with poor communities observed that apart from the occurrence of hazards and people's proximity and exposure to them, people also suffered because of prevailing socio-economic and political conditions that made them vulnerable to these natural hazards. They realised that natural hazards were often inevitable but disasters were not. This insight led to an emphasis on detailed analysis of their situation by communities at risk. Work on such community-generated analysis and action led slowly to the formal development of an approach now known as community-based disaster risk reduction (CBDRR).

Practice of CBDRR emerged in the 1970s and was eventually formalised and widely promoted in the 1980s through the creation of national and international networks of non-governmental organisations (NGOs) and civil society organisations involved in grassroots activities. In the Philippines, a nationwide network of NGOs called the Citizens Disaster Response Network was created back in the 1980s. The network propagated people's participation and put a premium on the organisational capacity of the vulnerable sectors through the formation of grassroots disaster response organisations (Heijmans and Victoria 2001). In Latin America the *Red de Estudios Sociales en Prevención de Desastres en América Latina* (LA RED) was established in 1992 to build evidence from local disaster experiences in order to advocate change in national and international policies that dealt with relief, preparedness and scientific research (Heijmans 2009). In countries such as Peru a community-based approach had been used since the early 1980s (Maskrey 1989). A network of organisations and individuals was also established in South Asia in 1994, initiated by the Intermediate Technology Development Group (ITDG), which is now called Practical Action. The network Duryog Nivaran, which means 'disaster mitigation', aims to reduce local communities' exposure and vulnerability to disasters and conflict by promoting an 'alternative perspective' at the level of concept, policy and implementation on CBDRR and development programmes in South Asia (Ariyabandu 1999). In Africa, Partners Enhancing Resilience to People Exposed to Risks (Periperi) is an umbrella organisation that was created in 2006 for a number of risk-reducing initiatives in Africa concerning livelihoods and urban vulnerability.

Many governments have now come to recognise the importance of CBDRR. In the Philippines, the National Disaster Coordinating Council in the Philippines, in collaboration with the Center for Disaster Preparedness, organised a nationwide CBDRR conference in 2003. The government of Vietnam, with funding from the World Bank and Japan, piloted CBDRR in ten communes in three provinces and with additional funding from the Dutch government scaled-up the CBDRR process to 100 communes. With the support of various governments, banks and NGOs, the government aimed to reach 1,000 communes in 2010 and 6,000 communes in 2020. More recently, the Ministry of Civil Affairs of the People's Republic of China held a workshop on CBDRR: Policy and Practice in the Province of Guangyuan. The royal government of Bhutan has just recently started the adoption of CBDRR nationwide by introducing participatory planning at the village up to district levels. The results will be incorporated into the national disaster risk reduction plan. CBDRR is part of the country's national disaster management framework.

The United Nations Development Programme (UNDP) and other UN agencies have also supported CBDRR in their programme countries. Specifically, the Special Unit for South–South Cooperation (SUSSC) of the UNDP has started the 'South–South Citizenry Based Development Academy' to promote the sharing and replication of best CBDRR practices. The first Academy was held in Bhuj, India, and the second was in Banda Aceh, Indonesia. Training on CBDRR has helped in its promotion. In 1997 the Asian Disaster Preparedness Center (ADPC) provided a home for CBDRR by offering the first course in community-based disaster management. A topic on CBDRR is also included in a course on 'disaster risk reduction and sustainable local development' offered by the International Training Center of the International Labour Organization (ILO) in Turin, Italy, in partnership with the UN International Strategy for Disaster Reduction (UNISDR) and the SUSSC.

CBDRR: theory and practice

CBDRR consists of self-developed, culturally and socially acceptable, economically and politically feasible ways of coping with and avoiding crises related to natural hazards (e.g. Anderson and Woodrow 1989; Maskrey 1989). It enhances local resources and may provide mediation or brokerage that also allows access to outside resources and technical means for dealing with disaster risk without perpetuating a cycle of dependency. CBDRR relies on three crucial principles: people's participation and empowerment, development-oriented activities and a multi-stakeholder approach.

People's participation and empowerment

The communities that engage in DRR are not homogenous. They have varied and diverse interests, views, experiences, risk perception and background. Communities include various groups of people that may be differentiated according to class and caste, occupation, age, gender and, not uncommonly, ethnicity and religion. Multi-ethnic and multi-faith communities are more and more common in rural and urban places around the world due to accelerated migration, displacement and mobility. People in such communities can be further categorised into more- and less-vulnerable groups. Although all face the same hazards and, if they have an interest in DRR, should have a common goal of reducing disaster risks, some are more vulnerable to injury and loss than others. In practice, community-based activities often have a bias in favour of one group or another within a community: at one extreme the local elite may want to protect themselves; at the other there may be a deliberate focusing on the poor and most vulnerable. A good deal of negotiation and sophisticated understanding of local politics is often employed so that a community organisation can win the support of people who are less vulnerable in order to help the poorest sector of the community. In such situations a moral case may be made or, more commonly, DRR activities are wrapped up in and integrated with development activities that benefit all (see below).

People affected by the 1998 East Gippsland floods in Australia had very diverse needs. Traditional farmers were in need of solutions to dispose of dead stock, repair farm assets and replace lost stock. Conversely, some families that had chosen to live an alternative lifestyle based on subsistence farming had limited need for farm support. In urban areas, needs pertained to health and child care. There were also some significant differences between age groups and family origins. To assess those varied needs and provide adequate support the government engaged in a dialogue with the locals and some members of the community were actually enrolled as community development officers during the post-disaster recovery stage. As a consequence, the assistance provided to the communities affected by the floods matched local needs (Buckle 1998–99).

In CBDRR, goals, objectives and activities are rooted in the people's understanding of disaster risks and their priorities. When people are aware of their level of exposure to hazards, they have taken the first step. They still need to identify and implement their priority DRR measures. Implementation may require resources, information and knowledge or political access. Yet whatever the level of outside inputs to this process, the community remains the locus of activity. 'Community-based' means community ownership of activities, processes and outcomes. In essence, the intention of CBDRR is not merely projects and programmes that are located in a community, but risk reduction measures that are managed by the people themselves in a community. The process may have been facilitated by outside agencies, but the local people are the prime movers in reducing disaster risks in their community.

In Afghanistan, many communities of Kapisar and Kandahar that are prone to earthquakes, droughts, floods and landslides live in isolation up in the mountains (Moss 2009). Yet most of the households have a radio set, which has become the focal point for DRR programmes initiated by the NGO Tearfund in collaboration with the government and radio stations. Radio programmes are not only used to convey messages to reduce the risk of disasters, but are actually broadcast by the communities themselves. Radio clubs were organised and trained in different villages. These clubs include a number of women to ensure that their voice is not left unheard. Local communities have thus been able to craft radio programmes to answer their own needs and lobby local governments to support DRR. Such effort has resulted in a strong feeling of community ownership of the DRR activities.

Development-oriented activities

Beyond people's participation, CBDRR generally takes place in the context of development-oriented actions. In this respect, interventions do not simply seek relief from disasters. Neither do actions merely pursue a vicious cycle of restoration of 'normal conditions' present before disasters. These pre-existing socio-economic and political conditions produce the vulnerability that forms part of the equation that results in disaster in the first place (Wisner *et al.* 2004). CBDRR aims to address the root causes of vulnerabilities and increase the communities' capacities. The short- and medium-term goal is to minimise human, property and environmental losses, limit social and economic disruption and enjoy the benefits of a secure and safe environment. For this purpose, development is considered both a means and an end. As an end in itself, development moves the community toward the attainment of a just, prosperous and humane situation wherein people become capable and secure (Sen 1999).

Investing in small infrastructures such as irrigation facilities, access roads and sturdy housing, and providing basic services like water supply, health care, etc. to communities contributes to improving people's everyday life while reducing their vulnerability in facing natural hazards. For example, in the district of Phalombe in Malawi, local communities supported by national and international NGOs organised to channel spring water towards their fields where they grow maize and other crops. In complement, farmers built grain banks to store the surplus of cereals that resulted from increasing irrigated production. This simple irrigation system provides farmers with additional food and stocks, which helps them face frequent droughts, prevents food insecurity, and helps them cope with hoarding by capitalist traders (Moss 2009).

In northwestern Bangladesh, the NGO Practical Action initiated a livelihood diversification project, which strengthened community cohesion and social bonding. Members of the local community were involved in the decision-making and planning process. As a result, the livelihoods of community members have been strengthened and diversified (agriculture, fisheries and livestock) and communities have increased capacity to make informed decisions regarding their own well-being and increased self-confidence among the poorest and most vulnerable families. Henceforth, stronger livelihoods mean less vulnerability and greater capacity to face the impact of natural hazards (Practical Action 2007).

As these examples suggest, CBDRR ideally strengthens people's livelihoods and makes them more sustainable, resistant and diverse (see Chapter 14).

CBDRR as a multi-stakeholder approach

Implementation of CBDRR on a large scale requires the support of local and national governments. Since local government often provides basic services to communities and primary assistance in time of disaster, it is crucial that local authorities are involved in both the assessment of the vulnerabilities and needs of communities. In addition, a directive approach from the top by government is necessary to enforce laws and regulations that protect the natural environment, promote public health and oppose discriminatory practices – without which people's vulnerability to hazards usually increases. Decision- and policy-makers should welcome the CBDRR successes and provide support mechanisms to scale them up, such as investing in the communities' social resources, local DRR planning, appropriate management structures, and implementation and co-ordination mechanisms.

Around the UK, people in communities affected by, or with the possibility of being affected by, flooding have set up locally run flood action groups under a variety of names, to understand, communicate and advise on flood topics. Actions have included lobbying politicians, informing residents about flood topics, taking legal action against flood-related decisions with which they disagree and providing support to flooded people dealing with their insurance company or other organisations. Such local actions have been scaled up to the national level through the National Flood Forum. Created as a charity by people affected by flooding, the

Box 59.1 From disaster response to poverty alleviation: a community initiative to reduce the risk of disasters in the Philippines

Noli Abinales
Buklod Tao, San Mateo, The Philippines

In May 1996 the community-based organisation Buklod Tao emerged victorious against a construction company that planned to establish a cement batching plant on agricultural land near the village of Banaba, San Mateo, Rizal Province. However, damage to the environment had already been done, and as a result the community became more exposed to flooding. The members of Buklod Tao needed to search for a way to reduce their vulnerability to flooding and other hazards. The community-based organisation submitted a modest project proposal to the Royal Netherlands Embassy in Manila, with a component in establishing a community-based disaster management (CBDM) programme.

In June 1997, using the first tranche of the project funding, Buklod Tao conducted its own seminar workshop on CBDM. In the course of that forum the community was organised into three CBDM teams by geographical areas: (1) the north Libis area; (2) the south Libis area; and (3) a back-up team. The community also fabricated three fibreglass rescue boats and purchased rescue ropes, flash lights, first aid kits and mega-phones for early warning communications. The grassroots disaster-response mechanism was tested in August 1997 when Typhoon Ibiang brought heavy rain, and local rivers overflowed. Fortunately, all families were saved through the efforts of the local CBDM teams.

Through alliance building with local NGO partners, Buklod Tao was eventually able to pursue disaster risk reduction along the river bank of one local river. To prevent soil erosion the community placed sandbags along the river bank in October and November 2003, and in May and June 2005 bamboo seedlings were planted. Since January 2010 Buklod Tao has been establishing gabion boxes to form a gabion wall to prevent soil erosion along the riverbank, work carried out in collaboration with students of De La Salle University in Manila. The community has also increased the number of rescue boats to eight and accomplished repairs of the old boats. These rescue boats saved many families when Typhoon Ketsana struck the Philippines badly on 26 September 2009.

During a two-day workshop conducted in 2007 and intended to address core problems and find solutions to disaster risk, the community arrived at a consensus that poverty alleviation should become a focus of their work. Buklod Tao thus initiated five community-based business programmes, namely tetra pots production, organic compost production, urban container gardening, fibreglass fabrication and green charcoal trading. Some 1,314 people eventually benefited from these poverty-alleviation initiatives. In May 2010 Buklod Tao also started a social enterprise capital augmentation programme (SECAP), or microcredit scheme, which provides 218 individual business beneficiaries with capital assistance (with an average loan of US$120). Such income-generating activities contribute to reducing people's vulnerability in facing typhoons and other natural hazards, and help them earn and save money so that, someday, there will be an option for them to move to safer places. Thus from its modest beginnings and struggle with flood risk, Buklod Tao now engages in a holistic and fully integrated community developmental programme.

Flood Forum raises awareness, supports communities, assists in setting up local groups, brings organisations together and runs events to disseminate information regarding flooding and flood-risk reduction.

In Japan, local governments serve as a catalyst for the organisation of *Jishubo* or 'autonomous organisations for disaster risk reduction'. *Jishubo* are community-based organisations in charge of emergency management within neighbourhoods, which, to date, have proved to be relatively efficient in mobilising local people and resources. Since the 1995 Kobe earthquake, the Japanese national government has been strongly encouraging the creation of *Jishubo*, while local authorities facilitate the organisation of such associations and provide them with disaster preparedness training (Bajek *et al.* 2008).

Scientists and outside professionals may also provide additional knowledge which CBDRR may successfully combine with local people's knowledge of hazards and potential preventative measures (see Chapter 9). This is essential when communities face hazards that have not occurred for a very long period of time and of which local knowledge may be lacking. In that sense, scientific knowledge may also be useful in assisting communities with long-term climate change adaptation.

For example, the province of East Nusa Tenggara in southeastern Indonesia has a three-month rainy season and a nine-month drought season. Hunger is commonly experienced during the long dry season. The PMPB-Community Association for Disaster Management and Yayasan Pikul initiated a project that combined outside and local knowledge to build a community early warning system to prevent food shortages and increase community capacity to face drought. As a result, local farmers have acquired increased capacity to manage dry lands and the community early warning system has been established through locally developed monitoring indicators for food security and livelihood (PMPB-Community Association for Disaster Management 2007).

CBDRR strategies

CBDRR first requires a strong community organisation to ensure sustainability. An organisation or a committee that will carry out the CBDRR process, which includes participatory risk assessment, planning, implementation, monitoring and evaluation of community-based risk reduction programmes, is indeed important. CBDRR entails social mobilisation. It brings together community stakeholders for DRR to expand its resource base. The local community level links up with the intermediate, national and even international levels to address the complexity of disaster risks. Networking and building partnership at all levels is crucial in social mobilisation. Advocacy, lobbying and campaigning for favourable policy formulation and legislation on CBDRR are necessary activities for mobilising various sectors. CBDRR also fosters public awareness and disaster risk communication. Public awareness of disaster is a significant step in organising a community. Awareness-raising and skills development should go together in order for the people to fully participate in the CBDRR process.

CBDRR further draws on a three-step analysis–action–reflection framework. Before implementing a plan, a thorough analysis of the situation is indeed undertaken. During and after implementation, people reflect on what went wrong and what went well in the process. Lessons drawn from practice are always considered to improve performance. Lessons learned continue to feed and improve CBDRR theory.

CBDRR should also be flexible and adaptive enough to adjust to changing physical and social environment. This is a crucial dimension of sustainability. People's vulnerabilities, resources and needs constantly evolve over time in accordance with changing threats and

Box 59.2 Community-based disaster risk reduction through 'town-watching' in Japan

Etsuko Tsunozaki
SEEDS Asia

The Japanese have learned that, in order to effectively advance DRR, each and every individual needs to be prepared to save their family and neighbours. The concept of 'self-help' and 'mutual-help', as opposed to 'public-help', has developed and spread in recent years. Many activities are being encouraged and carried out by various stakeholders at different levels, such as national and local governments, schools, universities, private companies and NGOs.

One of the effective activities to raise people's awareness about disasters and to promote DRR through active participation of community members is the 'town-watching for DRR' exercise, also known as 'DIG (disaster imagination game)' or 'exploration for disaster prevention'. The 'town-watching' exercise includes four steps:

Step 1 consists of a lecture to discuss natural hazards, past disasters in the area and the 'town-watching' method. It is important that all people are aware of potential natural hazards, past disasters in the area and how people then recovered. The lecture may be given by a school teacher or an expert in DRR, using text books, photos and videos.

Step 2 consists of walking around the neighbourhood, taking photographs and getting to know the local issues related to disasters, including positive and negative aspects. This step allows the participants to appraise and understand what is currently being done in terms of disaster preparedness (e.g. measures undertaken to prevent disasters, evacuation routes and shelters) and what the gaps are. Interviewing local people, government officials, police officers or fire fighters also helps people learn more about their neighbourhood.

Step 3 consists of developing maps with the information collected during the participatory walking exercise. Participants plot hazard and vulnerability data on a large-scale map to create a community-based risk map. Photographs taken during the walk are attached to the map.

Step 4 consists of community-based action planning. Participants discuss local issues, define potential risk reduction measures, identify stakeholders in charge of implementing these measures, and draft action plans.

Many municipalities throughout Japan are promoting 'town-watching' as a regular exercise, the outcomes of which are incorporated into official DRR maps. The 'town-watching' method may be used in any community, including rural, mountainous or coastal areas, not only for DRR but also for environmental management and city/rural planning. In Saijo, Ehime Prefecture, 'town-watching' was used as a sustainable DRR tool. A 'teachers' association for disaster education' was established and a 'kids' disaster prevention club' consisting of students, teachers, parents and local community members was set up. A forum for DRR called 'kids' DRR summit' is held three times a year, gathering almost 1,000 students of twelve years old to discuss what they have learned and make a DRR declaration.

Manuals/handbooks available in English:

- Guidance information on the 'exploration for disaster prevention' programme by the General Insurance Association of Japan (GIAJ): www.sonpo.or.jp/en/about/activity/pdf/edp_guidance.pdf.
- Town-watching handbook for disaster education – enhancing experimental learning, by the European Union, UNISDR and Kyoto University: www.preventionweb.net/english/professional/publications/v.php?id=12062, www.crid.or.cr/digitalizacion/pdf/eng/doc18044/doc18044-a.pdf.

economic opportunities. CBDRR plans and actions should adjust to such trends through frequent community self-assessment.

As an example of both reflectivity and flexibility, the communities of the Rimac Valley in Peru used to think that the solution to the flooding threat they face on a seasonal basis lay in large infrastructure that only the central government with experts and engineers would have been able to implement (Maskrey 1989). With the assistance of a local NGO called Predes they eventually realised that they had a significant role to contribute to DRR. They built flood defences made of local materials and using the local labour force. As a result, the occurrence of disasters in the area was significantly reduced. The project eventually grew bigger and after two years included other areas and organisations as the initial community built-up knowledge of the context and problems it faced. Such local and incremental evolution of DRR actions often turns out to be difficult when programmes are managed by centralised governmental structures.

Issues and challenges in CBDRR

Forty years of CBDRR practice and twenty-five years of reflection on its strategies and results highlight a series of issues and challenges. First, CBDRR practitioners have varying degrees of understanding and practice. Some so-called CBDRR programmes fall short of the ideal. Some CBDRR initiatives are simply located in a community (area-based), but the process and outcomes are not owned by the community. For example, the monitoring team of an international NGO reported that one of its well-intentioned CBDRR projects in Southeast Asia was not truly appreciated by the community where the project was based. The project involved the provision of a water supply, the installation of sanitary facilities, and drainage improvement. Though this project was implemented with the local people, there was vandalism of small monuments that recorded the date of the project implemented by the international NGO and funded by a specific funder. The monuments did not mention the community where the project was based. The community did not feel a sense of ownership and accomplishment in the project, feeling instead that it had been implemented by an outside agency.

Second, there are challenges related to funding and sustainability. On the side of human resources, there is a danger that CBDRR projects are too dependent on external facilitators' expertise. NGO workers who implement the project often fail to adequately empower members of the community with skills and resources which would enable them to ensure the sustainability of CBDRR plans and actions on their own. As regards financial resources, most

CBDRR projects rely upon short-term funding. Yet CBDRR often requires more time than the usual 'project cycle'. Participatory risk assessment may have been used, but this is only the beginning of the CBDRR process. Social preparation alone takes time. It takes a tremendous effort for mitigation and preparedness actions to be fully integrated in the daily lives of the people who are responsible for protecting themselves and ensuring their community's safety.

A third challenge pertains to participation and empowerment being concepts that are easier to promote than to implement. Although people are often eager to take part in any undertaking that will advance their interests, they need to understand entirely the disaster issues in order to participate fully. This requires general awareness, integration with the people and empowering them to take control of their future plans and actions. Facilitating this process is an arduous task for CBDRR facilitators. In many rural and urban places people have been systematically taught by government officials for generations that they are 'ignorant' and incapable of making 'big' decisions. 'Participation' is often an alien notion to people with such a history of oppression (see Chapter 8 and Chapter 9). A related set of issues concerns power relations within the community. 'Who participates?' is a critical question. Power relations within the community may constitute a serious challenge to be faced by CBDRR practitioners. For example, it proves very difficult to work in a multi-caste context where political, economic and social resources are concentrated in the hand of the most powerful, with whom it is often necessary to meet in order to get in touch with other segments of the community (see Chapter 38). This is often difficult in contexts where patronage politics are predominant (Luna 2011). Community organisation leaders may change in accordance with political turnover, thus leading to difficulties in ensuring project sustainability.

Fourth, accountability has emerged as a crucial issue (Twigg 1999–2000). For a long time, accountability was in just one direction – towards the source of the money. CBDRR designers and implementers failed to account to the communities whose plight they articulated in the proposal's design. Time seems ripe to account to the communities how much money is raised, how much goes to the project and how much goes to the facilitating agency. Responsible accounting is necessary; transparency is essential in all CBDRR operations.

Conclusions

Despite these issues and challenges, CBDRR is gaining ground worldwide. While there are significant gains, the ideal features of CBDRR are not yet fully captured in practice. In addition, CBDRR continues to be promoted mainly by outside agencies in more accessible communities. Though there may seem to be a proliferation of CBDRR projects, their application is still limited to selected districts or provinces in a country. There is a need for official guidelines and budgetary allocations.

Furthermore, there is a need for a thorough evaluation of current CBDRR practices in order to determine how faithful they are to the principles of CBDRR as initially promoted. Lessons in CBDRR practices need to be studied and examined for possible widespread replication, not only in the global South, but also in countries in the North.

At present, the assessment of CBDRR activities is often short-term and conducted by the same organisations that implemented the projects under scrutiny. This is a significant shortcoming of current CBDRR practice. The impact of CBDRR programmes on local communities needs to be evaluated by communities themselves in collaboration with independent organisations. This should be part of the accountability process.

Such initiatives should rely on recent efforts to provide criteria to evaluate the outcomes of CBDDR programmes. This is a difficult task, as a successful project is a project that results in the absence of evidence, i.e. no disaster. In that direction Twigg (2007) provides a long list of thematic criteria of strong communities and enabling environments.

Finally, there is a need for further institutional and legal recognition of the importance of CBDRR. Large-scale, sustainable progress in reducing the risk of disasters will only occur if national governments endorse local initiatives; otherwise CBDRR will be limited to valuable but small, temporary and rapid initiatives.

60

Civil society and disaster

Martha Thompson

UNITARIAN UNIVERSALIST SERVICE COMMITTEE, BOSTON, USA

Introduction

Beginning in the 1980s, many kinds of non-state actors have been active in disaster risk reduction (DRR). They range from large international non-governmental organisations (NGOs) financially organised and registered as charities, to local community-based organisations. Their movement into DRR came on the basis of several decades of development activity by this broad class of non-state actors, called 'civil society'. In the 1990s and 2000s civil society has become more and more central to DRR.

Definition and history of civil society

Civil society is a term that can mean quite different things to different people, depending on their political stance or their understanding of how the world functions. Edwards, in his book *Civil Society*, weaves the different strands of thinking into three main themes: associational life, the movement towards a good society, and the public sphere (Edwards 2005). He then weaves those ideas together, defining civil society as the way people organise themselves to seek 'good society' in the public sphere.

Civil society refers to the way in which citizens organise themselves to achieve an agreed-upon purpose to improve, regulate or change society. The resulting array of organisations – ranging from huge international NGOs, to lobby groups, to women's village organisations – are called civil society organisations (CSOs). They are not managed by governments and are not oriented towards profit. CSOs can provide services, emergency assistance, charity, infrastructure, non-profit loans, technical assistance, training, education, information, mobilisation around rights or specific issues, and advocacy. Civil society organisations provide networks where groups can come together around common goals and frameworks through which they can influence government. They reflect a spectrum of social, cultural, religious, political and economic viewpoints, but they all work to change or better society according to their particular vision and goal.

Civil society has existed for centuries in different forms in different parts of the world, from the Lollards, a grassroots religious reform movement in England in the fourteenth century, to the Luddites, nineteenth-century textile workers put out of work by the industrial revolution, to the abolitionist movement and the women's suffragist movement in nineteenth-century Britain and the USA. In the twentieth century, mass-organisations in the colonised countries of Africa and Asia played a fundamental role in liberating their countries from colonisation (Tandon 1991: 5). The civil rights movement in the USA, the land-reform movement in Latin America and the anti-apartheid movement are all examples of civil society organisations making fundamental changes in the political landscape.

In the last twenty years, there has been growing divergence of opinion about the role civil society can and should play within the paradigm of state, civil society and the marketplace. Mercer's (2002) literature review lays out these differences. Authors such as Diamond (1994: 5) see civil society primarily as social movements, students' and women's groups, farmers' organisations, trade unions and human rights organisations exerting change through mass-mobilisation.

The break-up of the Soviet Union produced a new understanding of or function for civil society. After the Soviet-backed government in Poland fell to a broad front of social organisations, capitalist democracies saw civil society actors as key players in the fall of the Soviet bloc (Aksartova 2006: 1). For many analysts in the West, the subsequent global imposition of a neo-liberal economic framework created political triumphalism around democracy and neo-liberal capitalism as the dominant paradigm (Howell and Pearce 2001: 4). Neo-liberal market policy and the brand of capitalist democracy that it promoted limited the role of the state, reduced state sovereignty and decentralised power and responsibility without decentralising resources. Neo-liberal thinking assigns civil society organisations a significant role in service and infra-structure, replacing government. In this analysis, civil society's function is to legitimise the status quo, not challenge it (Mercer 2002: 10). The evolution of this analysis was accompanied by massive growth in the NGO sector fuelled by Western governments putting considerable funding into developing civil society, in both former Soviet bloc countries and in the global South (Aksartova 2006: 3).

The debate about the role of civil society continues today. The neo-liberal definition of the role of civil society is a motivating force behind the funding of NGOs and other private service institutions that serve that role. At the same time, new social movements seeking to re-appropriate power from the state, effect structural changes and transform the situation of the marginalised continue to emerge and define themselves as part of civil society. It is important to bear in mind that civil society is not monolithic, and that different organisations within it have sharply differing ideas of the desired paradigm, their purpose within it, other civil society actors and the role of civil society itself.

Types of civil society organisations

The brief typology below is intended to give an idea of the range and diversity of civil society organisations involved in DRR. These are not exclusive types, and there is overlap; for example, international or national NGOs can also consider themselves to be rights organisations or include rights work in their mandate. The typology of civil society organisations responding in the Indian Ocean tsunami in Tamil Nadu (Srinivasan et al. 2005: 10) demonstrates the different roles that each of these sectors plays in a given situation, in this case crisis response.

It is important to note the growing role of the NGO sector within civil society. The number of NGOs dramatically increased in the 1990s. There were 220 NGOs registered in Nepal in 1990 and 1,210 in 1993. In Tanzania, NGOs grew from 137 in 1986 to an estimated 1,500 in 2005. In the USA, national non-profit organisations grew from 10,299 in 1968 to 23,000 in 1997

(Kajimbwa 2006: 2). The dramatic growth of NGOs, and their increased role in sustainable development due to government funding, has sharply increased the role of civil society organisations in DRR.

It can be misleading to equate NGOs with civil society. Some academics equate the presence of many NGOs, not social movements, with a healthy civil society (Mercer 2002: 6). This is a limited view, since NGOs are one sector within civil society, albeit a powerful and well-funded one. Their relationship with other members of civil society can be complex. Many social movements see NGOs' role as increasingly influenced by the neo-liberal forces acting through donors, states and multilateral institutions. Mercer (2002) cites extensive literature to illuminate the tensions in the relationship between NGOs and social movements.

The following is a rough typology of civil society organisations:

- *International NGOs* seek funding from individuals, governments or foundations to develop programmes in the global South depending on their particular vision of development, emergency relief and, more recently, disaster risk reduction. They either provide direct service, contract national organisations to carry out work, or develop and fund projects with national organisations. They operate with a great deal of latitude, with limited accountability to donors and less to beneficiaries. They can also provide technical assistance, resources, training, and work on advocacy and policy.

- *National NGOs* can range from huge, multi-activity groups such as the Bangladesh Rural Advancement Committee (BRAC), which now employs over 100,000 people, to small, single-focus organisations such as Komisyon Famn Viktim pou Viktim in Haiti, which works on behalf of women who are victims of sexual violence. They usually depend on outside sources such as bilateral institutions, international NGOs and foundations for their funds. Grassroots organisations sometimes contest the role of national NGOs, claiming that they are not accountable to the groups for which they often speak (Christoplos 2001). National NGOs can provide services that were previously seen as the responsibility of the government, such as Fundación Salvadorena para el Desarrollo y Vivienda Mínima (FUNDASAL) in El Salvador, which builds low-cost housing for marginalised communities and has developed earthquake-resistant designs. Among national-level NGOs, the national societies of the International Federation of Red Cross and Red Crescent Societies (IFRC) have particular characteristics, being national NGOs that belong to an international federation, thus being both national and international.

- *Membership organisations* organise around an identity or common goal to which their members subscribe. They can range from women's groups to land rights groups to unions to mass-organisations. The leadership requires active participation by members to help achieve goals and is accountable to the members. Membership confers legitimacy of representation.

- *Community-based organisations (CBOs)* can be based in one community or represent a group of related communities working together on a common goal for their own interests. They can form organically or be fostered by an NGO. Even small rural communities are not one-dimensional, having complex layers of power that are not immediately apparent to outsiders. It should not be automatically assumed that a CBO represents the whole spectrum of interests in a community.

- *Technical assistance and training organisations* provide specific knowledge and skills to NGOs, membership and community organisations or communities themselves. They are sometimes called development support organisations. For example, Architecture for Humanity is an organisation that provides technical assistance in solving design difficulties (architecture-forhumanity.org). Asociación Latinoamericana de Organizaciones para la Promoción

(ALOP) provides technical assistance and training on non-formal education to NGOs in Latin America (www.alop.or.cr).

- *Rights organisations* defend the rights of particular segments of society. For instance, People's Watch, a human rights group in Tamil Nadu, monitors the violation of human rights of marginalised groups, holding the government accountable and using the available legal tools (www.pwtn.org). They provide human rights education in Tamil Nadu. Advocates for Environmental Human Rights in Louisiana is an example of a rights group that provides legal aid, community support and campaigns to advance the human right to a healthy environment (www.ehumanrights.org).
- *Research and educational organisations* include organisations such as the Asian Pacific Forum on Women, Law and Development, which adds advocacy to their research, rights and education work on women and their access to the law. There are many for-profit research organisations which can do good work but do not necessarily identify themselves as actors for social change. University research centres can take on an activist role supporting grassroots communities, such as the Center for Hazards Response and Technology (CHART) at New Orleans University, which uses participatory action research as a tool to increase the resilience of indigenous communities (chart.uno.edu) (see Chapter 61).
- *Advocacy and campaign organisations* can be single or multi-issue. The Nigerian organisation Environmental Rights Action mobilises people around a single issue, the impact of resource extraction on the livelihoods of the people of the Niger Delta (www.eraction.org). Some international and national NGOs, such as Oxfam and Action Aid, have strong advocacy and campaign components as part of their strategy for change.

Box 60.1 provides a sketch of an international network of many of these kinds of civil society organisations that are pooling their efforts towards DRR.

Box 60.1 The Global Network of Civil Society Organisations for Disaster Reduction (GNDR)

Marcus Oxley
Co-ordinator, GNDR, London, UK

The Global Network of Civil Society Organisations for Disaster Reduction (GNDR, www.globalnetwork-dr.org) is an international network of civil society organisations working to influence and implement disaster risk reduction policies and practice around the world. It relies on the commitment, diversity of skills, knowledge and extensive reach of its membership at all levels (particularly local) across virtually every region of the world. Members work for a broad range of organisations including large national and international NGOs, local or community-based organisations, academic and/or research institutions. Many within the network serve on their own national and regional networks, alliances and associations. The GNDR includes 600 people from 300 organisations in 90 countries.

The goal of the GNDR is to harness the potential of global civil society to engage strategically in disaster risk reduction policy and practice, placing the interests and concerns of vulnerable people at the heart of policy formation and implementation.

The need for the GNDR was identified by a number of civil society organisations at the World Conference for Disaster Reduction in 2005. It was initiated with the close support of the United Nations International Strategy for Disaster Reduction (UNISDR) Secretariat, in collaboration with the Special Unit for South–South Cooperation (SUSSC) of the United Nations Development Programme (UNDP). It was officially launched in Geneva during the first session of the Global Platform for Disaster Risk Reduction (GP-DRR) in June 2007.

The GNDR works closely with civil society organisations, government bodies, UN agencies and international institutions, with the intention of amplifying the voice and influence of disaster-prone communities at the national, regional and international levels, through sharing of learning and experiences, building consensus and supporting collaborative approaches and joint actions.

A key project based on these principles is the 'Views from the Frontline' (VFL) participatory monitoring project (www.globalnetwork-dr.org/index.php?option=com_content&view=article&id=57&Itemid=65), which was undertaken by the GNDR in 2009 and 2011 and was to be every two years subsequently. The VFL project involves network members in a face-to-face survey of stakeholders involved in DRR at the local level, in order to establish what progress had been made in terms of the UN 'Hyogo Framework for Action' (HFA) on DRR, which was established in 2005. The first survey generated data from more than 7,000 respondents at local level, including local government officials, civil society members and community members. There was a special focus on women and on young people. The second survey focuses specifically on the performance of local government in DRR and increased the sample size to 50,000, experimenting with the use of SMS-based interviewing.

The first VFL report was presented at the second session of the GP-DRR in June 2009, with the aim of promoting stronger linkages between policy established at the national level based on the HFA, and practical implementation and resourcing at the local level. The first report was called 'Clouds but little rain ... Views from the frontline' and is available at www.preventionweb.net/files/9822_9822VFLfullreport06091.pdf.

Civil society and disaster risk reduction

The evolution of civil society's role in DRR

From the 1970s, some practitioners and academics argued the need to analyse and address economic, political and social causes of 'natural disasters' (O'Keefe et al. 1976). The UK-based organisation War on Want broke with tradition and used some of the money for immediate relief through a common appeal by the Disasters Emergency Committee to fund action research on farmers' coping strategies on the Mauritanian side of the Senegal River (Bradley et al. 1977). Cuny's 1983 book *Disasters and Development* called on international NGOs and governments to look at the impact that development has on disasters, not just vice versa. Anderson and Woodrow (1989) in their book, *Rising from the Ashes*, argued that enhancing local capacities was key to disaster response and avoiding future disasters. The thinking and initiatives that came out of the 1970s and 1980s laid the foundations for current disaster risk reduction work by civil society organisations. Thus, in 2000 a team from Practical Action published the results of action research in southern Zimbabwe that is strikingly similar to what War on Want had done more than two decades earlier (Murwira et al. 2000).

A multi-level role for CSOs: beyond the grassroots

The HFA on DRR (see Chapter 50) exhorts three main actors – government, NGOs and communities – to work together in five programme areas during 2005–15:

- Governance;
- Risk assessment, monitoring and warning;
- Knowledge and education;
- Underlying risk factors; and
- Disaster preparedness and response.

Moss (2007) argues that civil society must play a multi-level, multi-sector role in DRR that goes beyond the work at the grassroots level in order to have a substantial impact. She argues the need to simultaneously work on policy, legislation and multi-sectoral approaches in order to complement community-centred DRR work (Moss 2007: 3–4).

The GNDR report, 'Clouds but little rain …', lays out an impressive argument for civil society's role at multiple levels. The GNDR's own three main objectives clearly demonstrate a broad vision for multi-sector and multi-level work for CSOs in DRR; these are:

- Influencing policy formulation;
- Increasing public accountability for the HFA; and
- Raising resources and political will for the HFA.

Box 60.2 Abridged summary of civil society work on DRR

Civil society organisations, particularly NGOs, international NGOs and CBOs, have worked on aspects of disaster preparedness for many years. The major famines in the 1970s in Bangladesh and Biafra triggered a qualitatively different response from the international community, including large NGOs, which brought more concerted focus on prevention, early warning systems, stockpiling food for famine-prone countries, and facilitation of food transfer from highly productive countries to famine-prone ones (Cuny and Hill 1999: 3). Different international NGOs such as Oxfam began to include building livelihood resiliency into famine response.

Community preparedness work at the grassroots level began to grow in the late 1980s and early 1990s in countries prone to multiple hazards with large civil societies such as Bangladesh, the Philippines and Nicaragua. The rapid growth of NGOs in the 1990s and the growing linkages between sustainable development, disaster recovery and risk reduction increased the number and variety of civil society actors involved in disaster preparedness at the grassroots level.

In the early 1990s, three regional networks formed around community preparedness for education, advocacy, training and education. In 1992 the Network for Social Studies on Disaster Prevention in Latin America (LA RED, www.desenredando.org) was established. It was followed by Duryog Nivaran, the South Asian network promoting community-focused alternative perspectives on disaster management, in 1994 (www.duryognivaran.org), and Partners Enhancing Resilience to People Exposed to Risks (Periperi) in South Africa (IFRC 2002: 22). Beginning in the late 1990s and accelerating in the 2000s, community-based disaster preparedness gained importance as a viable alternative disaster preparedness model. A growing group of civil society organisations

---

gained a range of experience in DRR at the grassroots level, particularly in situations of mega-disasters of extraordinary impact, such as Hurricane Mitch in Central America (1998), the Gujarat earthquake (2001), the Indian Ocean tsunami (2004), Hurricane Katrina (2005) and the Haiti earthquake (2010).

During 2000–05 the UNISDR provided a framework for network and discussion that raised the profile of disaster preparedness in the NGO development world. Disaster preparedness became an important bridge between humanitarian and development communities. More and more NGOs made the link between vulnerability to disaster and marginalisation and pressure livelihoods.

In 2005 at the World Conference for Disaster Reduction (WCDR) in Kobe, Japan, numerous civil society organisations raised the need for a global network to increase co-operation and co-ordination (Wisner and Walker 2005). The marked failures in government response to mega-disasters such as Hurricane Mitch and Hurricane Katrina suggested the need for an expanded role for civil society. In the post-Mitch period, NGOs and CSOs from Central America demanded and received a role in developing the internationally funded plan for recovery (Wisner 2001a).

Thus by 2005 the situation was ripe for the emergence of the GNDR. It was officially launched in Geneva during the first session of the GP-DRR in June 2007. The GNDR began an independent review of progress in the implementation of the HFA, the international framework for risk reduction approved by 168 governments in 2005. The initial report, 'Clouds but little rain … ' (2009) was presented at the Second Global Platform for DRR. The study was the result of more than 7,000 interviews in 48 countries (GNDR 2009).

This report found that civil society organisations are the most dynamic, active actors in implementing and promoting the five programme areas of HFA at the local level. It stimulated literally thousands of conversations among civil society organisations, local governments and communities about HFA and led to co-operation on multiple initiatives.

When CSOs begin to work on policy, accountability and political will, whether they are grassroots, advocacy, international NGOs or research institutions, they have to address both governance and causes of vulnerability. International human rights institutions (see Chapter 6) assert that a state has the obligation to respect, protect and fulfil the human right to life of its citizens, and that this extends to protecting them from disasters. The GNDR's objectives recognise the responsibility of NGOs to hold governments accountable, influence policy formulation, claim resources and stimulate political will. Some governments will see these actions as too critical (Wisner and Haghebaert 2006). The more polarised a society, the more chance that the government will see civil society efforts to change the status quo as inimical and destabilising.

The HFA requires signatory states to fulfil their obligation to protect their citizens by developing a holistic approach to DRR policies, structures and programmes. It specifically recognises that implementation of DRR is fundamentally linked to good governance and addresses the underlying risks that create/increase vulnerability.

Civil society's relations with governments and communities

Vulnerability and governance

Disasters have greater impacts on poor and marginalised populations and communities (Wisner *et al.* 2004). The growing acceptance in the 1990s that natural disasters are strongly influenced by complex and political factors increased understanding that the uneven results of disasters triggered by natural hazards are embedded in the social, political and economic relationships in a particular country (Christoplos and Rocha 2001: 241–2).

Moss points out the linkage between vulnerability and governance:

> The root causes of people's vulnerability to disasters can often be found in national and global political, social and economic structures and trends: weak planning and building codes, inadequate policies governing civil protection and disaster response, inadequate international policies on greenhouse gas reduction and climate change, a lack of national welfare system or safety nets, indebtedness and aid dependency.
>
> *(Moss 2007: 2)*

As a remedy to the picture Moss paints above, the IFRC has drawn up a list of the elements of good governance for risk reduction, including the following:

- Social cohesion and solidarity;
- Trust between authorities and civil society;
- Investment in economic development;
- Investment in human development;
- Investment in social capital;
- Good co-ordination and information sharing among institutions for DRR;
- Attention to the most vulnerable;
- Attention to lifeline infrastructure;
- Effective risk communication system;
- Political commitment to risk reduction; and
- Laws, regulations and directives to support all of the above.

(IFRC 2002: 28)

If good governance for risk reduction means changing the vulnerability of a marginalised population, it leads to the question: What makes these groups vulnerable in the first place (see Chapter 5 and Chapter 38)? That question inevitably challenges the interests of any group that benefits from inequality in society. If a government is not committed to reducing vulnerability of the marginalised population, they can be threatened by organisations that champion the marginalised.

Governments, governance and civil society actors

Marginalised populations are created by specific economic, political and social policies and targeted restrictions, which determine how different groups in society can or cannot access services, rights, knowledge and resources (Wisner *et al.* 2004). In many cases, these policies are created or tolerated by the same governments that have signed the HFA. When civil society advances vulnerable communities' claims to rights and resources or disputes policies it sees as inequitable, its own work can be impacted. CSOs may not be regarded as neutral organisations by their governments, nor will the organisations that fund them or train them. Individual government response depends on a country's political stability, the overall exercise of civil rights and the degree of existing socio–economic inequality.

Examples of extreme responses by government include the banning of many international NGOs in Zimbabwe, especially after they spoke out concerning the role of government policy in creating the preconditions for an epidemic of cholera. After Hurricane Mitch in 1998 in

Nicaragua, 350 CSOs formed the Civil Coordinator for Emergency and Reconstruction Platform to co-ordinate emergency relief and use reconstruction to transform vulnerability. Their well-researched criticism of the government response and successful lobbying of international aid donors created tensions with the Nicaraguan government. The government reacted by cancelling a meeting of the World Bank-led consultative group in February 2000 and challenged the legitimacy of this civil society movement (IFRC 2002: 157–8).

Ideally, a government should see a vibrant civil society as an important vehicle for citizen participation and local ownership of development processes. A government that upholds free speech, freedom of the press and of assembly will see civil society's activities as legitimate. However, many governments fall short of this ideal.

There are, though, also instances of positive response by governments. For example, in Nepal the Lumanti Shelter Society noticed that there was little understanding of the HFA among local government officials. They invited local government staff to a workshop about the HFA to educate them and invite collaboration with NGOs. The government was willing to respond and agreed to a joint workshop and planning (GNDR 2009: 24). In Zambia, the district government of Sinozonwe partnered with the Zambian Red Cross in a participatory study of DRR (Zambia Red Cross Society 2003).

Failing governance and civil society organisations

A government in crisis will be less open to engage with CSOs bent on change and may act against them. Haiti is a good example of what happens when a government sees the organisation of marginalised people as threatening. Haiti has thousands of CSOs, ranging from small rural community organisations to mass-peasant organisations with 50,000 members, such as the Peasant Movement of Papaye (MPP), or a network of over thirty CSOs such as the Haitian Platform to Advocate Alternative Development (PAPDA). Many of these organisations were born out of struggle against different repressive government regimes for a most just distribution of resources and political power. Under 'Baby Doc' Duvalier almost all the peasant and union organising had to be clandestine. Now it is overt, and the widespread peasant movement is key to promoting DRR as linked to soil enrichment, water conservation and reforestation. Logically, a government concerned with the need for these things would work with such organisations. However, because the peasant movements' agenda also includes land reform and changes in the power structure, successive governments in Haiti, including the one backed by the military that took power in September 1991, sought to crush this popular movement. They tortured and killed peasant leaders and intimidated them into leaving the political arena (Ridgeway 1994: 64–9). Tension over land reform and the distribution of power continues in the country following the January 2010 earthquake, and as a result civil society was virtually excluded from planning earthquake recovery (Caistor 2010; CEPR 2010).

Government–CSO relations in civil war

Civil society organisations working on DRR in situations of institutionalised discrimination or civil war (see Chapter 7) can run the risk of attracting harassment, punitive action, defamation and repressive actions when they criticise the government. For example, the Sri Lankan government was suspicious of any civil society actors providing aid to Tamils in Sri Lanka during the war, and saw them as politically suspect. The French organisation Action Contre la Faim (ACF, Action Against Hunger) provided relief aid to Tamils in a contested area on the east coast after the 1994 tsunami. They had received permission from the government to have an office in the contested area. On 5 August 2006 the Sri Lankan military attacked their compound and killed seventeen ACF workers, mostly Tamils (Asian Center for Human Rights 2006: 1–2).

Organisations in India that worked with the Dalit community (formerly untouchables) after the Indian Ocean tsunami often found themselves at odds with the Indian government. After the tsunami, they had to challenge the government to extend aid allocation to the Dalits (Siscawati 2009). On paper, the Indian government is in agreement with implementing the five programme areas of the HFA. Addressing vulnerability would certainly include working on DRR with the Dalits. However, in reality the Indian government ends up in conflict with watchdog organisations such as People's Watch, precisely because this CSO champions the Dalits' rights. For example, on 17 August 2010 five Dalit human rights workers in training organised by People's Watch in Tamil Nadu visited the local police station as part of their training. They were all arrested arbitrarily. When People's Watch protested their unjust detention, Henry Tiphagne, People's Watch Director in Tamil Nadu, was charged for absconding from arrest although he was not even present when the five Dalits went to the police station (AWID 2010).

The particular risks for community-based organisations

CBOs that use analysis of power to claim the rights of groups that are either ignored by the government or suffer institutionalised discrimination openly confront established power structures. They often suffer for this. For example, in El Salvador the communities that made up National Co-ordination for Repopulation (CNR) in the 1980s during the civil war used their strong internal cohesion and support from NGOs to return as communities to their places of origin in conflict areas, against the will of the government. Even after the war ended in 1992, the Salvadorian government saw them as politically suspect. When Hurricane Mitch struck in 1999, the government decided to open dams in the Lempa River watershed in order to release flood waters. The engineering decision would put one group of the CNR communities in Usulutan and San Vicente along the Lempa River underwater. Rather than mounting assistance to evacuate the communities that were going to be flooded, the government did not even inform them of the imminent flooding. When people in these communities saw the floodwaters rising, they used their existing level of organisation to help each other to evacuate. Due to their level of organisation, not one life was lost (Thompson and Gaviria 2004: 52).

Grassroots organisations like the CNR communities and the Dalit rights workers who were arrested are more vulnerable to repressive actions by their government than mainstream NGOs. They have fewer legal or political resources to defend themselves. Governments pay a lower political price for any actions against these organisations, which have little access to the formal power structure.

Civil society, representation and voice

The ability of CSOs to carry out DRR work is conditioned by the relationships they have with the communities with which they work. Power dynamics amongst and between civil society organisations are as complex as those of the communities in which they work (Hauschildt and Lybaek 2006: 8–10). There is a power discrepancy between organisations based in the global North, such as Oxfam, CARE, Agence d'Aide à la Coopération Technique et au Développement (ACTED) and the local or national organisations that they fund. Ultimately, it is the quality of the relationship, the power analysis and the understanding of partnership between the northern organisation and the southern one which determines how that power difference affects agendas, activities, agency and priorities.

Representation is another issue. Who speaks for whom? International and national NGOs often see their role as advocating for the populations with which they work. However, those same populations do not necessarily want the NGOs to represent them. People's Watch in Tamil Nadu, working with the Dalits after the 1994 tsunami, addressed that power differential in an empowering way. Rather than continue to speak for the Dalits in discussions with the government, they helped the Dalits to form their own membership organisation to advance their own agenda (Siscawati 2009: 47).

In Latin America, particularly, there are serious questions around the discrepancy between NGOs' ideal and actions (Christoplos and Rocha 2001: 246). NGOs often have far greater resources and staff than the grassroots and membership organisations that they serve.

Class and urban/rural discrepancies, power dynamics, divergent concepts of authority and representation, and organisational culture can all play a role in creating distrust between NGOs, CBOs and the population they are supposedly serving.

Successes and challenges

How successful have civil society organisations been in mobilising communities for local hazard mapping and vulnerability assessment and action to reduce risk? Box 60.3 contains experiences from Central America that demonstrate success.

Factors influencing success of CSOs' DRR work at the community level

There are many negative examples of CSO work in DRR at the community level given the huge range of differences in approach and culture among CSO organisations. There is nothing magic about being a civil society organisation; it doesn't automatically confer cultural sensitivity or effectiveness. However, there is now a sufficient body of evidence to identify key factors that affect the success or failure of CSOs' work in DRR at the grassroots:

- Issues of power and participation: successful CSOs show an ability to listen, respect the community, recognise and tap local resources and experience, engage the community, incorporate their ideas and suggestions and enable them to 'own the plan'.
- Mobilisation of additional resources, the management of which is co-ordinated with or led by the community: the example of Oxfam in Box 60.3 is an excellent demonstration of a successful multi-faceted approach, including additional resources to enhance those of the community.
- Linkages with government policy and activities: a CSO is not omnipotent. Maximising impact, enhancing work and sustaining actions to reduce risk can depend on their ability to link into larger government plans and frameworks.
- Understanding of the cultural, economic, political and social fault lines in society that create marginalisation: if a CSO does not use a power analysis to understand why the population they work with is vulnerable, they cannot adequately address the causes of that vulnerability and therefore their solutions will be flawed.
- Technical expertise: the technical aspects of hazard mapping, tools for assessment and ability to plan for action are essential. CSOs with good skills develop more trust and credibility among those with whom they work.

Five key factors that can diminish a CSO's success are:

Box 60.3 Successful DRR work by CSOs in Central America

Honduras: a coalition of international NGOs funding, training and developing local community emergency committees in DRR

Action of Churches Together (ACT) in Honduras trained and organised local emergency committees in forty highly vulnerable communities in Naranjito and Azacualpa, Santa Bárbara Department; Nueva Arcadia, Copán Department; and La Unión, Lempira Department. The emergency committees created risk maps and contingency plans to facilitate evacuation and rescue, installed an early warning system to alert the population to signs of danger and thresholds for evacuation, and carried out small-scale mitigation projects to reduce risk in the most vulnerable communities (USAID/OFDA 2008: 4).

An important element of this project was that ACT worked with communities that had strong community organisations and previous collaborative experience with local church organisations. ACT was able to build on that experience and bring in new knowledge and skills such as mapping and contingency planning. The high levels of organisation in the involved communities maximised the impact of the knowledge and skills brought in by ACT.

El Salvador: Oxfam in San Vicente, El Salvador – a multi-faceted approach to DRR

When Hurricane Ida hit San Vicente, El Salvador on 7 November 2009, the communities were prepared, evacuation was organised, people arrived safely at shelters and relief was distributed due to a multi-level disaster preparedness response that Oxfam America developed over several years. They trained twenty local professionals on water and sanitation in emergencies and thirty-six local leaders on humanitarian response. They formed a network of eight partner organisations and significantly developed their capacity for disaster response. This network managed a common warehouse stocked for emergencies and rehabilitated five shelters (McGovern 2009).

Oxfam America provided training and resources, but also paid attention to power dynamics, gave control of resources to local groups they had trained to use them, and made sure that technical support teams were trained and readied as well as grassroots leaders.

- The politics of a government's attitude toward a particular organisation or population: this does not have to diminish success. The CNR communities in the Rio Lempa area in El Salvador developed excellent work in DRR that saved them during Hurricane Mitch, although the government at the time was inimical to these communities. However, a government's hostile attitude towards an organisation or population can prevent that group from accessing necessary resources or from full participation.
- Government indifference to the HFA and lack of progress on DRR: this by no means precludes success at a local level, but government indifference can produce a 'bonsai effect' – perfect small-scale successes that never evolve into larger-scale efforts.
- Competition among or between civil society actors over representation, trust, work modalities and resources.
- Lack of co-ordination for any of the above reasons.

- Lack of a gender lens and a children's rights lens will inevitably lead to the lack of participation on the part of some of the community, and lack of adequate analysis and solutions.

Conclusions

Civil society can play a very positive key role in the implementation of DRR. In the process of developing its 2009 report, 'Clouds but little rain … ', the global CSO network (GNDR) created an important forum for networking, joint discussion and reflection among civil society organisations. Through this, over 600 Northern and Southern CSOs engaged in DRR learned from each other and developed common analysis and common strategy. Their research found the role of CSOs to be extremely positive. According to one national participating CSO:

> Throughout all the priorities, it is evident that civil society organisations are scoring higher than the government. This shows the importance of using the capacities of the civil society organisations in order to strengthen the government's action in DRR. Therefore, partnerships at the local level between civil society and the local government and governmental bodies are the best way to optimise the available resources and capacities to build the resilience of the community.
>
> *(GNDR 2009: 23)*

The core message of the GNDR report was that governments must engage vulnerable and marginal groups. Engagement means listening to their ideas in the formulation of national policy, development of national institutions and implementation of DRR. The report underscored the key role civil society organisations can play in bridging the gap between the government and marginalised populations

However, civil society represents a broad spectrum. Many factors condition the ability of CSOs to play a positive role in DRR. Most obviously, a CSO's own ability, technical skill, organisational ability and grasp of DRR will affect the impact of their work. Less tangible but fundamental factors also include their relationships with communities and with government, and the CSO's understanding of power dynamics and the causes of vulnerability.

The relationships that CSOs have with the government where they work determine how easily they can access the local populations and what resources are available. The relationships that CSOs develop with the people with whom they work determine how they envision the work, the degree to which people take ownership and agency, and ultimately the real ability to reduce risk.

Vulnerable people are vulnerable because they have less access to power and resources in a society. If a CSO tries to provide technical assistance without understanding or addressing those key power issues, their work will have less impact on mitigating vulnerability.

In the end, civil society activity is about having a voice, the need of citizens to have their own voice apart from the government apparatus and the market. Through that voice the marginalised can be heard in the halls of power.

Communication and participation

61

University research's role in reducing disaster risk

Dorothea Hilhorst

DISASTER STUDIES, WAGENINGEN UNIVERSITY, THE NETHERLANDS

Annelies Heijmans

DISASTER STUDIES, WAGENINGEN UNIVERSITY, THE NETHERLANDS

Introduction

This chapter is about the history and current trends and challenges in making academic research more relevant to the policy and practice of disaster risk reduction (DRR). Although it promotes a broad definition of DRR, this chapter will mainly focus on university research addressing disaster in the context of development. It presents different paradigms of viewing disaster as they have appeared historically, moving from hazard-centred approaches to vulnerability approaches and approaches that centre on the mutuality between disasters and social and ecological processes. The point is that these can still be recognised today in different approaches to disaster studies: technocratic, structural and comprehensive approaches. It could even be stated that all these different approaches are necessary and complementary.

However, in reality, university research on disasters is fragmented, meaning that it results in parallel contributions rather than that it is organised in an interdisciplinary manner or through close collaboration. This chapter identifies factors that fragment the field, while discussing initiatives that aim to resolve the resulting gaps. It further delineates different roles of university research for better practice.

Scope of DRR and university research

DRR is often associated with the so-called prevention and mitigation phases of a 'disaster cycle'. Such a disaster cycle consists of different phases, including prevention, mitigation, impact, relief, rehabilitation and development. The notion of a cycle has been criticised because the different phases are not distinct in practice and are to some extent occurring together. Moreover, the scope for DRR is much broader than prevention and mitigation. DRR concerns measures to curb disaster losses, through minimising the hazard, reducing vulnerability and enhancing coping and adaptive capacity. Today, risk reduction is seen as an activity that permeates the full range of risk- and disaster-related phases from prevention through to reconstruction. Quite contrary to how many people in the field perceive DRR, this includes relief activities in the emergency phase. In the words of Lavell (2000: 20): 'All activities developed in the Emergency Phase are, or should be, risk reducing by nature'. The search for and rescue of disaster victims, the control of disease vectors and malnutrition, the guaranteeing of food-stuffs and potable water, the rapid stimulation of economic recovery and employment, security in temporary shelters and the provision of temporary housing, etc., are all risk-reducing activities that operate in the context of new post-impact risk scenarios. Examples of approaches aiming to integrate DRR in the response phase are: using relief to restore livelihood assets and rebuild livelihoods like cash- and food-for-work; building on survivors' capacities; using participatory approaches; and avoiding the reproduction of risk and pre-disaster vulnerability levels.

It is important to note that this notion of DRR encompasses all kinds of activities by all kinds of actors. It is not restricted to specific interventions or policies. Affected people take measures against disaster and build disaster preparedness into their social, cultural and economic institutions. Policies and practices of governments and other authorities in different fields have ramifications for disaster risks. This means that national governments, non-governmental organisations (NGOs), international agencies, the private sector, community-based organisations and people all take part in arenas of DRR. This arena is intricately linked to other arenas addressing poverty reduction, climate change and environment degradation. This remark is important, as many international NGOs associate DRR strictly with community-based approaches. Unwarranted and restrictive notions of the scope of DRR are one of the reasons why the disaster research community is fragmented.

The broad scope of DRR implies that every kind of university research into disasters may be relevant for DRR. In this chapter, however, we will especially focus on those university research endeavours that deal with disaster in the context of development. This means that we focus on those areas where the linkage between disasters and development is an issue.

Academic understanding of disasters and DRR

Until WWII, DRR, mainly through water management, was hazard-oriented and seen as a matter for engineering. Although there have been centuries and in some regions thousands of years of management of water-related disasters, in pre-modern times most rivers went untamed and people adjusted to disasters (Fagan 2000). In the twentieth century optimistic modernism led to the 'hydraulic mission' (Turton and Ohlsson 1999). The state controlled water resources and 'developed' the basin with mega-structures for flood protection and water retention. Embankments and barriers were built in order to tame the rivers and safeguard life and livelihoods of the population behind the dikes.

A social science perspective on disasters did not gain momentum until the 1960s. Gilbert White (1960) proposed that people should be given a greater range of options, which would make them consider leaving or avoiding high-risk areas. This so-called behavioural paradigm came to dominate disaster studies. It coupled a hazard-centred interest in the geo-physical processes underlying disaster with the conviction that people had to be taught to anticipate it. It is a technocratic paradigm dominated by geologists, seismologists, meteorologists and other scientists who can monitor and predict the hazards, while social scientists are brought in to explain people's behaviour in response to risk and disaster, and develop early warning mechanisms and disaster preparedness schemes.

In the 1970s anthropologists, sociologists and geographers increasingly began to challenge the technocratic, hazard-centred approach to disaster. This culminated in the 1983 landmark publication of *Interpretations of Calamity from the Viewpoint of Human Ecology*, edited by Kenneth Hewitt. The authors in this volume argued that structural factors such as increasing poverty and related social processes accounted for people and societies' vulnerability to disaster. The recognition of social vulnerability touched at the heart of understanding disaster. Whereas disasters used to be practically equated to natural hazards, they now became understood as the interaction between hazard and vulnerability (O'Keefe *et al.* 1976). DRR would then entail the transformation of social and political structures that breed poverty and the social dynamics that serve to perpetuate it (Heijmans and Victoria 2001), from the local to the international level (Wisner *et al.* 2004). It brought along increasing attention to the role of disaster-affected people and their communities in DRR. In 1989 Anderson and Woodrow developed the influential Capacities and Vulnerabilities analytical framework that sought to build disaster response on local strengths and capacities (Anderson and Woodrow 1989). Community-based DRR became a widely advocated approach.

In the 1990s the understanding of disasters shifted to emphasise the mutuality of hazard and vulnerability to disaster due to complex interactions between nature and society. The mutuality or complexity paradigm takes a step further than the structural analysis of disaster as a function of mal-development. While structural theory mainly looked at society to explain people's vulnerability to disaster, the mutuality paradigm looks at the mutual constitution of society and environment. People, in this view, are not just vulnerable to hazards, but hazards are increasingly the result of human activity. This has the important implication that vulnerability might not just be understood as how people are susceptible to hazards, but can also be considered as a measure of the impact of society on the environment (Oliver-Smith and Hoffman 2003). This is also associated with the notion of complexity since it recognises that different risks affect each other at different levels, in ways that escape simple cause and effect relations (Hilhorst 2004). This way of thinking implies a different outlook on DRR as well. The complex nature of disaster calls for more integrated responses. We increasingly see a shift from top-down interventionist forms of governance to governance as a quality of interacting social–political systems, such as international communities, national states, cities and localities, as well as in sectors such as agriculture, fisheries and domestic water use (Kooiman *et al.* 1997; Warner *et al.* 2003). This has led to co-governance arrangements such as public–private–NGO water partnerships, and can be recognised in the DRR model proposed in the UN's Hyogo Framework for Action (HFA) where DRR aims to involve different societal stakeholders.

While the different paradigms have partly evolved over time, it is important to realise that new paradigms do not replace the older ones completely. It is more accurate to think of them as existing alongside each other, and in some cases partly layered one over the other and influencing one another. Although not very pronounced and blurred in their boundaries, we can still distinguish different 'schools of thought' in disaster studies. Technocratic approaches, with a focus on a single hazard or a focus on a technocratic response mechanism, continue to be central to many disaster disciplines. Structural approaches that frame disasters in terms of vulnerability and development represent a large body of scholarship and are, for example, prominent in Radical Interpretations of Disaster and Radical Solutions, the RADIX website (www.radixonline.org). More systemic approaches emphasise the relationship between disaster and the wider system in which they occur, and are currently especially booming in work that addresses the nexus between disaster and climate change. In addition, there is a body of disaster studies that is interested in the symbolic and cultural aspects of disasters, and emphasises the socially constructed nature of disasters.

Box 61.1 Bradford Disaster Research Unit, 1973–77

Ilan Kelman
Center for International Climate and Environmental Research–Oslo, Norway

Interdisciplinary, cutting-edge disaster research based at universities has a long history. One of the earlier groups was the Bradford Disaster Research Unit (BDRU), founded at the Project Planning Centre at the University of Bradford in England. It put together a team that was specifically aimed at dealing with the problem of 'natural disasters' before they happened. It deliberately did not take any special disciplinary perspective or focus on post-disaster response.

Co-founded by Michael Gane, who headed the Project Planning Centre, and James Lewis, an architect supported by the Leverhulme Trust, its team was built up to include a geographer (Phil O'Keefe), a recent graduate of geography and economics (Ken Westgate) and an anthropologist (Alec Baird). With Lewis responsible for managing the unit, work included theoretical reviews, field studies and teaching, aiming to understand and solve disaster-related problems from a pre-disaster standpoint.

The unit produced one of the earlier textbooks devoted to interdisciplinary disaster studies, *Natural Disasters: An Intermediate Text*, by Phil O'Keefe and Ken Westgate, and a series of Occasional Papers, which included a compilation of disaster research references to date and a discussion of defining 'disaster'. Field-orientated reports included forestry planning for hurricanes in Fiji, analysis of disaster planning in the Bahamas commissioned by the League of Red Cross Societies, and an examination of the root causes of vulnerability to natural hazards on other Caribbean islands. Given the Centre where the unit was based, and as a reflection of its time, much of the work included planning perspectives, in terms of planning for disaster and planning to avoid disaster. All publications are today available at www.ilankelman.org/bdru.html.

Following Michael Gane's sudden resignation from the university, BDRU was discontinued by the university in 1977. Its lasting legacy was to contribute to examining long-term processes that create and maintain vulnerability leading to disaster, as well as combining knowledge and techniques from many academic and practitioner fields to better understand and tackle disasters. Its work became the basis of subsequent activities for all of its principal members.

Communities of research?

Although general trends in disaster studies can be outlined, the field of disaster studies is also characterised by gaps. There are a large number of specialised academic institutions on natural hazards and disasters that mainly focus on the technical sciences, although they may also incorporate social science. One of these is the Natural Hazards Center at the University of Colorado at Boulder, which has been prominent in the field, amongst others, by the annual conference it has been organising since 1975. The Aon Benfield University College London (UCL) Hazard Research Centre is another prominent institution that studies floods, earthquakes and volcanoes, and combines this with social science research on such themes as disability and disaster, corporate social responsibility and the development of indicators of community disaster resilience. In Japan the Disaster Prevention Research Institute at Kyoto University has more than forty professors studying most aspects of natural hazards and such social issues as risk communication and community-based DRR. A comparable institute at Beijing Normal University is the Institute for Disaster Prevention and Public Safety, where a dozen senior specialists study in a comprehensive and policy-oriented manner hazards including drought, dust storms, floods and earthquakes.

Nonetheless, in disaster studies that focus on the relation between disasters and development, the more general pattern is one of individual experts, often surrounded by small groups of PhD candidates, who are embedded in disciplinary or more general university departments. Many of the authors of this *Handbook* have been consistently working on the issue of disasters from within different university positions they have held throughout the years. The result is that part of the disaster studies community is more a network of people than of institutions.

As this *Handbook* demonstrates, natural hazards and disasters are the subject of dozens of disciplines, comprising the entire gamut of natural and social sciences. Inevitably, this means that the field consists of a whole range of sub-communities. The International Sociological Association, for example, has an international research committee on disasters. The committee publishes the *International Journal of Mass Emergencies and Disasters*. The International Geographical Union has for many years had a series of commissions on natural hazards and disasters, while the International Union of Geological Sciences encourages its members to study earthquakes and volcanoes. At the apex level, the International Council of Scientific Unions – now the International Council for Science (ICSU) – has recently developed a programme on integrated research on disaster risk, and its 'Strategic Action Plan 2006–11' identifies natural and human-induced hazards as one of the major research-led issues for ICSU over the planning period.

This broad multi-disciplinary interest in disasters is a result of the complex character of hazard and disaster phenomena. It also leads, however, to a lack of linkages between disciplines, where overspecialisation may lead to non-communicating vessels of parallel knowledge systems (Alexander 1997). A plethora of specialised and generalised conferences prevents any single scholar from following the field entirely. For example, the 'upcoming conferences' rubric of the website of the Natural Hazard Center alone lists around seven conferences for every month (www.colorado.edu/hazards/resources/conferences.html).

Another distinction within disaster studies that hampers cross-fertilisation is between disasters in high-income countries and disasters in low- and medium-income countries. For a long time, disaster studies in high-income countries tended towards a managerial outlook on disasters, where the possibility that processes within society could contribute to the cause of disaster is simply not conceivable. Disaster management in this perspective rested firmly within the domain of the state that is expected to take care of the risks that threaten its citizens. This was contrasted with studies that concentrated on disasters in low-income countries, where disasters were more explicitly framed in relation to vulnerability and social causation and where the role of the state in disaster response has been subject to debate and some doubt, both in terms of legitimacy and capacity. A less prominent role for the state left space for community-based disaster response, where non-state actors were considered by researchers to be legitimate and knowledgeable stakeholders in disaster response. There also has emerged in such development-oriented disaster studies considerable attention to international aid architecture, agencies and programmes – both in the fields of humanitarian aid and DRR.

At present, there are some trends towards overcoming this largely artificial distinction between disaster studies in high- versus low-income countries. There is increasing attention to state-based DRR in some low-income and middle-income countries. Some countries have made vast improvements over the last decades that enhance their disaster response capacity. For example, state performance in Cuba and countries like Mozambique and Vietnam has been impressive. So attention has shifted to multi-stakeholder governance of disaster. Meanwhile, scholars in high-income countries are becoming interested in analysing disaster in relation to society and in questioning the central role of the state and its capability to handle risk. Beck's landmark publication, *Risk Society*, directed more attention to the risks created by control-oriented risk management regimes (Beck 1992). Climate change has brought about a realisation that rich countries may be more vulnerable to disaster than hitherto assumed. Hurricane Katrina demonstrated that the differential vulnerability of the poor and elderly, and the role of race in access to state assistance in disaster, is pertinent in rich countries too.

A final set of fragmentising gaps can be seen between what we may call neighbouring fields. Disasters find their origin in the complex interplay of natural hazards with development (vulnerability) and environmental degradation. Currently, there is much attention to two other contributing factors to disasters: climate change and conflict.

Although the exact nature of the relationships between climate change and disasters is still subject to academic debate, it is generally accepted that such relationships exist and disasters have increasingly become a topic of interest to the climate change community, especially the one focusing on adaptation to climate change. In the 1990s scholars started taking up the study of natural hazards in the framework of climate change. However, the two communities continued diverging by working with different interpretations of concepts and developing different platforms for exchange and publications (Helmer and Hilhorst 2006; UNISDR 2007d). It is only in recent years that events have addressed both communities simultaneously. One initiative that explicitly aimed to bring scholars together working in the two artificially separated fields was the international conference on Climate Change and DRR organised by the Red Cross/Red Crescent Climate Centre in The Hague in June 2005 (see Chapter 18).

Attention to the interplay between disasters and conflict is also increasing. Between 1998 and 2003, Buchanan-Smith and Christoplos (2004) identified at least 140 incidents where natural hazards struck countries experiencing conflict or post-conflict situations. The co-occurrence of conflict and disaster is not a coincidence: conflict tends to negatively affect both the vulnerability and response capacity to natural hazards, while natural hazards like floods and prolonged drought can be one of the contributing factors to conflict, especially in harsh environments. Nonetheless, research and bodies of literature on disaster and risk reduction have developed rather separately from those on conflict (see Chapter 7). The Asian tsunami in Aceh and Sri Lanka and the earthquake in Kashmir triggered renewed interest in how to address the interplay between disasters and conflict. Some of these new efforts focus on 'disaster diplomacy' at the national and international level, studying how a disaster can be an opportunity to achieve peace (see www.disasterdiplomacy.org), while other efforts explore the interplay of disaster and conflict at the local level and study how people and communities cope with the multiple insecurities that result from the conflict–disaster nexus (Heijmans *et al.* 2009).

Another gap exists that separates researchers who work on questions of prevention and preparedness and those who study disaster response after the fact. Research into humanitarian action has been mainly focused on conflict-related humanitarian work and the provision of relief. An important exception is the work on famine. Since the 1970s famine research started to study the interplay between conflict, drought and relief policies in creating the slow-onset famine crises. Research on DRR before the disaster event has tended to undervalue the possibility of integrating DRR into the provision of humanitarian assistance. As a result, the discussions about DRR have tended to become separate from those concerning humanitarian action. One recent attempt to bridge these gaps among neighbouring fields was the formation of the International Humanitarian Studies Association during the first World Conference of

Humanitarian Studies in 2009 (www.ihsa.info). The association aims to bring together scholars addressing humanitarian crises from different angles including natural hazards, conflict, DRR, humanitarian action and climate change adaptation.

In general, research on disasters and DRR is still mono-disciplinary or at best multi-disciplinary in nature, meaning that it results in parallel contributions, rather than in integrated, innovative studies. If joint problem-solving were the aim of disaster studies, then the perspectives of the varying disciplines would be integrated in a manner that identifies and formulates a shared problem. Such collaboration would further require that researchers understand the language of each other's disciplines. However, in practice it appears very difficult for physical, earth or climate scientists to work together with historians, economists, political scientists and other social scientists in disaster studies, and vice versa. Why? One big obstacle is a differing perception among varying disciplines of the nature of knowledge and the nature of science. Natural sciences tend to view science as neutral, objective and separate theory construction from application. Many social scientists, on the other hand, view science as value-laden, subjective, connecting theory and practice in a dialogical manner. Different assumptions about knowledge and the nature of science are closely related to differing worldviews and how one defines concepts like disaster, risk or vulnerability. These worldviews in turn correspond to the different paradigms discussed earlier in this chapter. Methodologies across disciplines differ – either emphasising quantitative or qualitative methods – and these are valued differently, instead of appreciating them as complementary. This causes distorted communication and confusion.

Finally, there may be obstacles related to funding. Academic grants usually favour mono-disciplinary work and specific calls to study DRR are rare. Donor countries do occasionally fund research, depending on the priority they extend to DRR. For example, the European Union (EU) has given some attention, and hence funding, to evidence-based policy for DRR, particularly within its humanitarian programmes and development assistance programmes. An interesting development is the contribution of some insurance companies to academic research. The United Nations University's Institute for Environment and Human Security is partly sponsored by Munich Re, while the Aon Benfield reinsurance company sponsors the Benfield UCL Hazard Research Centre in London and, partially, the Risk Frontiers centre in Australia. Insurance companies are concerned with reducing risks of property losses and, therefore, they likely invest in research on specific hazards such as earthquakes, floods or volcanic eruptions affecting more affluent areas.

University research, policy and practice

University research has different roles to play in relation to policy and practice. It can inform policy and practice, it can be critical of policy and practice, and it can engage directly with policy and practice. Each role, though, has its challenges.

The first way in which university research can enhance policy and practice is by producing knowledge and making this available and applicable for policy. A prime example is formed by the disaster database EM-DAT, which is produced by the Center for Research on the Epidemiology of Disaster of the Catholic University of Louvain (www.cred.be and www.emdat.be). The database underlies much research and forms the basis of the yearly overview of disasters in the *World Disaster Report* of the International Federation of Red Cross and Red Crescent Societies (see Chapter 16). A problem with this function of university research is the fragmented and voluminous nature of research, which makes it difficult for policy and implementing actors to grasp what is important. In addition, the findings of university research often require translation to specific contexts to become applicable.

745

Box 61.2 LA RED: a pioneering research and action network in Latin America

Allan Lavell
Latin American Social Science Faculty–FLACSO, Costa Rica, and La Red de Estudios Sociales en Prevención de Desastres en America Latina

The Network for the Social Study of Disaster Prevention in Latin America – LA RED – was created by a group of fifteen people from seven countries in August 1992, in the Caribbean coastal town of Limon, Costa Rica. This small group of professionals sought to provide a collective social science perspective on disaster and DRR although they came from diverse disciplinary backgrounds, including engineering and geology as well as the social sciences. They met at a time when the orientation of the newly created International Decade for Natural Disaster Reduction (IDNDR) was still uncertain, but fears existed that it would be overly dominated by physical science and technocratic approaches.

Founding members of LA RED came from an array of professional and institutional backgrounds including academia, development and environmental NGOs, and government disaster organisations. All held a firm belief that disaster risk was essentially caused by skewed or failed economic and social development, and that DRR must be based on equitable and sustainable development. The varied backgrounds of founding members allowed for free flow of ideas that were nevertheless practical and pragmatic. Between 1992 and the present, LA RED has embraced directly or indirectly thousands of academics and practitioners looking for alternative ways of analysing and responding to disaster risk, and has become a major influence throughout the Latin American region. It has organised dozens of conferences and seminars, and published a new disaster journal called *Desastres y Sociedad* (Disasters and Society) and thirteen books. It has also developed new methodologies for risk and disaster analysis and training, and has reached tens of thousands of people regionally and worldwide through its website (www.desenredando.org). Moreover, its work has been fundamental in inspiring the creation of sister organisations, Periperi in Africa and Duryog Nivaran in Asia.

LA RED and its members have advised and worked closely with regional intergovernmental disaster organisations and many development assistance and disaster agencies as well as civil society organisations. It has produced two influential manifestos that helped shape international discourse and policy – one prior to the Yokohama World Conference marking the mid-point of the IDNDR (www.desenredando.org/lared/bitacora/cartagena1994/index.html) and one prior to the 2005 World Conference on Disaster Reduction in Kobe, Japan (www.unisdr.org/wcdr/preparatory-process/inputs/Declaration-Manizales-eng.pdf).

Major international contributions of LA RED include the DesInventar disaster database, which registers hazard events associated with damage and loss based on the premise that small and medium-sized disasters are as significant an aspect of the national and local risk profile as large-scale disasters (www.desinventar.org). LA RED also pioneered local-level training methods and tools for DRR, which subsequently informed many other institutional and organisational training programmes from the late 1990s onwards throughout the region.

Different university institutes have specialised in providing training to policy-makers and practitioners. For example, the International Institute for Geo-Information Science and Earth Observation (ITC) in Enschede, the Netherlands, has trained hundreds of DRR professionals from 170 countries since 1950 (www.itc.nl). The alumnae form an active network that further enhances research and training in the field. The translation of research to practice can also be facilitated by creating venues for exchange between academics and policy/practice actors. There are a number of networks where the different groups are joined together. To be mentioned are LA RED, a network founded in 1992 of academics and practitioners in Latin America (www. desenredando.org), and Duryog Nivaran in South Asia, which publishes the *South Asia Disaster Report* (www.duryognivaran.org). In Africa, Periperi is a university partnership to reduce disaster risks in Africa, originally led by University of Cape Town but now based at Stellenbosch University. It set out as an educational initiative but is now moving to the formation of centres of excellence to initiate research programmes. The Asian Disaster Preparedness Center (ADPC) serves a similar training function throughout its region (www.adpc.net).

The biannual Global Platform for DRR in Geneva, organised by the United Nations International Strategy for Disaster Reduction (UNISDR), brings together academics and other researchers, policy-makers, practitioners and key players from governments, international organisations, civil society, media and the private sector to exchange knowledge in the areas of risk management and reduction. The ProVention Consortium initiative played a major role in brokering between research and policy and practice communities between 2000 and 2010, and its website still unlocks vast resources from the different communities (www. proventionconsortium.org).

A second role of university research is to develop critical analysis of policy and practice. Although this leads to seemingly 'academic' research that is not easily translated into policy or practice, this gadfly function is an indispensable quality of academic research to advance DRR in the long term. It can be combined with the other roles of academic research, but because it requires a healthy distance between researchers and the field of DRR it needs to be consciously cherished. PhD research is very suitable for this kind of project as it allows scholars to do in-depth case studies and reflect away from the hectic everyday pressures of the field. It is, therefore, important to find venues to bring PhD findings to the field. This critical research explores, among others, fields like the politics of disasters. Disaster politics is an area long neglected by many researchers and practitioners 'who essentially believe that there shouldn't be a politics of disaster' (Olson 2000: 265). Research of this nature focuses on state–society relationships before and after disasters happen, and investigates how and when disasters have the potential to threaten the powerful elite, where elites capture the space created after the disaster to further their vested interests or in what circumstances disasters may lead to change towards more inclusion or protection of vulnerable people (Drury and Olson 1998; Pelling and Dill 2009a). Such critical research also focuses on institutional failure and weak governance as root causes for people's vulnerability to disasters (Ahrens and Rudolph 2006) (see Chapter 5).

Critical research may also take up the theme of the everyday politics of disaster response, asking how disaster response outcomes get shaped, *who* decides, *whose* risks are prioritised and *which* risk reduction measures will be implemented (Christoplos et al. 2001). These scholars view DRR as a political process and concentrate on how interventions are conceptualised, how choices are made and implemented. They are concerned that risk often is not reduced but re-allocated to the poorer, marginalised and socially excluded segments of society (Lebel et al. 2006). Importantly, this research also addresses the growing body of practice around community-based DRR. Further, it has a task in identifying and testing the implicit assumptions in DRR practices of government, NGOs and other agencies.

A third role of university research may be to join in teams together with DRR practitioners and policy-makers. It is a way to make research more accessible and relevant to the questions and dilemmas confronting organisations working in this field, and thus to better-informed and more sensitive policy and practice of DRR interventions. Collaborative or interactive forms of research break through the boundaries of disciplines, but also through the distinction between scientific knowledge, practitioners' expertise and possibly practical wisdom of disaster-affected populations. It is assumed that research in real-time settings will lead to a better match between new theories and relevant practice, particularly in disaster and conflict areas. Most of this kind of research, however, is short-term in nature and consultancy-based. Long-term, systematic inter-active research involving academic institutions, government officials and, for instance, aid agencies is still rare. There are specific barriers and challenges. One is the difficulty to synchro-nise academic theorising with social engagement, and to balance scientific quality with the practical relevance of the research. Another important barrier is the academic performance rating system which follows the American-based ISI system (Institute for Scientific Informa-tion). This system does not recognise policy reports, consultancy products or other written outcomes produced through collaborative research. This has no scientific status and therefore it may negatively impact on the academic career prospective. Academics who want to keep their university position, and who wish to do interactive research, often do not get the full support of their institutional boards. A strong lobby is needed to get societal relevance included in the definition of 'good science'.

In a similar vein it could be argued that a specialised code of ethics should be developed for disaster-related research to ensure that researchers do not inadvertently 'do harm' and that they remain open to opportunities to 'do some good' (Kelman 2005). Commonly used standards for 'human subjects research' provide a starting point involving such normal issues as informed consent, anonymity of informants, etc. In situations of distress, false hopes can easily arise when aims of the research are not clearly articulated. Unrealistic expectations may be avoided when researchers work closely with operational agencies to ensure that findings will be considered in subsequent aid interventions, provided that communication between researcher and aid agency is open and clear to not negatively affect NGO–community relations. 'Do some good' can imply that research counters myths and stereotypes, or gives voice to the marginalised segment of the affected population. In the field of DRR research, as more generally in development studies, many believe that there should be a commitment on the part of researchers to provide the affected communities with information and knowledge products as acknowledgement of local 'ownership' and an orientation by researchers towards changing the world and not just understanding it.

Conclusions

The importance of DRR slowly finds policy recognition and, as a result, university research that explicitly aims to enhance the effectiveness of DRR is also increasing. Policy analyses of hazards have become more sensitive to the complexities involved, which further leads to specialised research.

The time is ripe for more innovative ways of conducting research for DRR. In this chapter, we have argued that research must become more interdisciplinary – or transdisciplinary as some argue – and that research must become more interactive with and accountable to stakeholders in the field. The organisation of academia is unfortunately not well geared to this need, and continues to be biased towards mono-disciplinary studies. An important need, therefore, is lobbying within universities and research institutions to create an infrastructure for interdisciplinary DRR research that values collaborative research with policy-makers and practitioners.

Acknowledgements

We kindly thank the editors of this *Handbook* for their valuable contributions to this chapter, and thank Bruno Haghebaert, Sabine Maresch and Jeroen Warner for their comments.

62

Education and disaster

Emmanuel M. Luna

COLLEGE OF SOCIAL WORK AND COMMUNITY DEVELOPMENT, UNIVERSITY OF THE PHILIPPINES

Introduction

It is imperative to ensure the rights of people to education and to produce education that sustains life. While protecting the school children and adult students from any harm and disasters, we recognise too the role of education in creating knowledge and awareness that will contribute to disaster risk reduction (DRR) in all aspects of human lives, in all places, at all times, at the personal level, family, community, workplaces and society in general. There is a vast reserve of resources that can be mobilised for DRR in the educational system, such as the students, teachers, researchers, staff, parents and community association, the school facilities and knowledge itself.

Education is a basic universal need, a passage to a better life. Education enables one to be prepared and to contribute fruitfully to society. Education is a means of emancipating the poor from poverty and an agent for social mobility. As one gains access to education and becomes productive, one also gains prestige and better social status in the community. Similarly, society develops with the advancement in knowledge and technology and economic growth. Growth, though, is not always for the better but is accompanied by conditions that can perpetuate sufferings of people and depletion of the world in which we live. The issue at hand is not just environmental sustainability but survival of humanity as humans become a co-determinant of what used to be a natural phenomenon such as climatic conditions. Disasters taking place can no longer be attributed solely to nature, and education needs a re-examination of its values, particularly on how it contributes to environmental disharmony and how it can be an instrument for DRR and global sustainability.

Disasters as barriers to rights to education

Schools and school children are often affected by disasters. For example, in May 2008, approximately 7,000 students were killed in school buildings in Sichuan, China in the 2008 earthquake. The earthquake in the Spitak area of northern Armenia in 1988 resulted in 25,000 deaths, two-thirds of whom were children and adolescents. In one school alone, 285 out of 302 children died. The 1999 Chi-Chi Earthquake damaged 700 schools nationwide, with 43 schools in the Nantou and Taiching area completely destroyed (UNCRD 2009: 7–9).

A school's buildings and facilities might not be directly affected by disasters. However, they may have a role in providing a place for care and shelter for the general public who have been displaced by disasters (Risk RED for Earthquake County Alliance 2009: 14). In the first place, some of those affected may be students and their families. There are public schools mandated by national and local authorities to serve as evacuation centres. Private schools also volunteer and open their facilities as a refuge. In times of disasters and emergencies, classes are suspended, adversely affecting the education of the children.

A study conducted in the Philippines (Luna *et al.* 2008) found that in times of disaster, school administrators experienced problems with the disruption of classes, cleanliness of the school premises and disturbance among the students affecting their concentration in their studies. Students had to attend classes even on Saturdays and Sundays. Some missed classes and had lower attention span and interest in their studies. In cases where the classes were held in tents, the set-up was not conducive for learning and was uncomfortable for the students. Teachers were also adversely affected by disasters since they were also mandated to assist during emergencies. They had additional tasks like managing the evacuation centre, helping the family evacuees, acting as an information officer, care-giver and room co-ordinator. They also had difficulties in adjusting to the school environment and had to bear with the discomforts of the tents used as classrooms. The evacuees occupying the classrooms further forgot to clean up after themselves and caused damage to facilities.

Reducing disaster risks during education

Safe school locations for reducing disaster risk

Knowingly or unknowingly, school campuses are established in locations that are vulnerable to natural hazards. The school site could be flooded, eroded if along the river, or washed away by a tsunami. Unaware of the risks, schools could be located in the path of lava or lahar flow. Some river channels are unstable and schools that used to be distant from the river bank suddenly find themselves in mid-water if the river changes its course. In other cases, the school could be located on a steep slope or under an unstable slope in the path of a potential landslide. An example of this was the case of Guinsaugon in the Philippines. The whole village was totally engulfed by a landslide in mid-morning of 17 February 2006, killing about 1,000 people, and burying an elementary school with around 350 children, teachers and women from a non-governmental organisation (NGO) having a meeting in an open auditorium. It was not only the school that was poorly sited, but the whole community that was at risk due to its location, lack of risk assessment and unawareness of the hazards. An established settlement is not an indicator that the area is safe.

A school site can also be dependent on donation of a building site, regardless of its safety. Taking this cost-cutting opportunity, government agencies have built schools on donated but unsafe locations, jeopardising the lives of the children (Luna *et al.* 2008).

Reducing risk though safe buildings

A major cause of casualties in school is faulty building design and construction. The collapse of school buildings that have been improperly constructed, often because of corruption, and the death of school children in El Salvador and India (2001), Italy (2002) and Turkey (2003), have led to an increasing outcry both locally and internationally (Wisner *et al.* 2004: 317).

Earthquakes are the most common hazard causing the destruction of school buildings. In Algeria, ninety per cent of its 30 million people live in a band of land located on an African and Eurasian tectonic plate boundary, which experiences moderate-to-strong earthquakes. There were five destructive earthquakes from 1980 to 2003. Reports indicate school damage of twenty per cent for the 1989 magnitude 5.7 earthquake, and ninety-five per cent in 1980 for a magnitude 7.3 earthquake. The typical damage to school buildings reported included rupture of staircases, destruction of joints or of short columns, damage to masonry and 'pancake' collapse. While there were seismic building codes, the school buildings constructed since 1983 had adopted one design that could be duplicated across the country, but which was far from up to the standards prescribed by Algeria's own seismic codes (Benouar 1994; Benouar and Laradi 1996).

School buildings must have designs resistant to and able to withstand hazards, and with protective measures to reduce the risks to children and save lives. One of the initiatives to demonstrate this was through the United Nations (UN) Center for Regional Development project, 'Reducing Vulnerability of School Children to Earthquake', in the Asia-Pacific region, specifically in the Fiji islands, India, Indonesia and Uzbekistan. The project aimed to 'ensure that school children living in seismic regions have earthquake resilient schools and that local communities build capacities to cope with earthquake disasters. The project includes retrofitting of some school buildings in a participatory way with the involvement of local communities, local governments and resource institutions, training on safer construction practices to technicians, and disaster education in school and communities' (UNCRD 2009: 8–14). The project resulted in country-specific guidelines on earthquake safe construction which incorporate solutions to practical problems during school retrofitting.

In Nepal, the National Society for Earthquake Technology–Nepal also published 'Protection of Educational Buildings Against Earthquake: A Manual for Designers and Builders' (Bothara *et al.* 2002). Other resources for school protection that can be used in building new schools or retrofitting existing structures were listed by Wisner (2006a).

Reducing risk in the immediate external school environment

Children could also be at risk in the school's immediate environment. As the children come to school, they might not be aware of the hazards they might face along the way. The failure of local authorities and the school administration to create such awareness can result in child fatalities, as shown in the following case. Heavy rain is common in the Philippines, even if there is no typhoon. One day in January 2008, an eight-year-old boy was drowned when he was swept away by the swelling river on his way home from school. He was previously able to cross the river because there was no flood. There was no typhoon, hence no warning signal was issued, although the locality had been experiencing days of heavy rains (Mosqueda 2008: 21). This incident shows the need for protection of children beyond the school premises; the need for children, teachers, parents and community leaders to be aware of the risks and the urgency to respond to risky conditions, especially by local authorities. It also shows the need for more effective warning systems.

Protecting school children in times of emergency

In times of disaster or emergency, the primary role of the school is the protection of students by providing shelter, meals and health care until the students are released to their parents or previously authorised adults (Risk RED for Earthquake County Alliance 2009: 14). This is a complex task that requires familiarity with the procedures, routine, communications, equipment and materials to be used, and roles to be played by students, teachers and other school staff, local authorities, fire and security personnel, rescue groups, health and medical teams, and other key players. The case of the boy who was carried away by the flood on his way home could have been prevented if the school administrators were conscious of their protective role in times of emergency.

Familiarising oneself with these tasks requires a routine, which is something that is practiced regularly. This can be done through drills. One myth says that 'a drill now and then is enough', but human retention is short, hence regular drills must be undertaken, at least every six months, so that the pattern can be ingrained (Burnstein 2006: 50). Considering the frequency and magnitude of disaster events, disaster drills are becoming a necessity for preparation for protection in schools. In California, which experienced a devastating magnitude 6.7 earthquake in 1994, the Standard Emergency Management System requires both public and private schools from kindergarten to grade twelve to have school disaster planning and drills. With 3.6 million children enrolled in 262 public districts in seven counties, and 7,537 school buildings constructed before 1978 that are of questionable safety, a major earthquake could result in an unprecedented catastrophe (Risk RED for Earthquake County Alliance 2009).

To determine the state of school disaster prevention and preparedness in California and its implications for school disaster management worldwide, Risk RED for Earthquake County Alliance conducted participatory action research that culminated in a simultaneous drill called the 'Great Southern California ShakeOut'. This was participated in by almost 4 million children and adults, 207 school districts, 95 additional public schools and 650 private schools in 8 counties. The earthquake preparedness activities during the drill ranged 'from "Drop, Cover and Hold on", to school building evacuation, school emergency response coordination, and full response simulation drill using standard emergency response system' (Risk RED for Earthquake County Alliance 2009: 9). The research reported tremendous learning, which benefited millions more people as the children brought home the lessons gained from the school drill.

Ensuring rights to education during emergencies

Individuals should not forfeit their rights to education during emergencies. Neither should education remain outside mainstream humanitarian discussions, planning and resource allocation. It should be seen as a priority. Steps are required to ensure a minimum level of quality, access and accountability for education in crisis situations (Inter-Agency Network for Education in Emergencies 2004: 6).

The Inter-Agency Network for Education in Emergencies (INEE), an open network of UN agencies, NGOs, donors, practitioners, researchers and individuals from affected populations, has been working to ensure the right to education in emergencies and post-crisis reconstruction. 'The network is responsible for gathering and disseminating good practices, tools and research, promoting the rights to education for people affected by emergencies through advocacy, and ensuring the regular exchange of information among its members and partners' (Inter-Agency Network for Education in Emergencies 2004: 6).

Education in an emergency situation requires gender-responsive measures that ensure that girls and boys have equal access to education. Schools and learning spaces should be located near to the learners' homes and away from sources of danger and risk. The community members could be involved to ensure safe travel to and from school. Proper timing of school should be kept to enable the students to attend to their responsibilities. Child care facilities could be provided for female teachers and students who have children. Similarly, the schools should provide sanitary facilities and a meal programme or take-home rations for all children. The students could also help to prepare a 'missing-out map', or map of the children in the community who are currently not in school. They could be involved in designing a gender responsive education programme to reach out-of-school children. Furthermore, education must be empowering and protective for girls and boys (UNESCO 2006: 8–9).

Box 62.1 Comprehensive school safety

Marla Petal
Risk RED

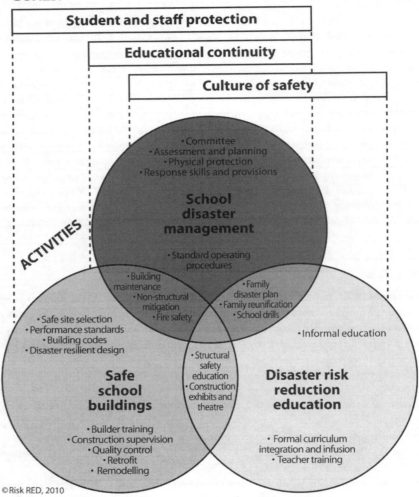

Figure 62.1 Goals and activities of comprehensive school safety

The goals of comprehensive school safety in the face of expected natural and man-made hazards (Figure 62.1) are:

- Student and staff protection;
- Educational continuity; and
- A culture of safety.

These three goals are accomplished through three overlapping spheres of activity: safe school buildings, school disaster management and disaster risk reduction (DRR) education. Each of these is typically addressed through distinct people and processes.

- Safe school buildings: Safe site selection, building code compliance, DRR performance standards and design are often the responsibility of central education, construction and planning authorities. No matter who is responsible, in order to assure safe buildings, the key actors in this sphere must also follow through to be sure that construction workers have adequate training and supervision and that there is a process of quality control. Retrofit may be an option, and remodelling should also be seen as an opportunity to improve safety. While the maintenance of school facilities to sustain building safety typically rests with the administration and authorities to whom the building is handed over, technical and material resources are required to assure sustained safety.
- School disaster and emergency management: Each and every local school community is at the heart of ensuring its own safety and educational continuity. This takes place through assessment and planning, physical and environmental risk reduction, development of response skills and provisions and advocacy. An ongoing committee with strong support and leadership from the school administration will need to involve teachers, staff, parents, students and community representatives in developing and implementing standardised policies and procedures, contingency plans and staff skills. This group leads in addressing physical safety through oversight of building maintenance, non-structural risk reduction and fire safety. Plans are tested and improved through regular school drills. Families are critical to success through their own family disaster planning and through joint family reunification planning.
- DRR education: The development of a culture of safety rests heavily upon what children (and by extension their families) learn at school. This in turn rests on school teachers and the training and support they receive to implement both formal and informal DRR education. The content of this education must move beyond hazard identification and awareness, to include practical life skills for early warning system design and use, methods of risk reduction, preparedness measures and response skills. This includes the use of the school building and other constructions as an educational opportunity to teach the core elements of structural safety.

A wide and growing selection of resources is available through the UNISDR PreventionWeb educational materials collection, at www.preventionweb.net/go/edu-materials.

Education for disaster risk reduction

Experiences and cases show how schools can be an instrument for DRR (Hayashi 2010; UNCRD 2009; UNISDR 2007b; Wisner 2006a) by doing simple yet participatory risk and vulnerability analysis, conducting disaster risk awareness, mainstreaming DRR in the lessons and curriculum, serving as an evacuation centre and supporting advocacy for policy changes that would ensure safety and protection from disasters. The significant role of education in DRR is well articulated in the Hyogo Framework for Action (HFA), Priority Action 3. In 2006–07 the

UNISDR launched a campaign, 'Disaster Risk Reduction Begins at School'. The campaign came out with a compilation of good practices and lessons on DRR in school, showcasing experiences from around the world (UNISDR 2007b). Wisner (2006a) also made a comprehensive review of the role of education and knowledge in DRR.

The activities at the school and local level are concrete expressions and manifestations of real actions on the ground. When educational activities are affected by disasters, it is imperative for school children to return to school as soon as possible. In cases where the school facilities are destroyed, then school construction provides an opportunity to 'build back better'. This principle of reducing risks in schools through rehabilitation and construction is illustrated by the programme of ActionAid helping the school children who have been displaced because of the earthquake in Haiti. The first stage planned to April 2010 was for emergency measures such as the creation of child-friendly spaces that could be used for non-formal schools, provision of scholarships and educational materials, and establishment of temporary schools. The second stage to December 2010 was designed to assess the safety of schools, the inclusion of children with disabilities and promote disaster preparedness in schools. The third phase in 2011–12 aimed to prepare Haiti for a better future by reviewing the curriculum and advocacy work on DRR and the extension of the educational programme (ActionAid 2010).

It is ironic that just a few days after the UN Educational, Scientific and Cultural Organisation (UNESCO) in Santiago facilitated a conference on risk reduction in the education sector among the education ministries in Chile and in the region, a magnitude 8.8 earthquake struck Chile in February 2010. The Ministry of Education was faced with the challenge of putting into action the good practices discussed during the conference. The earthquake resulted in twenty-three per cent of the 4,432 educational institutions being completely unusable and fourteen per cent being partially operational. There was a swift response by the Chilean government, with the Ministry of Education seeking the assistance of UNESCO Santiago in rapid needs assessment, and provision of temporary learning spaces to help restart primary and secondary schooling. By 26 April 2010, it was announced by the Ministry of Education that all students in the affected areas would return to the classroom. This was considered a milestone achievement. It was also pointed out that 'the early reactivation of the education system is

---

Box 62.2 A spare-time university for disaster risk reduction education

Michael H. Glantz
Consortium for Capacity Building, University of Colorado, USA

Ilan Kelman
Center for International Climate and Environmental Research–Oslo, Norway

Educational initiatives use various forms of technology. The Spare Time University (STU) is one conceptual as well as practical approach for applying technology to education. The goal of STU is to empower anyone to have access to knowledge and information, according to the level of technology that they use already, without having any specific level of formal education.

The basis of STU is information delivery by any technological means so that it matches users' needs. The building blocks are the packaging of information in different ways, depending on the technology available and user information needs. That means using emerging as well as traditional technologies and techniques to provide free educational information in formats that users request through media to which they already have access. STU thus becomes an information exchanger and stimulator, not just an information disseminator, with users encouraged to become active in spreading STU's information and in developing STU's material. In the case of disaster risk, this approach is perfectly adapted to exchange and co-production of knowledge concerning hazards, vulnerability, impacts and potential risk reduction actions.

Examples are brief information 'nuggets', sent as mobile phone text messages or as text-only emails, along with information modules, audio only for radios and MP3 players, and multimedia and interactive for high-bandwidth internet and MP4 players. The smaller packages, applying to mobile phones and low-bandwidth email, could be a few paragraphs written or recorded with basic facts or action items. Examples of simple messages for smaller nuggets are: 'In an earthquake in places with enforced seismic building codes, you should drop, cover and hold on', and 'If the road is flooded, then don't drive, walk, ride, or cycle through the water'.

The larger packages could be multimedia, interactive lectures, seminars or podcasts accessible online with high bandwidth or viewed on a DVD player, television or computer. They might be downloaded in people's homes, viewed in internet cafés or displayed in a room of a community centre or library for anyone to attend. These can be discussed and reactions and comments posted on a wide variety of social networking forums. For larger packages, both theory and practice could be covered. Theoretical modules might cover soil erosion from landslides and drought, root causes of vulnerability, explosive volcanic eruptions or how to communicate DRR effectively to different audiences.

Practical topics range from implementing community-based DRR by combining local and non-local knowledge forms or how emergency managers could assist the survival of occupants in buildings hit by volcanic ash or mud flows. Mountain tourist areas might use avalanche or ski safety nuggets sent to televisions as part of routine guest check-in procedure, while Caribbean islands might exchange hurricane modules with typhoon-prone Pacific islands to share each other's experiences on vulnerability reduction.

STU should complement and enhance, not replace, already-existing structures and initiatives, including face-to-face learning, teaching and exchange. Those who engage in volunteer teaching could access STU material and then pass it on to others in day-to-day conversations with neighbours and friends or by holding community gatherings. This also ensures that DRR information becomes integrated into people's daily lives.

essential to instilling a sense of normality ... thereby bridging the gap between the calamity and reconstruction ... getting back into school is one of the best ways to help youth overcome negative experiences and regain hope' (UNESCO 2010).

Local and national efforts require policies and guidelines that could help direct resources and initiatives for education and DRR. Box 62.1 shows the priorities, links and tasks of education, with sample case illustrations from practice in the field.

There are gaps that have to be addressed pertaining to education in DRR. Wisner (2006a) identified gaps and opportunities concerning primary and secondary education, tertiary education, training, protecting educational infrastructures, community–based DRR (see Chapter 59), media (see Chapter 63), communication and risk awareness, scientific knowledge, and the research and knowledge network (see Chapter 9). In the three years following the 2005 World Conference on Disaster Reduction and the formulation of the HFA, there has been an 'explosion' in the production of new educational materials on DRR in various forms. However, 'the widespread practice of public and school-based education for disaster risk reduction is in its infancy ... most of the educational efforts directed towards the public and children have neither been systematically conceived or tested ... and disaster risk reduction education is not yet an integral part of disaster risk management policy, planning and implementation' (Petal 2009: 285–6). Raising awareness of people even in areas where there are established warning systems and high-tech preparedness measures appears to be indispensable (Morin *et al.* 2008: 433).

Mainstreaming DRR in primary and secondary curricula

Mainstreaming of DRR in the school curriculum has been widely propagated and supported by several UN agencies, international funding institutions and humanitarian and development organisations.

There are many ways of mainstreaming DRR in formal education at the various levels. For primary and secondary levels, DRR can be integrated into subjects such as sciences, social sciences, languages, mathematics and the like. In sciences, DRR can be easily integrated since many of the hazards are natural and can be covered by earth and physical sciences. In language classes, the students may be asked to write essays and do exercises about disasters, their experiences and views. In social studies, DRR can be included as a topic and can blend well with sessions on geography, history, social relations and psychology. In mathematics or physics, exercises may be given using disaster-related situations. In all these subjects, school projects could be made a requirement in order to promote greater DRR awareness, such as scrap books about disasters in the community, poster-making or song- and poem-writing with DRR as the theme. It could also consist of forming a student group that could prepare for disaster relief, or a forum or community training.

The school could use participatory methods to do hazard, vulnerability and capacity analysis through actual survey of the campus premises and its vicinity. They could make community maps showing hazardous areas. The students could share these assessments with their parents and the community by displaying the outcome or conducting seminars and workshops with residents. For example, PLAN International has supported projects in El Salvador in which children and youth have made such maps and have shown that young people are capable of full participation in community-based DRR (UNISDR 2007b: 12–6).

Mainstreaming DRR in education requires teachers who are equipped to teach DRR in a more specialised way. More often than not, DRR was not part of the curriculum that the teachers underwent in their tertiary education. For them to be able to teach DRR, they must first undergo teacher training in DRR. The challenge here is the need to cascade the training to a broader scale (Wisner 2006a: 67).

Professional DRR training in tertiary education

One of the recommendations of the HFA is to 'encourage universities to develop degree programmes specific to disaster management and risk reduction issues' (UNISDR 2007c: 64). Professional training and formal education as an aspect of risk reduction have not yet been studied. However, it has been noted that there is an increasing number of courses on risk and disaster management. Most of these university courses are still in Europe and North America (Twigg 2004: 183), though there are some universities in Latin America and the Caribbean region that offer diploma and masters programmes (Wisner 2006a: 30). In many academic disciplines, subjects exist that are directly related to disaster concerns, such as those offered in geology, geography, sociology, psychology, engineering, medicine, public health and the like (see Chapter 61).

Advocacy and mobilisation for education and DRR

South–South collaboration

One support mechanism for education in promoting DRR is to establish a network among individuals, groups and institutions for a specific purpose. One of these is the Training and Learning Circle (TLC), a network of training institutions and universities organised to re-examine, strengthen and facilitate the crucial interface between training and education for community-based DRR. For example, this is being promoted and facilitated by the All India Disaster Mitigation Institute (AIDMI), the Center for Disaster Preparedness in the Philippines, the Asian Disaster Preparedness Center and the Special Unit for South–South Cooperation of the United Nations Development Programme. The TLC aims to strengthen the capacity of training institutions and universities by reviewing existing and developing new learner-centred learning materials and methodologies. The TLC enhances learning through South–South knowledge and solution exchanges, with a focus on addressing systemic gaps and topics in training and education. For example, the AIDMI published course material on making schools safer which can be used by trainers across the region (AIDMI 2009b: ii).

Teacher volunteer mobilisation

Teachers can be mobilised for DRR and response in times of disasters. They have varied skills and these can be tapped to carry out different functions. A good illustration of this is the Japan Hyogo Teachers' Union. In January 1995 the Kobe earthquake killed more than 6,000 people and damaged about 700,000 houses and buildings. The basic lifelines were cut off and the evacuees needing shelter surpassed 300,000, including 180,000 people who sought shelter in a nearby school. Though many teachers were affected by the disaster themselves, they volunteered and provided services such as managing the evacuation centre and attending to the needs of the evacuees. At the same time, they ensured the safety of their pupils and prepared for the early opening of the school. The Japan Hyogo Teachers' Union was able to dispatch 7,567 members who worked two days and one night each time in groups of three. This continued for 182 days, until 31 July. The teachers generated funds through street donations amounting to 215,108,000 yen. The Union was able to support 2,490 pupils for the next fifteen years (Izumi 2010: 1).

Political advocacy

Sometimes it is necessary for parents, teachers and students to work together in order to call attention to hazardous school conditions and generate political will among leaders. A very successful example is the group 'Families for School Seismic Safety' (fsssbc.org) in Vancouver, British Columbia, Canada. Parents, teachers and students became aware of the potential danger posed by hundreds of nineteenth-century multi-storey brick school buildings in the city of Vancouver, as a large earthquake to the south in Seattle, Washington, USA reminded them of the fact that the Pacific Coast of Canada, where Vancouver is located, is prone to large earthquakes. They teamed up with a seismic engineer from the University of British Columbia, educated themselves and carried out a public advocacy campaign that finally resulted in funding from city, province and federal levels in Canada for repairs and in some cases demolition and the building of new schools.

Conclusions

The education sector, schools and children face real dangers and risks due to natural hazards. The losses from previous disasters in terms of lives, assets, relationships and socio-psychological well-being are astonishingly depressing and difficult to imagine. Global concern and commitment to education to reduce the risks of disaster is becoming more concrete through policy formulation and support for innovative programmes in institutionalising DRR in education policies, main-streaming it in curricula, producing informational, educational and campaign materials, devel-oping new building designs for schools, constructing and re-constructing safe and resilient school buildings, organising and mobilising support groups for DRR in education and mainstreaming human rights and gender rights in education.

As the innovations in education and DRR programmes are tested and implemented in a few selected areas, the exigent calls for regularising these programmes need to be extended in scale and reach. Time is not the only element to consider here but resources as well. Funds are needed for new school buildings, training of teachers, production of educational materials, dis-semination of knowledge and technology. While the less-developed nations could mobilise their own resources as their contribution to the global strategy for participation in development, there are simply very limited resources on which to draw. Deep political, institutional and cultural reasons for the inability to support DRR programmes for education need to be tackled by political advocacy and legislation, and administrative reforms that reduce the 'silo mentality' and competition among different government departments (see Chapter 51). Education in DRR programmes entails greater collaboration, horizontally and vertically, for the sharing of appropriate and low-cost technology, and for mobilising various stakeholders at the macro, meso and micro levels, including local schools and communities, for collective action.

As an emerging field in academia, DRR and management must be conceptualised to pass the rigorous requirements of an academic discipline so as to effectively facilitate the mainstreaming of DRR in education, with professional training in tertiary education. Research and knowledge development on DRR must be fully supported, as these will 'kill two birds with one stone': the application that results in risk reduction and the advancement of the field of disaster risk reduction and management.

Building the capacities of schools and those in academia, and knowledge production would, it is hoped, help to reduce risks and respond to disasters. Schools, communities, governments and other supporting institutions are building blocks to ensure safety in schools and to make education an effective tool for DRR.

63

Media, communication and disasters

Tim Radford

JOURNALIST, BASED IN EASTBOURNE, UK

Ben Wisner

AON-BENFIELD UCL HAZARD RESEARCH CENTRE, UNIVERSITY COLLEGE LONDON, UK

Introduction

Until 8 May 1902 Marius Hurard was the editor of a local paper called *Les Colonies*. When all about him in Martinique the skies darkened and citizens began to react, he interviewed a science teacher at the local *lycée* (secondary or high school), a certain Gaston Landes (Scarth 2002). The interview ended with the conclusion, 'Mont Pelée offers no more danger to the inhabitants of Saint-Pierre than Vesuvius does to those of Naples'. Not content with glossing the science teacher's faulty information (M. Landes believed that Pompeii 'had been evacuated in time'), M. Hurard then in the same edition commented on the fourfold increase in the number of people taking the steam ship to safety and asked, rhetorically, 'Was there a better place to be than Saint-Pierre?'

History supplies the answer: yes, absolutely anywhere, especially at 8 am on 8 May, when a *nuée ardente* (pyroclastic flow) sent a wall of flaming rock dust and gas at temperatures of up to 450°C racing from the mountain at around 500 km/hr (see Chapter 28). Within two minutes, Saint-Pierre had been incinerated and more than 27,000 people scorched to death. An editor always has the final say, and by that time M. Hurard had certainly published his last word.

History may judge him harshly, but many newspapermen would have a sneaking empathy. Newspapers have an investment in the community they serve. They rely on advertising, reader loyalty and community support to cover overheads and pay reporters. A newspaper – Benjamin Franklin's *Pennsylvania Gazette*, Lenin's *Iskra*, H.L. Mencken's *The Baltimore Sun*, C.P. Scott's *The Manchester Guardian* – is an expression of optimism, of the belief that the people who read it today will be here to read it tomorrow. Disaster risk reduction and management was a phrase absent from M. Hurard's vocabulary, quite possibly because it rarely crossed the lips of any of his readers. Quite possibly, nobody ever bought a newspaper hoping or expecting to read the words 'disaster risk reduction' (DRR). Since newspapers are successful because they accurately reflect the needs and preoccupations of their readers, the reasonable conclusion is that they are not consciously interested in DRR.

Several questions arise from these observations and are the focus of this chapter. If DRR is not part of the media agenda or the readership's, should it be? Can it be?

The case for media involvement with DRR

The answer to each question is yes. The first is a simple moral issue: a free society is inseparable from a free press, a phrase that must include broadcasting and internet media as well as printed publications. People read media for information and for diversion, but, ultimately, they expect – and would certainly prefer – media to alert them to any threats to their lives and to their freedom, and therefore to any folly, corruption and incompetence in the governments that claim to uphold that freedom. Newspaper titles often incorporate the idea of vigilance and warning: Argus, Sentinel, Observer, Guardian, Beacon, Messenger, Clarion, Tribune, Spectator and so on, an implicit recognition of a newspaper's responsibility to monitor a community's sins of commission and omission (Pilger 2005).

Journalism contains people as disparate as any other human profession, trade or vocation, but the evidence from direct conversations with journalist colleagues from Russia both before and after 1990, and from China, India and Africa, suggests that journalists everywhere seem to have instinctively similar attitudes to authority and to the readership. After the Sichuan earthquake of May 2008, China's journalists asked the same kinds of questions that might have been expected from US, British or Japanese news media (Watts 2008). They did so because the readers they represented needed answers. The answers they got were unsatisfactory, but they asked. So much for 'should'.

Now, the second question: could DRR ever become a topic at a morning editorial conference, or a radio phone-in programme, or an evening news bulletin 'special'? The answer is, yes of course, but it is most likely to do so in the wake of an event. This, of course, is invariably the moment at which it becomes clear that there has been insufficient attention to disaster risk. There is a widespread assumption within the media that readers and viewers are simply not interested in disasters until after they have happened. In democracies, the material normally found in the media tends to reflect popular pre-occupation: football, economic depression, political scandal, fashion, celebrity misbehaviour and so on. People are moved, and shocked, by the disasters that overtake others, but unless constantly reminded of danger, most people will not regard hazards as frequent enough, or sufficiently probable, to be an urgent threat to themselves. Earthquakes, until they happen, are invisible, unpredictable events. Most people do not live in the shadow of a volcano, and the story of the editor of *Les Colonies* reminds us that even those close to violent eruption can persuade themselves that they are safe. There is well established psychological and sociological research that has ascertained the general disconnect between 'objective' measures of risk and the 'subjective' assessments and rankings given by the general population (Gardner 2008).

So does that mean that DRR is a niche topic: a no-go area, a subject to be marked only by a few pious but perfunctory editorial sentiments on World Disaster Reduction Day? (UNISDR 2009d) The answer is no. Attitudes can be changed, and journalists have traditionally been part of the machinery of change. The mistake is to think that these things can be done overnight. It took three decades to establish the widespread but still incomplete public acceptance of the link between smoking and health. It has taken so far two decades to persuade most political leaders to accept that climate change represents a political problem so serious that they must actually commit to effective political action at some future point. In both of these instances, the media played an important and perhaps decisive role, but in both of these instances the media responded to energetic and persuasive pressure from informed campaigners (Siegel 1998). So if DRR ever becomes a normal part of the news desk agenda – that is, becomes a routine theme at daily editorial conferences, as such themes as environment and health are – it will do so because engineers and scientists have made the subject seem compelling, exciting and urgent (see Box 63.1).

Potentials for media engagement with DRR

The press and broadcasters have, on occasion, seen trouble coming and said so. After the human tragedy, political farrago and civic fiasco that followed Hurricane Katrina, it became obvious that some experts and some reporters had warned of the threat to New Orleans (McQuaid and Schleifstein 2002; Laska 2004). Hindsight always provides perfect focus, and sadly most elected officials and other authorities became conscious of these warnings after the event. Therefore Hurricane Katrina offers a bleak lesson for politicians, lobbyists, reporters and newspaper readers concerned with disaster risk: to say something only once or twice is almost the same as not saying it at all. People need to be convinced. To convince people, you have to persuade, and then go on and on persuading. Information involves a campaign of attrition. What the media can, could and should do is play a role in this process. It helps to be able to tell the same story – and reiterate the same message – in as many different ways as possible.

Following up on the same story later is also effective, but it is seldom done. For example, farming techniques in the mountains of Nicaragua and locally run warning systems have made these rural people safer than they were when Hurricane Mitch struck in 1998 (see Chapter 14). However, it is generally hard to convince editors that a disaster that does not occur is news. Not only editors, but donors, development professionals, government officials and politicians find that the bad thing that has been avoided is less salient than the good thing achieved. Even the risk-bearers themselves, in this case the farmers on the steep volcano slopes, would have been far less likely to take up the soil and water management technologies that prevent landslides if they had not seen increased production and income. Livelihoods are foremost in their minds and more important, generally, than risk, just as for publishers it is the sale of newspapers, for politicians it is votes, or for broadcasters it is listeners and viewers.

One must therefore ask, what is in it for the media? In the first place, good stories; in the second place, a new and interesting audience; and in the third, an imprecise but not insignificant personal satisfaction for those reporters, broadcasters and editors who believe that journalism is more than just a form of daily entertainment, and that in a democracy, journalists have an obligation to support, defend, inform and, above all, alert society. Press censorship has been shown to create silences in which disasters such as famines can go unnoticed (Article 19 1990). Sen (1990) wrote that there had never been a famine in a country with a free press, and elsewhere that 'a free press and an active political opposition constitute the best early-warning system a country threatened by famines can have' (Sen 1999: 222). The situation is more complicated, however. Myhrvold-Hanssen (2003) points out that the literacy rate among the population is of great importance; writing about Kerala state in India he notes that high life expectancy is correlated with high literacy rate. He continues:

> What conclusions can we draw from this correlation? It is not unreasonable to state that a more educated people will have more ability to express their views and current living conditions. Moreover, the news media will have more incentives for producing reliable *and*

important news about the social and economic status of a community, since this news will certainly be read and criticised.

(Myhrvold-Hanssen 2003: 6)

Opportunities for such good stories come along almost every day. On average over the past decade, a disaster precipitated by the combination of natural hazard and lack of human preparedness has been more than a daily event. For journalists, gross statistics tend to conceal practical reality rather than highlight it. Every one of those deaths was potentially a powerful story; every one of those survivors a potential reader, a potential news source or a potential friend; and every one of those dollars a depleted investment in communities that deserved better from national government, better from civic authorities and better from the media. The bottom line is that disasters occur on average more than once a day, and there are perfectly good reasons why newspaper reporters and broadcasters should be prepared, should have done their homework on DRR.

The reporter who has taken the trouble to learn a little about earthquakes and earthquake engineering, about meteorology and storm damage, about the practical response to floods, avalanches, ice storms and volcanic eruptions is a reporter who can make the most of the awful moment when the ground begins to shake, or the river bursts its banks. Simply from a self-centred, venal career point of view, a journalist's investment in the practicalities of disaster hazard can be a rewarding one.

The reporter who knows what questions to ask, which places and communities must be most at risk, which authority is most immediately responsible and which sources will be the most immediately helpful is not only doing his or her job well. She or he is reminding the readers and listeners why a free press is, ultimately, a matter of life and death for a community. For journalists, modest intimacy with DRR ought to be straightforward: they can do their jobs, multiply their audiences and quite possibly save human lives, all at the same time.

Constraints to media engagement with DRR

Why has this not happened already? Why do media not already have DRR correspondents, in the way that they have technology and show-business reporters, political commentators and economics editors? One possible answer is that mainstream media organisations are, like farmers, essentially conservative, in that they tend to focus on what delivered a harvest of listeners and readers (and advertisers) the last time, and the time before that. Social observers frequently express alarm at the unhappy mix of populist political conservatism, disproportionate power and shameless irresponsibility that characterises the modern Western corporate media as well as the influence that economic interests have over what gets covered (Chomsky and Herman 2002). Because of lack of space this chapter cannot delve deeply into the political economy of media. The question will have to remain open as to whether similar constraints face media workers in the non-profit sector (for example, National Public Radio in the USA or the BBC in the UK) and also what influences there are on the presentation of DRR where the press is controlled (to one degree or another) by the state (e.g. Cuba, Iran or China).

This mix is not modern, however, and nor is the unhappiness. The British politician, Stanley Baldwin, eighty years ago coined a memorable phrase, when he accused two press barons, the Lords Beaverbrook and Rothermere, of 'power without responsibility, the prerogative of the harlot through the ages'. Newspapers and broadcasters have always been conservative, but that does not mean the media is unresponsive to pressure or incapable of change.

There is also a great tradition of campaigning journalism, and a great tradition of investigative reporting. Inertia can be overcome. A loose conspiracy of the informed and the concerned, the expert and the interested, the scientist and the journalist, could make DRR as amusing or as provoking as the misadventures of a Britney Spears or a Lady Gaga; as exhilarating as the presidential campaign of Barack Obama; as urgent and demanding as the sudden apparent worldwide collapse of the banking system.

This would be enough to put DRR on the political and media agenda within a fortnight. Could it be done? Probably not in a fortnight, but it could be done, by addressing the issues of DRR whenever there is an opportunity, and in as many imaginative ways as possible. In the course of the last four decades, journalists in Britain, Europe and the USA have become unevenly, but increasingly, interested in the politics and economics of the developing world (the journalists in Africa, Asia and South America, of course, were already on the case) (George 2009; PANOS 2010). There has been over the same decades a parallel growth of interest in the state of the global environment, and swift recognition that the two issues – development and environment – are intimately linked. DRR connects these issues even more intimately. Disasters interfere with, often setting back, sustainable development (see Chapter 14). Quite separately, the intensity and frequency of some extreme meteorological events could increase with global warming (see Chapter 18). There is a loop simply waiting to be closed by media workers as well as others.

In some respects this will not represent a change of direction. It is the duty, as well as the delight, of journalists to expose folly, tardiness, inconsistency and dishonesty in national and local government. Some 168 nations have signed up to the Hyogo Framework for Action, an international instrument that promised action on DRR, so there will potentially be 168 governments that are doing things inadequately or not at all (see Chapter 50 and Chapter 60). Newspapers, broadcasters and bloggers are – although they are properly wary of conceding such a thing – part of the public education programme, and over the decades have played interesting and sometimes decisive roles in influencing public attitudes to health behaviour, nuclear energy and space research (see Chapter 41).

SARS, or severe acute respiratory syndrome, provides an example. It threatened to become a deadly pandemic in 2002 (WHO 2003). By 2003, the drama had more or less ended. Thousands had been infected and hundreds had died, but the infection was contained by urgent government and international action made considerably more urgent by worldwide news coverage. While more controversial, the media frenzy around the H1N1 virus in 2009 may also have played a role in international and national response that contained the epidemic (Council of the European Union 2010).

In democracies, politicians and public officials respond to public alarm, and the media both raise the alarm and provide a useful index of the intensity of this alarm (Sen 1999). In countries that are not democracies, governments and autocracies are usually uneasily aware of the clamour from the free press in the neighbouring democracies, and will often respond anyway.

Journalists have already played an important part in promoting awareness of the forthcoming energy crisis, of environmental pollution and of climate change. Questions of DRR are only extensions of these larger programmes. In a wider sense, they are part of the media diet, because natural hazard and disaster risk encompass all the great news themes: sudden violence on a colossal scale, tragic death, widespread suffering, poignant human drama, sensational pictures and revelations of gross governmental failure. The challenge is to get them onto the news pages before the death and the suffering happen. This requires a certain amount of bridge building between the professional disaster community and the media. However, there is the challenge: an analysis of 200 English-language newspapers found that the 2004 Indian Ocean tsunami

Tim Radford and Ben Wisner

generated more column inches in six weeks than the total for the world's top ten 'neglected' or 'forgotten' disasters in the previous year. Advice on how scientists can best communicate with journalists (another link in the communication chain) is provided in Box 63.1 below.

Box 63.1 How scientists can best communicate with journalists

Tim Radford

Think of an encounter between Dr Grant Application, of Sanctimony College Milwaukee, and Ruthless Griller, scourge of news bulletins. The scientist is at home among caveats, uncertainties and very long answers, while the radio reporter prefers rasping questions to be answered in a single sound bite. However, Dr Application knows the answers, and the reporter does not, so he must use the air time constructively.

RG: 'But Professor Application, do you think the ground is really going to open up and swallow Oberlin, Ohio?'

GA: 'Stranger and more terrible things have happened. I think right now that earthquake hazards in Oberlin are not great, but in dozens of cities around the world people live in daily risk of dreadful destruction. Cities in the shadow of a volcano may be engulfed by fire, whole communities on the river deltas could be swept away overnight by sudden, overwhelming storm surges, and we can all imagine the terrible consequences of a powerful quake in Tokyo, or San Francisco ... '

Just think: you have sidestepped the silly question, and got to the issue of disaster risk management just by adopting the form 'Yes, but ... ' Your audience will have started to listen to you, not to the urbane Griller. *Remember:* nobody listens to long words. Make it clear and vivid.

RG: 'Surely an earthquake of this magnitude is quite mild?'

GA: 'Imagine hundreds of you crowded into a jerry-built apartment block. Seismic waves race through the city at hundreds of miles an hour. First the ground moves up and down. A second or two later, another set of waves shake the building from side to side. If your home is just breeze-blocks stacked one on top of another, then you are about to become a statistic. The whole place is likely to fall upon you ... '

Just think: you have painted a picture. Listeners have begun to see the world as more dangerous because of corrupt government and shameful building standards. Did you use phrases such as 'longitudinal oscillation subsequently impacted by shear wave radiation'? No, you have them hanging on your words. *Remember:* you know what you want to say, so get the message across before he can interrupt.

RG: But professor, could people ever protect themselves against volcanic eruption?'

GA: 'If you see a 200-foot-high wave of superheated gas and molten rock racing towards you at thirty-three metres a second, as people did in Saint-Pierre in Martinique in 1902, there is not much you can do, except wish you'd spent more on research,

---

766

more on a master plan, more on ways to evacuate a city. The trick is not to be there when the mountain erupts. That is why we are pushing the government to think about disasters before they happen … '

Just think: you told them about a disaster that really did happen, and you told them about an initiative to make cities safer. Go on talking like that, and maybe cities will become safer.

Tim Large, editor of Thompson–Reuter's AlertNet, suggests the following 'tricks of the trade' for journalists to bring attention to neglected disasters (Wynter 2005):

- *Invest in media relations*, communications training and expertise, down to the local level.
- *Keep up a dialogue with the media*: provide background material on complex emergencies, but not fifteen minutes before the deadline.
- *Put a number on it*: death tolls give journalists pegs on which to hang their stories, and they go some way towards quantifying the unimaginable.
- *Bring in the big names*: it is controversial, but enlisting celebrities can work. The press follows the famous face and ends up reporting on the cause.
- *Make it visual*: nothing sells a story like a good picture. In disasters, aid agencies may have the only photos available.
- *Be creative and proactive*: tell the bigger story through the eyes of individuals. Fit what you're doing into the news agenda. Organise trips for reporters.
- *Never give up*: in this game, persistence really does pay off.

News organisations, corporate or otherwise, are not monolithic states; they are loose and sometimes slightly anarchic republics, and each republic contains a number of ambitious individuals on the alert for good stories from unexpected sources. The disaster community in turn contains a rich mix of United Nations (UN) propagandists, national civil servants, government geophysicists, hospital medical officers, university social scientists and development charity chiefs with a flair for handling the media.

Mainstream, routine coverage of DRR will not happen in a hurry and, at a guess, when it does, it won't involve phrases such as 'disaster risk management' or 'disaster risk reduction'. Such polysyllabic labels somehow obscure meaning. Truly effective DRR may very well have to begin with a change of language, a switch to emotional and vivid words and phrases that will fall somewhere between the overdramatic and the sensationally lurid. Academics have no great fondness for emotional language, and journalists are often accused of hyperbole, but what could be more lurid than 50,000 deaths in an earthquake, or more hyperbolic than a tsunami that sweeps a quarter of a million people from the beaches, ports, shores and marinas of the Indian Ocean?

Support for the journalist willing to try

Protection of the journalist

Media workers are killed and imprisoned every year. As the lines continue to blur between civilians and combatants in conflict, and as natural hazards continue to threaten conflict zones, the journalists covering a disaster may become more insecure (see Chapter 7). Harassment and possible imprisonment is also a risk to media workers who ask awkward questions about disasters. In China, for example, despite some early increase in media openness (Gang 2009), journalists have been imprisoned for pursuing the story of collapsed schools during the Sichuan earthquake in 2008 (Branigan 2010) (see Chapter 5).

Technical assistance for the journalist

Once upon a time, journalists had to go looking for experts to consult. Things have changed. National bodies such as the British Geological Survey (BGS) and the United States Geological Survey (USGS) have been open, helpful and innovative, prompt to announce and above all swift to help. Thomson-Reuters launched the quick-off-the-mark disaster website AlertNet. A number of organisations collaborate in producing an excellent multi-purpose portal managed by the UN Office for the Coordination of Humanitarian Affairs, Reliefweb. The UN Space Agency builds up satellite data from disaster-hit areas and makes it available on the web. The UN funds an International Secretariat for Disaster Reduction which tends to respond to big events, inevitably a little more slowly. The UN, through the World Health Organization, also backs CRED, at the Catholic University of Louvain. This database provides a lot of useful statistical information. Insurance giants such as Munich Re and Swiss Re have a clear stake in DRR, and have produced annual tallies of the cost, in lives, human suffering and economic toll, along with some thoughtful analysis.

Journals such as *Nature* and *Science*, *Scientific American* and *New Scientist* have over the decades built up a huge corpus of quickly accessible material about disasters. Francophone readers enjoy *Pour la Science*, *Science et Vie*, *Sciences Humaines* and *La Recherche*. Contributors have then gone on to write valuable popular books that should be available in media libraries. There is no shortage of general material for those who start looking. The problem is that although all disasters are likely to follow one of a set of broad general possible patterns, no existing website or encyclopaedic library can help with all the immediate questions about a precise catastrophic event, which has only just happened, in a particular place on the map.

Some media institutions have themselves attempted to provide support for improved disaster journalism. AlertNet, a major resource funded in-house as a non-profit subsidiary by Thomson-Reuters, was mentioned above. AlertNet has a tool called MediaBridge which specifically targets working journalists (www.alertnet.org/mediabridge/index.htm). In the USA, the CBS television network offers a portal to many sources of data and analysis; so many, however, that one needs some prior experience to use it well (www.cbsnews.com/digitaldan/disaster/disasters.shtml).

Box 63.2 Examples of research resources for journalists

Ben Wisner
Aon-Benfield UCL Hazard Research Centre, University College London, UK

- BGS: www.bgs.ac.uk/research/earth_hazards.html
- USGS: www.usgs.gov/hazards and www.usgs.gov/emergency
- AlertNet: www.alertnet.org
- Reliefweb: www.reliefweb.int/rw/dbc.nsf/doc100?OpenForm
- UN Space Agency: www.un-spider.org
- Munich Re: www.munichre-foundation.org/StiftungsWebsite/Projects/DisasterPrevention
- Swiss Re: www.swissre.com/rethinking
- CRED: www.cred.be

Also, networks of civil society organisations and international non-governmental organisations (INGOs), as well as bloggers, now provide journalists with accessible ways in which to learn of success stories in avoiding disaster and fast-breaking impressions of sites of disaster impact (see Box 63.3). Blogs are a fascinating and little-studied phenomenon. They range from those produced by highly qualified scientists (e.g. daveslandslideblog.blogspot.com/2009/04/british-geological-survey-landslide-web.html), to lay people whose grasp of the issues, perhaps even reality, is suspect. So a good deal of care and discrimination is required in utilising this burgeoning resource. All of these developments are manifestations of the revolution in communication made possible by information and computer technology (Bulkley 2010).

Box 63.3 ICT, social media and bloggers

Terry Gibson
Global Network of Civil Society Organisations for Disaster Reduction, London, UK

Information communication technologies (ICTs) are changing the media landscape in ways that may help to shift popular and political attitudes to disaster risk reduction. The coverage of the 2004 Asian tsunami sits on the dividing line between old and new media perspectives. It predated now familiar channels such as Facebook, YouTube and Twitter, and yet it revealed a cultural contrast between the distant 'helicopter' perspective of the mass media and the grounded social media perspective of SMS (text messages) and blogs. The BBC news website opened its pages to blogs which made both stark and fascinating reading, including text-to-blog messages from people at the sharp end, in devastated communities in Sri Lanka, for example. The then head of BBC news, Richard Sambrook, described this as part of a 'sea change' in the locus of power of news delivery – particularly of fast-moving topical events. The Sichuan earthquake in 2008 was first broken on Twitter, and Ushahidi's SMS-monitoring and campaigning platform was born that same year out of the social turmoil of the Kenyan elections as the first large-scale example of SMS-based social campaigning. Just after the earthquake in 2010 in Haiti, Ushahidi linked survivors, responders and response co-ordinators via mobile phone texting and text-to-map analysis. YouTube broadcast eyewitness material from Haiti hours after the earthquake struck.

In the Philippines, since the early 2000s people have been using these sites for social networking, especially Friendster before Facebook emerged. It has been particularly used among overseas workers to maintain social relationships beyond borders among relatives, friends, alumni, etc., as a new medium of strengthening transnational communities. Now it is being used as a way of mobilising people in time of crisis. During a series of typhoons that battered the country during September 2009, there were stories of people mobilising resources among relatives and friends locally and across borders through such networks. In a country where remittances are crucial to both the daily economy and to coping with crises, Friendster and Facebook turn out to be powerful new additional media.

Two transitions are seen here: one from top-down mass media to narrowcast social media, and another from messaging to social mobilisation. Both offer possibilities for raising awareness and securing action for better disaster risk reduction. The shift from top-down to social media signifies a transformation of news representation through social media channels. Campaigning INGOs such as Greenpeace have been swift to grasp this opportunity – for example, with their 'Save the Turtle' campaign leading to

100,000 cyberactivists pressurising Indian corporation TATA to back down on wrecking a habitat with a port-building project. NGOs will increasingly recognise the opportunity to connect their collaborators – the excluded and the vulnerable – to wide audiences using these technologies.

The shift from messaging to social mobilisation emphasises the direct voice of groups – whether local or global – rather than the managing of messages by media organisations. For example, HablaHonduras.com was born out of the 2009 political turmoil in Honduras, and overnight it mobilised news and citizen action through Twitter, Facebook, Skype, Flickr and SMS. This was a response to a political rather than a natural crisis, but illustrates the way in which grassroots social movements are able to seize the initiative. The global social movement Global Network for Disaster Reduction focuses specifically on disaster risk reduction and has used a digitally orchestrated international monitoring project, combining blogs, Google Earth and social mapping, to create bottom-up pressure for change within the UN system.

This is a very different world from that of the traditional mass media. As with any innovations, many will fail to stand the test of time. However, the opportunity for social groupings and movements to secure a more powerful voice cannot be ignored. In a world where the excluded and vulnerable suffer disproportionately from the impact of disasters, the shift in the media landscape may create new opportunities for changed attitudes and action at a popular and political level.

Experts can help with answers to those questions that occur, spontaneously, as one follows the latest headlines. For example, is it true that someone could have foreseen this earthquake, that volcanic eruption or the once-in-100-year flood? Is it true that authorities were repeatedly warned about the vulnerability of this district, on that hillside; or the weakness of that levee, under these cyclone conditions? Is it true that the government formally committed to a disaster emergency plan, but failed ever to implement it? From the journalist's point of view, an expert in DRR is a rich resource: someone who is familiar with the mechanics of death and destruction, and someone who also knows what should have been done, could have been done and was not done: as we journalists say rather callously, someone who knows where the bodies are buried.

Studying the media

The area of media studies is vast and asks many questions that are germane to disaster and public perception of them; however, these are beyond the scope of this chapter. One might ask about the role of media in shaping public attitudes toward risks and toward natural hazards in particular. Another set of questions concerns how individual and group behaviour itself is influenced by media. Whole academic industries have grown up around the analysis of biases and ideological assumptions in the media (Bassett *et al.* 2000). All the questions can be asked of the treatment of hazard, risk and disaster.

There are also some more descriptive studies of importance. Some centres monitor the coverage of various events including disasters and their reception by various publics. The Pew Center Project for Excellence in Journalism is one such centre (Pew 2008). Similar research work is sponsored by the Annenberg. For example, a major study during 1993–94 looked at 'Media, Disaster Relief and Images of the Developing World' (Cate 2010) (see Chapter 11). In 2005 the International Federation for Red Cross and Red Crescent Societies' *World Disaster*

Report found that media coverage of disasters was not proportional to human suffering and need, but biased in various ways (Wynter 2005). A number of scholars have found evidence of bias and of 'neglected disasters' (Shah 2005; Wisner and Lavell 2006). Critical scholars have deconstructed the stereotypes and images of disaster 'victims' (Fordham and Ketteridge 1998) (see Chapter 35).

Conclusions

What could be more shameful than a set of civil and national authorities that were made aware of the risks, but did nothing; or a guardian of liberty, an Argus, a Sentinel or an Observer that pounced upon the folly of a drug-addicted actress or a philandering footballer, that denounced venality in a minor politician or applauded bigotry in a conceited bishop, but failed to see, and warn of, the headlong approach of death, destruction and appalling suffering on an epic scale? M. Marius Hurard, editor of *Les Colonies*, should not be held up as a figure of fun, or journalistic ignominy. He cannot be accused of failing to address the question of danger to the town of Saint-Pierre. He did address it. He made precisely the wrong judgement at a critical moment in history, and when editors do such things, they tend to lose their jobs. It seems heartless under the circumstances to say so, but M. Hurard paid for his misjudgement, and was fired on the spot.

Some recommendations follow. Media workers should make more of an effort to inform themselves about the resources available to make disaster risk reduction an exciting and engaging topic for their audiences. Many resources do, in fact, exist. Scientists and humanitarian workers, for their part, need to make an effort to communicate with journalists in a way that makes it easier for the latter to put together good stories. Finally, energy and resources need to go into the perennial struggle against state censorship and exposure of special economic interests that manipulate 'the news'. One should never forget that powerful interests may not want disaster risk discussed, and that the media workers who do so need to be protected by international networks of people who care.

64

Participatory action research and disaster risk

Michael K. McCall

ITC FACULTY, UNIVERSITY OF TWENTE, ENSCHEDE, THE NETHERLANDS, AND
UNIVERSIDAD NACIONAL AUTÓNOMA DE MÉXICO (UNAM), MORELIA, MÉXICO

Graciela Peters-Guarin

ANTHROPOLOGY DEPARTMENT, VANDERBILT UNIVERSITY, NASHVILLE, TN, USA

Introduction

The focus of this chapter is participatory research for action that incorporates and legitimates the knowledge and capacities of local communities (see Chapter 59), households, authorities and other local actors engaged in disaster risk reduction (DRR). The purpose of participatory action research (PAR) is to bring local knowledge and capacities into the same arena as the formally accepted knowledge of scientific researchers, policy-makers and planners, and to help establish a dialogue and co-production of 'hybrid' knowledge that combines local and outside specialist knowledge. Wadsworth (1998) defines PAR as 'research which involves all relevant parties ... by critically reflecting on the historical, political, cultural, economic, geographic and other contexts ... [I]t aims to be active co-research, by and for those to be helped'.

The 'A' in PAR is important. The goal is action intended to reduce disaster risk, i.e. how communities, researchers and local authorities can work together and make use of local capacities and activities in all stages of DRR. PAR is problem-driven and activist; as Kindon *et al.* put it, PAR is a process involving 'researchers and participants working together to examine a problematic situation or action to change it for the better' (Kindon *et al.* 2007: 1). Not all participatory research (PR) is focused directly on action. For example, PR has been used for national agenda-setting where feedback from local people is vital (Wisner 2008; Sierra Leone Red Cross 2004).

Origins and rationale of participatory action research

Participatory research and PAR are grounded in the philosophical writings of the Brazilian adult educator, Paulo Freire, with strong empirical evidence of its effectiveness provided by Creighton

(2005) among others. Earlier origins may be traced to 'action research' inspired by educational philosopher John Dewey in the 1930s (Wisner *et al.* 1991). PAR's methodological and episte-mological origins are in pragmatic, functional learning in management and development where people's participation has many concrete benefits, besides 'participation' as an end in itself. PAR has evolved out of rapid rural appraisal (RRA) and participatory rural appraisal (PRA) methods (Kindon *et al.* 2007), participatory technology development (PTD) and participatory learning and action (PLA) (Chambers 2007a). The PAR approach represents a higher intensity of participation compared with RRA and PRA, as summarised in Figure 64.1 developed for this paper based on the concept of 'ladders of participation'.

Intensities/types of participation	Actors - Local insiders (community members, local technicians) and outsiders (researchers, government, NGOs)	Overlap of research approaches			Some methods
		RRA	PRA	PAR	
***Transformation* Initiating research**	Self-mobilisation for research. Invite outsiders to assist with research			↑	Community's recognition of research needs. Community invitation to research
***Collaboration* in decision-making re. methodology**	All actors decide on all issues including methods, 'at all stages'. Interactive		↑		Participatory research design. Joint scenario setting. Joint assessment
***Consultation* on results and (interim) findings**	Locals refine and prioritise external ideas on findings	↑			Citizens juries/ feedback/ validation workshops
***Consultation* on topics and issues**	Locals refine or prioritise external ideas on topics and on information source				Pre-research workshop/ joint reconnaissance
***Information* sharing**	Two-way communication after the results, between insiders and outsiders		↓	↓	Presentations/ reporting of results session
***Exploitation* 'Acquire Information and Leave' = no participation (very common)**	One-way collection of information. Ousiders 'return' some edited information to local people	↓			Standard questionnaires/ censuses/ observation

Figure 64.1 Intensities of participation in research

It is important to distinguish between the intensities of (community) participation, and the underlying intentions of the agencies (e.g. government departments, international organisations, non-governmental organisations (NGOs) or academic researchers) that are using participation as a strategy. When engaging in PAR, one should analyse the intentions and ideologies – instrumental or transformative (Wisner 1988) – behind participation and PAR (McCall 2003; Pelling 2007). The goal of participation may be to facilitate external research, to lubricate the external interventions and co-opt communities into supporting them or to pass on a share of the cost burden of a project. The goal may also be empowerment. In such a case the research process is oriented towards transformative goals such as social redistribution and providing voice and access for weak groups. Alternatively, the goal may be empowerment through collaborative PAR that builds links between 'external' purpose and 'internal' demands. If successful, this third approach increases effectiveness, builds local capacities and redirects external research towards local needs and solutions. This mediation strategy recognises that PAR can serve either master, mitigating shorter-term risks or furthering longer-term empowerment.

PAR methods for DRR planning and implementation

Participatory methods are at the heart of community-based DRR (CBDRR). They are used for acquiring and improving understanding of field observations and local knowledge about hazard characteristics, vulnerability of people and structures, coping mechanisms and adaptation, post-disaster mitigation and preparedness, and for conducting monitoring and evaluation. PAR

Box 64.1 Use of PAR for empowerment and autonomy

- In Bolivia, rural Aymara communities around Lake Titicaca in the farmer organisation PROSUKO worked with yapuchiris, local specialist vocational farmers dedicated to agricultural learning. In this initiative, traditional agricultural and climatic knowledge was consolidated in yapuchiri groups, supported by intercooperation, to sell technological and financial services to local farmers. The project significantly reduced crop losses from drought, hail, frost and flooding, and stabilised market access for local crops (UNISDR 2008a).
- The 'Girls In Risk Reduction Leadership' (GIRRL) project in Sonderwater, a poor neighbourhood of Ikageng (South Africa), aimed to reduce the social vulnerability of marginalised adolescent girls. Practical capacity-building initiatives incorporated girls and their perspectives into decision-making processes towards increased resilience. The project, implemented by the African Centre for Disaster Studies through the ProVention Consortium, fostered girls' involvement in designing plans to reduce the impact of disasters and extreme events. Girls have been empowered to voice their opinions, and have gained confidence and respect, and improved their livelihoods, families and communities (UNISDR 2008a).
- After the devastating 2001 earthquake in Gujarat, the Sustainable Environment and Ecological Development Society (SEEDS) incorporated DRR features into rehabilitation programmes. One initiative trained local masons in earthquake-resistant construction. In five years 200 of them were certified in construction skills at international standards. The masons now work in their local communities and educate fellow masons in similar earthquake-risk regions for shelter reconstruction and capacity building (UNISDR 2007a).

---

methods also contribute to analysing and communicating the elicited information in transparent, local community-accessible media.

Tools for understanding the situation

PRA and PAR have developed and adapted many specialised participatory descriptive methods for understanding local risk situations: tools such as mobility profiles, problem trees, wealth ranking and conflict and stakeholder analyses. More holistic tools used in community-based DRR include situational analysis, role-playing games, theatre (e.g. the Venezuela case in Wisner *et al.* 2008; Kindon *et al.* 2007) and story-telling (e.g. El Salvador, Guatemala, Peru and Venezuela cases in Wisner *et al.* 2008; Kindon *et al.* 2007). Working with photography and video sets up intensive participation, both in the making of images and their subsequent discussion. Examples include participatory video imaging of women's perception of safe spaces in Belfast (McIntyre 2003) and in Guatemala (in Wisner *et al.* 2008), and with children in Nepal and Malawi (both in Reid *et al.* 2009). The same 'double' participation of involvement in creating and in explaining knowledge is achieved when using P3DM (participatory 3-D modelling), as for instance by Gaillard and Maceda (2009) in the Philippines.

PAR and spatial information tools for DRR

DRR requires locational information and local spatial knowledge, such as the sites of specific events, locations of people in terms of exposure, vulnerability, distance from risk sources and from potential coping capacity and resources, proximity to shelter and support and transport networks. Many spatial thresholds are significant for DRR, such as area subject to floods or earth movements, boundary conditions for air or water pollution, and the social and culturally-determined perceptions of secure movement or safe places in socially divided cities or hazardous environments. Social networks have spatial as well as a-spatial relationships because of the physical proximity necessary for communication and support mechanisms for family, kin and friends.

There are specific mapping tools for working with locations and spaces. These include simple sketching of people's mental maps, standard survey maps, aerial photos and satellite images and Web2 as mash-ups with Google Earth (www.google.com/earth) or OpenStreetMap (www.openstreetmap.org). The products may be 'counter-maps', which identify and record the spatial knowledge and priorities of excluded groups that are less powerful, less articulate and less integrated. By providing alternative visions as 'people's maps', they are essential for participatory risk mapping for vulnerable and marginalised groups, such as in 'gendered risk maps' (e.g. Honduras case in Wisner *et al.* 2008). Local people know very well the distortions that any externally sourced map can display; therefore, participatory mapping, using ephemeral maps (temporary in sand or mud), sketch maps or overlays, has been a mainstay for obtaining local spatial knowledge.

In addition, such maps serve as vital 'talking points' to stimulate discussion. Photos, drawings or three-dimensional models have a similar effect, but a map or aerial photo gives reference to a larger area. PAR in DRR applications engages individuals and communities with topographic maps, aerial photos, anaglyphs (3-D images) and satellite images. People mark objects of interest (landmarks and locations, hazards, routes, etc.) on a transparent plastic sheet laid over the base image. People generally respond better to aerial photos and high resolution satellite images, including Google Earth, than to regular maps as the base source, because of the verisimilitude of the image. These technologies still have weaknesses associated with ownership and control, lack of multi-temporal information and various deficiencies in technical quality.

In the past decade, more sophisticated spatial tools have entered the DRR repertoire, especially the array of earth observation (remote sensing) products and (participatory) geographic information systems (GIS). Mobile GIS use platforms such as global positioning system (GPS) receivers or GPS-enabled Smartphones, iPads and inexpensive hand-held computers (iPaq) or ruggedised tablet laptops. Mobile GIS also may use internet functionalities like Google Earth and other 'virtual globes' and field-mapping and recording software such as CyberTracker (www.cybertracker.org). The visual output of this spatial information is greatly enhanced through the use of graphics software tools and dynamic (web-based) GIS.

The development of these tools into a PGIS (participatory GIS) approach releases their enormous potential as cheap, user-friendly, sufficiently detailed and relatively transparent tools to acquire, analyse and present spatial information from a community point of view. These technologies are described on the Participatory Avenues website (www.iapad.org).

Williams and Dunn (2003) combined PGIS with local people's first-hand knowledge of landmine locations in Cambodia. Other DRR applications include community landslide risk mapping utilising indigenous languages in central Mexico (in Wisner *et al.* 2008), and the Jaringan Kerja Pemetaan Partisipatif (JKPP, Indonesia Community Mapping Network) and Yayasan Rumpun Bambu Indonesia (YRBI, Center for People's Economic Development) community video of the post-tsunami situation in Aceh (JKPP and YRBI 2006). Kienberger (2008) produced hazard maps in flood-prone rural Mozambique, as did Peters-Guarin *et al.* (2011) in urban Philippines.

PAR and temporal information tools for DRR

Historical information holds a special place in the PAR framework. Frequently it is only available from individual or collective memory. The historical knowledge of hazards is always richer and probably more convincing than is statistical data. This is especially the case for details of dynamic socio–economic phenomena such as differential household vulnerability and traditional coping strategies, but it is often true also for knowledge of the hazards themselves: the locations, frequency, intensity and magnitude of storms, inundations or landslides, etc., where physical evidence has been covered by re-vegetation or human alterations. PAR helps to uncover the significance of seasonal, weekly and diurnal changes for hazard impacts – for example, the devastating impact that even small floods may have on livelihoods of the most vulnerable, when they occur during peak periods of employment or marketing (Peters-Guarin *et al.* 2011). Something as simple as drawing a timeline on paper, noting commonly recognised dates (e.g. salient political events, the millennium, etc.) and using this as a focus for group discussion among elders can be richly rewarding.

Many PAR tools depend on the memories of elders, individual key informants or community and even folk legends as the starting points for discussion. Significant sites, perhaps now abandoned, may also be visited and discussed (see Chapter 4 and Chapter 10). Visual clues with active interpretation by local actors are important; for instance, elders' memories of rare cold snaps in Peru (in Wisner *et al.* 2008) and flood heights in Indonesia and the Philippines. Besides conversations with key informants, there are more formal PAR tools for analyses of change, such as seasonal calendars, personal diaries, place histories, content analysis, critical event analysis, and historical transects and trend lines (see Box 64.2).

Respect for local knowledge

Ideally PAR incorporates the specialised local knowledge of often previously excluded actors – the poor, women, children, landless, subordinate castes, etc. – as well as that of recognised local

Box 64.2 PAR tools applicable for DRR

Participation facilitation

Purposes of these tools are to promote participation as a process for its own ends; specifically, how to form groups, to develop organisational skills, strengthen confidence in negotiations, etc. The tools therefore support the initiation of research by local communities.

Tools: Group dynamics; team-building; group formation; leadership skills; DEMOCS (deliberative meetings organised by citizens); SARAR (self-esteem, associative strengths, resourcefulness, action-planning, responsibility); RAAKS – participatory action research; social mobilisation and animation; citizens' juries.

Key sources: AFPP 2006; Creighton 2005; Kindon *et al.* 2007; Wisner *et al.* 2008; World Bank 2007a; web.worldbank.org/wbsite/external/topics/extsocialdevelopment/extpceng/0,menuPK:410312~pagePK:149018~piPK:149093~theSitePK:410306,00.html.

Action learning tools – participatory research

These tools have similarities with the participation facilitation tools above, but they are more specifically focused on instigating and promoting participatory learning and collaborative research. Applications are for participatory assessment, knowledge dissemination, implementation.

Tools: Participatory theatre, participatory video and photography; games, gaming, music; RAAKS – participatory action research; contextual analysis; mind mapping; scenario development and imagineering; participant observation.

Key sources: AFPP 2006; Creighton 2005; ActionAid 2005; Kindon *et al.* 2007; Reid *et al.* 2009.

Participatory research – integrated disaster risk

These tools are compilations of specific data collection and data analysis tools which are dedicated to hazard measures, vulnerability and coping capacity analyses and risk assessment.

Tools: Participatory disaster risk assessment; PVA (participatory vulnerability analysis); VCA (vulnerability and capacity assessment); RA (risk assessment).

Key sources: ActionAid 2005; Heijmans and Victoria 2001; Falk 2005; UNISDR 2008a; ADPC 2004; ProVention www.proventionconsortium.org.

Data collection: knowledge acquisition

These tools are mainly derived from well-developed RRA and PRA tools. The basis is to use local and indigenous knowledge for richer descriptions, and as input to analysis – to identify hazardous past events, hazard-prone areas, profiling of vulnerable groups, etc.

Tools: Aerial inspection; checklists; direct measurement; participant observation; transect walks; photograph interpretation; folktales; digital camera and video, multimedia; VGI (voluntary geographical information); human sensor networks; crowd sourcing.

Key sources: Pincha 2008; ADPC 2004; Heijmans and Victoria 2001; Laituri and Kodrich 2008; JKPP and YRBI 2006; Wisner *et al.* 2008; World Bank web.worldbank. org/wbsite/external/topics/extsocialdevelopment/extpceng/0,menuPK:410312~pagePK: 149018~piPK:149093~theSitePK:410306,00.html.

Verbal information

Tools based on interviews, discussions, conversations.

Tools: Interviewing (key informants; semi-structured interviewing); checklists; focus groups; life histories; questionnaires; serendipitous meetings.

Key sources: Wisner *et al.* 2008.

Description of current situation

There is some overlap between the knowledge acquisition tools, these descriptive tools and the analytical tools.

Tools: Beneficiary assessment; conflict analysis; citizens' diaries; gender analysis; impact flows; participatory livelihood analysis; mobility diagrams; problem trees; resource ownership and access; social relations matrix; spidergrams; stakeholder analysis; systems diagrams; visualisation and graphics.

Key sources: AFPP 2006; Pincha 2008; ADPC 2004; Oxfam 2002; CIDA; ProVention www.proventionconsortium.org; IDRC www.idrc.ca/en/ev-80983-201-1-DO_TOPIC. html; World Bank web.worldbank.org/wbsite/external/topics/extsocialdevelopment/ extpceng/0,menuPK:410312~pagePK:149018~piPK:149093~theSitePK:410306,00.html.

Participatory mapping, PGIS, spatial analysis

The purpose of these tools is the interactive acquisition of local spatial knowledge of the locations and spatial extent of, for example, hazard-prone areas and their evolution over time, vulnerable infrastructure and groups of people, perceptions of risk, accessibility and mobility.

Tools: Participatory mapping with: sketch maps, standard maps, aerial photos or satellite images; gender mapping; counter-mapping; time–space profiles; risk maps; participatory 3-D modelling; visualisation and graphics; participatory GIS; dynamic GIS (web-based GIS); mobile GIS, GPS, CyberTracker.

Key sources: AFPP 2006; Reid *et al.* 2009; ADPC 2004; Oxfam 2002; Falk 2005; ProVention www.proventionconsortium.org.

Analysis of change – histories

Tools to identify patterns and explanations in the dynamics of hazards, risk and vulnerability.

Tools: Biographical analysis; content analysis; narrative inquiry; critical event analysis; local diaries and histories; historical transects; timelines and trend lines; seasonal calendars of livelihood, land use, stresses, hazards and conflicts; visualisation and graphics.

Key sources: ADPC 2004; Heijmans and Victoria 2001; ProVention www.provention consortium.org.

Eliciting local and indigenous knowledge

There are specialised tools and approaches for acquiring and handling local (and indigenous) knowledge, including spatial knowledge, which are different from simple data collection. An important aspect is the imperative for ethical participatory procedures.

Tools: Critical event analysis; diaries; expert systems; chains of interviews; key informants; life histories; photograph interpretation; analysis of folktales, songs and dances; analysis of indigenous technical knowledge (ITK).

Key sources: IDRC www.idrc.ca/en/ev-80983-201-1-DO_TOPIC.html; World Bank Indigenous Technical Knowledge www.worldbank.org/afr/ik/basic.htm.

Analytical tools: preferences and evaluations

Besides complex preference analyses, there are many simple (and some participatory) tools for analysing people's attitudes, preferences, objectives and priorities, etc.

Tools: Attitude scales; content analysis; discourse analysis; Delphi technique; Guttman and Likert scales; motives, interests and objectives tables; goals achievement matrix; problem and decision trees; options assessment; preference matrix; pair-wise ranking; multi-criteria analysis, rating and scoring; repertory grid; situational analysis; wealth and well-being ranking; citizens' juries.

Key sources: Oxfam 2002; Pincha 2008; ADPC 2004; ProVention www.proventioncon sortium.org.

Futures planning tools

PAR in DRR eventually leads to the participatory design and analysis of improved futures, for using these tools.

Tools: SWOT (strengths, weaknesses, opportunities, threats) analysis; scenario development; imagineering; visioning and pathways; transformation matrix; brainstorming; visualisation and graphics; games and gaming; mind mapping.

Key sources: AFPP 2006; Kindon *et al.* 2007; Reid *et al.* 2009; ProVention www.proven tionconsortium.org.

experts and key informants. PAR should especially involve the less articulate, less connected and less powerful actors. Other key informants often overlooked are local technical staff, such as extension agents, welfare officers, midwives, teachers or local police officers. Experts often see them as scientifically ignorant and lacking initiative, and NGOs may consider them as not 'speaking for the people'. In reality, they are often the first doorkeeper between external scientific and technological and local experiential knowledge.

PAR activities have revealed the impressive range and depth of local technical and spatial knowledge about hazards, vulnerabilities, capacities and risks. Amongst numerous examples drawn from the ProVention Consortium case studies (Wisner *et al.* 2008) are local knowledge of: volcano and lahar behaviour in Vanuatu; earthquake preparedness and coping strategies in Nepal, Pakistan and Venezuela; a focus on women's earthquake knowledge in Turkey; and responses to flood behaviour in the Philippines. Elsewhere one finds documentation of the significance of social capital and relationships in Bangladesh and Mozambique flood plains, and community warning systems for landslides in the Karakorams and Nepal (Dekens 2007).

In most cultures, women have special knowledge of risk reduction possibilities. Similarly so do children and youth because they notice different things than adults and visit other locations in the course of schooling, play and carrying out chores (e.g. herding, running errands, etc.). Urban risks in play areas, town centres and the journey to school have been researched in Sweden, the UK, Uganda, the USA and Nepal (e.g. Reid *et al.* 2009).

Applications of PAR in DRR

PAR for hazard analysis and mapping

PAR may elicit people's local knowledge of hazard characteristics such as magnitude, frequency, return periods, duration, seasonality, spatial extent and shape, hotspots, safety areas, development rates, decline rates, and relations between intensity, duration and extent. PAR also works to cross-check the local parameters of locations and duration, etc., with external expert knowledge.

Floods are a common focus for PAR studies, appropriately so because they are often relatively mild, manageable and sufficiently frequent to enable the build-up of local spatial knowledge. PAR has been used to interpret local understandings of the different characteristics of floods, flash floods and storm surges (see Chapter 21), e.g. in Haarlem in South Africa, Bangladesh, Laos, Nepal, Bihar and Andhra Pradesh in India, Honduras, and in El Salvador, where three types of flood hazard were assessed (in Wisner *et al.* 2008). Urban inhabitants in the Philippines differentiated among floodwater depths at ankle, knee and waist height, and their consequences (Peters-Guarin *et al.* 2011).

PAR has also tapped local knowledge of earth movements, mudslides, avalanches and landslips (e.g. in Nepal and Mexico), and the specific characteristics and impacts of forest and bush fires, township and urban fires (e.g. in South Africa, Belize and Guatemala). There are PAR applications to bio-hazards and disease outbreaks like plague and malaria (e.g. in Trinidad) and pest infestations such as mosquitoes, locusts and rats. These cases are described in Wisner *et al.* (2008).

Rare hazard events generate less accumulated local experience; however, if there has been no external monitoring, local people may hold the only information. PAR has been applied to knowledge of the behaviour of volcanoes, pyroclastic flows and lahars in the Philippines and Vanuatu, earthquakes in Turkey and Venezuela, and cold waves and blizzards in Peru (all cases in Wisner *et al.* 2008; also see Box 64.3)

Participatory research is also valuable for analysing slow-onset hazards – drought, sea-level rise and diffusion of pests and diseases. Climate predictions are made at macro–meso scales, whilst PAR interprets how people are already experiencing and responding to extreme weather. PAR can provide 'ground truth' for climate models and explain the actual impacts and adaptation strategies of local people. For instance, PAR methods were used to identify women's local knowledge of climate impacts in northern Mexico (Buechler 2009); drought damage in Zimbabwe, Laos and South Africa; forest degradation and biodiversity losses in El Salvador; and the spread of farm pests and diseases and food security issues in Zambia and Zimbabwe (all cases in Wisner *et al.* 2008).

PAR for vulnerability and capacity assessment (VCA)

Vulnerability assessment in DRR refers to the analysis of vulnerabilities and capacities of different categories of people under different circumstances. VCA may be researched in relation to specific

Box 64.3 Participatory three-dimensional mapping for reducing the risk of disasters

JC Gaillard
School of Environment, The University of Auckland, New Zealand

Jake Rom Cadag
Université de Montpellier III, France

Participatory three-dimensional mapping (P3DM) consists of building stand-alone scaled relief maps made of locally available and cheap materials (e.g. carton, paper, cork) over which are overlapped thematic layers of geographic information. P3DM enables the plotting of landforms and topographic landmarks, land cover and use, and anthropogenic features, which are depicted by push-pins (points), yarn (lines) and paint (polygons). It has been extensively used for natural resource management and land conflict resolution, now being applied to disaster risk reduction (DRR) (www.p3dmfordrr.com).

P3DM for DRR follows a five-step methodology that blends mapping activities with other participatory tools for assessing and reducing disaster risks. Local community members start by stacking up layers of cartons or other materials which have been previously cut following the contour lines of a reference map. The scale of this three-dimensional map is large, varying from 1:500 to 1:1,000, to enable the mapping of assets at the household level.

Participants plot land use and other geo-referenced features threatened by natural hazards based on the history of disasters in the community. These features usually include fishing and hunting grounds, agricultural fields, settlements, roads, houses and public or private buildings. Pins of different shapes, sizes and colours enable differentiating building materials and locating the most vulnerable people in the community. It is also possible to use marker pens to label the number of people living in the different houses, the different resources that form the household livelihoods, land tenure and power relationships within the community. Noteworthy is that P3DM only partially addresses social vulnerability/capacities (e.g. client–patron relationships, gender-related inequalities and social networks) and variation of vulnerability and capacities in time according to people's mobility.

Members of the community eventually delineate hazard-prone areas and locate local resources to face these threats. It is then easy and quick to evaluate disaster risk based on hazards, threatened assets, vulnerabilities and capacities.

The next step consists of planning disaster risk reduction measures based on multi-stakeholder group discussions over the map, helping to find consensus among participants. P3DM fosters the participation of a large range of stakeholders, especially the collaboration amongst scientists, government officials and local communities, thus enabling the integration of bottom-up and top-down risk reduction measures. P3DM provides a tangible tool where the most marginalised people, including the illiterate who may have a poor understanding of scientific concepts, can discuss disaster risk reduction with scientists, who on the other hand may have a poor understanding of the local context.

All stakeholders can contribute their knowledge on the same tool and in the same forum. P3DM is credible to both locals, who build the map and plot most of the information, and to scientists and local government representatives who can easily overlap their own data and plans on scaled and geo-referenced maps. In the process, NGO partners serve as facilitators and moderators. The integration of bottom-up and top-down actions is further facilitated when P3DM data are integrated into geographic information systems (GIS) to make use of people's knowledge beyond the community that built the map.

hazards or as composite vulnerability facing the multiple hazards to which a place is exposed. Vulnerability assessment involves an extensive range of factors, such as demographics, family status, age, ethnicity and gender, people's poverty/wealth status and livelihoods, especially the strength of resource ownership and access, political connections, as well as individual lifestyle and risk attitudes. Vulnerability is heavily influenced by temporal, e.g. seasonal and diurnal, and locational specificities (see Chapter 3).

The array of participatory research tools shown in Box 64.2 are equally applicable to vulnerability. Vulnerability and capacity are explored in PAR by means of seasonal calendars, risk maps and priority rankings, among many techniques. Several agencies have developed guidelines for participatory vulnerability assessment: ADPC (2004); ActionAid (2005); Benson et al. (2007); Pincha (2008) on vulnerability of women; and Kienberger (2008) with examples from Mozambique.

PAR to understand coping strategies

Participatory assessment of people's coping strategies and capacities (see Chapter 58) is guided by the concepts of impact mitigation, adaptation and manageability of risks. Productive and diversified livelihoods turn out to be a key driver of coping strategies (see Box 64.4). People's capacities and coping strategies vary amongst social groups, over time and in different locations.

Physical characteristics of the hazards likewise affect coping strategies. For instance, in higher-frequency hazards people can develop adequate coping over time; in the Philippines, urban poor conceptualised flood 'manageability' as a combination of household and community responses to the intensity (depth) and the duration of inundation (Peters-Guarin et al. 2011).

Coping mechanisms may be thought of in relation to event periods: (1) immediately before the hazard event, when 'coping' is closely related to 'preparedness'; (2) during the event; (3) immediately after; and (4) long-term preparedness for, and adaptation to, expected future events. PAR examples for the first period include transfer to safe places of valuable, moveable items like foods, livestock, motorbikes and TVs, and traditional early flood warning systems such as sounding gongs in Java (kentongan) and Laos (in Wisner et al. 2008) and signal fires in Pakistan (Dekens 2007). The second and third periods have been addressed by PAR real-time investigations of coping resources, during and after hazard events, and include such practices as protecting or adapting livelihoods, locating water points and famine foods, and identifying emergency secure buildings and safe sites.

Examples of the fourth period, long-term preparedness, are various 'insurance' schemes based on social relations and family networks that share or spread risk. Using diverse crop mixes and spatial distribution of crops, livestock and storage are examples.

PAR in post-disaster response and disaster recovery

Many NGOs and agencies are engaged in rapid-response and post-disaster work. Response is less participatory because the imperative is for urgently acquiring and disseminating information to

Box 64.4 Using participatory action research for disaster risk reduction in Jakarta, Indonesia

P. Texier
University of Lyon 3–Jean Moulin, CNRS UMR 5600, France

Jakarta slums are highly prone to floods. The impact of flooding is aggravated by uncontrolled urbanisation, poor waste management and bad sewage. To reduce the vulnerability of poor and marginalised communities living along the Ciliwung River, a PAR project was conducted in 2007–08 in collaboration with the local NGO, Ciliwung Merdeka, in Bukit Duri. It was meant to raise the community environmental awareness and strengthen waste management through community-based activities.

The project fostered collaboration between the community, the NGO and the researcher as an initiator and mediator of participatory activities. It was a two-way process that strengthened the community and provided scientific outputs at the same time.

The project started by building rapport with local stakeholders and defining each one's role. Afterwards, a participatory survey enabled assessment of the volume of waste produced, to spot areas of particular concern, to learn about recycling and commercial options, to identify vacant spaces where recycling facilities could be developed and to appraise local constraints and weaknesses in pursuing alternative management.

The second phase of the project consisted of the actual design and set-up of an alternative waste management system based on the systematic collection of waste and the production and commercialisation of compost. Although some participants were reluctant to take on responsibilities, local stakeholders agreed to create a community-based organisation tasked with implementing these activities. Participation of the locals was crucial to the sustainability of the activities and demonstrated their creativity and knowledge.

The project eventually thrived, leading to a significant reduction in the amount of waste dumped into the river and therefore preventing the occurrence of floods. It further contributed to the improvement of people's daily life through a cleaner and healthier environment. It shows that PAR is appropriate for both DRR and development-related activities. The collaboration between the local community and outside stakeholders such as the NGO and the researcher was instrumental as it helped to integrate local and scientific knowledge.

disaster workers and local administrations. However, even under these conditions the value of PAR is being recognised – for example, in the use of local knowledge to identify health hazards, security of people and property, and damage assessment. There are specific emergency mapping needs for emergency shelter, pollution hazards and services, especially water. This PAR always benefits from local involvement and knowledge, but the latter may be limited because people have been displaced to new locations by the emergency.

Participation is vital in the design of new housing and in planning services and new settlement layout, when relocation cannot be avoided (see Chapter 46). To the extent that the exhortation to 'build back better' is actually implemented, PAR is vital in using the recent disaster as a 'learning moment' and working with communities to map likely future hazards, assess people's vulnerabilities and capacities and develop plans to mitigate the next hazard. An example is site selection for the flood-displaced population in the Nepal terai (in Wisner *et al.* 2008).

Preparedness in DRR – planning for long-term adaptation

Long-term preparedness planning begins with mitigation, i.e. endeavouring to reduce the impacts of future hazards by lowering vulnerability and raising coping capacity. PAR can elicit local spatial knowledge of significant items such as geological risks, social characteristics and deficient infrastructure. Preparedness planning may use PAR for selecting appropriate designs of early warning systems and locating escape routes, stores and shelter. PAR studies found that the siting of some flood shelters in Bangladesh was influenced by local elites (Dekens 2007). Van Aalst *et al.* (2008) assessed participatory DRR methodologies for understanding adaptation and preparedness under climate changes. Others have developed participatory scenarios of alternative futures and farmers' expectations of climate change (in Reid *et al.* 2009). Common tools for PAR approaches to longer-term mitigation and preparedness are analysis of 'strengths, weaknesses, opportunities and threats' (SWOT), and the 'transformation matrix' summarising mitigation and adaptation mechanisms, as employed in Costa Rica and Guatemala (in Wisner *et al.* 2008).

Future of PAR in DRR – promises and warnings

The value of local knowledge and PAR

PAR elicits, represents and validates local and indigenous knowledge which is rarely available or even excluded in official documents, statistics, maps or conventional plans. This may be considered the most significant and valuable contribution of PAR. PAR helps to uncover specific information about local perceptions, values and priorities because it is driven by and focused on local actors and interests. PAR involves multiple sources and multiple processes of people's participation in knowledge identification and selection, providing many opportunities for cross-checking and alternative validations.

PAR legitimises endogenous local knowledge (such as indigenous or gendered knowledge) that does not necessarily conform to official assumptions and assertions about people, places and risks (see Chapter 8 and Chapter 9). It places local knowledge on equal footing with the exogenous Western scientific knowledge. PAR uses other media such as visual images as narratives; since pictures are rich in information and shared understanding, they can provide the 'conviction' factor for both inside and outside partners.

Participatory research is not easy to scale up from the community level, because that is not the basic motivation behind PAR; however, where PAR results are acceptably representative, they may form a sound basis for scaling-up. Lessons learnt from successful community PAR can be transferred to municipality or district level, and higher decision-makers can be made aware of the policy-relevant lessons learned. For example, community PAR results were transmitted to municipality level in South Africa; national government institutions were specifically involved in PAR activities in Laos; in Peru, community and local government efforts were integrated by PAR; and in Turkey, local authorities connected with local NGOs (all cases in Wisner *et al.* 2008).

Another value added by PAR is training and capacity building. A keystone of PAR is 'training by doing'. This is well illustrated by valuable practical field exercises in Turkey on vulnerability and capacity assessment for local NGOs, particularly for earthquake hazards; and by multi-hazard training workshops for local government officials in the Philippines, Honduras and Peru (cases in Wisner *et al.* 2008).

Inclusivity in participatory research

PAR seeks social inclusivity to represent the interests and values of communities as well as individuals. Achieving consensus from PAR can be problematic because of the range of needs, priorities and opinions among stakeholders. There is also unequal access to power and influence in the research process (Pelling 2007). How do the inarticulate, the ultra-poor, elderly, children or disabled make their voices heard? Despite their good intentions, participation methods have been criticised for co-opting the powerless and oppressed and weakening their resolve to change basic structures (Cooke and Kothari 2001).

An opposite danger also exists. There can be high-level external political resistance to local empowerment and PAR, and internal local holders of power will not easily share it. The inconvenient truths thrown up by participatory research and the consequent actions which should follow may provoke resistance, particularly among traditionalist or patriarchal social systems, as seen in PAR case studies from Lesotho and the Solomon Islands (in Wisner et al. 2008). Existing power relations might not be altered; indeed they may be reinforced by PAR activities (McCall 2003; Pelling 2007). Participatory mapping of traditional lands, for instance, easily results in the formalisation and privatisation of previously communally owned and managed property.

There are serious time constraints to any participatory research process. PAR demands engagement over long periods, both from the local actors and the promoting agencies. Local actors may not think it is worth their time and effort to participate when they calculate these costs against anticipated benefits. A practical question common in PAR and other participatory research is whether to provide some incentives along the way, such as small culturally appropriate gifts. Outright pay for participants' time is a contested issue, which can only be resolved through reviewing the specific cultural context. What is certain is that the situation of participants needs to be addressed – for example, child care may be needed to facilitate women's participation, and travel and day expenses always reimbursed.

Ethics in participatory research

All participatory research raises a series of ethical questions concerning the transparency, legitimacy and ultimately the ownership of the whole process (McCall 2003). Who initiated the PAR activity? Who will design and operate it? Who will own or control the outputs? Who will take it further?

There are also issues concerning the intentions of the external actors (NGOs, government agencies or researchers). What is their commitment to the community? What are their goals? Are they seeking objectivity and value neutrality, and if they are, how does that affect their 'position' in a community that may be at loggerheads with government or a poor group of landless who have grievances against a landlord? Are there signs of critical reflexivity and self-awareness in the external actors as the process develops?

As the PAR becomes operational, one needs to ask about the breadth and the depth of community participation. In principle, participation extends from the study design and choice of methods, through the initiation of activities, to field activities, analysis, storage and dissemination of the outputs. Here there are two issues. First, what respect is shown for local spaces, local agendas, temporality and terminology during the whole process? Acceptance of PAR is more likely when outside agencies adapt to local culture, such as communicating in the vernacular language (a Mexican example, in Wisner et al. 2008). Are insights 'lost in translation' if the researchers use their own technical language to communicate findings to government officials or other non-locals? Second, what about the 'A' in PAR? Is the research team committed to following up with an action plan and implementation? What if the budget for this work does not include funding for such follow-up?

Finally, there are procedural ethics in ensuring 'fully informed consent' (so far as that is possible), usually providing anonymity to informants and allowing for withdrawal, and for appropriate control and dissemination of results. Indeed, beyond procedural issues, a major ethical responsibility is to protect the community and individuals from negative backlash by powerful people or groups (governmental or non-governmental), which might perceive their interests threatened by what is revealed (see Chapter 60).

Conclusions

PAR can bridge many gaps: between Western scientific knowledge and local vernacular (traditional) knowledge; between theoretical and practical DRR; and between top-down interventions and bottom-up initiatives. At an operational level, PAR can integrate disparate elements of local DRR knowledge, the fields and sources of which fall under disconnected government agencies and civil society organs like NGOs. PAR strengthens opportunities for community actors to benefit from exogenous scientific methods and technology (e.g. IT) for handling DRR issues at local level. Despite these advantages and the large body of practice and experience that has accumulated, PAR also has limitations, and therefore needs to be applied as just one among many approaches to providing a knowledge base for DRR.

Part V
Conclusion

65

Challenging risk – has the left foot stepped forward?

The Editors

The left foot's main themes

While the title of this *Handbook* emphasises hazard, risk and disaster, the key aspect that emerges from the foregoing chapters is the need to address vulnerability and the capacities required to overcome it. That does not mean neglecting risk, hazard and disaster. It means recognising that hazards, risk and disaster tend to be products of human action and inaction, of decisions about the allocation of and access to resources and locations, and the power and lack of power to make these decisions (Mueller-Mann 2011).

The journey through this *Handbook* has been broad and deep. Across dozens of chapters, boxes and authors from all inhabited continents and numerous, disparate disciplines, as well as non-disciplinary approaches, we now use our position as editors to extract and emphasise the 'lucky thirteen' main themes that are important to us. These themes indicate the challenges and opportunities inherent in hazards, vulnerabilities, risks and disasters – and dealing with all of them in tandem through disaster risk reduction (DRR). The themes may read as excessively general, at times even philosophical; however, we believe that adequate understanding of disaster risk and how to meet its challenge calls forth fundamental questions about human existence and the Earth as the home of humanity amongst other species.

Our lucky thirteen left foot themes are:

1 *Complexity*. The world is complex. Life is complex. That should not be feared, but should be grasped, and it should inform policy and practice for DRR. Much human action takes place in the face of complexity. Consequences of actions or inaction are not always foreknown or immediately evident. We must nonetheless act through DRR. Many tools and a good deal of practical experience exist to help us 'muddle through' complexity, from applications of the so-called precautionary principle (Chapter 14) to engineering models (Chapter 26) and local-scale participatory exercises such as group discussions of strengths, weaknesses, opportunities and threats or constraints (SWOT/SWOC analysis) (Chapter 64).

2 *Interconnectedness*. Key sub-systems must be considered in DRR – politics, culture, technology and the built environment, and economics and social relations – as well as the

interactions of these human sub-systems with the natural world around them. That natural world also has sub-systems, divided amongst categories such as ecosystems, geological systems, water systems such as rivers and seas, and taxonomic levels of flora and fauna. Each of these may manifest as hazard or opportunity – for human use or perhaps only to contemplate their beauty. Much research and policy is focused too narrowly on hazards and not on the web of relations and cascade of consequences that characterise the interactions among these sub-systems. Most of all, political and economic power and differential access to choices and resources are often downplayed or sidelined. Connections may be denied because they are too complex to untangle. The chapters here assist with identifying and working with interconnectedness, such as in emergencies (Chapter 42) or applying financial mechanisms for DRR (Chapter 54). Many hazards display interconnectedness with human activities, such as soil degradation (Chapter 29), while vulnerabilities and capacities can emerge from the connected cultural characteristics such as caste and ethnicity (Chapter 38).

3   *Time and space.* To incorporate complexity and key sub-system interactions into DRR analysis and action, multiple scales across space and time need to be linked. This may be difficult because institutions, households and other entities (natural and human-made) function on different temporal and spatial scales. Political electoral cycles, business cycles, life cycles and fission of households as grandchildren come into the world – all of these have a rhythm, as do crop and livestock production, epidemics, climatic seasons and ENSO-related ocean changes. Some people never leave the town or village where they are born, yet global power relations, and decisions made in national capitals thousands of kilometres distant, affect them in the form of information flows, price fluctuations and geopolitics creating or reducing conflict. Witness the complex knock-on effects of converting farm land to production of bio-fuels, as this combined with many other factors to cause a dramatic rise in food prices in 2007–08 (Adam 2008; Bello 2009).

4   *Multiple stakeholders.* Various participants, all with their own interests and perspectives, need to be involved and understood. That includes their motivations, perceptions, interactions and the effects of decisions and actions on others. Much work still tends to focus on either larger-scale, institutional stakeholders, or on micro-scale households or individual stakeholders. This conventional research and practice has institutional and behavioural biases that can miss out the influences of mediating institutions such as local government (Chapter 52), civil society (Chapter 60), the private sector (especially small and medium-sized enterprises) and informal or emergent groups (Chapter 59).

5   *Knowledge and wisdom.* Knowledge and wisdom are not unitary or universal. Many forms of knowledge are relevant to understanding hazards, vulnerabilities, capacities, risks, disasters and DRR. Local knowledge and practice are as important to take into consideration as outside specialist, technical (often Western) knowledge. Yet there are numerous obstacles to establishing a fruitful dialogue between local and non-local. Furthermore, just the possession of knowledge is not sufficient. Wisdom means having the power to discern and judge which types of knowledge to use, and how to use them, under which circumstances, for which purposes and in whose interest or benefit. Wise use of knowledge is a matter of cultural judgement. Various cultural systems including religions – and world views more generally – contribute in their own ways to knowledge and wisdom, inhibiting or spurring on desire to address disasters. All knowledge and wisdom approaches have limitations and advantages.

6   *Learning and teaching.* Institutions and systems have been created that synthesise and evaluate disaster and DRR experience. International standards and guidelines abound. This *Handbook*

covers an extensive range of examples of this systematic study and diffusion of accumulated experience. This *Handbook* also misses many topics that are touched upon lightly or that are explicitly stated as not being included. Understanding and covering all such topics is not easy, leading to the need to continually exchange with others through co-learning. If this *Handbook* has taught you something, pass it on and use it to teach the editors and authors what you know. Feedback of this kind is vital. **Post your questions, comments, objections and insights through the Radix network and email list at www.radixonline.org**.

7   *Capacity and choice.* Numerous *Handbook* chapters make it clear that people rarely end up in harm's way due to stupidity, ignorance, apathy or fatalism. Conversely, when people are given options, resources and choices, they frequently show initiative with regards to hazards, reducing vulnerabilities, building capacities and managing risks to avoid disasters. Day-to-day life for many does not permit them to use their capacities to do so, while power, governance, knowledge and access structures may deny opportunities to use those capacities. Making resources and options available, and removing discriminatory practices that block some people's access to resources and options, are essential for DRR. Expelled from a newly established national park together with their livestock, Maasai in Kisiwani, Same District in northern Tanzania are struggling to find drought reserve pasture in a narrow strip of land between the national park and Pare mountain escarpment. These Maasai know very well how to keep their animals healthy during a drought but are denied access to resources and options (Wisner 2010c).

8   *Livelihoods and access.* For people to achieve dignified lives and to support the well-being of their families and communities, they need stable and productive livelihoods. The focus should be access to the resources and locations needed to achieve these ends without harming others now or future generations. A focus on livelihoods and access to resources for those livelihoods must be a baseline and long-term context for DRR. 'Baseline' means that the additional productivity and stability of diverse livelihoods and expanded access to resources allow some surplus time and money to be invested by households and communities in DRR, as an investment for the future. 'Long-term context' means that DRR practices and recurrent investments at the local level will become routine and part of normal life and livelihoods (Chapters 14, 22, 31, 57 and 58). Focusing on livelihoods further enables blending DRR and development as it strengthens people's ability to sustain their everyday needs while preventing their suffering from the harmful effect of natural hazards.

9   *Development and sustainability.* Many *Handbook* chapters reveal that good governance, good planning and sustainable development must take risk into account. They are each, to a large degree, about risk management and risk choices. The converse is also true. Effective DRR cannot be achieved in a vacuum, but needs to use tangible means and relationships within wider development and sustainability contexts and processes. Just as DRR needs to be integrated with enhanced, diversified and sustainable livelihoods at the local scale, national development policy and programmes should make DRR central and explicit. National governments and other authorities thereby 'insure' their investments in development against loss in disaster (World Bank 2010e) (Chapters 50, 51, 54 and 55).

10 *Implementation.* The reader might have been struck by the gap that still exists between what humanity collectively knows and what actually gets funded and implemented. Several sub-themes run through the *Handbook* that partly explain this gap: conflict, misuse of power, corruption, poor governance and lack of political interest or will loom large among the determinants (Chapters 5 and 7), alongside more benign, yet no less obstructive, 'silo' thinking and competition among different sectors, interests and actors at different scales (Chapter 50 and 51). A balance of top-down and bottom-up initiatives is required for

successful implementation; however, a persistent problem yet to be solved is establishing local 'ownership' and routinising DRR in ways that last – that is, that become interwoven into the patterns of daily life and livelihoods (Chapters 45, 46 and 58).

11 *Conflict*. People have different material interests. These interests often conflict, and nowhere as clearly as concerning access to land and other resources and advantageous locations. As part of good DRR and development, such conflicts can and should be negotiated non-violently. One has only to think of the situation of the urban poor forced by the price of land to live in dangerous locations (Chapter 53) or the Nicaraguan farmers who died because they had nowhere else to farm except the hazardous slopes of a volcano (Chapter 14). Conflict of a different sort may be manifested when established bureaucratic and other territorial and vested interests refuse to co-operate on an inter-departmental programme of DRR. Such resistance should also be recognised and overcome through negotiation.

12 *Consciousness*. Notions of hazard, vulnerability, capacity, risk and disaster populate human history, cultural production, religion and worldviews (Chapters 4, 10, 11, 12 and 38). However, given the complexity of the relations involved and the chains of cause and effect that run throughout all these, most people are not aware of how interwoven with day-to-day life is the creation, acceptance and management of risk. Effective DRR depends, in part, on people recognising these complexities and realising that they can and should take action, i.e. people being self-conscious of themselves as an interest group, which is a step toward self-organisation and action on behalf of themselves. Whether it is for psycho-social support after a disaster (Chapter 47) or stopping a tornado from becoming a disaster (Chapter 20), people's consciousness of what they can achieve for themselves is part of DRR.

13 *Daily life*. The *Handbook*'s topics demonstrate that dealing with disasters should be a part of usual day-to-day life. Over the long term, vulnerability can be tackled only by treating it as a condition built up over time, a process rooted in life and livelihood routines and the way in which long-standing systems of social, economic and political power either provide resources and options or constrain lives and livelihoods. Solutions are not one-off exercises and are not about short-term interventions and 'deliverables'.

From disaster to DRR

The *Handbook*'s work cannot stop here. It provides just one beginning, a basis from which to move forward – and, we hope, a basis for moving forward from disaster, towards DRR. Many possibilities exist for building on the *Handbook* and the knowledge and experience on which the *Handbook* is built. Some directions are provided here for the readers to forge their own paths according to their own interests and needs.

On the practical side, an impressive array of educational curricula continues to expand. In addition to professional training, undergraduate courses and primary and secondary school units related to disasters and DRR are becoming more popular (Chapter 62) (Wisner 2006a). The topic is also starting to be incorporated throughout regular curricula rather than always being viewed as apart and separate. Both tracks are needed. As one example from amongst many of such work being scaled up, millions of people have been participating in 'Shake Out' drills in California in which an earthquake is assumed to strike at a specific time and everyone is meant to respond appropriately.

On the theoretical side, it is encouraging to see in examples covered by the *Handbook* how disaster-related research is moving further into the human and social context of vulnerability and capacities without neglecting the natural environment as a locus of hazards as well as live-lihood opportunities. Neither hazard nor vulnerability should be denied for DRR.

Many techniques exist to follow the DRR path, as laid out throughout the *Handbook*. No single person can possess the knowledge and skills to map out and successfully implement DRR. Teams are required that provide many kinds of knowledge. Such teamwork needs to be complemented with partnership among communities and local groups, non-governmental organisations, sub-national and national governments, regional and international organisations, and the private sector.

Understanding of hazards and disaster risk reduction is never complete. Please send your comments, additions, objections and questions to www.radixonline.org.

Bibliography

ABC News (2009) *Cairo–Pig cull protests turn violent*, New York: ABC News. Online www.abc.net.au/news/stories/2009/05/03/2559580.htm (accessed 28 April 2010).

Abdulai, A., Barrett, C.B. and Hoddinnott, J. (2005) 'Does food aid really have disincentive effects?' *World Development* 33, 10, 1689–704.

Abolghasemi, H., Radfar, M.H., Khatami, M., Nia, M.S., Amid, A. and Briggs, S.M. (2006) 'International medical response to a natural disaster: Lessons learned from the Bam earthquake experience', *Prehospital and Disaster Medicine* 21, 3: 141–47.

Acharya, K.P. (2005) 'Private, collective, and centralized institutional arrangements for managing "forest commons" in Nepal', *Mountain Research and Development* 25: 269–77.

ACHR (Asian Center for Human Rights) (2006) 'Sri Lanka's proposed international commission of inquiry: OHCHR, eminent organizations and person, watch out', *ACHR Weekly Review* 137/06.

ACT Government (2003) *The Report of the Bushfire Recovery Taskforce Australian Capital Territory October 2003*, Canberra: ACT (Australian Capital Territory) Government.

ActionAid (2005) *Participatory Vulnerability Analysis: A step-by-step guideline for field staff*, London/Johannesburg: ActionAid International.

—— (2010) *Getting Children Back to School in Haiti*, Somerset: ActionAid. Online www.actionaid.org.uk/102326/getting_children_back_to_school_in_haiti.html (accessed 6 April 2010).

Adam, B. and Allan, S. (eds) (1995) *Theorizing Culture: An interdisciplinary critique after postmodernism*, London: University College London Press.

Adam, D. (2008) 'Food prices threaten global security – UN', *The Guardian*, 9 April. Online www.guardian.co.uk/environment/2008/apr/09/food.unitednations (accessed 21 December 2010).

Adams, J., Maslin, M. and Thomas, E. (1999) 'Sudden climate changes during the Quaternary', *Progress in Physical Geography* 23, 1: 1–36.

Adams, W.M. (2001) *Green Development: Environment and sustainability in the third world*, London: Routledge.

ADB (2007) *Report and Recommendation of the President to the Board of Directors: Proposed Program Loan and Technical Assistance Grant Republic of the Philippines: Local Government Financing and Budget Reform Program Cluster (Subprogram 1)*, Project Number 39516, Manila: Asian Development Bank. Online www.adb.org/Documents/RRPs/PHI/39516-PHI-RRP.pdf (accessed 29 December 2009).

Adeola, F.O. (2000) 'Cross-national environmental injustice and human rights issues', *American Behavioral Scientist* 43, 686–706.

ADPC (Asian Disaster Preparedness Center) (no date a) *India: Technological hazard mitigation in Baroda and metropolitan Calcutta*, Asian Urban Disaster Management Programme (AUDMP). Online www.adpc.net/audmp/pprofile.html (accessed 7 January 2010).

—— (no date b) *Child Focused Disaster Risk Reduction*, Bangkok: ADPC. Online ineesite.org/uploads/documents/store/doc_1_Child_Focused_Disaster_Risk_Reduction.pdf (accessed 4 January 2010).

—— (2004) *Field Practitioners' Handbook*, Pathumthani, Thailand: ADPC. Online www.adpc.net/pdr-sea/publications/12Handbk.pdf (accessed 18 December 2010).

—— (2007) *Child Focused Disaster Risk Reduction*, Bangkok: ADPC.

—— (2008) *Community Based Early Warning System and Evacuation: Planning, development and testing protecting people's lives and properties from flood risks in Dagupan City, Philippines*, Bangkok: ADPC.

—— (2009) *Government of India-UNDP Disaster Management Programme, 2002–2009: Evaluation and Review of Lessons Learnt*, Bangkok: ADPC. Online www.undp.org.in/sites/default/files/reports_publication/DRM-Report_0.pdf (accessed 17 August 2010).

Advameg, Inc. (2010) 'Australian mythology', *Myth Encyclopedia*. Online www.mythencyclopedia.com/Ar-Be/Australian-Mythology.html (accessed 25 May 2010).

AFAC (Australasian Fire Authorities Council) (2005) *Position Paper on Bushfires and Community Safety*, East Melbourne: AFAC Limited.

AFPP (Alliance Frontiers Prevention Project) (2006) *Tools Together Now! 100 Participatory Tools to Mobilise Communities for HIV/AIDS*, Brighton: International HIV/AIDS Alliance. Online www.aidsalliance.org/graphics/secretariat/publications/Tools_Together_Now.pdf (accessed 14 December 2010).

Aguirre, B. (2005) 'Cuba's disaster management model: Should it be emulated?', *International Journal of Mass Emergencies and Disasters* 23, 3: 55–71.

Agyeman, J. and Evans, T. (2003) 'Toward just sustainability in urban communities: Building equity rights with sustainable solutions', *Annals of the American Academy of Political and Social Science* 590: 35–53.

Ahrens, J. and Rudolph, P.M. (2006) 'The importance of governance in risk reduction and disaster management', *Journal of Contingencies and Crisis Management* 14, 4: 207–20.

AIDMI (All India Disaster Mitigation Institute) (2009a) *Case Study for the First South-South Citizenry Based Development Academy on Community Based Disaster Risk Reduction*, Gujarat: All India Disaster Mitigation Institute/Special Unit for South-South Cooperation, United Nations Development Programme.

—— (2009b) *Making Schools Safer*, Ahmedabad: All India Disaster Mitigation Institute.

—— (2009c) *Proceeding of the First South–South Citizenry Based Development Academy on Community Based Disaster Risk Reduction*, Gujarat: All India Disaster Mitigation Institute.

Akasoy, A. (2009) 'Interpreting earthquakes in Medieval Islamic texts', in C. Mauch and C. Pfister C. (eds), *Natural Disaster, Cultural Responses: Case studies towards a global environmental history*, Lanham, MD: Lexington Books, pp. 183–96.

Akhand, M. (2003) 'Disaster management and cyclone warning systems in Bangladesh', in J. Zschau and A. Kuppers (eds), *Early Warning Systems for Natural Disaster Reduction*, Berlin: Springer Verlag, pp. 49–64.

Aklilu, Y. (2008) *Livestock Marketing in Kenya and Ethiopia: A review of policies and practice*, Addis Ababa and Medford, MA: Feinstein International Center.

Aklilu, Y. and Wekesa, M. (2002) *Drought, Livestock and Livelihoods: Lessons from the 1999–2001 emergency response in the pastoral sector in Kenya*, London: Humanitarian Practice Network. Online livestock-emergency.net/userfiles/file/general/Aklilu-Wekesa-2002.pdf (accessed 25 March 2010).

Aksartova, S. (2006) 'Why NGOs? How American donors embraced civil society after the Cold War', *International Journal of Not-for-Profit Law* 8, 3: 16–21.

Alexander, B. (2008) 'Sustainagility', *Proceedings of the International Conference-Workshop on Climate Change and Biodiversity: Mitigation and adaptation*, 18–20 February 2008, Manila, Philippines.

Alexander, D. (1987) *The 1982 Urban Landslide Disaster at Ancona, Italy*, Working Paper No. 57, Boulder, CO: Natural Hazards Center.

—— (1993) *Natural Disasters*, London: UCL Press.

—— (1997) 'The study of natural disasters, 1977–97: Some reflections on a changing field of knowledge', *Disasters* 21, 4: 284–304.

—— (2000) *Confronting Catastrophe: New perspectives on natural disasters*, Oxford: Oxford University Press.

—— (2002a) 'From civil defence to civil protection – and back again', *Disaster Prevention and Management* 11, 3: 209–13.

—— (2002b) *Principles of Emergency Planning and Management*, New York: Oxford University Press.

Algermissen, S. and Perkins, D. (1976) *A Probabilistic Estimate of the Maximum Ground Acceleration in Rock in the Contiguous United States*, USGS Open File Rept. 76–416, Denver, CO: USGS.

ALNAP (Active Learning Network for Humanitarian Practice) (2008) *Responding to Earthquakes 2008: Learning from earthquake relief and recovery operations: Lessons learned*. Online www.alnap.org/pool/files/ALNAPLessonsEarthquakes.pdf (accessed 9 October 2010).

—— (2010) *State of the Humanitarian System*, London: ALNAP/ODI. Online www.alnap.org/pool/files/alnap-sohs-final.pdf (accessed 30 November 2010).

ALNAP and ProVention Consortium (2009) *Responding to Urban Disasters: Learning from previous relief and recovery operations*, London: ALNAP.

Altman, D. (1988) 'Legitimation through disaster: AIDS and the gay movement', in E. Fee and D. Fox (eds), *AIDS: The burdens of history*, Berkeley, CA: University of California Press.

Alvarez, W. (1997) *T-Rex and the Crater of Doom*, Princeton: Princeton University Press.

Ambler, J., Pandolfelli, L., Kramer, A. and Meinzen-Dick, R. (2007) *Strengthening Women's Assets and Status: Programs improving poor women's lives*, 2020 Focus Brief on the World's Poor and Hungry People, Washington, DC: International Food Policy Research Institute.

American Society of Civil Engineers (ASCE) (2009) 'TCLEE 2009: Lifeline earthquake engineering in a multihazard environment', *2009 ASCE Technical Council on Lifeline Earthquake Engineering Conference*, Oakland, CA, 28 June–1 July 2009.

Analitis, A., Katsouyanni, K., Biggeri, A., Baccini, M., Forsberg, B., Bisanti, L., Kirchmayer, U., Ballester, F., Cadum, E. and Goodman, P.G. (2008) 'Effects of cold weather on mortality: Results from 15 European cities within the PHEWE project', *American Journal of Epidemiology* 168: 1397.

Anderskov, C. (2004) *Anthropology and Disaster*, Special Thesis, Aarhus University, Denmark. Online www.anthrobase.com/Txt/A/Anderskov_C_03.htm (accessed 11 November 2010).

Anderson, M.B. (1999) *Do No Harm: How aid can support peace – or war*, Boulder, CO: Lynne Rienner.

Anderson, M.B. and Woodrow, P. (1989) *Rising from the Ashes: Development strategies in times of disasters*, Boulder: Westview Press.

Anderson, W.A. (1969) 'Disaster warning and communication processes in two communities', *Journal of Communication* 19: 92–104.

—— (2005) 'Bringing the children into focus on the social science disaster research Agenda', *International Journal of Mass Emergencies and Disasters* 23, 3: 59–175.

Anderson-Berry, L., Iroi, C. and Rangi, A. (2003) *The Environmental and Societal Impacts of Cyclone Zoe and the Effectiveness of the Tropical Cyclone Warning Systems in Tikopia and Anuta*, Townsville: James Cook University Centre for Disaster Studies.

Andrienko, G. and Andrienko, N. (2008) 'Spatio-temporal aggregation for visual analysis of movements', *IEEE Symposium on Visual Analytics Science and Technology (VAST)*, 2008: 51–58.

Annan, K.A. (2002) *Prevention of Armed Conflict: Report of the Secretary-General*, New York: United Nations.

Ansal, A. and Slejko, D. (2001) 'The long and winding road from earthquakes to damage', *Soil Dynamics and Earthquake Engineering* 21, 5: 369–75.

Ansell, C. and Gingrich, J. (2003) 'Trends in decentralization', in B. Cain, R. Dalton and S. Scarrow (eds), *Democracy Transformed? Expanding Political Opportunities in Advanced Industrial Democracies*, New York: Oxford University Press, pp. 140–63.

Anton, R.J. (1990) 'Combining singing and psychology', *Hispania* 73, 4: 1166–70.

Antonopoulos, J. (1992) 'The great Minoan eruption of Thera Volcano and the ensuing tsunami in the Greek archipelago', *Natural Hazards* 5: 153–68.

Aoo, K., Butters, S., Lamhauge, N., Napier-Moore, R. and Ono, Y. (2007) 'Whose (transformative) reality counts? A critical review of the transformative social protection framework', *IDS Bulletin* 38, 3: 29–33.

Ariyabandu, M. (1999) *Defeating Disasters: Ideas for action*, Colombo: ITDG/Duryog Nivaran.

Ariza, E., Jiménez, J.A. and Sardá, R. (2008) 'A critical assessment of beach management on the Catalan coast', *Ocean Coastal Management* 51: 141–60.

Armstrong, M.J. (2000) 'The political economy of hazards', *Environmental Hazards* 2: 53–55.

Arnold, J.L., Dembry, L.M, Tsai, M.C., Dainiak, N., Rodoplu, U., Schonfeld, D.J., Paturas, J., Cannon, C. and Selig, S. (2005) 'Recommended modifications and applications of the hospital emergency incident command system for hospital emergency management', *Prehospital and Disaster Medicine* 20, 5: 290–300.

Arriens, W.T. and Benson, C. (1999) *Post Disaster Rehabilitation: The experience of the Asian Development Bank*. Online www.eird.org/estrategias/pdf/eng/doc11509/doc11509-contenido.pdf (accessed 17 January 2010).

Article 19 (1990) *Starving in Silence: A report on famine and censorship*, Article 19, London. Online www.reliefweb.int/rw/lib.nsf/db900sid/OCHA-6NMTSW/$file/art-gen-apr90.pdf?openelement (accessed 14 December 2010).

Arunotai, N. (2008) 'Saved by an old legend and a keen observation: The case of Moken sea nomads in Thailand', in R. Shaw, N. Uy and J. Baumwoll (eds), *Indigenous Knowledge for Disaster Risk Reduction: Good Practices and Lessons Learnt from the Asia-Pacific Region*, Bangkok: UNISDR Asia and Pacific, pp. 73–78.

ASCE (American Society of Civil Engineers) (2006) *Minimum Design Loads for Buildings and Other Structures*, Reston: ASCE.

Ashmore, J., Babister, E., Corsellis, T., Fowler, J., Kelman, I., Manfield, P., McRobie, A., Spence, R., Vitale, A., Battilana, R. and Crawford, K. (2003) 'Diversity and adaptation of shelters in transitional settlements for IDPs in Afghanistan', *Disasters* 27, 4: 273–87.

Asian Center for Human Rights (2006) 'Sri Lanka's proposed International Commission of Inquiry: OHCHR, eminent organizations and person, watch out', *ACHR Weekly Review* 137, 6.

Askartova, S. (2006) 'Why NGO's? How American donors embraced civil society after the Cold War', *International Journal of Not-for Profit Law* 8(3). Online www.icnl.org/knowledge/ijnl/vol8iss3/special_4 (accessed 14 June 2010).

Askew, A.J. (1997) 'Water in the International Decade for Natural Disaster Reduction', *Proceedings of Destructive Water: Water-Caused Natural Disasters, their Abatement and Control Conference* held at Anaheim, California, June 1996. IAHS 239. Online iahs.info/redbooks/a239/iahs_239_0003.pdf (accessed 3 October 2010).

AUDMP (no date) *India: Technological hazard mitigation in Baroda and Metropolitan Calcutta.* Online www.adpc.net/audmp/India.html (accessed 3 October 2010).

Auf der Heide, E. (1989) *Disaster Response: Principles of preparation–coordination*, Chicago: Mosby.

Austin, J.D. and Bruch, C.E. (eds) (2000) *The Environmental Consequences of War: Legal, economic, and scientific perspectives*, Cambridge: Cambridge University Press.

AWID (Association for Women's Rights in Development) (2010) 'Take action: Front line: India: Arrest and detention of five human rights defenders investigating Dalit Human Rights Violations'. Online www.awid.org/.../Take-Action-Front-Line-India-Arrest-and-detention-of-five-human-rights-defenders-investigating-Dalit-human-rights-violation (accessed August 20 2010).

A'zami, A. (2005) 'The Badgir in traditional Iranian architecture', *International Conference: Passive and Low Energy Cooling for the Built Environment*, Santorini, Greece. Online www.inive.org/members_area/medias/pdf/Inive%5Cpalenc%5C2005%5CAzami2.pdf (accessed 5 October 2010).

Babugura, A.A. (2008) 'Vulnerability of children and youth in drought disasters: A case study of Botswana', *Children, Youth and Environments* 18, 1: 126–57.

Back, E., Cameron, C. and Tanner, T. (2009) *Children and Disaster Risk Reduction: Taking stock and moving forward*, Brighton: Institute of Development Studies.

Bail, R. (2007) 'Nicaragua exports its poor', *Le Monde Diplomatique*, English edn, January. Online mondediplo.com/2007/01/12nicaragua (accessed 19 December 2010).

Bailey, G. and Walker, J. (2007) 'Heat related disasters', in D. Hogan and J. Burstein (eds), *Disaster Medicine*, 2nd edn, Philadelphia: Lippincott, Williams & Wilkins, pp. 256–65.

Baillie, M.G.L. (2007a) 'The case for significant numbers of extraterrestrial impacts through the Late Holocene', *Journal of Quaternary Science* 22: 101–9.

—— (2007b) 'Tree-rings indicate global environmental downturn that could have been caused by comet debris', in P. Bobrowsky and H. Rickman (eds), *Comet/Asteroid Impacts and Human Society*, Berlin: Springer, pp. 105–22.

Baird, A., O'Keefe, P., Westgate, K. and Wisner, B. (1975) *Towards an Explanation and Reduction in Disaster Proneness*, Bradford Disaster Research Unit Occasional Paper 11, Bradford: University of Bradford.

Bajek, R., Matsuda, Y. and Okada, N. (2008) 'Japan's jishu-bosai-soshiki community activities: Analysis of its role in participatory community disaster risk management', *Natural Hazards* 44, 2: 281–29.

Baker, J.E., Sedney, M.A. and Gross, E. (1992). 'Psychological tasks for bereaved children', *American Journal of Orthopsychiatry* 62, 1: 105–16.

Baldi, B. (1999) *Beyond the Federal–Unitary Dichotomy*, Working Paper 99–7, Institute of Governmental Studies, University of California, Berkeley. Online igs.berkeley.edu/publications/working_papers/99–7.pdf (accessed 16 August 2010).

Bankoff, G. (2001) 'Rendering the world unsafe: "Vulnerability" as western discourse', *Disasters* 25, 1: 19–35.

—— (2007a) 'Fire and quake in the construction of old Manila', *Medieval History Journal* 10, 1–2: 411–27.

—— (2007b) 'The dangers of going it alone: Social capital and the origins of community resilience in the Philippines', *Continuity and Change* 22, 2: 327–55.

Barash, D.P. (2009) *Approaches to Peace*, 2nd edn, Oxford: Oxford University Press.

Barbaro, P. (2009) 'Explanation of the earth's features and origin in pre-Meiji Japan', in M. Kölbl-Ebert (ed.), *Geology and Religion: A history of harmony and hostility*, London: The Geological Society of London, pp. 25–36.

Barber, C.V., Miller, K.R. and Boness, M. (eds) (2004) *Securing Protected Areas in the Face of Global Change: Issues and strategies*, Gland and Cambridge: IUCN.

Barbier, C. (2010) 'Extreme La Niña brings misery and illness to Peru', *The Guardian*, 24 August. Online www.guardian.co.uk/world/2010/aug/24/peru-cold-weather-temperature-el-ninia (accessed 8 October 2010).

Barbina, G. (1993) 'Les communautés ethno-linguistiques et la conscience de leur territoire', in A.-L. Sanguin (ed.), *Les Minorités Ethniques en Europe*, Paris: L'Harmattan, pp. 55–60.

Barenstein, J.D. (2006) *Housing Reconstruction in Post-Earthquake Gujarat: A comparative analysis*, Network Paper No. 54, London: Humanitarian Practice Network.

Bari, F. (1998) 'Gender, disaster and empowerment: A case study from Pakistan', in E. Enarson and B. Hearn-Morrow (eds), *The Gendered Terrain of Disaster: Through women's eyes*, London: Praeger, pp. 125–32.

Barnes, L.R., Gruntfest, E.C., Hayden, M.H., Schultz, D.M. and Benight, C. (2007) 'False alarms and close calls: A conceptual model of warning accuracy', *Journal of the American Meteorological Society* 22, 5: 1140–47.

Barnett, B.J., Barrett, C.B. and Skees, J.R. (2008) 'Poverty traps and index-based risk transfer products', *World Development* 36: 1766–85.

Baron, N. (2002). 'Southern Sudanese refugees: In exile forever?', in R.C.-Y. Fred Bemak (ed.), *Counseling Refugees: A psychosocial approach to innovative multicultural interventions*, New York: Praeger, pp. 188–207.

Barrett, C.B. and Maxwell, D.G. (2005) *Food Aid After Fifty Years: Recasting its role*, New York: Routledge.

Barrett, T., Pastoret, P. and Taylor, W.P. (2006) *Rinderpest and Peste des Petits Ruminants*, Holland: Elsevier.

Barrientos, A. and Hulme, D. (2009) 'Social protection for the poor and poorest in developing countries: Reflections on a quiet revolution', *Oxford Development Studies* 37, 4: 439–56.

Barrientos, A., Hulme, D. and Shepherd, A. (2005) 'Can social protection tackle chronic poverty?', *European Journal of Development Research* 17, 1: 8–23.

Barth, F. (1969) 'Introduction', in F. Barth (ed.), *Ethnic Groups and Boundaries: The social organization of culture difference*, Boston: Little Brown and Company, pp. 9–38.

Bassett, C., Thornham, S. and Marris, P. (2000) *Media Studies*, 2nd edn, New York: New York University Press.

Bates, F.L., Fogleman, C.W., Parenton, V.J., Pittman, R.H. and Tracy, G.H. (1963) *The Social and Psychological Consequences of a Natural Disaster: A longitudinal study of Hurricane Audrey*, Publication No. 1081, Washington, DC: National Academy of Sciences – National Research Council.

Bates, F.L and Peacock, W.G. (1987) 'Disasters and social change', in R.R. Dynes, B. de Marchi and C. Pelanda (eds), *The Sociology of Disasters*, Milan, Italy: Franco Angeli Press, pp. 291–330.

Batjargal, Z. (2001) 'Lessons learnt from consecutive *dzud* disaster of 1999–2000 in Mongolia in Asia', in *Abstracts of Open Symposium on Change and Sustainability of Pastoral Land Use Systems in Temperate and Central Asia*, Mongolia: Ulaanbaatar.

Baxter, P., Kapila, M. and Mfonfu, D. (1989) 'Lake Nyos disaster, Cameroon, 1986: The medical effects of large scale emission of carbon dioxide?' *British Medical Journal* 298: 1437–41.

BBC (2009) 'Huge Bolivian glacier disappears', *BBC News Online*, 12 May. Online news.bbc.co.uk/1/hi/8046540.stm (accessed 12 April 2010).

—— (2010a) 'Himalayan glaciers' mixed picture', *BBC News Online*. Online news.bbc.co.uk/2/hi/8355837.stm (accessed 12 April 2010).

—— (2010b) '"No alert" in Indonesian tsunami', BBC Online, 27 October. Online www.bbc.co.uk/news/world-asia-pacific-11635714 (accessed 4 December 2010).

—— (2010c) 'Haiti cholera: UN peacekeepers to blame, report says', BBC Online, 8 December. Online www.bbc.co.uk/news/world-latin-america-11943902 (accessed 22 December 2010).

Beall, J. and Kanji, N. (1999) 'Households, livelihoods and urban poverty', Background paper for the ESCOR commissioned research project *Urban Development: Urban governance, partnership and poverty*. Online www.ucl.ac.uk/dpu-projects/drivers_urb_change/urb_society/pdf_liveli_vulnera/DFID_ESCOR_Beall_Households_Livelihoods.pdf (accessed 17 December 2010).

Beam, A.R., de Caceres, L. and Moroney Jr, M.J. (1999) 'Restoration of maritime navigation systems in Central American ports', *Oceans Conference Record (IEEE)* 3: 1317.

Beck, T. (2005) *South Asia Earthquake 2005: Learning from previous recovery operations*, ALNAP and ProVention Consortium. Online www.proventionconsortium.org/themes/default/pdfs/ALNAP-ProVention_SAsia_Quake_Lessonsb.pdf (accessed 28 March 2010).

Beck, U. (1992) *Risk Society: Towards a new modernity*, London: Sage.

Begum, R. (1993) 'Women in environmental disasters: The 1991 cyclone in Bangladesh', *Gender and Development* 1, 1: 34–39.

Bello, W. (2009) *The Food Wars*, New York: Verso.

Below, R, Grover-Kopec, E. and Dilley, M. (2007) 'Documenting drought-related disasters: A global reassessment', *Journal of Environment and Development* 16, 3: 328–44.

Benouar, D. (1994) 'Materials for the investigation of the seismicity of Algeria and adjacent regions during the twentieth century', *Annali di Geofisica* 37, 4: 459–860.

Benouar, D. and Laradi, N. (1996) 'A reappraisal of the seismicity of the Maghreb countries: Algeria, Morocco, Tunisia', *Natural Hazards* 13: 275–96.

Benson, C. (2009) *Mainstreaming Disaster Risk Reduction into Development: Challenges and experience in the Philippines*, Geneva: ProVention Consortium. Online www.proventionconsortium.org/?pageid=37& publicationid=161#161 (accessed 30 December 2009).

Benson, C., Arnold, M. and Christoplos, I. (2009a) 'Disaster risk financing consultative brief', unpublished draft, Geneva: ProVention Consortium.

Benson, C. and Clay, E. (2000) 'The economic dimensions of drought in sub-Saharan Africa', in D. Wilhite (ed.), *Drought: A Global Assessment*, London: Routledge, vol. 1, pp. 287–311.

Benson, C., Gyanwaly, R.P. and Regmi, H. (2009b) *Economic and Financial Decision Making in Disaster Risk Reduction: Nepal case study*, Geneva: Bureau of Conflict Prevention and Recovery, United Nations Development Programme.

Benson, C. and Mangani, R. (2008) *Economic and Financial Decision Making in Disaster Risk Reduction: Malawi case study*, Geneva and Lilongwe: United Nations Development Programme and Department of Disaster Management Affairs, Office of the President and Cabinet, Government of Malawi.

Benson, C., Twigg, J. and Rossetto, T. (2007) *Tools for Mainstreaming Disaster Risk Reduction: Guidance notes for development organisations*, Geneva: ProVention Consortium. Online www.proventionconsortium.org/?pageid=37&publicationid=132#132 (accessed 14 December 2010).

Benson, L. and Bugge, J. (2007) *Child-led Disaster Risk Reduction: A practical guide*, Sweden: Save the Children Sweden.

Bentley, S.P. and Siddle, H.J. (1996) 'Landslide research in the South Wales coalfield', *Engineering Geology* 43: 65–80.

Berger, G. and Wisner, B. (2008) 'Pouring oil on troubled water? How disaster capitalism blackmails flood victims', Geneva: United Nations International Strategy for Disaster Reduction. Online www.preventionweb.net/english/professional/news/v.php?id=2321 (accessed 5 April 2010).

Bergin, A. (2009) 'Defending the home front is top priority', *The Age*, 21 February. Online www.theage.com.au/opinion/defending-the-home-front-is-top-priority-20090220-8dl2.html (accessed 4 June 2010).

Berkovitch, J. and Jackson, R. (1997) *International Conflict: A chronological encyclopedia of conflicts and their management 1945–1995*, Ann Arbor, MI: University of Michigan Press.

Berman, M. (1988) *All That's Solid Melts into Air: The experience of modernity*, New York: Penguin.

Bermejo, P. (2006) 'Preparation and response in case of natural disasters: Cuban programs and experience', *Journal of Public Health Policy* 27: 13–21.

Bernard, S. and McGeehin, M.A. (2004) 'Municipal heat wave response plans', *American Journal of Public Health* 94: 1520–22.

Betts, R. (2003) 'The missing links in community warning systems: Findings from two Victorian community warning system projects', *Australian Journal of Emergency Management* 18, 3: 37–45.

Bezlova, A. (2007) 'Environment-China: Three gorges dam may displace millions more', IPS-Inter Press Service, Washington, DC: IPS-International Federation of Environmental Journalists. Online ipsnews.net/news.asp?idnews=39621 (accessed 10 September 2010).

Bhalla, N. (2009) 'Traffickers prey on disaster-hit children in India – agencies', *AlertNet*, 23 March. Online www.alertnet.org/db/an_art/55867/2009/02/23-145911-1.htm (accessed 3 January 2010).

Bibbee, A., Gonenc, R., Jacobs, S., Konvitz, J. and Price, R. (2000) *Economic Effects of the 1999 Turkish Earthquakes: An interim report*, Economics Department Working Papers No. 247, Paris: OECD. Online www.oecd.org/dataoecd/15/7/1885266.pdf (accessed 25 March 2010).

Bicknell, J., Dodman, D. and Satterthwaite, D. (eds) (2009) *Adapting Cities to Climate Change: Understanding and addressing the development challenges*, London: Earthscan.

Binzel, R.P. (2000) 'The Torino impact hazard scale', *Planetary and Space Science* 48: 297–303.

Binzel, R.P., Rivkin, A.S., Thomas, C.A., Vernazza, P., Burbine, T.H., DeMeo, F.E., Bus, S.J., Tokunaga, A.T. and Birlan, M. (2009) 'Spectral properties and composition of potentially hazardous asteroid (99942) Apophis', *Icarus* 200: 480–85.

Biodiversity in Development Project (2001) *Guiding Principles for Biodiversity in Development: Lessons from field projects*, Brussels, Gland and Cambridge: European Commission and IUCN.

Birch, E.L. and Wachter, S.M. (eds) (2006) *Rebuilding Urban Places After Disaster: Lessons from hurricane Katrina*, Philadelphia: University of Pennsylvania Press.

Birkmann, J. (2006) 'Measuring vulnerability to promote disaster-resilient societies: Conceptual frameworks and definitions', in J. Birkmann (ed.), *Measuring Vulnerability to Natural Hazards*, Tokyo, New York and Paris: United Nations University Press, pp. 9–54.

Bisson, J.I. and Lewis, C. (2009) *Systematic Review of Psychological First Aid*, Geneva: World Health Organization.

Biswas, S.R., Mallik, A., Choudhury, J. and Nishat, A. (2009) 'A unified framework for restoration of southeast Asian mangroves: Bridging ecology, society and economics', *Wetlands Ecology Management* 17: 365–83.

Bjerrum, L. (1967) 'Progressive failure in slopes of overconsolidated plastic clay and clay shales', *Journal of the Soil Mechanics and Foundations* 93: 1–49.

Black, R. (1998) *Refugees, Environment and Development*, Harlow, Essex: Longman.

Blackburn, J., Chambers, R. and Gaventa, J. (2002) 'Mainstreaming participation in development', in N. Hanna and R. Picciotto (eds), *Making Development Work*, New Brunswick, NJ: Transaction Publishers, pp. 61–82.

Blaikie, P. and Brookfield, H. (1987) *Land Degradation and Society*, London: Routledge.

Blaikie, P., Cannon, T., Davis, I. and Wisner, B. (1994) *At Risk*, London: Routledge.

Blair, H. (2000) 'Participation and accountability at the periphery: Democratic local governance in six countries', *World Development* 28, 1: 21–39.

Blancou, J. (2003) *History of Surveillance and Control of Transmissible Animal Diseases*, France: Office International des Epizooties.

Blank, R.M. (1994) *Social Protection versus Economic Flexibility: Is there a trade-off?* Chicago: University of Chicago Press.

Blench, R.M. (2001) *Pastoralism in the New Millennium*, Animal Health and Production Series, No. 150, Rome: Food and Agriculture Organization of the United Nations.

Blondet, M., Villa-Garcia, G. and Brzev, S. (2003) 'Earthquake-resistant construction of adobe buildings: A tutorial', EERI/IAEE World Housing Encyclopedia. Online www.world-housing.net (accessed 4 June 2009).

Blong, R.J. (1984) *Volcanic Hazards: A sourcebook on the effects of eruptions*, Sydney: Academic Press.

Blong, R. (2003) 'Building damage in Rabaul, Papua New Guinea, 1994', *Bulletin of Volcanology* 65, 1: 43–54.

BNPB (Badan Nasional Penanggulangan Bencana) and UNDP (United Nations Development Programme) (2008) *Lessons Learned: Disaster management legal reform*, Jakarta: United Nations Development Programme. Online www.undp.or.id/pubs/docs/Lessons%20Learned%20Disaster%20Management%20Legal%20Reform.pdf (accessed 30 December 2009).

Bobrowsky, P.T. and Rickman, H. (eds) (2007) *Comet/Asteroid Impacts and Human Society*, Berlin, Heidelberg, New York: Springer.

Bodley, J.H. (1998) *Victims of Progress*, 4th edn, Palo Alto: Mayfield Publishing.

Boff, L. and Boff, C. (1987) *Introducing Liberation Theology*, New York: Orbis Books.

Bolin, B. (2007) 'Race, class, ethnicity, and disaster vulnerability', in H. Rodríguez, E.L. Quarantelli and R.R. Dynes (eds), *The Handbook of Disaster Research*, New York: Springer, pp. 113–29.

Bolin, R. and Bolton, P. (1986) *Race, Religion and Ethnicity in Disaster Recovery*, Monograph No. 42, Natural Hazard Center, Boulder: University of Colorado.

Bolin, R. and Stanford, L. (1998) *The Northridge Earthquake: Vulnerability and disaster*, London and New York: Routledge.

Bommer, J., Spence, R., Erdik, M., Tabuchi, S., Aydinoglu, N., Booth, E., del Re, D. and Peterken, O. (2002) 'Development of an earthquake loss model for Turkish catastrophe insurance', *Journal of Seismology* 6: 431–46.

Bompangue, D., Giraudoux, P., Piarroux, M., Mutombo, G., Shamavu, R., Bertrand, S., Mutombo, A., Mondonge, V. and Piarroux, R. (2009) 'Cholera epidemics, war and disasters around Goma and Lake Kivu: An eight-year survey', *PLoS Neglected Tropical Diseases* 3, 5: e436.

Bor, J. (2007) 'The political economy of AIDS leadership in developing countries: An exploratory analysis', *Social Science and Medicine* 64, 8: 1585–99.

Bosher, L. (2007) *Social and Institutional Elements of Disaster Vulnerability: The case of South India*, Bethesda, MD: Academica Press.

Boslough, M.B.E. and Crawford, D.A. (2008) 'Low-altitude airbursts and the impact threat', *International Journal of Impact Engineering* 35: 1441–48.

Boswell, J. (1822) *The of Life of Samuel Johnson, LL.D.*, 9th edn, London: T. Cadell.

Bothara, J.K., Guragain, R. and Dixit, A. (2002) *Protection of Educational Buildings Against Earthquakes: A manual for designers and builders*, Katmandu: National Society for Earthquake Technology–Nepal.

Bouchon, S., Di Mauro, C., Logtmeijer, C., Nordvik, J.P., Pride, R., Schupp, B. and Thornton, M. (2008) *Non-Binding Guidelines for Application of the Council Directive on the Identification and Designation of European Critical Infrastructure and the Assessment of the Need to Improve their Protection*, European Commission, Joint Research Centre, Ispra, Italy: EUR 23665 EN-2008.

Boudreau, T. (2009) 'Solving the risk equation: People-centred disaster risk assessment in Ethiopia', *Humanitarian Practice Network* (HPN), Network Paper, No. 66, London: Overseas Development Institute.

Bourgeois, J., Hansen, T.A., Wiberg, P.L. and Kauffman, E.G. (1988) 'A tsunami deposit at the Cretaceous–Tertiary boundary in Texas', *Science* 29: 567–70.

Bowker, J. (1970) *Problems of Suffering in the Religions of the World*, Cambridge: Cambridge University Press.

Bradley, D., Raynaud, C. and Torrealba, J. (1977) *Guidimakha*, London: War on Want.

Bradshaw, S. and Víquez, A.Q. (2008) 'Women beneficiaries or women bearing the cost? A gendered analysis of the *Red de Protección Social* in Nicaragua', *Development and Change* 39, 5: 823–44.

Bradt, D.A. (2009) *Evidence-Based Decision-Making in Humanitarian Assistance*, HPN Network Paper No. 67, London: Overseas Development Institute.

Branigan, T. (2010) 'China jails investigator into Sichuan earthquake schools', *The Guardian*, 9 February. Online www.guardian.co.uk/world/2010/feb/09/china-eathquake-schools-activist-jailed (accessed 14 December 2010).

Brauer, M. and Hisham-Hashim, J. (1998) 'Fires in Indonesia: Crisis and reaction', *Environmental Science and Technology* 32, 17: 404A–407A.

Bray, D.B. and Merino-Pérez, L. (2004) *La Experiencia de las Comunidades Forestales en México: Veinticinco años de silvicultura y construcción de empresas forestales comunitarias*, Mexico City: Instituto Nacional de Ecología.

Brendon, P. (2000) *The Dark Valley: A panorama of the 1930s*, New York: Knopf.

Brennan, R.J., Keim, M.E., Sharp, T.W., Wetterhall, S.F., Williams, R.J., Baker, E.L., Cantwell, J.D. and Lillibridge, S.R. (1997) 'Medical and Public Health Services at the 1996 Atlanta Olympic Games: An overview', *Medical Journal of Australia* 167: 595–98.

Bretton Woods Project (2010) 'World Bank admits "land grab" risk, proceeds anyway', Bretton Woods Project, 30 September. Online www.brettonwoodsproject.org/art-566651 (accessed 4 October 2010).

Brewer, C.A. (2005) *Designing Better Maps: A guide for GIS users*, Redlands, CA: ESRI Press.

Brewin, C., Andrews, B. and Velntine, J. (2000). 'Meta-analysis of risk factors for post-traumatic stress disorder in trauma-exposed adults', *Journal of Consulting and Clinical Psychology* 68, 5: 748–66.

Briggs, C.L. (2004) 'Theorizing modernity conspiratorially: Science, scale, and the political economy of public discourse in explanations of a cholera epidemic', *American Ethnologist* 31, 2: 164–87.

Briggs, J. and Sharp, J. (2004) 'Indigenous knowledge and development: A postcolonial caution', *Third World Quarterly* 25, 4: 661–76.

Bright, A.D., Don Carlos, A.W., Vaske, J.J. and Absher, J.D. (2007) 'Source credibility and the effectiveness of firewise information', *Proceedings of the 2006 Northeastern Recreation Research Symposium*, General Technical Report NRS-P-14, Newtown Square, PA: US Department of Agriculture, Forest Service, Northern Research Station.

Bristol, G. (2010) 'Surviving the second tsunami: Land rights in the face of buffer zones, land grabs and development', in G. Lizarralde, C. Johnson and C.H. Davidson (eds), *Rebuilding after Disasters: From emergency to sustainability*, Abingdon: Spon Press, pp. 133–48.

Brokensha, D., Warren, D.M. and Werner, O. (1980) *Indigenous Knowledge Systems and Development*, Lanham, MD: University Press of America.

Bromet, E.J. and Havenaar, J.M. (2002) 'Mental health consequences of disasters', in N. Sartorius, W. Gaebel, J.J. Lopez-Ibor and M. Maj (eds), *Psychiatry in Society*, New York: Wiley, pp. 241–62.

Brookfield, H. (2001) *Exploring Agrodiversity*, New York: Columbia University Press.

Brookfield, H., Parsons, H. and Brookfield, M. (2003) *Agrobiodiversity: Learning from farmers across the world*, Tokyo: United Nations University Press.

Brooks, E. and Emel, J. (1995) 'The Llano Estacado of the American southern high plains', in J.X. Kasperson, R.E. Kasperson and B.L. Turner II (eds), *Regions at Risk: Comparisons of threatened environments*, Tokyo: United Nations University Press, pp. 255–303.

Bruijnzeel, L.A. (2001) 'Hydrology of tropical montane cloud forests: A reassessment', *Land Use and Water Resources Research* 1: 1.1–1.18.

Brundtland, G. (1987) *Our Common Future*, New York: Oxford University Press.

Brunkard, J., Namulanda, G. and Ratard, R. (2008) 'Hurricane Katrina deaths, Louisiana, 2005', *Disaster Medicine and Public Health Preparedness* 2: 215–23.

Brunner, K. (2007) *Participatory Disaster Risk Management and Food Security in the Rio San Pedro Watershed*, GTZ. Online www.preventionweb.net/files/2646_gtz20080076enriskmanagementbolivia.pdf (accessed 11 September 2010).

Brusset, E., Bhatt, M., Bjornestad, K., Cosgrave, J., Davies, A., Deshmukh, Y., Haleem, J., Hidalgo, S., Immajati, Y., Jayasundere, R., Mattsson, A., Muhaimin, N., Polastro, R. and Wu, T. (2009) *A Ripple in*

Development? Long term perspectives on the response to the Indian Ocean tsunami 2004, Stockholm: SIDA. Online www.preventionweb.net/files/9667_SIDA52010enwebArippleindevelopment.pdf (accessed 16 August 2010).

Bryan, R.E. (2000) 'Soil erodibility and processes of water erosion on hillslope', *Geomorphology* 32: 385–415.

Bryceson, D.F. (2009) 'Sub-saharan Africa's vanishing peasantries and the specter of a global food crisis', Monthly Review, July–August. Online www.monthlyreview.org/090720bryceson.php (accessed 2 October 2010).

Brzev, S. (2007) *Earthquake-Resistant Confined Masonry Construction*, New Delhi: National Information Center of Earthquake Engineering (NICEE), Indian Institute of Technology Kanpur (India).

Buchanan-Smith, M. and Christoplos, I. (2004) 'Natural disasters amid complex political emergencies', *Humanitarian Exchange* 27: 36–38.

Buchanan-Smith, M. and Maxwell, S. (1994) *Linking Relief and Development: An introduction and overview*, London: Overseas Development Institute.

Buckle, P. (1998–99) 'Re-defining community and vulnerability in the context of emergency management', *Australian Journal of Emergency Management*, Summer: 21–26.

Budayeva and others v. Russia (2008) 'European Court of Human Rights applications nos. 15339/02, 21166/02, 20058/02, 11673/02 and 15343/02', judgment 20 March 2008.

Buechler, S. (2009) 'Gender, water, and climate change in Sonora, Mexico: Implications for policies and programmes on agricultural income-generation', *Gender and Development* 17, 1: 51–66.

Building Seismic Safety Council and Applied Technology Council (1997) *Guidelines for the Seismic Rehabilitation of Buildings*, Federal Emergency Management Agency Report FEMA-273, Washington, DC: FEMA.

Bulkley, K. (2010) 'Technology takes centre stage in disaster relief', *The Guardian*, 18 June. Online www.guardian.co.uk/activate/mobile-technology-disaster-relief (accessed 14 December 2010).

Bullard, R.D. (1994) *Unequal Protection: Environmental justice and communities of color*, San Francisco: Sierra Club Books.

Burby, R. (ed.) (1998) *Cooperating with Nature: Confronting natural hazards with land-use planning for sustainable communities*, Washington, DC: Joseph Henry Press.

Burke, E.M. (1968) 'Citizen participation strategies', *Journal of the American Planning Association* 34, 5: 287–94.

Burnstein, J.E. (2006) 'The myth of disaster education', *Annals of Emergency Medicine* 47, 1: 50–52.

Burton, I., Kates, R. and White, G. (1993) *Environment as Hazard*, 2nd edn, New York: Guilford Press.

Butry, D.T., Mercer, E.D., Prestemon, J.P., Pye, J.M. and Holmes, T.P. (2001) 'What is the price of catastrophic wildfire?', *Journal of Forestry* 99, 11: 9–17.

Bwenge, C. (2010a) Personal communication, 12 November 2010, concerning Tanzanian field research.

—— (2010b) Personal communication, 23 April 2010, concerning Tanzanian field research.

Byun, N. and Wilhite, D. (1999) 'Objective quantification of drought severity and duration', *Journal of Climate* 12: 2747–56.

Cabieses, F. (2010) 'The potato: Treasure of the Andes', International Potato Center. Online www.cipotato.org/publications/books/potato_treasure_andes_online/14_timeless_story.asp (accessed 4 October 2010).

Cabinet Office (UK) (2003) *Dealing with Disaster*, revised 3rd edn, Liverpool: Brodie Publishing.

—— (2009) *Civil Protection Lexicon: A developing single point of reference for UK civil protection terminology*, London: Cabinet Office, UK Government.

Cai, W. and Cowan, T. (2008) 'Evidence of impacts from rising temperature on inflows to the Murray-Darling basin', *Geophysical Research Letters* 35: 7.

Caistor, N. (2010) 'Haiti's history of misery', *BBC Online*, 13 January. Online news.bbc.co.uk/2/hi/8456728.stm (accessed 18 December 2010).

Campbell, D. (1999) 'Response to drought among farmers and herders in southern Kajiado District, Kenya: A comparison of 1972–76 and 1994–95', *Human Ecology* 27, 3: 377–416.

Campbell, J.R. (1990) 'Disasters and development in historical context: Tropical cyclone response in the Banks Island, Northern Vanuatu', *International Journal of Mass Emergencies* 8, 3: 401–24.

—— (2010) 'An overview of natural hazard planning in the Pacific Island region', *Australasian Journal of Disaster and Trauma Studies* 1. Online www.massey.ac.nz/~trauma/issues/2010–11/campbell.htm (accessed 27 November 2010).

Camper, F. (2000) *Volcano Girl*. Online www.fredcamper.com/Film/Rossellini.html (accessed 4 September 2010).

Campese, J., Sunderland, T., Greiber, T. and Oviedo, G. (2009) *Rights-based Approaches: Exploring issues and opportunities for conservation*, Bogor: CIFOR and IUCN.

Cannon, T. (2002) 'Gender and climate hazards in Bangladesh', *Gender & Development* 10, 2: 45–50.

—— (2008) *Reducing People's Vulnerability to Natural Hazards*, Research Paper 2008/34, Helsinki: WIDER.

Carballo, M., Daita, S. and Hernandez, M. (2005) 'Impact of the tsunami on health care systems', *Journal of the Royal Society of Medicine* 98, 9: 390–5.

Cardenas, V. (2008) 'Potential financial solutions for developing countries', in *Mechanisms to Manage Financial Risks from Direct Impacts of Climate Change in Developing Countries*, Technical paper, FCCC/TP/2008/9, Geneva: United Nations Office at Geneva, pp. 69–86. Online unfccc.int/4159.php (accessed 7 April 2010).

Cardenas, V., Hochrainer, S., Mechler, R., Pflug, G. and Linnerooth-Bayer, J. (2007) 'Sovereign financial disaster risk management: The case of Mexico', *Environmental Hazards* 7: 40–53.

Cardona, O.D. (2008) *Indicators of Disaster Risk and Risk Management*, Washington, DC: Inter-American Development Bank. Online www.manizales.unal.edu.co/ProyectosEspeciales/bid2/documentos/Indicators INEMarch282008English.pdf (accessed 16 December 2010).

CARE Bangladesh (2007) *Urban Policy Environment in Bangladesh: Making CARE's response effective*, Dakha: CARE Bangladesh.

—— (2008) *Marginalized Groups in Urban Areas: CARE Bangladesh impact statement*, Dhaka: CARE Bangladesh.

CARE Bolivia (2008) 'Community preparedness for disaster helps reduce poverty', in ISDR (ed.), *Linking Disaster Risk Reduction with Poverty Reduction*, pp. 6–10. Online www.unisdr.org/eng/about_isdr/isdr-publications/14_Linking_Disaster_Risk_Reduction_Poverty_Reduction/Linking_Disaster_Risk_Reduction_Poverty_Reduction.pdf (accessed 12 September 2010).

Caribbean Development Bank (2004) *Sourcebook for the Integration of Natural Hazards into the Environmental Impact Assessment Process*. Online www.caribank.org/titanweb/cdb/webcms.nsf/AllDoc/05463F8ADC0B 8BA904257539005DA94D/$File/Source%20Book5.pdf (accessed 7 November 2010).

Carmalt, J. (2011) 'Human rights, care ethics and situated universal norms', *Antipode*, 43, 2: 296–325.

Carminati, E. and Martinelli, G. (2002) 'Subsidence rates in the Po Plain, northern Italy: The relative impact of natural and anthropogenic causation', *Engineering Geology* 66: 241–55.

Carney, D. (ed.) (1998) *Sustainable Rural Livelihoods: What contribution can we make?* London: Department for International Development.

Carney, J. (1991) 'Indigenous soil and water management in Senegambian rice farming systems', *Agriculture and Human Values* 8: 37–58.

—— (2010) 'Why Haiti's horrific earthquake won't ravage insurance companies', *Business Insider*. Online www.businessinsider.com/haitis-horrific-earthquake-wont-ravage-insurance-companies-2010-1 (accessed 23 May 2010).

Carrasco, E. and Ro, J. (1999) 'Remittances and development', University of Iowa Center for International Finance and Development. Online www.uiowa.edu/ifdebook/ebook2/ebook.shtml (accessed 7 April 2010).

Carson, R. (1962) *Silent Spring*, New York: Fawcett World Library.

Carter, M.R., Little, P.D., Mogues, T. and Negatu, W. (2007) 'Poverty traps and natural disasters in Ethiopia and Honduras', *World Development* 35, 5: 835–56.

Castaños, H. and Lomnitz, C. (2011) *Disasters: A holistic approach*, New York: Springer.

Cataldo, A. (1995) 'Japan industry weighing Kobe earthquake impact: Infrastructure damage may pose problem', *Electronic News* 41: 2.

Cate, F. (2010) 'Media, disaster relief and images of the developing world: Strategies for rapid, accurate, and effective coverage of complex stories from around the globe'. Online www.annenberg.northwestern. edu/pubs/disas/disas10.htm (accessed 14 December 2010).

Caughey, J.L. (2006) *Negotiating Cultures and Identities: Life history issues, methods and readings*, Lincoln: University of Nebraska Press.

Cavallo, E.A., Powell, A. and Becerra, O. (2010) *Estimating the Direct Economic Damage of the Earthquake in Haiti*, Washington: Inter-American Development Bank. Online www.iadb.org/research/pub_hits.cfm? pub_id=35074108 (accessed 25 March 2010).

CCAD (Comisión Centroamericana de Ambiente y Desarrollo) (2002) *Desarrollo de un Plan Estratégico Regional para el Manejo de Incendios y Plagas Forestales*, Antiguo Cuscatlán: CCAD. Online www.sica.int/ busqueda/Noticias.aspx?IDItem=7932&IDCat=3&IdEnt=2&Idm=1&IdmStyle=1 (accessed 20 April 2011).

CCOP (Technical Secretariat of the Coordinating Committee for Geoscience Programmes in East and Southeast Asia) (2009) *Tsunami Risk Assessment and Mitigation in S & SE Asia: Evaluation tsunami hazard in Sri Lanka*, Bangkok: CCOP.

CDC (United States Centers for Disease Control and Prevention) (2005) 'Heat-related mortality: Arizona, 1993–2002 and United States, 1979–2002', *Morbidity & Mortality Weekly Report* 54: 628–30.

CENESTA (Centre for Sustainable Development) (2004) *The Role of Local Institutions in Reducing Vulnerability to Recurrent Natural Disasters and in Sustainable Livelihoods Development in Iran – Case study: The role of Qashqai nomadic communities in reducing vulnerability torRecurrent drought and sustainabl livelihoods development in Iran*, Rome: FAO. Online www.unisdr.org/eng/library/Literature/7762.pdf (accessed 25 March 2010).

Center for Global Education (2010) 'Bolivia', *Better by the Year*. Online www.betterbytheyear.org/bolivia/bolivia_health.htm (accessed 10 September 2010).

CEPR (Center for Economic and Policy Research) (2010) 'Haitian NGOs decry total exclusion from donor conferences on Haitian reconstruction', 19 March. Online www.cepr.net/index.php/relief-and-reconstruction-watch/qhaitian-ngos-decry-total-exclusion-from-donors-conferences-on-haitian-reconstr uctionq (accessed 18 December 2010).

CEPREDENAC (Central American Coordination Center for Natural Disaster Prevention) (2004), *Mitch + 5 Regional Forum Report, Where do we stand? Where are we heading?* Annex B-4: Thematic Session 'D': Research, Information and Early Warning Systems. Online www.undp.org/cpr/disred/documents/regions/america/m5regforum/report.pdf (accessed 22 March 2010).

Cernea, M. (1997) 'The risks and reconstruction model for resettling displaced populations', *World Development* 25, 10: 1569–87.

CESR (Centre for Economic and Social Rights) (2003) *The Human Costs of War in Iraq*, Cambridge: CESR. Online www.cesr.org/downloads/Human%20Costs%20of%20War%20in%20Iraq.pdf (accessed 7 December 2010).

CGAP (Consultative Group to Assist the Poor) (2005) 'Protecting microfinance borrowers', Washington, DC. Online www.cgap.org/gm/document-1.9.2864/Portfolio_03.pdf (accessed 8 April 2010).

CGIAR (Consultative Group on International Agricultural Research) (2010) 'Sorghum'. Online www.cgiar.org/impact/research/sorghum.html (accessed 5 October 2010).

Chagnon, S. (2006) 'Thunderstorm impacts: A mix of curses and blessings', paper presented at the *14th Joint Conference on the Applications of Air Pollution Meteorology with the Air and Waste Management Association*. Online ams.confex.com/ams/Annual2006/techprogram/paper_105612.htm (accessed 12 March 2010).

Chambers, R. (1983) *Rural Development: Putting the last first*, London: Longman.

—— (1995) 'Poverty and livelihoods: Whose reality counts?' *Environment and Urbanization* 7: 173–204.

—— (2007a) 'From PRA to PLA to pluralism: Practice and theory', *IDS Working Paper* No. 286, Brighton: IDS Sussex.

—— (2007b) *Who Counts? The Quiet Revolution of Participation and Numbers*, Brighton: Institute of Development Studies at the University of Sussex.

—— (2008) *Revolutions in Development Inquiry*, London: Earthscan.

Chambers, R. and Conway, G. (1991) *Sustainable rural livelihoods: Practical concepts for the 21st century*, IDS discussion paper 296, Brighton: Institute of Development Studies.

Chambote, R.M. and Boaventura, S.V. (2008) *Reassentamento pela Metade no Vale de Zambeze. O caso de Mutarara*, Maputo: Oxfam.

Chan, E. (2008) 'The untold stories of the Sichuan earthquake', *The Lancet* 372: 359–62.

Chan, Y.F., Alagappan, K., Gandhi, A., Donovan, C., Tewari, M. and Zaets, S.B. (2006) 'Disaster management following the Chi-Chi earthquake in Taiwan', *Prehospital and Disaster Medicine* 21, 3: 196–202.

Chapman, C.R. (2007) 'The asteroid impact hazard and interdisciplinary issues', in P. Bobrowsky and H. Rickman (eds), *Comet/Asteroid Impacts and Human Society*, Berlin and Heidelberg: Springer, pp. 145–62.

Chardon, A.-C. (1999) 'A geographic approach of the global vulnerability in urban area: Case of Manizales, Colombian Andes', *Geojournal* 49, 2: 197–212.

Charlton, R. and McKinnon, R. (2001) *Pensions in Development*, Aldershot: Ashgate.

Charvériat, C. (2000) *Natural Disasters in Latin America and the Caribbean: An overview of risk*, Washington, DC: Inter-American Development Bank, Research Department. Online ideas.repec.org/p/idb/wpaper/4233.html (accessed 9 April 2010).

Chávez, H. (2010) 'Pueblo y Gobierno Unidos', *Las Líneas de Chávez* 97: 2, 5 December.

Chavez-Demoulin, V. and Roehrl, A. (2004) 'Extreme value theory can save your neck', *Approximity GmbH*. Online www.approximity.com/papers/evt_wp.pdf (accessed 6 June 2010).

Chen, M. (2008) 'Informality and social protection: Theories and realities', *IDS Bulletin* 39, 2: 18–27.

Chester, D.K. and Duncan, A.M. (2008) 'Geomythology, theodicy and the continuing relevance of religious worldviews on responses to volcanic eruptions', in J. Grattan and Torrence, R. (eds), *Living Under the Shadow*, Walnut Creek, CA: Left Coast Press, pp. 203–24.

—— (2009) 'The Bible, theodicy and Christian responses to historic and contemporary earthquakes and volcanic eruptions', *Environmental Hazards* 8: 304–32.

Chester, D.K., Duncan, A.M. and Dibben, C.R.J. (2008) 'The importance of religion in shaping volcanic risk perceptions in Italy, with special reference to Vesuvius and Etna', *Journal of Volcanology and Geothermal Research* 172, 3–4: 216–28.

China Planning Group of Post-Wenshuan Earthquake Restoration and Reconstruction, NDRC (National Development and Reform Committee) and MOHURD (The People's Government of Sichuan Province, Ministry of Housing and Urban–Rural Development) (2008) *The State Overall Planning for Post-Wenchuan Earthquake Restoration and Reconstruction*, Public Opinion Soliciting Draft. Online www.recovery platform.org/assets/publication/china%20re construction%20plan.pdf (accessed 16 March 2010).

China.org.cn (2010) 'China's animal welfare advocates' voice finally heard'. Online www.china.org.cn/ opinion/2010–02/17/content_19431499.htm (accessed 28 April 2010).

Chizmar, C. (2009) *Plantas Comestibles de Centroamerica*, Santo Domingo de Heredia: Instituto Nacional de Biodiversidad.

Choi, B.H., Pelinovsky, E., Kim, K.O. and Lee, J.S. (2003) 'Simulation of the trans-oceanic tsunami propagation due to the 1883 Krakatau volcanic eruption', *Natural Hazards and Earth System Sciences* 3: 321–32.

Chomsky, N. and Herman, E. (2002) *Manufacturing Consent: The political economy of the mass media*, 2nd edn, New York: Pantheon.

Chowdhury, A.M., Mushtaque, R., Bhuyia, A.U., Choudhury, A.Y. and Sen, R. (1993) 'The Bangladesh cyclone of 1991: Why so many people died?' *Disasters* 17, 4: 291–304.

Christen, H. and Maniscalco, P. (1998) *The EMS Incident Management System: Operations for mass casualty and high impact incidents*, Englewood Cliffs, NJ: Prentice-Hall.

Christensen, K.M., Blair, M.E. and Holt, J.M. (2007) 'The built environment, evacuations and individuals with disabilities: A guiding framework for disaster policy and preparation', *Journal of Disability Policy Studies* 17: 249–54.

Christie, F. and Hanlon, J. (2001) *Mozambique and the Great Flood of 2000*, Oxford: James Currey.

Christoplos, I. (2001) 'Extension, poverty and vulnerability in Nicaragua', Overseas Development Institute (ODI) Working Paper, London: ODI.

—— (2006a) *Links Between Relief, Rehabilitation and Development in the Tsunami Response*, London: Tsunami Evaluation Coalition.

—— (2006b) *Out of Step? Agricultural Policy and Afghan Livelihoods*, Issues Paper, Kabul: Afghanistan Research and Evaluation Unit.

—— (2007) 'Between the CAPs: Agricultural policies, programming and the market in Bosnia and Herzegovina', HPG Background Paper, London: Overseas Development Institute.

Christoplos, I., Anderson, S., Arnold, M., Galaz, V., Hedger, M. and Klein, R. (2009) *The Human Dimension of Climate Adaptation: The importance of local and institutional issues*, Stockholm: Commission on Climate Change and Development.

Christoplos, I., Aysan, Y. and Galperin, A. (2005) *External Evaluation of the Inter-Agency Secretariat of the International Strategy for Disaster Reduction*, Geneva: UNISDR.

Christoplos, I., Liljelund, A. and Mitchell, J. (2001) 'Re-framing risk: The changing context of disaster mitigation and preparedness', *Disasters* 25, 3: 185–98.

Christoplos, I. and Rocha, J. (2001) 'Disaster mitigation and preparedness on the Nicaraguan post-Mitch agenda', *Disasters* 25, 3: 240–50.

Christoplos, I., Rodríguez, T., Schipper, L., Narvaez, E.A., Bayres Mejia, K.M., Buitrago, R., Gómez, L. and Pérez, F.J. (2010) 'Learning from recovery after Hurricane Mitch', *Disasters* 34, 2: 202–19.

CIDA Canada (2010) *Vibrant Communities, Gender and Poverty Project*, Gender Analysis Tools. Online tamarackcommunity.ca/downloads/gender/Tools.pdf (accessed 14 December 2010).

CIR (Center for International Rehabilitation) (2005) *International Disability Rights Monitor: Disability and early tsunami relief efforts in India, Indonesia and Thailand*, Chicago: International Disability Network.

CIS (Comite d'Information Sahel) (1973) *Que se nourrit de la famine en Afrique?* Paris: Maspero.

City Administrative Officer (1994) *City of Los Angeles Northridge Earthquake After-Action Report*, report presented to the Emergency Operations Board, City of Los Angeles, CA: 3 June.

Clinch, J.P. and Healy, J.D. (2000) 'Housing standards and excess winter mortality', *British Medical Journal* 54: 719.

Clinton, W.J. (2006) *Lessons Learned from Tsunami Recovery: Key propositions for building back better*, New York: Office of the UN Secretary General's Special Envoy for Tsunami Recovery.

Cloke, H. and Pappenberger, F. (2009) 'Ensemble flood forecasting: A review', *Journal of Hydrology* 375: 613–26.

Coburn, A., Hughes, R., Spence, R. and Pomonis, A. (1995) *Technical Principles of Building for Safety*, London: IT Publications.

Coburn, A.W. and Spence, R.J.S. (2002) *Earthquake Protection*, London: John Wiley & Sons.

Cochard, R., Ranamukharachchi, S.L., Shivakoti, G., Shipin, O., Edwards, P.J. and Seeland, K.T. (2008) 'The 2004 tsunami in Aceh and southern Thailand: A review on coastal ecosystems, wave hazards and vulnerability', *Perspectives in Plant Ecology, Evolution and Systematics* 10: 3–40.

Cohen, J. (2001) *Shots in the Dark: The wayward search for an AIDS vaccine*, London: W.W. Norton & Company.

Cohn Jr, S.K. (2008) 'Epidemiology of the Black Death and successive waves of plague', *Medical History Supplement* 27: 74–100.

COHRE (Centre on Housing Rights and Evictions) (2000) *Legal Resources for Housing Rights*, Sources No. 4, International and National Standards, Geneva, Switzerland: Centre on Housing Rights and Evictions.

Coleman, P.G., Perry, B.D. and Woolhouse, M.E.J. (2001) 'Endemic stability: A veterinary idea applied to human public health', *The Lancet* 357: 1284–86.

Coleridge, P. (1993) *Disability, Liberation and Development*, Oxford: Oxfam Publications.

Collier, P. (2009) 'Africa's organic peasantry: Beyond romanticism', *Harvard International Review* 31, 2. Online hir.harvard.edu/agriculture/africa-s-organic-peasantry (accessed 16 December 2010).

Colten, C. (2007) 'Environmental justice in a landscape of tragedy', *Technology and Society* 29, 2: 173–79.

Comfort, L., Boin, A. and Demchak, C. (eds) (2010) *Designing Resilience: Preparing for extreme events*, Pittsburgh, PA: University of Pittsburgh Press.

Commission Report (2004) *European Union Solidarity Fund Annual Report 2002–2003 and Report on the Experience Gained after One Year of Applying the New Instrument, COM (2004) 397 final*, Brussels, 26 May 2004.

Conticini, A. (2005) 'On the streets of Dhaka: Urban livelihoods from children's perspectives: Protecting and promoting assets' *Environment and Urbanization* 17: 69–81.

Cook, B.I., Miller, R.L. and Seager, R. (2009) 'Amplification of the North American "Dust Bowl" drought through human-induced land degradation', *Proceedings of the National Academy of Sciences* 106: 4997–5001.

Cooke, B. and Kothari, U. (2001) *Participation: The new tyranny*, London: Zed Books.

Cooper, M.M. (2000) 'A fifth mechanism of lightning injury', paper presented at the *International Lightning Detection Conference*, November 2000, Tuscon, Arizona.

Coppola, D. and Maloney, E. (2009) *Communicating Emergency Preparedness: Strategies for creating a disaster resilient public*, Boca Raton, FL: CRC Press.

Corsellis, T. and Vitale, A. (2005) *Guidelines for the Transitional Settlement of Displaced Populations*, Oxford: Oxfam.

Council of the European Union (2010) 'Council conclusions on lessons learned from the A/H1N1 pandemic', 13 September. Online www.consilium.europa.eu/uedocs/cms_data/docs/pressdata/en/lsa/116478.pdf (accessed 14 December 2010).

Cox, D. and Fafchamps, M. (2006) 'Extended family and kinship networks: Economic insights and evolutionary directions'. Online www.economics.ox.ac.uk/members/marcel.fafchamps/homepage/Cox Fafchamps_FirstDraft.pdf (accessed 7 April 2010).

Crabbé, P. and Robin, M. (2006) 'Institutional adaptation of water resource infrastructures to climate change in eastern Ontario', *Climatic Change* 78: 103–33.

Craddock, S. (2000) *City of Plagues*, Minneapolis, MN: University of Minnesota Press.

—— (2007) 'Market incentives, human lives, and AIDS vaccines', *Social Science and Medicine* 64, 5: 1042–56.

Creighton, J. (2005) *The Public Participation Handbook: Making better decisions through citizen involvement*, San Francisco: Jossey-Bass.

Crilly, B. (2008) 'UN food aid debate: Give cash, not food?', *Christian Science Monitor*, 4 June. Online www.csmonitor.com/World/Africa/2008/0604/p01s02-woaf.html (accessed 4 December 2010).

Cronin, S.J., Gaylord, D.R., Charley, D., Alloway, B.V., Wallez, S. and Esau, J.W. (2004) 'Participatory methods of incorporating scientific with traditional knowledge for volcanic hazard management on Ambae Island, Vanuatu', *Bulletin of Volcanology* 66: 652–68.

Crosby, A. (2004) 'Ecological imperialism', in S. Krech, J.R. McNeill and C. Merchant (eds), *Encyclopedia of World Environmental History: A–E*, London: Routledge, pp. 368–71.

Cross, J.A. (2001) 'Megacities and small towns: Different perspectives on hazard vulnerability', *Environmental Hazards* 3, 2: 63–80.

Cuny, F. (1983) *Disaster and Development*, New York: Oxford University Press.

Cuny, F.C. (1987) 'Sheltering the urban poor: Lessons and strategies of the Mexico City and San Salavador earthquakes', *Open House International* 12, 3: 16–20.

Cuny, F. and Hill, R. (1999) *Famine, Conflict and Response*, Connecticut: Kumarian Press.

CUR (2003) '50 years after the storm disaster of 1953 – past and future', in K. d'Angremond and J.K. Vrijling (eds), *Symposium Proceedings*, Gouda, Netherlands: CUR Bouw & Infra.

Cutter, S. (2006) 'Vulnerability to environmental hazards', in S. Cutter (ed.), *Hazards, Vulnerability and Environmental Justice*, London: Earthscan, pp. 71–82.

da Silva, J. (2010) *Lessons from Aceh: Key considerations in post-disaster reconstruction*, Rugby: Practical Action Publishing.

Dahdouh-Guebas, F., Jayatissa, L.P., Di Nitto, D., Bosire, J.O., Lo Seen, D. and Koedam, N. (2005) 'How effective were mangroves as a defence against the recent tsunami?' *Current Biology* 15, 12: R 443–47.

Dai, A. (2001) 'Global precipitation and thunderstorm frequencies, Part I: Seasonal and interannual variations', *Journal of Climate* 14: 1092–111.

Dalberg Global Development Advisors (2010) *Secretariat Evaluation: Final report*, Geneva: Dalberg. Online www.unisdr.org/preventionweb/files/12659_UNISDRevaluation2009finalreport.pdf (accessed 1 December 2010).

Daly, M., Poutasi, N., Nelson, F. and Kohlhase, J. (2010) 'Reducing the climate vulnerability of coastal communities in Samoa', *Journal of International Development* 22: 265–81.

Daneshvaran, S. and Morden, R.E. (2007) 'Tornado risk analysis in the United States', *Journal of Risk Finance* 8, 12: 97–111.

Dartmouth (2010) *Space-based Measurement of Surface Water for Research, Educational, and Humanitarian Applications*, Dartmouth College, Hanover, NH. Online www.dartmouth.edu/~floods (accessed 5 April 2010).

Davidson, R. and Lambert, K. (2001) 'Comparing the hurricane disaster risk of coastal counties in the U.S.', *Natural Hazards Review* 3, 3: 132–42.

Davidson, R. and Shah, H. (1997) 'Risk classification of megacities', paper presented at *Proceedings of the First International Workshop on Earthquakes and Mega-Cities*, Frankfurt, Germany, 1–4 September 1997.

Davies, A. (2008) 'Is humanitarian reform improving IDP protection and assistance?' *Forced Migration Review* 29: 16.

Davies, J., Sandstrom, S., Shorrocks, A. and Wolff, E. (2009) 'The Global distribution of personal wealth', *Journal of International Development* 21: 1111–24.

Davies, M., Oswald, K., Mitchell, T. and Tanner, T. (2008) *Climate Change Adaptation, Disaster Risk Reduction and Social Protection: Briefing note*, Brighton: Institute for Development Studies.

Davies, S. (2007) 'Remittances as insurance for idiosyncratic and covariate shocks in Malawi: The importance of distance and relationship'. Online mpra.ub.uni-muenchen.de/4463 (accessed 7 April 2010).

Davis, I. (1977) 'Emergency shelter', *Disasters* 1, 1: 23–40.

—— (1978) *Shelter After Disaster*, Oxford: Oxford Polytechnic Press.

—— (1981) 'Disasters and settlements: Towards an understanding of the key issues', *Habitat International* 5, 5/6: 723–40.

Davis, M. (1999) *Ecology of Fear*, New York: Vintage.

—— (2001) *Late Victorian Holocausts: El Niño famines and the making of the Third World*, London and New York: Verso.

—— (2006) *Planet of Slums*, New York: Verso.

Dawkins, W. (1995) 'Corporate Japan shakes in after-shock of quake – some companies' losses from the Kobe tragedy may prove competitors' gains', *Financial Times*, 24 February, p.30.

Day, W., Pirie, A. and Roys, C. (2007) *Strong and Fragile: Learning from older people in emergencies*, London: HelpAge International.

de Cordier, B. (2009) 'Faith-based aid, globalisation and the humanitarian frontline: An analysis of Western-based Muslim aid organisations', *Disasters* 33, 4: 608–28.

de Jong, K., Prosser, S. and Ford, N. (2005) 'Addressing the psychosocial needs in the aftermath of the tsunami', *PLoS Medicine* 2, 6: 486–88.

de Roo, A.P.J, Gouweleeuw, B., Thieelen, J., Bartholmes, J., Bongioannini, P., Todini, E., Bates, P.D., Horritt, M.S., Hunter, N., Beven, K.J., Pappenberger, F., Heise, E., Rivin, G., Hils, M., Hollingsworth, A., Holst, B., Kwadijk, J., Reggiani, P., Van Dijk, M., Sattler, K. and Sprokkereef, E. (2003) 'Development of a European flood forecasting system', *International Journal of River Basin Management* 1, 1: 49–59.

de Schutter, O. (2009) 'Agribusiness and the right to food: Report of the Special Rapporteur on the right to food, Olivier De Schutter', United Nations General Assembly, Human Rights Council, A/HRC/13/33.

de Silva, P. (2006) 'The tsunami and its aftermath in Sri Lanka: Explorations of a Buddhist perspective', *International Review of Psychiatry* 18, 3: 281–87.

de Ville de Goyet, C. (2004) 'Epidemics caused by dead bodies: A disaster myth that does not want to die', *Pan American Journal of Public Health* 15, 5: 297–99.

de Ville de Goyet, C., del Cid, E., Romero, A., Jeannee, E. and Lechat, M. (1976) 'Earthquake in Guatemala: Epidemiologic evaluation of the relief effort', *Bulletin of the Pan American Health Organization* 10, 2: 95–109.

de Waal, A. (2005) *Famine that Kills: Dafur, Sudan*, 2nd edn, New York: Oxford University Press.

—— (2010) 'The humanitarian's tragedy: Escapable and inescapable cruelties', *Disasters* 34, S2: 130–37.

Dean, D.R. (1979) 'The influence of geology on American literature and thought', in C.J. Schneer (ed.), *Two hundred years of Geology in America*, Durham, NH: University Press of New England, pp. 289–303.

DEFRA (Department for Environment, Food and Rural affairs) (2004) *Working Together for a Better Flood Response: Exercise Triton 04: Overview report of lessons identified*, Bristol: Environment Agency.

Degg, M. and Homan, J. (2005) 'Earthquake vulnerability in the Middle East', *Geography* 90, 1: 54–66.

Dekens, J. (2007) *Local Knowledge for Disaster Preparedness: A literature review*. Kathmandu: International Centre for Integrated Mountain Development (ICIMOD). Online books.icimod.org/index.php/search/author/614 (accessed 14 December 2010).

Delaney, P.L. and Shrader, E. (2000) *Gender and Post-Disaster Reconstruction: The case of Hurricane Mitch in Honduras and Nicaragua*, Washington, DC: The World Bank.

Delfin, F.G. and Gaillard, J.C. (2008) 'Extreme versus quotidian: Addressing temporal dichotomies in Philippine disaster management', *Public Administration and Development* 28, 3: 190–99.

Delica, Z. (1998) 'Balancing vulnerability and capacity: Women and children in the Philippines', in E. Enarson and B. Morrow (eds), *The Gendered Terrain of Disaster: Through women's eyes*, Santa Barbara: Praeger Publisher, pp. 110–13.

Delica-Willison, Z. and Willison, R. (2004) 'Vulnerability reduction: A task for the vulnerable people themselves', in G. Bankoff, G. Frerks and D. Hilhorst (eds), *Mapping Vulnerability: Disasters, development and people*, London: Earthscan, pp. 145–58.

Denevan, W.M. and Lovell, W.G. (1992) *The Native Population of the Americas in 1492*, Madison: University of Wisconsin Press.

Dengler, L. (2005) 'The role of education in the National Tsunami Hazard Mitigation Program', *Natural Hazards* 35: 141–53.

DeRose, L., Messer, E. and Millman, S. (eds) (1999) *Who's Hungry and How Do We Know? Food shortage, Poverty and Deprivation*, Tokyo: United Nations University Press.

Devereux, S. (2001) 'Sen's entitlement approach: Critiques and counter-critiques', *Oxford Development Studies* 29, 3: 245–63.

DFID (Department for International Development) (2006) 'Social protection in poor countries', Social Protection Briefing Note Series No. 1. Online www.gsdrc.org/docs/open/SP17.pdf (accessed 25 July 2010).

Diagne, K. and Ndiaye, A. (2009) 'History, governance and the millennium development goals: Flood risk reduction in Saint-Louis, Senegal', in M. Pelling and B. Wisner (eds), *Disaster Risk Reduction: Cases from urban Africa*, London: Earthscan, pp. 147–67.

Diamond, L. (1994) 'Rethinking civil society: Towards democratic consolidation', *Journal of Democracy* 5: 4–18.

Diario de Los Andes (2005), regional Venezuelan newspaper, February–March.

Diario Frontera (2005), regional Venezuelan newspaper, online www.diariofrontera.com.

Dias, G. (1981) 'Famine and disease in the history of Angola c. 1830–1930', *Journal of African History* 22, 3: 349–78.

Díaz, J.L.C., Alberdi, J., Jordán, A., García, R. and Hernández, E. (2002) 'Heat waves in Madrid, 1986–97: Effects on the health of the elderly', *International Archives of Occupational and Environmental Health* 75: 163–70.

Dilley, M., Chen, R.S., Deichmann, U., Lerner-Lam, A.L., Arnold, M., Agwe, A., Buys, P., Oddvar, K., Lyon, B. and Yetman, G. (2005) *Natural Hazard Hotspots: Global risk analysis: Synthesis report*, Washington, DC: World Bank and Colombia University. Online sedac.ciesin.columbia.edu/hazards/hotspots/synthesisreport.pdf (accessed 30 November 2010).

Dimitrakopoulos, A.P. and Mitsopoulos, I.D. (2006) *Global Forest Resources Assessment 2005: Report on fires in the Mediterranean region*, Fire Management Working Papers, Rome: United Nations, Food and Agriculture Organization.

Disaster Risk Reduction Network Philippines (2010) *Praymer ng Disaster: Risk Reduction and Management (DRRM) Act of 2010*, Quezon City: Christian Aid.

Disaster Watch (2008) *Community Hazard Mapping: Learning exchange on resilience in Honduras July 18, 2008*, Trujillo, Honduras: GROOTS International Community Resilience Program/Comite de Emergencia Garifuna.

Dispatch Online (2010) 'Lightning kills 14 in a village in one month', *Daily Dispatch*, 17 May 2010. Online www.dispatch.co.za/dottydispatch/article.aspx?id=362744 (accessed 2 June 2010).

Djindil, S.N. and de Bruijn, M. (2009) 'The silent victims of humanitarian crises and livelihood (in)security: A case study among migrants in two Chadian towns', *Jàmbá: Journal of Disaster Risk Studies* 2, 3: 253–72.

Dodman, D., Mitlin, D. and Rayos Co, J. (2010) 'Victims to victors, disasters to opportunities: Community-driven responses to climate change in the Philippines', *International Development Planning Review* 32, 1: 1–26.

Dods, R.R. (2004) 'Knowing ways/ways of knowing: Reconciling science and tradition', *World Archaeology* 36, 4: 547–57.

Doherty, T. (2003) *Cold War, Cool Medium*, New York: Columbia University Press.

Domhoff, G. (2007) *Wealth, Income, and Power: Who rules America?* Online sociology.ucsc.edu/whorulesamerica/power/wealth.html (accessed on 24 January 2010).

Doppler, J.V. (2009) *Gender and Tsunami: Vulnerability and coping of Sinhalese widows and widowers on the southwest coast of Sri Lanka*, Magistra der Philosophie thesis, University of Vienna. Online othes.univie.ac.at/6775/1/2009-10-01_9949167.pdf (accessed 24 September 2010).

Doswell, C.A. (2001) 'Severe convective storms – an overview', in C.A. Doswell, III (ed.), *Severe Convective Storms*, AMS Meteorological Monograph Series, 28(50). Boston, Massachusetts: American Meteorological Society, pp. 1–26.

Dotzek, N. (2003) 'An updated estimate of tornado occurrence in Europe', *Atmospheric Research* 67–68: 153–61.

Douglas, M. and Wildavsky, A. (1982) *Risk and Culture: An essay on the selection of technical and environmental dangers*, Berkeley: University of California Press.

Douzinas, C. and Zizek, C. (eds) (2010) *The Idea of Communism*, London: Verso.

Dovers, S.R., Norton, T.W. and Handmer, J. (1996) 'Uncertainty, ecology, sustainability and policy', *Biodiversity and Conservation* 5: 1143–67.

Dowsell, C.A. (2003) 'Societal impacts of severe thunderstorms and tornadoes: Lessons learned and implications for Europe', *Atmospheric Research* 67–68: 135–52.

Drabek, T.E. (1986) *Human System Responses to Disaster: An inventory of sociological findings*, New York: Springer-Verlag.

Drèze, J. and Sen, A.K. (1989) *Hunger and Public Action*, Oxford: Oxford University Press.

Drury, A.C. and Olson, R.S. (1998) 'Disasters and political unrest: An empirical investigation', *Journal of Contingencies and Crisis Management* 6, 3: 153–61.

Dudley, N., Stolton, S., Belokurov, A., Krueger, L., Lopoukhine, N.P., MacKinnon, K., Sandwith, T. and Sekhran, N. (eds) (2010) *Natural Solutions: Protected areas helping people cope with climate change*, Gland, Washington DC and New York: IUCN-WCPA, TNC, UNDP, WCS, The World Bank and WWF.

Duggan, P., Deeny, P., Spelman, R. and Vitale, C. (2010) 'Perceptions of older people on disaster response and preparedness', *International Journal of Older People Nursing* 5: 71–76.

Dumeni, I. and Giorgis, D.W. (1993) 'Institutional mechanisms to deal with drought emergencies in Namibia', in NEPRU (Namibian Economic Policy Research Unit) (ed.), *Drought Impacts and Preparedness in Namibia*, Windhoek: NEPRU.

Dumont, L. (1980) *Homo Hierarchicus: The caste system and its implications*, Chicago: University of Chicago Press.

Duncan, A.M., Chester, D.K. and Guest, J.E. (1981) 'Mount Etna volcano: Environmental impact and problems of volcanic prediction', *The Geographical Journal* 147, 2: 164–78.

Dundes, A. (ed.) (1988) *The Flood Myth*, Berkeley and Los Angeles: University of California Press.

Dupree, H. and Roder, W. (1974) 'Coping with drought in a preindustrial, preliterate farming society', in G.F. White (ed.), *Natural Hazards: Local, national, global*, New York: Oxford University Press.

Duryog Nivaran (2009) *Disaster Risk and Poverty in South Asia: A contribution to the 2009 ISDR Global Assessment Report on Disaster Risk Reduction*, Colombo: Duryog Nivaran. Online www.preventionweb.net/english/hyogo/gar/background-papers/?pid:34& pil:1 (accessed 3 December 2009).

Dutta, D., Herath, S. and Musiake, K. (2003) 'A mathematical model for flood loss estimation', *Journal of Hydrology* 277: 24–49.

Dvorak, V.F. (1974) 'Tropical cyclone intensity analysis and forecasting from satellite imagery', *Monthly Weather Review* 103: 420–30.

Dyer, C., Regev, M., Burnett, J., Festa, N. and Cloyd, B. (2008) 'SWiFT: A rapid triage tool for vulnerable older adults in disaster situations', *Disaster Medicine and Public Health Preparedness* 2, S1: S45–S50.

Dyer, O. (2004) 'Infectious diseases increase in Iraq as public health service deteriorates', *British Medical Journal* 329, 7472: 940.

Dykes, J., MacEachren, A. and Kraak, M. (2005) *Exploring Geovisualization*, International Cartographic Association, Oxford: Pergamon.

Dymon, U.J. and Winter, N.L. (1993) 'Evacuation mapping: The utility of guidelines', *Disasters* 17: 12–24.

Easterling, W. and Mendelsohn, R. (2000) 'Estimating the economic impacts of drought on agriculture', in D. Wilhite (ed.), *Drought: A global assessment*, London: Routledge, vol. 1, pp. 256–68.

Eberwine, D. (2005) 'Disaster myths that just won't die', *Perspectives in Health: The magazine of the Pan American Health Organization* 10 (1): 2–7. Online www.paho.org/english/dd/pin/Number21_article01.htm (accessed 4 December 2010).

Ebi, K.L. and Meehl, G.A. (2007) *The Heat is On: Climate change & heatwaves in the Midwest*, Pew Center on Global Climate Change. Online www.pewclimate.org/docUploads/Regional-Impacts-Midwest.pdf (accessed 6 March 2010).

ECLAC (Economic Commission for Latin America and the Caribbean Programme Planning and Operations Division) (1991) *Manual for Estimating the Socio-economic Effects of Natural Disasters*, Santiago. Online www.eclac.org/publicaciones/xml/8/7818/partone.pdf (accessed 17 January 2010).

EDCflix (2010) 'Earthquake oreparedness', *YouTube*. Video clip, available www.youtube.com/watch?v=5C0eGOLPKrg&feature=PlayList&p=ACEC8EB00F11639A&playnext_from=PL&playnext=1 (accessed 28 May 2010).

Edgar, A. and Sedgwick, P. (eds) (1999) *Key Concepts in Cultural Theory*, London: Routledge.

Edwards, B., Gray, M. and Hunter, B. (2009) 'A sunburnt country: The economic and financial impact of drought on rural and regional families in Australia in an era of climate change', *Australian Journal of Labour Economics* 12, 1: 109–31.

Edwards, M. (2005) 'Civil society: The encyclopedia of informal education'. Online www.infed.org/association/civil_society.htm (accessed 2 June 2010).

Edwards, M. and Hulme, D. (eds) (1995) *Non-Governmental Organizations – Performance and Accountability: Beyond the magic bullet*, London: Earthscan.

Edwards, P. (2002) 'Infrastructure and modernity: Force, time, and social organization in the history of sociotechnical systems', in T. Misa, P. Brey and A. Feenberg (eds), *Modernity and Technology*, Cambridge, MA: MIT Press, pp. 185–225.

Eide, A. (2001) 'Economic, social and cultural rights as human rights', in A. Eide, C. Krause and A. Rosas (eds), *Economic, Social and Cultural Rights: A textbook*, 2nd edn, Dordrecht: Kluwer Law International (Martinus Nijhoff Publishers), pp. 9–28.

Eidinger, J.M. (2001) 'The Gujarat (Kutch) India earthquake of January 26, 2001: Lifeline performance', Technical Council on Lifeline Earthquake Engineering, *Monograph No. 19*, April.

Eisenman, D., Cordasco, K., Asch, S., Golden, J. and Glik, D. (2007) 'Disaster planning and risk communication with vulnerable populations: Lessons from hurricane Katrina', *American Journal of Public Health* 97, S1: S109–S115.

Eldridge, C. (2002) 'Why was there no famine following the 1992 Southern African drought? The contributions and consequences of household responses', *IDS Bulletin* 33, 4: 79–87.

Elliott, J. and Pais, J. (2006) 'Race, class, and Hurricane Katrina: Social differences in human responses to disaster', *Social Science Research* 35, 2: 295–321.

EMA (Emergency Management Australia) (1998) *Australian Emergency Management Glossary*, Canberra: Emergency Management Australia.

—— (2002) *Economic and Financial Aspects of Disaster Recovery*, Barton: Emergency Management Australia. Online www.preventionweb.net/english/professional/publications/v.php?id=1596 (accessed 25 March 2010).

—— (2005) *Evacuation Planning*, Canberra: Emergency Management Australia.

EM-DAT (Emergency events database) (2010) *OFDA/CRED International Disaster Database*, www.cred.be/emdat, Université Catholique de Louvain, Brussels.

Emergency Management New South Wales (2010) 'Natural disaster assistance schemes'. Online www.emergency.nsw.gov.au/content.php/501.html (accessed 2 March 2010).

Enarson, E. (2000) *Gender and Natural Disasters*, Geneva: Recovery and Reconstruction Department, International Labour Organization.

Enarson, E., Fothergill, A. and Peek, L. (2006) 'Gender and disaster: Foundation and directions', in H. Rodriguez, E.L. Quarantelli and R.R. Dynes (eds), *The Handbook of Disaster Research*, New York: Springer, pp. 130–46.

Enarson, E. and Morrow, B.H. (1998) *The Gendered Terrain of Disaster: Through women's eyes*, Westport: Praeger.

Enders, A. and Brandt, Z. (2007) 'Using geographic information system technology to improve emergency management and disaster response for people with disabilities', *Journal of Disability Policy Studies* 17: 223–29.

Endfield, G.H. and Nash, D.J. (2002) 'Missionaries and morals: Climatic discourse in nineteenth-century Central Southern Africa', *Annals of the Association of American Geographers* 92, 4: 727–42.

Engle, P.S.C. and Menon, P. (1996) 'Child development: Vulnerability and resilience', *Social Science and Medicine* 43, 5: 621–35.

Ensor, J. and Berger, G. (2009) *Understanding Climate Change Adaptation*, Rugby: Practical Action Publishing.

Environment Agency (2010) *HiFlows-UK*, Environment Agency (EA), UK. Online www.environment-agency.gov.uk/hiflowsuk (accessed 7 May 2010).

EPA (US Environmental Protection Agency) (2010) *Environmental Justice*, Office of Compliance and Enforcement. Online www.epa.gov/environmentaljustice (accessed 24 October 2010).

Erdik, M. (1998) 'Seismic vulnerability of megacities', in E. Booth (ed.), *Seismic Design Practice into the Next Century*, Research and Application, Rotterdam: Balkema, pp. 35–42.

Eriksson, J., Adelman, H., Borton, J., Christensen, H., Kumar, K., Suhrke, A., Tardif-Douglin, D., Villumstad, S., Wohlgemuth, L. and Millwood, D. (eds) (1996) *The International Response to Conflict and Genocide: Lessons from the Rwanda experience*, London: Steering Committee of the Joint Evaluation of Emergency Assistance to Rwanda.

ERM (Environmental Resources Management) (2005) *Natural Disaster and Disaster Risk Reduction Measures*, London: ERM, Department of International Development. Online www.unisdr.org/news/DFID-Economics-Study-for-DfID.pdf (accessed 28 March 2010).

Eskijian, M.L. (2006) 'Mitigation of Seismic and Meteorological Hazards to Marine Oil Terminals and other Pier and Wharf Structures in California', *Natural Hazards* 39, 2: 343–51.

Espinosa-Aranda, J.M., Jimenez, A., Ibarrola, G., Alcantara, F., Aguilar, A., Inostroza, M., Maldonado, S. and Higareda, R. (2003) 'The seismic alert system in Mexico City and the School Prevention Program', in J. Zschau and A. Küppers (eds), *Early Warning Systems for Natural Disaster Reduction*, Berlin and Heidelberg: Springer-Verlag, pp. 441–46.

Esser, I., Ferrarini, T., Nelson, K. and Sjöberg, O. (2009) 'A framework for comparing social protection in developing and developed countries: The example of child benefits', *International Social Security Review* 62: 91–115.

Essick, K. (2001) 'Guns, money and cell phones', *Global Issues*. Online www.globalissues.org/article/442/guns-money-and-cell-phones (accessed 4 October 2010).

Etkin, D. (1999). 'Risk transference and related trends: Driving forces towards more mega-disasters', *Environmental Hazards* 1: 69–75.

Etkin, D., Brun, S.E., Shabbar, A. and Joe, P. (2001) 'Tornado climatology of Canada revisited: Tornado activity during different phases of ENSO', *International Journal of Climatology* 21, 8: 915–38.

Evans, D. (2006) 'Bessie Smith's "Back-Water Blues": The story behind the song', *Popular Music* 26, 1: 97–116.

Evans, S.G., Guthrie, R.H., Roberts, N.J. and Bishop, N.F. (2007) 'The disastrous 17 February 2006 rockslide-debris avalanche on Leyte Island, Philippines: A catastrophic landslide in tropical mountain terrain', *Natural Hazards and Earth System Science* 7: 89–101.

Fagan, B. (2000) *Floods, Famines and Emperors: El Niño and the fate of civilizations*, London: Pimlico.

Falk, K. (ed.) (2005) *Preparing for Disaster – A Community-Based Approach*, Copenhagen: Danish Red Cross, 2nd edn, and Manila: Philippines National Red Cross. Online www.allindiary.org/pool/resources/preparing-for-disaster-a-community-based-approach.pdf (accessed 14 December 2010).

Fanuthor, M. (2008) 'After quake, China's elderly long for family', *The Washington Post*, 3 June: A01.

FAO (Food and Agriculture Organization of the United Nations) (2005) *Post-Earthquake Rapid Livelihoods Assessment*, Muzaffarabad: FAO.

—— (2006) *State of Food and Agriculture 2006: Food aid for food security?* Rome: FAO.

—— (2007) *Fire Management: Global assessment 2006*, Rome: FAO.

FAO/CIFOR (2005) *Forests and Floods: Drowning in fiction or thriving on facts?* Bogor: FAO/CIFOR.

FAP 24 (Flood Action Plan 24) (1996) *River Survey Project: Annex 3: Hydrology*, Dhaka: Flood Plan Coordination Organization and Commission of the European Communities.

Farmer, P. (2001) *Infections and Inequalities: The modern plagues*, 1st edn, Berkeley: University of California Press.

Farrer, A. (1966) *A Science of God*, London: Bles.

Farvar, T. and Milton, J. (1972) *The Careless Technology*, Garden City, NJ: The Natural History Press.

FEMA (US Federal Emergency Management Agency) (1999) *Hazus*. Online www.fema.gov/plan/prevent/hazus/index.shtm (accessed 6 June 2010).

—— (2004a) *Amendments to National Flood Insurance Maps*. Online www.dhs.gov/xlibrary/assets/privacy/privacy_pia_mip_apnd_a.pdf (accessed 9 October 2010).

—— (2004b) *Nonstructural Earthquake Mitigation Guidance Manual*. Online www.bchelpline.com/BCAToolkit/resource_files/tech_manuals/earthquake/Nonstructural_EQ_Tech_Manual.pdf (accessed 6 June 2010).

—— (2008) 'FEMA ongoing response to formaldehyde', *FEMA for the Record*. Online www.fema.gov/news/newsrelease.fema?id=42586 (accessed 20 October 2010).

FEMACT (2009) 'Tanzania: Loliondo report findings', 23 September, *Pambazuka News 449*. Online pambazuka.org/en/category/advocacy/58956 (accessed 7 August 2010).

Fernandes, P.M. and Botelho, H.S. (2003) 'A review of prescribed burning effectiveness in fire hazard reduction', *International Journal of Wildland Fire* 12: 117–28.

Fernandez, L.S., Byard, D., Lin, C.-C., Benson, S. and Barbera, J. (2002) 'Frail elderly as disaster victims: Emergency management strategies', *Prehospital and Disaster Medicine* 17: 67–74.

Fierro, E. and Perry, C. (2010) *Preliminary Reconnaissance Report: 12 January 2010 Haiti earthquke report*, Pacific Earthquake Research Center, Berkeley, CA. Online peer.berkeley.edu/publications/haiti_2010/related_events_haiti.html (accessed 6 October 2010).

Finch, J.D. (2010) 'Bangladesh and East India tornadoes background information'. Online bangladeshtornadoes.org/bengaltornadoes.html (accessed 13 March 2010).

Fiorillo, F. and Wilson, R.C. (2004) 'Rainfall induced debris flows in pyroclastic deposits, Campania (Southern Italy)', *Engineering Geology* 75: 263–89.

First People (2010) 'Lakota creation myth', *First People – the Legends*. Online www.firstpeople.us/FP-Html-Legends/LakotaCreationMyth-Lakota.html (accessed 1 June 2010).

Fjord, L. and Manderson, L. (eds) (2009) 'Anthropological perspectives on disasters and disability', *Human Organization* 68, 1: 64–112.

Flannigan, M.D., Bergeron, Y., Engelmark, O. and Wotton, B.M. (1998) 'Future wildfire in circumboreal forests in relation to global warming', *Journal of Vegetation Science* 9, 4: 469–76.

Flavier, J.M., De Jesus, A. and Navarro, C. (1995) 'The regional program for the promotion of indigenous knowledge in Asia', in D.M. Warren, L.J. Slikkerverr and D. Brokensha (eds), *The Cultural Dimension of Development: Indigenous Knowledge Systems*, London: Intermediate Technology Publications, pp. 479–87.

Fleming, G. (1871) *Animal Plagues: Their history, nature and prevention*, London: Chapman & Hall.

Flood, L. and Kamalani, A. (2002) 'Hawaiian Pele', *The Ocean Refuses No River* [CD], LionHeart Studios, Canada.

Floret, N., Viel, J.F., Mauny, F., Hoen, B. and Piarroux, R. (2006) 'Negligible risk for epidemics after geophysical disasters', *Emerging Infectious Diseases* 12 (4): 543–48.

Fordham, M. (1998) 'Making women visible in disasters: Problematising the private domain', *Disasters* 22, 2: 126–43.

—— (1998/1999) 'Participatory planning for flood mitigation: Models and approaches', *Australian Journal of Emergency Management*, Summer 1998/99: 27–34.

—— (2004) 'Gendering vulnerability analysis: Towards a more nuanced approach', in G. Bankoff, G. Frerks and D. Hilhorst (eds), *Mapping Vulnerability: Disasters, development and people*, London: Earthscan, pp. 174–82.

—— (2009) 'We can make things better for each other: Women and girls organize to reduce disasters in Central America', in E. Enarson and P.G. Dhar Chakrabarti (eds), *Women, Gender and Disaster: Global issues and initiatives*, Delhi: Sage, pp. 175–88.

Fordham, M., Ariyabandu, M., Gopalan, P. and Peterson, K. (2006) 'Please don't raise gender now – we're in an emergency!', in IFRC (International Federation of Red Cross and Red Crescent Societies) (ed.), *2006 World Disasters Report: Focus on neglected crises*, Geneva: IFRC, pp. 140–63.

Fordham, M. and Ketteridge, A.-M. (1998) 'Men must work and women must weep: Examining gender stereotypes', in E. Enarson and B. Morrow (eds), *Through Women's Eyes: The gendered terrain of disaster*, New York: Praeger, pp. 81–94.

Forsyth, F. (2007) *The Biafra Story*, London: Pen & Sword.

Fosbrooke, H. and Young, R. (1960) *Smoke in the Hills: Political tension in the Morogoro district of Tanganyika*, Evanston, IL: Northwestern University Press.

Fothergill, A. (1996) 'Gender, risk and disaster', *International Journal of Mass Emergencies and Disasters* 14: 33–56.

Fouillet, A., Rey, G., Wagner, V., Laaidi, K., Empereur-Bissonnet, P., Le Tertre, A., Frayssinet, P., Bessemoulin, P., Laurent, F., De Crouy-Chanel, P., Jougla, E. and Hemon, D. (2008) 'Has the impact of heat waves on mortality changed in France since the European heat wave of summer 2003? A study of the 2006 heat wave', *International Journal of Epidemiology* 37: 309–17.

Fox, F. (2001) 'New humanitarianism: Does it provide a moral banner for the 21st century?' *Disasters* 25, 4, 275–89.

Fox, M.H., White, G.W., Rooney, C. and Rowland, J.L. (2007) 'Disaster preparedness and response for persons with mobility impairments: Results from the University of Kansas Nobody Left Behind Project', *Journal of Disability Policy Studies* 17: 196–205.

Francis, P. and Oppenheimer, C. (2004) *Volcanoes*, Oxford: Oxford University Press.

Franke, R. and Chassin, B. (1980) *Seeds of Famine*, Montclair, NJ: Allen & Unwin.

Frankenberger, T. (1995) *Household Livelihood Security: A unifying conceptual framework for CARE programming*, Atlanta: CARE USA.

—— (2002) *CARE Household Livelihood Security Assessments: A toolkit for practitioners*, Tucson: TANGO International Inc.

Fraser, E. (2003) 'Social vulnerability and ecological fragility: Building bridges between social and natural sciences using the Irish potato famine as a case study', *Conservation Ecology* 7, 2. Online www.ecologyandsociety.org/vol7/iss2/art9 (accessed 15 November 2010).

Freeman, P.K., Keen, M. and Mani, M. (2003) *Dealing with Increased Risk of Natural Disasters: Challenges and options*, IMF Working Paper 03/197, Washington, DC: International Monetary Fund.

Freeth, S.J. (1993) 'On the problems of translation in the investigation of the Lake Nyos disaster', *Journal of Volcanology and Geothermal Research* 54: 353–56.

Freire, M.E. and Stren, R. (eds) (2001) *The Challenge of Urban Government*, Washington, DC: World Bank Institute.

Friedsam, H. (1961) 'Reactions of older persons to disaster-caused losses: A hypothesis of relative deprivation', *Gerontologist* 1: 34–37.

Froot, K. and O'Connell, P.G.J. (1999) 'The pricing of U.S. catastrophe reinsurance', in K. Froot (ed.), *The Pricing of U.S. Catastrophe Reinsurance*, Chicago: University of Chicago Press, pp. 195–232.

Frydenlund, M.M. (1993) *Lightning Protection for People and Property*, New York: Van Nostrand Reinhold.

Fu, K.W., White, J., Chan, Y.Y., Zhou, L., Zhang, Q. and Lu, Q. (2010) 'Enabling the disabled: Media use and communication needs of people with disabilities during and after the Sichuan earthquake in China', *International Journal of Emergency Management* 7, 1: 75–87.

Fujita, T.T. (1973) *Experimental Classification of Tornadoes in FPP Scale*, SMRP Report 98, Chicago: University of Chicago.

Fukuyama, F. (1993) *The End of History and the Last Man*, New York: Harper.

Fullen, M.A. and Catt, J.A. (2004) *Soil Management. Problems and Solutions*, London: Arnold.

Fuller, R.B. (1969) *Operating Manual for Spaceship Earth*, Carbondale, IL: Southern Illinois University Press.

Funari, P., Zarankin, A. and Salerno, M. (eds) (2009) *Memories of Darkness: Archaeology of repression and resistance in Latin America*, Berlin: Springer-Verlag.

Gabre-Madhin, E. (2002) *Making Markets Work in Malawi*, Washington: International Food Policy Research Institute. Online ocha-gwapps1.unog.ch/rw/rwb.nsf/db900sid/ACOS-64CTW7?OpenDocument (accessed 27 April 2010).

Gabriel, P. (2000) *Personal Communication from the Office of the Emergency Services Commissioner*, Melbourne: Government of Victoria.

Gaillard, J.C. (2007) 'Resilience of traditional societies in facing natural hazards', *Disaster Prevention and Management* 16, 4: 522–44.

—— (2008) 'Differentiated adjustment to the 1991 Mt Pinatubo Resettlement Program among lowland ethnic groups of the Philippines', *Australian Journal of Emergency Management* 23, 2: 31–39.

—— (2010a) Personal communication, based on field work conducted in Cape Verde in April 2009.

—— (2010b) Personal communication, based on field work conducted in Guadeloupe in April 2010.

Gaillard, J.C. and Cadag, J.R.D. (2009) 'From marginality to further marginalization: Experiences from the victims of the July 2000 Payatas trashslide in the Philippines', *Jàmbá: Journal of Disaster Risk Studies* 2, 3: 197–215.

Gaillard, J.C., Clavé, E. and Kelman, I. (2008a) 'Wave of peace? Tsunami disaster diplomacy in Aceh, Indonesia', *Geoforum* 39, 1: 511–26.

Gaillard, J.C., Clavé, E., Vibert, O., Azhari, Dedi, Denain, J.-C., Efendi, Y., Grancher, D., Liamzon, C. C., Sari, D.S.R. and Setiawan, R. (2008b) 'Ethnic groups' response to the 26 December 2004 earthquake and tsunami in Aceh, Indonesia', *Natural Hazards* 47, 1: 17–38.

Gaillard, J.C., Liamzon, C.C. and Villanueva, J.D. (2007) 'Natural disaster? A retrospect into the causes of the late-2004 typhoon disaster in Eastern Luzon, Philippines', *Environmental Hazards* 7, 4: 257–70.

Gaillard, J.C. and Maceda, E.A. (2009) 'Participatory three-dimensional mapping for disaster risk reduction', *Participatory Learning and Action* 60: 109–18.

Gaillard, J.C., Maceda, E.A., Stasiak, E., Le Berre, I. and Espaldon, M.A.O. (2009) 'Sustainable livelihoods and people's vulnerability in the face of coastal hazards', *Journal of Coastal Conservation* 13, 2–3: 119–29.

Gaillard, J.C., Wisner, B., Benouar, D., Cannon, T., Créton-Cazanave, L., Dekens, J., Fordham, J., Gilbert, C., Hewitt, K., Kelman, I., Morin, J., N'Diaye, A., O'Keefe, P., Oliver-Smith, A., Quesada, C., Revet, S., Sudmeier-Rieux, K., Texier, P. and Vallette, C. (2010) 'Alternatives pour une réduction durable des risques de catastrophe', *Human Geography* 3, 1: 66–88.

Gang, Q. (2009) 'Looking back on Chinese media reporting of school collapses', *Chine Media Project*, 7 May. Online cmp.hku.hk/2009/05/07/1599 (accessed 14 December 2010).

Gao, H. (2010) 'Scientific approach to natural disaster mitigation', presentation during EU-China Science and Technology Week, Shanghai, 18 June. Online ec.europa.eu/dgs/jrc/downloads/jrc_20100618_shanghai_expo_guo.pdf (accessed 9 October 2010).

Gardner, D. (2008) *The Science of Fear*, New York: Dutton.

Gardner, R.A.M. and Gerrard, A.J. (2003) 'Runoff and soil erosion on cultivated rainfed terraces in the middle hills of Nepal', *Applied Geography* 23: 23–45.

Gaventa, J. (2002) 'Introduction: Exploring citizenship, participation, and accountability', *IDS Bulletin* 33, 2: 1–11.

—— (2004) 'Toward participatory governance: Assessing the transformative possibilities', in S. Hickey and G. Mohan (eds), *Participation: From tyranny to transformation*, London: Zed Books, pp. 25–41.

—— (2007) *Powercube: Understanding power for social change*, Institute for Development Studies. Online www.powercube.net/analyse-power (accessed 3 October 2010).

Ge, G. (2009) 'The development and importance of pebble mulch in China', *Journal of Agricultural Science* 30, 4: 52–54.

Geertsema, M., Highland, L. and Vaugeouis, L. (2009) 'Environmental impact of landslides', in K. Sassa and P. Canuti (eds), *Landslide Disaster Risk Reduction*, Berlin: Springer-Verlag, pp. 588–608.

Geertz, C. (1973) *The Interpretation of Cultures*, New York: Basic Books.

Gell, A. (2010) 'Haiti's unnatural disaster', *Black Agenda Report*. Online blackagendareport.com/?q=content/haitis-unnatural-disasters (accessed 2 December 2010).

Gentilini, U. (2009) 'Social protection in the "real world": Issues, models and challenges', *Development Policy Review* 27, 2: 147–66.

George, S. (1976) *How the Other Half Dies*, New York: Penguin.

—— (2009) 'What is development journalism?', *The Guardian*, 23 November. Online www.guardian.co.uk/journalismcompetition/professional-what-is-development-journalism (accessed 14 December 2010).

Gerritsen, H. (2005) 'What happened in 1953? The big flood in the Netherlands in retrospect', *Philosophical Transactions of the Royal Society A* 363: 1271–91.

Gerrity, E.T. and Flynn, B.W. (1997) 'Mental health consequences of disasters', in E.K. Noji (ed.), *The Public Health Consequences of Disasters*, Oxford: Oxford University Press, pp. 102–3.

Gerulis-Darcy, M.L. (2008) 'Vulnerability and the social-production of disaster Hurricane Mitch in Posoltega, Nicaragua', unpublished PhD thesis, Northeastern University.

GFDRR (Global Facility for Disaster Reduction and Recovery) (2010) 'About the global facility'. Online www.gfdrr.org/gfdrr/node/1 (accessed 1 December 2010).

GIEWS (Global Information and Early Warning System) (2010) 'GIEWS country brief: Mongolia', Rome: GIEWS/FAO. Online www.reliefweb.int/rw/RWFiles2010.nsf/FilesByRWDocUnidFilename/EKIM-87B59U-full_report.pdf/$File/full_report.pdf (accessed 21 December 2010).

Girot, P. (2000) *Raíz y Vuelo: El uso de los recursos naturales vivientes en Mesoamérica*, San José: IUCN.

—— (2002) 'Scaling Up: Resilience to hazards and the importance of cross-scale linkages', paper presented at the UNDP Expert Group Meeting on Integrating Disaster Reduction and Adaptation to Climate Change, Havana, April 2002.

Giuffrida, A. (2005) 'Clerics, rebels and refugees: Mobility strategies and networks among the Kel Antessar', *Journal of North African Studies*, 10, 3–4: 529–43.

Glade, T. (2003) 'Vulnerability assessment in landslide risk analysis', *Die Erde* 134: 123–46.

Glade, T. and Crozier, M.J. (2005) 'The nature of landslide hazard impact', in T. Glade, M. Anderson and M.J. Crozier (eds), *Landslide Hazard and Risk*, London: John Wiley & Sons, pp. 43–74.

Glade, T. and Dikau, R. (2001) 'Gravitative massenbewegungen – vom naturereignis zur naturkatastrophe', *Petermanns Geographische Mitteilungen* 145: 42–55.

Glantz, M.H. (ed.) (1987) *Drought and Hunger in Africa: Denying famine a future*, Cambridge: Cambridge University Press.

—— (ed.) (1999) *Creeping Environmental Problems and Sustainable Development in the Aral Sea*, Cambridge: Cambridge University Press.

—— (ed.) (2001) *Once Burned, Twice Shy?* Tokyo: United Nations University Press.

—— (2007) 'How about a spare-time university?' *WMO Bulletin* 56, 2: 1–6.

Glavovic, B., Saunders, W. and Becker, J. (2010) 'Land-use planning for natural hazards in New Zealand', *Natural Hazards* 54, 3: 679–706.

Gliessman, S. (2006) *Agroecology: The ecology of sustainable food systems*, Boca Raton, FL: CRC Press.

GNDR (Global Network of Civil Society Organizations for Disaster Reduction) (2009) '*Clouds but little rain ...*' – *Views from the Frontline: A local perspective of progress towards implementation of the Hyogo Framework for Action*, Teddington: Global Network of Civil Society Organisations for Disaster Reduction. Online www.globalnetwork-dr.org/reports/VFLfullreport0609.pdf (accessed 2 April 2010).

—— (2011) *If We Do Not Join Hands: Views from the front line 2011*, Teddington: Global Network of Civil Society Organisations for Disaster Reduction. Online www.globalnetwork-dr.org/voices-from-the-frontline-2011.html (accessed 14 May 2011).

GOB (Government of Bangladesh) (2005) *Bangladesh: Unlocking the potential National Strategy for Accelerated Poverty Reduction*, Dhaka: General Economics Division, Planning Commission, Government of People's Republic of Bangladesh. Online siteresources.worldbank.org/INTPRS1/Resources/Bangladesh_PRSP (Oct-16-2005).pdf (accessed 29 December 2009).

—— (2009) *Steps Towards Change: National Strategy for Accelerated Poverty Reduction II FY 2009–11*, Dhaka: General Economics Division, Planning Commission, Government of People's Republic of Bangladesh. Online www.lcgbangladesh.org/prsp/docs/PRS%20Bangladesh%202010%20final.pdf (accessed 2 April 2010).

Gobat, J.-M., Aragno, M. and Matthey, W. (2004) *The Living Soil. Fundamentals of Soil Science and Soil Biology*, Enfield: Science Publishers.

Godshalk, D. (2003) 'Urban hazard mitigation: Creating resilient cities', *Natural Hazards Review* 4, 3: 136–43.

Godshalk, D.R., Kaiser, E.J. and Berke, P.R. (1998) 'Integrating hazard mitigation and local land use planning', in R. Burby (ed.), *Cooperating with Nature: Confronting natural hazards with land-use planning for sustainable communities*, Washington, DC: Joseph Henry Press, pp. 85–118.

GOI (Government of India) (2005) *Scheme for Constitution and Administration of the Calamity Relief Fund*, Delhi: Ministry of Finance, Government of India. Online finmin.nic.in/the_ministry/dept_expenditure/plan_finance/FCD/Guidelines-CRF.html (accessed 22 December 2009).

—— (2007) 'Revision of items and norms of assistance from the Calamity Relief Fund (CRF) and the National Calamity Contingency Fund (NCCF) for the period between 2005–2010', Communication No. 32–34/2005-NDM-I, 27 June 2007, to the Chief Secretaries of all States and the Relief Commissioners/Secretaries, Department of Disaster Management of all States. New Delhi: Ministry of Home Affairs (Disaster Management – I Division), Government of India.

Goldammer, J.G. and de Ronde, C. (2004) *Wildland Fire Management Handbook for Sub-Sahara Africa*, Freiburg, Germany: Global Fire Monitoring Center.

Goldstein, M. and Amin, S. (eds) (2008) *Data Against Natural Disasters Establishing Effective Systems for Relief, Recovery, and Reconstruction*, Washington, DC, World Bank. Online siteresources.worldbank.org/INTPOVERTY/Resources/335642-1130251872237/9780821374528.pdf (accessed 30 January 2010).

Gonzales Devant, S. (2008) 'Displacement in the 2006 Dili crisis', *Oxford Refugee Studies Centre Working Paper 45*, January. Online www.reliefweb.int/rw/RWFiles2008.nsf/FilesByRWDocUnidFilename/AMMF-7C9HVT-full_report.pdf/$File/full_report.pdf (accessed 26 December 2010).

Gonzalez, F.I., Bernard, E.N., Meinig, C., Eble, M.C., Mofjeld, H.O. and Stalin, S. (2005) 'The NTHMP tsunameter network', *Natural Hazards* 35: 25–39.

Goss, J. (2010) Personal communication, medical response to Haiti earthquake, 4 February.

Goudge, J., Russell, S., Gilson, L., Gumede, T., Tollman, S. and Mills, A. (2009) 'Illness-related impoverishment in rural South Africa: Why does some social protection work for some households but not others?' *Journal of International Development* 21: 231–51.

Government of Australia (2010) 'Disaster lesson plans', *Government of Australia: Emergency management for schools*. Online www.ema.gov.au/www/ema/schools.nsf/Page/TeachLesson_Plans (accessed 23 May 2010).

Government of Ethiopia (2004) *Ethiopia: National information on disaster reduction*, World Conference on Disaster Reduction, Kobe, Japan. Online www.unisdr.org/eng/country-inform/reports/Ethiopia-report.pdf (accessed 17 September 2010).

Government of Mongolia (2004) *Report to the World Conference on Disaster Reduction*, Kobe, Japan. Online www.unisdr.org/eng/mdgs-drr/national-reports/Mongolia-report.pdf (accessed 21 December 2010).

Government of Namibia (1997) *National Drought Policy and Strategy*, Windhoek, Namibia. Online www.mawf.gov.na/Documents/app. htm (accessed 5 November 2010).

Government of New Zealand (2005) *National Civil Defence Emergency Management Plan Order 2005*, Wellington: Government of New Zealand.

——— (2008) *Natural Hazards Guidance Note*, Ministry of Environment. Online www.qp.org.nz/plan-topics/natural-hazards.php#abstract (accessed 7 August 2010).

Government of Samoa (1997) *National Disaster Management Plan and Emergency Procedures*, Apia: Government of Samoa.

Government of the Republic of South Africa (2004) 'National Information', World Conference on Disaster Reduction, Kobe, Japan. Online www.unisdr.org/wcdr/preparatory-process/national-reports/South-Africa-report.pdf (accessed 15 August 2010).

Government of Zambia (1999) *Forests Act*, Lusaka: Government of Zambia.

Grace, D. (2003) *Rational Drug Use: Practical training for farmers*, Nairobi: International Livestock Research Institute, Nairobi, Kenya.

Grace, D., Randolph, T., Affognon, H., Dramane, D., Diall, O. and Clausen, P.-H. (2009) 'Characterisation and validation of farmers' knowledge and practice of cattle trypanosomosis management in the cotton zone of West Africa', *Acta Tropica* 111: 137–143.

Grameen Foundation (2008) 'Frequently asked questions about microfinance'. Online www.grameenfoundation.org/what-we-do/microfinance-basics (accessed 7 April 2010).

Grant, K.A. (2007) 'Tacit knowledge revisited – we can still learn from Polanyi', *Electronic Journal of Knowledge Management* 5, 2: 173–180.

Grayland, E. (1957) *New Zealand Disasters*, Wellington: A.H & A.W. Reed.

Griffin, D.W., Kellogg, C.A. and Shinn, E.A. (2001) 'Dust in the wind: Long range transport of dust in the atmosphere and its implications for global and public ecosystem health', *Global Change & Human Health* 2: 20–33.

Grover, J.Z. (1987) 'AIDS: Keywords', *October* 43: 17–30.

Gruntfest, E.C., Downing, T.E. and White, G.F. (1978) 'Big Thompson flood exposes need for better flood reaction system to save lives', *Civil Engineering ASCE February* 1978: 72–73.

GSDMA (Gujarat State Disaster Management Authority) (2001) *Gujarat Earthquake Reconstruction and Rehabiliation Policy*, Gujarat: GSDMA.

Guèye, C., Fall, A.S. and Tall, S.M. (2007) 'Climatic perturbation and urbanization in Senegal', *Geographical Journal* 173: 88–92.

Guijt, I. and Shah, M.K. (1998) *The Myth of Community: Gender issues in participatory development*, London: Intermediate Technology Publications.

Gunn, A. (2008) 'New Madrid, Missouri, earthquakes', in A. Gunn (ed.), *Encyclopedia of Disaster*, Westport, CT: Greenwood, vol. 1, pp. 90–94.

Gupta, M. (2008) 'Building inherent resilience', in S. Nicklin, B. Cornwell, J. Dodd, J. Griffiths and S. Townsend (eds), *Risk Wise*, Leicester: Tudor Rose.

Gurenko, E.N. (2004) 'Introduction', in E.N. Gurenko (ed.), *Catastrophe Risk and Reinsurance: A country risk management perspective*, London: Risk Books, pp. 3–16.

Gutiérrez, G. (1988) *A Theology of Liberation*, New York: Orbis Books.

Guzzetti, F. (2006) 'Landslide hazard and risk assessment', unpublished PhD, University of Bonn.

Hacking, I. (1990) *The Taming of Chance*, Cambridge: Cambridge University Press.

Hagman, G., Beer, H., Bendz, M. and Wijkman, A. (1984) *Prevention Better than Cure*, Stockholm: Swedish Red Cross.

Håkansson, I. and Voorhees, W.B. (1998) 'Soil compaction', in R. Lal, W.H. Blum, C. Valentine and B.A. Stewart (eds), *Methods for Assessment of Soil Degradation*, Boca Raton, FL: CRC Press, pp. 167–79.

Hallegatte, S. (2008) 'An adaptive regional input–output model and its application to the assessment of the economic cost of Katrina', *Risk Analysis* 28, 3: 779–99.

Hallegatte, S., Hourcade, J.-C. and Dumas, P. (2007) 'Why economic dynamics matter in assessing climate change damages: Illustration on extreme events', *Ecological Economics* 62, 2, 330–340.

Hamnett, M. and Anderson, C. (1999) *The Pacific ENSO Applications Center and the 1997–98 ENSO Warm Event in the US-Affiliated Micronesian Islands: Minimizing impacts through rainfall forecasts and hazard mitigation*, Honolulu: Pacific ENSO Applications Center. Online www.pacificrisa.org/pubs/PEACpaper1999.pdf (accessed 23 May 2010).

Handicap International (2005) *How to Include Disability Issues in Disaster Management following Floods 2004 in Bangladesh*, Dhaka: Handicap International.

Handmer, J. (2000) 'Are flood warnings futile? Risk communication in emergencies', *Australian Journal of Disaster and Trauma Studies* 2. Online www.massey.ac.nz/~trauma/issues/2000–2/handmer.htm (accessed 8 October 2010).

—— (2002) 'Flood warning reviews in North America and Europe: Statements and silence', *Australian Journal of Emergency Management* 17, 3: 17–24.

Handmer, J. and Dovers, S. (2007) *Handbook of Disaster and Emergency Policies and Institutions*, London: Earthscan.

Handmer, J. and Haynes, K. (eds) (2008) *Community Bushfire Safety*, Melbourne: CSIRO Publishing.

Handmer, J. and Tibbits, A. (2005) 'Is staying at home the safest option during wildfire? Historical evidence for an Australian approach', *Environmental Hazards* 6: 81–91.

Hans, A. and Mohanty, R. (2006) 'Disaster, disability and technology', *Development* 49, 4: 119–122.

Hans, A., Patel, A.M., Sharma, R.K., Prasad, D., Mahapatra, K. and Mohanty, R. (2008) *Mainstreaming Disability in Disaster Management: A tool kit*, New Delhi: United Nations Development Programme.

Hapuarachchi, B. (ed.) (2009) *South Asia Disaster Report 2008: Disaster and development in South Asia*, Colombo, Sri Lanka: Duryog Nivaran and Practical Action. Online www.duryognivaran.org/documents/SADR2008.pdf (accessed 13 September 2010).

Hardoy, J.E., Mitlin, D. and Satterthwaite, D. (2001) *Environmental Problems in an Urbanizing World*, London: Earthscan.

Hare, J. (2010) 'Children of the Gods', *Sacred Texts*. Online www.sacred-texts.com/pac/hm/hm13.htm (accessed 27 May 2010).

Harris, S.L. (2000) 'Archaeology and volcanism', in H. Sigurdsson, B. Houghton, S.R. McNutt, H. Rymer and J. Stix (eds), *The Encyclopedia of Volcanoes*, San Diego: Academic Press, pp. 1301–14.

Hartmann, D.L. (2002) 'Climate change: Tropical surprises', *Science* 295, 5556: 811–812.

Hartwell, W.T. (2007) 'The sky on the ground: Celestial objects and events in archaeology and popular culture', in P. Bobrowsky and H. Rickman (eds), *Comet/Asteroid Impacts and Human Society*, Berlin and Heidelberg: Springer, pp. 71–87.

Harvard University Medical School (2004) *Climate Change Futures: Confronting risks, emerging opportunities*, Draft report, Zurich: Swiss Re/UNDP.

Harvey, P. (2006) 'Editorial: Mini special issues on cash transfers', *Disasters* 30, 3: 273–6.

—— (2007) *Cash-based Responses in Emergencies*, HPG Report 24, London: Overseas Development Institute.

Hasan, A., Ranaweera, R., Saeed, A. and Sanderson, D. (2009), *ASPK-61 Appeal Earthquake Recovery and Rehabilitation – Pakistan Evaluation Report*, Geneva: Action by Churches Together (ACT).

Hauschildt, L.S. and Lybaek, R. (2006) 'The tyranny of participation: A discussion of potentials and pitfalls in the application of participatory and social capital approaches to development', University of Aarhus Denmark. Online www.ulandslaere.au.dk/.../LineSkou%20Hauschildt_RasmusLybaek.pdf (accessed 21 August 2010).

Hausmann, R., Tyson, L.D. and Zahidi, S. (2009) *The Global Gender Gap Report 2009*, Geneva: World Economic Forum. Online www.weforum.org/pdf/gendergap/report2009.pdf (accessed 24 August 2010).

Havnevik, K. (1993) *Tanzania: The limits to development from above*, Uppsala: Nordiska Afrikainstitutet.

Hawkins-Bell, L. and Rankin, J.T. (1994) 'Heat-related deaths: Philadelphia and United States, 1993–1994', *Morbidity and Mortality Weekly Report* 43: 453–55.

Hayashi, I. (2010) 'Building resilient culture to large scale earthquakes: Activities of voluntary associations in Kushimoto-cho of Wakayama prefecture, Japan', paper presented at International Symposium on Best Education Practice of Disaster Prevention in Southeast Asia and Japan, University of the Philippines Diliman, February 2010.

Hayes, M.J. (2010) *Drought Indices*, National Drought Mitigation Center, University of Nebraska, Lincoln. Online www.drought.unl.edu/whatis/Indices.pdf (accessed 9 September 2009).

Haynes, K., Barclay, J. and Pidgeon, N.F. (2007) 'An evaluation of volcanic hazard maps as a communication tool on the Caribbean island of Montserrat', *Bulletin of Volcanology* 70, 2: 123–38.

—— (2008a) 'Whose reality counts? Factors affecting the perception of volcanic risk', *Journal of Volcanology and Geothermal Research* 172, 3–4: 259–72.

—— (2008b) 'The issue of trust and its influence on risk communication during a volcanic crisis', *Bulletin of Volcanology* 70, 5: 605–21.

Healey, M. (2002) 'The fragility of the moment: Politics and class in the aftermath of the 1944 Argentine earthquake', *International Labor and Working-Class History* 62: 50–9.

Heffes, E.M. (2008) 'Finance execs have a "What, Me Worry?" attitude towards disasters', *Financial Executive* 24, 8: 9.

Heijmans, A. (2009) *The Social Life of Community-Based Disaster Risk Management: Origins, politics and framing policies*, Working Paper No. 20, London: Aon Benfield UCL Hazard Research Centre.

Heijmans, A., Okechukwu, I., Schuller, A., Pearsum, T. and Skarubowiz, R. (2009) 'A grassroots perspective on risks stemming from disasters and conflict', *Humanitarian Exchange* 44: 34–5.

Heijmans, A. and Victoria, L.P. (2001) *Citizenry-Based and Development Oriented Disaster Response: Experiences and practices in disaster management of the citizens' disaster response network in the Philippines*, Quezon City: Center for Disaster Preparedness.

Helmer, M. and Hilhorst, D. (2006) 'Natural disasters and climate change', *Disasters* 30, 1: 1–3.

HelpAge International (2000) *Older People in Disasters and Humanitarian Crisis: Guidelines for best practice*, London: HelpAge International.

—— (2002) *State of the World's Older People 2002*, London: HelpAge International.

—— (2005) *The Impact of the Indian Ocean Tsunami on Older People: Issues and recommendations*, London: HelpAge International.

—— (2007) *Older People's Associations in Community Disaster Risk Reduction: A resource book on good practice*, London: HelpAge International.

—— (2009) *Witness to Climate Change: Learning from older people's experience*, London: HelpAge International.

Henderson, A. (2008) 'Points of origin: The social impact of the 1906 San Francisco earthquake and fire', paper presented at *Conference on Flammable Cities: Fire, Urban Environment, and Culture in History*, Washington, DC: German Historical Institute.

Herman, E.S. and Chomsky, N. (2002) *Manufacturing Consent*, 2nd edn, New York: Pantheon.

Hermelin, M. and Bedoya, G. (2008) 'Community participation in natural risk prevention: Case histories from Colombia', *Geological Society of London, Special Publications* 305: 39–51.

Hess, U. and Syroka, J. (2005) 'Weather-based insurance in Southern Africa: The case of Malawi', *Agriculture and Rural Development Discussion Paper* 13, Washington, DC: The World Bank.

Hess, U., Wiseman, W. and Robertson, T. (2006) 'Ethiopia: Integrated risk financing to protect livelihoods and foster development', Working Paper, Rome: World Food Programme.

Hewitt, K. (1983a) 'The idea of calamity in a technocratic age', in K. Hewitt (ed.), *Interpretations of Calamity from the Viewpoint of Human Ecology*, Boston: Allen & Unwin, pp. 3–32.

—— (1983b) *Interpretations of Calamity from the Viewpoint of Human Ecology*, Boston: Allen & Urwin.

—— (1995) 'Sustainable disasters? Perspectives and powers in the discourse of calamity', in J. Crush (ed.), *Power of Development*, London: Routledge, pp. 115–28.

—— (1997) *Regions of Risk: A geographical introduction to disaster*, New York: Addison Wesley Longman.

—— (2007) 'Preventable disasters: Addressing social vulnerability, institutional risk, and civil ethics', *Geographisches Rundschau: International Edition* 3, 1: 43–52.

Hewitt, K. and Burton, I. (1971) *The Hazardousness of a Place: A regional ecology of damaging events*, Toronto: University of Toronto Press.

Heylighten, F. (1996) *What is complexity?* Principia Cybernetica Web. Online pespmc1.vub.ac.be/COMPLEXI.html (accessed 14 December 2010).

Hilhorst, D. (2004) 'Complexity and diversity: Unlocking domains of disaster response', in G. Bankoff, G. Frerks and D. Hilhorst (eds), *Mapping Vulnerability: Disaster, Development and People*, London: Earthscan.

Hilson, G. (2006) 'Abatement of mercury pollution in the small-scale gold mining industry: Restructuring the policy and research agendas', *Science of the Total Environment* 362: 1–14.

Hinshaw, R.E. (2006) *Hurricane Stan Response in Guatemala*, Quick Response Report No. 182, Boulder: Natural Hazards Center.

Hobsbawm, E. and Ranger, T. (eds) (1983) *The Invention of Tradition*, Cambridge: Cambridge University Press.

Hochrainer, S., Linnerooth-Bayer, J. and Mechler, R. (2010) 'The European Union Solidarity Fund: Its legitimacy, viability and efficiency', *Mitigation and Adaptation Strategies for Global Change* 15, 7: 797–810.

Hochrainer, S., Patnaik, U., Kull, D., Wajih, S. and Singh, P. (forthcoming) 'Disaster risk financing for poor households: Realities in northern India', forthcoming publication submitted to the *Journal of Disaster Studies, Policy and Management*.

Hochrainer, S. and Pflug, G. (2010) 'Natural disaster risk bearing ability of governments: Consequences of kinked utility', *Journal of Natural Disaster Science* 31, 1: 11–21.

Hochschild, A. (1999) *King Leopold's Ghost*, New York: Mariner.

Hocke, I. and O'Brien, A. (2003) 'Strengthening the capacity of remote indigenous communities through emergency management', *Australian Journal of Emergency Management* 18, 2: 62–70.

Hodgetts, T.J. and Porter, C. (2002) *Major Incident Management System*, London: BMJ Books.

Holden, J. and Wright, A. (2004) 'UK Tornado climatology and the development of simple prediction tools', *Quarterly Journal of Royal Meteorological Society* 130: 1009–21.

Holle, R.L. (2008) 'Annual rates of lightning fatalities by country', paper presented at the 20th International Lightning Detection Conference, 24–25 April 2008, Tucson, Arizona.

Holloway, A. (2000) 'Drought emergency, yes … drought disaster, no: Southern Africa 1991–1993', *Cambridge Review of International Affairs* 14, 1: 254–76.

Holt-Gimenez, E. (2002) 'Measuring farmers' agroecological resistance after Hurricane Mitch in Nicaragua: A case study in participatory, sustainable land management impact monitoring', *Agriculture Ecosystems and Environment* 93: 87–105.

Holzmann, R. (ed.) (2009) *Social Protection and Labor at the World Bank, 2001–2008*, Washington, DC: World Bank.

Holzmann, R. and Jorgensen, S. (2001) 'Social risk management: A new conceptual framework for social protection and beyond', *International Tax and Public Finance* 8: 529–56.

Homan, J. (2004) 'Seismic cultures: Myth or reality', paper presented at Second International Conference on Post-Disaster Reconstruction: Planning for Reconstruction, Coventry: Coventry University.

Homewood, K.M. and Rodgers, W.A. (1991) *Maasailand Ecology: Pastoralist development and wildlife conservation in Ngorongoro*, Cambridge: Cambridge University Press.

—— (1999) *Maasailand Ecology: Pastoralist Development and Wildlife Conservation in Ngorongoro, Tanzania*, Cambridge: Cambridge University Press.

Hoodbhoy, P.A. (2007) 'Science and the Islamic world: The quest for rapprochement', *Physics Today*, August: 49–55.

Hoover, B. (2009) 'In Mongolia: Democracy's roots grow deeper with the presidential election', The Asia Foundation. Online asiafoundation.org/in-asia/2009/06/03/in-mongolia-democracy%E2%80%99s-roots-grow-deeper-with-the-presidential-election/(accessed 21 December 2010).

Hopson, T.M. and Webster, P.J. (2009) 'A 1–10 day ensemble forecasting scheme for the major river basins of Bangladesh: Forecasting severe floods of 2003–2007', *Journal of Hydrometeorology* 11: 618–41.

Horowitz, M. (1995) 'Dams, cows, and vulnerable people: Anthropological contributions to sustainable development', *The Pakistan Development Review* 34, 4: 481–508.

Houghton, J.T., Ding, Y., Griggs, D.J., Noguer, M., van der Linden, P.J., Dai, X., Maskell, K. and Johnson, C.A. (eds) (2001) *Climate Change 2001: The scientific basis*, Contribution of the Working Group I to the Third Assessment Report of the Intergovernmental Panel on Climate Change (IPCC), Cambridge: Cambridge University Press.

Howell, B. (1990) *An Introduction to Seismological Research: History and development*, Cambridge: Cambridge University Press.

Howell, J. and Pearce, J. (2001) *Civil Society and Development: A critical exploration*, Boulder: Lynne Rienner.

Hsu, S.A. (1988) *Coastal Meteorology*, San Diego: Academic Press.

Hsu, S.S. (2010) 'FEMA's sale of Katrina trailers sparks criticism', *The Washington Post*, 13 March. Online www.washingtonpost.com/wp-dyn/content/article/2010/03/12/AR2010031202213.html (accessed 20 October 2010).

Human Rights Center (2005) *After the Tsunami: Human rights of vulnerable populations*, Berkeley: East–West Center, University of California. Online hrc.berkeley.edu/pdfs/tsunami_full.pdf (accessed 11 June 2010).

Hurley-Glowa, S. (2010) 'Reworking the Santiago sound: A cultural history of Badiu roots music in the Cape Verdean diaspora'. Online www.prio.no/private/jorgen/cvmd/papers/CVMD_Hurley-Glowa_Susan.pdf (accessed 27 May 2010).

Hussein, K. (2002) *Livelihoods Approaches Compared: A multi-agency review of current practice*, London: Overseas Development Institute.

Hutton, D. (2008) *Older People in Emergencies: Considerations for action and policy development*, Geneva: World Health Organization.

Huynen, M.M., Martens, P., Schram, D., Weijenberg, M.P. and Kunst, A.E. (2001) 'The impact of heat waves and cold spells on mortality rates in the Dutch population', *Environmental Health Perspectives* 109: 463–70.

IASC (2002) *Improving the Quality, Coverage and Accuracy of Disaster Data: A comparative analysis of global and national datasets*, Task Force On Disaster Reduction Working Group No. 3-IASC-TFDR-WG3, Geneva: UNISDR.

—— (2007) *IASC Guidelines on Mental Health and Psychosocial Support in Emergency Settings*, Geneva: IASC.

IATO (International Art Therapy Organization) (2010) *Art Therapy and Disaster Relief*. Online www.internationalarttherapy.org/disaster.html (accessed 20 October 2010).

IBC (Insurance Bureau of Canada) (2008) 'FACTS of the general insurance industry in Canada, 2008', *IBC Fact Book*. Online www.tcim.ca/documents/Facts_Book_may14_08.pdf (accessed 18 June 2010).

—— (2009) *International Building Code*, Illinois: International Code Council.

ICCPR (International Covenant on Civil and Political Rights) (1966) United Nations General Assembly Resolution 2200A (XXI).

ICERD (International Covenant on the Elimination of All Forms of Racial Discrimination) (1965) United Nations General Assembly Resolution 2106 (XX).

ICESCR (International Covenant on Economic, Social and Cultural Rights) (1966) General Assembly Resolution 2200A (XXI).

ICL (International Campaign to Ban Landmines) (2010) 'Major findings', *Landmine and Cluster Munitions Monitor*. Online www.the-monitor.org/index.php/publications/display?url=lm/2010/es/Major_Findings.html (accessed 7 December 2010).

IDB (2008) *Disaster Risk Management Policy Guidelines*, Washington, DC: Inter-American Development Bank, March. Online idbdocs.iadb.org/wsdocs/getdocument.aspx?docnum=360026 (accessed 29 December 2009).

IDEA (Institute of Environmental Studies, National University, Manizales, Colombia) (2004) *Results of Application of the System of Indicators on Twelve Countries of the Americas*, Information and Indicators Program for Disaster Risk Management – Execution of Component II, Indicators for Disaster Risk Management Operation ATN/JF-7907-RG, Manizales and Washington, DC: Instituto de Estudios Ambientales, Universidad Nacional de Colombia and Inter-American Development Bank, Sustainable Development Department.

IDL (2003) *Community Based Animal Health Workers: Threat or opportunity?* Crewkerne: IDL Group.

IDMC (Internal Displacement Monitoring Centre) (2009) 'Colombia: New displacement continues, response still ineffective'. Online www.internal-displacement.org/countries/colombia (accessed 5 December 2010).

IDNDR-EWP (International Decade for Natural Disaster Reduction – Early Warning Programme) (1997) *Guiding Principles for Effective Early Warning*, Geneva: IDNDR.

IEG (Independent Evaluation Group) (2006) *Hazards of Nature: Risks to Development*, Washington, DC: World Bank. Online lnweb90.worldbank.org/oed/oeddoclib.nsf/DocUNIDViewForJavaSearch/F0FCEB17632CB93485257155005081BE/$file/natural_disasters_evaluation.pdf (accessed 10 April 2011).

—— (2008) *Disaster Risk Management: Taking lessons from evaluation*, IEG Working Paper 2008/5, Proceedings of the Conference on Evaluation, Paris, November 2006, Washington, DC: World Bank. Online lnweb90.worldbank.org/oed/oeddoclib.nsf/DocUNIDViewForJavaSearch/50CB58565D1096B8852574A6006EFD73/$file/disaster_risk_man.pdf (accessed 1 December 2010).

Iezzoni, L. (2010) 'Disability legacy of the Haitian earthquake', *Annals of Internal Medicine*, 15 March. Online www.annals.org (accessed 27 July 2010).

Iezzoni, L. and Ronan, L. (2010) 'Disability legacy of the Haitian earthquake', *Annals of Internal Medicine* 152, 12: 812–14.

IFRC (International Federation of Red Cross and Red Crescent Societies) (2000) *Introduction to Disaster Preparedness*, Geneva: International Federation of Red Cross and Red Crescent Societies.

—— (2001) *Psychological Support: Best practices from Red Cross and Red Crescent programmes*, Geneva: IFRC.

—— (2002) *World Disasters Report: Focus on reducing risk*, Bloomfield: Connecticut Kumarian Press.

—— (2003a) *Ethiopian Droughts: Reducing the risk to livelihoods through cash transfers*, Geneva: IFRC. Online www.ifrc.org/docs/pubs/disasters/reduction/ethiopia-droughts-en.pdf (accessed 25 March 2010).

—— (2003b) *2003 World Disasters Report: Focus on ethics in aid*, Geneva: IFRC. Online www.ifrc.org/publicat/wdr2003 (accessed 27 April 2010).

—— (2004) *2004 World Disaster Report: Focus on community resilience*, Geneva: IFRC.

—— (2005) 'Data or Dialogue? The Role of Information in Disasters', in *World Disaster Report 2005*, Geneva: IFRC.

—— (2006) *World Disaster Report 2006*, Geneva: IFRC.

—— (2007) *2007 World Disaster Report: Focus on discrimination*, Geneva: IFRC.

—— (2008a) *Angola: Cholera*. Online www.ifrc.org/cgi/pdf_appeals.pl?06/MDRAO001final.pdf (accessed 5 October 2010).

—— (2008b) *Rebuilding Homes and Livelihoods in Grenada after Hurricane Ivan*, Geneva: IFRC. Online www.recoveryplatform.org/assets/publication/rebuilding_in_grenada_after_hurricane_ivan.pdf (accessed 25 March 2010).

—— (2008c) *Recovery and Risk Reduction through Livelihood Support in Timor-Leste*, Geneva: IFRC. Online www.ifrc.org/Docs/pubs/disasters/resources/reducing-risks/cs-timor-leste-2-en.pdf (accessed 25 March 2010).

—— (2009) *World Disasters Report 2009: Focus on early warning, early action*, Geneva: IFRC.

—— (2010a) *Community Preparedness*. Online www.ifrc.org/what/disasters/preparing/community.asp (accessed 9 June 2010).

—— (2010b) *Haiti from Tragedy to Opportunity*, Geneva: IFRC.

—— (2010c) *Revised Plan 2011: Mongolia*, Geneva: IFRC. Online www.reliefweb.int/rw/RWFiles2010. nsf/FilesByRWDocUnidFilename/EDIS-8ATN9P-full_report.pdf/$File/full_report.pdf (accessed 21 December 2010).

IFRC and Johns Hopkins University (2008) *The Johns Hopkins and Red Cross Red Crescent Public Health Guide in Emergencies*, Geneva: IFRC.

IFRC Reference Centre for Psychosocial Support (2009) *Psychosocial Interventions*, Copenhagen: IFRC.

Ignatieff, M. (2005) 'The broken contract', *New York Times*, 25 September.

Ikeda, S., Sato, T. and Fukuzono, T. (2008) 'Towards an integrated management framework for emerging disaster risks in Japan', *Natural Hazards* 44: 267–80.

ILO (International Labour Organization) (2005) 'After the tsunami: In the wake of disaster, ILO helps rebuild lives and livelihoods', *World of Work Magazine* 53: 15–16.

IMC (International Medical Corps) (2006) *Displaced in America – Health Status Among Internally Displaced Persons in Louisiana and Mississippi Travel Trailer Parks: A Global Perspective*, Santa Monica, CA: IMC.

Inbar, M., Ostera, H.A. and Parica, C.A. (1995) 'Environmental assessment of 1991 Hudson volcano eruption ash fall effects on southern Patagonia region, Argentina', *Environmental Geology* 25: 119–25.

Ingalsbee, T. (2006) 'The war on wildfire: Firefighting and the militarization of forest fire management', in G. Wuerthner (ed.), *Wildfire: A century of failed forest policy*, Washington: Island Press, pp. 223–31.

Inglis, F. (2004) *Key Concepts: Culture*, Cambridge: Polity Press.

Inter-Agency Network for Education in Emergencies (2004) *Minimum Standards for Education in Emergencies, Chronic Crises and Early Reconstruction*, London: Inter-Agency Network for Education in Emergencies.

Inter-American Development Bank (2007) 'Survey finds lower percentage of Mexican migrants sending money home from the United States'. Online www.iadb.org/news-releases/2007–08/english/survey-finds-lower-percentage-of-mexican-migrants-sending-money-home-from-the-un-3985.html (accessed 7 April 2010).

International Geosphere-Biosphere Programme (2001) *Environmental Variability and Climate Change*, PAGES Project, IGBP Science no. 3, Stockholm: IGBP.

International Rivers (2006) *Fizzy Science: Loosening the hydro industry's grip on reservoir greenhouse gas emissions research*, Berkeley, CA: International Rivers. Online www.internationalrivers.org/node/1349 (accessed 12 February 2010).

—— (2010) *Questions and Answers about Big Dams*, International Rivers Network. Online www.international rivers.org/en/node/570 (accessed 6 October 2010).

International Society for Mangrove Ecosystems. Online www.mangroverestoration.com/MBC_Code_AAA_WB070803_TN.pdf (accessed 10 November 2010).

IPCC (Intergovernmental Panel on Climate Change) (2007a) *Climate Change 2007: The physical science basis: Contribution of Working Group I to the Fourth Assessment Report of the Intergovernmental Panel on Climate Change*, Cambridge: Cambridge University Press.

—— (2007b) *Climate Change 2007: Impacts, adaptation and vulnerability: Working Group II contribution to the Fourth Assessment Report of the Climate Change Intergovernmental Panel on Climate Change*, Cambridge: Cambridge University Press.

—— (2007c) *Climate Change 2007: Mitigation of climate change: Working Group III contribution to the Fourth Assessment Report of the Climate Change Intergovernmental Panel on Climate Change*, Cambridge: Cambridge University Press.

IPET (Interagency Performance Evaluation Taskforce) (2007) *Performance Evaluation of the New Orleans and South East Louisiana Hurricane Protection System, Volume VII – The Consequences*, Final Report of the Interagency Performance Evaluation Task Force, 26 March.

IRIN (2009) *GLOBAL: How to measure vulnerability to climate change?* Geneva: United Nations Office for the Coordination of Humanitarian Affairs (OCHA).

Irwin, R.L. (1989) 'The incident command system', in E. Auf der Heide (ed.), *Disaster Response: Principles of preparation–coordination*, Chicago: Mosby, pp. 133–63.

IUCN (International Union for the Conservation of Nature) (2010) 'Nature's backbone at risk'. Online www.iucnredlist.org/news/vertebrate-story (accessed 30 October 2010).

Ivers, L and Ryan, E. (2006) 'Infectious diseases of severe weather-related and flood-related natural disasters', *Current Opinion on Infectious Diseases* 19, 5: 408–14.

Izumi, Y. (2010) 'Fifteen years of educational activities learned from lessons in the Kobe Earthquake', paper presented at International Symposium on Best Education Practice of Disaster Prevention in Southeast Asia and Japan, University of the Philippines Diliman, February 2010.

Izzadeen, A. (2005) 'No war, no peace', WWW Virtual Library Sri Lanka. Online www.lankalibrary.com/pol/no-war_no-peace.htm (accessed 5 December 2010).

Jabeen, H., Johnson, C. and Allen, A. (2010) 'Built-in resilience: Learning from grassroots coping strategies for climate variability', *Environment and Urbanization* 22, 2: 415–31.

Jabry, A. (2005) *After the Cameras Have Gone: Children in disasters*, London: Plan International.

Jacobs, G.A. (2007) 'The development and maturation of humanitarian psychology', *American Psychologist* 62, 8: 932–41.

Jacobs, G.A. and Meyer, D.L. (2006) 'Psychological first aid: Clarifying the concept', in L. Barbanel and R.J. Sternberg (eds), *Psychological Interventions in Times of Crisis*, Berlin: Springer, pp. 57–71.

Jakes, P., Kruger, L., Monroe, M., Nelson, K. and Sturtevant, V. (2007) 'Improving wildfire preparedness: Lessons from bommunities across the U.S.', *Human Ecology Review* 14, 2: 188–97.

Jeffrey, P. (2000) *Organization Saved Lives in Caracas Slum*, New York: UMCOR (United Methodist Committee On Relief).

Jeggle, T. (2003) 'Bringing early warning to the people – public and partnership responsibilities for early warning', in J. Zschau and A. Küppers (eds), *Early Warning Systems for Natural Disaster Reduction*, Berlin and Heidelberg: Springer-Verlag, pp. 13–14.

—— (2010) Personal communication, International DRR Consultant, Pittsburgh, PA, USA.

Jenkins, R. (2008) *Rethinking Ethnicity: Arguments and explorations*, 2nd edn, London: Sage.

Jenks, A.L. (2010) *The Perils of Progress: Environmental disasters in the 20th century*, New York: Prentice-Hall.

Jennings, A.A. (2008) 'Analysis of worldwide regulatory guidance for surface soil contamination', *Journal of Environmental Engineering Science* 7: 597–615.

Jha, A.K., Barenstein, J.D., Phelps, P.M., Pittet, D. and Sena, S. (2010) *Safer Homes, Stronger Communities: Handbook for reconstructing after natural disasters*, Washington: Global Facility for Disaster Reduction and Recovery, The World Bank.

Jigyasu, R. (2001) 'From natural to cultural disaster: Consequences of the post-earthquake rehabilitation process on the cultural heritage in Marathwada region, India', *Bulletin of the New Zealand Society for Earthquake Engineering* 34, 3: 237–42.

—— (2002) *Reducing Disaster Vulnerability through Local Knowledge and Capacity: The case of earthquake prone rural communities in India and Nepal*, DrIng thesis, Norwegian University of Science and Technology, Trondheim.

—— (2010) 'Appropriate technology for post disaster reconstruction', in G. Lizarralde, C. Johnson and C. Davidson (eds), *Rebuilding after Disasters: From emergency to sustainability*, London and New York: Spon Press, pp. 49–69.

Jitendra, K.B., Ramesh, G. and Dixit, A. (2002) *Protection of Educational Buildings Against Earthquakes: A manual for designers and builders*, Katmandu: National Society for Earthquake Technology–Nepal.

JKPP (Jaringan Kerja Pemetaan Partisipatif – Indonesia Community Mapping Network) and YRBI (Yayasan Rumpun Bambu Indonesia – Center for People's Economic Development) (2006) 'Community mapping in tsunami affected areas in Aceh, Indonesia', Video. Online www.iapad.org/aceh.htm (accessed 14 December 2010).

Johnson, C. (2007) 'Strategic planning for post-disaster temporary housing', *Disasters* 31, 4: 435–58.

—— (2010) 'Urban disaster trends', in International Federation of Red Cross and Red Crescent Societies (IFRC) *World Disasters Report 2010*, Geneva: IFRC, pp. 29–49.

Johnson, C., Tunstall, S. and Penning-Rowsell, E. (2005) 'Floods as catalysts for policy change: Historical lessons from England and Wales', *Water Resources Development* 21, 4: 561–75.

Johnson, K., Olson, E.A. and Manandhar, S. (1982) 'Environmental knowledge and response to natural hazards in mountainous Nepal', *Mountain Research and Development* 2: 175–88.

Johnson, S., Satu, R., Jadon, N. and Duca, C. (2009) *Contamination of Soil and Water Inside and Outside the Union Carbide India Limited, Bhopal*, New Delhi: Centre for Science and Environment.

Jones-DeWeever, A. (2008) *Women in the Wake of the Storm: Examining the post-Katrina realities of the women of New Orleans and the Gulf Coast*, Washington, DC: The Institute for Women's Policy Research. Online www.iwpr.org/pdf/D481.pdf (accessed 23 August 2010).

Jonientz-Trisler, C., Simmons, R.S., Yanagi, B.S., Crawford, G.L., Darienzo, M., Eisner, R.K., Petty, E. and Priest, G.R. (2005) 'Planning for tsunami-resilient communities', *Natural Hazards* 35: 121–39.

Jonkman, S.N. (2005) 'Global perspectives on loss of human life caused by flood', *Natural Hazards* 34: 151–7.

—— (2007) 'Loss of life estimation in flood risk assessment: Theory and applications', unpublished PhD thesis, Delft University.

Jonkman, S.N. and Kelman, I. (2005) 'An analysis of the causes and circumstances of flood disaster deaths', *Disasters* 29, 1: 75–97.

Jonkman, S.N., Maaskant, B., Boyd, E. and Levitan, M.L. (2009) 'Loss of life caused by the flooding of New Orleans after Hurricane Katrina: Analysis of the relationship between flood characteristics and mortality', *Risk Analysis* 29, 5: 676–98.

Jung, C.G. (1989) 'Late thoughts', footnote by editor A. Jaffé, in *Memories, Dreams and Reflections*, Boston: Beacon [originally published 1961, New York: Random House].

Kailes, J.I. (2002) *Evacuation Preparedness – Taking Responsibility for Your Safety: A guide for people with disabilities and other activity limitations*, Pomona, California: Center for Disability Issues and the Health Profession, Western University of Health Sciences.

Kailes, J.I. and Enders, A. (2007) 'Moving beyond "Special Needs": A function based framework for emergency management and planning', *Journal of Disability Policy Studies* 17: 230–7.

Kaimowitz, D. (2000) *Useful Myths and Intractable Truths: The politics of the link between forests and water in Central America*, Bogor: CIFOR.

Kajimbwa, M. (2006) 'NGOs and their role in the global South', *International Journal of Not-for-Profit Law* 9, 1. Online www.icnl.org/knowledge/ijnl/vol9iss1/art_7.htm (accessed 24 August 2010).

Kälin, W. (2005) 'Protection of internally displaced persons in situations of natural disasters – a working visit to Asia of the Representative of the UN Secretary General on the Human Rights of Internally Displaced Persons', 27 February to 5 March.

Kalnay, E. and Cai, M. (2003) 'The impact of urbanization and land-use change on climate', *Nature* 423, 29: 528–31.

Kaniasty, K. and Norris, F. (1995) 'In search of altruistic community: Patterns of social support mobilization following Hurricane Hugo', *American Journal of Community Psychology* 23, 4: 447–77.

Kaufmann, S.H.E. (2009) *The New Plagues: Pandemics and poverty in a globalized world*, London: Haus Publishing.

Kawachi, I. and Berkman, L. (2000) 'Social cohesion, social capital, and health', in I. Kawachi and L. Berkman (eds), *Social Epidemiology*, New York: Oxford University Press, pp. 174–90.

Kearns, C. and Lowe, B. (2007) 'Disasters and people with disabilities', *Journal of Emergency Management* 5: 35–40.

Keegan, N. (2010) *The National Commission on Children and Disasters: Overview and issues*, Washington, DC: Congressional Research Service.

Keen, D. (1994) *The Benefits of Famine: A political economy of famine and relief in southwestern Sudan 1983–1989*, Princeton, NJ: Princeton University Press.

Keen, M., Brown, V. and Dyball, R. (2005) 'Social learning: A new approach to environmental management', in M. Keen, V. Brown and R. Dyball (eds), *Learning in Environmental Management: Towards a sustainable future*, London: Earthscan, pp. 2–21.

Keim, M. (2010a) 'Flood disasters', in K. Koenig and C. Schultz (eds), *Disaster Medicine: Comprehensive Principles and Practices*, New York: Cambridge University Press, pp. 529–42.

—— (2010b) 'Sea-level rise disaster in Micronesia: Sentinel event for climate change?' *Disaster Medicine and Public Health Preparedness* 4, 1: 81–7.

—— (2010c) 'Environmental disasters', in H. Frumkin (ed.), *Environmental Health: From Global to Local*, San Francisco: Wiley, pp. 843–75.

Keller, B. (1988) 'As hope dies, quake rescuers pull out', *New York Times*, 16 December.

Kelman, I. (2005) 'Operational ethics for disaster research', *International Journal of Mass Emergencies and Disasters* 23, 4: 141–58.

—— (2006) 'Warning for the 26 December 2004 tsunamis', *Disaster Prevention and Management* 15, 1: 178–89.

Kelman, I. and Gaillard, J.C. (2008) 'Placing climate change within disaster risk reduction', *Disaster Advances* 1, 3: 3–5.

Kelman, I. and Mather, T.A. (2008) 'Living with volcanoes: The sustainable livelihoods approach for volcano-related opportunities', *Journal of Volcanology and Geothermal Research* 172, 3–4: 189–98.

Kelman, I. and Spence, R. (2004) 'An overview on flood actions on buildings', *Engineering Geology* 73: 297–309.

Kennedy, J., Ashmore, J., Babister, E. and Kelman, I. (2008) 'The meaning of "build back better": Evidence from post-tsunami Aceh and Sri Lanka', *Journal of Contingencies and Crisis Management* 16, 1: 24–36.

Kennett, D.J., Kennet, J.P., West, A., Mercer, C., Que Hee, S.S., Bement, L., Bunch, T.E., Sellers, M. and Wolbach, W.S. (2009) 'Nanodiamonds in the younger Dryas boundary sediment layer', *Science* 323: 94.

Kenny, C. (2009) 'Why do people die in earthquakes? The costs, benefits and institutions of disaster risk reduction in developing countries', Washington, DC: World Bank Policy Research Working Paper Series, 4823. Online ssrn.com/abstract = 1334526 (accessed 17 January 2010).

Kentucky (2010) *National Flood Insurance Program*, State of Kentucky, Division of Water. Online www.water.ky.gov/floodplainmanagement/nationalfloodinsuranceprogram (accessed 3 October 2010).

Kett, M., Stubbs, S., Yeo, R., Deshpande, S. and Cordeiro, V. (2005) *Disability in Conflict and Emergency Situations: Focus on tsunami-affected areas*, Brussels: IDDC Research Report, International Disability and Development Consortium.

Khan, S. and Haque, E. (2010) 'Wetland resource management in Bangladesh', *Environmental Hazards* 9, 1: 54–73.

Khazai, B. and Sitar, N. (2003) 'Evaluation of factors controlling earthquake-induced landslides caused by Chi-Chi earthquake and comparison with the Northridge and Loma Prieta events', *Engineering Geology* 71: 79–95.

Kienberger, S. (2007) *Field Trip Report – Vulnerability Mapping Búzi*. Online projects.stefankienberger.at/vulmoz/wpcontent/uploads/2007/12/fieldtripreport_kienberger_102007.pdf (accessed 3 May 2010).

Kienberger, S. (ed.) (2008) *Toolbox and Manual. Mapping the Vulnerability of Communities: Example from Búzi, Mozambique*, Salzburg: Salzburg University, Centre for Geoinformatics (Z_GIS), INGC, Maputo; CIG-UCM, Beira. Online projects.stefankienberger.at/vulmoz/wp-content/uploads/2008/08/Toolbox_CommunityVulnerabilityMapping_V1.pdf (accessed 14 December 2010).

Killick, T. (1995) 'Structural adjustment and poverty alleviation: An interpretative survey', *Development and Change* 26, 2: 305–30.

Kim, D. (2006) 'The natural environment control system of traditional Korean architecture', *Building and Environment* 41, 12: 1905–12.

Kim, J.Y., Millen, J.V., Irwin, A. and Gersham, J. (eds) (2000) *Dying for Growth: Global inequality and the health of the poor*, Monroe: Common Courage Press.

Kindon, S., Pain, R. and Kesby, M. (eds) (2007) *Participatory Action Research Approaches and Methods: Connecting people, participation and place*, London: Routledge.

Kiunsi, R. and Lupala, J. with Lerise, F., Manoris, M., Malele, B., Namangaya, A. and Mchome, E. (2009) 'Building disaster-resistant communities: Dar es Salaam, Tanzania', in M. Pelling and B. Wisner (eds), *Disaster Risk Reduction: Cases from urban Africa*. London: Earthscan, pp. 127–46.

Kjekshus, H. (1996) *Ecology Control and Economic Development in East Africa*, London: James Curry [first edn 1976].

Klein, N. (2005) 'Power to the victims of New Orleans', *The Guardian*, 9 September. Online www.guardian.co.uk/world/2005/sep/09/hurricanekatrina.usa4 (accessed 27 April 2010).

—— (2007) *The Shock Doctrine: The rise of disaster capitalism*, London: Penguin.

Kleinman, P.J.A., Pimentel, D. and Bryant, R.B. (1995) 'The ecological sustainability of slash-and-burn agriculture', *Agriculture, Ecosystems and Environment* 52: 235–49.

Klinenberg, E. (2002) *Heatwave: A social autopsy of disaster in Chicago*, Chicago: University of Chicago Press.

Kling, G.W., Evans, W.C., Tanyileke, G., Kusakabe, M., Ohba, T., Yoshida, Y. and Hell, J.V. (2005) 'Degassing Lakes Nyos and Monoun: Defusing certain disaster', *Proceedings of the National Academy of Sciences* 102, 40: 14185–90.

Knight, B., Heller, K. and Bengston, V. (2000) 'Age and emotional response to the Northridge earthquake: A longitudinal analysis', *Psychology and Aging* 15, 4: 627–34.

Kohler, G. and Chaves, E.J. (eds) (2003) *Globalization: Critical perspectives*, New York: Nova Science Publishers.

Kohn, R., Levav, I., Garcia, I., Machuca, M. and Tamashiro, R. (2005) 'Prevalence, risk factors and aging vulnerability for psychopathology following a natural disaster in a developing country', *International Journal of Geriatric Psychiatry* 20: 835–41.

Kolawole, A. (2007) 'Responses to natural and man-made hazards in Borno, northeast Nigeria', *Disasters* 11, 1: 59–66.

Kolmannskog, V. (2009) *Climate Change, Disaster, Displacement and Migration: Initial evidence from Africa*, Research Paper No. 180, Geneva: United Nations High Commissioner for Refugees.

Kondo, T., Orbeta Jr, A., Dingcong, C. and Infantado, C. (2008) 'Impact of microfinance on rural households in the Philippines', *IDS Bulletin* 39, 1: 51–68.

Kooiman, J., Van Vliet, M. and Jentoft, S. (1997) *Creative Governance: Opportunities for fisheries in Europe*, Aldershot: Ashgate.

Kovacs, P. and Hallak, A. (2007) 'Insurance coverage of meteorite, asteroid and comet impacts – issues and options', in P. Bobrowsky and H. Rickman (eds), *Comet/Asteroid Impacts and Human Society*, Berlin and Heidelberg: Springer, pp. 469–78.

Krausmann, E., Cruz, A.M. and Affeltranger, B. (2010) 'The impact of the 12 May 2008 Wenchuan earthquake on industrial facilities', *Journal of Loss Prevention in the Process Industries* 23: 242–8.

Kreimer, A., Arnold, M., Barham, C., Freeman, P., Gilbert, R., Krimgold, F., Lester, R., Pollner, J.D. and Vogt, T. (1999) *Managing Disaster Risk in Mexico: Market incentives for mitigation investment*, Washington, DC: World Bank.

Kreimer, A. and Echeverría, E. (1991) 'Case study: Housing reconstruction in Mexico City', in A. Kreimer and M. Munasinghe (eds), *Managing Natural Disasters and the Environment*, Washington, DC: World Bank.

Kreimer, A. and Munasinghe, M. (eds) (1992) *Environmental Management and Urban Vulnerability*, World Bank Discussion Paper 168, Washington, DC: World Bank.

Krings, M. (2007) 'Black Titanic. African-American and African appropriations of the White Star liner', *Arbeitspapiere/Working Papers* 81: 1–24.

Krinitzsky, E. (1993) 'Earthquake probability in engineering – Part 1: The use and misuse of expert opinion', *Engineering Geology* 33: 257–88.

Kroll-Smith, J.S. and Crouch, S.R. (1990) *The Real Disaster is Above Ground: A mine fire and social conflict*, Lexington: University Press of Kentucky.

Kuban, R. and MacKenzie-Carey, H. (2001) *Community-Wide Vulnerability and Capacity Assessment (CVCA)*, Ottawa: Office of Critical Infrastructure Protection and Emergency Preparedness.

Kull, C.A. (2004) *Isle of Fire: The political ecology of landscape burning in Madagascar*, London: University of Chicago Press.

Kull, D. (2006) 'Financial services for disaster risk management for the poor', in P.G.D. Chakrabarti and M.R. Bhatt (eds), *Micro-Finance and Disaster Risk Reduction*, Delhi: National Institute of Disaster Management/Knowledge World, pp. 39–64.

Kumar, S. (1998) 'India's heat wave and rains result in massive death toll', *Lancet* 351: 1869.

Kunkel, K.E., Pielke, R.A. and Changnon, S.A. (1999) 'Temporal fluctuations in weather and climate extremes that cause economic and human health impacts: A review', *Bulletin of the American Meteorological Society* 8: 1077–98.

Kurita, T., Arakida, M. and Colombage, S. (2007) 'Regional characteristics of tsunami risk perception among the tsunami affected countries in the Indian Ocean', *Journal of Natural Disaster Science* 29, 1: 29–38.

La Red (2010) *Antecedentes*, Desenredando. Online www.desenredando.org/lared/antecedentes.html (accessed 20 October 2010).

Laffaille, J., Ferrer, C. and Dugarte, M. (2005) 'Evaluación de campo: Estudio preliminar de los efectos geomorfológicos de evento meteorológico observado el día 5 de Febrero del ano 2005', Mérida, Venezuela: Fundación para Prevención de los Riesgos Sísmicos de Estado Mérida (FUNDAPRIS-Mérida).

Laituri, M. and Kodrich, K. (2008.) 'On line disaster response community: People as sensors of high magnitude disasters using Internet GIS', *Sensors* 8, 3037–55. Online www.mdpi.net/sensors/papers/s8053037.pdf (accessed 14 December 2010).

Lal, R. (1995) 'Erosion crop-productivity relationships for soils in Africa', *Soil Science Society of America Journal* 59: 661–7.

Lamb, H. (1977) *Climate: Present, past and future*, 2 vols, London: Methuen.

Langenbach, R. (1989) 'Bricks, mortar and earthquakes: Historic preservation vs. earthquake safety', *Journal of the Association for Preservation Technology* 21, 3–4: 3–43.

Laska, S. (2004) 'What if Hurricane Ivan had not missed New Orleans?' *Natural Hazards Observer*, 5–6 November. Online chart.uno.edu/publications/docs/nov4revised.pdf (accessed 14 December 2010).

Lathrop, D. (1994) 'Disaster! If you have a disability, the forces of nature can be meaner to you than anyone else. But you can fight back. Be prepared', *Mainstream*, November. Online www.accessiblesociety.org/topics/independentliving/disaster.htm (accessed 20 November 2009).

Lavell, A. (1999) 'The impact of disasters on development gains: Clarity or controversy', Paper presented at the IDNDR Programme Forum, Geneva, 5–9 July. Online www.desenredando.org/public/articulos/1999/iddg/IDDG1999_mar-1-2002.pdf (accessed 14 December 2010).

—— (2000) *An Approach to Concept and Definition in Risk Management Terminology and Practice*, Lima: La Red de Estudios Sociales en Prevencion de Desastres en America Latina. Online www.desenredando.org/public/articulos/2000/acdrmtp/ACDRMTP2000_mar-4-2002.pdf (accessed 5 July 2010).

—— (2004a) 'The Lower Lempa River valley, El Salvador: Risk reduction and development project', in G. Bankoff, G. Frerks and D. Hilhorst (eds), *Mapping Vulnerability*, London: Earthscan, pp. 67–82.

—— (2004b) *La Gestión Local del Riesgo: Nociones y precisiones en torno al concepto y la práctica*, Quito, Ecuador: CEPREDENAC-PNUD.

—— (2008) *Programme for Mitigation of Flood Disasters in the Lower Lempa River Basin*, ProVention Consortium CRA Tool Kit. Online www.proventionconsortium.org/themes/default/pdfs/CRA/El_Salvador.pdf (accessed 18 October 2010).

Lavell, A. and Franco, E. (eds) (1996) *Estado, Sociedad y Gestión de los Desastres: En búsqueda del paradigma perdida*, Lima, Peru: LA RED, FLACSO and ITDG.

Lavigne, F., De Coster, B., Juvin, N., Flohic, F., Gaillard, J.-C., Texier, P., Morin, J. and Sartohadi, J. (2008) 'People's behaviour in the face of volcanic hazards: Perspectives from Javanese communities, Indonesia', *Journal of Volcanology and Geothermal Research* 172, 3–4: 273–87.

Lavigne, F. and Gunnel, Y. (2006) 'Land cover change and abrupt environmental impacts on Javan volcanoes, Indonesia: A long-term perspective on recent events', *Regional Environmental Change* 6: 86–100.

Lawry, L. and Burkle, F. (2010) 'Measuring the true human cost of natural disasters', *Disaster Medicine and Public Health Preparedness* 2, 4: 208–10.

Lebel, L., Nikitina, E., Kotov, V. and Manuta, J. (2006) 'Assessing institutionalized capacities and practices to reduce risk of flood disaster', in J. Birkmann (ed.), *Measuring Vulnerability to Natural Hazards: Towards disaster resilient societies*, Tokyo: United Nations University, pp. 359–79.

Lee, W.H.K and Espinosa-Aranda, J.M. (2003) 'Earthquake early warning systems: Current status and perspectives', in J. Zschau and A. Küppers (eds), *Early Warning Systems for Natural Disaster Reduction*, Berlin, Heidelberg: Springer-Verlag, pp. 409–23.

Leigh, R. (2007) 'Hail storm – one of the costliest natural hazards', Paper presented at the Coastal Cities Natural Disasters Conference, 20–21 February 2007, Sydney, Australia.

Leitmann, J. (2007) 'Cities and calamities: Learning from post-disaster response in Indonesia', *Journal of Urban Health: Bulletin of the New York Academy of Medicine* 84, 1: 144–53.

The Lens (2010) 'Eastern New Orleans and the Lower Ninth Ward demand a full accounting of recovery spending. So do we', *The Lens*, 5 August. Online thelensnola.org/2010/08/05/on-the-list-of-priorities-for-new-orleanians-transparent-accounting-of-recovery-dollars-ranks-high-along-with-less-blight-safer-streets-and-supermarkets (accessed 20 August 2010).

Leon, E., Kelman, I., Kennedy, J. and Ashmore, J. (2009) 'Capacity building lessons from a decade of transitional settlement and shelter', *International Journal of Strategic Property Management* 13, 3: 247–65.

Leonard, M. (2004) 'GPS geodetic monitoring in regions of high intra-plate seismic activity in Australia', *Proceedings of the 2004 New Zealand Society of Earthquake Engineering 9NZSEE Conference*, Geoscience Australia.

LESLP (London Emergency Services Liaison Panel) (2007) *Major Incident Procedure Manual*, 7th edn, London: Her Majesty's Stationery Office (HMSO).

Lester, R. (1999) *The Changing Risk Landscape: Implications for Insurance Risk Management*, The World Bank and Natural Catastrophe Funding, Proceedings of a conference sponsored by Aon Group Australia Ltd., Sydney.

Levick, J.J. (1859) 'Remarks on sunstroke', *American Journal of Medical Science* 73: 40.

Levine, J.S., Bobbe, T., Ray, N., Singh, A. and Witt, R.G. (1999) *Wildland Fires and the Environment: A global synthesis*, UNEP/DEIAEW/TR.99-1, Nairobi: UNEP (United Nations Environment Programme).

Levine, S. and Chastre, C. (2004) 'Missing the point: An analysis of food security interventions in the Great Lakes', HPN Network Paper Number 47, London: Overseas Development Institute.

Levy, B.R., Slade, M.D. and Ranasinghe, P. (2009) 'Casual thinking after a tsunami wave: Karma beliefs, pessimistic explanatory style and health among Sri Lankan Survivors', *Journal of Religion and Health* 48: 38–45.

Lewis, J. (1999) *Development in Disaster-Prone Places: Studies of vulnerability*, London: Intermediate Technology Publications.

—— (2005) 'Earthquake destruction: Corruption on the fault line', in Transparency International (ed.), *Global Corruption Report 2005: Corruption in construction and post-conflict reconstruction*. Berlin: Transparency

International, pp. 23–30. Online datum.gn.apc.org/PDFs/Transparency%20Int%20Corruption%20&%
20%20Earthquakes.pdf (accessed 12 November 2010).

—— (2009) 'An island characteristic: Derivative vulnerabilities to indigenous and exogenous hazards', *Shima: The International Journal of Research into Island Cultures* 3, 1: 3–15.

Lewis, J. and Kelman, I. (2010) 'Places, people and perpetuity: Community capacities in ecologies of catastrophe', *ACME: An International E-Journal for Critical Geographies* 9, 2: 191–220.

Lightfoot, D.R. (1994) 'Morphology and ecology of lithic-mulch agriculture', *The Geographical Review* 84: 172–84.

Lin, M., Huang, W., Huang, C., Hwang, H. and Tsai, L. (2002) 'The impact of the Chi-Chi earthquake on quality of life among elderly survivors in Taiwan: A before and after study', *Quality of Life Research* 11: 379–88.

Lin, W.T., Lin, C.Y. and Chou, W.C. (2006) 'Assessment of vegetation recovery and soil erosion at landslides caused by a catastrophic earthquake: A case study in Central Taiwan', *Ecological Engineering* 28: 79–89.

Lin, W.T., Lin, C.Y., Tsai, J.S. and Huang, P.H. (2008) 'Eco-environmental changes assessment at the Chiufenershan landslide area caused by catastrophic earthquake in central Taiwan', *Ecological Engineering* 33: 220–32.

Linayo, A. (2006) 'Pautas para la implementación de un SAT en el Valle del rió Mocotíes, Mérida, Venezuela', Presentation at the Third International Conference on Early Warning, Bonn. Online www.ewc3.org/upload/downloads/Forum_Earth_06_Venezuela.pdf (accessed 27 December 2010).

Lindberg-Falk, M. (2010) 'Recovery and Buddhist practices in the aftermath of the tsunami in southern Thailand', *Religion* 40, 2: 96–103.

Lindell, M.K. and Perry, R.W. (1997) 'Hazardous materials releases in the Northridge earthquake: Implications for seismic risk assessment', *Risk Analysis* 17, 2: 147–56.

Linnerooth-Bayer, J. (2008) 'Non-insurance mechanisms for managing climate-related risks: Mechanisms to manage financial risks from direct impacts of climate change in developing countries', *Technical paper, FCCC/TP/2008/9*, United Nations Office at Geneva. Online unfccc.int/4159.php (accessed 23 May 2010).

Linnerooth-Bayer, J., Bals, M.J. and Mechler, R. (2010) 'Insurance as part of a climate adaptation strategy', in M. Hulme and H. Neufeldt (eds), *Making Climate Change Work for Us: European perspectives on adaptation and mitigation strategies*, Cambridge: Cambridge University Press, pp. 340–66.

Linnerooth-Bayer, J. and Mechler, R. (2007) 'Disaster safety nets for developing countries: Extending public–private partnerships', *Environmental Hazards* 7, 1: 54–61.

Linnerooth-Bayer, J., Mechler, R. and Pflug, G. (2005) 'Refocusing disaster aid', *Science* 309: 1044–6.

Linthicum, K. (2009) 'Hurricane victims get chance to buy trailers for as little as 1$', *Los Angeles Times*, 4 June. Online articles.latimes.com/2009/jun/04/nation/na-katrina-trailers4 (accessed 20 October 2010).

LIPI (2010) *The Tsunami of 2004 and How the Smong Story Developed*. Online www.jtic.org/jtic/images/en/dlPDF/Lipi_CBDP/reports/SMGChapter5.pdf (accessed 1 June 2010).

Litman, T. (2006) 'Lessons from Katrina and Rita: What major disasters can teach transportation planners', *Journal of Transportation Engineering* 132, 1: 11–18.

Little, P.D., Mahmoud, H. and Layne Coppock, D. (2001) 'When deserts flood: Risk management and climatic processes among East African pastoralists', *Climate Research* 19: 149–59.

Livengood, A. (2011) 'Enabling participatory planning with GIS: A case study of settlement mapping in Cuttack, India', *Environment and Urbanization* 23: in press.

Liverman, D. (2000) 'Adaptation to drought in Mexico', in D.A. Wilhite (ed.), *Drought: A global assessment*, London: Routledge, vol. 2, pp. 35–45.

Lizarralde, G., Johnson, C. and Davidson, C. (eds) (2009) *Rebuilding after Disasters: From emergency to sustainability*, London: Taylor & Francis.

Loane, I.T. and Gould, J.S. (1985) *Aerial Suppression of Bushfires: Cost–benefit study for Victoria*, Canberra: National Bush Fire Research Unit CSIRO.

Loescher, G. (1993) *Beyond Charity: International cooperation and the global refugee crisis*, New York: Oxford University Press.

Longley, C., Christoplos, I., Slaymaker, T. and Meseka, S. (2007) *Rural Recovery in Fragile States: Agricultural support in countries emerging from conflict*, Natural Resource Perspectives 105, London: Overseas Development Institute.

Longley, C. and Sperling, L. (eds) (2002) *Beyond Seeds and Tools: Effective support to farmers in emergencies*, Special issue of *Disasters* 26, 4.

Longo, G. (2007) 'The Tunguska event', in P. Bobrowsky and H. Rickman (eds), *Comet/Asteroid Impacts and Human Society*, Berlin: Springer, pp. 303–30.

Lopes, R. (1992) *Public Perception of Disaster Preparedness Presentations using Disaster Damage Images*, Working Paper #79, Boulder, Colorado: Natural Hazards Research and Applications Information Center, University of Colorado.

Lopez-Caressi, A. (2011) 'Disaster myths', in B. Wisner and A. Lopez-Caressi (eds), *Disaster Management: International Lessons in Risk Reduction, Response and Recovery*, London: Earthscan.

Luhmann, N. (2002) *Risk: A sociological theory*, New Brunswick, NJ: Aldine Transaction.

Luis Rocha, J. (1999) *Posoltega: Unresolved property problems and continuing vulnerability*, Revista Envio Digital. Online www.envio.org.ni/articulo/2258 (accessed 10 October 2010).

Luna, E.M. (2007) 'Mainstreaming community-based disaster risk management in local development planning', Paper presented at the *Forum on Framework-Building for Investigation of Local Government Settlement Planning Responses to Disaster Mitigation*, 17 January, Alternative Planning Initiatives (ALTERPLAN), Quezon City.

—— (2009) *The Institutionalization of Disaster Risk Reduction in Community Development Education*, Quezon City: Center for Disaster Preparedness.

—— (2011) 'Community self help and partnership', in B. Wisner and A. Lopez-Carresi (eds), *Disaster Management: International lessons in risk reduction, response and recovery*, London: Earthscan.

Luna, E., Bautista, M.L. and De Guzman, M. (2008) 'The impact of disasters on the education sector', in Center for Disaster Preparedness (ed.), *Mainstreaming Disaster Risk Reduction in the Education Sector in the Philippines*, Quezon City: Center for Disaster Preparedness, pp. 32–118.

Luneta, M. and Molina, J.G. (2008) 'Community-based early warning system and evacuation: Planning, development and testing – protecting peoples' lives and properties from flood risks in Dagupan City, Philippines', *Safer Cities* 20: 1–8.

Luseno, W.K, McPeak, J.G., Barrett, C.B., Little, P.D. and Gebru, G. (2003) 'Assessing the value of climate forecast information for pastoralists: Evidence from southern Ethiopia and northern Kenya', *World Development* 31, 9: 1477–94.

Ly, Z.-S. and Chen, B.-Y. (1955) 'Research on gravel mulch fields in Gansu Province', *Journal of Agriculture* 6, 3: 299–312.

Maastricht Guidelines (1997) *Maastricht Guidelines on Violations of Economic, Social and Cultural Rights*, Maastricht, 22–26 January 1997. Online www1.umn.edu/humanrts/instree/Maastrichtguidelines_.html (accessed 16 October 2010).

Maathai, W. (2009) *The Challenge for Africa*, London: Arrow Books.

Mabogunje, A. and Kates, R.W. (2004) 'Sustainable development in Ijebu-Ode, Nigeria: The role of social capital, participation, and science and technology', Center for International Development Working Paper 201, Harvard University. Online www.hks.harvard.edu/var/ezp_site/storage/fckeditor/file/pdfs/centers-programs/centers/cid/publications/faculty/wp/102.pdf (accessed 6 September 2010).

McAdoo, B. (2009) 'Indigenous knowledge and the near field population response during the 2007 Solomon Islands tsunami', *Natural Hazards* 48: 73–82.

McAdoo, B.G., Fritz, H., Jackson, K., Kalligeris, N., Kruger, J., Bonte-Grapentin, M., Moore, A., Rafiau, W., Billy, D. and Tiano, B. (2008) 'Solomon Islands tsunami: One year later', *EOS, Transactions, American Geophysical Union* 89, 18: 169–70.

McAlpin, M.B. (1987) 'Famine relief policy in India: Six lessons for Africa', in M.H. Glantz (ed.), *Drought and Hunger in Africa: Denying famine a future*, Cambridge: Cambridge University Press, pp. 391–414.

McCabe, T. (2004) *Cattle Bring Us to Our Enemies: Turkana ecology, politics, and raiding in a disequilibrium system*, Ann Arbor: University of Michigan Press.

McCall, M.K. (2003) 'Seeking good governance in participatory-GIS: A review of processes and governance dimensions in applying GIS to participatory spatial planning', *Habitat International* 27: 549–73.

McCauley, M. (1976) *Khrushchev and the Development of Soviet Agriculture: The Virgin Land Programme 1953–1964*, New York: Holmes & Meier.

McCorkle, C., Mathias, E. and Schillhorn Van Veen, T. (1996) *Ethnoveterinary Research and Development*, London: IT Books.

McGovern, L. (2009) 'Oxfam America tests its disaster preparedness', Washington, DC: Business Civic Leadership Center. Online bclc.chamberpost.com/2009/11/oxfam-america-tests-its-disaster-preparedness.html (accessed 15 June 2010).

McGranahan, G., Balk, D. and Anderson, B. (2008) 'Risks of climate change for urban settlements in low elevation coastal zones', in G. Martine, G. McGranahan, M. Montgomery and R. Fernández-Castilla (eds), *The New Global Frontier: Urbanization, poverty and environment in the 21st century*, London and Sterling, VA: Earthscan, pp. 165–81.

McGuire, W.J. (1998) 'Volcanic hazards and their mitigation', in J.G. Maund and M. Eddleston (eds), *Geohazards in Engineering Geology*, Engineering Geology Special Publications, London: Geological Society, pp. 79–95.

—— (2006a) 'Global risk from extreme geophysical events: Threat identification and assessment', *Philosophical Transactions of the Royal Society A* 364: 1889–909.

—— (2006b) 'Lateral collapse and tsunamigenic potential of marine volcanoes', in C. Troise, G. De Natale and C.R.J. Kilburn (eds), *Mechanisms of Activity and Unrest at Large Calderas*, Geological Society, London, Special Publication 269, pp. 121–40.

McGuire, L.C., Ford, E.S. and Okoro, C.A. (2007) 'Natural disasters and older US adults with disabilities: Implications for evacuation', *Disasters* 31: 49–56.

McGurty, E. (2007) *Transforming Environmentalism: Warren County, PCBs, and the origins of environmental justice*, New Brunswick: Rutgers University Press.

MacInnes, B., Bourgeois, J., Pinegina, T. and Kravchunovskaya, E. (2009) 'Tsunami geomorphology: Erosion and deposition from the 15 November 2006 Kuril island tsunami', *Geology* 37, 11: 995–8.

Macintosh, D. and Ashton, E. (2003) *Draft Code of Conduct for the Sustainable Management of Mangrove Ecosystems*, Aarhus: Center for Tropical Ecosystems Research and World Bank.

McIntyre, A. (2003) 'Through the eyes of women: Photovoice and participatory research as tools for reimagining place', *Gender, Place and Culture* 10, 1: 47–66.

MacMillan, S. and Brickenden, J. (1993) Liner notes, *On Celtic Mass for the Sea* [CD], Halifax, Nova Scotia: Scojen Music Productions.

McNutt, S.R., Rymer, H. and Stix, J. (2000) 'Synthesis of volcano monitoring', *Encyclopedia of Volcanoes*, Burlington, MA: Academic Press, pp. 1165–83.

McQuaid, J. and Schleifstein, M. (2002) 'Washing away', *Times Picayune*, 23–27 June.

Maddisson, A. (2003) *The World Economy: Historical statistics*, Paris: OECD.

Mahoney, J., Chandra, V., Gambheera, H., Silva, T. and Suveendran, T. (2006) 'Responding to the mental health and psychosocial needs of the people of Sri Lanka in disasters', *International Review of Psychiatry* 18, 6: 593–7.

Mainhardt-Gibbs, H. (2009) *World Bank Energy Sector Lending: Encouraging the world's addiction to fossil fuels*, Washington, DC: World Bank Information Center. Online www.bicusa.org/en/Article.11033.aspx (accessed 2 November 2010).

Makary, A. (2008) 'Cairo disaster could happen again'. Online english.aljazeera.net/focus/2008/09/2008914213410998175.html (accessed 9 May 2010).

Malilay, J. (1997) 'Tropical cyclones', in E. Noji (ed.), *The Public Health Consequences of Disasters*, New York: Oxford University Press, pp. 207–27.

Mankiller, W. (2009) 'Being indigenous in the 21st century', *Cultural Survival Quarterly* 32, 1: 32–7.

Mann, J., Drucker, E., Tarantola, D. and McCabe, M.P. (1994) 'Bosnia: The war against public health', *Medicine and Global Survival* 1: 130–46.

Manuel-Navarrete, D., Pelling, M. and Redclift, M. (2009) 'Coping, governance, and development: The climate change adaptation triad', Environment, Politics and Development Working Paper Series No. 18, Department of Geography, King's College London. Online www.kcl.ac.uk/content/1/c6/03/95/42/WorkingPaperTriad.pdf (accessed 8 November 2010).

Marchand, M. (2008) 'Differential vulnerability in coastal communities: Evidences and lessons learned from two deltas', in D. Proverbs, C.A. Brebbia and E. Penning-Rowsell (eds), *Proceedings of the Conference on Flood Recovery, Innovation and Response*, London, 2–3 July 2008, Southampton: WIT Press, pp. 283–93.

Maret, I. and Amdal, J. (2010) 'Stakeholder participation in post-disaster reconstruction programmes – New Orleans' lakeview: A case study', in G. Lizarralde, C. Johnson and C.H. Davidson (eds), *Rebuilding after Disasters: From emergency to sustainability*, Abingdon: Spon Press, pp. 110–32.

Marglin, S. and Marglin, F. (1990) *Dominating Discourse: Development, culture and resistance*, Helsinki and Tokyo: United Nations University–World Institute for Development Economics Research.

Marks, G. and Beatty, W.K. (1976) *Epidemics*, New York: Scribner.

Martin, S. (2005) *Must Boys Be Boys? Ending sexual exploitation and abuse in UN Peacekeeping Missions*, Washington, DC: Refugees International. Online www.refintl.org/sites/default/files/MustBoysbeBoys.pdf (accessed 24 August 2010).

Marx, K. and Engels, F. (1848) *The Communist Manifesto*, New York: Signet Classics, chapter 1. Online www.marxists.org/archive/marx/works/1848/communist-manifesto/ch01.htm#007 (accessed 7 December 2010).

Maskrey, A. (1989) *Disaster Mitigation: A community-based approach*, Oxford: Oxfam.

—— (1997) *Report on National and Local Capabilities for Early Warning*, Geneva: IDNDR (International Decade for Natural Disaster Reduction) Secretariat.

Mason, B.G., Pyle, D.M. and Oppenheimer, C. (2004) 'The size and frequency of the largest explosive eruptions on Earth', *Bulletin of Volcanology* 66: 735–48.

Masse, W.B. (2007) 'The archaeology and anthropology of Quaternary Period cosmic impact', in P. Bobrowsky and H. Rickman (eds), *Comet/Asteroid Impacts and Human Society*, Berlin and Heidelberg: Springer, pp. 26–70.

Mathias, E. (2007) 'Ethnoveterinary medicine in the era of evidence-based medicine: Mumbo-jumbo, or a valuable resource?' *The Veterinary Journal* 173, 2: 241–2.

Mauelshagen, F. (2009) 'Disaster and political culture in Germany since 1500', in C. Mauch and C. Pfister (eds), *Natural Disaster, Cultural Responses: Case studies towards a global environmental history*, Lanham, MD: Lexington Books, pp. 41–75.

Mayer, T. (2008) 'Five years after the Bam earthquake: When victims become volunteers', *IFRC News*, 22 December. Online www.ifrc.org/Docs/News/08/08122202/index.asp (accessed 8 June 2010).

Mayr, E. (2004) *Après Darwin: La biologie, une science pas comme les autres*, Paris: DUNOD.

Mazda, Y., Magi, M., Kogo, M. and Hong, P.N. (1997) 'Mangrove on coastal protection from waves in the Tong King Delta, Vietnam', *Mangroves and Salt Marshes* 1: 127–35.

Mazoyer, M. and Roudart, L. (1997) *Histoire des Agricultures du Monde: Du néolithique à la crise contemporaine*, Paris: Éditions du Seuil.

Mbaiwa, J.E. (2006) 'The effects of veterinary fences on wildlife populations in Okavango Delta, Botswana', *International Journal of Wilderness* 12, 3: 17–41.

MCDEM (Ministry of Civil Defence and Emergency Management) (2008) *Mass Evacuation Planning Director's Guidelines for Civil Defence Emergency Management (CDEM) Groups [DGL 07/08]*, Wellington: MCDEM.

MCEER (Multidisciplinary Center for Earthquake Engineering Research) (2010) *Major Iranian Earthquakes of the 20th Century*. Online mceer.buffalo.edu/infoservice/reference_services/major-iran-earthquake.asp (accessed 20 October 2010).

Mechler, R. (2003) *Natural Disaster Risk Management and Financing Disaster Losses in Developing Countries*, Karlsruhe: Verlag Versicherungswirtsch.

Meister, R. (2002) *Lawinen: Warnung, rettung, prävention*, Switzerland: International Commission for Alpine Rescue.

Menoni, S. (2001) 'Chains of damages and failures in a metropolitan environment: Some observations on the Kobe Earthquake in 1995', *Journal of Hazardous Materials* 86: 101–19.

Mercer, C. (2002) 'NGOs, civil society and democratization', *Progress in Development Studies* 2: 15–22.

Mercer, J., Dominey-Howes, D., Kelman, I. and Lloyd, K. (2007) 'The potential for combining indigenous and western knowledge in reducing vulnerability to environmental hazards in small island developing states', *Environmental Hazards* 7, 4: 245–56.

Mercer, J., Kelman, I., Lloyd, K. and Suchet-Pearson, S. (2008) 'Reflections on use of participatory research for disaster risk reduction', *Area* 40, 2: 172–83.

Mercer, J., Kelman, I., Taranis, L. and Suchet-Pearson, S. (2010) 'Framework for integrating indigenous and scientific knowledge for disaster risk reduction', *Disasters* 34: 214–39.

Merino, L. and Robson, J. (eds) (2005) *Managing the Commons: Conservation of biodiversity*, Consejo Civil Mexicano para la Silvicultura Sostenible, Mexico City: The Christensen Fund, Ford Foundation, SEMARNAT and Instituto Nacional de Ecología.

Meron, T. (1986) 'On a hierarchy of international human rights', *American Journal of International Law* 80: 1–23.

Merrill, S., Alfred, D., Black, L., Fryrear, D.W., Saleh, A., Zobeck, T.M., Halvorson, A.D. and Tanaka, D.L. (1999) 'Soil wind erosion hazard of spring wheat-fallow as affected by long-term climate and tillage', *Soil Science Society of America Journal* 63: 1768–77.

Metzger, P., D'Ercole, R. and Sierra, A. (1999) 'Political and scientific uncertainties in volcanic risk management: The yellow alert in Quito in October 1998', *Geojournal* 49, 2: 213–21.

Meyers, N. (1989) 'The future of forests', in L. Friday and R.A. Laskey (eds), *The Fragile Environment: The Darwin College Lectures*, Cambridge: Cambridge University Press, pp. 22–40.

Michael-Leiba, M., Andrews, K. and Blong, R. (1997) 'Impact of landslides in Australia', *Australian Journal of Emergency Management* 49, 1: 23–5.

Midgley, J. and Tang, K. (2010) *Social Policy and Poverty in East Asia*, London: Routledge.

Mignan, A. (2009) *Comet and Asteroid Risk: An analysis of the 1908 Tunguska event*, RMS Special Report, London: Research Management Solutions.

Mileti, D.S. and Sorensen, J.H. (1990) *Communication of Emergency Public Warnings: A social science perspective and state-of-the-art assessment*, ORNL-6609, Oak, Ridge, TN: Oak Ridge National Laboratory.

Millennium Ecosystem Assessment (2005) *Ecosystems and Human Well-Being: Biodiversity synthesis*, Washington, DC: World Resources Institute.

Miller, J. (1960) 'Giant waves in Lituya Bay, Alaska', *US Geological Survey Professional Paper* 354-C: 51–86, Washington, DC: USGS.

Minear, L. (1991) *Operation Lifeline Sudan*, Trenton, NJ: Red Sea Press.

Mirza, M.M., Patwardhan, Q.A., Attz, M., Marchand, M., Ghimire, M., Hanson, R. and Norgaard, R. (2005) 'Flood and storm control', in K. Chopra, R. Leemans, P. Kumar and H. Simons (eds), *Ecosystems and Human Well-Being: Policy responses*, vol. 3, Findings of the Responses Working Group of the Millennium Ecosystem Assessment, Washington, DC: Island Press, pp. 337–52.

Mishra, P.K., Samarth, R.M., Pathak, N., Jain, S.K., Banerjee, S. and Maudar, K.K. (2009) 'Bhopal gas tragedy: Review of clinical and experimental findings after 25 years', *International Journal of Occupational Medicine and Environmental Health* 22: 193–202.

Mitchell, A., Glavovic, B.C., Hutchinson, B., MacDonald, G., Roberts, M. and Goodland, J. (2010) 'Community-based civil defence emergency management planning in Northland, New Zealand', *Australasian Journal of Disaster and Trauma Studies* 2010–1. Online www.massey.ac.nz/~trauma/issues/2010–1/mitchell.htm (accessed 19 August 2010).

Mitchell, J.K. (ed.) (1999) *Crucibles of Hazard: Mega-cities and disasters in transition*, Tokyo: UNU Press.

Mitchell, T., Tanner, T. and Haynes, K. (2009) *Children as Agents of Change for Disaster Risk Reduction: Lessons from El Salvador and the Philippines*, Brighton: Institute of Development Studies.

Mitchell, T.C. (2006) 'Building a disaster resilient future: Lessons from participatory research in St. Kitts and Montserrat', unpublished PhD thesis, University College London.

Mitlin, D. and Muller, A. (2004) 'Windhoek, Namibia: Towards progressive urban land policies in southern Africa', *International Development Planning Review* 26, 2: 167–86.

Mlay, W. (1985) 'Environmental implications of land-use patterns in the new villages in Tanzania', in J. Arntzen, L. Ngcongco and S. Turner (eds), *Proceedings of a Conference on Land Policy and Agricultural Production in Eastern and Southern Africa*, Tokyo: United Nations University. Online www.unu.edu/unupress/unupbooks/80604e/80604E0e.htm (2 October 2010).

Moen, J.E.T. and Ale, B.J.M. (1998) 'Risk maps and communication', *Journal of Hazardous Materials* 61: 271–8.

Mohamed, A.B., van Duivenbooden, N. and Abdoussallam, S. (2002) 'Impact of climate change on agricultural production in the Sahel – Part 1: Methodological approach and case study for millet in Niger', *Climatic Change* 54, 3: 327–48.

Mohan, G., Brown, E., Milward, B. and Zack-Williams, A. (2000) *Structural Adjustment: Theory, practice and impacts*, London: Routledge.

Monecke, K., Finger, W., Klarer, D., Kongko, W., McAdoo, B.G., Moore, A.L. and Sudrajat, S.U. (2008) 'A 1,000-year sediment record of tsunami recurrence in northern Sumatra', *Nature* 455: 1232–4.

Monk, T. (2005) 'The importance of community involvement on the road to school seismic safety in British Columbia', *Families for School Seismic Safety*. Online info.worldbank.org/etools/docs/library/229567/Session%203/Case%20Study%203b%20-%20Monk%20re%20School%20Safety.pdf (accessed 6 June 2010).

Monmonier, M. (1996) *How to Lie with Maps*, Chicago: University of Chicago Press.

Moore, G. and Tymowski, W. (2005) *Explanatory Guide to the International Treaty on Plant Genetic Resources for Food and Agriculture*, Gland and Cambridge: IUCN.

Moore, T. (2008) *Disaster and Emergency Management Systems*, London: BSI.

Moore, T. and Lakha, R. (2006) *Tolley's Handbook of Disaster and Emergency Management*, Amsterdam: Elsevier.

Mora, S. (2010) 'Natural hazards have been known to affect Hispanola since long time', *Understanding Risk: Haiti: 12 January and beyond discussions*, World Bank. Online community.understandrisk.org/group/haitijanuary12thandbeyond/forum/topics/natural-hazards-have-been (accessed 4 December 2010).

Moran, E. (1981) *Developing the Amazon*, Bloomington, IN: Indiana University Press.

Morgan, O. (2004) 'Infectious disease risks from dead bodies following natural disasters', *Revista Panamericana de Salud Pública* 15, 5: 307–12.

Morgan, O. and de Ville de Goyet, C. (2005) 'Dispelling disaster myths about dead bodies and disease: The role of scientific evidence and the media', *Revista Panamericana de Salud Pública* 18, 1: 33–6.

Morin, J., De Coster, B., Paris, R., Flohic, F., Le Floch, D. and Lavigne, F. (2008) 'Tsunami-resilient communities' development in Indonesia through educative action', *Disaster Prevention and Management* 17, 3: 430–46.

Moriondo, M., Good, P., Durao, R., Bindi, M., Giannakopoulos, C. and Corte-Real, J. (2006) 'Potential impact of climate change on fire risk in the Mediterranean area', *Climate Research* 31: 85–95.

Morris, A. (2003) 'Understandings of catastrophe: The landslide at La Josefina, Ecuador', in M. Pelling (ed.), *Natural Disaster, Development and Global Change*, London: Routledge, pp. 157–69.

Morris, M. (2000) *CADAM Concerted Action on Dambreak Modelling: Final report*, Tech. Rep. SR 571: HR Wallingford Ltd.

Morris, S.S., Neidecker-Gonzales, O., Carletto, C., Munguia, M., Medina, J.M. and Wodon, Q. (2002) 'Hurricane Mitch and the livelihoods of the rural poor in Honduras', *World Development* 30, 1: 49–60.

Morrison, D. (2006) 'Asteroid and comet impacts: The ultimate environmental catastrophe', *Philosophical Transactions of the Royal Society A* 364: 2041–54.

Moser, C. (1998) 'The asset vulnerability framework: Reassessing urban poverty reduction strategies', *World Development* 26, 1: 1–9.

Mosqueda Jr, M.W. (2008) 'Pupil dies in Cebu flood', *Manila Bulletin*, 13 January.

Moss, S. (2009) *Local Voices, Global Choices: For successful disaster risk reduction*, A Collection of Case Studies about Community-Centre Partnerships for DRR, London: BOND Disaster Risk Reduction Group.

—— (2007) 'Christian Aid and disaster risk reduction', *Overseas Development Humanitarian Practice Network*, ODIHPN, Issue 38. Online www.odihpn.org/report.asp?id=2892 (accessed 24 March 2010).

Mossman, B. (2009) 'Water', *Baker's Dozen* [CD], 709 Studios: Honolulu, Hawai'i.

Moteff, J., Copeland, C. and Fischer, J. (2003) *Critical Infrastructures: What makes an infrastructure critical?* Washington, DC: Congressional Research Service.

Mourits, B. (2010) 'Dust bowl ballads or How to get from Oklahoma to California'. Online www.univie. ac.at/Anglistik/easyrider/data/CoodDsBl.htm (accessed 25 May 2010).

Msilimba, G.G. (2010) 'The socioeconomic and environmental effects of the 2003 landslides in the Rumphi and Ntcheu Districts (Malawi)', *Natural Hazards* 53: 347–60.

Mueller-Mann, D. (ed.) (2011) *The Spatial Dimension of Risk*, London: Earthscan.

Mukerjee, M. (2009) *Churchill's Secret War: The British Empire and the ravaging of India during World War II*, New York: Basic Books.

Mullin, J.R. (1992) 'The reconstruction of Lisbon following the earthquake of 1755: A study in despotic planning', *Journal of the International History of City Planning Association* 7, 2: 64–88.

Munich Re (2004) *Megacities-Megarisks: Trends and challenges for insurance and risk management*, Knowledge Series: Munich Re.

—— (2005a) *Topics: Natural disasters 2004 annual review of natural disasters Munich*, Munich Reinsurance Group.

—— (2005b) *Natural Disasters According to Country Income Groups 1980–2004*, NatCatSERVICE, Munich Re, Munich.

Murray, B. (1998) *People with Disabilities in Lao PDR – Training for Employment and Income Generation: Key issues and potential strategies*, Gladnet Collection Paper No. 135, Ithaca, New York: Cornell University.

Murray, C.J., Lopez, A.D., Chin, B., Feehan, D. and Hill, K.H. (2007) 'Estimation of potential global pandemic influenza mortality on the basis of vital registry data from the 1918–20 pandemic: A quantitative analysis', *The Lancet* 368, 9554: 2211–18.

Murphy, F.J. (2005) 'Unknowable world; Solving the problem of natural evil', *Religious Studies* 41: 343–6.

Murray, C., King, G., Lopez, A., Tomijima, N. and Krug, E. (2002) 'Armed conflict as a public health problem', *British Medical Journal* 324: 346–9.

Murwira, K., Wedgwood, H., Watson, C., Win, E.J. and Tawney, C. (2000) *Beating Hunger: The chivi experience*, London: IT Press. Online www.proventionconsortium.org/themes/default/pdfs/CRA/Zimbabwe. pdf (accessed 24 August 2010).

Musso, M. and Pejon, O. (2006) *Expansive Soils Engineering Geological Mapping: Applied method in clayey soils of Montevideo, Uruguay*, IAEG2006 Paper number 59, London: The Geological Society of London.

Mustafa, D. (1998) 'Structural causes of vulnerability to flood hazard in Pakistan', *Economic Geography* 47, 3: 289–305.

Myers, D. and Wee, D. (2005) *Disaster Mental Health Services*, London: Brunner-Routledge.

Myhrvold-Hanssen, T. (2003) 'Democracy, news media and famine prevention: Amartya Sen and the Bihar famine of 1966–67', *Disaster Diplomacy*. Online www.disasterdiplomacy.org/MyhrvoldHanssenBihar Famine.rtf (accessed 14 December 2010).

Nadim, F., Kjekstad, O., Peduzzi, P., Herold, C. and Jaedicke, C. (2006) 'Global landslide and avalanche hotspots', *Landslides* 3, 2: 159–73.

Nagel, J. (1994) 'Constructing ethnicity: Creating and recreating ethnic identity and culture', *Social Problems* 41, 1: 152–76.

Nakagawa, Y. and Shaw, R. (2004) 'Social capital: A missing link to disaster recovery', *International Journal of Mass Emergencies and Disasters* 22, 1: 5–34.

Narayan, D., Chambers, R., Shah, M.K. and Petesch, P. (2000) *Voices of the Poor: Crying out for change*, Washington DC/New York: World Bank/Oxford University Press.

Narayan, D. and Petesch, P. (2000) *Voices of the Poor: Crying out for a change*, Washington, DC: World Bank.

NAS (National Academy of Sciences) (2003) *Critical Issues in Weather Modification Research*, Washington, DC: Committee on the Status and Future Directions in US Weather Modification Research and Operations, National Research Council of the National Academy, NAS.

—— (2008) *Severe Space Weather Events: Understanding societal and economic impacts*, Washington, DC: NAS.

—— (2009) *Near-Earth Object Surveys and Hazard Mitigation Strategies: Interim report*, National Academy of Sciences, Washington, DC: NAS.

Nash, R. (1985) 'Sorry, Bambi, but man must enter the forest: Perspectives on the old wilderness and the new', in J.E. Lotan, B.M. Kilgore, W.C. Fischer and R.W. Mutch (eds), *Proceedings of symposium and workshop on wilderness fire*, 1983 November 15–18, Missoula, Montana. General Technical Report INT-182, Forest Service, Intermountain Forest and Range Experiment Station,Ogden, Utah: US Department of Agriculture, pp. 264–8.

Nathan, F. (2008) 'Risk perception, risk management and vulnerability to landslides in the hill slopes in the city of La Paz, Bolivia: A preliminary statement', *Disasters* 32, 3: 337–57.

National Disability Authority (2008) *Promoting Safe Egress and Evacuation for People with Disabilities*, Dublin: National Disability Authority of Ireland.

National Labor Committee (2001) *El Salvador Earthquake: 'From poverty to misery'*, Pittsburgh: National Labor Committee. Online www.nlcnet.org (accessed 5 January 2010).

National Research Council (2003) *Living on an Active Earth: Perspectives on earthquake science*, Committee on the Science of Earthquakes, Washington, DC: National Academies Press.

Naudé, W., Santos-Paulino, A. and McGillivray, M. (2009a) *Vulnerability in Developing Countries*, Tokyo: UNU Press.

Naudé, W., Santos-Paulino, A.U. and McGillivray, M. (2009b) 'Measuring vulnerability: An overview and introduction', *Oxford Development Studies* 37, 3: 183–91.

Naughton, M.P., Henderson, A., Mirabelli, M.C., Kaiser, R., Wilhelm, J.L., Kieszak, S.M., Rubin, C.H. and Mcgeehin, M.A. (2002) 'Heat-related mortality during a 1999 heat wave in Chicago', *American Journal of Preventive Medicine* 22: 221–27.

NCD (US National Council on Disability) (2005) *Saving Lives: Including people with disabilities in emergency planning*, Washington, DC: US-NCD.

NDMC (National Disaster Management Committee, Marshall Islands) (1997a) *Hazard Mitigation Plan*, Majuro: NDMC.

NDMC (National Disaster Management Committee, Samoa) (1997b) *National Disaster Management Plan and Emergency Procedures*, Apia: NDMC.

NDMC (National Drought Mitigation Center) (2006) *Understanding and Defining Drought*, Lincoln: NDMC. Online www.drought.unl.edu/whatis/concept.htm (accessed 7 April 2011).

Neal, J., Bates, P., Fewtrell, T., Hunter, N.M., Wilson, M. and Horritt, M. (2009) 'Hydrodynamic modelling of the Carlisle 2005 urban flood event and comparison with validation data', *Journal of Hydrology* 368: 42–55.

Neumayer, E. and Plümper, T. (2007) 'The gendered nature of natural disasters: The impact of catastrophic events on the gender gap in life expectancy, 1981–2002', *Annals of the Association of American Geographers* 97, 3: 551–66.

Newhall, C.G., Hendley, J.W. and Stauffer, P.H. (1997) 'Benefits of volcano monitoring far outweigh costs – the case of Mount Pinatubo', *US Geological Survey Fact Sheet*, 115–97.

Newhall, C.G. and Punongbayan, R.S. (1996) *Fire and Mud: Eruptions and lahars of Mount Pinatubo, Philippines*, Quezon City and Seattle, WA: Philippine Institute of Volcanology and Seismology and University of Washington Press.

Newhall, C.G. and Self, S. (1982) 'The volcanic explosivity index (VEI): An estimate of explosive magnitude for historical volcanism', *Journal of Geophysical Research* 87, C2: 1231–8.

Ngo, E. (2001) 'When disasters and age collide: Reviewing vulnerability of the elderly', *Natural Hazards Review* 2, 2: 80–89.

Nguyen, K. (2010) 'Haitian women lose out in post-quake "Survival of the Strongest"', *Alertnet*, 29 January. Online www.alertnet.org/db/an_art/57964/2010/00/29-162402-1.htm (accessed 24 August 2010).

NHWG (Natural Hazard Working Group) (2005) *The Role of Science in Physical Natural Hazard Assessment*, Report to the UK Government by the Natural Hazard Working Group, London: HMSO and Department of Trade and Industry.

Nicholls, N. (2003) 'Continued anomalous warming in Australia', *Geophysical Research Letters* 30: 1370.

Ninkovich, D., Sparks, R.S.J. and Ledbetter, M. (1978) 'The exceptional magnitude and intensity of the Toba eruption, Sumatra: An example of the use of deep-sea tephra layers as a geological tool', *Bulletin of Volcanology* 41, 3: 286–98.

Nisbet, E.G. and Piper, D.J.W. (1998), 'Giant submarine landslides', *Nature* 392: 329–30.

Nixdorf-Miller, A., Hunsaker, D.M. and Hunsaker, J.C. (2006) 'Hypothermia and hyperthermia medico-legal investigation of morbidity and mortality from exposure to environmental temperature extremes', *Archives of Pathology and Laboratory Medicine* 130: 1297–304.

Noji, E. (2005) 'Public health issues in disasters', *Critical Care Medicine* 33, Supplement 1: 29–33.

Nord, D.P. (2006) *Communities of Journalism: A history of American newspapers and their readers*, Champaign-Urbana, Illinois: University of Illinois Press.

Norris, F.H. and Elrod, C.L. (2006) 'Psychosocial consequences of disaster: A review of past research', in F.H. Norris, S. Galea, M.J. Friedman and P.J. Watson (eds), *Methods for Disaster Mental Health Research*, New York: Guilford Press, pp. 20–43.

Northridge, M.E., Sclar, E.D. and Biswas, P. (2003) 'Sorting out the connections between the built environment and health: A conceptual framework for navigating pathways and planning healthy cities', *Journal of Urban Health* 80: 556–68.

Norton, A., Conway, T. and Forster, M. (2000) 'Social protection concepts and approaches – implication for policy and practice in international development', in T. Conway, A. de Haan and A. Norton (eds), *Social Protection: New directions of donor agencies*, London: DFID, pp. 5–18.

NSET (National Society for Earthquake Technology) (2008) *National Strategy for Disaster Risk Management in Nepal*, Kathmandu: NSET.

O'Brien, G. (2006) 'UK emergency preparedness: A step in the right direction?', *Journal of International Affairs* 59, 2: 63–85.

O'Brien, G., O'Keefe, P., Rose, J. and Wisner, B. (2008) 'Climate change and disaster management' *Disasters* 30, 1: 64–80.

OECD (Organisation of Economic Co-operation and Development) (2009a) *Integrating Climate Change Adaptation into Development Co-operation*, Paris: Organisation of Economic Co-operation and Development. Online www.oecd.org/dataoecd/45/45/44887764.pdf (accessed 29 December 2009).

—— (2009b) *OECD Development Assistance Committee Tracks Aid in Support of Climate Change Mitigation and Adaptation – Information note – December 2009*, Paris: OECD. Online www.oecd.org/dataoecd/32/31/44275379.pdf (accessed 29 January 2010).

Ofrin, R. and Nelwan, I (2009) 'Disaster risk reduction through strengthened primary health care', *World Health Organization Southeast Asia Regional Health Forum* 13, 1: 29–34.

Ogawa, Y., Fernandez, A.L. and Yoshimura, T. (2005) 'Town watching as a tool for citizen participation in developing countries: Applications in disaster training', *International Journal of Mass Emergencies and Disasters* 23, 2: 5–36.

O'Hare, G. (2001) 'Hurricane 07B in the Godavari Delta, Andhra Pradesh, India: Vulnerability, mitigation and the spatial impact', *The Geographical Journal* 167: 23–38.

O'Keefe, P., Westgate, K. and Wisner, B. (1976) 'Taking the "naturalness" out of "natural" disasters', *Nature* 260, 5552: 566–67.

Oliver-Smith, A. (1979) 'Post-disaster consensus and conflict in a traditional society: The 1970 avalanche of Yungay, Peru', *Mass Emergencies* 4: 39–52.

—— (1986) *The Martyred City: Death and rebirth in the Andes*, Albuquerque: University of New Mexico Press.

—— (1991) 'Success and failures in post-disaster resettlement', *Disasters* 14, 1: 7–19.

—— (1994) 'Peru's five hundred year earthquake: Vulnerability in historical context', in A. Varley (ed.), *Disasters, Development and Environment*, Chichester: John Wiley & Sons, pp. 31–48.

—— (1999) 'Peru's five hundred year earthquake: Vulnerability in historical context', in A. Oliver-Smith and S.M. Hoffman (eds), *The Angry Earth: Disaster in anthropological perspectives*, New York: Routledge, pp. 74–88.

—— (2006) 'Disasters and forced migration in the 21st century', in Social Science Research Council (ed.), *Understanding Katrina*, New York: Social Science Research Council. Online understandingkatrina.ssrc. org/Oliver-Smith (accessed 11 November 2010).

Oliver-Smith, A.S. and Hoffman, S.M. (eds) (2003) *The Angry Earth: Disaster in anthropological perspective*, New York: Routledge.

Olson, R.S. (2000) 'Toward a politics of disaster: Losses, values, agendas, and blame', *International Journal of Mass Emergencies and Disasters* 18, 2: 265–88.

—— (2010) 'From disaster event to political crisis', *International Studies Perspectives* 11: 205–21.

Omenugha, K.A. (2004) 'The Nigerian press and the politics of difference', in J. Hands and E. Siapera (eds), *At the Interface: Continuity and transformation in culture and politics*, Amsterdam: Rodopi B.V.

O'Neill, M.S., Zanobetti, A. and Schwartz, J. (2005) 'Disparities by race in heat-related mortality in four US cities: The role of air conditioning prevalence', *Journal of Urban Health* 82: 191–7.

Öneryildiz v. Turkey (2004) European Court of Human Rights, Application 48939/99, Judgment of 30 November 2004.

Øni, I.O., Adamou, A. and Schulz, E. (2009) 'Landmines, drugs and justice. The recent history of two Saharan mountains (Adrar des Iforas/Mali and Air Mts./Niger)', in R. Baumhauer and J. Runge (eds), *Holocene Palaeoenvironmental History of the Central Sahara*, Boca Raton, FL: CRC Press, pp. 221–38.

Open CRS (2008) 'Hurricane Katrina: Insurance losses and national capacities for financing disaster risks'. Online opencrs.com/document/RL33086/2008-01-31 (accessed 23 May 2010).

Organization of American States (1988) *Incorporating Natural Hazard Mitigation into Project Preparation*, Committee of International Development Institutions on the Environment, CIDIE Publication Series 2, Nairobi, Kenya: CIDIE.

Osofsky, J.D., Osofsky, H.J. and Harris, W.W. (2007) 'Katrina's children: Social policy consideration for children in disasters', *Social Policy Report* 21, 1: 3–18.

Oxfam (2006) 'Making the case: A national drought contingency fund for Kenya', Oxfam Briefing Paper No. 89, Oxford: Oxfam.

—— (2002) *Participatory Capacities and Vulnerabilities Assessment: Finding the link between disasters and development*, Quezon City: Oxfam Great Britain – Philippines Programme. Online www.proventionconsortium. org/themes/default/pdfs/CRA/PCVA_2002_meth.pdf (accessed 19 December 2010).

Özdemir, A. and Delikanli, M. (2009) 'A geotechnical investigation of the retrogressive Yaka landslide and the debris flow threatening the town of Yaka (Isparta, SW Turkey)', *Natural Hazards* 49: 113–36.

Özerdem, A. and Barakat, S (2000) 'After the Mamara earthquake: Lessons for avoiding short cuts to disasters', *Third World Quarterly* 21, 93: 425–39.

PAHO (Pan American Health Organization) (2004) *Safe Hospitals: A collective responsibility – a global measure of disaster reduction*, Washington, DC: PAHO.

—— (2008) *Hospital Safety Index: Guide for evaluators*, Washington, DC: PAHO. Online www.prevention web.net/english/professional/publications/v.php?id=8974 (accessed 18 October 2010).

PAHO-WHO (World Health Organization) (2000) *Natural Disasters: Protecting the public's health*, Washington, DC: PAHO/WHO. Online www.paho.org/English/Ped/sp575.htm (accessed 17 January 2010).

—— (2003) *Guidelines for an Effective Use of the Foreign Field Hospitals*. Online publications.paho.org/product.php? productid=878 (accessed 17 January 2010).

Palermo, M.P. and Engle, D. (2010) 'Rio rescuers scour for new mudslide victims; 173 dead', *Reuters*. Online www.reuters.com/article/idUSTRE6375KZ20100408 (accessed 1 May 2010).

Pampanin, S. (2008) 'Development in seismic design and retrofit of structures: Modern technology built on "ancient wisdom"', in L. Bosher (ed.) *Hazards and the Built Environment: Attaining built-in resilience*, London and New York: Routledge, pp. 96–123.

PANAP (2010) 'Rice: The grain that shapes cultures, traditions, and rituals'. Online www.panap.net/ uploads/media/ricecult.pdf (accessed 27 May 2010).

Pandya, M. (ed.) (2007) *Tsunami Evaluation Coalition Synthesis Report*, ColomboCo: United Nations Development Programme – Disaster Management Programme.

PANOS (2010) Panos Institute. Online www.panos.org.

Parnell, S., Simon, D. and Vogel, C. (2007) 'Global environmental change: Conceptualising the growing challenge for cities in poor countries', *Area* 39, 3: 357–69.

Parr, A.R. (1987) 'Disasters and disabled persons: An examination of the safety needs of a neglected minority', *Disasters* 11: 148–59.

Parry, M., Evans, A., Rosegrant, M.W. and Wheeler, T. (2009) *Climate Change and Hunger: Responding to the challenge*, Rome: World Food Programme.

Pasos, R. (1994) *El Ultimo Despale: La frontera agrícola Centroamericana*, San José: FUNDESCA, European Union, GRET and UNEP.

Passioura, J.B. (1996) 'Drought and drought tolerance', *Plant Growth Regulation* 20: 79–83.

Pattiaratchi, C. (2005) 'Tsunami impacts on Sri Lanka: Lessons for disaster reduction on coasts', Paper presented at International Symposium on Disaster Reduction on Coasts, Scientific-Sustainable-Holistic-Accessible, Melbourne, November. Online civil.eng.monash.edu.au/drc/symposium-papers/drc128-pattiaratchi.pdf (accessed 11 June 2010).

Pattullo, P. (2000) *Fire from the Mountain: The tragedy of Montserrat and the betrayal of its people*, London: Constable & Robinson.

Paudel, G.S. (2002) 'Coping with land scarcity: Farmers' changing land-use and management practices in two mountain watersheds of Nepal', *Norwegian Journal of Geography* 56: 21–31.

Paul, B.K. and Bhuiyan, R.H. (2009) 'Urban earthquake hazard: Perceived seismic risk and preparedness in Dhaka City, Bangladesh', *Disasters* 34, 2: 289–591.

PDRI (Participatory Research and Development Initiative) (2009) *Front-liners' Views on Disaster Risk Reduction in Bangladesh*, Draft Report, Dhaka: Participatory Research and Development Initiative.

Peacock, W.G. and Girard, C. (1997) 'Ethnic and racial inequalities in hurricane damage and insurance settlements', in W.G. Peacock, B.H. Morrow and H. Gladwin (eds), *Hurricane Andrew: Ethnicity, gender and the sociology of disasters*, London: Routledge, pp. 171–90.

Peckley, D.C., Bagtang, E.T. and Zarco, M.A.H. (2010) 'Development of a non-expert tool for site specific evaluation of landslide susceptibility', in D.W. Eka Putra and W. Wilopo (eds), *Protecting Life from Geo-Disaster and Environmental Hazards*, Proceeding of International Symposium on Geo-Disaster Mitigation in ASEAN and the 2nd AUN/SEED-Net, Bali, Indonesia, pp. 165–72.

Peduzzi, P., Dao, H., Herold, C. and Mouton, F. (2009) 'Assessing global exposure and vulnerability towards natural hazards: The disaster risk index', *Natural Hazards and Earth System Sciences* 9: 1149–59.

Peek, L. (2008) 'Children and disasters: Understanding vulnerability, developing capacities, and promoting resilience: An introduction', *Children, Youth and Environments* 18, 1: 1–29.

Peek, L. and Stough, L.M. (2010) 'Children with disabilities in the context of disaster: A social vulnerability perspective', *Child Development* 81, 4: 1260–70.

Peel, S. (1977). 'Practical relief and preventive methods', *Disasters* 1, 2: 179–97.

Pei, M. (2007) *Corruption Threatens China's Future*, Policy Brief No. 55, Washington, DC: Carnegie Endowment for International Peace. Online www.carnegieendowment.org/publications/index.cfm?fa=view&id=19628&prog=zch (accessed 17 December 2010).

Pekovic, V., Seff, L. and Rothman, M. (2007) 'Planning for and responding to special needs of elders in natural disasters', *Generations* 31: 37–41.

Pelling, M. (1999) 'The political ecology of flood hazard in urban Guyana', *Geoforum* 30: 249–61.

—— (2003) *The Vulnerability of Cities: Natural disasters and social resilience*, London: Earthscan.

—— (2004) *Visions of Risk: A review of international indicators of disaster risk and its management*, A report for the ISDR Inter-Agency Task Force on Disaster Reduction, Working Group 3: Risk, Vulnerability and Disaster Impact Assessment, Geneva: International Strategy for Disaster Reduction and UNDP.

—— (2007) 'Learning from others: The scope and challenges for participatory disaster risk management', *Disasters* 31, 4: 373–85.

Pelling, M. and Dill, K. (2009a) 'Disaster politics: Tipping points for change in the adaptation of socio-political regimes', *Progress in Human Geography* 34, 5: 1–17.

—— (2009b) *'Natural' disasters as catalysts of political action*, Chatham House ISP/NSC Briefing Paper 06/01, London: Chatham House.

Pelling, M. and Holloway, A. (2006) *Legislation for Mainstreaming Disaster Risk Reduction*, Teddington: Tearfund. Online www.tearfund.org/webdocs/website/Campaigning/Policy%20and%20research/DRR%20legislation.pdf (accessed 22 December 2009).

Pelling, M. and Uitto, J. (2001) 'Small island developing states: Natural disaster vulnerability and global change', *Environmental Hazards* 3, 2: 49–62.

Pelling, M. and Wisner, B. (eds) (2009) *Disaster Risk Reduction: Cases from urban Africa*, London: Earthscan.

Perrow, C. (1998) *Normal Accidents*, 2nd edn, Princeton, NJ: Princeton University Press.

—— (2006) *The Next Catastrophe*, Princeton, NJ: Princeton University Press.

Perry, B.D., Randolph, T.F., Ashley, S., Chimedza, R., Forman, T., Morrison, J., Poulton, C., Sibanda, L., Stevens, C., Nebele, T. and Yngström, I. (2003) *The Impact and Poverty Reduction Implications of Foot and Mouth Disease Control in Southern Africa with Special Reference to Zimbabwe*, Nairobi: International Livestock Research Institute.

Perry, R. and Quarantelli, E. (2005) *What is a Disaster?* New York: Xlibris.

Petal, M. (2009) 'Education in disaster risk reduction', in R. Shaw and R.R. Krishnamurthy (eds), *Disaster Management Global Challenges and Local Solutions*, India: University Press, pp. 285–320.

Petal, M., Green, R., Kelman, I., Shaw, R. and Dixit, A. (2008) 'Community-based construction for disaster risk reduction', in L. Bosher (ed.), *Hazards and the Built Environment: Attaining built-in resilience*, London: Routledge, pp. 191–217.

Peters, M. (1996) 'Hospitals respond to water loss during the Midwest Floods of 1993: Preparedness and improvisation', *Journal of Emergency Medicine* 14, 3: 345–50.

Peters-Guarin, G., McCall, M.K. and van Westen, C. (2011) 'Coping strategies and manageability: How participatory geographical information systems can transform local knowledge into better policies for disaster risk management', *Disasters*, in press.

Petley, D. (2008) 'International symposium of landslides: Wenchuan (Sichuan) earthquake landslides', *Dave's landslide blog*, International Landslide Centre, Durham University. Online daveslandslideblog.blogspot.com/2008/06/international-symposium-of-landslides.html (accessed 6 June 2010).

Pew (Pew Research Center) (2008) 'Foreign disasters attract interest despite modest coverage', Pew Research Center. Online people-press.org/report/423/foreign-disasters-attract-interest-despite-modest-coverage (accessed 14 December 2010).

Phifer, J., Kaniasty, K. and Norris, F. (1988) 'The impact of natural disaster on the health of older adults: A multiwave prospective study', *Journal of Health and Social Behavior* 29: 65–78.

PIANC (2001) *Seismic Design Guidelines for Port Structures*, Working Group No. 34 of the Maritime Navigation Commission, International Navigation Association, Lisse, Belgium: A.A. Balkema Publishers.

Pilger, J. (ed.) (2005) *Tell Me No Lies: Investigative journalism that changed the world*, New York: Basic Books.

Pincha, C. (2008) *Gender Sensitive Disaster Management: A toolkit for practitioners*, Mumbai: Earthworm Books, for Oxfam America & NANBAN Trust. Online thinkbeyondboundaries.org/index2.php?option=com_docman&task=doc_view&gid=7&Itemid=38 (accessed 14 December 2010).

Pincha, C. and Krishna, N.H. (2009) 'Post disaster death ex gratia payments and their gendered impact: Implications for practice and policies', *Regional Development Dialogue* 30, 1: 95–105.

Pinder, C. (2008) 'Including the sustainable livelihood approach in international and national development frameworks', Livelihoods Connect Discussion Paper, Brighton: Institute for Development Studies.

Pisani, E. (2008) *The Wisdom of Whores: Bureaucrats, brothels, and the business of AIDS*, New York: WW Norton.

Plan International and World Vision (2009) *Children on the Front Line: Children and young people in disaster risk reduction*, London and Federal Way, WA: Plan International and World Vision.

PMPB-Community Association for Disaster Management (2007) 'Combining science and indigenous knowledge to build a community early warning system', in UNISDR and UNDP (eds), *Building Disaster Resilient Communities: Good practices and lessons learned*, Geneva: UNISDR, pp. 26–8.

Polanyi, M. (1958) *Personal Knowledge: Towards a post critical philosophy*, Chicago: Chicago University Press.

Pollet, J. and Omi, P.N. (2002) 'Effect of thinning and prescribed burning on crown fire severity in ponderosa pine forests', *International Journal of Wildland Fire* 11, 1: 1–10.

Polman, L. (2010) *The Crisis Caravan: What's wrong with humanitarian aid*, New York: Metropolitan.

Popescu, M.H. and Sasahara, K. (2009) 'Engineering measures for landslide disaster mitigation', in K. Sassa and P. Canuti (eds), *Landslides: Disaster risk reduction*, Berlin: Springer-Verlag, pp. 609–31.

Porphyrios, T. (1971) 'Traditional earthquake-resistant construction on a Greek island', *Journal of the Society of Architectural Historians* 30, 1: 31–9.

Poumadere, M., Mays, C., Le Mer, S. and Blong, R. (2005) 'The 2003 heat wave in France: Dangerous climate change here and now', *Risk Analysis* 25: 1483–94.

PPEW-ISDR (Platform for the Promotion of Early Warning – International Strategy for Disaster Reduction) (2006a) 'Statement made by former President William B. Clinton on the occasion of the Third International Early Warning Conference', Bonn: PPEW-ISDR. Online www.ewc3.org/upload/downloads/Clinton.pdf (accessed 18 November 2009).

—— (2006b) *Global Survey of Early Warning Systems: An assessment of capacities, gaps, and opportunities towards building a comprehensive global early warning system for all natural hazards*, Bonn: PPEW-ISDR. Online www.unisdr.org/ppew/info-resources/ewc3/Global-Survey-of-Early-Warning-Systems.pdf (accessed 23 November 2009).

—— (2006c) *Developing Early Warning Systems: A checklist*, Bonn: PPEW-ISDR. Online www.unisdr.org/ppew/inforesources/ewc3/checklist/English.pdf (accessed 23 November 2009).

Practical Action (Bangladesh) (2007) 'Voluntary formation of community organizations to implement DRR', in UNISDR and UNDP (eds), *Building Disaster Resilient Communities: Good practices and lessons learned*, Geneva: United National International Strategy for Disaster Reduction, pp. 6–8.

Pradesh, U. (2001) *Why Some Village Water and Sanitation Committees are Better than Others*, New Delhi: Water and Sanitation Program – South Asia.

Pratt, N., Bird, J., Taylor, R. and Carter, R. (1997) 'Estimating areas of land under small-scale irrigation using satellite imagery and ground data for a study area in N.E. Nigeria', *The Geographical Journal* 163, 1: 65–77.

Pretty, J. and Ward, H. (2001) 'Social capital and the environment', *World Development* 29: 209–27.

Preuss, J. (2005) 'Europe's flood disaster of 2002: Vienna's evolving flood mitigation projects', *Quick Response Research Report 175*, Natural Hazards Center, Boulder, CO. Online www.colorado.edu/hazards/research/qr/qr175/qr175.html (accessed 9 October 2006).

ProVention Consortium (2010) 'Community disaster resilience fund'. Online www.proventionconsortium.org/?pageid=32&projectid=34 (accessed 1 December 2010).

Provincial Emergency Program (2000) *British Columbia Emergency Response Management System Overview, Interim, Based on Operations and Management Standard 1000*, Victoria: Provincial Emergency Program, British Columbia Ministry of Public Safety and Solicitor General.

Prucha, F.P (1984) *The Great Father: The United States government and the American Indians*, Lincoln: University of Nebraska Press.

The Psychosocial Working Group (2004) *Considerations in Planning Psychosocial Programmes*, Edinburgh: Queen Margaret University College.

Quarantelli, E.L. (1983) *Delivery of Emergency Medical Services in Disasters: Assumptions and realities*, New York: Irvington.

—— (1986–87) 'Le jour où le désastre frappera vous serez admirable', *Le Temps Stratégique*, Winter: 75–80.

—— (1995) 'Patterns of shelter and housing in US disasters', *Disaster Prevention and Management* 4, 3: 43–53.

—— (ed.) (1998) *What is a Disaster?* New York: Routledge.

Quarantelli, E.L. and Dynes, R.R. (1972) 'When disaster strikes: It isn't much like what you've heard and read about', *Psychology Today* 5, 9: 66–70.

Quinn, T. (2008) *Flu: A social history of influenza*, London: New Holland Publishers.

Raby, R.L. (2006) *Democracy and Revolution: Latin America and socialism today*. London: Pluto.

Rahimi, M. (1993) 'An examination of behaviour and hazards faced by physically disabled people during the Loma Prieta earthquake', *Natural Hazards* 7: 59–82.

Ramirez Gomez, F. and Cardona, O.D. (1996) 'El sistema nacional de prevención y atención de desastres de Colombia', in A. Lavell and Franco, E. (eds), *Estado, Sociedad y Gestión de los Desastres: En Búsqueda del Paradigma Perdida*, Lima: Peru: LA RED, FLACSO, ITDG, pp. 214–55.

Ramsey, D. (2002) 'The role of music in environmental education: Lessons from the cod fishery crisis and the dust bowl days', *Canadian Journal of Environmental Education* 7, 1: 183–98.

Rappaport, E.N. (2000) 'Loss of life in the United States associated with recent tropical Atlantic cyclones', *Bulletin of American Meteorological Society* 81, 9: 2065–73.

Rashid, S.F. (2000) 'The urban poor in Dhaka City: Their struggles and coping strategies during the floods of 1998', *Disasters* 24, 3: 240–53.

Rawal, V. 'Intermediate semi-permanent shelters: An active precursor to owner driven post-disaster reconstruction', Paper presented at National Conference on Owner Driven Reconstruction, New Delhi, January 2009.

Razal, R.A., Tolentino, E.L.T., Carandang, W.M., Nghia, N.H., Hao, P.S. and Luoma-Aho, T. (2005) *Status of Genetic Resources of Pinus merkusii (Jungh et De Vriese) and Pinus kesiya (Royle ex Gordon) in Southeast Asia*. Online www.apforgen.org/Edited%20final%20pine%20project%20report.PDF (accessed 5 May 2010).

RCC (Regional Consultative Committee) (2007) *Integrating Disaster Risk Reduction into School Curriculum*, Bangkok: Asian Disaster Preparedness Center.

Redclift, M.R. (2006) 'Sustainable development (1987–2005): An oxymoron comes of age', *Horizontes Antropológicos* 12: 65–84.

Reddy, A., Sharma, V.S. and Chitoor, M. (2000) *Cyclones in Andhra Pradesh: A multidisciplinary study to profile cyclone response in coastal Andhra Pradesh, India*, Hyderabad: Booksline.

Regalsky, P. and Hosse, T. (2008) *Andean Indigenous Small Farmer Strategies for Climate Risk Reduction: State-of-the-art and progress of research in the Bolivian Andes*, London: CENDA-CAFOD.

Regel, S. (2007) 'Resilience in trauma and disaster', in B. Monroe and D. Olivere (eds), *Resilience in Palliative Care: Achievement in adversity*, Oxford: Oxford University Press, pp. 261–80.

Reid, H., Alam, M., Berger, R., Cannon, T. and Milligan, A. (eds) (2009) *Community-Based Adaptation to Climate Change*, Participatory Learning and Action PLA No. 60, London: IIED. Online www.iied.org/pubs/display.php?o=14573IIED&n=2&l=445&c=part (accessed 14 December 2010).

Reij, C., Scoones, I. and Toulmin, C. (eds) (1996) *Sustaining the Soil: Indigenous soil and water conservation in Africa*, London: Earthscan.

Resilient Earth (The) (2010) 'Himalayan glaciers not melting'. Online www.theresilientearth.com/?q=content/himalayan-glaciers-not-melting (accessed 11 April 2010).

Ribot, J. (2003) 'Democratic decentralization of natural resources: Institutional choice and discretionary power transfers in sub-Saharan Africa', *Public Administration and Development* 23, 1: 53–65.

Rich, A. (1980) 'Compulsory heterosexuality and lesbian existence', *Signs* 5, 4: 631–60.

Richards, P. (1986) *Coping with Hunger: Hazard and experiment in an African rice-farming system*, London: Allen & Unwin.

—— (1996) *Fighting for the Forest: War, youth and resources in Sierra Leone*, Oxford: James Currey.

Ridgeway, J. (1994) *The Haiti Files*, Washington: Essential Books/Azul Edition.

Riehl, H. (1979) *Climate and Weather in the Tropics*, London: Academic Press.

Ripley, A. (2008) *The Unthinkable: Who survives when disaster strikes – and why*, New York: Crown Publishers.

Risk RED for Earthquake County Alliance (2009) *School Disaster Readiness: Lessons from the first great southern California shakeout*, Los Angeles: Risk RED for Earthquake County Alliance.

Rist, S. and Dahdouh-Guebas, F. (2006) 'Ethnosciences: A step towards the integration of scientific and indigenous forms of knowledge in the management of natural resources for the future', *Environment, Development and Sustainability* 8, 4: 467–93.

Roberts, D. (2008) 'Thinking globally, acting locally: Institutionalizing climate change at the local government level in Durban, South Africa', *Environment and Urbanization* 20, 2: 521–37.

Roberts, L., Lafta, R., Garfield, R., Khudhairi, J. and Burnham, G. (2004) 'Mortality before and after the 2003 invasion of Iraq: Cluster sample survey', *The Lancet* 364, 9448: 1857–64.

Roberts, L. and Toole, M. (1995) 'Cholera deaths in Goma', *The Lancet* 346, 8987: 1431.

Robertson, G., Fernandez, R., Fisher, D, Lee, B. and Stasko, J. (2008) 'Effectiveness of animation in trend visualization', *IEEE Transactions on Visualization and Computer Graphics* 14, 6: 1325–32.

Robinson, M. (1994) 'Governance, democracy and conditionality: NGOs and the new policy agenda', in A. Clayton (ed.), *Governance, Democracy and Conditionality: What role for NGOs?*, Oxford: INTRAC, pp. 35–52.

—— (1997) 'Privatizing the voluntary sector: NGOs as public service contractors?' in D. Hulme and M. Edwards (eds), *NGOs, States and Donors: Too close for comfort?* London: McMillan, pp. 59–78.

Rodríguez, H. and Dynes, R. (2006) 'Finding and framing Katrina: The social construction of disaster', in Social Science Research Council (ed.), *Understanding Katrina*, New York: Social Science Research Council. Online understandingkatrina.ssrc.org/Dynes_Rodriguez (accessed 11 November 2010).

Rodriguez, S.R., Tocco, J.S., Mallonee, S., Smithee, L., Cathey, T. and Bradley, K. (2006) 'Rapid needs assessment of Hurricane Katrina evacuees – Oklahoma, September 2005'. *Prehospital and Disaster Medicine* 21, 6: 390–95.

Roeder, P.L. and Taylor, W.P. (2007) 'Mass vaccination and herd immunity: Cattle and buffalo', *Revue Scientifique et Technique (International Office of Epizootics)* 26, 1: 253–63.

Rooney, C. and White, G.W. (2007) 'Consumer perspective: A narrative analysis of a disaster preparedness and emergency response survey from persons with mobility impairments', *Journal of Disability Policy Studies* 17: 206–15.

Ropelewski, C.F. and Folland, C.F. (2000) 'Prospects for the prediction of meteorological drought', in D.A. Wilhite (ed.), *Drought: A global assessment – Vol. 1*, London: Routledge, pp. 21–41.

Rosenberg, C.E. (1992) 'Framing disease: Illness, society and history', in C.E. Rosenberg (eds), *Explaining Epidemics and Other Stories in the History of Medicine*, Cambridge: Cambridge University Press, pp. 305–18.

Rothstein, B. (2005) *Social Traps and the Problem of Trust*, Cambridge: Cambridge University Press.

Rowland, J.L., Fox, M.H., White, G.W. and Rooney, C. (2007) 'Emergency response training practices for people with disabilities: Analysis of some current practices and recommendations for future training programs', *Journal of Disability Policy Studies* 17: 216–22.

Royal Society of Edinburgh (2002) *Inquiry Into Foot and Mouth Disease in Scotland*. Online www.rse.org.uk/enquiries/footandmouth/fm_mw.pdf (accessed 28 April 2010).

Sabates-Wheeler, R. and Devereux, S. (2007) 'Social protection for transformation', *IDS Bulletin* 38, 3: 23–8.

Sagala, S.A.H. (2009) 'Systems analysis of social resilience against volcanic risks: Case studies of Mt. Merapi, Indonesia and Mt. Sakurajima, Japan', unpublished PhD thesis, Department of Urban Management, Graduate school of Engineering, Kyoto University.

Saito, K. (2009) 'High-resolution optical satellite images for post-earthquake damage assessment', unpublished PhD thesis, Department of Architecture, University of Cambridge.

Salem-Pickartz, J. (2009) 'Surviving in Gaza', *JO Magazine*, 1 March.

Sánchez-Rodríguez, R. (2008) 'Urban sustainability and global environmental change: Reflections for an urban agenda', in G. Martine, G. McGranahan, M. Montgomery and R. Fernández-Castilla (eds), *The New Global Frontier: Urbanization, poverty and environment in the 21st century*, London: Earthscan, pp. 149–63.

Sánchez-Rodríguez, R., Seto, K.C., Simon, D., Solecki, W.D., Kraas, F. and Laumann, G., (2005) *Science Plan, Urbanization and Global Environmental Change*, IHDP Report 15, Bonn: International Human Dimensions Programme on Global Environmental Change. Online www.ugec.org.

Sandberg, A. (2010) 'Institutional challenges to robustness of delta and floodplain agricultural systems', *Environmental Hazards* 9: 284–300.

Sanderson, D. (2000) 'Cities, disasters and livelihoods', *Environment and Urbanization* 12, 2: 93–102.

—— (2009) 'Integrating development and disaster management concepts to reduce vulnerability in low income urban settlements', unpublished PhD thesis, Oxford Brookes University, Oxford.

Sanderson, L. (1997) 'Fires', in E. Noji (ed.), *The Public Health Consequences of Disasters*, New York: Oxford University Press, pp. 373–96.

San Mateo County Health Services Agency (1989) *HEICS: The Hospital Emergency Incident Command System*, San Mateo County Health Services Agency – Emergency Medical Services. Online www.heics.com/index.html (accessed 5 October 2010).

Santiago III, C., Francia, F.F. and Brusola, M.M. (2009) *Evacuation Skills Showcased in Calabanga Barangay*. Online www.preventionweb.net/english/professional/news/v.php?id=10640 (accessed 3 March 2010).

Sardá, R., Mora, J. and Avila, C. (2005) 'Tourism development in the Costa Brava (Girona, Spain): How integrated coastal zone management may rejuvenate its lifecycle', in J. Vermaat, W. Salomons, L. Bouwer and K. Turner (eds), *Managing European Coasts: Past, present and future*, Berlin: Springer, pp. 291–314.

Sathirathai, S. and Barbier, E. (2001) 'Valuing mangrove conservation in southern Thailand', *Contemporary Economic Policy* 19, 2: 109–22.

Satterthwaite, D. (2007) 'The transition to a predominantly urban world and its underpinnings', *Human Settlements Discussion Paper Series*, Urban Change – 4, London: International Institute for Environment and Development. Online pubs.iied.org/pdfs/10550IIED.pdf (accessed 6 November 2010).

Sauer, C.O. (1952) *Agricultural Origin and Dispersals*, The American Geographical Society, Series Two, New York: George Grady Press.

Saul, M. and Austen, R. (2010) *Viewing African Film in the Twentieth-Century: Art films and the Nollywood video revolution*, Columbus, OH: University of Ohio Press.

Saunders, W.S.A., Forsyth, J., Johnston, D.J. and Becker, J. (2007) 'Strengthening linkages between land-use planning and emergency management in New Zealand', *Australian Journal of Emergency Management* 22, 1: 36–43.

Saunders, W.S.A. and Glassey, P. (2007) *Guidelines for Assessing Planning Policy and Consent Requirements for Landslide Prone Land*, GNS Science Miscellaneous Series 7, Lower Hutt, New Zealand: GNS Science.

Savage, K. and Harvey, P. (2007) 'Remittances during crises: Implications for humanitarian response', London: Oversees Development Institute. Online www.odi.org.uk/resources/download/228.pdf (accessed 18 March 2010).

Save the Children (2008) *In the Face of Disaster: Children and climate change*, London: Save the Children.

Sawada, T. (1992) 'A report from Tent City: eExperience in medical relief', in H. Shimizu (ed.), *After the Eruption: Pinatubo Aetas at the crisis of their survival*, Tokyo: Foundation for Human Rights in Asia, pp. 63–73.

Scarpa, R. and Tilling, R.I. (1996) *Monitoring and Mitigation of Volcanic Hazards*, Berlin: Springer-Verlag.

Scarth, A. (2002) *La Catastrophe: Mont Pelée and the destruction of Saint-Pierre*, London: Terra Publishing.

Scawthorn, C. (2008) 'A brief history of seismic risk assessment', in A. Bostrom, S.P. French and S. Gottlieb (eds), *Risk Assessment, Modeling and Decision Support*, Berlin: Springer, pp. 5–82.

Scha, R. (2005) 'Natural disasters', *Radical Art Info*. Online radicalart.info/destruction/NaturalDisasters/index.html (accessed 20 October 2010).

Schacher, T. (2009) *Confined Masonry for One and Two Storey Buildings in Low-Tech Environments: A guidebook for technicians and artisans*, National Information Centre for Earthquake Engineering, Indian Institute of Technology, Kanpur, India.

Schaetzl, R. and Anderson, S. (2005) *Soils: Genesis and geomorphology*, Cambridge: Cambridge University Press.

Schipper, L. and Pelling, M. (2006) 'Disaster risk, climate change and international development: Scope for, and challenges to, integration', *Disasters* 30, 1: 19–38.

Schmuck-Widmann, H. (1996) *Living with the Floods: Survival strategies of char-dwellers in Bangladesh*, Berlin: FDCL.

—— (2001) *Facing the Jamuna River: Indigenous and engineering knowledge in Bangladesh*, Dhaka: Bangladesh Centre for Indigenous Knowledge.

Schoch-Spana, M. (2005) 'Public response to extreme events: Top 5 disaster myths', Resources for the Future, Washington, DC, Homeland Security, Environment and the Public, First Wednesday Seminar, 5 October. Online www.paho.org/english/dd/pin/Number21_article01.htm (accessed 4 December 2010).

Schoeder, J.M. and Polusny, M.A. (2004) 'Risk factors for adolescent alcohol use following a natural disaster', *Prehospital & Disaster Medicine* 19, 1: 122–7.

Schoenbaum, S.C. (2001) 'The impact of pandemic influenza, with special reference to 1918', *International Congress Series* 1219, October: 43–51.

Schreier, H., Shah, P.B., Lavkulich, L.M. and Brown, S. (1994) 'Maintaining soil fertility under increasing land use pressure in the middle mountains of Nepal', *Soil Use and Management* 10: 137–42.

Schroeder, R., Martin, K.S., Wilson, B. and Sen, D. (2008) 'Third world environmental justice', *Society and Natural Resources* 21: 547–55.

Schulte, P., Alegret, L., Arenillas, I., Arz, J.A., Barton, P.J., Bown, P.R., Bralower, T.J., Christeson, G.L., Claeys, P., Cockell, C.S., Collins, G.S., Deutsch, A., Goldin, T.J., Goto, K., Grajales-Nishimura, J.M., Grieve, R.A.F., Gulick, S.P.S., Johnson, K.R., Kiessling, W., Koeberl, C., Kring, D.A., MacLeod, K.G., Matsui, T., Melosh, J., Montanari, A., Morgan, J.V., Neal, C.R., Nichols, D.J., Norris, R.D., Pierazzo, E., Ravizza, G., Rebolledo-Vieyra, M., Reimold, W.U., Robin, E., Salge, T., Speijer, R.P., Sweet, A.R., Urrutia-Fucugauchi, J., Vajda, V., Whalen, M.T. and Willumsen, P.S. (2010) 'The Chicxulub asteroid impact and mass extinction at the Cretaceous–Paleogene boundary', *Science* 327: 1214–18.

Schultz, J.M., Russell, J. and Espinel, Z. (2005) 'Epidemiology of tropical cyclones: The dynamics of disaster, disease, and development', *Epidemiologic Reviews* 27: 21–35.

Schumann, G., Bates, P.D., Horritt, M.S., Matgen, P. and Pappenberger, F. (2009) 'Progress in integration of remote sensing-derived flood extent and stage data and hydraulic models', *Reviews of Geophysics* 47: RG4001.

Schuster, R. and Highland, L.M. (2007) 'The third Hans Cloos lecture: Urban landslides: Socioeconomic impacts and overview of mitigative strategies', *Bulletin of Engineering Geology and the Environment* 66: 1–27.

Schuster, R.L., Salcedo, D.A. and Valenzuela, L. (2002) 'Overview of catastrophic landslides of South America in the twentieth century', in S.G. Evans and J.V. de Graff (eds), *Catastrophic Landslides: Effects, occurrence, and mechanisms*, Reviews in Engineering Geology No. 15, Boulder, CO: Geological Society of America, pp. 1–34.

Scoones, I. (1998) 'Sustainable rural livelihoods: A framework for analysis', Working Paper 72. Brighton: Institute of Development Studies

—— (2001) 'Transforming soils: The dynamics of soil-fertility management in Africa', in I. Scoones (ed.), *Dynamics and Diversity: Soil fertility and farming livelihoods in Africa*, London: Earthscan, pp. 1–44.

Scott, J.C. (1998) *Seeing Like a State: How certain schemes to improve the human condition have failed*, New Haven: Yale University Press.

SEARO (WHO Regional Office in Southeast Asia) (2006) *Guidelines for Seismic Vulnerability Assessment Hospitals*. Online www.searo.who.int/LinkFiles/Nepal-EPR_Publications_NonStructural_Vulnerability_Assessment.pdf, full guidelines source, www.searo.who.int/en/Section23/Section1108/Section2019/Section2021.htm (accessed 6 June 2010).

Secretariat of the Convention on Biological Diversity (2009) *Connecting Biodiversity and Climate Change Mitigation and Adaptation: Report of the second ad-hoc technical expert group on biodiversity and climate change*, Montreal: Secretariat of the Convention on Biological Diversity.

Seed, D. (1999) *American Science Fiction and the Cold War*, New York: Routledge.

SEEP (Small Enterprise Education and Promotion) Network (2009) *Minimum Standards for Economic Recovery after a Crisis*, Washington, DC: SEEP Network.

Seicshnaydre, S. (2010) 'Postcards from post-Katrina New Orleans: Why government-assisted housing seems destined to perpetuate racial segregation and what can be done about it', written testimony to the US House Sub-Committee on Constitution, Civil Rights and Civil Liberties, 29 July. Online judiciary.house.gov/hearings/pdf/Seicshnaydre100729.pdf (accessed 20 October 2010).

Sellström, T., Wohlgemuth, L., Dupont, P. and Schiebe, K.A. (1996) 'The international response to conflict and genocide: Lessons from the Rwanda experience', *Journal of Humanitarian Assistance*, 14 April. Online www.reliefweb.int/library/nordic/book1/pb020.html (accessed 1 December 2010).

Sen, A. (1981) *Poverty and Famines: An essay on entitlement and deprivation*, Oxford: Clarendon Press.

—— (1990) 'Individual freedom as a social commitment', *New York Review of Books*, 14 June. Online www.nybooks.com/articles/archives/1990/jun/14/individual-freedom-as-a-social-commitment/(accessed 14 December 2010).

—— (1999) *Development as Freedom*, New York: Anchor Books.

Sengezer, B. and Koç, E. (2005) 'A critical analysis of earthquakes and urban planning in Turkey', *Disasters* 29, 2: 171–94.

Shackleton, C.M. and Scholes, R.J. (2000) 'Impact of fire frequency on woody community structure and soil nutrients in the Kruger National Park', *Koedoe* 43, 1: 75–81.

Shah, A. (2005) 'Difference in media coverage', *Global Issues*. Online www.globalissues.org/article/568/media-and-natural-disasters#Differencesinmediacoverage (accessed 14 December 2010).

—— (2010) 'Democratic Republic of Congo', *Global Issues*. Online www.globalissues.org/article/87/the-democratic-republic-of-congo (accessed 3 September 2010).

Sharp, T., Brennan, R., Keim, M., Williams, R., Eitzen, E. and Lillibridge, S. (1998) 'Medical preparedness for a terrorist incident during the Atlanta Olympics', *Annals of Emergency Medicine* 32, 2: 214–23.

Shaw, R., Uy, N. and Baumwoll, J. (eds) (2008) *Indigenous Knowledge for Disaster Risk Reduction: Good practices and lessons learnt from the Asia-Pacific region*, Bangkok: UNISDR.

Shelter Centre (2008) *Transitional Settlement and Reconstruction after Natural Disasters: Field edition*, Geneva: Shelter Centre.

Sheridan, S. (2007) 'A survey of public perception and response to heat warnings across four North American cities: An evaluation of municipal effectiveness', *International Journal of Biometeorology* 52: 3–15.

Shi, P, (2008) 'China Wenchuan earthquake disaster (May 12, 2008) and its loss assessment', Paper presented at the *8th IIASA-DPRI Conference on Integrated Disaster Risk Management*, Induno-Olona, Varese, Italy, 1–2 September.

Shiva, V. (2000) *Stolen Harvest: The hijacking of the global food supply*, Cambridge, MA: South End Press.

—— (2002) *Water Wars: Privatization, pollution and profit*, Toronto: Between the Lines.

Shrestha, R.K. (1992) 'Agro-ecosystem of the mid-hills', in J.B. Abington (ed.), *Sustainable Livestock Production in the Mountain Agro-Ecosystems of Nepal*, Rome: FAO. Online www.fao.org/docrep/004/t0706e/T0706E02.htm#ch2 (accessed 5 October 2010).

Sian, T.A. (2000) 'Malaria control in Malaysia', *Mekong Malaria Forum: Information exchange on malaria control in Southeast Asia* 5: 6–9.

Siapera, E. and Hands, J. (2004) 'Introduction', in J. Hands and E. Siapera (eds), *At the Interface: Continuity and transformation in culture and politics*, Amsterdam: Rodopi B.V., pp. ix–xii.

Siddique, A.K., Salam, A., Islam, M.S., Akram, K., Majumdar, R.N., Zaman, K., Fronczak, N. and Laston, S. (1995) 'Why treatment centres failed to prevent cholera deaths among Rwandan refugees in Goma, Zaire', *The Lancet* 345, 8946: 359–61.

Sidel, V., Onel, E. and Geiger, H. (1992) 'Public health responses to natural and human-made disasters', in J.M. Last and R.B. Wallace (eds), *Maxcy-Rosenau-Lastublic Health and Preventative Medicine*, 13th edn, Norwalk, CT: Appleton & Lange, pp. 118–34.

Sidner, S. (2009) 'Desperate farmers sell wives to pay debts in rural India', *CNN.com*. Online www.cnn.com/2009/WORLD/asiapcf/10/24/intl.india.farmers.selling.wives/index.html (accessed 18 March 2010).

Siegel, M. (1998) 'Mass media anti smoking campaigns: A powerful tool for health promotion', *Annals of Internal Medicine* 129, 2: 128–32.

Sierra Leone Red Cross (2004) *Vulnerability and Capacity Assessment (VCA) for 19 Communities in Kono and Tonkolili Districts*, Community Reintegration and Development Project (CRDP), Sierra Leone Red Cross, Freetown, Sierra Leone. Online www.proventionconsortium.org/themes/default/pdfs/CRA/Sierra_Leone.pdf (accessed 16 October 2010).

Sigurdsson, H. (1999) *Melting the Earth: The history of ideas on volcanic eruptions*, Oxford: Oxford University Press.

Sigurdsson, H., Houghton, B.F., McNutt, S.R., Rymer, H. and Stix, J. (eds) (2000) *Encyclopedia of Volcanoes*, San Diego: Academic Press.

Sillitoe, P. (1998) 'The development of indigenous knowledge', *Current Anthropology* 39, 2: 223–52.

Silverstein, L.M. (2008) 'Guidelines for gender-sensitive disaster management: A revolutionary document', by Asia Pacific Forum on Women, Law and Development, *Reproductive Health Matters* 16, 31: 153–8.

Simon, D. (2007) 'Cities and global environmental change: Exploring the links', *The Geographical Journal* 173: 75–9.

—— (2010) 'The challenges of global invironmental change for urban Africa', *Urban Forum* 21, 3: 235–48.

Simpson, D. (2006) *Indicator Issues and Proposed Framework for a Disaster Preparedness Index (DPi)*, University of Louisville: Center for Hazards Research and Policy Development.

Sims, H. and Vogelmann, K. (2002) 'Popular mobilization and disaster management in Cuba', *Public Administration and Development* 22: 389–400.

Singh, B.R. (1998) 'Soil pollution and contamination', in R. Lal, W.H. Blum, C. Valentine and B.A. Stewart (eds), *Methods for Assessment of Soil Degradation*, Boca Raton, FL: CRC Press, pp. 279–99.

Sirohi, S. and Michaelowa, A. (2007) 'Sufferer and cause: Indian livestock and climate change', *Climatic Change* 85: 285–98.

Siscawati, M. (2009) 'Evaluation report on the Tsunami Response Program prepared for the Unitarian Universalist Service Committee'. Online www.uusc.org/files/Full%20Tsunami%20Evaluation%20Report. pdf (accessed 10 March 2010).

Skees, J.R. and Enkh-Amgalan, A. (2002) 'Examining the feasibility of livestock insurance in Mongolia', *World Bank Working Paper 2886*, 17 September, Washington, DC: World Bank.

Skempton, A.W. (1966) 'Bedding-plane slip, residual strength and the Vaiont landslide', *Geotechnique* 16: 82–4.

Skupin, A. and Fabrikant, S.I. (2003) 'Spatialization methods: A cartographic research agenda for non-geographic information visualization', *Cartography and Geographic Information Science* 30: 99–119.

Slingsby, A., Dykes, J. and Wood, J. (2009) 'Configuring hierarchical questions to address research questions', *IEEE Transactions on Visualization and Computer Graphics* 15, 6: 977–84.

Smart, J. (2009) *Disability, Society and the Individual*, 2nd edn, Austin, TX: Pro-ed.

Smil, V. (1993) *Global Ecology: Environmental change and social flexibility*, London: Routledge.

Smillie, I. (1998) 'Relief and development: The struggle for synergy', Thomas J. Watson Jr Institute for International Studies, Occasional Paper 33, Brown University. Online www.watsoninstitute.org/pub/OP33.pdfBuildingBridgesBetweenReliefandDevelopment (accessed 7 December 2010).

Smith, K. and Petley, D. (2009) *Environmental Hazards: Assessing risk and reducing disaster*, 5th edn, London: Routledge.

Smith, P.C. and Simpson, D.M. (2005) 'The role of mobile emergency tactical communication systems for disaster response', Center for Hazards Research and Policy Development, Working Paper 06–05, Louisville, Kentucky: University of Louisville.

Smithson, M.J. (1989) *Ignorance and Uncertainty: Emerging paradigms*, New York: Springer-Verlag.

Smithsonian Institution (2008) 'Chaitén', *Weekly Reports of the Global Volcanism Network*, 30 April–6 May, Washington, DC: Smithsonian Institution.

Smolinski, M.S.,Hamburg, M.A. and Lederberg, J. (2003) *Microbial Threats to Health: Emergence, detection and response*, Washington, DC: National Academies Press.

Smolka, A. (2006) 'Natural disasters and the challenge of extreme events: Risk management from an insurance perspective', *Philosophical Transactions of the Royal Society A* 364: 2147–65.

Smoller, L.A. (2000) 'Of earthquakes, hail, frogs and geography: Plague and the investigation of the Apolcalypse in the Later Middle Ages', in C.W. Bynum and P. Freeman (eds), *Last Things. Death and the Apocalypse in the Middle Ages*, Philadelphia, University of Pensylvania Press, pp. 156–85.

Smoyer-Tomic, K.E. and Rainham, D.G.C. (2001) 'Beating the heat: Development and evaluation of a Canadian hot weather health-response plan', *Environmental Health Perspectives* 109: 1241–48.

Smucker, T. (2003) *Land Tenure Reform and Changes in Land-Use and Land Management in Semi-Arid Tharaka, Kenya*, Working Paper Number No. 11, Nairobi: International Livestock Research Institute.

Smucker, T. and Wisner, B. (2008) 'Changing household responses to drought in Tharaka, Kenya: Vulnerability, persistence, and challenge', *Disasters* 32, 2: 190–215.

Snyder, B. (2000) *Music and Memory: An introduction*, Boston: MIT Press.

Sobrino, J. (2004) *Where is God? Earthquake, terrorism, barbarity and hope*, New York: Orbis Books.

Solesbury, W. (2003) *Sustainable Livelihoods: A case study of the evolution of DFID policy*, Working Paper 217, London: Overseas Development Institute.

Solnit, R. (2009) *A Paradise Built in Hell: The extraordinary communities that arise in disaster*, New York: Viking.

Sommer, A. and Mosley, W.H. (1971) *Survey of 1970 Cyclone-Affected Region of East Pakistan, February-March 1971: Final report*, Atlanta: Center for Disease Control.

—— (1972) 'East Bengal cyclone of November, 1970', *The Lancet* 299, 7759: 1030–6.

Songsore, J., Nabila, J.S, Yangyuoru, Y., Avle, S., Bosque–Hamilton, E.K., Amponsah, P.E. and Alhassan, O. (2009) 'Integrated disaster risk and environmental health monitoring: Greater Accra Metropolitan Area, Ghana', in M. Pelling and B. Wisner (eds), *Disaster Risk Reduction: Cases from urban Africa*, London and Sterling, VA: Earthscan, pp. 65–85.

Sontag, S. (1965) 'The imagination of disaster', *Commentary* 40: 42–8.

SOPAC (Pacific Islands Applied Geoscience Commission) (2005) *A Framework for Action 2005–2015*, SOPAC Miscellaneous Report 613, Suva, Fiji.

Sorensen, J.H. (2000) 'Hazard warning systems: Review of 20 years of progress', *Natural Hazards Review* 1, 2: 119–25.

SPC (Secretariat of the Pacific Community) (2008) *Land Use Planning in the Pacific*, Policy Brief No. 3, Suva, Fiji: Secretariat of the Pacific Community.

Sphere Project (2004) *Humanitarian Charter and Minimum Standards in Disaster Response*, Geneva: Sphere Project.

—— (2010) *Humanitarian Charter and Minimum Standards in Disasters*, 3rd edn, Geneva: Sphere Project.

Spiker, E.C and Gori, P.L. (2003) 'National landslides hazards mitigation strategy: A framework for loss reduction', *US Geological Survey Circular*, Issue 1244: 1–54.

Squires, G. and Hartman, C. (2006) *There is No Such Thing as a Natural Disaster: Race, class and Katrina*, London: Routledge.

Srinivasan, K., Nagaraj, K. and Venkatesh, V. (2005) 'The state and civil society in disaster response: An analysis of the Tamil Nadu tsunami experience', Tata Institute of Social Sciences. Online www.eSocialSciences.com/data/articles/Document12872009370.7873499.pdf (accessed 20 August 2010).

Srivastava, A.K., Dandekar, M.M., Kshirsagar, S.R. and Dikshit, S.K. (2007) 'Is summer becoming more uncomfortable at Indian cities?', *MAUSAM* 58: 335–44.

Steel, D.I., Asher, D.J., Napier, W.M. and Clube, S.V.M. (1994) 'Are impacts correlated in time?', in T. Gehrels, M.S. Matthews and A. Schumann (eds), *Hazards Due to Comets and Asteroids*, Space Science Series, Tucson: University of Arizona Press, pp. 463–78.

Stehlik, D., Lawrence, G. and Gray, I. (2000) 'Gender and drought: Experiences of Australian women in the drought of the 1990s', *Disasters* 24, 1: 38–53.

Stein, G. (2010) 'Reports accuse WHO of exaggerating H1N1 threat, possible ties to drug makers', *Washington Post*, 4 June. Online www.washingtonpost.com/wp-dyn/content/article/2010/06/04/AR2010060403034.html (accessed 22 December 2010).

Stein, S. and Mazzotti, S. (2007) *Continental Intraplate Earthquakes: Science, hazard and policy issues*, Geological Society of America Special Paper, 403, Washington, DC: Geological Society of America.

Steinberg, L.J. and Cruz, A.M. (2004) 'When natural and technological disasters collide: Lessons from the Turkey earthquake of August 17, 1999', *Natural Hazards Review* 5, 3: 121–30.

Steinberg, T. (2000) *Acts of God: The unnatural history of natural disaster in America*, New York: Oxford University Press.

Stern, G. (2007a) *Can God Intervene? How Religion Explains Natural Disasters*, Westport, CN: Praeger.

Stern, N. (2007b) *The Economics of Climate Change: The Stern review*, Cambridge: Cambrige University Press.

Stevens, M.R., Berke, P.R. and Song, T. (2010) 'Public participation in local government review of development proposals in hazardous locations: Does it matter and what do local government planners have to do with it?' *Environmental Management* 45: 320–35.

Stoddard, A. (2003) 'Humanitarian NGOs: Challenges and trends', HPG Briefing No. 12, London: Overseas Development Institute.

Stokes, G.H. (2003) 'Study to determine the feasibility of extending the search for near Earth objects to smaller limiting diameters', Report of the NASA NEO Science Definition Team. Online neo.jpl.nasa.gov/neo/neoreport030825.pdf (accessed 3 March 2010).

Strand, A. and Borchgrevink, K. (2006) *Review of Norwegian Earthquake Assistance to Pakistan 2005 and 2006*, Norway: Chr. Michelsen Institute. Online www.cmi.no/publications/file/2449-review-of-norwegian-earthquake-assistance-to.pdf (accessed 27 September 2010).

Strand, R.T., Fernandes Dias, L., Bergström, S. and Andersson, S. (2007) 'Unexpected low prevalence of HIV among fertile women in Luanda, Angola: Does war prevent the spread of HIV?' *International Journal of STD & AIDS* 18, 7: 467–71.

Streefland, P.H. (1996) 'Enhancing sustainable vaccination programmes in an unstable world: A social science perspective', *Journal of Epidemiology and Community Health* 50, 5: 601–4.

Street, A. (2007) 'The UN cluster approach in the Pakistan earthquake response: An NGO perspective', *Humanitarian Exchange Magazine*, London: ODI, 37 (March). Online www.odihpn.org/report.asp?ID=2880 (accessed 30 November 2010).

Stulman, D. and Warnke, N. (1990) 'The disaster film at the turn of the century', *Film History* 4, 2: 101–11.

Suarez, P. and Linnerooth-Bayer, J. (2010) 'Micro-insurance for local adaptation', *Climate Change* 1, 2: 271–8.

Sudmeier-Rieux, K. and Ash, N. (2009) *Environmental Guidance Note for Disaster Risk Reduction: Healthy ecosystems for human security*, Ecosystem Management Series No. 8, Commission on Ecosystem Management, revised edn, Gland: IUCN.

Sudmeier-Rieux, K., Masundire, H., Rizvi, A. and Rietbergen, S. (eds) (2006) *Ecosystems, Livelihoods and Disasters: An integrated approach to disaster risk management*, Gland: IUCN.

Sudmeier-Rieux, K., Qureshi, R.A., Peduzzi, P., Nessi, J., Breguet, A., Dubois, Jaboyedoff, M., Jaubert, R., Rietbergen, S., Klaus, R. and Cheema, M.A. (2007) *Disaster Risk, Livelihoods and Natural Barriers, Strengthening Decision-Making Tools for Disaster Risk Reduction: A case study from Northern Pakistan*, Gland: The World Conservation Union.

Sunstein, C. (2005) *Laws of Fear: Beyond the precautionary principle*, Cambridge: Cambridge University Press.

Susman, P., O'Keefe, P. and Wisner, B. (1983) 'Global disasters: A radical interpretation', in K. Hewitt (ed.), *Interpretations of Calamity from the Viewpoint of Human Ecology*, London: Allen & Unwin, pp. 263–83.

Sutter, D. and Simmons, K. (2010) 'Tornado fatalities and mobile homes in the United States', *Natural Hazards* 53, 1: 125–37.

Svensen, H. (2009) *The End is Nigh: A history of natural disasters*, London: Reaktion Press.

Swift, J. (1973) 'Disaster and a Sahelian nomad economy', in D. Dalby and R.J.H. Church (eds), *Drought in Africa*, London: School of Oriental and African Studies, pp. 71–8.

—— (1989) 'Why are rural people vulnerable to famine?' *IDS Bulletin* 20, 2: 8–15.

Swiss Re (2000) *Space Weather: Hazard to the Earth?* Zurich: Swiss Re.

—— (2006) 'Natural catastrophes and man-made disasters in 2005', *Sigma 2/2006*. Online media.swissre.com/documents/sigma2_2006_en.pdf (accessed 9 October 2010).

—— (2010) 'Natural catastrophes and man-made disasters in 2009', Sigma 1/2010. Online media.swissre.com/documents/sigma1_2010_en.pdf (accessed 9 October 2010).

Szabolcs, I. (1998) 'Salt buildup as a factor of soil degradation', in R. Lal, W.H. Blum, C. Valentine and B.A. Stewart (eds), *Methods for Assessment of Soil Degradation*, Boca Raton, FL: CRC Press, pp. 253–64.

Tanaka, K. (2005) 'The impact of disaster education on public preparation and mitigation for earthquakes: A cross-country comparison between Fukui, Japan and the San Francisco Bay Area, California, USA', *Applied Geography* 25, 3: 201–25.

Tandon, R. (1991) 'Civil society, the state and roles of NGOs', *Institute for Development Research Reports* 8, 3. Online www.worlded.org/docs/Publications/idr/pdf/8-3.pdf (accessed 18 December 2010).

Tang, A.K. (ed.) (2000) *Izmit (Kocaeli), Earthquake of August 17, 1999 including Duzce Earthquake of November 12, 1999: Lifeline performance*, Virginia: Technical Council on Lifeline Earthquake Engineering, Monograph No. 17, ASCE.

—— (2009) *Preliminary ASCE Technical Council on Lifeline Earthquake Engineering (TCLEE) Reconnaissance Report: Padang, Sumatra, Indonesia Earthquake of September 30, 2009*. Online asce.org/files/pdf/instfound/Preliminary_Padang_Indonesia_Reconnaissance_Report.pdf (accessed 5 February 2010).

Tanida, N. (1996) 'What happened to elderly people in the Great Hanshin Earthquake?', *Behavioral Medical Journal* 313: 1133–5.

Tanner, M. and Vlassoff, C. (1998) 'Treatment-seeking behaviour for malaria: A typology based on endemicity and gender', *Social Science and Medicine* 46, 4–5: 523–32.

Tanner, T., Mitchell, T. and Haynes, K. (2009) 'Children as agents of change for disaster risk reduction: Lessons from El Salvador and the Philippines', Children in a Changing Climate Research Working Paper No. 1, Brighton: Institute of Development Studies.

Tappin, D.R., Matsumota, T., Watts, P., Satake, K., McMurty, G.M., Matsuyama, M., Lafoy, Y., Tsuji, Y., Kanamatsu, T., Lus, W., Iwabuchi, Y., Yeh, H., Matsumotu, Y., Nakamura, M., Mahoi, M., Hill, P., Crook, K., Anton, L. and Walsh, J.P. (1999) 'Sediment slump likely caused the 1998 Papua New Guinea tsunami', *Eos, Transactions American Geophysical Union* 80, 30: 329–40.

TEC (Tsunami Evaluation Commission) (2006a) *Joint Evaluation of the International Response to the Indian Ocean Tsunami: Synthesis report*, London: Tsunami Evaluation Commission. Online www.alnap.org/pool/files/synthrep(1).pdf (accessed 24 August 2010).

—— (2006b) *The Role of Needs Assessment in the Tsunami Response*, London: Tsunami Evaluation Coalition. Online www.alnap.org/pool/files/needs-assessment-final-report.pdf (accessed 7 April 2011).

—— (2006c) *Funding the Tsunami Response*, London: Tsunami Evaluation Coalition. Online www.alnap.org/initiatives/tec/thematic/fundingresponse.aspx (accessed 17 January 2010).

Terry, F. (2002) *Condemned to Repeat*, Cornell, NY: Cornell University Press.

Thacker, M., Lee, R., Sabogal, R. and Henderson, A. (2008) 'Overview of deaths associated with natural events, United States, 1979–2004', *Disasters* 32, 2: 303–15.

Thomalla, F. and Schmuck, H. (2004) 'We all knew that a cyclone was coming: Disaster preparedness and the cyclone of 1999 in Orissa, India', *Disasters* 4, 28: 373–87.

Thomas, J. and Cook, K. (2005) *Illuminating the Path: The research and development agenda for visual analytics*, Richland: National Visualization and Analytics Center.

Thomas Jr, W.L. (ed.) (1956) *Man's Role in Changing the Face of the Earth*, Chicago: University of Chicago Press.

Thompson, M. and Gaviria, I. (2004) *Weathering the Storm: Lessons in risk reduction from Cuba*, Oxfam International. Online www.proventionconsortium.org/themes/default/pdfs/CRA/Cuba.pdf (accessed 20 August 2010).

Thornton, W.E. and Voigt, L. (2007) 'Lydia disaster rape: Vulnerability of women to sexual assaults during Hurricane Katrina', *Journal of Public Management & Social Policy* 13, 2: 23–49.

Thorsheim, P. (2006) *Inventing Pollution: Coal, smoke and pollution in Britain since 1800*, Columbus, OH: University of Ohio.

Thrupp, L.A. (1998) *Cultivating Diversity: Agrobiodiversity and food security*, Washington, DC: World Resources Institute.

Tibby, J., Lane, M.B. and Gell, P.A. (2007) 'Local knowledge and environmental management: A cautionary tale from Lake Ainsworth, New South Wales, Australia', *Environmental Conservation* 34, 4: 334–41.

Ticehurst, S., Webster, R., Carr, V. and Lewin, T. (1996) 'The psychosocial impact of an earthquake on the elderly', *International Journal of Geriatric Psychiatry* 11: 943–51.

Tiki, W. and Oba, G. (2009) 'Ciinna – the Borana Oromo narration of the 1890s great rinderpest epizootic in North Eastern Africa', *Journal of Eastern African Studies* 3, 3: 479–508.

To Waninara, C.G. (2000) *The 1994 Rabaul Volcanic Eruption: Human sector impacts on the Tolai displaced communities*, Goroka: Melanesian Research Institute.

Tobin, G.A. (1995) 'The levee love affair: A stormy relationship', *Water Resources Bulletin* 31, 3: 359–67.

Tobin, G.A. and Whiteford, L.M. (2002) 'Community resilience and volcano hazard: The eruption of Tungurahua and evacuation of the Faldas in Ecuador', *Disasters* 26, 1: 28–48.

Tobriner, S. (1983) 'La casa baraccata: Earthquake-resistant construction in 18th-century Calabria', *Journal of the Society of Architectural Historians* 42, 2: 131–8.

Todes, A., Karam, A., Klug, N. and Malaza, N. (2010) 'Beyond master planning? New approaches to spatial planning in Ekurhuleni, South Africa', *Habitat International* 34, 4: 414–20.

Tolfree, D. (2003) *Community Based Care for Separated Children*, Sweden: Save the Children Sweden.

Toole, M. (1997) 'Communicable disease and disease control', in E. Noji (ed.), *The Public Health Consequences of Disasters*, New York: Oxford University Press, pp. 79–100.

Tran, P., Marincioni, F., Shaw, R., Sarti, M. and Van An, L. (2008) 'Flood risk management in Central Vietnam: Challenges and potentials', *Journal of Natural Hazards* 46, 1: 119–38.

Transparency International (2005) *Global Corruption Report 2005*. Online www.transparency.org/publications/gcr/gcr_2005 (accessed 6 June 2010).

Trivelli, C., Yancari, J. and De los Ríos, C. (2009) *Crisis y pobreza rural en América Latina*, Documento de Trabajo 37, Programa Dinámicas Territoriales Rurales, Santiago: RIMISP.

Tsunozaki, E. (2007) *Capacity Building and Awareness Training for Disaster Reduction through Formal Education: Lessons learned from the Indian Ocean tsunami*. Online 150.217.73.85/wlfpdf/07_Tsunozaki.pdf (accessed 27 May 2010).

Tu, D., Geheb, K., Susumu, U. and Vitoon (2004) 'Rice is the life and culture of the people of the Lower Mekong Basin Region', Paper presented at the *Mekong Rice Conference*, Mekong River Commission Secretariat, Ho Chi Minh City, 15–17 October 2004.

Tuan, Y.-F. (1979) *Landscapes of Fear*, New York: Pantheon.

Turner, B.L., Kasperson, R., Matson, P., McCarthy, J., Cortell, R., Christensen, L., Eckley, N., Kasperson, J., Luers, A., Martello, M., Polsky, C., Pulsipher, A. and Schiller, A. (2003) 'A framework for assessing vulnerability in sustainability science', *Proceedings of the National Academy of Sciences* 100, 14: 8074–79.

Turner, K. and Freedman, B. (2004) 'Music and environmental studies', *Journal of Environmental Education* 36, 1: 45–52.

Turner, M.D. (2010) 'Climate change and social resilience: "Adaptive" conflict in the Sahel', Paper presented at the Berkeley Environmental Politics Workshop, Institute for International Studies, University of California, Berkeley, CA. Online globetrotter.berkeley.edu/bwep/colloquium/index.php (accessed 10 November 2010).

Turton, A.R. and Ohlsson, L. (1999) *Water Scarcity and Social Stability: Towards a deeper understanding of key concepts needed to manage water scarcity in developing countries*, Occasional Paper No. 17, London: School of Oriental and African Studies.

Tversky, A. and Kahneman, D. (1974) 'Judgment under uncertainty: Heuristics and biases', *Science* 185, 4157: 1124–31.

Twigg, J. (1999–2000) 'The age of accountability? Future community involvement in disaster reduction', *Australian Journal of Emergency Management*, Summer: 51–8.

—— (2001) *Sustainable Livelihoods and Vulnerability to Disasters*, London: Benfield Greig Hazard Research Centre.

—— (2003) 'The human factor in early warnings. Risk perception and appropriate communications', in J. Zschau and A.N. Küppers (eds), *Early Warning Systems for Natural Disaster Reduction*, Berlin: Springer, pp. 19–27.

—— (2004) *Disaster Risk Reduction: Mitigation and preparedness in developing and emergency programming*, Good Practice Review No. 9, London: Humanitarian Practice Network.

—— (2007) *Characteristics of a Disaster-Resilient Community: A guidance note*, London: DFID Disaster Risk Reduction Interagency Coordination Group.

UDHR (Universal Declaration of Human Rights) (1948) United Nations General Assembly Resolution 217A (III), UN Doc. A/810 at 71.

Uganda Red Cross (2010) *Bududa Landslides: One month down the road, what has been done?* Uganda Red Cross, Kampala.

Uhernik, J.A. and Husson, M.A. (2009) 'Psychological first aid: An evidence informed approach for acute disaster behavioral health response', in G. Walz, J. Bleuer and R. Yep (eds), *Compelling Counseling Interventions: VISTAS 2009*, Alexandria: American Counseling Association, pp. 271–80.

Uitto, J.I. (1998) 'The geography of disaster vulnerability in megacities: A theoretical framework', *Applied Geography* 8, 1: 7–16.

Ullberg, S. (2009) 'Watermarks: Flood and memoryscape in Santa Fe, Argentina', Paper presented at the 9th Conference of the European Sociological Association. Online esa.abstractbook.net/abstract.php?aID=353 (accessed 20 October 2010).

UNCCD (United Nations Convention to Combat Desertification) (no date) *Expanatory Leaflet*. Online www.unccd.int/convention/text/leaflet.php (accessed 14 May 2010).

UNCCPR (United Nations Human Rights Committee) (1982) General Comment No. 06: The right to life (art. 6). UN Document No. HRI/GEN/1/Rev.9 (Vol.I).

—— (2004) General Comment No. 31 [80]: Nature of the general legal obligation imposed on states parties to the covenant. UN Document No. CCPR/C/21/Rev.1/Add.13.

—— (2006) Concluding Observations of the Human Rights Committee: United States of America, UN Document No. CCPR/C/USA/CO/3/Rev.1.

UNCESCR (United Nations Committee on Economic, Social and Cultural Rights) (2000) General Comment No. 14: The right to the highest attainable standard of health (Article 12 of the International Covenant on Economic, Social and Cultural Rights), UN Doc. E/C.12/2000/4.

UNCRD (United Nations Center for Regional Development) (2009) *Reducing Vulnerability of School Children to Earthquakes*, Nagoya: UNCRD.

Underwood, R.J., Sneeuwjagt, R.J. and Styles, H.G. (1985) *The Contribution of Prescribed Burning to Forest Fire Control in Western Australia: Case studies*, WAIT Environmental Studies Group Report No 14. Perth: Western Australian Institute of Technology.

UNDG (United Nation Development Group) (no date) *Integrating Disaster Risk Reduction into the CCA and UNDAF: A Guide for UN Country Teams*, United Nations Development Group. Online www.undg.org/docs/9866/UNDG-DRR-Guidance-Note-2009_DUP_08-07-2009_11-43-02-734_AM.PDF (accessed 29 December 2009).

UNDP (United Nation Development Programme) (1997) *Governance for Sustainable Human Development: A UNDP policy document*, New York: UNDP.

—— (2004) *Reducing Disaster Risk: A challenge for development*, Geneva and New York: UNDP.

—— (2007) *Human Development Report 2007/2008: Fighting climate change – human solidarity in a divided world*, New York: UNDP.

—— (2010a) 'Early recovery', *Crisis Prevention and Recovery*. Online www.undp.org/cpr/we_do/early_recovery.shtml (accessed 5 December 2010).

—— (2010b) 'Flagship programme for disaster risk management as proposed by the Nepal Risk Reduction Consortium: Concepts, outlines and expected outcomes', Presentation prepared by UNDP DRM Unit on behalf of the Nepal Risk Reduction Consortium for National Symposium on 'Experiences in Disaster Risk Reduction & Response', on the Occasion of 12th Earthquake Safety Day 2010, United Nations Development Programme, Kathmandu. Online www.nset.org.np/nset/html/esd/2010/presentation/Flagship-Concept-Outline-Expected-Outcomes-Jan12–2010%20YC%20%5BCompatibility%20Mode%5D.pdf (accessed 4 March 2010).

—— (2010c) *Assessment of Development Results, Indonesia*, Evaluation Office, March. Online www.undp.org/evaluation/documents/ADR/ADR_Reports/Indonesia/ADR-Indonesia.pdf (accessed 5 September 2010).

UNDP – Afghanistan (2010) *National Disaster Management Project (NDMP): First Quarter Project Progress Report 2010*, Kabul: United Nations Development Programme – Afghanistan.

UNDRO (United Nations Disaster Relief Organisation) (1982a) *Disaster and the Disabled*, New York: UNDRO.

—— (1982b) *Shelter after Disaster*, New York: UNDRO.

UNESCO (United Nations Educational, Scientific and Cultural Organization) (2006) *Education in Emergencies: The gender implications*, Bangkok: UNESCO.

—— (2007) *Natural Preparedness and Education for Sustainable Development*, Bangkok: UNESCO.

—— (2009) Report by the World Commission on the Ethics of Scientific Knowledge and Technology (COMEST) on the Ethical Implications of Global Climate Change, 182 EX/INF.16.

—— (2010) *Earthquakes and Tsunamis: OREALC/UNESCO Santiago and disaster risk reduction in Latin America*, Santiago, Chile: UNESCO. Online portal.unesco.org/geography/en/ev.php-URL_ID=12577&URL_DO=DO_TOPIC&URL_SECTION=201.html (accessed 14 April 2011).

Ungar, M. (2006) 'Resilience across cultures', *British Journal of Social Work Advance Access*. Online bjsw.oxfordjournals.org/cgi/content/short/bcl343v1 (accessed 20 July 2010).

UN-HABITAT (2003) *Global Report on Human Settlements, 2003: The challenge of the slums*, London: Earthscan.

—— (2007) *Global Report on Human Settlements: Enhancing urban safety and security*, London: Earthscan.

—— (2008) *Secure Land Rights for All*, Nairobi: *UN-HABITAT*. Online www.unhabitat.org/categories.asp?catid=283 (accessed 24 August 2010).

—— (2009) *Global Report on Human Settlements: Planning sustainable cities: Policy directions*, London: Earthscan.

UNHCR (United Nations High Commission for Refugees) (2010) *Global Trends 2009*. Online www.unhcr.org/4c11f0be9.pdf (accessed 7 December 2010).

UNHRC (United Nations Human Rights Committee) (1984) General Comment 14: Article 6 (Right to life). HRI/GEN/1/Rev.9 (Vol.I).

UNICEF (United Nations International Children's Emergency Fund) (2007a) *Promoting the Rights of Children with Disabilities*, Florence, Italy: UNICEF Innocenti Research Center.

—— (2007b) *The Participation of Children and Young People in Emergencies: A guide for relief agencies, based largely on experiences in the Asian tsunami response*, Geneva: UNICEF.

—— (2010) *Humanitarian Action: Partnering for children in emergencies*, Geneva: UNICEF.

UNIDNDR (United Nations International Decade for Natural Disaster Reduction) (1994) *Yokohama Strategy and Plan of Action for a Safer World: Guidelines for natural disaster prevention, preparedness and mitigation*, Geneva: UNIDNDR.

UNISDR (United Nations International Strategy for Disaster Reduction) (2004) *Living with Risk: A global review of disaster reduction initiatives*, Geneva: ISDR.

—— (2005a) *National Reports*, World Conference on Disaster Reduction 2005. Online www.unisdr.org/eng/mdgs-drr/national-reports.htm (accessed 27 November 2010).

—— (2005b) *Hyogo Framework for Action 2005–2015: Building the resilience of nations and communities to disasters*, Geneva: UNISDR. Online www.unisdr.org/eng/hfa/docs/Hyogo-framework-for-action-english.pdf (accessed 22 December 2009).

—— (2006) *Lessons for a Safer Future: Drawing on the experience of the Indian Ocean tsunami disaster*, Geneva: ISDR.

—— (2007a) *Building Disaster Resilient Communities, Good Practices and Lessons Learned*. Online www.unisdr.org/eng/about_isdr/isdr-publications/06-ngos-good-practices/ngos-good-practices.pdf (accessed 14 December 2010).

—— (2007b) *Towards a Culture of Prevention: Disaster risk reduction begins at school*, Geneva: UNISDR.

—— (2007c) *Words Into Action: A guide for implementing the Hyogo Framework*, Geneva: UNISDR.

—— (2007d) *On Better Terms: Key climate change and DRR Concepts*, Geneva: UNISDR.

—— (2008a) *Aprendamos a Prevenir los Desastres: Los niños y las niñas también participamos en la reducción de riesgos – Juegos y Proyectos*, San José, Costa Rica: Unidad Regional para América Latina y el Caribe, UNICEF Costa Rica.

—— (2008b) *Indicators of Progress: Guidance on measuring the reduction of disaster risks and the implementation of the Hyogo Framework for Action*, Geneva: UNISDR. Online www.preventionweb.net/files/2259_IndicatorsofProgressHFA.pdf (accessed 30 December 2009).

—— (2008c) *Indigenous Knowledge. Policy note*, Geneva: UNISDR.

—— (2008d) *Indigenous Knowledge for Disaster Risk Reduction: Good practices and lessons learned from experiences in the Asia-Pacific region*, Geneva: UNISDR.

—— (2008e) *Climate Change and Disaster Risk Reduction*, Briefing Note 1, Geneva: ISDR.

—— (2009a) *Global Assessment Report on Disaster Risk Reduction 2009*, Geneva: UNISDR. Online www.preventionweb.net/english/hyogo/gar/report/index.php?id=9413 (accessed 10 November 2010).

—— (2009b) *Drought Risk Reduction Framework and Practices: Contributing to the implementation of the Hyogo Framework for Action*, Geneva: UNISDR. Online www.preventionweb.net/files/11541_DroughtRisk Reduction2009library.pdf (accessed 23 December 2010).

—— (2009c) *UNISDR Terminology on Disaster Risk Reduction (2009)*, Geneva: UNISDR. Online www.unisdr.org/eng/terminology/terminology-2009-eng.html (accessed 3 March 2010).

—— (2009d) *International Day for Disaster Reduction*, Geneva: UNISDR. Online www.unisdr.org/eng/public_aware/world_camp/2008–2009/wdrc-2008-2009.html (accessed 18 December 2010).

—— (2010a) 'UN development group to seize opportunity to build back better in Haiti', UNISDR News Briefs, 28 January. Online www.unisdr.org/news/v.php?id=12466 (accessed 4 December 2010).

—— (2010b) 'Mission and objectives', Geneva: UNISDR. Online www.unisdr.org/eng/about_isdr/isdr-mission-objectives-eng.htm (accessed 1 December 2010).

—— (2010c) 'ISDR system partners', Geneva: UNISDR. Online www.preventionweb.net/english/hyogo/isdr/partners/?pid:21& pil:1 (accessed 1 December 2010).

United Nations (1945) 'Charter of the United Nations', in *The United Nations and Human Rights 1945–1995*, New York: United Nations Department of Public Information.

—— (1989a) *Convention on the Rights of the Child*, Geneva: United Nations.

—— (1989b) 'International decade for natural disaster reduction', A/RES/44/236, 85th Plenary Meeting, 22 December, New York: UN General Assembly. Online www.un.org/documents/ga/res/44/a44r236.htm (accessed 14 December 2010).

—— (1989c) CCPR General Comment No. 18: Non-discrimination, 10/11/89. HRI/GEN/1/Rev.7, New York: United Nations Human Rights Committee.

—— (1998) Guiding Principles on Internal Displacement, UN Doc. E/CN.4/1998/53/Add.2. Online www.icrc.org/web/eng/siteeng0.nsf/html/57jpg1 (15 October 2010).

—— (2000) *Millennium Development Goals*, New York: United Nations. Online www.un.org/millenniumgoals.

—— (2002a) *The Madrid International Plan of Action on Ageing*, Geneva: United Nations.

—— (2002b) *Report of the World Summit on Sustainable Development*, Johannesburg, South Africa, 26 August–4 September 2002, A/CONF.199/20, New York: United Nations.

—— (2005) *Know Risk*, London: Tudor Rose.

—— (2010) 'Top UN health official refutes conflict of interest claims in handling of H1N1 pandemic', UN News Centre, 8 June. Online www.un.org/apps/news/story.asp?NewsID=34954&Cr=world+health+organization&Cr1= (accessed 6 September 2010).

United Nations Radio (2010) 'Cash for work helps reconstruction in Haiti and incomes too', Geneva/New York: United Nations Radio. Online www.unmultimedia.org/radio/english/detail/91089.html (accessed on 27 April 2010).

United Nations Special Rapporteur on Adequate Housing (2005) *Report on the Central-Asia/Eastern Europe Regional Consultation on Women's Right to Adequate Housing: The interlinkages between multiple discrimination and women's right to adequate housing*, Geneva: Office of the High Commissioner for Human Rights.

United Nations University (2010) *Resilient Bangladesh: Songs for a changing world*, OurWorld2.0, 26 March 2010. Online ourworld.unu.edu/en/resilient-bangladesh-songs-for-a-changing-world/(accessed 28 May 2010).

Universidade Eduardo Mondlane (2009) *Estudo sócio-antropológico sobre reassentamento pós-cheias no vale do Zambeze – 2008: Tete, Manica, Sofala e Zambézia*, Elaborado pelo Departamento de Arqueologia e Antropologia da Universidade Maputo, Mozambique: Eduardo Mondlane Para o INGC.

UNOCHA (United Nations Office of the Coordinator for Humanitarian Assistance) (2003) 'Bangladesh – cold wave OCHA Situation Report No 1', Geneva: UNOCHA. Online www.reliefweb.int/rw/rwb. nsf/AllDocsByUNID/360004bb373b026085256ca90060a1e6 (accessed 4 October 2010).

—— (2007) Swaziland Drought Flash Appeal 2007, Geneva: UNOCHA. Online www.reliefweb.int/rw/ RWB.NSF/db900SID/SHES-75ARAC?OpenDocument (accessed 20 November 2010).

—— (2009) 'Indonesia – Earthquake Situation Report No. 14', Geneva: UNOCHA. Online www. reliefweb.int/rw/rwb.nsf/db900SID/EDIS-7WSKEP (accessed 7 February 2010).

—— (2010a) 'Syria: Over a million people affected by drought', Geneva: UNOCHA. Online www. reliefweb.int/rw/rwb.nsf/db900SID/ACIO-82RJVU?OpenDocument (accessed 17 February 2010).

—— (2010b) ' Cluster approach', Geneva: UNOCHA. Online oneresponse.info/Coordination/Cluster Approach/Pages/Cluster%20Approach.aspx (accessed 30 November 2010).

Ursano, R., Fullerton, C. and McCaughey, B. (1994) 'Trauma and disaster', in R. Ursano, C. Fullerton and B. McCaughey (eds), Individual and Community Responses to Trauma and Disaster: The structure of human chaos, Cambridge: Cambridge University Press, pp. 241–56.

US Congress (2009) Far from Home: Deficiencies in federal housing assistance after Hurricanes Katrina and Rica and recommendations for improvement, Washington, DC: Government Printing Office. Online biotech.law.lsu. edu/blaw/FEMA/DisasterHousingInves.pdf (accessed 20 October 2010).

USAID (United States Agency for International Development) and OFDA (Office of US Foreign Disaster Assistance) (2008) Latin America and the Caribbean – Disaster Preparedness and Mitigation Program Fact Sheet #1, Washington, DC: USAID-OFDA. Online www.reliefweb.int/rw/RWFiles2008.nsf/FilesByRW DocUnidFilename/MUMA-7N63E6-full_report.pdf/$File/full_report.pdf (accessed 28 August 2010).

Usamah, M. (2010) Personal communication to S. Jenkins and K. Haynes.

USGS (United States Geological Survey) (2000) 'Implications for earthquake risk reduction in the United States from the Kocaeli, Turkey, earthquake of August 17, 1999', US Geological Survey Circular 1193, US Geological Survey, United States Government Printing Office.

—— (2010a) 'Ring of Fire', Earthquake Glossary, Earthquake Hazard Program. Online earthquake.usgs. gov/learn/glossary/?termID=150 (accessed 17 May 2010).

—— (2010b) 'Earthquakes with 50,000 or more deaths', Hazards Program. Online earthquake.usgs.gov/ earthquakes/world/most_destructive.php (accessed 9 October 2010).

—— (2010c) ShakeMaps, USGS. Online earthquake.usgs.gov/earthquakes/shakemap (accessed 4 July 2010).

—— (2010d) USGS Water Data for the Nation, USGS. Online waterdata.usgs.gov (accessed 4 July 2010).

Utting, P. (2008) 'The struggle for corporate accountability', Development and Change 39, 6: 959–75.

Vaillant, G. (1995) Adaptation to Life, Cambridge, MA: Harvard University Press.

Vakis, R. (2006) Complementing Natural Disaster Management: The role of social protection, Social Protection Discussion Paper no. 0543, Washington, DC: World Bank.

Vale, L.J. and Campanella, T.J. (2005) The Resilient City: How cities recover from disaster, Oxford: Oxford University Press.

Valodia, I. (2008) 'Informal employment, labour markets and social protection: Some considerations based on South African estimates', IDS Bulletin 39, 2: 57–62.

van Aalst, M.K., Cannon, T. and Burton, I. (2008) 'Community level adaptation to climate change: The potential role of participatory community risk assessment', Global Environmental Change 18: 165–79.

Van Dijk, S. (2007) Environmental Impact Assessment: Tsunami Indonesia, United Nations, Office for the Coordination of Humanitarian Affaires, Banda Aceh, Aceh, Indonesia. Online humanitarianinfo.org/ sumatra/reference/assessments/doc/other/report_def_draft_send_2601.pdf (accessed March 2008).

Van Emmerik, A., Kamphuis, J., Hulsbosch, A. and Emmelkamp, P. (2002) 'Single session debriefing after psychological trauma: A meta-analysis', The Lancet 360, 9335: 766–71.

Van Ommeren, M., Saxena, S. and Saraceno, B. (2005) 'Mental and social sealth during and after acute emergencies: Emerging consensus?' Bulletin of the World Health Organization 83, 1: 71–4.

Van Veen, J. (1962) Dredge, Drain and Reclaim, 5th edn, The Hague: Martinus Nijhoff.

Van Willigen, M., Edwards, T., Edwards, B. and Hessee, S. (2002) 'Riding out the storm: Experiences of the physically disabled during Hurricanes Bonnie, Dennis, and Floyd', Natural Hazards Review 3: 98–106.

Vavilov, N.I. (1926) 'The origin, variation, immunity and breeding of cultivated plants', trans. K. Starr Chester, Chronica Botanica 13, 1–6: 1949–50.

Velasquez, G.T., Uitto, J.I., Wisner, B. and Takahashi, S. (1999) 'A new approach to disaster mitigation and planning in mega-cities: The pivotal role of social vulnerability in disaster risk management', in T. Inoguchi, E. Newman and G. Paoletto (eds), Cities and the Environment: New approaches to eco-societies, Tokyo: United Nations University Press, pp. 161–84.

Velthuis, A.G.J., Saatkamp, H.W., Mourits, M.C.M., de Koeijer, A.A. and Elbers, A.R.W. (2010) 'Financial consequences of the Dutch bluetongue serotype 8 epidemics of 2006 and 2007', *Preventive Veterinary Medicine* 93, 4: 294–304.

Venugopal, V. and Yilmaz, S. (2010) 'Decentralization in Tanzania', *Public Administration and Development* 30: 215–31.

Verchick, R. (2010) *Facing Catastrophe: Environmental action for a post-Katrina world*, Cambridge, MA: Harvard University Press.

Verdin, J., Funk, C., Senay, G. and Choularton, R. (2005) 'Climate science and famine early warning', *Philosophical Transactions of the Royal Society B* 360: 2155–68.

Veski, S., Heinsalu, A., Lang, V., Kestlane, Ü. and Possnert, G. (2004) 'The age of the Kaali meteorite craters and the effect of the impact on the environment and man: Evidence from inside the Kaali craters, island of Saaremaa, Estonia', *Vegetation History and Archaeobotany* 13: 197–206.

Vie, J.-C., Hilton-Taylor, C. and Stuart, S. (eds) (2008) *Wildlife in a Changing World: An analysis of the 2008 Red List of threatened species*, Gland and Cambridge: IUCN.

Villagrán de León, J.C. (2001) *Community-Operated Early Warning Systems in Central America*. Online www. eird.org/eng/revista/No4_2001/pagina11.htm (accessed 5 March 2010).

—— (2003) *SATs: Sistemas de Alerta Temprana para Emergencias de Inundaciones en Centro América*, UNICEF and CEPREDENAC. Online www.crid.or.cr/cd/CD_GERIMU06/pdf/spa/doc14297/doc14297-a. pdf (accessed 7 March 2010).

—— (2005) 'Precursores indígenas en el contexto de sistemas de alerta temprana de Guatemala', in HEI-ILO Research Programme on Strengthening Employment in Response to Crises, *Volume III: Strengthening Crisis Prevention Through Early Warning Systems*. Online www.ilo.org/wcmsp5/groups/public/-ed_emp/-emp_ent/-ifp_crisis/documents/publication/wcms_116544.pdf (accessed 10 December 2009).

—— (2008) 'GIRO: The integral risk management framework: An overview', UNU-EHS Working Paper No. 6/2008 Bonn: UNU-EHS. Online www.ehs.unu.edu/file.php?id=485 (accessed 10 December 2009).

Viratkapan, V. and Perera, R. (2006) 'Slum relocation projects in Bangkok: What has contributed to their success or failure?' *Habitat International* 30: 157–74.

Vlahov, D., Quinn, A., Putnam, S., Proietti, F. and Caiaffa, W.T. (2007) 'Urban health: Latin America and the Caribbean', Proceedings from Rockefeller Foundation Global Urban Summit, 1–20 July 2007, Bellagio, Italy. Online csud.ei.columbia.edu/?id=projects_urbansummit (accessed 15 November 2010).

Vogel, C.H. and Drummond, J.H. (1993) 'Dimensions of drought: South African case studies', *GeoJournal* 30, 1: 93–8.

VOICE (Voluntary Organizations in Cooperation in Emergencies) (2010) *Voice Position Paper on UN-Lead Humanitarian Reform Process*. Online www.ngovoice.org/documents/VOICE%20position%20paper%20on%20UN-led%20%20Humanitarian%20Reform%20process-FINAL.pdf (accessed 30 November 2010).

Von Einsiedel, N., Bendimerad, F., Rodil, A.S. and Deocariza, M. (2010) 'The challenge of urban redevelopment in disaster-affected communities', *Environment and Urbanization ASIA* 1, 1: 27–44.

Von Kotze, A. and Holloway, A. (1999) *Living with Drought: Drought mitigation for sustainable livelihoods*, London: Intermediate Technology Publications.

Von Schreeb, J., Riddez, L., Sammegard, H. and Rosling, H. (2008) 'Foreign field hospitals and the recent sudden-impact disasters in Iran, Haiti, Indonesia and Pakistan', *Prehospital and Disaster Medicine* 23, 2: 144–51.

Wachtendorf, T., Brown, B. and Nickle, M. (2008) 'Big bird, disaster masters and high school students taking charge: the social capacities of children in disaster education', *Children, Youth and Environments* 18, 1: 456–69.

Wadsworth, Y. (1998) 'What is participatory action research?', Southern Cross Institute of Action Research, Action Research International Paper No. 2. Online www.scu.edu.au/schools/gcm/ar/ari/p-ywadsworth98.html (accessed 14 December 2010).

Wald, P., (1997) 'Cultures and carriers: "Typhoid Mary" and the science of social control', *Social Text* 52–53: 181–214.

Walden, B. (2009) *The Food Wars*, London: Verso.

Walden, V.M., Mwangulube, K. and Makhumula-Nkhoma, P. (1999) 'Measuring the impact of a behaviour changeIntervention for commercial sex workers and their potential clients in Malawi', *Health Education Research* 14, 4: 545–54.

Walker, C. (2008) *Shaky Colonialism: The 1746 earthquake-tsunami in Lima, Peru, and its long aftermath*, Durham, NC: Duke University Press.

Walker, P. (1989) *Famine Early Warning Systems: Victims and destitution*, London: Earthscan.

—— (2010) 'The origins, development and future of the international humanitarian system: Containment, compassion and crusades', unpublished paper, Feinstein International Center, Tufts University. Online www.allacademic.com//meta/p_mla_apa_research_citation/2/5/3/5/6/pages253562/p253562–1. php (accessed 14 December 2010).

Walker, P. and Maxwell, D.G. (2008) *Shaping the Humanitarian World*, London: Routledge.

Walker, P., Wisner, B., Leaning, J. and Minear, L. (2005) 'Smoke and mirrors: Deficiencies in disaster funding', *British Medical Journal*, 29 January, 330, 7485: 247–50. Online www.hsph.harvard.edu/fxbcenter/Smoke%20and%20Mirrors.pdf (accessed 22 December 2010).

Wallin, S. (2005) 'Disasters in art'. Online sarah.macmanx.com/documents/disastersinart.pdf (accessed 20 October 2010).

Walmsley, J. (2006) 'The nature of community', *Dialogue* 25, 1: 5–12.

Walsh, D. (2010) 'Floodwaters could sweep Punjab's feudals away', *The Guardian Weekly*, 15 October, pp. 26–27.

Wamsler, C. (2006) 'Integrating risk reduction, urban planning and housing: Lessons from El Salvador', *Open House International* 31, 1: 71–83.

—— (2007a) 'Bridging the gaps: Stakeholder-based strategies for risk reduction and financing for the urban poor', *Environment and Urbanization* 19, 1: 115–42.

—— (2007b) 'Coping strategies in urban slums', in L. Starke (ed.), *Worldwatch Institute: State of the world*, London: Earthscan, p. 124.

—— (2007c) 'Mainstreaming risk reduction in urban planning and housing: A challenge for international aid organizations', Paper presented at the International conference on Housing Growth and Regeneration (European Network of Housing Research), 2–6 July 2004, Cambridge, UK.

—— (2008) 'Planning ahead: Adapting settlements before disasters strike', in L. Bosher (ed.), *Hazards and the Built Environment*, London: Taylor & Francis, pp. 317–54.

Wangui, E.E. (2008) 'Development interventions, changing livelihoods, and the making of female Maasai pastoralists', *Agriculture and Human Values* 25, 3: 365–78.

Wanzala, W., Zessin, K.H., Kyule, N.M., Baumann, M.P.O., Mathias, E. and Hasanali, A. (2005) 'Ethnoveterinary medicine: A critical review of its evolution, perception, understanding and the way forward', *Livestock Research for Rural Development* 17, 11: 1–31.

Warner, J., Waalewijn, P. and Hilhorst, D. (2003) 'Public participation in disaster-prone watersheds: Time for multi-stakeholder platforms?' Paper for the Water and Climate Dialogue No. 6, Wageningen: Disaster Studies – Wageningen University.

Warner, K., Ranger, N., Surminksi, S., Arnold, M., Linnerooth-Bayer, J., Michel-Kerjan, E., Kovacs, P. and Herweijer, C. (2009) *Adapting to Climate Change: Linking disaster risk reduction and insurance*, Geneva: International Strategy for Disaster Reduction.

Watters, E. (2007) 'Suffering differently', *New York Times Magazine*, 12 August. Online www.nytimes.com/2007/08/12/magazine/12wwln-idealab-t.html (accessed 2 January 2011).

Watts, J. (2008) 'China earthquake: Regime cordons off destroyed schools and bans media', *The Guardian*, 12 June. Online www.guardian.co.uk/world/2008/jun/12/chinaearthquake.chinathemedia (accessed 14 December 2010).

Watts, M.J. (1983) *Silent Violence: Food, famine, and peasantry in Northern Nigeria*, Berkeley: University of California Press.

WCC (Welsh Consumer Council) (1992) *In Deep Water: A study of consumer problems in Towyn and Kinmell Bay after the 1990 floods*, Cardiff: Welsh Consumer Council.

WCED (World Commission on Environment and Development) (1987) *Our Common Future*, Geneva: WCED.

Weber, M. (2004) *The Vocation Lectures*, trans. A. Owen and T.R. Strong (eds), Livingstone, NY: Hackett.

Wedgwood, H., Watson, C., Win, E., Tawney, C. and Muwira, K. (2001) *Beating Hunger: The Chivi experience*, London: IT Books.

Weichselgartner, J. and Obersteiner, M. (2002) 'Knowing sufficient and applying more: Challenges in hazard management', *Environmental Hazards* 4, 2–3: 73–7.

Weiland, S.K., Husing, A., Strachan, D.P., Rzehak, P. and Pearce, N. (2004) 'Climate and the prevalence of symptoms of asthma, allergic rhinitis and atopic eczema in children', *Occupational and Environmental Medicine* 61: 609–15.

Weiss, T.G. (2005) *Military–Civilian Interactions: Humanitarian crises and the responsibility to protect*, Maryland: Rowman & Littlefield.

Weisskopf, M., Anderson, H., Foldy, S., Hanrahan, L., Blair, K., Torok, T. and Rumm, P. (2002) 'Heat wave morbidity and mortality, Milwaukee, Wis., 1999 vs. 1995: An improved response?', *American Journal of Public Health* 92: 830–33.

Wells, J. (2005) 'Protecting and assisting older people in emergencies', Network Paper No. 53, London: Humanitarian Practice Network.

Wenger, D.E. and James, T.F. (1994) 'The convergence of volunteers in a consensus crisis: The case of the 1985 Mexico City Earthquake', in R.R. Dynes and K.J. Tierney (eds), *Disasters, Collective Behavior and Social Organization*, Newark: University of Delaware Press, pp. 229–43.

Westerling, A.L., Hidalgo, H.G., Cayan, D.R. and Swetnam, T.W. (2006) 'Warming and earlier spring increased western U.S. forest wildfire activity', *Science* 313, 5789: 940–43.

Westing, A.H. (1999) 'Conflict versus cooperation in a regional setting: Lessons from Eritrea', in M. Suliman (ed.), *Ecology, Politics and Violent Conflicts*, London: Zed Books, pp. 286–90.

Westley, K. and Mikhalev, V. (2002) 'The use of participatory methods for livelihood assessment in situations of political instability: A case study from Kosovo', Working Paper 190, London: Overseas Development Institute.

Westley, K. and Sanderson, D. (1999) *Kosovo Participatory Livelihoods Assessment*, Briefing Note, London: CARE UK.

WFP (World Food Programme) (2006) *Emergency Food Security Assessment: Dili, East Timor, 9–14 June*, Bangkok: WFP. Online www.reliefweb.int/rw/RWFiles2006.nsf/FilesByRWDocUNIDFileName/VBOL-6SCESY-wfp-tls-14jun.pdf/$File/wfp-tls-14jun.pdf (accessed 26 December 2010).

Whelan, D. (1999) *Gender and HIV/AIDS: Taking stock of research and programmes*, Geneva: UNAIDS.

White, G.F. (1945) *Human Adjustment to Floods: A geographical approach to the flood problem in the United States*, Research Paper Number 29, Chicago: Department of Geography, University of Chicago.

—— (1960) 'Strategic aspects of urban floodplain occupance', *Journal of the Hydraulics Division* 86, 2: 89–102.

White, G.F., Kates, R.W. and Burton, I. (2001) 'Knowing better and losing even more: The use of knowledge in hazards management', *Environmental Hazards* 3, 3–4: 81–92.

White, G.W. (2007) 'Katrina and other disasters: lessons learned and lessons to teach', *Journal of Disability Policy Studies* 17: 194–5.

Whiteside, K.H. (2006) *Precautionary Politics: Principle and practice in confronting environmental risk*, London and Cambridge, MA: MIT Press.

Whitfield Åslund, M.L., Zeeb, B.A., Rutter, A. and Reimer, K.J. (2007) 'In situ phytoextraction of polychlorinated biphenyl (PCB) contaminated soil', *Science of the Total Environment* 374: 1–12.

WHO (World Health Organization) (2003) 'Severe acute respiratory syndrome (SARS)', Geneva: WHO. Online www.who.int/csr/media/sars_wha.pdf (accessed 14 December 2010).

—— (2005) *Communicable Disease Control in Emergencies: A field manual*, Geneva: WHO.

—— (2008) *The World Health Report 2008: Primary health care: Now more than ever*, Geneva: WHO.

—— (2009a) 'West Africa flood-related crisis: Situation report no. 1–15 September 2009', Geneva: WHO. Online www.who.int/hac/crises/international/wafrica/sitreps/15september2009/en/index.html (accessed 19 June 2010).

—— (2009b) *Health Cluster Guide: A practical guide for country-level implementation of the health cluster*, Provisional Version, Geneva: WHO. Online www.who.int/hac/global_health_cluster/guide/en/index.html (accessed 1 June 2010).

—— (2009c) 'WHO takes lead on health as UN tackles crises', *Bulletin of the World Health Organization* 87, 4. Online www.scielosp.org/scielo.php?pid=S0042–96862009000400005&script=sci_arttext&tlng=en (accessed 1 June 2010).

—— (2009d) 'Typhoon in the Philippines', Manila: WHO. Online www.wpro.who.int/NR/rdonlyres/C0F8C9E8-A383-47D5-A86B-2511CD6AA80E/0/RECAP_vol_5_December_2009.pdf (accessed 4 October 2010).

WHO, UNFPA (United Nations Fund for Population Activities) and UNHCR (United Nations High Commissioner for Refugees) (1999) *Reproductive Health in Refugee Situations*, Geneva: WHO.

Widyastuti, E., Silaen, G., Pricesca, A., Handoko, A., Blanton, C., Handzel, T., Brennan, M. and Mach, O. (2006) 'Assessment of health-related needs after tsunami and earthquake: Three districts, Aceh Province, Indonesia, July–August 2005', *Morbidity and Mortality Weekly Report* 55, 4: 93–7.

Wiest, R.E., Mocellin, J.S.P. and Motsisi, D.T. (2004) *The Needs of Women in Disasters and Emergencies*, Winnipeg: University of Manitoba – Disaster Research Institute.

Wilches-Chaux, G. (1984) 'The reconstruction programme developed in Popayan by a professional training institution', Paper presented at International Conference on Disaster Mitigation Programme Implementation, Jamaica, November.

—— (1993) 'La vulnerabilidad global', in A. Maskrey (ed.), *Los Desastres no Son Naturales*, Bogotá: LA RED, p. 23.

Wiles, P., Selvester, K., Fidalgo, L. (2005) *Learning Lessons from Disaster Recovery: The case of Mozambique*, Washington, DC: World Bank.

Wilhite, D.A. (1993) *Drought Mitigation Technologies in the United States: With future policy recommendations*, International Drought Information Center Technical Report Series 93–1, Lincoln: University of Nebraska.

—— (2000) 'Drought as a natural hazard: Concepts and definitions', in D.A. Wilhite (ed.), *Drought: A global assessment*, London, UK: Routledge, vol. 1, pp. 3–18.

Wilhite, D.A., Svoboda, M.D. and Hayes, M.J. (2007) 'Understanding the complex impacts of drought: A key to enhancing drought mitigation and preparedness', *Water Resources Management* 21: 763–74.

Wilhite, D.A. and Vanyarkho, O. (2000) 'Drought: Pervasive impacts of a creeping phenomenon', in D.A. Wilhite (ed.), *Drought: A global assessment*, London: Routledge, vol. 2, pp. 245–55.

Wilkinson, L. (1992) *Animals and Disease: An introduction to the history of comparative medicine*, Cambridge: Cambridge University Press.

Will, P.-E. and Wong, R. (1991) *Nourish the People: The state civilian granary system in China, 1650–1850*, Ann Arbor: University of Michigan Press.

Williams, C. and Dunn, C.E. (2003) 'GIS in participatory research: Assessing the impact of landmines on communities in northwest Cambodia', *Transactions in GIS* 7, 3: 393–410.

Williams, K. (2009) *Read All About It! A History of the British Newspaper*, London: Taylor & Francis.

Wilson, N. (2010) '9th Ward residents "Call for Action"', *The Louisiana Weekly*, 8 March. Online www.louisianaweekly.com/news.php?viewStory=2485 (accessed 21 August 2010).

Wilson, T.M. (2009) 'Vulnerability of pastoral farming systems to volcanic ashfall hazards', unpublished PhD thesis, Natural Hazard Research Centre, University of Canterbury.

Wilson, T., Daly, M. and Johnston, D. (2009) 'Review of impacts of volcanic ash on electricity distribution systems, broadcasting and communication networks', *Auckland Engineering Lifelines Group* 79.

Win, E. (1996) *Our Community Ourselves: A search for food security by Chivi's farmers*, Harare, Zimbabwe: Intermediate Technology Development Group.

Winchester, P. (1992) *Power, Choice and Vulnerability: A case study in disaster mismanagement in South India 1977–1988*, London: James & James.

Wisner, B. (1975) 'Famine relief and people's war', *Review of African Political Economy* 2, 3: 77–83.

—— (1978) 'The human ecology of drought in eastern Kenya', unpublished PhD thesis, Graduate School of Geography, Clark University.

—— (1979) 'Flood prevention and mitigation in the People's Republic of Mozambique', *Disasters* 3, 3: 293–306.

Wisner, B. (1988) *Power and Need in Africa: Basic human needs and development strategies*, London: Earthscan.

—— (1993) 'Disaster vulnerability: Scale, power, and daily life', *Geojournal* 30 (2): 127–40.

—— (1995) 'Bridging "expert" and "local" knowledge for counter-disaster planning in urban South Africa', *GeoJournal* 37, 3: 335–48.

—— (1998) 'Marginality and vulnerability: Why the homeless of Tokyo don't "count" in disaster preparations', *Applied Geography* 18, 1: 25–33.

—— (2001a) 'Risk and the neoliberal state: Why post-Mitch lessons didn't reduce El Salvador's earthquake losses', *Disasters* 25, 3: 251–68.

—— (2001b) 'Notes on social vulnerability: Categories, situations, capabilities, and circumstances', Paper presented at the annual meeting of the Association of American Geographers. Online www.radixonline.org/resources/vulnerability-aag2001.rtf (accessed 21 December 2011).

—— (2002) 'Disability and disaster: Victimhood and agency in earthquake risk reduction', *RADIX*. Online www.radixonline.org/disability.html (accessed 27 July 2010).

—— (2006a) *Let Our Children Teach Us*, UN-ISDR and ActionAid. Online www.unisdr.org/eng/task%20force/working%20groups/knowledge-education/docs/Let-our-Children-Teach-Us.pdf (accessed 18 October 2010).

—— (2006b) 'Self-assessment of coping capacity: Participatory, proactive and qualitative engagement of communities in their own risk management', in J. Birkmann (ed.), *Measuring Vulnerability to Natural Hazard: Towards disaster resilient societies*, Tokyo: UNU, pp. 328–40.

—— (2008) 'El Salvador guidance note', *ProVention Consortium CRA Tool Kit*. Online www.provention consortium.org/themes/default/pdfs/CRA/El_Salvador_GN.pdf (accessed 18 October 2010).

—— (2010a) 'Untapped potential of the world's religious communities for disaster reduction in an age of accelerated climate change: An epilogue and prologue', *Religion* 40: 128–31.

—— (2010b) 'Climate change and cultural diversity', *International Social Science Journal*, 61, 199: 131–40.

—— (2010c) Personal communication, based on field work conducted in Tanzania in July 2010.

—— (2010d) 'Marginality', in P. Bobrowky (ed.), *Springer Encyclopedia of Natural Hazards*, Berlin: Springer, entry 226.

Wisner, B. and Adams, J. (2003) *Environmental Health in Emergencies and Disasters*, Geneva: World Health Organization.

Wisner, B., Blaikie, P., Cannon, T. and Davis, I. (2004) *At Risk: Natural hazards, people's vulnerability and disasters*, London: Routledge.

Wisner, B. and Gaillard, J.C. (2009) 'An introduction to neglected disasters', *Jàmbá: Journal of Disaster Risk Studies* 2, 3: 151–8.

Wisner, B. and Haghebaert, B. (2006) 'Fierce friends and friendly enemies: State/civil society relations in disaster risk reduction', Concept Paper, ProVention Consortium Forum, Bangkok, 2–3 February. Online www.proventionconsortium.org/themes/default/pdfs/Forum06/Forum06_Session4_State-CommunityAction.pdf (accessed 18 December 2010).

Wisner, B. and Lavell, A. (2006) 'Neglected crises: Partial response perpetuates suffering', in IFRC, *World Disaster Report 2006*, Geneva: IFRC, pp. 10–41. Online www.ifrc.org/publicat/wdr2006/summaries.asp (accessed 14 December 2010).

Wisner, B., Lavell, A., Ruiz, V. and Meyereles, L. (2005) 'Run tell your neighbour: Early warning in the 2004 Caribbean hurricane season', in IFRC, *World Disaster Report 2005*, Geneva: IFRC, summary, pp. 38–59. Online www.ifrc.org/PUBLICAT/wdr2005/chapter2.asp (accessed 4 December 2010).

Wisner, B. and Mbithi, P. (1974) 'Drought in eastern Kenya: Nutritional status and farmer activity', in G. White (ed.), *Natural Hazards: Local, national, global*, New York: Oxford University Press, pp. 87–97.

Wisner, B., O'Keefe, P. and Westgate, K. (1977) 'Global systems and local disasters: The untapped power of peoples' Science', *Disasters* 1, 1: 47–57.

Wisner, B., Schaerer, M., Haghebert, B. and Arnold, M. (2008) 'Compendium of case studies', *Community Risk Assessment Tool Kit*, ProVention Consortium. Online www.proventionconsortium.org/?pageid=43 (accessed 16 October 2010).

Wisner, B., Stea, D. and Kruks, S. (1991) 'Participatory and action research methods', in E. Zube and G. Moore (eds), *Advances in Environment, Behavior and Design*, New York: Plenum, pp. 271–96.

Wisner, B., Toulmin, C. and Chitiga, R. (2006), 'Introduction', in B. Wisner, C. Toulin and R. Chitiga (eds), *Towards a New Map of Africa*, London: Earthscan, pp. 1–50.

Wisner, B. and Uitto, J.I. (2009) 'Life on the edge: Urban social vulnerability and decentralized, citizen-based disaster risk reduction in four large cities of the Pacific Rim', in H.-G. Brauch, U.O. Spring, J. Grin, C. Mesjasz, P. Kameri-Mbote, N.C. Behera, B. Chourou and H. Krummenache (eds), *Facing Global Environmental Change: Environmental, human, energy, food, health and water security concepts*, Heidelberg: Springer, pp. 215–31.

Wisner, B. and Walker, P. (2005) *Beyond Kobe*. Online www.unisdr.org/wcdr/thematic-sessions/Beyond-Kobe-may-2005.pdf (accessed 16 October 2010).

Wisner, B., Westgate, K. and O'Keefe, P. (1977) 'Global systems and local disasters: The untapped power of people's science', *Disasters* 1, 1: 47–57.

Witham, C.S. (2005) 'Volcanic disasters and incidents: A new database', *Journal of Volcanology and Geothermal Research* 148, 3–4: 191–233.

Wittfogel, K. (1964) *Le Despotisme Oriental*, Paris: Éditions de Minuit.

WMO (World Meteorological Association) (2006) *Drought Monitoring and Early Warning: Concepts, progress and future challenges*, WMO-No. 1006, Geneva: WMO.

—— (1995) *Meeting of Experts to Review the Present Status of Hail Suppression*, WMO Report No. 26, WMO/TD No. 764, Geneva: World Meteorological Organization.

Wolfe, N.D., Dunavan, C.P. and Diamond, J. (2007) 'Origins of major human infectious diseases', *Nature* 447, 7142: 279–83.

Wolfenstein, M. (1957) *Disaster: A psychological essay*, Glencoe: Free Press.

Women's Refugee Commission (2005) *Assessment of the Minimum Initial Service Package in Tsunami-Affected Areas in Indonesia*. Online www.womensrefugeecommission.org/docs/id_misp_eng.pdf (accessed 19 June 2010).

Wood, D. and Lenné, J.M. (eds) (1999) *Agrobiodiversity: Characterization, utilization and management*, Wallingford: CABI Publishing.

Woodhouse, C. (2003) 'Droughts of the past: Implications for the future?', in S.L. Smith (ed.), *The Future of the Southern Plains*, Norman: University of Oklahoma Press, pp. 95–113.

World Bank (2001) *World Development Report 2000–2001: Attacking poverty*, London: Oxford University Press.

—— (2003) *Financing Rapid Onset Natural Disaster Losses in India: A risk management approach, Report No. 26844*, Washington, DC: World Bank.

—— (2005) 'Tsunami and earthquake reconstruction', Washington, DC: World Bank. Online www.worldbank.org/id/tsunami.

—— (2006a) *Ethiopia: Managing water resources to maximize sustainable growth: A World Bank water resources assistance strategy for Ethiopia*, Report No 36000-ET, Washington, DC: Agriculture and Rural Development Department, World Bank. Online www-wds.worldbank.org/external/default/WDSContentServer/WDSP/IB/2009/04/21/000334955_20090421081602/Rendered/PDF/360000REVISED01final1text1and1cover.pdf (accessed 30 December 2009).

—— (2007a) World Bank Participation Sourcebook, Appendix I. Methods and Tools, Washington, DC: World Bank. Online gametlibrary.worldbank.org/FILES/226_Reference%20Guide%20to%20participation%20methods%20and%20tools%20World%20Bank.pdf (accessed 14 December 2010).

—— (2007b) *World Development Report 2008: Agriculture for development*, Washington, DC: World Bank

—— (2008) *Catastrophe Risk Deferred Drawdown Option (DDO) or CAT DDO: Background note*, Washington, DC: World Bank.

—— (2009) *World Development Report 2010: Development and climate change*, Washington, DC: World Bank.

—— (2010a) *Rising Global Interest in Farm Land: Can it yield sustainable and equitable benefits?* Washington, DC: World Bank. Online siteresources.worldbank.org/INTARD/Resources/ESW_Sept7_final_final.pdf (accessed 4 October 2010).

—— (2010b) 'Haiti earthquake recovery', Washington, DC: World Bank. Online www.reliefweb.int/rw/RWFiles2010.nsf/FilesByRWDocUnidFilename/EGUA-89JQTL-full_report.pdf/$File/full_report.pdf (accessed 9 October 2010).

—— (2010c) 'Country papers: Sub-Saharan Africa', Washington, DC: World Bank. Online web.worldbank.org/WBSITE/EXTERNAL/TOPICS/EXTPOVERTY/EXTPRS/0,contentMDK:20195487~menuPK:421515~pagePK:148956~piPK:216618~theSitePK:384201,00.html (accessed 25 August 2010).

—— (2010d) *PRSP Sourcebook*, Washington, DC: World Bank. Online web.worldbank.org/WBSITE/EXTERNAL/TOPICS/EXTPOVERTY/EXTPRS/0,contentMDK:22404376~pagePK:210058~piPK:210062~theSitePK:384201~isCURL:Y,00.html (accessed 17 August 2010).

—— (2010e) *Natural Hazards, UnNatural Disasters: The economics of effective prevention*, Washington, DC: World Bank and United Nations. Online issuu.com/world.bank.publications/docs/9780821380505 (accessed 1 December 2010).

World Bank and ADB (Asian Development Bank) (2005) *Preliminary Damage and Needs Assessment*, Washington, DC: World Bank.

World Bank, UNISDR and Central Asia Regional Economic Cooperation (CAREC) (2009) *Central Asia and Caucasus Disaster Risk Management Initiative (CAC DRMI)*, Washington DC, Geneva and Manila: World Bank and UNISDR.

World Bank – Urban Development Unit South Asia Region (2006) *Dhaka: Improving living conditions for the urban poor*, Report No. 35824-BD, Washington, DC: World Bank.

World Commission on Dams (2000) *The World Dams Report*, London and Sterling, VA: Earthscan.

World Vision (2010) *Protecting Children Post-Disasters*, Federal Way, WA: World Vision.

Wrong, D. (1995) *Power: Its forms, bases, and uses*, 2nd edn, New Brunswick: Transaction Publishers.

WTO (World Trade Organization) (2006) *World Trade Report: Exploring the links between subsidies, trade and the WTO*, Geneva: WTO.

Wu, J., Huang, J. and Han, X. (2004) 'The Three Gorges Dam: An ecological perspective', *Frontiers in Ecology and the Environment* 2: 241–48.

Wucker, M. (2010) 'Haiti: A historical perspective', in *Newsweek*, 15 January 2010. Online www.newsweek.com/2010/01/15/haiti-a-historical-perspective.html#.

WWAP (World Water Assessment Programme) (2009) *The United Nations World Water Development Report 3: Water in a changing world*, Paris and London: UNESCO and Earthscan.

Wynter, A. (2005) 'Humanitarian media coverage in the digital age', in IFRC, *World Disaster Report 2004*, Geneva: IFRC, pp. 126–49. Online www.ifrc.org/publicat/wdr2005/chapter6.asp (accessed 14 December 2010).

Xanthopoulos, G. (2000) 'Fire situation in Greece', *International Forest Fire News (IFFN)*, 23: 76–84.

Yang, T.C., Wu, P.C., Chen, V.Y.J. and Su, H.J. (2009) 'Cold Surge: A sudden and spatially varying threat to health?', *Science of the Total Environment* 407: 3421–4.

Yates, L. and Anderson-Berry, L. (2004) 'The societal and environmental impacts of Cyclone Zoë and the effectivness of the tropical cyclone warning system in Tikopia and Anuta Solomon Islands: December 26–29, 2002', *Australian Journal of Emergency Management* 19, 1: 16–20.

Yonder, A. with Akcar, S. and Gopalan, P. (2005) *Women's Participation in Disaster Relief and Recovery*, Pamphlet of the Population Council No. 22, New York: Population Council.

Yonder, A. and Turkoglu, H. (2011) 'Case study: Post-1999 developments in disaster management in Turkey', in C. Johnson (ed.), *Recent Experience of Land Use Plans and Building Codes to Reduce Disaster Risk: Background Paper for Global Assessment Report on Disaster Risk Reduction*, in press, Geneva: UNISDR.

Yusuf, S. (2009) 'A star is born', in S. Yusuf, A. Deaton, K. Dervis, W. Easterly, T. Ito and J.E. Steiglitz (eds), *Development Economics through the Decades: A critical look at 30 years of the World Development Report*, Washington, DC: World Bank, pp. 1–18.

Zack, N. (2010) 'Race, class and money in disaster', *Southern Journal of Philosophy* 47, S1: 84–103.

Zambia Red Cross Society (2003) 'Vulnerability and capacity assessment: Sinazongwe district' in B. Wisner, B. Haghebaert, M. Schaerer and M. Arnold (eds), *CRA Toolkit*, Geneva: ProVention Consortium. Online www.proventionconsortium.org/themes/default/pdfs/CRA/Zambia.pdf (accessed 18 December 2010).

Zhao, Y., Li, C.-J., Kang, J.-H., Wu, H.-L. and Xu, Q. (2009) 'Development of sandy fields and application of research in Ningxia', *Journal of Agricultural Science* 30, 4: 35–8.

Zimmerer, K.S. (ed.) (2006) *Globalization and New Geographies of Conservation*, Chicago: University of Chicago Press.

Zimmerman, E. (1951) *World Resources and Industries*, revised edn, New York: Harper & Row.

Zipse, D.W. (1994) 'Lightning protection systems: Advantages and disadvantages', *IEEE Transactions on Industry Applications* 30: 1351–61.

Zonn, I., Glantz, M.H. and Rubinstein, A. (1994) 'The virgin lands scheme in the former Soviet Union', in M.H. Glantz (ed.), *Drought Follows the Plow: Cultivating marginal areas*, Cambridge: Cambridge University Press, pp. 135–50.

Index